Lecture Notes in Computer Science 12357

More information about this series at http://www.springer.com/series/7412

Andrea Vedaldi · Horst Bischof ·
Thomas Brox · Jan-Michael Frahm (Eds.)

Computer Vision – ECCV 2020

16th European Conference
Glasgow, UK, August 23–28, 2020
Proceedings, Part XII

 Springer

Editors
Andrea Vedaldi 🆔
University of Oxford
Oxford, UK

Horst Bischof 🆔
Graz University of Technology
Graz, Austria

Thomas Brox 🆔
University of Freiburg
Freiburg im Breisgau, Germany

Jan-Michael Frahm
University of North Carolina at Chapel Hill
Chapel Hill, NC, USA

ISSN 0302-9743 ISSN 1611-3349 (electronic)
Lecture Notes in Computer Science
ISBN 978-3-030-58609-6 ISBN 978-3-030-58610-2 (eBook)
https://doi.org/10.1007/978-3-030-58610-2

LNCS Sublibrary: SL6 – Image Processing, Computer Vision, Pattern Recognition, and Graphics

This Springer imprint is published by the registered company Springer Nature Switzerland AG
The registered company address is: Gewerbestrasse 11, 6330 Cham, Switzerland

Foreword

Hosting the European Conference on Computer Vision (ECCV 2020) was certainly an exciting journey. From the 2016 plan to hold it at the Edinburgh International Conference Centre (hosting 1,800 delegates) to the 2018 plan to hold it at Glasgow's Scottish Exhibition Centre (up to 6,000 delegates), we finally ended with moving online because of the COVID-19 outbreak. While possibly having fewer delegates than expected because of the online format, ECCV 2020 still had over 3,100 registered participants.

Although online, the conference delivered most of the activities expected at a face-to-face conference: peer-reviewed papers, industrial exhibitors, demonstrations, and messaging between delegates. In addition to the main technical sessions, the conference included a strong program of satellite events with 16 tutorials and 44 workshops.

Furthermore, the online conference format enabled new conference features. Every paper had an associated teaser video and a longer full presentation video. Along with the papers and slides from the videos, all these materials were available the week before the conference. This allowed delegates to become familiar with the paper content and be ready for the live interaction with the authors during the conference week. The live event consisted of brief presentations by the oral and spotlight authors and industrial sponsors. Question and answer sessions for all papers were timed to occur twice so delegates from around the world had convenient access to the authors.

As with ECCV 2018, authors' draft versions of the papers appeared online with open access, now on both the Computer Vision Foundation (CVF) and the European Computer Vision Association (ECVA) websites. An archival publication arrangement was put in place with the cooperation of Springer. SpringerLink hosts the final version of the papers with further improvements, such as activating reference links and supplementary materials. These two approaches benefit all potential readers: a version available freely for all researchers, and an authoritative and citable version with additional benefits for SpringerLink subscribers. We thank Alfred Hofmann and Aliaksandr Birukou from Springer for helping to negotiate this agreement, which we expect will continue for future versions of ECCV.

August 2020

Vittorio Ferrari
Bob Fisher
Cordelia Schmid
Emanuele Trucco

Preface

Welcome to the proceedings of the European Conference on Computer Vision (ECCV 2020). This is a unique edition of ECCV in many ways. Due to the COVID-19 pandemic, this is the first time the conference was held online, in a virtual format. This was also the first time the conference relied exclusively on the Open Review platform to manage the review process. Despite these challenges ECCV is thriving. The conference received 5,150 valid paper submissions, of which 1,360 were accepted for publication (27%) and, of those, 160 were presented as spotlights (3%) and 104 as orals (2%). This amounts to more than twice the number of submissions to ECCV 2018 (2,439). Furthermore, CVPR, the largest conference on computer vision, received 5,850 submissions this year, meaning that ECCV is now 87% the size of CVPR in terms of submissions. By comparison, in 2018 the size of ECCV was only 73% of CVPR.

The review model was similar to previous editions of ECCV; in particular, it was double blind in the sense that the authors did not know the name of the reviewers and vice versa. Furthermore, each conference submission was held confidentially, and was only publicly revealed if and once accepted for publication. Each paper received at least three reviews, totalling more than 15,000 reviews. Handling the review process at this scale was a significant challenge. In order to ensure that each submission received as fair and high-quality reviews as possible, we recruited 2,830 reviewers (a 130% increase with reference to 2018) and 207 area chairs (a 60% increase). The area chairs were selected based on their technical expertise and reputation, largely among people that served as area chair in previous top computer vision and machine learning conferences (ECCV, ICCV, CVPR, NeurIPS, etc.). Reviewers were similarly invited from previous conferences. We also encouraged experienced area chairs to suggest additional chairs and reviewers in the initial phase of recruiting.

Despite doubling the number of submissions, the reviewer load was slightly reduced from 2018, from a maximum of 8 papers down to 7 (with some reviewers offering to handle 6 papers plus an emergency review). The area chair load increased slightly, from 18 papers on average to 22 papers on average.

Conflicts of interest between authors, area chairs, and reviewers were handled largely automatically by the Open Review platform via their curated list of user profiles. Many authors submitting to ECCV already had a profile in Open Review. We set a paper registration deadline one week before the paper submission deadline in order to encourage all missing authors to register and create their Open Review profiles well on time (in practice, we allowed authors to create/change papers arbitrarily until the submission deadline). Except for minor issues with users creating duplicate profiles, this allowed us to easily and quickly identify institutional conflicts, and avoid them, while matching papers to area chairs and reviewers.

Papers were matched to area chairs based on: an affinity score computed by the Open Review platform, which is based on paper titles and abstracts, and an affinity score computed by the Toronto Paper Matching System (TPMS), which is based on the paper's full text, the area chair bids for individual papers, load balancing, and conflict avoidance. Open Review provides the program chairs a convenient web interface to experiment with different configurations of the matching algorithm. The chosen configuration resulted in about 50% of the assigned papers to be highly ranked by the area chair bids, and 50% to be ranked in the middle, with very few low bids assigned.

Assignments to reviewers were similar, with two differences. First, there was a maximum of 7 papers assigned to each reviewer. Second, area chairs recommended up to seven reviewers per paper, providing another highly-weighed term to the affinity scores used for matching.

The assignment of papers to area chairs was smooth. However, it was more difficult to find suitable reviewers for all papers. Having a ratio of 5.6 papers per reviewer with a maximum load of 7 (due to emergency reviewer commitment), which did not allow for much wiggle room in order to also satisfy conflict and expertise constraints. We received some complaints from reviewers who did not feel qualified to review specific papers and we reassigned them wherever possible. However, the large scale of the conference, the many constraints, and the fact that a large fraction of such complaints arrived very late in the review process made this process very difficult and not all complaints could be addressed.

Reviewers had six weeks to complete their assignments. Possibly due to COVID-19 or the fact that the NeurIPS deadline was moved closer to the review deadline, a record 30% of the reviews were still missing after the deadline. By comparison, ECCV 2018 experienced only 10% missing reviews at this stage of the process. In the subsequent week, area chairs chased the missing reviews intensely, found replacement reviewers in their own team, and managed to reach 10% missing reviews. Eventually, we could provide almost all reviews (more than 99.9%) with a delay of only a couple of days on the initial schedule by a significant use of emergency reviews. If this trend is confirmed, it might be a major challenge to run a smooth review process in future editions of ECCV. The community must reconsider prioritization of the time spent on paper writing (the number of submissions increased a lot despite COVID-19) and time spent on paper reviewing (the number of reviews delivered in time decreased a lot presumably due to COVID-19 or NeurIPS deadline). With this imbalance the peer-review system that ensures the quality of our top conferences may break soon.

Reviewers submitted their reviews independently. In the reviews, they had the opportunity to ask questions to the authors to be addressed in the rebuttal. However, reviewers were told not to request any significant new experiment. Using the Open Review interface, authors could provide an answer to each individual review, but were also allowed to cross-reference reviews and responses in their answers. Rather than PDF files, we allowed the use of formatted text for the rebuttal. The rebuttal and initial reviews were then made visible to all reviewers and the primary area chair for a given paper. The area chair encouraged and moderated the reviewer discussion. During the discussions, reviewers were invited to reach a consensus and possibly adjust their ratings as a result of the discussion and of the evidence in the rebuttal.

After the discussion period ended, most reviewers entered a final rating and recommendation, although in many cases this did not differ from their initial recommendation. Based on the updated reviews and discussion, the primary area chair then made a preliminary decision to accept or reject the paper and wrote a justification for it (meta-review). Except for cases where the outcome of this process was absolutely clear (as indicated by the three reviewers and primary area chairs all recommending clear rejection), the decision was then examined and potentially challenged by a secondary area chair. This led to further discussion and overturning a small number of preliminary decisions. Needless to say, there was no in-person area chair meeting, which would have been impossible due to COVID-19.

Area chairs were invited to observe the consensus of the reviewers whenever possible and use extreme caution in overturning a clear consensus to accept or reject a paper. If an area chair still decided to do so, she/he was asked to clearly justify it in the meta-review and to explicitly obtain the agreement of the secondary area chair. In practice, very few papers were rejected after being confidently accepted by the reviewers.

This was the first time Open Review was used as the main platform to run ECCV. In 2018, the program chairs used CMT3 for the user-facing interface and Open Review internally, for matching and conflict resolution. Since it is clearly preferable to only use a single platform, this year we switched to using Open Review in full. The experience was largely positive. The platform is highly-configurable, scalable, and open source. Being written in Python, it is easy to write scripts to extract data programmatically. The paper matching and conflict resolution algorithms and interfaces are top-notch, also due to the excellent author profiles in the platform. Naturally, there were a few kinks along the way due to the fact that the ECCV Open Review configuration was created from scratch for this event and it differs in substantial ways from many other Open Review conferences. However, the Open Review development and support team did a fantastic job in helping us to get the configuration right and to address issues in a timely manner as they unavoidably occurred. We cannot thank them enough for the tremendous effort they put into this project.

Finally, we would like to thank everyone involved in making ECCV 2020 possible in these very strange and difficult times. This starts with our authors, followed by the area chairs and reviewers, who ran the review process at an unprecedented scale. The whole Open Review team (and in particular Melisa Bok, Mohit Unyal, Carlos Mondragon Chapa, and Celeste Martinez Gomez) worked incredibly hard for the entire duration of the process. We would also like to thank René Vidal for contributing to the adoption of Open Review. Our thanks also go to Laurent Charling for TPMS and to the program chairs of ICML, ICLR, and NeurIPS for cross checking double submissions. We thank the website chair, Giovanni Farinella, and the CPI team (in particular Ashley Cook, Miriam Verdon, Nicola McGrane, and Sharon Kerr) for promptly adding material to the website as needed in the various phases of the process. Finally, we thank the publication chairs, Albert Ali Salah, Hamdi Dibeklioglu, Metehan Doyran, Henry Howard-Jenkins, Victor Prisacariu, Siyu Tang, and Gul Varol, who managed to compile these substantial proceedings in an exceedingly compressed schedule. We express our thanks to the ECVA team, in particular Kristina Scherbaum for allowing open access of the proceedings. We thank Alfred Hofmann from Springer who again serve as the publisher. Finally, we thank the other chairs of ECCV 2020, including in particular the general chairs for very useful feedback with the handling of the program.

August 2020

Andrea Vedaldi
Horst Bischof
Thomas Brox
Jan-Michael Frahm

Organization

General Chairs

Vittorio Ferrari   Google Research, Switzerland
Bob Fisher   University of Edinburgh, UK
Cordelia Schmid   Google and Inria, France
Emanuele Trucco   University of Dundee, UK

Program Chairs

Andrea Vedaldi   University of Oxford, UK
Horst Bischof   Graz University of Technology, Austria
Thomas Brox   University of Freiburg, Germany
Jan-Michael Frahm   University of North Carolina, USA

Industrial Liaison Chairs

Jim Ashe   University of Edinburgh, UK
Helmut Grabner   Zurich University of Applied Sciences, Switzerland
Diane Larlus   NAVER LABS Europe, France
Cristian Novotny   University of Edinburgh, UK

Local Arrangement Chairs

Yvan Petillot   Heriot-Watt University, UK
Paul Siebert   University of Glasgow, UK

Academic Demonstration Chair

Thomas Mensink   Google Research and University of Amsterdam, The Netherlands

Poster Chair

Stephen Mckenna   University of Dundee, UK

Technology Chair

Gerardo Aragon Camarasa   University of Glasgow, UK

Tutorial Chairs

Carlo Colombo	University of Florence, Italy
Sotirios Tsaftaris	University of Edinburgh, UK

Publication Chairs

Albert Ali Salah	Utrecht University, The Netherlands
Hamdi Dibeklioglu	Bilkent University, Turkey
Metehan Doyran	Utrecht University, The Netherlands
Henry Howard-Jenkins	University of Oxford, UK
Victor Adrian Prisacariu	University of Oxford, UK
Siyu Tang	ETH Zurich, Switzerland
Gul Varol	University of Oxford, UK

Website Chair

Giovanni Maria Farinella	University of Catania, Italy

Workshops Chairs

Adrien Bartoli	University of Clermont Auvergne, France
Andrea Fusiello	University of Udine, Italy

Area Chairs

Lourdes Agapito	University College London, UK
Zeynep Akata	University of Tübingen, Germany
Karteek Alahari	Inria, France
Antonis Argyros	University of Crete, Greece
Hossein Azizpour	KTH Royal Institute of Technology, Sweden
Joao P. Barreto	Universidade de Coimbra, Portugal
Alexander C. Berg	University of North Carolina at Chapel Hill, USA
Matthew B. Blaschko	KU Leuven, Belgium
Lubomir D. Bourdev	WaveOne, Inc., USA
Edmond Boyer	Inria, France
Yuri Boykov	University of Waterloo, Canada
Gabriel Brostow	University College London, UK
Michael S. Brown	National University of Singapore, Singapore
Jianfei Cai	Monash University, Australia
Barbara Caputo	Politecnico di Torino, Italy
Ayan Chakrabarti	Washington University, St. Louis, USA
Tat-Jen Cham	Nanyang Technological University, Singapore
Manmohan Chandraker	University of California, San Diego, USA
Rama Chellappa	Johns Hopkins University, USA
Liang-Chieh Chen	Google, USA

Yung-Yu Chuang	National Taiwan University, Taiwan
Ondrej Chum	Czech Technical University in Prague, Czech Republic
Brian Clipp	Kitware, USA
John Collomosse	University of Surrey and Adobe Research, UK
Jason J. Corso	University of Michigan, USA
David J. Crandall	Indiana University, USA
Daniel Cremers	University of California, Los Angeles, USA
Fabio Cuzzolin	Oxford Brookes University, UK
Jifeng Dai	SenseTime, SAR China
Kostas Daniilidis	University of Pennsylvania, USA
Andrew Davison	Imperial College London, UK
Alessio Del Bue	Fondazione Istituto Italiano di Tecnologia, Italy
Jia Deng	Princeton University, USA
Alexey Dosovitskiy	Google, Germany
Matthijs Douze	Facebook, France
Enrique Dunn	Stevens Institute of Technology, USA
Irfan Essa	Georgia Institute of Technology and Google, USA
Giovanni Maria Farinella	University of Catania, Italy
Ryan Farrell	Brigham Young University, USA
Paolo Favaro	University of Bern, Switzerland
Rogerio Feris	International Business Machines, USA
Cornelia Fermuller	University of Maryland, College Park, USA
David J. Fleet	Vector Institute, Canada
Friedrich Fraundorfer	DLR, Austria
Mario Fritz	CISPA Helmholtz Center for Information Security, Germany
Pascal Fua	EPFL (Swiss Federal Institute of Technology Lausanne), Switzerland
Yasutaka Furukawa	Simon Fraser University, Canada
Li Fuxin	Oregon State University, USA
Efstratios Gavves	University of Amsterdam, The Netherlands
Peter Vincent Gehler	Amazon, USA
Theo Gevers	University of Amsterdam, The Netherlands
Ross Girshick	Facebook AI Research, USA
Boqing Gong	Google, USA
Stephen Gould	Australian National University, Australia
Jinwei Gu	SenseTime Research, USA
Abhinav Gupta	Facebook, USA
Bohyung Han	Seoul National University, South Korea
Bharath Hariharan	Cornell University, USA
Tal Hassner	Facebook AI Research, USA
Xuming He	Australian National University, Australia
Joao F. Henriques	University of Oxford, UK
Adrian Hilton	University of Surrey, UK
Minh Hoai	Stony Brooks, State University of New York, USA
Derek Hoiem	University of Illinois Urbana-Champaign, USA

Timothy Hospedales	University of Edinburgh and Samsung, UK
Gang Hua	Wormpex AI Research, USA
Slobodan Ilic	Siemens AG, Germany
Hiroshi Ishikawa	Waseda University, Japan
Jiaya Jia	The Chinese University of Hong Kong, SAR China
Hailin Jin	Adobe Research, USA
Justin Johnson	University of Michigan, USA
Frederic Jurie	University of Caen Normandie, France
Fredrik Kahl	Chalmers University, Sweden
Sing Bing Kang	Zillow, USA
Gunhee Kim	Seoul National University, South Korea
Junmo Kim	Korea Advanced Institute of Science and Technology, South Korea
Tae-Kyun Kim	Imperial College London, UK
Ron Kimmel	Technion-Israel Institute of Technology, Israel
Alexander Kirillov	Facebook AI Research, USA
Kris Kitani	Carnegie Mellon University, USA
Iasonas Kokkinos	Ariel AI, UK
Vladlen Koltun	Intel Labs, USA
Nikos Komodakis	Ecole des Ponts ParisTech, France
Piotr Koniusz	Australian National University, Australia
M. Pawan Kumar	University of Oxford, UK
Kyros Kutulakos	University of Toronto, Canada
Christoph Lampert	IST Austria, Austria
Ivan Laptev	Inria, France
Diane Larlus	NAVER LABS Europe, France
Laura Leal-Taixe	Technical University Munich, Germany
Honglak Lee	Google and University of Michigan, USA
Joon-Young Lee	Adobe Research, USA
Kyoung Mu Lee	Seoul National University, South Korea
Seungyong Lee	POSTECH, South Korea
Yong Jae Lee	University of California, Davis, USA
Bastian Leibe	RWTH Aachen University, Germany
Victor Lempitsky	Samsung, Russia
Ales Leonardis	University of Birmingham, UK
Marius Leordeanu	Institute of Mathematics of the Romanian Academy, Romania
Vincent Lepetit	ENPC ParisTech, France
Hongdong Li	The Australian National University, Australia
Xi Li	Zhejiang University, China
Yin Li	University of Wisconsin-Madison, USA
Zicheng Liao	Zhejiang University, China
Jongwoo Lim	Hanyang University, South Korea
Stephen Lin	Microsoft Research Asia, China
Yen-Yu Lin	National Chiao Tung University, Taiwan, China
Zhe Lin	Adobe Research, USA

Haibin Ling	Stony Brooks, State University of New York, USA
Jiaying Liu	Peking University, China
Ming-Yu Liu	NVIDIA, USA
Si Liu	Beihang University, China
Xiaoming Liu	Michigan State University, USA
Huchuan Lu	Dalian University of Technology, China
Simon Lucey	Carnegie Mellon University, USA
Jiebo Luo	University of Rochester, USA
Julien Mairal	Inria, France
Michael Maire	University of Chicago, USA
Subhransu Maji	University of Massachusetts, Amherst, USA
Yasushi Makihara	Osaka University, Japan
Jiri Matas	Czech Technical University in Prague, Czech Republic
Yasuyuki Matsushita	Osaka University, Japan
Philippos Mordohai	Stevens Institute of Technology, USA
Vittorio Murino	University of Verona, Italy
Naila Murray	NAVER LABS Europe, France
Hajime Nagahara	Osaka University, Japan
P. J. Narayanan	International Institute of Information Technology (IIIT), Hyderabad, India
Nassir Navab	Technical University of Munich, Germany
Natalia Neverova	Facebook AI Research, France
Matthias Niessner	Technical University of Munich, Germany
Jean-Marc Odobez	Idiap Research Institute and Swiss Federal Institute of Technology Lausanne, Switzerland
Francesca Odone	Universita di Genova, Italy
Takeshi Oishi	The University of Tokyo, Tokyo Institute of Technology, Japan
Vicente Ordonez	University of Virginia, USA
Manohar Paluri	Facebook AI Research, USA
Maja Pantic	Imperial College London, UK
In Kyu Park	Inha University, South Korea
Ioannis Patras	Queen Mary University of London, UK
Patrick Perez	Valeo, France
Bryan A. Plummer	Boston University, USA
Thomas Pock	Graz University of Technology, Austria
Marc Pollefeys	ETH Zurich and Microsoft MR & AI Zurich Lab, Switzerland
Jean Ponce	Inria, France
Gerard Pons-Moll	MPII, Saarland Informatics Campus, Germany
Jordi Pont-Tuset	Google, Switzerland
James Matthew Rehg	Georgia Institute of Technology, USA
Ian Reid	University of Adelaide, Australia
Olaf Ronneberger	DeepMind London, UK
Stefan Roth	TU Darmstadt, Germany
Bryan Russell	Adobe Research, USA

Mathieu Salzmann	EPFL, Switzerland
Dimitris Samaras	Stony Brook University, USA
Imari Sato	National Institute of Informatics (NII), Japan
Yoichi Sato	The University of Tokyo, Japan
Torsten Sattler	Czech Technical University in Prague, Czech Republic
Daniel Scharstein	Middlebury College, USA
Bernt Schiele	MPII, Saarland Informatics Campus, Germany
Julia A. Schnabel	King's College London, UK
Nicu Sebe	University of Trento, Italy
Greg Shakhnarovich	Toyota Technological Institute at Chicago, USA
Humphrey Shi	University of Oregon, USA
Jianbo Shi	University of Pennsylvania, USA
Jianping Shi	SenseTime, China
Leonid Sigal	University of British Columbia, Canada
Cees Snoek	University of Amsterdam, The Netherlands
Richard Souvenir	Temple University, USA
Hao Su	University of California, San Diego, USA
Akihiro Sugimoto	National Institute of Informatics (NII), Japan
Jian Sun	Megvii Technology, China
Jian Sun	Xi'an Jiaotong University, China
Chris Sweeney	Facebook Reality Labs, USA
Yu-wing Tai	Kuaishou Technology, China
Chi-Keung Tang	The Hong Kong University of Science and Technology, SAR China
Radu Timofte	ETH Zurich, Switzerland
Sinisa Todorovic	Oregon State University, USA
Giorgos Tolias	Czech Technical University in Prague, Czech Republic
Carlo Tomasi	Duke University, USA
Tatiana Tommasi	Politecnico di Torino, Italy
Lorenzo Torresani	Facebook AI Research and Dartmouth College, USA
Alexander Toshev	Google, USA
Zhuowen Tu	University of California, San Diego, USA
Tinne Tuytelaars	KU Leuven, Belgium
Jasper Uijlings	Google, Switzerland
Nuno Vasconcelos	University of California, San Diego, USA
Olga Veksler	University of Waterloo, Canada
Rene Vidal	Johns Hopkins University, USA
Gang Wang	Alibaba Group, China
Jingdong Wang	Microsoft Research Asia, China
Yizhou Wang	Peking University, China
Lior Wolf	Facebook AI Research and Tel Aviv University, Israel
Jianxin Wu	Nanjing University, China
Tao Xiang	University of Surrey, UK
Saining Xie	Facebook AI Research, USA
Ming-Hsuan Yang	University of California at Merced and Google, USA
Ruigang Yang	University of Kentucky, USA

Kwang Moo Yi   University of Victoria, Canada
Zhaozheng Yin   Stony Brook, State University of New York, USA
Chang D. Yoo   Korea Advanced Institute of Science and Technology,
                           South Korea
Shaodi You   University of Amsterdam, The Netherlands
Jingyi Yu   ShanghaiTech University, China
Stella Yu   University of California, Berkeley, and ICSI, USA
Stefanos Zafeiriou   Imperial College London, UK
Hongbin Zha   Peking University, China
Tianzhu Zhang   University of Science and Technology of China, China
Liang Zheng   Australian National University, Australia
Todd E. Zickler   Harvard University, USA
Andrew Zisserman   University of Oxford, UK

Technical Program Committee

Sathyanarayanan
  N. Aakur
Wael Abd Almgaeed
Abdelrahman
  Abdelhamed
Abdullah Abuolaim
Supreeth Achar
Hanno Ackermann
Ehsan Adeli
Triantafyllos Afouras
Sameer Agarwal
Aishwarya Agrawal
Harsh Agrawal
Pulkit Agrawal
Antonio Agudo
Eirikur Agustsson
Karim Ahmed
Byeongjoo Ahn
Unaiza Ahsan
Thalaiyasingam Ajanthan
Kenan E. Ak
Emre Akbas
Naveed Akhtar
Derya Akkaynak
Yagiz Aksoy
Ziad Al-Halah
Xavier Alameda-Pineda
Jean-Baptiste Alayrac

Samuel Albanie
Shadi Albarqouni
Cenek Albl
Hassan Abu Alhaija
Daniel Aliaga
Mohammad
  S. Aliakbarian
Rahaf Aljundi
Thiemo Alldieck
Jon Almazan
Jose M. Alvarez
Senjian An
Saket Anand
Codruta Ancuti
Cosmin Ancuti
Peter Anderson
Juan Andrade-Cetto
Alexander Andreopoulos
Misha Andriluka
Dragomir Anguelov
Rushil Anirudh
Michel Antunes
Oisin Mac Aodha
Srikar Appalaraju
Relja Arandjelovic
Nikita Araslanov
Andre Araujo
Helder Araujo

Pablo Arbelaez
Shervin Ardeshir
Sercan O. Arik
Anil Armagan
Anurag Arnab
Chetan Arora
Federica Arrigoni
Mathieu Aubry
Shai Avidan
Angelica I. Aviles-Rivero
Yannis Avrithis
Ismail Ben Ayed
Shekoofeh Azizi
Ioan Andrei Bârsan
Artem Babenko
Deepak Babu Sam
Seung-Hwan Baek
Seungryul Baek
Andrew D. Bagdanov
Shai Bagon
Yuval Bahat
Junjie Bai
Song Bai
Xiang Bai
Yalong Bai
Yancheng Bai
Peter Bajcsy
Slawomir Bak

Mahsa Baktashmotlagh
Kavita Bala
Yogesh Balaji
Guha Balakrishnan
V. N. Balasubramanian
Federico Baldassarre
Vassileios Balntas
Shurjo Banerjee
Aayush Bansal
Ankan Bansal
Jianmin Bao
Linchao Bao
Wenbo Bao
Yingze Bao
Akash Bapat
Md Jawadul Hasan Bappy
Fabien Baradel
Lorenzo Baraldi
Daniel Barath
Adrian Barbu
Kobus Barnard
Nick Barnes
Francisco Barranco
Jonathan T. Barron
Arslan Basharat
Chaim Baskin
Anil S. Baslamisli
Jorge Batista
Kayhan Batmanghelich
Konstantinos Batsos
David Bau
Luis Baumela
Christoph Baur
Eduardo
    Bayro-Corrochano
Paul Beardsley
Jan Bednavr'ik
Oscar Beijbom
Philippe Bekaert
Esube Bekele
Vasileios Belagiannis
Ohad Ben-Shahar
Abhijit Bendale
Róger Bermúdez-Chacón
Maxim Berman
Jesus Bermudez-cameo

Florian Bernard
Stefano Berretti
Marcelo Bertalmio
Gedas Bertasius
Cigdem Beyan
Lucas Beyer
Vijayakumar Bhagavatula
Arjun Nitin Bhagoji
Apratim Bhattacharyya
Binod Bhattarai
Sai Bi
Jia-Wang Bian
Simone Bianco
Adel Bibi
Tolga Birdal
Tom Bishop
Soma Biswas
Mårten Björkman
Volker Blanz
Vishnu Boddeti
Navaneeth Bodla
Simion-Vlad Bogolin
Xavier Boix
Piotr Bojanowski
Timo Bolkart
Guido Borghi
Larbi Boubchir
Guillaume Bourmaud
Adrien Bousseau
Thierry Bouwmans
Richard Bowden
Hakan Boyraz
Mathieu Brédif
Samarth Brahmbhatt
Steve Branson
Nikolas Brasch
Biagio Brattoli
Ernesto Brau
Toby P. Breckon
Francois Bremond
Jesus Briales
Sofia Broomé
Marcus A. Brubaker
Luc Brun
Silvia Bucci
Shyamal Buch

Pradeep Buddharaju
Uta Buechler
Mai Bui
Tu Bui
Adrian Bulat
Giedrius T. Burachas
Elena Burceanu
Xavier P. Burgos-Artizzu
Kaylee Burns
Andrei Bursuc
Benjamin Busam
Wonmin Byeon
Zoya Bylinskii
Sergi Caelles
Jianrui Cai
Minjie Cai
Yujun Cai
Zhaowei Cai
Zhipeng Cai
Juan C. Caicedo
Simone Calderara
Necati Cihan Camgoz
Dylan Campbell
Octavia Camps
Jiale Cao
Kaidi Cao
Liangliang Cao
Xiangyong Cao
Xiaochun Cao
Yang Cao
Yu Cao
Yue Cao
Zhangjie Cao
Luca Carlone
Mathilde Caron
Dan Casas
Thomas J. Cashman
Umberto Castellani
Lluis Castrejon
Jacopo Cavazza
Fabio Cermelli
Hakan Cevikalp
Menglei Chai
Ishani Chakraborty
Rudrasis Chakraborty
Antoni B. Chan

Kwok-Ping Chan
Siddhartha Chandra
Sharat Chandran
Arjun Chandrasekaran
Angel X. Chang
Che-Han Chang
Hong Chang
Hyun Sung Chang
Hyung Jin Chang
Jianlong Chang
Ju Yong Chang
Ming-Ching Chang
Simyung Chang
Xiaojun Chang
Yu-Wei Chao
Devendra S. Chaplot
Arslan Chaudhry
Rizwan A. Chaudhry
Can Chen
Chang Chen
Chao Chen
Chen Chen
Chu-Song Chen
Dapeng Chen
Dong Chen
Dongdong Chen
Guanying Chen
Hongge Chen
Hsin-yi Chen
Huaijin Chen
Hwann-Tzong Chen
Jianbo Chen
Jianhui Chen
Jiansheng Chen
Jiaxin Chen
Jie Chen
Jun-Cheng Chen
Kan Chen
Kevin Chen
Lin Chen
Long Chen
Min-Hung Chen
Qifeng Chen
Shi Chen
Shixing Chen
Tianshui Chen

Weifeng Chen
Weikai Chen
Xi Chen
Xiaohan Chen
Xiaozhi Chen
Xilin Chen
Xingyu Chen
Xinlei Chen
Xinyun Chen
Yi-Ting Chen
Yilun Chen
Ying-Cong Chen
Yinpeng Chen
Yiran Chen
Yu Chen
Yu-Sheng Chen
Yuhua Chen
Yun-Chun Chen
Yunpeng Chen
Yuntao Chen
Zhuoyuan Chen
Zitian Chen
Anchieh Cheng
Bowen Cheng
Erkang Cheng
Gong Cheng
Guangliang Cheng
Jingchun Cheng
Jun Cheng
Li cheng
Ming-Ming Cheng
Yu Cheng
Ziang Cheng
Anoop Cherian
Dmitry Chetverikov
Ngai-man Cheung
William Cheung
Ajad Chhatkuli
Naoki Chiba
Benjamin Chidester
Han-pang Chiu
Mang Tik Chiu
Wei-Chen Chiu
Donghyeon Cho
Hojin Cho
Minsu Cho

Nam Ik Cho
Tim Cho
Tae Eun Choe
Chiho Choi
Edward Choi
Inchang Choi
Jinsoo Choi
Jonghyun Choi
Jongwon Choi
Yukyung Choi
Hisham Cholakkal
Eunji Chong
Jaegul Choo
Christopher Choy
Hang Chu
Peng Chu
Wen-Sheng Chu
Albert Chung
Joon Son Chung
Hai Ci
Safa Cicek
Ramazan G. Cinbis
Arridhana Ciptadi
Javier Civera
James J. Clark
Ronald Clark
Felipe Codevilla
Michael Cogswell
Andrea Cohen
Maxwell D. Collins
Carlo Colombo
Yang Cong
Adria R. Continente
Marcella Cornia
John Richard Corring
Darren Cosker
Dragos Costea
Garrison W. Cottrell
Florent Couzinie-Devy
Marco Cristani
Ioana Croitoru
James L. Crowley
Jiequan Cui
Zhaopeng Cui
Ross Cutler
Antonio D'Innocente

Rozenn Dahyot
Bo Dai
Dengxin Dai
Hang Dai
Longquan Dai
Shuyang Dai
Xiyang Dai
Yuchao Dai
Adrian V. Dalca
Dima Damen
Bharath B. Damodaran
Kristin Dana
Martin Danelljan
Zheng Dang
Zachary Alan Daniels
Donald G. Dansereau
Abhishek Das
Samyak Datta
Achal Dave
Titas De
Rodrigo de Bem
Teo de Campos
Raoul de Charette
Shalini De Mello
Joseph DeGol
Herve Delingette
Haowen Deng
Jiankang Deng
Weijian Deng
Zhiwei Deng
Joachim Denzler
Konstantinos G. Derpanis
Aditya Deshpande
Frederic Devernay
Somdip Dey
Arturo Deza
Abhinav Dhall
Helisa Dhamo
Vikas Dhiman
Fillipe Dias Moreira
    de Souza
Ali Diba
Ferran Diego
Guiguang Ding
Henghui Ding
Jian Ding

Mingyu Ding
Xinghao Ding
Zhengming Ding
Robert DiPietro
Cosimo Distante
Ajay Divakaran
Mandar Dixit
Abdelaziz Djelouah
Thanh-Toan Do
Jose Dolz
Bo Dong
Chao Dong
Jiangxin Dong
Weiming Dong
Weisheng Dong
Xingping Dong
Xuanyi Dong
Yinpeng Dong
Gianfranco Doretto
Hazel Doughty
Hassen Drira
Bertram Drost
Dawei Du
Ye Duan
Yueqi Duan
Abhimanyu Dubey
Anastasia Dubrovina
Stefan Duffner
Chi Nhan Duong
Thibaut Durand
Zoran Duric
Iulia Duta
Debidatta Dwibedi
Benjamin Eckart
Marc Eder
Marzieh Edraki
Alexei A. Efros
Kiana Ehsani
Hazm Kemal Ekenel
James H. Elder
Mohamed Elgharib
Shireen Elhabian
Ehsan Elhamifar
Mohamed Elhoseiny
Ian Endres
N. Benjamin Erichson

Jan Ernst
Sergio Escalera
Francisco Escolano
Victor Escorcia
Carlos Esteves
Francisco J. Estrada
Bin Fan
Chenyou Fan
Deng-Ping Fan
Haoqi Fan
Hehe Fan
Heng Fan
Kai Fan
Lijie Fan
Linxi Fan
Quanfu Fan
Shaojing Fan
Xiaochuan Fan
Xin Fan
Yuchen Fan
Sean Fanello
Hao-Shu Fang
Haoyang Fang
Kuan Fang
Yi Fang
Yuming Fang
Azade Farshad
Alireza Fathi
Raanan Fattal
Joao Fayad
Xiaohan Fei
Christoph Feichtenhofer
Michael Felsberg
Chen Feng
Jiashi Feng
Junyi Feng
Mengyang Feng
Qianli Feng
Zhenhua Feng
Michele Fenzi
Andras Ferencz
Martin Fergie
Basura Fernando
Ethan Fetaya
Michael Firman
John W. Fisher

Matthew Fisher
Boris Flach
Corneliu Florea
Wolfgang Foerstner
David Fofi
Gian Luca Foresti
Per-Erik Forssen
David Fouhey
Katerina Fragkiadaki
Victor Fragoso
Jean-Sébastien Franco
Ohad Fried
Iuri Frosio
Cheng-Yang Fu
Huazhu Fu
Jianlong Fu
Jingjing Fu
Xueyang Fu
Yanwei Fu
Ying Fu
Yun Fu
Olac Fuentes
Kent Fujiwara
Takuya Funatomi
Christopher Funk
Thomas Funkhouser
Antonino Furnari
Ryo Furukawa
Erik Gärtner
Raghudeep Gadde
Matheus Gadelha
Vandit Gajjar
Trevor Gale
Juergen Gall
Mathias Gallardo
Guillermo Gallego
Orazio Gallo
Chuang Gan
Zhe Gan
Madan Ravi Ganesh
Aditya Ganeshan
Siddha Ganju
Bin-Bin Gao
Changxin Gao
Feng Gao
Hongchang Gao

Jin Gao
Jiyang Gao
Junbin Gao
Katelyn Gao
Lin Gao
Mingfei Gao
Ruiqi Gao
Ruohan Gao
Shenghua Gao
Yuan Gao
Yue Gao
Noa Garcia
Alberto Garcia-Garcia
Guillermo
   Garcia-Hernando
Jacob R. Gardner
Animesh Garg
Kshitiz Garg
Rahul Garg
Ravi Garg
Philip N. Garner
Kirill Gavrilyuk
Paul Gay
Shiming Ge
Weifeng Ge
Baris Gecer
Xin Geng
Kyle Genova
Stamatios Georgoulis
Bernard Ghanem
Michael Gharbi
Kamran Ghasedi
Golnaz Ghiasi
Arnab Ghosh
Partha Ghosh
Silvio Giancola
Andrew Gilbert
Rohit Girdhar
Xavier Giro-i-Nieto
Thomas Gittings
Ioannis Gkioulekas
Clement Godard
Vaibhava Goel
Bastian Goldluecke
Lluis Gomez
Nuno Gonçalves

Dong Gong
Ke Gong
Mingming Gong
Abel Gonzalez-Garcia
Ariel Gordon
Daniel Gordon
Paulo Gotardo
Venu Madhav Govindu
Ankit Goyal
Priya Goyal
Raghav Goyal
Benjamin Graham
Douglas Gray
Brent A. Griffin
Etienne Grossmann
David Gu
Jiayuan Gu
Jiuxiang Gu
Lin Gu
Qiao Gu
Shuhang Gu
Jose J. Guerrero
Paul Guerrero
Jie Gui
Jean-Yves Guillemaut
Riza Alp Guler
Erhan Gundogdu
Fatma Guney
Guodong Guo
Kaiwen Guo
Qi Guo
Sheng Guo
Shi Guo
Tiantong Guo
Xiaojie Guo
Yijie Guo
Yiluan Guo
Yuanfang Guo
Yulan Guo
Agrim Gupta
Ankush Gupta
Mohit Gupta
Saurabh Gupta
Tanmay Gupta
Danna Gurari
Abner Guzman-Rivera

JunYoung Gwak
Michael Gygli
Jung-Woo Ha
Simon Hadfield
Isma Hadji
Bjoern Haefner
Taeyoung Hahn
Levente Hajder
Peter Hall
Emanuela Haller
Stefan Haller
Bumsub Ham
Abdullah Hamdi
Dongyoon Han
Hu Han
Jungong Han
Junwei Han
Kai Han
Tian Han
Xiaoguang Han
Xintong Han
Yahong Han
Ankur Handa
Zekun Hao
Albert Haque
Tatsuya Harada
Mehrtash Harandi
Adam W. Harley
Mahmudul Hasan
Atsushi Hashimoto
Ali Hatamizadeh
Munawar Hayat
Dongliang He
Jingrui He
Junfeng He
Kaiming He
Kun He
Lei He
Pan He
Ran He
Shengfeng He
Tong He
Weipeng He
Xuming He
Yang He
Yihui He

Zhihai He
Chinmay Hegde
Janne Heikkila
Mattias P. Heinrich
Stéphane Herbin
Alexander Hermans
Luis Herranz
John R. Hershey
Aaron Hertzmann
Roei Herzig
Anders Heyden
Steven Hickson
Otmar Hilliges
Tomas Hodan
Judy Hoffman
Michael Hofmann
Yannick Hold-Geoffroy
Namdar Homayounfar
Sina Honari
Richang Hong
Seunghoon Hong
Xiaopeng Hong
Yi Hong
Hidekata Hontani
Anthony Hoogs
Yedid Hoshen
Mir Rayat Imtiaz Hossain
Junhui Hou
Le Hou
Lu Hou
Tingbo Hou
Wei-Lin Hsiao
Cheng-Chun Hsu
Gee-Sern Jison Hsu
Kuang-jui Hsu
Changbo Hu
Di Hu
Guosheng Hu
Han Hu
Hao Hu
Hexiang Hu
Hou-Ning Hu
Jie Hu
Junlin Hu
Nan Hu
Ping Hu

Ronghang Hu
Xiaowei Hu
Yinlin Hu
Yuan-Ting Hu
Zhe Hu
Binh-Son Hua
Yang Hua
Bingyao Huang
Di Huang
Dong Huang
Fay Huang
Haibin Huang
Haozhi Huang
Heng Huang
Huaibo Huang
Jia-Bin Huang
Jing Huang
Jingwei Huang
Kaizhu Huang
Lei Huang
Qiangui Huang
Qiaoying Huang
Qingqiu Huang
Qixing Huang
Shaoli Huang
Sheng Huang
Siyuan Huang
Weilin Huang
Wenbing Huang
Xiangru Huang
Xun Huang
Yan Huang
Yifei Huang
Yue Huang
Zhiwu Huang
Zilong Huang
Minyoung Huh
Zhuo Hui
Matthias B. Hullin
Martin Humenberger
Wei-Chih Hung
Zhouyuan Huo
Junhwa Hur
Noureldien Hussein
Jyh-Jing Hwang
Seong Jae Hwang

Sung Ju Hwang
Ichiro Ide
Ivo Ihrke
Daiki Ikami
Satoshi Ikehata
Nazli Ikizler-Cinbis
Sunghoon Im
Yani Ioannou
Radu Tudor Ionescu
Umar Iqbal
Go Irie
Ahmet Iscen
Md Amirul Islam
Vamsi Ithapu
Nathan Jacobs
Arpit Jain
Himalaya Jain
Suyog Jain
Stuart James
Won-Dong Jang
Yunseok Jang
Ronnachai Jaroensri
Dinesh Jayaraman
Sadeep Jayasumana
Suren Jayasuriya
Herve Jegou
Simon Jenni
Hae-Gon Jeon
Yunho Jeon
Koteswar R. Jerripothula
Hueihan Jhuang
I-hong Jhuo
Dinghuang Ji
Hui Ji
Jingwei Ji
Pan Ji
Yanli Ji
Baoxiong Jia
Kui Jia
Xu Jia
Chiyu Max Jiang
Haiyong Jiang
Hao Jiang
Huaizu Jiang
Huajie Jiang
Ke Jiang

Lai Jiang
Li Jiang
Lu Jiang
Ming Jiang
Peng Jiang
Shuqiang Jiang
Wei Jiang
Xudong Jiang
Zhuolin Jiang
Jianbo Jiao
Zequn Jie
Dakai Jin
Kyong Hwan Jin
Lianwen Jin
SouYoung Jin
Xiaojie Jin
Xin Jin
Nebojsa Jojic
Alexis Joly
Michael Jeffrey Jones
Hanbyul Joo
Jungseock Joo
Kyungdon Joo
Ajjen Joshi
Shantanu H. Joshi
Da-Cheng Juan
Marco Körner
Kevin Köser
Asim Kadav
Christine Kaeser-Chen
Kushal Kafle
Dagmar Kainmueller
Ioannis A. Kakadiaris
Zdenek Kalal
Nima Kalantari
Yannis Kalantidis
Mahdi M. Kalayeh
Anmol Kalia
Sinan Kalkan
Vicky Kalogeiton
Ashwin Kalyan
Joni-kristian Kamarainen
Gerda Kamberova
Chandra Kambhamettu
Martin Kampel
Meina Kan

Christopher Kanan
Kenichi Kanatani
Angjoo Kanazawa
Atsushi Kanehira
Takuhiro Kaneko
Asako Kanezaki
Bingyi Kang
Di Kang
Sunghun Kang
Zhao Kang
Vadim Kantorov
Abhishek Kar
Amlan Kar
Theofanis Karaletsos
Leonid Karlinsky
Kevin Karsch
Angelos Katharopoulos
Isinsu Katircioglu
Hiroharu Kato
Zoltan Kato
Dotan Kaufman
Jan Kautz
Rei Kawakami
Qiuhong Ke
Wadim Kehl
Petr Kellnhofer
Aniruddha Kembhavi
Cem Keskin
Margret Keuper
Daniel Keysers
Ashkan Khakzar
Fahad Khan
Naeemullah Khan
Salman Khan
Siddhesh Khandelwal
Rawal Khirodkar
Anna Khoreva
Tejas Khot
Parmeshwar Khurd
Hadi Kiapour
Joe Kileel
Chanho Kim
Dahun Kim
Edward Kim
Eunwoo Kim
Han-ul Kim

Hansung Kim
Heewon Kim
Hyo Jin Kim
Hyunwoo J. Kim
Jinkyu Kim
Jiwon Kim
Jongmin Kim
Junsik Kim
Junyeong Kim
Min H. Kim
Namil Kim
Pyojin Kim
Seon Joo Kim
Seong Tae Kim
Seungryong Kim
Sungwoong Kim
Tae Hyun Kim
Vladimir Kim
Won Hwa Kim
Yonghyun Kim
Benjamin Kimia
Akisato Kimura
Pieter-Jan Kindermans
Zsolt Kira
Itaru Kitahara
Hedvig Kjellstrom
Jan Knopp
Takumi Kobayashi
Erich Kobler
Parker Koch
Reinhard Koch
Elyor Kodirov
Amir Kolaman
Nicholas Kolkin
Dimitrios Kollias
Stefanos Kollias
Soheil Kolouri
Adams Wai-Kin Kong
Naejin Kong
Shu Kong
Tao Kong
Yu Kong
Yoshinori Konishi
Daniil Kononenko
Theodora Kontogianni
Simon Korman

Adam Kortylewski
Jana Kosecka
Jean Kossaifi
Satwik Kottur
Rigas Kouskouridas
Adriana Kovashka
Rama Kovvuri
Adarsh Kowdle
Jedrzej Kozerawski
Mateusz Kozinski
Philipp Kraehenbuehl
Gregory Kramida
Josip Krapac
Dmitry Kravchenko
Ranjay Krishna
Pavel Krsek
Alexander Krull
Jakob Kruse
Hiroyuki Kubo
Hilde Kuehne
Jason Kuen
Andreas Kuhn
Arjan Kuijper
Zuzana Kukelova
Ajay Kumar
Amit Kumar
Avinash Kumar
Suryansh Kumar
Vijay Kumar
Kaustav Kundu
Weicheng Kuo
Nojun Kwak
Suha Kwak
Junseok Kwon
Nikolaos Kyriazis
Zorah Lähner
Ankit Laddha
Florent Lafarge
Jean Lahoud
Kevin Lai
Shang-Hong Lai
Wei-Sheng Lai
Yu-Kun Lai
Iro Laina
Antony Lam
John Wheatley Lambert

Xiangyuan lan
Xu Lan
Charis Lanaras
Georg Langs
Oswald Lanz
Dong Lao
Yizhen Lao
Agata Lapedriza
Gustav Larsson
Viktor Larsson
Katrin Lasinger
Christoph Lassner
Longin Jan Latecki
Stéphane Lathuilière
Rynson Lau
Hei Law
Justin Lazarow
Svetlana Lazebnik
Hieu Le
Huu Le
Ngan Hoang Le
Trung-Nghia Le
Vuong Le
Colin Lea
Erik Learned-Miller
Chen-Yu Lee
Gim Hee Lee
Hsin-Ying Lee
Hyungtae Lee
Jae-Han Lee
Jimmy Addison Lee
Joonseok Lee
Kibok Lee
Kuang-Huei Lee
Kwonjoon Lee
Minsik Lee
Sang-chul Lee
Seungkyu Lee
Soochan Lee
Stefan Lee
Taehee Lee
Andreas Lehrmann
Jie Lei
Peng Lei
Matthew Joseph Leotta
Wee Kheng Leow

Gil Levi
Evgeny Levinkov
Aviad Levis
Jose Lezama
Ang Li
Bin Li
Bing Li
Boyi Li
Changsheng Li
Chao Li
Chen Li
Cheng Li
Chenglong Li
Chi Li
Chun-Guang Li
Chun-Liang Li
Chunyuan Li
Dong Li
Guanbin Li
Hao Li
Haoxiang Li
Hongsheng Li
Hongyang Li
Houqiang Li
Huibin Li
Jia Li
Jianan Li
Jianguo Li
Junnan Li
Junxuan Li
Kai Li
Ke Li
Kejie Li
Kunpeng Li
Lerenhan Li
Li Erran Li
Mengtian Li
Mu Li
Peihua Li
Peiyi Li
Ping Li
Qi Li
Qing Li
Ruiyu Li
Ruoteng Li
Shaozi Li

Sheng Li
Shiwei Li
Shuang Li
Siyang Li
Stan Z. Li
Tianye Li
Wei Li
Weixin Li
Wen Li
Wenbo Li
Xiaomeng Li
Xin Li
Xiu Li
Xuelong Li
Xueting Li
Yan Li
Yandong Li
Yanghao Li
Yehao Li
Yi Li
Yijun Li
Yikang LI
Yining Li
Yongjie Li
Yu Li
Yu-Jhe Li
Yunpeng Li
Yunsheng Li
Yunzhu Li
Zhe Li
Zhen Li
Zhengqi Li
Zhenyang Li
Zhuwen Li
Dongze Lian
Xiaochen Lian
Zhouhui Lian
Chen Liang
Jie Liang
Ming Liang
Paul Pu Liang
Pengpeng Liang
Shu Liang
Wei Liang
Jing Liao
Minghui Liao

Renjie Liao
Shengcai Liao
Shuai Liao
Yiyi Liao
Ser-Nam Lim
Chen-Hsuan Lin
Chung-Ching Lin
Dahua Lin
Ji Lin
Kevin Lin
Tianwei Lin
Tsung-Yi Lin
Tsung-Yu Lin
Wei-An Lin
Weiyao Lin
Yen-Chen Lin
Yuewei Lin
David B. Lindell
Drew Linsley
Krzysztof Lis
Roee Litman
Jim Little
An-An Liu
Bo Liu
Buyu Liu
Chao Liu
Chen Liu
Cheng-lin Liu
Chenxi Liu
Dong Liu
Feng Liu
Guilin Liu
Haomiao Liu
Heshan Liu
Hong Liu
Ji Liu
Jingen Liu
Jun Liu
Lanlan Liu
Li Liu
Liu Liu
Mengyuan Liu
Miaomiao Liu
Nian Liu
Ping Liu
Risheng Liu

Sheng Liu
Shu Liu
Shuaicheng Liu
Sifei Liu
Siqi Liu
Siying Liu
Songtao Liu
Ting Liu
Tongliang Liu
Tyng-Luh Liu
Wanquan Liu
Wei Liu
Weiyang Liu
Weizhe Liu
Wenyu Liu
Wu Liu
Xialei Liu
Xianglong Liu
Xiaodong Liu
Xiaofeng Liu
Xihui Liu
Xingyu Liu
Xinwang Liu
Xuanqing Liu
Xuebo Liu
Yang Liu
Yaojie Liu
Yebin Liu
Yen-Cheng Liu
Yiming Liu
Yu Liu
Yu-Shen Liu
Yufan Liu
Yun Liu
Zheng Liu
Zhijian Liu
Zhuang Liu
Zichuan Liu
Ziwei Liu
Zongyi Liu
Stephan Liwicki
Liliana Lo Presti
Chengjiang Long
Fuchen Long
Mingsheng Long
Xiang Long

Yang Long
Charles T. Loop
Antonio Lopez
Roberto J. Lopez-Sastre
Javier Lorenzo-Navarro
Manolis Lourakis
Boyu Lu
Canyi Lu
Feng Lu
Guoyu Lu
Hongtao Lu
Jiajun Lu
Jiasen Lu
Jiwen Lu
Kaiyue Lu
Le Lu
Shao-Ping Lu
Shijian Lu
Xiankai Lu
Xin Lu
Yao Lu
Yiping Lu
Yongxi Lu
Yongyi Lu
Zhiwu Lu
Fujun Luan
Benjamin E. Lundell
Hao Luo
Jian-Hao Luo
Ruotian Luo
Weixin Luo
Wenhan Luo
Wenjie Luo
Yan Luo
Zelun Luo
Zixin Luo
Khoa Luu
Zhaoyang Lv
Pengyuan Lyu
Thomas Möllenhoff
Matthias Müller
Bingpeng Ma
Chih-Yao Ma
Chongyang Ma
Huimin Ma
Jiayi Ma

K. T. Ma
Ke Ma
Lin Ma
Liqian Ma
Shugao Ma
Wei-Chiu Ma
Xiaojian Ma
Xingjun Ma
Zhanyu Ma
Zheng Ma
Radek Jakob Mackowiak
Ludovic Magerand
Shweta Mahajan
Siddharth Mahendran
Long Mai
Ameesh Makadia
Oscar Mendez Maldonado
Mateusz Malinowski
Yury Malkov
Arun Mallya
Dipu Manandhar
Massimiliano Mancini
Fabian Manhardt
Kevis-kokitsi Maninis
Varun Manjunatha
Junhua Mao
Xudong Mao
Alina Marcu
Edgar Margffoy-Tuay
Dmitrii Marin
Manuel J. Marin-Jimenez
Kenneth Marino
Niki Martinel
Julieta Martinez
Jonathan Masci
Tomohiro Mashita
Iacopo Masi
David Masip
Daniela Massiceti
Stefan Mathe
Yusuke Matsui
Tetsu Matsukawa
Iain A. Matthews
Kevin James Matzen
Bruce Allen Maxwell
Stephen Maybank

Helmut Mayer
Amir Mazaheri
David McAllester
Steven McDonagh
Stephen J. Mckenna
Roey Mechrez
Prakhar Mehrotra
Christopher Mei
Xue Mei
Paulo R. S. Mendonca
Lili Meng
Zibo Meng
Thomas Mensink
Bjoern Menze
Michele Merler
Kourosh Meshgi
Pascal Mettes
Christopher Metzler
Liang Mi
Qiguang Miao
Xin Miao
Tomer Michaeli
Frank Michel
Antoine Miech
Krystian Mikolajczyk
Peyman Milanfar
Ben Mildenhall
Gregor Miller
Fausto Milletari
Dongbo Min
Kyle Min
Pedro Miraldo
Dmytro Mishkin
Anand Mishra
Ashish Mishra
Ishan Misra
Niluthpol C. Mithun
Kaushik Mitra
Niloy Mitra
Anton Mitrokhin
Ikuhisa Mitsugami
Anurag Mittal
Kaichun Mo
Zhipeng Mo
Davide Modolo
Michael Moeller

Pritish Mohapatra
Pavlo Molchanov
Davide Moltisanti
Pascal Monasse
Mathew Monfort
Aron Monszpart
Sean Moran
Vlad I. Morariu
Francesc Moreno-Noguer
Pietro Morerio
Stylianos Moschoglou
Yael Moses
Roozbeh Mottaghi
Pierre Moulon
Arsalan Mousavian
Yadong Mu
Yasuhiro Mukaigawa
Lopamudra Mukherjee
Yusuke Mukuta
Ravi Teja Mullapudi
Mario Enrique Munich
Zachary Murez
Ana C. Murillo
J. Krishna Murthy
Damien Muselet
Armin Mustafa
Siva Karthik Mustikovela
Carlo Dal Mutto
Moin Nabi
Varun K. Nagaraja
Tushar Nagarajan
Arsha Nagrani
Seungjun Nah
Nikhil Naik
Yoshikatsu Nakajima
Yuta Nakashima
Atsushi Nakazawa
Seonghyeon Nam
Vinay P. Namboodiri
Medhini Narasimhan
Srinivasa Narasimhan
Sanath Narayan
Erickson Rangel
    Nascimento
Jacinto Nascimento
Tayyab Naseer

Lakshmanan Nataraj
Neda Nategh
Nelson Isao Nauata
Fernando Navarro
Shah Nawaz
Lukas Neumann
Ram Nevatia
Alejandro Newell
Shawn Newsam
Joe Yue-Hei Ng
Trung Thanh Ngo
Duc Thanh Nguyen
Lam M. Nguyen
Phuc Xuan Nguyen
Thuong Nguyen Canh
Mihalis Nicolaou
Andrei Liviu Nicolicioiu
Xuecheng Nie
Michael Niemeyer
Simon Niklaus
Christophoros Nikou
David Nilsson
Jifeng Ning
Yuval Nirkin
Li Niu
Yuzhen Niu
Zhenxing Niu
Shohei Nobuhara
Nicoletta Noceti
Hyeonwoo Noh
Junhyug Noh
Mehdi Noroozi
Sotiris Nousias
Valsamis Ntouskos
Matthew O'Toole
Peter Ochs
Ferda Ofli
Seong Joon Oh
Seoung Wug Oh
Iason Oikonomidis
Utkarsh Ojha
Takahiro Okabe
Takayuki Okatani
Fumio Okura
Aude Oliva
Kyle Olszewski

Björn Ommer
Mohamed Omran
Elisabeta Oneata
Michael Opitz
Jose Oramas
Tribhuvanesh Orekondy
Shaul Oron
Sergio Orts-Escolano
Ivan Oseledets
Aljosa Osep
Magnus Oskarsson
Anton Osokin
Martin R. Oswald
Wanli Ouyang
Andrew Owens
Mete Ozay
Mustafa Ozuysal
Eduardo Pérez-Pellitero
Gautam Pai
Dipan Kumar Pal
P. H. Pamplona Savarese
Jinshan Pan
Junting Pan
Xingang Pan
Yingwei Pan
Yannis Panagakis
Rameswar Panda
Guan Pang
Jiahao Pang
Jiangmiao Pang
Tianyu Pang
Sharath Pankanti
Nicolas Papadakis
Dim Papadopoulos
George Papandreou
Toufiq Parag
Shaifali Parashar
Sarah Parisot
Eunhyeok Park
Hyun Soo Park
Jaesik Park
Min-Gyu Park
Taesung Park
Alvaro Parra
C. Alejandro Parraga
Despoina Paschalidou

Nikolaos Passalis
Vishal Patel
Viorica Patraucean
Badri Narayana Patro
Danda Pani Paudel
Sujoy Paul
Georgios Pavlakos
Ioannis Pavlidis
Vladimir Pavlovic
Nick Pears
Kim Steenstrup Pedersen
Selen Pehlivan
Shmuel Peleg
Chao Peng
Houwen Peng
Wen-Hsiao Peng
Xi Peng
Xiaojiang Peng
Xingchao Peng
Yuxin Peng
Federico Perazzi
Juan Camilo Perez
Vishwanath Peri
Federico Pernici
Luca Del Pero
Florent Perronnin
Stavros Petridis
Henning Petzka
Patrick Peursum
Michael Pfeiffer
Hanspeter Pfister
Roman Pflugfelder
Minh Tri Pham
Yongri Piao
David Picard
Tomasz Pieciak
A. J. Piergiovanni
Andrea Pilzer
Pedro O. Pinheiro
Silvia Laura Pintea
Lerrel Pinto
Axel Pinz
Robinson Piramuthu
Fiora Pirri
Leonid Pishchulin
Francesco Pittaluga

Daniel Pizarro
Tobias Plötz
Mirco Planamente
Matteo Poggi
Moacir A. Ponti
Parita Pooj
Fatih Porikli
Horst Possegger
Omid Poursaeed
Ameya Prabhu
Viraj Uday Prabhu
Dilip Prasad
Brian L. Price
True Price
Maria Priisalu
Veronique Prinet
Victor Adrian Prisacariu
Jan Prokaj
Sergey Prokudin
Nicolas Pugeault
Xavier Puig
Albert Pumarola
Pulak Purkait
Senthil Purushwalkam
Charles R. Qi
Hang Qi
Haozhi Qi
Lu Qi
Mengshi Qi
Siyuan Qi
Xiaojuan Qi
Yuankai Qi
Shengju Qian
Xuelin Qian
Siyuan Qiao
Yu Qiao
Jie Qin
Qiang Qiu
Weichao Qiu
Zhaofan Qiu
Kha Gia Quach
Yuhui Quan
Yvain Queau
Julian Quiroga
Faisal Qureshi
Mahdi Rad

Filip Radenovic
Petia Radeva
Venkatesh
　B. Radhakrishnan
Ilija Radosavovic
Noha Radwan
Rahul Raguram
Tanzila Rahman
Amit Raj
Ajit Rajwade
Kandan Ramakrishnan
Santhosh
　K. Ramakrishnan
Srikumar Ramalingam
Ravi Ramamoorthi
Vasili Ramanishka
Ramprasaath R. Selvaraju
Francois Rameau
Visvanathan Ramesh
Santu Rana
Rene Ranftl
Anand Rangarajan
Anurag Ranjan
Viresh Ranjan
Yongming Rao
Carolina Raposo
Vivek Rathod
Sathya N. Ravi
Avinash Ravichandran
Tammy Riklin Raviv
Daniel Rebain
Sylvestre-Alvise Rebuffi
N. Dinesh Reddy
Timo Rehfeld
Paolo Remagnino
Konstantinos Rematas
Edoardo Remelli
Dongwei Ren
Haibing Ren
Jian Ren
Jimmy Ren
Mengye Ren
Weihong Ren
Wenqi Ren
Zhile Ren
Zhongzheng Ren

Zhou Ren
Vijay Rengarajan
Md A. Reza
Farzaneh Rezaeianaran
Hamed R. Tavakoli
Nicholas Rhinehart
Helge Rhodin
Elisa Ricci
Alexander Richard
Eitan Richardson
Elad Richardson
Christian Richardt
Stephan Richter
Gernot Riegler
Daniel Ritchie
Tobias Ritschel
Samuel Rivera
Yong Man Ro
Richard Roberts
Joseph Robinson
Ignacio Rocco
Mrigank Rochan
Emanuele Rodolà
Mikel D. Rodriguez
Giorgio Roffo
Grégory Rogez
Gemma Roig
Javier Romero
Xuejian Rong
Yu Rong
Amir Rosenfeld
Bodo Rosenhahn
Guy Rosman
Arun Ross
Paolo Rota
Peter M. Roth
Anastasios Roussos
Anirban Roy
Sebastien Roy
Aruni RoyChowdhury
Artem Rozantsev
Ognjen Rudovic
Daniel Rueckert
Adria Ruiz
Javier Ruiz-del-solar
Christian Rupprecht

Chris Russell
Dan Ruta
Jongbin Ryu
Ömer Sümer
Alexandre Sablayrolles
Faraz Saeedan
Ryusuke Sagawa
Christos Sagonas
Tonmoy Saikia
Hideo Saito
Kuniaki Saito
Shunsuke Saito
Shunta Saito
Ken Sakurada
Joaquin Salas
Fatemeh Sadat Saleh
Mahdi Saleh
Pouya Samangouei
Leo Sampaio
　Ferraz Ribeiro
Artsiom Olegovich
　Sanakoyeu
Enrique Sanchez
Patsorn Sangkloy
Anush Sankaran
Aswin Sankaranarayanan
Swami Sankaranarayanan
Rodrigo Santa Cruz
Amartya Sanyal
Archana Sapkota
Nikolaos Sarafianos
Jun Sato
Shin'ichi Satoh
Hosnieh Sattar
Arman Savran
Manolis Savva
Alexander Sax
Hanno Scharr
Simone Schaub-Meyer
Konrad Schindler
Dmitrij Schlesinger
Uwe Schmidt
Dirk Schnieders
Björn Schuller
Samuel Schulter
Idan Schwartz

William Robson Schwartz
Alex Schwing
Sinisa Segvic
Lorenzo Seidenari
Pradeep Sen
Ozan Sener
Soumyadip Sengupta
Arda Senocak
Mojtaba Seyedhosseini
Shishir Shah
Shital Shah
Sohil Atul Shah
Tamar Rott Shaham
Huasong Shan
Qi Shan
Shiguang Shan
Jing Shao
Roman Shapovalov
Gaurav Sharma
Vivek Sharma
Viktoriia Sharmanska
Dongyu She
Sumit Shekhar
Evan Shelhamer
Chengyao Shen
Chunhua Shen
Falong Shen
Jie Shen
Li Shen
Liyue Shen
Shuhan Shen
Tianwei Shen
Wei Shen
William B. Shen
Yantao Shen
Ying Shen
Yiru Shen
Yujun Shen
Yuming Shen
Zhiqiang Shen
Ziyi Shen
Lu Sheng
Yu Sheng
Rakshith Shetty
Baoguang Shi
Guangming Shi

Hailin Shi
Miaojing Shi
Yemin Shi
Zhenmei Shi
Zhiyuan Shi
Kevin Jonathan Shih
Shiliang Shiliang
Hyunjung Shim
Atsushi Shimada
Nobutaka Shimada
Daeyun Shin
Young Min Shin
Koichi Shinoda
Konstantin Shmelkov
Michael Zheng Shou
Abhinav Shrivastava
Tianmin Shu
Zhixin Shu
Hong-Han Shuai
Pushkar Shukla
Christian Siagian
Mennatullah M. Siam
Kaleem Siddiqi
Karan Sikka
Jae-Young Sim
Christian Simon
Martin Simonovsky
Dheeraj Singaraju
Bharat Singh
Gurkirt Singh
Krishna Kumar Singh
Maneesh Kumar Singh
Richa Singh
Saurabh Singh
Suriya Singh
Vikas Singh
Sudipta N. Sinha
Vincent Sitzmann
Josef Sivic
Gregory Slabaugh
Miroslava Slavcheva
Ron Slossberg
Brandon Smith
Kevin Smith
Vladimir Smutny
Noah Snavely

Roger
  D. Soberanis-Mukul
Kihyuk Sohn
Francesco Solera
Eric Sommerlade
Sanghyun Son
Byung Cheol Song
Chunfeng Song
Dongjin Song
Jiaming Song
Jie Song
Jifei Song
Jingkuan Song
Mingli Song
Shiyu Song
Shuran Song
Xiao Song
Yafei Song
Yale Song
Yang Song
Yi-Zhe Song
Yibing Song
Humberto Sossa
Cesar de Souza
Adrian Spurr
Srinath Sridhar
Suraj Srinivas
Pratul P. Srinivasan
Anuj Srivastava
Tania Stathaki
Christopher Stauffer
Simon Stent
Rainer Stiefelhagen
Pierre Stock
Julian Straub
Jonathan C. Stroud
Joerg Stueckler
Jan Stuehmer
David Stutz
Chi Su
Hang Su
Jong-Chyi Su
Shuochen Su
Yu-Chuan Su
Ramanathan Subramanian
Yusuke Sugano

Masanori Suganuma
Yumin Suh
Mohammed Suhail
Yao Sui
Heung-Il Suk
Josephine Sullivan
Baochen Sun
Chen Sun
Chong Sun
Deqing Sun
Jin Sun
Liang Sun
Lin Sun
Qianru Sun
Shao-Hua Sun
Shuyang Sun
Weiwei Sun
Wenxiu Sun
Xiaoshuai Sun
Xiaoxiao Sun
Xingyuan Sun
Yifan Sun
Zhun Sun
Sabine Susstrunk
David Suter
Supasorn Suwajanakorn
Tomas Svoboda
Eran Swears
Paul Swoboda
Attila Szabo
Richard Szeliski
Duy-Nguyen Ta
Andrea Tagliasacchi
Yuichi Taguchi
Ying Tai
Keita Takahashi
Kouske Takahashi
Jun Takamatsu
Hugues Talbot
Toru Tamaki
Chaowei Tan
Fuwen Tan
Mingkui Tan
Mingxing Tan
Qingyang Tan
Robby T. Tan

Xiaoyang Tan
Kenichiro Tanaka
Masayuki Tanaka
Chang Tang
Chengzhou Tang
Danhang Tang
Ming Tang
Peng Tang
Qingming Tang
Wei Tang
Xu Tang
Yansong Tang
Youbao Tang
Yuxing Tang
Zhiqiang Tang
Tatsunori Taniai
Junli Tao
Xin Tao
Makarand Tapaswi
Jean-Philippe Tarel
Lyne Tchapmi
Zachary Teed
Bugra Tekin
Damien Teney
Ayush Tewari
Christian Theobalt
Christopher Thomas
Diego Thomas
Jim Thomas
Rajat Mani Thomas
Xinmei Tian
Yapeng Tian
Yingli Tian
Yonglong Tian
Zhi Tian
Zhuotao Tian
Kinh Tieu
Joseph Tighe
Massimo Tistarelli
Matthew Toews
Carl Toft
Pavel Tokmakov
Federico Tombari
Chetan Tonde
Yan Tong
Alessio Tonioni

Andrea Torsello
Fabio Tosi
Du Tran
Luan Tran
Ngoc-Trung Tran
Quan Hung Tran
Truyen Tran
Rudolph Triebel
Martin Trimmel
Shashank Tripathi
Subarna Tripathi
Leonardo Trujillo
Eduard Trulls
Tomasz Trzcinski
Sam Tsai
Yi-Hsuan Tsai
Hung-Yu Tseng
Stavros Tsogkas
Aggeliki Tsoli
Devis Tuia
Shubham Tulsiani
Sergey Tulyakov
Frederick Tung
Tony Tung
Daniyar Turmukhambetov
Ambrish Tyagi
Radim Tylecek
Christos Tzelepis
Georgios Tzimiropoulos
Dimitrios Tzionas
Seiichi Uchida
Norimichi Ukita
Dmitry Ulyanov
Martin Urschler
Yoshitaka Ushiku
Ben Usman
Alexander Vakhitov
Julien P. C. Valentin
Jack Valmadre
Ernest Valveny
Joost van de Weijer
Jan van Gemert
Koen Van Leemput
Gul Varol
Sebastiano Vascon
M. Alex O. Vasilescu

Subeesh Vasu
Mayank Vatsa
David Vazquez
Javier Vazquez-Corral
Ashok Veeraraghavan
Erik Velasco-Salido
Raviteja Vemulapalli
Jonathan Ventura
Manisha Verma
Roberto Vezzani
Ruben Villegas
Minh Vo
MinhDuc Vo
Nam Vo
Michele Volpi
Riccardo Volpi
Carl Vondrick
Konstantinos Vougioukas
Tuan-Hung Vu
Sven Wachsmuth
Neal Wadhwa
Catherine Wah
Jacob C. Walker
Thomas S. A. Wallis
Chengde Wan
Jun Wan
Liang Wan
Renjie Wan
Baoyuan Wang
Boyu Wang
Cheng Wang
Chu Wang
Chuan Wang
Chunyu Wang
Dequan Wang
Di Wang
Dilin Wang
Dong Wang
Fang Wang
Guanzhi Wang
Guoyin Wang
Hanzi Wang
Hao Wang
He Wang
Heng Wang
Hongcheng Wang

Hongxing Wang
Hua Wang
Jian Wang
Jingbo Wang
Jinglu Wang
Jingya Wang
Jinjun Wang
Jinqiao Wang
Jue Wang
Ke Wang
Keze Wang
Le Wang
Lei Wang
Lezi Wang
Li Wang
Liang Wang
Lijun Wang
Limin Wang
Linwei Wang
Lizhi Wang
Mengjiao Wang
Mingzhe Wang
Minsi Wang
Naiyan Wang
Nannan Wang
Ning Wang
Oliver Wang
Pei Wang
Peng Wang
Pichao Wang
Qi Wang
Qian Wang
Qiaosong Wang
Qifei Wang
Qilong Wang
Qing Wang
Qingzhong Wang
Quan Wang
Rui Wang
Ruiping Wang
Ruixing Wang
Shangfei Wang
Shenlong Wang
Shiyao Wang
Shuhui Wang
Song Wang

Tao Wang
Tianlu Wang
Tiantian Wang
Ting-chun Wang
Tingwu Wang
Wei Wang
Weiyue Wang
Wenguan Wang
Wenlin Wang
Wenqi Wang
Xiang Wang
Xiaobo Wang
Xiaofang Wang
Xiaoling Wang
Xiaolong Wang
Xiaosong Wang
Xiaoyu Wang
Xin Eric Wang
Xinchao Wang
Xinggang Wang
Xintao Wang
Yali Wang
Yan Wang
Yang Wang
Yangang Wang
Yaxing Wang
Yi Wang
Yida Wang
Yilin Wang
Yiming Wang
Yisen Wang
Yongtao Wang
Yu-Xiong Wang
Yue Wang
Yujiang Wang
Yunbo Wang
Yunhe Wang
Zengmao Wang
Zhangyang Wang
Zhaowen Wang
Zhe Wang
Zhecan Wang
Zheng Wang
Zhixiang Wang
Zilei Wang
Jianqiao Wangni

Anne S. Wannenwetsch
Jan Dirk Wegner
Scott Wehrwein
Donglai Wei
Kaixuan Wei
Longhui Wei
Pengxu Wei
Ping Wei
Qi Wei
Shih-En Wei
Xing Wei
Yunchao Wei
Zijun Wei
Jerod Weinman
Michael Weinmann
Philippe Weinzaepfel
Yair Weiss
Bihan Wen
Longyin Wen
Wei Wen
Junwu Weng
Tsui-Wei Weng
Xinshuo Weng
Eric Wengrowski
Tomas Werner
Gordon Wetzstein
Tobias Weyand
Patrick Wieschollek
Maggie Wigness
Erik Wijmans
Richard Wildes
Olivia Wiles
Chris Williams
Williem Williem
Kyle Wilson
Calden Wloka
Nicolai Wojke
Christian Wolf
Yongkang Wong
Sanghyun Woo
Scott Workman
Baoyuan Wu
Bichen Wu
Chao-Yuan Wu
Huikai Wu
Jiajun Wu

Jialin Wu
Jiaxiang Wu
Jiqing Wu
Jonathan Wu
Lifang Wu
Qi Wu
Qiang Wu
Ruizheng Wu
Shangzhe Wu
Shun-Cheng Wu
Tianfu Wu
Wayne Wu
Wenxuan Wu
Xiao Wu
Xiaohe Wu
Xinxiao Wu
Yang Wu
Yi Wu
Yiming Wu
Ying Nian Wu
Yue Wu
Zheng Wu
Zhenyu Wu
Zhirong Wu
Zuxuan Wu
Stefanie Wuhrer
Jonas Wulff
Changqun Xia
Fangting Xia
Fei Xia
Gui-Song Xia
Lu Xia
Xide Xia
Yin Xia
Yingce Xia
Yongqin Xian
Lei Xiang
Shiming Xiang
Bin Xiao
Fanyi Xiao
Guobao Xiao
Huaxin Xiao
Taihong Xiao
Tete Xiao
Tong Xiao
Wang Xiao

Yang Xiao
Cihang Xie
Guosen Xie
Jianwen Xie
Lingxi Xie
Sirui Xie
Weidi Xie
Wenxuan Xie
Xiaohua Xie
Fuyong Xing
Jun Xing
Junliang Xing
Bo Xiong
Peixi Xiong
Yu Xiong
Yuanjun Xiong
Zhiwei Xiong
Chang Xu
Chenliang Xu
Dan Xu
Danfei Xu
Hang Xu
Hongteng Xu
Huijuan Xu
Jingwei Xu
Jun Xu
Kai Xu
Mengmeng Xu
Mingze Xu
Qianqian Xu
Ran Xu
Weijian Xu
Xiangyu Xu
Xiaogang Xu
Xing Xu
Xun Xu
Yanyu Xu
Yichao Xu
Yong Xu
Yongchao Xu
Yuanlu Xu
Zenglin Xu
Zheng Xu
Chuhui Xue
Jia Xue
Nan Xue

Tianfan Xue
Xiangyang Xue
Abhay Yadav
Yasushi Yagi
I. Zeki Yalniz
Kota Yamaguchi
Toshihiko Yamasaki
Takayoshi Yamashita
Junchi Yan
Ke Yan
Qingan Yan
Sijie Yan
Xinchen Yan
Yan Yan
Yichao Yan
Zhicheng Yan
Keiji Yanai
Bin Yang
Ceyuan Yang
Dawei Yang
Dong Yang
Fan Yang
Guandao Yang
Guorun Yang
Haichuan Yang
Hao Yang
Jianwei Yang
Jiaolong Yang
Jie Yang
Jing Yang
Kaiyu Yang
Linjie Yang
Meng Yang
Michael Ying Yang
Nan Yang
Shuai Yang
Shuo Yang
Tianyu Yang
Tien-Ju Yang
Tsun-Yi Yang
Wei Yang
Wenhan Yang
Xiao Yang
Xiaodong Yang
Xin Yang
Yan Yang

Yanchao Yang
Yee Hong Yang
Yezhou Yang
Zhenheng Yang
Anbang Yao
Angela Yao
Cong Yao
Jian Yao
Li Yao
Ting Yao
Yao Yao
Zhewei Yao
Chengxi Ye
Jianbo Ye
Keren Ye
Linwei Ye
Mang Ye
Mao Ye
Qi Ye
Qixiang Ye
Mei-Chen Yeh
Raymond Yeh
Yu-Ying Yeh
Sai-Kit Yeung
Serena Yeung
Kwang Moo Yi
Li Yi
Renjiao Yi
Alper Yilmaz
Junho Yim
Lijun Yin
Weidong Yin
Xi Yin
Zhichao Yin
Tatsuya Yokota
Ryo Yonetani
Donggeun Yoo
Jae Shin Yoon
Ju Hong Yoon
Sung-eui Yoon
Laurent Younes
Changqian Yu
Fisher Yu
Gang Yu
Jiahui Yu
Kaicheng Yu

Ke Yu
Lequan Yu
Ning Yu
Qian Yu
Ronald Yu
Ruichi Yu
Shoou-I Yu
Tao Yu
Tianshu Yu
Xiang Yu
Xin Yu
Xiyu Yu
Youngjae Yu
Yu Yu
Zhiding Yu
Chunfeng Yuan
Ganzhao Yuan
Jinwei Yuan
Lu Yuan
Quan Yuan
Shanxin Yuan
Tongtong Yuan
Wenjia Yuan
Ye Yuan
Yuan Yuan
Yuhui Yuan
Huanjing Yue
Xiangyu Yue
Ersin Yumer
Sergey Zagoruyko
Egor Zakharov
Amir Zamir
Andrei Zanfir
Mihai Zanfir
Pablo Zegers
Bernhard Zeisl
John S. Zelek
Niclas Zeller
Huayi Zeng
Jiabei Zeng
Wenjun Zeng
Yu Zeng
Xiaohua Zhai
Fangneng Zhan
Huangying Zhan
Kun Zhan

Xiaohang Zhan
Baochang Zhang
Bowen Zhang
Cecilia Zhang
Changqing Zhang
Chao Zhang
Chengquan Zhang
Chi Zhang
Chongyang Zhang
Dingwen Zhang
Dong Zhang
Feihu Zhang
Hang Zhang
Hanwang Zhang
Hao Zhang
He Zhang
Hongguang Zhang
Hua Zhang
Ji Zhang
Jianguo Zhang
Jianming Zhang
Jiawei Zhang
Jie Zhang
Jing Zhang
Juyong Zhang
Kai Zhang
Kaipeng Zhang
Ke Zhang
Le Zhang
Lei Zhang
Li Zhang
Lihe Zhang
Linguang Zhang
Lu Zhang
Mi Zhang
Mingda Zhang
Peng Zhang
Pingping Zhang
Qian Zhang
Qilin Zhang
Quanshi Zhang
Richard Zhang
Rui Zhang
Runze Zhang
Shengping Zhang
Shifeng Zhang

Shuai Zhang
Songyang Zhang
Tao Zhang
Ting Zhang
Tong Zhang
Wayne Zhang
Wei Zhang
Weizhong Zhang
Wenwei Zhang
Xiangyu Zhang
Xiaolin Zhang
Xiaopeng Zhang
Xiaoqin Zhang
Xiuming Zhang
Ya Zhang
Yang Zhang
Yimin Zhang
Yinda Zhang
Ying Zhang
Yongfei Zhang
Yu Zhang
Yulun Zhang
Yunhua Zhang
Yuting Zhang
Zhanpeng Zhang
Zhao Zhang
Zhaoxiang Zhang
Zhen Zhang
Zheng Zhang
Zhifei Zhang
Zhijin Zhang
Zhishuai Zhang
Ziming Zhang
Bo Zhao
Chen Zhao
Fang Zhao
Haiyu Zhao
Han Zhao
Hang Zhao
Hengshuang Zhao
Jian Zhao
Kai Zhao
Liang Zhao
Long Zhao
Qian Zhao
Qibin Zhao

Qijun Zhao
Rui Zhao
Shenglin Zhao
Sicheng Zhao
Tianyi Zhao
Wenda Zhao
Xiangyun Zhao
Xin Zhao
Yang Zhao
Yue Zhao
Zhichen Zhao
Zijing Zhao
Xiantong Zhen
Chuanxia Zheng
Feng Zheng
Haiyong Zheng
Jia Zheng
Kang Zheng
Shuai Kyle Zheng
Wei-Shi Zheng
Yinqiang Zheng
Zerong Zheng
Zhedong Zheng
Zilong Zheng
Bineng Zhong
Fangwei Zhong
Guangyu Zhong
Yiran Zhong
Yujie Zhong
Zhun Zhong
Chunluan Zhou
Huiyu Zhou
Jiahuan Zhou
Jun Zhou
Lei Zhou
Luowei Zhou
Luping Zhou
Mo Zhou
Ning Zhou
Pan Zhou
Peng Zhou
Qianyi Zhou
S. Kevin Zhou
Sanping Zhou
Wengang Zhou
Xingyi Zhou

Yanzhao Zhou
Yi Zhou
Yin Zhou
Yipin Zhou
Yuyin Zhou
Zihan Zhou
Alex Zihao Zhu
Chenchen Zhu
Feng Zhu
Guangming Zhu
Ji Zhu
Jun-Yan Zhu
Lei Zhu
Linchao Zhu
Rui Zhu
Shizhan Zhu
Tyler Lixuan Zhu

Wei Zhu
Xiangyu Zhu
Xinge Zhu
Xizhou Zhu
Yanjun Zhu
Yi Zhu
Yixin Zhu
Yizhe Zhu
Yousong Zhu
Zhe Zhu
Zhen Zhu
Zheng Zhu
Zhenyao Zhu
Zhihui Zhu
Zhuotun Zhu
Bingbing Zhuang
Wei Zhuo

Christian Zimmermann
Karel Zimmermann
Larry Zitnick
Mohammadreza
   Zolfaghari
Maria Zontak
Daniel Zoran
Changqing Zou
Chuhang Zou
Danping Zou
Qi Zou
Yang Zou
Yuliang Zou
Georgios Zoumpourlis
Wangmeng Zuo
Xinxin Zuo

Additional Reviewers

Victoria Fernandez
   Abrevaya
Maya Aghaei
Allam Allam
Christine
   Allen-Blanchette
Nicolas Aziere
Assia Benbihi
Neha Bhargava
Bharat Lal Bhatnagar
Joanna Bitton
Judy Borowski
Amine Bourki
Romain Brégier
Tali Brayer
Sebastian Bujwid
Andrea Burns
Yun-Hao Cao
Yuning Chai
Xiaojun Chang
Bo Chen
Shuo Chen
Zhixiang Chen
Junsuk Choe
Hung-Kuo Chu

Jonathan P. Crall
Kenan Dai
Lucas Deecke
Karan Desai
Prithviraj Dhar
Jing Dong
Wei Dong
Turan Kaan Elgin
Francis Engelmann
Erik Englesson
Fartash Faghri
Zicong Fan
Yang Fu
Risheek Garrepalli
Yifan Ge
Marco Godi
Helmut Grabner
Shuxuan Guo
Jianfeng He
Zhezhi He
Samitha Herath
Chih-Hui Ho
Yicong Hong
Vincent Tao Hu
Julio Hurtado

Jaedong Hwang
Andrey Ignatov
Muhammad
   Abdullah Jamal
Saumya Jetley
Meiguang Jin
Jeff Johnson
Minsoo Kang
Saeed Khorram
Mohammad Rami Koujan
Nilesh Kulkarni
Sudhakar Kumawat
Abdelhak Lemkhenter
Alexander Levine
Jiachen Li
Jing Li
Jun Li
Yi Li
Liang Liao
Ruochen Liao
Tzu-Heng Lin
Phillip Lippe
Bao-di Liu
Bo Liu
Fangchen Liu

Hanxiao Liu
Hongyu Liu
Huidong Liu
Miao Liu
Xinxin Liu
Yongfei Liu
Yu-Lun Liu
Amir Livne
Tiange Luo
Wei Ma
Xiaoxuan Ma
Ioannis Marras
Georg Martius
Effrosyni Mavroudi
Tim Meinhardt
Givi Meishvili
Meng Meng
Zihang Meng
Zhongqi Miao
Gyeongsik Moon
Khoi Nguyen
Yung-Kyun Noh
Antonio Norelli
Jaeyoo Park
Alexander Pashevich
Mandela Patrick
Mary Phuong
Bingqiao Qian
Yu Qiao
Zhen Qiao
Sai Saketh Rambhatla
Aniket Roy
Amelie Royer
Parikshit Vishwas
    Sakurikar
Mark Sandler
Mert Bülent Sarıyıldız
Tanner Schmidt
Anshul B. Shah

Ketul Shah
Rajvi Shah
Hengcan Shi
Xiangxi Shi
Yujiao Shi
William A. P. Smith
Guoxian Song
Robin Strudel
Abby Stylianou
Xinwei Sun
Reuben Tan
Qingyi Tao
Kedar S. Tatwawadi
Anh Tuan Tran
Son Dinh Tran
Eleni Triantafillou
Aristeidis Tsitiridis
Md Zasim Uddin
Andrea Vedaldi
Evangelos Ververas
Vidit Vidit
Paul Voigtlaender
Bo Wan
Huanyu Wang
Huiyu Wang
Junqiu Wang
Pengxiao Wang
Tai Wang
Xinyao Wang
Tomoki Watanabe
Mark Weber
Xi Wei
Botong Wu
James Wu
Jiamin Wu
Rujie Wu
Yu Wu
Rongchang Xie
Wei Xiong

Yunyang Xiong
An Xu
Chi Xu
Yinghao Xu
Fei Xue
Tingyun Yan
Zike Yan
Chao Yang
Heran Yang
Ren Yang
Wenfei Yang
Xu Yang
Rajeev Yasarla
Shaokai Ye
Yufei Ye
Kun Yi
Haichao Yu
Hanchao Yu
Ruixuan Yu
Liangzhe Yuan
Chen-Lin Zhang
Fandong Zhang
Tianyi Zhang
Yang Zhang
Yiyi Zhang
Yongshun Zhang
Yu Zhang
Zhiwei Zhang
Jiaojiao Zhao
Yipu Zhao
Xingjian Zhen
Haizhong Zheng
Tiancheng Zhi
Chengju Zhou
Hao Zhou
Hao Zhu
Alexander Zimin

Contents – Part XII

Generative Low-Bitwidth Data Free Quantization

Shoukai Xu[1,2], Haokun Li[1], Bohan Zhuang[3], Jing Liu[1], Jiezhang Cao[1], Chuangrun Liang[1], and Mingkui Tan[1(✉)]

[1] South China University of Technology, Guangzhou, China
{sexsk,selihaokun,seliujing,secaojiezhang,selcr}@mail.scut.edu.cn,
mingkuitan@scut.edu.cn
[2] PengCheng Laboratory, Shenzhen, China
[3] Monash University, Melbourne, Australia
bohan.zhuang@monash.edu

Abstract. Neural network quantization is an effective way to compress deep models and improve their execution latency and energy efficiency, so that they can be deployed on mobile or embedded devices. Existing quantization methods require original data for calibration or fine-tuning to get better performance. However, in many real-world scenarios, the data may not be available due to confidential or private issues, thereby making existing quantization methods not applicable. Moreover, due to the absence of original data, the recently developed generative adversarial networks (GANs) cannot be applied to generate data. Although the full-precision model may contain rich data information, such information alone is hard to exploit for recovering the original data or generating new meaningful data. In this paper, we investigate a simple-yet-effective method called Generative Low-bitwidth Data Free Quantization (GDFQ) to remove the data dependence burden. Specifically, we propose a knowledge matching generator to produce meaningful fake data by exploiting classification boundary knowledge and distribution information in the pre-trained model. With the help of generated data, we can quantize a model by learning knowledge from the pre-trained model. Extensive experiments on three data sets demonstrate the effectiveness of our method. More critically, our method achieves much higher accuracy on 4-bit quantization than the existing data free quantization method. Code is available at https://github.com/xushoukai/GDFQ.

Keywords: Data free compression · Low-bitwidth quantization · Knowledge matching generator

---

S. Xu, H. Li and B. Zhuang—Authors Contributed Equally.

Electronic supplementary material The online version of this chapter (https://doi.org/10.1007/978-3-030-58610-2_1) contains supplementary material, which is available to authorized users.

A. Vedaldi et al. (Eds.): ECCV 2020, LNCS 12357, pp. 1–17, 2020.
https://doi.org/10.1007/978-3-030-58610-2_1

1   Introduction

Deep neural networks (DNNs) have achieved great success in many areas, such as computer vision [15,16,37,40,43] and natural language processing [12,28,36]. However, DNNs often contain a considerable number of parameters, which makes them hard to be deployed on embedded/edge devices due to unbearable computation costs for inference. Network quantization, which aims to reduce the model size by quantizing floating-point values into low precision(e.g., 4-bit), is an effective way to improve the execution latency and energy efficiency. Existing quantization methods [10,24,48,49,51] generally require training data for calibration or fine-tuning. Nevertheless, in many real-world applications in medical [46], finance [47] and industrial domains, the training data may not be available due to privacy or confidentiality issues. Consequently, these methods are no longer applicable due to the absence of training data. Therefore, the data-free quantization is of great practical value.

To address this issue, one possible way is to directly sample random inputs from some distribution and then use these inputs to quantize the model so that the output distributions of the full-precision model and quantized model are as close as possible. Unfortunately, since random inputs contain no semantic information and are far away from the real data distribution, this simple method suffers from huge performance degradation. Alternatively, one can use generative adversarial networks (GANs) to produce data. Due to the absence of training data, GANs cannot be applied in the data-free quantization. In practice, the quantization performance highly depends on the quality of the input data. Therefore, the process of constructing meaningful data to quantize models is very challenging.

To generate meaningful data, it is important and necessary to exploit data information from a pre-trained model. A recent study [44] revealed that a well-trained over-parameterized model maintains sufficient information about the entire data set. Unfortunately, what information exists and how to exploit such information are still unknown. In the training, a neural network uses batch

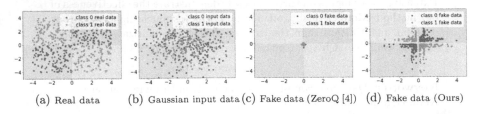

(a) Real data   (b) Gaussian input data (c) Fake data (ZeroQ [4])   (d) Fake data (Ours)

Fig. 1. The comparisons of generated fake data. ZeroQ generates fake data by gradient updating; however, since it neglects the inter-class information, it is hard to capture the original data distribution. Meanwhile, the knowledge matching generator produces fake data with label distribution, and it is more likely to produce data distributed near the classifier boundaries.

normalization [19] to stabilize the data distribution and learns a classification boundary to divide data into different classes (See Fig. 1(a)). In this sense, some information about the classification boundary and data distribution is hidden in the pre-trained model. However, these kinds of information are disregarded by existing data-free quantization methods [4,30], which only focus on a single sample or network parameters. For example, ZeroQ [4] exploits the information of a single sample instead of entire data, causing the constructed distribution to be far away from the real data distribution (See Fig. 1(c)). To address these issues, how to construct meaningful data by fully exploiting the classification boundary and distribution information from a pre-trained model remains an open question.

In this paper, we propose a simple-yet-effective method called Generative Low-bitwidth Data Free Quantization to achieve completely data-free quantization. Without original data, we aim to learn a good generator to produce meaningful data by mining the classification boundary knowledge and distribution information in batch normalization statistics (BNS) from the pre-trained full-precision model.

The main contributions of this paper are summarized as follows:

- We propose a scheme called Generative Low-bitwidth Data Free Quantization (GDFQ), which performs 4-bitwidth quantization without any real data. To our knowledge, this is the first low-bitwidth data free quantization method.
- We propose an effective knowledge matching generator to construct data by mining knowledge from the pre-trained full-precision model. The generated data retain classification boundary knowledge and data distribution.
- Extensive experiments on image classification data sets demonstrate the superior performance of our method compared to the existing data-free quantization method.

2   Related Work

Model Quantization. Model quantization targets to quantize weights, activations and even gradients to low-precision, to yield highly compact models, where the expensive multiplication operations can be replaced by additions or bit-wise operations. According to the trade-off between accuracy and computational complexity, quantization can be roughly categorized into binary neural networks (BNNs) [18,35,52] and fixed-point quantization [13,49,51]. Moreover, quantization studies mainly focus on tackling two core bottlenecks, including designing accurate quantizers to fit the categorical distribution to the original continuous distribution [5,21,45], and approximating gradients of the non-differential quantizer during back-propagation [26,41,50]. In addition, the first practical 4-bit post-training quantization approach was introduced in [3]. To improve the performance of neural network quantization without retraining, outlier channel splitting (OCS) [48] proposed to move affected outliers toward the center of the distribution. Besides, many mainstream deep learning frameworks support 16-bit half-precision or 8-bit fixed-point quantization, such as TensorFlow [1], PyTorch [34], PaddleSlim, etc. In particular, these platforms provide both post-training quantization and quantization-aware training. However, all of them require the original data. In short, whether in scientific studies or practical applications, if the original data is unavailable, it is hard for quantization methods to work normally.

Data-Free Model Compression. Recently, the researches about data-free had included more model compression methods, such as knowledge distillation [8,31], low-rank approximation [42] and model quantization [48]. A novel framework [8] exploited generative adversarial networks to perform data-free knowledge distillation. Another work [25] reconstructed a new data set based solely on the trained model and the activation statistics, and finally distilled the pre-trained "teacher" model into the smaller "student" network. Zero-shot knowledge distillation [31] synthesized pseudo data by utilizing the prior information about the underlying data distribution. Specifically, the method extracted class similarities from the parameters of the teacher model and modeled the output space of the teacher network as a Dirichlet distribution. KEGNET [42] proposed a novel data-free low-rank approximation approach assisted by knowledge distillation. This method contained a generator that generated artificial data points to replace missing data and a decoder that aimed to extract the low-dimensional representation of artificial data. In Adversarial Belief Matching [27], a generator generated data on which the student mismatched the teacher, and then the student network was trained using these data.

Quantization also faces the situation without original data while previous quantization methods generally need original data to improve their performance. However, in some instances, it is difficult to get the original data. Recently, some studies focused on data-free model quantization. DFQ [30] argued that data-free quantization performance could be improved by equalizing the weight ranges of each channel in the network and correcting biased quantization error. ZeroQ [4], a novel zero-shot quantization framework, enabled mixed-precision quantization without any access to the training or validation data. However, these data-free quantization methods work well for 8-bit, but got poor performance in aggressively low bitwidth regimes such as 4-bit.

3   Problem Definition

Data-Free Quantization Problem. Quantization usually requires original data for calibration or fine-tuning. In many practical scenarios, the original data may not be available due to private or even confidential issues. In this case, we cannot use any data; thus, the general scheme of the network quantization will lose efficacy and even fail to work completely, resulting in a quantized model with inferior performance. Given a full-precision model M, data-free quantization aims to construct fake data $(\hat{\mathbf{x}}, y)$ and meanwhile quantize a model Q from the model M_θ. Specifically, to compensate for the accuracy loss from quantization, training-aware quantization can fine-tune the quantized model by optimizing the following problem:

$$\min_{Q,\hat{\mathbf{x}}} \mathbb{E}_{\hat{\mathbf{x}},y} \left[\ell(Q(\hat{\mathbf{x}}), y) \right], \tag{1}$$

where $\hat{\mathbf{x}}$ is a generated fake sample, y is the corresponding label, and $\ell(\cdot, \cdot)$ is a loss function, such as cross-entropy loss and mean squared error.

Challenges of Constructing Data. Because of the absence of original data, one possible way is to construct data by exploiting information from a pre-trained full-precision model. Although the full-precision model may contain rich data information, such latent information alone is hard to exploit for recovering the original data. In practice, the performance of quantization highly depends on the quality of constructed data. With the limited information of the pre-trained model, constructing meaningful data is very challenging.

Recently, one data-free quantization method (ZeroQ) [4] constructs fake data by using a linear operator with gradient update information. With the help of constructed data, ZeroQ is proposed to improve the quantization performance. However, ZeroQ has insufficient information to improve the performance of quantization with the following two issues. First, it constructs fake data without considering label information, and thus neglects to exploit the classification boundary knowledge from the pre-trained model. Second, it enforces the batch normalization statistics of a single data point instead of the whole data, leading to being far away from the real data distribution. To address these issues, one can construct a generator G to produce fake data by considering label information and using powerful neural networks,

$$\hat{\mathbf{x}} = G(\mathbf{z}|y), \quad \mathbf{z} \sim p(\mathbf{z}), \tag{2}$$

where $\mathbf{z}$ is a random vector drawn from some prior distribution $p(\mathbf{z})$, *e.g.*, Gaussian distribution or uniform distribution. By using the generator, we are able to construct fake data to improve quantization performance. However, what knowledge in the pre-trained model can be exploited and how to learn a good generator remain to be answered.

4   Generative Low-Bitwidth Data Free Quantization

In many practical scenarios, training data are unavailable; thus, an existing method [4] performs data-free quantization by constructing data. However, without exploiting knowledge from a pre-trained model, directly constructing data lacks label information and leads to being far away from the real data distribution. To address this issue, we aim to design a generator to produce fake data. Then, we are able to perform supervised learning to improve the performance of quantization. The overall framework is shown in Fig. 2.

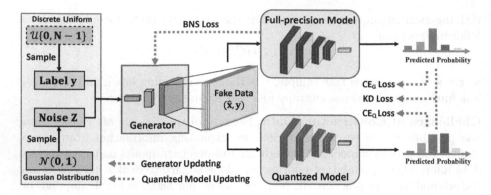

Fig. 2. An overview of the proposed method. Given Gaussian noise and the label as input, the generator creates fake data and feeds them into both the full-precision model and the quantized model. The fixed full-precision model provides knowledge for updating the generator. The quantized model learns latent knowledge from the generator and the full-precision model.

4.1 Knowledge Matching Generator

When training a deep neural network, it captures sufficient data information to make a decision [44]. In this sense, the pre-trained DNN contains some knowledge information of the training data, e.g., classification boundary information and distribution information. We therefore elaborate how to effectively construct informative data from the pre-trained model.

Classification Boundary Information Matching. The pre-trained model contains classification boundary information. Unfortunately, such information is difficult to exploit to recover the data near to the classification boundary. Recently, generative adversarial networks (GANs) [6,7,14] have achieved considerable success in producing data. Since the real data are unavailable, the discriminator in a GAN cannot work in the data-free quantization task. Without the discriminator, learning a generator to produce meaningful data is difficult.

In this paper, we propose a generator to produce fake data for the data-free quantization task. For this task, although the original data cannot be observed, we are able to easily confirm the number of categories of original data by the last layer in a pre-trained model. As shown in Fig. 1(a), different categories of data should be distributed in different data spaces. To generate fake data, we introduce a noise vector $\mathbf{z}$ conditioned on a label y. Here, we sample noise from Gaussian distribution $\mathcal{N}(0,1)$ and uniformly sample a label from $\{0, 1, ..., n-1\}$. Then, the generator maps a prior input noise vector $\mathbf{z}$ and the given label y to the fake data $\hat{\mathbf{x}}$. Formally, we define the knowledge matching generator as follows:

$$\hat{\mathbf{x}} = G(\mathbf{z}|y), \quad \mathbf{z} \sim \mathcal{N}(0,1). \tag{3}$$

To improve the quantization performance, the generator should have the ability to generate data that are effective to fine-tune a quantized model. To this end, given the Gaussian noise $\mathbf{z}$ and the corresponding label y, the generated fake data should be classified to the same label y by the full-precision pre-trained model M. Therefore, we introduce the following cross-entropy loss $\mathrm{CE}(\cdot, \cdot)$ to train the generator G:

$$\mathcal{L}_{\mathrm{CE}}^{G}(G) = \mathbb{E}_{\mathbf{z},y}\left[\mathrm{CE}(M(G(\mathbf{z}|y)), y)\right]. \tag{4}$$

Distribution Information Matching. In addition, the pre-trained model contains the distribution information of training data. Such distribution information can be captured by batch normalization [19], which is used to control the change of the distribution. While training the full-precision model, the batch normalization statistics (BNS), $i.e.$, the mean and variance, are computed dynamically. For every batch, batch normalization (BN) layers only compute statistics on the current mini-batch inputs and accumulate them with momentum. Finally, the exponential moving average (EMA) batch normalization statistics will be obtained and then they will be used in network validation and test.

To retain the BNS information, the mean and variance of the generated distribution should be the same as that of the real data distribution. To this end, we learn a generator G with the BNS information ($\boldsymbol{\mu}_l$ and $\boldsymbol{\sigma}_l$) encoded in the l-th BN layer of the pre-trained model using $\mathcal{L}_{\mathrm{BNS}}$:

$$\mathcal{L}_{\mathrm{BNS}}(G) = \sum_{l=1}^{L} \|\boldsymbol{\mu}_l^r - \boldsymbol{\mu}_l\|_2^2 + \|\boldsymbol{\sigma}_l^r - \boldsymbol{\sigma}_l\|_2^2, \tag{5}$$

where $\boldsymbol{\mu}_l^r$ and $\boldsymbol{\sigma}_l^r$ are the mean and variance of the fake data's distribution at the l-th BN layer, respectively, and $\boldsymbol{\mu}_l$ and $\boldsymbol{\sigma}_l$ are the corresponding mean and variance parameters stored in the l-th BN layer of the pre-trained full-precision model, respectively. In this way, we are able to learn a good generator to keep the distribution information from the training data.

4.2   Fake-Data Driven Low-Bitwidth Quantization

With the help of the generator, the data-free quantization problem can be turned to a supervised quantization problem. Thus, we are able to quantize a model using produced meaningful data. However, transferring knowledge from a pre-trained model to a quantized model is difficult. To address this, we introduce a fake-data driven quantization method and solve the optimization problem by exploiting knowledge from the pre-trained model.

Quantization. Network quantization maps full-precision (32-bit) weights and activations to low-precision ones, $e.g.$, 8-bit fixed-point integer. We use a simple-yet-effective quantization method refer to [20] for both weights and activations. Specifically, given full-precision weights θ and the quantization precision k, we quantize θ to θ_q in the symmetric k-bit range:

$$\theta_q = \begin{cases} -2^{k-1}, & \text{if } \theta' < -2^{k-1} \\ 2^{k-1}-1, & \text{if } \theta' > 2^{k-1}-1 \\ \theta', & \text{otherwise,} \end{cases} \tag{6}$$

where θ' are discrete values mapped by a linear quantization, *i.e.*, $\theta' = \lfloor \Delta \cdot \theta - b \rceil$, $\lfloor \cdot \rceil$ is the round function, and Δ and b can be computed by $\Delta = \frac{2^k - 1}{u - l}$ and $b = l \cdot \Delta + 2^{k-1}$. Here, l and u can be set as the minimum and maximum of the floating-point weights θ, respectively.

Optimization Problem. When the real training data are unavailable, quantization may suffer from some limitations. First, direct quantization from a full-precision model may result in severe performance degradation. To address this issue, we aim to train the quantized model to approximate the full-precision model through the fine-tuning process. To this end, a well fine-tuned quantized model Q and the pre-trained model M should classify the fake data correctly. For this purpose, we use the cross-entropy loss function $\text{CE}(\cdot, \cdot)$ to update Q:

$$\mathcal{L}_{\text{CE}}^Q(Q) = \mathbb{E}_{\hat{\mathbf{x}}, y} \left[\text{CE}(Q(\hat{\mathbf{x}}), y) \right]. \tag{7}$$

By minimizing Eq. (7), the quantization model can be trained with the generated data to perform multi-class classification.

Second, the traditional fine-tuning process with a common classification loss function is insufficient because the data are fake. However, with the help of fake data, we are able to apply knowledge distillation [17] to improve the quantization performance. Specifically, given the same inputs, the outputs of a quantized model and full-precision model should be close enough to guarantee that the quantized model is able to achieve nearly performance compared with the full-precision model. Therefore, we utilize knowledge distillation to recover the performance of the quantized model by using fake data $\hat{\mathbf{x}}$ to simulate the training data. Then, the quantized model can be fine-tuned using the following Kullback-Leibler loss function $\text{KL}(\cdot, \cdot)$:

$$\mathcal{L}_{\text{KD}}(Q) = \mathbb{E}_{\hat{\mathbf{x}}} \left[\text{KL}(Q(\hat{\mathbf{x}}), M(\hat{\mathbf{x}})) \right]. \tag{8}$$

By optimizing the loss in (8), the quantization model can learn knowledge from the full-precision model.

Fine-Tuning with Fixed BNS. To stabilize the training, we fix the batch normalization statistics (BNS) in the quantized model during fine-tuning. In this way, the BNS in the quantized model are corresponding to the statistics of the real data distribution. With the help of fixed BNS, the quantized model always maintains the real data information to improve quantization performance.

4.3   Training Process

We propose a low-bitwidth quantization algorithm to alternately optimize the generator G and the quantized model Q. In our alternating training strategy, the generator is able to generate different data with each update. By increasing the diversity of data, the quantized model Q can be trained to improve the performance. In addition, to make the fine-tuning of Q more stable, we firstly train G solely several times as a warm-up process. The overall process is shown in

---

Algorithm 1. Generative Low-bitwidth Data Free Quantization

Input: Pre-trained full-precision model M.
Output: Generator G, quantized model Q.
   Update G several times as a warm-up process.
   Quantize model M using Eq. (6), get quantized model Q.
   Fix the batch normalization statistics in all BN layers of quantized model Q.
   for $t = 1, \ldots, T_{fine-tune}$ **do**
      Obtain random noise $\mathbf{z} \sim \mathcal{N}(0, 1)$ and label y.
      Generate fake image $\hat{\mathbf{x}}$ using Eq. (3).
      Update generator G by minimizing Loss (9).
      Update quantized model Q by minimizing Loss (10).
   end for

Algorithm 1. In contrast, one can train the generator G with a full-precision model, and then fix the generator to train the quantized model Q until convergence. In this separate training strategy, when the diversity of the generated data is poor, the quantized model has a limitation to improve the quantization performance.

Training Generator G. First, we randomly sample a Gaussian noise vector $\mathbf{z} \sim \mathcal{N}(0, 1)$ with a label y. Then we use G to generate fake data from the distribution and update G. The final generator loss $\mathcal{L}_1(G)$ is formulated as follows:

$$\mathcal{L}_1(G) = \mathcal{L}_{\mathrm{CE}}^G(G) + \beta \mathcal{L}_{\mathrm{BNS}}(G), \tag{9}$$

where β is a trade-off parameter.

Training Quantized Model Q. We quantize the model according to Eq. (6). Then, we replace the BNS in the quantized model with the fixed batch normalization statistics (FBNS) as described in Sect. 4.2. So far, the quantized model has inherited the information contained in BNS and a part of latent knowledge from the parameters of the pre-trained model. In the fine-tuning process, we train the G and Q alternately in every epoch. Based on the warmed up G, we obtain fake samples and use them to optimize the quantized model Q. The final quantized model loss function $\mathcal{L}_2(Q)$ is formulated as follows:

$$\mathcal{L}_2(Q) = \mathcal{L}_{\mathrm{CE}}^Q(Q) + \gamma \mathcal{L}_{\mathrm{KD}}(Q), \tag{10}$$

where γ is a trade-off parameter. We do not stop updating G because if we have a better G, the fake data will be more similar to real training data and the upper limit of optimizing Q will be improved. Note that we keep the pre-trained full-precision model fixed at all times.

5 Experiments

5.1 Data Sets and Implementation Details

We evaluate the proposed method on well-known data sets including CIFAR-10 [23], CIFAR-100 [23], and ImageNet [11]. CIFAR-10 consists of 60k images

from 10 categories, with 6k images per category. There are 50k images for training and 10k images for testing. CIFAR-100 has 100 classes and each class contains 500 training images and 100 testing images. ImageNet is one of the most
challenging and largest benchmark data sets for image classification, which has
around 1.2 million real-world images for training and 50k images for validation.

Based on the full-precision pre-trained models from pytorchcv[1], we quantize ResNet-20 [16] on CIFAR-10/100 and ResNet-18 [16], BN-VGG-16 [38] and
Inception v3 [40] on ImageNet. In all experiments, we quantize all layers including the first and last layers of the network following [4] and the activation clipping
values are per-layer granularity. All implementations are based on PyTorch.

For CIFAR-10/100, we construct the generator following ACGAN [33] and
the dimension of noise is 100. During training, we optimize the generator and
quantized model using Adam [22] and SGD with Nesterov [32] respectively, where
the momentum term and weight decay in Nesterov are set to 0.9 and 1×10^{-4}.
Moreover, the learning rates of quantized models and generators are initialized
to 1×10^{-4} and 1×10^{-3}, respectively. Both of them are decayed by 0.1 for
every 100 epochs. In addition, we train the generator and quantized model for
400 epochs with 200 iterations per epoch. To obtain a more stable clip range
for activation, we calculate the moving average of activation's range in the first
four epochs without updating the quantized models and then fix this range for
subsequent training. For $\mathcal{L}_1(G)$ and $\mathcal{L}_2(Q)$, we set $\beta = 0.1$ and $\gamma = 1$ after
a simple grid search. For ImageNet, we replace the generator's standard batch
normalization layer with the categorical conditional batch normalization layer
for fusing label information following SN-GAN [29] and set the initial learning
rate of the quantized model as 1×10^{-6}. Other training settings are the same as
those on CIFAR-10/100.

5.2   Toy Demonstration for Classification Boundary Matching

To evaluate that the fake data generated by our G are able to match the classification boundary information, we design a toy experiment. The results are shown
in Fig. 1. First, we create a toy binary classification data set by uniform sampling from -4 to $+4$, and the label is shown in Fig. 1(a). Second, we construct a
simple neural network T, which is composed of several linear layers, BN layers,
and ReLU layers, and we train it using the toy data. The classification boundaries are shown in each subfigure. To simulate the process of our method, we
sample noises from Gaussian distribution and every noise has a random label of
0 or 1 (Fig. 1(b)). Then, we generate fake data from noises by learning from the
pre-trained model T. Figure 1(c) and Fig. 1(d) show the fake data generated by
the ZeroQ method and our method, respectively. The data generated by ZeroQ
do not capture the real data distribution since it neglects the inter-class information; while our method is able to produce fake data that not only have label
information but also match the classification boundaries.

---

[1] https://pypi.org/project/pytorchcv/.

Table 1. Comparisons on CIFAR-10/100 and ImageNet. We report the average and standard deviation of our method to show that our method is stable. We quantize both the weights and activations of the models to 4-bits and report the top1 accuracy.

Data Set	Model	Real data		Data free	
		FP32	FT	ZeroQ [4]	Ours
CIFAR-10	ResNet-20	94.03	93.11	79.30	**90.25 ± 0.30**
CIFAR-100	ResNet-20	70.33	68.34	45.20	**63.58 ± 0.23**
ImageNet	BN-VGG16	74.28	68.83	1.15	**67.10 ± 0.29**
	ResNet-18	71.47	67.84	26.04	**60.60 ± 0.15**
	Inception v3	78.80	73.80	26.84	**70.39 ± 0.20**

5.3 Comparison of the Results

To further evaluate the effectiveness of our method, we include the following methods for study. **FP32:** the full-precision pre-trained model. **FT:** we use real training data instead of fake data to fine-tune the quantized model by minimizing $\mathcal{L}_2$. **ZeroQ:** a data-free post-training quantization method. We obtain the result from the publicly released code of ZeroQ [4].

We quantize both weights and activations to 4-bit and report the comparison results in Table 1. For CIFAR-10, our method achieves much higher accuracy than that of ZeroQ [4]. When the number of categories increases in CIFAR-100, our method suffers a much smaller degradation in accuracy compared with that of ZeroQ. The main reason is that, our method gains more prior knowledge from the full-precision model. These results demonstrate the effectiveness of our method on simple data sets with 4-bit quantization. For large scale and categories data set, such as ImageNet, existing data-free quantization methods suffer from severe performance degradation. However, our generated images contain abundant category information and similar distribution with real data. As a result, our method recovers the accuracy of quantized models significantly with the help of generated fake data and knowledge distillation on three typical networks. More experiments on different models and methods can be found in the supplementary material.

5.4 Ablation Studies

In this section, we first evaluate the effectiveness of each component in $\mathcal{L}_1$ and $\mathcal{L}_2$. Second, we explore how fixed BNS affects our method. Then, we compare our method with different quantization methods. Last, we further study the effect of different stopping conditions. All the ablation experiments are conducted on the CIFAR-10/100 data sets.

Effect of Different Losses. To verify the effectiveness of different components in our method, we conduct a series of ablation experiments on CIFAR-100 with

Table 2. Effect of different loss functions of generator G. We quantize both the weights and activations of the models to 4-bits and report the top1 accuracy on CIFAR-100.

Model	CE loss	BNS loss	Acc. (%)
ResNet-20 (4-bit)	×	×	30.70
	✓	×	54.51
	×	✓	44.40
	✓	✓	**63.91**

Table 3. Effect of different loss functions of Q. We keep the weights and activations of the models to 4-bits and report the top1 accuracy on CIFAR-100.

Model	CE loss	KD loss	Acc. (%)
ResNet-20 (4-bit)	✓	×	55.55
	×	✓	62.98
	✓	✓	**63.91**

ResNet-20. Table 2 reports the top-1 accuracy of quantized models with different components of $\mathcal{L}_1$. In this ablation experiment, we fine-tune quantized models with complete $\mathcal{L}_2$. Since we do not use both CE loss and BNS loss, we have no way to optimize G, which means we use the fake data generated from the initial G to fine-tune the quantized model. In this case, the distribution of the fake data is far away from that of original data because the generator receives no guidance from the full-precision model. Therefore, the quantized model suffers from a large performance degradation. To utilize the knowledge in the full-precision model, we use CE loss to optimize G and achieve a better quantized model. In this case, the generator produces fake data that can be classified with high confidence by the full-precision model. Last, we combine CE loss and BNS loss with a coefficient and achieve the best result. The BNS loss encourages the generator to generate fake data that match the statistics encoded in full-precision model's BN layers so that these fake data have a much similar distribution with real data. In summary, both CE loss and BNS loss contribute to better performance of the quantized model.

We further conduct ablation experiments to analyze the effectiveness of each component in $\mathcal{L}_2$. Table 3 reports the top-1 accuracy of quantized models with different components of $\mathcal{L}_2$. In this experiment, we optimize the generator with complete $\mathcal{L}_1$. When only introducing KD loss, the quantized model receives knowledge from the full-precision model's prediction and achieves 62.98% on top-1 accuracy. To use the additional label information, we combine BNS loss with CE loss. The resultant model achieves a 0.93% improvement on top-1 accuracy.

Effect of the Fixed BNS. To verify the effectiveness of fixing batch normalization, we conduct ablation studies with ResNet-20 on CIFAR-10/100. The results are shown in Table 4. When we fix batch normalization statistics during

Table 4. Ablation experiments on the fixed BNS (FBNS). We keep the weights and activations of the models to be 4-bits and report the top1 accuracy on CIFAR-10/100. We use "w/o FBNS" to represent that we use fake data to fine-tune the quantized models without fixed BNS. Similarly, we use "w/ FBNS" to represent the fine-tuning process with fixed BNS.

Data Set	w/o FBNS	w/ FBNS
CIFAR-10	89.21	**90.23**
CIFAR-100	61.12	**63.91**

Table 5. Comparison of different post-training quantization methods. We use real data sets as calibration sets and report the accuracy of the quantized model without fine-tuning.

Data Set	Model	Method	W8A8	W6A6	W5A5	W4A4
CIFAR-10	ResNet-20 (94.03)	MSE [39]	93.86	93.10	91.38	81.59
		ACIQ [2]	93.69	92.22	86.66	61.87
		KL [9]	93.72	92.32	90.71	80.05
		Ours	**93.92**	**93.38**	**92.39**	**85.20**
CIFAR-100	ResNet-20 (70.33)	MSE [39]	70.11	66.87	60.49	27.11
		ACIQ [2]	69.29	63.21	48.21	8.72
		KL [9]	70.15	67.65	57.55	15.83
		Ours	**70.29**	**68.63**	**64.03**	**43.12**
ImageNet	ResNet-18 (71.47)	MSE [39]	71.01	66.96	54.23	15.08
		ACIQ [2]	68.78	61.15	46.25	7.19
		KL [9]	70.69	61.34	56.13	16.27
		Ours	**71.43**	**70.43**	**64.68**	**33.25**

fine-tuning, we narrow the statistics gap between the quantized model and the full-precision model. As a result, we achieve a much higher top-1 accuracy than that with standard batch normalization.

5.5 Further Experiments

Comparisons with Different Post-training Quantization Methods. We compare different post-training quantization methods [2,9,39] on CIFAR-10/100 and ImageNet and show the results in Table 5. In this experiment, we use images from real data sets as calibration sets for quantized models with different post-training quantization methods. Specifically, we compare our quantization method with MSE(mean squared error), ACIQ, and KL(Kullback-Leibler), which are popular methods to decide the clip values of weight and activation. Our method shows much better performance than that of the other methods, which means it is more suitable in low-bitwidth quantization. Furthermore, the

Table 6. Comparison of separate training and alternating training of G and Q.

Training Strategy	Acc. (%)
Separate training	61.81
Alternating training	**63.91**

Table 7. Effect of different thresholds in the stopping condition of G.

Threshold η (%)	90.00	95.00	99.00	99.50	w/o stopping condition
Acc. (%)	57.20	57.56	59.09	59.67	**63.91**

experimental results show that as the data set gets larger, the accuracy decreases. With the decrease of precision, all the quantization methods behave more poorly. Specifically, when the precision drops to 4-bit, the accuracy declines sharply.

Effect of Two Training Strategies. We investigate the effect of two kinds of training strategies. 1) Training generator and quantized model in two steps. We first train the generator by minimizing the loss (9) until convergence. Then, we train the quantized model by minimizing the loss (10). 2) Training the generator and quantized model alternately in each iteration following Algorithm 1. From the results in Table 6, alternating training performs significantly better than separate training. Therefore, we use alternating training in other experiments.

Effect of Different Thresholds in Stopping Conditions. In this experiment, we stop the training of the generator if the classification accuracy of the full-precision model on fake data is larger than a threshold η. Table 7 reports the results of different thresholds η in stopping condition. When increasing the threshold, the generator is trained with quantized models for more epochs, and we get a better fine-tuning result. We achieve the best performance when we do not stop optimizing the generator. These results demonstrate that optimizing the generator and quantized model simultaneously increases the diversity of data, which is helpful for fine-tune quantized models.

6   Conclusion

In this paper, we have proposed a Generative Low-bitwidth Data Free Quantization scheme to eliminate the data dependence of quantization methods. First, we have constructed a knowledge matching generator to produce fake data for the fine-tuning process. The generator is able to learn the classification boundary knowledge and distribution information from the pre-trained full-precision model. Next, we have quantized the full-precision model and fine-tuned the quantized model using the fake data. Extensive experiments on various image classification data sets have demonstrated the effectiveness of our data-free method.

Acknowledgements. This work was partially supported by the Key-Area Research and Development Program of Guangdong Province 2018B010107001, Program for Guangdong Introducing Innovative and Entrepreneurial Teams 2017ZT07X183, Fundamental Research Funds for the Central Universities D2191240.

References

1. Abadi, M., et al.: Tensorflow: a system for large-scale machine learning. In: 12th {USENIX} Symposium on Operating Systems Design and Implementation ({OSDI} 16), pp. 265–283 (2016)
2. Banner, R., Nahshan, Y., Hoffer, E., Soudry, D.: Aciq: analytical clipping for integer quantization of neural networks (2018)
3. Banner, R., Nahshan, Y., Soudry, D.: Post training 4-bit quantization of convolutional networks for rapid-deployment. In: Proceedings of Advance Neural Information Processing System (2019)
4. Cai, Y., Yao, Z., Dong, Z., Gholami, A., Mahoney, M.W., Keutzer, K.: Zeroq: a novel zero shot quantization framework. In: Proceedings of IEEE Conference on Computer Vision Pattern Recognition (2020)
5. Cai, Z., He, X., Sun, J., Vasconcelos, N.: Deep learning with low precision by half-wave gaussian quantization. In: Proceedings of IEEE Conference on Computer Vision Pattern Recognition (2017)
6. Cao, J., Guo, Y., Wu, Q., Shen, C., Tan, M.: Adversarial learning with local coordinate coding. In: Proceedings of International Conference Machine Learning (2018)
7. Cao, J., Mo, L., Zhang, Y., Jia, K., Shen, C., Tan, M.: Multi-marginal wasserstein gan. In: Proceedings of Advances in Neural Information Processing Systems (2019)
8. Chen, H., et al.: Data-free learning of student networks. In: Proceedings of the IEEE International Conference on Computer Vision (2019)
9. Chen, T., et al.: Mxnet: a flexible and efficient machine learning library for heterogeneous distributed systems. arXiv preprint arXiv:1512.01274 (2015)
10. Choi, J., Wang, Z., Venkataramani, S., Chuang, P.I.J., Srinivasan, V., Gopalakrishnan, K.: Pact: Parameterized clipping activation for quantized neural networks. arXiv preprint arXiv:1805.06085 (2018)
11. Deng, J., Dong, W., Socher, R., Li, L.J., Li, K., Fei-Fei, L.: Imagenet: a large scale hierarchical image database. In: Proceedings of IEEE Conference on Computer Vision and Pattern Recognition (2009)
12. Devlin, J., Chang, M.W., Lee, K., Toutanova, K.: Bert: pre-training of deep bidirectional transformers for language understanding. arXiv preprint arXiv:1810.04805 (2018)
13. Esser, S.K., McKinstry, J.L., Bablani, D., Appuswamy, R., Modha, D.S.: Learned step size quantization. In: Proceedings of International Conference on Learning Representations (2020)
14. Goodfellow, I., et al.: Generative adversarial nets. In: Proceedings of Advances in Neural Information Processing Systems (2014)
15. Guo, Y., et al.: Nat: neural architecture transformer for accurate and compact architectures. In: Proceedings of Advances in Neural Information Processing Systems (2019)
16. He, K., Zhang, X., Ren, S., Sun, J.: Deep residual learning for image recognition. In: Proceedings of IEEE Conference on Computer Vision and Pattern Recognition (2016)

17. Hinton, G.E., Vinyals, O., Dean, J.: Distilling the knowledge in a neural network. arXiv:1503.02531 (2015)
18. Hubara, I., Courbariaux, M., Soudry, D., El-Yaniv, R., Bengio, Y.: Binarized neural networks. In: Proceedings of Advances in Neural Information Processing Systems (2016)
19. Ioffe, S., Szegedy, C.: Batch normalization: accelerating deep network training by reducing internal covariate shift. arXiv preprint arXiv:1502.03167 (2015)
20. Jacob, B., et al.: Quantization and training of neural networks for efficient integer-arithmetic-only inference. In: Proceedings of IEEE Conference on Computer Vision and Pattern Recognition (2018)
21. Jung, S., et al.: Learning to quantize deep networks by optimizing quantization intervals with task loss. In: Proceedings of IEEE Conference on Computer Vision and Pattern Recognition (2019)
22. Kingma, D.P., Ba, J.: Adam: a method for stochastic optimization. In: Bengio, Y., LeCun, Y. (eds.) Proceedings of International Conference on Learning Representations (2015)
23. Krizhevsky, A., Hinton, G., et al.: Learning multiple layers of features from tiny images (2009)
24. Lin, J., Gan, C., Han, S.: Defensive quantization: when efficiency meets robustness. arXiv preprint arXiv:1904.08444 (2019)
25. Lopes, R.G., Fenu, S., Starner, T.: Data-free knowledge distillation for deep neural networks. arXiv preprint arXiv:1710.07535 (2017)
26. Louizos, C., Reisser, M., Blankevoort, T., Gavves, E., Welling, M.: Relaxed quantization for discretized neural networks. In: Proceedings of International Conference on Learning Representations (2019)
27. Micaelli, P., Storkey, A.: Zero-shot knowledge transfer via adversarial belief matching. arXiv:1905.09768 (2019)
28. Mikolov, T., Karafiát, M., Burget, L., Černocký, J., Khudanpur, S.: Recurrent neural network based language model. In: Conference of the International Speech Communication Association (ISCA) (2010)
29. Miyato, T., Kataoka, T., Koyama, M., Yoshida, Y.: Spectral normalization for generative adversarial networks. arXiv preprint arXiv:1802.05957 (2018)
30. Nagel, M., Baalen, M.v., Blankevoort, T., Welling, M.: Data-free quantization through weight equalization and bias correction. In: Proceedings of the IEEE International Conference on Computer Vision (2019)
31. Nayak, G.K., Mopuri, K.R., Shaj, V., Babu, R.V., Chakraborty, A.: Zero-shot knowledge distillation in deep networks. In: Proceedings of the International Conference on Machine Learning (2019)
32. Nesterov, Y.E.: A method for solving the convex programming problem with convergence rate o $(1/k^2)$. In: Proceedings of the USSR Academy of Sciences, vol. 269, pp. 543–547 (1983)
33. Odena, A., Olah, C., Shlens, J.: Conditional image synthesis with auxiliary classifier GANs. In: Proceedings of International Conference on Machine Learning (2017)
34. Paszke, A., Gross, S., Chintala, S., Chanan, G.: Pytorch: tensors and dynamic neural networks in python with strong GPU acceleration. PyTorch: Tensors Dynamic Neural Networks in Python with strong GPU Acceleration, 6 (2017)
35. Rastegari, M., Ordonez, V., Redmon, J., Farhadi, A.: XNOR-Net: imagenet classification using binary convolutional neural networks. In: Leibe, B., Matas, J., Sebe, N., Welling, M. (eds.) ECCV 2016. LNCS, vol. 9908, pp. 525–542. Springer, Cham (2016). https://doi.org/10.1007/978-3-319-46493-0_32

36. Sak, H., Senior, A.W., Beaufays, F.: Long short-term memory recurrent neural network architectures for large scale acoustic modeling. In: Conference of the International Speech Communication Association (ISCA), pp. 338–342 (2014)
37. Sandler, M., Howard, A., Zhu, M., Zhmoginov, A., Chen, L.C.: Mobilenetv 2: inverted residuals and linear bottlenecks. In: Proceedings of IEEE Conference on Computer Vision and Pattern Recognition (2018)
38. Simonyan, K., Zisserman, A.: Very deep convolutional networks for large-scale image recognition. In: Bengio, Y., LeCun, Y. (eds.) Proceedings of International Conference on Learning Representations (2015)
39. Sung, W., Shin, S., Hwang, K.: Resiliency of deep neural networks under quantization. arXiv preprint arXiv:1511.06488 (2015)
40. Szegedy, C., Vanhoucke, V., Ioffe, S., Shlens, J., Wojna, Z.: Rethinking the inception architecture for computer vision. In: Proceedings of IEEE Conference on Computer Vision and Pattern Recognition (2016)
41. Yang, J., et al.: Quantization networks. In: Proceedings of IEEE Conference on Computer Vision and Pattern Recognition (2019)
42. Yoo, J., Cho, M., Kim, T., Kang, U.: Knowledge extraction with no observable data. In: Proceedings of Advances in Neural Information Processing Systems, pp. 2701–2710 (2019)
43. Zeng, R., et al.: Graph convolutional networks for temporal action localization. In: Proceedings of the IEEE International Conference on Computer Vision (2019)
44. Zhang, C., Bengio, S., Hardt, M., Recht, B., Vinyals, O.: Understanding deep learning requires rethinking generalization. In: Proceedings of International Conference on Learning Representations (2017)
45. Zhang, D., Yang, J., Ye, D., Hua, G.: LQ-Nets: learned quantization for highly accurate and compact deep neural networks. In: Ferrari, V., Hebert, M., Sminchisescu, C., Weiss, Y. (eds.) ECCV 2018. LNCS, vol. 11212, pp. 373–390. Springer, Cham (2018). https://doi.org/10.1007/978-3-030-01237-3_23
46. Zhang, Y., et al.: From whole slide imaging to microscopy: deep microscopy adaptation network for histopathology cancer image classification. In: Shen, D., et al. (eds.) MICCAI 2019. LNCS, vol. 11764, pp. 360–368. Springer, Cham (2019). https://doi.org/10.1007/978-3-030-32239-7_40
47. Zhang, Y., Zhao, P., Wu, Q., Li, B., Huang, J., Tan, M.: Cost-sensitive portfolio selection via deep reinforcement learning. IEEE Trans. Knowl. Data Eng. (2020)
48. Zhao, R., Hu, Y., Dotzel, J., De Sa, C., Zhang, Z.: Improving neural network quantization without retraining using outlier channel splitting. In: Proceedings of the International Conference on Machine Learning (2019)
49. Zhou, S., Wu, Y., Ni, Z., Zhou, X., Wen, H., Zou, Y.: Dorefa-net: training low bitwidth convolutional neural networks with low bitwidth gradients. arXiv preprint arXiv:1606.06160 (2016)
50. Zhuang, B., Liu, L., Tan, M., Shen, C., Reid, I.: Training quantized neural networks with a full-precision auxiliary module. In: Proceedings of IEEE Conference on Computer Vision and Pattern Recognition (2020)
51. Zhuang, B., Shen, C., Tan, M., Liu, L., Reid, I.: Towards effective low-bitwidth convolutional neural networks. In: Proceedings of IEEE Conference on Computer Vision and Pattern Recognition (2018)
52. Zhuang, B., Shen, C., Tan, M., Liu, L., Reid, I.: Structured binary neural networks for accurate image classification and semantic segmentation. In: Proceedings of IEEE Conference on Computer Vision and Pattern Recognition (2019)

Local Correlation Consistency
for Knowledge Distillation

Xiaojie Li[1], Jianlong Wu[2,3](✉), Hongyu Fang[4], Yue Liao[5],
Fei Wang[1], and Chen Qian[1]

[1] SenseTime Research, Beijing, China
`{lixiaojie,wangfei,qianchen}@sensetime.com`
[2] School of Computer Science and Technology, Shandong University, Qingdao, China
`jlwu1992@sdu.edu.cn`
[3] Zhejiang Laboratory,Hangzhou, China
[4] School of Electronics Engineering and Computer Science, Peking University,
Beijing, China
`fanghongyu@pku.edu.cn`
[5] School of Computer Science and Engineering, Beihang University, Beijing, China
`liaoyue.ai@gmail.com`

Abstract. Sufficient knowledge extraction from the teacher network plays a critical role in the knowledge distillation task to improve the performance of the student network. Existing methods mainly focus on the consistency of instance-level features and their relationships, but neglect the local features and their correlation, which also contain many details and discriminative patterns. In this paper, we propose the local correlation exploration framework for knowledge distillation. It models three kinds of local knowledge, including intra-instance local relationship, inter-instance relationship on the same local position, and the inter-instance relationship across different local positions. Moreover, to make the student focus on those informative local regions of the teacher's feature maps, we propose a novel class-aware attention module to highlight the class-relevant regions and remove the confusing class-irrelevant regions, which makes the local correlation knowledge more accurate and valuable. We conduct extensive experiments and ablation studies on challenging datasets, including CIFAR100 and ImageNet, to show our superiority over the state-of-the-art methods.

Keywords: Knowledge distillation · Local correlation consistency · Class-aware attention

1 Introduction

Convolutional Neural Networks have achieved great successes in the vision community, significantly facilitating the development of many practical tasks,

Electronic supplementary material The online version of this chapter (https:// doi.org/10.1007/978-3-030-58610-2_2) contains supplementary material, which is available to authorized users.

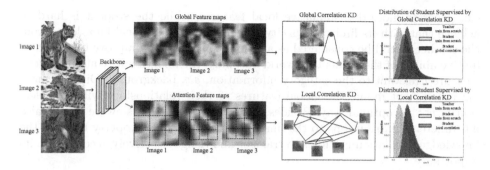

Fig. 1. Comparison between traditional global correlation methods and our proposed method. Instead of directly using the global feature maps to construct the relationship matrix, we first select the class-relevant regions by a class-aware attention module and then construct the local correlation matrix based on the selected local parts to guide the learning of the student network. Feature distributions comparison in the rightmost figure shows that the proposed method can help the student model to better mimic the teacher model than the global correlation method

such as image classification [7,14,23] and face recognition [19,24,28]. Currently, many complicated neural networks with deeper and wider architectures have been proposed to pursuit high performance [25,34]. However, these networks cost plenty of parameters and computations, which limits their deployments on computationally limited platforms, such as mobile devices, embedded systems. Towards this issue, model compression and acceleration become popular research topics recently. Typical methods include network pruning [4,6,17], compact architecture design [9,40], network quantization [5,11,31], knowledge distillation [1,8,26,35,36,41], and so on. Among them, knowledge distillation has been validated as a very effective approach to improving the performance of a light-weight network, *i.e.* student, with the guidance of a pre-trained large deep network, *i.e.* teacher. It encourages the student to learn the teacher's knowledge by applying some consistency based regularization between teacher and student.

The essential point of knowledge distillation is to extract sufficient knowledge from a teacher network to guide a student network. Conventional methods mostly focused on instance-level feature learning, which aims to mimic output activations [1,2,8,41] or transfer the correlation in feature space [16,18,20,27]. The instance-level based methods have achieved good performance, but they still suffer from the following limitations. Firstly, it is hard for a student to thoroughly understand the transferred knowledge from the teacher only based on global supervision. We observe that the local features are also important for the network to understand and recognize an object. As can be seen in Fig. 1, the teacher network can make the right predictions for different categories of objects with similar appearance based on those distinguishing local regions, such as the head, the streaks of the body, or the foot appearance, but the student network may fail. We consider that the teacher network with more learn-able parameters can generate more discriminative local features, while the student is hard to achieve that with its limited capacity. Therefore, learning local knowledge from the teacher should be considered as an important factor to improve the discriminative ability of the student network. Secondly, the images may contain regions that are irrelevant to the category information, *e.g.* background. Directly making the student mimic the global features or their relationships without selection is not an optimal way. Besides, each pixel of the class-aware region also has different contributions to the final classification. This property requires the knowledge distillation methods to transfer knowledge selectively according to its importance.

To resolve the above limitations, a novel local correlation exploration framework is proposed for knowledge distillation, which models sufficient relationships of those class-aware local regions. For the first limitation, we greatly enrich the family of network knowledge by proposing three different kinds of local relationships: (1) the local intra-instance relationship across different positions; (2) the local inter-instance relationship in the same position; (3) the local inter-instance relationship across different positions. Based on the above local relationships, we represent the intermediate feature maps using a more concise and structural form. Further, we hope the correlations computed by the teacher network could be well preserved by the student network. Therefore, we define the consistency regularization to minimize their difference between the teacher and student models. For the second limitation, to transfer the knowledge of those valuable class-aware regions and reduce the influence of invalid class-irrelevant information, we propose a novel class-aware attention module to generate the attention maps before the construction of the local correlation matrices.

We conduct extensive experiments on typical datasets to validate the effectiveness of the proposed framework as well as the local relationships. As shown in the rightmost figure of Fig. 1, we allocated a set of feature maps from the middle layer of a set of models and draw the $cos(\theta)$ similarity distributions between the local patches of those feature maps. The red one is the distribution of the teacher. The yellow one is from the student trained from scratch. The blue one is from the student supervised by global correlation. And the green one is from the student supervised by our local correlation. The higher the coincidence between the histograms of the student and teacher, the more knowledge the student learns from the teacher. This graph shows that the student supervised by our local correlation achieves higher distribution coincidence with the teacher as well as higher accuracy than the student supervised by global correlation.

Our main contributions are summarized as follows:

1) We make the first attempt to explore local relationships in knowledge distillation and propose a novel local correlation consistency based framework. Instead of the traditional global feature based relationship, we mainly focus on the local correlation knowledge, which contains more details and discriminative patterns. By thoroughly investigating three kinds of local relationships, the student network in our framework can sufficiently preserve the important knowledge of the large teacher network.

2) To make the local correlation knowledge more accurate and valuable, we propose a novel class-aware attention module to generate attention masks for valuable class-relevant regions, which can reduce the influence of invalid class-irrelevant regions, highlight the contribution of important pixels, and improve the performance as well.

3) Extensive experiments and ablation studies conducted on CIFAR100 [13] and ImageNet [3] show the superiority of the proposed method and effectiveness of each proposed module.

2 Related Works

The concept of knowledge distillation (KD) with neural networks is first presented by Hinton et al. in 2015 [8], where they come up with the teacher-student framework. Since then, many works have been proposed to improve its applicability and generalization ability. According to the types of knowledge to transfer, existing KD methods can be divided into three categories, including the feature representation learning based methods, attention based methods, and graph learning based methods. We briefly introduce them in this section.

Feature learning based methods mainly aim to train the student to mimic output activations of individual data examples represented by the teacher. Zhang et al. [41] learn a projection matrix to project the teacher-level knowledge and its visual representations from an intermediate layer of teacher network to an intermediate layer of student network. Yim et al. [33] construct the flow of solution procedure matrix across two different layers and minimize the difference between that matrix of teacher and student. Aguilar et al. [1] adopt both the activations and internal representations of the teacher network to guide the learning of the student network and achieve good performance on text classification. Chung et al. [2] try to capture the consistent feature map of intermediate layers by the adversarial learning. Similarly, Shu et al. [22] incorporate the intermediate supervision under the adversarial training framework. To better learn discriminative feature representation, Tian et al. [26] come up with the contrastive learning framework. Lan et al. [15] construct a multi-branch network, whose ensemble predictions are taken as supervision for the training of every single branch.

Attention mechanisms have been widely used in computer vision [10,29,32] and have been successfully applied in the field of KD [12,38,39]. Zagoruyko et al. [38] first show that attention transfer can significantly improve the performance of convolutional neural networks. Zhang et al. [39] present the self distillation framework to distill knowledge within the network itself. Kim et al. [12] make use of the output errors for self-attention based KD models.

Correlation learning [30] receives much attention for KD recently. Instead of directly teach the student to fit the instance features of the teacher, it transfers the correlation among training samples from the teacher network to the student network. Liu et al. [16] construct the instance relationship matrix, which takes the instance features, instance relationships, and feature space transformation into consideration to transfer sufficient knowledge. Park et al. [18] propose distance-wise and angle-wise distillation losses to penalize structural differences in relations. Both Tung et al. [27] and Peng et al. [20] hope to preserve the pair-wise similarity based on the correlation consistency.

Our method focuses on correlation learning and introduces a class-aware attention module. Compared with existing work, our differences mainly lie in two aspects. First, we are the first to explore local correlation during knowledge transfer, while previous methods mainly use the global features to compute the correlations among instances. Second, our class-aware attention module learns the soft attention mask under the supervision of the ground-truth label, which can strengthen the class-aware regions and weaken the class-irrelevant regions during knowledge transfer. To our knowledge, the above attention mechanism is new in the knowledge distillation area.

3   Methods

In this section, we first summarize the basic framework of traditional global embedding based feature learning and correlation learning KD methods. Then we describe our local relationship based KD framework, and introduce the class-aware attention module to filter the semantic-irrelevant knowledge from the feature maps before the correlation construction. Finally, we come up with the overall loss function to supervise the training of the student network.

3.1   Problem Formulation

Given a teacher model T, a student model S and N training samples $\mathcal{X} = \{x_i\}_{i=1}^{N}$, we denote $f^T(x_i)$ and $f^S(x_i)$ as the outputs of teacher and student network for sample x_i, which can be the final outputs after softmax or intermediate feature maps from the middle layers. In the preliminary stage, the conventional KD methods mainly focus on transferring individual outputs from teacher to student. For example, the milestone of KD proposed by Hinton et al. [8] makes the student mimic the teacher's behavior by minimizing the Kullback-Leibler divergence between predictions of student and teacher:

$$\mathcal{L}_{KD} = \frac{1}{n} \sum_{x_i \in \mathcal{X}} \text{KL}(\text{softmax}(\frac{f^T(x_i)}{\tau}), \text{softmax}(\frac{f^S(x_i)}{\tau})), \qquad (1)$$

where τ is a relaxation hyperparameter referred to as temperature in [8]. Recently, many methods have started to take the relationships among instances as a new kind of knowledge for transfer. Based on the outputs of the network, they construct the instance correlations and minimizes the following objective function:

$$\mathcal{L}_{GKD} = \frac{1}{n^2} \mathcal{D}(G(f^T(x_1), f^T(x_2), ...f^T(x_n)), G(f^S(x_1), f^S(x_2), ...f^S(x_n))), \qquad (2)$$

where $\mathcal{D}(\cdot)$ is a loss function that penalizes the difference between correlations of teacher and student, and $G(\cdot)$ is the function to construct the similarity correlation, which in this paper is represented by a correlation matrix. Given feature

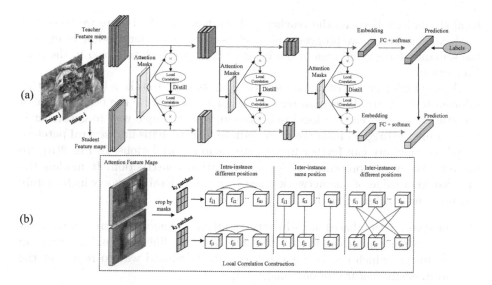

(a)

(b)

Fig. 2. Overall framework of the proposed method. (a) We supervise the training of the student network by the local correlation consistency losses. Class-aware attention module is trained using the teacher's feature maps to extract those class-specific foreground information before constructing the local correlation matrices of the teacher or student network. (b) We crop the main region of the attention feature maps based on the thresholding attention mask and split it into k^2 patches to investigate the proposed three kinds of local relationships

vectors of n samples, the (i, j) element of similarity correlation G is computed as follows:

$$G_{ij} = \varphi(f(x_i), f(x_j)), \; G \in \mathbb{R}^{n \times n}, \tag{3}$$

where φ can be any function that calculates the similarity between two examples, such as the cosine similarity [18] and the Gaussian kernel based similarity [20].

The above similarity correlation among instances has been validated as an effective knowledge for transfer. However, existing methods usually utilize the global features to construct the relationship correlation and neglect the discrimination power implied in local image regions. To make full use of the discriminative local information, we propose our local correlation based knowledge distillation framework.

3.2   Local Correlation Construction

The overall architecture of the proposed method is shown in Fig. 2(a). Specifically, we divide both the teacher and the student networks into several stages according to the resolution of the feature maps. For each stage, based on its corresponding feature maps, we investigate three different kinds of local relationships and construct the similarity matrix, after which we minimize the difference of local similarities between the teacher and student models. Before we construct the local correlation, we propose a novel class-aware attention module to extract the semantic foreground area of the image, which will be introduced in the next subsection.

As local information contains more details, we hope to take advantage of this information to learn discriminative correlations and improve the performance of the student model. For the local information, it is a simple way to construct it by dividing the original image or intermediate feature maps into several patches, based on which we can further investigate various correlations. In Fig. 2(b), we present the overall procedure of local correlation construction. It models the distilled knowledge of one network stage in a more detailed way, which mainly contains three different kinds of relationships:

(1) Local correlation based intra-instance relationship across different local positions: it corresponds to the relationship between different spatial regions in one image, which can be regarded as a more relaxed way to represent the intermediate features of one image.
(2) Local correlation based inter-instance relationship on the same position: it corresponds to the relationship between regions at the same position among images in one mini-batch, which is a more strict way than the global correlation method to achieve the correlation consistency.
(3) Local correlation based inter-instance relationship across different positions: it corresponds to the relationship between regions at different positions among images in one mini-batch, which contains more abundant knowledge compared with the second relationship and explores more knowledge between local regions without the limitation of position.

For each mini-batch with n images, we compute the correlation matrix of the local regions based on the output feature maps of the teacher and student network. We denote the activation maps produced by the teacher network at l-th stage as $f_l^T \in \mathbb{R}^{n \times c \times h \times w}$, where c, h, w are the size of the channel, height and width, respectively. The corresponding activation maps of the student network can be represented by $f_{l'}^S \in \mathbb{R}^{n \times c' \times h' \times w'}$. Note that c does not necessarily have to equal c' in our method since our correlation-based knowledge transfer method only needs to compute the correlation among features of the same model. For the feature maps f_l^T or $f_{l'}^S$ of each stage, we split it into $k \times k$ patches for each sample and get nk^2 patches for the whole mini-batch, where each patch has the shape of $c \times \frac{h}{k} \times \frac{w}{k}$ or $c' \times \frac{h'}{k} \times \frac{w'}{k}$ (to simplify, here we suppose that h, h', w and w' can be fully divided by k). For the j-th patch from image x_i, we denote $f_l^T(x_{i,j})$ and $f_{l'}^S(x_{i,j})$ as the corresponding local patch features of the teacher and student networks, respectively. After reshaping the features of each patch to a vector, we compute the local correlations we introduced before.

For the first kind of local relationship, it models the intra-instance relationship across different local positions. For the l-th stage, we denote $F_{l,intra}(x_i) = \{f_l(x_{i,1}), ..., f_l(x_{i,k^2})\}$ as the collection of k^2 local features for sample x_i. Then we can define the corresponding loss function in a mini-batch with n samples as:

$$\mathcal{L}_{intra} = \sum_{i=1}^{n}\sum_{l=1}^{L}\|G(F_{l,intra}^{T}(x_i)) - G(F_{l',intra}^{S}(x_i))\|_{F}^{2}, \tag{4}$$

where $G(\cdot)$ is the function defined in Eq. (3) to construct the similarity matrix, and L is the total number of stages. The permutations of k^2 local features in $F_{l,intra}^{T}(x_i)$ and $F_{l,intra}^{S}(x_i)$ are the same. We adopt the Frobenius norm $\|.\|_F$ to penalize the distance between local correlation matrices computed by student and teacher. For the similarity matrix construction, we use cosine similarity to compute the correlation between the embeddings of two local patches to penalize angular differences.

The second one is the inter-instance relationship on the same local position. Similarly, we denote $F_{l,inter-s}(i) = \{f_l(x_{1,i}), ..., f_l(x_{n,i})\}$ as the collection of local features of the l-th stage, corresponding to the i-th local patch ($i \in [1, 2, \cdots, k^2]$) for n samples of the mini-batch. Then we can define the corresponding loss function as:

$$\mathcal{L}_{inter-s} = \sum_{i=1}^{k^2}\sum_{l=1}^{L}\|G(F_{l,inter-s}^{T}(i)) - G(F_{l',inter-s}^{S}(i))\|_{F}^{2}. \tag{5}$$

Similarly, the loss function for the third relationship that explores inter-instance relationship across different positions can be defined by:

$$\mathcal{L}_{inter-d} = \sum_{p,q=1,p\neq q}^{k^2} \sum_{i,j=1,i\neq j}^{n} \sum_{l=1}^{L} \left(\varphi\left(f_l^{T}(x_{i,p}), f_l^{T}(x_{j,q})\right) - \varphi\left(f_{l'}^{S}(x_{i,p}), f_{l'}^{S}(x_{j,q})\right)\right)^{2}, \tag{6}$$

where $\varphi(\cdot)$ is the function to compute cosine similarity between two feature vectors.

Based on the above loss functions for the above three local relationships, we combine them to get the following overall loss function:

$$\mathcal{L}_{LKD} = \mathcal{L}_{intra} + \mathcal{L}_{inter-s} + \mathcal{L}_{inter-d}. \tag{7}$$

The local correlation based relationships we explored mainly have two advantages. On the one hand, the local features contain more detailed information about this category, which can introduce some discriminative knowledge to facilitate the distillation. For example, many classes in ImageNet belong to a large category. The difference only lies in small local regions, while other regions are very similar. Our local feature based method can well capture and transfer these local patterns, while previous global feature based methods may ignore it. On the other hand, our method investigates various kinds of correlations, which are much more sufficient than previous methods. While the key challenge of knowledge distillation lies in extracting moderate and sufficient knowledge for guidance [16], our method can better guide the learning of the student network.

3.3   Class-Aware Attention

In the previous subsection, we divide the feature map into several non-overlapped patches as the local information. However, the original images also contain a part of unrelated information, which contributes less to the final prediction and may even have a negative influence on the quality of local patches as well as the local correlation. To solve this issue and extract these high related semantic regions, we introduce a class-aware attention module (CAAT) to filter out the invalid information.

The module consists of two parts: a mask generator and an auxiliary classifier. Supervised by the ground-truth label, CAAT can generate the pixel-level attention mask, which can identify the importance of each pixel and its correlation with the final prediction of the teacher. Given the feature maps of the teacher model $f_l^T \in \mathbb{R}^{n \times c \times h \times w}$, the generated spatial masks $M \in \mathbb{R}^{n \times h \times w}$ can be computed by:

$$M = \mathcal{G}(f_l^T),\tag{8}$$

where $\mathcal{G}(\cdot)$ denotes the mask generator network, which is constructed by a stack of conv-bn-relu blocks followed by the Sigmoid thresholding layer so that each value in the mask is a continuous value between 0 and 1. $M(i, :, :)$ $(i \in [1, n])$ corresponds to the mask for the feature maps of i-th image in the mini-batch. Each value in $M(i, :, :)$ reflects the contribution of the corresponding location to the final prediction of the teacher network. For the same position of different channels, we assign the same mask information. By repeating the mask M along the channel dimension, we can make the mask have the same shape as the feature map f_l^T and $f_{l'}^S$. Then we can get the class-aware attention feature map $\tilde{f}_l^T$ and $\tilde{f}_{l'}^S$ by the following element-wise product:

$$\tilde{f}_l^T = \mathcal{O}_{repeat}(M) \otimes f_l^T, \tilde{f}_{l'}^S = \mathcal{O}_{repeat}(M) \otimes f_{l'}^S,\tag{9}$$

where $\mathcal{O}_{repeat}(.)$ denotes the repeat operation.

To guide the training of network $\mathcal{G}$, we further introduce an auxiliary classifier network $\mathcal{C}$, which takes $\tilde{f}^T$ as input and is supervised by the ground truth label. This sub-network consists of a sequence of bottleneck blocks and utilizes a fully-connected layer for final classification. By minimizing the softmax loss, the auxiliary classifier $\mathcal{C}$ forces the generated mask to pay more attention to informative regions and ignore helpless information like background.

We get the attention feature maps of teacher and student by applying the class-aware attention mask to the original feature maps to highlight those important pixels and weaken those class-irrelevant pixels. Furthermore, we generate a bounding box of the main part of the feature maps based on the thresholding attention mask (the value that larger than threshold $\mathcal{H}$ will be set to 1. The opposite will be set to 0). The top-left point and the right-down point of the bounding box are decided by the boundaries of the thresholding attention mask. We crop the main part of the attention feature maps based on the generated bounding box and divide it into several patches like the way we introduced in the last subsection. Finally, we resize the patches to the same size as the original patch by bilinear interpolation and calculated the local correlation we introduced in the last section. In this part, we modify the proposed losses $\mathcal{L}_{LKD}$ in Eq. (7) by replacing the original local features with the cropped masked local features and then get $\widetilde{\mathcal{L}}_{LKD}$, which is formulated as follows:

$$\widetilde{\mathcal{L}}_{LKD} = \widetilde{\mathcal{L}}_{intra} + \widetilde{\mathcal{L}}_{inter-s} + \widetilde{\mathcal{L}}_{inter-d}. \tag{10}$$

3.4   The Overall Model and Optimization

By combining the cross-entropy loss $\mathcal{L}_{CE}$ supervised by the ground truth labels, the classic KD loss $\mathcal{L}_{KD}$, and the proposed local correlation based consistency loss $\widetilde{\mathcal{L}}_{LKD}$, we come up with the final overall loss function:

$$\mathcal{L} = (1 - \alpha)\mathcal{L}_{CE} + \alpha\mathcal{L}_{KD} + \beta\widetilde{\mathcal{L}}_{LKD}, \tag{11}$$

where α, β are hyper-parameters to balance contributions of different terms.

During training, we first optimize network $\mathcal{G}$ and $\mathcal{C}$ by minimizing the softmax loss. Then we fix the parameters of mask generator $\mathcal{G}$, and train the student network by minimizing the overall loss function in Eq. (11).

3.5   Complexity Analysis

We present the computational complexity in training a mini-batch. The computational complexities of Eqs. (4), (5) and (6) for l-th stage of teacher are $\mathcal{O}(nk^2chw)$, $\mathcal{O}(nchw)$ and $\mathcal{O}(n^2k^2chw)$, respectively. Therefore, the total computational complexity of our method is $\mathcal{O}(n^2k^2chw)$. For comparison, SP [27] has $\mathcal{O}(n^2chw)$ complexity and CCKD [20] has $\mathcal{O}(n^2pd)$ complexity, where p and d correspond to the p-order Taylor-series and dimension of feature embedding. In fact, k is very small in our method. For example, we set k to 4 on CIFAR100, and 3 on ImageNet. In this case, the complexity of our method is comparable and in the same order with these conventional KD methods. Besides, the complexity of Eq. (4) is much smaller than SP and CCKD. With only Eq. (4) as the loss function, the accuracy of our method is also better than SP and CCKD, which will be proved by the ablation study. Therefore, when the computation resources are limited, you can only use this term as the loss function.

4   Experiments

In this section, we conduct several experiments to demonstrate the effectiveness of our proposed local graph supervision as well as the class-aware attention module. We first compare the results on CIFAR100 [13] and ImageNet [3] with four knowledge distillation methods, including Hinton's traditional knowledge distillation (KD) [8], attention transfer (AT) [38], similarity-preserving knowledge distillation (SP) [27], and correlation congruence knowledge distillation (CC) [20]. Besides, cross-entropy (CE) loss is also chosen as a baseline. Then we perform ablation studies to evaluate the effect of different modules.

Table 1. Comparison of classification accuracy on CIFAR100. The best results of the student network are highlighted in bold

Teacher Net.	Student Net.	CE	KD	AT	SP	CC	LKD	Teacher
ResNet110	ResNet14	67.45	69.78	69.51	69.59	69.77	**70.48**	75.76
ResNet110	ResNet20	69.47	71.47	71.8	71.42	71.78	**72.63**	75.76
WRN-40-2	WRN-16-1	66.79	66.74	66.75	66.4	66.76	**67.72**	75.61
WRN-40-2	WRN-16-2	73.1	74.89	75.15	74.69	75.05	**75.44**	75.61

4.1 Evaluation on CIFAR100

The CIFAR100 dataset contains 100 classes. For each class, there are 500 images in the training set and 100 images in the testing set. Similar to the settings in [20], we randomly crop 32×32 image from zero-padded 40×40 image, and apply random horizontal flipping for data augmentation. SGD is used to optimize the model with batch size 64, momentum 0.9, and weight decay $5e^{-4}$. For the class-aware attention module, we train the mask generators and auxiliary classifiers for 60 epochs with learning rate starting from 0.05 and multiplied by 0.1 at 30, 40, 50 epochs. The threshold $\mathcal{H}$ is set to 0 because most of the images in CIFAR100 are occupied by the main object. For the extraction of the local features, we set $k = 4$ for all the stages to split the feature maps to 16 patches. Then we train the student network for 200 epochs with the learning rate starting from 0.1 and multiplied by 0.1 at 80, 120, 160 epochs. For CE, we set $\alpha = 0$ in Eq. (11). For traditional KD, AT, CC, SP and our methods, we set $\alpha = 1$ and $\tau = 4$ following the CIFAR100 experiments in [38]. For a fair comparison, we carefully tune the loss weight of all the methods by grid-search for each teacher-student pair and report the average accuracy over 3 runs with the chosen loss weight. $\beta \in [0.001, 0.1]$ works reasonably well for our methods.

We also test the performance under four combinations of teacher and student networks using ResNet [7] and Wide ResNet (WRN) [37]. For the teacher network of ResNet110, the accuracy is 75.76%, and we adopt ResNet14 and ResNet20 as two different student networks. For the teacher network of WRN-40-2, the accuracy is 75.61%, and we adopt WRN-16-1 and WRN-16-2 as two different student networks.

In Table 1, we show the results of different methods on CIFAR100. We can see that our proposed LKD method achieves the best performance under all these four different settings of the teacher and student networks, which can demonstrate the effectiveness and robustness of our method. Based on the results, we also have the following observations. First, our method substantially surpasses the baseline methods KD and AT by a large margin. While these two methods mainly minimize the distance between instance features of the teacher and student models, our improvement can verify that mimicking the correlation between local regions of the feature maps is a more effective way. Second, we find that compared with these methods with global feature based correlation, including SP and CC, our local features based correlation consistency shows the superiority, which can be attributed to the sufficient details and discriminative patterns that local features contain.

Table 2. Comparison of classification accuracy on ImageNet. The best results of the student network are highlighted in bold

Accuracy	CE	KD	AT	SP	CC	LKD	Teacher
Top-1	70.58	71.34	71.33	71.38	71.45	**71.54**	73.27
Top-5	89.45	90.27	90.26	90.28	90.26	**90.30**	91.27

4.2 Evaluation on ImageNet

After successfully demonstrating our method's superiority on the relatively small CIFAR100 dataset, we move to validate its effectiveness on the large-scale ImageNet dataset, which contains 128k training images and 50k testing images. The resolution of input images after pre-processing in ImageNet is 224×224, which is much larger than that in CIFAR100. With more images and larger resolution, classification on ImageNet is more challenging than that on CIFAR100.

Following the setting in AT [38], we adopt ResNet34 as the teacher network and ResNet18 as the student network. Mask generators and auxiliary classifiers are trained for 48 epochs with learning rate starting from 0.8 and multiplied by 0.1 at 36, 44 epochs. The threshold $\mathcal{H}$ is set to 0.1 for the cropping of the attention feature maps. The local relationships based loss function $\mathcal{L}_{LKD}$ is added on the last stage of the network following the implementation of SP [27] with the loss scale $\beta = 0.5$. The patch number k is set to 3. The student network is trained for 120 epochs with mini-batch size 1024 (on 16 GPUs, each with batch size 64 and weight decay $4e-4$). The learning rate starts from 0.4 and is multiplied by 0.1 at 40, 72, and 96 epochs. The α is set to 1 with temperature $\tau = 2$.

In Table 2, we compare the classification accuracy with other methods on ImageNet. We can see that our method continuously outperforms the competing methods on both Top-1 and Top-5 accuracy. Because the ImageNet dataset is very challenging, our small improvement is also very hard. The above result further demonstrates the effectiveness of our LKD on the large-scale and high-resolution dataset.

4.3 Ablation Study

To verify the effectiveness of each of the three kinds of local relationships based knowledge and the class-aware attention module in our method, we conduct ablation studies on CIFAR100 with ResNet110 as the teacher network and ResNet14, ResNet20 as the student networks. Results are shown in Table 3. By adding each of these three local relationship based loss functions into the baseline KD method, the result can be stably improved. By combining these three loss functions, it can achieve a much better result. Based on $\mathcal{L}_{LKD}$, our class-aware attention module can further improve the performance. The above results can sufficiently show the effectiveness of each local correlation based knowledge as well as the attention module. Besides, we can observe similar results with both

Table 3. Ablation study on CIFAR100. *intra*, *inter-same* and *inter-diff* denote three local relationships introduced in Sect. 3. CAAT is the class-aware attention module

Methods	Local Relationships			CAAT	Top1 accuracy	
	intra	*inter-same*	*inter-diff*		ResNet14	ResNet20
$\mathcal{L}_{KD}$					69.78	71.47
$\mathcal{L}_{KD} + \mathcal{L}_{intra}$	✓				70.00	72.04
$\mathcal{L}_{KD} + \mathcal{L}_{inter-s}$	✓				70.20	71.96
$\mathcal{L}_{KD} + \mathcal{L}_{inter-d}$			✓		70.03	72.10
$\mathcal{L}_{KD} + \mathcal{L}_{LKD}$	✓	✓	✓		70.37	72.31
$\mathcal{L}_{KD} + \widetilde{\mathcal{L}}_{LKD}$	✓	✓	✓	✓	**70.48**	**72.63**

Table 4. Results on CIFAR100 with different number of k, which denotes how many patches that we divide the feature map into along each axis

Student	LKD $(k=1)$	LKD $(k=2)$	LKD $(k=4)$
ResNet14	69.98	70.09	70.37
ResNet20	71.82	71.86	72.31

ResNet14 and ResNet20, which also demonstrates the robustness and generalization ability of our contributions.

4.4   Sensitivity Analysis

Influence of the Parameter k. To extract local features, recall that we split the foreground feature map of each image into $k \times k$ patches. In the above experiments on CIFAR100, we simply set $k = 4$ on CIFAR100. In this part, we purely evaluate the performance of the student network with different k. For simplification, we only add the local correlation based loss on ResNet14 and ResNet20 and do not add the class-aware attention module. The results are presented in Table 4. We can observe that with the increase of k, the performance is improved gradually. The results with $k = 4$ obviously surpasses that of $k = 1$ and $k = 2$. The reason is that the larger k we use to extract the local features, the more sufficient knowledge we will extract from the teacher to transfer, which can bring the performance improvement in return.

Effect of Class-Aware Attention. In this part, we evaluate the effect of our class-aware attention module and show whether it can filter out the invalid information. We conduct experiments on CIFAR100 with several different attention methods, including the activation-based attention in AT [38] and Grad-CAM [21]. In all experiments, the grid number k is set to 4, and the baseline experiment is conducted without the attention module. For a fair comparison, we utilize a sigmoid function to normalize the attention masks obtained by all the attention methods mentioned above.

Table 5. Top-1 accuracy on CIFAR100 for LKD with different attention methods

Student	LKD	LKD+AT	LKD+Grad-CAM	LKD+CAAT
ResNet14	70.37	69.88	70.41	**70.48**
ResNet20	72.31	72.34	72.18	**72.63**

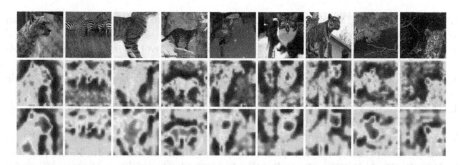

Fig. 3. Visualization of the attention maps. First row: images sampled from ImageNet. Second row: original feature maps generated by the teacher model. Third row: corresponding attention masks generated by CAAT module at the third stage of teacher network. High value is shown in red and low value in blue (Color figure online)

The results are summarized in Table 5. We can see that our proposed CAAT module works much better than other attention methods as well as the baseline.

We also visualize the attention masks of some sample images in Fig. 3. We can find that the informative regions are assigned relatively high value while the confusing background regions are on the contrary. The mask generated by CAAT can well filter out the background that has less contribution to the classification task. And more importantly, it helps the student network to focus on those class-relevant regions and ignore these confusing regions in images, such as the messy background of all the images in Fig. 3.

5   Conclusions

In this paper, we proposed the local correlation consistency: a novel form of knowledge distillation that aims to represent the relationships of local regions in the feature space. By minimizing the local correlation matrices of teacher and student, we could make the student generate more discriminative local features. Furthermore, we applied a class-aware attention mask to both the teacher and the student's feature maps before constructing the local correlation matrices. We trained the class-aware attention module using teacher's feature maps to highlight those informative and class-relevant regions and weaken the effect of those confusing regions. Our Experiments on CIFAR100 and ImageNet demonstrate the effectiveness of the proposed local correlation consistency knowledge distillation and the class-aware attention module.

Acknowledgment. Jianlong Wu is the corresponding author, who is supported by the Fundamental Research Funds and the Future Talents Research Funds of Shandong University.

References

1. Aguilar, G., Ling, Y., Zhang, Y., Yao, B., Fan, X., Guo, E.: Knowledge distillation from internal representations. In: AAAI (2020)
2. Chung, I., Park, S., Kim, J., Kwak, N.: Feature-map-level online adversarial knowledge distillation. In: ICML (2020)
3. Deng, J., Dong, W., Socher, R., Li, L.J., Li, K., Fei-Fei, L.: ImageNet: a large-scale hierarchical image database. In: IEEE CVPR, pp. 248–255 (2009)
4. Dong, X., Yang, Y.: Network pruning via transformable architecture search. In: NeurIPS, pp. 759–770 (2019)
5. Han, S., Mao, H., Dally, W.J.: Deep compression: compressing deep neural networks with pruning, trained quantization and Huffman coding. In: ICLR (2016)
6. Han, S., Pool, J., Tran, J., Dally, W.: Learning both weights and connections for efficient neural network. In: NeurIPS, pp. 1135–1143 (2015)
7. He, K., Zhang, X., Ren, S., Sun, J.: Deep residual learning for image recognition. In: IEEE CVPR, pp. 770–778 (2016)
8. Hinton, G., Vinyals, O., Dean, J.: Distilling the knowledge in a neural network. In: NeurIPS Deep Learning Workshop (2014)
9. Howard, A.G., et al.: Mobilenets: efficient convolutional neural networks for mobile vision applications. arXiv preprint arXiv:1704.04861 (2017)
10. Hu, J., Shen, L., Sun, G.: Squeeze-and-excitation networks. In: IEEE CVPR, pp. 7132–7141 (2018)
11. Hubara, I., Courbariaux, M., Soudry, D., El-Yaniv, R., Bengio, Y.: Quantized neural networks: training neural networks with low precision weights and activations. J. Mach. Learn. Res. **18**(1), 6869–6898 (2017)
12. Kim, H.G., et al.: Knowledge distillation using output errors for self-attention end-to-end models. In: ICASSP, pp. 6181–6185 (2019)
13. Krizhevsky, A., Hinton, G., et al.: Learning multiple layers of features from tiny images. Technical report (2009)
14. Krizhevsky, A., Sutskever, I., Hinton, G.E.: Imagenet classification with deep convolutional neural networks. In: NeurIPS, pp. 1097–1105 (2012)
15. Lan, X., Zhu, X., Gong, S.: Knowledge distillation by on-the-fly native ensemble. In: NeurIPS, pp. 7528–7538 (2018)
16. Liu, Y., et al.: Knowledge distillation via instance relationship graph. In: IEEE CVPR, pp. 7096–7104 (2019)
17. Molchanov, P., Tyree, S., Karras, T., Aila, T., Kautz, J.: Pruning convolutional neural networks for resource efficient inference. In: ICLR (2017)
18. Park, W., Kim, D., Lu, Y., Cho, M.: Relational knowledge distillation. In: IEEE CVPR, pp. 3967–3976 (2019)
19. Parkhi, O.M., Vedaldi, A., Zisserman, A.: Deep face recognition. In: BMVC (2015)
20. Peng, B., et al.: Correlation congruence for knowledge distillation. In: IEEE ICCV, pp. 5007–5016 (2019)
21. Selvaraju, R.R., Cogswell, M., Das, A., Vedantam, R., Parikh, D., Batra, D.: Grad-cam: Visual explanations from deep networks via gradient-based localization. In: IEEE ICCV, pp. 618–626 (2017)

22. Shu, C., Li, P., Xie, Y., Qu, Y., Dai, L., Ma, L.: Knowledge squeezed adversarial network compression. arXiv preprint arXiv:1904.05100 (2019)
23. Simonyan, K., Zisserman, A.: Very deep convolutional networks for large-scale image recognition. In: ICLR (2015)
24. Sun, Y., Chen, Y., Wang, X., Tang, X.: Deep learning face representation by joint identification-verification. In: NeurIPS, pp. 1988–1996 (2014)
25. Tan, M., Le, Q.V.: Efficientnet: rethinking model scaling for convolutional neural networks. arXiv preprint arXiv:1905.11946 (2019)
26. Tian, Y., Krishnan, D., Isola, P.: Contrastive representation distillation. In: ICLR (2020)
27. Tung, F., Mori, G.: Similarity-preserving knowledge distillation. In: IEEE ICCV, pp. 1365–1374 (2019)
28. Wang, F., et al.: The devil of face recognition is in the noise. In: Ferrari, V., Hebert, M., Sminchisescu, C., Weiss, Y. (eds.) ECCV 2018. LNCS, vol. 11213, pp. 780–795. Springer, Cham (2018). https://doi.org/10.1007/978-3-030-01240-3_47
29. Wang, F., et al.: Residual attention network for image classification. In: IEEE CVPR, pp. 3156–3164 (2017)
30. Wu, J., et al.: Deep comprehensive correlation mining for image clustering. In: IEEE ICCV, pp. 8150–8159 (2019)
31. Wu, J., Leng, C., Wang, Y., Hu, Q., Cheng, J.: Quantized convolutional neural networks for mobile devices. In: IEEE CVPR, pp. 4820–4828 (2016)
32. Yang, L., Song, Q., Wu, Y., Hu, M.: Attention inspiring receptive-fields network for learning invariant representations. IEEE TNNLS **30**(6), 1744–1755 (2018)
33. Yim, J., Joo, D., Bae, J., Kim, J.: A gift from knowledge distillation: fast optimization, network minimization and transfer learning. In: IEEE CVPR, pp. 7130–7138 (2017)
34. You, S., Huang, T., Yang, M., Wang, F., Qian, C., Zhang, C.: GreedyNAS: towards fast one-shot NAS with greedy supernet. In: IEEE CVPR, pp. 1999–2008 (2020)
35. You, S., Xu, C., Xu, C., Tao, D.: Learning from multiple teacher networks. In: KDD, pp. 1285–1294 (2017)
36. You, S., Xu, C., Xu, C., Tao, D.: Learning with single-teacher multi-student. In: AAAI, pp. 4390–4397 (2018)
37. Zagoruyko, S., Komodakis, N.: Wide residual networks. In: BMVC (2016)
38. Zagoruyko, S., Komodakis, N.: Paying more attention to attention: improving the performance of convolutional neural networks via attention transfer. In: ICLR (2017)
39. Zhang, L., Song, J., Gao, A., Chen, J., Bao, C., Ma, K.: Be your own teacher: improve the performance of convolutional neural networks via self distillation. In: IEEE ICCV, pp. 3713–3722 (2019)
40. Zhang, X., Zhou, X., Lin, M., Sun, J.: Shufflenet: an extremely efficient convolutional neural network for mobile devices. In: IEEE CVPR, pp. 6848–6856 (2018)
41. Zhang, Z., Ning, G., He, Z.: Knowledge projection for deep neural networks. arXiv preprint arXiv:1710.09505 (2017)

Perceiving 3D Human-Object Spatial Arrangements from a Single Image in the Wild

Jason Y. Zhang[1]($\boxtimes$), Sam Pepose[2], Hanbyul Joo[2], Deva Ramanan[1,3], Jitendra Malik[2,4], and Angjoo Kanazawa[4]

[1] Carnegie Mellon University, Pittsburgh, USA
jasonyzhang@cmu.edu
[2] Facebook AI Research, Menlo Park, USA
[3] Argo AI, Pittsburgh, USA
[4] UC Berkeley, Berkeley, USA

Abstract. We present a method that infers spatial arrangements and shapes of humans and objects in a globally consistent 3D scene, all from a single image in-the-wild captured in an uncontrolled environment. Notably, our method runs on datasets without any scene- or object-level 3D supervision. Our key insight is that considering humans and objects jointly gives rise to "3D common sense" constraints that can be used to resolve ambiguity. In particular, we introduce a scale loss that learns the distribution of object size from data; an occlusion-aware silhouette re-projection loss to optimize object pose; and a human-object interaction loss to capture the spatial layout of objects with which humans interact. We empirically validate that our constraints dramatically reduce the space of likely 3D spatial configurations. We demonstrate our approach on challenging, in-the-wild images of humans interacting with large objects (such as bicycles, motorcycles, and surfboards) and handheld objects (such as laptops, tennis rackets, and skateboards). We quantify the ability of our approach to recover human-object arrangements and outline remaining challenges in this relatively unexplored domain. The project webpage can be found at https://jasonyzhang.com/phosa.

1 Introduction

Tremendous strides have been made in estimating the 2D structure of in-the-wild scenes in terms of their constituent objects. While recent work has also demonstrated impressive results in estimating 3D structures, particularly human bodies, the focus is often on bodies [29,36] and objects [10,16] imaged in isolation or

---

J. Y. Zhang and S. Pepose—Equal contribution.

Electronic supplementary material The online version of this chapter (https://doi.org/10.1007/978-3-030-58610-2_3) contains supplementary material, which is available to authorized users.

© Springer Nature Switzerland AG 2020
A. Vedaldi et al. (Eds.): ECCV 2020, LNCS 12357, pp. 34–51, 2020.
https://doi.org/10.1007/978-3-030-58610-2_3

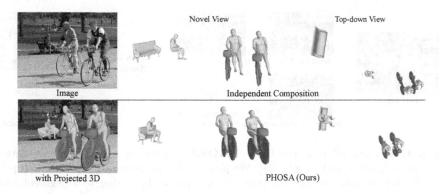

Fig. 1. We present PHOSA, Perceiving Human-Object Spatial Arrangements, an approach that recovers the spatial arrangements of humans and objects in 3D space from a single image by reasoning about their intrinsic scale, human-object interaction, and depth ordering. Given the input image (top left), we show two possible interpretations of the 3D scene that have similar 2D projections (bottom left). Using priors of humans, objects, and their interactions, our approach is able to recover the more reasonable interpretation (bottom row).

in controlled lab conditions [22,52]. To enable true 3D in-the-wild scene understanding, we argue that one must look at the *holistic* 3D scene, where objects and bodies can provide contextual cues for each other so as to correct local ambiguities. Consider the task of understanding the image in Fig. 1. Independently estimated 3D poses of humans and objects are not necessarily consistent in the spatial arrangement of the 3D world of the scene (top row). When processed holistically, one can produce far more plausible 3D arrangements by exploiting contextual cues, such as the fact that humans tend to sit on park benches and ride bicycles rather than float mid-air.

In this paper, we present a method that can similarly recover the 3D spatial arrangement and shape of humans and objects in the scene from a single image. We demonstrate our approach on challenging, in-the-wild images containing multiple and diverse human-object interactions. We propose an optimization framework that relies on automatically predicted 2D segmentation masks to recover the 3D pose, shape, and location of humans along with the 6-DoF pose and intrinsic scale of key objects in the scene. Per-instance intrinsic scale allows one to convert each instance's local 3D coordinate system to coherent world coordinates, imbuing people and objects with a consistent notion of metric size.

There are three significant challenges to address. First is that the problem is inherently ill-posed as multiple 3D configurations can result in the same 2D projection. It is attractive to make use of data-driven priors to resolve such ambiguities. But we immediately run into the second challenge: obtaining training data with 3D supervision is notoriously challenging, particularly for entire 3D scenes captured in-the-wild. Our key insight is that considering humans and objects jointly gives rise to 3D scene constraints that reduce ambiguity. We make use

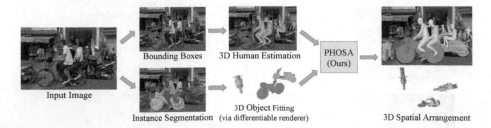

Fig. 2. Overview of our method, PHOSA. Given an image, we first detect instances of humans and objects [24]. We predict the 3D pose and shape of each person [27] and optimize for the 3D pose of each object by fitting to a segmentation mask [31]. Then, we convert each 3D instance in its own local coordinate frame into world coordinates using an intrinsic scale. Using our Human-Object Spatial Arrangement optimization, we produce a globally consistent output, as shown here. Our framework produces plausible reconstructions that capture realistic human-object interaction, preserve depth ordering, and obey physical constraints.

of physical 3D constraints including a prior on the typical size of objects within a category. We also incorporate spatial constraints that encode typical modes of interactions with humans (e.g. humans typically interact with a bicycle by grabbing its handlebars). Our final challenge is that while there exists numerous mature technologies supporting 3D understanding of humans (including shape models and keypoint detectors), the same tools do not exist for the collective space of all objects. In this paper, we take the first step toward building such tools by learning the natural size distributions of object categories without any supervision. Our underlying thesis, bourne out by experiment, is that contextual cues arising from holistic processing of human-object arrangements can still provide enough information to understand objects in 3D.

We design an optimization-based framework, where we first reconstruct the humans and objects *locally* in each detected bounding box. For humans, we make use of state-of-the-art 3D human reconstruction output [27]. For objects, we solve for the 6-DoF parameters of a category-specific 3D shape exemplar that fits the local 2D object instance segmentation mask [31]. We then use a per-instance intrinsic scale to convert each local 3D prediction into a world coordinate frame by endowing metric size to each object and define a *global* objective function that scores different 3D object layouts, orientations, and shape exemplars. We operationalize constraints through loss terms in this objective. We make use of gradient-based solvers to optimize for the globally consistent layout. Although no ground truth is available for this task, we evaluate our approach qualitatively and quantitatively on the COCO-2017 dataset [40], which contains challenging images of humans interacting with everyday objects obtained in uncontrolled settings. We demonstrate the genericity of our approach by evaluating on objects from 8 categories of varying size and interaction types: baseball bats, bicycles, laptops, motorcycles, park benches, skateboards, surfboards, and tennis rackets.

2   Related Work

3D Human Pose and Shape from a Single Image. Recovering the 3D pose and shape of a person from a single image is a fundamentally ambiguous task. As such, most methods employ statistical 3D body models with strong priors on shape learned from large-scale 3D scans and with known kinematic structure to model the articulation [4,28,41,47,64]. Seminal works in this area [3,17,53] fit the parameters of a 3D body model to manual annotation such as silhouettes and keypoints obtained from users interaction [19,53,64]. Taking advantage of the progress in 2D pose estimation, [7] proposes a fully automatic approach where the parameters of the SMPL body model [41] are fit to automatically detected 2D joint locations in combination with shape and pose priors. More recent approaches employ expressive human models with faces and fingers [47,60]. Another line of work develops a learning based framework, using a feed-forward model to directly predict the parameters of the body model from a single image [29,42,45,50,56,58]. More recent approaches combine human detection with 3D human pose and shape prediction [20]. Much of the focus in such approaches is training on in-the-wild images of humans without any paired 3D supervision. [29] employ an adversarial prior on pose, [49] explore using ordinal supervision, and [48] use texture consistency. More recently, [27,36] have proposed hybrid approaches that combine feed-forward networks with optimization-based methods to improve 2D keypoint fit, achieving state-of-the-art results. In this work, we use the 3D regression network from [27] to recover the 3d pose and shape of humans.

Note that all of these approaches consider the human in isolation. More related to our work are methods that recover 3D pose and shape of multiple people [26,62,63]. These approaches use collision constraints to avoid intersection, and use bottom-up grouping or ordinal-depth constraints to resolve ambiguities. We take inspiration from these works for the collision loss and the depth ordering loss from [26], but our focus is on humans and objects in this work.

3D Objects from a Single Image. There has also been significant literature in single-view 3D object reconstruction. Earlier methods optimize a deformable shape model to image silhouettes [5,25,30,32,39]. Recent approaches train a deep network to predict the 3D shape from an image [9,10,12,16,18,43,46]. Most of these approaches require 3D supervision or multi-view cues and are trained on synthetic datasets such as [54,59]. Many of these approaches reason about 3D object shape in isolation, while in this work we focus on their spatial arrangements. There are several works [15,38,55] that recover the 3D shape of multiple objects but still reason about them independently. More recently, [37] proposes a graph neural network to reason about the relationships between object to infer their layout, trained on a synthetic dataset with no humans. In this work we explore 3D spatial arrangements of humans and objects in the wild. As there is no 3D supervision for these images, we take the traditional category-based model-fitting approach to get the initial 6-DoF pose of the objects and refine their spatial arrangements in relation to humans and other objects in the scene.

3D Human-to-Object Interaction. Related to our work are earlier approaches that infer about the 3D geometry and affordances from observing people interacting with scenes over time [11,13,21]. These approaches are similar to our work in spirit in that they use the ability to perceive humans to understand the 3D properties of the scene. The majority of the recent works rely on a pre-captured 3D scene to reason about 3D human-object interaction. [52] use RGB-D sensors to capture videos of people interacting with indoor scenes and use this data to learn a probabilistic model that reasons about how humans interact with its environment. Having access to 3D scenes provides scene constraints that improve 3D human pose perception [35,51,61]. The recent PROX system [22] demonstrates this idea through an optimization-based approach to improve 3D human pose estimation conditioned on a known 3D scene captured by RGB-D sensors. While we draw inspiration from their contact terms to model human object interaction, we critically do not assume that 3D scenes are available. We experiment on single images captured in an uncontrolled *in-the-wild* environment, often outdoors.

More related to our approach are those that operate on images. There are several hand-object papers that recover both 3D object and 3D hand configurations [23]. In this work we focus on 3D spatial arrangements of humans and objects. Imapper [44] uses priors built from RGB-D data [52] to recover a plausible global 3D human motion and a global scene layout from an indoor video. Most related to our work is [8], who develop an approach that recovers a parse graph that represents the 3D human pose, 3D object, and scene layout from a single image. They similarly recover the spatial arrangements of humans and objects, but rely on synthetic 3D data to (a) learn priors over human-object interaction and (b) train 3D bounding box detectors that initialize the object locations. In this work we focus on recovering 3D spatial arrangements of humans and objects in the wild where no 3D supervision is available for 3D objects and humans and their layout. Due to the reliance on 3D scene capture and/or 3D synthetic data, many previous work on 3D human object interaction focus on indoor office scenes. By stepping out of this supervised realm, we are able to explore and analyze how 3D humans and objects interact in the wild.

3   Method

Our method takes a single RGB image as input and outputs humans and various categories of objects in a common 3D coordinate system. We begin by separately estimating 3D humans and 3D objects in each predicted bounding box provided by an object detector [24]. We use a state-of-the art 3D human pose estimator [27] to obtain 3D humans in the form of a parametric 3D human model (SMPL [41]) (in Sect. 3.1), and use a differentiable renderer to obtain 3D object pose (6-DoF translation and orientation) by fitting 3D mesh object models to predicted 2D segmentation masks [34] (in Sect. 3.2). The core idea of our method is to exploit the interaction between humans and objects to spatially arrange them in a common 3D coordinate system by optimizing for the per-instance *intrinsic scale*, which specifies their metric size (in Sect. 3.3). In particular, our method can also improve the performance of 3D pose estimation for objects by exploiting cues from the estimated 3D human pose. See Fig. 2 for the overview of our method.

3.1 Estimating 3D Humans

Given a bounding box for a human provided by a detection algorithm [24], we estimate the 3D shape and pose parameters of SMPL [41] using [27]. The 3D human is parameterized by pose $\theta \in \mathbb{R}^{72}$ and shape $\beta \in \mathbb{R}^{10}$, as well as a weak-perspective camera $\Pi = [\sigma, t_x, t_y] \in \mathbb{R}^3$ to project the mesh into image coordinates. To position the humans in the 3D space, we convert the weak-perspective camera to the perspective camera projection by assuming a fixed focal length f for all images, where the distance of the person is determined by the reciprocal of the camera scale parameter σ. Thus, the 3D vertices of the SMPL model for the i-th human is represented as,

Fig. 3. **Ambiguity in scale.** Recovering a 3D scene from a 2D image is fundamentally ambiguous because multiple 3D interpretations can have the same 2D projection. Consider the photo of the surfer on the left. A large surfboard far away (c) and a small one closer (a) have the same 2D projection (second panel) as the correct interpretation (b). In this work, we aim to resolve this scale ambiguity by exploiting cues from human-object interaction, essentially using the human as a "ruler."

$$V_h^i = \mathcal{M}(\beta^i, \theta^i) + \left[t_x^i \; t_y^i \; f/\sigma^i \right] \qquad (1)$$

where $\mathcal{M}$ is the differentiable SMPL mapping from pose and shape to a human mesh with 6890 vertices in meters. The SMPL shape parameter β controls the height and size of the person. In practice, this is difficult to reliably estimate from an image since a tall, far-away person and a short, closeby person may project to similar image regions (Fig. 1 and Fig. 6). To address this ambiguity, we fix the estimated SMPL pose and shape and introduce an additional per-human intrinsic scale parameter $s_i \in \mathbb{R}$ that changes the size and thus depth of the human in world coordinates: $V_h^{i*} = s_i V_h^i$. While the shape parameter β also captures size, we opt for this parameterization as it can also be applied to objects (described below) and thus optimized to yield a globally consistent layout of the scene.

3.2 Estimating 3D Objects

We consider each object as a rigid body mesh model and estimate the 3D location $\mathbf{t} \in \mathbb{R}^3$, 3D orientation $\mathcal{R} \in SO(3)$, and an intrinsic scale $s \in \mathbb{R}$. The intrinsic scale converts the local coordinate frame of the template 3D mesh to the world frame. We consider single or multiple exemplar mesh models for each object category, pre-selected based on the shape variation within each category. For example, we use a single mesh for skateboards but four meshes for motorcycle. The mesh models are obtained from [1,2,38] and are pre-processed to have fewer faces (about 1000) to make optimization more efficient. See Fig. 5 for some examples of mesh models and the supplementary for a full list. The 3D state of the jth object is represented as,

$$V_o^j = s^j \mathcal{R}^j \mathcal{O}(c^j, k^j) + \mathbf{t}^j, \tag{2}$$

where $\mathcal{O}(c^j, k^j)$ specifies the k^j-th exemplar mesh for category c^j. Note that the object category c^j is provided by the object detection algorithm [24], and k^j is automatically determined in our optimization framework (by selecting the exemplar that minimizes reprojection error).

Our first goal is to estimate the 3D pose of each object independently. However, estimating 3D object pose in the wild is challenging because (1) there are no existing parametric 3D models for target objects; (2) 2D keypoint annotations or 3D pose annotations for objects in the wild images are rare; and (3) occlusions are common in cluttered scenes, particularly those with humans. We propose an optimization-based approach using a differentiable renderer [31] to fit the 3D object to instance masks from [24] in a manner that is robust to partial occlusions. We began with an pixel-wise L2 loss over rendered silhouettes S versus predicted masks M, but found that it ignored boundary details that were important for reliable pose estimation. We added a symmetric chamfer loss [14] which focuses on boundary alignment, but found it computationally prohibitive since it required recomputing a distance transform of S at each gradient iteration. We found good results with L2 mask loss augmented with a *one-way* chamfer loss that computes the distance of each silhouette boundary pixel to the nearest mask boundary pixel, which requires computing a *single* distance transform once for the mask M. Given an no-occlusion indicator I (0 if pixel only corresponds to mask of different instance, 1 else), we write our loss as follows:

$$L_{\text{occ-sil}} = \sum (I \circ S - M)^2 + \sum_{p \in E(I \circ S)} \min_{\hat{p} \in E(M)} \|p - \hat{p}_2\| \tag{3}$$

where $E(M)$ computes the edge map of mask M. Note that this formulation can handle partial occlusions by object categories for which we do not have 3D models, as illustrated in Fig. 4. We also add an offscreen penalty to avoid degenerate solutions when minimizing the chamfer loss. To estimate the 3D object pose, we minimize the occlusion-aware silhouette loss:

$$\{\mathcal{R}^j, \mathbf{t}^j\}^* = \operatorname*{argmin}_{\mathcal{R}, \mathbf{t}} L_{\text{occ-sil}}\left(\Pi_{\text{sil}}(V_o^j), M^j\right), \tag{4}$$

where Π_{sil} is the silhouette rendering of a 3D mesh model via a perspective camera with a fixed focal length (same as f in (1)) and M^j is a 2D instance mask for the j-th object. We use PointRend [34] to compute the instance masks. See Fig. 4 for a visualization and the supplementary for more implementation details. While this per-instance optimization provides a reasonable 3D pose estimate, the mask-based 3D object pose estimation is insufficient since there remains a fundamental ambiguity in determining the global location relative to other objects or people, as shown in Fig. 3. In other words, reasoning about instances in isolation cannot resolve ambiguity in the intrinsic scale of the object.

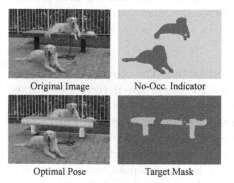

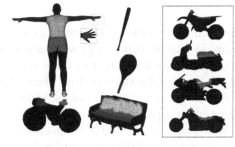

Fig. 4. Occlusion-Aware Silhouette Loss for optimizing object pose. Given an image, a 3D mesh model, and instance masks, our occlusion-aware silhouette loss finds the 6-DoF pose that most closely matches the target mask (bottom right). To be more robust to partial occlusions, we use a no-occlusion indicator (top right) to ignore regions that correspond to other object instances, including those for which we do not have 3D mesh models (e.g. dogs).

Fig. 5. Part labels for fine-grained interaction. To model human-object interactions, we label each object mesh with interaction regions corresponding to parts of the human body. Each color-coded region on the person (top left) interacts with the matching colored region for each object. The interaction loss pulls pairs of corresponding parts closer together. To better capture variation in shape, we can use multiple mesh instances for the same category (e.g. the motorcycles shown on the right). See the supplementary for all mesh models and interaction correspondences. (Color figure online)

3.3 Modeling Human-Object Interaction for 3D Spatial Arrangement

Reasoning about the 3D poses of humans and objects independently may produce inconsistent 3D scene arrangements. In particular, objects suffer from a fundamental depth ambiguity: a large object further away can project to the same image coordinates as a small object closer to the camera (see Fig. 3).

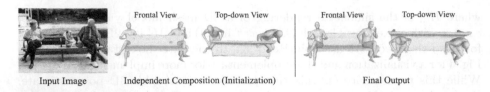

Fig. 6. Recovering realistic human-object spatial arrangements by reasoning about depth ordering and interpenetration. Given the image on the left as input, we first initialize the spatial arrangement by independently estimating the 3D human and object poses (Independent Composition). By incorporating physical priors such as avoiding mesh inter-penetration as well as preserving the depth ordering inferred from the predicted segmentation mask, we can produce the much more plausible spatial arrangement shown on the right.

As such, the absolute 3D depth cannot be estimated. The interactions between humans and objects can provide crucial cues to reason about the relative spatial arrangement among them. For example, knowing that two people are riding on the same bike suggests that they should have similar depths in Fig 2. This pair-wise interaction cue can be propagated to determine the spatial arrangement of multiple humans and objects together. Furthermore, given the fact that a wide range of 2D and 3D supervision exists for human pose, we can leverage 3D human pose estimation to further adjust the orientation of the 3D object. For instance, knowing that a person is sitting down on the bench can provide a strong prior to determine the 3D orientation of the bench. Leveraging this requires two important steps: (1) identifying a human and an object that are interacting and (2) defining an objective function to correctly adjust their spatial arrangements.

Identifying Human-Object Interaction. We hypothesize that an interacting person and object must be nearby in the world coordinates. In our formulation, we solve for the 6-DoF object pose and the intrinsic scale parameter s^j which places the object into world coordinates and imbues the objects with metric size. We use 3D bounding box overlap between the person and object to determine whether the object is interacting with a person. The size of the per-category 3D bounding box in world coordinates is set larger for larger object categories. See supplementary for a full list of 3D box sizes. Endowing a reasonable initial scale is important for identifying human-object interaction because if the object is scaled to be too large or too small in size, it will not be nearby the person. We first initialize the scale using common sense reasoning, via an internet search to find the average size of objects (e.g. baseball bats and bicycles are $\sim$0.9 m $\sim$2 m long respectively). Through our proposed method, the per-instance intrinsic scales change during optimization. From the final distribution of scales obtained over the test set, we compute the empirical mean scale and repeat this process using this as the new initialization (Fig. 7).

Objective Function to Optimize 3D Spatial Arrangements. Our objective includes multiple terms to provide constraints for interacting humans and objects:

$$L = \lambda_1 L_{\text{occ-sil}} + \lambda_2 L_{\text{interaction}} + \lambda_3 L_{\text{depth}} + \lambda_4 L_{\text{collision}} + \lambda_5 L_{\text{scale}}. \quad (5)$$

We optimize (5) using a gradient-based optimizer [33] w.r.t. intrinsic scale $s^i \in \mathbb{R}$ for the i-th human and intrinsic scale $s^j \in \mathbb{R}$, rotation $\mathcal{R}^j \in SO(3)$, and translation $\mathbf{t}^j \in \mathbb{R}^3$ for the j-th object instance jointly. The object poses are initialized from Sect. 3.2. $L_{\text{occ-sil}}$ is the same as (4) except without the chamfer loss which didn't help during joint optimization. We define the other terms below.

Interaction loss: We first introduce a coarse, instance-level interaction loss to pull the interacting object and person close together:

$$L_{\text{coarse inter}} = \sum_{h \in \mathcal{H}, o \in \mathcal{O}} \mathbb{1}(h, o) \| C(h) - C(o) \|_2, \quad (6)$$

where $\mathbb{1}(h, o)$ identifies whether human h and object o are interacting according to the 3D bounding box overlap criteria described before.

Humans generally interact with objects in specific ways. For example, humans hold tennis rackets by the handle. This can be used as a strong prior for human-object interaction and adjust their spatial arrangement. To do this, we annotate surface regions on the SMPL mesh and on our 3D object meshes where there is likely to be interaction, similar to PROX [22]. These include the hands, feet, and back of a person or the handlebars and seat of a bicycle, as shown in Fig. 5. To encode spatial priors about human-object interaction (e.g. people grab bicycle handlebars by the hand and sit on the seat), we enumerate pairs of object and human part regions that interact (see supplementary for a full list). We incorporate a fine-grained, parts-level interaction loss by using the part-labels (Fig. 5) to pull the interaction regions closer to achieve better alignment:

$$L_{\text{fine inter}} = \sum_{h \in \mathcal{H}, o \in \mathcal{O}} \sum_{\substack{\mathcal{P}_h, \mathcal{P}_o \in \\ \mathcal{P}(h,o)}} \mathbb{1}(\mathcal{P}_h, \mathcal{P}_o) \| C(\mathcal{P}_h) - C(\mathcal{P}_o) \|_2, \quad (7)$$

where $\mathcal{P}_h$ and $\mathcal{P}_o$ are the interaction regions on the person and object respectively. Note that we define the parts interaction indicator $\mathbb{1}(\mathcal{P}_h, \mathcal{P}_o)$ using the same criteria as instances, i.e. 3D bounding box overlap. The interactions are recomputed at each iteration. Finally, $L_{\text{interaction}} = L_{\text{coarse inter}} + L_{\text{fine inter}}$.

Scale loss: We observe that there is a limit to the variation in size within a category. Thus, we incorporate a Gaussian prior on the intrinsic scales of instances in the same category using a category-specific mean scale:

$$L_{\text{scale}} = \sum_{c} \sum_{j \in [|\mathcal{O}_c|]} \| s_j - \bar{s}_c \|_2. \quad (8)$$

We initialize the intrinsic scale of all objects in category c to $\bar{s}_c$. The mean object scale $\bar{s}_c$ is initially set using common sense estimates of object size. In Fig. 7, we visualize the final distribution of object sizes learned for the COCO-2017 [40] test set after optimizing for human interaction. We then repeat the process with the empirical mean as a better initialization for $\bar{s}_c$. We also incorporate the scale loss for the human scales s_i with a mean of 1 (original size) and a small variance.

Ordinal Depth loss: The depth ordering inferred from the 3D placement should match that of the image. While the correct depth ordering of people and objects would also minimize the occlusion-aware silhouette loss, we posit that the ordinal depth loss introduced in Jiang et al. [26] can help recover more accurate depth orderings from the modal masks. Using an ordinal depth can give smoother gradients to both the occluder and occluded object. Formally, for each pair of instances, we compare the pixels at the intersections of the silhouettes with the segmentation mask. If at pixel p, instance i is closer than instance j but the segmentation masks at p show j and not i, then we apply a ranking loss on the depths of both instances at pixel p:

$$L_{\text{depth}} = \sum_{o_i \in \mathcal{H} \cup \mathcal{O}} \sum_{o_j \in \mathcal{H} \cup \mathcal{O}} \sum_{\substack{p \in \text{Sil}(o_i) \\ \cap \text{Sil}(o_j)}} \mathbb{1}(p, o_i, o_j) \log \left(1 + \exp(D_{o_j}(p) - D_{o_i}(p))\right), \quad (9)$$

where $\text{Sil}(o)$ is the rendered silhouette of instance o, $D_o(p)$ is the depth of instance o at pixel p, and $\mathbb{1}(p, o_i, o_j)$ is 1 if the segmentation label at pixel p is o_j but $D_{o_i}(p) < D_{o_j}(p)$. See [26] for more details.

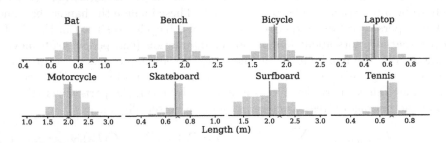

Fig. 7. Learned size distribution from human interaction: Here, we visualize the distribution of object sizes across the COCO-2017 [40] test set at the end of optimization. The red caret denotes the size resulting from the hand-picked scale used for initialization. The blue line denotes the size produced by the empirical mean scale of all category instances at the end of optimization. We then use the empirical mean as the new initialization for intrinsic object scale.

Collision loss: Promoting proximity between people and objects can exacerbate the problem of instances occupying the same 3D space. To address this, we penalize poses that would human and/or object interpenetration using the collision loss $L_{\text{collision}}$ introduced in [6,57]. We use a GPU implementation based on [47] which detects colliding mesh triangles, computes a 3D distance field, and penalizes based on the depth of the penetration. See [47] for more details.

4   Evaluation

In this section, we provide quantitative and qualitative analysis on the performance of our method on the COCO-2017 [40] dataset. We focus our evaluation on 8 categories: baseball bats, benches, bicycles, laptops, motorcycles, skateboards, surfboards, and tennis rackets. These categories cover a significant variety in size, shape, and types of interaction with humans.

4.1   Quantitative Analysis

Since 3D ground truth annotations for both humans and objects do not exist for in-the-wild images, we used a forced-choice evaluation procedure on COCO-2017 [40] images. To test the contribution of the holistic processing of human and object instances, we evaluate our approach against an "independent composition," which uses our approach from Sect. 3.1 and Sect. 3.2 to independently estimate the human and object poses. To make the independent composition competitive, we set the intrinsic scale to be the empirical mean per-category scale learned over the test set by our proposed method in Sect. 3.3. This actually gives the independent composition global information through human-object interaction. This is the best that can be done without considering all instances holistically.

We collected a random subset of images from the COCO 2017 test set in which at least one person and object overlap in 2D. For each of our object categories, we randomly sample 50 images with at least one instance of that category. For each image, annotators see the independent composition and the result of our proposed method in random order, marking whether our result looks better than, equal to, or worse than the independent composition (see supplementary for screenshots of our annotating tool). We compute the average percentage of images for which our method performs better (treating equal as 50) in Table 1. Overall, we find that our method safely outperforms the independent composition.

To evaluate the importance of the individual loss terms, we run an ablative study. We run the same forced choice test for the full proposed method compared with dropping a single loss term in Table 1. We find that omitting the occlusion-aware silhouette loss and the interaction loss has the most significant effect. Using the silhouette loss during global optimization ensures that the object poses continue to respect image evidence, and the interaction loss encodes the spatial arrangement of the object relative to the person. We did observe that the interaction loss occasionally pulls the object too aggressively toward the person for the laptop category. The scale loss appears to have a positive effect for most categories. Note that because we initialized the scale to the empirical mean, the effects of the scale loss are not as pronounced as they would be if initialized to something else. The depth ordering loss gave a moderate boost while the collision loss had a less pronounced effect. We attribute this to the collision loss operating only on the surface triangles and thus prone to

Fig. 8. Qualitative results of our method on test images from COCO 2017.
Our method, PHOSA, recovers plausible human-object spatial arrangements by explicitly reasoning about human interaction. We evaluate the importance of modeling interaction by comparing with independently estimated human and object poses also using our method (Independent Composition). The intrinsic scale for the independent composition is set to the per-category empirical mean scale learned by our method.

Table 1. Percentage of images for which our proposed method performs better on a subset of COCO 2017 test set. In this table, we evaluate our approach against an independent composition and ablations of our method. The independent composition estimates human and object pose independently using Sect. 3.1 and Sect. 3.2 and sets the intrinsic scale to the empirical mean category scale learned by our method. The ablations each drop one loss term from our proposed method. In each row, we compute the average percentage of images for which our method performs better across a random sample of COCO 2017 test set images [40]. A number greater than 50 implies that our proposed method performs better than the independent composition or that the ablated loss term is beneficial for that category.

Ours vs.	Bat	Bench	Bike	Laptop	Motor.	Skate.	Surf.	Tennis
Indep. Comp.	83	74	73	61	74	71	82	82
No $L_{occ\text{-}sil}$	79	74	87	76	70	96	80	77
No $L_{interaction}$	82	59	57	46	71	71	76	68
No L_{scale}	77	49	54	51	54	55	55	56
No L_{depth}	50	55	55	55	52	50	51	50
No $L_{collision}$	52	40	51	51	50	52	50	50

getting stuck in local minima in which objects get embedded inside the person (especially for benches).

4.2 Qualitative Analysis

In Fig. 8, we demonstrate the generality of our approach on the COCO 2017 dataset by reconstructing the spatial arrangement of multiple people and objects engaged in a wide range of 3D human-object interactions. We find that our method works on a variety of everyday objects with which people interact, ranging from handheld objects (baseball bats, tennis rackets, laptops) to full-sized objects (skateboards, bicycles, motorcycles) to large objects (surfboards, benches). In the middle column of Fig. 8, we visualize the spatial arrangement produced by the independent composition introduced in Sect. 4. We find that independently estimating human and object poses is often insufficient for resolving fundamental ambiguities in scale. Explicitly reasoning about human-object interaction produces more realistic spatial arrangements. Please refer to the Supplementary Materials for discussion of failure modes and significantly more qualitative results.

5 Discussion

In summary, we have found that 2D and 3D technologies for understanding objects and humans have advanced considerably. Armed with these advances, we believe the time is right to start tackling broader questions of holistic 3D scene-understanding—and moving such questions from the lab to uncontrolled in-the-wild imagery! Our qualitative analysis suggests that 3D human understanding has particularly matured, even for intricate interactions with objects in the wild. Detailed recovery of 3D object shape is still a challenge, as illustrated by our rather impoverished but surprisingly effective exemplar-based shape model. It would be transformative to learn statistical models (a "SMPL-for-objects"), and we take the first step by learning the intrinsic scale distribution from data.

A number of conclusions somewhat surprised us. First, even though object shape understanding lacks some tools compared to its human counterpart (such as statistical 3D shape models and keypoint detectors), 2D object instance masks combined with a differentiable renderer and a 3D shape library proves to be a rather effective initialization for 3D object understanding. Perhaps even more remarkable is the scalability of such an approach. Adding new objects and defining their modes of interactions is relatively straightforward, because it is far easier to "paint" annotations on 3D models than annotate individual image instances. Hence 3D shapes provide a convenient coordinate frame for *meta*-level supervision. This is dramatically different from the typical supervised pipeline, in which adding a new object category is typically quite involved.

While the ontology of objects will likely be diverse and continue to grow and evolve over time, humans will likely remain a consistent area of intense focus and targeted annotation. Because of this, we believe it will continue to be fruitful to pursue approaches that leverage contextual constraints from humans that act as "rulers" to help reason about objects. In some sense, this philosophy harkens back to Protagoras's quote from Ancient Greece—"man is the measure of all things"!

Acknowledgements. We thank Georgia Gkioxari and Shubham Tulsiani for insightful discussion and Victoria Dean and Gengshan Yang for useful feedback. We also thank Senthil Purushwalkam for deadline reminders. This work was funded in part by the CMU Argo AI Center for Autonomous Vehicle Research.

References

1. 3d Warehouse. https://3dwarehouse.sketchup.com
2. Free3d. https://free3d.com
3. Agarwal, A., Triggs, B.: Recovering 3D human pose from monocular images. TPAMI **28**(1), 44–58 (2006)
4. Anguelov, D., Srinivasan, P., Koller, D., Thrun, S., Rodgers, J., Davis, J.: SCAPE: shape completion and animation of people. In: SIGGRAPH (2005)
5. Aubry, M., Maturana, D., Efros, A.A., Russell, B.C., Sivic, J.: Seeing 3d chairs: exemplar part-based 2d-3d alignment using a large dataset of cad models. In: CVPR (2014)
6. Ballan, L., Taneja, A., Gall, J., Van Gool, L., Pollefeys, M.: Motion capture of hands in action using discriminative salient points. In: Fitzgibbon, A., Lazebnik, S., Perona, P., Sato, Y., Schmid, C. (eds.) ECCV 2012. LNCS, vol. 7577, pp. 640–653. Springer, Heidelberg (2012). https://doi.org/10.1007/978-3-642-33783-3_46

7. Bogo, F., Kanazawa, A., Lassner, C., Gehler, P., Romero, J., Black, M.J.: Keep it SMPL: automatic estimation of 3D human pose and shape from a single image. In: Leibe, B., Matas, J., Sebe, N., Welling, M. (eds.) ECCV 2016. LNCS, vol. 9909, pp. 561–578. Springer, Cham (2016). https://doi.org/10.1007/978-3-319-46454-1_34

8. Chen, Y., Huang, S., Yuan, T., Qi, S., Zhu, Y., Zhu, S.C.: Holistic++ scene understanding: single-view 3D holistic scene parsing and human pose estimation with human-object interaction and physical commonsense. In: CVPR (2019))

9. Chen, Z., Zhang, H.: Learning implicit fields for generative shape modeling. In: CVPR (2019)

10. Choy, C.B., Xu, D., Gwak, J.Y., Chen, K., Savarese, S.: 3D-R2N2: a unified approach for single and multi-view 3D object reconstruction. In: Leibe, B., Matas, J., Sebe, N., Welling, M. (eds.) ECCV 2016. LNCS, vol. 9912, pp. 628–644. Springer, Cham (2016). https://doi.org/10.1007/978-3-319-46484-8_38

11. Delaitre, V., Fouhey, D.F., Laptev, I., Sivic, J., Gupta, A., Efros, A.A.: Scene semantics from long-term observation of people. In: Fitzgibbon, A., Lazebnik, S., Perona, P., Sato, Y., Schmid, C. (eds.) ECCV 2012. LNCS, vol. 7577, pp. 284–298. Springer, Heidelberg (2012). https://doi.org/10.1007/978-3-642-33783-3_21

12. Fan, H., Su, H., Guibas, L.J.: A point set generation network for 3D object reconstruction from a single image. In: CVPR (2017)

13. Fouhey, D.F., Delaitre, V., Gupta, A., Efros, A.A., Laptev, I., Sivic, J.: People watching: human actions as a cue for single view geometry. Int. J. Comput. Vision 110(3), 259–274 (2014). https://doi.org/10.1007/s11263-014-0710-z

14. Gavrila, D.M.: Pedestrian detection from a moving vehicle. In: Vernon, D. (ed.) ECCV 2000. LNCS, vol. 1843, pp. 37–49. Springer, Heidelberg (2000). https://doi.org/10.1007/3-540-45053-X_3

15. Gkioxari, G., Malik, J., Johnson, J.: Mesh r-cnn. In: ICCV (2019)

16. Girdhar, R., Fouhey, D.F., Rodriguez, M., Gupta, A.: Learning a predictable and generative vector representation for objects. In: Leibe, B., Matas, J., Sebe, N., Welling, M. (eds.) ECCV 2016. LNCS, vol. 9910, pp. 484–499. Springer, Cham (2016). https://doi.org/10.1007/978-3-319-46466-4_29

17. Grauman, K., Shakhnarovich, G., Darrell, T.: Inferring 3D structure with a statistical image-based shape model. In: ICCV (2003)

18. Groueix, T., Fisher, M., Kim, V.G., Russell, B.C., Aubry, M.: A papier-mâché approach to learning 3D surface generation. In: CVPR (2018)

19. Guan, P., Weiss, A., Balan, A.O., Black, M.J.: Estimating human shape and pose from a single image. In: ICCV (2009)

20. Guler, R.A., Kokkinos, I.: Holopose: holistic 3D human reconstruction in-the-wild. In: CVPR (2019)

21. Gupta, A., Satkin, S., Efros, A.A., Hebert, M.: From 3D scene geometry to human workspace. In: CVPR (2011)

22. Hassan, M., Choutas, V., Tzionas, D., Black, M.J.: Resolving 3D human pose ambiguities with 3D scene constraints. In: ICCV (2019)

23. Hasson, Y., et al.: Learning joint reconstruction of hands and manipulated objects. In: CVPR (2019)

24. He, K., Gkioxari, G., Dollár, P., Girshick, R.: Mask r-cnn. In: ICCV (2017)

25. Izadinia, H., Shan, Q., Seitz, S.M.: Im2cad. In: CVPR (2017)

26. Jiang, W., Kolotouros, N., Pavlakos, G., Zhou, X., Daniilidis, K.: Coherent reconstruction of multiple humans from a single image. In: CVPR, pp. 5579–5588 (2020)

27. Joo, H., Neverova, N., Vedaldi, A.: Exemplar fine-tuning for 3D human pose fitting towards in-the-wild 3D human pose estimation. arXiv preprint arXiv:2004.03686 (2020)

28. Joo, H., Simon, T., Sheikh, Y.: Total capture: a 3D deformation model for tracking faces, hands, and bodies. In: CVPR (2018)
29. Kanazawa, A., Black, M.J., Jacobs, D.W., Malik, J.: End-to-end recovery of human shape and pose. In: CVPR (2018)
30. Kar, A., Tulsiani, S., Carreira, J., Malik, J.: Category-specific object reconstruction from a single image. In: CVPR (2015)
31. Kato, H., Ushiku, Y., Harada, T.: Neural 3D mesh renderer. In: CVPR (2018)
32. Kholgade, N., Simon, T., Efros, A., Sheikh, Y.: 3D object manipulation in a single photograph using stock 3D models. ACM Trans. Graph. (TOG) **33**(4), 1–12 (2014)
33. Kingma, D.P., Ba, J.: Adam: a method for stochastic optimization. arXiv preprint arXiv:1412.6980 (2014)
34. Kirillov, A., Wu, Y., He, K., Girshick, R.: Pointrend: image segmentation as rendering. In: Proceedings of the IEEE/CVF Conference on Computer Vision and Pattern Recognition, pp. 9799–9808 (2020)
35. Kjellström, H., Kragić, D., Black, M.J.: Tracking people interacting with objects. In: CVPR (2010)
36. Kolotouros, N., Pavlakos, G., Black, M.J., Daniilidis, K.: Learning to reconstruct 3D human pose and shape via model-fitting in the loop. In: ICCV (2019)
37. Kulkarni, N., Misra, I., Tulsiani, S., Gupta, A.: 3D-relnet: joint object and relational network for 3D prediction. In: CVPR (2019)
38. Kundu, A., Li, Y., Rehg, J.M.: 3D-rcnn: Instance-level 3D object reconstruction via render-and-compare. In: CVPR (2018)
39. Lim, J.J., Pirsiavash, H., Torralba, A.: Parsing ikea objects: fine pose estimation. In: ICCV (2013)
40. Lin, T.-Y., et al.: Microsoft COCO: common objects in context. In: Fleet, D., Pajdla, T., Schiele, B., Tuytelaars, T. (eds.) ECCV 2014. LNCS, vol. 8693, pp. 740–755. Springer, Cham (2014). https://doi.org/10.1007/978-3-319-10602-1_48
41. Loper, M., Mahmood, N., Romero, J., Pons-Moll, G., Black, M.J.: SMPL: a skinned multi-person linear model. In: SIGGRAPH Asia (2015)
42. Mehta, D., et al.: Vnect: real-time 3D human pose estimation with a single RGB camera. In: SIGGRAPH (2017)
43. Mescheder, L., Oechsle, M., Niemeyer, M., Nowozin, S., Geiger, A.: Occupancy networks: learning 3D reconstruction in function space. In: CVPR (2019)
44. Monszpart, A., Guerrero, P., Ceylan, D., Yumer, E., Mitra, N.J.: imapper: interaction-guided scene mapping from monocular videos. ACM Trans. Graph. (TOG) **38**(4), 1–15 (2019)
45. Omran, M., Lassner, C., Pons-Moll, G., Gehler, P.V., Schiele, B.: Neural body fitting: unifying deep learning and model-based human pose and shape estimation. In: 3DV (2018)
46. Park, J.J., Florence, P., Straub, J., Newcombe, R., Lovegrove, S.: Deepsdf: learning continuous signed distance functions for shape representation. In: CVPR (2019)
47. Pavlakos, G., et al.: Expressive body capture: 3D hands, face, and body from a single image. In: CVPR (2019)
48. Pavlakos, G., Kolotouros, N., Daniilidis, K.: TexturePose: supervising human mesh estimation with texture consistency. In: ICCV (2019)
49. Pavlakos, G., Zhou, X., Daniilidis, K.: Ordinal depth supervision for 3D human pose estimation. In: CVPR (2018)
50. Pavlakos, G., Zhu, L., Zhou, X., Daniilidis, K.: Learning to estimate 3D human pose and shape from a single color image. In: CVPR (2018)
51. Rosenhahn, B., Schmaltz, C., Brox, T., Weickert, J., Cremers, D., Seidel, H.P.: Markerless motion capture of man-machine interaction. In: CVPR (2008)

52. Savva, M., Chang, A.X., Hanrahan, P., Fisher, M., Nießner, M.: Pigraphs: learning interaction snapshots from observations. ACM Trans. Graph. (TOG) **35**(4), 1–12 (2016)
53. Sigal, L., Balan, A., Black, M.J.: Combined discriminative and generative articulated pose and non-rigid shape estimation. In: NeurIPS (2008)
54. Song, S., Yu, F., Zeng, A., Chang, A.X., Savva, M., Funkhouser, T.: Semantic scene completion from a single depth image. In: CVPR (2017)
55. Tulsiani, S., Gupta, S., Fouhey, D., Efros, A.A., Malik, J.: Factoring shape, pose, and layout from the 2D image of a 3D scene. In: CVPR (2018)
56. Tung, H.Y.F., Tung, H.W., Yumer, E., Fragkiadaki, K.: Self-supervised learning of motion capture. In: NeurIPS (2017)
57. Tzionas, D., Ballan, L., Srikantha, A., Aponte, P., Pollefeys, M., Gall, J.: Capturing hands in action using discriminative salient points and physics simulation. Int. J. Comput. Vision **118**(2), 172–193 (2016). https://doi.org/10.1007/s11263-016-0895-4
58. Varol, G., et al.: Bodynet: volumetric inference of 3D human body shapes. In: ECCV (2018)
59. Wu, Z., et al.: 3D shapenets: a deep representation for volumetric shapes. In: CVPR (2015)
60. Xiang, D., Joo, H., Sheikh, Y.: Monocular total capture: posing face, body, and hands in the wild. In: CVPR (2019)
61. Yamamoto, M., Yagishita, K.: Scene constraints-aided tracking of human body. In: CVPR (2000)
62. Zanfir, A., Marinoiu, E., Sminchisescu, C.: Monocular 3D pose and shape estimation of multiple people in natural scenes the importance of multiple scene constraints. In: CVPR (2018)
63. Zanfir, A., Marinoiu, E., Zanfir, M., Popa, A.I., Sminchisescu, C.: Deep network for the integrated 3D sensing of multiple people in natural images. In: NeurIPS (2018)
64. Zhou, S., Fu, H., Liu, L., Cohen-Or, D., Han, X.: Parametric reshaping of human bodies in images. In: SIGGRAPH (2010)

Sep-Stereo: Visually Guided Stereophonic Audio Generation by Associating Source Separation

Hang Zhou$^{(\boxtimes)}$, Xudong Xu, Dahua Lin, Xiaogang Wang, and Ziwei Liu

CUHK - SenseTime Joint Lab, The Chinese University of Hong Kong, Hong Kong, China
zhouhang@link.cuhk.edu.hk,
{xx018,dhlin}@ie.cuhk.edu.hk,
xgwang@ee.cuhk.edu.hk,
zwliu.hust@gmail.com

Abstract. Stereophonic audio is an indispensable ingredient to enhance human auditory experience. Recent research has explored the usage of visual information as guidance to generate binaural or ambisonic audio from mono ones with stereo supervision. However, this fully supervised paradigm suffers from an inherent drawback: the recording of stereophonic audio usually requires delicate devices that are expensive for wide accessibility. To overcome this challenge, we propose to leverage the vastly available mono data to facilitate the generation of stereophonic audio. Our key observation is that the task of visually indicated audio separation also maps independent audios to their corresponding visual positions, which shares a similar objective with stereophonic audio generation. We integrate both stereo generation and source separation into a unified framework, **Sep-Stereo**, by considering source separation as a particular type of audio spatialization. Specifically, a novel associative pyramid network architecture is carefully designed for audio-visual feature fusion. Extensive experiments demonstrate that our framework can improve the stereophonic audio generation results while performing accurate sound separation with a shared backbone (Code, models and demo video are available at https://hangz-nju-cuhk.github.io/projects/Sep-Stereo.).

1 Introduction

Sight and sound are both crucial components of human perceptions. Sensory information around us is inherently multi-modal, mixed with both pixels and

---

H. Zhou and X. Xu—Equal Contribution.

Electronic supplementary material The online version of this chapter (https://doi.org/10.1007/978-3-030-58610-2_4) contains supplementary material, which is available to authorized users.

© Springer Nature Switzerland AG 2020
A. Vedaldi et al. (Eds.): ECCV 2020, LNCS 12357, pp. 52–69, 2020.
https://doi.org/10.1007/978-3-030-58610-2_4

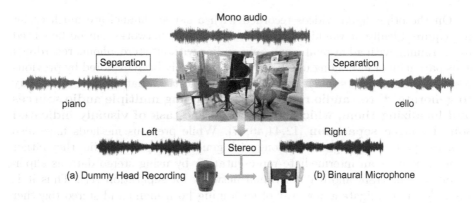

Fig. 1. We propose to integrate audio source separation and stereophonic generation into one framework. Since both tasks build associations between audio sources and their corresponding visual objects in videos. Below is the equipment for recording stereo: (a) dummy head recording system, (b) Free Space XLR Binaural Microphone of 3Dio. The equipment is not only expensive but also not portable. This urges the need for leveraging mono data in stereophonic audio generation

vocals. More importantly, the stereophonic or spatial effect of sound received by two ears gives us the superiority to roughly reconstruct the layout of the environment, which complements the spatial perception in the vision system. This spatial perception of sound makes it appealing for content creators to create audio information with more than one channel. For example, the user's experience will be greatly promoted if stereo music instead of mono is provided when watching a recording of a concert.

However, it is still inconvenient for portable devices to record stereophonic audio. Normally, cell phones and cameras have only mono or line array microphones that can not record real binaural audio. To achieve such goals, dummy head recording systems or fake auricles need to be employed for creating realistic 3D audio sensations that humans truly perceive. This kind of system requires two microphones attached to artificial ears (head) for mimicking a human listener in the scenery, as shown in Fig. 1. Due to the cost and weight of the devices, such recorded binaural audio data is limited, particularly the ones associated with visual information. Therefore, developing a system to generate stereophonic audio automatically from visual guidance is highly desirable.

To this end, Gao and Grauman [17] contribute the FAIR-Play dataset collected with binaural recording equipment. Along with this dataset, they propose a framework for recovering binaural audio with the mono input. Nevertheless, their method is built in a data-driven manner, fusing visual and audio information with a simple feature concatenation, which is hard to interpret. Moreover, the data is valuable yet insufficient to train a network that can generalize well on all scenarios. Similarly, Morgado et al. [30] explore to use 360° videos uploaded on youtube for generating ambisonic audio.

On the other hand, videos recorded with a single channel are much easier to acquire. Ideally, it would be a great advantage if networks can be benefited from training with additional mono-only audios without stereophonic recordings. This paradigm of leveraging unlabeled data has rarely been explored by previous research because of its inherent challenge. We observe that **an essential way to generate stereo audio requires disentangling multiple audio sources and localizing them, which is similar to the task of visually indicated sound source separation** [12,44,46,47]. While previous methods have also explored performing separation along with generating spatial audio, they either implement it as an intermediate representation by using stereo data as supervision [30], or train another individual network for separation [17]. Thus it is valuable to investigate a new way of leveraging both mono and stereo together for solving both problems. Moreover, it is particularly desirable if the two tasks can be solved within one unified framework with a shared backbone as illustrated in Fig. 1. However, there are considerable differences between these two tasks and great challenges to overcome. The details are explained in Sect. 3.1.

Our key insight is to **regard the problem of separating two audios as an extreme case of creating binaural audio**. More specifically, we perform duet audio separation with the hypothesis that the two sources are only visible at the edges of human sight. And no visual information can be provided about the whole scene. Based on this assumption and our observation above, we propose to explicitly associate different local visual features to different responses in spectrograms of audios. In this manner, the intensity of sound can be represented on both the image domain visually and the audio domain auditorily. A novel **associative pyramid network** architecture is proposed for better fusing the information within the two modalities. The whole learning process can be divided into two parts, namely **separative learning** and **stereophonic learning**. We perform multi-task training with two different sets of data: mono ones (MUSIC [47]) for separation, and binaural ones (FAIR-Play [17]) for stereophonic audio generation. Following traditional multi-task learning settings, the two learning stages share a same backbone network but learn with different heads with the same architecture. This framework is uniformly called **Sep-Stereo**.

Extensive experiments regarding stereophonic audio generation have validated the effectiveness of our proposed architecture and learning strategy. At the same time, the trained model can preserve competitive results on the task of audio separation simultaneously. Moreover, we show that with the aid of mono data, our framework is able to achieve generalization under a low data regime using only a small amount of stereo audio supervision.

Our **contributions** are summarized as follows: **1)** We unify audio source separation and stereophonic audio generation into a principled framework, **Sep-Stereo**, which performs joint **separative** and **stereophonic** learning. **2)** We propose a novel **associative pyramid network** architecture for coupling audio-visual responses, which enables effective training of both tasks simultaneously with a shared backbone. **3)** Our **Sep-Stereo** framework has a unique advantage of leveraging mono audio data into stereophonic learning. Extensive experiments demonstrate that our approach is capable of producing more realistic binaural audio while preserving satisfying source separation quality.

2   Related Works

2.1   Joint Audio-Visual Learning

The joint learning of both audio and visual information has received growing attention in recent years [15,19,23,35,53]. By leveraging data within the two modalities, researchers have shown success in learning audio-visual self-supervision [2–4,22,25,31], audio-visual speech recognition [21,39,45,48], localization [34,37,38,47], event localization (parsing) [40,41,43], audio-visual navigation [5,13], cross-modality generation between the two modalities [6–9,42,48–52] and so on. General representation learning across the two modalities is normally conducted in a self-supervised manner. Relja *et al.* [2,3] propose to learn the association between visual objects and sound, which supports localizing the objects that sound in an image. Owens *et al.* [31] and Korbar *et al.* [25] train neural networks to predict whether video frames and audios are temporally aligned. Researchers have also explored the possibility of directly generating sound according to videos [8,32,51], which is more related to our task. Different from their aims, our Sep-Stereo framework exploits visual-audio correspondence to improve the generation of stereophonic audios.

2.2   Audio Source Separation and Spatialization

Source Separation. Source separation with visual guidance has been an interest of research for decades [1,10–12,16,29,33,44,46,47]. Compared with audio-only source separation, visual information could provide rich clues about the types and movements of audio sources. Thus the performance of audio separation is expected to improve with the guidance of vision. Recently, deep learning has been widely applied into this filed of research. For separating speech segments, Afouras *et al.* [1] propose to leverage mouth movements as guidance, and Ephrat *et al.* [10] use cropped human faces. Owens *et al.* [31] do not crop faces and modify their pipeline from learning synchronization. On the other hand, instrumental music [16,18,44,46,47] is the field that we care more about. Gao *et al.* [16] propose to combine non-negative matrix factorization (NMF) with audio features. Zhao *et al.* [47] use a learnable U-Net instead, and match feature activations with different audio channels. Based on this work, motion information is merged into the main framework to achieve better performance in [46]. In [18], object detection and instrument labels are leveraged to co-separating sound. In our work, we will not model motion explicitly, thus adopt a similar setting as [47].

Spatialization. Visually guided audio spatialization has received relatively less attention [14,17,26,28,30] compared with separation. Recently, Li *et al.* [26] leverage synthesised early reverberation and measured late reverberation tail to generate stereo sound in a specific room, which cannot generalize well to other scenarios. With the assistance of deep learning, Morgado *et al.* [30] propose to generate ambisonic for 360° videos using recorded data as self-supervision. The work mostly related to ours is [17]. They contribute a self-collected binaural audio dataset, and propose a U-Net based framework for mono-to-binaural generation on normal field of view (NFOV) videos. These works all only leverage the limited stereophonic audio. In this paper, we propose to boost the spatialization performance with additional mono data.

3 Our Approach

Our proposed framework, **Sep-Stereo**, is illustrated in Fig. 2. This whole pipeline consists of two parts: (a) stereophonic learning and (b) separative learning. We will first introduce the overall framework of visually guided stereophonic audio generation and source separation (Sect. 3.1). Then we demonstrate how our proposed network architectures can effectively associate and integrate audio and visual features into a unified network with a shared backbone.

3.1 Framework Overview

Stereophonic Learning. The whole process of stereophonic learning is depicted in the lower part in Fig. 2. In the setting of stereophonic learning, we care for the scenario that human perceives and has access to binaural data. The visual information V_s corresponds to its audio recording of the left ear $a_l(t)$ and the right-ear one $a_r(t)$. Notably, all spatial information is lost when they are averaged to be a mono clip $a_{mono} = (a_l + a_r)/2$, and our goal is to recover left and right given the mono and video. We operate in the Time-Frequency (TF) domain by transferring audio to spectrum using Short-Time Fourier Transformation (STFT) as a common practice. Here we use S_l^t and S_r^t to denote the STFT of the ground truth left and right channels, with t here represents "target". The input of our network is the mono audio by averaging the STFT of the two audio channels:

$$S_{mono} = (S_l^t + S_r^t)/2 = \text{STFT}(a_{mono}). \tag{1}$$

This can be verified by the property of the Fourier Transformation. Please note that due to the complex operation of STFT, each spectrum $S = S_R + j * S_I$ is a complex matrix that consists of the real S_R and imagery part S_I. So the input size of our audio network is $[T, F, 2]$ by stacking the real and imagery channels.

Separative Learning. The task of separation is integrated for its ability to leverage mono data. Our separative learning follows the Mix-and-Separate training procedure [47], where we elaborately mix two independent audios as input and manage to separate them using ground truth as supervision. It is illustrated at the top of Fig. 2. Given two videos V_A and V_B with only mono audios accompanied, the input of the separative phase is the mixture of two mono audios

$a_{mix} = (a_A + a_B)/2$. They can be represented in the STFT spectrum domain as S_A^t, S_B^t and S_{mix}. Aiming at disentangling the mixed audio, separative learning targets at recovering two mono audios with the guidance of corresponding videos.

Connections and Challenges. Apart from our observation that both tasks connect salient image positions with specific audio sources, they all take mono audio clips as input and attempt to split them into two channels. One can easily find a mapping from stereo to separation as: $\{S_{mono} \Rightarrow S_{mix}, S_l^t \Rightarrow S_A^t, S_r^t \Rightarrow S_B^t\}$. From this point of view, the two tasks are substantially similar. However, the goals of the two tasks are inherently different. While each separated channel should contain the sound of one specific instrument, both sources should be audible, *e.g.* in the task of stereo for a scene shown in Fig. 1. Also, the spatial effect would exist if there is only one source, but separation is not needed in such a case. As for the usage of visual information, the separation task aims at finding the most salient area correctly while the stereo one is affected by not only the sources' positions but also the environment's layout. Thus neither an existing stereo framework [17] nor a separation one [47] is capable of handling both tasks.

3.2   Associative Neural Architecture

Backbone Network. Our audio model is built upon Mono2Binaural [17], which is a conditional U-Net [36]. This audio backbone is denoted as Net_a. It consists of skip-connected encoder Net_a^E and decoder Net_a^D. The visual features are extracted using a visual encoder with ResNet18 [20] architecture called Net_v. The input video clip with input dimension $[T, W_v, H_v, 3]$ is encoded into a feature map F_v of size $[w_v, h_v, C_v]$, by conducting a temporal max pooling operation. We assume that the feature map would correspond to each spatial part in the original video with high responses on salient positions.

Associative Pyramid Network. Based on the backbone network, we propose a novel Associative Pyramid Network (APNet) for both learning stages. It is inspired by PixelPlayer [47] that maps one vision activation with one source feature map, but with a different formulation and underlying motivation. Our key idea is to associate different intensities of audio sources with different vision activations in the whole scene with feature map re-scheduling. As illustrated in Fig. 2 and 3, it works as a side-way network along-side the backbone in a coarse-to-fine manner.

We operate on each layer of the decoder Net_a^D in the U-Net after the upsample deconvolutions. Suppose the ith deconv layer's feature map F_a^i is of shape $[W_a^i, H_a^i, C_a^i]$, we first reshape F_v to $[(w_v \times h_v), C_v]$ and multiply it by a learned weight with size $[C_v, C_a^i]$ to be K_v^i with dimension $[1, 1, C_a^i, (h_v \times w_v)]$. This is called the kernel transfer operation. Then K_v^i operates as a 1×1 2D-convolution kernel on the audio feature map F_a^i, and renders an entangle audio-visual feature

$F_{ap}^{i'}$ of size $[W_a^i, H_a^i, (h_v \times w_v)]$. This process can be formulated as:

$$F_{ap}^{i'} = \underset{K_v^i}{\text{Conv2d}}(F_a^i). \tag{2}$$

Note that each $[1, 1, C_a^i]$ sub-kernel of K_v^i corresponds to one area in the whole image space. So basically, audio channels and positional vision information are associated through this learned convolution. This operation is named Associative-Conv. The output feature $F_{ap}^{i'}$ can be regarded as the stack of audio features associated with different visual positions.

The $F_{ap}^{1'}$ is the first layer of APNet F_{ap}^1. When $i > 1$, the $(i-1)$th feature map F_{ap}^{i-1} will be upsampled through a deconvolution operation to be of the same size as $F_{ap}^{i'}$. Then a new feature map F_{ap}^i can be generated by a concatenation:

$$F_{ap}^i = \text{Cat}([\text{DeConv}(F_{ap}^{i-1}), F_{ap}^{i'}]). \tag{3}$$

In this way, the side-way APNet can take advantage of the pyramid structure by coarse-to-fine tuning with both low-level and high-level information.

The goal of APNet is to predict the target left and right channels' spectrum in both learning stages. Thus, two parallel final convolutions are applied to the last layer of F_{ap}^i, and map it to two outputs with real and imagery channels. As discussed before, each channel in the APNet is specifically associated with one visual position, the final convolution acts also as a reconfiguration to different source intensities.

3.3   Learning Sep-Stereo

Directly predicting spectrum is difficult due to the large dynamic range of STFTs, so we predict the complex masks $M = \{M_R, M_I\}$ following [17] as our objectives. Suppose the input spectrum is $S_{mono} = S_{R(mono)} + j * S_{I(mono)}$, then a prediction can be written as:

$$S^p = (S_{R(mono)} + j * S_{I(mono)})(M_R + j * M_I). \tag{4}$$

The outputs of our networks are all in the form of complex masks, and the predictions are made by the complex multiplication stated above.

Stereophonic Learning. The base training objective of the backbone network for stereo is to predict the subtraction of the two spectrums $S_D^t = (S_l^t - S_r^t)/2$ as proposed in [17]. The left and right ground truth can be written as:

$$S_r^t = S_{mono} + S_D^t, \quad S_l^t = S_{mono} - S_D^t. \tag{5}$$

The backbone network is to predict the difference spectrum S_D^p, the training objective is:

$$L_D = ||S_D^t - S_D^p||_2^2. \tag{6}$$

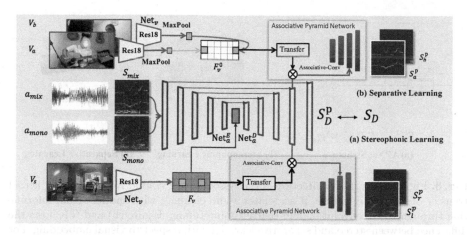

Fig. 2. The whole pipeline of our Sep-Stereo framework. It aims at learning the associations between visual activation and the audio responses. The framework consists of a U-Net backbone whose weights are shared, and our proposed Associative Pyramid Network (APNet). The visual features are firstly extracted and fused with audio features through a multi-scale feature mapping network (APNet). To perform multitask training, (a) stereophonic learning and (b) separative learning stages are leveraged to tackle this problem. One set of APNet parameters are trained for each task

The APNet's outputs are the left and right spectrums, the loss fuction is:

$$L_{rl} = ||S_l^t - S_l^p||_2^2 + ||S_r^t - S_r^p||_2^2. \tag{7}$$

Separative Learning. While the backbone network and APNet seem to be suitable for learning binaural audio, it has the advantage of handling separation by manually modifying features. Our key insight is to regard separation as an extreme case of stereo, that the visual information is only available at the left and right edges. Normally, the visual feature map F_v is a global response which contains salient and non-salient regions. During the separation stage, we manually create the feature map.

Specifically, we adopt max-pooling to the visual feature map F_v to be of size $[1, 1, C_v]$. The feature vectors for video A and B are denoted as F_A and F_B respectively. Then we create an all-zero feature map F_v^0 of the same size as $[w_v, h_v, C_v]$ to serve as a dummy visual map. Then the max-pooled vectors are sent to the left and right most positions as illustrated in Fig. 3. It can be written as:

$$F_v^0(\lceil H/2 \rceil, 1) = F_A, \quad F_v^0(\lceil H/2 \rceil, W) = F_B. \tag{8}$$

Then we replace the F_v with F_v^0. This process is called the **rearrangement** for visual feature.

The intuition for the separative learning to work is based on our design that each channel of the APNet layers corresponds to one visual position. While with

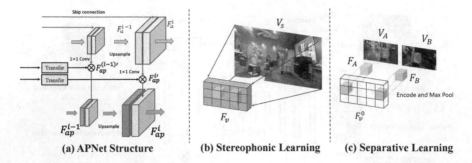

(a) APNet Structure   (b) Stereophonic Learning   (c) Separative Learning

Fig. 3. Figure (a) is the architecture of the Associative Pyramid Network. After kernel transfer of the visual feature, it associates audio channels with spatial visual information through 1×1 convolution and coarse-to-fine tuning. Figure (b) and (c) shows the difference between stereo and separative learning with respect to visual embedding. For (b) stereophonic learning, a visual feature map F_v is directly extracted, with each bin corresponds to a region in the image. While in (c), the visual feature is max-pooled to a 1×1 bin and manually sent to an all-zero feature map F_v^0. This process is the *rearrangement* of visual feature

separative learning, take the left-ear for instance. Most information correlates to the left-ear is zero, thus the left-ear spectrum should correspond to only the left-most visual information. Training the separation task provides especially the backbone network with more audio spectrum and vision information instead of only overfitting to the limited binaural data. Besides, it is also assumed that the non-salient visual features also help our APNet understand the sound field. Thus, without the environment information, we can expect the network to implicitly ignore the distribution of sound around the space but focus on the two sides.

At the separative learning stage, only the APNet predicts the masks for audios A and B, as the left and right channels in the same way as Eq. 4. The predicted spectrums can be represented as S_a^p and S_b^p. So the training objective is:

$$L_{ab} = ||S_a^t - S_a^p||_2^2 + ||S_b^t - S_b^p||_2^2. \tag{9}$$

Final Objective. The final objective of our network is the combination of all the losses for training stereo and separation.

$$L_{all} = L_D + \lambda_1 L_{rl} + \lambda_2 L_{ab} \tag{10}$$

where λ_1 and λ_2 are loss weights that are empirically set to 1 in the experiments through cross-validation.

4 Experiments

4.1 Implementation Details

Preprocessing. We fix all of our audio sampling rate to $16\,\mathrm{kHz}$ and clip the raw audios to ensure their values are between -1 and 1. For performing STFT, our window size is 512, and the hop length is 160. During stereophonic training, we sample a 0.63 s clip randomly from the whole 10 s video. Thus can lead to an STFT map with the size of [257, 64]. Separative learning samples a 0.63 s clip from each individual video as well, and mixes them up as inputs. Other configurations are the same as [17]. The length of the sliding window for testing is 0.1 s. The videos are extracted to frames at 10 fps. At each training time step, the center frame is used as the input of the visual embedding network.

Model Configurations. The backbone audio U-Net Net_a is borrowed from [17], which consists of 5 downsample convolution and 5 de-convolution layers with 4 skip connections between feature maps of the same scale. The Associative Pyramid Network consists of 4 Associative-Conv which couples visual features with audio features. Additionally, there are 3 upsampling operations in APNet. The visual embedding network Net_v is adopted from [47], which is a modified ResNet18 network [20]. The final pooling and fully-connected layers are removed from this network, and the dilation of the network's kernels is 2. Thus F_v is of size [14, 7, 512], where 512 is its channel size.

Training Details. The networks are trained using Adam [24] optimizer with learning rate at 5e−4 and batch size 144. For stereophonic learning, we use the same data augmentation diagram as Mono2Binaural [17]. For separative learning, the amplitude of selected audio is augmented with a random scale disturb of 0.5 to 1.5. The separative learning part is firstly trained, then both data of stereo and separation are sent into the network at the same time. Our original design is to share the backbone and APNet parameters through both learning stages. However, it requires careful tuning for both tasks to converge simultaneously. In our final version, the parameters of Net_a and Net_v are **shared** across two learning stages while **different** sets of APNet parameters are trained for different stages. As the backbone takes up most of the parameters, sharing it with separative learning is the key for improving stereophonic learning. Moreover, visual information is also fused into the backbone, thus our insights all stand even without sharing the APNet parameters.

4.2 Datasets and Evaluation Metrics

In the sense of improving audiences' experiences, videos with instrumental music are the most desired scenario for stereophonic audio. Thus in this paper, we choose music-related videos and audios as a touchstone. Our approach is trained and evaluated on the following datasets:

FAIR-Play. The FAIR-Play dataset is proposed by Gao and Grauman [17]. It consists of 1,871 video clips. The train/val/test has already been split by the authors. We follow the same split and evaluation protocol of [17].

YT-MUSIC. This is also a stereophonic audio dataset that contains video recordings of music performances in 360° view. It is the most challenging dataset collected in paper [30]. As the audios are recorded in first-order ambisonics, we

Table 1. Comparisons between different approaches on FAIR-Play and YT-Music dataset with the evaluation metric of STFT distance and envelop distance. The lower the score the better the results. The training data types for each method are also listed. It can be seen from the results that each component contributes to the network

Method	Training Data		FAIR-Play		YT-Music	
	Stereo	Mono	$STFT_D$	ENV_D	$STFT_D$	ENV_D
Mono2Binaural [17]	✓	✗	0.959	0.141	1.346	0.179
Baseline (MUSIC)	✓	✓	0.930	0.139	1.308	0.175
Assoicative-Conv	✓	✗	0.893	0.137	1.147	0.150
APNet	✓	✗	0.889	0.136	1.070	0.148
Sep-Stereo (Ours)	✓	✓	**0.879**	**0.135**	**1.051**	**0.145**

convert them into binaural audios using an existing converter and also follow the protocol of [17].

MUSIC. We train and evaluate the visually indicated audio source separation on the solo part of MUSIC dataset [47]. Note that in our comparing paper [46], this dataset is enriched to a version with 256 videos for testing named MUSIC21. We follow this setting and use an enriched version with 245 videos for testing, so the comparisons are basically fair. Please be noted that our whole Sep-Stereo model with separative learning is trained on this dataset.

Stereo Evaluation Metrics. We evaluate the performance of audio spatialization using similar metrics used in Mono2Binaural [17] and Ambisonics [30].

- *STFT Distance* ($STFT_D$). As all existing methods are trained in the form of STFT spectrum, it is natural to evaluate directly using the training objective on the test set.
- *Envelope Distance* (ENV_D). As for evaluations on raw audios, we use the envelope distance. It is well-known that direct comparison on raw audios is not informative enough due to the high-frequency nature of audio signals. So we follow [30] to use differences between audio envelopes as a measurement.

Separation Evaluation Metrics. We use these source separation metrics following [47]: Signal-to-Distortion Ratio (SDR), Signal-to-Interference Ratio (SIR), and Signal-to-Artifact Ratio (SAR). The units are dB.

4.3   Evaluation Results on Stereo Generation

As the model of Mono2Binaural is also the baseline of our model, we re-produce Mono2Binaural with our preprocessing and train it carefully using the authors' released code. Besides, we perform extensive ablation studies on the effect of each component in our framework brings on the FAIR-Play and MUSIC dataset. Our modification to the original Mono2Binaural are basically the following modules:

Table 2. Source separation results on MUSIC dataset. The units are dB. †Note that DDT results are directly borrowed from the original paper [46]. It uses additional motion information, while other methods only use static input

Metric	Baseline(MUSIC)	Associative-Conv	PixelPlayer	Sep-Stereo(Ours)	DDT†
SDR	5.21	5.79	7.67	8.07	8.29
SIR	6.44	6.87	14.81	10.14	14.82
SAR	14.44	14.49	11.24	15.51	14.47

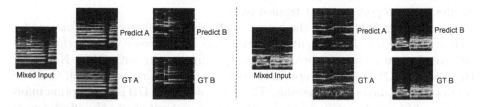

Fig. 4. The visualization of separation results from spectrums. Our results are very similar to the ground truth (GT)

1) **Associative-Conv.** While our APNet associates the visual and audio features at multiple scales, we conduct an additional experiment with Associative-Conv operating at only the outmost layer. This module aims to validate the effectiveness of the associative operation.

2) **APNet.** Then we perform the whole process of stereo learning with the complete version of APNet. Four layers of associative mappings are utilized to perform the coarse-to-fine tuning of the stereo learning.

3) **Baseline (MUSIC).** It is not possible for Mono2Binaural to use mono data in its original setting. Nevertheless, we manage to integrate our separative learning into the baseline by using our **rearrangement** module for the visual feature illustrated in Fig. 3(c). The other parts of the network remain the same. This model can also validate the advantage of our proposed separative learning over Mono2Binaural.

4) **Sep-Stereo (Ours).** Finally, we add the data from MUSIC dataset for training separation. In this model, the separative learning and stereo learning are working together towards more accurate stereophonic audio generation.

The results of the experiments tested on FAIR-Play and YT-MUSIC dataset are listed in Table 1. The "Training Data" column shows whether these models are trained on MUSIC. Due to different preprocessing and sliding window sizes for testing, the results reported in paper [17] in not directly comparable. So we use our re-produced results for comparison. It can be seen that step by step adding our module can lead to better stereo results. Particularly, adding Associative-Conv can shorten the STFT distance by a large margin, which proves that this procedure can efficiently merge audio and visual data. Then improvements can be seen when expanding it to be APNet. Finally, integrating separative learning into our framework gives the network more diverse data for training which leads to be the best outcome.

4.4   Evaluation Results on Source Separation

Our competing methods are the baseline (MUSIC) trained directly for separation, ablation of Associative-Conv, the results of self-implemented PixelPlayer [47] and DDT [46] which are originally designed for the separation task. With or without training on FAIR-Play for predicting stereo has little influence on our separation results, so we report the duet trained ones.

As shown in Table 2 that the baseline (MUSIC) and Associative-Conv model cannot achieve satisfying results. However, our Sep-Stereo can outperform PixelPlayer by two metrics and can keep competitive results with DDT. Note that the results of DDT are the reported ones from the original paper [46], thus the results are not directly comparable. The state-of-the-art DDT uses motion information which we do not leverage. Reaching such a result shows the effectiveness of APNet and the value of our model. There is no doubt that we have the potential for further improvements. We visualize two cases of separation results with duet music in Fig. 4. It can be observed that our method can mostly disentangle the two individual spectrums from the mixed one.

4.5   Further Analysis

User Study. We conduct user studies to verify the stereophonic quality of our results. We show the users a total of 10 cases from the FAIR-Play [17] dataset. Four are results selected from Mono2Binaural's video and six are results generated by our own implementation. A total number of 15 users with normal hearing are asked to participate in the study, and a monitoring-level earphone Shure SE846 is used for conducting the study in a quiet place.

The users are asked to listen to the audio and watch the video frames at the same time. They will listen to the generated audios first and listen to the ground truth. They are responsible for telling their preferences over (1) the audio-visual **matching** quality; which of the two audios better matches the video. And (2) **similarity** to the ground truth; which of the two audios are closer to the ground truth. The users can listen to the clips multiple times. One *Neutral* option is provided if it is really difficult to tell the difference. The results show the users' preferences in Fig. 5(a). The final results are averaged per video and per user. The table shows the ratio between selections. It can be seen that it is hard to tell the differences for certain untrained users without the ground truth. However, more people prefer our results than Mono2Binaural under both the two evaluations. The confusion is less when the ground truth is given. It can be inferred that our results are more similar to the ground truth.

Audio-Based Visual Localization. We illustrate the visually salient areas learned from our model in Fig. 6. The way is to filter intense responses in feature

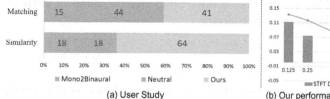
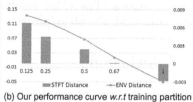

(a) User Study   (b) Our performance curve *w.r.t* training partition

Fig. 5. (a) User study for their preferences between ours and Mono2Binaural [17]. The **matching** stands for audio-visual matching quality and **similarity** represents their similarity with ground truth. The **Neutral** selection means the users cannot tell which one is better. The results are shown in percentage (%). It can be seen that ours are more preferred by users. (b) The curve of our relative performances *w.r.t* the percentage of training data we use on FAIR-Play. The X-axis is the fraction of training data used. Axis on the left represents STFT distance and right is ENV. The lower the better. The curve is drawn in a relative setting, where the performance of Mono2Binaural serves as the reference (zero in both metrics). It can be observed that our framework can reach their full performance using only 67% of the training data

Fig. 6. The visualization of the visual responses according to audio-visual associative learning. The bright places are the salient regions in the feature map, which correspond to intense audio information. The images are selected from MUSIC and FAIR-Play

F_v back to the image space. It can be seen that our network focuses mostly on instruments and humans, which are undoubtedly the potential sound sources.

Generalization Under Low Data Regime. We show the curve of our relative performance gains *w.r.t* the percentage of training data used on FAIR-Play in Fig. 5(b). The curve is drawn in a relative setting, where the performance of Mono2Binaural serves as the reference (zero in both metrics). It can be observed that our framework can reach their full performance by using only 67% of the training data, which is an inherent advantage that our separative and stereophonic learning with mono data brings under low data regime [27].

We also highlight our model's ability of generalization to unseen scenarios by leveraging separative learning. Previous methods such as Mono2Binaural can only be trained with stereo data. It is difficult for them to handle out-of-distribution data if no supervision can be provided. While our method is naturally trained on mono ones, by additional training on only a small portion of stereophonic data with supervision, our method can generalize to in-the-wild mono scenarios. The video results and comparisons can be found at https://hangz-nju-cuhk.github.io/projects/Sep-Stereo.

5   Conclusion

In this work, we propose to integrate the task of stereophonic audio generation and audio source separation into a unified framework namely **Sep-Stereo**. We introduce a novel perspective of regarding separation as a particular type of stereo audio generation problem through manual manipulation on visual feature maps. We further design Associative Pyramid Network (APNet) which associates the visual features and the audio features with a learned Associative-Conv operation. Our proposed Sep-Stereo has the following appealing properties that are rarely achieved before: **1)** Rich mono audio clips can be leveraged to assist the learning of binaural audios. **2)** The task of audio separation and spatialization can be solved with a shared backbone with different heads, thus additional parameters for an entire extra network can be removed. **3)** Stereophonic generation can be generalized to low data regime with the aid of mono data. Extensive evaluation, analysis and visualization demonstrate the effectiveness of our proposed framework.

Acknowledgements. This work is supported by SenseTime Group Limited, the General Research Fund through the Research Grants Council of Hong Kong under Grants CUHK14202217, CUHK14203118, CUHK14205615, CUHK14207814, CUHK14208619, CUHK14213616, CUHK14203518, and Research Impact Fund R5001-18.

References

1. Afouras, T., Chung, J.S., Zisserman, A.: The conversation: deep audio-visual speech enhancement. In: Proceedings Interspeech 2018 (2018)
2. Arandjelovic, R., Zisserman, A.: Look, listen and learn. In: Proceedings of the IEEE International Conference on Computer Vision (ICCV) (2017)
3. Arandjelovic, R., Zisserman, A.: Objects that sound. In: Proceedings of the European Conference on Computer Vision (ECCV) (2018)
4. Aytar, Y., Vondrick, C., Torralba, A.: Soundnet: learning sound representations from unlabeled video. In: Advances in Neural Information Processing Systems (NeurIPS) (2016)
5. Chen, C., et al.: Audio-visual embodied navigation. In: Proceedings of the European Conference on Computer Vision (ECCV) (2020)
6. Chen, L., Li, Z., Maddox, R.K., Duan, Z., Xu, C.: Lip movements generation at a glance. In: Proceedings of the European Conference on Computer Vision (ECCV) (2018)
7. Chen, L., Maddox, R.K., Duan, Z., Xu, C.: Hierarchical cross-modal talking face generation with dynamic pixel-wise loss. In: Proceedings of the IEEE Conference on Computer Vision and Pattern Recognition (CVPR) (2019)
8. Chen, L., Srivastava, S., Duan, Z., Xu, C.: Deep cross-modal audio-visual generation. In: Proceedings of the on Thematic Workshops of ACM Multimedia (2017)
9. Chung, J.S., Jamaludin, A., Zisserman, A.: You said that? In: BMVC (2017)

10. Ephrat, A., et al.: Looking to listen at the cocktail party: a speaker-independent audio-visual model for speech separation. ACM Trans. Graph. (TOG) (2018)
11. Fisher III, J.W., Darrell, T., Freeman, W.T., Viola, P.A.: Learning joint statistical models for audio-visual fusion and segregation. In: Advances In Neural Information Processing Systems (NeurIPS) (2001)
12. Gan, C., Huang, D., Zhao, H., Tenenbaum, J.B., Torralba, A.: Music gesture for visual sound separation. In: IEEE/CVF Conference on Computer Vision and Pattern Recognition (CVPR) (2020)
13. Gan, C., Zhang, Y., Wu, J., Gong, B., Tenenbaum, J.B.: Look, listen, and act: towards audio-visual embodied navigation. In: ICRA (2020)
14. Gan, C., Zhao, H., Chen, P., Cox, D., Torralba, A.: Self-supervised moving vehicle tracking with stereo sound. In: Proceedings of the IEEE International Conference on Computer Vision (2019)
15. Gao, R., Chen, C., Al-Halah, Z., Schissler, C., Grauman, K.: Visualechoes: spatial image representation learning through echolocation. In: Proceedings of the European Conference on Computer Vision (ECCV) (2020)
16. Gao, R., Feris, R., Grauman, K.: Learning to separate object sounds by watching unlabeled video. In: Proceedings of the European Conference on Computer Vision (ECCV) (2018)
17. Gao, R., Grauman, K.: 2.5 D visual sound. In: Proceedings of the IEEE Conference on Computer Vision and Pattern Recognition (CVPR) (2019)
18. Gao, R., Grauman, K.: Co-separating sounds of visual objects. In: Proceedings of the IEEE International Conference on Computer Vision (ICCV) (2019)
19. Gao, R., Oh, T.H., Grauman, K., Torresani, L.: Listen to look: action recognition by previewing audio. In: Proceedings of the IEEE/CVF Conference on Computer Vision and Pattern Recognition (CVPR) (2020)
20. He, K., Zhang, X., Ren, S., Sun, J.: Deep residual learning for image recognition. In: Proceedings of the IEEE Conference on Computer Vision and Pattern Recognition (CVPR) (2016)
21. Hu, D., Li, X., Lu, X.: Temporal multimodal learning in audiovisual speech recognition. In: The IEEE Conference on Computer Vision and Pattern Recognition (CVPR) (2016)
22. Hu, D., Nie, F., Li, X.: Deep multimodal clustering for unsupervised audiovisual learning. In: The IEEE Conference on Computer Vision and Pattern Recognition (CVPR) (2019)
23. Huang, Q., Xiong, Y., Rao, A., Wang, J., Lin, D.: Movienet: A holistic dataset for movie understanding. In: Proceedings of the European Conference on Computer Vision (ECCV) (2020)
24. Kingma, D.P., Ba, J.: Adam: a method for stochastic optimization. arXiv preprint arXiv:1412.6980 (2014)
25. Korbar, B., Tran, D., Torresani, L.: Cooperative learning of audio and video models from self-supervised synchronization. In: Advances in Neural Information Processing Systems (NeurIPS) (2018)
26. Li, D., Langlois, T.R., Zheng, C.: Scene-aware audio for 360 videos. ACM Trans. Graph. (TOG) **37**(4), 1–12 (2018)
27. Liu, Z., Miao, Z., Zhan, X., Wang, J., Gong, B., Yu, S.X.: Large-scale long-tailed recognition in an open world. In: Proceedings of the IEEE Conference on Computer Vision and Pattern Recognition (2019)
28. Lu, Y.D., Lee, H.Y., Tseng, H.Y., Yang, M.H.: Self-supervised audio spatialization with correspondence classifier. In: 2019 IEEE International Conference on Image Processing (ICIP) (2019)

29. Maganti, H.K., Gatica-Perez, D., McCowan, I.: Speech enhancement and recognition in meetings with an audio-visual sensor array. IEEE Trans. Audio Speech Lang. Process. **15**(8), 2257–2269 (2007)
30. Morgado, P., Nvasconcelos, N., Langlois, T., Wang, O.: Self-supervised generation of spatial audio for 360 video. In: Advances in Neural Information Processing Systems (NeurIPS) (2018)
31. Owens, A., Efros, A.A.: Audio-visual scene analysis with self-supervised multisensory features. In: European Conference on Computer Vision (ECCV) (2018)
32. Owens, A., Isola, P., McDermott, J., Torralba, A., Adelson, E.H., Freeman, W.T.: Visually indicated sounds. In: Proceedings of the IEEE Conference on Computer Vision and Pattern Recognition (CVPR) (2016)
33. Parekh, S., Essid, S., Ozerov, A., Duong, N.Q., Pérez, P., Richard, G.: Motion informed audio source separation. In: 2017 IEEE International Conference on Acoustics, Speech and Signal Processing (ICASSP) (2017)
34. Qian, R., Hu, D., Dinkel, H., Wu, M., Xu, N., Lin, W.: Learning to visually localize multiple sound sources via a two-stage manner code. In: Proceedings of the European Conference on Computer Vision (ECCV) (2020)
35. Rao, A., et al.: A local-to-global approach to multi-modal movie scene segmentation. In: Proceedings of the IEEE/CVF Conference on Computer Vision and Pattern Recognition (2020)
36. Ronneberger, O., Fischer, P., Brox, T.: U-Net: convolutional networks for biomedical image segmentation. In: Navab, N., Hornegger, J., Wells, W.M., Frangi, A.F. (eds.) MICCAI 2015. LNCS, vol. 9351, pp. 234–241. Springer, Cham (2015). https://doi.org/10.1007/978-3-319-24574-4_28
37. Rouditchenko, A., Zhao, H., Gan, C., McDermott, J., Torralba, A.: Self-supervised audio-visual co-segmentation. In: IEEE International Conference on Acoustics, Speech and Signal Processing (ICASSP) (2019)
38. Senocak, A., Oh, T.H., Kim, J., Yang, M.H., So Kweon, I.: Learning to localize sound source in visual scenes. In: Proceedings of the IEEE Conference on Computer Vision and Pattern Recognition (CVPR) (2018)
39. Son Chung, J., Senior, A., Vinyals, O., Zisserman, A.: Lip reading sentences in the wild. In: The IEEE Conference on Computer Vision and Pattern Recognition (CVPR) (2017)
40. Tian, Y., Li, D., Xu, C.: Unified multisensory perception: weakly-supervised audio-visual video parsing. In: Proceedings of the European Conference on Computer Vision (ECCV) (2020)
41. Tian, Y., Shi, J., Li, B., Duan, Z., Xu, C.: Audio-visual event localization in unconstrained videos. In: Proceedings of the European Conference on Computer Vision (ECCV) (2018)
42. Wen, Y., Raj, B., Singh, R.: Face reconstruction from voice using generative adversarial networks. In: Advances in Neural Information Processing Systems (NeurIPS) (2019)
43. Wu, Y., Zhu, L., Yan, Y., Yang, Y.: Dual attention matching for audio-visual event localization. In: Proceedings of the IEEE International Conference on Computer Vision (ICCV) (2019)
44. Xu, X., Dai, B., Lin, D.: Recursive visual sound separation using minus-plus net. In: Proceedings of the IEEE International Conference on Computer Vision (ICCV) (2019)
45. Yu, J., et al.: Audio-visual recognition of overlapped speech for the lrs2 dataset. In: ICASSP 2020–2020 IEEE International Conference on Acoustics, Speech and Signal Processing (ICASSP) (2020)

46. Zhao, H., Gan, C., Ma, W.C., Torralba, A.: The sound of motions (2019)
47. Zhao, H., Gan, C., Rouditchenko, A., Vondrick, C., McDermott, J., Torralba, A.: The sound of pixels. In: Proceedings of the European Conference on Computer Vision (ECCV) (2018)
48. Zhou, H., Liu, Y., Liu, Z., Luo, P., Wang, X.: Talking face generation by adversarially disentangled audio-visual representation. In: Proceedings of the AAAI Conference on Artificial Intelligence (AAAI) (2019)
49. Zhou, H., Liu, Z., Xu, X., Luo, P., Wang, X.: Vision-infused deep audio inpainting. In: Proceedings of the IEEE International Conference on Computer Vision (ICCV) (2019)
50. Zhou, Y., Li, D., Han, X., Kalogerakis, E., Shechtman, E., Echevarria, J.: Makeittalk: Speaker-aware talking head animation. arXiv preprint arXiv:2004.12992 (2020)
51. Zhou, Y., Wang, Z., Fang, C., Bui, T., Berg, T.L.: Visual to sound: generating natural sound for videos in the wild. In: Proceedings of the IEEE Conference on Computer Vision and Pattern Recognition (CVPR) (2018)
52. Zhu, H., Huang, H., Li, Y., Zheng, A., He, R.: Arbitrary talking face generation via attentional audio-visual coherence learning. In: International Joint Conference on Artificial Intelligence (IJCAI) (2020)
53. Zhu, H., Luo, M., Wang, R., Zheng, A., He, R.: Deep audio-visual learning: a survey. arXiv preprint arXiv:2001.04758 (2020)

CelebA-Spoof: Large-Scale Face Anti-spoofing Dataset with Rich Annotations

Yuanhan Zhang[1], ZhenFei Yin[2], Yidong Li[1], Guojun Yin[2], Junjie Yan[2], Jing Shao[2(✉)], and Ziwei Liu[3]

[1] Beijing Jiaotong University, Beijing, China
{18120454,yidongli}@bjtu.edu.cn
[2] SenseTime Group Limited, Hong Kong, China
{yinzhenfei,yinguojun,yanjunjie,shaojing}@sensetime.com
[3] The Chinese University of Hong Kong, Hong Kong, China
zwliu@ie.cuhk.edu.hk

Abstract. As facial interaction systems are prevalently deployed, security and reliability of these systems become a critical issue, with substantial research efforts devoted. Among them, face anti-spoofing emerges as an important area, whose objective is to identify whether a presented face is live or spoof. Though promising progress has been achieved, existing works still have difficulty in handling complex spoof attacks and generalizing to real-world scenarios. The main reason is that current face anti-spoofing datasets are limited in both quantity and diversity. To overcome these obstacles, we contribute a large-scale face anti-spoofing dataset, **CelebA-Spoof**, with the following appealing properties: *1) Quantity:* CelebA-Spoof comprises of 625,537 pictures of 10,177 subjects, significantly larger than the existing datasets. *2) Diversity:* The spoof images are captured from 8 scenes (2 environments * 4 illumination conditions) with more than 10 sensors. *3) Annotation Richness:* CelebA-Spoof contains 10 spoof type annotations, as well as the 40 attribute annotations inherited from the original CelebA dataset. Equipped with CelebA-Spoof, we carefully benchmark existing methods in a unified multi-task framework, **Auxiliary Information Embedding Network (AENet)**, and reveal several valuable observations. Our key insight is that, compared with the commonly-used binary supervision or mid-level geometric representations, rich semantic annotations as auxiliary tasks can greatly boost the performance and generalizability of face anti-spoofing across a wide range of spoof attacks. Through comprehensive studies, we show that CelebA-Spoof serves as an effective training data source. Models trained on CelebA-Spoof (without fine-tuning) exhibit state-of-the-art performance on standard benchmarks such as CASIA-MFSD. The datasets are available at https://github.com/Davidzhangyuanhan/CelebA-Spoof.

Electronic supplementary material The online version of this chapter (https://doi.org/10.1007/978-3-030-58610-2_5) contains supplementary material, which is available to authorized users.

Keywords: Face anti-spoofing · Large-scale dataset

1 Introduction

Face anti-spoofing is an important task in computer vision, which aims to facilitate facial interaction systems to determine whether a presented face is live or spoof. With the successful deployments in phone unlock, access control and e-wallet payment, facial interaction systems already become an integral part in the real world. However, there exists a vital threat to these face interaction systems. Imagine a scenario where an attacker with a photo or video of you can unlock your phone and even pay his bill using your e-wallet. To this end, face anti-spoofing has emerged as a crucial technique to protect our privacy and property from being illegally used by others.

Most modern face anti-spoofing methods [8,14,31] are fueled by the availability of face anti-spoofing datasets [4,5,18,24,29,32,34], as shown in Table 1. However, there are several limitations with the existing datasets: 1) *Lack of Diversity.* Existing datasets suffer from lacking sufficient subjects, sessions and input sensors (*e.g.* mostly less than 2000 subject, 4 sessions and 10 input sensors). 2) *Lack of Annotations.* Existing datasets have only annotated the type of spoof type. Face anti-spoof community lacks a densely annotated dataset covering rich attributes, which can further help researchers to explore face anti-spoofing task with diverse attributes. 3) *Performance Saturation.* The classification performance on several face anti-spoofing datasets has already saturated, failing to evaluate the capability of existing and future algorithms. For example, the recall under FPR $= 0.5\%$ on SiW and Oulu-NPU datasets using vanilla ResNet-18 has already reached 100.0% and 99.0%, respectively (Fig. 1).

To address these shortcomings in existing face anti-spoofing dataset, in this work we propose a large-scale and densely annotated dataset, **CelebA-Spoof**. Besides the standard *Spoof Type* annotation, CelebA-Spoof also contains annotations for *Illumination Condition* and *Environment*, which express more information in face anti-spoofing, compared to categorical label like *Live/Spoof*. Essentially, these dense annotations describe images by answering questions like "Is the people in the image Live or Spoof?", "What kind of spoof type is this?", "What kind of illumination condition is this?" and "What kind of environment in the background?". Specifically, all live images in CelebA-Spoof are selected from CelebA [20], and all Spoof images are collected and annotated by skillful annotators. CelebA-Spoof has several appealing properties. 1) **Large-Scale.** CelebA-Spoof comprises of a total of 10177 subjects, 625537 images, which is the largest dataset in face anti-spoofing. 2) **Diversity.** For collecting images, we use more than 10 different input tensors, including phones, pads and personal computers (PC). Besides, we cover images in 8 different sessions. 3) **Rich Annotations.** Each image in CelebA-Spoof is defined with 43 different attributes: 40 types of *Face Attribute* defined in CelebA [20] plus 3 attributes of face anti-spoofing, including: *Spoof Type, Illumination Condition* and *Environment*. With

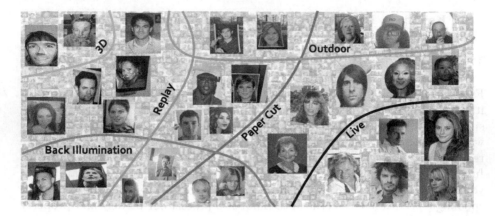

Fig. 1. A quick glance of CelebA-Spoof face anti-spoofing dataset with its attributes. Hypothetical space of scenes are partitioned by attributes and Live/Spoof. In reality, this space is much higher dimensional and there are no clean boundaries between attributes presence and absence

rich annotations, we can comprehensively investigate face anti-spoofing task from various perspectives.

Equipped with CelebA-Spoof, we design a simple yet powerful network named **A**uxiliary information **E**mbedding **N**etwork (**AENet**), and carefully benchmark existing methods within this unified multi-task framework. Several valuable observations are revealed: **1)** We analyze the effectiveness of auxiliary **geometric information** for different spoof types and illustrate the sensitivity of geometric information to special illumination conditions. Geometric information includes *depth map* and *reflection map*. **2)** We validate auxiliary **semantic information**, including face attribute and spoof type, plays an important role in improving classification performance. **3)** We build three CelebA-Spoof benchmarks based on this two auxiliary information. Through extensive experiments, we demonstrate that our large-scale and densely annotated dataset serves as an effective data source in face anti-spoofing to achieve state-of-the-art performance. Furthermore, models trained with auxiliary semantic information exhibit great generalizability compared to other alternatives.

In summary, the **contributions** of this work are three-fold:**1)** We contribute a large-scale face anti-spoofing dataset, **CelebA-Spoof**, with 625,537 images from 10,177 subjects, which includes 43 rich attributes on face, illumination, environment and spoof types. **2)** Based on these rich attributes, we further propose a simple yet powerful multi-task framework, namely **AENet**. Through AENet, we conduct extensive experiments to explore the roles of semantic information and geometric information in face anti-spoofing. **3)** To support comprehensive evaluation and diagnosis, we establish three versatile benchmarks to evaluate the performance and generalization ability of various methods under different carefully-designed protocols. With several valuable observations revealed,

Table 1. The comparison of CelebA-Spoof with existing datasets of face anti-spoofing. Different illumination conditions and environments make up different sessions, (V means video, I means image; Ill. Illumination condition, Env. Environment; - means this information is not annotated)

Dataset	Year	Modality	#Subjects	#Data(V/I)	#Sensor	#Face Attribute	#Semantic Attribute Spoof type	#Session (Ill.,Env.)
Replay-Attack [5]	2012	RGB	50	1,200 (V)	2		1 Print, 2 Replay	1 (-,-)
CASIA-MFSD [35]	2012	RGB	50	600 (V)	3		1 Print, 1 Replay	3 (-,-)
3DMAD [7]	2014	RGB/Depth	14	255 (V)	2		1 3D mask	3 (-,-)
MSU-MFSD [30]	2015	RGB	35	440 (V)	2		1 Print, 2 Replay	1 (-,-)
Msspoof [10]	2015	RGB/IR	21	4,704 (I)	2		1 Print	7 (-,7)
HKBU-MARs V2 [17]	2016	RGB	12	1,008 (V)	7		2 3D masks	6 (6,-)
MSU-USSA [25]	2016	RGB	1,140	10,260 (I)	2		2 Print, 6 Replay	1 (-,-)
Oulu-NPU [4]	2017	RGB	55	5,940 (V)	6		2 Print, 2 Replay	3 (-,-)
SiW [19]	2018	RGB	165	4,620 (V)	2		2 Print, 4 Replay	4 (-,-)
CASIA-SURF [33]	2018	RGB/IR/Depth	1,000	21,000 (V)	1		5 Paper Cut	1 (-,-)
CSMAD [1]	2018	RGB/IR/Depth/LWIR	14	246 (V),17 (I)	1		1 silicone mask	4 (4,-)
HKBU-MARs V1+ [16]	2018	RGB	12	180(v)	1		1 3D mask	1 (1,-)
SiW-M [20]	2019	RGB	493	1,628 (V)	4		1 Print, 1 Replay 5 3D Mask, 3 Make Up, 3 Partial	3 (-,-)
CelebA-Spoof	2020	RGB	10,177	625,537 (I)	>10	40	3 Print, 3 Replay 1 3D, 3 Paper Cut	8 (4,2)

we demonst rate the effectiveness of CelebA-Spoof and its rich attributes which can significantly facilitate future research.

2   Related Work

Face Anti-spoofing Datasets. Face anti-spoofing community mainly has three types of datasets. First, the multi-modal dataset: 3DMAD [7], Msspoof [6], CASIA-SURF [32] and CSMAD [1]. However, since widespread used mobile phones are not equipped with suitable modules, such datasets cannot be widely used in the real scene. Second is the single-modal dataset, such as Replay Attack [5], CASIA-MFSD [34], MSU-MFSD [29], MSU-USSA [24] and HKBU-MARS V2 [16]. But these datasets have been collected for more than three years. With the rapid development of electronic equipment, the acquisition equipment of these datasets is completely outdated and cannot meet the actual needs. SiW [18], Oulu-NPU [4] and HKBU-MAR V1+ [15] are relatively up-to-date. However, the limited number of subjects, spoof types, and environment (Only indoors) in these datasets does not guarantee for the generalization capability required in the real application. Third, SiW-M [19] is mainly used for Zero-Shot face anti-spoofing tasks. CelebA-Spoof datasets have 625537 pictures from 10177 subjects, 8 scenes (2 environments * 4 illumination conditions) with rich annotations. The characteristic of Large-scale and diversity can further fill the gap between face anti-spoofing dataset and real scenes. With rich annotations we can better analyze face anti-spoofing task. All datasets mentioned above are listed in Table 1.

Face Anti-spoofing Methods. In recent years, face anti-spoofing algorithms have seen great progress. Most traditional algorithms focus on handcrafted features, such as LBP [5,21,22,30], HoG [21,25,30] and SURF [2]. Other works also focused on temporal features such as eye-blinking [23,27] and lips motion

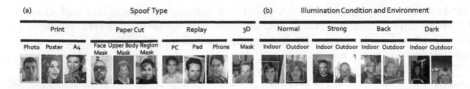

Fig. 2. Representative examples of the semantic attributes (*i.e.* spoof type, illumination and environment) defined upon spoof images. In detail, (a) 4 macro-types and 11 micro-types of spoof type and (b) 4 illumination and 2 types of environmental conditions are defined

[12]. In order to improve the robustness to light changes, some researchers have paid attention to different color spaces, such as HSV [3], YCbcR [2] and Fourier spectrum [13]. With the development of the deep learning model, researchers have also begun to focus on Convolutional Neural Network based methods. [8,14] considered the face PAD problem as binary classification and perform good performance. The method of auxiliary supervision is also used to improve the performance of binary classification supervision. Atoum *et al.* let the full convolutional network to learn the depth map and then assist the binary classification task. Liu *et al.* [15,17] proposed remote toplethysmography (rPPG signal)-based methods to foster the development of 3D face anti-spoofing. Liu *et al.* [18] proposed to leverage depth map combined with rPPG signal as the auxiliary supervision information. Kim *et al.* [11] proposed using depth map and reflection map as the Bipartite auxiliary supervision. Besides, Yang *et al.* [31] proposed to combine the spatial information with the temporal information in the video stream to improve the generalization of the model. Amin *et al.* [10] solved the problem of face anti-spoofing by decomposing a spoof photo into a Live photo and a Spoof noise pattern. These methods mentioned above are prone to over-fitting on the training data, the generalization performance is poor in real scenarios. In order to solve the poor generalization problem, Shao *et al.* [26] adopted transfer learning to further improve performance. Therefore, a more complex face anti-spoofing dataset with large-scale and diversity is necessary. From extensive experiments, CelebA-Spoof has been shown to significantly improve generalization of basic models, In addition, based on auxiliary semantic information method can further achieve better generalization.

3   CelebA-Spoof Dataset

Existing face anti-spoofing datasets cannot satisfy the requirements for real scenario applications. As shown in Table 1, most of them contain fewer than 200 subjects and 5 sessions, meanwhile they are only captured indoor with fewer than 10 types of input sensors. On the contrary, our proposed CelebA-Spoof dataset provides 625, 537 pictures and 10, 177 subjects, therefore offering a superior comprehensive dataset for the area of face anti-spoofing. Furthermore, each image

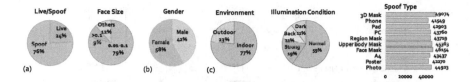

Fig. 3. The statistical distribution of CelebA-Spoof dataset. (a) Overall live and spoof distribution as well as the face size statistic. (b) An exemplar of live attribute, *i.e.* "gender". (c) Three types of spoof attributes

is annotated with 43 attributes. This abundant information enrich the diversity and make face anti-spoofing more illustrative. To our best knowledge, our dataset surpasses all the existing datasets both in scale and diversity.

In this section, we describe our CelebA-Spoof dataset and analyze it through a variety of informative statistics. The dataset is built based on CelebA [20], where all the live people in this dataset are from CelebA. We collect and annotate Spoof images of CelebA-Spoof.

3.1 Semantic Information Collection

In recent decades, studies in attribute-based representations of objects, faces, and scenes have drawn large attention as a complement to categorical representations. However, rare works attempt to exploit semantic information in face anti-spoofing. Indeed, for face anti-spoofing, additional semantic information can characterize the target images by attributes rather than discriminated assignment into a single category, *i.e.* "live" or "spoof".

Semantic for Live - Face Attribute $\mathcal{S}^f$. In our dataset, we directly adopt 40 types of face attributes defined in CelebA [20] as "live" attributes. Attributes of "live" faces always refer to gender, hair color, expression and *etc*. These abundant semantic cues have shown their potential in providing more information for face identification. It is the first time to incorporate them into face anti-spoofing. Extensive studies can be found in Sect. 5.1.

Semantic for Spoof - Spoof Type $\mathcal{S}^s$, Illumination $\mathcal{S}^i$, and Environment $\mathcal{S}^e$. Differs to "live" face attributes, "spoof" images might be characterized by another bunch of properties or attributes as they are not only related to the face region. Indeed, the material of spoof type, illumination condition and environment where spoof images are captured can express more semantic information in "spoof" images, as shown in Fig. 2. Note that the combination of illumination and environment forms the "session" defined in the existing face anti-spoofing dataset. As shown in Table 1, the combination of four illumination conditions and two environments forms 8 sessions. To our best knowledge, CelebA-Spoof is the first dataset covering spoof images in outdoor environment.

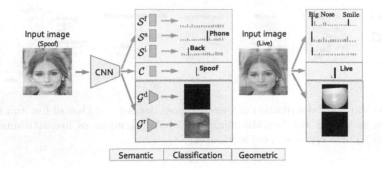

Fig. 4. Auxiliary information Embedding Network (AENet). We use two $\mathrm{Conv}_{3\times3}$ after CNN and upsample to size 14×14 to learn the geometric information. Besides, we use three FC layers to learn the semantic information. The prediction score of $\mathcal{S}^f$ of spoof image should be very low and the prediction result of $\mathcal{S}^s$ and $\mathcal{S}^i$ of live image should be "No illumination" and "No attack" which belongs to the first label in $\mathcal{S}^s$ and $\mathcal{S}^i$ (Color figure online)

3.2 Statistics on CelebA-Spoof Dataset

The CelebA-Spoof dataset is constructed with a total of $625,537$ images. As shown in Fig. 3(a), the ratio of live and spoof is $1 : 3$. Face size in all images is mainly between 0.01 million pixels to 0.1 million pixels. We split the CelebA-Spoof dataset into training, validation, and test sets with a ratio of $8 : 1 : 1$. Note that all three sets are guaranteed to have no overlap on subjects, which means there is no case of a live image of one certain subject in the training set while its counterpart spoof image in the test set. The distribution of live images in three splits is the same as that defined in the CelebA dataset.

The semantic attribute statistics are shown in Fig. 3(c). The portion of each type of attack is almost the same to guarantee a balanced distribution. It is easy to collect data under normal illumination in an indoor environment where most existing datasets adopt. Besides such easy cases, in CelebA-Spoof dataset, we also involve 12% dark, 11% back, and 19% strong illumination. Furthermore, both indoor and outdoor environments contain all illumination conditions.

4   Auxiliary Information Embedding Network

Equipped with CelebA-Spoof dataset, in this section, we design a simple yet effective network named **A**uxiliary information **E**mbedding **N**etwork (AENet), as shown in Fig. 4. In addition to the main binary classification branch (in green), we **1)** Incorporate the *semantic* branch (in orange) to exploit the auxiliary capacity of rich annotated semantic attributes in the dataset, and **2)** Benchmark the existing *geometric* auxiliary information within this unified multi-task framework.

AENet$_{C,S}$. Refers to the multi-task jointly learn auxiliary "semantic" attributes and binary "classification" labels. Such auxiliary semantic attributes defined in our dataset provide complement cues rather than discriminated assignment into a single category. The semantic attributes are learned via the backbone network followed by three FC layers. In detail, given a batch of n images, based on AENet$_{C,S}$, we learn live/spoof class $\{C_k\}_{k=1}^n$ and semantic information, *i.e.* live face attributes $\{S_k^f\}_{k=1}^n$, spoof type $\{S_k^s\}_{k=1}^n$ and illumination conditions $\{S_k^i\}_{k=1}^n$ simultaneously[1]. The loss function of our AENet$_{C,S}$ is

$$\mathcal{L}_{c,s} = \mathcal{L}_C + \lambda_f \mathcal{L}_{S^f} + \lambda_s \mathcal{L}_{S^s} + \lambda_i \mathcal{L}_{S^i}, \tag{1}$$

where $\mathcal{L}_{S^f}$ is binary cross entropy loss. $\mathcal{L}_C$, $\mathcal{L}_{S^s}$ and $\mathcal{L}_{S^i}$ are softmax cross entropy losses. We set the loss weights $\lambda_f = 1$, $\lambda_s = 0.1$ and $\lambda_i = 0.01$, λ values are empirically selected to balance the contribution of each loss.

AENet$_{C,G}$. Besides the semantic auxiliary information, some recent works claim some geometric cues such as *reflection map* and *depth map* can facilitate face anti-spoofing. As shown in Fig. 4 (marked in blue), spoof images exhibit even and the flat surfaces which can be easily distinguished by the depth map. The reflection maps, on the other hand, may display reflection artifacts caused by reflected light from flat surface. However, rare works explore their pros and cons.

AENet$_{C,G}$ also learn auxiliary geometric information in a multi-task fashion with live/spoof classification. Specifically, we concate a Conv_3 × 3 after the backbone network and upsample to 14 × 14 to output the geometric maps. We denote depth and reflection cues as $\mathcal{G}^d$ and $\mathcal{G}^r$ respectively. The loss function is defined as

$$\mathcal{L}_{c,g} = \mathcal{L}_c + \lambda_d \mathcal{L}_{\mathcal{G}^d} + \lambda_r \mathcal{L}_{\mathcal{G}^r}, \tag{2}$$

where $\mathcal{L}_{\mathcal{G}^d}$ and $\mathcal{L}_{\mathcal{G}^r}$ are mean squared error losses. λ_d and λ_r are set to 0.1. In detail, refer to [11], the ground truth of the depth map of live image is generated by PRNet [9] and the ground truth of the reflection map of the spoof image is generated by the method in [33]. Besides, the ground truth of the depth map of the spoof image and the ground truth of the reflection map of the live images are zero.

5   Ablation Study on CelebA-Spoof

Based on our rich annotations in CelebA-Spoof and the designed AENet, we conduct extensive experiments to analyze semantic information and geometric information. Several valuable observations have been revealed: **1)** We validate that S^f and S^s can facilitate live/spoof classification performance greatly. **2)** We analyze the effectiveness of geometric information on different spoof types and find that depth information is particularly sensitive to dark illumination.

---

[1] Note that we do not learn environments S^e since we take face image as input where environment cues (*i.e.* indoor or outdoor) cannot provide more valuable information yet illumination influences much.

Table 2. Different settings in ablation study. For Baseline, we use softmax score of $\mathcal{C}$ for classification (a) For AENet$_\mathcal{S}$, we use the average softmax score of $\mathcal{S}^f$, $\mathcal{S}^s$ and $\mathcal{S}^i$ for classification. AENet$_{\mathcal{S}^f}$, AENet$_{\mathcal{S}^s}$ and AENet$_{\mathcal{S}^i}$ refer to each single spoof semantic attribute respectively. Based on AENet$_{\mathcal{C},\mathcal{S}}$, w/o $\mathcal{S}^f$, w/o $\mathcal{S}^s$, w/o $\mathcal{S}^i$ mean AENet$_{\mathcal{C},\mathcal{S}}$ discards $\mathcal{S}^f$, $\mathcal{S}^s$ and $\mathcal{S}^i$ respectively. (b) For AENet$_{\mathcal{G}^d}$, we use $\|\mathcal{G}^d\|_2$ for classification. Based on AENet$_{\mathcal{C},\mathcal{G}}$, w/o $\mathcal{G}^d$, w/o $\mathcal{G}^r$ mean AENet$_{\mathcal{C},\mathcal{G}}$ discards $\mathcal{G}^d$ and $\mathcal{G}^r$ respectively

(a)	Baseline	AENet$_\mathcal{S}$	AENet$_{\mathcal{S}^f}$	AENet$_{\mathcal{S}^s}$	AENet$_{\mathcal{S}^i}$	AENet$_{\mathcal{C},\mathcal{S}}$ w/o $\mathcal{S}^f$	AENet$_{\mathcal{C},\mathcal{S}}$ w/o $\mathcal{S}^s$	AENet$_{\mathcal{C},\mathcal{S}}$ w/o $\mathcal{S}^i$	AENet$_{\mathcal{C},\mathcal{S}}$	(b)	Baseline	AENet$_{\mathcal{G}^d}$	AENet$_{\mathcal{C},\mathcal{G}}$ w/o $\mathcal{G}^r$	AENet$_{\mathcal{C},\mathcal{G}}$ w/o $\mathcal{G}^d$	AENet$_{\mathcal{C},\mathcal{G}}$
Live/Spoof	✓					✓	✓	✓	✓	Live/Spoof	✓		✓	✓	✓
Face Attribute		✓	✓				✓	✓	✓	Reflection Map				✓	✓
Spoof Type		✓		✓		✓		✓	✓	Depth map		✓	✓		✓
Illumination Conditions		✓			✓	✓	✓		✓						

5.1   Study of Semantic Information

In this subsection, we explore the role of different semantic informations annotated in CelebA-Spoof on face anti-spoofing. Based on AENet$_{\mathcal{C},\mathcal{S}}$, we design eight different models in the Table 2(a). The *key* observations are:

Binary Supervision is Indispensable. As shown in Table 3(a), Compared to baseline, AENet$_\mathcal{S}$ which only leverages three semantic attributes to do the auxiliary job cannot surpass the performance of baseline. However, as shown in 3(b), AENet$_{\mathcal{C},\mathcal{S}}$ which jointly learns auxiliary semantic attributes and binary classification significantly improves the performance of baseline. Therefore we can infer that even such rich semantic information cannot fully replace live/spoof information. But live/spoof with semantic attributes as auxiliary information can be more effective. This is because the semantic attributes of an image cannot be included completely, and a better classification performance cannot be achieved only by relying on several annotated semantic attributes. However, semantic attributes can help the model pay more attention to cues in the image, thus improving the classification performance of the model.

Semantic Attribute Matters. From Table 3(c), we study the impact of different individual semantic attributes on AENet$_{\mathcal{C},\mathcal{S}}$. As shown in this table, AENet$_{\mathcal{C},\mathcal{S}}$ w/o $\mathcal{S}^s$ achieves the worst APCER. Since APCER reflects the classification ability of spoof images, it shows that compared to other semantic attributes, *spoof types* would significantly affect the performance of the spoof images classification of AENet$_{\mathcal{C},\mathcal{S}}$. Furthermore, we list detail information of AENet$_{\mathcal{C},\mathcal{S}}$ in Fig. 5(a). As shown in this figure, AENet$_{\mathcal{C},\mathcal{S}}$ without *spoof types* gets the 5 worst APCER$_{\mathcal{S}^s}$ out of 10 APCER$_{\mathcal{S}^s}$ and we show up these 5 values in this figure. Besides, in Table 3(b), AENet$_{\mathcal{C},\mathcal{S}}$ w/o $\mathcal{S}^f$ gets the highest BPCER. And we also obtain the BPCER$_{\mathcal{S}^f}$ of each face attribute. As shown in Fig. 5(b), among 40 face attributes, BPCER$_{\mathcal{S}^f}$ of AENet$_{\mathcal{C},\mathcal{S}}$ w/o $\mathcal{S}^f$ occupies 25 worst scores. Since BPCER reflects the classification ability of live images, it demonstrate $\mathcal{S}^f$ plays an important role in the classification of live images.

Qualitative Evaluation. Success and failure cases on live/spoof and semantic attributes predictions are shown in Fig. 6. For *live* examples, the first example

Table 3. Semantic information study results in Sect. 5.1. (a) AENet$_\mathcal{S}$ which only depends on semantic attributes for classification cannot surpass the performance of baseline. (b) AENet$_{\mathcal{C},\mathcal{S}}$ which leverages all semantic attributes achieve the best result. **Bolds** are the best results; ↑ means bigger value is better; ↓ means smaller value is better

	Model	Recall (%)↑			AUC ↑	EER (%) ↓	APCER (%) ↓	BPCER (%) ↓	ACER (%) ↓
		FPR = 1%	FPR = 0.5%	FPR = 0.1%					
(a)	Baseline	97.9	95.3	85.9	0.9984	1.6	6.1	1.6	3.8
	AENet$_\mathcal{S}$	98.0	96.0	80.4	0.9981	1.4	6.89	1.44	4.17
(b)	AENet$_{\mathcal{C},\mathcal{S}}$	**98.8**	**97.4**	**90.0**	**0.9988**	**1.1**	**4.62**	**1.09**	**2.85**
(c)	AENet$_{\mathcal{C},\mathcal{S}}$ w/o $\mathcal{S}^i$	98.1	96.5	86.4	0.9982	1.3	**4.62**	1.35	2.99
	AENet$_{\mathcal{C},\mathcal{S}}$ w/o $\mathcal{S}^s$	98.2	96.5	89.4	0.9986	1.3	5.31	1.25	3.28
	AENet$_{\mathcal{C},\mathcal{S}}$ w/o $\mathcal{S}^f$	97.8	95.4	83.6	0.9979	1.3	5.19	1.37	3.28

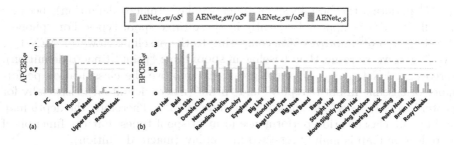

Fig. 5. Representative examples of dropping partial semantic attributes on AENet$_{\mathcal{C},\mathcal{S}}$ performance. In detail, higher APCER$_{\mathcal{S}^s}$ and BPCER$_{\mathcal{S}^f}$ are worse results. (a) Spoof types where AENet$_{\mathcal{C},\mathcal{S}}$ w/o $\mathcal{S}^s$ achieve the worst APCER$_{\mathcal{S}^s}$. (b) Face attributes where AENet$_{\mathcal{C},\mathcal{S}}$ w/o $\mathcal{S}^f$ achieve the worst BPCER$_{\mathcal{S}^f}$

in Fig. 6(a-i) with "glasses" and "hat" help AENet$_{\mathcal{C},\mathcal{S}}$ to pay more attention to the clues of the live image and further improve the performance of prediction of live/spoof. Besides, the first example in Fig. 6(a-ii). AENet$_{\mathcal{C},\mathcal{S}}$ significantly improve the classification performance of live/spoof comparing to baseline. This is because spoof semantic attributes including "back illumination" and "phone" help AENet$_{\mathcal{C},\mathcal{S}}$ recognize the distinct characteristics of spoof image. Note that the prediction of the second example in Fig. 6(b-i) is mistaken.

5.2 Study of Geometric Information

Based on AENet$_{\mathcal{C},}$ under different settings, we design four models as shown in Table 2(b) and use semantic attributes we annotated to analyze the usage of geometric information in face anti-spoofing task. The *key* observations are:

Depth Maps are More Versatile. As shown in Table 4(a), geometric information is insufficient to be the unique supervision for live/spoof classification. However, it can boost the performance of the baseline when it serves as an auxiliary supervision. Besides, we study the impact of different individual geometric information on AENet$_{\mathcal{C},\mathcal{G}}$ performance. As shown in Fig. 7(a), AENet$_{\mathcal{C},\mathcal{G}}$

Table 4. Geometric information study results in Sect. 5.2. (a) AENet$_{\mathcal{G}^d}$ which only depends on the depth map for classification performs worst than baseline. (b) AENet$_{\mathcal{C},\mathcal{G}}$ which leverages all semantic attributes achieve the best result. **Bolds** are the best results; ↑ means bigger value is better; ↓ means smaller value is better

	Model	Recall (%)↑			AUC↑	EER (%)↓	APCER (%)↓	BPCER (%)↓	ACER (%)↓
		FPR = 1%	FPR = 0.5%	FPR = 0.1%					
(a)	Baseline	97.9	95.3	85.9	0.9984	1.6	6.1	1.6	3.8
	AENet$_{\mathcal{G}^d}$	97.8	96.2	87.0	0.9946	1.6	7.33	1.68	4.51
(b)	AENet$_{\mathcal{C},\mathcal{G}}$	**98.4**	**96.8**	86.7	**0.9985**	1.2	**5.34**	**1.19**	**3.26**
(c)	AENet$_{\mathcal{C},\mathcal{G}}$ w/o $\mathcal{G}^d$	98.3	96.1	**87.7**	0.9976	1.2	5.91	1.27	3.59
	AENet$_{\mathcal{C},\mathcal{G}}$ w/o $\mathcal{G}^r$	97.9	95.7	84.1	0.9973	1.3	5.71	1.38	3.55

w/o $\mathcal{G}^d$ performs the best in spoof type: "replay" (macro definition), because the reflect artifacts appear frequently in these three spoof types. For "phone", AENet$_{\mathcal{C},\mathcal{G}}$ w/o $\mathcal{G}^d$ improves 56% comparing to the baseline. However AENet$_{\mathcal{C},\mathcal{G}}$ w/o $\mathcal{G}^d$ gets worse result than baseline in spoof type: "print" (macro definition). Moreover, AENet$_{\mathcal{C},\mathcal{G}}$ w/o $\mathcal{G}^r$ helps greatly to improve the classification performance of baseline in both "replay" and "print"(macro definition). Especially for "poster", AENet$_{\mathcal{C},\mathcal{G}}$ w/o $\mathcal{G}^r$ improves baseline by 81%. Therefore, the depth map can improve classification performance in most spoof types, but the function of the reflection map is mainly reflected in "replay"(macro definition).

Sensitive to Illumination. As shown in Fig. 7(a), in spoof type "print"(macro definition), the performance of the AENet$_{\mathcal{C},\mathcal{G}}$ w/o $\mathcal{G}^r$ on "A4" is much worse than "poster" and "photo", although they are both in "print" spoof type. The main reason for the large difference in performance among these three spoof types for AENet$_{\mathcal{C},\mathcal{G}}$ w/o $\mathcal{G}^r$ is that the learning of the depth map is sensitive to dark illumination, as shown in Fig. 7(b). When we calculate APCER under other illumination conditions: normal, strong and back, AENet$_{\mathcal{C},\mathcal{G}}$ w/o $\mathcal{G}^r$ achieves almost the same results among "A4", "poster" and "photo".

6   Benchmarks

In order to facilitate future research in the community, we carefully build three different benchmarks to investigate face anti-spoofing algorithms. Specifically, for a comprehensive evaluation, besides ResNet-18, we also provide the corresponding results based on a heavier backbone, *i.e.* Xception. Detailed information of the results based on Xception are shown in the supplementary material.

6.1   Intra-dataset Benchmark

Based on this benchmark, models are trained and evaluated on the whole training set and testing set of CelebA-Spoof. This benchmark evaluates the overall capability of the classification models. According to different input data types, there are two kinds of face anti-spoof methods, *i.e.* " video-driven methods"

Fig. 6. Success and failure cases. The row(i) present the live image and row(ii) present the spoof image. For each image, the first row is the highest score of live/spoof prediction of baseline and others are the highest live/spoof and the highest semantic attributes predictions of AENet$_{c,s}$. Blue indicates correctly predicted results and orange indicates the wrong results. In detail, we list the top three prediction scores of face attributes in the last three rows of each image (Color figure online)

Table 5. Intro-dataset Benchmark results on CelebA-Spoof. AENet$_{c,s,g}$ achieved the best result. **Bolds** are the best results; ↑ means bigger value is better; ↓ means smaller value is better. * Model 2 defined in *Auxiliary* can be used as "image driven method"

Model	Backbone	Parm. (MB)	Recall (%)↑			AUC↑	EER (%)↓	APCER (%)↓	BPCER (%)↓	ACER (%)↓
			FPR = 1%	FPR = 0.5%	FPR = 0.1%					
Auxiliary* [18]	–	22.1	97.3	95.2	83.2	0.9972	1.2	5.71	1.41	3.56
BASN [11]	VGG16	569.7	98.9	**97.8**	**90.9**	**0.9991**	1.1	4.0	1.1	2.6
AENet$_{c,s,g}$	ResNet-18	42.7	**98.9**	97.3	87.3	0.9989	**0.9**	**2.29**	**0.96**	**1.63**

and "image-driven methods". Since the data in CelebA-Spoof are image-based, we benchmark state-of-the-art "image-driven methods" in this subsection. As shown in Table 5, AENet$_{c,s,g}$ which combines geometric and semantic information has achieved the best results on CelebA-Spoof. Specifically, our approach outperforms the state-of-the-art by 38% with much fewer parameters.

6.2 Cross-Domain Benchmark

Since face anti-spoofing is an open-set problem, even though CelebA-Spoof is equipped with diverse images, it is impossible to cover all spoof types, environments, sensors, *etc.* that exist in the real world. Inspired by [4,18], we carefully design two protocols for CelebA-Spoof based on real-world scenarios. In each protocol, we evaluate the performance of trained models under controlled domain

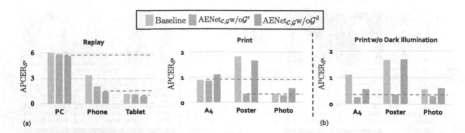

Fig. 7. Representative examples of the effectiveness of geometric information. Higher $APCER_{S^s}$ is worse. (a) $AENet_{C,G}$ w/o $\mathcal{G}^d$ perform the best in spoof type: "replay"(macro definition) and $AENet_{C,G}$ w/o $\mathcal{G}^r$ perform the best in spoof type: "print"(macro definition). (b) The performance of $AENet_{C,G}$ w/o $\mathcal{G}^r$ improve largely on spoof type: "A4", if we only calculate APCER under illumination conditions: "normal", "strong" and "back"

Table 6. Cross-domain benchmark results of CelebA-Spoof. **Bolds** are the best results; ↑ means bigger value is better; ↓ means smaller value is better

Protocol	Model	Recall (%) ↑			AUC↑	EER (%)↓	APCER (%)↓	BPCER (%)↓	ACER (%)↓
		FPR = 1%	FPR = 0.5%	FPR = 0.1%					
1	Baseline	93.7	86.9	69.6	**0.996**	2.5	5.7	2.52	4.11
	$AENet_{C,G}$	93.3	88.6	**74.0**	0.994	2.5	5.28	2.41	3.85
	$AENet_{C,S}$	93.4	89.3	71.3	**0.996**	2.4	5.63	2.42	4.04
	$AENet_{C,S,G}$	**95.0**	**91.4**	73.6	0.995	**2.1**	**4.09**	**2.09**	**3.09**
2	Baseline	#	#	#	0.998 ±0.002	1.5±0.8	8.53±2.6	1.56±0.81	5.05±1.42
	$AENet_{C,G}$	#	#	#	0.995±0.003	1.6±4.5	8.95±1.07	1.67±0.9	5.31±0.95
	$AENet_{C,S}$	#	#	#	0.997±0.002	1.2±0.7	**4.01±2.9**	1.24±0.67	3.96±1.79
	$AENet_{C,S,G}$	#	#	#	**0.998±0.002**	1.3±0.7	4.94±3.42	**1.24±0.73**	**3.09±2.08**

shifts. Specifically, we define two protocols. **1)** *Protocol 1* - Protocol 1 evaluates the cross-medium performance of various spoof types. This protocol includes 3 macro types of spoof, where each covers 3 micro types of spoof. These three macro types of spoof are "print", "repay" and "paper cut". In detail, in each macro type of spoof, we choose 2 of their micro type of spoof for training, and the others for testing. Specifically, "A4", "face mask" and "PC" are selected for testing. **2)** *Protocol 2* - Protocol 2 evaluates the effect of input sensor variations. According to imaging quality, we split input sensors into three groups: *low-quality sensor*, *middle-quality sensor* and *high-quality sensor*[2]. Since we need to test on three different kinds of sensor and the average performance of FPR-Recall is hard to measure, we do not include FPR-Recall in the evaluation metrics of protocol 2. Table 6 shows the performance under each protocol.

6.3   Cross-Dataset Benchmark

In this subsection, we perform cross-dataset testing on CelebA-Spoof and CASIA-MFSD dataset to further construct the cross-dataset benchmark. On the one hand, we offer a quantitative result to measure the quality of our

---

[2] Please refer to supplementary for the detailed input sensors information.

Table 7. Cross-dataset benchmark results. AENet$_{c,s,g}$ based on ResNet-18 achieves the best generalization performance. **Bolds** are the best results; ↑ means bigger value is better; ↓ means smaller value is better

Model	Training	Testing	HTER (%) ↓
FAS-TD-SF [28]	SiW	CASIA-MFSD	39.4
FAS-TD-SF [28]	CASIA-SURF	CASIA-MFSD	37.3
AENet$_{c,s,g}$	SiW	CASIA-MFSD	27.6
Baseline	CelebA-Spoof	CASIA-MFSD	14.3
AENet$_{c,g}$	CelebA-Spoof	CASIA-MFSD	14.1
AENet$_{c,s}$	CelebA-Spoof	CASIA-MFSD	12.1
AENet$_{c,s,g}$	CelebA-Spoof	CASIA-MFSD	**11.9**

dataset. On the other hand, we can evaluate the generalization ability of different methods according to this benchmark. The current largest face anti-spoofing dataset CASIA-SURF [32] adopted *FAS-TD-SF* [28] (which is trained on SiW or CASIA-SURF and tested on CASIA-MFSD) to demonstrate the quality of CASIA-SURF. Following this setting, we first train AENet$_{c,g}$, AENet$_{c,s}$ and AENet$_{c,s,g}$ based on CelebA-Spoof and then test them on CASIA-MFSD to evaluate the quality of CelebA-Spoof. As shown in Table 7, we can conclude that: **1)** The diversity and large quantities of CelebA-Spoof drastically boosts the performance of vanilla model; a simple ResNet-18 achieves state-of-the-art cross-dataset performance. **2)** Comparing to geometric information, semantic information equips the model with better generalization ability.

7 Conclusion

In this paper, we construct a large-scale face anti-spoofing dataset, **CelebA-Spoof**, with 625,537 images from 10,177 subjects, which includes 43 rich attributes on face, illumination, environment and spoof types. We believe CelebA-Spoof would be a significant contribution to the community of face anti-spoofing. Based on these rich attributes, we further propose a simple yet powerful multi-task framework, namely **AENet**. Through AENet, we conduct extensive experiments to explore the roles of semantic information and geometric information in face anti-spoofing. To support comprehensive evaluation and diagnosis, we establish three versatile benchmarks to evaluate the performance and generalization ability of various methods under different carefully-designed protocols. With several valuable observations revealed, we demonstrate the effectiveness of CelebA-Spoof and its rich attributes which can significantly facilitate future research.

Acknowledgments. This work is supported in part by SenseTime Group Limited, in part by National Science Foundation of China Grant No. U1934220 and 61790575, and the project "Safety data acquisition equipment for industrial enterprises No.134". The corresponding author is Jing Shao. The contributions of Yuanhan Zhang and Zhenfei Yin are Equal.

References

1. Bhattacharjee, S., Mohammadi, A., Marcel, S.: Spoofing deep face recognition with custom silicone masks. In: Proceedings of IEEE 9th International Conference on Biometrics: Theory, Applications, and Systems (BTAS) (2018)
2. Boulkenafet, Z., Komulainen, J., Hadid, A.: Face antispoofing using speeded-up robust features and fisher vector encoding. IEEE Signal Process. Lett. **24**(2), 141–145 (2016)
3. Boulkenafet, Z., Komulainen, J., Hadid, A.: Face spoofing detection using colour texture analysis. TIFS **11**(8), 1818–1830 (2016)
4. Boulkenafet, Z., Komulainen, J., Li, L., Feng, X., Hadid, A.: Oulu-npu: a mobile face presentation attack database with real-world variations. In: FG, pp. 612–618. IEEE (2017)
5. Chingovska, I., Anjos, A., Marcel, S.: On the effectiveness of local binary patterns in face anti-spoofing. In: BIOSIG, pp. 1–7. IEEE (2012)
6. Chingovska, I., Erdogmus, N., Anjos, A., Marcel, S.: Face recognition systems under spoofing attacks. In: Bourlai, T. (ed.) Face Recognition Across the Imaging Spectrum, pp. 165–194. Springer, Cham (2016). https://doi.org/10.1007/978-3-319-28501-6_8
7. Erdogmus, N., Marcel, S.: Spoofing 2D face recognition systems with 3D masks. In: BIOSIG, pp. 1–8. IEEE (2013)
8. Feng, L., Po, L.M., Li, Y., Xu, X., Yuan, F., Cheung, T.C.H., Cheung, K.W.: Integration of image quality and motion cues for face anti-spoofing: a neural network approach. J. Visual Commun. Image Represent. **38**, 451–460 (2016)
9. Feng, Y., Wu, F., Shao, X., Wang, Y., Zhou, X.: Joint 3D face reconstruction and dense alignment with position map regression network. In: ECCV, pp. 534–551 (2018)
10. Jourabloo, A., Liu, Y., Liu, X.: Face de-spoofing: anti-spoofing via noise modeling. In: ECCV, pp. 290–306 (2018)
11. Kim, T., Kim, Y., Kim, I., Kim, D.: BASN: enriching feature representation using bipartite auxiliary supervisions for face anti-spoofing. In: ICCV Workshops (2019)
12. Kollreider, K., Fronthaler, H., Faraj, M.I., Bigun, J.: Real-time face detection and motion analysis with application in "liveness" assessment. TIFS **2**(3), 548–558 (2007)
13. Li, J., Wang, Y., Tan, T., Jain, A.K.: Live face detection based on the analysis of fourier spectra. In: Biometric Technology for Human Identification, vol. 5404, pp. 296–303. International Society for Optics and Photonics (2004)
14. Li, L., Feng, X., Boulkenafet, Z., Xia, Z., Li, M., Hadid, A.: An original face anti-spoofing approach using partial convolutional neural network. In: IPTA, pp. 1–6. IEEE (2016)
15. Liu, S.Q., Lan, X., Yuen, P.C.: Remote photoplethysmography correspondence feature for 3D mask face presentation attack detection. In: ECCV, September 2018

16. Liu, S., Yang, B., Yuen, P.C., Zhao, G.: A 3D mask face anti-spoofing database with real world variations. In: CVPR Workshops, pp. 1551–1557, June 2016

17. Liu, S., Yuen, P.C., Zhang, S., Zhao, G.: 3D mask face anti-spoofing with remote photoplethysmography. In: Leibe, B., Matas, J., Sebe, N., Welling, M. (eds.) ECCV 2016. LNCS, vol. 9911, pp. 85–100. Springer, Cham (2016). https://doi.org/10.1007/978-3-319-46478-7_6

18. Liu, Y., Jourabloo, A., Liu, X.: Learning deep models for face anti-spoofing: binary or auxiliary supervision. In: CVPR, pp. 389–398 (2018)

19. Liu, Y., Stehouwer, J., Jourabloo, A., Liu, X.: Deep tree learning for zero-shot face anti-spoofing. In: CVPR, pp. 4680–4689 (2019)

20. Liu, Z., Luo, P., Wang, X., Tang, X.: Deep learning face attributes in the wild. In: ICCV (2015)

21. Määttä, J., Hadid, A., Pietikäinen, M.: Face spoofing detection from single images using texture and local shape analysis. IET Biom. 1(1), 3–10 (2012)

22. Ojala, T., Pietikainen, M., Maenpaa, T.: Multiresolution gray-scale and rotation invariant texture classification with local binary patterns. TPAMI 24(7), 971–987 (2002)

23. Pan, G., Sun, L., Wu, Z., Lao, S.: Eyeblink-based anti-spoofing in face recognition from a generic webcamera. In: ICCV, pp. 1–8. IEEE (2007)

24. Patel, K., Han, H., Jain, A.K.: Secure face unlock: spoof detection on smartphones. TIFS 11(10), 2268–2283 (2016)

25. Schwartz, W.R., Rocha, A., Pedrini, H.: Face spoofing detection through partial least squares and low-level descriptors. In: 2011 International Joint Conference on Biometrics (IJCB), pp. 1–8. IEEE (2011)

26. Shao, R., Lan, X., Li, J., Yuen, P.C.: Multi-adversarial discriminative deep domain generalization for face presentation attack detection. In: CVPR (2019)

27. Sun, L., Pan, G., Wu, Z., Lao, S.: Blinking-based live face detection using conditional random fields. In: Lee, S.-W., Li, S.Z. (eds.) ICB 2007. LNCS, vol. 4642, pp. 252–260. Springer, Heidelberg (2007). https://doi.org/10.1007/978-3-540-74549-5_27

28. Wang, Z., et al.: Exploiting temporal and depth information for multi-frame face anti-spoofing. arXiv (2018)

29. Wen, D., Han, H., Jain, A.K.: Face spoof detection with image distortion analysis. TIFS 10(4), 746–761 (2015)

30. Yang, J., Lei, Z., Liao, S., Li, S.Z.: Face liveness detection with component dependent descriptor. In: ICB, pp. 1–6. IEEE (2013)

31. Yang, X., et al.: Face anti-spoofing: model matters, so does data. In: CVPR, pp. 3507–3516 (2019)

32. Zhang, S., et al.: A dataset and benchmark for large-scale multi-modal face anti-spoofing. In: CVPR, pp. 919–928 (2018)

33. Zhang, X., Ng, R., Chen, Q.: Single image reflection separation with perceptual losses. In: ICCV, pp. 4786–4794 (2018)

34. Zhang, Z., et al.: A face antispoofing database with diverse attacks. In: ICB, pp. 26–31. IEEE (2012)

Thinking in Frequency: Face Forgery Detection by Mining Frequency-Aware Clues

Yuyang Qian[1,2] ![ID], Guojun Yin[1(✉)] ![ID], Lu Sheng[3(✉)] ![ID], Zixuan Chen[1,4] ![ID], and Jing Shao[1] ![ID]

[1] SenseTime Research, Hong Kong, China
{yinguojun,shaojing}@sensetime.com
[2] University of Electronic Science and Technology of China, Chengdu, China
qyy@std.uestc.edu.cn
[3] College of Software, Beihang University, Beijing, China
lsheng@buaa.edu.cn
[4] Northwestern Polytechnical University, Xi'an, China
zixuan.sean.chen@hotmail.com

Abstract. As realistic facial manipulation technologies have achieved remarkable progress, social concerns about potential malicious abuse of these technologies bring out an emerging research topic of face forgery detection. However, it is extremely challenging since recent advances are able to forge faces beyond the perception ability of human eyes, especially in compressed images and videos. We find that mining forgery patterns with the awareness of frequency could be a cure, as frequency provides a complementary viewpoint where either subtle forgery artifacts or compression errors could be well described. To introduce frequency into the face forgery detection, we propose a novel Frequency in Face Forgery Network (F^3-Net), taking advantages of two different but complementary frequency-aware clues, 1) frequency-aware decomposed image components, and 2) local frequency statistics, to deeply mine the forgery patterns via our two-stream collaborative learning framework. We apply DCT as the applied frequency-domain transformation. Through comprehensive studies, we show that the proposed F^3-Net significantly outperforms competing state-of-the-art methods on all compression qualities in the challenging FaceForensics++ dataset, especially wins a big lead upon low-quality media.

Keywords: Face forgery detection · Frequency · Collaborative learning

---

Y. Qian and Z. Chen—This work was done during the internship of Yuyang Qian and Zixuan Chen at SenseTime Research.
The first two authors contributed equally.

A. Vedaldi et al. (Eds.): ECCV 2020, LNCS 12357, pp. 86–103, 2020.
https://doi.org/10.1007/978-3-030-58610-2_6

1   Introduction

Rapid development of deep learning driven generative models [8,11,26,34,35] enables an attacker to create, manipulate or even forge the media of a human face (*i.e.*, images and videos, etc.) that cannot be distinguished even by human eyes. However, malicious distribution of forged media would cause security issues and even crisis of confidence in our society. Therefore, it is supremely important to develop effective face forgery detection methods.

Various methods [3,30,33,43,45,46,58,60] have been proposed to detect the forged media. A series of earlier works relied on hand-crafted features *e.g.*, local pattern analysis [21], noise variances evaluation [47] and steganalysis features [13, 24] to discover forgery patterns and magnify faint discrepancy between real and forged images. Deep learning introduces another pathway to tackle this challenge, recent learning-based forgery detection methods [12,14] tried to mine the forgery patterns in feature space using convolutional neural networks (CNNs), having achieved remarkable progresses on public datasets, *e.g.*, FaceForensics++ [50].

Current state-of-the-art face manipulation algorithms, such as DeepFake [1], FaceSwap [2], Face2Face [56] and NeuralTextures [55], have been able to conceal the forgery artifacts, so that it becomes extremely difficult to discover the flaws of these refined counterfeits, as shown in Fig. 1(a). What's worse, if the visual quality of a forged face is tremendously degraded, such as compressed by JPEG or H.264 with a large compression ratio, the forgery artifacts will be contaminated by compression error, and sometimes cannot be captured in RGB domain any more. Fortunately, these artifacts can be captured in frequency domain, as many prior studies suggested [19,32,38,57,58], in the form of unusual frequency distributions when compared with real faces. However, how to involve frequency-aware clues into the deeply learned CNN models? This question also raises alongside. Conventional frequency domains, such as FFT and DCT, do not match the shift-invariance and local consistency owned by nature images, thus vanilla CNN structures might be infeasible. As a result, CNN-compatible frequency representation becomes pivotal if we would like to leverage the discriminative representation power of learnable CNNs for frequency-aware face forgery detection. To this end, we would like to introduce two frequency-aware forgery clues that are compatible with the knowledge mining by deep convolutional networks.

From one aspect, it is possible to decompose an image by separating its frequency signals, while each decomposed image component indicates a certain band of frequencies. The first frequency-aware forgery clue is thus discovered by the intuition that we are able to identify subtle forgery artifacts that are somewhat salient (*i.e.*, in the form of unusual patterns) in the decomposed components with higher frequencies, as the examples shown in the middle column of Fig. 1(b). This clue is compatible with CNN structures, and is surprisingly robust to compression artifacts. From the other aspect, the decomposed image components describe the frequency-aware patterns in the spatial domain, but not explicitly render the frequency information directly in the neural networks. We suggest the second frequency-aware forgery clue as the local frequency statistics.

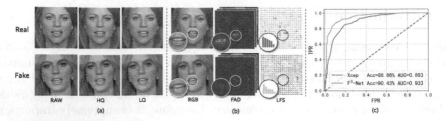

Fig. 1. Frequency-aware tampered clues for face forgery detection. (a) RAW, high quality (HQ) and low quality (LQ) real and fake images with the same identity, manipulation artifacts are barely visible in low quality images. (b) Frequency-aware forgery clues in low quality images using the proposed *Frequency-aware Decomposition (FAD)* and *Local Frequency Statistics (LFS)*. (c) ROC Curve of the proposed **Frequency in Face Forgery Network (F³-Net)** and baseline (*i.e.*, Xception [12]). The proposed F³-Net wins the Xception with a large margin. Best viewed in color. (Color figure online)

In each densely but regularly sampled local spatial patch, the statistics is gathered by counting the mean frequency responses at each frequency band. These frequency statistics re-assemble back to a multi-channel spatial map, where the number of channels is identical to the number of frequency bands. As shown in the last column of Fig. 1(b), the forgery faces have distinct local frequency statistics than the corresponding real ones, even though they look almost the same in the RGB images. Moreover, the local frequency statistics also follows the spatial layouts as the input RGB images, thus also enjoy effective representation learning powered by CNNs. Meanwhile, since the decomposed image components and local frequency statistics are complementary to each other but both of them share inherently similar frequency-aware semantics, thus they can be progressively fused during the feature learning process.

Therefore, we propose a novel **Frequency in Face Forgery Network (F³-Net)**, that capitalizes on the aforementioned frequency-aware forgery clues. The proposed framework is composed of two frequency-aware branches, one aims at learning subtle forgery patterns through *Frequency-aware Image Decomposition (FAD)*, and the other would like to extract high-level semantics from *Local Frequency Statistics (LFS)* to describe the frequency-aware statistical discrepancy between real and forged faces. These two branches, are further gradually fused through a cross-attention module, namely *MixBlock*, which encourages rich interactions between the aforementioned FAD and LFS branches. The whole face forgery detection model is learned by the cross-entropy loss in an end-to-end manner. Extensive experiments demonstrate that the proposed F³-Net significantly improves the performance over low-quality forgery media with a thorough ablation study. We also show that our framework largely exceeds competing state-of-the-arts on all compression qualities in the challenging Face-Forensics++ [50]. As shown in Fig. 1(c), the effectiveness and superiority of the proposed frequency-aware F³-Net is obviously demonstrated by comparing the

ROC curve with Xception [12](baseline, previous state-of-the-art seeing in Sect. 4). Our contributions in this paper are summarized as follows:

1) Frequency-aware Decomposition (FAD) aims at learning frequency-aware forgery patterns through frequency-aware image decomposition. The proposed FAD module adaptively partitions the input image in the frequency domain according to learnable frequency bands and represents the image with a series of frequency-aware components.

2) Local Frequency Statistics (LFS) extracts local frequency statistics to describe the statistical discrepancy between real and fake faces. The localized frequency statistics not only reveal the unusual statistics of the forgery images at each frequency band, but also share the structure of natural images, and thus enable effective mining through CNNs.

3) The proposed framework collaboratively learns the frequency-aware clues from FAD and LFS, by a cross-attention (a.k.a MixBlock) powered two-stream networks. The proposed method achieves the state-of-the-art performance on the challenging FaceForensics++ dataset [50], especially wins a big lead in the low quality forgery detection.

2   Related Work

With the development of computer graphics and neural networks especially generative adversarial networks (GANs) [8,11,26,34,35], face forgery detection has gained more and more interest in our society. Various attempts have been made for face forgery detection and achieved remarkable progress, but learning-based generation methods such as NeuralTextures [55] are still difficult to detect because they introduce only small-scale subtle visual artifacts especially in low quality videos. To address the problem, various additional information is used to enhance performance.

Spatial-Based Forgery Detection. To address face forgery detection tasks, a variety of methods have been proposed. Most of them are based on the spatial domain such as RGB and HSV. Some approaches [9,16] exploit specific artifacts arising from the synthesis process such as color or shape cues. Some studies [27, 37,44] extract color-space features to classify fake and real images. For example, ELA [27] uses pixel-level errors to detect image forgery. Early methods [6,14] use hand-crafted features for shallow CNN architectures. Recent methods [3,30, 33,43,45,46,58,60] use deep neural networks to extract high-level information from the spatial domain and get remarkable progress. MesoInception-4 [3] is a CNN-based Network inspired by InceptionNet [54] to detect forged videos. GANs Fingerprints Analysis [58] introduces deep manipulation discriminator to discover specific manipulation patters. However, most of them use only spatial domain information and therefore are not sensitive to subtle manipulation clues that are difficult to detect in color-space. In our works, we take advantage of frequency cues to mine small-scale detailed artifacts that are helpful especially in low-quality videos.

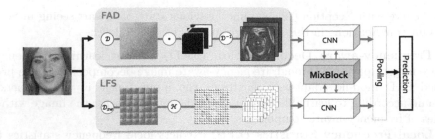

Fig. 2. Overview of the F³-Net. The proposed architecture consists of three novel methods: *FAD* for learning subtle manipulation patterns through frequency-aware image decomposition; *LFS* for extracting local frequency statistics and *MixBlock* for collaborative feature interaction.

Frequency-Based Forgery Detection. Frequency domain analysis is a classical and important method in image signal processing and has been widely used in a number of applications such as image classification [23,52,53], steganalysis [10,17], texture classification [22,25,28] and super-resolution [31,39]. Recently, several attempts have been made to solve forgery detection using frequency cues. Some studies use Wavelet Transform (WT) [7] or Discrete Fourier Transform (DFT) [19,57,59] to convert pictures to frequency domain and mine underlying artifacts. For example, Durall *et al.* [19] extracts frequency-domain information using DFT transform and averaging the amplitudes of different frequency bands. Stuchi *et al.* [53] uses a set of fixed frequency domain filters to extract different range of information followed by a fully connected layer to get the output. Besides, filtering, a classic image signal processing method, is used to refine and mine underlying subtle information in forgery detection, which leverages existing knowledge of the characteristics of fake images. Some studies use high-pass filters [15,29,48,57], Gabor filters [10,22] etc. to extract features of interest (*e.g.* edge and texture information) based on features regarding with high frequency components. Phase Aware CNN [10] uses hand-crafted Gabor and high-pass filters to augment the edge and texture features. Universal Detector [57] finds that significant differences can be obtained in the spectrum between real and fake images after high-pass filtering. However, the filters used in these studies are often fixed and hand-crafted thus fail to capture the forgery patterns adaptively. In our work, we make use of frequency-aware image decomposition to mine frequency forgery cues adaptively.

3   Our Approach

In this section, we introduce the proposed two kinds of frequency-aware forgery clue mining methods, *i.e.*, frequency-aware decomposition (in Sect. 3.1) and local frequency statistics (in Sect. 3.2), and then present the proposed cross-attention two-stream collaborative learning framework (in Sect. 3.3).

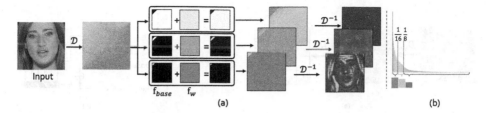

Fig. 3. (a) The proposed *Frequency-aware Decomposition (FAD)* to discover salient frequency components. $\mathcal{D}$ indicates applying Discrete Cosine Transform (DCT). $\mathcal{D}^{-1}$ indicates applying Inversed Discrete Cosine Transform (IDCT). Several frequency band components can be concatenated together to extract a wider range of information. (b) The distribution of the DCT power spectrum. We flatten 2D power spectrum to 1D by summing up the amplitudes of each frequency band. We divide the spectrum into 3 bands with roughly equal energy.

3.1   FAD: Frequency-Aware Decomposition

Towards the frequency-aware image decomposition, former studies usually apply hand-crafted filter banks [10,22] in the spatial domain, thus fail to cover the complete frequency domain. Meanwhile, the fixed filtering configurations make it hard to adaptively capture the forgery patterns. To this end, we propose a novel frequency-aware decomposition (FAD), to adaptively partition the input image in the frequency domain according to a set of learnable frequency filters. The decomposed frequency components can be inversely transformed to the spatial domain, resulting in a series of frequency-aware image components. These components are stacked along the channel axis, and then inputted into a convolutional neural network (in our implementation, we employ an Xception [12] as the backbone) to comprehensively mine forgery patterns.

To be specific, we manually design N binary base filters $\{\mathbf{f}_{base}^i\}_{i=1}^N$ (or called masks) that explicitly partition the frequency domain into low, middle and high frequency bands. And then we add three learnable filters $\{\mathbf{f}_w^i\}_{i=1}^N$ to these base filters. The frequency filtering is a dot-product between the frequency response of the input image and the combined filters $\mathbf{f}_{base}^i + \sigma(\mathbf{f}_w^i), i = \{1,\ldots,N\}$, where $\sigma(x) = \frac{1-\exp(-x)}{1+\exp(-x)}$ aims at squeezing x within the range between -1 and $+1$. Thus, to an input image $\mathbf{x}$, the decomposed image components are obtained by

$$\mathbf{y}_i = \mathcal{D}^{-1}\{\mathcal{D}(\mathbf{x}) \odot [\mathbf{f}_{base}^i + \sigma(\mathbf{f}_w^i)]\}, \quad i = \{1,\ldots,N\}. \tag{1}$$

$\odot$ is the element-wise product. We apply $\mathcal{D}$ as the Discrete Cosine Transform (DCT) [4], according to its wide applications in image processing, and its nice layout of the frequency distribution, *i.e.*, low-frequency responses are placed in the top-left corner, and high-frequency responses are located in the bottom-right corner. Moreover, recent compression algorithms, such as JPEG and H.264, usually apply DCT in their frameworks, thus DCT-based FAD will be more compatible towards the description of compression artifacts out of the forgery patterns. Observing the DCT power spectrum of natural images, we find that the

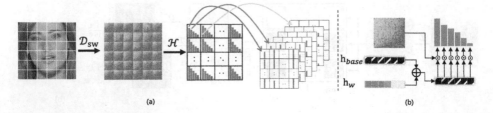

(a)   (b)

Fig. 4. (a) The proposed *Local Frequency Statistics (LFS)* to extract local frequency domain statistical information. SWDCT indicates applying Sliding Window Discrete Cosine Transform and $\mathcal{H}$ indicates gathering statistics on each grid adaptively. (b) Extracting statistics from a DCT power spectrum graph, $\oplus$ indicates element-wise addition and $\odot$ indicates element-wise multiplication.

spectral distribution is non-uniform and most of the amplitudes are concentrated in the low frequency area. We apply the base filters $\mathbf{f}_{base}$ to divide the spectrum into N bands with roughly equal energy, from low frequency to high frequency. The added learnable $\{\mathbf{f}_w^i\}_{i=1}^N$ provides more adaptation to select the frequency of interest beyond the fixed base filters. Empirically, as shown in Fig. 3(b), the number of bands $N = 3$, the low frequency band $\mathbf{f}_{base}^1$ is the first $1/16$ of the entire spectrum, the middle frequency band $\mathbf{f}_{base}^2$ is between $1/16$ and $1/8$ of the spectrum, and the high frequency band $\mathbf{f}_{base}^3$ is the last $7/8$.

3.2   LFS: Local Frequency Statistics

The aforementioned FAD has provided frequency-aware representation that is compatible with CNNs, but it has to represent frequency-aware clues back into the spatial domain, thus fail to directly utilize the frequency information. Also knowing that it is usually infeasible to mine forgery artifacts by extracting CNN features directly from the spectral representation, we then suggest to estimate local frequency statistics (LFS) to not only explicitly render frequency statistics but also match the shift-invariance and local consistency that owned by natural RGB images. These features are then inputted into a convolutional neural network, *i.e.*, Xception [12], to discover high-level forgery patterns.

As shown in Fig. 4(a), we first apply a Sliding Window DCT (SWDCT) on the input RGB image (*i.e.*, taking DCTs densely on sliding windows of the image) to extract the localized frequency responses, and then counting the mean frequency responses at a series of learnable frequency bands. These frequency statistics reassemble back to a multi-channel spatial map that shares the same layout as the input image. This LFS provides a localized aperture to detect detailed abnormal frequency distributions. Calculating statistics within a set of frequency bands allows a reduced statistical representation, whilst yields a smoother distribution without the interference of outliers.

To be specific, in each window $\mathbf{p} \in \mathbf{x}$, after DCT, the local statistics is gathered in each frequency band, which is constructed similarly as the way used in FAD (see Sect. 3.1). In each band, the statistics become

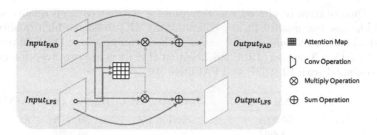

Fig. 5. The proposed *MixBlock*. $\otimes$ indicates matrix multiplication and $\oplus$ indicates element-wise addition.

$$\mathbf{q}_i = \log_{10} \|\mathcal{D}(\mathbf{p}) \odot [\mathbf{h}_{base}^i + \sigma(\mathbf{h}_w^i)]\|_1, \quad i = \{1, \ldots, M\}, \tag{2}$$

Note that $\log_{10}$ is applied to balance the magnitude in each frequency band. The frequency bands are collected by equally partitioning the spectrum in to M parts, following the order from low frequency to high frequency. Similarly, $\mathbf{h}_{base}^i$ is the base filter, $\mathbf{h}_w^i$ is the learnable filter, $i = \{1, \ldots, M\}$. The local frequency statistics $\mathbf{q}$ for a window $\mathbf{p}$ is then transposed as a $1 \times 1 \times M$ vector. These statistics vectors gathered from all windows are re-assembled into a matrix with downsampled spatial size of the input image, whose number of channels is equal to M. This matrix will act as the input to the later convolutional layers.

Practically in our experiments, we empirically adopt the window size as 10, the sliding stride as 2, and the number of bands as $M = 6$, thus the size of the output matrix will be $149 \times 149 \times 6$ if the input image is of size $299 \times 299 \times 3$.

3.3 Two-Stream Collaborative Learning Framework

As mentioned in Sect. 3.1 and Sect. 3.2, the proposed FAD and LFS modules mine the frequency-aware forgery clues from two different but inherently connected aspects. We argue that these two types of clues are different but complementary. Thus, we propose a collaborative learning framework that powered by cross-attention modules, to gradually fuse two-stream FAD and LFS features. To be specific, the whole network architecture of our F^3-Net is composed of two branches equipped with Xception blocks [12], one is for the decomposed image components generated by FAD, and the other is for local frequency statistics generated by LFS, as shown in Fig. 2.

We propose a cross-attention fusion module for the feature interaction and message passing every several Xception blocks. As shown in Fig. 5, different from the simple concatenation widely used in previous methods [20,32,60], we firstly calculate the cross-attention weight using the feature maps from the two branches. The cross-attention matrix is adopted to augment the attentive features from one stream to another.

In our implementation, we use Xception network [12] pretrained on the ImageNet [18] for both branches, each of which has 12 blocks. The newly-introduced layers and blocks are randomly initialized. The cropped face is adopted as the

Table 1. Quantitative results on FaceForensics++ dataset with all quality settings, *i.e.* LQ indicates low quality (heavy compression), HQ indicates high quality (light compression) and RAW indicates raw videos without compression. The bold results are the best. Note that Xception+ELA and Xception-PAFilters are two Xception baselines that are equipped with ELA [27] and PAFilters [10].

Methods	Acc (LQ)	AUC (LQ)	Acc (HQ)	AUC (HQ)	Acc (RAW)	AUC (RAW)
Steg.Features [24]	55.98%	-	70.97%	-	97.63%	-
LD-CNN [14]	58.69%	-	78.45%	-	98.57%	-
Constrained Conv [6]	66.84%	-	82.97%	-	98.74%	-
CustomPooling CNN [49]	61.18%	-	79.08%	-	97.03%	-
MesoNet [3]	70.47%	-	83.10%	-	95.23%	-
Face X-ray [40]	-	0.616	-	0.874	-	-
Xception [12]	86.86%	0.893	95.73%	0.963	99.26%	0.992
Xception-ELA [27]	79.63%	0.829	93.86%	0.948	98.57%	0.984
Xception-PAFilters [10]	87.16%	0.902	-	-	-	-
F^3-Net (Xception)	**90.43%**	**0.933**	**97.52%**	**0.981**	**99.95%**	**0.998**
Optical Flow [5]	81.60%	-	-	-	-	-
Slowfast [20]	90.53%	0.936	97.09%	0.982	99.53%	0.994
F^3-Net(Slowfast)	**93.02%**	**0.958**	**98.95%**	**0.993**	**99.99%**	**0.999**

input of the framework after resized as 299×299. Empirically, we adopt MixBlock after block 7 and block 12 to fuse two types of frequency-aware clues according to their mid-level and high-level semantics. We train the F^3-Net by the cross entropy loss, and the whole system can be trained in an end-to-end fashion.

4 Experiment

4.1 Setting

Dataset. Following previous face forgery detection methods [5,19,40,51], we conduct our experiments on the challenging FaceForensics++ [50] dataset. Face-Forensics++ is a face forgery detection video dataset containing 1,000 real videos, in which 720 videos are used for training, 140 videos are reserved for validation and 140 videos for testing. Most videos contain frontal faces without occlusions and were collected from YouTube with the consent of the subjects. Each video undergoes four manipulation methods to generate four fake videos, therefore there are 5,000 videos in total. The number of frames in each video is between 300 and 700. The size of the real videos is augmented four times to solve category imbalance between the real and fake data. 270 frames are sampled from each video, following the setting as in FF++ [50]. Output videos are generated with different quality levels, so as to create a realistic setting for manipulated videos, *i.e.*, RAW, High Quality (HQ) and Low Quality (LQ), respectively.

We use the face tracking method proposed by Face2Face [56] to crop the face and adopt a conservative crop to enlarge the face region by a factor of 1.3 around the center of the tracked face, following the setting in [50].

Table 2. Quantitative results (Acc) on FaceForensics++ (LQ) dataset with four manipulation methods, *i.e.* DeepFakes (DF) [1], Face2Face (F2F) [56], FaceSwap (FS) [2] and NeuralTextures (NT) [55]). The bold results are the best.

Methods	DF [1]	F2F [56]	FS [2]	NT [55]
Steg.Features [24]	67.00%	48.00%	49.00%	56.00%
LD-CNN [14]	75.00%	56.00%	51.00%	62.00%
Constrained Conv [6]	87.00%	82.00%	74.00%	74.00%
CustomPooling CNN [49]	80.00%	62.00%	59.00%	59.00%
MesoNet [3]	90.00%	83.00%	83.00%	75.00%
Xception [12]	96.01%	93.29%	94.71%	79.14%
F³-Net(Xception)	**97.97%**	**95.32%**	**96.53%**	**83.32%**
Slowfast [20]	97.53%	94.93%	95.01%	82.55%
F³-Net(Slowfast)	**98.62%**	**95.84%**	**97.23%**	**86.01%**

Evaluation Metrics. We apply the Accuracy score (Acc) and Area Under the Receiver Operating Characteristic Curve (AUC) as our evaluation metrics. (1) **Acc.** Following FF++ [50], we use the accuracy score as the major evaluation metric in our experiments. This metric is commonly used in face forgery detection tasks [3,13,46]. Specifically, for single-frame methods, we average the accuracy scores of each frame in a video. (2) **AUC.** Following face X-ray [40], we use AUC score as another evaluation metric. For single-frame methods, we also average the AUC scores of each frame in a video.

Implementation Details. We use Xception [12] pretrained on the ImageNet [18] as backbone for the proposed F³-Net. The newly-introduced layers and blocks are randomly initialized. The networks are optimized via SGD. We set the base learning rate as 0.002 and use Cosine [41] learning rate scheduler. The momentum is set as 0.9. The batch size is set as 128. We train for about $150k$ iterations.

Some studies [5,51] use videos as the input of the face forgery detection system. To demonstrate the generalization of the proposed methods, we also plug LFS and FAD into existing video-based methods, *i.e.* Slowfast-R101 [20] pre-trained on Kinetics-400 [36]. The networks are optimized via SGD. We set the base learning rate as 0.002. The momentum is set as 0.9. The batch size is set as 64. We train the model for about $200k$ iterations.

4.2   Comparing with Previous Methods

In this section, on the FaceForensics++ dataset, we compare our method with previous face forgery detection methods.

Evaluations on Different Quality Settings. The results are listed in Table 1. The proposed F³-Net outperforms all the reference methods on all quality settings, *i.e.*, LQ, HQ and RAW, respectively. According to the low-quality (LQ)

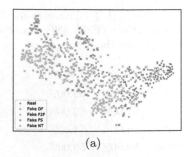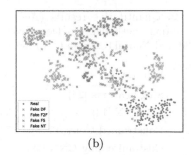

(a)   (b)

Fig. 6. The t-SNE embedding visualization of the baseline (a) and F^3-Net (b) on FF++ [50] low quality (LQ) task. Red color indicates the real videos, the rest colors represent data generated by different manipulation methods. Best viewed in color. (Color figure online)

setting, the proposed F^3-Net achieves 90.43% in Acc and 0.933 in AUC respectively, with a remarkable improvement comparing to the current state-of-the-art methods, *i.e.*, about 3.5% performance gain on Acc score against the best performed reference method (*i.e.*, Xception-PAFilters with 87.16% *v.s.* F^3-Net with 90.43%). The performance gains mainly benefit from the information mining from frequency-aware FAD and LFS clues, which helps the proposed F^3-Net more capable of detecting subtle manipulation artifacts as well as robust to heavy compression errors than plain RGB-based networks. It is worth noting that some methods [10,19,27,45] also try to employ complementary information from other domains, and try to take advantages of prior knowledge. For example, Steg.Features [24] employs hand-crafted steganalysis features and PAFilters [10] tries to augment the edge and texture features by hand-crafted Gabor and high-pass filters. Different from these methods, the proposed F^3-Net makes good use of CNN-friendly and adaptive mechanism to augment the FAD and LFS module, thus significantly boost the performance by a considerable margin.

Towards Different Manipulation Types. Furthermore, we evaluate the proposed F^3-Net on different face manipulation methods listed in [50]. The models are trained and tested exactly on the low quality videos from one face manipulation methods. The results are shown in Table 2. Of the four manipulation methods, the videos generated by NeuralTextures (NT) [55] is extremely challenging due to its excellent generation performance in synthesizing realistic faces without noticeable forgery artifacts. The performance of our proposed method is particularly impressive when detecting forged faces by NT, leading to an improvement of about 4.2% on the Acc score, against the baseline method Xception [12].

Furthermore, we also showed the t-SNE [42] feature spaces of data in FaceForensics++ [50] low quality (LQ) task, by the Xception and our F^3-Net, as shown in Fig. 6. Xception cannot divide the real data and NT-based forged data since their features are cluttered in the t-SNE embedding space, as shown in Fig. 6(a). However, although the feature distances between real videos and NT-based forged videos are closer than the rest pairs in the feature space of F^3-Net,

ID	FAD	LFS	MixBlock	Acc	AUC
1	-	-	-	86.86%	0.893
2	√	-	-	87.95%	0.907
3	-	√	-	88.73%	0.920
4	√	√	-	89.89%	0.928
5	√	√	√	**90.43%**	**0.933**

(a)

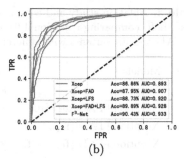

(b)

Fig. 7. (a) Ablation study of the proposed F³-Net on the low quality task(LQ). We compare F³-Net and its variants by removing the proposed FAD, LFS and MixBlock step by step. (b) ROC Curve of the models in our ablation studies.

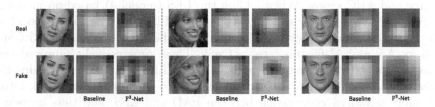

Fig. 8. The visualization of the feature map extracted by baseline (*i.e.*, Xception) and the proposed F³-Net respectively.

they are still much farther away than those in the feature space of Xception. It, from another viewpoint, proves that the proposed F³-Net can mine effective clues to distinguish the real and forged media.

Video-Based Extensions. Meanwhile, there are also several studies [5,51] using multiple frames as the input. To evaluate the generalizability of our methods, we involve the proposed LFS and FAD into Slowfast-R101 [20] due to its excellent performance for video classification. The results are shown in Table 1 and Table 2. More impressively, our F³-Net (Slowfast) achieves the better performances than the baseline using Slowfast only, *i.e.*, 93.02% and 0.958 of Acc and AUC scores in comparison to 90.53% and 0.936, in low quality (LQ) task, as shown in Table 1. Slowfast-F³-Net also wins over 3% on the NT-based manipulation, as shown in Table 2, not to mention the rest three manipulation types. These excellent performances further demonstrate the effectiveness of our proposed frequency-aware face forgery detection method.

4.3 Ablation Study

Effectiveness of LFS, FAD and MixBlock. To evaluate the effectiveness of the proposed LFS, FAD and MixBlock, we quantitatively evaluate F³-Net and its variants: 1) the baseline (Xception), 2) F³-Net w/o LFS and MixBlock, 3) F³-Net w/o FAD and MixBlock, 4) F³-Net w/o MixBlock.

Table 3. Ablation study and component analysis on FAD in FF++ low quality (LQ) tasks. Left: comparing traditional fixed filters with the proposed FAD. Right: comparing FAD and its variants with different kinds of frequency components.

Models	Acc	AUC	Models	Acc	AUC
Xception	86.86%	0.893	FAD-Low	86.95%	0.901
Xception+PAFilters [10]	87.16%	0.902	FAD-Mid	87.57%	0.904
Xception+FAD (f_{base})	87.12%	0.901	FAD-High	87.77%	0.906
Xception+FAD ($f_{base} + f_w$)	87.95%	0.907	FAD-All	87.95%	0.907

The quantitative results are listed in Fig. 7(a). By comparing model 1 (baseline) and model 2 (Xception with FAD), the proposed FAD consistently improves the Acc and AUC scores. When adding the LFS (model 4) based on model 2, the Acc and AUC scores become even higher. Plugging MixBlock (model 5) into the two branch structure (model 4) gets the best performance, 90.43% and 0.933 for Acc and AUC scores, respectively. These progressively improved performances validate that the proposed FAD and LFS module indeed helps the forgery detection, and they are complementary to each other. MixBlock introduce more advanced cooperation between FAD and LFS, and thus introduce additional gains. As shown in the ROC curves in Fig. 7(b), F^3-Net receives the best performance at lower false positive rate (FPR), while low FPR rate is a most challenging scenario to forgery detection system. To better understand the effectiveness of the proposed methods, we visualize the feature maps extracted by the baseline (Xception) and F^3-Net, respectively, as shown in Fig. 8. The discriminativeness of these feature maps is obviously improved by the proposed F^3-Net, e.g., there are clear differences between real and forged faces in the feature distributions of F^3-Net, while the corresponding feature maps generated by Xception are similar and indistinguishable.

Ablation Study on FAD. To demonstrate the benefits of adaptive frequency decomposition on complete frequency domain in FAD, we evaluate the proposed FAD and its variants by removing or replacing some components, i.e., 1) Xception (baseline), 2) a group of hand-drafted filters used in Phase Aware CNN [10], denoted as Xception + PAFilters, 3) proposed FAD without learnable filters, denoted as Xception + FAD (f_{base}), and 4) Xception with the full FAD, denoted as Xception+FAD ($f_{base} + f_w$). All the experiments are under the same hyper-parameters for fair comparisons. As shown in the left part of Table 3, the performance of Xception is improved by a considerable margin on the Acc and AUC scores after applying the FAD, in comparison with other methods using fixed filters (Xception + PAFilters). If the learnable filters are removed, there will also be a sudden performance drop.

We further demonstrate the importance of extracting complete information from complete frequency domain by quantitatively evaluating FAD with different kinds of frequency components, i.e., 1) FAD-Low, FAD with low frequency band components, 2) FAD-Mid, FAD with middle frequency band components, 3)

Table 4. Ablation study on LFS. Here we use only LFS branch and add components step by step. SWDCT indicates using Sliding Window DCT instead of traditional DCT, Stat indicates adopting frequency statistics and D-Stat indicates using our proposed adaptive frequency statistics.

SWDCT	Stat	D-Stat	Acc (LQ)	AUC (LQ)	Acc (HQ)	AUC (HQ)	Acc (RAW)	AUC (RAW)
-	-	-	76.16%	0.724	90.12%	0.905	95.28%	0.948
√	-	-	82.47%	0.838	93.85%	0.940	97.02%	0.964
√	√	-	84.89%	0.865	94.12%	0.936	97.97%	0.975
√	√	√	86.16%	0.889	94.76%	0.951	98.37%	0.983

FAD-High, FAD with high frequency band components and 4) FAD-All, FAD with all frequency bands components. The quantitative results are listed in the right part of Table 3. By comparing FAD-Low, FAD-Mid and FAD-High, the model with high frequency band components achieves the best scores, which indicates that high frequency clues are indubitably helpful for forgery detection. It is because high-frequent clues are usually correlated with forgery-sensitive edges and textures. After making use of all three kinds of information (*i.e.*, FAD-All), we achieved the highest result. Since the low frequency components preserve the global picture, the middle and high frequency reveals the small-scale detailed information, concatenating them together helps to obtain richer frequency-aware clues and is able to mine forgery patterns more comprehensively.

Ablation study on LFS. To demonstrate the effectiveness of SWDCT and dynamic statistical strategy in the proposed LFS introduced in Sect. 3.2, we take the experiments (Xception as backbone) on the proposed LFS and its variants, 1) Baseline, of which the frequency spectrum of the full image by traditional DCT; 2) SWDCT, adopting the localized frequency response by SWDCT; 3) SWDCT+Stat, adopting the general statistical strategy with filters h_{base}; 4) SWDCT+Stat+D-Stat, the proposed FAD consisted of SWDCT and the adaptive frequency statistics with learnable filters h_w. The results are shown in Table 4. Comparing with traditional DCT operation on the full image, the proposed SWDCT significantly improves the performance by a large margin since it is more sensitive to the spatial distributions of the local statistics, and letting the Xception back capture the forgery clues. The improvement of using the statistics is significant and local statistics are more robust to unstable or noisy spectra, especially when optimized by adding the adaptive frequency statistics.

5 Conclusions

In this paper, we propose an innovative face forgery detection framework that can make use of frequency-aware forgery clues, named as F³-Net. The proposed framework is composed of two frequency-aware branches, one focuses on mining subtle forgery patterns through frequency components partition, and the other aims at extracting small-scale discrepancy of frequency statistics between real and forged images. Meanwhile, a novel cross-attention module is applied for two-stream collaborative learning. Extensive experiments demonstrate the effectiveness and significance of the proposed F^3-Net on FaceForencis++ dataset, especially in the challenging low quality task.

Acknowledgements. This work is supported by SenseTime Group Limited, in part by key research and development program of Guangdong Province, China, under grant 2019B010154003. The contribution of Yuyang Qian and Guojun Yin are Equal.

References

1. Deepfakes github. https://github.com/deepfakes/faceswap
2. Faceswap. https://github.com/MarekKowalski/FaceSwap/
3. Afchar, D., Nozick, V., Yamagishi, J., Echizen, I.: Mesonet: a compact facial video forgery detection network. In: 2018 IEEE International Workshop on Information Forensics and Security (WIFS), pp. 1–7. IEEE (2018)
4. Ahmed, N., Natarajan, T., Rao, K.R.: Discrete cosine transform. IEEE Trans. Comput. **100**(1), 90–93 (1974)
5. Amerini, I., Galteri, L., Caldelli, R., Del Bimbo, A.: Deepfake video detection through optical flow based CNN. In: Proceedings of the IEEE International Conference on Computer Vision Workshops (2019)
6. Bayar, B., Stamm, M.C.: A deep learning approach to universal image manipulation detection using a new convolutional layer. In: Proceedings of the 4th ACM Workshop on Information Hiding and Multimedia Security, pp. 5–10 (2016)
7. Bentley, P.M., McDonnell, J.: Wavelet transforms: an introduction. Electron. Commun. Eng. J. **6**(4), 175–186 (1994)
8. Brock, A., Donahue, J., Simonyan, K.: Large scale GAN training for high fidelity natural image synthesis. arXiv preprint arXiv:1809.11096 (2018)
9. Carvalho, T., Faria, F.A., Pedrini, H., Torres, R.D.S., Rocha, A.: Illuminant-based transformed spaces for image forensics. IEEE Trans. Inf. Forensics Secur. **11**(4), 720–733 (2015)
10. Chen, M., Sedighi, V., Boroumand, M., Fridrich, J.: JPEG-phase-aware convolutional neural network for steganalysis of JPEG images. In: Proceedings of the 5th ACM Workshop on Information Hiding and Multimedia Security, pp. 75–84 (2017)
11. Choi, Y., Choi, M., Kim, M., Ha, J.W., Kim, S., Choo, J.: StarGAN: unified generative adversarial networks for multi-domain image-to-image translation. In: Proceedings of the IEEE Conference on Computer Vision and Pattern Recognition, pp. 8789–8797 (2018)
12. Chollet, F.: Xception: deep learning with depthwise separable convolutions. In: Proceedings of the IEEE Conference on Computer Vision and Pattern Recognition, pp. 1251–1258 (2017)
13. Cozzolino, D., Gragnaniello, D., Verdoliva, L.: Image forgery localization through the fusion of camera-based, feature-based and pixel-based techniques. In: 2014 IEEE International Conference on Image Processing (ICIP), pp. 5302–5306. IEEE (2014)
14. Cozzolino, D., Poggi, G., Verdoliva, L.: Recasting residual-based local descriptors as convolutional neural networks: an application to image forgery detection. In: Proceedings of the 5th ACM Workshop on Information Hiding and Multimedia Security, pp. 159–164 (2017)

15. D'Avino, D., Cozzolino, D., Poggi, G., Verdoliva, L.: Autoencoder with recurrent neural networks for video forgery detection. Electron. Imaging **2017**(7), 92–99 (2017)
16. De Carvalho, T.J., Riess, C., Angelopoulou, E., Pedrini, H., de Rezende Rocha, A.: Exposing digital image forgeries by illumination color classification. IEEE Trans. Inf. Forensics Secur. **8**(7), 1182–1194 (2013)
17. Denemark, T.D., Boroumand, M., Fridrich, J.: Steganalysis features for content-adaptive JPEG steganography. IEEE Trans. Inf. Forensics Secur. **11**(8), 1736–1746 (2016)
18. Deng, J., Dong, W., Socher, R., Li, L.J., Li, K., Fei-Fei, L.: Imagenet: a large-scale hierarchical image database. In: 2009 IEEE Conference on Computer Vision and Pattern Recognition, pp. 248–255. IEEE (2009)
19. Durall, R., Keuper, M., Pfreundt, F.J., Keuper, J.: Unmasking deepfakes with simple features. arXiv preprint arXiv:1911.00686 (2019)
20. Feichtenhofer, C., Fan, H., Malik, J., He, K.: Slowfast networks for video recognition. In: Proceedings of the IEEE International Conference on Computer Vision, pp. 6202–6211 (2019)
21. Ferrara, P., Bianchi, T., De Rosa, A., Piva, A.: Image forgery localization via fine-grained analysis of CFA artifacts. IEEE Trans. Inf. Forensics Secur. **7**(5), 1566–1577 (2012)
22. Fogel, I., Sagi, D.: Gabor filters as texture discriminator. Biol. Cybern. **61**(2), 103–113 (1989)
23. Franzen, F.: Image classification in the frequency domain with neural networks and absolute value DCT. In: Mansouri, A., El Moataz, A., Nouboud, F., Mammass, D. (eds.) ICISP 2018. LNCS, vol. 10884, pp. 301–309. Springer, Cham (2018). https://doi.org/10.1007/978-3-319-94211-7_33
24. Fridrich, J., Kodovsky, J.: Rich models for steganalysis of digital images. IEEE Trans. Inf. Forensics Secur. **7**(3), 868–882 (2012)
25. Fujieda, S., Takayama, K., Hachisuka, T.: Wavelet convolutional neural networks for texture classification. arXiv preprint arXiv:1707.07394 (2017)
26. Goodfellow, I., et al.: Generative adversarial nets. In: Advances in Neural Information Processing Systems, pp. 2672–2680 (2014)
27. Gunawan, T.S., Hanafiah, S.A.M., Kartiwi, M., Ismail, N., Za'bah, N.F., Nordin, A.N.: Development of photo forensics algorithm by detecting photoshop manipulation using error level analysis. Indonesian J. Electr. Eng. Comput. Sci. (IJEECS) **7**(1), 131–137 (2017)
28. Haley, G.M., Manjunath, B.: Rotation-invariant texture classification using a complete space-frequency model. IEEE Trans. Image Process. **8**(2), 255–269 (1999)
29. Hsu, C.C., Hung, T.Y., Lin, C.W., Hsu, C.T.: Video forgery detection using correlation of noise residue. In: 2008 IEEE 10th Workshop on Multimedia Signal Processing, pp. 170–174. IEEE (2008)
30. Hsu, C.C., Lee, C.Y., Zhuang, Y.X.: Learning to detect fake face images in the wild. In: 2018 International Symposium on Computer, Consumer and Control (IS3C), pp. 388–391. IEEE (2018)
31. Huang, H., He, R., Sun, Z., Tan, T.: Wavelet-SRNet: A wavelet-based CNN for multi-scale face super resolution. In: Proceedings of the IEEE International Conference on Computer Vision, pp. 1689–1697 (2017)
32. Huang, Y., Zhang, W., Wang, J.: Deep frequent spatial temporal learning for face anti-spoofing. arXiv preprint arXiv:2002.03723 (2020)
33. Jeon, H., Bang, Y., Woo, S.S.: FDFtNet: Facing off fake images using fake detection fine-tuning network. arXiv preprint arXiv:2001.01265 (2020)

34. Karras, T., Aila, T., Laine, S., Lehtinen, J.: Progressive growing of GANs for improved quality, stability, and variation. arXiv preprint arXiv:1710.10196 (2017)
35. Karras, T., Laine, S., Aila, T.: A style-based generator architecture for generative adversarial networks. In: Proceedings of the IEEE Conference on Computer Vision and Pattern Recognition, pp. 4401–4410 (2019)
36. Kay, W., et al.: The kinetics human action video dataset. arXiv preprint arXiv:1705.06950 (2017)
37. Li, H., Li, B., Tan, S., Huang, J.: Detection of deep network generated images using disparities in color components. arXiv preprint arXiv:1808.07276 (2018)
38. Li, J., Wang, Y., Tan, T., Jain, A.K.: Live face detection based on the analysis of Fourier spectra. In: Biometric Technology for Human Identification, vol. 5404, pp. 296–303. International Society for Optics and Photonics (2004)
39. Li, J., You, S., Robles-Kelly, A.: A frequency domain neural network for fast image super-resolution. In: 2018 International Joint Conference on Neural Networks (IJCNN), pp. 1–8. IEEE (2018)
40. Li, L., et al.: Face X-ray for more general face forgery detection. arXiv preprint arXiv:1912.13458 (2019)
41. Loshchilov, I., Hutter, F.: SGDR: Stochastic gradient descent with warm restarts. arXiv preprint arXiv:1608.03983 (2016)
42. van der Maaten, L., Hinton, G.: Visualizing data using t-SNE. J. Mach. Learn. Res. **9**(Nov), 2579–2605 (2008)
43. Marra, F., Gragnaniello, D., Cozzolino, D., Verdoliva, L.: Detection of GAN-generated fake images over social networks. In: 2018 IEEE Conference on Multimedia Information Processing and Retrieval (MIPR), pp. 384–389. IEEE (2018)
44. McCloskey, S., Albright, M.: Detecting GAN-generated imagery using color cues. arXiv preprint arXiv:1812.08247 (2018)
45. Nguyen, H.H., Fang, F., Yamagishi, J., Echizen, I.: Multi-task learning for detecting and segmenting manipulated facial images and videos. arXiv preprint arXiv:1906.06876 (2019)
46. Nguyen, H.H., Yamagishi, J., Echizen, I.: Use of a capsule network to detect fake images and videos. arXiv preprint arXiv:1910.12467 (2019)
47. Pan, X., Zhang, X., Lyu, S.: Exposing image splicing with inconsistent local noise variances. In: 2012 IEEE International Conference on Computational Photography (ICCP), pp. 1–10. IEEE (2012)
48. Pandey, R.C., Singh, S.K., Shukla, K.K.: Passive forensics in image and video using noise features: a review. Digit. Invest. **19**, 1–28 (2016)
49. Rahmouni, N., Nozick, V., Yamagishi, J., Echizen, I.: Distinguishing computer graphics from natural images using convolution neural networks. In: 2017 IEEE Workshop on Information Forensics and Security (WIFS), pp. 1–6. IEEE (2017)
50. Rossler, A., Cozzolino, D., Verdoliva, L., Riess, C., Thies, J., Nießner, M.: Faceforensics++: learning to detect manipulated facial images. In: Proceedings of the IEEE International Conference on Computer Vision, pp. 1–11 (2019)
51. Sabir, E., Cheng, J., Jaiswal, A., AbdAlmageed, W., Masi, I., Natarajan, P.: Recurrent convolutional strategies for face manipulation detection in videos. Interfaces (GUI) **3**, 1 (2019)
52. Sarlashkar, A., Bodruzzaman, M., Malkani, M.: Feature extraction using wavelet transform for neural network based image classification. In: Proceedings of Thirtieth Southeastern Symposium on System Theory, pp. 412–416. IEEE (1998)
53. Stuchi, J.A., et al.: Improving image classification with frequency domain layers for feature extraction. In: 2017 IEEE 27th International Workshop on Machine Learning for Signal Processing (MLSP), pp. 1–6. IEEE (2017)

54. Szegedy, C., Ioffe, S., Vanhoucke, V., Alemi, A.A.: Inception-v4, inception-ResNet and the impact of residual connections on learning. In: Thirty-First AAAI Conference on Artificial Intelligence (2017)
55. Thies, J., Zollhöfer, M., Nießner, M.: Deferred neural rendering: image synthesis using neural textures. ACM Trans. Graph. (TOG) **38**(4), 1–12 (2019)
56. Thies, J., Zollhofer, M., Stamminger, M., Theobalt, C., Nießner, M.: Face2face: real-time face capture and reenactment of RGB videos. In: Proceedings of the IEEE Conference on Computer Vision and Pattern Recognition, pp. 2387–2395 (2016)
57. Wang, S.Y., Wang, O., Zhang, R., Owens, A., Efros, A.A.: CNN-generated images are surprisingly easy to spot... for now. arXiv preprint arXiv:1912.11035 (2019)
58. Yu, N., Davis, L.S., Fritz, M.: Attributing fake images to GANs: Learning and analyzing GAN fingerprints. In: Proceedings of the IEEE International Conference on Computer Vision, pp. 7556–7566 (2019)
59. Zhang, X., Karaman, S., Chang, S.F.: Detecting and simulating artifacts in GAN fake images. arXiv preprint arXiv:1907.06515 (2019)
60. Zhou, P., Han, X., Morariu, V.I., Davis, L.S.: Two-stream neural networks for tampered face detection. In: 2017 IEEE Conference on Computer Vision and Pattern Recognition Workshops (CVPRW), pp. 1831–1839. IEEE (2017)

Weakly-Supervised Cell Tracking via Backward-and-Forward Propagation

Kazuya Nishimura[1]($\boxtimes$), Junya Hayashida[1], Chenyang Wang[2],
Dai Fei Elmer Ker[2], and Ryoma Bise[1]

[1] Kyushu University, Fukuoka, Japan
{kazuya.nishimura,bise}@human.ait.kyushu-u.ac.jp
[2] The Chinese University of Hong Kong, Sha Tin, Hong Kong

Abstract. We propose a weakly-supervised cell tracking method that can train a convolutional neural network (CNN) by using only the annotation of "cell detection" (*i.e.*, the coordinates of cell positions) without association information, in which cell positions can be easily obtained by nuclear staining. First, we train co-detection CNN that detects cells in successive frames by using weak-labels. Our key assumption is that co-detection CNN implicitly learns association in addition to detection. To obtain the association, we propose a backward-and-forward propagation method that analyzes the correspondence of cell positions in the outputs of co-detection CNN. Experiments demonstrated that the proposed method can associate cells by analyzing co-detection CNN. Even though the method uses only weak supervision, the performance of our method was almost the same as the state-of-the-art supervised method. Code is publicly available in https://github.com/naivete5656/WSCTBFP.

Keywords: Cell tracking · Weakly-supervised learning · Multi-object tracking · Cell detection · Tracking · Weakly-supervised tracking

1   Introduction

Cell behavior analysis plays an important role in biology and medicine. To create quantitative cell-behavior metrics, cells are often captured with time-lapse images by using phase-contrast microscopy, which is a non-invasive imaging technique, and then hundreds of cells over thousands of frames are tracked in populations. However, it is time-consuming to track a large number of cells manually. Thus, automatic cell tracking is required.

Cell tracking in phase-contrast microscopy has several difficulties compared with general object tracking. First, cells have similar appearances and their shapes may be severely deformed. Second, cells often touch each other and have

Electronic supplementary material The online version of this chapter (https://doi.org/10.1007/978-3-030-58610-2_7) contains supplementary material, which is available to authorized users.

blurry intercellular boundaries. Third, a cell may divide into two cells (cell mitosis); this is very different from general object tracking. These aspects make it difficult to track cells by using only shape similarity and proximity of cells.

To address such difficulties, the positional relationship of nearby cells is important information to identify the association. The recently proposed CNN-based methods that use such context [13,31] have outperformed the conventional image-processing-based methods. However, learning-based methods require enough training data including individual cell positions in each frame and their correspondences in successive frames (*i.e.*, cell location and motion). In addition, since the appearance and behaviors of a cell often change the appearance and behaviors of a cell often change may often change depending on condition (*e.g.*, growth-factors, type of microscope), we usually have to prepare a training dataset for each individual case.

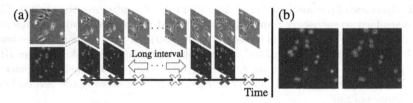

Fig. 1. (a) **Top**: Phase-contrast image sequence. **Bottom**: Fluorescent images with the nucleus stain cell; these were only captured with longer intervals (red crosses) due to phototoxicity. Fluorescent images were not captured on the white crosses. (b) Rough centroid positions (in red) can be easily identified in fluorescent images. (Color figure online)

On the other hand, there are invasive imaging techniques such as fluorescent imaging to facilitate observation of cells. If we can obtain fluorescent images showing cells whose nuclei are stained (Fig. 1) in addition to the phase-contrast images, the rough centroid positions can be easily detected by using simple image processing techniques. However, because fluorescent imaging damages cells, these images can be only captured for training, not for testing. Moreover, fluorescent images cannot be captured frequently over a long period, since phototoxicity may affect the shapes and migration of the cells. Instead, we can capture fluorescent images only several times in enough long period (Fig. 1) since cells can recover from the damage during the non-invasive imaging period. From such sequences, we can automatically obtain point labels for detection [28]. Although these labels do not include the correspondence information between frames, they can be considered as weak-labels for the tracking task.

In this paper, we propose a weakly supervised cell tracking that can obtain the correspondences from training data for the detection tasks (without association). In order to obtain the association information, we designed a method that has three steps as shown in Fig. 2: (1) Our co-detection CNN is trained to detect cells in successive frames by using the rough cell centroid positions,

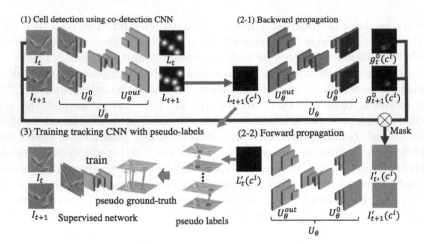

Fig. 2. Overview of our method. (1) co-detection CNN U_θ estimates the position likelihood maps for two successive frames. (2-1) Backward-propagation estimates relevance maps $g_t^0(c^i)$, $g_{t+1}^0(c^i)$ of the cell of interest c^i (red). (2-2) Forward-propagation estimates the cell position likelihood map $L_t'(c^i)$ (blue) with inputting the masked images $I_t'(c^i)$, $I_{t+1}'(c^i)$ which are generated using $g_t^0(c^i)$, $g_{t+1}^0(c^i)$. The pseudo-labels are generated using this estimated regions. (3) The tracking CNN is trained using the pseudo-labels. (Color figure online)

which are weak-labels for the tracking task but it can be used as supervision for detection. Our key assumption is that co-detection CNN implicitly learns the association. (2) The proposed method performs backward-and-forward propagation to extract associations from co-detection CNN without any ground-truth. When we focus on a particular detection response in the output layer L_{t+1} (e.g., the red region in the left image of Fig. 2 (2-1)), the association problem can be considered to be one of finding the position corresponding with the cell of interest (blue region in Fig. 2 (2-2)) from L_t. (3) Using the detection results (1) and association results (2), we can generate the pseudo-training data for the tracking task. We train the cell tracking method [13] with pseudo-training data and a masked loss function that ignores the loss from the false-negative regions, in which we can know the false-negative regions where the cell of interest cannot be associated with any cells in the second step. It is expected that the trained tracking network has better tracking performance compared with the pseudo-labels.

Our main contributions are summarized as follows:

- We propose a weakly-supervised tracking method that can track multiple cells by only using training data for detection. Our method obtains cell association information from co-detection CNN. The association information is used as pseudo-training data for cell tracking CNN.
- We propose a novel network analysis method that can estimate the relation between output and output in two outputs CNN. Our method can extract the

positional correspondences of cells from two successive frames by analyzing co-detection CNN.
- We demonstrated the effectiveness of our method using open data and realistic data. In realistic data, we do not use any human annotations. Our method outperformed current methods that do not require training data. In addition, the performance of our method was almost the same as the state-of-the-art supervised method by using only weak supervision.

2   Related Work

Cell Tracking: Many cell tracking methods have been proposed, which is particle filters [30,37], active contour [22,45,46,51], and detection-and-association [8–10,17,35,40,51]. The detection-and-association methods, which first detect cells in each frame and then solve associations between successive frames, are the most popular tracking paradigm due to the good quality of detection algorithms that use CNNs in the detection step [3,25,32,33]. To associate the detected cells, many methods use hand-crafted association scores based on proximity and shape similarity [10,17,35,40,51]. To extract the similarity features from images, Payer *et al.* [31] proposed a recurrent hourglass network that not only extracts local features but also memorizes inter-frame information. Hayashida *et al.* [13,14] proposed a cell motion field that represents the cell association between successive frames and it can be estimated by a CNN. These methods outperform ones that use hand-crafted association scores. However, they require sufficient training data for both detection and association.

Unsupervised or Weakly-Supervised Tracking for General Objects: Recently, several unsupervised or weakly-supervised tracking methods have been proposed. To track a single object, correspondence learning have been proposed with several weakly-supervision [49] or unsupervised scenarios [42–44]. Zhong *et al.* [49] proposed a tracking method that combines the outputs of multiple trackers to improve tracking accuracy in order to address noisy labels (weak labels). The weak label scenario is different from ours. These methods assumed for tracking a single object, and thus these are short to our problem. Several methods have been proposed for multi-object tracking [16,29]. Nwoye *et al.* [29] proposed a weakly-supervised tracking of surgical tools appearing in endoscopic video. The weak label in this case is a class label of the tool type, and they assumed that one tool appears for each tool type even though several different types of tools appear at a frame. Huang *et al.* [16] tackled a similar problem of semantic object tracking. He *et al.* [15] proposed an unsupervised tracking-by-animation framework. This method uses shape and appearance information for updating the track states of multiple-objects. It assumes that the target objects have different appearances. The above methods may become confused if there are many similar appearance objects in the image; such is the case in cell tracking.

Relevant Pixel Analysis: Visualization methods have been proposed for analyzing relevant pixels for classification in CNNs [4,12,20,27,36,38,39,48,50].

Layer-wise relevance propagation (LRP) [4,27] and guided backpropagation [39] back-propagate signals from the output layer to the input layer on the basis of the weights and signals in the forward-propagation for inference. Methods based on class activation mapping (CAM) [12,36,50], such as Grad-CAM [12], produces the relevance map from CNN using the semantic features right before the fully connected layer for classification. There are several methods that uses such backward operation in a network for instance segmentation [28] and object tracking [23]. For example, Li *et al.* [23] propose a Gradient-Guided Network (GradNet) for a single object tracking that exploits the information in the gradient for template update. Although this method uses the backward operation for guided calculation, the purpose is totally different from ours (analysis of the relevance of the two output layers). These methods assume that they analyze the relationship between the input and output layers but not for two output layers in a multi-branch network.

Unlike the above methods, our method can obtain correspondences between objects in successive frames without training data for the association by analyzing the co-detection CNN, in spite of the challenging conditions wherein many cells having similar appearances migrate.

3   Weakly-Supervised Cell Tracking

3.1   Overview

Figure 2 shows an overview of the proposed method. The method consists of three parts: 1) cell detection using co-detection CNN that jointly detects cells at successive frames using weak labels (cell position label): it is expected to implicitly learn the association between the frames in addition to detection; 2) backward-and-forward propagation for extracting the association from co-detection CNN: 3) training the cell tracking network using pseudo-labels that were generated using the estimated results in (2); The details of each step are explained as follows.

3.2   Co-detection CNN

In the co-detection task [6], the detection results in frame t can facilitate to detect the corresponding cell in frame $t + 1$ and vice versa. Based on this key observation, we designed co-detection CNN U_θ for jointly detecting cells at the successive frames, in which θ indicates the network parameters. In our problem setup, the rough cell centroid position is obtained from the fluorescent images as training data, but the nuclei position may shift from the ground-truth of the centroid position. Therefore, we follow the cell detection network [28] that mitigates this gap by representing cell positions as a position likelihood map. The position likelihood map can be generated from the rough cell centroid position, where a given cell position becomes an intensity peak and the intensity value gradually decreases away from the peak in accordance with a Gaussian distribution [28]. In contrast to [28] (U-Net [34] architecture), our network has two

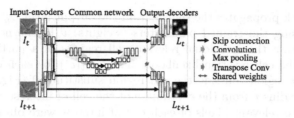

Fig. 3. Architecture of co-detection CNN.

input-encoders, a common network, and two output-decoders to simultaneously estimates the detection results in successive frames as shown in Fig. 3. The two input-encoders have shared weights, and these extract the cell appearance features from the inputted successive images I_t, I_{t+1}. The features are concatenated and input into a common network that has a U-Net architecture. We consider that the common network performs co-detection and it implicitly learns the cell association. Finally, the output-decoders decode the extracted co-detection features into the cell position likelihood maps; the layers of the input and output networks have skip connections to adjust the local positions similar to U-Net. The loss function $Loss_{CD}$ for co-detection CNN is the sum of the Mean Square Errors (MSE) of the likelihood maps of the two frames:

$$Loss_{CD} = MSE(L_t - \hat{L}_t) + MSE(L_{t+1} - \hat{L}_{t+1}), \tag{1}$$

where $\hat{L}_t$, $\hat{L}_{t+1}$ are the ground-truths of the cell position likelihood map of each frame and L_t, L_{t+1} are the estimated maps. In the inference, the peaks in the estimated map are the detected cell positions.

3.3   Backward-and-Forward Propagation

Next, in accordance with our assumption that co-detection CNN implicitly learns the association, we extract the cell association information from co-detection CNN U_θ. Here, we will focus on a particular detection response in the output layer L_{t+1} (e.g., the red regions in Fig. 2 (2-1)). The association problem can be considered to be one of finding the position (blue region) corresponding to the cell of interest from L_t. We propose the following backward-and-forward propagation for this task.

Backward Propagation: Figure 2 (2-1) illustrates the backward-propagation process on U_θ for the cell of interest c^i that is selected from frame $t + 1$. In this step, we extract the relevance maps that are expected to relevant to producing the detection response of interest by using guided backpropagation (GB) [39]. For this process, we modified the weakly-supervised instance segmentation method proposed by Nishimura [28], which extracts individual relevant cell regions of a particular cell in U-Net for a single image. Different from [28], in our case, two relevance maps $g_t^0(c^i)$, $g_{t+1}^0(c^i)$ are extracted from a single output $L_{t+1}(c^i)$.

The GB back propagates the signals from the output layer U_θ^{out} to the input layer U_θ^0 by using the trained parameters (weights) θ in the network. In our method, to obtain the individual relevant cell regions of a particular cell, we first initialize the cell position likelihood map $L_{t+1}(c^i)$ for each cell of interest c^i, in which all regions outside the cell region substitute 0 (Fig. 2 (2-1)). The region within radius r from the coordinate of the cell of interest c^i is defined as $S(c^i)$. Then, the relevant pixels of each cell of interest were obtained by back-propagation from $S(c^i)$. The backpropagating signals are propagated to both layers at the branch of the input-encoders. The red nodes of the intermediate layers in the network in Fig. 2 (2-1) show the illustration of the back-propagation process. This backward process is performed for each cell $i = 1, ..., N$, where N is the number of cells.

It is expected that the corresponding cell regions have positive values in the relevance maps. However, regions of the outside the cell of interest may also have values in the relevance map. This adversely affects the process of extracting the cell association. Therefore, we compare the pixel values of $g_{p,t}^0(c^i)$ ($i = \{1, ..., N\}$) for all cells, where pixel p corresponds to c^i if it takes the maximum value among the cells The maximum projection of the p-th pixel for c^i at frame t can be formalized as:

$$g'_{p,t}(c^i) = \psi_p(t, i, \arg \max_k g_{p,t}^0(c^k)), \qquad (2)$$

$$\psi_p(t, i, k) = \begin{cases} g_{p,t}^0(c^i) & if\ (k = i), \\ 0 & otherwise, \end{cases} \qquad (3)$$

where, p is the p-th pixel on the relevance map. $g'_{t+1}(c^i)$ is calculated by same manner. By applying maximum projection to all cells, we get the maximum projection relevance maps $g'_t(c^i)$ and $g'_{t+1}(c^i)$ for each cell.

Forward Propagation: This step estimates the corresponding cell position likelihood map $L'_t(c^i)$ by using the relevance maps $g'_t(c^i)$ and $g'_{t+1}(c^i)$ (Fig. 2 (2-2)). The high value pixels in $g'_t(c^i)$, $g'_{t+1}(c^i)$ show the pixels that contribute to detect c^i. It indicates that co-detection CNN U_θ is able to detect the corresponding cell position from only these relevant pixels in the input images.

We generate masked images $I'_t(c^i)$, $I'_{t+1}(c^i)$ that only have values at the high relevance pixels of only the cell of interests c^i. In order to generate it, we first initialize the images so that it has the background intensities of the input image. Then, we set the pixel values in the initialized image as the input image intensity if the value of the p-th pixel of $g'_t(c^i)$ is larger than a threshold th.

$$I'_{p,t}(c^i) = \begin{cases} I_{p,t}(c^i) & if\ g'_{p,t}(c^i) > th, \\ B_{p,t}(c^i) & otherwise, \end{cases} \qquad (4)$$

where the background image at frame t B_t is estimated by using quadratic curve fitting [47] and p is the p-th pixel on I'. $I'_{t+1}(c^i)$ is also made by the same manner. The corresponding cell position likelihood map $L'_t(c^i) = U_\theta(I'_t(c^i), I'_{t+1}(c^i))$ can be obtained by inputting the masked images $I'_t(c^i)$, $I'_{t+1}(c^i)$ to U_θ.

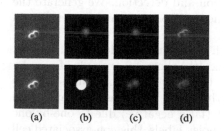

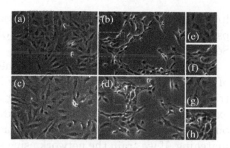

Fig. 4. Top: Example without using the masked loss. **Bottom:** Example with using the masked loss. (a) phase-contrast image, (b) pseudo-training data, (c) output of trained network, and (d) overlapping images. The white region in (b) indicates the region that ignores the loss.

Fig. 5. Example images on four culture conditions. (a) Control, (b) FGF2, (c) BMP2, (d) FGF2+BMP2, (e)–(h) are enlarged images of the red box in (a)–(d). (Color figure online)

The estimated map $L'_t(c^i)$ indicates the i-th detection response at t that corresponds to the detected cell at frame t. It is expected that the high intensity region in $L'_t(c^i)$ is expected to appears on the same region of either cell detection result in L_t. To obtain the cell position at t corresponding to the i-th cell position at $t+1$, we perform one-by-one matching by using linear programming between these two maps, in which we use a simple MSE of intensities for the matching score. This one-by-one matching may correspond either of the cell position even if the estimated response signal is too small. We omit the low confidence associations in order to keep the precision of the pseudo-labels high enough. If the matching score is less than the threshold th_{conf}, we define it as the low confidence association. By omitting low confidence associations, the result includes some false-negative. One of the interesting points is that we can know where the low confidence region at $t+1$ is since we explicitly give the region of the cell of interest $S_{t+1}(c^i)$. If the cell c^i is not associated with any cell, we add the pixels of the cell $S_{t+1}(c^i)$ to the set of unassociated cell region Γ. Finally, we obtain the cell position and association in successive frames and there will use as pseudo-labels in the next step.

3.4   Training Tracking CNN Using Pseudo-labels

It is known that the generalization performance and estimation speed can be improved by training a CNN using pseudo-labels in weakly-supervised segmentation tasks [1,2,5,19,24]; we took this approach for our tracking task. We train a state-of-the-art cell tracking network called MPM-Net [14] with the pseudo-labels. The MPM-Net estimates Motion and Position Map (MPM) that simultaneously represents the cell positions and their association between frames from inputting the images at successive frames. The advantage of this method is that it can extract the features of the spatio-temporal context about nearby cells from the inputted entire images for association and detection. We generate the pseudo-training data for MPM using obtained cell position and association in Sect. 3.3.

In the previous step, we obtained the high confidence pseudo-labels and a set of unassociated regions Γ. The top row in Fig. 4(b) shows an example of pseudo-training data directly generated using only the high confidence labels. In this example, a cell divides two cells. However, the mother cell was associated with one of the child cells and the other was not (false negative) due to one-by-one matching. If we train the network using such noisy labels, this non-associated cell region affects the learning. Indeed, the non-associated region was not detected due to over-fitting. Figure 4(c) shows the output of the trained network that only detects one cell by over-fitting. To avoid this problem, we train the network with the masked loss function that ignores the loss from the false-negative regions where the cell did not correspond to any detection responses due to its low confidence. The masked loss is formulated as:

$$Loss_{mask} = \begin{cases} 0 & \text{if } p \in \Gamma, \\ Loss_{ori} & \text{otherwise,} \end{cases} \tag{5}$$

where Γ is the set of the ignoring regions that contain the unassociated cell regions, p is a pixel in Γ, $Loss_{ori}$ is the original loss for MPM-Net [14]. As shown in the white circle in bottom row of Fig. 4(b), the false-negative region is not calculated in the masked loss. This effectively avoid over-fitting and correctly estimate the cell position likelihood map as shown in the bottom row of Fig. 4(c).

4   Experiment

4.1   Data Set and Experimental Setup

We evaluated our method on an open data set [18] that contains time-lapse sequences captured phase-contrast microscopy[1]. In the data set, the mybolast cells were cultured under four growth factor conditions: (a) Control, (b) FGF2, (c) BMP2, and (d) FGF2+BMP2 (Fig. 5). Each sequence consists of 780 frames, with a 5 min interval between consecutive frames. The resolution of each image is 1392×1040 pixels. There are four sequences for each condition and the total number of sequences is 16. The rough cell centroid positions are annotated with the cell ID. In one of the BMP2 sequence, all cells are annotated. For the other sequences, three cells were randomly selected at the beginning of the sequence and then their descendants were annotated. The total number of annotated cells in the 16 sequences is 135859. We used one of the BMP2 sequence as the training data for co-detection CNN and the other sequences were used as the test data. In the training process, we only used the cell position coordinates as weak labels.

---

[1] The data [18] is more challenging as a tracking task compared with ISBI Cell Tracking Challenging [26,41] that more focused on segmentation task, since the cells often partially overlapped and the boundary of cells is ambiguous.

The task was challenging because the training was only weakly-supervised and the appearances of the cells in the test data differed from those in the training data (see Fig. 5). To train co-detection CNN and MPM, we used Adam [21] optimizer with learning rate 10^{-3}. We set the threshold th in Eq. 4 to 0.01, the low confidence association threshold th_{conf} to 0.5, and $r = 18$ in all experiments; these parameters were decided using validation data and were not sensitive.

4.2 Performance of Cell Tracking on Open Data Set

We compared our method with five other methods by using the open data set [18]. Since our method only requires weak-supervision, we selected three methods that do not use association information; 1) asymmetric graphcut (A-Graph) [7] that segments cell regions using asymmetric graph-cut: it was trained with the small amount of additional ground-truth for segmentation; 2) Fogbank [11] that segments cell regions using image processing: the hyper-parameters were tuned using the validation data; 3) global data association (GDA) [10] that segments cell regions by physical-model-based method [47] and then performs spatial-temporal global data association: the hyper-parameters were tuned using validation data. In addition, in order to show that the performance of our method is comparable with the SOTA (state-of-the-art), we evaluated two supervised tracking methods that require the ground-truth of the cell position and association; 4) cell motion field (CMF) [13] that estimates the cell motion and position separately; 5) motion and position map (MPM) [14] that estimates the motion and position map, which achieved the SOTA performance. In addition, to confirm the effectiveness of the masked loss, we also compared with our method without the masked loss (Ours w/o ml).

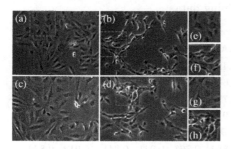

Fig. 6. Examples of tracking results of (a) GDA, (b) CMF, (c) MPM, and (d) ours. The horizontal axis indicates the time.

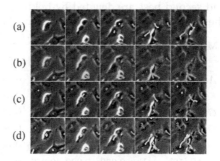

Fig. 7. Examples of tracking results under each conditions: (a) Control, (b)FGF2, and (c) FGF2+BMP2.

We used the association accuracy and target effectiveness as following the paper that proposed the MPM [14]. Each target was first assigned to a track (estimation) for each frame. The association accuracy indicates the number of true positive associations divided by the number of true positive associations in the ground-truth. If cell A switches into B, and B into A, we count two false-positive (A→B, B→A) and two false-negatives (no A→A, B→B). The target effectiveness was computed as the number of the assigned track observations over the total number of frames of the target after assigning each target to a track that contains the most observations from that ground-truth. It indicates how many frames of targets are followed by computer-generated tracks. This metric is a stricter than the association accuracy. If a switching error occurs in the middle of the trajectory, the target effectiveness is 0.5.

Table 1 shows the results of the performance comparison. Our method outperformed the other weakly or unsupervised methods (A-Graph [7], Fogbank [11], GDA [10]) and achieved comparable results with state-of-the-art supervised methods (MPM [14]). Even though our method used only weak-supervision, it outperformed that of the supervised MPM in FGF2. We consider that the MPM may be over-fitted to the condition of the training data (BMP2), and thus its performance may decrease since the cell appearance in FGF2 is different from that in BMP2 as shown in Fig. 5. In addition, the results show that the masked loss slightly improved the performance compared with 'Ours w/o ml'. Figure 6 (c) shows examples of tracking results under BMP2[2]. Although GDA did not detect the brighter cell and CMF did not identify the newly born cells after cell mitosis, our method successfully tracked almost all the cells as the same with MPM.

Table 1. Tracking performance in terms of association accuracy (AA) and target effectiveness (TE) on open data set [18]. Su. indicates the condition of the training data: weak-supervision (W), un-supervision (U), and fully-supervision (F). The best and second best are denoted by boldface and the best one is underlined. '*' indicates the culture condition is the same as in the training data.

Method	Su.	*BMP2		FGF2		Control		FGF2+BMP2		Ave.	
		AA	TE	AA	TE	AA	TE	AA	TE	AA	TE
A-Graph [7]	W	0.801	0.621	0.604	0.543	0.499	0.448	0.689	0.465	0.648	0.519
Fogbank [11]	U	0.769	0.691	0.762	0.683	0.650	0.604	0.833	0.587	0.753	0.641
GDA [10]	U	0.855	0.788	0.826	0.733	0.775	0.710	0.942	0.633	0.843	0.771
Ours w/o ml	W	0.979	0.960	0.950	0.861	0.917	0.786	0.972	0.880	0.954	0.873
Ours	W	**0.982**	**0.970**	**0.955**	**0.869**	**0.926**	**0.806**	**0.976**	**0.911**	**0.960**	**0.881**
CMF [13]	F	0.958	0.939	0.866	0.756	0.884	0.761	0.941	0.841	0.912	0.822
MPM [14]	F	**0.991**	0.958	**0.947**	0.803	**0.952**	**0.829**	**0.987**	**0.911**	**0.969**	0.875

---

[2] Since the tracking results of A-Graph and Fogbank were very worse, we omitted their results on these figures due to the page limitation.

Table 2. Detection performance.

Method	*BMP2 sparse	*BMP2 medium	*BMP2 dense	Control	FGF2	FGF2+BMP2
Nishimura [28]	0.998	0.978	0.977	0.922	0.924	**0.962**
Ours (5 min. int.)	**0.999**	0.983	**0.980**	0.923	**0.928**	0.945
Ours (25 min. int.)	0.998	**0.984**	0.978	**0.926**	0.911	0.946

4.3   Ablation Study

Next, we performed an ablation study to evaluate the performance of the co-detection CNN, backward-and-forward propagation, and re-training individually. In this ablation study, in order to confirm the robustness in various conditions, we additionally added annotations for other three conditions (Control, FGF2, FGF2+BMP2) since only some of the cells were annotated under these three conditions in the original data. Then, we evaluated each step of our method under these three conditions.

Co-detection CNN: We first evaluated our co-detection CNN against the method proposed by Nishimura *et al.* [28] that estimates the cell position likelihood map of a single image. In addition, in order to demonstrate the robustness for cell migration speed since the speed is depending on the cell types and time-interval, we also evaluated two intervals (5 and 25 min), in which the speed in 25 min is much faster than that in 5 min. We used F1-score as the detection performance metric. Table 2 shows the results. Our co-detection CNN performed almost as well as the state-of-the-art method under all conditions (BMP2-sparse,

Table 3. Association performance of backward-and-forward propagation.

Interval	Metrics	*BMP2 sparse	*BMP2 medium	*BMP2 dense	Control	FGF2	FGF2+BMP2
5 min.	Precision	0.999	0.989	0.992	0.971	0.966	0.964
	Recall	0.997	0.976	0.957	0.849	0.844	0.900
	F1-score	0.998	0.982	0.974	0.906	0.901	0.931
25 min.	Precision	0.998	0.982	0.974	0.906	0.901	0.931
	Recall	0.961	0.975	0.960	0.849	0.753	0.902
	F1-score	0.979	0.981	0.976	0.904	0.830	0.930

Table 4. Comparison of backward-and-forward propagation (BF) with MPM trained by backward-and-forward propagation (T). AA: association accuracy, TE: target effectiveness.

Met.	*BMP2 sparse		*BMP2 medium		*BMP2 dense		Control		FGF2		FGF2+BMP2	
	AA	TE	AA	TE	AA	TE	AA	TE	AA	TE	AA	TE
BF	0.983	0.968	0.973	0.962	0.840	0.914	0.826	0.765	**0.794**	**0.672**	0.955	0.945
T	**0.993**	**0.976**	**0.980**	**0.974**	**0.970**	**0.969**	**0.858**	**0.800**	0.773	0.640	**0.982**	**0.970**

BMP2-medium, BMP2-dense, Control, FGF2, FGF2+FGF2). In addition, the results show that our method was robust to the different cell migration speeds, since the performances of both interval conditions are almost the same. Figure 8 shows examples in which co-detection CNN improved the detection results. In the upper case, the cell shape is ambiguous at t but it is more clear at $t + 1$, co-detection CNN uses these two image and it may facilitate to detect the cell. In the bottom case, a tips of cell (noise) appears at both frames. Since the noise traveled a large distance, co-detection CNN may reduce over-detections.

Backward-and-Forward Propagation: Next, we evaluated the association performance of Backward-and-Forward propagation (BF-prop). We used precision, recall, F1-score of association accuracy as the performance metrics. Figure 9 shows examples. L indicates the output of co-detection CNN given two input images I at t and $t + 1$. I' indicates the masked image generated using the relevance map produced by backward propagation. L' indicates the estimated likelihood map by forward propagation by inputting I'. In both cases, the backward propagation could obtain the target cell regions and forward propagation successfully estimated the detection map of the corresponding cell. Under all conditions, backward-and-forward propagation performed association accurately (over 90% in the terms of F1-score as shown in Table 3). As discussed in Sect. 3.3, precision is more important than recall when using the pseudo-labels. BF-prop achieved higher precision than recall on all data-sets. In addition, we conducted evaluations using the different intervals (5 and 25 min.). The results for 5 min were slightly better than those in 25 min, but not significantly.

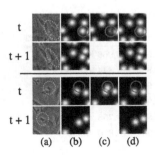

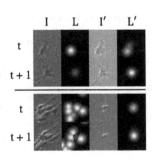

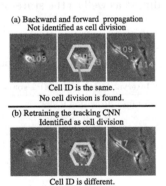

Fig. 8. Example results of detection. (a) phase-contrast, (b) ground-truth, (c) Nishimura [28], and (d) Ours.

Fig. 9. Example results of BF-prop. The red regions are the cell region of interest, and the blue regions are the estimated corresponding cell. (Color figure online)

Fig. 10. Example results from (a) BF-prop, (b) retrained MPM-net.

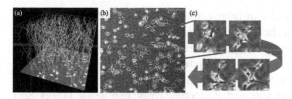

Fig. 11. Examples of tracking result on realistic dat. (a) 3D view of estimated cell trajectories. The z-axis is the time, and each color indicates the trajectory of a cell. (a) Entire image. (c) Sequence of enlarged images at the red box in (b). (Color figure online)

Table 5. Quantitative evaluation. AA: association accuracy, TE: target effectiveness

Method	AA	TE
A-Graph [7]	0.216	0.169
Fogbank [11]	0.695	0.321
GDA [10]	0.773	0.527
Ours	**0.857**	**0.804**

Training Tracking CNN: To show the effectiveness of retraining the tracking CNN, we compared the trained CNN with the results of BF-prop in terms of the same tracking metrics (AA and TE). As shown in Table 4, the retraining improved the performance under almost conditions except FGF2. The important thing is that although the BF-prop only tracks one of the cell when a cell divided two cells, the retraining could identify a cell division since the masked loss helped to detect the another cell of the divided two cells. Figure 10 shows the example of the cell division case. The two new cells were successfully identified by MPM-Net and new IDs were assigned to them. In contrast, BF-prop tracked only one of them continuously and did not identify cell division.

4.4   Cell Tracking Without Any Human Annotation

In this section, we consider a more realistic scenario when pairs of phase-contrast and fluorescent microscopy images for training and the test image sequence that was captured by only phase-contrast microscopy were provided by biologists without any human annotation. In order to confirm that our method can perform such a realistic scenario, we also prepared a data-set with this problem setup. In this experiment, 86 pairs of phase-contrast and fluorescent images were given as the training data, and the 95 images was given as the test data, in which the cell appearance is different from the open data set we used in the previous section.

In this setting, our method could perform tracking in four steps. (1) We trained the detection CNN with phase-contrast images using the ground-truth of detection automatically generated from the given fluorescent image(the procedure was the same as that of Nishimura [28].). Then, we generated co-detection pseudo labels. (2) We trained co-detection CNN with the generated detection pseudo labels, and (3) generated the pseudo-labels for association by BF-prop. (4) We trained the MPM-net using the pseudo-labels and applied it to the test data. In the evaluation, we also compared our method with un-supervised and weakly-supervised tracking methods; A-Graph [7], Fogbank [11], GDA [10]. Here, we could not compare with the supervised method on this scenario since there was no supervised training data.

Table 5 shows the tracking results in terms of association accuracy (AA) and target effectiveness (TE). Our method outperformed the other methods on both metrics. Our method achieved an 8% improvement in association accuracy and 28% improvement in target effectiveness compared with the second best. As shown in Fig. 11, our method can track many cells without supervised annotation. These results show that our method can effectively use weak labels and obtain good tracking results.

5 Conclusion

We proposed a weakly-supervised tracking method that can track multiple cells by using only training data for detection. The method first trains co-detection CNN that detects cells in successive frames by using weak supervision. Then, the method estimates association from co-detection CNN by using our novel backward-and-forward propagation method on the basis of the key assumption that co-detection CNN implicitly learns the association. The association is used as pseudo-labels for a state-of-the-art tracking network (MPM-net). Our method outperformed the compared methods and achieved comparable results to those of supervised state-of-the-art methods. In addition, we demonstrated the effectiveness of our method in a realistic scenario in which the tracking network was trained without any human annotations.

Acknowledgement. This work was supported by JSPS KAKENHI Grant Number 20H04211.

References

1. Ahn, J., Cho, S., Kwak, S.: Weakly supervised learning of instance segmentation with inter-pixel relations. In: CVPR, pp. 2209–2218 (2019)
2. Ahn, J., Kwak, S.: Learning pixel-level semantic affinity with image-level supervision for weakly supervised semantic segmentation. In: CVPR, pp. 4981–4990 (2018)
3. Akram, S.U., Kannala, J., Eklund, L., Heikkilä, J.: Joint cell segmentation and tracking using cell proposals. In: ISBI, pp. 920–924 (2016)
4. Bach, S., Binder, A., Montavon, G., Klauschen, F., Müller, K.R., Samek, W.: On pixel-wise explanations for non-linear classifier decisions by layer-wise relevance propagation. PloS One **10**(7), e0130140 (2015)
5. Bansal, A., Chen, X., Russell, B., Gupta, A., Ramanan, D.: Pixelnet: Representation of the pixels, by the pixels, and for the pixels. arXiv:1702.06506 (2017)
6. Bao, S.Y., Xiang, Y., Savarese, S.: Object co-detection. In: Fitzgibbon, A., Lazebnik, S., Perona, P., Sato, Y., Schmid, C. (eds.) ECCV 2012. LNCS, vol. 7572, pp. 86–101. Springer, Heidelberg (2012). https://doi.org/10.1007/978-3-642-33718-5_7
7. Bensch, R., Olaf, R.: Cell segmentation and tracking in phase contrast images using graph cut with asymmetric boundary costs. In: ISBI, pp. 1220–1223 (2015)
8. Bise, R., Li, K., Eom, S., Kanade, T.: Reliably tracking partially overlapping neural stem cells in DIC microscopy image sequences. In: International Conference on Medical Image Computing and Computer-Assisted Intervention Workshop (MICCAIW), pp. 67–77 (2009)

9. Bise, R., Maeda, Y., Kim, M.H., Kino-oka, M.: Cell tracking under high conflu-ency conditions by candidate cell region detection-based-association approach. In: Biomedical Engineering, pp. 1004–1010 (2013)
10. Bise, R., Yin, Z., Kanade, T.: Reliable cell tracking by global data association. In: ISBI, pp. 1004–1010 (2011)
11. Chalfoun, J., Majurski, M., Dima, A., Halter, M., Bhadriraju, K., Brady, M.: Lineage mapper: a versatile cell and particle tracker. Sci. Rep. **6**, 36984 (2016)
12. Chattopadhay, A., Sarkar, A., Howlader, P., Balasubramanian, V.N.: GRAD-CAM++: generalized gradient-based visual explanations for deep convolutional networks. In: WACV, pp. 839–847 (2018)
13. Hayashida, J., Bise, R.: Cell tracking with deep learning for cell detection and motion estimation in low-frame-rate. In: Shen, D., et al. (eds.) MICCAI 2019. LNCS, vol. 11764, pp. 397–405. Springer, Cham (2019). https://doi.org/10.1007/978-3-030-32239-7_44
14. Hayashida, J., Nishimura, K., Bise, R.: MPM: joint representation of motion and position map for cell tracking. In: CVPR, pp. 3823–3832 (2020)
15. He, Z., Li, J., Liu, D., He, H., Barber, D.: Tracking by animation: Unsupervised learning of multi-object attentive trackers. In: CVPR, pp. 1318–1327 (2019)
16. Huang, K., Shi, Y., Zhao, F., Zhang, Z., Tu, S.: Multiple instance deep learning for weakly-supervised visual object tracking. Sig. Process.: Image Commun. 115807 (2020)
17. Kanade, T., et al.: Cell image analysis: algorithms, system and applications. In: WACV, pp. 374–381 (2011)
18. Ker, E., et al.: Phase contrast time-lapse microscopy datasets with automated and manual cell tracking annotations. Sci. Data **5**, 180237 (2018). https://doi.org/10.1038/sdata.2018.237
19. Khoreva, A., Benenson, R., Hosang, J., Hein, M., Schiele, B.: Simple does it: weakly supervised instance and semantic segmentation. In: CVPR, pp. 876–885 (2017)
20. Kindermans, P.J., et al.: Learning how to explain neural networks: Patternnet and patternattribution. In: International Conference on Learning Representations (2018)
21. Kingma, D., Ba, J.: Adam: a method for stochastic optimization. In: ICLR (2015)
22. Li, K., Miller, E.D., Chen, M., Kanade, T., Weiss, L.E., Campbell, P.G.: Cell population tracking and lineage construction with spatiotemporal context. Med. Image Anal. **12**(5), 546–566 (2008)
23. Li, P., Chen, B., Ouyang, W., Wang, D., Yang, X., Lu, H.: GradNet: Gradient-guided network for visual object tracking. In: ICCV, pp. 6162–6171 (2019)
24. Li, Q., Arnab, A., Torr, P.H.S.: Weakly- and semi-supervised panoptic segmenta-tion. In: Ferrari, V., Hebert, M., Sminchisescu, C., Weiss, Y. (eds.) ECCV 2018. LNCS, vol. 11219, pp. 106–124. Springer, Cham (2018). https://doi.org/10.1007/978-3-030-01267-0_7
25. Lux, F., Matula, P.: DIC image segmentation of dense cell populations by combin-ing deep learning and watershed. In: ISBI, pp. 236–239 (2019)
26. Maška, M., et al.: A benchmark for comparison of cell tracking algorithms. Bioin-formatics **30**(11), 1609–1617 (2014)
27. Montavon, G., Lapuschkin, S., Binder, A., Samek, W., Müller, K.R.: Explaining nonlinear classification decisions with deep Taylor decomposition. Pattern Recogn. **65**, 211–222 (2017)

28. Nishimura, K., Ker, D.F.E., Bise, R.: Weakly supervised cell instance segmentation by propagating from detection response. In: Shen, D., et al. (eds.) MICCAI 2019. LNCS, vol. 11764, pp. 649–657. Springer, Cham (2019). https://doi.org/10.1007/978-3-030-32239-7_72

29. Nwoye, C.I., Mutter, D., Marescaux, J., Padoy, N.: Weakly supervised convolutional lstm approach for tool tracking in laparoscopic videos. Int. J. Comput. Assist. Radiol. Surg. **14**(6), 1059–1067 (2019)

30. Okuma, K., Taleghani, A., de Freitas, N., Little, J.J., Lowe, D.G.: A boosted particle filter: multitarget detection and tracking. In: Pajdla, T., Matas, J. (eds.) ECCV 2004. LNCS, vol. 3021, pp. 28–39. Springer, Heidelberg (2004). https://doi.org/10.1007/978-3-540-24670-1_3

31. Payer, C., Stern, D., Neff, T., Bischof, H., Urschler, M.: Instance segmentation and tracking with cosine embeddings and recurrent hourglass networks. In: Frangi, A.F., Schnabel, J.A., Davatzikos, C., Alberola-López, C., Fichtinger, G. (eds.) MICCAI 2018. LNCS, vol. 11071, pp. 3–11. Springer, Cham (2018). https://doi.org/10.1007/978-3-030-00934-2_1

32. Rempfler, M., Kumar, S., Stierle, V., Paulitschke, P., Andres, B., Menze, B.H.: Cell lineage tracing in lens-free microscopy videos. In: Descoteaux, M., Maier-Hein, L., Franz, A., Jannin, P., Collins, D.L., Duchesne, S. (eds.) MICCAI 2017. LNCS, vol. 10434, pp. 3–11. Springer, Cham (2017). https://doi.org/10.1007/978-3-319-66185-8_1

33. Rempfler, M., et al.: Tracing cell lineages in videos of lens-free microscopy. Med. Image Anal. **48**, 147–161 (2018)

34. Ronneberger, O., Fischer, P., Brox, T.: U-Net: convolutional networks for biomedical image segmentation. In: Navab, N., Hornegger, J., Wells, W.M., Frangi, A.F. (eds.) MICCAI 2015. LNCS, vol. 9351, pp. 234–241. Springer, Cham (2015). https://doi.org/10.1007/978-3-319-24574-4_28

35. Schiegg, M., Hanslovsky, P., Kausler, B.X., Hufnagel, L., Hamprecht, F.A.: Conservation tracking. In: ICCV, pp. 2928–2935 (2013)

36. Selvaraju, R.R., Cogswell, M., Das, A., Vedantam, R., Parikh, D., Batra, D.: Grad-CAM: visual explanations from deep networks via gradient-based localization. In: ICCV, pp. 618–626 (2017)

37. Smal, I., Niessen, W., Meijering, E.: Bayesian tracking for fluorescence microscopic imaging. In: ISBI, pp. 550–553 (2006)

38. Smilkov, D., Thorat, N., Kim, B., Viégas, F., Wattenberg, M.: Smoothgrad: removing noise by adding noise. arXiv:1706.03825 (2017)

39. Springenberg, J., Dosovitskiy, A., Brox, T., Riedmiller, M.: Striving for simplicity: the all convolutional net. In: ICLRW (2015)

40. Su, H., Yin, Z., Huh, S., Kanade, T.: Cell segmentation in phase contrast microscopy images via semi-supervised classification over optics-related features. Med. Image Anal. **17**(7), 746–765 (2013)

41. Ulman, V., et al.: An objective comparison of cell-tracking algorithms. Nat. Methods **14**(12), 1141 (2017)

42. Vondrick, C., Shrivastava, A., Fathi, A., Guadarrama, S., Murphy, K.: Tracking emerges by colorizing videos. In: Ferrari, V., Hebert, M., Sminchisescu, C., Weiss, Y. (eds.) ECCV 2018, pp. 391–408. Springer, Heidelberg (2018). https://doi.org/10.1007/978-3-030-01261-8_24

43. Wang, N., Song, Y., Ma, C., Zhou, W., Liu, W., Li, H.: Unsupervised deep tracking. In: CVPR, pp. 1308–1317 (2019)

44. Wang, X., Jabri, A., Efros, A.A.: Learning correspondence from the cycle-consistency of time. In: CVPR, pp. 2566–2576 (2019)

45. Wang, X., He, W., Metaxas, D., Mathew, R., White, E.: Cell segmentation and tracking using texture-adaptive snakes. In: ISBI, pp. 101–104 (2007)
46. Yang, F., Mackey, M.A., Ianzini, F., Gallardo, G., Sonka, M.: Cell segmentation, tracking, and mitosis detection using temporal context. In: Duncan, J.S., Gerig, G. (eds.) MICCAI 2005. LNCS, vol. 3749, pp. 302–309. Springer, Heidelberg (2005). https://doi.org/10.1007/11566465_38
47. Yin, Z., Kanade, T., Chen, M.: Understanding the phase contrast optics to restore artifact-free microscopy images for segmentation. Med. Image Anal. 16(5), 1047–1062 (2012)
48. Zhang, J., Bargal, S.A., Lin, Z., Brandt, J., Shen, X., Sclaroff, S.: Top-down neural attention by excitation backprop. Int. J. Comput. Vis. 126(10), 1084–1102 (2018)
49. Zhong, B., Yao, H., Chen, S., Ji, R., Chin, T.J., Wang, H.: Visual tracking via weakly supervised learning from multiple imperfect oracles. Pattern Recogn. 47(3), 1395–1410 (2014)
50. Zhou, B., Khosla, A., Lapedriza, A., Oliva, A., Torralba, A.: Learning deep features for discriminative localization. In: CVPR, pp. 2921–2929 (2016)
51. Zhou, Z., Wang, F., Xi, W., Chen, H., Gao, P., He, C.: Joint multi-frame detection and segmentation for multi-cell tracking. In: Zhao, Y., Barnes, N., Chen, B., Westermann, R., Kong, X., Lin, C. (eds.) ICIG 2019. LNCS, vol. 11902, pp. 435–446. Springer, Cham (2019). https://doi.org/10.1007/978-3-030-34110-7_36

SeqHAND: RGB-Sequence-Based 3D Hand Pose and Shape Estimation

John Yang[1], Hyung Jin Chang[2], Seungeui Lee[1], and Nojun Kwak[1(✉)]

[1] Seoul National University, Seoul, South Korea
{yjohn,seungeui.lee,nojunk}@snu.ac.kr
[2] University of Birmingham, Birmingham, UK
h.j.chang@bham.ac.uk

Abstract. 3D hand pose estimation based on RGB images has been studied for a long time. Most of the studies, however, have performed frame-by-frame estimation based on independent static images. In this paper, we attempt to not only consider the appearance of a hand but incorporate the temporal movement information of a hand in motion into the learning framework, which leads to the necessity of a large-scale dataset with sequential RGB hand images. We propose a novel method that generates a synthetic dataset that mimics natural human hand movements by re-engineering annotations of an extant static hand pose dataset into *pose-flows*. With the generated dataset, we train a newly proposed recurrent framework, exploiting visuo-temporal features from sequential synthetic hand images and emphasizing smoothness of estimations with temporal consistency constraints. Our novel training strategy of detaching the recurrent layer of the framework during domain finetuning from synthetic to real allows preservation of the visuo-temporal features learned from sequential synthetic hand images. Hand poses that are sequentially estimated consequently produce natural and smooth hand movements which lead to more robust estimations. Utilizing temporal information for 3D hand pose estimation significantly enhances general pose estimations by outperforming state-of-the-art methods in our experiments on hand pose estimation benchmarks.

Keywords: 3D hand pose estimations · Pose-flow generation · Synthetic-to-real domain gap reduction · Synthetic hand motion dataset

1 Introduction

Since expressions of hands reflect much of human behavioral features in a daily basis, hand pose estimations are essential for many human-computer interactions,

---

J. Yang and H. J. Chang—Equal Contribution.

Electronic supplementary material The online version of this chapter (https://doi.org/10.1007/978-3-030-58610-2_8) contains supplementary material, which is available to authorized users.

A. Vedaldi et al. (Eds.): ECCV 2020, LNCS 12357, pp. 122–139, 2020.
https://doi.org/10.1007/978-3-030-58610-2_8

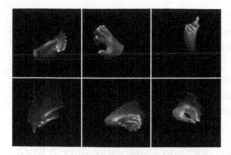

Fig. 1. Illustrations of sequential 2D images of hand pose-flows that are generated by the proposed method.

Fig. 2. Each frame of sequential hand motion videos is composed of varying poses and moving backgrounds.

such as augmented reality (AR), virtual reality (VR) [16] and computer vision tasks that require gesture tracking [8]. Hand pose estimations conventionally struggle from an extensive space of pose articulations and occlusions including self-occlusions. Some recent 3D hand pose estimators that take sequential depth image frames as inputs have tried to enhance their performance considering temporal information of hand motions [14, 22, 26, 43]. Motion context provides temporal features for narrower search space, hand personalizing, robustness to occlusion and refinement of estimations. In this paper we focus on the hand pose estimation considering its movements using only RGB image sequences for better inference of 3D spatial information.

Although the problem of estimating a hand pose in a single RGB image is an ill-posed problem, its performance is rapidly improving due to the development of various deep learning networks [3, 11, 24]. However, most studies have focused on accurately estimating 3D joint locations for each image without considering motion tendency. Pose of hands changes very quickly and in many cases contains more information on the movements of the successive poses than on the momentary ones. In addition, the current pose is greatly affected by the pose from the previous frames. Until now, there has been a lack of research on the estimation network considering the continuous changes of poses. The main reason that conventional RGB-based deep 3D hand pose estimators [1, 3, 24, 49] have only proposed frameworks with per-frame pose estimation approaches is that any large scale RGB sequential hand image dataset has not been available unlike the datasets with static images of hand poses. The diversity and the authenticity of hand motions along with generalization over skin colors, backgrounds and occlusions is a challenging factor for a dataset to be assured.

In this paper, we present a novel perspective on hand pose and shape estimation tasks and propose to consider temporal movements of hands as well as their appearances for more accurate 3D estimations of hand poses based on RGB image inputs. In order to train a framework that exploits visuo-temporal features to manage successive hand pose images, we are required to have sufficient pose data samples that are sequentially correlated. We thus propose a new generation method of dataset, SeqHAND dataset, with sequential synthetic RGB images of natural hand movements, re-engineering extant static hand pose annotations of BigHand2.2M dataset [46]. To effectively test our generated dataset, we extend the framework of [3] with a recurrent layer based on empirical validity of its structure. Also since it is widely accepted that models trained with synthetic images perform poorly on real images [24], we present a new training pipeline to preserve pre-trained image-level temporal mapping during synthetic-real domain transition. Our contributions to this end are as follows:

- We design a new generation method for sequential RGB image dataset with realistic hand motions that allows 3D hand pose and shape estimators to learn the dynamics of hand pose variations (See Fig. 1) by proposing a pose-flow generation procedure.
- We propose a new recurrent framework with convolution-LSTM layer to directly exploit visuo-temporal information from hand pose and shape variations in image space and map to 3D space.
- We present a novel training pipeline of preserving extracted spatio-temporal features from sequential RGB hand images during domain finetuning from synthetic to real.
- Our approach achieves not only state-of-the-art performance in standard 3D hand pose estimation dataset benchmarks, but also smooth human-like 3D pose fittings for the image sequences.

To the best of our knowledge, we propose the first deep-learning based 3D hand pose and shape estimator without any external 2D pose estimator that exploits temporal information directly from sequential RGB images.

2   Related Works

Many approaches of hand pose estimation (HPE) have been actively studied. To acquire hand information, the literature of single hand 3D pose estimation has been mainly based on visual inputs of depth sensors and/or RGB cameras.

Per-Frame RGB-Based 3D HPE. As views of a single 3D scene in multiple perspectives are correlated, efforts of 3D estimation based on multiple RGB images of a hand have also been introduced [6,12,27,34,37]. Multi-view camera setups allow refinements against occlusions, segmentation enhancements and better sense of depth. In the work of [34], bootstrapping pose estimations among images from multiple perspectives help the estimator to retrain badly annotated data samples and refine against occlusions. A pair of stereo images provides similar effects in a more limited setting. Integration of paired stereo images has yielded better 3D hand pose estimations through manipulations of disparity between paired images [28,30,32,48].

Monocular RGB-only setup is even more challenging because it only provides visual 2D vision of hand poses. With deep learning methods that have allowed successful achievements of hand detection [13,19], deep pose estimators have recently been able to concentrate on per-frame hand 3D pose estimation problems [51]. To overcome the lack of 3D spatial information from the 2D inputs, there are needs of constraints and guidance to infer 3D hand postures [29]. Most recently, works of [1,3,49] employ a prior hand model of MANO [31] and have achieved significant performance improvement in the RGB-only setup.

Temporal Information in 3D HPE. Considering temporal features of depth maps, sequential data of hand pose depth images [25,26,46,48] have been trained with hand pose estimators. The temporal features of hand pose variations are used for encoding temporal variations of hand poses with recurrent structure of a model [14,43], modeling of hand shape space [17], and refinement of current estimations [22,26]. With sequential monocular RGB-D inputs, Taylor et al. [41] optimize surface hand shape models, updating subdivision surfaces on corresponding 3D hand geometric models. Temporal feature exploitation has not been done for deep-learning based 3D hand pose estimators that take color images as inputs because large scale sequential RGB hand pose datasets have not been available in the literature. We share the essential motivation with the work of [5], but believe that, even without the assistance of 2D pose estimation results, sequential RGB images provide sufficient temporal information and spatial constraints for better 3D hand pose inference with robustness to occlusions.

Synthetic Hand Data Generations. Since RGB images also consist of background noise and color diversity of hands that distract pose estimations, synthetic RGB data samples are generated from the hand model to incite the robustness of models [3,4,11,25]. In [35,45], cross-modal data is embedded in a latent space, which allows 3D pose labeling of unlabeled samples generated from (disentangled) latent factor traverses. Mueller et al. [24] had applied cycleGAN [50] for realistic appearances of generated synthetic samples to reduce the synthetic-real domain gap. While there have been recent attempts to solve an issue of lacking reliable RGB datasets through generations of hand images [3,4,24,35,51], most of the works have focused on generation of realistic appearances of hands that are not in motions. To strictly imitate human perception of hand poses, it is critical for RGB-based hand pose estimators to understand the dynamics of pose variations in a spatio-temporal space. We further consider that synthetic hand pose dataset in realistic motions provides efficient information for pose estimations as much as appearances.

3   SeqHAND Dataset

3.1   Preliminary: MANO Hand Model

MANO hand model [31] is a mesh deformation model that takes two low-dimensional parameters θ and β as inputs for controlling the pose and the shape, respectively, of the 3D hand mesh outputs. With a given mean template $\bar{T}$, the rigid hand mesh is defined as:

$$M(\theta, \beta) = W(T(\theta, \beta, \bar{T}), J(\beta), \theta, \omega) \tag{1}$$

where $T(\cdot)$ defines the overall shape for the mesh model based on pre-defined deformation criteria with pose and shape, and $J(\cdot)$ yields 3D joint locations using a kinematic tree. $W(\cdot)$ represents the linear blend skinning function that is applied with blend weights ω. MANO model may take up to 45-dimensional pose parameters θ and 10-dimensional shape parameters β while the original MANO framework uses 6-dimensional PCA (principal component analysis) subspace of θ for computational efficiency.

2D Reprojection of MANO Hands: The location of joints $J(\beta)$ can be globally rotated based on the pose θ, denoted as R_θ, to obtain a hand posture P with corresponding 3D coordinates of 21 joints:

$$P = J(\theta, \beta) = R_\theta(J(\beta)).$$ (2)

After 3D estimations for mesh vertices $M(\theta, \beta)$ and joints $J(\theta, \beta)$ are computed by MANO model, in [3], 3D estimations are re-projected to 2D image plane with a weak-perspective camera model to acquire 2D estimations with a given rotation matrix $R \in SO(3)$, a translation $t \in \mathbb{R}^2$ and a scaling factor $s \in \mathbb{R}^+$:

$$M_{2D} = s\Pi R M(\theta, \beta) + t$$ (3)

$$J_{2D} = s\Pi R J(\theta, \beta) + t$$ (4)

where Π represents orthographic projections. Hand mesh $M(\theta, \beta)$ is composed of 1,538 mesh faces and defined by 3D coordinates of 778 vertices, and joint locations $J(\theta, \beta)$ are represented by 3D coordinates of 21 joints. The re-projected 2D coordinates of M_{2D} and J_{2D} are represented in 2D locations in the image coordinates. We have utilized MANO hand model in both synthetic hand motion data generation and the proposed pose and shape estimator.

3.2   Generation of SeqHAND Dataset

Although the potential of temporal features have been shown promising results for 3D HPE tasks [5,26,41], large scale RGB sequential hand image datasets have not been available during recent years in the literature of RGB-based 3D HPE. In this section, we describe a new generation method of hand motions that consist of sequential RGB frames of synthetic hands.

To generate sequential RGB image data with human-like hand motions, all poses during the variation from an initial pose to a final pose need to be realistic. We thus utilize BigHand2.2M (BH) [46] for sequential hand motion image dataset generation. BH dataset consists of 2.2 million pose samples with 3D annotations for joint locations acquired from 2 hour-long hand motions collected from 10 real subjects. With BH datasets, the generated samples are expected to inherit the manifold of its real human hand articulation space and kinematics of real hand postures. As 3D mapping of BH samples using t-SNE [21] in Fig. 3 shows, BH is known that the pose samples are densely and widely collected.

Pose-Flow Generation: We firstly define a *pose-flow*, a set of poses at each time step during the variation. Putting gradually changing poses in a sequential

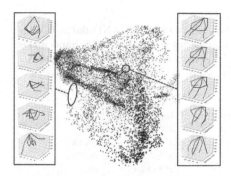
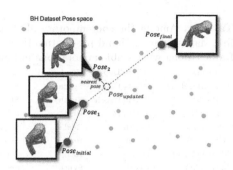

Fig. 3. 3D t-SNE visualization of 10,000 of BH data samples randomly selected. BH dataset completes a pose space that covers previously reported datasets, having a dense pool of related neighboring poses.

Fig. 4. An illustration of the pose-flow generation procedure. All poses per pose-flow are selected from the annotations of BH dataset. At each frame, a current pose is updated by the difference between the previous pose and the final pose. The pose nearest to the updated pose is then selected for the frame.

manner, we newly propose a *pose-flow* generation method. For each pose-flow generation, an initial and a final poses, $P_{initial}$ and P_{final}, are independently and randomly selected from BH dataset. While varying from the initial to the final pose during n frames, the coordinates of joints are updated by α/n of the difference between the current coordinates and the ones of the final pose.[1] The update size α is empirically chosen for the desirable speed of pose variations. A pose P_i^{BH} from BH dataset that is the nearest to the updated pose in terms of Euclidean distance is then newly selected as the current pose for the k-th frame:

$$P_0 = P_{initial}^{BH} \tag{5}$$

$$P_{updated} = P_{k-1} - \frac{\alpha}{n}(P_{k-1} - P_{final}^{BH}) \tag{6}$$

$$P_k = P_i^{BH} \quad \text{s.t.} \quad \min_i ||P_{updated} - P_i^{BH}||. \tag{7}$$

The overall procedure of the Pose-flow generation is summarized in Fig. 4. The intermediate pose ($P_{updated}$) is calculated as stochastic update. Such stochasticity of our pose updates helps avoiding strict updates of pose gradients and encourages wandering more within the pose space. Pose selections from the BH annotations, again, allows assurance on the authencity of hand poses during the variation.

To generate RGB images for a pose, a shallow four-layer network with which takes inputs of 3D coordinates for joints of BH annotations is trained to output corresponding pose parameters θ for MANO hand model. For each pose at a

---

[1] Note that direct random samplings from continuous pose parameter space $\theta \in \mathbb{R}$ does not assure diversity and authenticity of poses [31].

Table 1. Among the contemporary 3D hand pose datasets, SeqHAND dataset is the first dataset for 3D hand pose estimations that provides sequential RGB hand image frames along with stable annotations in both 3rd-person and egocentric perspectives.

Datasets	RGB/Depth	Real/Synth	Static/Sequential	3rd/Ego view	# of frames
SynthHands [25]	RGB+Depth	Synth	Static	Ego	63k
RHD [51]	RGB+Depth	Synth	Static	3rd	43.7k
FHAD [10]	RGB+Depth	Real	Sequential	Ego	100k
NYU [42]	Depth	Real	Sequential	3rd	80k
ICVL [40]	Depth	Real	Sequential	3rd	332.5k
MSRA15 [38]	Depth	Real	Sequential	3rd	76,375
MSRC [33]	Depth	Synth	Sequential	3rd+Ego	100k
SynHand5M [23]	Depth	Synth	Sequential	3rd	5M
GANerated [24]	RGB	Synth	Static	Ego	330k
SeqHAND	RGB	Synth	Sequential	3rd+Ego	410k

frame, we feed corresponding 21 joint location coordinates to the this network to acquire a hand mesh model in the desired pose, which is then re-projected to an image plane. As done in [3], we assign each vertex in a mesh the RGB value of predefined color templates of hands to create appearances of hands. Sampled hand shape parameter $\beta \in [-2,2]^{10}$ and selected color template are set unchanged along per flow. Camera parameters of rotation R, scale s and translation t factors are independently sampled for initial and final poses and updated at each frame in the same way as the poses are. All frames are in the size of w and h. Figure 1 depicts illustrations of our generated pose-flows.

Further mimicking images of hand motions in the wild, we sample two (initial and ending) random patches from VOC2012 data [9] with the size of w and h and move the location of the patch for backgrounds along the frames. As Table 1 denotes, the generated SeqHAND dataset provides not only both 3rd-person and egocentric viewpoints of hand postures but also sequential RGB images of hand poses that firstly allow data-hungry neural networks to exploit visuo-temporal features directly from RGB inputs.[2]

4   SeqHand-Net for Visuo-Temporal Feature Exploitation

With SeqHAND dataset, we are able to overcome the scarcity of sequential RGB dataset which limits conventional RGB-based 3D HPE methods from exploiting temporal image features. Motivated by [3], we design sequential hand pose and shape estimation network (SeqHAND-Net). On top of the encoder network of [3], we incorporate convolution-LSTM (ConvLSTM) layer [44] to capture sequential relationship between consecutive hand poses. Our method does not consider additional hand 2D joint locations as inputs, and purely performs 3D hand pose estimation based on sequentially streaming RGB images in an effort to overcome the dependency on external 2D pose estimators. We also propose, in this section,

---

[2] Although we can generate as many synthetic data as we want, our SeqHand dataset contains 400K/10K samples used for training/validation.

a training pipeline for domain adaptation from synthetic to real, adapting low-level features with real hand images while preserving high-level visuo-temporal features of hand motions.

From each frame, a cropped hand image is fed into SeqHAND-Net as illustrated in Fig. 5. Our problem scope is to better perform hand pose estimations on streaming cropped frames that are unseen by the estimator. The encoder of our SeqHAND-Net has the backbone structure of ResNet-50 [13] and, for training, expects sequential inputs with k frames. A single ConvLSTM is implemented right before the last layer as a recurrent visual feature extractor so that the dynamics of hand motions are embedded in the highest-level latent space. Learning of hand motion sequential dynamics in the high-level space is important since low-level visual features are changed with the ConvLSTM layer fixed during finetuning for real hand images. After the recurrent layer, a simple linear mapping

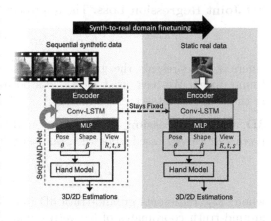

Fig. 5. Our training strategy for preservation of temporal features during domain adaptation synth-to-real. SeqHAND-Net is created from [3] with an extra visuo-temporal feature exploitation layer for sequential RGB inputs. During finetuning, SeqHAND-Net considers a static real data as 1-frame-long sequence. The temporal high-level feature encoding is preserved while low-level image feature encoding layers are finetuned.

layer from hidden features to the output vector is set. The encoder's resultant vector consists of parameters for pose $\theta \in \mathbb{R}^{10}$, shape $\beta \in \mathbb{R}^{10}$, scale $s \in \mathbb{R}^{+}$, translation $t \in \mathbb{R}^2$ and rotation $r \in \mathbb{R}^3$ which turns into a matrix $R \in SO(3)$ through Rodrigues rotation formula for Eqs (3) and (4).

Synth-to-Real Domain Transfer with Preservation of Temporal Features. As mentioned earlier, many recent researches have used synthetic hand images for pre-training and finetuned into real domain to overcome the scarcity of real hand images. While finetuning into real domain may allow faster training convergence, further training with a smaller dataset not only causes overfitting and may result in catastrophic forgetting [18]. To preserve visuo-temporal features learned from synthetic hand motions of SeqHAND dataset, we exclude the ConvLSTM layer of SeqHAND-Net from domain transfer to real hand images, allowing the network to only finetune low-level image features. Only the 'Encoder' and 'MLP' layers from Fig. 5 are finetuned with a real static hand image dataset (e.g. FreiHand [52]). SeqHAND-Net is therefore trained, considering each image sample as 1-frame-long sequential image during domain transition to real.

Training Objectives. The followings are the types of criteria used for training our proposed framework to consider visuo-temporal features and emphasize the temporal smoothness of estimations:

2D Joint Regression Loss: The re-projected 2D joint loss is represented as:

$$L_{2D}^J = ||J_{2D} - x_{2D}^J||_1, \tag{8}$$

where x_{2D} represents the ground-truth 2D locations of hand joints within a frame image. We have used the L1 loss because of inaccuracies in annotations in the training datasets.

3D Joint Regression Loss: The ground-truth joint locations and the ones predicted are regressed to be the same using the following loss:

$$L_{3D}^J = ||RJ(\theta, \beta) - x_{3D}^J||_2^2, \tag{9}$$

where x_{3D}^J represents ground-truth 3D joint coordinates. If a dataset provides ground-truth coordinates of 3D vertex points (e.g. FreiHand dataset), the 3D coordinates of each vertex predicted and the ones of ground-truth is minimized as done for 3D joint loss, based on the following loss:

$$L_{3D}^M = ||RM(\theta, \beta) - x_{3D}^M||_2^2 \tag{10}$$

where x_{3D}^M represents ground-truth 3D mesh vertex coordinates.

Hand Mask Fitting Loss: The hand mask loss is proposed in [3] to fit the shape and pose predictions in the binary mask of hands in the image plane. This loss ensures predicted coordinates of mesh vertices to be inside of a hand region when re-projected:

$$L_{mask} = 1 - \frac{1}{N}\sum_i H(M_{2D}^i), \quad H(x) = \begin{cases} 1, & \text{if } x \text{ inside a hand region.} \\ 0, & \text{otherwise.} \end{cases} \tag{11}$$

where H is a hand mask indicator function that tells if vertex point x is inside the hand region or not. The loss represents the percentage of vertices that are outside the region.

Temporal Consistency Loss: For pre-training on SeqHAND dataset, our method needs to be constrained with temporal consistency to ensure smoothness of pose and shape predictions. Similar to [5], we have adopted the temporal consistency loss for smoothness of temporal variation of poses:

$$L_{temp} = ||\beta_{t-1} - \beta_t||_2^2 + \lambda_{temp}^\theta ||\theta_{t-1} - \theta_t||_2^2. \tag{12}$$

While finely penalizing current estimations with the previous ones, this loss allows the reduction of search space and natural 3D hand motion estimations.

Camera Parameter Regression Loss: During training with SeqHAND dataset where all ground-truths for pose, shape and viewpoint parameters

$\{\theta, \beta, r, t, s\}$ are available, our model is trained with L2-norm loss between predictions and the ground-truth.

$$L_{cam} = \sum_{i \in \{\theta, \beta, r, t, s\}} ||\hat{i} - i||_2^2 \tag{13}$$

where $\hat{i}$ and i respectively refer to predicted and ground-truth parameters for pose, shape and viewpoint.

5 Experiments

Datasets for Training. For visuo-temporal feature encodings of sequential RGB hand images, we pretrain SeqHAND-Net with our SeqHAND dataset. We have generated 40,000 sequences for training and 1,000 for validate samples each of which is 10-frames-long. All images are generated in the size of 224×224 for ResNet-50 input size. SeqHAND data samples are exemplified in Fig. 1 and 2.

To finetune SeqHAND-Net for synthetic-real domain gap reduction, we have used STB (Stereo Hand Pose Tracking Benchmark) [48] and FR (FreiHand) [52] datasets. STB dataset consists of real hand images captured in a sequential manner during 18,000 frames with 6 different lighting conditions and backgrounds. Each frame image is labeled with 2D and 3D annotations of 21 joints. Since STB dataset has annotations for joint locations of palm centers instead of wrist, we have interpolated related mesh vertices of MANO hand model to mach the annotation of STB dataset. The dataset is divided into training and testing sets as done in [3].

FR dataset has 130,240 data samples that are made up of 32,560 non-sequential real hand images with four different backgrounds. Since the dataset has hands that are centered within the image planes, we have modified each sample by re-positioning the hand randomly within the image for more robust training results. FR dataset provides MANO-friendly annotations of 21 joint 3D/2D locations along with 778 vertex ground-truth 2D/3D coordinates with hand masks.

We have finetuned the SeqHAND-Net pretrained on SeqHAND dataset with real-hand image datasets mentioned above in a non-sequential manner while conserving hand motion dynamic features detached from further learning.

Datasets for Evaluation. We evaluate various framework structures that consider temporal features on the validation set of SeqHAND dataset for the logical framework choice. For the comparison against other state-of-the-art methods, we have selected standard hand pose estimation datasets of the splitted test set of STB, EgoDexter(ED)[25] and Dexter+Obeject(DO)[36] in which there exists temporal relations among data samples since our network requires sequential RGB inputs for fair comparisons. While STB and DO datasets consist of real hand images in 3rd-person viewpoints, ED dataset has samples that are in egocentric perspective. For all datasets, our method is evaluated on every frame of input sequences.

Table 2. Ablation study results of various structures of the framework proposed in [3] for sequential inputs.

Frameworks	AUC		Error (px/mm)		# params
	2D	3D	2D	3D	
ResNet50-Encoder (baseline) [3]	0.855	0.979	3.44	7.85	28.8M
ResNet101-Encoder [3]	0.861	0.981	3.31	7.54	47.8M
I3D-Encoder [7]	0.831	0.967	4.19	9.24	31.5M
MFNet-Encoder [20]	0.818	0.912	5.48	10.54	41.7M
ResNet50-Encoder+LSTM	0.826	0.956	4.64	9.63	39.3M
ResNet50-Encoder+ConvLSTM	**0.873**	**0.986**	**3.17**	**7.18**	43.2M

Table 3. Performances of differently (partially) trained models on ED, DO, STB datasets.

Methods	AUC			Avg. 3D Error (mm)		
	ED	DO	STB	ED	DO	STB
Encoder + Train(SynthHAND)	0.350	0.095	0.140	52.11	100.84	68.86
Encoder + Train(SynthHAND) + Train(FH + STB)	0.397	0.516	**0.985**	49.18	33.12	**9.80**
Encoder + ConvLSTM + Train(SeqHAND)	0.373	0.151	0.121	52.18	81.51	71.10
Encoder + ConvLSTM + Train(SeqHAND) + Train(FH + STB)	0.444	0.581	0.981	40.94	29.41	9.82
Encoder + ConvLSTM + Train(SeqHAND) + Train$_C$(FH + STB)	**0.766**	**0.843**	0.978	**17.16**	**18.12**	9.87

Metrics. For evaluation results, we measure the percentage of correct key-points for 3D joint locations (3D-PCK) along with the area under the curve (AUC) of various thresholds. In addition, we provide average Euclidean distance error for all 2D/3D joint key-points so that more absolute comparisons can be made.

Hand Localizations. For all experiments, we have used MobileNet+SSD version of hand detection implementation [19] trained with a hand segmentation dataset [2] for providing sequential cropped hand images to SeqHAND-Net. For localized hands with tight bounding rectangular boxes, we choose the longer edge with a length size l and crop the region based on the center point of boxes so that the cropped images have a square ratio with width and height size of $2.2 * l$, as done in [3].

Table 4. Average 3D joint distance (mm) to ground-truth for RGB Sequence datasets hand pose benchmarks.

	Avg. 3D Error (mm)		
	ED	DO	STB
Our Method	**17.16**	**18.12**	9.87
Bouk. et al. (RGB)	51.87	33.16	**9.76**
Bouk. et al. (Best)	45.33	25.53	**9.76**
Spurr et al.	56.92	40.20	-
Zimmer. et al.	52.77	34.75	-

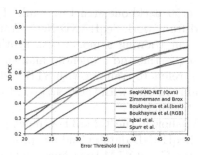

Fig. 6. 3D PCK for ED

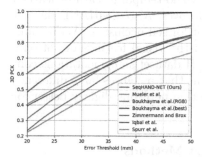

Fig. 7. 3D PCK for DO

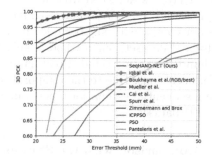

Fig. 8. 3D PCK for STB

5.1 Ablation Study

Framework Selection: To show the logic behind the selection of the proposed framework, we evaluate various forms of extended baseline model [3] shown in Table 2 for managing sequential inputs on our newly generated SeqHAND dataset. The extended versions of baseline encoder (ResNet-50) include the baseline model with a LSTM layer [39], the baseline model with a ConvLSTM layer [44], the baseline encoder with the structure of I3D [7] and the baseline encoder with the structure of MF-Net [20]. Both I3D-Encoder and MFNet-Encoder represent methods that incorporate sequential inputs with 3D convolutional neural network. For I3D, we have changed few features from the original form of I3D so that its structure fits into the hand pose estimation task. The original backbone structure of I3D with Inception modules have changed into ResNet-50 for a fair comparison. MFNet is another examplary 3D convolution network proposed specifically for motion feature extractions. Of the candidates, the encoder with a ConvLSTM layer has performed the best.

Effectiveness of *SeqHAND-Net* and *SeqHAND* Dataset: To clarify the effectiveness of our proposed framework and our generated dataset, variations of the proposed method and the baseline model are investigated. We report AUCs of 3D PCK curves and average 3D joint location errors for ED, DO and the evaluation set of STB datasets. In Table 3, 'Encoder' denotes the baseline model with

ResNet50 backbone structure while 'Encoder + ConvLSTM' denotes our proposed framework SeqHAND-Net. 'Train(SynthHAND)' and 'Train(SeqHAND)' represent training a model with synthetic hand image dataset respectively in non-sequential and sequential manner. 'Train(FH + STB)' and 'Train$_C$(FH + STB)' refers to training with STB and FreiHand datasets for the synthetic-real domain transfer with the ConvLSTM layer, respectively, attached and detached from finetuning.

We show in the Table 3 how much performance enhancement can be obtained with SeqHAND dataset and our proposed domain adaptation strategy. Encoder with the ConvLSTM layer finetuned to real domain consequently performs similar to the encoder that does not consider visuo-temporal correlations. If the ConvLSTM layer is detached from finetuning and visuo-temporal features learned are preserved, the performance significantly improves. Also, SeqHAND dataset does not consist with any occluded hands except for self-occlusions. With training for FH dataset, our method is able to learn the visual features of not only real hands but also occluded real hands since FH dataset's augmentations consist of occlusions. Due to the temporal constraint that penalizes large difference among sequential estimations, per-frame estimation performs slightly better for the STB dataset.

5.2 Comparison with the State-of-the-art Methods

In Figs. 6, 7 and 8, we have plotted 3D-PCK graph with various thresholds for STB, ED and DO datasets. For STB dataset, deep-learning based works of [3,4,15,24,35,51] and approaches from [29,47] are compared. Many previous methods have reached near the maximum performance for STB dataset. With our temporal constraints and fixing the ConvLSTM layer during finetuning, our method reaches a competitive performance. For both ED and DO datasets, our method outperforms other methods. For ED dataset, contemporary works of [3, 15,35,51] are compared to our method. The best performance of our baseline [3] is reached with inputs of RGB and 2D pose estimations provided by an external 2D pose estimator. Our method results in outstanding performance against other compared methods [3,15,24,35,51] for DO dataset with heavy occlusions, which shows that the learning of pose-flow continuity enhances robustness to occlusions. Temporal information exploitation from sequential RGB images affect our model to be robust against dynamically moving scene. For more absolute comparisons, we provide our average 3D error of joint location in Table 4.

We provide qualitative results in Fig. 9 for visual comparison against a frame-by-frame 3D pose estimator, our reproduced work of [3]. All images in the figure are sequentially inputted to both estimators from left to right. Per-frame estimations that fit postures at each frame result in unnatural 3D hand posture changes over a sequence while our method's leaning trajectories biased by previous frames produces natural hand motions and robust estimations to frames that lack visual information of hand postures. Our method models estimated hand shapes as consistent as possible per sequence. During the qualitative evaluation on a RGB image sequence of a single real hand, our method's average

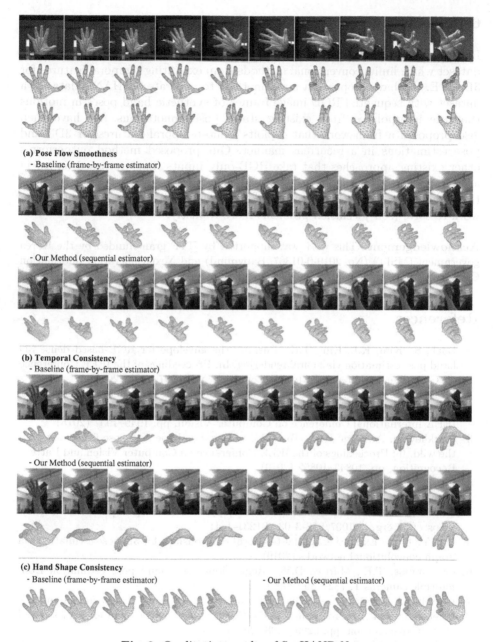

(a) Pose Flow Smoothness
  - Baseline (frame-by-frame estimator)

  - Our Method (sequential estimator)

(b) Temporal Consistency
  - Baseline (frame-by-frame estimator)

  - Our Method (sequential estimator)

(c) Hand Shape Consistency
  - Baseline (frame-by-frame estimator)   - Our Method (sequential estimator)

Fig. 9. Qualitative results of SeqHAND-Net.

difference among temporal changes of shape parameters $\beta_{t-1} - \beta_t$ is $4.16e^{-11}$ while that of the frame-by-frame estimator is $2.38e^{-5}$. The average difference among temporal changes of the pose parameters $\theta_{t-1} - \theta_t$ are $1.88e^{-6}$ for our method and $6.90e^{-6}$ for the other.

6 Conclusion

In this paper, we have addressed and tackled the scarcity of sequential RGB dataset which limits conventional methods from exploiting temporal features for 3D HPE. We have proposed a novel method to generate SeqHAND dataset, a dataset with sequential RGB image frames of synthetic hand poses in motions that are interpolated from existing static pose annotations. We have then also proposed a framework that exploits visuo-temporal features for 3D hand pose estimations in a recurrent manner. Our proposed method outperforms other existing approaches that take RGB-only inputs that are based on solely appearance-based methods, and consequently produces pose-flow estimations that mimick natural movements of human hands. We plan to enable the framework to solve (self-)occlusion problems more robustly.

Acknowledgement. This work was supported by IITP grant funded by the Korea government (MSIT) (No. 2019-0-01367, Babymind) and Next-Generation Information Computing Development Program through the NRF of Korea (2017M3C4A7077582).

References

1. Baek, S., Kim, K.I., Kim, T.K.: Pushing the envelope for RGB-based dense 3D hand pose estimation via neural rendering. In: Proceedings of the IEEE Conference on Computer Vision and Pattern Recognition, pp. 1067–1076 (2019)
2. Bambach, S., Lee, S., Crandall, D.J., Yu, C.: Lending a hand: detecting hands and recognizing activities in complex egocentric interactions. In: Proceedings of the IEEE International Conference on Computer Vision, pp. 1949–1957 (2015)
3. Boukhayma, A., Bem, R.D., Torr, P.H.: 3D hand shape and pose from images in the wild. In: Proceedings of the IEEE Conference on Computer Vision and Pattern Recognition, pp. 10843–10852 (2019)
4. Cai, Y., Ge, L., Cai, J., Yuan, J.: Weakly-supervised 3D hand pose estimation from monocular RGB images. In: Ferrari, V., Hebert, M., Sminchisescu, C., Weiss, Y. (eds.) ECCV 2018. LNCS, vol. 11210, pp. 678–694. Springer, Cham (2018). https://doi.org/10.1007/978-3-030-01231-1_41
5. Cai, Y., et al.: Exploiting spatial-temporal relationships for 3D pose estimation via graph convolutional networks (2019)
6. de Campos, T.E., Murray, D.W.: Regression-based hand pose estimation from multiple cameras 1, 782–789 (2006)
7. Carreira, J., Zisserman, A.: Quo vadis, action recognition? A new model and the kinetics dataset. In: Proceedings of the IEEE Conference on Computer Vision and Pattern Recognition, pp. 6299–6308 (2017)
8. Chang, H.J., Garcia-Hernando, G., Tang, D., Kim, T.K.: Spatio-temporal hough forest for efficient detection-localisation-recognition of fingerwriting in egocentric camera. Comput. Vis. Image Understand. 148, 87–96 (2016)
9. Everingham, M., Van Gool, L., Williams, C.K.I., Winn, J., Zisserman, A.: The PASCAL Visual Object Classes Challenge 2012 (VOC2012) Results. http://www.pascal-network.org/challenges/VOC/voc2012/workshop/index.html

10. Garcia-Hernando, G., Yuan, S., Baek, S., Kim, T.K.: First-person hand action benchmark with RGB-D videos and 3D hand pose annotations. In: Proceedings of the IEEE Conference on Computer Vision and Pattern Recognition, pp. 409–419 (2018)

11. Ge, L., et al.: 3D hand shape and pose estimation from a single RGB image. In: Proceedings of the IEEE Conference on Computer Vision and Pattern Recognition, pp. 10833–10842 (2019)

12. Gomez-Donoso, F., Orts-Escolano, S., Cazorla, M.: Large-scale multiview 3D hand pose dataset. arXiv preprint arXiv:1707.03742 (2017)

13. He, K., Zhang, X., Ren, S., Sun, J.: Deep residual learning for image recognition. In: Proceedings of the IEEE Conference on Computer Vision and Pattern Recognition, pp. 770–778 (2016)

14. Hu, Z., Hu, Y., Liu, J., Wu, B., Han, D., Kurfess, T.: A CRNN module for hand pose estimation. Neurocomputing **333**, 157–168 (2019)

15. Iqbal, U., Molchanov, P., Breuel Juergen Gall, T., Kautz, J.: Hand pose estimation via latent 2.5 d heatmap regression In: Ferrari, V., Hebert, M., Sminchisescu, C., Weiss, Y. (eds.) ECCV 2018. LNCS, vol. 11215, pp. 118–134. Springer, Cham (2018). https://doi.org/10.1007/978-3-030-01252-6_8

16. Jang, Y., Noh, S., Chang, H.J., Kim, T., Woo, W.: 3D finger cape: clicking action and position estimation under self-occlusions in egocentric viewpoint. IEEE Trans. Vis. Comput. Graph. **21**(4), 501–510 (2015)

17. Khamis, S., Taylor, J., Shotton, J., Keskin, C., Izadi, S., Fitzgibbon, A.: Learning an efficient model of hand shape variation from depth images, pp. 2540–2548 (2015)

18. Kirkpatrick, J., et al.: Overcoming catastrophic forgetting in neural networks. Proc. Natl. Acad. Sci. **114**(13), 3521–3526 (2017)

19. Le, T.H.N., Quach, K.G., Zhu, C., Duong, C.N., Luu, K., Savvides, M.: Robust hand detection and classification in vehicles and in the wild. In: 2017 IEEE Conference on Computer Vision and Pattern Recognition Workshops (CVPRW), pp. 1203–1210. IEEE (2017)

20. Lee, M., Lee, S., Son, S., Park, G., Kwak, N.: Motion feature network: fixed motion filter for action recognition. In: Ferrari, V., Hebert, M., Sminchisescu, C., Weiss, Y. (eds.) ECCV 2018. LNCS, vol. 11214, pp. 392–408. Springer, Cham (2018). https://doi.org/10.1007/978-3-030-01249-6_24

21. van der Maaten, L., Hinton, G.: Visualizing data using t-SNE. J. Mach. Learn. Res. **9**(Nov), 2579–2605 (2008)

22. Madadi, M., Escalera, S., Carruesco, A., Andujar, C., Baró, X., Gonzàlez, J.: Top-down model fitting for hand pose recovery in sequences of depth images. Image Vis. Comput. **79**, 63–75 (2018)

23. Malik, J., et al.: DeepHPS: end-to-end estimation of 3D hand pose and shape by learning from synthetic depth. In: 2018 International Conference on 3D Vision (3DV), pp. 110–119. IEEE (2018)

24. Mueller, F., et al.: Ganerated hands for real-time 3D hand tracking from monocular RGB, pp. 49–59 (2018)

25. Mueller, F., Mehta, D., Sotnychenko, O., Sridhar, S., Casas, D., Theobalt, C.: Real-time hand tracking under occlusion from an egocentric RGB-D sensor, pp. 1284–1293 (2017)

26. Oberweger, M., Riegler, G., Wohlhart, P., Lepetit, V.: Efficiently creating 3D training data for fine hand pose estimation, pp. 4957–4965 (2016)

27. Oikonomidis, I., Kyriazis, N., Argyros, A.A.: Full DOF tracking of a hand interacting with an object by modeling occlusions and physical constraints, pp. 2088–2095 (2011)

28. Panteleris, P., Argyros, A.: Back to RGB: 3D tracking of hands and hand-object interactions based on short-baseline stereo, pp. 575–584 (2017)
29. Panteleris, P., Oikonomidis, I., Argyros, A.: Using a single RGB frame for real time 3D hand pose estimation in the wild. In: 2018 IEEE Winter Conference on Applications of Computer Vision (WACV), pp. 436–445. IEEE (2018)
30. Remilekun Basaru, R., Slabaugh, G., Alonso, E., Child, C.: Hand pose estimation using deep stereovision and Markov-chain Monte Carlo, pp. 595–603 (2017)
31. Romero, J., Tzionas, D., Black, M.J.: Embodied hands: modeling and capturing hands and bodies together. ACM Trans. Graph. (TOG) **36**(6), 245 (2017)
32. Rosales, R., Athitsos, V., Sigal, L., Sclaroff, S.: 3D hand pose reconstruction using specialized mappings, vol. 1, pp. 378–385 (2001)
33. Sharp, T., et al.: Accurate, robust, and flexible real-time hand tracking. In: Proceedings of the 33rd Annual ACM Conference on Human Factors in Computing Systems, pp. 3633–3642 (2015)
34. Simon, T., Joo, H., Matthews, I., Sheikh, Y.: Hand keypoint detection in single images using multiview bootstrapping, pp. 1145–1153 (2017)
35. Spurr, A., Song, J., Park, S., Hilliges, O.: Cross-modal deep variational hand pose estimation, pp. 89–98 (2018)
36. Sridhar, S., Mueller, F., Zollhöfer, M., Casas, D., Oulasvirta, A., Theobalt, C.: Real-time joint tracking of a hand manipulating an object from RGB-D input. In: Leibe, B., Matas, J., Sebe, N., Welling, M. (eds.) ECCV 2016. LNCS, vol. 9906, pp. 294–310. Springer, Cham (2016). https://doi.org/10.1007/978-3-319-46475-6_19
37. Sridhar, S., Rhodin, H., Seidel, H.P., Oulasvirta, A., Theobalt, C.: Real-time hand tracking using a sum of anisotropic Gaussians model **1**, 319–326 (2014)
38. Sun, X., Wei, Y., Liang, S., Tang, X., Sun, J.: Cascaded hand pose regression. In: Proceedings of the IEEE Conference on Computer Vision and Pattern Recognition, pp. 824–832 (2015)
39. Sundermeyer, M., Schlüter, R., Ney, H.: LSTM neural networks for language modeling. In: Thirteenth Annual Conference of the International Speech Communication Association (2012)
40. Tang, D., Jin Chang, H., Tejani, A., Kim, T.K.: Latent regression forest: Structured estimation of 3D articulated hand posture, pp. 3786–3793 (2014)
41. Taylor, J., et al.: User-specific hand modeling from monocular depth sequences. In: Proceedings of the IEEE Conference on Computer Vision and Pattern Recognition, pp. 644–651 (2014)
42. Tompson, J., Stein, M., Lecun, Y., Perlin, K.: Real-time continuous pose recovery of human hands using convolutional networks. ACM Trans. Graph. (ToG) **33**(5), 169 (2014)
43. Wu, Y., Ji, W., Li, X., Wang, G., Yin, J., Wu, F.: Context-aware deep spatiotemporal network for hand pose estimation from depth images. IEEE Trans. Cybern. (2018)
44. Xingjian, S., Chen, Z., Wang, H., Yeung, D.Y., Wong, W.K., Woo, W.C.: Convolutional LSTM network: a machine learning approach for precipitation nowcasting. In: Advances in Neural Information Processing Systems, pp. 802–810 (2015)
45. Yang, L., Yao, A.: Disentangling latent hands for image synthesis and pose estimation. In: Proceedings of the IEEE Conference on Computer Vision and Pattern Recognition. pp. 9877–9886 (2019)
46. Yuan, S., Ye, Q., Stenger, B., Jain, S., Kim, T.K.: Bighand2. 2m benchmark: Hand pose dataset and state of the art analysis, pp. 4866–4874 (2017)
47. Zhang, J., Jiao, J., Chen, M., Qu, L., Xu, X., Yang, Q.: 3D hand pose tracking and estimation using stereo matching. arXiv preprint arXiv:1610.07214 (2016)

48. Zhang, J., Jiao, J., Chen, M., Qu, L., Xu, X., Yang, Q.: A hand pose tracking benchmark from stereo matching, pp. 982–986 (2017)
49. Zhang, X., Li, Q., Zhang, W., Zheng, W.: End-to-end hand mesh recovery from a monocular RGB image. arXiv preprint arXiv:1902.09305 (2019)
50. Zhu, J.Y., Park, T., Isola, P., Efros, A.A.: Unpaired image-to-image translation using cycle-consistent adversarial networks. In: Proceedings of the IEEE International Conference on Computer Vision, pp. 2223–2232 (2017)
51. Zimmermann, C., Brox, T.: Learning to estimate 3D hand pose from single RGB images, pp. 4903–4911 (2017)
52. Zimmermann, C., Ceylan, D., Yang, J., Russell, B., Argus, M., Brox, T.: Freihand: a dataset for markerless capture of hand pose and shape from single RGB images. In: Proceedings of the IEEE International Conference on Computer Vision, pp. 813–822 (2019)

Rethinking the Distribution Gap of Person Re-identification with Camera-Based Batch Normalization

Zijie Zhuang[1]([✉]), Longhui Wei[2,4], Lingxi Xie[2], Tianyu Zhang[2], Hengheng Zhang[3], Haozhe Wu[1], Haizhou Ai[1], and Qi Tian[2]

[1] Tsinghua University, Beijing, China
jayzhuang42@gmail.com, wuhz1997@163.com, ahz@tsinghua.edu.cn
[2] Huawei Inc., Shenzhen, China
weilh2568@gmail.com, 198808xc@gmail.com, tianyu1949@gmail.com,
tian.qi1@huawei.com
[3] Hefei University of Technology, Hefei, China
imhmhm@gmail.com
[4] University of Science and Technology of China, Hefei, China

Abstract. The fundamental difficulty in person re-identification (ReID) lies in learning the correspondence among individual cameras. It strongly demands costly inter-camera annotations, yet the trained models are not guaranteed to transfer well to previously unseen cameras. These problems significantly limit the application of ReID. This paper rethinks the working mechanism of conventional ReID approaches and puts forward a new solution. With an effective operator named Camera-based Batch Normalization (CBN), we force the image data of all cameras to fall onto the same subspace, so that the distribution gap between any camera pair is largely shrunk. This alignment brings two benefits. First, the trained model enjoys better abilities to generalize across scenarios with unseen cameras as well as transfer across multiple training sets. Second, we can rely on intra-camera annotations, which have been undervalued before due to the lack of cross-camera information, to achieve competitive ReID performance. Experiments on a wide range of ReID tasks demonstrate the effectiveness of our approach. The code is available at https://github.com/automan000/Camera-based-Person-ReID.

Keywords: Person re-identification · Distribution gap · Camera-based batch normalization

1 Introduction

Person re-identification (ReID) aims at matching identities across disjoint cameras. Generally, it is achieved by mapping images from the same and different

Electronic supplementary material The online version of this chapter (https://doi.org/10.1007/978-3-030-58610-2_9) contains supplementary material, which is available to authorized users.

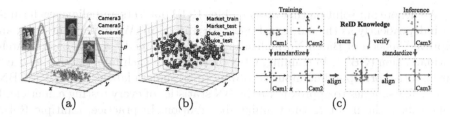

(a)   (b)   (c)

Fig. 1. (a) We visualize the distributions of several cameras in Market-1501. Each curve corresponds to an approximated marginal density function. Curves of different cameras demonstrate the differences between the corresponding distributions. (b) The Barnes-Hut t-SNE [39] visualization of the distribution inconsistency among datasets. (c) Illustration of the proposed camera-based formulation. Note that **Cam1**, **Cam2**, and **Cam3** could come from any ReID datasets. This figure is best viewed in color. (Color figure online)

cameras into a feature space, where features of the same identity are closer than those of different identities. To learn the relations between identities from all cameras, there are two different objectives: learning the relations between identities in the same camera and learning identity relations across cameras.

However, there is an inconsistency between these two objectives. As shown in Fig. 1(a), due to the large appearance variation caused by illumination conditions, camera views, *etc.*, images from different cameras are subject to distinct distributions. Handling the distribution gap between cameras is crucial for inter-camera identity matching, yet learning within a single camera is much easier. As a consequence, the conventional ReID approaches mainly focus on associating different cameras, which demands costly inter-camera annotations. Besides, after learning on a training set, part of the learned knowledge is strongly correlated to the connections among these particular cameras, making the model generalize poorly on scenarios consisting of unseen cameras. As shown in Fig. 1b, the ReID model learned on one dataset often has a limited ability of describing images from other datasets, *i.e.*, its generalization ability across datasets is limited. For simplicity, we denote this formulation neglecting within-dataset inconsistencies as the **dataset-based formulation**. We emphasize that lacking the ability to bridge the distribution gap between all cameras from all datasets leads to two problems: the unsatisfying generalization ability and the excessive dependence on inter-camera annotations. To tackle these problems simultaneously, we propose to align the distribution of all cameras explicitly. As shown in Fig. 1(c), we eliminate the distribution inconsistency between all cameras, so the ReID knowledge can always be learned, accumulated, and verified in the same input distribution, which facilitates the generalization ability across different ReID scenarios. Moreover, with the aligned distributions among all cameras, intra- and inter-camera annotations can be regarded as the same, *i.e.*, labeling the image relations under the same input distribution. This allows us to approximate the effect of inter-camera annotations with only intra-camera annotations. It may relieve the exhaustive human labor for the costly inter-camera annotations.

We denote our solution that disassembles ReID datasets and aligns each camera independently as the **camera-based formulation**. We implement it via an improved version of Batch Normalization (BN) [9] named Camera-based Batch Normalization (CBN). In training, CBN disassembles each mini-batch and standardizes the corresponding input according to its camera labels. In testing, CBN utilizes few samples to approximate the BN statistics of every testing camera and standardizes the input to the training distribution. In practice, multiple ReID tasks benefit from our work, such as *fully-supervised learning* [1,36,51,53,54,58], *direct transfer* [8,21], *domain adaptation* [3,4,33,41,52,57], and *incremental learning* [12,15,28]. Extensive experiments indicate that our method improves the performance of these tasks simultaneously, such as 0.9%, 5.7%, and 14.2% averaged Rank-1 accuracy improvements on *fully-supervised learning*, *domain adaptation*, and *direct transfer*, respectively, and 9.7% less forgetting on Rank-1 accuracy for *incremental learning*. Last but not least, even without inter-camera annotations, a *weakly-supervised* pipeline [60] with our formulation can achieve competitive performance on multiple ReID datasets, which demonstrates that the value of intra-camera annotations may have been undervalued in the previous literature. To conclude, our contribution is three-fold:

- In this paper, we emphasize the importance of aligning the distribution of all cameras and propose a camera-based formulation. It can learn discriminative knowledge for ReID tasks while excluding training-set-specific information.
- We implement our formulation with Camera-based Batch Normalization. It facilitates the generalization and transfer ability of ReID models across different scenarios and makes better use of intra-camera annotations. It provides a new solution for ReID tasks without costly inter-camera annotations.
- Experiments on *fully-supervised, weakly-supervised, direct transfer, domain adaptation*, and *incremental learning* tasks validate our method, which confirms the universality and effectiveness of our camera-based formulation.

2   Related Work

Our formulation aligns the distribution per camera. In training, it eliminates the distribution gap between all cameras. ReID models can treat intra- and inter-camera annotations equally and make better use of them, which benefits both *fully-supervised* and *weakly-supervised* ReID tasks. It also guarantees that the distribution of each testing camera is aligned to the same training distribution. Thus, the knowledge can better generalize and transfer across datasets. It helps *direct transfer, domain adaptation*, and *incremental learning*. In this section, we briefly categorize and summarize previous works on the above ReID topics.

Supervision. The supervision in ReID tasks is usually in the form of identity annotations. Although there are many outstanding unsupervised methods [44–47] that do not need annotations, it is usually hard for them to achieve competitive performance as the supervised ReID methods. For better performance, lots of previous methods [1,11,36,42,51,53,54,58] utilized *fully-supervised learning*, in which identity labels are annotated manually across all training cameras.

Many of them designed spatial alignment [34,37,49], visual attention [13,19], and semantic segmentation [11,31,38] for extracting accurate and fine-grained features. GAN-based methods [10,20,23] were also utilized for data augmentation. However, although these methods achieved remarkable performance on ReID tasks, they required costly inter-camera annotations. To reduce the cost of human labor, ReID researchers began to investigate *weakly-supervised learning*. SCT [48] presumes that each identity appears in only one camera. In ICS [60], an intra-camera supervision task is studied in which an identity could have different labels under different cameras. In [17,18], pseudo labels are used to supervised the ReID model.

Generalization. The generalization ability in ReID tasks denotes how well a trained model functions on unseen datasets, which is usually examined by *direct transfer* tasks. Researchers found that many fully-supervised ReID models perform poorly on unseen datasets [3,32,41]. To improve the generalization ability, various strategies were adopted as additional constraints to avoid over-fitting, such as label smoothing [21] and sophisticated part alignment approaches [8].

Transfer. The transfer ability in ReID tasks corresponds to the capability of ReID models transferring and preserving the discriminative knowledge across multiple training sets. There are two related tasks. *Domain adaptation* transfers knowledge from labeled source domains to unlabeled target domains. One solution [3,41,57] bridged the domain gap by transferring source images to the target image style. Other solutions [4,6,16,33,40] utilized the knowledge learned from the source domain to mine the identity relations in target domains. *Incremental learning* [12,15,28] also values the transfer ability. Its goal is to preserve the previous knowledge and accumulate the common knowledge for all seen datasets. A recent ReID work that relates to incremental learning is MASDF [43], which distilled and incorporated the knowledge from multiple datasets.

3   Methodology

3.1   Conventional ReID: Learning Camera-Related Knowledge

ReID is a task of retrieving identities according to their appearance. Given a training set consisting of disjoint cameras, learning a ReID model on it requires two types of annotations: inter-camera annotations and intra-camera annotations. The conventional ReID formulation regards a ReID dataset as a whole and learns the relations between identities as well as the connections between training cameras. Given an image $\mathbf{I}_i^{\mathcal{D}_j}$ from any training set $\mathcal{D}_j$, the training goal of this formulation is:

$$\arg\min \mathbb{E}\left[\mathbf{y}_i^{\mathcal{D}_j} - \mathbf{g}^{\mathcal{D}_j}\left(\mathbf{f}^{\mathcal{D}_j}\left(\mathbf{I}_i^{\mathcal{D}_j}\right)\right)\right], \left(\mathbf{I}_i^{\mathcal{D}_j}, \mathbf{y}_i^{\mathcal{D}_j}\right) \in \mathcal{D}_j, \qquad (1)$$

where $\mathbf{f}^{\mathcal{D}_j}\left(\cdot\right)$ and $\mathbf{g}^{\mathcal{D}_j}\left(\cdot\right)$ are the corresponding feature extractor and classifier for $\mathcal{D}_j$, respectively. $\mathbf{y}_i^{\mathcal{D}_j}$ denotes the identity label of the image $\mathbf{I}_i^{\mathcal{D}_j}$.

In our opinion, this formulation has three drawbacks. First, images from different cameras, even of the same identity, are subject to distinct distributions. To associate images across cameras, conventional approaches strongly demand the costly inter-camera annotations. Meanwhile, the intra-camera annotations are less exploited since they provide little information across cameras. Second, such learned knowledge not only discriminates the identities in the training set but also encodes the connections between training cameras. These connections are associated with the particular training cameras and hard to generalize to other cameras, since the learned knowledge may not apply to the distribution of previously unseen cameras. For example, when transferring a ReID model trained on Market-1501 to DukeMTMC-reID, it produces a poor Rank-1 accuracy of 37.0% without fine-tuning. Third, the learned knowledge is hard to preserve when being fine-tuned. For instance, after fine-tuning the aforementioned model on DukeMTMC-reID, the Rank-1 accuracy drops 14.2% on Market-1501, because it turns to fit the relations between the cameras in DukeMTMC-reID. We analyze these three problems and find that the particular relations between training cameras are the primary cause of them. Thus, we believe that the conventional method of handling these camera-related relations may need a re-design.

3.2 Our Insight: Towards Camera-Independent ReID

We rethink the relations between cameras. More specifically, we believe that the exclusive knowledge for bridging the distribution gap between the particular training cameras should be suppressed during training. Such knowledge is associated to the cameras in the training set and sacrifices the discriminative and generalization ability on unseen scenarios.

To this end, we propose to align the distribution of all cameras explicitly, so that the distribution gap between all cameras is eliminated, and much less camera-specific knowledge will be learned during training. We denote this formulation as the **camera-based formulation**. To align the distribution of each camera, we estimate the raw distribution of each camera and standardize images from each camera with the corresponding distribution statistics. We use $\eta\left(\cdot\right)$ to denote the estimated statistics related to the distribution of a camera. Then, given a related image $\mathbf{I}_i^{(c)}$, aligning the camera-wise distribution will transform this image as:

$$\tilde{\mathbf{I}}_i^{(c)} = \mathbf{DA}\left(\mathbf{I}_i^{(c)}; \eta\left(c\right)\right), \tag{2}$$

where $\mathbf{DA}\left(\cdot\right)$ represents a distribution alignment mechanism, $\tilde{\mathbf{I}}_i^{(c)}$ denotes the aligned $\mathbf{I}_i^{(c)}$ and $\eta\left(c\right)$ is the estimated alignment parameters for camera c. For any training set $\mathcal{D}_j$, we can now learn the ReID knowledge from this aligned distribution by replacing $\mathbf{I}_i^{\mathcal{D}_j}$ in Eq. 1 with $\tilde{\mathbf{I}}_i^{(c)}$.

With the distributions of all cameras aligned by $\mathbf{DA}\left(\cdot\right)$, images from all these cameras can be regarded as distributing on a "standardized camera". By learning on this "standardized camera", we eliminate the distribution gap between cameras, so the raw learning objectives within the same and across different cameras can be treated equally, making the training procedure more efficient and effective. Besides, without the disturbance caused by the training-camera-related connections, the learned knowledge can generalize better across various ReID scenarios. Last but not least, now that the additional knowledge for associating diverse distributions is much less required, our formulation can make better use of the intra-camera annotations. It may relieve human labor for the costly inter-camera annotations, and provides a solution for ReID in a large-scale camera network with fewer demands of inter-camera annotations.

3.3   Camera-Based Batch Normalization

In practice, a possible solution for aligning camera-related distributions is to conduct batch normalization in a camera-wise manner. We propose the Camera-based Batch Normalization (CBN) for aligning the distribution of all training and testing cameras. It is modified from the conventional Batch Normalization [9], and estimates camera-related statistics rather than dataset-related statistics.

Batch Normalization Revisited. The Batch Normalization [9] is designed to reduce the internal covariate shifting. In training, it standardizes the data with the mini-batch statistics and records them for approximating the global statistics. During testing, given an input $\mathbf{x}_i$, the output of the BN layer is:

$$\hat{\mathbf{x}}_i = \gamma \frac{\mathbf{x}_i - \hat{\mu}}{\sqrt{\hat{\sigma}^2 + \epsilon}} + \beta, \tag{3}$$

where $\mathbf{x}_i$ is the input and $\hat{\mathbf{x}}_i$ is the corresponding output. $\hat{\mu}$ and $\hat{\sigma}^2$ are the global mean and variance of the training set. γ and β are two parameters learned during training. In ReID tasks, BN has significant limitations. It assumes and requires that all testing images are subject to the same training distribution. However, this assumption is satisfied only when the cameras in the testing set and training set are exactly the same. Otherwise, the standardization fails.

Batch Normalization within Cameras. Our Camera-based Batch Normalization (CBN) aligns all training and testing cameras independently. It guarantees an invariant input distribution for learning, accumulating, and verifying the ReID knowledge. Given images or corresponding intermediate features $\mathbf{x}_m^{(c)}$ from camera c, CBN standardizes them according to the camera-related statistics:

$$\mu_{(c)} = \frac{1}{M} \sum_{m=1}^{M} \mathbf{x}_m^{(c)}, \quad \sigma_{(c)}^2 = \frac{1}{M} \sum_{m=1}^{M} \left(\mathbf{x}_m^{(c)} - \mu_{(c)} \right)^2, \quad \hat{\mathbf{x}}_m = \gamma \frac{\mathbf{x}_m - \mu_{(c)}}{\sqrt{\sigma_{(c)}^2 + \epsilon}} + \beta, \tag{4}$$

where $\mu_{(c)}$ and $\sigma_{(c)}^2$ denote the mean and variance related to this camera c. During training, we disassemble each mini-batch and calculate the camera-related mean and variance for each involved camera. The camera with only one sampled images is ignored. During testing, before employing the learned ReID model to extract features, the above statistics have to be renewed for every testing camera.

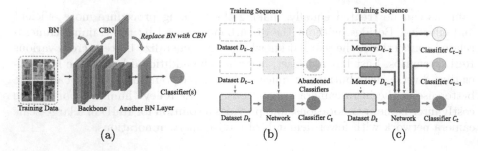

Fig. 2. Demonstrations of our bare-bones baseline network and two *incremental learning* settings involved in this paper. (a) Given an arbitrary backbone with BN layers, we simply replace all BN layers with our CBN layers. (b) **Data-Free**. (c) **Replay**.

In short, we collect several unlabeled images and calculate the camera-related statistics per testing camera. Then, we employ these statistics and the learned weights to generate the final features.

3.4 Applying CBN to Multiple ReID Scenarios

The proposed CBN is generic and nearly cost-free for existing methods on multiple ReID tasks. To demonstrate its superiority, we setup a bare-bones baseline, which only contains a deep neural network, an additional BN layer as the bottleneck, and a fully connected layer as the classifier. As shown in Fig. 2(a), our camera-based formulation can be implemented by simply replacing all BN layers in a usual convolutional network with CBN layers.

With a modified network mentioned above, our camera-based formulation can be applied to many popular tasks, such as *fully-supervised learning, weakly-supervised learning, direct transfer,* and *domain adaptation*. Apart from them, we also evaluate a rarely discussed ReID task, *i.e., incremental learning*. It studies the problem of learning knowledge incrementally from a sequence of training sets while preserving and accumulating the previously learned knowledge. As shown in Fig. 2, we propose two settings. (1) **Data-Free**: once we finish the training procedure on a dataset, the training data along with the corresponding classifier are abandoned. When training the model on the subsequent training sets, the old data will never show up again. (2) **Replay**: unlike Data-Free, we construct an exemplar set from each old training set. The exemplar set and the corresponding classifier are preserved and used during the entire training sequence.

3.5 Discussions

Bridging ReID Tasks. We briefly demonstrate our understandings of the relations between ReID tasks and how we bridge these tasks. Different ReID tasks handle different combinations of training and testing sets. Since datasets have distinct cameras, previous methods have to learn exclusive relations between particular training cameras and adapt them to specific testing camera sets. Our formulation aligns the distribution of all cameras for learning and testing ReID knowledge, and suppresses the exclusive training-camera relations. It may reveal the latent connections between ReID tasks. First, by aligning the distribution of seen and unseen cameras, *fully-supervised learning* and *direct transfer* are united since training and testing distributions are always aligned in a camera-wise manner. Second, since there is no need to learn relations between distinct camera-related distributions, intra- and inter-camera annotations can be treated almost equally. Knowledge is better shared among cameras which helps *fully-* and *weakly-supervised learning*. Third, with the aligned training and testing distributions, it is more efficient to learn, accumulate, and preserve knowledge across datasets. It offers an elegant solution to preserve old knowledge (*incremental learning*) and absorb new knowledge (*domain adaptation*) in the same model.

Relationship to Previous Works. There are two types of previous works that closely relate to ours: camera-related methods and BN variants. Camera-related methods such as CamStyle [57] and CAMEL [45] noticed the camera view discrepancy inside the dataset. CamStyle augmented the dataset by transferring the image style in a camera-to-camera manner, but still learned ReID models in the dataset-based formulation. Consequently, transferring across datasets is still difficult. CAMEL [45] is the most similar work with ours, which learned camera-related projections and mapped camera-related distributions into an implicit common distribution. However, these projections are associated with the training cameras, limiting its ability to transfer across datasets. BN variants such as AdaBN also inspire us. AdaBN aligned the distribution of the entire dataset. It neither eliminated the camera-related relations in training nor handled the camera-related distribution gap in testing. Unlike them, CBN is specially designed for our camera-based formulation. It is much more general and precise for ReID tasks. More comparisons will be provided in Sects. 4.2 and 4.3.

4   Experiments

4.1   Experiment Setup

Datasets. We utilize three large scale ReID datasets, including Market-1501 [50], DukeMTMC-reID [52], and MSMT17 [41]. Market-1501 dataset has 1,501 identities in total. 751 identities are used for training and the rest for testing. The training set contains 12,936 images and the testing set contains 15,913 images. DukeMTMC-reID dataset contains 16,522 images of 702 identities for training, and 1,110 identities with 17,661 images are used for testing. MSMT17 dataset is the current largest ReID dataset with 126,441 images of 4,101 identities from 15 cameras. For short, we denote Market-1501 as Market, DukeMTMC-reID as Duke, and MSMT17 as MSMT in the rest of this paper. *It is worth noting that in these datasets, the training and testing subsets contain the same camera combinations. It could be the reason that previous dataset-based methods create remarkable fully-supervised performance but catastrophic direct transfer results.*

Implementation Details. In this paper, all experiments are conducted with PyTorch. The image size is 256×128 and the batch size is 64. In training, we sample 4 images for each identity. The baseline network presented in Sect. 3.4 uses ResNet-50 [7] as the backbone. To train this network, we adopt SGD optimizer with momentum [27] of 0.9 and weight decay of 5×10^{-4}. The initial learning rate is 0.01, and it decays after the $40th$ epoch by a factor of 10. For all experiments, the training stage will end up with 60 epochs. For incremental learning, we include a warm-up stage. In this stage, we freeze the backbone and only fine-tune the classifier(s) to avoid damaging the previously learned knowledge. During testing, our framework will first sample a few unlabeled images from each camera and use them to approximate the camera-related statistics. Then, these statistics are fixed and employed to process the corresponding testing images. Following the conventions, mean Average Precision (mAP) and Cumulative Matching Characteristic (CMC) curves are utilized for evaluations.

4.2 Performance on Different ReID Tasks

We evaluate our proposed method on five types of ReID tasks, *i.e.*, *fully-supervised learning*, *weakly-supervised learning*, *direct transfer*, *domain adaptation*, and *incremental learning*. The corresponding experiments are organized as follows. First, we demonstrate the importance of aligning the distribution of all cameras from all datasets, and simultaneously conduct *fully-supervised learning* and *direct transfer* on multiple ReID datasets. Second, we demonstrate that it is possible to learn discriminative knowledge with only intra-camera annotations. We utilize the network architecture in Sect. 3.4 to compare the *fully-supervised learning* and *weakly-supervised learning*. To evaluate the generalization ability, *direct transfer* is also conducted for these two settings. Third, we evaluate the transfer ability of our method. This part of experiments includes *domain adaptation*, *i.e.*, transferring the knowledge from the old domain to new domains, and *incremental learning*, *i.e.*, preserving the old knowledge and accumulating the common knowledge for all training sets.

Note that, for simplicity, we denote the results of training and testing the model on the same dataset with fully annotated data as the *fully-supervised learning results*. For similar experiments that only use the intra-camera annotations, we denote their results as the *weakly-supervised learning results*.

Supervisions and Generalization. In this section, we evaluate and analyze the supervisions and the generalization ability in ReID tasks. For all experiments in this section, the testing results on both the training domain and other unseen testing domains are always obtained by the same learned model. We first conduct experiments on *fully-supervised learning* and *direct transfer*. As shown in Table 1, our proposed method shows good advantages, *e.g.*, there is an averaged 1.1% improvement in Rank-1 accuracy for the *fully-supervised learning* task. Meanwhile, without bells and whistles, there is an average 13.6% improvement in Rank-1 accuracy for the *direct transfer* task. We recognize that our method has to collect a few unlabeled samples from each testing camera for estimating the camera-related statistics. However, this process is fast and nearly cost-free.

Table 1. Results of the baseline method with our formulation and the conventional formulation. The fully-supervised learning results are in *italics*.

Training set	Testing set	Market		Duke		MSMT	
	Formulation	Rank-1	mAP	Rank-1	mAP	Rank-1	mAP
Market	Conventional	*90.2*	*74.0*	37.0	20.7	17.1	5.5
	Ours	*91.3*	*77.3*	**58.7**	**38.2**	**25.3**	**9.5**
Duke	Conventional	53.2	25.1	*81.5*	*66.6*	27.2	9.1
	Ours	**72.7**	**43.0**	*82.5*	*67.3*	**35.4**	**13.0**
MSMT	Conventional	58.1	30.8	57.8	38.4	*71.5*	*42.3*
	Ours	**73.7**	**45.0**	**66.2**	**46.7**	*72.8*	*42.9*

Our method can also boost previous methods. Take BoT [21], a recent state-of-the-art method, as an example. We integrate our proposed CBN into BoT and conduct experiments with almost the same settings as in the original paper, including the network architecture, objective functions, and training strategies. The only difference is that we disable Random Erasing [54] due to its constant negative effects on *direct transfer*. The results of the *fully-supervised learning* on Market and Duke are shown in Table 2. It should be pointed out that in *fully-supervised learning*, training and testing subsets contain the same cameras. Therefore, there is no significant shift among the BN statistics of the training set and the testing set, which favors the conventional formulation. Even so, our method still improves the performance on both Market and Duke. We believe that both aligning camera-wise distributions and better utilizing all annotations contribute to these improvements. Moreover, we also present results on *direct transfer* in Table 4. It is clear that our method improves BoT significantly, *e.g.*, there is a 15.3% Rank-1 improvement when training on Duke but testing on Market. These improvements on both *fully-supervised learning* and *direct transfer* demonstrate the advantages of our camera-based formulation.

Weak Supervisions. As we demonstrated in Sect. 3.1, the conventional ReID formulation strongly demands the inter-camera annotations for associating identities under distinct camera-related distributions. Since our method eliminates the distribution gap between cameras, the intra-camera annotations can be better used for learning the appearance features. We compare the performance of using all annotations (*fully-supervised learning*) and only intra-camera annotations (*weakly-supervised learning*). The results are in Table 3. For weakly-supervised experiments, we follow the same settings in MT [60]. Since there are no inter-camera annotations, the identity labels of different cameras are independent, and we assign each individual camera with a separate classifier. Each of these classifiers is supervised by the corresponding intra-camera identity labels. Surprisingly, even without inter-camera annotations, the *weakly-supervised learning* achieves competitive performance. According to these results, we believe that the importance of intra-camera annotations is significantly undervalued.

Table 2. Results of the state-of-the-art fully-supervised learning methods. BoT* denotes our results with the official BoT code. In BoT*, Random Erasing is disabled due to its negative effect on direct transfer. Unless otherwise stated, the **baseline** method in the following sections refers to the network described in Sect. 3.4.

Method	Market				Duke			
	Rank-1	Rank-5	Rank-10	mAP	Rank-1	Rank-5	Rank-10	mAP
CamStyle [57]	88.1	-	-	68.7	75.3	-	-	53.5
MLFN [2]	90.0	-	-	74.3	81.0	-	-	62.8
SCPNet [5]	91.2	97.0	-	75.2	80.3	89.6	-	62.6
HA-CNN [13]	91.2	-	-	75.7	80.5	-	-	63.8
PGFA [24]	91.2	-	-	76.8	82.6	-	-	65.5
MVP [35]	91.4	-	-	80.5	83.4	-	-	70.0
SGGNN [30]	92.3	96.1	97.4	82.8	81.1	88.4	91.2	68.2
SPReID [11]	92.5	97.2	98.1	81.3	84.4	91.9	93.7	71.0
BoT* [21]	93.6	97.6	98.4	82.2	84.3	91.9	94.2	70.1
PCB+RPP [37]	93.8	97.5	98.5	81.6	83.3	90.5	92.5	69.2
OSNet [59]	94.8	-	-	84.9	88.6	-	-	73.5
VA-reID [61]	**96.2**	**98.7**	-	**91.7**	**91.6**	**96.2**	-	**84.5**
Baseline	90.2	96.7	97.9	74.0	81.5	91.4	94.0	66.6
Ours+Baseline	91.3	97.1	98.4	77.3	82.5	91.7	94.1	67.3
Ours+BoT*	94.3	97.9	98.7	83.6	84.8	92.5	95.2	70.1

Transfer. In this section, we evaluate the ability to transfer ReID knowledge between the old and new datasets. First, we evaluate the ability to transfer previous knowledge to new domains. The related task is *domain adaptation*, which usually involves a labeled source training set and another unlabeled target training set. We integrate our formulation into a recent state-of-the-art method ECN [56]. The results are shown in Table 4. By aligning the distributions of source labeled images and target unlabeled images, the performance of ECN is largely boosted, *e.g.*, when transferring from Duke to Market, the Rank-1 accuracy and mAP are improved by 6.6% and 9.0%, respectively. Meanwhile, compared to other methods that also utilize camera labels, such as CamStyle [57] and CASCL [44], our method outperforms them significantly. These improvements demonstrate the effectiveness of our camera-based formulation in *domain adaptation*.

Second, we evaluate the ability to preserve old knowledge as well as accumulate common knowledge for all seen datasets when being fine-tuned. *Incremental learning*, which fine-tunes a model on a sequence of training sets, is used for this evaluation. Experiments are designed as follows. Given three large-scale ReID datasets, there are in total six training sequences of length 2, such as (Market→Duke) and six sequences of length 3, such as (Market→Duke→MSMT). We use the baseline method described in Sect. 3.4 and train it on all sequences separately. After training on each dataset of every

Table 3. The comparisons of fully- and weakly-supervised learning. Results of training and testing on the same domain are in *italics*. MT [60] is our baseline. Except for the camera-based formulation, our weakly-supervised model follows all its settings.

Training set	Testing set	Market		Duke		MSMT	
	Supervision	Rank-1	mAP	Rank-1	mAP	Rank-1	mAP
Market	MT [60]	*78.4*	*52.1*	–	–	–	–
	Weakly	*83.3*	*60.4*	48.9	29.7	**26.8**	**9.6**
	Fully	***91.3***	***77.3***	**58.7**	**38.2**	25.3	9.5
Duke	MT	–	–	*65.2*	*44.7*	–	–
	Weakly	68.4	37.7	*73.9*	*54.4*	33.7	11.9
	Fully	**72.7**	**43.0**	***82.5***	***67.3***	**35.4**	**13.0**
MSMT	MT	–	–	–	–	*39.6*	*15.9*
	Weakly	68.3	37.2	59.2	38.2	*49.4*	*21.5*
	Fully	**73.7**	**45.0**	**66.2**	**46.7**	***72.8***	***42.9***

Table 4. The results of testing ReID models across datasets. ‡ marks methods that only use the source domain data for training, *i.e.*, direct transfer. Other methods listed in this table utilize both the source and target training data, *i.e.*, domain adaptation.

Method	Duke to market				Market to duke			
	Rank-1	Rank-5	Rank-10	mAP	Rank-1	Rank-5	Rank-10	mAP
UMDL [26]	34.5	52.6	59.6	12.4	18.5	31.4	37.6	7.3
PTGAN [41]	38.6	–	66.1	–	27.4	–	50.7	–
PUL [4]	45.5	60.7	66.7	20.5	30.0	43.4	48.5	16.4
SPGAN [3]	51.5	70.1	76.8	22.8	41.1	56.6	63.0	22.3
BoT*‡ [21]	53.3	69.7	76.4	24.9	43.9	58.8	64.9	26.1
MMFA [16]	56.7	75.0	81.8	27.4	45.3	59.8	66.3	24.7
TJ-AIDL [40]	58.2	74.8	81.1	26.5	44.3	59.6	65.0	23.0
CamStyle [57]	58.8	78.2	84.3	27.4	48.4	62.5	68.9	25.1
HHL [55]	62.2	78.8	84.0	31.4	46.9	61.0	66.7	27.2
CASCL [44]	64.7	80.2	85.6	35.6	51.5	66.7	71.7	30.5
ECN [56]	75.1	87.6	91.6	43.0	63.3	75.8	80.4	40.4
Baseline‡	53.2	70.0	76.0	25.1	37.0	52.6	58.9	20.7
Ours+BoT*‡	68.6	82.5	87.7	39.0	60.6	74.0	78.5	39.8
Ours+Baseline‡	72.7	85.8	90.7	43.0	58.7	74.1	78.1	38.2
Ours+ECN	**81.7**	**91.9**	**94.7**	**52.0**	**68.0**	**80.0**	**83.9**	**44.9**

sequence, we evaluate the latest model on the first dataset of the corresponding sequence and record the performance decreases. Both the **Data-Free** and **Replay** settings are tested. For the Replay settings, the exemplars are selected by randomly sampling one image for each identity. Compared to the original

Table 5. Results of ReID models on incremental learning tasks. Each result denotes the percentage of the performance preserved on the first dataset after learning on new datasets. § marks the Data-Free settings. † corresponds to the Replay settings.

Testing set		Market		Duke		MSMT	
Seq Length	Formulation	Rank-1	mAP	Rank-1	mAP	Rank-1	mAP
1	–	100%	100%	100%	100%	100%	100%
2	Conventional§	82.2%	62.5%	80.2%	68.8%	55.5%	38.7%
	Ours§	88.3%	71.2%	89.3%	83.2%	74.5%	58.9%
	Conventional†	92.5%	84.1%	90.9%	84.7%	81.7%	70.1%
	Ours†	**95.0%**	**85.7%**	**94.3%**	**91.1%**	**91.6%**	**84.6%**
3	Conventional§	74.8%	52.2%	75.2%	63.0%	38.9%	24.7%
	Ours§	85.8%	66.0%	85.8%	77.4%	56.6%	39.4%
	Conventional†	86.5%	74.0%	84.1%	76.4%	74.3%	60.9%
	Ours†	**94.4%**	**83.1%**	**91.5%**	**87.6%**	**86.4%**	**76.0%**

training sets, the size of the exemplar set for Market, Duke, and MSMT is only 5.5%, 4.2%, and 3.4%, respectively. Note that in Replay settings, the old classifiers will also be updated in training. The corresponding results are shown in Table 5. To better demonstrate our improvements, we report the averaged results of the sequences that are of the same length and share the same initial dataset, *e.g.*, averaging the results of testing Market on the sequences Market→Duke and Market→MSMT. In short, our formulation outperforms the dataset-based formulation in all experiments. These results further demonstrate the effectiveness of our formulation.

4.3 Ablation Study

The experiments above demonstrate that our camera-based formulation boosts all the mentioned tasks. Now, we conduct more ablation studies to validate CBN.

Comparisons Between CBN and Other BN Variants. We compare CBN with three types of BN variants. (1) BN [9] and IBN [25] correspond to the methods that use training-set-specific statistics to normalize all testing data. (2) AdaBN [14] is a dataset-wise adaptation that utilizes the testing-set-wise statistics to align the entire testing set. (3) The combination of BN and our CBN is to verify the importance of training ReID models with CBN. As shown in Table 6, training and testing the ReID model with CBN achieves the best performance in both *fully-supervised learning* and *direct transfer*.

Table 6. Results of combining different normalization strategies in fully-supervised learning and direct transfer. In this table, BN and IBN correspond to the training-set-specific normalization methods. AdaBN adapts the dataset-wise normalization statistics. CBN follows our camera-based formulation and aligns each camera independently.

Training method	Testing method	Duke to duke		Duke to market	
		Rank-1	mAP	Rank-1	mAP
BN	BN	81.5	66.6	53.2	25.1
IBN [25]	IBN	77.6	57.0	61.7	29.5
BN	AdaBN [14]	81.2	66.2	55.8	28.1
BN	Our CBN	80.2	63.7	69.5	40.6
Our CBN	Our CBN	**82.5**	**67.3**	**72.7**	**43.0**

Table 7. The mAP of our method on fully-supervised learning and direct transfer. We repeat each experiment 10 times and calculate the mean and variance of all results.

# Batches	Market to market		Market to duke	
	Mean	Variance	Mean	Variance
1	76.29	0.032	37.34	0.047
5	77.21	0.010	38.08	0.017
10	77.33	0.007	38.19	0.008
20	77.37	0.005	38.18	0.002
50	**77.39**	**0.001**	**38.21**	**0.001**

Samples Required for CBN Approximation. We conduct experiments for approximating the camera-related statistics with different numbers of samples. Note that if a camera contains less than the required number of images, we simply use all available images rather than duplicate them. We repeat all experiments 10 times and list the averaged results in Table 7. As demonstrated, the performance is better and more stable when using more samples to estimate the camera-related statistics. Besides, results are already good enough when only utilizing very few samples, e.g., 10 mini-batches. For the balance of simplicity and performance, we adopt 10 mini-batches for approximation in all experiments.

Compatibility with Different Backbones. Apart from ResNet [7] used in the above experiments, we further evaluate the compatibility of CBN. We embed CBN with other commonly used backbones: MobileNet V2 [29] and ShuffleNet V2 [22], and evaluate their performance on *fully-supervised learning* and *direct transfer*. As shown in Table 8, the performance is also boosted significantly.

Table 8. Results of combining our camera-based formulation with different convolutional backbones. The fully-supervised learning results are in *italics*.

Backbone	Training set	Testing set	Market		Duke	
		Formulation	Rank-1	mAP	Rank-1	mAP
MobileNet V2 [29]	Market	Conventional	*87.7*	*69.2*	34.7	18.9
		Ours	***89.8***	***73.7***	**54.4**	**34.0**
	Duke	Conventional	51.4	22.6	*79.8*	*60.2*
		Ours	**70.7**	**39.0**	*79.9*	*62.4*
ShuffleNet V2 [22]	Market	Conventional	*82.6*	*58.4*	34.6	18.4
		Ours	***85.9***	***65.8***	**53.8**	**33.8**
	Duke	Conventional	48.1	20.3	*74.7*	*52.8*
		Ours	**70.0**	**38.9**	***77.1***	***58.6***

5 Conclusions

In this paper, we advocate for a novel camera-based formulation for person re-identification and present a simple yet effective solution named camera-based batch normalization. With only a few additional costs, our approach shrinks the gap between intra-camera learning and inter-camera learning. It significantly boosts the performance on multiple ReID tasks, regardless of the source of supervision, and whether the trained model is tested on the same dataset or transferred to another dataset. Our research delivers two key messages. **First**, it is crucial to align *all* camera-related distributions in ReID tasks, so the ReID models can enjoy better abilities to generalize across different scenarios as well as transfer across multiple datasets. **Second**, with the aligned distributions, we unleash the potential of intra-camera annotations, which may have been undervalued in the community. With promising performance under the weakly-supervised setting (only intra-camera annotations are available), our approach provides a practical solution for deploying ReID models in large-scale, real-world scenarios.

Acknowledgements. This work was supported by National Science Foundation of China under grant No. 61521002.

References

1. Almazan, J., Gajic, B., Murray, N., Larlus, D.: Re-id done right: towards good practices for person re-identification. arXiv preprint arXiv:1801.05339 (2018)
2. Chang, X., Hospedales, T.M., Xiang, T.: Multi-level factorisation net for person re-identification. In: CVPR. IEEE (2018)
3. Deng, W., Zheng, L., Ye, Q., Kang, G., Yang, Y., Jiao, J.: Image-image domain adaptation with preserved self-similarity and domain-dissimilarity for person re-identification. In: CVPR. IEEE (2018)

4. Fan, H., Zheng, L., Yan, C., Yang, Y.: Unsupervised person re-identification: Clustering and fine-tuning. ACM Trans. Multimedia Comput. Commun. Appl. (TOMM) **14**(4), 83 (2018)

5. Fan, X., Luo, H., Zhang, X., He, L., Zhang, C., Jiang, W.: SCPNet: spatial-channel parallelism network for joint holistic and partial person re-identification. In: Jawahar, C.V., Li, H., Mori, G., Schindler, K. (eds.) ACCV 2018. LNCS, vol. 11362, pp. 19–34. Springer, Cham (2019). https://doi.org/10.1007/978-3-030-20890-5_2

6. Fu, Y., Wei, Y., Wang, G., Zhou, Y., Shi, H., Huang, T.S.: Self-similarity grouping: a simple unsupervised cross domain adaptation approach for person re-identification. In: ICCV. IEEE (2019)

7. He, K., Zhang, X., Ren, S., Sun, J.: Deep residual learning for image recognition. In: CVPR. IEEE (2016)

8. Huang, H., et al.: Eanet: Enhancing alignment for cross-domain person re-identification. arXiv preprint arXiv:1812.11369 (2018)

9. Ioffe, S., Szegedy, C.: Batch normalization: Accelerating deep network training by reducing internal covariate shift. arXiv preprint arXiv:1502.03167 (2015)

10. Jiao, J., Zheng, W.S., Wu, A., Zhu, X., Gong, S.: Deep low-resolution person re-identification. In: AAAI (2018)

11. Kalayeh, M.M., Basaran, E., Gökmen, M., Kamasak, M.E., Shah, M.: Human semantic parsing for person re-identification. In: CVPR. IEEE (2018)

12. Kirkpatrick, J., et al.: Overcoming catastrophic forgetting in neural networks. Proc. Natl. Acad. Sci. **114**(13), 3521–3526 (2017)

13. Li, W., Zhu, X., Gong, S.: Harmonious attention network for person re-identification. In: CVPR. IEEE (2018)

14. Li, Y., Wang, N., Shi, J., Liu, J., Hou, X.: Revisiting batch normalization for practical domain adaptation. arXiv preprint arXiv:1603.04779 (2016)

15. Li, Z., Hoiem, D.: Learning without forgetting. IEEE Trans. Pattern Anal. Mach. Intell. **40**(12), 2935–2947 (2018)

16. Lin, S., Li, H., Li, C.T., Kot, A.C.: Multi-task mid-level feature alignment network for unsupervised cross-dataset person re-identification. In: BMVC (2018)

17. Lin, Y., Dong, X., Zheng, L., Yan, Y., Yang, Y.: A bottom-up clustering approach to unsupervised person re-identification. In: AAAI (2019)

18. Lin, Y., Xie, L., Wu, Y., Yan, C., Tian, Q.: Unsupervised person re-identification via softened similarity learning. In: CVPR. IEEE (2020)

19. Liu, H., Feng, J., Qi, M., Jiang, J., Yan, S.: End-to-end comparative attention networks for person re-identification. IEEE Trans. Image Process. **26**(7), 3492–3506 (2017)

20. Liu, J., Ni, B., Yan, Y., Zhou, P., Cheng, S., Hu, J.: Pose transferrable person re-identification. In: CVPR. IEEE (2018)

21. Luo, H., Gu, Y., Liao, X., Lai, S., Jiang, W.: Bag of tricks and a strong baseline for deep person re-identification. In: CVPRW. IEEE (2019)

22. Ma, N., Zhang, X., Zheng, H.-T., Sun, J.: ShuffleNet V2: practical guidelines for efficient CNN architecture design. In: Ferrari, V., Hebert, M., Sminchisescu, C., Weiss, Y. (eds.) Computer Vision – ECCV 2018. LNCS, vol. 11218, pp. 122–138. Springer, Cham (2018). https://doi.org/10.1007/978-3-030-01264-9_8

23. Mao, S., Zhang, S., Yang, M.: Resolution-invariant person re-identification. arXiv preprint arXiv:1906.09748 (2019)

24. Miao, J., Wu, Y., Liu, P., Ding, Y., Yang, Y.: Pose-guided feature alignment for occluded person re-identification. In: ICCV. IEEE (2019)

25. Pan, X., Luo, P., Shi, J., Tang, X.: Two at once: enhancing learning and generalization capacities via IBN-Net. In: Ferrari, V., Hebert, M., Sminchisescu, C., Weiss, Y. (eds.) ECCV 2018. LNCS, vol. 11208, pp. 484–500. Springer, Cham (2018). https://doi.org/10.1007/978-3-030-01225-0_29
26. Peng, P., et al.: Unsupervised cross-dataset transfer learning for person re-identification. In: CVPR. IEEE (2016)
27. Qian, N.: On the momentum term in gradient descent learning algorithms. Neural Netw. **12**(1), 145–151 (1999)
28. Rannen, A., Aljundi, R., Blaschko, M.B., Tuytelaars, T.: Encoder based lifelong learning. In: ICCV. IEEE (2017)
29. Sandler, M., Howard, A., Zhu, M., Zhmoginov, A., Chen, L.C.: Mobilenetv 2: inverted residuals and linear bottlenecks. In: CVPR. IEEE (2018)
30. Shen, Y., Li, H., Yi, S., Chen, D., Wang, X.: Person re-identification with deep similarity-guided graph neural network. In: Ferrari, V., Hebert, M., Sminchisescu, C., Weiss, Y. (eds.) ECCV 2018. LNCS, vol. 11219, pp. 508–526. Springer, Cham (2018). https://doi.org/10.1007/978-3-030-01267-0_30
31. Song, C., Huang, Y., Ouyang, W., Wang, L.: Mask-guided contrastive attention model for person re-identification. In: CVPR. IEEE (2018)
32. Song, J., Yang, Y., Song, Y.Z., Xiang, T., Hospedales, T.M.: Generalizable person re-identification by domain-invariant mapping network. In: CVPR. IEEE (2019)
33. Song, L., et al.: Unsupervised domain adaptive re-identification: theory and practice. Pattern Recogn. (2020)
34. Suh, Y., Wang, J., Tang, S., Mei, T., Lee, K.M.: Part-aligned bilinear representations for person re-identification. In: Ferrari, V., Hebert, M., Sminchisescu, C., Weiss, Y. (eds.) Computer Vision – ECCV 2018. LNCS, vol. 11218, pp. 418–437. Springer, Cham (2018). https://doi.org/10.1007/978-3-030-01264-9_25
35. Sun, H., Chen, Z., Yan, S., Xu, L.: MVP matching: a maximum-value perfect matching for mining hard samples, with application to person re-identification. In: ICCV. IEEE (2019)
36. Sun, Y., Zheng, L., Deng, W., Wang, S.: Svdnet for pedestrian retrieval. In: ICCV. IEEE (2017)
37. Sun, Y., Zheng, L., Yang, Y., Tian, Q., Wang, S.: Beyond part models: person retrieval with refined part pooling (and a strong convolutional baseline). In: Ferrari, V., Hebert, M., Sminchisescu, C., Weiss, Y. (eds.) ECCV 2018. LNCS, vol. 11208, pp. 501–518. Springer, Cham (2018). https://doi.org/10.1007/978-3-030-01225-0_30
38. Tian, M., et al.: Eliminating background-bias for robust person re-identification. In: CVPR. IEEE (2018)
39. Van Der Maaten, L.: Accelerating t-SNE using tree-based algorithms. JMLR **15**(1), 3221–3245 (2014)
40. Wang, J., Zhu, X., Gong, S., Li, W.: Transferable joint attribute-identity deep learning for unsupervised person re-identification. In: CVPR. IEEE (2018)
41. Wei, L., Zhang, S., Gao, W., Tian, Q.: Person transfer GAN to bridge domain gap for person re-identification. In: CVPR. IEEE (2018)
42. Wei, L., Zhang, S., Yao, H., Gao, W., Tian, Q.: Glad: global-local-alignment descriptor for pedestrian retrieval. In: ACMMM. ACM (2017)
43. Wu, A., Zheng, W.S., Guo, X., Lai, J.H.: Distilled person re-identification: towards a more scalable system. In: CVPR. IEEE (2019)
44. Wu, A., Zheng, W.S., Lai, J.H.: Unsupervised person re-identification by camera-aware similarity consistency learning. In: ICCV. IEEE (2019)

45. Yu, H.X., Wu, A., Zheng, W.S.: Cross-view asymmetric metric learning for unsupervised person re-identification. In: ICCV. IEEE (2017)
46. Yu, H.X., Wu, A., Zheng, W.S.: Unsupervised person re-identification by deep asymmetric metric embedding. TPAMI (2018)
47. Yu, H.X., Zheng, W.S., Wu, A., Guo, X., Gong, S., Lai, J.H.: Unsupervised person re-identification by soft multilabel learning. In: CVPR (2019)
48. Zhang, T., Xie, L., Wei, L., Zhang, Y., Li, B., Tian, Q.: Single camera training for person re-identification. In: AAAI (2020)
49. Zhang, X., et al.: Alignedreid: surpassing human-level performance in person re-identification. arXiv preprint arXiv:1711.08184 (2017)
50. Zheng, L., Shen, L., Tian, L., Wang, S., Wang, J., Tian, Q.: Scalable person re-identification: a benchmark. In: ICCV. IEEE (2015)
51. Zheng, Z., Zheng, L., Yang, Y.: A discriminatively learned CNN embedding for person reidentification. ACM Trans. Multimedia Comput. Commun. Appl. **14**(1), 13 (2017)
52. Zheng, Z., Zheng, L., Yang, Y.: Unlabeled samples generated by GAN improve the person re-identification baseline in vitro. In: ICCV. IEEE (2017)
53. Zhong, Z., Zheng, L., Cao, D., Li, S.: Re-ranking person re-identification with k-reciprocal encoding. In: CVPR. IEEE (2017)
54. Zhong, Z., Zheng, L., Kang, G., Li, S., Yang, Y.: Random erasing data augmentation. In: AAAI (2020)
55. Zhong, Z., Zheng, L., Li, S., Yang, Y.: Generalizing a person retrieval model hetero- and homogeneously. In: Ferrari, V., Hebert, M., Sminchisescu, C., Weiss, Y. (eds.) ECCV 2018. LNCS, vol. 11217, pp. 176–192. Springer, Cham (2018). https://doi.org/10.1007/978-3-030-01261-8_11
56. Zhong, Z., Zheng, L., Luo, Z., Li, S., Yang, Y.: Invariance matters: exemplar memory for domain adaptive person re-identification. In: CVPR. IEEE (2019)
57. Zhong, Z., Zheng, L., Zheng, Z., Li, S., Yang, Y.: Camera style adaptation for person re-identification. In: CVPR. IEEE (2018)
58. Zhou, J., Yu, P., Tang, W., Wu, Y.: Efficient online local metric adaptation via negative samples for person reidentification. In: ICCV. IEEE (2017)
59. Zhou, K., Yang, Y., Cavallaro, A., Xiang, T.: Omni-scale feature learning for person re-identification. In: ICCV. IEEE (2019)
60. Zhu, X., Zhu, X., Li, M., Murino, V., Gong, S.: Intra-camera supervised person re-identification: a new benchmark. In: ICCVW. IEEE (2019)
61. Zhu, Z., et al.: Viewpoint-aware loss with angular regularization for person re-identification. In: AAAI (2020)

AMLN: Adversarial-Based Mutual Learning Network for Online Knowledge Distillation

Xiaobing Zhang[1] , Shijian Lu[2(✉)] , Haigang Gong[1] , Zhipeng Luo[2,3] ,
and Ming Liu[1]

[1] University of Electronic Science and Technology of China, Sichuan, China
zhangxiaobing@std.uestc.edu.cn, {hggong,csmliu}@uestc.edu.cn
[2] Nanyang Technological University, Singapore, Singapore
shijian.lu@ntu.edu.sg, zhipeng001@e.ntu.edu.sg
[3] Sensetime Research, Singapore, Singapore

Abstract. Online knowledge distillation has attracted increasing interest recently, which jointly learns teacher and student models or an ensemble of student models simultaneously and collaboratively. On the other hand, existing works focus more on outcome-driven learning according to knowledge like classification probabilities whereas the distilling processes which capture rich and useful intermediate features and information are largely neglected. In this work, we propose an innovative adversarial-based mutual learning network (AMLN) that introduces process-driven learning beyond outcome-driven learning for augmented online knowledge distillation. A block-wise training module is designed which guides the information flow and mutual learning among peer networks adversarially throughout different learning stages, and this spreads until the final network layer which captures more high-level information. AMLN has been evaluated under a variety of network architectures over three widely used benchmark datasets. Extensive experiments show that AMLN achieves superior performance consistently against state-of-the-art knowledge transfer methods.

Keywords: Mutual learning network · Adversarial-based learning strategy · Online knowledge transfer and distillation

1 Introduction

Deep neural networks (DNNs) have been widely studied and applied in various fields such as image classification [7,36], object detection [11,26], semantic segmentation [10,27], etc. One direction pursues the best accuracy which tends to introduce over-parameterized models [24,26] and demands very high computation and storage resources that are often not available for many edge computing devices. This has triggered intensive research in developing lightweight yet competent network models in recent years, typically through four different

© Springer Nature Switzerland AG 2020
A. Vedaldi et al. (Eds.): ECCV 2020, LNCS 12357, pp. 158–173, 2020.
https://doi.org/10.1007/978-3-030-58610-2_10

approaches: 1) network pruning [14,15,17,21,28], 2) network quantization [9,29], 3) building efficient small networks [6,19,20], and 4) knowledge transfer (KT) [4,5,12,18,30]. Among the four approaches, KT works in a unique way by pre-training a large and powerful teacher network and then distilling features and knowledge to a compact student network. Though compact yet powerful student networks can be trained in this manner, the conventional distillation is usually a multi-stage complex offline process requiring extra computational costs and memory.

Online knowledge distillation [3,22,25,37] has attracted increasing interest in recent years. Instead of pre-training a large teacher network in advance, it trains two or more student models simultaneously in a cooperative peer-teaching manner. In other words, the training of the teacher and student networks is merged into a one-phase process, and the knowledge is distilled and shared among peer networks. This online distilling paradigm can generalize better without a clear definition of teacher/student role, and it has achieved superior performance as compared to offline distillation from teacher to student networks. On the other hand, this online distillation adopts an outcome-driven distillation strategy in common which focuses on minimizing the discrepancy among the final predictions. The rich information encoded in the intermediate layers from peer networks is instead largely neglected which has led to various problems such as limited knowledge transfer in deep mutual learning [37], constrained coordination in on-the-fly native ensemble [25], etc.

In this work, we propose a novel adversarial-based mutual learning network (AMLN) that includes both process-driven and outcome-driven learning for optimal online knowledge distillation. Specifically, AMLN introduces a block-wise learning module for process-driven distillation that guides peer networks to learn the intermediate features and knowledge from each other in an adversarial manner as shown in Fig. 1. At the same time, the block-wise module also learns from the final layer of the peer networks which often encodes very useful high-level features and information. In addition, the softened class posterior of each network is aligned with the class probabilities of its peer, which works together with a conventional supervised loss under the outcome-driven distillation. By incorporating supervision from both intermediate and final network layers, AMLN can be trained in an elegant manner and the trained student models also produce better performance than models trained from scratch in a conventional supervised learning setup. Further, AMLN outperforms state-of-the-art online or offline distillation methods consistently. More details will be described in Experiments and Analysis sections.

The contributions of this work are thus threefold. First, it designs an innovative adversarial-based mutual learning network AMLN that allows an ensemble of peer student networks to transfer knowledge and learn from each other collaboratively. Second, it introduces a block-wise module to guide the peer networks to learn intermediate features and knowledge from each other which augments the sole outcome-driven peer learning greatly. Third, AMLN does not require pre-training a large teach network, and extensive experiments over several public

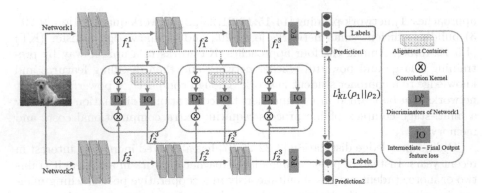

Fig. 1. Overview of the proposed adversarial-based mutual learning network (AMLN): AMLN achieves process-driven mutual distillation by dividing each peer network into same blocks and employing a discriminator to align the block-wise learned features adversarially. Additionally, the intermediate features are also guided by the peer's final output for learning high-level features. The outcome-driven learning instead employs the conventional cross-entropy loss (with one-hot labels) and Kullback-Leibler (KL) loss (with softened labels). Note this pipeline focuses on the distillation from Network2 to Network1. For distillation from Network1 to Network2, a similar pipeline applies as highlighted by the dashed lines.

datasets show that it achieves superior performance as compared to state-of-the-art online/offline knowledge transfer methods.

2   Related Work

2.1   Knowledge Transfer

Knowledge transfer (KT) is one of the most popular methods used in model compression. The early KT research follows a teacher-student learning paradigm in an offline learning manner [5,12,23,30,34]. In recent years, online KT is developed to strengthen the student's performance without a pre-trained teacher network [3,25,33,37]. Our work falls into the online KT learning category.

Offline KT aims to enforce the efficiency of the student's learning from scratch by distilling knowledge from a pre-trained powerful teacher network. Cristian et al. [5] first uses soft-labels for knowledge distillation, and this idea is further improved by adjusting the temperature of softmax activation function to provide additional supervision and regularization on the higher entropy soft-targets [12]. Recently, various new KT systems have been developed to enhance the model capabilities by transferring intermediate features [23,30,34] or by optimizing the initial weights of student networks [8,18].

Online KT trains a student model without the requirement of training a teacher network in advance. With online KT, the networks teach each other mutually by sharing their distilled knowledge and imitating the peer network's

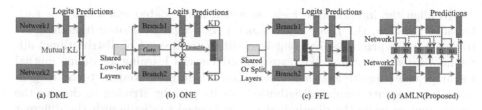

Fig. 2. Four different mutual learning networks: The architectures in (a), (b) and (c) perform mutual learning from the predictions or features of peer networks. The deep mutual learning (DML) [37] in (a) uses the distilled softened prediction of the peer network. The on-the-fly native ensemble (ONE) [25] in (b) creates a teacher with the gating mechanism for the peer network training. The feature fusion learning (FFL) [22] in (c) applies mutual knowledge learning between peer networks and fused classifier. Unlike these outcome-driven learning architectures, our adversarial-based mutual learning network (AMLN) in (d) uses mutual knowledge distillation between block-wise output features and final generated predictions, which enhances the performance of each peer network by distilling more multifarious features from peers.

performance during the training process. Deep mutual learning (DML) [37] and on-the-fly native ensemble (ONE) [25] are the two representative online KT methods that have demonstrated very promising performance as illustrated in Fig. 2. DML proposes to train the students by mutually exchanging the softened classification information using the Kullback-Leibler(KL) divergence loss. Similar to [37], Rohan et al. [3] introduces the codistillation method that forces student networks to maintain diversity longer by using the distillation loss after enough *burn in steps*. Rather than mutually distilling between peer networks, ONE generates a gated ensemble logit of the networks during training and adopts it as a target to guide each network. In addition, feature fusion learning (FFL) [22] uses a fusion module to combine the feature maps from sub-networks, aiming for enhancing the performance of each sub-network.

All the above methods adopt an outcome-driven distillation approach where the distillation during the intermediate network layers is largely neglected. AMLN addresses this issue by further incorporating process-driven distillation which guides the sharing and transfer of intermediate knowledge beyond the knowledge from the final outputs. Unlike ONE [25], AMLN also has better applicability which can work with peer networks with the same or different architecture.

2.2 Adversarial Learning

Generative Adversarial Learning [13] is proposed to create realistic-looking images from random noise. An adversarial training scheme is proposed which consists of a generator network G and a discriminator network D. Specifically, G learns to synthesize images to fool D, meanwhile, D is trained to distinguish the real images in the dataset from the fake images generated by G.

To align the intermediate features which are updated continually at each training iteration, the L_1 or L_2 distance is not applicable since it is designed to evaluate the pixel-level or point-level difference instead of distributional differences between features. We introduce adversarial learning for online mutual learning among multiple student networks, where each student tends to generate features with similar distributions as its peer by striving to deceive the discriminators while the discriminators are trained to distinguish the different distributions of the generated features from multiple peer student networks.

3   Proposed Method

In this section, we describe how to effectively guide the peer-teaching student networks to learn collaboratively with the proposed Adversarial-based Mutual Learning Network (AMLN). Unlike existing online KT methods, AMLN takes into account not only the distillation based on the final prediction, but also the intermediate mutual supervision between the peer networks. We start by giving the architecture overview in Subsect. 3.1, and introduce our novel online process-driven mutual knowledge distillation in Subsect. 3.2. In Subsect. 3.3, we give an explanation of the outcome-driven mutual learning method. Finally, the whole optimization pipeline is presented in Subsect. 3.4.

3.1   The Architecture of AMLN

We formulate our proposed method by considering two peer networks S_1 and S_2. As illustrated in Fig. 1, S_1 and S_2 could adopt identical of different architectures, but should have the same number of blocks for intermediate feature alignment. During the training, the process-driven mutual knowledge distillation is implemented with a proposed block-wise module that contains a discriminator and an alignment container. Specifically, each network is trained to fool its corresponding block-wise discriminators so that it can produce similar feature maps to mimic that from its peer network. The alignment container is employed to align the block-wise outputs to the peer network's final feature maps for high-level information distillation. On the other hand, the outcome-driven mutual knowledge distillation is realised by minimizing the peer model's softened output distributions, which encodes higher entropy as extra supervision. Moreover, ground truth labels are used as a conventional supervision for the task-specific features learning.

3.2   Process-Driven Mutual Knowledge Distillation

Given N samples $X = \{x_i\}_{i=1}^N$ from M classes, we denote the corresponding label set as $Y = \{y_i\}_{i=1}^N$ with $y_i \in \{1, 2, ..., M\}$. As can be seen in Fig. 1, the backbone networks are first divided into the same blocks according to their depth. Suppose that the block-wise generated feature is defined as f_j^b, where j and b indicate the network number and block number respectively, i.e. $j = 1, 2$ and $b = 1, 2, 3$. Each block is followed with a block-wise training module, including a discriminator D_j^b and an alignment container C_j^b. The discriminator D_j^b is formed by three convolution layers with ReLU operation, where the last layer with two neurons is responsible for identifying the network number j of the injected feature f_j^b. For each alignment container C_j^b, it applies depthwise convolution and pointwise convolution to align the block-wise generated feature f_j^b with the peer's final output f_{3-j}^3 for high-level knowledge distillation. Therefore, there are two loss items for the process-driven mutual learning, one of which is the adversarial-based distilling loss defined as follows:

$$L_D^j = \min_{f_j^b} \max_D \sum_{b=1}^3 E_{f_j^b \sim P_{S_j}}[1 - D_j^b(\sigma(f_j^b))] + E_{f_{3-j}^b \sim P_{S_{3-j}}}[D_j^b(\sigma(f_{3-j}^b))] \quad (1)$$

Here, σ denotes the convolution kernel, which is utilized to reduce the number of channels of f_j^b. P_{S_j} corresponds to the logits distribution of the network S_j.

Another loss works by evaluating the distance between the block-wise distilled feature and the peer's final generated feature, which can be computed as:

$$L_F^j = \sum_{b=1}^3 d(C_j^b(f_j^b), f_{3-j}^3) \quad (2)$$

where C_j^b denotes the alignment container that transforms f_j^b into the same shape as f_{3-j}^3, and the distance metric d is adopted with L_2 method consistently.

The overall process-driven mutual distillation loss function is then formulated with the weight balance parameter β as:

$$L_{S_j^P} = L_D^j + \beta L_F^j \quad (3)$$

3.3 Outcome-Driven Mutual Knowledge Distillation

For outcome-driven distillation, two evaluation items are employed where one is the conventional cross-entropy (CE) loss and the other is the Kullback Leibler (KL) loss between the softened predicted outputs. Suppose that the probability of class m for sample x_i given by S_j is computed as:

$$p_j^m(x_i) = \frac{exp(z_j^m)}{\sum_{m=1}^M exp(z_j^m)} \quad (4)$$

where z_j^m is the predicted output of S_j. Thus, the CE loss between the predicted outputs and one-hot labels for S_j can be evaluated as:

$$L_C^j = -\sum_{i=1}^N \sum_{m=1}^M u(y_i, m)log(p_j^m(x_i)) \quad (5)$$

Here, u is an indicator function, which returns 1 if $y_i = m$ and 0 otherwise.

Algorithm 1. Adversarial-based Mutual Learning Network (AMLN)

Require:

Training set X, label set Y;

Ensure:

Iteration = 0; Initialize S_1 and S_2 to different conditions;

1: Compute intermediate feature maps f_1^b, predicted probabilities p_1 and softened predictions ρ_1, b=1,2,3;

2: Compute the total loss L_{S_1} (Equ. 9);

3: Update the parameters of network S_1 by the SGD algorithm;

4: Compute intermediate feature maps f_2^b, predicted probabilities p_2 and softened predictions ρ_2, b=1,2,3;

5: Compute the total loss L_{S_2} (Equ. 9);

6: Update the parameters of network S_2 by the SGD algorithm;

7: Iteration = Iteration + 1; Begin with Step 1.

8: **return** Both converged models S_1 and S_2.

To improve the generalization performance of sub-networks on the test data, we apply the peer network to generate softened probability with a temperature term T. Given z_j, the softened probability is defined as:

$$\rho_j^m(x_i, T) = \frac{exp(z_j^m/T)}{\sum_{m=1}^{M} exp(z_j^m/T)} \tag{6}$$

when $T = 1$, ρ_j^m is the same as p_j^m. As the temperature term T increases, it generates a softened probability distribution where the probability of each class distributes more evenly and less dominantly. Same as [22,37], we use $T = 3$ consistently during our experiments.

KL divergence is then used to quantify the alignment of the peer networks' softened predictions as:

$$L_{KL}^j(\rho_j||\rho_{3-j}) = \sum_{i=1}^{N} \sum_{m=1}^{M} \rho_j^m(x_i) log \frac{\rho_j^m(x_i)}{\rho_{3-j}^m(x_i)} \tag{7}$$

The overall outcome-driven distillation loss function $L_{S_j^R}$ is formulated as:

$$L_{S_j^R} = L_C^j + T^2 \times L_{KL}^j \tag{8}$$

Since the scale of the gradient produced by the softened distribution is $1/T^2$ of the original value, we multiply T^2 according to the KD recommendations [12] to ensure that the relative contributions of the ground-truth and the softened peer prediction remain roughly unchanged.

3.4 Optimization

Combining both process-driven and outcome-driven distillation loss, the overall loss for each sub-network S_j is as follows:

$$L_{S_j} = L_{S_j^P} + L_{S_j^R} \tag{9}$$

The mutual learning strategy in AMLN works in such a way that the peer networks are closely guided and optimized jointly and collaboratively. At each training iteration, we compute the generated features and predictions of the two peer networks, and update both models' parameters according to Eq. 9. The optimization details are summarized in Algorithm 1.

4   Experimental Results and Analysis

4.1   Datasets and Evaluation Setups

AMLN is evaluated over three datasets that have been widely used for evaluations of knowledge transfer methods. **CIFAR10** [1] and **CIFAR100** [2] are two publicly accessible datasets that have been widely used for the image classification studies. The two datasets have 50,000 training images and 10,000 test images of 10 and 100 image classes, respectively. All images in the two datasets are in RGB format with an image size of 32×32 pixels. **ImageNet** [31] refers to the LSVRC 2015 classification dataset which consists of 1.2 million training images and 50,000 validation images of 1,000 image classes.

Evaluation Metrics: We use the Top-1 and Top-5 mean classification accuracy (%) for evaluations, the former is calculated for all studied datasets while the latter is used for the ImageNet only. To measure the computation cost in model inference stage, we apply the criterion of floating point operations (FLOPs) and the inference time of each image for efficiency comparison.

Networks: The evaluation networks in our experiments include ResNet [16] as well as Wide ResNet(WRN) [35] of different network depths. Table 1 shows the number of parameters of different AMLN-trained network models that are evaluated over the dataset CIFAR100.

4.2   Implementation Details

All experiments are implemented by PyTorch on NVIDIA GPU devices. On the CIFAR dataset, the initial learning rate is 0.1 and is multiplied by 0.1 every 200 epochs. We used SGD as the optimizer with Nesterov momentum 0.9 and weight decay 1e−4, respectively. Mini-batch size is set to 128. For ImageNet, we use SGD with a weight decay of 10^{-4}, a mini-batch size of 128, and an initial learning rate of 0.1. The learning rate is decayed every 30 epochs by a factor of 0.1 and we train for a total of 90 epochs.

Table 1. The number of parameters in Millions over CIFAR100 dataset.

Network Types	WRN-40-2	WRN-16-2	ResNet110	ResNet32
Parameters	2.27M	0.72M	1.74M	0.48M
Network Types	WRN-40-1	WRN-16-1	ResNet56	ResNet20
Parameters	0.57M	0.18M	0.86M	0.28M

Table 2. Comparison with online distillation methods DML [37], ONE [25] and FFL [22] over CIFAR10 in (a) and CIFAR100 in (b) with the same network architecture. '↑' denotes accuracy increases over 'vanilla', 'Avg denotes the average accuracy of Net1 and Net2, and '*' indicates the reported accuracies in [22] under the same network setup.

(a) Top-1 accuracy(%) with the same architecture networks on CIFAR10.

Network Types		vanilla	DML [37]		ONE [25]		FFL [22]		AMLN	
Net1	Net2		Avg	↑	Avg*	↑	Avg*	↑	Avg	↑
ResNet32	ResNet32	93.10	93.15	0.05	93.76	0.66	93.81	0.71	94.25	1.15
ResNet56	ResNet56	93.79	94.19	0.40	94.38	0.59	94.43	0.64	94.68	0.89
WRN-16-2	WRN-16-2	93.58	93.72	0.14	93.76	0.18	93.79	0.21	94.39	0.81
WRN-40-2	WRN-40-2	94.71	95.03	0.32	95.06	0.35	95.17	0.46	95.21	0.50

(b) Top-1 accuracy(%) with the same architecture networks on CIFAR100.

Network Types		vanilla	DML [37]		ONE [25]		FFL [22]		AMLN	
Net1	Net2		Avg	↑	Avg*	↑	Avg*	↑	Avg	↑
ResNet32	ResNet32	69.71	70.98	1.27	72.57	2.86	72.97	3.26	74.69	4.98
ResNet56	ResNet56	71.76	74.13	2.37	74.58	2.82	74.78	3.02	75.77	4.01
WRN-16-2	WRN-16-2	71.41	73.27	1.86	73.95	2.54	74.17	2.76	75.56	4.15
WRN-40-2	WRN-40-2	74.47	76.49	2.02	77.63	3.16	77.77	3.30	77.97	3.50

4.3 Comparisons with the Online Methods

Comparisons over CIFAR: This section presents the comparison of AMLN with state-of-the-art mutual learning methods DML [37], ONE [25] and FFL [22] over CIFAR10 and CIFAR100. Since ONE cannot work for peer networks with different architectures, we evaluate both scenarios when peer networks have the same and different architectures. Tables 2 and Table 3 show experimental results, where 'vanilla' denotes the accuracy of backbone networks that are trained from scratch with classification loss alone, 'Avg' shows the averaged accuracy of the two peer networks Net 1 and Net 2, and the column highlighted with '*' represents the values as extracted from [22] under the same setup.

Case 1: Peer Networks with the Same Architecture. Tables 2(a) and 2(b) show the Top-1 accuracy over the datasets CIFAR10 and CIFAR100, respectively, when peer networks have the same architecture. As Table 2 shows, ONE, DML, and FFL all outperform the 'vanilla' consistently though ONE and FFL achieve larger margins in performance improvement. In addition, AMLN outperforms all three state-of-the-art methods consistently under different network architectures and different datasets. Specifically, the average accuracy improvements (across the four groups of peer networks) over DML, ONE and FFL are up to 0.61%, 0.39% and 0.33% for CIFAR10 and 2.28%, 1.32% and 1.08% for CIFAR100, respectively. Further, it can be observed that the performance improvement over the more challenging CIFAR100 is much larger than that over CIFER10, demonstrating the good scalability and generalizability of ALMN when applied to complex datasets with more image classes.

Table 3. Comparison with online distillation methods DML [37], ONE [25] and FFL [22] over CIFAR10 in (a) and CIFAR100 in (b) with different network architectures.

(a) Top-1 accuracy(%) with different architecture networks on CIFAR10.

Network Types		vanilla		DML		FFL		AMLN	
Net1	Net2	Net1	Net2	Net1	Net2	Net1	Net2	Net1	Net2
WRN-16-2	ResNet32	93.58	93.10	93.91	93.39	94.01	93.99	94.37	94.35
WRN-40-2	ResNet56	94.71	93.79	94.87	93.87	94.89	94.05	94.94	94.39

(b) Top-1 accuracy(%) with different architecture networks on CIFAR100.

Network Types		vanilla*		DML*		FFL*		AMLN	
Net1	Net2	Net1	Net2	Net1	Net2	Net1	Net2	Net1	Net2
WRN-16-2	ResNet32	71.41	69.71	73.55	71.69	74.07	72.94	75.88	74.65
WRN-40-2	ResNet56	74.47	71.76	76.67	73.25	76.94	73.77	76.76	75.29

Case 2: Peer Networks with Different Architectures. This experiment evaluates the peer networks with different architectures WRN-16-2/ResNet32 and WRN-40-2/ResNet56, where the former pair has relatively lower depths. Table 3 shows experimental results. As Table 3(a) shows, the AMLN-trained Net1 and Net2 outperform the same networks trained by 'DML' and 'FFL' consistently on CIFAR10. For CIFAR100, AMLN-trained Net2 achieves significant improvements of 1.71% (WRN-16-2/ResNet32) and 1.52% (WRN-40-2/ResNet56) over the state-of-the-art method FFL as shown in Table 3(b). The good performance is largely attributed to the complementary knowledge distillation with both process-driven learning and outcome-driven learning which empower the peer networks to learn and transfer more multifarious and meaningful features from each other.

Comparisons over ImageNet: To demonstrate the potential of AMLN to transfer more complex information, we conduct a large-scale experiment over the ImageNet LSVRC 2015 classification task. For a fair comparison, we choose the same peer networks of ResNet34 as in ONE [25] and FFL [22]. Table 4 shows experimental results. As Table 4 shows, ONE and FFL achieve similar performance as what is observed over the CIFAR datasets. Our AMLN method performs better consistently, with 1.09% and 1.06% improvements in the Top-1 accuracy as compared with ONE and FFL, respectively. The consistent strong performance over the large-scale dataset ImageNet further demonstrates the scalability of our proposed method.

4.4   Comparisons with the Offline Methods

Several experiments have been carried out to compare AMLN with state-of-the-art offline knowledge transfer methods including AT [34], KD [12], FT [23], as well as the combinations of AT+KD and FT+KD. Among all compared methods,

Table 4. Comparison of Top-1/Top-5 accuracy(%) with online methods ONE [25] and FFL [22] on the ImageNet dataset with the same network architecture (ResNet34). #FLOPs and inference time of each image are also provided.

Method	Top-1(%)	Top-5(%)	#FLOPs	Inference time(per/image)
Vanilla	73.31	91.42	3.67B	1.13×10^{-2} s
ONE	74.39	92.04	4.32B	1.13×10^{-2} s
FFL	74.42	92.05	4.35B	1.13×10^{-2} s
AMLN	75.48	92.54	3.72B	1.13×10^{-2} s

Table 5. Comparison results with offline knowledge transfer methods AT [34], KD [12], FT [23], as well as their hybrid methods AT+KD and FT+KD over CIFAR10 (a) and CIFAR100 (b). The results shown in the last 7 columns are from Table 3 of [23], where the 'vanilla' column represents the performance of the backbone network trained from scratch and the last five columns are the Top-1 accuracy of Net2 under the guidance of Net1.

(a) Comparison results of Top-1 accuracy(%) on CIFAR10.

Network Types		AMLN		vanilla		AT	KD	FT	AT+KD	FT+KD
Net1	Net2	Net1	Net2	Net1	Net2					
WRN-40-1	ResNet20	94.42	93.48	93.16	92.22	92.66	92.91	93.15	93.00	93.05
WRN-16-2	WRN-16-1	94.16	92.92	93.73	91.38	91.90	92.36	92.36	92.48	92.41

(b) Comparison results of Top-1 accuracy(%) on CIFAR100.

Network Types		AMLN		vanilla		AT	KD	FT	AT+KD	FT+KD
Net1	Net2	Net1	Net2	Net1	Net2					
ResNet110	ResNet20	76.12	72.44	73.09	68.76	68.96	66.86	70.92	65.22	67.81
ResNet110	ResNet56	76.74	74.79	73.09	71.06	72.72	72.04	74.38	71.99	73.07

KD adopts an outcome-driven learning strategy and AT and FT adopt process-driven learning strategy.

Tables 5(a) and 5(b) show experimental results over CIFAR10 and CIFAR100, respectively, where Net1 serves as the teacher to empower the student Net2. Three points can be observed from the experimental results: 1) AMLN-trained student Net2 outperforms that trained by all other offline distillation methods consistently for both CIFAR10 and CIFAR100, regardless of whether Net1 and Net2 are of different types (WRN-40-1/ResNet20), having different widths (WRN-16-2/WRN-16-1) or depths (ResNet110/ResNet20, ResNet110/ResNet56); 2) Compared to the 'vanilla' teacher Net1 trained from scratch, AMLN-trained teacher Net1 (mutually learnt with the student Net2) obtains significantly better performance with 0.43%-1.26% and 3.03%-3.65% improvements on CIFAR10 and CIFAR100, respectively. This shows that small networks with fewer parameters or smaller depths can empower larger networks effectively through distilling useful features; and 3) AMLN-trained student Net2 even achieves higher accuracy than its corresponding

Table 6. Ablation study of AMLN with the same peer network ResNet32.

Case	Outcome-driven loss		Process-driven loss		CIFAR10	CIFAR100
	L_C	L_{KL}	L_D	L_F		
A	✓				93.10	69.71
B	✓	✓			93.15	70.98
C	✓	✓		✓	93.89	73.24
D	✓	✓	✓		94.01	74.16
E	✓	✓	✓	✓	94.25	74.69

teacher Net1 in 'vanilla'. Specifically, AMLN-trained ResNet56 (0.86M parameters) produces a better classification accuracy with an improvement of 1.70% than the teacher ResNet110 (1.74M parameters) trained from scratch (in the ResNet110/ResNet56 setup). This shows that a small network trained with proper knowledge distillation could have the same or even better representation capacity than a large network.

4.5 Ablation Study

In AMLN, we have moved one step forward from previous researches by introducing the block-wise module which consists of mutual adversarial learning (MDL) and intermediate-final feature learning (MFL). We perform ablation studies to demonstrate the effectiveness of the proposed method on the datasets CIFAR10 and CIFAT100 by using two identical peer networks ResNet32. Table 6 shows experimental results.

As Table 6 shows, Cases A and E refer to the models trained from scratch and from AMLN, respectively. Case B refers to the network when only the outcome-driven losses L_C (Equ. 5) and L_{KL} (Equ. 7) are included. By including MFL(L_F) in Case C, the averaged accuracy is improved by 0.74% and 2.26% over datasets CIFAR10 and CIFAR100, respectively, as compared with case B. The further inclusion of MDL(L_D) on top of the outcome-driven losses in Case D introduces significant improvements of 0.86% and 3.18% over the datasets CIFAR10 and CIFAR100, respectively. The improvements indicate that MDL has a greater impact on the model performance, which is largely attributed to the convolutional structure of the discriminator that can interpret the spatial information in block-wise intermediate features and map the peer model's features to a similar probability distribution. As expected, AMLN performs the best when both outcome-driven loss and process-driven loss are included for mutual learning. This demonstrates that the two learning strategies are actually complementary to each other in achieving better knowledge distillation and transfer between the collaboratively learning peer networks.

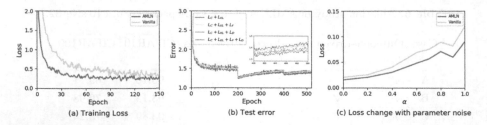

Fig. 3. Analysis of AMLN: The graph in (a) shows the training loss in the first 150 epoch of AMLN and a vanilla model. The graph in (b) shows the testing error under the guidance of different transferring losses. The graph in (c) shows the loss fluctuation when adding parameter noise α during the training of AMLN and a vanilla model.

4.6   Discussion

Benefits of Intermediate Supervision. To evaluate the benefit of combining outcome-driven and process-driven learning in the training procedure, we visualize the training loss (in the first 150 epoch) and test error with the peer networks of ResNet32 on CIFAR100. As illustrated in Fig. 3, our model (the purple line) converges faster than the fully trained vanilla model in Fig. 3(a). Compared to other loss combinations, AMLN (the $L_C + L_{KL} + L_F + L_D$ case) has a relatively lower testing error, especially after 400 epoch. See the zoom-in window for details in Fig. 3(b). In addition, we compare the training loss of the learned models before and after adding Gaussian noise α to model parameters. As shown in Fig. 3(c), the training loss of AMLN increases much less than the independent model after adding the perturbation. These clearly indicate that process-driven learning could improve the model stability and AMLN provides better generalization performance.

Qualitative Analysis. To provide insights on how AMLN contributes to the improved performance consistently, we visualize the heatmaps of learned features after the last convolution layer from four different networks AMLN, FFL, ONE and the vanilla model. We use the Grad-CAM [32] algorithm which works by visualizing the important regions where the network has focused on to discover how our model is taking advantage of the features. Figure 4 shows the Grad-CAM visualizations from each network with the highest probability and the corresponding predicted class. From the first two columns where all the evaluated models predict the correct class, it shows that our AMLN detects the object better with higher rate of confidence. In addition, the last four columns are the cases where AMLN predicts the correct answer but others do not. It again demonstrates the superior performance of our proposed online distillation method AMLN, in which both process-driven and outcome-driven learning effectively complement with each other for multifarious and discriminative feature distillation.

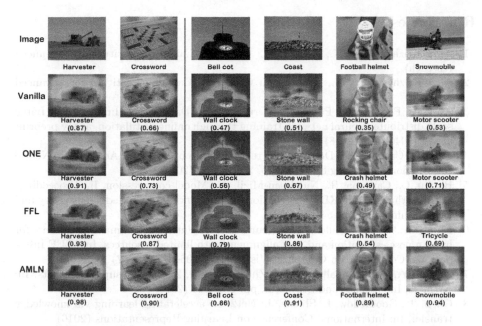

Fig. 4. The comparison of Grad-CAM [32] visualizations of the proposed AMLN with state-of-the-art methods FFL and ONE as well as the vanilla model where the peer networks use the same architecture ResNet32. The label under each heatmap is the corresponding predicted class with the highest prediction probability in the parenthesis.

5   Conclusion

In this paper, a novel online knowledge distillation method is proposed, namely the adversarial-based mutual learning network (AMLN). Unlike existing methods, AMLN employs both process-driven and outcome-driven mutual knowledge distillation, where the former is conducted by the proposed block-wise module with a discriminator and an alignment container for intermediate supervision from the peer network. Extensive evaluations of our proposed AMLN method are conducted on three challenging image classification datasets, where a clear outperformance over the state-of-the-art knowledge transfer methods is achieved. In our future work, we will investigate how to incorporate different tasks to train the peer networks cooperatively, not limited to using the same dataset while mutually training the peer networks as in this work.

Acknowledgements. This work is supported in part by National Science Foundation of China under Grant No. 61572113, and the Fundamental Research Funds for the Central Universities under Grants No. XGBDFZ09.

References

1. Alex Krizhevsky, V.N., Hinton, G.: Cifar-10 (Canadian institute for advanced research)
2. Alex Krizhevsky, V.N., Hinton, G.: Cifar-100 (Canadian institute for advanced research)
3. Anil, R., Pereyra, G., Passos, A., Ormandi, R., Dahl, G.E., Hinton, G.E.: Large scale distributed neural network training through online distillation. arXiv preprint arXiv:1804.03235 (2018)
4. Ba, L.J., Caruana, R.: Do deep nets really need to be deep? In: Advances in Neural Information Processing Systems, pp. 2654–2662 (2013)
5. Bucilu, C., Caruana, R., Niculescu-Mizil, A.: Model compression. In: Proceedings of the 12th ACM SIGKDD International Conference on Knowledge Discovery and Data Mining, pp. 535–541 (2006)
6. Bulat, A., Tzimiropoulos, G.: Binarized convolutional landmark localizers for human pose estimation and face alignment with limited resources. In: IEEE International Conference on Computer Vision, pp. 3706–3714 (2017)
7. Lee, C.-Y., Xie, S., Gallagher, P., Zhang, Z., Tu, Z.: Deeply supervised nets. In: Artificial Intelligence and Statistics, pp. 562–570 (2015)
8. Chen, T., Goodfellow, I., Shlens, J.: Net2net: accelerating learning via knowledge transfer. In: International Conference on Learning Representations (2016)
9. Courbariaux, M., Hubara, I., Soudry, D., Ran, E.Y., Bengio, Y.: Binarized neural networks: Training deep neural networks with weights and activations constrained to $+1$ or -1. arXiv preprint arXiv:1602.02830 (2016)
10. Dai, J., He, K., Sun, J.: Instance-aware semantic segmentation via multi-task network cascades. In: Proceedings of IEEE Conference on Computer Vision and Pattern Recognition, pp. 3150–3158 (2016)
11. Felzenszwalb, P.F., Girshick, R.B., Mcallester, D., Ramanan, D.: Object detection with discriminatively trained part-based models. IEEE Trans. Pattern Anal. Mach. Intell. 1627–1645 (2010)
12. Hinton, G., Vinyals, O., Dean, J.: Distilling the knowledge in a neural network. arXiv preprint arXiv:1503.02531 (2014)
13. Goodfellow, I., et al.: Generative adversarial nets. In: Advances in Neural Information Processing Systems (2014)
14. Han, S., Pool, J., Tran, J., Dally, W.J.: Learning both weights and connections for efficient neural networks. In: Advances in Neural Information Processing Systems, pp. 1135–1143 (2015)
15. Hao, L., Kadav, A., Durdanovic, I., Samet, H., Graf, H.P.: Pruning filters for efficient convnets. arXiv preprint arXiv:1608.08710 (2016)
16. He, K., Zhang, X., Ren, S., Jian, S.: Deep residual learning for image recognition, pp. 770–778 (2016)
17. He, Y., Zhang, X., Jian, S.: Channel pruning for accelerating very deep neural networks. In: IEEE International Conference on Computer Vision, pp. 1389–1397 (2017)
18. Heo, B., Lee, M., Yun, S., Choi, J.Y.: Knowledge transfer via distillation of activation boundaries formed by hidden neurons. In: Proceedings of AAAI Conference on Artificial Intelligence, pp. 3779–3787 (2019)
19. Howard, A.G., et al.: Mobilenets: Efficient convolutional neural networks for mobile vision applications. arXiv preprint arXiv:1704.04861 (2017)

20. Iandola, F.N., Han, S., Moskewicz, M.W., Ashraf, K., Dally, W.J., Keutzer, K.: Squeezenet: Alexnet-level accuracy with 50x fewer parameters and <0.5 mb model size. arXiv preprint arXiv:1602.07360 (2016)
21. Luo, J.-H., Wu, J., Lin, W.: Thinet: a filter level pruning method for deep neural network compression. In: IEEE International Conference on Computer Vision, pp. 5058–5066 (2017)
22. Kim, J., Hyun, M., Chung, I., Kwak, N.: Feature fusion for online mutual knowledge distillation. In: Proceedings of IEEE Conference on Computer Vision and Pattern Recognition (2019)
23. Kim, J., Park, S., Kwak, N.: Paraphrasing complex network: network compression via factor transfer. In: Advances in Neural Information Processing Systems, pp. 2760–2769 (2018)
24. Krizhevsky, A., Sutskever, I., Hinton, G.: Imagenet classification with deep convolutional neural networks. In: Advances in Neural Information Processing Systems, pp. 1097–1105 (2012)
25. Lan, X., Zhu, X., Gong, S.: Knowledge distillation by on-the-fly native ensemble. In: Advances in Neural Information Processing Systems, pp. 7528–7538 (2018)
26. Liu, W., et al.: SSD: single shot multibox detector. In: Leibe, B., Matas, J., Sebe, N., Welling, M. (eds.) ECCV 2016. LNCS, vol. 9905, pp. 21–37. Springer, Cham (2016). https://doi.org/10.1007/978-3-319-46448-0_2
27. Long, J., Shelhamer, E., Darrell, T.: Fully convolutional networks for semantic segmentation. In: Proceedings of IEEE Conference on Computer Vision and Pattern Recognition, pp. 3431–3440 (2015)
28. Molchanov, P., Tyree, S., Karras, T., Aila, T., Kautz, J.: Pruning convolutional neural networks for resource efficient transfer learning. arXiv preprint arXiv:1611.06440 (2016)
29. Rastegari, M., Ordonez, V., Redmon, J., Farhadi, A.: XNOR-Net: imagenet classification using binary convolutional neural networks. In: Leibe, B., Matas, J., Sebe, N., Welling, M. (eds.) ECCV 2016. LNCS, vol. 9908, pp. 525–542. Springer, Cham (2016). https://doi.org/10.1007/978-3-319-46493-0_32
30. Romero, A., Ballas, N., Kahou, S.E., Chassang, A., Bengio, Y.: Fitnets: Hints for thin deep nets. arXiv preprint arXiv:1412.6550 (2014)
31. Russakovsky, O., et al.: Imagenet large scale visual recognition challenge. Int. J. Comput. Vis. 211–252 (2015)
32. Selvaraju, R.R., Cogswell, M., Das, A., Vedantam, R., Parikh, D., Batra, D.: Grad-CAM: visual explanations from deep networks via gradient-based localization. In: Proceedings of IEEE International Conference on Computer Vision, pp. 618–626 (2019)
33. Song, G., Chai, W.: Collaborative learning for deep neural networks. In: Advances in Neural Information Processing Systems, pp. 1837–1846 (2018)
34. Zagoruyko, S., Komodakis, N.: Paying more attention to attention: Improving the performance of convolutional neural networks via attention transfer. arXiv preprint arXiv:1612.03928 (2016)
35. Zagoruyko, S., Komodakis, N.: Wide residual networks. arXiv preprint arXiv:1605.07146 (2016)
36. Zhang, X., Gong, H., Dai, X., Yang, F., Liu, N., Liu, M.: Understanding pictograph with facial features: end-to-end sentence-level lip reading of Chinese. In: Proceedings of the AAAI Conference on Artificial Intelligence, pp. 9211–9218 (2019)
37. Zhang, Y., Xiang, T., Hospedales, T.M., Lu, H.: Deep mutual learning. In: Proceedings of IEEE Conference on Computer Vision and Pattern Recognition, pp. 4320–4328 (2018)

Online Multi-modal Person Search
in Videos

Jiangyue Xia[1], Anyi Rao[2(✉)], Qingqiu Huang[2], Linning Xu[2], Jiangtao Wen[1],
and Dahua Lin[2]

[1] Department of Computer Science and Technology, Tsinghua University,
Beijing, China
`xiajy16@mails.tsinghua.edu.cn,jtwen@tsinghua.edu.cn`
[2] CUHK-SenseTime Joint Lab, The Chinese University of Hong Kong,
Hong Kong, China
`{anyirao,hq016,dhlin}@ie.cuhk.edu.hk,`
`linningxu@link.cuhk.edu.cn`

Abstract. The task of searching certain people in videos has seen
increasing potential in real-world applications, such as video organization and editing. Most existing approaches are devised to work in an
offline manner, where identities can only be inferred after an entire video
is examined. This working manner precludes such methods from being
applied to online services or those applications that require real-time
responses. In this paper, we propose an online person search framework,
which can recognize people in a video on the fly. This framework maintains a multi-modal memory bank at its heart as the basis for person
recognition, and updates it dynamically with a policy obtained by reinforcement learning. Our experiments on a large movie dataset show that
the proposed method is effective, not only achieving remarkable improvements over online schemes but also outperforming offline methods.

Keywords: Online person search · Multi-modality · Dynamic memory
bank · Uncertain instance cache · Reinforcement learning

1 Introduction

Person identification in videos can be specified into different forms and tasks.
Among them, *person search with one portrait* is especially related to real-
world applications, such as "intelligent fast forwards" on online video platforms and multimedia-oriented web search, and can further benefit video summarization and story understanding. This task is very challenging compared
with other person identification problems such as *person Re-ID* [6,8,41] and
person recognition in photo album [19,43], as the appearance, pose, and clothing of the characters may vary dramatically through the videos. To overcome
this difficulty, the research community has explored the use of various modalities [1,3,5,11,16,25,34], such as face, lip motion, body, audio, subtitle, and
screenplay.

© Springer Nature Switzerland AG 2020
A. Vedaldi et al. (Eds.): ECCV 2020, LNCS 12357, pp. 174–190, 2020.
https://doi.org/10.1007/978-3-030-58610-2_11

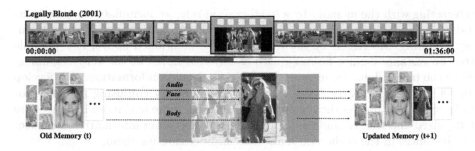

Fig. 1. Illustration of the memory updating scheme of human movie watching experience. We select out instances of *Elle Woods* in movie *Legally Blonde* (2001) and demonstrate how we update our memory about actress *Reese Witherspoon* with them. The multi-modal memory stores face, body and audio information, which are closely related to human identities

However, those methods are mainly offline, *i.e.* an instance is compared with the rest to determine its identity, which leads to high computational complexity. Additionally, for scenarios such as suspect discovery in real-time surveillance videos and story understanding in live broadcasting, the offline approaches cannot recognize the identities immediately. In this paper, we work on *online* person search to meet the emerging requirement of timely inference.

Online search is very challenging, as decisions need to be made on the fly based on limited memory. The key to this problem is to effectively update the memory so that it can adapt to the changes as the video proceeds. Think about how human tackle with online person search. Suppose we are watching the movie *Legally Blonde* (2001), as shown in Fig. 1. When we see the figure of *Elle Woods*, we compare it with previous images stored in our memories to infer the actress's name. There are two possibilities. 1) If the instance appears to be very similar to *Reese Witherspoon*, we recognize her name immediately, and update the impression of *Reese Witherspoon* in our memories with the current looking of *Elle Woods*. Similar processes are also carried out for other cast. When another new instance comes, we continue to compare it with our dynamically updated memory to judge his/her identity. 2) The other possible reaction is that we cannot confirm her identity since her looking is quite different from any cast that exists in our memories. In this case, we stay confused until she appears again and again. We gradually build up our memories on *Elle Woods* and may be capable of recognizing her as *Reese Witherspoon* in the future.

Inspired by this cognitive process, we propose an *online multi-modal searching machine* (OMS). Specifically, to mimic how human recognize characters and store representations in memory, a *dynamic memory bank* is developed to store *face*, *body* and *audio* features of each cast. These multi-modal feature representations are closely related to human identities. The memory bank is dynamically updated to capture the latest changes to the cast's features as new instances come in. To adapt to diverse movie contents and appearance changes, instead of

176 J. Xia et al.

interacting with the memory by a hand-crafted rule, we formulate the process as a decision making problem and design a controller to learn the strategy of memory updating. Motivated by the second case we mentioned above, it is possible that an instance cannot be recognized as any cast in list at the very beginning, since the initial dynamic memory bank lacks adequate information. We develop an *uncertain instance cache* to keep these temporarily confusing instances for judgments later on. As the online process goes on, more and more instances are recognized and the dynamic memory bank becomes more informative, we select out instances in the cache and make a second decision for them.

Experiments are conducted on *Cast Search in Movies* dataset [16] to verify the effectiveness of our online multi-modal searching method. Thanks to the adaptive multi-modal feature integration and reinforcement learning based memory updating strategy, our approach raises the mAP from 61.24% to 69.08% and outperforms all the online methods. Surprisingly, it achieves better results than offline methods and declines computational cost at the same time.

2 Related Work

Person Identification in Videos. In order to identify characters in videos, frameworks using diverse features have been proposed. What commonly used are visual features of face [1] and body [16,17], audio features of speaking voice [25], text features of subtitle [3,11] and screenplay [5,34], and contextual features of scene and social relation [15]. In [5,34], with the alignment of subtitles and screenplay, time-stamped annotations are acquired to provide supervision of character naming. Nagrani *et al.* [25] train face and voice classifiers in a joint framework to recognize characters. With face and body features, Huang *et al.* [16,17] propagate identity labels through visual and temporal links between the instances. However, most previous studies work on an offline manner, *i.e.* all the instances are compared with each other, and the corresponding identities are inferred after an entire video is examined, which increases computational complexity. In this paper, we propose an online framework that dynamically updates the memory with features of newly identified instances to enable real-time inference. Since text information such as subtitles and screenplay is more difficult to acquire compared with the internal features, we utilize face, body and audio features to infer identities.

Multi-modal Fusion. In person identification methods, fusion of visual and audio features can be classified into two categories: late integration [4,28] and early integration [13,14,30,47]. Late integration methods design a specific classifier for each modality and combine decisions by voting or scoring, while early integration merges features from different modalities by concatenation, weighted summation, or learning joint presentations, etc., before decision. Erzin *et al.* [4] determine the reliable modality combinations with a cascade of classifiers. Hu *et al.* [14,30] propose a cross-modality weight sharing LSTM to capture correlation of face and audio features for speaker identification. In this paper, the strategy of multi-modal fusion is learnt implicitly in the decision making process.

Memory Modelling. To strengthen the ability of conventional neural networks in modelling long-range temporal dependencies, several memory models are proposed. Graves *et al.* [9] design a Neural Turing Machine (NTM) which holds an external memory to interact with the neural networks through attentional reading and writing operations. While NTM focuses on problems of sorting, copying and recall, Memory Networks [39] utilize large long-term static external memory and target to language and reasoning tasks. Sukhbaatar *et al.* [35] extend the model to a continuous form to enable end-to-end training, making it more generally applicable to tasks with less supervision. These memory models have also been modified to different structures [18,33] and adopted in video-related researches such as summarization [7,38], captioning [37], visual question answering [24] and object tracking [40]. In this paper, we utilize a dynamic memory bank to store updated multi-modal features of cast in movies.

Reinforcement Learning. Reinforcement learning (RL) is a technique for solving decision making problems, aiming at learning a policy for the sequence of state-action pairs to obtain maximal rewards [36]. In recent years, RL has been applied in person Re-ID [20,26,44] and face recognition [29]. In [26,29,44], RL is used to find the most representative frames in video sequences, while in [20], RL guides an agent to select informative training samples which are used to finetune a pre-trained Re-ID model. In this paper, we formulate the updating of memory as a decision making problem, where we learn the strategy with RL to maximize recognition accuracy.

3 Online Multi-modal Search

Given the portraits of a list of cast, our goal is to search them in a sequential movie with an online fashion following the human behaviors. To tackle this challenging problem, we propose a novel *online multi-modal searching machine* (OMS) as shown in Fig. 2. There are four key components in OMS, *i.e.* multi-modal feature representations (MFR), a dynamic memory bank (DMB), an uncertain instance cache (UIC) and a controller. Each instance is a tracklet and is represented by multi-modal features. It is compared with the cast stored in the memory bank to judge its identity. The controller then determines whether this instance should be used to update memory or put into the uncertain instance cache for later comparisons. The memory bank and the uncertain instance cache are dynamically updated over time, with a strategy operated by the controller. All these components together build an "intelligent machine" to watch a movie and gradually recognize the characters like humans do.

3.1 Multi-modal Feature Representations

When watching a movie, we can identify a person based on various cues, *e.g.* facial appearance, clothing, and even speech. These modalities are complementary to each other. Therefore, it is necessary for us to capture the representations of different modalities for each instance in the movie. Specifically, we take face, body

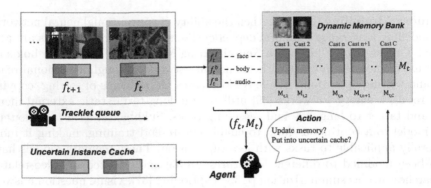

Fig. 2. Pipeline of inference in our proposed OMS. A dynamic memory bank stores the multi-modal feature representations of each actor/actress. When a new instance comes, we compare it with each candidate cast, then the trained agent decides whether to update his memory with this instance or to put it into the uncertain instance cache

and audio information into consideration in our framework. Given an instance x, we represent it with three feature vectors $(f^f(x), f^b(x), f^a(x))$. Here $f^f \in \mathbb{R}^d$ is the face feature that comes from a face recognition model, $f^b \in \mathbb{R}^d$ is the body feature obtained by a Re-ID network, and $f^a \in \mathbb{R}^d$ is the audio feature acquired by a speech recognition model. These feature vectors are concatenated to form a holistic representation $f = [f^f, f^b, f^a] \in \mathbb{R}^{3d}$.

3.2 Dynamic Memory Bank

A simple way to search cast is to calculate the similarity between the given portrait and the detected instances by their face features. However, as the movie proceeds, the appearance of a cast may change dramatically, and a clear face is missing in many cases where the body is partially occluded or even blurred. Humans can tackle this problem easily with the help of memory. Imagine that when you watch a movie, you may not be able to recognize some of the people at the beginning. However, with the playing of the video, you become more and more familiar with the characters as more identified instances enter the memory.

Inspired by the above observation, we construct a *dynamic memory bank* (DMB) $\mathcal{M}_t \in \mathbb{R}^{C \times 3d}$ to store the most representative features of each person. Here $t \in [1, \cdots, N]$ represents the time when the t-th instance appears, N is the total number of instances in a movie and C denotes the number of cast in list. The memory bank is initialized with the features of the provided portrait of each actor/actress. When an instance x_t comes, we search for it in our memory and then predict its identity. The procedure can be formulated as Eq. 1, where f_t is the multi-modal feature representation of x_t.

$$p_t = \mathcal{M}_t \cdot f_t^T \tag{1}$$

As the movie goes by, the DMB keeps updating, with the strategy shown as Eq. 2. Here $\mu \in [0, 1]$ is a pre-defined updating factor. $\mathcal{G}_{t,j}^1 \in \{0, 1\}$ is a gate of the controller, the details of which will be introduced in Sect. 3.4, and $j \in [1, \cdots, C]$ represents the j-th cast.

$$\mathcal{M}_{t+1,j} = (1 - \mu \mathcal{G}_{t,j}^1)\mathcal{M}_{t,j} + \mu \mathcal{G}_{t,j}^1 f_t \tag{2}$$

3.3   Uncertain Instance Cache

At the beginning of a movie, we are not familiar with the characters. Therefore, it may be quite hard for us to recognize some of the tough samples. For example, if a man appears in the first frame of the movie without a visible face, it is impossible for us to identify him at that time. However, as the movie goes on, we begin to know more about the story and the people. We may suddenly recall the uncertain instance before and recognize him with our stronger knowledge.

Motivated by the fact described above, we build a novel module in our machine to store the uncertain instances temporarily, which is named as *uncertain instance cache* (UIC). We denote the cache as $\mathcal{C} \in \mathbb{R}^{k \times 3d}$. k is the size of the cache, which dynamically changes as time goes on. Whether to place an instance x_t into the cache or not is also represented by a gate of the controller, denoted as $\mathcal{G}_t^2 \in \{0, 1\}$, which will be introduced in Sect. 3.4. The updating strategy can be formulated as Eq. 3.

$$\mathcal{C}_k = f_t, \quad k \leftarrow k + 1 \qquad \text{if } \mathcal{G}_t^2 = 1 \tag{3}$$

Whenever the DMB updates, we recall all the instances in the UIC to make new predictions. Specifically, we compare each instance x_i in the cache with the updated memory bank $\mathcal{M}_t$, as shown in Eq. 4. $\mathcal{C}_i$ ($i \in [1, \cdots, k]$) is the multi-modal feature representation of x_i.

$$p_i = \mathcal{M}_t \cdot \mathcal{C}_i^T \tag{4}$$

The p_i here is not the final prediction of the uncertain instance x_i. Whether x_i can be confidently identified and removed from the cache is controlled by the third gate $\mathcal{G}_i^3 \in \{0, 1\}$, the details of which will also be introduced in Sect. 3.4.

3.4   Controller

As we mentioned before, there are three gates, *i.e.* $\mathcal{G}_{t,j}^1, \mathcal{G}_t^2, \mathcal{G}_i^3 \in \{0, 1\}$, in our framework. The three gates determine "whether to update the memory with instance x_t", "whether to put x_t into the uncertain cache", and "whether to remove x_i from the cache", respectively. In this section, we will provide details on how to construct a controller with all these three gates.

A Manual Controller. A simple way is to design the gates by setting thresholds for the prediction, *i.e.* the similarity. Equation 5 shows such a manual controller, where α, β and γ are three pre-defined thresholds. $\mathcal{F}(\Delta t) = \tau \Delta t$ is a regularization function to control the size of the cache. Here Δt is the duration that an instance is stored in the cache and τ is the weight.

$$
\begin{cases}
\mathcal{G}_{t,j}^1 = \mathrm{sgn}(p_{t,j} - \alpha) \\
\mathcal{G}_t^2 = \displaystyle\prod_{j=1}^{C} \mathrm{sgn}(\beta - p_{t,j}) \\
\mathcal{G}_i^3 = 1 - \displaystyle\prod_{j=1}^{C} \mathrm{sgn}(\gamma - \mathcal{F}(\Delta t)p_{i,j})
\end{cases}
, \quad
\mathrm{sgn}(x) = \begin{cases} 1, & \text{if } x >= 0, \\ 0, & otherwise \end{cases}
\tag{5}
$$

A Learnable Controller. Designing the gates according to some manually designed rules will highly reduce the generality. Also, it is hard for us to search for an optimal value of all the hyper parameters. To make our approach more adaptable, we resort to reinforcement learning (RL) to get a learnable controller. RL is characterized by an agent that continuously interacts and learns from the environment through *trial-and-error* games. Its key characteristics include: 1) lack of supervisor, 2) delayed feedback, 3) sequential decisions, and 4) actions affect states, which accord with the peculiarities of our online memory learning setting. Specifically, at each time step, we do not know if updating memory can earn long-term benefits; we observe the instances and make the judgments sequentially; and the updating of our memory will influence future judgments. RL has the potential to find a better policy to replace naive threshold-based strategy. Here, we take $\mathcal{G}_{t,j}^1$ as an example for analysis.

Problem Formulation. The game we teach our agent to play is learning a policy $\mu_\theta(s)$ to decide whether to update the memory bank. For a new instance x_t with feature representation f_t in the sequential movie, we compare it with $\mathcal{M}_t$, and repeat this procedure for each cast $j \in \{1, \cdots, C\}$.

State. State space here is formulated as $\mathcal{S}_t = (\mathcal{M}_t, f_t)$.

Action. Action space here is a one-dimensional discrete space $\{0, 1\}$. If action 1 is taken, we update the memory as Eq. 2.

Reward. Denote the recognition reward at time step t as r_t. If the action is matched with the ground truth label, *i.e.* if x_t is indeed the person j and action 1 is taken, or x_t is not j and action 0 is taken, then the recognition reward at the current time step is $r_t = 1$. Since the effect of the update can only be reflected in future decisions, we define the long-term reward for each action as the cumulative recognition reward in the near future, $R_t = \sum_{m=t}^{t+T} r_m$. We use deep Q-learning network (DQN) to find the improved policy.

The formulation of a learnable $\mathcal{G}_t^2$ is similar to $\mathcal{G}_{t,j}^1$. Note that we do not employ a learnable $\mathcal{G}_i^3$ here. The reason is that $\mathcal{G}_i^3$ is dependent on the samples in the UIC, yet the cache size is quite small and unstable, with which we are not able to train an agent. Through our study, we find that the manual $\mathcal{G}_i^3$ can work well with the other two learned gates. An extensive analysis on the parameters of $\mathcal{G}_i^3$ is provided in the experiment section.

4   Experiments

4.1   Experimental Settings

Data. To validate the effectiveness of our approach, we conduct experiments on the state-of-the-art large-scale *Cast Search in Movies* (CSM) dataset [16]. Extracted from 192 movies, CSM consists of a *query* set that contains the portraits of $1,218$ cast (the actors and actresses) and a *gallery* set that contains $127K$ instances (tracklets). The movies in CSM are split into training, validation and testing sets without any overlap of cast. The training set contains 115 movies with 739 cast and $79K$ instances, while the testing set holds 58 movies with 332 cast and $32K$ instances, and the rest 19 movies are in the validation set.

Evaluation. Given a query with the form of a portrait, our method should present a ranking of all the instances in the gallery to suggest the corresponding possibilities that the instances and the query share a same identity. Therefore, we use *mean Average Precision* (mAP) to evaluate the performance. The training, validation and testing are under the setting of **"per movie"**, *i.e.* given a query, a ranking of instances from only the specific movie will be returned, which is in accordance with real-world applications such as "intelligent fast forwards". Among the 192 movies in CSM, the average size of query and gallery for each movie is 6.4 and 560.5, respectively.

4.2   Implementation Details

Multi-modal Feature Representations. For each instance in CSM, we collect face, body and audio features to facilitate multi-modal person search. The face and body features are extracted for each frame, and averaged to produce the instance-level descriptors. For body feature, we utilize the IDE descriptor [45] extracted by a ResNet-50 [12], which is pre-trained on ImageNet [32] and fine-tuned on the training set of CSM. We detect face region [27,42] and extract face feature with a ResNet-101 trained on MS-Celeb-1M [10]. NaverNet [2] pre-trained with AVA-ActiveSpeaker dataset [31] is applied on the instances to align the characters with their speech audio, which distinguishes a character's voice with the others' as well as background noises. With the proper setting of sampling rate and Mel-frequency cepstral coefficients (MFCC) [21] to reduce the noises, ultimately, each speaking instance is assigned with an audio feature.

Memory Initialization and Update. Recall that the multi-modal memory bank is $\mathcal{M} = \{M_f, M_b, M_a\}$, where M_f is initialized with the face features

Table 1. Person search results on CSM under "per movie" setting

Methods	Online	mAP (%, ↑)	Complexity * (↓)
Face matching	✓	61.24	$\mathcal{O}(NC)$
TwoStep [22] (face+body)		64.79	$\mathcal{O}(NC)$
TwoStep [22] (face+body+audio)		64.40	$\mathcal{O}(NC)$
LP [46]		9.33	$\mathcal{O}(NC + N^2)$
PPCC [16]		67.99	$\mathcal{O}(NC + N^2)$
OMS (DMB w/ manual updating rule)	✓	63.83	$\mathcal{O}(NC)$
OMS-R (DMB w/ RLC)	✓	64.39	$\mathcal{O}(NC)$
OMS-RM (DMB w/ RLC+MFR)	✓	66.42	$\mathcal{O}(NC)$
OMS-RMQ (DMB w/ RLC+MFR+UIC)	✓	**69.08**	$\mathcal{O}(NC + \hat{k}NC)$

* N: number of instances; C: number of cast; $\hat{k}$: average size of UIC

extracted from the IMDb portrait of each actor/actress in the movie, and M_b, M_a are void. The optimal μ in Eq. 2 is set to 0.01 through grid search.

RL Training. The DQN mentioned above is instantiated by a two-layer fully-connected network. The training epoch is 100 with learning rate 0.001. Each epoch is run on the whole movie list, with each movie taking 200 Q-learning iterations. The future reward length is set to be 30. We run the framework on a desktop with a TITAN X GPU.

4.3 Quantitative Results

We compare our method with five baselines: **1) Face matching (online):** The instances are sequentially compared with the cast portraits by face feature similarity, without memory updating. **2) TwoStep (face+body):** After comparisons between face features, instances with high recognition confidence are assigned with identity labels, then a round of body feature comparisons is conducted. **3) TwoStep (face+body+audio):** The second step of comparisons in 2) is based on the combination of body and audio features. **4) LP:** The identities of labeled nodes are propagated to the unlabeled nodes with conventional linear diffusion [46] through multi-modal feature links, where a node updates its probability vector by taking a linear combination of vectors from the neighbors [16]. In addition to face features, the body and audio features are combined for matching, where the weights are 0.9 and 0.1, respectively. **5) PPCC [16]:** Based on the combination of visual and temporal links, the label propagation scheme only spreads identity information when there is high certainty.

Moreover, four variants of our OMS method are compared to validate the influences of different modules. For *DMB with manual updating rule*, only face features are compared between instances and cast in the memory. When the face similarity exceeds a fixed threshold, the memory is updated with the newly recognized instance. The RL-based controller, multi-modal feature representations and UIC are added sequentially to form the other three variants.

We compare different approaches in three aspects: (1) feasibility of online inference; (2) effectiveness measured by mAP; and (3) computational complexity.

The results are presented in Table 1, from which we can see that: 1) Almost all the previous works tackle this problem in an offline manner except for the simple face matching baseline, while OMS can handle the online scenarios. 2) OMS is quite effective, which can even outperform the offline methods significantly. 3) The computational cost of OMS is low. Without UIC, OMS is as efficient as the matching-based methods, *e.g.* face matching. Note that the cache size $\hat{k}$ is usually smaller than N and C is less than 10 here. Therefore, even for the complete version of OMS, *i.e.* OMS-RMQ, the complexity is still lower than the popular propagation-based methods [16,46]. 4) The gradually added components of OMS can continuously raise the performances, which proves the effectiveness of the design for each module. All these results demonstrate that OMS is an effective and efficient framework for person search.

4.4   Ablation Studies

What Is the Behavior of the Framework Along the Time? To discover the behavior of our online framework, we study the development of UIC along the time in our OMS-RMQ method. We select the movies with above 600 instances and record their varied cache sizes at each time step during testing, and average the results among all the movies. The result is presented in Fig. 3 (a). Each time step represents that a decision is made to an instance, and the first 600 steps are shown. The "total size" denotes the cumulative number of uncertain instances that have been put into the cache, while "current size" indicates the number of existing instances therein. It is observed that as time goes, total cache size increases gradually. After processing 600 instances, there are around 100 instances that have ever been put into the UIC. The current cache size raises at the very beginning. After around 350 time steps, it drops gradually to zero. It demonstrates that our DMB becomes better after absorbing informative features to assist recognition, thus more and more uncertain instances get a confident result and are popped out of the cache.

Additionally, we record the cumulative recall of instance identification results, namely R@k, along the time. R@k means the fraction of instances where the correct identity is listed within the top k results. The performance improves gradually with time, as shown in Fig. 3 (b). The R@1 raises from 59% at the beginning to 67%, which proves the effectiveness of our online design.

What Does RL Learn? Recall that in the manual rule setting, we update the memory if the similarity between the memory and the instance is higher than a given threshold. With RL, whether to update or not is decided by the trained agent. To have a deeper understanding of how the RL agent makes the decision and why the RL-trained strategy performs better than manual rules, during testing with OMS-RMQ method, we record the similarity scores on different modalities when an instance is used to update the memory. After regressing the data points into Gaussian distributions as shown in Fig. 3 (c), the mean and

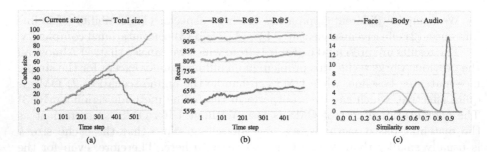

Fig. 3. Ablation studies. (a) The variation of cache size along the time. (b) The development of R@k performance along the time. (c) The distribution of similarity scores of the instances that update the DMB

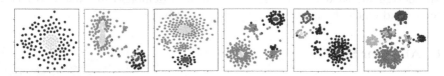

Fig. 4. t-SNE plot of instance features and evolution of memory in 6 movies. Each cluster represents a cast. The "remembered" features in the memory are plotted by light-colored dots, while the instances of a cast are in dark colors. We notice that all the "remembered" features lie at the center of the spread instances. This indicates that the memory absorbs reliable features and well represents the cast's peculiarities (Color figure online)

standard deviation of similarity scores on face, body and audio are 0.89, 0.025 (face), 0.64, 0.063 (body), and 0.46, 0.089 (audio), respectively. The RL agent implicitly adjusts the thresholds of updating memory. Interestingly, the mean values are almost the same with the thresholds we carefully designed before that achieve the highest performance in the manual rule setting.

What Does Memory Learn? To prove the effectiveness of the DMB, we visualize the features of a cast's memory and all his/her ground-truth instances in our OMS-RMQ method using t-SNE [23]. Figure 4 shows cast from 6 movies who have at least 15 memory updates, where each cluster represents a cast. We observe that the updated memory features lie at the center of the instance cluster, which indeed provide typical representations of the cast. This shows that our DMB can accurately capture the characteristics of all his/her possible lookings.

How Do Different Modalities Work? To study how different modalities contribute to the online multi-modal search, *OMS using DMB with RLC and UIC* is taken as the baseline. The results are shown in Table 2. The performance improves when we gradually add a new modality information in. We observe that the introduction of body and audio features brings 3% and 0.5% improvement to the baseline, respectively. With all these modalities together, OMS achieves a

Table 2. Performances of OMS (DMB w/ RLC+UIC) based on different modalities

Method	Face	Body	Audio	mAP (%, ↑)
Face matching (*online*, w/o DMB)	✓			61.24
OMS (DMB w/ RLC+UIC)	✓			64.91
OMS (DMB w/ RLC+UIC)	✓	✓		67.93
OMS (DMB w/ RLC+UIC)	✓		✓	65.39
OMS (DMB w/ RLC+UIC)	✓	✓	✓	**69.08**

Table 3. Performances of OMS (DMB w/ RLC+MFR+UIC) with different cache sizes

Weight τ	0	0.04	0.08	0.12	0.16	0.20
mAP (%, ↑)	66.84	68.92	**69.08**	68.13	65.67	63.69
Mean cache size	199	158	96	63	40	16

4.2% enhancement in recognition precision, which validates that all the modalities are complementary to each other and are informative to the online search.

What Is the Effect of Different UIC Sizes? As we mentioned above in $\mathcal{G}_i^3$, $\mathcal{F}(\Delta t) = \tau \Delta t$ is the regularization function to control the cache size and τ is the weight. A larger weight leads to a smaller cache, and vice versa. We select different weights and show the corresponding performances provided by OMS-RMQ in Table 3. Under each setting, we record the mean cache size of each movie and average the values among all the movies. The average of mean cache size drops from 199 to 16 as the weight raises from 0 to 0.20. The mAP achieves the maximum 69.08% when the weight is 0.08. When the cache is too small, the uncertain instances are not able to benefit from the gradually absorbed knowledge, which causes inferior performance. Since a character is likely to appear again in the movie before long, a medium-sized cache encourages the uncertain instances to match with a neighboring confidently recognized one. Thus, when the mean cache size is 96, the framework achieves the best result.

4.5 Qualitative Results

Which Instances Contribute to the Memory/Are Sent into the UIC? In Fig. 5, we present some sample instances and the corresponding actions given by the agent during inference. The samples demonstrate that *person search with one portrait* is extremely challenging due to varied illumination, sizes, expressions, poses and clothing. During inference, the trained agent successfully selects informative instances which are mostly easier to recognize to update the memory bank, while the instances that contain profile faces, back figures and occlusions are sent into the UIC for later comparisons when more information is acquired.

Method Comparison. In the "per movie" setting, given a portrait as a query, instances are ranked in descending order according to their similarity to the

(a)

(b)

Fig. 5. The sample instances and their corresponding decision making results given by the trained agent. Samples shown above the dash line with green boxes are well-recognized and used to update memory, while those below the dash line with yellow boxes are temporarily put into the UIC. (a) Movie IMDb ID: tt0250494, cast IMDb ID: nm0205127. (b) Movie IMDb ID: tt0250494, cast IMDb ID: nm0000702 (Color figure online)

cast. In Fig. 6, we show some searching results provided by our OMS-R and OMS-RMQ methods. The green bounding boxes represent correct recognition, while the red ones are mistakenly identified. It is shown that with the introduction of UIC and multi-modal features, the recognition accuracy is evidently improved, which is in accordance with the quantitative result that the mAP raises from 64.39% to 69.08%. Even though the rankings of the samples presented are approaching the length of ground-truth instance list, *i.e.* 11-20/22 and 71-80/109, where instances are harder to recognize due to varied poses and face sizes, OMS-RMQ still provides satisfying results.

Fig. 6. Samples searched by different methods, ranked in descending order according to similarity. The green bounding boxes represent correct recognition, and the red ones are mistakenly identified. (a) The 11th–20th searching results of the actor's portrait. Movie IMDb ID: tt0072684, cast IMDb ID: nm0578527. (b) The 71th–80th searching results of the actress's portrait. Movie IMDb ID: tt0129387, cast IMDb ID: nm0000139 (Color figure online)

5 Conclusion

In this paper, we systematically study the challenging problem of *person search in videos with one portrait*. To meet the demand of timely inference in real-world video-related applications, we propose an *online multi-modal searching machine*. Inspired by the cognitive process in movie watching experience, we construct a dynamic memory bank to store multi-modal feature representations of the cast, and develop a controller to determine the strategy of memory updating. An uncertain instance cache is also introduced to temporarily keep unrecognized instances for further comparisons. Experiments show that our method provides remarkable improvements over online schemes and outperforms offline methods.

Acknowledgment. This work is partially supported by the SenseTime Collaborative Grant on Large-scale Multi-modality Analysis (CUHK Agreement No. TS1610626 & No. TS1712093), the General Research Fund (GRF) of Hong Kong (No. 14203518 & No. 14205719), and Innovation and Technology Support Program (ITSP) Tier 2, ITS/431/18F.

References

1. Arandjelovic, O., Zisserman, A.: Automatic face recognition for film character retrieval in feature-length films. In: 2005 IEEE Conference on Computer Vision and Pattern Recognition (CVPR), pp. 860–867 (2005)
2. Chung, J.S.: Naver at ActivityNet challenge 2019-task B active speaker detection (AVA). arXiv preprint arXiv:1906.10555 (2019)
3. Cour, T., Sapp, B., Nagle, A., Taskar, B.: Talking pictures: temporal grouping and dialog-supervised person recognition. In: 2010 IEEE Conference on Computer Vision and Pattern Recognition (CVPR), pp. 1014–1021 (2010)
4. Erzin, E., Yemez, Y., Tekalp, A.M.: Multimodal speaker identification using an adaptive classifier cascade based on modality reliability. IEEE Trans. Multimedia **7**(5), 840–852 (2005)
5. Everingham, M., Sivic, J., Zisserman, A.: "Hello! my name is... Buffy"-automatic naming of characters in TV video. In: 2006 British Machine Vision Conference (BMVC), pp. 899–908 (2006)
6. Farenzena, M., Bazzani, L., Perina, A., Murino, V., Cristani, M.: Person re-identification by symmetry-driven accumulation of local features. In: 2010 IEEE Conference on Computer Vision and Pattern Recognition (CVPR), pp. 2360–2367 (2010)
7. Feng, L., Li, Z., Kuang, Z., Zhang, W.: Extractive video summarizer with memory augmented neural networks. In: 2018 ACM International Conference on Multimedia (MM), pp. 976–983 (2018)
8. Gheissari, N., Sebastian, T.B., Hartley, R.: Person reidentification using spatiotemporal appearance. In: 2006 IEEE Conference on Computer Vision and Pattern Recognition (CVPR), pp. 1528–1535 (2006)
9. Graves, A., Wayne, G., Danihelka, I.: Neural turing machines. arXiv preprint arXiv:1410.5401 (2014)
10. Guo, Y., Zhang, L., Hu, Y., He, X., Gao, J.: MS-Celeb-1M: a dataset and benchmark for large-scale face recognition. In: Leibe, B., Matas, J., Sebe, N., Welling, M. (eds.) ECCV 2016. LNCS, vol. 9907, pp. 87–102. Springer, Cham (2016). https://doi.org/10.1007/978-3-319-46487-9_6
11. Haurilet, M., Tapaswi, M., Al-Halah, Z., Stiefelhagen, R.: Naming TV characters by watching and analyzing dialogs. In: 2016 IEEE Winter Conference on Applications of Computer Vision (WACV), pp. 1–9 (2016)
12. He, K., Zhang, X., Ren, S., Sun, J.: Deep residual learning for image recognition. In: 2016 IEEE Conference on Computer Vision and Pattern Recognition (CVPR), pp. 770–778 (2016)
13. Hu, D., Li, X., Lu, X.: Temporal multimodal learning in audiovisual speech recognition. In: 2016 IEEE Conference on Computer Vision and Pattern Recognition (CVPR), pp. 3574–3582 (2016)

14. Hu, Y., Ren, J.S., Dai, J., Yuan, C., Xu, L., Wang, W.: Deep multimodal speaker naming. In: 2015 ACM International Conference on Multimedia (MM), pp. 1107–1110 (2015)
15. Huang, Q., Xiong, Y., Lin, D.: Unifying identification and context learning for person recognition. In: 2018 IEEE Conference on Computer Vision and Pattern Recognition (CVPR), pp. 2217–2225 (2018)
16. Huang, Q., Liu, W., Lin, D.: Person search in videos with one portrait through visual and temporal links. In: Ferrari, V., Hebert, M., Sminchisescu, C., Weiss, Y. (eds.) ECCV 2018. LNCS, vol. 11217, pp. 437–454. Springer, Cham (2018). https://doi.org/10.1007/978-3-030-01261-8_26
17. Huang, Q., Xiong, Y., Rao, A., Wang, J., Lin, D.: MovieNet: a holistic dataset for movie understanding. In: 2020 European Conference on Computer Vision (ECCV) (2020)
18. Li, D., Kadav, A.: Adaptive memory networks. In: 2018 International Conference on Learning Representations Workshop (ICLRW) (2018)
19. Lin, D., Kapoor, A., Hua, G., Baker, S.: Joint people, event, and location recognition in personal photo collections using cross-domain context. In: Daniilidis, K., Maragos, P., Paragios, N. (eds.) ECCV 2010. LNCS, vol. 6311, pp. 243–256. Springer, Heidelberg (2010). https://doi.org/10.1007/978-3-642-15549-9_18
20. Liu, Z., Wang, J., Gong, S., Lu, H., Tao, D.: Deep reinforcement active learning for human-in-the-loop person re-identification. In: 2019 IEEE International Conference on Computer Vision (ICCV), pp. 6121–6130 (2019)
21. Logan, B.: Mel frequency cepstral coefficients for music modeling. In: 2000 International Symposium on Music Information Retrieval (ISMIR) (2000)
22. Loy, C.C., et al.: Wider face and pedestrian challenge 2018: methods and results. arXiv preprint arXiv:1902.06854 (2019)
23. Maaten, L.V.D., Hinton, G.: Visualizing data using t-SNE. J. Mach. Learn. Res. 9(Nov), 2579–2605 (2008)
24. Na, S., Lee, S., Kim, J., Kim, G.: A read-write memory network for movie story understanding. In: 2017 IEEE International Conference on Computer Vision (ICCV), pp. 677–685 (2017)
25. Nagrani, A., Zisserman, A.: From Benedict Cumberbatch to Sherlock Holmes: character identification in TV series without a script. In: 2017 British Machine Vision Conference (BMVC), pp. 107.1–107.13 (2017)
26. Ouyang, D., Shao, J., Zhang, Y., Yang, Y., Shen, H.T.: Video-based person re-identification via self-paced learning and deep reinforcement learning framework. In: 2018 ACM International Conference on Multimedia (MM), pp. 1562–1570 (2018)
27. Rao, A., et al.: A unified framework for shot type classification based on subject centric lens. In: 2020 European Conference on Computer Vision (ECCV) (2020)
28. Rao, A., et al.: A local-to-global approach to multi-modal movie scene segmentation. In: 2020 IEEE Conference on Computer Vision and Pattern Recognition (CVPR), pp. 10146–10155 (2020)
29. Rao, Y., Lu, J., Zhou, J.: Attention-aware deep reinforcement learning for video face recognition. In: 2017 IEEE International Conference on Computer Vision (ICCV), pp. 3951–3960 (2017)
30. Ren, J.S.J., et al.: Look, listen and learn - a multimodal LSTM for speaker identification. In: 2016 AAAI Conference on Artificial Intelligence (AAAI), pp. 3581–3587 (2016)
31. Roth, J., et al.: AVA-active speaker: an audio-visual dataset for active speaker detection. arXiv preprint arXiv:1901.01342 (2019)

32. Russakovsky, O., et al.: ImageNet large scale visual recognition challenge. Int. J. Comput. Vis. **115**(3), 211–252 (2015)
33. Shen, Y., Tan, S., Hosseini, A., Lin, Z., Sordoni, A., Courville, A.C.: Ordered memory. In: Advances in Neural Information Processing Systems, pp. 5037–5048 (2019)
34. Sivic, J., Everingham, M., Zisserman, A.: "Who are you?" - learning person specific classifiers from video. In: 2009 IEEE Conference on Computer Vision and Pattern Recognition (CVPR), pp. 1145–1152 (2009)
35. Sukhbaatar, S., Szlam, A., Weston, J., Fergus, R.: End-to-end memory networks. In: Advances in Neural Information Processing Systems, pp. 2440–2448 (2015)
36. Sutton, R.S., Barto, A.G.: Reinforcement learning: an introduction. IEEE Trans. Neural Netw. **9**(5), 1054–1054 (1998)
37. Wang, J., Wang, W., Huang, Y., Wang, L., Tan, T.: Hierarchical memory modelling for video captioning. In: 2018 ACM International Conference on Multimedia (MM), pp. 63–71 (2018)
38. Wang, J., Wang, W., Wang, Z., Wang, L., Feng, D., Tan, T.: Stacked memory network for video summarization. In: 2019 ACM International Conference on Multimedia (MM), pp. 836–844 (2019)
39. Weston, J., Chopra, S., Bordes, A.: Memory networks. In: 2015 International Conference on Learning Representations (ICLR) (2015)
40. Yang, T., Chan, A.B.: Learning dynamic memory networks for object tracking. In: Ferrari, V., Hebert, M., Sminchisescu, C., Weiss, Y. (eds.) ECCV 2018. LNCS, vol. 11213, pp. 153–169. Springer, Cham (2018). https://doi.org/10.1007/978-3-030-01240-3_10
41. Zajdel, W., Zivkovic, Z., Krose, B.J.A.: Keeping track of humans: have I seen this person before? In: 2005 IEEE International Conference on Robotics and Automation (ICRA), pp. 2081–2086 (2005)
42. Zhang, K., Zhang, Z., Li, Z., Qiao, Y.: Joint face detection and alignment using multitask cascaded convolutional networks. IEEE Signal Process. Lett. **23**(10), 1499–1503 (2016)
43. Zhang, N., Paluri, M., Taigman, Y., Fergus, R., Bourdev, L.: Beyond frontal faces: improving person recognition using multiple cues. In: 2015 IEEE Conference on Computer Vision and Pattern Recognition (CVPR), pp. 4804–4813 (2015)
44. Zhang, W., He, X., Lu, W., Qiao, H., Li, Y.: Feature aggregation with reinforcement learning for video-based person re-identification. IEEE Trans. Neural Netw. Learn. Syst. **30**(12), 3847–3852 (2019)
45. Zheng, L., et al.: MARS: a video benchmark for large-scale person re-identification. In: 2016 European Conference on Computer Vision (ECCV), pp. 868–884 (2016)
46. Zhou, D., Bousquet, O., Lal, T.N., Weston, J., Schölkopf, B.: Learning with local and global consistency. In: Advances in Neural Information Processing Systems, pp. 321–328 (2003)
47. Zhou, H., Liu, Z., Xu, X., Luo, P., Wang, X.: Vision-infused deep audio inpainting. In: 2019 IEEE International Conference on Computer Vision (ICCV), pp. 283–292 (2019)

Single Image Super-Resolution via a Holistic Attention Network

Ben Niu[1], Weilei Wen[2,3], Wenqi Ren[3], Xiangde Zhang[1], Lianping Yang[1(✉)],
Shuzhen Wang[2], Kaihao Zhang[5], Xiaochun Cao[3,4], and Haifeng Shen[6]

[1] Northeastern University, Shenyang, China
yanglp@mail.neu.edu.cn
[2] Xidian University, Xi'an, China
[3] SKLOIS, IIE, CAS, Beijing, China
[4] Peng Cheng Laboratory, Cyberspace Security Research Center, Shenzhen, China
[5] ANU, Canberra, Australia
[6] AI Labs, Didi Chuxing, Beijing, China

Abstract. Informative features play a crucial role in the single image super-resolution task. Channel attention has been demonstrated to be effective for preserving information-rich features in each layer. However, channel attention treats each convolution layer as a separate process that misses the correlation among different layers. To address this problem, we propose a new holistic attention network (HAN), which consists of a layer attention module (LAM) and a channel-spatial attention module (CSAM), to model the holistic interdependencies among layers, channels, and positions. Specifically, the proposed LAM adaptively emphasizes hierarchical features by considering correlations among layers. Meanwhile, CSAM learns the confidence at all the positions of each channel to selectively capture more informative features. Extensive experiments demonstrate that the proposed HAN performs favorably against the state-of-the-art single image super-resolution approaches.

Keywords: Super-resolution · Holistic attention · Layer attention · Channel-spatial attention

1 Introduction

Single image super-resolution (SISR) is an important task in computer vision and image processing. Given a low-resolution image, the goal of super-resolution (SR)

B. Niu and W. Wen—Equal contribution.

The original version of this chapter was revised: City and country of the first affiliation was corrected from "Boston, USA" to "Shenyang, China". The correction to this chapter is available at https://doi.org/10.1007/978-3-030-58610-2_47

Electronic supplementary material The online version of this chapter (https://doi.org/10.1007/978-3-030-58610-2_12) contains supplementary material, which is available to authorized users.

© Springer Nature Switzerland AG 2020, corrected publication 2020
A. Vedaldi et al. (Eds.): ECCV 2020, LNCS 12357, pp. 191–207, 2020.
https://doi.org/10.1007/978-3-030-58610-2_12

is to generate a high-resolution (HR) image with necessary edge structures and texture details. The advance of SISR will immediately benefit many application fields, such as video surveillance and pedestrian detection.

SRCNN [3] is an unprecedented work to tackle the SR problem by learning the mapping function from LR input to HR output using convolutional neural networks (CNNs). Afterwards, numerous deep CNN-based methods [26,27] have been proposed in recent years and generate a significant progress. The superior reconstruction performance of CNNs based methods are mainly from deep architecture and residual learning [7]. Networks with very deep layers have larger receptive fields and are able to provide a powerful capability to learn a complicated mapping between the LR input and the HR counterpart. Due to the residual learning, the depth of the SR networks are going to deeper since residual learning could efficiently alleviate the gradient vanishing and exploding problems.

Though significant progress have been made, we note that the texture details of the LR image often tend to be smoothed in the super-resolved result since most existing CNN-based SR methods neglect the feature correlation of intermediate layers. Therefore, generating detailed textures is still a non-trivial problem in the SR task. Although the results obtained by using channel attention [2,40] retain some detailed information, these channel attention-based approaches struggle in preserving informative textures and restoring natural details since they treat the feature maps at different layers equally and result in lossing some detail parts in the reconstructed image.

To address these problems, we present a novel approach termed as holistic attention network (HAN) that is capable of exploring the correlations among hierarchical layers, channels of each layer, and all positions of each channel. Therefore, HAN is able to stimulate the representational power of CNNs. Specifically, we propose a layer attention module (LAM) and a channel-spatial attention module (CSAM) in the HAN for more powerful feature expression and correlation learning. These two sub-attention modules are inspired by channel attention [40] which weighs the internal features of each layer to make the network pay more attention to information-rich feature channels. However, we notice that channel attention cannot weight the features from multi-scale layers. Especially the long-term information from the shallow layers are easily weakened. Although the shallow features can be recycled via skip connections, they are treated equally with deep features across layers after long skip connection, hence hindering the representational ability of CNNs. To solve this problem, we consider exploring the interrelationship among features at hierarchical levels, and propose a layer attention module (LAM). On the other hand, channel attention neglects that the importance of different positions in each feature map varies significantly. Therefore, we also propose a channel-spatial attention module (CSAM) to collaboratively improve the discrimination ability of the proposed SR network.

Our contributions in this paper are summarized as follows:

- We propose a novel super-resolution algorithm named Holistic Attention Network (HAN), which enhances the representational ability of feature representations for super-resolution.

- We introduce a layer attention module (LAM) to learn the weights for hierarchical features by considering correlations of multi-scale layers. Meanwhile, a channel-spatial attention module (CSAM) is presented to learn the channel and spatial interdependencies of features in each layer.
- The proposed two attention modules collaboratively improve the SR results by modeling informative features among hierarchical layers, channels, and positions. Extensive experiments demonstrate that our algorithm performs favorably against the state-of-the-art SISR approaches.

2   Related Work

Numerous algorithms and models have been proposed to solve the problem of image SR, which can be roughly divided into two categories. One is the traditional algorithm [11,12,35], the other one is the deep learning model based on neural network [4,15,16,19,22,30,31,41]. Due to the limitation of space, we only introduce the SR algorithms based on deep CNN.

Deep CNN for Super-Resolution. Dong et al. [3] proposed a CNN architecture named SRCNN, which was the pioneering work to apply deep learning to single image super-resolution. Since SRCNN successfully applied deep learning network to SR task, various efficient and deeper architectures have been proposed for SR. Wang et al. [33] combined the domain knowledge of sparse coding with a deep CNN and trained a cascade network to recover images progressively. To alleviate the phenomenon of gradient explosion and reduce the complexity of the model, DRCN [16] and DRRN [30] were proposed by using a recursive convolutional network. Lai et al. [19] proposed a LapSR network which employs a pyramidal framework to progressively generate ×8 images by three sub-networks. Lim et al. [22] modified the ResNet [7] by removing batch normalization (BN) layers, which greatly improves the SR effect.

In addition to above MSE minimizing based methods, perceptual constraints are proposed to achieve better visual quality [28]. SRGAN [20] uses a generative adversarial networks (GAN) to predict high-resolution outputs by introducing a multi-task loss including a MSE loss, a perceptual loss [14], and an adversarial loss [5]. Zhang et al. [42] further transferred textures from reference images according to the textural similarity to enhance textures. However, the aforementioned models either result in the loss of detailed textures in intermediate features due to the very deep depth, or produce some unpleasing artifacts or inauthentic textures. In contrast, we propose a holistic attention network consists of a layer attention and a channel-spatial attention to investigate the interaction of different layers, channels, and positions.

Attention Mechanism. Attention mechanisms direct the operational focus of deep neural networks to areas where there is more information. In short, they help the network ignore irrelevant information and focus on important information [8,9]. Recently, attention mechanism has been successfully applied into deep CNN based image enhancement methods. Zhang et al. [40] proposed a residual channel attention network (RCAN) in which residual channel attention

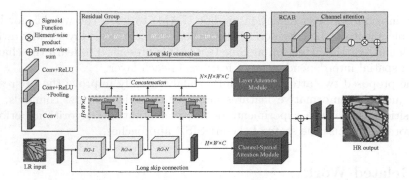

Fig. 1. Network architecture of the proposed holistic attention network (HAN). Given a low-resolution image, the first convolutional layer of the HAN extracts a set of shallow feature maps. Then a series of residual groups further extract deeper feature representations of the low-resolution input. We propose a layer attention module (LAM) to learn the correlations of each output from RGs and a channel-spatial attention module (CSAM) to investigate the interdependencies between channels and pixels. Finally, an upsampling block produces the high-resolution image

blocks (RCAB) allow the network to focus on the more informative channels. Woo et al. [34] proposed channel attention (CA) and spatial attention (SA) modules to exploit both inter-channel and inter-spatial relationship of feature maps. Kim et al. [17] introduced a residual attention module for SR which is composed of residual blocks and spatial channel attention for learning the inter-channel and intra-channel correlations. More recently, Dai et al. [2] presented a second-order channel attention (SOCA) module to adaptively refine features using second-order feature statistics.

However, these attention based methods only consider the channel and spatial correlations while ignore the interdependencies between multi-scale layers. To solve this problem, we propose a layer attention module (LAM) to exploit the nonlinear feature interactions among hierarchical layers.

3   Holistic Attention Network (HAN) for SR

In this section, we first present the overview of HAN network for SISR. Then we give the detailed configurations of the proposed layer attention module (LAM) and channel-spatial attention module (CSAM).

3.1   Network Architecture

As shown in Fig. 1, our proposed HAN consists of four parts: feature extraction, layer attention module, channel-spatial attention module, and the final reconstruction block.

Features Extraction. Given a LR input I_{LR}, a convolutional layer is used to extract the shallow feature F_0 of the LR input

$$F_0 = \mathrm{Conv}(I_{LR}). \tag{1}$$

Then we use the backbone of the RCAN [40] to extract the intermediate features F_i of the LR input

$$F_i = H_{RB_i}(F_{i-1}), \quad i = 1, 2, ..., N, \tag{2}$$

where H_{RB_i} represents the i-th residual group (RG) in the RCAN, N is the number of the residual groups. Therefore, except F_N is the final output of RCAN network backbone, all other feature maps are intermediate outputs.

Holistic Attention. After extracting hierarchical features F_i by a set of residual groups, we further conduct a holistic feature weighting, which includes: i) layer attention of hierarchical features, and ii) channel-spatial attention of the last layer of RCAN.

The proposed layer attention makes full use of features from all the preceding layers and can be represented as

$$F_L = H_{LA}(\text{concatenate}(F_1, F_2, ..., F_N)), \tag{3}$$

where H_{LA} represents the LAM which learns the feature correlation matrix of all the features from RGs' output and then weights the fused intermediate features F_i capitalized on the correlation matrix (see Sect. 3.2). As a results, LAM enables the high contribution feature layers to be enhanced and the redundant ones to be suppressed.

In addition, channel-spatial attention aims to modulate features for adaptively capturing more important information of inter-channel and intra-channel for the final reconstruction, which can be written as

$$F_{CS} = H_{CSA}(F_N), \tag{4}$$

where H_{CSA} represents the CSAM to produce channel-spatial attention for discriminately abtaining feature information, F_{CS} denotes the filtered features after channel-spatial attention (details can be found in Sect. 3.3). Although we can filter all the intermediate features of F_i using CSAM, we only modulate the last feature layer of F_N as a trade-off between accuracy and speed.

Image Reconstruction. After obtaining features from both LAM and CSAM, we integrate the layer attention and channel-spatial attention units by element-wise summation. Then, we employ the sub-pixel convolution [29] as the last upsampling module, which converts the scale sampling with a given magnification factor by pixel translation. We perform the sub-pixel convolution operation to aggregate low-resolution feature maps and simultaneously impose projection to high dimensional space to reconstruct the HR image. We formulate the process as follows

$$I_{SR} = U_{\uparrow}(F_0 + F_L + F_{CS}), \tag{5}$$

Where $U_{\uparrow}$ represents the operation of sub-pixel convolution, and I_{SR} is the reconstructed SR result. The long skip connection is introduced in HAN to stabilize the training of the proposed deep network, $i.e.$, the sub-pixel upsampling block takes $F_0 + F_L + F_{CS}$ as input.

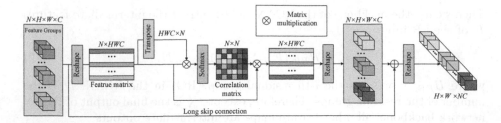

Fig. 2. Architecture of the proposed layer attention module

Loss Function. Since we employ the RCAN network as the backbone of the proposed method, only L_1 distance is selected as our loss function as in [40] for a fair comparison

$$L(\Theta) = \frac{1}{m}\sum_{i=1}^{m}\left\|H_{HAN}(I_{LR}^i) - I_{HR}^i\right\|_1 = \frac{1}{m}\sum_{i=1}^{m}\left\|I_{SR}^i - I_{HR}^i\right\|_1, \qquad (6)$$

Where H_{HAN}, Θ, and m denote the function of the proposed HAN, the learned parameter of the HAN, and the number of training pairs, respectively. Note that we do not use other sophisticated loss functions such as adversarial loss [5] and perceptual loss [14]. We show that simply using the naive image intensity loss $L(\Theta)$ can already achieve competitive results as demonstrated in Sect. 4.

3.2   Layer Attention Module

Although dense connections [10] and skip connections [7] allow shallow information to be bypassed to deep layers, these operations do not exploit interdependencies between the different layers. In contrast, we treat the feature maps from each layer as a response to a specific class, and the responses from different layers are related to each other. By obtaining the dependencies between features of different depths, the network can allocate different attention weights to features of different depths and automatically improve the representation ability of extracted features. Therefore, we propose an innovative LAM that learns the relationship between features of different depths, which automatically improve the feature representation ability.

The structure of the proposed layer attention is shown in Fig. 2. The input of the module is the extracted intermediate feature groups FGs, with the dimension of $N \times H \times W \times C$, from N residual groups. Then, we reshape the feature groups FGs into a 2D matrix with the dimension of $N \times HWC$, and apply matrix multiplication with the corresponding transpose to calculate the correlation $W_{la} = w_{i,j=1}^N$ between different layers

$$w_{i,j} = \delta(\varphi(FG)_i \cdot (\varphi(FG))_j^T), \quad i,j = 1,2,...,N, \qquad (7)$$

where $\delta(\cdot)$ and $\varphi(\cdot)$ denote the softmax and reshape operations, $w_{i,j}$ represents the correlation index between i-th and j-th feature groups. Finally, we multiply

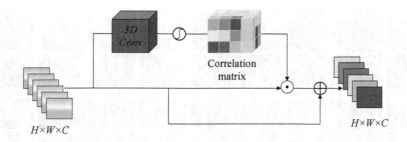

Fig. 3. Architecture of the proposed channel-spatial attention module

the reshaped feature groups FGs by the predicted correlation matrix with a scale factor α, and add the input features FGs

$$F_{L_j} = \alpha \sum_{i=1}^{N} w_{i,j} FG_i + FG_j, \tag{8}$$

where α is initialized to 0 and is automatically assigned by the network in the following epochs. As a result, the weighted sum of features allow the main parts of network to focus on more informative layers of the intermediate LR features.

3.3 Channel-Spatial Attention

The existing spatial attention mechanisms [17,34] mainly focuse on the scale dimension of the feature, with little uptake of channel dimension information, while the recent channel attention mechanisms [2,40,41] ignore the scale information. To solve this problem, we propose a novel channel-spatial attention mechanism (CSAM) that contains responses from all dimensions of the feature maps. Note that although we can perform the CSAM for all the feature groups FGs extracted from RCAN, we only modulate the last feature group of F_N for a trade-off between accuracy and speed as shown in Fig. 1.

The architecture of the proposed CSAM is shown in Fig. 3. Given the last layer feature maps $F_N \in R^{H \times W \times C}$, we feed F_N to a 3D convolution layer [13] to generate correlation matrix by capturing joint channel and spatial features. We operate the 3D convolution via convolving 3D kernels with the cube constructed from multiple neighboring channels of F_N. Specifically, we perform 3D convolutions with kernel size of $3 \times 3 \times 3$ with step size of 1 (*i.e.*, three groups of consecutive channels are convolved with a set of 3D kernels respectively), resulting in three groups of channel-spatial correlation matrix W_{csa}. By doing so, our CSAM can extract powerful representations to describe inter-channel and intra-channel information in continuous channels.

In addition, we perform element-wise multiplication with the correlation matrix W_{csa} and the input feature F_N. Finally, multiply the weighted result by a scale factor β, and then add the input feature F_N to obtain the weighted features

$$F_{CS} = \beta \sigma(W_{csa}) \odot F_N + F_N, \tag{9}$$

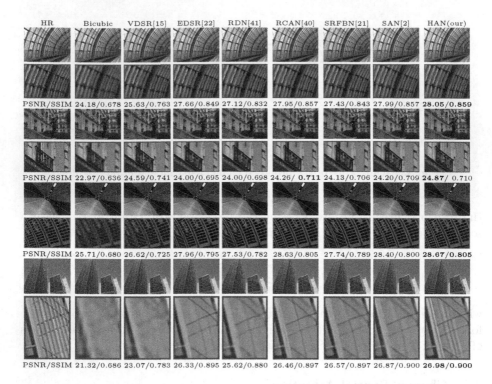

Fig. 4. Visual comparison for 4× SR with BI degradation model on the Urban100 datasets. The best results are highlighted. Our method obtains better visual quality and recovers more image details compared with other state-of-the-art SR methods

where $\sigma(\cdot)$ is the sigmoid function, $\odot$ is the element-wise product, the scale factor β is initialized as 0 and progressively improved in the follow iterations. As a results, F_{CS} is the weighted sum of all channel-spatial position features as well as the original features. Compared with conventional spatial attention and channel attention, our CSAM adaptively learns the inter-channel and intra-channel feature responses by explicitly modelling channel-wise and spatial feature interdependencies.

4   Experiments

In this section, we first analyze the contributions of the proposed two attention modules. We then compare our HAN with state-of-the-art algorithms on five benchmark datasets. The implementation code will be made available to the public. Results on more images can be found in the supplementary material.

Table 1. Effectiveness of the proposed LAM and CSAM for image super-resolution

	Baseline	w/o CSAM	w/o LAM	Ours
PSNR/SSIM	31.22/0.9173	31.38/0.9175	31.28/0.9174	**31.42/0.9177**

Table 2. Ablation study about using different numbers of RGs

	Set5	Set14	B100	Urban100	Manga100
RCAN	32.63	28.87	27.77	26.82	31.22
HAN 3RGs	32.63	28.89	27.79	26.82	31.40
HAN 6RGs	**32.64**	**28.90**	27.79	26.84	**31.42**
HAN 10RGs	**32.64**	**28.90**	**27.80**	**26.85**	**31.42**

4.1 Settings

Datasets. We selecte DIV2K [32] as the training set as like in [2,22,40,41]. For the testing set, we choose five standard datasets: Set5 [1], Set14 [36], B100 [23], Urban100 [11], and Manga109 [24]. Degraded data was obtained by bicubic interpolation and blur-downscale degradation model. Following [40], the reconstruct RGB results by the proposed HAN are first converted to YCbCr space, and then we only consider the luminance channel to calculate PSNR and SSIM in our experiments.

Implementation Details. We implement the proposed network using PyTorch platform and use the pre-trained RCAN ($\times$2), ($\times$3), ($\times$4), ($\times$8) model to initialize the corresponding holistic attention networks, respectively. In our network, patch size is set as 64×64. We use ADAM [18] optimizer with a batch size 16 for training. The learning rate is set as 10^{-5}. Default values of β_1 and β_2 are used, which are 0.9 and 0.999, respectively, and we set $\epsilon = 10^{-8}$. We do not use any regularization operations such as batch normalization and group normalization in our network. In addition to random rotation and translation, we do not apply other data augmentation methods in the training. The input of the LAM is selected as the outputs of all residual groups of RCAN, we use $N = 10$ residual groups in out network. For all the results reported in the paper, we train the network for 250 epochs, which takes about two days on an Nvidia GTX 1080Ti GPU.

4.2 Ablation Study About the Proposed LAM and CSAM

The proposed LAM and CSAM ensure that the proposed SR method generate the feature correlations between hierarchical layers, channels, and locations. One may wonder whether the LAM and CSAM help SISR. To verify the performance of these two attention mechanisms, we compare the method without using LAM and CSAM in Table 1, where we conduct experiments on the Manga109 dataset with the magnification factor of $\times$4.

Table 3. Quantitative results with BI degradation model. The best and second best results are highlighted in **bold** and <u>underlined</u>

Methods	Scale	Set5 PSNR	Set5 SSIM	Set14 PSNR	Set14 SSIM	B100 PSNR	B100 SSIM	Urban100 PSNR	Urban100 SSIM	Manga109 PSNR	Manga109 SSIM
Bicubic	×2	33.66	0.9299	30.24	0.8688	29.56	0.8431	26.88	0.8403	30.80	0.9339
SRCNN [3]	×2	36.66	0.9542	32.45	0.9067	31.36	0.8879	29.50	0.8946	35.60	0.9663
FSRCNN [4]	×2	37.05	0.9560	32.66	0.9090	31.53	0.8920	29.88	0.9020	36.67	0.9710
VDSR [15]	×2	37.53	0.9590	33.05	0.9130	31.90	0.8960	30.77	0.9140	37.22	0.9750
LapSRN [19]	×2	37.52	0.9591	33.08	0.9130	31.08	0.8950	30.41	0.9101	37.27	0.9740
MemNet [31]	×2	37.78	0.9597	33.28	0.9142	32.08	0.8978	31.31	0.9195	37.72	0.9740
EDSR [22]	×2	38.11	0.9602	33.92	0.9195	32.32	0.9013	32.93	0.9351	39.10	0.9773
SRMDNF [38]	×2	37.79	0.9601	33.32	0.9159	32.05	0.8985	31.33	0.9204	38.07	0.9761
D-DBPN [6]	×2	38.09	0.9600	33.85	0.9190	32.27	0.9000	32.55	0.9324	38.89	0.9775
RDN [41]	×2	38.24	0.9614	34.01	0.9212	32.34	0.9017	32.89	0.9353	39.18	0.9780
RCAN [40]	×2	38.27	0.9614	34.12	0.9216	32.41	0.9027	33.34	0.9384	39.44	<u>0.9786</u>
SRFBN [21]	×2	38.11	0.9609	33.82	0.9196	32.29	0.9010	32.62	0.9328	39.08	0.9779
SAN [2]	×2	<u>38.31</u>	**0.9620**	34.07	0.9213	<u>32.42</u>	<u>0.9028</u>	33.10	0.9370	39.32	0.9792
HAN(ours)	×2	38.27	0.9597	<u>34.16</u>	<u>0.9217</u>	32.41	0.9027	<u>33.35</u>	0.9385	<u>39.46</u>	0.9785
HAN+(ours)	×2	**38.33**	<u>0.9617</u>	**34.24**	**0.9224**	**32.45**	**0.9030**	**33.53**	**0.9398**	**39.62**	**0.9787**
Bicubic	×3	30.39	0.8682	27.55	0.7742	27.21	0.7385	24.46	0.7349	26.95	0.8556
SRCNN [3]	×3	32.75	0.9090	29.30	0.8215	28.41	0.7863	26.24	0.7989	30.48	0.9117
FSRCNN [4]	×3	33.18	0.9140	29.37	0.8240	28.53	0.7910	26.43	0.8080	31.10	0.9210
VDSR [15]	×3	33.67	0.9210	29.78	0.8320	28.83	0.7990	27.14	0.8290	32.01	0.9340
LapSRN [19]	×3	33.82	0.9227	29.87	0.8320	28.82	0.7980	27.07	0.8280	32.21	0.9350
MemNet [31]	×3	34.09	0.9248	30.00	0.8350	28.96	0.8001	27.56	0.8376	32.51	0.9369
EDSR [22]	×3	34.65	0.9280	30.52	0.8462	29.25	0.8093	28.80	0.8653	34.17	0.9476
SRMDNF [38]	×3	34.12	0.9254	30.04	0.8382	28.97	0.8025	27.57	0.8398	33.00	0.9403
RDN [41]	×3	34.71	0.9296	30.57	0.8468	29.26	0.8093	28.80	0.8653	34.13	0.9484
RCAN [40]	×3	34.74	0.9299	30.65	0.8482	29.32	0.8111	29.09	0.8702	34.44	0.9499
SRFBN [21]	×3	34.70	0.9292	30.51	0.8461	29.24	0.8084	28.73	0.8641	34.18	0.9481
SAN [2]	×3	34.75	<u>0.9300</u>	30.59	0.8476	<u>29.33</u>	<u>0.8112</u>	28.93	0.8671	34.30	0.9494
HAN(ours)	×3	<u>34.75</u>	0.9299	<u>30.67</u>	<u>0.8483</u>	29.32	0.8110	<u>29.10</u>	<u>0.8705</u>	<u>34.48</u>	<u>0.9500</u>
HAN+(ours)	×3	**34.85**	**0.9305**	**30.77**	**0.8495**	**29.39**	**0.8120**	**29.30**	**0.8735**	**34.80**	**0.9514**
Bicubic	×4	28.42	0.8104	26.00	0.7027	25.96	0.6675	23.14	0.6577	24.89	0.7866
SRCNN [3]	×4	30.48	0.8628	27.50	0.7513	26.90	0.7101	24.52	0.7221	27.58	0.8555
FSRCNN [4]	×4	30.72	0.8660	27.61	0.7550	26.98	0.7150	24.62	0.7280	27.90	0.8610
VDSR [15]	×4	31.35	0.8830	28.02	0.7680	27.29	0.0726	25.18	0.7540	28.83	0.8870
LapSRN [19]	×4	31.54	0.8850	28.19	0.7720	27.32	0.7270	25.21	0.7560	29.09	0.8900
MemNet [31]	×4	31.74	0.8893	28.26	0.7723	27.40	0.7281	25.50	0.7630	29.42	0.8942
EDSR [22]	×4	32.46	0.8968	28.80	0.7876	27.71	0.7420	26.64	0.8033	31.02	0.9148
SRMDNF [38]	×4	31.96	0.8925	28.35	0.7787	27.49	0.7337	25.68	0.7731	30.09	0.9024
D-DBPN [6]	×4	32.47	0.8980	28.82	0.7860	27.72	0.7400	26.38	0.7946	30.91	0.9137
RDN [41]	×4	32.47	0.8990	28.81	0.7871	27.72	0.7419	26.61	0.8028	31.00	0.9151
RCAN [40]	×4	32.63	0.9002	28.87	0.7889	27.77	0.7436	26.82	0.8087	31.22	0.9173
SRFBN [21]	×4	32.47	0.8983	28.81	0.7868	27.72	0.7409	26.60	0.8015	31.15	0.9160
SAN [2]	×4	32.64	<u>0.9003</u>	<u>28.92</u>	0.7888	27.78	0.7436	26.79	0.8068	31.18	0.9169
HAN(ours)	×4	<u>32.64</u>	0.9002	28.90	<u>0.7890</u>	<u>27.80</u>	<u>0.7442</u>	<u>26.85</u>	<u>0.8094</u>	<u>31.42</u>	<u>0.9177</u>
HAN+(ours)	×4	**32.75**	**0.9016**	**28.99**	**0.7907**	**27.85**	**0.7454**	**27.02**	**0.8131**	**31.73**	**0.9207**
Bicubic	×8	24.40	0.6580	23.10	0.5660	23.67	0.5480	20.74	0.5160	21.47	0.6500
SRCNN [3]	×8	25.33	0.6900	23.76	0.5910	24.13	0.5660	21.29	0.5440	22.46	0.6950
FSRCNN [4]	×8	20.13	0.5520	19.75	0.4820	24.21	0.5680	21.32	0.5380	22.39	0.6730
SCN [33]	×8	25.59	0.7071	24.02	0.6028	24.30	0.5698	21.52	0.5571	22.68	0.6963
VDSR [15]	×8	25.93	0.7240	24.26	0.6140	24.49	0.5830	21.70	0.5710	23.16	0.7250
LapSRN [19]	×8	26.15	0.7380	24.35	0.6200	24.54	0.5860	21.81	0.5810	23.39	0.7350
MemNet [31]	×8	26.16	0.7414	24.38	0.6199	24.58	0.5842	21.89	0.5825	23.56	0.7387
MSLapSRN[19]	×8	26.34	0.7558	24.57	0.6273	24.65	0.5895	22.06	0.5963	23.90	0.7564
EDSR [22]	×8	26.96	0.7762	24.91	0.6420	24.81	0.5985	22.51	0.6221	24.69	0.7841
D-DBPN [6]	×8	27.21	0.7840	25.13	0.6480	24.88	0.6010	22.73	0.6312	25.14	0.7987
RCAN [40]	×8	27.31	0.7878	25.23	<u>0.6511</u>	24.88	0.6058	<u>23.00</u>	<u>0.6452</u>	25.24	<u>0.8029</u>
SAN [2]	×8	27.22	0.7829	25.14	0.6476	24.88	0.6011	22.70	0.6314	24.85	0.7906
HAN(ours)	×8	<u>27.33</u>	<u>0.7884</u>	<u>25.24</u>	0.6510	<u>24.98</u>	<u>0.6059</u>	22.98	0.6437	25.20	0.8011
HAN+(ours)	×8	**27.47**	**0.7920**	**25.39**	**0.6552**	**25.04**	**0.6075**	**23.20**	**0.6518**	**25.54**	**0.8080**

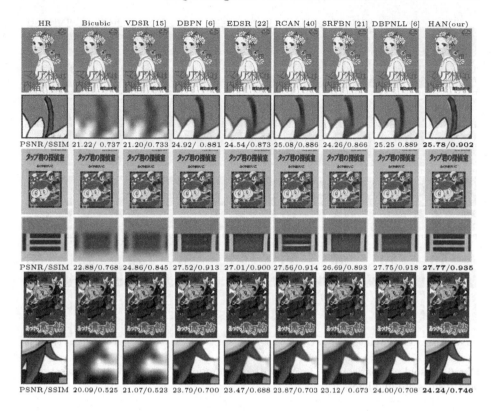

Fig. 5. Visual comparison for 8× SR with BI model on the Manga109 dataset. The best results are highlighted

Table 1 shows the quantitative evaluations. Compared with the baseline method which is identical to the proposed network except for the absence of these two modules LAM and CSAM. CSAM achieves better results by up to 0.06 dB in terms of PSNR, while LAM promotes 0.16 dB on the test dataset. In addition, the improvement of using both LAM and CSAM is significant as the proposed algorithm improves 0.2 dB, which demonstrates the effectiveness of the proposed layer attention and channel-spatial attention blocks. Figure 4 further shows that using the LAM and CSAM is able to generate the results with clearer structures and details.

4.3 Ablation Study About the Number of Residual Group

We conduct an ablation study about feeding different numbers of RGs to the proposed LAM. Specifically, we apply severally three, six, and ten RGs to the LAM, and we evaluate our model on five standard datasets. As shown in Table 2, we compare our three models with RCAN, although using fewer RGs, our algorithm still generates higher PSNR values than the baseline of RCAN. This ablation study demonstrates the effectiveness of the proposed LAM.

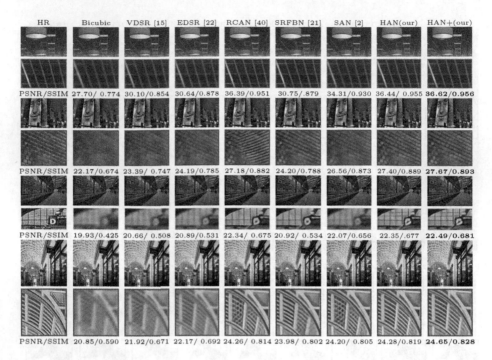

Fig. 6. Visual comparison for 3× SR with BD model on the Urban100 dataset. The best results are highlighted

4.4 Ablation Study About the Number of CSAM

In the paper, the channel-spatial attention module (CSAM) can extract powerful representations to describe inter-channel and intra-channel information in continuous channels. We conduct an ablation study about using different numbers of CSAM. We use one, three, five, and ten CSAMs in RGs. As shown in Table 4, with the increase of CSAM, the values of PSNR are increasing on the testing datasets. This ablation study demonstrates the effectiveness of the proposed CSAM.

Table 4. Ablation study about using different numbers of CSAMs

	Set5	Set14	B100	Urban100	Manga100
HAN(1 CSAM)	32.64	28.90	27.80	26.85	31.42
HAN(3 CSAM)	32.67	28.91	27.80	26.89	**31.46**
HAN(5 CSAM)	**32.69**	28.91	27.80	26.89	31.43
HAN(10 CSAM)	32.67	**28.91**	**27.80**	**26.89**	31.43

Table 5. Quantitative results with BD degradation model. The best and second best results are highlighted in **bold** and underlined

Method	Scale	Set5 PSNR	Set5 SSIM	Set14 PSNR	Set14 SSIM	B100 PSNR	B100 SSIM	Urban100 PSNR	Urban100 SSIM	Manga109 PSNR	Manga109 SSIM
Bicubic	×3	28.78	0.8308	26.38	0.7271	26.33	0.6918	23.52	0.6862	25.46	0.8149
SPMSR [25]	×3	32.21	0.9001	28.89	0.8105	28.13	0.7740	25.84	0.7856	29.64	0.9003
SRCNN [3]	×3	32.05	0.8944	28.80	0.8074	28.13	0.7736	25.70	0.7770	29.47	0.8924
FSRCNN [4]	×3	26.23	0.8124	24.44	0.7106	24.86	0.6832	22.04	0.6745	23.04	0.7927
VDSR [15]	×3	33.25	0.9150	29.46	0.8244	28.57	0.7893	26.61	0.8136	31.06	0.9234
IRCNN [37]	×3	33.38	0.9182	29.63	0.8281	28.65	0.7922	26.77	0.8154	31.15	0.9245
SRMDNF [38]	×3	34.01	0.9242	30.11	0.8364	28.98	0.8009	27.50	0.8370	32.97	0.9391
RDN [41]	×3	34.58	0.9280	30.53	0.8447	29.23	0.8079	28.46	0.8582	33.97	0.9465
RCAN [40]	×3	34.70	0.9288	30.63	0.8462	29.32	0.8093	28.81	0.8647	34.38	0.9483
SRFBN [21]	×3	34.66	0.9283	30.48	0.8439	29.21	0.8069	28.48	0.8581	34.07	0.9466
SAN [2]	×3	34.75	0.9290	30.68	0.8466	29.33	0.8101	28.83	0.8646	34.46	0.9487
HAN(ours)	×3	34.76	0.9294	30.70	0.8475	29.34	0.8106	28.99	0.8676	34.56	0.9494
HAN+(ours)	×3	**34.85**	**0.9300**	**30.79**	**0.8487**	**29.41**	**0.8116**	**29.21**	**0.8710**	**34.87**	**0.9509**

4.5 Results with Bicubic (BI) Degradation Model

We compare the proposed algorithm with 11 state-of-the-art methods: SRCNN [3], FSRCNN [4], VDSR [15], LapSRN [19], MemNet [31], SRMDNF [38], D-DBPN [6], RDN [41], EDSR [22], SRFBN [21] and SAN [2]. We provide more comparisons in supplementary material. Following [2,22,40], we also propose self-ensemble model and donate it as HAN+.

Quantitative Results. Table 3 shows the comparison of $2\times$, $3\times$, $4\times$, and $8\times$ SR quantitative results. Compared to existing methods, our HAN+ performs best on all the scales of reconstructed test datasets. Without using self-ensemble, our network HAN still obtains great gain compared with the recent SR methods. In particular, our model is much better than SAN which also uses the same backbone network of RCAN and has more computationally intensive attention module. Specifically, when we compare the reconstruction results at $\times 8$ scale on the Set5 dataset, the proposed HAN advances 0.11 dB in terms of PSNR than the competitive SAN.

To further evaluate the proposed HAN, we conduct experiments on the large test sets of B100, Urban100, and Manga109. Our algorithm still performs favorably against the state-of-the-art methods. For example, the super-resolved results by the proposed HAN is 0.06 dB and 0.35 dB higher than the very recent work of SAN for the $4\times$ and $8\times$ scales, respectively.

Visual Results. We also show visual comparisons of various methods on the Urban100 dataset for $4\times$ SR in Fig. 4. As shown, most compared SR networks cannot recover the grids of buildings accurately and suffer from unpleasant blurring artifacts. In contrast, the proposed HAN obtains clearer details and reconstructs sharper high-frequency textures.

Take the first and fourth images in Fig. 4 as example, VDSR and EDSR fail to generate the clear structures. The results generated by the recent work of RCAN, SRFBN, and SAN still contain noticeable artifacts caused by spatial aliasing. In contrast, our approach effectively suppresses such artifacts through the proposed two attention modules. As shown, our method accurately reconstructs the grid patterns on windows in the first row and the parallel straight lines on the building in the fourth image.

For 8× SR, we also show the super-resolved results by different SR methods in Fig. 5. As show, it is challenging to predict HR images from bicubic-upsampled input by VDSR and EDSR. Even the state-of-the-art methods of RCAN and SRFBN cannot super-resolve the fine structures well. In contrast, our HAN reconstructs high-quality HR images for 8× results by using cross-scale layer attention and channel-spatial attention modules on the limited information.

4.6   Results with Blur-Downscale Degradation (BD) Model

Quantitative Results. Following the protocols of [37,38,41], we further compare the SR results on images with blur-downscale degradation model. We compare the proposed method with nine state-of-the-art super-resolution methods: SPMSR [25], SRCNN [3], FSRCNN [4], VDSR [15], IRCNN [37], SRMD [39], RDN [41], RCAN [40],SRFBN [21] and SAN [2]. Quantitative results on the 3× SR are reported in Table 5. As shown, both the proposed HAN and HAN+ perform favorably against existing methods. In particular, our HAN+ yields the best quantitative results and HAN obtains the second best scores for all the datasets, 0.06–0.2 dB PSNR better than the attention-based methods of RCAN and SAN and 0.2–0.8 dB better than the recently proposed SRFBN.

Visual Quality. In Fig. 6, we show visual results on images from the Urban 100 dataset with blur-downscale degradation model by a scale factor of 3. Both the full images and the cropped regions are shown for comparison. We find that our proposed HAN is able to recover structured details that were missing in the LR image by properly exploiting the layer, channel, and spatial attention in the feature space.

As shown, VDSR and EDSR suffer from unpleasant blurring artifacts and some results even are out of shape. RCAN alleviate it to a certain extent, but still misses some details and structures. SRFBN and SAN also fail to recover these structured details. In contrast, our proposed HAN effectively suppresses artifacts and exploits the scene details and the internal natural image statistics to super-resolve the high-frequency contents.

5   Conclusions

In this paper, we propose a holistic attention network for single image super-resolution, which adaptively learns the global dependencies among different depths, channels, and positions using the self-attention mechanism. Specifically, the layer attention module captures the long-distance dependencies among hierarchical layers. Meanwhile, the channel-spatial attention module incorporates the channel and contextual information in each layer. These two attention modules are collaboratively applied to multi-level features and then more informative features can be captured. Extensive experimental results on benchmark datasets demonstrate that the proposed model performs favorably against the state-of-the-art SR algorithms in terms of accuracy and visual quality.

Acknowledgements. This work is supported by the National Key R&D Program of China under Grant 2019YFB1406500, National Natural Science Foundation of China (No. 61971016, U1605252, 61771369), Fundamental Research Funds of Central Universities (Grant No. N160504007), Beijing Natural Science Foundation (No. L182057), Peng Cheng Laboratory Project of Guangdong Province PCL2018KP004, and the Shaanxi Provincial Natural Science Basic Research Plan (2019JM-557).

References

1. Bevilacqua, M., Roumy, A., Guillemot, C., Alberi-Morel, M.L.: Low-complexity single-image super-resolution based on nonnegative neighbor embedding. In: BMVC (2012)
2. Dai, T., Cai, J., Zhang, Y., Xia, S.T., Zhang, L.: Second-order attention network for single image super-resolution. In: CVPR (2019)
3. Dong, C., Loy, C.C., He, K., Tang, X.: Learning a deep convolutional network for image super-resolution. In: Fleet, D., Pajdla, T., Schiele, B., Tuytelaars, T. (eds.) ECCV 2014. LNCS, vol. 8692, pp. 184–199. Springer, Cham (2014). https://doi.org/10.1007/978-3-319-10593-2_13
4. Dong, C., Loy, C.C., Tang, X.: Accelerating the super-resolution convolutional neural network. In: Leibe, B., Matas, J., Sebe, N., Welling, M. (eds.) ECCV 2016. LNCS, vol. 9906, pp. 391–407. Springer, Cham (2016). https://doi.org/10.1007/978-3-319-46475-6_25
5. Goodfellow, I., et al.: Generative adversarial nets. In: NIPS (2014)
6. Haris, M., Shakhnarovich, G., Ukita, N.: Deep back-projection networks for super-resolution. In: CVPR (2018)
7. He, K., Zhang, X., Ren, S., Sun, J.: Deep residual learning for image recognition. In: CVPR (2016)
8. Hu, J., Shen, L., Sun, G.: Squeeze-and-excitation networks. In: CVPR (2018)
9. Hu, Y., Li, J., Huang, Y., Gao, X.: Channel-wise and spatial feature modulation network for single image super-resolution. IEEE Trans. Circ. Syst. Video Technol. (2019)
10. Huang, G., Liu, Z., Van Der Maaten, L., Weinberger, K.Q.: Densely connected convolutional networks. In: CVPR (2017)
11. Huang, J.B., Singh, A., Ahuja, N.: Single image super-resolution from transformed self-exemplars. In: CVPR (2015)
12. Huang, S., Sun, J., Yang, Y., Fang, Y., Lin, P., Que, Y.: Robust single-image super-resolution based on adaptive edge-preserving smoothing regularization. TIP **27**(6), 2650–2663 (2018)
13. Ji, S., Xu, W., Yang, M., Yu, K.: 3D convolutional neural networks for human action recognition. TPAMI **35**(1), 221–231 (2012)
14. Johnson, J., Alahi, A., Fei-Fei, L.: Perceptual losses for real-time style transfer and super-resolution. In: Leibe, B., Matas, J., Sebe, N., Welling, M. (eds.) ECCV 2016. LNCS, vol. 9906, pp. 694–711. Springer, Cham (2016). https://doi.org/10.1007/978-3-319-46475-6_43

15. Kim, J., Kwon Lee, J., Mu Lee, K.: Accurate image super-resolution using very deep convolutional networks. In: CVPR (2016)
16. Kim, J., Kwon Lee, J., Mu Lee, K.: Deeply-recursive convolutional network for image super-resolution. In: CVPR (2016)
17. Kim, J.H., Choi, J.H., Cheon, M., Lee, J.S.: Ram: Residual attention module for single image super-resolution. arXiv preprint arXiv:1811.12043 (2018)
18. Kingma, D.P., Ba, J.: Adam: a method for stochastic optimization. arXiv preprint arXiv:1412.6980 (2014)
19. Lai, W.S., Huang, J.B., Ahuja, N., Yang, M.H.: Deep Laplacian Pyramid Networks for fast and accurate super-resolution. In: CVPR (2017)
20. Ledig, C., et al.: Photo-realistic single image super-resolution using a generative adversarial network. In: CVPR (2017)
21. Li, Z., Yang, J., Liu, Z., Yang, X., Jeon, G., Wu, W.: Feedback network for image super-resolution. In: CVPR (2019)
22. Lim, B., Son, S., Kim, H., Nah, S., Mu Lee, K.: Enhanced deep residual networks for single image super-resolution. In: CVPR (2017)
23. Martin, D., Fowlkes, C., Tal, D., Malik, J.: A database of human segmented natural images and its application to evaluating segmentation algorithms and measuring ecological statistics. In: ICCV (2001)
24. Matsui, Y., et al.: Sketch-based manga retrieval using manga109 dataset. Multimedia Tools Appl. **76**(20), 21811–21838 (2016). https://doi.org/10.1007/s11042-016-4020-z
25. Peleg, T., Elad, M.: A statistical prediction model based on sparse representations for single image super-resolution. TIP **23**(6), 2569–2582 (2014)
26. Ren, W., Yang, J., Deng, S., Wipf, D., Cao, X., Tong, X.: Face video deblurring using 3D facial priors. In: Proceedings of the IEEE International Conference on Computer Vision, pp. 9388–9397 (2019)
27. Ren, W., et al.: Deep non-blind deconvolution via generalized low-rank approximation. In: Advances in Neural Information Processing Systems, pp. 297–307 (2018)
28. Sajjadi, M.S., Scholkopf, B., Hirsch, M.: EnhanceNet: single image super-resolution through automated texture synthesis. In: ICCV (2017)
29. Shi, W., et al.: Real-time single image and video super-resolution using an efficient sub-pixel convolutional neural network. In: CVPR (2016)
30. Tai, Y., Yang, J., Liu, X.: Image super-resolution via deep recursive residual network. In: CVPR (2017)
31. Tai, Y., Yang, J., Liu, X., Xu, C.: MemNet: a persistent memory network for image restoration. In: ICCV (2017)
32. Timofte, R., Agustsson, E., Van Gool, L., Yang, M.H., Zhang, L.: NTIRE 2017 challenge on single image super-resolution: methods and results. In: CVPRW (2017)
33. Wang, Z., Liu, D., Yang, J., Han, W., Huang, T.: Deep networks for image super-resolution with sparse prior. In: ICCV (2015)
34. Woo, S., Park, J., Lee, J.Y., So Kweon, I.: CBAM: convolutional block attention module. In: ECCV (2018)
35. Yang, J., Wright, J., Huang, T., Ma, Y.: Image super-resolution as sparse representation of raw image patches. In: CVPR (2008)
36. Zeyde, Roman, Elad, Michael, Protter, Matan: On single image scale-up using sparse-representations. In: Boissonnat, J.-D., et al. (eds.) Curves and Surfaces 2010. LNCS, vol. 6920, pp. 711–730. Springer, Heidelberg (2012). https://doi.org/10.1007/978-3-642-27413-8_47
37. Zhang, K., Zuo, W., Gu, S., Zhang, L.: Learning deep CNN denoiser prior for image restoration. In: CVPR (2017)

38. Zhang, K., Zuo, W., Zhang, L.: Learning a single convolutional super-resolution network for multiple degradations. In: CVPR (2018)
39. Zhang, L., Wu, X.: An edge-guided image interpolation algorithm via directional filtering and data fusion. TIP **15**(8), 2226–2238 (2006)
40. Zhang, Y., Li, K., Li, K., Wang, L., Zhong, B., Fu, Y.: Image super-resolution using very deep residual channel attention networks. In: ECCV (2018)
41. Zhang, Y., Tian, Y., Kong, Y., Zhong, B., Fu, Y.: Residual dense network for image super-resolution. In: CVPR (2018)
42. Zhang, Z., Wang, Z., Lin, Z., Qi, H.: Image super-resolution by neural texture transfer. In: CVPR (2019)

Can You Read Me Now? Content Aware Rectification Using Angle Supervision

Amir Markovitz$^{(\boxtimes)}$, Inbal Lavi, Or Perel, Shai Mazor, and Roee Litman

Amazon Web Services, Seattle, USA
{amirmak,ilavi,orperel,smazor,rlit}@amazon.com

Abstract. The ubiquity of smartphone cameras has led to more and more documents being captured by cameras rather than scanned. Unlike flatbed scanners, photographed documents are often folded and crumpled, resulting in large local variance in text structure. The problem of document rectification is fundamental to the Optical Character Recognition (OCR) process on documents, and its ability to overcome geometric distortions significantly affects recognition accuracy. Despite the great progress in recent OCR systems, most still rely on a pre-process that ensures the text lines are straight and axis aligned. Recent works have tackled the problem of rectifying document images taken in-the-wild using various supervision signals and alignment means. However, they focused on global features that can be extracted from the document's boundaries, ignoring various signals that could be obtained from the document's content.

We present CREASE: Content Aware Rectification using Angle Supervision, the first learned method for document rectification that relies on the document's content, the location of the words and specifically their orientation, as hints to assist in the rectification process. We utilize a novel pixel-wise angle regression approach and a curvature estimation side-task for optimizing our rectification model. Our method surpasses previous approaches in terms of OCR accuracy, geometric error and visual similarity.

1 Introduction

Documents are a common way to share information and record transactions between people. In order to digitize mass amounts of printed documents, the hard copies are scanned and text is extracted automatically by Optical Character Recognition (OCR) systems, such as [11,12]. In the past, most documents were scanned in flatbed scanners. However, the past few years have seen a rise in the use of smartphones, and with it the use of the smartphone camera as a document

---

A. Markovitz and I. Lavi—Equal Contribution.

Electronic supplementary material The online version of this chapter (https://doi.org/10.1007/978-3-030-58610-2_13) contains supplementary material, which is available to authorized users.

© Springer Nature Switzerland AG 2020
A. Vedaldi et al. (Eds.): ECCV 2020, LNCS 12357, pp. 208–223, 2020.
https://doi.org/10.1007/978-3-030-58610-2_13

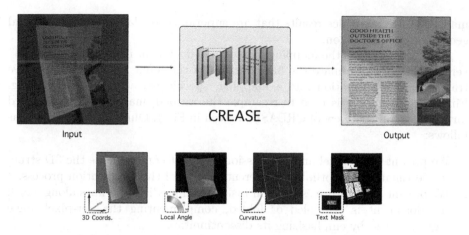

Fig. 1. Overview of CREASE. CREASE is a document rectification method that learns content based signals in addition to the document's 3D structure in order to estimate a transformation used for rectifying the image. On the left is a synthetically generated input image, and on the right is the image rectified using the transformation that CREASE predicted. The bottom images are the supervision signals, from left to right: The 3D coordinates of the warped document, the angle deformation map, the curvature map, and the text mask.

scanner. Camera captured documents such as receipts are often folded, curved, or crumpled, and vary greatly in camera angles, lighting and texture conditions. This makes the OCR task much more challenging compared to scanned images.

Recent OCR methods have had great success in recognizing text in very challenging scenarios. One example is scene text recognition [1,18] which aims to recognize text in natural images. The text is often sparse, and may also be rotated or curved. Another scenario is retrieving the content of a document with *dense* text, that poses the challenge of detecting and recognizing many words that are closely located.

While recognizing dense text, and similarly detecting sparse curved text, had been studied thoroughly, the combined problem of both dense and warped text detection and recognition has received significantly less attention. Many text detectors assume axis-aligned text and struggle with deformed lines [8,26], while text recognition systems struggle with fine deformations on the character level. Taking this into account, a line of works proposed to rectify the document as a pre-process to the recognition phase. Recent methods harnessed the power of deep learning to solve this task [6,13], but put more emphasis on the page boundaries and less emphasis on the contents of the document.

In this paper, we present CREASE: *Content Aware Rectification using Angle Supervision*. This method performs document rectification by relying on both global and local hints, with an emphasis on content. Our method predicts the 3D structure globally while simultaneously optimizing for the local structure of both the text orientation, the location of folds and creases, and the output backward map. CREASE provides results that are superior in readability, similarity and geometric reconstruction.

CREASE predicts the mapping of a warped document image to its "flatbed" version. First, we estimate the 3D structure of the input document. Then, we transform this estimation into a mapping, specifically, a backward mapping. Finally, the mapping is used to resample the warped image into the flattened form. A general overview of CREASE is given in Fig.1. Our contributions are as follows:

1. We present a per-pixel angle regression loss that complements the 3D structure estimation by optimizing different aspects of the rectification process.
2. We present a curvature estimation task, which predicts the lines along which the document is crumpled or folded, complementing the per-pixel angle regression loss by emphasizing its discontinuities.
3. The losses are learned as side tasks, focusing on the areas of the document that contain strong signals regarding the text orientation, and are optimized alongside the 3D structure estimation in an end-to-end optimization process.
4. We reduce the relative OCR error in a challenging warped document dataset by 20.2% and the relative geometric error by 14.1%, compared to the state-of-the-art method.

We train CREASE using synthetic data, which provides us with intricate details regarding each document in our training set without requiring manual annotation: the ground-truth transformation for every pixel, the 3D coordinates, angles, curvature values for every pixel, and the text segmentation mask.

We present visual and quantitative results and comparisons on both synthetic and real evaluation datasets. We also present a detailed study of the contribution of each individual model component.

2   Background

Many works have addressed the problem of extracting text from documents captured in challenging scenarios. These works have focused on different elements that improve OCR accuracy, such as illumination and noise correction [2,16], resolution enhancement [27], and document rectification [6,7,13]. This work focuses on the last problem of rectifying a warped document from a single image by prediction of the 3D model.

Early document rectification methods used hand crafted features to detect the structure of a document. These methods usually made strong assumptions on the deformation process such as smoothness [4,10] and folding structure [7]. Several works utilized special equipment to capture the 3D model of the document [3], or reconstructed the 3D model from multi-view images [25].

More recently, approaches such as [6,13,16] have used deep learning to rectify single-image, camera-captured documents, and were designed to solve the rectification problem by directly predicting the document warp. These works placed the focus on the aforementioned warp, and gave no special treatment to the content of the document, i.e., the data we wish to ultimately recognize.

The first in this line of works is DocUnet [13], which used a stacked hourglass architecture to predict the original 2D coordinates of each pixel in the warped document. This prediction, in essence, gives the forward mapping from the rectified image to the warped one, which can be inverted to get the final result.

A followup work presented DewarpNet [6], which added three learned postprocessing components for calculating the backward map, the surface normals and the shading, each as a separate hourglass network. Additionally, the original 2D forward map of [13] was replaced by a prediction of its 3D counterpart, and the stacked hourglass was substituted by a single hourglass network for this prediction. This method mostly relied on the document boundary, and did not explicitly address the document's content in the rectification module.

Another work by Li et al. [16] focused mainly on uneven background illumination, however it did this by predicting the document warp. This work computed a forward map in a method similar to [13], but divided the prediction into three phases. First, a local, patch-based network predicted the gradients of the forward map. Then, a graph-cut model stitched these patch-predictions into a global warp. Finally, the un-warped image underwent an illumination correction. The local and global level prediction allowed the warp estimation to take into account both the document boundaries and areas in the center of the document, but this pipeline required an expensive patch-stitching process and worked best on input documents with minimal background. Additionally, this method was not end-to-end trainable and did not take into account the document content.

One of the key points in our approach is the importance of predicting text orientation in the warped image. The notion of text angle prediction was previously explored in scene text recognition at the word (or, object) level [17, 19, 28], as opposed to our pixel-level approach.

The EAST text detector by Zhou et al. [28] and FOTS detector by Liu et al. [17] both predicted the angle for each word detection candidate in conjunction with other parameters, like bounding box size and quadrangle coordinates. Ma et al. [19] extended upon the Faster-RCNN [20] architecture by adding rotated anchors to accommodate for arbitrarily oriented text. It is important to stress that scene text methods deal with a sparse set of words, and moreover, each word is rectified separately. Documents, on the other hand, benefit more from a rectification process before word localization, due to the denser text but also due to a stronger prior on the structure thereof. To make use of this prior, CREASE applies an angle regression loss at the pixel level, focused on the salient text areas in the document, and optimized in an end-to-end manner over the predicted backward map.

3   Method

We design CREASE to exploit a document's content and geometry on both local and global levels. CREASE addresses different aspects of the input, such as global

212   A. Markovitz et al.

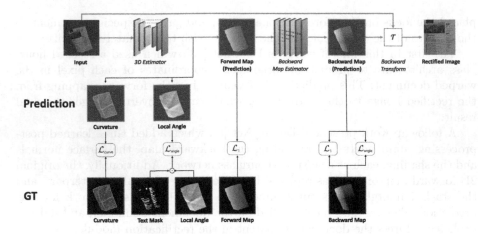

Fig. 2. Architecture of CREASE. The rectification process (in orange arrows) contains two steps: a 3D estimator predicts the 3D coordinates of the document in the image, and a backward map estimator that infers the backward map from the 3D estimation. The input image is rectified using the backward map. Red and blue frames denote ground truth supervisions and predictions, accordingly. Black arrows denote the training process, and the losses used for optimization. Training is performed first on the 3D estimation model, and is then fine-tuned in an end-to-end fashion. (Color figure online)

structure, creases and fold lines, and per-pixel angular deformation. This allows it to capture an accurate mapping for the entire document globally, and for fine-grained features such as characters and words locally (without detecting them explicitly).

First, we present the general architecture of our model (Subsect. 3.1). Next, we present the properties of documents that CREASE relies on: the flow field angles (Subsect. 3.2), and the curvature estimation (Subsect. 3.3). Finally, we present the optimization objective tying the various signals together (Subsect. 3.4).

3.1   Architecture

CREASE is comprised of a two-stage network, illustrated in Fig. 2. The first stage is used for estimating the location of each pixel in a normalized 3D coordinate system, the warp field angle values, and the curvature in each pixel. The 3D estimation module is followed by a backward mapping network. This network transforms the estimated 3D coordinate image into a backward map that can be used for rectifying the input image.

3D Estimation. The first stage provides per-pixel estimation for 3D coordinates (based on [6]) along with the angle and curvature outputs, used as side-tasks. A Unet [21] based architecture is used for mapping an input image into the angle, curvature and 3D coordinate maps. The three maps are used for supervision, and the 3D coordinate map also functions as the input for the backward mapping stage.

Backward Mapping. The second network stage transforms the 3D coordinate map outputted from the first stage to the backward mapping of the image. This mapping describes the transformation from the warped image to the rectified result. In other words, the backward map determines for every pixel in the rectified (output) image domain, its location in the input image.

The authors of [13] used a straightforward implementation of deducing the backward map by inverting their UV forward map prediction. This inversion is done by 'placing' the pixels of the forward map in the rectified image based on their values, and performing interpolation over the resulting non-regular grid[1]. This inversion is very sensitive to noise in the forward map, e.g., if two neighboring pixels swap their predictions. The same task was addressed in [16] using a parallel iterative method that isn't applicable for an end-to-end differential solution.

We rely on the DenseNet [9] based model provided by [6] for the backward mapper in our solution. The work of [6] introduced a learned model for this warp and trained it in a manner independent of the input texture. We rely on their work for transforming our 3D coordinate maps into backward maps. The use of a differentiable backward mapper allows for end-to-end training of the model using a combined objective, optimizing the 3D estimation and backward mapping networks jointly using a combined objective, as discussed in Subsect. 3.4.

3.2 Angle Supervision

While the 3D estimation network learns the global document structure, we wish the network will also be aware of the local angular deformation that each point of the document has undergone during the warping. Angular deformation estimation complements the 3D regression used in [6] because it is more sensitive to small deformations that might warp parts of words. To calculate this value, we warp a local Cartesian system from the source to the target image. We use angular deformation estimation in two places in our framework.

Angle from Backward Map. The first place we estimate the angle is the backward map, where in each pixel we create two infinitesimal vectors ε_x and ε_y, respectively directed at the x and y directions. We then measure the *rotation* that ε_x and ε_y undergo due to the warping process, and denote the resulting angles as θ_x and θ_y. This process is illustrated in Fig. 3. These angles capture the *rotation* and *shear* parts of a local affine transform, without the *translation* and *scale* counterparts that are captured by the coordinate regression.

---

[1] For more details see the Matlab code in [13].

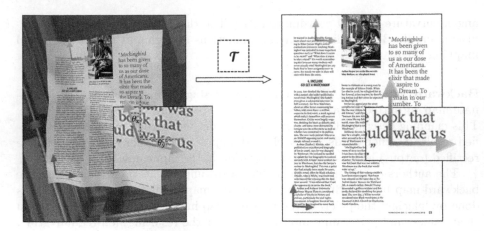

Fig. 3. Per-Pixel Angle Calculation. Illustration of the local warp angle calculation in three locations on the document. The green point is enlarged to provide better details, showing the resulting angles per axis, denoted by θ_x and θ_y. This process is used to penalize for angle prediction errors in the the backward map. A similar process is done for the forward map, by selecting points from the warped image and transferring them using τ. (Color figure online)

Auxiliary Angle Prediction. In addition to deriving the angles from the backward map, we predict them directly from the 3D estimator network as two auxiliary prediction maps. These are learned in parallel to the 3D coordinates prediction, as shown in Fig. 2, to better guide the training, and are not used during test time. Specifically, each of the two angles θ_x and θ_y is derived from its own pair of channels, followed by a Cartesian-to-polar conversion (see supplementary material for details). This conversion yields, in addition to angles θ_x and θ_y, corresponding magnitude values denoted ρ_x and ρ_y. We use the magnitude values as 'angle-confidence' to penalize the angle loss proportionally. This is beneficial since predictions that have small magnitude are more sensitive to small perturbations.

Angle Estimation Loss. We employ a per-pixel angle penalty on the two aforementioned predictions: one as derived from the backward map, and the other as an auxiliary prediction. The per-pixel prediction provided by the 3D estimation network is masked by a binary text segmentation map. Often, the strongest deformations appear around the borders, and far from content. Masking content-less areas allows the loss to target areas of interest, and avoid bias towards the highly deformed boundaries. Our loss minimizes the smallest angle (modulo 2π) between each of the predicted angles $\{\theta_x, \theta_y\}$ and their ground truth counterparts $\{\hat{\theta}_x, \hat{\theta}_y\}$. The per-pixel loss for angles is therefore:

$$L_{angle}(\boldsymbol{\theta}, \hat{\boldsymbol{\theta}}, \hat{\boldsymbol{\rho}}) = \sum_{i \in \{x,y\}} \hat{\rho}_i \odot (\|\theta_i - \hat{\theta}_i\| - \pi) \mod 2\pi, \qquad (1)$$

where $\odot$ denotes the Hadamard product. In the backward map angles the loss is without the confidence values $\hat{\rho}$, that are set to **1**, because the angles are derived from the backward map and are not predicted as an auxiliary.

3.3 Curvature Estimation

A key observation we utilize in this work is that the surface of a crumpled document behaves in manner similar to a 2D piecewise-planar surface. Each interface between two approximately-planar sections introduces a section of higher curvature, and higher local distortion. We wish to give the network a supervision signal that indicates the presence of such high curvature.

Intuitively, the more crumpled the paper, the more creases or discontinuities the warp function exhibits. The *curvature map* highlights non-planar areas of the paper, where 3D and angle regression might be less accurate. A point in the middle of a plane would have zero curvature, while a point at the tip of a needle would have the maximal curvature value. To generate this signal, we utilize the 3D mesh used to generate each document image.

Formally, for a paper mesh $\mathcal{M}$ we calculate a curvature map $H(\mathcal{M})$ using the Laplace-Beltrami operator, as defined for meshes in [23]. The mean curvature per mesh vertex $\mathbf{v}_i \in \mathbb{R}^3$ is obtained by:

$$H(\mathcal{M})_i = || \sum_{j \in N_i} (\mathbf{v}_i - \mathbf{v}_j)||_2. \tag{2}$$

The maps used for supervision are created by thresholding the curvature, to avoid noise and slight perturbations while emphasizing the actual lines defining the global deformation for the paper. The maps are used as supervision and are predicted as an additional segmentation mask by the 3D estimation network.

3.4 Optimization

The optimization of our model consists of two stages: An initial training stage for the 3D estimation network using the side-tasks, followed by an end-to-end fine tuning stage in which the network is optimized w.r.t. a combined loss term.

3D Estimation Model. Initially, we optimize the 3D estimation model using a loss objective that includes the 3D coordinate estimation loss and the aforementioned auxiliary losses, described in Eq. (3). We denote the predicted and ground-truth normalized world coordinates $\mathbf{C}$ and $\hat{\mathbf{C}}$. The first loss term is the L_1 loss over coordinate error, similarly to [6]. The second term is the angle loss term presented in Eq. (1), masked by the binary text segmentation mask $\hat{\mathbf{D}}$, averaged over all text containing pixels. The last term is the curvature estimation L_2 loss. The 3D estimation loss is:

$$L_{3D} = ||\mathbf{C} - \hat{\mathbf{C}}||_1 + \hat{\mathbf{D}} \odot L_{angle} + ||\mathbf{H} - \hat{\mathbf{H}}||_2. \tag{3}$$

End-to-end Fine Tuning. The first stage of our model may either be trained individually, or as part of an end-to-end architecture. When training the model end-to-end, the backward map B is inferred and used for penalizing the predicted 3D coordinates by the final result. We penalize the resulting backward map $\hat{B}$ by the L_1 loss as was done in [6], and additionally using our angle loss from Eq. (1). We append these penalty terms to the one in Eq. (3), resulting in the following combined end-to-end loss:

$$L_{combined} = L_{3D} + ||\mathbf{B} - \hat{\mathbf{B}}||_1 + L_{angle}. \tag{4}$$

4 Experiments

We evaluate CREASE on a new evaluation set comprised of 50 high resolution synthetic images, as well as on real images from the evaluation set proposed by [13]. The synthetic dataset is generated with both warp and text annotations, useful for OCR based evaluations and for evaluating the individual stages of our model. We provide geometric, visual, and OCR based metrics, as well as qualitative evaluations. We compare our results to Dewarpnet [6] trained on our training set, using the code and parameters that were published by the authors. As the method of Li et al. [16] isn't directly applicable to the task at hand, it is not evaluated in this section. Discussion and comparison to [16] are provided in the supplementary material.

All models were trained on 15,000 high resolution images rendered using an extension of the rendering pipeline provided by [6]. Our extensions include the generation of our supervision signals: text, curvature and angles in addition to the 3D coordinates provided by the original rendering pipeline. These added signals come at negligible cost and have no affect in test time. Further details regarding dataset generation are provided in the supplementary material.

4.1 Evaluation Metrics

OCR Based Metric. To correctly evaluate any word related metric, we must first obtain a set of aligned word location pairs, i.e., a matching polygon for each ground-truth bounding box in the predicted rectified image domain. Given the density and small scale of words in documents, a naive coordinate matching scheme is likely to fail, as a small global shift is to be expected even in the best case scenario.

During evaluation, we rectify an input image twice: using the network's predicted backward map, and using the ground-truth map. We then use an OCR engine for extracting words and bounding boxes from the rectified images.

To properly match bounding boxes, we perform the matching stage in the input image domain, visualized in Fig. 4. Each bounding box extracted from a rectified image is warped back and becomes a polygon in the input (warped) image domain.

We define polygon intersection as our distance metric and match pairs using the Hungarian algorithm [14]. With the paired prediction and ground-truth word

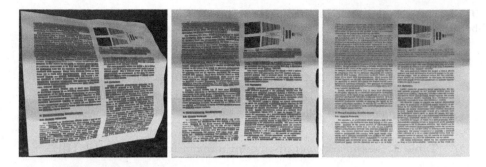

Fig. 4. OCR Polygon matching. Frames overlayed with OCR bounding boxes. Left: Input image overlayed with the warped polygons from images rectified using a predicted transformation (center) and the ground-truth transformation (right). Purple areas denote a correct match, blue and red areas were detected only in the ground-truth and prediction rectified images, respectively. (Color figure online)

boxes we can evaluate the *Levenshtein distance* [15], or edit distance, denoted by E_d. We first calculate the edit distance for each word in each document, then calculate the average edit distance over all the words in the dataset.

Following [6], we use an off-the-shelf OCR engine (Tesseract 4.0 [22]). This engine is quite basic, and does not reflect the advances and robustness of more modern OCR models. However, the vast majority of recent OCR methods are targeted at scene-text, with the number of proposed text instance detections often limited to 100-200. Thus, they are not suited to handle *dense* document text. As an alternative, there are a few commercial products designed to handle dense text recognition [11,12] that are far more advanced than Tesseract. We choose one of them, [11], for an additional evaluation. Results are presented in Tables 1 and 2.

Geometric and Visual Metrics. In addition to an OCR-based evaluation, we use two metrics for evaluating the geometric correctness and visual similarity of our results, *End Point Error (EPE)* and Multi-Scale Structural Similarity (MS-SSIM). The EPE metric is used to evaluate the calculated rectification warps and compare them to ground truth. Following [16], we include evaluation for this metric in our benchmark.

The MS-SSIM [24] metric quantifies how visually similar are the output images to the ground truth. Given that a small amount of shift is expected and is not considered an error, a naive evaluation using L_1 or L_2 metrics is not suited for our evaluation. Therefore, following [6] we use the MS-SSIM metric which focuses on statistical measures rather than per-pixel color accuracy. Evaluating statistics rather than per-pixel accuracy also has its limitations, as character level rectification is a fine-grained task and improvements on this scale are not always manifested in this metric. In fact, SSIM is much more sensitive to small visual deformations in documents containing large amounts of text or

Table 1. Benchmark comparison using Tesseract OCR [22]. For E_d and EPE, lower is better, while or SSIM, higher is better.

	$\downarrow E_d$	$\downarrow$ EPE	$\uparrow$ SSIM
DewarpNet [6]	0.223 ± 0.014	0.051 ± 0.001	0.403 ± 0.004
Ours	0.178 ± 0.003	0.043 ± 0.002	0.411 ± 0.002
Improvement	+20.2%	+14.1%	+1.4%

Table 2. Benchmark comparison using a commercial OCR model [11]

	$\downarrow E_d$
DewarpNet [6]	0.109 ± 0.005
Ours	0.103 ± 0.001
Improvement	+5.1%

sharp edges. Thus, we only use it to complement our finer-grained, OCR based metrics. For further discussion regarding the SSIM metric, see supplementary.

4.2   Implementation Details

Models are trained by first optimizing the 3D estimation network using 3D coordinates, text masks, curvature masks and local angle supervision signals to convergence. Starting from the converged 3D estimation models, we fine-tune our model in an end-to-end manner by using a fixed, pre-trained, differentiable backward mapper. We calculate the L_1 and angle losses over the output backward maps and back-propagate the losses to the 3D estimation network. Training is conducted using 15,000 high-resolution images rendered in Blender [5] using over 8,000 texture images. Further details regarding data generation are provided in the supplementary material.

4.3   Comparison to DewarpNet [6]

The first result we present is a comparison to the prior state-of-the-art trained on our training set, using the Tesseract [22] engine.

We show mean and standard deviation values over 5 experiments in Table 1. Our method improves the edit distance metric over the previous method by 4.5% absolute and 20.2% relative. We also see improvements in EPE and SSIM metrics, and a reduction in standard deviation for all three. The use of both angle regression and curvature estimation improves performance and stabilizes the optimization process, reducing the sensitivity to model initialization.

Next, we evaluate our method using the public online API of [11]. Results are presented in Table 2. In this case, our model still provides a 5.1% relative improvement. The commercial model [11] is superior to [22], reducing the mean

Table 3. Angle loss evaluation for the 3D estimation model.

Model	$\downarrow E_d$	$\downarrow$ EPE	$\uparrow$ SSIM
Vanilla	0.324 ± 0.169	0.066 ± 0.029	0.390 ± 0.024
Angles	0.246 ± 0.017	0.049 ± 0.002	0.403 ± 0.005
+ Mask	0.244 ± 0.014	0.052 ± 0.001	0.398 ± 0.005
+ Conf.	0.216 ± 0.017	0.049 ± 0.001	0.400 ± 0.005

Table 4. Ablation study.

	$\downarrow E_d$	$\downarrow$ EPE	$\uparrow$ SSIM
Vanilla	0.324 ± 0.169	0.066 ± 0.029	0.390 ± 0.024
Angles	0.216 ± 0.017	0.049 ± 0.001	0.400 ± 0.005
Angles + Curvature	0.187 ± 0.005	0.043 ± 0.001	0.409 ± 0.007
E2E	0.223 ± 0.014	0.051 ± 0.001	0.403 ± 0.004
E2E + Angles	0.204 ± 0.015	0.051 ± 0.002	0.402 ± 0.005
E2E + Angles + Curvature	0.178 ± 0.003	0.043 ± 0.002	0.411 ± 0.002

edit distance from 0.178 to 0.103, yet CREASE still maintains a significant gap over DewarpNet of 0.6% absolute and 5.1% relative.

4.4 Evaluation Using Real World Images

Figure 5 depicts a qualitative comparison between our rectification method and [6] on the real images provided by [13]. Notice how the text lines rectified using CREASE are better aligned and easier to read than the other method's outputs, especially for text near document edges. Additional examples are included in the supplementary material.

4.5 Angle Loss Evaluation

We show the contribution of the different elements of our angle-based loss presented in Sect. 3.2 for our metrics and for the OCR metric in particular in Table 3. 'Angles' refers to models trained with the angle loss applied to all image pixels, instead of only to those that contain text. '+ Mask' refers to applying the text mask over the loss, i.e., taking the loss only in text-containing pixels, using the mask denoted by $\hat{D}$ in Eq. (3). '+ Conf.' represents the use of the angle confidence values (denoted ρ in Eq. (1)). When not used, we set ρ to 1 for all pixels. We report results averaged over 5 experiments each, as well as the standard deviation. For this experiment, the curvature estimation term was omitted. Our contributions show a consistent improvement over the vanilla 3D estimation network and, in addition, a much more stable training framework with consistent results over multiple initializations.

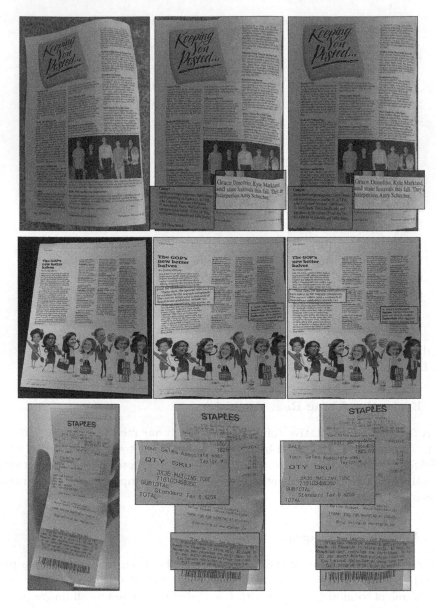

Fig. 5. Visual Examples. Results from the real image dataset of [13]. Left to right: input images, rectification of the input image according to output of [6] trained using our data, rectification of the input image according to our model's output transformation.

4.6   Ablation Study

Table 4 shows the effect of each component of our method. Models trained using angle and curvature estimation are compared to vanilla models. We compare both models trained end-to-end (denoted *E2E*) and models trained separately. As seen before, the improvement in results is also accompanied by a decrease in standard deviation, especially for models trained using curvature estimation.

We evaluate the contribution of end-to-end training of our model using a fixed, differentiable backward mapper and losses derived from its results, i.e., the backward map and angle prediction errors (shown in Table 4). The top three rows refer to models that were not trained in an end-to-end fashion, while the three rows below (starting with 'E2E') refer to models trained end-to-end. 'Angles' and 'Curvature' denote the use of each of our two added auxiliary predictions.

The dual usage of the angle loss, in both the 3D estimation model and the end-to-end training, as well as the curvature estimation, result in much more readable rectification and a more stable training scheme than the previous state-of-the-art.

5 Conclusion

We presented CREASE, a content aware document rectification method which optimizes a per-pixel angle regression loss, a curvature estimation loss and a 3D coordinate estimation loss for providing image rectification maps.

Our method rectifies folded and creased documents using hints found in both local and global scale properties of the document, and provides a significant improvement in OCR performance, geometry and visual similarity based metrics. In our proposed two stage model, the first stage is used for predicting 3D structure, angles and curvature, while the second stage predicts the backward map. We utilize a pixel-level angle regression loss that is shown to be a beneficial side-task in both the 3D estimation and the end-to-end training. Furthermore, our 3D estimation model learns the angle side-task specifically on the words in the document, thus optimizing for readability in the rectified image, while the curvature estimation side-task complements the angle regression by mapping its discontinuities.

Extensive testing and comparisons show our method's superior performance over diverse inputs, using both real and synthetic evaluation data. We show an increase in OCR performance, geometry and similarity metrics that is consistent over all experiments and on a variety of documents.

References

1. Baek, J., et al.: What is wrong with scene text recognition model comparisons? dataset and model analysis. In: International Conference on Computer Vision (ICCV) (2019, to appear)
2. Bajjer Ramanna, V.K., Bukhari, S.S., Dengel, A.: Document image dewarping using deep learning. In: The 8th International Conference on Pattern Recognition Applications and Methods, International Conference on Pattern Recognition Applications and Methods (ICPRAM-2019), 19–21 February Prague, Czech Republic. Insticc (2019)

3. Brown, M.S., Seales, W.B.: Image restoration of arbitrarily warped documents. IEEE Trans. Pattern Anal. Mach. Intell. **26**, 1295–1306 (2004)
4. Burden, A., Cote, M., Albu, A.B.: Rectification of camera-captured document images with mixed contents and varied layouts. In: 2019 16th Conference on Computer and Robot Vision (CRV), pp. 33–40. IEEE (2019)
5. Community B.O.: Blender - a 3D modelling and rendering package. Blender Foundation, Stichting Blender Foundation, Amsterdam (2018). http://www.blender.org
6. Das, S., Ma, K., Shu, Z., Samaras, D., Shilkrot, R.: DewarpNet: single-image document unwarping with stacked 3D and 2D regression networks. In: The IEEE International Conference on Computer Vision (ICCV), October 2019
7. Das, S., Mishra, G., Sudharshana, A., Shilkrot, R.: The common fold: utilizing the four-fold to dewarp printed documents from a single image. In: Proceedings of the 2017 ACM Symposium on Document Engineering, DocEng 2017, pp. 125–128. Association for Computing Machinery, New York (2017). https://doi.org/10.1145/3103010.3121030
8. Grüning, T., Leifert, G., Strauß, T., Michael, J., Labahn, R.: A two-stage method for text line detection in historical documents. Int. J. Docu. Anal. Recogn. (IJDAR) **22**(3), 285–302 (2019). https://doi.org/10.1007/s10032-019-00332-1
9. Huang, G., Liu, Z., Van Der Maaten, L., Weinberger, K.Q.: Densely connected convolutional networks. In: Proceedings of the IEEE Conference on Computer Vision and Pattern Recognition, pp. 4700–4708 (2017)
10. Huang, Z., Gu, J., Meng, G., Pan, C.: Text line extraction of curved document images using hybrid metric. In: 2015 3rd IAPR Asian Conference on Pattern Recognition (ACPR), pp. 251–255, November 2015. https://doi.org/10.1109/ACPR.2015.7486504
11. Amazon Inc.: Amazon textract. https://aws.amazon.com/textract
12. Google Inc.: Detect text in images. https://cloud.google.com/vision/docs/ocr
13. Ma, K., Shu, Z., Bai, X., Wang, J., Samaras, D.: DocUNet: document image unwarping via a stacked U-Net. In: Proceedings of IEEE Conference on Computer Vision and Pattern Recognition (2018)
14. Kuhn, H.W.: The Hungarian method for the assignment problem. Nav. Res. logist. Q. **2**(1–2), 83–97 (1955)
15. Levenshtein, V.I.: Binary codes capable of correcting deletions, insertions, and reversals
16. Li, X., Zhang, B., Liao, J., Sander, P.V.: Document rectification and illumination correction using a patch-based CNN. ACM Trans. Graph. (TOG) **38**(6), 1 (2019)
17. Liu, X., Liang, D., Yan, S., Chen, D., Qiao, Y., Yan, J.: FOTS: fast oriented text spotting with a unified network. In: 2018 IEEE/CVF Conference on Computer Vision and Pattern Recognition, June 2018
18. Lyu, P., Liao, M., Yao, C., Wu, W., Bai, X.: Mask TextSpotter: an end-to-end trainable neural network for spotting text with arbitrary shapes. In: Ferrari, V., Hebert, M., Sminchisescu, C., Weiss, Y. (eds.) Computer Vision – ECCV 2018. LNCS, vol. 11218, pp. 71–88. Springer, Cham (2018). https://doi.org/10.1007/978-3-030-01264-9_5
19. Ma, J., et al.: Arbitrary-oriented scene text detection via rotation proposals. IEEE Trans. Multimedia **20**(11), 3111–3122 (2018)
20. Ren, S., He, K., Girshick, R., Sun, J.: Faster R-CNN: towards real-time object detection with region proposal networks. In: Advances in Neural Information Processing Systems, vol. 28, pp. 91–99. Curran Associates Inc. (2015)

21. Ronneberger, O., Fischer, P., Brox, T.: U-Net: convolutional networks for biomedical image segmentation. In: Navab, N., Hornegger, J., Wells, W.M., Frangi, A.F. (eds.) MICCAI 2015. LNCS, vol. 9351, pp. 234–241. Springer, Cham (2015). https://doi.org/10.1007/978-3-319-24574-4_28
22. Smith, R.: An overview of the tesseract OCR engine. In: Ninth International Conference on Document Analysis and Recognition (ICDAR 2007), vol. 2, pp. 629–633. IEEE (2007)
23. Sorkine-Hornung, O.: Laplacian mesh processing. In: Eurographics (2005)
24. Wang, Z., Bovik, A.C., Sheikh, H.R., Simoncelli, E.P.: Image quality assessment: from error visibility to structural similarity. IEEE Trans. Image Process. **13**(4), 600–612 (2004)
25. You, S., Matsushita, Y., Sinha, S., Bou, Y.B., Ikeuchi, K.: Multiview rectification of folded documents. IEEE Trans. Pattern Anal. Mach. Intell. **40**, 505–511 (2016)
26. Yousef, M., Bishop, T.E.: OrigamiNet: weakly-supervised, segmentation-free, one-step, full page text recognition by learning to unfold. In: Proceedings of the IEEE/CVF Conference on Computer Vision and Pattern Recognition (2020)
27. Zheng, Y., Kang, X., Li, S., He, Y., Sun, J.: Real-time document image super-resolution by fast matting. In: 2014 11th IAPR International Workshop on Document Analysis Systems, pp. 232–236. IEEE (2014)
28. Zhou, X., et al.: East: an efficient and accurate scene text detector. In: 2017 IEEE Conference on Computer Vision and Pattern Recognition (CVPR), July 2017

Momentum Batch Normalization
for Deep Learning with Small Batch Size

Hongwei Yong[1,2], Jianqiang Huang[2], Deyu Meng[3,4], Xiansheng Hua[2],
and Lei Zhang[1,2(✉)]

[1] Department of Computing, The Hong Kong Polytechnic University,
Kowloon, Hong Kong, China
{cshyong,cslzhang}@comp.polyu.edu.hk
[2] DAMO Academy, Alibaba Group, Hangzhou, China
jianqiang.jqh@gmail.com, huaxiansheng@gmail.com
[3] Macau University of Science and Technology, Macau, China
[4] School of Mathematics and Statistics, Xi'an Jiaotong University, Xi'an, China
dymeng@mail.xjtu.edu.cn

Abstract. Normalization layers play an important role in deep network
training. As one of the most popular normalization techniques, batch
normalization (BN) has shown its effectiveness in accelerating the model
training speed and improving model generalization capability. The success of BN has been explained from different views, such as reducing
internal covariate shift, allowing the use of large learning rate, smoothing optimization landscape, etc. To make a deeper understanding of BN,
in this work we prove that BN actually introduces a certain level of noise
into the sample mean and variance during the training process, while
the noise level depends only on the batch size. Such a noise generation
mechanism of BN regularizes the training process, and we present an
explicit regularizer formulation of BN. Since the regularization strength
of BN is determined by the batch size, a small batch size may cause the
under-fitting problem, resulting in a less effective model. To reduce the
dependency of BN on batch size, we propose a momentum BN (MBN)
scheme by averaging the mean and variance of current mini-batch with
the historical means and variances. With a dynamic momentum parameter, we can automatically control the noise level in the training process.
As a result, MBN works very well even when the batch size is very small
(e.g., 2), which is hard to achieve by traditional BN.

Keywords: Batch normalization · Small batch size · Noise ·
Momentum

Electronic supplementary material The online version of this chapter (https://
doi.org/10.1007/978-3-030-58610-2_14) contains supplementary material, which is
available to authorized users.

A. Vedaldi et al. (Eds.): ECCV 2020, LNCS 12357, pp. 224–240, 2020.
https://doi.org/10.1007/978-3-030-58610-2_14

1   Introduction

During the past decade, deep neural networks (DNNs) have achieved remarkable success in a variety of applications, such as image classification [14], object detection [13,31], speech recognition [1], natural language processing [28] and computer games [29,36], etc. The success of DNNs comes from the advances in higher computing power (e.g., GPUs), large scale datasets [10], and learning algorithms [11,20,37]. In particular, advanced network architecture [14,15] and optimization techniques [20,22] have been developed, making the training of very deep networks from a large amount of training data possible.

One of the key issues in DNN training is how to normalize the training data and intermediate features. It is well-known that normalizing the input data makes training faster [22]. The widely used batch normalization (BN) technique [19] naturally extends this idea to the intermediate layers within a deep network by normalizing the samples in a mini-batch during the training process. It has been validated that BN can accelerate the training speed, enable a bigger learning rate, and improve the model generalization accuracy [14,15]. BN has been adopted as a basic unit in most of the popular network architectures such as ResNet [14] and DenseNet [15]. Though BN has achieved a great success in DNN training, how BN works remains not very clear. Researchers have tried to explain the underlying working mechanism of BN from different perspectives. For example, it is argued in [19] that BN can reduce internal covariate shift (ICS). However, it is indicated in [33] that there is no clear link between the performance gain of BN and the reduction of ICS. Instead, it is found that BN makes the landscape of the corresponding optimization problem smoother so that it allows larger learning rates, while stochastic gradient decent (SGD) with a larger learning rate could yield faster convergence along the flat direction of the optimization landscape so that it is less likely to get stuck in sharp minima [5].

Apart from better convergence speed, another advantage of BN is its regularization capability. Because the sample mean and variance are updated on mini-batches during training, their values are not accurate. Consequently, BN will introduce a certain amount of noise, whose function is similar to dropout. It will, however, increase the generalization capability of the trained model. This phenomenon has been empirically observed from some experimental results in [43,44]. Teye et al. [27,38] tried to give a theoretical explanation of the generalization gain of BN from a Bayesian perspective; however, it needs additional assumptions and priors, and the explanation is rather complex to understand.

In this paper, we present a simple noise generation model to clearly explain the regularization nature of BN. Our explanation only assumes that the training samples are independent and identically distributed (i.i.d.), which holds well for the randomly sampled mini-batches in the DNN training process. We prove that BN actually introduces a certain level of noise into the sample mean and variance, and the noise level only depends on the batch size. When the training batch size is small, the noise level becomes high, increasing the training difficulty. We consequently propose a momentum BN (MBN) scheme, which can automatically control the noise level in the training process. MBN can work stably for different mini-batch sizes, as validated in our experiments on benchmark datasets.

2 Related Work

Batch Normalization: BN [19] was introduced to address the internal covariate shift (ICS) problem by performing normalization along the batch dimension. For a layer with d-dimensional input $\mathbf{x} = (x^{(1)}, x^{(2)}, ..., x^{(d)})$ in a mini-batch $\mathcal{X}_B$ with size m, BN normalizes each dimension of the input samples as:

$$\widehat{x}^{(k)} = \frac{x^{(k)} - \mu_{\mathcal{B}}^k}{\sqrt{\sigma_{\mathcal{B}}^{k^2} + \epsilon}} \tag{1}$$

where $\mu_{\mathcal{B}}^k = \frac{1}{m} \sum_{i=1}^{m} x_i^{(k)}$, $\sigma_{\mathcal{B}}^k = \frac{1}{m} \sum_{i=1}^{N} (x_i^{(k)} - \mu_{\mathcal{B}}^k)^2$, and ϵ is a small positive constant. And for inference step, the mean and variance of mini-batch are replaced with that of population, often estimated by moving average.

BN has achieved remarkable performance in terms of improving training speed and model generalization ability for many applications [14,45]. In the case of small batch size, unfortunately, the sample mean and variance can be very different from those of the population. Consequently, BN may not perform well with a small batch size. To address this problem, batch renormalization (BReN) [18] was proposed by constraining the range of estimated mean and variance of a batch. However, it is very hard to tune the hyper-parameters in BReN, which limits its application to different tasks.

Other Normalization Methods: Besides BN, other normalization methods [3,23,30,32,39,43] have been proposed to normalize data along other dimensions. For example, layer normalization (LN) [23] normalizes all activations or feature maps along feature dimension, instance normalization (IN) [39] performs normalization for each feature map of each sample and group normalization (GN) [43] normalizes feature maps for each input sample in a divided group. Although these methods depend less on training batch size, BN still outperforms them in many visual recognition tasks. Switchable normalization [25,26,34] uses a weighted combination of BN, LN and IN, which introduces more parameters and costs more computation. There are also some variants of BN, such as Batch Kalman Norm (BKN) [41], L1 BN [42], Decorrelated BN (DBN) [16], Riemannian BN [8], Iterative BN [17], Memorized BN [12] etc. Instead of operating on features, Weight Normalization (WN) [32] normalizes the filter weights. WN can also accelerate training but cannot outperform BN.

3 The Regularization Nature of BN

3.1 Noise Generation of BN

Several previous works [43,44] have indicated that the BN layer can enhance the generalization capability of DNNs experimentally; however, little work has

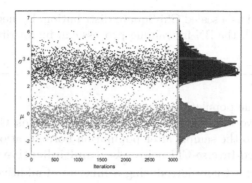

Fig. 1. The mean and variance of mini-batches vs. iterations. The mean (green points) and variance (blue points) are from one channel of the first BN layer of ResNet18 in the last epoch when training with batch size 16 on CIFAR100. The histograms of batch mean and variance are plotted on the right, which can be well fitted by Gaussian distribution and Chi-square distribution, respectively. (Color figure online)

been done on the theoretical analysis about why BN has this capability. The only work we can find is [38], where Teye et al. tried to give a theoretical illustration for the generalization gain of BN from a Bayesian perspective with some additional priors. In the work [27], Luo et al. presented a regularization term based on the result of [38]. Shekhovtsov et al. [35] gave an interpretation of BN from the perspective of noise generation. However, it is assumed that the input activations follows strictly i.i.d. Gaussian distribution and there is no further theoretical analysis on how the noise affects the training process. In this section, we theoretically shows that BN can be modeled as process of noise generation.

Let's first assume that one activation input in a layer follows the Gaussian distribution $\mathcal{N}(x|\mu, \sigma^2)$, where μ and σ^2 can be simply estimated by population mean $\mu_{\mathcal{P}}$ and variance $\sigma_{\mathcal{P}}$ of training data. This assumption can be extended to more general cases other than Gaussian distribution, as we will explain later. In stochastic optimization [6,7], randomly choosing a mini-batch of training samples can be considered as a sample drawing process, where all samples x_i in a mini-batch $\mathcal{X}_b = \{x_i\}_{i=1}^m$ are i.i.d., and follows $\mathcal{N}(x|\mu, \sigma^2)$. For the mini-batch $\mathcal{X}_b$ with mean $\mu_B = \frac{1}{m}\sum_{i=1}^m x_i$ and variance $\sigma_B^2 = \frac{1}{m}\sum_{i=1}^m (x_i - \mu_B)^2$, we can define two random variables ξ_μ and ξ_σ as follows [9]:

$$\xi_\mu = \frac{\mu - \mu_B}{\sigma} \sim \mathcal{N}(0, \frac{1}{m}), \quad \xi_\sigma = \frac{\sigma_B^2}{\sigma^2} \sim \frac{1}{m}\chi^2(m-1) \tag{2}$$

where χ^2 denotes the Chi-squared distribution and ξ_σ follows a Scaled-Chi-squared distribution with $E(\xi_\sigma) = \frac{m-1}{m}$ and $Var(\xi_\sigma) = \frac{2(m-1)}{m^2}$.

In Fig. 1 we plot the means and variances of mini-batches computed at the first BN layer of ResNet18 in the last training epoch when training with batch size 16 on CIFAR100 dataset. One can see that these means and variances are distributed like biased random noise. Specifically, the histogram of mean values can be well modeled as a Gaussian distribution, while the histogram of variances can be well modeled as a scaled Chi-Square distribution. By neglecting the small constant ϵ in Eq. (1), the BN in training process can be rewritten as

$$\widehat{x} = \frac{x - \mu_B}{\sigma_B} = \frac{x - \mu + (\mu - \mu_B)}{\sigma \frac{\sigma_B}{\sigma}} = \frac{\frac{x-\mu}{\sigma} + \xi_\mu}{\sqrt{\xi_\sigma}} = \frac{\widetilde{x} + \xi_\mu}{\sqrt{\xi_\sigma}} \tag{3}$$

where $\widetilde{x} = \frac{x-\mu}{\sigma}$ is the population normalized formula.

From Eq. (3), we can see that BN actually first adds Gaussian noise ξ_μ (**additive noise**) to the sample after population normalization, and then multiplies with a Scaled-Inverse-Chi noise $\frac{1}{\sqrt{\xi_\sigma}}$ (**multiplicative noise**). That is, training with BN is actually introducing a mixture of additive and multiplicative noise. With the introduced additive noise ξ_μ and multiplicative noise ξ_σ, the output variable $\widehat{x}$ follows $Nt(\widetilde{x}, m-1)$, which is a noncentral t-distribution [24], and its probability density function is very complex. Fortunately, we can still get the mean and variance of $\widehat{x}$ as follows:

$$E[\widehat{x}] = \widetilde{x}\sqrt{\frac{m-1}{2}}\frac{\Gamma((m-2)/2)}{\Gamma((m-1)/2)}, \quad Var[\widehat{x}] = \frac{1}{m}(\frac{m-1}{m-3}(1+\widetilde{x}^2) - E[\widehat{x}]^2) \tag{4}$$

When m is very large, $E[\widehat{x}] \approx \widetilde{x}$ and $Var[\widehat{x}] \approx 0$. However, when m is small, the noise generated by BN depends on not only the statistics of entire training data $\mathcal{X}$ (e.g., mean μ and variance σ^2) but also the batch size m.

With the above analyses, we can partition BN into three parts: a normalizer part (i.e., $\widetilde{x} = \frac{x-\mu}{\sigma}$); a noise generator part (i.e., $\widehat{x} = \frac{\widetilde{x}+\xi_\mu}{\sqrt{\xi_\sigma}}$) ; and an affine transformation part (i.e., $y = \gamma\widehat{x} + \beta$). In the training stage, only the noise generator part is related to batch size m. In the inference stage, the batch mean and variance are replaced with population mean and variance, and thus BN only has the normalizer part and the affine transformation part. It should be emphasized that μ and σ are unknown in training, and they also vary during the training process. At the end of training and when statistics for activations of all samples are stable, they can be viewed as fixed.

Now we have shown that the BN process actually introduces noises ξ_μ and ξ_σ into the BN layer in the training process. When the batch size is small, the variances of both additive noise ξ_μ and multiplicative noise ξ_σ become relatively large, making the training process less stable. In our above derivation, it is assumed that the activation input follows the Gaussian distribution. However, in practical applications the activations may not follow exactly the Gaussian distribution. Fortunately, we have the following theorem.

Theorem 1: *Suppose samples x_i for $i = 1, 2, ..., m$ are i.i.d. with $E[x] = \mu$ and $Var[x] = \sigma^2$, ξ_μ and ξ_σ are defined in Eq. (2), we have:*

$$\lim_{m \to \infty} p(\xi_\mu) \to \mathcal{N}(0, \frac{1}{m}), \quad \lim_{m \to \infty} p(\xi_\sigma) \to \frac{1}{m}\chi^2(m-1).$$

Theorem 1 can be easily proved by the central limit theorem. Please refer to the **Supplementary Materials** for the detailed proof. In particular, when m

is larger than 5, ξ_μ and ξ_σ nearly meet the distribution assumptions. As for the i.i.d. assumption on the activations of samples in a mini-batch, it generally holds because the samples are randomly drawn from the pool in training.

3.2 Explicit Regularization Formulation

It has been verified in previous works [43,44] that introducing a certain amount of noise into training data can increase the generalization capability of the neural network. However, there lacks a solid theoretical analysis on how this noise injection operation works. In this section, we aim to give a clear formulation.

Additive Noise: We first take additive noise ξ_μ into consideration. Let $l(t, f(x))$ (abbreviate as $l(x)$ in the following development) denote the loss w.r.t. one activation input x, where t is the target, $f(\cdot)$ represents the network and $l(\cdot)$ is the loss function. When additive noise ξ_μ is added to the activation, the loss becomes $l(x + \xi_\mu)$. By Taylor expansion [4], we have

$$E_{\xi_\mu}[l(x + \xi_\mu)] = l(x) + R^{add}(x), \quad R^{add}(x) = \sum_{n=1}^{\infty} \frac{E[\xi_\mu^n]}{n!} \frac{d^n l(x)}{dx^n}. \tag{5}$$

where $E(\cdot)$ is the expectation and R^{add} is the additive noise residual term, which is related to the n-th order derivative of loss function w.r.t. activation input and the n-th order moment of noise distribution.

According to [2], by considering only the major term in R^{add}, it can be shown that $R^{add}(x) \approx \frac{E[\xi_\mu^2]}{2} \left|\frac{\partial f(x)}{\partial x}\right|^2$ for mean square-error loss; and $R^{add}(x) \approx \frac{E[\xi_\mu^2]}{2} \frac{f(x)^2 - 2tf(x) + t}{f(x)^2(1-f(x))^2} \left|\frac{\partial f(x)}{\partial x}\right|^2$ for cross-entropy loss. This indicates that R^{add} regularizes the smoothness of the network function, while the strength of smoothness is mainly controlled by the second order moment of the distribution of noise ξ_μ (i.e., $\frac{1}{m}$), which is only related to training batch size m. In Fig. 2, we illustrate the influence of additive noise on learning a classification hyperplane with different noise levels. The yellow points and blue points represent samples from two classes, and the Gaussian noise is added to the samples for data augmentation. By increasing the noise level σ to a proper level (e.g., $\sigma = 0.5$), the learned classification hyperplane becomes smoother and thus has better generalization capability. However, a too big noise level (e.g., $\sigma = 1$) will over-smooth the classification boundary and decrease the discrimination ability.

Multiplicative Noise: For multiplicative noise ξ_σ, we can use a simple logarithmic transformation $l(\frac{x}{\sqrt{\xi_\sigma}}) = l(e^{\log|x| - \frac{1}{2}\log \xi_\sigma} \text{sign}(x))$ to transform it into the form of additive noise. Then according to our analyses of additive noise:

$$E_{\xi_\sigma}[l(\frac{x}{\sqrt{\xi_\sigma}})] = l(x) + R^{mul}(x), \quad R^{mul}(x) = \sum_{n=1}^{\infty} \sum_{k=1}^{d} I(x \neq 0) \frac{E[\log^n \xi_\sigma]}{(-2)^n n!} \frac{d^n l(x)}{(d \log |x|)^n}. \tag{6}$$

where $I(x \neq 0)$ is an indicator function and $R^{mul}(x)$ is the residual term of Taylor expansion for multiplicative noise. Similar to the residual term of addictive

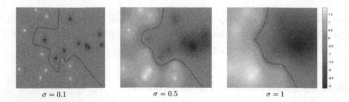

$\sigma = 0.1$ $\sigma = 0.5$ $\sigma = 1$

Fig. 2. The influence of noise injection on classification hyperplane (shown as red curve) learning with different noise levels. The yellow points and blue points represent samples from two classes, and the Gaussian noise with variance σ^2 is added to the samples for data augmentation. We can see that by increasing the noise level σ to a proper level (e.g., $\sigma = 0.5$), the learned classification hyperplane becomes smoother and thus has better generalization capability. However, a too big noise level (e.g., $\sigma = 1$) will over-smooth the classification boundary and decrease the discrimination ability. (Color figure online)

noise $R^{add}(x)$, the major term of $R^{mul}(x)$ can also be viewed as a regularizer to $\left| \frac{\partial f(x)}{\partial \log |x|} \right|^2$, which controls the smoothness of network on log-scale, and $E[\log^2 \xi_\sigma]$ is related to the strength of the regularizer.

Compound Noise: In Sect. 3.1 we have shown that BN will introduce both addictive noise and multiplicative noise into the normalized activation input, i.e., $\hat{x} = \frac{\tilde{x} + \xi_\mu}{\sqrt{\xi_\sigma}}$. In the following theorem, we present the joint residual formulation for the compound of addictive noise and multiplicative noise.

Theorem 2: *If the infinite derivative of $l(x)$ exists for any x, given two random variables ξ_μ and ξ_σ (> 0), then we have the Taylor expansion for $l(\frac{x+\xi_\mu}{\sqrt{\xi_\sigma}})$:*

$$E_{\xi_\mu, \xi_\sigma}[l(\frac{x+\xi_\mu}{\sqrt{\xi_\sigma}})] = l(x) + R^{add}(x) + R^{mul}(x) + R(x), \quad R(x) = \sum_{n=1}^{\infty} \frac{E[\xi_\mu{}^n]}{n!} \frac{d^n R^{mul}(x)}{dx^n}$$

(7)

where $R^{add}(x)$ and $R^{mul}(x)$ are defined in Eq. (5) and (6), respectively.

The proof of Theorem 2 can be found in the **Supplementary Material**. From Theorem 2, we can see the Taylor expansion residual can be divided into three parts: a residual term $R^{add}(x)$ for addictive noise, a residual term $R^{mul}(x)$ for multiplicative noise and a cross residual term $R(x)$. When the noise level is small, $R(x)$ can be ignored. Particularly, the distributions of ξ_μ and ξ_σ are give in Eq. (2) so that the regularizer strength parameters $E[\xi_\mu^2]$ and $E[\log^2 \xi_\sigma]$ can be easily calculated, which are only determined by training batch size m. The noise is injected into the normalized data $\tilde{x} = \frac{x-\mu}{\sigma}$. If the introduced noise by BN is strong (e.g., when batch size is small), the training forward propagation through the DNN may accumulate and amplify noise, which leads to undesirable model performance. Therefore, it is crucial to choose a suitable batch size for training to make BN keep a proper noise level and ensure a favorable regularization function. However, in some situations of limited memory and computing resources, we can only use a small batch size for training. It is hence important to find an approach to control the noise level of BN with small batch size, which will be investigated in the next section.

4   Momentum Batch Normalization

As proved in Sect. 3, the batch size m directly controls the strength of the regularizer in BN so that BN is sensitive to batch size. In most previous literature [18,43], the batch size m is set around 64 by experience. However, in some applications the batch size may not be set big enough due to the limited memory and large size of input. How to stably train a network with small batch size in BN remains an open problem. Owe to our theoretical analyses in Sect. 3, we propose a simple solution to alleviate this problem by introducing a parameter to control the strength of regularizer in BN. Specifically, we replace the batch means and variances in BN by their momentum or moving average:

$$\mu_M^{(n)} = \lambda\mu_M^{(n-1)} + (1-\lambda)\mu_B, \quad \left(\sigma_M^{(n)}\right)^2 = \lambda\left(\sigma_M^{(n-1)}\right)^2 + (1-\lambda)\sigma_B^2, \qquad (8)$$

where λ is the momentum parameter to controls the regularizer strength, and n refers to the number of batches (or iterations). We name our new BN method as Momentum Batch Normalization (MBN), which can make the noise level generated by using a small batch size almost the same as that by using a large batch size when the training stage ends.

4.1   Noise Estimation

At the end of the training process, all statistics of variables tend to be converged. According to Eq. (8), it can be derived that

$$\mu_M^{(n)} = (1-\lambda)\sum_{i=1}^{n}\lambda^{n-i}\mu_B, \quad \left(\sigma_M^{(n)}\right)^2 = (1-\lambda)\sum_{i=1}^{n}\lambda^{n-i}\sigma_B^2 \qquad (9)$$

When n is very large, let μ_M and σ_M denote the final momentum mean and variance, we can derive that

$$\xi_\mu = \frac{\mu - \mu_M}{\sigma} \sim \mathcal{N}(0, \frac{1-\lambda}{m}) \qquad (10)$$

$\xi_\sigma = \frac{\sigma_M^2}{\sigma^2}$ follows Generalized-Chi-Squared distribution, whose expectation is $E[\xi_\sigma] = \frac{m-1}{m}$ and variance is $Var[\xi_\sigma] = \frac{1-\lambda}{1+\lambda}\frac{2(m-1)}{m^2}$.

We can see that the variances of ξ_μ and ξ_σ approach to zero when λ is close to 1, MBN degenerates into standard BN when λ is zero. This implies that the noise level can be controlled by momentum parameter λ. A larger value λ will weaken the regularization function of MBN, and vice versa. Even when the batch size m is very small, we are still able to reduce the noise level through adjusting λ. This is an important advantage of MBN over conventional BN. For instance, if we want to make MBN with batch size 4 have similar noise level with batch size 16, the momentum parameter λ can be set as 3/4 to make their variances of ξ_μ similar, $(\frac{1-\frac{3}{4}}{4} = \frac{1}{16})$, and the multiplicative noise ξ_σ will also be reduced.

Algorithm 1. [MBN] Momentum Batch Normalization

Input: Values of x over a training mini-batch $\mathcal{X}_b$; parameters γ, β; current training moving mean μ and variance σ^2; current inference moving mean μ_{inf} and variance σ_{inf}^2; momentum parameters λ for training and τ for inference.

Output: $\{y_i = \text{MBN}(x_i)\}$; updated μ and σ^2;

updated μ_{inf} and σ_{inf}^2

Training step:

$\mu_B = \frac{1}{n}\sum_{i=1}^{m} x_i$

$\sigma_B^2 = \frac{1}{m}\sum_{i=1}^{m}(x_i - \mu_b)^2$

$\mu \leftarrow \lambda\mu + (1-\lambda)\mu_B$

$\sigma^2 \leftarrow \lambda\sigma^2 + (1-\lambda)\sigma_B^2$

$\widehat{x}_i = \frac{x_i - \mu}{\sqrt{\sigma^2 + \epsilon}}$

$y_i = \gamma\widehat{x}_i + \beta$

$\mu_{inf} \leftarrow \tau\mu_{inf} + (1-\tau)\mu_B$

$\sigma_{inf}^2 \leftarrow \tau\sigma_{inf}^2 + (1-\tau)\sigma_B^2$

Inference step: $y_i = \gamma\frac{x_i - \mu_{inf}}{\sqrt{\sigma_{inf}^2 + \epsilon}} + \beta$

4.2 Momentum Parameter Setting

Dynamic Momentum Parameter for Training: Since the momentum parameter λ controls the final noise level, we need to set a proper momentum parameter to endow the network a certain generalization ability. Please note that our noise analysis in Sect. 4.1 holds only when network statistics are stable at the end of training. In the beginning of training, we cannot directly use the moving average of batch mean and variance, because the population mean and variance also change significantly. Therefore, we hope that in the beginning of training the normalization is close to BN, while at the end of it tends to be MBN. To this end, we propose a dynamic momentum parameter as follows:

$$\lambda^{(t)} = \rho^{\frac{T}{T-1}\max(T-t,0)} - \rho^T, \qquad \rho = \min(\frac{m}{m_0}, 1)^{\frac{1}{T}} \tag{11}$$

where t refers to the t-th iteration epoch, T is the number of the total epochs, m is the actual batch size and m_0 is a large enough batch size (e.g., 64).

We use the same momentum parameter within one epoch. $\lambda^{(t)}$ starts from zero. When $\frac{m}{m_0}$ is small, $\lambda^{(t)}$ tends to be a number close to 1 at the end of the training. If m is equal to or larger than m_0, $\lambda^{(t)}$ is always equal to zero, and then MBN degenerates into BN. The dynamic setting of momentum parameter ensures that at the beginning of training process, the normalization is similar to standard BN, while at the end of the training the normalization approaches to MBN with a noise level similar to that of BN with batch size m_0.

Momentum Parameter for Inference: For inference step, we also need to set a momentum parameter. For the clarity of description, here we use τ to denote this momentum parameter to differentiate it from the momentum parameter λ in the training stage. One can straightforwardly set τ as a constant, e.g. $\tau = 0.9$, which is independent of batch size. However, this setting is not very reasonable because it cannot reflect the final noise level when training is ended, which is related to batch size m. Therefore, we should set τ to be adaptive to batch size m. Denote by τ_0 the desired momentum value for an ideal batch size m_0, to make the inference momentum have the same influence on the last sample, we take $\tau^{\frac{N}{m}}$ as a reference to determine the value of τ for batch size m as follows:

$$\tau^{\frac{N}{m}} = \tau_0^{\frac{N}{m_0}} \quad \Rightarrow \quad \tau = \tau_0^{\frac{m}{m_0}} \tag{12}$$

where N is the number of samples, m_0 is an ideal batch size and τ_0 is its corresponding momentum parameter. In most of our experiments, we set $m_0 = 64$ and $\tau_0 = 0.9$ for the inference step. One can see that when the training batch size m is small, a larger inference momentum parameter τ will be used, and consequently the noise in momentum mean and variance will be suppressed.

4.3 Algorithm

The back-propagation (BP) process of MBN is similar to that of traditional BN. During training, the gradients of loss w.r.t. to activations and model parameters are calculated and back-propagated. The formulas of BP are listed as follows:

$$
\begin{aligned}
&\frac{\partial L}{\partial \widehat{x}_i} = \frac{\partial L}{\partial \widehat{y}_i}\gamma, \quad \frac{\partial L}{\partial \gamma} = \sum_{i=1}^{m} \frac{\partial L}{\partial \widehat{y}_i}\widehat{x}_i, \quad \frac{\partial L}{\partial \beta} = \sum_{i=1}^{m} \frac{\partial L}{\partial \widehat{y}_i} \\
&\frac{\partial L}{\partial \sigma_B^2} = \sum_{i=1}^{m} \frac{\partial L}{\partial \widehat{x}_i}(\widehat{x}_i - \mu_M)\frac{-1}{2}(\sigma_M^2 + \epsilon)^{-\frac{3}{2}}(1 - \lambda) \\
&\frac{\partial L}{\partial \mu_B} = (\sum_{i=1}^{m} \frac{\partial L}{\partial \widehat{x}_i}\frac{\lambda - 1}{\sqrt{\sigma_M^2 + \epsilon}}) + \frac{\partial L}{\partial \sigma_B^2}\frac{\sum_{i=1}^{m} -2(x_i - \mu_B)}{m} \\
&\frac{\partial L}{\partial x_i} = \frac{\partial L}{\partial \widehat{x}_i}\frac{1}{\sqrt{\sigma_M^2 + \epsilon}} + \frac{\partial L}{\partial \sigma_B^2}\frac{2(x_i - \mu_B)}{m} + \frac{\partial L}{\partial \mu_B}\frac{1}{m}
\end{aligned}
\tag{13}
$$

Since the current moving averages of μ_M and σ_M^2 are related to the mean μ_B and variance σ_B^2 of the current mini-batch, they also contribute to the gradient, while the previous μ_M and σ_M^2 can be viewed as two constants for the current mini-batch. The training and inference of MBN are summarized in Algorithm 1.

5 Experimental Results

5.1 Datasets and Experimental Setting

To evaluate MBN, we apply it to image classification tasks and conduct experiments on CIFAR10, CIFAR100 [21] and Mini-ImageNet100 datasets [40].

Datasets. CIFAR10 consists of 50k training images from 10 classes, while CIFAR100 consists of 50k training and 10k testing images from 100 classes. The resolution of sample images in CIFAR10/100 is 32×32. Mini-ImageNet is a subset of the well-known ImageNet dataset. It consists of 100 classes with 600 images each class, and the image resolution is 84×84. We use the first 500 images from each class as training data, and the rest 100 images for testing, i.e., 50k images for training and 10k images for testing.

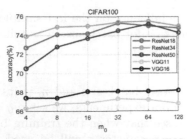

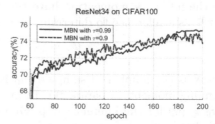

(a) Recognition rates of MBN with different m_0 on CIFAR100 by using different DNNs, including ResNet18, ResNet34, ResNet50, VGG11 and VGG16.

(b) Testing accuracy curves on CIFAR100 of BN and MBN with different τ for inference and training batch size 2 per GPU.

Fig. 3. Parameters tuning of MBN.

Experimental Setting. We use SGD with momentum 0.9 and weight decay 0.0001, employ standard data augmentation and preprocessing techniques, and nd decrease the learning rate when learning plateaus occur. The model is trained for 200 epochs and 100 epochs for CIFAR and Mini-ImageNet-100, respectively. We start with a learning rate of $0.1 * \frac{m}{64}$ both for CIFAR10 and CIFAR100 and $0.1 * \frac{m}{128}$ for Mini-ImageNet-100, and divide it by 10 for every 60 epochs and 30 epochs, respectively. We mainly employ ResNet [14] as our backbone network, and use similar experimental settings to the original ResNet paper. All the experiments are conducted on Pytorch1.0 framework.

5.2   Parameters Setting

There are two hyper-parameters in our proposed MBN, m_0 and τ_0, which are used to determine the momentum parameters λ and τ for training and inference.

The Setting for m_0: We first fix τ_0 (e.g., 0.9) to find a proper m_0. We adopt ResNet18 as the backbone and train it with batch size 8 and 16 on 4 GPUs, i.e., batch size 2 and 4 per GPU, to test the classification accuracy with different m_0. Particularly, we let m_0 be 4, 8, 16, 32, 64, 128 in MBN. Figure 3(a) shows the accuracy curves on CIFAR100. We can see that if m_0 is too small (e.g., 4), MBN will be close to BN, and the performance is not very good. The accuracies of MBN are very close for m_0 from 16 to 128, which shows that MBN is not very sensitive to parameter m_0. Considering that if m_0 is too large (e.g., 128), the momentum parameter λ may change too quickly so that the training may not converge, we set it to 32 in all the experiments.

The Setting for τ_0: We then fix m_0 as 32 and find a proper τ_0 based on Eq. (12). Figure 3(b) shows the testing accuracy curves for MBN with different values of τ. $\tau = 0.9$ is the original BN setting, and $\tau = 0.99$ is our setting based on Eq. (12) with $\tau_0 = 0.85$. We can see that when τ is small the testing accuracy curves of

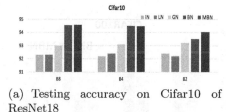
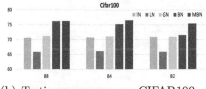

(a) Testing accuracy on Cifar10 of ResNet18

(b) Testing accuracy on CIFAR100 of ResNet18

Fig. 4. Testing accuracy on CIFAR10 and CIFAR100 of ResNet18 with training batch size (BS) 8, 4, and 2 per GPU.

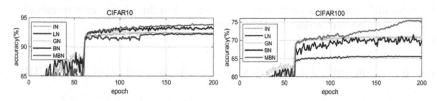

Fig. 5. Comparison of accuracy curves for different normalization methods with a batch size of 2 per GPU. We show the test accuracies vs. the epoches on CIFAR10 (left) and CIFAR100 (right). The ResNet18 is used.

both BN and MBN have big fluctuations; while τ is large, the accuracy curves become more stable and the final accuracies can be improved. We set $\tau_0 = 0.85$ in the following experiments.

5.3 Results on CIFAR10/100

We first conduct experiments on the CIFAR10 and CIFAR100 datasets [21]. We first use ResNet18 as the backbone network to evaluate MBN with different batch sizes, and then test the performance of MBN with more networks.

Training with Different Batch Size: To testify whether MBN is more robust than BN with small batch size, we train Resnet18 on CIFAR10 and CIFAR100 by setting the batch size m as 8, 4, 2 per GPU, respectively. We also compare the behaviors of other normalization methods, including IN [39], LN [23] and GN [43], by replacing the BN layer with them. For GN, we use 32 groups as set in [43]. And we set $T = 180$ in Eq. (11) for MBN.

Figure 4 shows the results for different normalization methods. We can see that on both CIFAR10 and CIFAR100, when the batch size is relatively large (e.g., 8), the accuracy of MBN is similar to BN. This is in accordance to our theoretical analysis in Sects. 3 and 4. However, when training batch size becomes small (e.g., 2), the accuracy of BN drops largely, while the accuracy of MBN decreases slightly. This shows that MBN is more robust than BN for training with small batch size. Meanwhile, MBN works much better than IN, LN and GN.

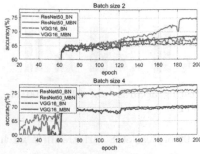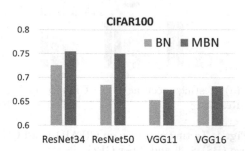

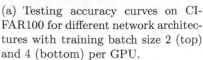

(a) Testing accuracy curves on CI-FAR100 for different network architectures with training batch size 2 (top) and 4 (bottom) per GPU.

(b) Testing accuracy on CIFAR100 for different network architectures with training batch size 2 per GPU.

Fig. 6. Training with different DNN models with BN and MBN on CIFAR100.

Figure 5 shows the training and testing accuracy curves vs. epoch of ResNet18 with batch size 2. We can see that at the last stage of training when all statistics become stable, MBN can still achieve certain performance gain. This is because with MBN the momentum mean and variance approach to the population mean and variance, and hence the noise becomes small. Consequently, MBN can still keep improving though other methods are saturated.

On More Network Architectures: We further test MBN with different network architectures, including ResNet34, ResNet50, VGG11 and VGG16, by using batch size 2 per GPU on CIFAR100. Figure 6(a) shows the training and testing accuracy curves vs. epochs, and Fig. 6(b) shows the final testing accuracies. We can have the following observations. First, on all the four networks, MBN always outperforms BN. Second, under such a small batch size, the accuracy of deeper network ResNet50 can be lower than its shallower counterpart ResNet34. That is because the deeper networks have more BN layers, and each BN layer would introduce relatively large noise when batch size is small. The noise is accumulated so that the benefit of more layers can be diluted by the accumulated noise. However, with MBN the performance drop from ResNet50 to ResNet34 is very minor, where the drop by BN is significant. This again validates that MBN can suppress the noise effectively in training.

5.4　Results on Mini-ImageNet-100

On Small Batch Size: On Mini-imageNet, we use ResNet50 as our backbone network. The input size is the same as image size 84×84. The settings for MBN are the same as Sect. 5.3. Figure 7(a) compares the testing accuracies of IN, LN, GN, BN and MBN with batch sizes 16, 8, 4 and 2 per GPU. Figure 7(b) shows their testing accuracy curves with training batch size 2 per GPU. We can see that BN and MBN achieve better results than other normalization methods

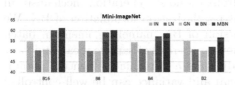

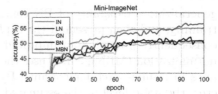

(a) Testing accuracy on Mini-ImageNet of IN, LN, GN, BN and MBN with training batch size 16, 8, 4 and 2 per GPU.

(b) Testing accuracy curves on Mini-ImageNet of IN, LN, GN, BN and MBN with training batch size 2 per GPU.

Fig. 7. Comparison of different normalization methods on Mini-ImageNet.

when batch size is larger than 2, while other normalization methods, such as IN, LN and GN, usually work not very well on Mini-imageNet. But the performance of BN drops significantly when batch size is 2, even worse that IN, while MBN still works well when batch size is 2. This clearly demonstrates the effectiveness of MBN. Furthermore, we also compare MBN with BN on full ImageNet using ResNet50 with 64 GPUs and 4 batch size per GPU. It is found that find MBN outperforms BN by 2.5% in accuracy on the validation set.

Comparison with BreN: BreN [18] was also proposed to make BN work for training with small batch size. It adopts a heuristic clipping strategy to control the influence of current moving average on the normalizer. Though BreN and our proposed MBN have similar goals, they are very different in theoretical development and methodology design. First, the dynamic momentum setting in MBN makes it easy to analyze the noise level in the final training stage, while in BreN it is hard to know the noise level with the heuristic clipping strategy. Second, the hyper-parameters m_0 and τ_0 are very easy to be tuned and fixed in MBN (we fixed them in all our experiments on all datasets), while the hyper-parameters (clipping bounds) in BreN are very difficult to set. Although a strategy to set the clipping bound was given in [18], we found that this setting usually leads unsatisfactory performance when the dataset or training batch size changes. We have tried various parameter settings for BreN on Mini-ImageNet when training batch size is 2, but found that in most cases the results are even worse. So we report the best result of BreN on Mini-ImageNet with $r_{max} = 1.5$ and $d_{max} = 0.5$, which is 55.47%, lower than the performance of MBN (56.50%).

6   Conclusion

Batch normalization (BN) is a milestone technique in deep learning and it largely improves the effectiveness and efficiency in optimizing various deep networks. However, the working mechanism of BN is not fully revealed yet, while the performance of BN drops much when the training batch size is small because of the inaccurate batch statistics estimation. In this work, we first revealed that the generalization capability of BN comes from its noise generation mechanism in training, and then presented the explicit regularization formulation of BN. We consequently presented an improved version of BN, namely momentum batch normalization (MBN), which uses the moving average of sample mean and variance in a mini-batch for training. By adjusting a dynamic momentum parameter, the noise level in the estimated mean and variance can be well controlled in MBN. The experimental results demonstrated that MBN can work stably for different batch sizes. In particular, it works much better than BN and other popular normalization methods when the batch size is small.

References

1. Amodei, D., et al.: Deep speech 2: end-to-end speech recognition in English and Mandarin. In: International Conference on Machine Learning, pp. 173–182 (2016)
2. An, G.: The effects of adding noise during backpropagation training on a generalization performance. Neural Comput. 8(3), 643–674 (1996)
3. Arpit, D., Zhou, Y., Kota, B.U., Govindaraju, V.: Normalization propagation: a parametric technique for removing internal covariate shift in deep networks. arXiv preprint arXiv:1603.01431 (2016)
4. Bishop, C.M.: Training with noise is equivalent to tikhonov regularization. Neural Comput. 7(1), 108–116 (1995)
5. Bjorck, J., Gomes, C., Selman, B., Weinberger, K.Q.: Understanding batch normalization, pp. 7694–7705 (2018)
6. Bottou, L.: Stochastic gradient learning in neural networks. Proc. Neuro-Nımes 91(8), 12 (1991)
7. Bottou, L.: Large-scale machine learning with stochastic gradient descent. In: Lechevallier, Y., Saporta, G. (eds.) Proceedings of COMPSTAT 2010, pp. 177–186. Springer (2010). https://doi.org/10.1007/978-3-7908-2604-3_16
8. Cho, M., Lee, J.: Riemannian approach to batch normalization. In: Advances in Neural Information Processing Systems, pp. 5225–5235 (2017)
9. Crocker, L., Algina, J.: Introduction to classical and modern test theory. ERIC (1986)
10. Deng, J., Dong, W., Socher, R., Li, L.J., Li, K., Fei-Fei, L.: ImageNet: a large-scale hierarchical image database. In: 2009 IEEE Conference on Computer Vision and Pattern Recognition, pp. 248–255. IEEE (2009)
11. Duchi, J., Hazan, E., Singer, Y.: Adaptive subgradient methods for online learning and stochastic optimization. J. Mach. Learn. Res. 12(Jul), 2121–2159 (2011)
12. Guo, Y., Wu, Q., Deng, C., Chen, J., Tan, M.: Double forward propagation for memorized batch normalization. In: Thirty-Second AAAI Conference on Artificial Intelligence (2018)
13. He, K., Gkioxari, G., Dollár, P., Girshick, R.: Mask R-CNN. In: Proceedings of the IEEE International Conference on Computer Vision, pp. 2961–2969 (2017)
14. He, K., Zhang, X., Ren, S., Sun, J.: Deep residual learning for image recognition. In: Proceedings of the IEEE Conference on Computer Vision and Pattern Recognition, pp. 770–778 (2016)
15. Huang, G., Liu, Z., Van Der Maaten, L., Weinberger, K.Q.: Densely connected convolutional networks. In: Proceedings of the IEEE Conference on Computer Vision and Pattern Recognition, pp. 4700–4708 (2017)

16. Huang, L., Yang, D., Lang, B., Deng, J.: Decorrelated batch normalization. In: Proceedings of the IEEE Conference on Computer Vision and Pattern Recognition, pp. 791–800 (2018)
17. Huang, L., Zhou, Y., Zhu, F., Liu, L., Shao, L.: Iterative normalization: beyond standardization towards efficient whitening. In: Proceedings of the IEEE Conference on Computer Vision and Pattern Recognition, pp. 4874–4883 (2019)
18. Ioffe, S.: Batch renormalization: towards reducing minibatch dependence in batch-normalized models. In: Advances in Neural Information Processing Systems, pp. 1945–1953 (2017)
19. Ioffe, S., Szegedy, C.: Batch normalization: accelerating deep network training by reducing internal covariate shift. arXiv preprint arXiv:1502.03167 (2015)
20. Kingma, D.P., Ba, J.: Adam: a method for stochastic optimization. arXiv preprint arXiv:1412.6980 (2014)
21. Krizhevsky, A., Hinton, G., et al.: Learning multiple layers of features from tiny images. Technical report, Citeseer (2009)
22. LeCun, Y.A., Bottou, L., Orr, G.B., Müller, K.-R.: Efficient backprop. In: Montavon, G., Orr, G.B., Müller, K.-R. (eds.) Neural Networks: Tricks of the Trade. LNCS, vol. 7700, pp. 9–48. Springer, Heidelberg (2012). https://doi.org/10.1007/978-3-642-35289-8_3
23. Lei Ba, J., Kiros, J.R., Hinton, G.E.: Layer normalization. arXiv preprint arXiv:1607.06450 (2016)
24. Lenth, R.V.: Cumulative distribution function of the non-central T distribution. J. Roy. Stat. Soc. Ser. C (Appl. Stat.) 38(1), 185–189 (1989)
25. Luo, P., Peng, Z., Ren, J., Zhang, R.: Do normalization layers in a deep convnet really need to be distinct? arXiv preprint arXiv:1811.07727 (2018)
26. Luo, P., Ren, J., Peng, Z.: Differentiable learning-to-normalize via switchable normalization. arXiv preprint arXiv:1806.10779 (2018)
27. Luo, P., Wang, X., Shao, W., Peng, Z.: Towards understanding regularization in batch normalization (2018)
28. Luong, M.T., Pham, H., Manning, C.D.: Effective approaches to attention-based neural machine translation. arXiv preprint arXiv:1508.04025 (2015)
29. Mnih, V., et al.: Human-level control through deep reinforcement learning. Nature 518(7540), 529 (2015)
30. Ren, M., Liao, R., Urtasun, R., Sinz, F.H., Zemel, R.S.: Normalizing the normalizers: comparing and extending network normalization schemes. arXiv preprint arXiv:1611.04520 (2016)
31. Ren, S., He, K., Girshick, R., Sun, J.: Faster R-CNN: towards real-time object detection with region proposal networks. In: Advances in Neural Information Processing Systems, pp. 91–99 (2015)
32. Salimans, T., Kingma, D.P.: Weight normalization: a simple reparameterization to accelerate training of deep neural networks. In: Advances in Neural Information Processing Systems, pp. 901–909 (2016)
33. Santurkar, S., Tsipras, D., Ilyas, A., Madry, A.: How does batch normalization help optimization? (no, it is not about internal covariate shift), pp. 2483–2493 (2018)
34. Shao, W., et al.: SSN: learning sparse switchable normalization via sparsestmax. arXiv preprint arXiv:1903.03793 (2019)
35. Shekhovtsov, A., Flach, B.: Stochastic normalizations as bayesian learning. In: Jawahar, C.V., Li, H., Mori, G., Schindler, K. (eds.) ACCV 2018. LNCS, vol. 11362, pp. 463–479. Springer, Cham (2019). https://doi.org/10.1007/978-3-030-20890-5_30

36. Silver, D., et al.: Mastering the game of go with deep neural networks and tree search. Nature **529**(7587), 484 (2016)
37. Sutskever, I., Martens, J., Dahl, G., Hinton, G.: On the importance of initialization and momentum in deep learning. In: International Conference on Machine Learning, pp. 1139–1147 (2013)
38. Teye, M., Azizpour, H., Smith, K.: Bayesian uncertainty estimation for batch normalized deep networks. arXiv preprint arXiv:1802.06455 (2018)
39. Ulyanov, D., Vedaldi, A., Lempitsky, V.: Instance normalization: the missing ingredient for fast stylization. arXiv preprint arXiv:1607.08022 (2016)
40. Vinyals, O., Blundell, C., Lillicrap, T., Wierstra, D., et al.: Matching networks for one shot learning. In: Advances in Neural Information Processing Systems, pp. 3630–3638 (2016)
41. Wang, G., Peng, J., Luo, P., Wang, X., Lin, L.: Batch kalman normalization: towards training deep neural networks with micro-batches. arXiv preprint arXiv:1802.03133 (2018)
42. Wu, S., et al.: L1-norm batch normalization for efficient training of deep neural networks. IEEE Trans. Neural Netw. Learn. Syst. **30**(7), 2043–2051 (2018)
43. Wu, Y., He, K.: Group normalization. In: Proceedings of the European Conference on Computer Vision (ECCV), pp. 3–19 (2018)
44. Zhang, C., Bengio, S., Hardt, M., Recht, B., Vinyals, O.: Understanding deep learning requires rethinking generalization. arXiv preprint arXiv:1611.03530 (2016)
45. Zhang, K., Zuo, W., Chen, Y., Meng, D., Zhang, L.: Beyond a gaussian denoiser: Residual learning of deep CNN for image denoising. IEEE Trans. Image Process. **26**(7), 3142–3155 (2017)

AdvPC: Transferable Adversarial Perturbations on 3D Point Clouds

Abdullah Hamdi$^{(\boxtimes)}$, Sara Rojas, Ali Thabet, and Bernard Ghanem

King Abdullah University of Science and Technology (KAUST),
Thuwal, Saudi Arabia
{abdullah.hamdi,sara.rojasmartinez,ali.thabet,
bernard.ghanem}@kaust.edu.sa

Abstract. Deep neural networks are vulnerable to adversarial attacks, in which imperceptible perturbations to their input lead to erroneous network predictions. This phenomenon has been extensively studied in the image domain, and has only recently been extended to 3D point clouds. In this work, we present novel data-driven adversarial attacks against 3D point cloud networks. We aim to address the following problems in current 3D point cloud adversarial attacks: they do not transfer well between different networks, and they are easy to defend against via simple statistical methods. To this extent, we develop a new point cloud attack (dubbed AdvPC) that exploits the input data distribution by adding an adversarial loss, after Auto-Encoder reconstruction, to the objective it optimizes. AdvPC leads to perturbations that are resilient against current defenses, while remaining highly transferable compared to state-of-the-art attacks. We test AdvPC using four popular point cloud networks: PointNet, PointNet++ (MSG and SSG), and DGCNN. Our proposed attack increases the attack success rate by up to 40% for those transferred to unseen networks (transferability), while maintaining a high success rate on the attacked network. AdvPC also increases the ability to break defenses by up to 38% as compared to other baselines on the ModelNet40 dataset. The code is available at https://github.com/ajhamdi/AdvPC.

1 Introduction

Deep learning has shown impressive results in many perception tasks. Despite its performance, several works show that deep learning algorithms can be susceptible to adversarial attacks. These attacks craft small perturbations to the inputs that push the network to produce incorrect outputs. There is significant progress made in 2D image adversarial attacks, where extensive work shows diverse ways to attack 2D neural networks [2,4,6–8,11,18,23,35]. In contrast, there is little focus on their 3D counterparts [25,31,37,38]. 3D point clouds captured by 3D

Electronic supplementary material The online version of this chapter (https://doi.org/10.1007/978-3-030-58610-2_15) contains supplementary material, which is available to authorized users.

A. Vedaldi et al. (Eds.): ECCV 2020, LNCS 12357, pp. 241–257, 2020.
https://doi.org/10.1007/978-3-030-58610-2_15

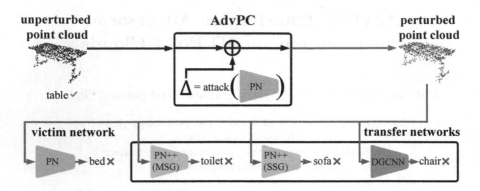

Fig. 1. Transferable Adversarial Perturbations on 3D point clouds: Generating adversarial attacks to fool PointNet [21] (PN) by perturbing a Table point cloud. The perturbed 3D object not only forces PointNet to predict an incorrect class, but also induces misclassification on other unseen 3D networks (PointNet++ [22], DGCNN [29]) that are not involved in generating the perturbation. Fooling unseen networks poses a threat to 3D deep vision models.

sensors like LiDAR are now widely processed using deep networks for safety-critical applications, including but not limited to self-driving [3,27]. However, as we show in this paper, 3D deep networks tend to be vulnerable to input perturbations, a fact that increases the risk of using them in such applications. In this paper, we present a novel approach to attack deep learning algorithms applied to 3D point clouds with a primary focus on attack transferability between networks.

The concept of attack transferability has been extensively studied in the 2D image domain [17,19,20]. Transferability allows an adversary to fool any network, without access to the network's architecture. Clearly, transferable attacks pose a serious security concern, especially in the context of deep learning model deployment. In this work, the goal is to generate adversarial attacks with network-transferability, *i.e.* the attack to a given point cloud is generated using a single and accessible *victim* network, and the perturbed sample is directly applied to an unseen and inaccessible *transfer* network. Accessibility here refers to whether the parameters and architecture of the network are known, while optimizing the attack (white-box). Figure 1 illustrates the concept of transferability. The perturbation generated by our method for a 3D point cloud not only flips the class label of a victim network to a wrong class (*i.e.* it is adversarial), but it also induces a misclassification for the transfer networks that are not involved in generating the perturbation (*i.e.* it is transferable).

Very few adversarial attacks have been developed for 3D point clouds. The first method was introduced by Xiang *et al.* [31] and it proposes point perturbation and adversarial point generation as two attack modes. More recently, Tsai *et al.* [25] proposed to make point cloud attacks more smooth and natural by incorporating a K-Nearest Neighbor (KNN) loss on the points, thus making the attacks physically realizable. We identify two main shortcomings in current 3D adversarial perturbations methods [25,31]. First, their attacks are unsuccessful in the presence of simple defenses, such as Statistical Outlier Removal [38]. Second, they are limited to the victim network and do not transfer well to other networks [31]. In contrast, our work not only focuses on adversarial perturbations that are significantly more resilient against currently available point cloud defenses, but also on those that transfer well between different point cloud networks.

To generate more transferable attacks, we use a point cloud Auto-Encoder (AE), which can effectively reconstruct the unperturbed input after it is perturbed, and then add a data adversarial loss. We optimize the perturbation added to the input to fool the classifier *before* it passes through the AE (regular adversarial loss) and *after* it passes through the AE (data adversarial loss). In doing so, the attack tends to be less dependent on the victim network, and generalizes better to different networks. Our attack is dubbed "AdvPC", and our full pipeline is optimized end-to-end from the classifier output to the perturbation. The AE learns the natural distribution of the data to generalize the attack to a broader range of unseen classifiers [26], thus making the attack more dangerous. Our attacks surpass state-of-the-art attacks [25,31] by a large margin (up to 40%) on point cloud networks operating on the standard ModelNet40 dataset [30] and for the same maximum allowed perturbation norms (norm-budgets).

Contributions. Our contributions are two-fold. **(1)** We propose a new pipeline and loss function to perform transferable adversarial perturbations on 3D point clouds. By introducing a data adversarial loss targeting the victim network after reconstructing the perturbed input with a point cloud AE, our approach can be successful in both attacking the victim network and transferring to unseen networks. Since the AE is trained to leverage the point cloud data distribution, incorporating it into the attack strategy enables better transferability to unseen networks. To the best of our knowledge, we are the first to introduce network-transferable adversarial perturbations for 3D point clouds. **(2)** We perform extensive experiments under constrained norm-budgets to validate the transferability of our attacks. We transfer our attacks between four point cloud networks and show superiority against the state-of-the-art. Furthermore, we demonstrate how our attacks outperform others when targeted by currently available point cloud defenses.

2   Related Work

2.1   Deep Learning for 3D Point Clouds

PointNet [21] paved the way as the first deep learning algorithm to operate directly on 3D point clouds. PointNet computes point features independently, and aggregates them using an order invariant function like max-pooling. An update to this work was PointNet++ [22], where points are aggregated at different 3D scales. Subsequent works focused on how to aggregate more local context [5] or on more complex aggregation strategies like RNNs [9,33]. More recent methods run convolutions across neighbors of points, instead of using point-wise operations [12–15,15,24,28,29]. Contrary to PointNet and its variants, these works achieve superior recognition results by focusing on local feature representation. In this paper and to evaluate/validate our adversarial attacks, we use three point-wise networks, PointNet [21] and PointNet++ [22] in single-scale (SSG) and multi-scale (MSG) form, and a Dynamic Graph convolutional Network, DGCNN [29]. We study the sensitivity of each network to adversarial perturbations and show the transferability of AdvPC attacks between the networks.

2.2 Adversarial Attacks

Pixel-Based Adversarial Attacks. The initial image-based adversarial attack was introduced by Szegedy *et al.* [23], who cast the attack problem as optimization with pixel perturbations being minimized so as to fool a trained classifier into predicting a wrong class label. Since then, the topic of adversarial attacks has attracted much attention [4,6,11,16,18]. More recent works take a learning-based approach to the attack [19,20,36]. They train a neural network (adversary) to perform the attack and then use the trained adversary model to attack unseen samples. These learning approaches [19,20,36] tend to have better transferability properties than the optimizations approaches [4,6,11,16,18], while the latter tend to achieve higher success rates on the victim networks. As such, our proposed AdvPC attack is a *hybrid* approach, in which we leverage an AE to capture properties of the data distribution but still define the attack as an optimization for each sample. In doing so, AdvPC captures the merits of both learning *and* optimization methods to achieve high success rates on the victim networks as well as better transferability to unseen networks.

Adversarial Attacks in 3D. Several adversarial attacks have moved beyond pixel perturbations to the 3D domain. One line of work focuses on attacking image-based CNNs by changing the 3D parameters of the object in the image, instead of changing the pixels of the image [2,7,8,32,35]. Recently, Xiang *et al.* [31] developed adversarial perturbations on 3D point clouds, which were successful in attacking PointNet [21]; however, this approach has two main shortcomings. First, it can be easily defended against by simple statistical operations [38]. Second, the attacks are non-transferable and only work on the attacked network [31,38]. In contrast, Zheng *et al.* [37] proposed dropping points from the point cloud using a saliency map, to fool trained 3D deep networks. As compared to [37], our attacks are modeled as an optimization on the additive perturbation variable with a focus on point perturbations instead of point removal. As compared to [31], our AdvPC attacks are significantly more successful against available defenses and more transferable beyond the victim network, since AdvPC leverages the point cloud data distribution through the AE. Concurrent to our work is the work of Tsai *et al.* [25], in which the attack is crafted with KNN loss to make smooth and natural shapes. The motivation of their work is to craft natural attacks on 3D point clouds that can be 3D-printed into real objects. In comparison, our novel AdvPC attack utilizes the data distribution of point clouds by utilizing an AE to generalize the attack.

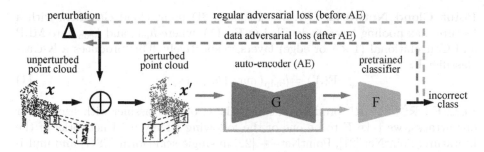

Fig. 2. AdvPC Attack Pipeline: We optimize for the constrained perturbation variable Δ to generate the perturbed sample $\mathcal{X}' = \mathcal{X} + \Delta$. The perturbed sample fools a trained classifier $\mathbf{F}$ (*i.e.* $\mathbf{F}(\mathcal{X}')$ is incorrect), and at the same time, if the perturbed sample is reconstructed by an Auto-Encoder (AE) $\mathbf{G}$, it too fools the classifier (*i.e.* $\mathbf{F}(\mathbf{G}(\mathcal{X}'))$ is incorrect). The AdvPC loss for network $\mathbf{F}$ is defined in Eq. (6) and has two parts: network adversarial loss (*purple*) and data adversarial loss (*green*). Dotted lines are gradients flowing to the perturbation variable Δ. (Color figure online)

Defending Against 3D Point Cloud Attacks. Zhou *et al.* [38] proposed a Statistical Outlier Removal (SOR) method as a defense against point cloud attacks. SOR uses KNN to identify and remove point outliers. They also propose DUP-Net, which is a combination of their SOR and a point cloud up-sampling network PU-Net [34]. Zhou *et al.* also proposed removing unnatural points by Simple Random Sampling (SRS), where each point has the same probability of being randomly removed. Adversarial training on the attacked point cloud is also proposed as a mode of defense by [31]. Our attacks surpass state-of-the-art attacks [25,31] on point cloud networks by a large margin (up to 38%) on the standard ModelNet40 dataset [30] against the aforementioned defenses [38].

3   Methodology

The pipeline of AdvPC is illustrated in Fig. 2. It consists of an Auto-Encoder (AE) $\mathbf{G}$, which is trained to reconstruct 3D point clouds and a point cloud classifier $\mathbf{F}$. We seek to find a perturbation variable Δ added to the input $\mathcal{X}$ to fool $\mathbf{F}$ before *and* after it passes through the AE for reconstruction. The setup makes the attack less dependent on the victim network and more dependent on the data. As such, we expect this strategy to generalize to different networks. Next, we describe the main components of our pipeline: 3D point cloud input, AE, and point cloud classifier. Then, we present our attack setup and loss.

3.1   AdvPC Attack Pipeline

3D Point Clouds ($\mathcal{X}$). We define a point cloud $\mathcal{X} \in \mathbb{R}^{N \times 3}$, as a set of N 3D points, where each point $\mathbf{x}_i \in \mathbb{R}^3$ is represented by its 3D coordinates (x_i, y_i, z_i).

Point Cloud Networks (F). We focus on 3D point cloud classifiers with a feature max pooling layer as detailed in Eq. (1), where h_{mlp} and h_{conv} are MLP and Convolutional (1×1 or edge) layers, respectively. This produces a K-class classifier **F**.

$$\mathbf{F}(\mathcal{X}) = h_{\mathrm{mlp}}(\max_{\mathbf{x}_i \in \mathcal{X}} \{h_{\mathrm{conv}}(\mathbf{x}_i)\}) \tag{1}$$

Here, $\mathbf{F} : \mathbb{R}^{N \times 3} \rightarrow \mathbb{R}^K$ produces the logits layer of the classifier with size K. For our attacks, we take **F** to be one of the following widely used networks in the literature: PointNet [21], PointNet++ [22] in single-scale form (SSG) and multi-scale form (MSG), and DGCNN [29]. Section 5.2 delves deep into the differences between them in terms of their sensitivities to adversarial perturbations.

Point Cloud Auto-Encoder (G). An AE learns a representation of the data and acts as an effective defense against adversarial attacks. It ideally projects a perturbed point cloud onto the natural manifold of inputs. Any AE architecture in point clouds can be used, but we select the one in [1] because of its simple structure and effectiveness in recovering from adversarial perturbation. The AE **G** consists of an encoding part, $\mathbf{g}_{\mathrm{encode}} : \mathbb{R}^{N \times 3} \rightarrow \mathbb{R}^q$ (similar to Eq. (1)), and an MLP decoder, $\mathbf{g}_{\mathrm{mlp}} : \mathbb{R}^q \rightarrow \mathbb{R}^{N \times 3}$, to produce a point cloud. It can be described formally as: $\mathbf{G}(.) = \mathbf{g}_{\mathrm{mlp}}(\mathbf{g}_{\mathrm{encode}}(.))$. We train the AE with the Chamfer loss as in [1] on the same data used to train **F**, such that it can reliably encode and decode 3D point clouds. We freeze the AE weights during the optimization of the adversarial perturbation on the input. Since the AE learns how naturally occurring point clouds look like, the gradients updating the attack, which is also tasked to fool the reconstructed sample after the AE, actually become more dependent on the data and less on the victim network. The enhanced data dependency of our attack results in the success of our attacks on unseen transfer networks besides the success on the victim network. As such, the proposed composition allows the crafted attack to successfully attack the victim classifier, as well as, fool transfer classifiers that operate on a similar input data manifold.

3.2 AdvPC Attack Loss

Soft Constraint Loss. In AdvPC attacks, like the ones in Fig. 3, we focus solely on perturbations of the input. We modify each point $\mathbf{x}_i$ by a an addictive perturbation variable δ_i. Formally, we define the perturbed point set $\mathcal{X}' = \mathcal{X} + \mathbf{\Delta}$, where $\mathbf{\Delta} \in \mathbb{R}^{N \times 3}$ is the perturbation parameter we are optimizing for. Consequently, each pair $(\mathbf{x}_i, \mathbf{x}_i')$ are in correspondence. Adversarial attacks are commonly formulated as in Eq. (2), where the goal is to find an input perturbation $\mathbf{\Delta}$ that successfully fools **F** into predicting an incorrect label t', while keeping $\mathcal{X}'$ and $\mathcal{X}$ close under distance metric $\mathcal{D}: \mathbb{R}^{N \times 3} \times \mathbb{R}^{N \times 3} \rightarrow \mathbb{R}$.

$$\min_{\mathbf{\Delta}} \quad \mathcal{D}(\mathcal{X}, \mathcal{X}') \quad \text{s.t.} \quad \left[\arg\max_i \mathbf{F}(\mathcal{X}')_i\right] = t' \tag{2}$$

The formulation in Eq. (2) can describe targeted attacks (if t' is specified before the attack) or untargeted attacks (if t' is any label other than the

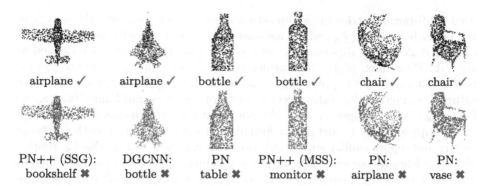

Fig. 3. Examples of AdvPC Attacks: Adversarial attacks are generated for victim networks PointNet, PointNet ++ (MSG/SSG) and DGCNN using AdvPC. The unperturbed point clouds are in black (*top*) while the perturbed examples are in blue (*bottom*). The network predictions are shown under each point cloud. The wrong prediction of each perturbed point cloud matches the target of the AdvPC attack. (Color figure online)

true label of $\mathcal{X}$). We adopt the following choice of t' for untargeted attacks: $t' = [\arg\max_{i\neq \text{true}} \mathbf{F}(\mathcal{X}')_i]$. Unless stated otherwise, we primarily use untargeted attacks in this paper. As pointed out in [4], it is difficult to directly solve Eq. (2). Instead, previous works like [25,31] have used the well-known C&W formulation, giving rise to the commonly known soft constraint attack: $\min_{\Delta} f_{t'}(\mathbf{F}(\mathcal{X}')) + \lambda \mathcal{D}(\mathcal{X}, \mathcal{X}')$ where $f_{t'}(\mathbf{F}(\mathcal{X}'))$ is the adversarial loss function defined on the network $\mathbf{F}$ to move it to label t' as in Eq. (3).

$$f_{t'}(\mathbf{F}(\mathcal{X}')) = \max\left(\max_{i\neq t'}(\mathbf{F}(\mathcal{X}')_i) - \mathbf{F}(\mathcal{X}')_{t'} + \kappa, 0\right), \quad (3)$$

where κ is a loss margin. The 3D-Adv attack [31] uses ℓ_2 for $\mathcal{D}(\mathcal{X}, \mathcal{X}')$, while the KNN Attack [25] uses Chamfer Distance.

Hard Constraint Loss. An alternative to Eq. (2) is to put $\mathcal{D}(\mathcal{X}, \mathcal{X}')$ as a hard constraint, where the objective can be minimized using Projected Gradient Descent (PGD) [11,16] as follows.

$$\min_{\Delta} f_{t'}(\mathbf{F}(\mathcal{X}')) \quad s.t. \quad \mathcal{D}(\mathcal{X}, \mathcal{X}') \leq \epsilon \quad (4)$$

Using a hard constraint sets a limit to the amount of added perturbation in the attack. This limit is defined by ϵ in Eq. (4), which we call norm-budget in this work. Having this bound ensures a fair comparison between different attack schemes. We compare these schemes by measuring their attack success rate at different levels of norm-budget. Using PGD, the above optimization in Eq. (4) with ℓ_p distance $\mathcal{D}_{\ell_p}(\mathcal{X}, \mathcal{X}')$ can be solved by iteratively projecting the perturbation Δ onto the ℓ_p sphere of size ϵ_p after each gradient step such that: $\Delta_{t+1} = \Pi_p(\Delta_t - \eta \nabla_{\Delta_t} f_{t'}(\mathbf{F}(\mathcal{X}')), \epsilon_p)$. Here, $\Pi_p(\Delta, \epsilon_p)$ projects the perturbation Δ onto the ℓ_p sphere of size ϵ_p, and η is a step size. The two most commonly used ℓ_p distance metrics in the literature are ℓ_2, which measures the energy of the perturbation, and ℓ_∞, which measures the maximum point perturbation of each $\boldsymbol{\delta}_i \in \boldsymbol{\Delta}$. In our experiments, we choose to use the ℓ_∞ distance defined as $\mathcal{D}_{\ell_\infty}(\mathcal{X}, \mathcal{X}') = \max_i \|\boldsymbol{\delta}_i\|_\infty$, The projection of $\boldsymbol{\Delta}$ onto the ℓ_∞ sphere of size ϵ_∞ is: $\Pi_\infty(\boldsymbol{\Delta}, \epsilon_\infty) = \mathrm{SAT}_{\epsilon_\infty}(\boldsymbol{\delta}_i),\ \forall \boldsymbol{\delta}_i \in \boldsymbol{\Delta}$, where $\mathrm{SAT}_{\epsilon_\infty}(\boldsymbol{\delta}_i)$ is the element-wise saturation function that takes every element of vector $\boldsymbol{\delta}_i$ and limits its range to $[-\epsilon_\infty, \epsilon_\infty]$. Norm-budget ϵ_∞ is used throughout the experiments in this work.

In **supplement**, we detail our formulation when ℓ_2 is used as the distance metric and report similar superiority over the baselines just as the ℓ_∞ results. For completeness, we also show in the supplement the effect of using different distance metrics (ℓ_2, Chamfer, and Earth Mover Distance) as soft constraints on transferability and attack effectiveness.

Data Adversarial Loss. The objectives in Eq. (2, 4) focus solely on the network $\mathbf{F}$. We also want to add more focus on the data in crafting our attacks. We do so by fooling $\mathbf{F}$ using both the perturbed input $\mathcal{X}'$ and the AE reconstruction $\mathbf{G}(\mathcal{X}')$ (see Fig. 2). Our new objective becomes:

$$\min_{\boldsymbol{\Delta}}\ \mathcal{D}(\mathcal{X}, \mathcal{X}') \quad \text{s.t.}\ [\arg\max_i \mathbf{F}(\mathcal{X}')_i] = t';\ [\arg\max_i \mathbf{F}(\mathbf{G}(\mathcal{X}'))_i] = t'' \quad (5)$$

Here, t'' is any incorrect label $t'' \neq \arg\max_i \mathbf{F}(\mathcal{X})_i$ and t' is just like Eq. (2). The second constraint ensures that the prediction of the perturbed sample after the AE differs from the true label of the unperturbed sample. Similar to Eq. (2), this objective is hard to optimize, so we follow similar steps as in Eq. (4) and optimize the following objective for AdvPC using PGD (with ℓ_∞ as the distance metric):

$$\min_{\boldsymbol{\Delta}}\ (1 - \gamma)\ f_{t'}(\mathbf{F}(\mathcal{X}')) + \gamma\ f_{t''}(\mathbf{F}(\mathbf{G}(\mathcal{X}'))) \quad \text{s.t.}\ \mathcal{D}_{\ell_\infty}(\mathcal{X}, \mathcal{X}') \leq \epsilon_\infty \quad (6)$$

Here, f is as in Eq. (3), while γ is a hyper-parameter that trades off the attack's success before and after the AE. When $\gamma = 0$, the formulation in Eq. (6) becomes Eq. (4). We use PGD to solve Eq. (6) just like Eq. (4). We follow the same procedures as in [31] when solving Eq. (6) by keeping a record of any $\boldsymbol{\Delta}$ that satisfies the constraints in Eq. (5) and by trying different initializations for $\boldsymbol{\Delta}$.

4   Experiments

4.1   Setup

Dataset and Networks. We use ModelNet40 [30] to train the classifier network ($\mathbf{F}$) and the AE network ($\mathbf{G}$), as well as test our attacks. ModelNet40 contains 12,311 CAD models from 40 different classes. These models are divided into 9,843 for training and 2,468 for testing. Similar to previous work [31,37,38], we sample 1,024 points from each object. We train the $\mathbf{F}$ victim networks: PointNet[21], PointNet++ in both Single-Scale (SSG) and Multi-scale (MSG) [22] settings, and DGCNN [29]. For a fair comparison, we adopt the subset of ModelNet40

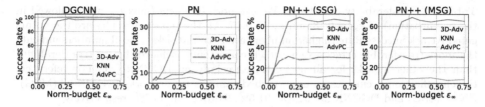

Fig. 4. Transferability Across Different Norm-Budgets: Here, the victim network is DGCNN [29] and the attacks are optimized using different ϵ_∞ norm-budgets. We report the attack success on DGCNN and on the transfer networks (PointNet, PointNet ++ MSG, and PointNet++ SSG). We note that our AdvPC transfers better to the other networks across different ϵ_∞ as compared to the baselines 3D-Adv[31] and KNN Attack [25]. Similar plots for the other victim networks are provided in the **supplement**.

detailed in [31] to perform and evaluate our attacks against their work (we call this the attack set). In the attack set, 250 examples are chosen from 10 ModelNet40 classes. We train the AE using the full ModelNet40 training set with the Chamfer Distance loss and then fix the AE when the attacks are being generated.

Adversarial Attack Methods. We compare AdvPC against the state-of-the-art baselines 3D-Adv [31] and KNN Attack [25]. For all attacks, we use Adam optimizer [10] with learning rate $\eta = 0.01$, and perform 2 different initializations for the optimization of Δ (as done in [31]). The number of iterations for the attack optimization for all the networks is 200. We set the loss margin $\kappa = 30$ in Eq. (3) for both 3D-Adv [31] and AdvPC and $\kappa = 15$ for KNN Attack [25] (as suggested in their paper). For other hyperparameters of [25,31], we follow what is reported in their papers. We pick $\gamma = 0.25$ in Eq. (6) for AdvPC because it strikes a balance between the success of the attack and its transferability (refer to Sect. 5.1 for details). In all of the attacks, we follow the same procedure as [31], where the best attack that satisfies the objective during the optimization is reported. We add the hard ℓ_∞ projection $\Pi_\infty (\Delta, \epsilon_\infty)$ described in Sect. 3 to all the methods to ensure fair comparison on the same norm-budget ϵ_∞. We report the best performance of the baselines obtained under this setup.

Transferability. We follow the same setup as [19,20] by generating attacks using the constrained ℓ_∞ metric and measure their success rate at different norm-budgets ϵ_∞ taken to be in the range $[0, 0.75]$. This range is chosen because it enables the attacks to reach 100% success on the victim network, as well as offer an opportunity for transferability to other networks. We compare AdvPC against the state-of-the-art baselines [25,31] under these norm-budgets (*e.g.* see Fig. 4 for attacking DGCNN). To measure the success of the attack, we compute the percentage of samples out of all attacked samples that the victim network misclassified. We also measure transferability from each victim network to the transfer networks. For each pair of networks, we optimize the attack on one network (victim) and measure the success rate of this optimized attack when

Table 1. Transferability of Attacks: We use norm-budgets (max ℓ_∞ norm allowed in the perturbation) of $\epsilon_\infty = 0.18$ and $\epsilon_\infty = 0.45$. All the reported results are the untargeted Attack Success Rate (higher numbers are better attacks). **Bold** numbers indicate the most transferable attacks. Our attack consistently achieves better transferability than the other attacks for all networks, especially on DGCNN [29]. For reference, the classification accuracies on unperturbed samples for networks PN, PN++(MSG), PN++(SSG) and DGCNN are 92.8%, 91.5%, 91.5%, and 93.7%, respectively.

		$\epsilon_\infty = 0.18$				$\epsilon_\infty = 0.45$			
Victim network	Attack	PN	PN++ (MSG)	PN ++ (SSG)	DGCNN	PN	PN++ (MSG)	PN++ (SSG)	DGCNN
PN	3D-Adv [31]	100	8.4	10.4	6.8	100	8.8	9.6	8.0
	KNN [25]	100	9.6	10.8	6.0	100	9.6	8.4	6.4
	AdvPC (Ours)	98.8	**20.4**	**27.6**	**22.4**	98.8	**18.0**	**26.8**	**20.4**
PN++	3D-Adv [31]	6.8	100	28.4	11.2	7.2	100	29.2	11.2
(MSG)	KNN [25]	6.4	100	22.0	8.8	6.4	100	23.2	7.6
	AdvPC (Ours)	**13.2**	97.2	**54.8**	**39.6**	**18.4**	98.0	**58.0**	**39.2**
PN++	3D-Adv [31]	7.6	9.6	100	6.0	7.2	10.4	100	7.2
(SSG)	KNN [25]	6.4	9.2	100	6.4	6.8	7.6	100	6.0
	AdvPC (Ours)	**12.0**	**27.2**	99.2	**22.8**	**14.0**	**30.8**	99.2	**27.6**
DGCNN	3D-Adv [31]	9.2	11.2	31.2	100	9.6	12.8	30.4	100
	KNN [25]	7.2	9.6	14.0	99.6	6.8	10.0	11.2	99.6
	AdvPC (Ours)	**19.6**	**46.0**	**64.4**	94.8	**32.8**	**48.8**	**64.4**	97.2

applied as input to the other network (transfer). We report these success rates for all network pairs. No defenses are used in the transferability experiment. All the attacks performed in this section are untargeted attacks (following the convention for transferability experiments [31]).

Attacking the Defenses. We also analyze the success of our attacks against point cloud defenses. We compare AdvPC attacks and the baselines [25,31] against several defenses used in the point cloud literature: SOR, SRS, DUP-Net [38], and Adversarial Training [31]. We also add a newly trained AE (different from the one used in the AdvPC attack) to this list of defenses. For SRS, we use a drop rate of 10%, while in SOR, we use the same parameters proposed in [38]. We train DUP-Net on ModelNet40 with an up-sampling rate of 2. For Adversarial Training, all four networks are trained using a mix of the training data of ModelNet40 and adversarial attacks generated by [31]. While these experiments are for untargeted attacks, we perform similar experiments under targeted attacks and report the results in **supplement** for reference and completeness.

4.2 Results

We present quantitative results that focus on two main aspects. First, we show the transferable power of AdvPC attacks to different point cloud networks. Second, we highlight the strength of AdvPC under different point cloud defenses.

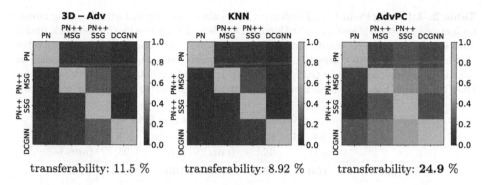

transferability: 11.5 %   transferability: 8.92 %   transferability: **24.9** %

Fig. 5. Transferability Matrix: Visualizing the overall transferability for 3D-Adv [31] (*left*), KNN Attack [25] (*middle*), and our AdvPC (*right*). Elements in the same row correspond to the same victim network used in the attack, while those in the same column correspond to the network that the attack is transferred to. Each matrix element measures the average success rate over the range of ϵ_∞ for the transfer network. We expect the diagonal elements of each transferability matrix (average success rate on the victim network) to have high values, since each attack is optimized on the same network it is transferred to. More importantly, brighter off-diagonal matrix elements indicate better transferability. We observe that our proposed AdvPC attack is more transferable than the other attacks and that DGCNN is a more transferable victim network than the other point cloud networks. The transferability score under each matrix is the average of the off-diagonal matrix values, which summarizes overall transferability for an attack.

Transferability. Table 1 reports transferability results for $\epsilon_\infty = 0.18$ and $\epsilon_\infty = 0.45$ and compares AdvPC with the baselines [25,31]. The value $\epsilon_\infty = 0.18$ is chosen, since it allows the DGCNN attack to reach maximum success (see Sect. 5.2), and the value $\epsilon_\infty = 0.45$ is arbitrarily chosen to be midway in the remaining range of ϵ_∞. It is clear that AdvPC attacks consistently beat the baselines when transferring between networks (up to 40%). Our method shows substantial gains in the case of DGCNN. We also report transferability results for a range of ϵ_∞ values in Fig. 4 when the victim network is DGCNN, and the attacks transferred to all other networks. In **supplement**, we show the same plots when the victim network is taken to be PN and PN++. To represent all these transferability curves compactly, we aggregate their results into a Transferability Matrix. Every entry in this matrix measures the transferability from the victim network (**row**) to the transfer network (**column**), and it is computed as the average success rate of the attack evaluated on the transfer network across all ϵ_∞ values. This value reflects how good the perturbation is at fooling the transfer network overall. As such, we advocate the use of the transferability matrix as a standard mode of evaluation for future work on network-transferable attacks. In Fig. 5, we show the transferability matrices for our attack and the baselines. AdvPC transfers better overall, since it leads to higher (brighter) off-diagonal values in the matrix. Using the average of off-diagonal elements in this matrix as a single scalar measure of transferability, AdvPC achieves 24.9% average transferability,

Table 2. Attacking Point Cloud Defenses: We evaluate untargeted attacks using norm-budgets of $\epsilon_\infty = 0.18$ and $\epsilon_\infty = 0.45$ with DGCNN [29] as the victim network under different defenses for 3D point clouds. Similar to before, we report attack success rates (**higher** indicates better attack). AdvPC consistently outperforms the other attacks [25,31] for all defenses. Note that both the attacks *and* evaluations are performed on DGCNN, which has an accuracy of 93.7% without input perturbations (for reference).

Defenses	$\epsilon_\infty = 0.18$			$\epsilon_\infty = 0.45$		
	3D-Adv [31]	KNN [25]	AdvPC (ours)	3D-Adv [31]	KNN [25]	AdvPC (ours)
No defense	100	99.6	94.8	100	99.6	97.2
AE (newly trained)	9.2	10.0	**17.2**	12.0	10.0	**21.2**
Adv Training [31]	7.2	7.6	**39.6**	8.8	7.2	**42.4**
SOR [38]	18.8	17.2	**36.8**	19.2	19.2	**32.0**
DUP Net [38]	28	28.8	**43.6**	28	31.2	**37.2**
SRS [38]	43.2	29.2	**80.0**	47.6	31.2	**85.6**

as compared to 11.5% for 3D-Adv [31] and 8.92% for KNN Attack [25]. We note that DGCNN [29] performs best in terms of transferability and is the hardest network to attack (for AdvPC and the baselines).

Attacking Defenses. Since DGCNN performs the best in transferability, we use it to evaluate the resilience of our AdvPC attacks under different defenses. We use the five defenses described in Sect. 4.1 and report their results in Table 2. Our attack is more resilient than the baselines against all defenses. We note that the AE defense is very strong against all attacks compared to other defenses [38], which explains why AdvPC works very well against other defenses and transfers well to unseen networks. We also observe that our attack is strong against simple statistical defenses like **SRS** (38% improvement over the baselines). We report results for other victim networks (PN and PN++) in the **supplement**, where AdvPC shows superior performance against the baselines under these defenses.

5   Analysis

We perform several analytical experiments to further explore the results obtained in Sect. 4.2. We first study the effect of different factors that play a role in the transferability of our attacks. We also show some interesting insights related to the sensitivity of point cloud networks and the effect of the AE on the attacks.

5.1   Ablation Study (Hyperparameter γ)

Here, we study the effect of γ used in Eq. (6) on the performance of our attacks. While varying γ between 0 and 1, we record the attack success rate on the victim network and report the transferability to all of the other three transfer networks

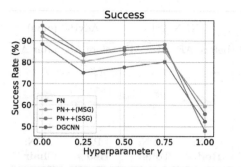
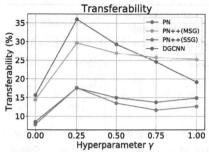

Fig. 6. Ablation Study: Studying the effect of changing AdvPC hyperparameter (γ) on the success rate of the attack (*left*) and on its transferability (*right*). The transferability score reported for each victim network is the average success rate on the transfer networks averaged across all different norm-budgets ϵ_∞. We note that as γ increases, the success rate of the attack on the victim network drops, and the transferability varies with γ. We pick $\gamma = 0.25$ in all of our experiments.

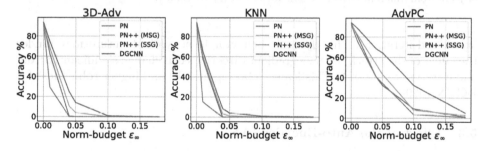

Fig. 7. Sensitivity of Architectures: We evaluate the sensitivity of each of the four networks for increasing norm-budget. For each network, we plot the classification accuracy under 3D-Adv perturbation [31] (*left*), KNN Attack [25] (*middle*), and our AdvPC attack (*right*). Overall, DGCNN [29] is affected the least by adversarial perturbation.

(average success rate on the transfer networks). We present averaged results over all norm-budgets in Fig. 6 for the four victim networks. One observation is that adding the AE loss with $\gamma > 0$ tends to deteriorate the success rate, even though it improves transferability. We pick $\gamma = 0.25$ in our experiments to balance success and transferability.

5.2 Network Sensitivity to Point Cloud Attacks

Figure 7 plots the sensitivity of the various networks when they are subject to input perturbations of varying norm-budgets ϵ_∞. We measure the classification accuracy of each network under our AdvPC attack ($\gamma = 0.25$), 3D-Adv [31], and KNN Attack [25]. We observe that DGCNN [29] tends to be the most robust to adversarial perturbations in general. This might be explained by the fact that the convolution neighborhoods in DGCNN are dynamically updated across

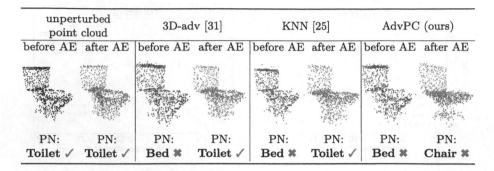

unperturbed point cloud		3D-adv [31]		KNN [25]		AdvPC (ours)	
before AE	after AE	before AE	after AE	before AE	after AE	before AE	after AE
PN: Toilet ✓	PN: Toilet ✓	PN: Bed ✗	PN: Toilet ✓	PN: Bed ✗	PN: Toilet ✓	PN: Bed ✗	PN: Chair ✗

Fig. 8. Effect of the Auto-Encoder (AE): The AE does not affect the unperturbed point cloud (classified correctly by PN before and after AE). The AE cleans the point cloud perturbed by 3D-Adv and KNN [25,31], which allows PN to predict the correct class label. However, our AdvPC attack can fool PN before and after AE reconstruction. Samples perturbed by AdvPC, if passed through the AE, transform into similar looking objects from different classes (Chair looks similar to Toilet).

layers and iterations. This dynamic behavior in network structure may hinder the effect of the attack because gradient directions can change significantly from one iteration to another. This leads to failing attacks and higher robustness for DGCNN [29].

5.3 Effect of the Auto-Encoder (AE)

In Fig. 8, we show an example of how AE reconstruction preserves the details of the unperturbed point cloud and does not change the classifier prediction. When a perturbed point cloud passes through the AE, it recovers a natural-looking shape. The AE's ability to reconstruct natural-looking 3D point clouds from various perturbed inputs might explain why it is a strong defense against attacks in Table 2. Another observation from Fig. 8 is that: when we fix the target t' and do not enforce a specific incorrect target t'' (*i.e.* untargeted attack setting) for the data adversarial loss on the reconstructed point cloud in the AdvPC attack (Eq. (6)), the optimization mechanism tends to pick t'' to be a *similar* class to the correct one. For example, a *Toilet* point cloud perturbed by AdvPC can be transformed into a *Chair* (similar in appearance to a toilet), if reconstructed by the AE. This effect is not observed for the other attacks [25,31], which do not consider the data distribution and optimize solely for the network. For completeness, we tried replacing the AE with other 3D generative models from [1] in our AdvPC attack, and we tried to use the learning approach in [19,20] instead of optimization, but the attack success was less than satisfactory in both cases (refer to **supplement**).

6 Conclusions

In this paper, we propose a new adversarial attack for 3D point clouds that utilizes a data adversarial loss to formulate network-transferable perturbations. Our attacks achieve better transferability to four popular point cloud networks than other 3D attacks, and they improve robustness against popular defenses. Future work would extend this attack to other 3D deep learning tasks, such as detection and segmentation, and integrate it into a robust training framework for point cloud networks.

Acknowledgments. This work was supported by the King Abdullah University of Science and Technology (KAUST) Office of Sponsored Research under Award No. RGC/3/3570-01-01.

References

1. Achlioptas, P., Diamanti, O., Mitliagkas, I., Guibas, L.: Learning representations and generative models for 3D point clouds (2018)
2. Alcorn, M.A., et al.: Strike (with) a pose: neural networks are easily fooled by strange poses of familiar objects. In: The IEEE Conference on Computer Vision and Pattern Recognition (CVPR) (2019)
3. Cao, Y., et al.: Adversarial objects against lidar-based autonomous driving systems. CoRR abs/1907.05418 (2019)
4. Carlini, N., Wagner, D.: Towards evaluating the robustness of neural networks. In: IEEE Symposium on Security and Privacy (SP) (2017)
5. Engelmann, F., Kontogianni, T., Hermans, A., Leibe, B.: Exploring spatial context for 3D semantic segmentation of point clouds. In: 2017 IEEE International Conference on Computer Vision Workshops (ICCVW), pp. 716–724, October 2017
6. Goodfellow, I., Shlens, J., Szegedy, C.: Explaining and harnessing adversarial examples. In: International Conference on Learning Representations (ICLR) (2015)
7. Hamdi, A., Ghanem, B.: Towards analyzing semantic robustness of deep neural networks. CoRR abs/1904.04621 (2019)
8. Hamdi, A., Muller, M., Ghanem, B.: SADA: semantic adversarial diagnostic attacks for autonomous applications. In: AAAI Conference on Artificial Intelligence (2020)
9. Huang, Q., Wang, W., Neumann, U.: Recurrent slice networks for 3D segmentation of point clouds. In: Proceedings of the IEEE Conference on Computer Vision and Pattern Recognition (CVPR), pp. 2626–2635 (2018)
10. Kingma, D.P., Ba, J.: Adam: a method for stochastic optimization. CoRR abs/1412.6980 (2014)
11. Kurakin, A., Goodfellow, I.J., Bengio, S.: Adversarial machine learning at scale. CoRR abs/1611.01236 (2016)
12. Landrieu, L., Boussaha, M.: Point cloud over segmentation with graph-structured deep metric learning, pp. 7440–7449 (2019)
13. Landrieu, L., Simonovsky, M.: Large-scale point cloud semantic segmentation with superpoint graphs. In: Proceedings of the IEEE Conference on Computer Vision and Pattern Recognition (CVPR), pp. 4558–4567 (2018)
14. Li, J., Chen, B.M., Hee Lee, G.: SO-Net: self-organizing network for point cloud analysis. In: Proceedings of the IEEE Conference on Computer Vision and Pattern Recognition (CVPR), pp. 9397–9406 (2018)

15. Li, Y., Bu, R., Sun, M., Wu, W., Di, X., Chen, B.: PointCNN: convolution on x-transformed points. In: Advances in Neural Information Processing Systems (NIPS), pp. 820–830 (2018)
16. Madry, A., Makelov, A., Schmidt, L., Tsipras, D., Vladu, A.: Towards deep learning models resistant to adversarial attacks. In: International Conference on Learning Representations (ICLR) (2018)
17. Moosavi-Dezfooli, S.M., Fawzi, A., Fawzi, O., Frossard, P.: Universal adversarial perturbations. In: The IEEE Conference on Computer Vision and Pattern Recognition (CVPR) (2017)
18. Moosavi-Dezfooli, S.M., Fawzi, A., Frossard, P.: DeepFool: a simple and accurate method to fool deep neural networks. In: The IEEE Conference on Computer Vision and Pattern Recognition (CVPR) (2016)
19. Naseer, M.M., Khan, S.H., Khan, M.H., Shahbaz Khan, F., Porikli, F.: Cross-domain transferability of adversarial perturbations. In: Advances in Neural Information Processing Systems (NeurIPS), pp. 12905–12915 (2019)
20. Poursaeed, O., Katsman, I., Gao, B., Belongie, S.: Generative adversarial perturbations. In: Proceedings of the IEEE Conference on Computer Vision and Pattern Recognition (CVPR), pp. 4422–4431 (2018)
21. Qi, C.R., Su, H., Mo, K., Guibas, L.J.: PointNet: deep learning on point sets for 3D classification and segmentation. In: Proceedings of the IEEE Conference on Computer Vision and Pattern Recognition (CVPR), pp. 652–660 (2017)
22. Qi, C.R., Yi, L., Su, H., Guibas, L.J.: PointNet++: deep hierarchical feature learning on point sets in a metric space. In: Advances in Neural Information Processing Systems (NIPS), pp. 5099–5108 (2017)
23. Szegedy, C., et al.: Intriguing properties of neural networks. CoRR abs/1312.6199 (2013)
24. Tatarchenko, M., Park, J., Koltun, V., Zhou, Q.Y.: Tangent convolutions for dense prediction in 3D. In: Proceedings of the IEEE Conference on Computer Vision and Pattern Recognition (CVPR), pp. 3887–3896 (2018)
25. Tsai, T., Yang, K., Ho, T.Y., Jin, Y.: Robust adversarial objects against deep learning models. In: AAAI Conference on Artificial Intelligence (2020)
26. Tu, C.C., et al.: Autozoom: autoencoder-based zeroth order optimization method for attacking black-box neural networks. In: Proceedings of the AAAI Conference on Artificial Intelligence, vol. 33, pp. 742–749 (2019)
27. Tu, J., et al.: Physically realizable adversarial examples for lidar object detection. In: Proceedings of the IEEE Conference on Computer Vision and Pattern Recognition (CVPR), pp. 13716–13725 (2020)
28. Wang, W., Yu, R., Huang, Q., Neumann, U.: SGPN: similarity group proposal network for 3D point cloud instance segmentation. In: Proceedings of the IEEE Conference on Computer Vision and Pattern Recognition (CVPR), pp. 2569–2578 (2018)
29. Wang, Y., Sun, Y., Liu, Z., Sarma, S.E., Bronstein, M.M., Solomon, J.M.: Dynamic graph CNN for learning on point clouds. ACM Trans. Graph. (TOG) 38, 1–12 (2019)
30. Wu, Z., et al.: 3D shapenets: a deep representation for volumetric shapes. In: 2015 IEEE Conference on Computer Vision and Pattern Recognition (CVPR), pp. 1912–1920 (2015)
31. Xiang, C., Qi, C.R., Li, B.: Generating 3D adversarial point clouds. In: Proceedings of the IEEE Conference on Computer Vision and Pattern Recognition (CVPR), pp. 9136–9144 (2019)

32. Xiao, C., Yang, D., Li, B., Deng, J., Liu, M.: MeshAdv: adversarial meshes for visual recognition. In: Proceedings of the IEEE Conference on Computer Vision and Pattern Recognition (CVPR), pp. 6898–6907 (2019)

33. Ye, X., Li, J., Huang, H., Du, L., Zhang, X.: 3D recurrent neural networks with context fusion for point cloud semantic segmentation. In: Ferrari, V., Hebert, M., Sminchisescu, C., Weiss, Y. (eds.) ECCV 2018. LNCS, vol. 11211, pp. 415–430. Springer, Cham (2018). https://doi.org/10.1007/978-3-030-01234-2_25

34. Yu, L., Li, X., Fu, C.W., Cohen-Or, D., Heng, P.A.: PU-Net: point cloud upsampling network. In: Proceedings of IEEE Conference on Computer Vision and Pattern Recognition (CVPR) (2018)

35. Zeng, X., et al.: Adversarial attacks beyond the image space. In: The IEEE Conference on Computer Vision and Pattern Recognition (CVPR) (2019)

36. Zhao, Z., Dua, D., Singh, S.: Generating natural adversarial examples. In: International Conference on Learning Representations (ICLR) (2018)

37. Zheng, T., Chen, C., Yuan, J., Li, B., Ren, K.: PointCloud saliency maps. In: The IEEE International Conference on Computer Vision (ICCV) (2019)

38. Zhou, H., Chen, K., Zhang, W., Fang, H., Zhou, W., Yu, N.: DUP-Net: denoiser and upsampler network for 3d adversarial point clouds defense. In: The IEEE International Conference on Computer Vision (ICCV) (2019)

Edge-Aware Graph Representation Learning and Reasoning for Face Parsing

Gusi Te[1], Yinglu Liu[2], Wei Hu[1](✉), Hailin Shi[2], and Tao Mei[2]

[1] Wangxuan Institute of Computer Technology, Peking University, Beijing, China
{tegusi,forhuwei}@pku.edu.cn
[2] JD AI Research, Beijing, China
{liuyinglu1,shihailin,tmei}@jd.com

Abstract. Face parsing infers a pixel-wise label to each facial component, which has drawn much attention recently. Previous methods have shown their efficiency in face parsing, which however overlook the correlation among different face regions. The correlation is a critical clue about the facial appearance, pose, expression, *etc.*, and should be taken into account for face parsing. To this end, we propose to model and reason the region-wise relations by learning graph representations, and leverage the edge information between regions for optimized abstraction. Specifically, we encode a facial image onto a global graph representation where a collection of pixels ("regions") with similar features are projected to each vertex. Our model learns and reasons over relations between the regions by propagating information across vertices on the graph. Furthermore, we incorporate the edge information to aggregate the pixel-wise features onto vertices, which emphasizes on the features around edges for fine segmentation along edges. The finally learned graph representation is projected back to pixel grids for parsing. Experiments demonstrate that our model outperforms state-of-the-art methods on the widely used Helen dataset, and also exhibits the superior performance on the large-scale CelebAMask-HQ and LaPa dataset. The code is available at https://github.com/tegusi/EAGRNet.

Keywords: Face parsing · Graph representation · Attention mechanism · Graph reasoning

1 Introduction

Face parsing assigns a pixel-wise label to each semantic component, such as facial skin, eyes, mouth and nose, which is a particular task in semantic segmentation. It has been applied in a variety of scenarios such as face understanding, editing, synthesis, and animation [1–3].

W. Hu—This work was in collaboration with JD AI Research during Gusi Te's internship there.

© Springer Nature Switzerland AG 2020
A. Vedaldi et al. (Eds.): ECCV 2020, LNCS 12357, pp. 258–274, 2020.
https://doi.org/10.1007/978-3-030-58610-2_16

The region-based methods have been recently proposed to model the facial components separately [4–6], and achieved state-of-the-art performance on the current benchmarks. However, these methods are based on the individual information within each region, and the correlation among regions is not exploited yet to capture long range dependencies. In fact, facial components present themselves with abundant correlation between each other. For instance, eyes, mouth and eyebrows will generally become more curvy when people smile; facial skin and other components will be dark when the lighting is weak, and so on.

The correlation between the facial components is the critical clue in face representation, and should be taken into account in the face parsing. To this end, we propose to learn graph representations over facial images, which model the relations between regions and enable reasoning over non-local regions to capture long range dependencies. To bridge the facial image pixels and graph vertices, we project a collection of pixels (a "region") with similar features to each vertex. The pixel-wise features in a region are aggregated to the feature of the corresponding vertex. In particular, to achieve accurate segmentation along the edges between different components, we propose the edge attention in the pixel-to-vertex projection, assigning larger weights to the features of edge pixels during the feature aggregation. Further, the graph representation learns the relations between facial regions, $i.e.$, the graph connectivity between vertices, and reasons over the relations by propagating information across all vertices on the graph, which is able to capture long range correlations in the facial image. The learned graph representation is finally projected back to the pixel grids for face parsing. Since the number of vertices is significantly smaller than that of pixels, the graph representation also reduces redundancy in features as well as computational complexity effectively.

Specifically, given an input facial image, we first encode the high-level and low-level feature maps by the ResNet backbone [7]. Then, we build a projection matrix to map a cluster of pixels with similar features to each vertex. The feature of each vertex is taken as the weighted aggregation of pixel-wise features in the cluster, where features of edge pixels are assigned with larger weights via an edge mask. Next, we learn and reason over the relations between vertices ($i.e.$, regions) via graph convolution [8,9] to further extract global semantic features. The learned features are finally projected back to a pixel-wise feature map. We test our model on Helen, CelebAMask-HQ and LaPa datasets, and surpass state-of-the-art methods.

Our main contributions are summarized as follows.

- We propose to exploit the relations between regions for face parsing by modeling on a region-level graph representation, where we project a collection of pixels with similar features to each vertex and reason over the relations to capture long range dependencies.
- We introduce edge attention in the pixel-to-vertex feature projection, which emphasizes on features of edge pixels during the feature aggregation to each vertex and thus enforces accurate segmentation along edges.

– We conduct extensive experiments on Helen, CelebAMask-HQ and LaPa datasets. The experimental results show our model outperforms state-of-the-art methods on almost every category.

2 Related Work

2.1 Face Parsing

Face parsing is a division of semantic segmentation, which assigns different labels to the corresponding regions on human faces, such as nose, eyes, mouth and *etc.*. The methods of face parsing could be classified into global-based and local-based methods.

Traditionally, hand crafted features including SIFT [10] are applied to model the facial structure. Warrell *et al.* describe spatial relationship of facial parts with epitome model [11]. Kae *et al.* combine Conditional Random Field (CRF) with a Restricted Boltzmann Machine (RBM) to extract local and global features [12]. With the rapid development of machine learning, CNN has been introduced to learn more robust and rich features. Liu *et al.* import CNN-based features into the CRF framework to model individual pixel labels and neighborhood dependencies [13]. Luo *et al.* propose a hierarchical deep neural network to extract multi-scale facial features [14]. Zhou *et al.* adopt adversarial learning approach to train the network and capture high-order inconsistency [15]. Liu *et al.* design a CNN-RNN hybrid model that benefits from both high quality features of CNN and non-local properties of RNN [6]. Zhou *et al.* present an interlinked CNN that takes multi-scale images as input and allows bidirectional information passing [16]. Lin *et al.* propose a novel RoI Tanh-Warping operator preserving central and peripheral information. It contains two branches with the local-based for inner facial components and the global based for outer facial ones. This method shows high performance especially on hair segmentation [4].

2.2 Attention Mechanism

Attention mechanism has been proposed to capture long-range information [17], and applied to many applications such as sentence encoding [18] and image feature extraction [19]. Limited by the locality of convolution operators, CNN lacks the ability to model global contextual information. Furthermore, Chen *et al.* propose Double Attention Model that gathers information spatially and temporally to improve complexity of traditional non-local modules [20]. Zhao *et al.* propose a point-wise spatial attention module, relaxing the local neighborhood constraint [21]. Zhu *et al.* also present an asymmetric module to reduce abundant computation and distillate features [22]. Fu *et al.* devise a dual attention module that applies both spatial and channel attention in feature maps [23]. To research underlying relationship between different regions, Chen *et al.* project original features into interactive space and utilize GCN to exploit high order relationship [24]. Li *et al.* devise a robust attention module that incorporates the Expectation-Maximization algorithm [25].

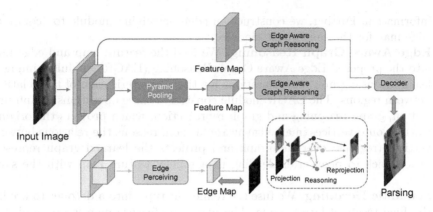

Fig. 1. The overview of the proposed face parsing framework.

2.3 Graph Reasoning

Interpreting images from the graph perspective is an interesting idea, since an image could be regarded as regular pixel grids. Chandra *et al.* propose Conditional Random Field (CRF) based method on image segmentation [26]. Besides, graph convolution network (GCN) is imported into image segmentation. Li *et al.* introduce graph convolution to the semantic segmentation, which projects features into vertices in the graph domain and applies graph convolution afterwards [27]. Furthermore, Lu *et al.* propose Graph-FCN where semantic segmentation is reduced to vertex classification by directly transforming an image into regular grids [28]. Pourian *et al.* propose a method of semi-supervised segmentation [29]. The image is divided into community graph and different labels are assigned to corresponding communities. Te *et al.* propose a computation-efficient and posture-invariant face representation with only a few key points on hypergraphs for face anti-spoofing beyond 2D attacks [30]. Zhang *et al.* utilize graph convolution both in the coordinate space and feature space [31].

3 Methods

3.1 Overview

As illustrated in Fig. 1, given an input facial image, we aim to predict the corresponding parsing label and auxiliary edge map. The overall framework of our method consists of three procedures as follows.

- **Feature and Edge Extraction.** We take ResNet as the backbone to extract features at various levels for multi-scale representation. The low-level features contain more details but lack semantic information, while the high-level features provide rich semantics with global information at the cost of image details. To fully exploit the global information in high-level features, we employ a spatial pyramid pooling operation to learn multi-scale contextual information. Further, we construct an edge perceiving module to acquire an edge map for the subsequent module.

- **Edge Aware Graph Reasoning.** We feed the feature map and edge map into the proposed Edge Aware Graph Reasoning (EAGR) module, aiming to learn intrinsic graph representations for the characterization of the relations between regions. The EAGR module consists of three operations: graph projection, graph reasoning and graph reprojection, which projects the original features onto vertices in an edge-aware fashion, reasons the relations between vertices (regions) over the graph and projects the learned graph representation back to pixel grids, leading to a refined feature map with the same size.

- **Semantic Decoding.** We fuse the refined features into a decoder to predict the final result of face parsing. The high-level feature map is upsampled to the same dimension as the low-level one. We concatenate both feature maps and leverage 1 × 1 convolution layer to reduce feature channels, predicting the final parsing labels.

3.2   Edge-Aware Graph Reasoning

Inspired by the non-local module [19], we aim to build the long-range interactions between distant regions, which is critical for the description of the facial structure. In particular, we propose edge-aware graph reasoning to model the long-range relations between regions on a graph, which consists of edge-aware graph projection, graph reasoning and graph reprojection.

Edge-Aware Graph Projection. We first revisit the typical non-local modules. Given a feature map $\mathbf{X} \in \mathbb{R}^{HW \times C}$, where H and W refer to the height and width of the input image respectively and C is the number of feature channels. A typical non-local module is formulated as:

$$\widetilde{\mathbf{X}} = \text{softmax} \left(\theta(\mathbf{X})\varphi^{\top}(\mathbf{X}) \right) \gamma(\mathbf{X}) = \mathbf{V}\gamma(\mathbf{X}), \tag{1}$$

where θ, φ and γ are convolution operations with 1×1 kernel size. $\mathbf{V} \in \mathbb{R}^{HW \times HW}$ is regarded as the attention maps to model the long-range dependencies. However, the complexity of computing $\mathbf{V}$ is $\mathcal{O}(H^2 W^2 C)$, which does not scale well with increasing number of pixels HW. To address this issue, we propose a simple yet effective edge-aware projection operation to eliminate the redundancy in features.

Given an input feature map $\mathbf{X} \in \mathbb{R}^{HW \times C}$ and an edge map $\mathbf{Y} \in \mathbb{R}^{HW \times 1}$, we construct a projection matrix $\mathbf{P}$ by mapping $\mathbf{X}$ onto vertices of a graph with $\mathbf{Y}$ as a prior. Specifically, we first reduce the dimension of $\mathbf{X}$ in the feature space via a convolution operation φ with 1×1 kernel size, leading to $\varphi(\mathbf{X}) \in \mathbb{R}^{HW \times T}$, $T < C$. Then, we duplicate the edge map Y to the same dimension of $\varphi(\mathbf{X})$ for ease of computation. We incorporate the edge information into the projection, by taking the Hadamard Product of $\varphi(\mathbf{X})$ and $\mathbf{Y}$. As the edge map $\mathbf{Y}$ encodes the probability of each pixel being an edge pixel, the Hadamard Product operation

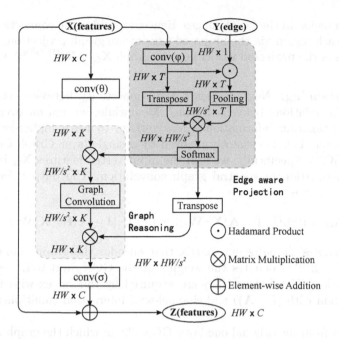

Fig. 2. Architecture of the Edge Aware Graph Reasoning module.

essentially assigns a weight to the feature of each pixel, with larger weights to features of edge pixels. Further, we introduce an average *pooling* operation $\mathcal{P}(\cdot)$ with stride s to obtain anchors of vertices. These anchors represent the centers of each region of pixels, and we take the multiplication of $\varphi(\mathbf{X})$ and anchors to capture the similarity between anchors and each pixel. We then apply a softmax function for normalization. Formally, the projection matrix takes the form (Fig. 2):

$$\mathbf{P} = \mathrm{softmax}\left(\mathcal{P}(\varphi(\mathbf{X}) \odot \mathbf{Y}) \cdot \varphi(\mathbf{X})^{\top}\right), \qquad (2)$$

where $\odot$ denotes the Hadamard product, and $\mathbf{P} \in \mathbb{R}^{HW/s^2 \times HW}$.

In Eq. (2), we have two critical operations: the edge attention and the pooling operation. The edge attention emphasizes the features of edge pixels by assigning larger weights to edge pixels. Further, we propose the pooling operation in the features, whose benefits are in twofold aspects. On one hand, the pooling leads to compact representations by averaging over features to remove the redundancy. On the other hand, by pooling with stride s, the computation complexity is reduced from $\mathcal{O}(H^2W^2C)$ in non-local modules to $\mathcal{O}(H^2W^2C/s^2)$.

With the acquired projection matrix $\mathbf{P}$, we project the pixel-wise features $\mathbf{X}$ onto the graph domain, *i.e.*,

$$\mathbf{X}_G = \mathbf{P}\theta(\mathbf{X}), \qquad (3)$$

where θ is a convolution operation with 1×1 kernel size so as to reduce the dimension of $\mathbf{X}$, resulting in $\theta(\mathbf{X}) \in \mathbb{R}^{HW \times K}$. The projection aggregates pixels with similar features as each anchor to one vertex, thus each vertex essentially represents a region in the facial images. Hence, we bridge the connection between pixels and each region via the proposed edge aware graph projection, leading to the features of the projected vertices on the graph $\mathbf{X}_G \in \mathbb{R}^{HW/s^2 \times K}$ via Eq. (3).

Graph Reasoning. Next, we learn the connectivity between vertices from $\mathbf{X}_G$, i.e., the relations between regions. Meanwhile, we reason over the relations by propagating information across vertices to learn higher-level semantic information. This is elegantly realized by a single-layer Graph Convolution Network (GCN). Specifically, we feed the input vertex features $\mathbf{X}_G$ into a first-order approximation of spectral graph convolution. The output feature map $\hat{\mathbf{X}}_G \in \mathbb{R}^{HW/s^2 \times K}$ is

$$\hat{\mathbf{X}}_G = \mathrm{ReLU}\left[(\mathbf{I} - \mathbf{A})\mathbf{X}_G\mathbf{W}_G\right] = \mathrm{ReLU}\left[(\mathbf{I} - \mathbf{A})\mathbf{P}\theta(\mathbf{X})\mathbf{W}_G\right], \qquad (4)$$

where $\mathbf{A}$ denotes the adjacent matrix that encodes the graph connectivity to learn, $\mathbf{W}_G \in \mathbb{R}^{K \times K}$ denotes the weights of the GCN, and ReLU is the activation function. The features $\hat{\mathbf{X}}_G$ are acquired by the vertex-wise interaction (multiplication with $(\mathbf{I} - \mathbf{A})$) and channel-wise interaction (multiplication with $\mathbf{W}_G$).

Different from the original one-layer GCN [32] in which the graph $\mathbf{A}$ is handcrafted, we randomly initialize $\mathbf{A}$ and learn from vertex features. Moreover, we add a residual connection to reserve features of raw vertices. Based on the learned graph, the information propagation across all vertices leads to the finally reasoned relations between regions. After graph reasoning, pixels embedded within one vertex share the same context of features modeled by graph convolution. We set the same number of output channels as the input to keep consistency, allowing the module to be compatible with the subsequent process.

Graph Reprojection. In order to fit into existing framework, we reproject the extracted vertex features in the graph domain to the original pixel grids. Given the learned graph representation $\hat{\mathbf{X}}_G \in \mathbb{R}^{HW/s^2 \times K}$, we aim to compute a matrix $\mathbf{V} \in \mathbb{R}^{HW \times HW/s^2}$ that maps $\hat{\mathbf{X}}_G$ to the pixel space. In theory, $\mathbf{V}$ could be taken as the inverse of the projection matrix $\mathbf{P}$. However, it is nontrivial to compute because $\mathbf{P}$ is not a square matrix. To tackle this problem, we take the transpose matrix $\mathbf{P}^\top$ as the reprojection matrix [27], in which $\mathbf{P}_{ij}^\top$ reflects the correlation between vertex i and pixel j. The limitation of this operation is that the row vectors in $\mathbf{P}^\top$ are not normalized.

After reprojection, we deploy a 1×1 convolution operation σ to increase the feature channels in consistent with the input features $\mathbf{X}$. Then, we take the summation of the reprojected refined features and the original feature map as the final features. The final pixel-wise feature map $\mathbf{Z} \in \mathbb{R}^{HW \times C}$ is thus computed by

$$\mathbf{Z} = \mathbf{X} + \sigma(\mathbf{P}^\top \hat{\mathbf{X}}_G). \qquad (5)$$

3.3 The Loss Function

To further strengthen the effect of the proposed edge aware graph reasoning, we introduce the boundary-attention loss (BA-Loss) inspired by [33] besides the traditional cross entropy loss for predicted parsing maps and edge maps. The BA-loss computes the loss between the predicted label and the ground truth only at edge pixels, thus improving the segmentation accuracy of critical edge pixels that are difficult to distinguish. Mathematically, the BA-loss is written as

$$\mathcal{L}_{\mathrm{BA}} = \sum_{i=1}^{HW} \sum_{j=1}^{N} [e_i = 1]\, y_{ij} \log p_{ij}, \tag{6}$$

where i is the index of pixels, j is the index of classes and N is the number of classes. e_i denotes the edge label, y_{ij} denotes the ground truth label of face parsing, and p_{ij} denotes the predicted parsing label. $[\cdot]$ is the Iverson bracket, which denotes a number that is 1 if the condition in the bracket is satisfied, and 0 otherwise.

The total loss function is then defined as follows:

$$\mathcal{L} = \mathcal{L}_{\mathrm{parsing}} + \lambda_1 \mathcal{L}_{\mathrm{edge}} + \lambda_2 \mathcal{L}_{\mathrm{BA}}, \tag{7}$$

where $\mathcal{L}_{\mathrm{parsing}}$ and $\mathcal{L}_{\mathrm{edge}}$ are classical cross entropy losses for the parsing and edge maps. λ_1 and λ_2 are two hyper-parameters to strike a balance among the three loss functions.

3.4 Analysis

Since non-local modules and graph-based methods have drawn increasing attention, it is interesting to analyze the similarities and differences between previous works and our method.

Comparison with Non-local Modules. Typically, a traditional non-local module models *pixel-wise* correlations by feature similarities. However, the high-order relationship between regions are not captured. In contrast, we exploit the correlation among distinct regions via the proposed graph projection and reasoning. The features of each vertex embed not only local contextual anchor aggregated by average pooling in a certain region but also global features from the overall pixels. We further learn and reason over the relations between regions by graph convolution, which captures high-order semantic relations between different facial regions.

Also, the computation complexity of non-local modules is expensive in general as discussed in Sect. 3.2. Our proposed edge-aware pooling addresses the issue by extracting significant anchors to replace redundant query points. Also, we do not incorporate pixels within each facial region during the sampling process while focusing on edge pixels, thus improving boundary details. The intuition is that pixels within each region tend to share similar features.

Comparison with Graph-Based Models. In comparison with other graph-based models, such as [24, 27], we improve the graph projection process by introducing locality in sampling in particular. In previous works, each vertex is simply represented by a weighted sum of image pixels, which does not consider edge information explicitly and brings ambiguity in understanding vertices. Besides, with different inputs of feature maps, the pixel-wise features often vary greatly but the projection matrix is fixed after training. In contrast, we incorporate the edge information into the projection process to emphasize on edge pixels, which preserves boundary details well. Further, we specify vertex anchors locally based on the average pooling, which conforms with the rule that the location of facial components keeps almost unchanged after face alignment.

4 Experiments

4.1 Datasets and Metrics

The Helen dataset includes 2,330 images with 11 categories: background, skin, left/right brow, left/right eye, upper/lower lip, inner mouth and hair. Specifically, we keep the same train/validation/test protocol as in [34]. The number of the training, validation and test samples are 2,000, 230 and 100, respectively. The CelebAMask-HQ dataset is a large-scale face parsing dataset which consists of 24,183 training images, 2,993 validation images and 2,824 test images. The number of categories in CelebAMask-HQ is 19. In addition to facial components, the accessories such as eyeglass, earring, necklace, neck, and cloth are also annotated in the CelebAMask-HQ dataset. The LaPa dataset is a newly released challenging dataset for face parsing, which contains 11 categories as Helen, covering large variations in facial expression, pose and occlusion. It consists of 18,176 training images, 2,000 validation images and 2,000 test images.

During training, we use the rotation and scale augmentation. The rotation angle is randomly selected from $(-30°, 30°)$ and the scale factor is randomly selected from $(0.75, 1.25)$. The edge mask is extracted according to the semantic label map. If the label of a pixel is different with its 4 neighborhoods, it is regarded as a edge pixel. For the Helen dataset, similar to [4], we implement face alignment as a pre-processing step and the results are re-mapped to the original image for evaluation.

We employ three evaluation metrics to measure the performance of our model: pixel accuracy, mean intersection over union (mIoU) and F1 score. Directly employing the accuracy metric ignores the scale variance amid facial components, while the mean IoU and F1 score are better for evaluation. To keep consistent with the previous methods, we report the overall F1-score on the Helen dataset, which is computed over the merged facial components: brows (left+right), eyes (left+right), nose, mouth (upper lip+lower lip+inner mouth). For the CelebAMask-HQ and LaPa datasets, the mean F1-score over all categories excluding background is employed.

Table 1. Ablation study on the Helen dataset.

Model	Baseline	Edge	Graph	Reasoning	BA-loss	mIoU	F1-score	Accuracy
1	✓					76.5	91.4	85.9
2	✓	✓				77.5	92.0	86.2
3	✓		✓	✓		77.3	92.3	85.8
4	✓	✓	✓	✓		77.8	92.4	84.6
5	✓	✓	✓		✓	77.3	92.3	86.7
6	✓	✓	✓	✓	✓	78.2	92.8	87.3

4.2 Implementation Details

Our backbone is a modified version of the ResNet-101 [7] excluding the average pooling layer, and the Conv1 block is changed to three 3×3 convolutional layers. For the pyramid pooling module, we follow the implementation in [35] to exploit global contextual information. The pooling factors are $\{1, 2, 3, 6\}$. Similar to [36], the edge perceiving module predicts a two-channel edge map based on the outputs of Conv2, Conv3 and Conv4 in ResNet-101. The outputs of Conv1 and the pyramid pooling serve as the low-level and high-level feature maps, respectively. Both of them are fed into the EAGR module separately for graph representation learning.

As for the EAGR module, we set the pooling size to 6×6. To pay more attention on the facial components, we just utilize the central 4×4 anchors for graph construction. The feature dimensions K and T are set to 128 and 64, respectively.

Stochastic Gradient Descent (SGD) is employed for optimization. We initialize the network with a pretrained model on ImageNet. The input size is 473×473 and the batch size is set to 28. The learning rate starts at 0.001 with the weight decay of 0.0005. The batch normalization is implemented with In-Place Activated Batch Norm [37].

4.3 Ablation Study

On Different Components. We demonstrate the effectiveness of different components in the proposed EAGR module. Specifically, we remove some components and train the model from scratch under the same initialization. The quantitative results are reported in Table 1. *Baseline* means the model only utilizes the ResNet backbone, pyramid pooling and multi-scale decoder without any EGAR module, and *Edge* represents whether edge aware pooling is employed. *Graph* represents the EAGR module, while *Reasoning* indicates the graph reasoning excluding graph projection and reprojection. We observe that *Edge* and *Graph* lead to improvement over the baseline by 1% in mIoU respectively. When both components are taken into account, we achieve even better performance. The boundary-attention loss (BA-loss) also leads to performance improvement.

Model 1   Model 2   Model 3   Ours   Ground Truth

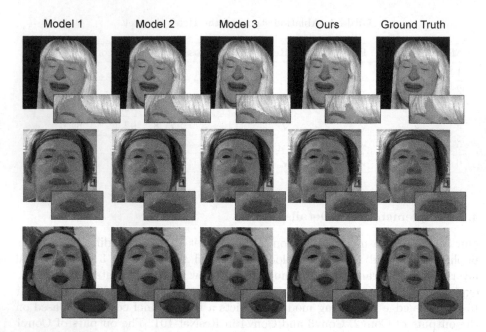

Fig. 3. Parsing results of different models on the Helen dataset. (Best viewed in color) (Color figure online)

Table 2. Performance comparison with different deployment of the EAGR module and pooling size.

Model	Deployment				Pooling size		
	0-module	1-module	2-modules	3-modules	4 × 4	6 × 6	8 × 8
mIoU	77.6	77.6	78.2	77.4	77.0	78.2	78.0
F1-score	92.0	92.5	92.8	92.3	92.1	92.8	92.6
Accuracy	85.5	86.0	87.3	85.4	87.4	87.3	87.0

We also provide subjective results of face parsing from different models in Fig. 3. Results of incomplete models exhibit varying degrees of deficiency around edges in particular, such as the edge between the hair and skin in the first row, the upper lips in the second row, and edges around the mouse in the third row. In contrast, our complete model produce the best results with accurate edges between face constitutes, which is almost the same as the ground truth. This validates the effectiveness of the proposed edge aware graph reasoning.

On the Deployment of the EAGR Module. We also conduct experiments on the deployment of the EAGR module with respect to the feature maps as well as pooling sizes. We take the output of Conv2 in the ResNet as the low-level

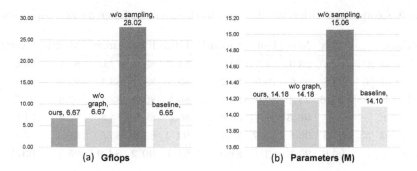

Fig. 4. Complexity comparison on the Helen dataset. We reset the start value of y-axis for better appearance.

feature map, and that of the pyramid pooling module as the high-level feature map. We compare four deployment schemes: 1) 0-module, where no EAGR module is applied; 2) 1-module, where the low-level and high-level feature maps are concatenated, and then fed into an EAGR module; 3) 2-modules, where the low-level and high-level feature maps are fed into one EAGR module respectively; 4) 3-moduels, which combines 2) and 3). As listed in Table 2, the scheme of 2-modules leads to the best performance, which is the one we finally adopt.

We also test the influence of the pooling size, where the number of vertices changes along with the pooling size. As presented in Table 2, the size of 6×6 leads to the best performance, while enlarging the pooling size further does not bring performance improvement. This is because more detailed anchors lead to the loss of integrity, which breaks the holistic semantic representation.

On the Complexity in Time and Space. Further, we study the complexity of different models in time and space in Fig. 4. We compare with three schemes: 1) a simplified version without the EAGR module, which we refer to as the *Baseline*; 2) a non-local module [19] employed without edge aware sampling (*i.e.*, pooling) as *Without sampling*; and 3) a version without graph convolution for reasoning as *Without graph*. As presented in Fig. 4, compared with the typical non-local module, our proposed method reduces the computation time by more than 4× in terms of flops. We also see that the computation and space complexity of our method is comparable to those of the *Baseline*, which indicates that most complexity comes from the backbone network. Using Nvidia P40, the time cost of our model for a single image is 89ms in the inference stage. This demonstrates that the proposed EAGR module achieves significant performance improvement with trivial computational overhead.

4.4   Comparison with the State-of-the-Art

We conduct experiments on the broadly acknowledged Helen dataset to demonstrate the superiority of the proposed model. To keep consistent with the previous

Table 3. Comparison with state-of-the-art methods on the Helen dataset (in F1 score).

Methods	Skin	Nose	U-lip	I-mouth	L-lip	Eyes	Brows	Mouth	Overall
Liu *et al.* [6]	92.1	93.0	74.3	79.2	81.7	86.8	77.0	89.1	88.6
Lin *et al.* [4]	94.5	95.6	79.6	86.7	89.8	89.6	83.1	95.0	92.4
Wei *et al.* [38]	**95.6**	95.2	80.0	86.7	86.4	89.0	82.6	93.6	91.7
Yin *et al.* [5]	-	**96.3**	82.4	85.6	86.6	89.5	84.8	92.8	91.0
Liu *et al.* [33]	94.9	95.8	**83.7**	89.1	**91.4**	89.8	83.5	**96.1**	93.1
Ours	94.6	96.1	83.6	**89.8**	91.0	**90.2**	**84.9**	95.5	**93.2**

Table 4. Experimental comparison on the CelebAMask-HQ dataset (in F1 score).

Methods	Face	Nose	Glasses	L-Eye	R-Eye	L-Brow	R-Brow	L-Ear	R-Ear	Mean
	I-Mouth	U-Lip	L-Lip	Hair	Hat	Earring	Necklace	Neck	Cloth	
Zhao *et al.* [35]	94.8	90.3	75.8	79.9	80.1	77.3	78	75.6	73.1	76.2
	89.8	87.1	88.8	90.4	58.2	65.7	19.4	82.7	64.2	
Lee *et al.* [1]	95.5	85.6	**92.9**	84.3	85.2	81.4	81.2	84.9	83.1	80.3
	63.4	88.9	90.1	86.6	**91.3**	63.2	26.1	**92.8**	68.3	
Ours	**96.2**	**94**	92.3	**88.6**	**88.7**	**85.7**	**85.2**	88	**85.7**	**85.1**
	95	**88.9**	**91.2**	**94.9**	87.6	**68.3**	**27.6**	89.4	**85.3**	

works[4–6,33,38], we employ the overall F1 score to measure the performance, which is computed by combining the merged eyes, brows, nose and mouth categories. As Table 3 shows, Our model surpasses state-of-the-art methods and achieves 93.2% on this dataset.

We also evaluate our model on the newly proposed CelebAMask-HQ [1] and LaPa [33] datasets, whose scales are about 10 times larger than the Helen dataset. Different from the Helen dataset, CelebAMask-HQ and LaPa have accurate annotation for hair. Therefore, mean F1-score (over all foreground categories) is employed for better evaluation. Table 4 and Table 5 give the comparison results of the related works and our method on these two datasets, respectively.

4.5   Visualization of Graph Projection

Further, we visualize the graph projection for intuitive interpretation. As in Fig. 5, given each input image (first row), we visualize the weight of each pixel that contributes to a vertex marked in a blue rectangle in the other rows, which we refer to as the response map. Darker color indicates higher response. We observe that the response areas are consistent with the vertex, which validates that our graph projection maps pixels in the same semantic component to the same vertex.

Table 5. Experimental comparison on the LaPa dataset (in F1 score).

Methods	Skin	Hair	L-Eye	R-Eye	U-lip	I-mouth	L-lip	Nose	L-Brow	R-Brow	Mean
Zhao *et al.* [35]	93.5	94.1	86.3	86.0	83.6	86.9	84.7	94.8	86.8	86.9	88.4
Liu *et al.* [33]	97.2	**96.3**	88.1	88.0	84.4	87.6	85.7	95.5	**87.7**	**87.6**	89.8
Ours	**97.3**	96.2	**89.5**	**90.0**	**88.1**	**90.0**	**89.0**	**97.1**	86.5	87.0	**91.1**

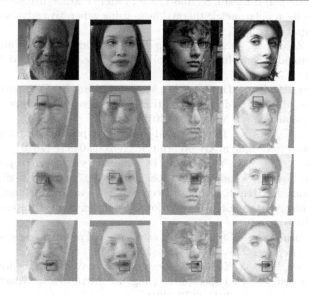

Fig. 5. Visualization of graph projection via response maps. The first row shows the input image, and the rest visualize response maps with respect to the vertex marked in a blue rectangle. Darker color indicates higher response. (Color figure online)

5   Conclusion

We propose a novel graph representation learning paradigm of edge aware graph reasoning for face parsing, which captures region-wise relations to model long-range contextual information. Edge cues are exploited in order to project significant pixels onto graph vertices on a higher semantic level. We then learn the relation between vertices (regions) and reason over all vertices to characterize the semantic information. Experimental results demonstrate that the proposed method sets the new state-of-the-art with low computation complexity, which efficiently reconstructs boundary details in particular. In future, we will apply the paradigm of edge aware graph reasoning to more segmentation applications, such as scene parsing.

Acknowledgement. This work was supported by National Natural Science Foundation of China [61972009], Beijing Natural Science Foundation [4194080] and Beijing Academy of Artificial Intelligence (BAAI).

References

1. Lee, C.H., Liu, Z., Wu, L., Luo, P.: MaskGAN: towards diverse and interactive facial image manipulation. In: Proceedings of the IEEE/CVF Conference on Computer Vision and Pattern Recognition, pp. 5549–5558 (2020)
2. Zhang, H., Riggan, B.S., Hu, S., Short, N.J., Patel, V.M.: Synthesis of high-quality visible faces from polarimetric thermal faces using generative adversarial networks. Int. J. Comput. Vis. **127**, 1–18 (2018)
3. Zhang, K., Zhang, Z., Li, Z., Qiao, Y.: Joint face detection and alignment using multitask cascaded convolutional networks. IEEE Sig. Process. Lett. **23**(10), 1499–1503 (2016)
4. Lin, J., Yang, H., Chen, D., Zeng, M., Wen, F., Yuan, L.: Face parsing with ROI tanh-warping. In: Proceedings of the IEEE Conference on Computer Vision and Pattern Recognition, pp. 5654–5663 (2019)
5. Yin, Z., Yiu, V., Hu, X., Tang, L.: End-to-end face parsing via interlinked convolutional neural networks. arXiv preprint arXiv:2002.04831 (2020)
6. Liu, S., Shi, J., Liang, J., Yang, M.H.: Face parsing via recurrent propagation. In: 28th British Machine Vision Conference, BMVC 2017, pp. 1–10 (2017)
7. He, K., Zhang, X., Ren, S., Sun, J.: Deep residual learning for image recognition. In: Proceedings of the IEEE International Conference on Computer Vision, pp. 770–778 (2016)
8. Henaff, M., Bruna, J., LeCun, Y.: Deep convolutional networks on graph-structured data. arXiv preprint arXiv:1506.05163 (2015)
9. Defferrard, M., Bresson, X., Vandergheynst, P.: Convolutional neural networks on graphs with fast localized spectral filtering. In: Advances in Neural Information Processing Systems, pp. 3844–3852 (2016)
10. Smith, B.M., Zhang, L., Brandt, J., Lin, Z., Yang, J.: Exemplar-based face parsing. In: Proceedings of the IEEE International Conference on Computer Vision, pp. 3484–3491 (2013)
11. Warrell, J., Prince, S.J.: Labelfaces: parsing facial features by multiclass labeling with an epitome prior. In: IEEE International Conference on Image Processing (ICIP), pp. 2481–2484 (2009)
12. Kae, A., Sohn, K., Lee, H., Learned-Miller, E.: Augmenting CRFs with Boltzmann machine shape priors for image labeling. In: Proceedings of the IEEE International Conference on Computer Vision, pp. 2019–2026 (2013)
13. Liu, S., Yang, J., Huang, C., Yang, M.H.: Multi-objective convolutional learning for face labeling. In: Proceedings of the IEEE International Conference on Computer Vision, pp. 3451–3459 (2015)
14. Luo, P., Wang, X., Tang, X.: Hierarchical face parsing via deep learning. In: Proceedings of the IEEE International Conference on Computer Vision, pp. 2480–2487 (2012)
15. Zhou, E., Fan, H., Cao, Z., Jiang, Y., Yin, Q.: Extensive facial landmark localization with coarse-to-fine convolutional network cascade. In: Proceedings of the IEEE International Conference on Computer Vision Workshops, pp. 386–391 (2013)
16. Zhou, Y., Hu, X., Zhang, B.: Interlinked convolutional neural networks for face parsing. In: Hu, X., Xia, Y., Zhang, Y., Zhao, D. (eds.) ISNN 2015. Lecture Notes in Computer Science, vol. 9377, pp. 222–231. Springer, Cham (2015). https://doi.org/10.1007/978-3-319-25393-0_25
17. Bahdanau, D., Cho, K., Bengio, Y.: Neural machine translation by jointly learning to align and translate. arXiv preprint arXiv:1409.0473 (2014)

18. Vaswani, A., et al.: Attention is all you need. In: Advances in neural information processing systems, pp. 5998–6008 (2017)
19. Wang, X., Girshick, R., Gupta, A., He, K.: Non-local neural networks. In: Proceedings of the IEEE Conference on Computer Vision and Pattern Recognition, pp. 7794–7803 (2018)
20. Chen, Y., Kalantidis, Y., Li, J., Yan, S., Feng, J.: A 2-nets: double attention networks. In: Advances in Neural Information Processing Systems, pp. 352–361 (2018)
21. Zhao, H., et al.: PSANet: point-wise spatial attention network for scene parsing. In: Proceedings of the European Conference on Computer Vision (ECCV), pp. 267–283 (2018)
22. Zhu, Z., Xu, M., Bai, S., Huang, T., Bai, X.: Asymmetric non-local neural networks for semantic segmentation. In: Proceedings of the IEEE International Conference on Computer Vision, pp. 593–602 (2019)
23. Fu, J., et al.: Dual attention network for scene segmentation. In: Proceedings of the IEEE Conference on Computer Vision and Pattern Recognition, pp. 3146–3154 (2019)
24. Chen, Y., Rohrbach, M., Yan, Z., Shuicheng, Y., Feng, J., Kalantidis, Y.: Graph-based global reasoning networks. In: Proceedings of the IEEE Conference on Computer Vision and Pattern Recognition, pp. 433–442 (2019)
25. Li, X., Zhong, Z., Wu, J., Yang, Y., Lin, Z., Liu, H.: Expectation-maximization attention networks for semantic segmentation. In: Proceedings of the IEEE International Conference on Computer Vision, pp. 9167–9176 (2019)
26. Chandra, S., Usunier, N., Kokkinos, I.: Dense and low-rank gaussian CRFs using deep embeddings. In: Proceedings of the IEEE International Conference on Computer Vision, pp. 5103–5112 (2017)
27. Li, Y., Gupta, A.: Beyond grids: learning graph representations for visual recognition. In: Advances in Neural Information Processing Systems, pp. 9225–9235 (2018)
28. Lu, Y., Chen, Y., Zhao, D., Chen, J.: Graph-FCN for image semantic segmentation. In: Lu, H., Tang, H., Wang, Z. (eds.) Advances in Neural Networks – ISNN 2019, ISNN 2019. Lecture Notes in Computer Science, vol. 11554, pp. 97–105. Springer, Cham (2019). https://doi.org/10.1007/978-3-030-22796-8_11
29. Pourian, N., Karthikeyan, S., Manjunath, B.S.: Weakly supervised graph based semantic segmentation by learning communities of image-parts. In: Proceedings of the IEEE International Conference on Computer Vision, pp. 1359–1367 (2015)
30. Te, G., Hu, W., Guo, Z.: Exploring hypergraph representation on face anti-spoofing beyond 2D attacks. In: 2020 IEEE International Conference on Multimedia and Expo (ICME), pp. 1–6. IEEE (2020)
31. Zhang, L., Li, X., Arnab, A., Yang, K., Tong, Y., Torr, P.H.: Dual graph convolutional network for semantic segmentation. arXiv preprint arXiv:1909.06121 (2019)
32. Kipf, T.N., Welling, M.: Semi-supervised classification with graph convolutional networks. In: 5th International Conference on Learning Representations, Conference Track Proceedings, OpenReview.net, ICLR 2017, Toulon, France, 24–26 April 2017 (2017)
33. Liu, Y., Shi, H., Shen, H., Si, Y., Wang, X., Mei, T.: A new dataset and boundary-attention semantic segmentation for face parsing. AAA I, 11637–11644 (2020)

274   G. Te et al.

34. Le, V., Brandt, J., Lin, Z., Bourdev, L., Huang, T.S.: Interactive facial feature localization. In: Fitzgibbon, A., Lazebnik, S., Perona, P., Sato, Y., Schmid, C. (eds.) Computer Vision – ECCV 2012, ECCV 2012. Lecture Notes in Computer Science, vol. 7574, pp. 679–692. Springer, Heidelberg (2012). https://doi.org/10.1007/978-3-642-33712-3_49
35. Zhao, H., Shi, J., Qi, X., Wang, X., Jia, J.: Pyramid scene parsing network. In: Proceedings of the IEEE International Conference on Computer Vision, pp. 2881–2890 (2017)
36. Ruan, T., Liu, T., Huang, Z., Wei, Y., Wei, S., Zhao, Y.: Devil in the details: towards accurate single and multiple human parsing. In: Proceedings of the AAAI Conference on Artificial Intelligence, vol. 33, pp. 4814–4821 (2019)
37. Rota Bulò, S., Porzi, L., Kontschieder, P.: In-place activated BatchNorm for memory-optimized training of DNNs. In: Proceedings of the IEEE Conference on Computer Vision and Pattern Recognition (2018)
38. Wei, Z., Liu, S., Sun, Y., Ling, H.: Accurate facial image parsing at real-time speed. IEEE Trans. Image Process. **28**(9), 4659–4670 (2019)

BBS-Net: RGB-D Salient Object Detection with a Bifurcated Backbone Strategy Network

Deng-Ping Fan[1], Yingjie Zhai[2], Ali Borji[3], Jufeng Yang[2(✉)],
and Ling Shao[1,4]

[1] Inception Institute of Artificial Intelligence, Abu Dhabi, UAE
dengpfan@gmail.com
[2] Nankai University, Tianjin, China
zhaiyingjie@mail.nankai.edu.cn, yangjufeng@nankai.edu.cn
[3] HCL America, Manhattan, NY, USA
aliborji@gmail.com
[4] Mohamed bin Zayed University of Artificial Intelligence, Abu Dhabi, UAE
ling.shao@ieee.org
https://github.com/zyjwuyan/BBS-Net

Abstract. Multi-level feature fusion is a fundamental topic in computer vision for detecting, segmenting and classifying objects at various scales. When multi-level features meet multi-modal cues, the optimal fusion problem becomes a hot potato. In this paper, we make the first attempt to leverage the inherent multi-modal and multi-level nature of RGB-D salient object detection to develop a novel cascaded refinement network. In particular, we 1) propose a bifurcated backbone strategy (BBS) to split the multi-level features into teacher and student features, and 2) utilize a depth-enhanced module (DEM) to excavate informative parts of depth cues from the channel and spatial views. This fuses RGB and depth modalities in a complementary way. Our simple yet efficient architecture, dubbed **B**ifurcated **B**ackbone **S**trategy **Net**work (**BBS-Net**), is backbone independent and outperforms 18 SOTAs on seven challenging datasets using four metrics.

Keywords: RGB-D saliency detection · Bifurcated backbone strategy

1 Introduction

Multi-modal and multi-level feature fusion [37] is essential for many computer vision tasks, such as object detection [8,21,26,42,70], semantic segmentation [29,30,32,67], co-attention tasks [19,72] and classification [38,40,53].

D.-P. Fan and Y. Zhai—Equal contributions.

Electronic supplementary material The online version of this chapter (https://doi.org/10.1007/978-3-030-58610-2_17) contains supplementary material, which is available to authorized users.

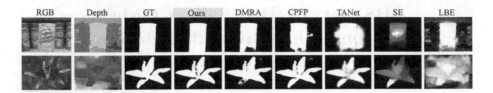

Fig. 1. Saliency maps of state-of-the-art (SOTA) CNN-based methods (*i.e.*, DMRA [52], CPFP [74], TANet [4], and our *BBS-Net*) and methods based on hand crafted features (*i.e.*, SE [27] and LBE [24]). Our method generates higher quality saliency maps and suppresses background distractors for challenging scenarios (first row: complex background; second row: depth with noise) more effectively.

Here, we attempt to utilize this idea for RGB-D salient object detection (SOD) [4,74], which aims at finding and segmenting the most visually prominent object(s) [2,75] in a scene according to the RGB and depth cues.

To efficiently integrate the RGB and depth cues for SOD, researchers have explored several multi-modal strategies [3,5], and have achieved encouraging results. Existing RGB-D SOD methods, however, still face the following challenges:

(1) **Effectively aggregating *multi-level* features.** As discussed in [44,63], *teacher features*[1] provide discriminative semantic information that serves as strong guidance for locating salient objects, while *student features* carry affluent details that are beneficial for refining edges. Therefore, previous RGB-D SOD algorithms focus on leveraging multi-level features, either via a progressive merging process [47,76] or by using a dedicated aggregation strategy [52,74]. However, these operations directly fuse multi-level features without considering level-specific characteristics, and thus suffer from the inherent noise often introduced by low-level features [4,65]. Thus, some methods tend to get distracted by the background (*e.g.*, first row in Fig. 1).

(2) **Excavating informative cues from the *depth modality*.** Previous methods combine RGB and depth cues by regarding the depth map as a fourth-channel input [13,51] or fusing RGB and depth modalities by simple summation [22,23] or multiplication [9,78]. These algorithms treat depth and RGB information the same and ignore the fact that depth maps mainly focus on the spatial relations among objects, whereas RGB information captures color and texture. Thus, such simple combinations are not efficient due to the modality difference. Besides, depth maps are sometimes low-quality, which may introduce feature noise and redundancy into the network. As an example, the depth map shown in the second row of Fig. 1 is blurry and noisy, and that is why many methods, including the top-ranked model (DMRA-iccv19 [52]), fail to detect the complete salient object.

---

[1] Note that we use the terms 'high-level features & low-level features' and 'teacher features & student features' interchangeably.

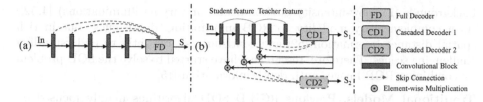

Fig. 2. (a) Existing multi-level feature aggregation methods for RGB-D SOD [3,4,47, 52,62,74,76]. (b) In this paper, we propose to adopt a bifurcated backbone strategy (BBS) to split the multi-level features into student and teacher features. The initial saliency map S_1 is utilized to refine the student features to effectively suppress distractors. Then, the refined features are passed to another cascaded decoder to generate the final saliency map S_2.

To address these issues, we propose a novel **Bifurcated Backbone Strategy Network** (***BBS-Net***) for RGB-D salient object detection. As shown in Fig. 2(b), *BBS-Net* consists of two cascaded decoder stages. In the first stage, teacher features are aggregated by a standard cascaded decoder $\mathbf{F}_{CD1}$ to generate an initial saliency map S_1. In the second stage, student features are refined by an element-wise multiplication with the initial saliency map S_1 and are then integrated by another cascaded decoder $\mathbf{F}_{CD2}$ to predict the final map S_2.

To the best of our knowledge, *BBS-Net* is the first work to explore the cascaded refinement mechanism for the RGB-D SOD task. Our main **contributions** are as follows:

(1) **We exploit multi-level features in a bifurcated backbone strategy (BBS)** to suppress distractors in the lower layers. This strategy is based on the observation that high-level features provide discriminative semantic information without redundant details [44,65], which may contribute significantly to eliminating distractors in lower layers.

(2) To fully capture the informative cues in the depth map and improve the compatibility of RGB and depth features, **we introduce a depth-enhanced module (DEM)**, which contains two sequential attention mechanisms: *channel attention* and *spatial attention*. The channel attention utilizes the inter-channel relations of the depth features, while the spatial attention aims to determine where informative depth cues are carried.

(3) We demonstrate that the proposed *BBS-Net* **exceeds 18 SOTAs on seven public datasets, by a large margin**. Our experiments show that **our framework has strong scalability** in terms of various backbones. This means that the bifurcated backbone strategy via a cascaded refinement mechanism is promising for multi-level and multi-modal learning tasks.

2   Related Works

Although RGB-based SOD has been thoroughly studied in recent years [7,39, 60,69,71], most of algorithms fail under complicated scenarios (*e.g.*, cluttered backgrounds [16], low-intensity environments, or varying illuminations) [4,52]. As a complementary modality to RGB information, depth cues contain rich spatial distance information [52] and contribute significantly to understanding challenging scenes. Therefore, researchers have started to solve the SOD problem by combining RGB images with depth information [15].

Traditional Models. Previous RGB-D SOD algorithms mainly focused on hand crafted features [9,78]. Some of these methods largely relied on contrast-based cues by calculating color, edge, texture and region contrast to measure the saliency in a local region. For example, [15] adopted the region based contrast to compute contrast strengths for the segmented regions. In [10], the saliency value of each pixel depended on the color contrast and surface normals. However, the local contrast methods focued on the boundaries of salient objects and were easily affected by high-frequency content [54]. Therefore, some algorithms, such as global contrast [11], spatial prior [9], and background prior [56], proposed to calculate the saliency by combining local and global information. To effectively combine saliency cues from the RGB and depth modalities, researchers have explored various fusion strategies. Some methods [13,51] regarded depth images as the fourth-channel inputs and processed the RGB and depth channes together (early fusion). This operation seems simple but disregards the differences between the RGB and depth modalities and thus cannot achieve reliable results. Therefore, to effectively extract the saliency information from the two modalities separately, some algorithms [22,78] first leveraged two backbones to predict saliency maps and then fused the saliency results (late fusion). Besides, considering that the RGB and depth modalities may positively influence each other, yet other methods [24,34] fused RGB and depth features in a middle stage and then predicted the saliency maps from the fused features (middle fusion). In fact, these three fusion strategies are also explored in the current deep models, and our model can be considered as a middle fusion.

Deep Models. Early deep algorithms [54,56] extracted hand crafted features first, and then fed them to CNNs to compute saliency confidence scores. However, these methods need to design low-level features first and cannot be trained in an end-to-end manner. More recently, researchers have exploited CNNs to extract RGB and depth features in a bottom-up way [28]. Compared with hand crafted features, deep features contain more semantic and contextual information that can better capture representations of the RGB and depth modalities and achieve encouraging performance. The success of these deep models [5,52] stems from two aspects of feature fusion. The first is the extracting of multi-scale features from different layers and then the effective fusion of these features. The second is the mechanism of fusing features from the two different modalities.

To effectively aggregate multi-scale features, researchers have designed various network architectures. For example, [47] fed a four-channel RGB-D image into a single backbone and then obtained saliency map outputs from each side-out features (single stream). Chen et al. [3] leveraged two networks to extract RGB features and depth features respectively, and then fused them in a progressive complementary way (double stream). Further, to exploit cross-modal

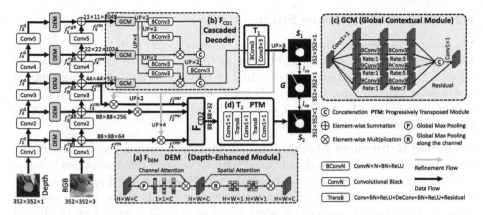

Fig. 3. **The architecture of *BBS-Net*.** Feature Extraction: '*Conv1*'∼'*Conv5*' denote different layers from ResNet-50 [31]. Multi-level features ($f_1^d \sim f_5^d$) from the depth branch are enhanced by the (a) DEM and then fused with features (*i.e.*, $f_1^{rgb} \sim f_5^{rgb}$) from the RGB branch. Stage 1: cross-modal teacher features ($f_3^{cm} \sim f_5^{cm}$) are first aggregated by the (b) cascaded decoder to produce the initial saliency map S_1. Stage 2: Then, student features ($f_1^{cm} \sim f_3^{cm}$) are refined by the initial saliency map S_1 and are integrated by another cascaded decoder to predict the final saliency map S_2.

complements in the bottom-up feature extraction process, Chen *et al.* [4] proposed a three-stream network that contains two modality-specific streams and a parallel cross-modal distillation stream to learn supplementary features (three streams). However, depth maps are often of low quality and thus may contain a lot of noise and misleading information. This greatly decreases the performance of SOD models. To address this problem, Zhao *et al.* [74] designed a contrast-enhanced network to improve the quality of depth maps by the contrast prior. Fan *et al.* [20] proposed a depth depurator unit that can evaluate the quality of the depth images and then filter out the low-quality maps automatically. Two recent famous works have also explored uncertainty [68], bilateral attention [73], graph neural network [48] and a joint learning strategy [25] and achieve good performance.

3 Proposed Method

3.1 Overview

Existing popular RGB-D SOD models directly aggregate multi-level features (Fig. 2(a)). As shown in Fig. 3, the network flow of our *BBS-Net* is different from the above mentioned models. We first introduce the bifurcated backbone strategy with the cascaded refinement mechanism in Sect. 3.2. To fully use informative cues in the depth map, we introduce a new depth-enhanced module (Sect. 3.3).

3.2   Bifurcated Backbone Strategy (BBS)

We propose to excavate the rich semantic information in high-level cross-modal features to suppress background distractors in a cascaded refinement way. We adopt a bifurcated backbone strategy (BBS) to divide the multi-level cross-modal features into two groups, i.e., $\mathbf{Q}_1 = \{Conv1,\ Conv2,\ Conv3\}$ and $\mathbf{Q}_2 = \{Conv3,\ Conv4,\ Conv5\}$, with the Conv3 as the split point. Each group still preserves the original multi-scale information.

Cascaded Refinement Mechanism. To effectively leverage the features of the two groups, the whole network is trained with a cascaded refinement mechanism. This mechanism first produces an initial saliency map with three cross-modal teacher features (i.e., $\mathbf{Q}_2$) and then improves the details of the initial saliency map S_1 with three cross-modal student features (i.e., $\mathbf{Q}_1$) refined by the initial saliency map itself. Using this mechanism, our model can iteratively refine the details in the low-level features. This is based on the observation that high-level features contain rich global contextual information which is beneficial for locating salient objects, while low-level features carry much detailed micro-level information that can contribute significantly to refining the boundaries. In other words, this strategy efficiently eliminates noise in low-level cross-modal features, by exploring the specialties of multi-level features, and predicts the final saliency map in a progressive refinement manner.

Specifically, we first compute cross-modal features $\{f_i^{cm}; i = 1, 2, ..., 5\}$ by merging RGB and depth features processed by the DEM (Fig. 3(a)). In stage one, the three cross-modality teacher features (i.e., $f_3^{cm}, f_4^{cm}, f_5^{cm}$) are aggregated by the first cascaded decoder, which is formulated:

$$S_1 = \mathbf{T}_1\big(\mathbf{F}_{CD1}(f_3^{cm}, f_4^{cm}, f_5^{cm})\big), \tag{1}$$

where S_1 is the initial saliency map, $\mathbf{F}_{CD1}$ is the first cascaded decoder and $\mathbf{T}_1$ represents two simple convolutional layers that change the channel number from 32 to 1. In stage two, the initial saliency map S_1 is leveraged to refine the three cross-modal student features, which is defined as:

$$f_i^{cm'} = f_i^{cm} \odot S_1, \tag{2}$$

where $f_i^{cm'}$ ($i \in \{1, 2, 3\}$) denotes the refined features and $\odot$ represents the element-wise multiplication. Then, the three refined student features are integrated by another decoder followed by a progressively transposed module (PTM), which is defined as,

$$S_2 = \mathbf{T}_2\Big(\mathbf{F}_{CD2}(f_1^{cm'}, f_2^{cm'}, f_3^{cm'})\Big), \tag{3}$$

where S_2 is the final saliency map. $\mathbf{T}_2$ represents the PTM module and $\mathbf{F}_{CD2}$ denotes the second cascaded decoder. Finally, we jointly optimize the two stages by defining the total loss:

$$\mathcal{L} = \alpha \ell_{ce}(S_1, G) + (1 - \alpha)\ell_{ce}(S_2, G), \tag{4}$$

in which ℓ_{ce} represents the widely used binary cross entropy loss and $\alpha \in [0, 1]$ controls the trade-off between the two parts of the losses. The ℓ_{ce} is computed as:

$$\ell_{ce}(S, G) = G \log S + (1 - G) \log(1 - S), \tag{5}$$

in which S is the predicted saliency map and G denotes the ground-truth binary saliency map.

Cascaded Decoder. Given the two groups of multi-level, cross-modal features ($\{f_i^{cm}, f_{i+1}^{cm}, f_{i+2}^{cm}\}, i \in \{1, 3\}$) fused by the RGB and depth features from different layers, we need to efficiently utilize the multi-level, multi-scale information in each group to carry out our cascaded refinement strategy. Thus, we introduce a light-weight cascaded decoder [65] to aggregate the two groups of multi-level, cross-modal features. As shown in Fig. 3(b), the cascaded decoder contains three global context modules (GCM) and a simple feature aggregation strategy. The GCM is refined from the RFB module [46] with an additional branch to enlarge the receptive field and a residual connection [31] to preserve the original information. Specifically, as illustrated in Fig. 3(c), the GCM module contains four parallel branches. For all of these branches, a 1×1 convolution is first applied to reduce the channel size to 32. For the kth branch ($k \in \{2, 3, 4\}$), a convolution operation with a kernel size of $2k - 1$ and dilation rate of 1 is applied. This is followed by another 3×3 convolution operation with a dilation rate of $2k - 1$. The goal here is to extract the global contextual information from the cross-modal features. Next, the outputs of the four branches are concatenated together and their channel number is reduced to 32 with a 1×1 convolution operation. Finally, the concatenated features form a residual connection with the input feature. The outputs of the GCM modules in the two cascaded decoders are defined by:

$$f_i^{gcm} = \mathbf{F}_{GCM}(f_i), \tag{6}$$

To further improve the representational ability of cross-modal features, we leverage a pyramid multiplication and concatenation feature aggregation strategy to integrate the cross-modal features ($\{f_i^{gcm}, f_{i+1}^{gcm}, f_{i+2}^{gcm}\}, i \in \{1, 3\}$). As shown in Fig. 3(b), first, each refined feature f_i^{gcm} is updated by multiplying it with all higher-level features:

$$f_i^{gcm'} = f_i^{gcm} \odot \Pi_{k=i+1}^{k_{max}} Conv\left(\mathbf{F}_{UP}\left(f_k^{gcm}\right)\right), \tag{7}$$

where $i \in \{1, 2, 3\}$, $k_{max} = 3$ or $i \in \{3, 4, 5\}$, $k_{max} = 5$. $Conv(\cdot)$ represents the standard 3×3 convolution operation, and $\mathbf{F}_{UP}$ denotes the upsampling operation if these features are not in the same scale. $\odot$ represents the element-wise multiplication. Second, the updated features are aggregated by a progressive concatenation strategy to generate the output:

$$S = \mathbf{T}\left(\left[f_k^{gcm'}; Conv\left(\mathbf{F}_{UP}\left[f_{k+1}^{gcm'}; Conv\left(\mathbf{F}_{UP}(f_{k+2}^{gcm'})\right)\right]\right)\right]\right), \tag{8}$$

where S is the generated salient map, $k \in \{1, 3\}$, and $[x; y]$ denotes the concatenation operation of x and y. In the first stage, $\mathbf{T}$ represents two sequential convolutional layers ($\mathbf{T}_1$), while it denotes the PTM module ($\mathbf{T}_2$) for the second stage. The output (88×88) of the second decoder is $1/4$ of the ground-truth resolution (352×352), so directly up-sampling the output to the ground-truth size will result in a loss of some details. To address this problem, we design a simple yet effective progressively transposed module (PTM, Fig. 3(d)) to predict the final saliency map (S_2) in a progressive upsampling way. It is composed of two sequential residual-based transposed blocks [33] and three sequential 1×1 convolutions. Each residual-based transposed block consists of a 3×3 convolution and a residual-based transposed convolution.

Note that our cascaded refinement mechanism is different from the recent refinement mechanisms R3Net [14], CRN [6], and RFCN [61] in its usage of multi-level features and the initial map. The obvious difference and superiority of our design is that we only need one round of saliency refinement to obtain a good performance, while R3Net, CRN, and RFCN all need more iterations, which will increase the training time and computational resources. In addition, our cascaded strategy is different from CPD [65] in that it exploits the details in low-level features and semantic information in high-level features, while suppressing the noise in low-level features, simultaneously.

3.3 Depth-Enhanced Module (DEM)

There are two main problems when trying to fuse RGB and depth features. One is the compatibility of the two due to the intrinsic modality difference, and the other is the redundancy and noise in low-quality depth features. Inspired by [64], we introduce a depth-enhanced module (DEM) to improve the compatibility of multi-modal features and to excavate the informative cues from the depth features.

Specifically, let f_i^{rgb}, f_i^d denote the feature maps of the ith ($i \in 1, 2, ..., 5$) side-out layer from the RGB and depth branches, respectively. As shown in Fig. 3, each DEM is added before each side-out feature map from the depth branch to improve the compatibility of the depth features. Such a side-out process enhances the saliency representation of depth features and preserves the multi-level, multi-scale information. The fusion process of the two modalities is formulated as:

$$f_i^{cm} = f_i^{rgb} + \mathbf{F}_{DEM}(f_i^d), \tag{9}$$

where f_i^{cm} represents the cross-modal features of the ith layer. As illustrated in Fig. 3(a), the DEM includes a sequential channel attention operation and a spatial attention operation. The operation of the DEM is defined as:

$$\mathbf{F}_{DEM}(f_i^d) = \mathbf{S}_{att}\Big(\mathbf{C}_{att}(f_i^d)\Big), \tag{10}$$

where $\mathbf{C}_{att}(\cdot)$ and $\mathbf{S}_{att}(\cdot)$ denote the spatial and channel attention, respectively. More specifically,

$$\mathbf{C}_{att}(f) = \mathbf{M}\Big(\mathbf{P}_{max}(f)\Big) \otimes f, \tag{11}$$

where $\mathbf{P}_{max}(\cdot)$ represents the global max pooling operation for each feature map, f denotes the input feature map, $\mathbf{M}(\cdot)$ is a multi-layer (two-layer) perceptron, and $\otimes$ denotes the multiplication by the dimension broadcast. The spatial attention is implemented as:

$$\mathbf{S}_{att}(f) = Conv\Big(\mathbf{R}_{max}(f)\Big) \odot f, \tag{12}$$

where $\mathbf{R}_{max}(\cdot)$ represents the global max pooling operation for each point in the feature map along the channel axis. Our depth enhanced module is different from previous RGB-D models. Previous models fuse the corresponding multi-level features from the RGB and depth branches by direct concatenation [3,4,76], enhance the depth map by contrast prior [74] or process the multi-level depth features by a simple convolutional layer [52]. To the best of our knowledge, we are the first to introduce the attention mechanism to excavate informative cues from depth features in multiple side-out layers. Our experiments (see Table 4 and Fig. 5) show the effectiveness of this approach in improving the compatibility of multi-modal features.

Moreover, the spatial and channel attention mechanisms are different from the operation proposed in [64]. We only leverage a single global max pooling [50] operation to excavate the most critical cues in the depth features and reduce the complexity of the module simultaneously, which is based on the intuition that SOD aims at finding the most important area in an image.

4 Experiments

4.1 Experimental Settings

Datasets. We tested seven challenging RGB-D SOD datasets, *i.e.*, NJU2K [34], NLPR [51], STERE [49], SIP [20], DES [9], LFSD [41], and SSD [77].

Training/Testing. Following the same training settings as in [52,74], we use 1,485 samples from the NJU2K dataset and 700 samples from the NLPR dataset as our training set. The remaining images in the NJU2K and NLPR datasets and the whole datasets of STERE, DES, LFSD, SSD, and SIP are used for testing.

Evaluation Metrics. We adopt four widely used metrics, including S-measure (S_α) [17], maximum E-measure (E_ξ) [18], maximum F-measure (F_β) [1], mean absolute error (MAE). Evaluation code: http://dpfan.net/d3netbenchmark/.

Contenders. We compare the proposed *BBS-Net* with ten models that use hand crafted features [9,12,24,27,34,43,51,55,58,78] and eight models [3–5,28,52,54, 62,74] based on deep learning. We train and test the above models using their default settings, as proposed in the original papers. For those models without released source codes, we used their published results for comparisons.

Inference Time. In terms of speed, *BBS-Net* 14 fps and 48 fps on a single GTX 1080Ti with a batch size of 1 and 10, respectively.

Table 1. Quantitative comparison of models using S-measure (S_α), max F-measure (F_β), max E-measure (E_ξ) and MAE (M) scores on seven datasets. $\uparrow$ ($\downarrow$) denotes that the higher (lower) the better. The best score in each row is highlighted in **bold**.

Dataset	Metric	Hand-crafted-Features-Based Models										CNNs-Based Models								BBS-Net
		LHM [51]	CDB [43]	DESM [9]	GP [55]	CDCP [78]	ACSD [34]	LBE [24]	DCMC [12]	MDSF [58]	SE [27]	DF [54]	AFNet [62]	CTMF [28]	MMCI [5]	PCF [3]	TANet [4]	CPFP [74]	DMRA [52]	Ours
NJU2K [34]	$S_\alpha\uparrow$	.514	.624	.665	.527	.669	.699	.695	.686	.748	.664	.763	.772	.849	.858	.877	.878	.879	.886	**.921**
	$F_\beta\uparrow$	.632	.648	.717	.647	.621	.711	.748	.715	.775	.748	.804	.775	.845	.852	.872	.874	.877	.886	**.920**
	$E_\xi\uparrow$	.724	.742	.791	.703	.741	.803	.803	.799	.838	.813	.864	.853	.913	.915	.924	.925	.926	.927	**.949**
	$M\downarrow$	.205	.203	.283	.211	.180	.202	.153	.172	.157	.169	.141	.100	.085	.079	.059	.060	.053	.051	**.035**
NLPR [51]	$S_\alpha\uparrow$	.630	.629	.572	.654	.727	.762	.724		.805	.756	.802	.799	.860	.856	.874	.886	.888	.899	**.930**
	$F_\beta\uparrow$	.622	.618	.640	.611	.645	.607	.745	.648	.793	.713	.778	.771	.825	.815	.841	.863	.867	.879	**.918**
	$E_\xi\uparrow$	.766	.791	.805	.723	.820	.780	.855	.793	.885	.847	.880	.879	.929	.913	.925	.941	.932	.947	**.961**
	$M\downarrow$	.108	.114	.146	.112	.179	.081	.117		.095	.091	.085	.058	.056	.059	.044	.041	.036	.031	**.023**
STERE [49]	$S_\alpha\uparrow$	.562	.615	.642	.588	.713	.692	.660	.731	.728	.708	.757	.825	.848	.873	.875	.871	.879	.835	**.908**
	$F_\beta\uparrow$	.683	.717	.700	.671	.664	.669	.633	.740	.719	.755	.757	.823	.831	.863	.860	.861	.874	.847	**.903**
	$E_\xi\uparrow$	.771	.823	.811	.743	.786	.806	.787	.819	.809	.846	.847	.887	.912	.927	.925	.923	.925	.911	**.942**
	$M\downarrow$	.172	.166	.295	.182	.149	.200	.250	.148	.176	.143	.141	.075	.086	.068	.064	.060	.051	.066	**.041**
DES [9]	$S_\alpha\uparrow$	.578	.645	.622	.636	.709	.728	.703	.707	.741	.741	.752	.770	.863	.848	.842	.858	.872	.900	**.933**
	$F_\beta\uparrow$	.511	.723	.765	.597	.631	.756	.788	.666	.746	.741	.766	.728	.844	.822	.804	.827	.846	.888	**.927**
	$E_\xi\uparrow$	.653	.830	.868	.670	.811	.850	.890	.773	.851	.856	.870	.881	.932	.928	.893	.910	.923	.943	**.966**
	$M\downarrow$	.114	.100	.299	.168	.115	.169	.208	.111	.122	.090	.093	.068	.055	.065	.049	.046	.038	.030	**.021**
LFSD [41]	$S_\alpha\uparrow$	.553	.515	.716	.635	.712	.727	.729	.753	.694	.692	.783	.738	.788	.787	.786	.801	.828	.839	**.864**
	$F_\beta\uparrow$	.708	.677	.762	.783	.702	.763	.722	.817	.779	.786	.813	.744	.787	.771	.775	.796	.826	.852	**.858**
	$E_\xi\uparrow$	.763	.766	.811	.824	.780	.829	.797	.856	.819	.832	.857	.815	.857	.839	.827	.847	.863	.893	**.901**
	$M\downarrow$	.218	.225	.253	.190	.172	.195	.214	.155	.197	.174	.145	.133	.127	.132	.119	.111	.088	.083	**.072**
SSD [77]	$S_\alpha\uparrow$	.566	.562	.602	.615	.603	.675	.621	.704	.673	.675	.747	.714	.776	.813	.841	.839	.807	.857	**.882**
	$F_\beta\uparrow$	.568	.592	.680	.740	.535	.682	.619	.711	.703	.710	.735	.687	.729	.781	.807	.810	.766	.844	**.859**
	$E_\xi\uparrow$	.717	.698	.769	.782	.700	.785	.736	.786	.779	.800	.828	.807	.865	.882	.894	.897	.852	.906	**.919**
	$M\downarrow$	.195	.196	.308	.180	.214	.203	.278	.169	.192	.165	.142	.118	.099	.082	.062	.063	.082	.058	**.044**
SIP [20]	$S_\alpha\uparrow$	.511	.557	.616	.588	.595	.732	.727	.683	.717	.628	.653	.720	.716	.833	.842	.835	.850	.806	**.879**
	$F_\beta\uparrow$	.574	.620	.669	.687	.505	.763	.751	.618	.698	.661	.657	.712	.694	.818	.838	.830	.851	.821	**.883**
	$E_\xi\uparrow$	.716	.737	.770	.768	.721	.838	.853	.743	.798	.771	.759	.819	.829	.897	.901	.895	.903	.875	**.922**
	$M\downarrow$	.184	.192	.298	.173	.224	.172	.200	.186	.167	.164	.185	.118	.139	.086	.071	.075	.064	.085	**.055**

Implementation Details. We perform our experiments using the PyTorch [59] framework on a single 1080Ti GPU. Parameters of the backbone network (ResNet-50 [31]) are initialized from the model pre-trained on ImageNet [36]. We discard the last pooling and fully connected layers of ResNet-50 and leverage each middle output of the five convolutional blocks as the side-out feature maps. The two branches do not share weights and the only difference between them is that the depth branch has the input channel number set to 1. The other parameters are initialized as the PyTorch default settings. We use the Adam algorithm [35] to optimize the proposed model. The initial learning rate is set to 1e−4 and is divided by 10 every 60 epochs. We resize the input RGB and depth images to 352×352 for both the training and test phases. All the training images are augmented using multiple strategies (*i.e.*, random flipping, rotating and border clipping). It takes about 10 h to train our model with a mini-batch size of 10 for 200 epochs.

4.2   Comparison with SOTAs

Quantitative Results. As shown in Table 1, our method performs favorably against all algorithms based on hand crafted features as well as SOTA CNN-based methods by a large margin, in terms of all four evaluation metrics. Performance gains over the best compared algorithms (ICCV'19 DMRA [52] and CVPR'19 CPFP [74]) are (2.5%–3.5%, 0.7%–3.9%, 0.8%–2.3%, 0.009–0.016) for the metrics (S_α, $maxF_\beta$, $maxE_\xi$, M) on seven challenging datasets.

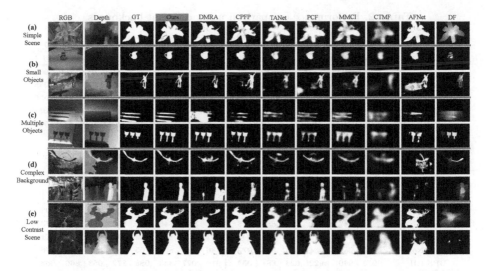

Fig. 4. Qualitative visual comparison of the proposed model versus 8 SOTAs.

Visual Comparison. Figure 4 provides sample saliency maps predicted by the proposed method and several SOTA algorithms. Visualizations cover simple scenes (a) and various challenging scenarios, including small objects (b), multiple objects (c), complex backgrounds (d), and low-contrast scenes (e). First, (a) is an easy example. The flower in the foreground is evident in the original RGB image, but the depth image is low-quality and contains some misleading information. The top two models, *i.e.*, DMRA and CPFP, fail to predict the whole extent of the salient object due to the interference from the depth map. Our method can eliminate the side-effects of the depth map by utilizing the complementary depth information more effectively. Second, two examples of small objects are shown in (b). Despite the handle of the teapot in the first row being tiny, our method can accurately detect it. Third, we show two examples with multiple objects in the image in (c). Our method locates all salient objects in the image. It segments the objects better and generates sharper edges compared to other algorithms. Even though the depth map in the first row of (c) lacks clear information, our algorithm predicts the salient objects correctly. Fourth, (d) shows two examples with complex backgrounds. Here, our method produces reliable results, while other algorithms confuse the background as a salient object. Finally, (e) presents two examples in which the contrast between the object and background is low. Many algorithms fail to detect and segment the entire extent of the salient object. Our method produces satisfactory results by suppressing background distractors and exploring the informative cues from the depth map.

5   Discussion

Scalability. There are three popular backbone architectures (*i.e.*, VGG-16 [57], VGG-19 [57] and ResNet-50 [31]) that are used in deep RGB-D models. To

Table 2. Performance comparison using different backbones.

Models	NJU2K [34]		NLPR [51]		STERE [49]		DES [9]		LFSD [41]		SSD [77]		SIP [20]	
	$S_\alpha \uparrow$	$M \downarrow$	$S_\alpha \uparrow$	$M \downarrow$	$S_\alpha \uparrow$	$M \downarrow$	$S_\alpha \uparrow$	$M \downarrow$	$S_\alpha \uparrow$	$M \downarrow$	$S_\alpha \uparrow$	$M \downarrow$	$S_\alpha \uparrow$	$M \downarrow$
CPFP [74]	.879	.053	.888	.036	.879	.051	.872	.038	.828	.088	.807	.082	.850	.064
DMRA [52]	.886	.051	.899	.031	.835	.066	.900	.030	.839	.083	.857	.058	.806	.085
BBS-Net(VGG-16)	.916	.039	.923	.026	.896	.046	.908	.028	.845	.080	.858	.055	.874	.056
BBS-Net(VGG-19)	.918	.037	.925	.025	.901	.043	.915	.026	.852	.074	.855	.056	.878	**.054**
BBS-Net(ResNet-50)	**.921**	**.035**	**.930**	**.023**	**.908**	**.041**	**.933**	**.021**	**.864**	**.072**	**.882**	**.044**	**.879**	.055

Table 3. Comparison of different feature aggregation strategies on seven datasets.

#	Settings	NJU2K [34]		NLPR [51]		STERE [49]		DES [9]		LFSD [41]		SSD [77]		SIP [20]	
		$S_\alpha \uparrow$	$M \downarrow$	$S_\alpha \uparrow$	$M \downarrow$	$S_\alpha \uparrow$	$M \downarrow$	$S_\alpha \uparrow$	$M \downarrow$	$S_\alpha \uparrow$	$M \downarrow$	$S_\alpha \uparrow$	$M \downarrow$	$S_\alpha \uparrow$	$M \downarrow$
1	Low3	.881	.051	.882	.038	.832	.070	.853	.044	.779	.110	.805	.080	.760	.108
2	High3	.902	.042	.911	.029	.886	.048	.912	.026	.845	.080	.850	.058	.833	.073
3	All5	.905	.042	.915	.027	.891	.045	.901	.028	.845	.082	.848	.060	.839	.071
4	BBS-NoRF	.893	.050	.904	.035	.843	.072	.886	.039	.804	.105	.839	.069	.843	.076
5	BBS-RH	.913	.040	.922	.028	.881	.054	.919	.027	.833	.085	.872	.053	.866	.063
6	BBS-RL (ours)	**.921**	**.035**	**.930**	**.023**	**.908**	**.041**	**.933**	**.021**	**.864**	**.072**	**.882**	**.044**	**.879**	**.055**

further validate the scalability of the proposed method, we provide performance comparisons using different backbones in Table 2. We find that our *BBS-Net* exceeds the SOTA methods (*e.g.*, CPFP [74], and DMRA [52]) with all of these popular backbones, showing the strong scalability of our framework.

Aggregation Strategies. We conduct several experiments to validate the effectiveness of our cascaded refinement mechanism. Results are shown in Table 3 and Fig. 5(a). 'Low3' means that we only integrate the student features (*Conv1~3*) using the decoder without the refinement from the initial map for training and testing. Student features contain abundant details that are beneficial for refining the object edges, but at the same time introduce a lot of background distraction. Integrating only low-level features produces unsatisfactory results and generates many distractors (*e.g.*, 1st–2nd row in Fig. 5(a)) or fails to locate the salient objects (*e.g.*, the 3rd row in Fig. 5(a)). 'High3' only integrates the teacher features (*Conv3~5*) using the decoder to predict the saliency map. Compared with student features, teacher features are 'sophisticated' and thus contain more semantic information. As a result, they help locate the salient objects and preserve edge information. Thus, integrating teacher features leads to better results. 'All5' aggregates features from all five levels (*Conv1~5*) directly using a single decoder for training and testing. It achieves comparable results with the 'High3' but may generate background noise introduced by the student features. 'BBS-NoRF' indicates that we directly remove the refinement flow of our model. This leads to poor performance. 'BBS-RH' can be seen as a reverse refinement strategy to our cascaded refinement mechanism, where teacher features (*Conv3~5*) are first refined by the initial map aggregated by student features (*Conv1~3*) and are then integrated to generate the final saliency map. It performs worse than our final mechanism (BBS-RL), because, with this reverse refinement strategy, noise

Table 4. Ablation study of our *BBS-Net*. 'BM' = base model. 'CA' = channel attention. 'SA' = spatial attention. 'PTM' = progressively transposed module.

#	BM	CA	SA	PTM	NJU2K [34] $S_\alpha\uparrow$	$M\downarrow$	NLPR [51] $S_\alpha\uparrow$	$M\downarrow$	STERE [49] $S_\alpha\uparrow$	$M\downarrow$	DES [9] $S_\alpha\uparrow$	$M\downarrow$	LFSD [41] $S_\alpha\uparrow$	$M\downarrow$	SSD [77] $S_\alpha\uparrow$	$M\downarrow$	SIP [20] $S_\alpha\uparrow$	$M\downarrow$
1	✓				.908	.045	.918	.029	.882	.055	.917	.027	.842	.083	.862	.057	.864	.066
2	✓	✓			.913	.042	.922	.027	.896	.048	.923	.025	.840	.086	.855	.057	.868	.063
3	✓		✓		.912	.045	.918	.029	.891	.054	.914	.029	.855	.083	.872	.054	.869	.063
4	✓	✓	✓		.919	.037	.928	.026	.900	.045	.924	.024	.861	.074	.873	.052	.869	.061
5	✓	✓	✓	✓	**.921**	**.035**	**.930**	**.023**	**.908**	**.041**	**.933**	**.021**	**.864**	**.072**	**.882**	**.044**	**.879**	**.055**

Table 5. Effectiveness analysis of the cascaded decoder.

Strategies	NJU2K [34] $S_\alpha\uparrow$	$M\downarrow$	NLPR [51] $S_\alpha\uparrow$	$M\downarrow$	STERE [49] $S_\alpha\uparrow$	$M\downarrow$	DES [9] $S_\alpha\uparrow$	$M\downarrow$	LFSD [41] $S_\alpha\uparrow$	$M\downarrow$	SSD [77] $S_\alpha\uparrow$	$M\downarrow$	SIP [20] $S_\alpha\uparrow$	$M\downarrow$
Element-wise sum	.915	.037	.925	.025	.897	.045	.925	.022	.856	.073	.868	.050	**.880**	**.052**
Cascaded decoder	**.921**	**.035**	**.930**	**.023**	**.908**	**.041**	**.933**	**.021**	**.864**	**.072**	**.882**	**.044**	.879	.055

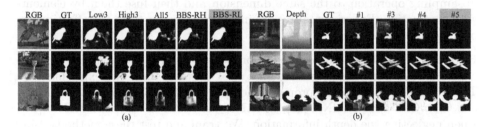

Fig. 5. (a): Visual comparison of different aggregation strategies, (b): Visual effectiveness of gradually adding modules. '#' denotes the corresponding row of Table 4.

in student features cannot be effectively suppressed. Besides, compared to 'All5', our method fully utilizes the features at different levels, and thus achieves significant performance improvement with fewer background distractors and sharper edges (*i.e.*, 'BBS-RL' in Fig. 5(a)).

Impact of Different Modules. As shown in Table 4 and Fig. 5(b), we conduct an ablation study to test the effectiveness of different modules in our *BBS-Net*. The base model (BM) is our *BBS-Net* without additional modules (*i.e.*, CA, SA, and PTM). Note that just the BM performs better than the SOTA methods over almost all datasets, as shown in Table 1 and Table 4. Adding the channel attention (CA) and spatial attention (SA) modules enhances performance on most of the datasets. See the results shown in the second and third rows of Table 4. When we combine the two modules (fourth row in Table 4), the performance is greatly improved on all datasets, compared with the BM. We can easily conclude from the '#3' and '#4' columns in Fig. 5(b) that the spatial attention and channel attention mechanisms in DEM allow the model to focus on the informative parts of the depth features, which results in better suppression of background distraction. Finally, we add a progressively transposed block before the second decoder to gradually upsample the feature map to the same resolution as the

Table 6. S_α comparison with SOTA RGB SOD methods on three datasets. 'w/o D' and 'w/ D' represent training and testing the proposed method without/with the depth.

Methods	CPD [65]		PoolNet [44]		PiCANet [45]		PAGRN [71]		R3Net [14]		Ours (w/o D)		Ours (w/ D)	
	$S_\alpha \uparrow$	$M \downarrow$	$S_\alpha \uparrow$	$M \downarrow$	$S_\alpha \uparrow$	$M \downarrow$	$S_\alpha \uparrow$	$M \downarrow$	$S_\alpha \uparrow$	$M \downarrow$	$S_\alpha \uparrow$	$M \downarrow$	$S_\alpha \uparrow$	$M \downarrow$
NJU2K [34]	.894	.046	.887	.045	.847	.071	.829	.081	.837	.092	.914	.038	**.921**	**.035**
NLPR [51]	.915	.025	.900	.029	.834	.053	.844	.051	.798	.101	.925	.026	**.930**	**.023**
DES [9]	.897	.028	.873	.034	.854	.042	.858	.044	.847	.066	.912	.025	**.933**	**.021**

ground truth. The results in the fifth row of Table 4 and the '#5' column of Fig. 5(b) show that the 'PTM' achieves impressive performance gains on all datasets and generates sharper edges with fine details.

To further analyze the effectiveness of the cascaded decoder, we experiment by changing the decoder to an element-wise summation mechanism. That is to say, we first change the features from different layers using 1×1 convolution and upsampling operation to the same dimension and then fuse them by element-wise summation. Experimental results in Table 5 demonstrate the effectiveness of the cascaded decoder.

Benefits of the Depth Map. To explore whether or not the depth information can really contribute to the performance of SOD, we conduct two experiments in Table 6: (i) We compare the proposed method with five SOTA RGB SOD methods (*i.e.*, CPD [66], PoolNet [44], PiCANet [45], PAGRN [71] and R3Net [14]) when neglecting the depth information. We train and test these methods using the same training and testing sets as our *BBS-Net*. It is shown that the proposed methods (*i.e.*, Ours (w/ D)) can significantly exceed SOTA RGB SOD methods due to the usage of depth information. (ii) We train and test the proposed method without using the depth information by setting the inputs of the depth branch to zero (*i.e.*, Ours (w/o D)). Comparing the results of Ours (w/ D) with Ours (w/o D), we find that the depth information can effectively improve the performance of the proposed model. The above two experiments together demonstrate the benefits of the depth information for SOD, since depth maps can be seen as prior knowledge that provides spatial-distance information and contour guidance to detect salient objects.

6   Conclusion

We presented a new multi-level multi-modality learning framework that demonstrates state-of-the-art performance on seven challenging RGB-D salient object detection datasets using several evaluation measures. Our *BBS-Net* is based on a novel bifurcated backbone strategy (BBS) with a cascaded refinement mechanism. Importantly, our simple architecture is backbone independent, making it promising for further research on other related topics, including semantic segmentation, object detection and classification.

Acknowledgments. This work was supported by the Major Project for New Generation of AI Grant (NO. 2018AAA0100403), NSFC (NO. 61876094, U1933114), Natural Science Foundation of Tianjin, China (NO. 18JCYBJC15400, 18ZXZNGX00110), the Open Project Program of the National Laboratory of Pattern Recognition (NLPR), and the Fundamental Research Funds for the Central Universities.

References

1. Achanta, R., Hemami, S., Estrada, F., Susstrunk, S.: Frequency-tuned salient region detection. In: CVPR, pp. 1597–1604 (2009)
2. Borji, A., Cheng, M.M., Jiang, H., Li, J.: Salient object detection: a benchmark. IEEE TIP **24**(12), 5706–5722 (2015)
3. Chen, H., Li, Y.: Progressively complementarity-aware fusion network for RGB-D salient object detection. In: CVPR, pp. 3051–3060 (2018)
4. Chen, H., Li, Y.: Three-stream attention-aware network for RGB-D salient object detection. IEEE TIP **28**(6), 2825–2835 (2019)
5. Chen, H., Li, Y., Su, D.: Multi-modal fusion network with multi-scale multi-path and cross-modal interactions for RGB-D salient object detection. IEEE TOC **86**, 376–385 (2019)
6. Chen, Q., Koltun, V.: Photographic image synthesis with cascaded refinement networks. In: CVPR, pp. 1511–1520 (2017)
7. Chen, S., Tan, X., Wang, B., Lu, H., Hu, X., Fu, Y.: Reverse attention-based residual network for salient object detection. IEEE TIP **29**, 3763–3776 (2020)
8. Cheng, G., Han, J., Zhou, P., Xu, D.: Learning rotation-invariant and fisher discriminative convolutional neural networks for object detection. IEEE TIP **28**(1), 265–278 (2018)
9. Cheng, Y., Fu, H., Wei, X., Xiao, J., Cao, X.: Depth enhanced saliency detection method. In: ICIMCS, pp. 23–27 (2014)
10. Ciptadi, A., Hermans, T., Rehg, J.M.: An in depth view of saliency. In: BMVC (2013)
11. Cong, R., Lei, J., Fu, H., Hou, J., Huang, Q., Kwong, S.: Going from RGB to RGBD saliency: a depth-guided transformation model. IEEE TOC, 1–13 (2019)
12. Cong, R., Lei, J., Zhang, C., Huang, Q., Cao, X., Hou, C.: Saliency detection for stereoscopic images based on depth confidence analysis and multiple cues fusion. IEEE SPL **23**(6), 819–823 (2016)
13. Cong, R., Lei, J., Fu, H., Huang, Q., Cao, X., Ling, N.: HSCS: hierarchical sparsity based co-saliency detection for RGBD images. IEEE TMM **21**(7), 1660–1671 (2019)
14. Deng, Z., et al.: R3Net: recurrent residual refinement network for saliency detection. In: IJCAI, pp. 684–690 (2018)
15. Desingh, K., Krishna, K., Rajanand, D., Jawahar, C.: Depth really matters: improving visual salient region detection with depth. In: BMVC, pp. 1–11 (2013)
16. Fan, D.P., Cheng, M.M., Liu, J.J., Gao, S.H., Hou, Q., Borji, A.: Salient objects in clutter: bringing salient object detection to the foreground. In: ECCV, pp. 186–202 (2018)
17. Fan, D.P., Cheng, M.M., Liu, Y., Li, T., Borji, A.: Structure-measure: a new way to evaluate foreground maps. In: ICCV, pp. 4548–4557 (2017)
18. Fan, D.P., Gong, C., Cao, Y., Ren, B., Cheng, M.M., Borji, A.: Enhanced-alignment measure for binary foreground map evaluation. In: IJCAI, pp. 698–704 (2018)
19. Fan, D.P., Lin, Z., Ji, G.P., Zhang, D., Fu, H., Cheng, M.M.: Taking a deeper look at co-salient object detection. In: CVPR, pp. 2919–2929 (2020)

20. Fan, D.P., Lin, Z., Zhang, Z., Zhu, M., Cheng, M.M.: Rethinking RGB-D salient object detection: models, datasets, and large-scale benchmarks. IEEE TNNLS (2020)
21. Fan, D.P., Wang, W., Cheng, M.M., Shen, J.: Shifting more attention to video salient object detection. In: CVPR, pp. 8554–8564 (2019)
22. Fan, X., Liu, Z., Sun, G.: Salient region detection for stereoscopic images. In: DSP, pp. 454–458 (2014)
23. Fang, Y., Wang, J., Narwaria, M., Le Callet, P., Lin, W.: Saliency detection for stereoscopic images. IEEE TIP **23**(6), 2625–2636 (2014)
24. Feng, D., Barnes, N., You, S., McCarthy, C.: Local background enclosure for RGB-D salient object detection. In: CVPR, pp. 2343–2350 (2016)
25. Fu, K., Fan, D.P., Ji, G.P., Zhao, Q.: JL-DCF: joint learning and densely-cooperative fusion framework for RGB-D salient object detection. In: CVPR, pp. 3052–3062 (2020)
26. Gao, S.H., Tan, Y.Q., Cheng, M.M., Lu, C., Chen, Y., Yan, S.: Highly efficient salient object detection with 100K parameters. In: ECCV (2020)
27. Guo, J., Ren, T., Bei, J.: Salient object detection for RGB-D image via saliency evolution. In: ICME, pp. 1–6 (2016)
28. Han, J., Chen, H., Liu, N., Yan, C., Li, X.: CNNs-based RGB-D saliency detection via cross-view transfer and multiview fusion. IEEE TOC **48**(11), 3171–3183 (2018)
29. Han, J., Yang, L., Zhang, D., Chang, X., Liang, X.: Reinforcement cutting-agent learning for video object segmentation. In: CVPR, pp. 9080–9089 (2018)
30. Han, Q., Zhao, K., Xu, J., Cheng, M.M.: Deep hough transform for semantic line detection. In: ECCV (2020)
31. He, K., Zhang, X., Ren, S., Sun, J.: Deep residual learning for image recognition. In: CVPR, pp. 770–778 (2016)
32. He, X., Yang, S., Li, G., Li, H., Chang, H., Yu, Y.: Non-local context encoder: robust biomedical image segmentation against adversarial attacks. In: AAAI 2019, pp. 8417–8424 (2019)
33. Hu, X., Yang, K., Fei, L., Wang, K.: ACNet: attention based network to exploit complementary features for RGBD semantic segmentation. In: ICIP, pp. 1440–1444 (2019)
34. Ju, R., Ge, L., Geng, W., Ren, T., Wu, G.: Depth saliency based on anisotropic center-surround difference. In: ICIP, pp. 1115–1119 (2014)
35. Kingma, D.P., Ba, J.: Adam: a method for stochastic optimization. In: ICLR (2015)
36. Krizhevsky, A., Sutskever, I., Hinton, G.E.: ImageNet classification with deep convolutional neural networks. In: NIPS, pp. 1106–1114 (2012)
37. Li, G., Yu, Y.: Visual saliency based on multiscale deep features. In: CVPR, pp. 5455–5463 (2015)
38. Li, G., Zhu, X., Zeng, Y., Wang, Q., Lin, L.: Semantic relationships guided representation learning for facial action unit recognition. In: AAAI, pp. 8594–8601 (2019)
39. Li, H., Chen, G., Li, G., Yu, Y.: Motion guided attention for video salient object detection. In: ICCV, pp. 7274–7283 (2019)
40. Li, J., et al.: Learning from large-scale noisy web data with ubiquitous reweighting for image classification. IEEE TPAMI (2019)
41. Li, N., Ye, J., Ji, Y., Ling, H., Yu, J.: Saliency detection on light field. In: CVPR, pp. 2806–2813 (2014)
42. Li, X., Yang, F., Cheng, H., Liu, W., Shen, D.: Contour knowledge transfer for salient object detection. In: ECCV, pp. 355–370 (2018)

43. Liang, F., Duan, L., Ma, W., Qiao, Y., Cai, Z., Qing, L.: Stereoscopic saliency model using contrast and depth-guided-background prior. Neurocomputing **275**, 2227–2238 (2018)
44. Liu, J.J., Hou, Q., Cheng, M.M., Feng, J., Jiang, J.: A simple pooling-based design for real-time salient object detection. In: CVPR, pp. 3917–3926 (2019)
45. Liu, N., Han, J., Yang, M.H.: PiCANet: learning pixel-wise contextual attention for saliency detection. In: CVPR, pp. 3089–3098 (2018)
46. Liu, S., Huang, D., Wang, Y.: Receptive field block net for accurate and fast object detection. In: ECCV, pp. 404–419 (2018)
47. Liu, Z., Shi, S., Duan, Q., Zhang, W., Zhao, P.: Salient object detection for RGB-D image by single stream recurrent convolution neural network. Neurocomputing **363**, 46–57 (2019)
48. Luo, A., Li, X., Yang, F., Jiao, Z., Cheng, H., Lyu, S.: Cascade graph neural networks for RGB-D salient object detection. In: ECCV (2020)
49. Niu, Y., Geng, Y., Li, X., Liu, F.: Leveraging stereopsis for saliency analysis. In: CVPR, pp. 454–461 (2012)
50. Oquab, M., Bottou, L., Laptev, I., Sivic, J.: Is object localization for free? - Weakly-supervised learning with convolutional neural networks. In: CVPR, pp. 685–694 (2015)
51. Peng, H., Li, B., Xiong, W., Hu, W., Ji, R.: RGBD salient object detection: a benchmark and algorithms. In: Fleet, D., Pajdla, T., Schiele, B., Tuytelaars, T. (eds.) ECCV 2014. Lecture Notes in Computer Science, vol. 8691, pp. 92–109. Springer, Cham (2014). https://doi.org/10.1007/978-3-319-10578-9_7
52. Piao, Y., Ji, W., Li, J., Zhang, M., Lu, H.: Depth-induced multi-scale recurrent attention network for saliency detection. In: ICCV, pp. 7254–7263 (2019)
53. Qiao, L., Shi, Y., Li, J., Wang, Y., Huang, T., Tian, Y.: Transductive episodic-wise adaptive metric for few-shot learning. In: ICCV, pp. 3603–3612 (2019)
54. Qu, L., He, S., Zhang, J., Tian, J., Tang, Y., Yang, Q.: RGBD salient object detection via deep fusion. IEEE TIP **26**(5), 2274–2285 (2017)
55. Ren, J., Gong, X., Yu, L., Zhou, W., Ying Yang, M.: Exploiting global priors for RGB-D saliency detection. In: CVPRW, pp. 25–32 (2015)
56. Shigematsu, R., Feng, D., You, S., Barnes, N.: Learning RGB-D salient object detection using background enclosure, depth contrast, and top-down features. In: ICCVW, pp. 2749–2757 (2017)
57. Simonyan, K., Zisserman, A.: Very deep convolutional networks for large-scale image recognition. arXiv preprint arXiv:1409.1556 (2014)
58. Song, H., Liu, Z., Du, H., Sun, G., Le Meur, O., Ren, T.: Depth-aware salient object detection and segmentation via multiscale discriminative saliency fusion and bootstrap learning. IEEE TIP **26**(9), 4204–4216 (2017)
59. Steiner, B., et al.: PyTorch: an imperative style, high-performance deep learning library. In: NIPS, pp. 8024–8035 (2019)
60. Su, J., Li, J., Zhang, Y., Xia, C., Tian, Y.: Selectivity or invariance: boundary-aware salient object detection. In: ICCV, pp. 3798–3807 (2019)
61. Wang, L., Wang, L., Lu, H., Zhang, P., Ruan, X.: Salient object detection with recurrent fully convolutional networks. IEEE TPAMI **41**(7), 1734–1746 (2018)
62. Wang, N., Gong, X.: Adaptive fusion for RGB-D salient object detection. IEEE Access **7**, 55277–55284 (2019)
63. Wang, T., et al.: Detect globally, refine locally: a novel approach to saliency detection. In: CVPR, pp. 3127–3135 (2018)
64. Woo, S., Park, J., Lee, J.Y., So Kweon, I.: CBAM: convolutional block attention module. In: ECCV, pp. 3–19 (2018)

65. Wu, Z., Su, L., Huang, Q.: Cascaded partial decoder for fast and accurate salient object detection. In: CVPR, pp. 3907–3916 (2019)
66. Wu, Z., Su, L., Huang, Q.: Stacked cross refinement network for edge-aware salient object detection. In: ICCV, pp. 7264–7273 (2019)
67. Zeng, Y., Zhuge, Y., Lu, H., Zhang, L.: Joint learning of saliency detection and weakly supervised semantic segmentation. In: ICCV, pp. 7223–7233 (2019)
68. Zhang, J., et al.: UC-Net: uncertainty inspired RGB-D saliency detection via conditional variational autoencoders. In: CVPR, pp. 8582–8591 (2020)
69. Zhang, L., Wu, J., Wang, T., Borji, A., Wei, G., Lu, H.: A multistage refinement network for salient object detection. IEEE TIP **29**, 3534–3545 (2020)
70. Zhang, Q., Huang, N., Yao, L., Zhang, D., Shan, C., Han, J.: RGB-T salient object detection via fusing multi-level CNN features. IEEE TIP **29**, 3321–3335 (2020)
71. Zhang, X., Wang, T., Qi, J., Lu, H., Wang, G.: Progressive attention guided recurrent network for salient object detection. In: CVPR, pp. 714–722 (2018)
72. Zhang, Z., Jin, W., Xu, J., Cheng, M.M.: Gradient-induced co-saliency detection. In: ECCV (2020)
73. Zhang, Z., Lin, Z., Xu, J., Jin, W., Lu, S.P., Fan, D.P.: Bilateral attention network for RGB-D salient object detection. arXiv preprint arXiv:2004.14582 (2020)
74. Zhao, J.X., Cao, Y., Fan, D.P., Cheng, M.M., Li, X.Y., Zhang, L.: Contrast prior and fluid pyramid integration for RGBD salient object detection. In: CVPR, pp. 3927–3936 (2019)
75. Zhao, J.X., Liu, J.J., Fan, D.P., Cao, Y., Yang, J., Cheng, M.M.: EGNet: edge guidance network for salient object detection. In: CVPR, pp. 8779–8788 (2019)
76. Zhu, C., Cai, X., Huang, K., Li, T.H., Li, G.: PDNet: prior-model guided depth-enhanced network for salient object detection. In: ICME, pp. 199–204 (2019)
77. Zhu, C., Li, G.: A three-pathway psychobiological framework of salient object detection using stereoscopic technology. In: ICCVW, pp. 3008–3014 (2017)
78. Zhu, C., Li, G., Wang, W., Wang, R.: An innovative salient object detection using center-dark channel prior. In: ICCVW, pp. 1509–1515 (2017)

G-LBM: Generative Low-Dimensional Background Model Estimation from Video Sequences

Behnaz Rezaei, Amirreza Farnoosh, and Sarah Ostadabbas[✉]

Augmented Cognition Lab, Electrical and Computer Engineering Department,
Northeastern University, Boston, USA
{brezaei,afarnoosh,ostadabbas}@ece.neu.edu
http://www.northeastern.edu/ostadabbas/

Abstract. In this paper, we propose a computationally tractable and theoretically supported non-linear low-dimensional generative model to represent real-world data in the presence of noise and sparse outliers. The non-linear low-dimensional manifold discovery of data is done through describing a joint distribution over observations, and their low-dimensional representations (i.e. manifold coordinates). Our model, called generative low-dimensional background model (G-LBM) admits variational operations on the distribution of the manifold coordinates and simultaneously generates a low-rank structure of the latent manifold given the data. Therefore, our probabilistic model contains the intuition of the non-probabilistic low-dimensional manifold learning. G-LBM selects the intrinsic dimensionality of the underling manifold of the observations, and its probabilistic nature models the noise in the observation data. G-LBM has direct application in the background scenes model estimation from video sequences and we have evaluated its performance on SBMnet-2016 and BMC2012 datasets, where it achieved a performance higher or comparable to other state-of-the-art methods while being agnostic to different scenes. Besides, in challenges such as camera jitter and background motion, G-LBM is able to robustly estimate the background by effectively modeling the uncertainties in video observations in these scenarios. (The code and models are available at: https://github.com/brezaei/G-LBM.)

Keywords: Background estimation · Foreground segmentation · Non-linear manifold learning · Deep neural network · Variational auto-encoding

---

Electronic supplementary material The online version of this chapter (https://doi.org/10.1007/978-3-030-58610-2_18) contains supplementary material, which is available to authorized users.

1   Introduction

Many high-dimensional real world datasets consist of data points coming from a lower-dimensional manifold corrupted by noise and possibly outliers. In particular, background in videos recorded by a static camera might be generated from a small number of latent processes that all non-linearly affect the recorded video scenes. Linear multivariate analysis such as robust principal component analysis (RPCA) and its variants have long been used to estimate such underlying processes in the presence of noise and/or outliers in the measurements with large data matrices [6,17,41]. However, these linear processes may fail to find the low-dimensional structure of the data when the mapping of the data into the latent space is non-linear. For instance background scenes in real-world videos lie on one or more non-linear manifolds, an investigation to this fact is presented in [16]. Therefore, a robust representation of the data should find the underlying non-linear structure of the real-world data as well as its uncertainties. To this end, we propose a generic probabilistic non-linear model of the background inclusive to different scenes in order to effectively capture the low-dimensional generative process of the background sequences. Our model is inspired by the classical background estimation methods based on the low-dimensional subspace representation enhanced with the Bayesian auto-encoding neural networks in finding the non-linear latent processes of the high-dimensional data. Despite the fact that finding the low-dimensional structure of data has different applications in real world [14,29,48], the main focus of this paper is around the concept of background scene estimation/generation in video sequences.

1.1   Video Background Model Estimation Toward Foreground Segmentation

Foreground segmentation is the primary task in a wide range of computer vision applications such as moving object detection [34], video surveillance [5], behavior analysis and video inspection [35], and visual object tracking [33]. The objective in foreground segmentation is separating the moving objects from the background which is mainly achieved in three steps of background estimation, background subtraction, and background maintenance.

The first step called background model estimation refers to the extracting a model which describes a scene without foreground objects in a video. In general, a background model is often initialized using the first frame or a set of training frames that either contain or do not contain foreground objects. This background model can be the temporal average or median of the consecutive video frames. However, such models poorly perform in challenging types of environments such as changing lighting conditions, jitters, and occlusions due to the presence of foreground objects. In these scenarios aforementioned simple background models require bootstrapping, and a sophisticated model is then needed to construct the first background image. The algorithms with highest overall performance applied to the SBMnet-2016 dataset, which is the largest public dataset on background modeling with different real world challenges are

Motion-assisted Spatio-temporal Clustering of Low-rank (MSCL) [16], Super-pixel Motion Detector (SPMD) [46], and LaBGen-OF [21], which are based on RPCA, density based clustering of the motionless superpixels, and the robust estimation of the median, respectively. Deep neural networks (DNNs) are suit-able for this type of tasks and several DNN methods have recently been used in this field. In Sect. 1.2, we give an overview of the DNN-based background model estimation algorithms.

Following the background model estimation, background subtraction in the second step consists of comparing the modeled background image with the cur-rent video frames to segment pixels as background or foreground. This is a binary classification task, which can be achieved successfully using a DNN. Dif-ferent methods for the background subtraction have been developed, and we refer the reader to look at [3,4] for comprehensive details on these methods. While we urge the background subtraction process to be unsupervised given the background model, the well performing methods are mostly supervised. The three top algorithms on the CDnet-2014 [42] which is the large-scale real-world dataset for background subtraction are supervised DNN-based methods, namely different versions of FgSegNet [22], BSPVGAN [49], cascaded CNN [43], followed by three unsupervised approaches, WisennetMD [18], PAWCS [37], IUTIS [1].

1.2 Related Work on Background Model Estimation

DNNs have been widely used in modeling the background from video sequences due to their flexibility and power in estimating complex models. Aside from prevalent use of convolutional neural networks (CNNs) in this field, successful methods are mainly designed based on the deep auto-encoder networks (DAE), and generative adversarial networks (GAN).

1. Model architectures based on convolutional neural networks (CNNs): FC-FlowNet model proposed in [13] is a CNN-based architecture inspired from the FlowNet proposed by Dosovitskiy et al. in [10]. FlowNet is a two-stage architecture developed for the prediction of the optical flow motion vectors: A contractive stage, composed of a succession of convolutional layers, and a refinement stage, composed of deconvolutional layers. FC-FlowNet modifies this architecture by creating a fully-concatenated version which combines at each convolutional layer multiple feature maps representing different high level abstractions from previous stages. Even though FC-FlowNet is able to model background in mild challenges of real-world videos, it fails to address challenges such as clutters, background motions, and illumination changes.
2. Model architectures based on deep auto-encoding networks (DAEs): One of the earliest works in background modeling using DAEs was presented in [45]. Their model is a cascade of two auto-encoder networks. The first network approximates the background images from the input video. Back-ground model is then learned through the second auto-encoder network. Qu et al. [31] employed context-encoder to model the motion-based back-ground from a dynamic foreground. Their method aims to restore the overall scene of a video by removing the moving foreground objects and learning

the feature of its context. Both aforementioned works have limited number of experiments to evaluate their model performance. More recently two other unsupervised models for background modeling inspired by the successful novel auto-encoding architecture of U-net [36] have been proposed in [25,39]. BM-Unet and its augmented version presented by Tao et al. [39] is a background modelling method based on the U-net architecture. They augment their baseline model with the aim of robustness to rapid illumination changes and camera jitter. However, they did not evaluate their proposed model on the complete dataset of SBMnet-2016. DeepPBM in [11] is a generative scene-specific background model based on the variational auto-encoders (VAEs) evaluated on the BMC2012 dataset compared with RPCA. Mondejar et al. in [25] proposed an architecture for simultaneous background modeling and subtraction consists of two cascaded networks which are trained together. Both sub-networks have the same architecture as U-net architecture. The first network, namely, background model network takes the video frames as input and results in M background model channel as output. The background subtraction sub-network, instead, takes the M background model channels plus the target frame channels from the same scene as input. The whole network is trained in a supervised manner given the ground truth. Their model is scene-specific and cannot be used for unseen videos.

3. Model architectures based on generative adversarial network (GAN): Considering the promising paradigm of GANs for unsupervised learning, they have been used in recent research of background modeling. Sultana et al. in [38] designed an unsupervised deep context prediction (DCP) for background initialization using hybrid GAN. DCP is a scene-specific background model which consists of four steps: (1) Object masking by creating the motion masks. (2) Evaluating the missing regions resulted from masking the motions using the context prediction hybrid GAN. (3) improving the fine texture details by scene-specific fine tuning of VGG-16 network. (4) Obtaining the final background model by applying modified Poisson blending technique. Their model is containing two different networks which are trained separately.

1.3   Our Contributions

Background models are utilized to segment the foreground in videos, generally regarded as object of interest towards further video processing. Therefore, providing a robust background model in various computer vision applications is an essential preliminary task. However, modeling the background in complex real-world scenarios is still challenging due to presence of dynamic backgrounds, jitter induced by unstable camera, occlusion, illumination changes. None of the approaches proposed so far could address all the challenges in their model. Moreover, current background models are mostly scene-specific. Therefore, DNN models need to be retrained adjusting their weights for each particular scene and non-DNN models require parameter tuning for optimal result on different video sequences, which makes them unable to extend to unseen scenes. According to the aforementioned challenges, we propose our generative low-dimensional background model (G-LBM) estimation approach that is applicable to different video sequences. Our main contributions in this paper are listed as follows:

- The G-LBM, our proposed background model estimation approach is the first generative probabilistic model which learns the underlying nonlinear manifold of the background in video sequences.
- The probabilistic nature of our model yields the uncertainties that correlate well with empirical errors in real world videos, yet maintains the predictive power of the deterministic counter part. This is verified with extensive experiments on videos with camera jitters, and background motions.
- The G-LBM is scene non-specific and can be extended to the new videos with different scenes.
- We evaluated the proposed G-LBM on large scale background model datasets SBMnet as well as BMC. Experiments show promising results for modeling the background under various challenges.

Our contributions are built upon the assumption that there is a low-dimensional non-linear latent space process that generates background in different videos. In addition, background from the videos with the same scene can be non-linearly mapped into a lower dimensional subspace. In different words the underlying non-linear manifold of the background in different videos is locally linear.

2   Generative Low-Dimensional Background Model

We designed an end-to-end architecture that performs a scene non-specific background model estimation by finding the low-dimensional latent process which generates the background scenes in video sequences. An overview of our generative model is presented in Fig. 1. As described in Sect. 2.1, the latent process $\mathbf{z}$ is

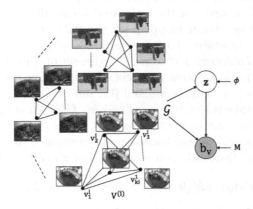

Fig. 1. Graphical representation of the generative process in G-LBM. Given the video dataset $\{\mathbf{v}_1, \ldots, \mathbf{v}_n\}$, we construct a neighbourhood graph $\mathcal{G}$ in which video frames from the same scene create a clique in the graph, as $V^{(i)} = [\mathbf{v}_1^i, \ldots, \mathbf{v}_{ki}^i]$. The distribution over latent process $\mathbf{z}$ is controlled by the graph $\mathcal{G}$ as well as the parameters of the non-linear mapping ϕ. The latent process $\mathbf{z}$ along with the motion mask M determine the likelihood of the background $\mathbf{b_v}$ in video frames $\mathbf{v} \in \{\mathbf{v}_1, \ldots, \mathbf{v}_n\}$.

estimated through a non-linear mapping from the corrupted/noisy observations
$\mathbf{v} = \{\mathbf{v_1}, ..., \mathbf{v_n}\}$ parameterized by ϕ.

Notation: In the following, a diagonal matrix with entries taken from vector
$\mathbf{x}$ is shown as $diag(\mathbf{x})$. Vector of n ones is shown as $\mathbf{1}_n$ and $n \times n$ identity
matrix is I_n. The Nuclear norm of a matrix B is $||B||_*$ and its l_1-norm is $||B||_1$.
The Kronecker product of matrices A and B is $A \otimes B$. Khatri–Rao product is
defined as $A * B = (A_{ij} \otimes B_{ij})_{ij}$, in which the ij-th block is the $m_i p_i \times n_j q_j$
sized Kronecker product of the corresponding blocks of A and B, assuming the
number of row and column partitions of both matrices is equal.

2.1   Nonlinear Latent Variable Modeling of Background in Videos

Problem Formulation: Suppose that we have n data points $\{\mathbf{v}_1, \mathbf{v}_2, ..., \mathbf{v}_n\} \subset$
$\mathbb{R}^m$, and a graph $\mathcal{G}$ with n nodes corresponding to each data point with the edge
set $\mathcal{E}_{\mathcal{G}} = \{(i, j)|\mathbf{v}_i$ and $\mathbf{v}_j$ are neighbours$\}$. In context of modeling the back-
ground $\mathbf{v}_i$ and $\mathbf{v}_j$ are neighbours if they are video frames from the same scene.
We assume that there is a low-dimensional (latent) representation of the high-
dimensional data $\{\mathbf{v}_1, \mathbf{v}_2, ..., \mathbf{v}_n\}$ with coordinates $\{\mathbf{z}_1, \mathbf{z}_2, ..., \mathbf{z}_n\} \subset \mathbb{R}^d$, where
$d \ll m$. It is helpful to concatenate data points from the same clique in the
graph to form $V^{(i)}$ and all the cliques to form the $\mathbf{V} = concat(V^{(1)}, ..., V^{(N)})$.

Assumptions: Our essential assumptions are as follows: (1) Latent space is
locally linear in a sense that neighbour data points in the graph $\mathcal{G}$ lie in a lower
dimensional subspace. In other words, mapped neighbour data points in latent
space $\{(\mathbf{z}_i, \mathbf{z}_j)|(i, j) \in \mathcal{E}_{\mathcal{G}}\}$ belong to a subspace with dimension lower than the
manifold dimension. (2) Measurement dataset is corrupted by sparse outliers
(foreground in the videos). Under these assumptions, we aim to find the non-
linear mapping from observed input data into a low-dimensional latent manifold
and the distribution over the latent process $\mathbf{z}$, $p(\mathbf{z}|\mathcal{G}, \mathbf{v})$, which best describes
the data such that samples of the estimated latent distribution can generate
the data through a non-linear mapping. In the following, we describe the main
components of our Generative Low-dimensional Background Model (G-LBM).

Adjacency and Laplacian matrices: The edge set of $\mathcal{G}$ for n data points specifies a $n \times n$ symmetric adjacency matrix $A_{\mathcal{G}}$. a_{ij} defined as i, jth element of
$A_{\mathcal{G}}$, is 1 if $\mathbf{v}_i$ and $\mathbf{v}_j$ are neighbours and 0 if not or $i = j$ (diagonal elements).
Accordingly, the Laplacian matrix is defined as: $L_{\mathcal{G}} = diag(A_{\mathcal{G}}\mathbf{1}_n) - A_{\mathcal{G}}$.

Prior distribution over $\mathbf{z}$: We assume that the prior on the latent variables
$\mathbf{z}_i$, $i \in \{1, ..., n\}$ is a unit variant Gaussian distribution $\mathcal{N}(\mathbf{0}, I_d)$. This prior as a
multivariate normal distribution on concatenated $\mathbf{z}$ can be written as:

$$p(\mathbf{z}|A_{\mathcal{G}}) = \mathcal{N}(\mathbf{0}, \Sigma), \quad \text{where} \quad \Sigma^{-1} = 2L_{\mathcal{G}} \otimes I_d. \tag{1}$$

Posterior distribution over $\mathbf{z}$: Under the locally linear dependency assumption on the latent manifold, the posterior is defined as a multivariate Gaussian
distribution given by Eq. (2). Manifold coordinates construct the expected value
Λ and covariance Π of the latent process variables corresponding to the neighbouring high dimensional points in graph $\mathcal{G}$.

$$p(\mathbf{z}|A_{\mathcal{G}}, \mathbf{v}) = \mathcal{N}(\Lambda, \Pi), \quad \text{where} \tag{2}$$

$$\Pi^{-1} = 2L_{\mathcal{G}} * [diag(f_\phi^\sigma(\mathbf{v}_1)), \dots, diag(f_\phi^\sigma(\mathbf{v}_n))]^T [diag(f_\phi^\sigma(\mathbf{v}_1)), \dots, diag(f_\phi^\sigma(\mathbf{v}_n))]$$

$$\Lambda = [f_\phi^\mu(\mathbf{v}_1)^T, \dots, f_\phi^\mu(\mathbf{v}_n)^T]^T \in \mathbb{R}^{nd},$$

where $f_\phi^\sigma(\mathbf{v}_i)$ and $f_\phi^\mu(\mathbf{v}_i)$, for $i = \{1, \dots, n\}$ are corresponding points on the latent manifold mapped from high-dimensional point $\mathbf{v}_i$ by nonlinear function $f_\phi(.)$. These points are treated as the mean and variance of the latent process respectively. Our aim is to infer the latent variables $\mathbf{z}$ as well as the non-linear mapping parameters ϕ in G-LBM. We infer the parameters by minimizing the reconstruction error when generating the original data points through mapping the corresponding samples of the latent space into the original high-dimensional space. Further details on finding the parameters of the non-linear mapping in G-LBM from video sequences are provided in Sect. 2.2.

2.2   Background Model Estimation in G-LBM Using VAE

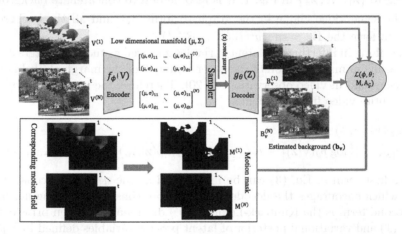

Fig. 2. Schematic illustration of the proposed G-LBM training procedure for background model estimation. Given a batch of input video clips $V^{(i)}$ consist of consecutive frames from the same scene, input videos $V^{(i)}$ are mapped through the encoder $f_\phi(.)$ to a low-dimensional manifold representing the mean and covariance (μ, Σ) of the latent process $\mathbf{z}$. The latent process $\mathbf{z}$ generates the estimated backgrounds for each video clip through decoder $g_\theta(.)$. Imposing the locally linear subspace is done by minimizing the rank of the manifold coordinates correspond to the video frames of the same clip $(f_\phi(V^{(i)}) := (\mu, \Sigma)^{(i)})$. Learning the parameters of the non-linear mappings ϕ, θ is done by incorporating the reconstruction error between input videos $V^{(i)}$ and estimated background $B_\mathbf{v}^{(i)}$ where motion mask value is zero to the final loss $\mathcal{L}(\phi, \theta; M, A_{\mathcal{G}})$.

Consider that backgrounds in video frames $\mathbf{v}$ belong to $\{\mathbf{v}_1, \dots, \mathbf{v}_n\}$, each of size $m = w \times h$ pixels, are generated from n underlying probabilistic latent process vectorized in $\mathbf{z} \in \mathbb{R}^d$ for $d \ll m$. Video frame $\mathbf{v}_i$ is interpreted as the corrupted background in higher dimension by sparse perturbations/outliers called foreground objects, and vector z_i is interpreted as the low-dimensional representation of the background in video frame v_i. The neighbourhood graph $\mathcal{G}$ represents the video frames recorded from the same scene as nodes of a clique in the graph. A variational auto-encoding considers the joint probability of the background in input video frames v and its representation z to define the underlying generative model as $p_\theta(v, z|A_\mathcal{G}) = p_\theta(v|z)p(z|A_\mathcal{G})$, where $p(z|A_\mathcal{G})$ is the Gaussian prior for latent variables z defined in Eq. (1), and $p_\theta(v|z)$ is the generative process of the model illustrated in Fig. 1.

In variational auto-encoding (VAE), the generative process is implemented by the decoder part $g_\theta(.)$ that is parameterized by a DNN with parameters θ as demonstrated in Fig. 2. In the encoder part of the VAE, the posterior distribution is approximated with a variational posterior $p_\phi(z|v, A_\mathcal{G})$ defined in Eq. (2) with parameters ϕ corresponding to non-linear mapping $f_\phi(.)$ which is also parameterized by a DNN. We assume that backgrounds in input video frames specified as b_v belong to $\{b_v^1, \ldots, b_v^n\}$ are generated by n underlying process specified as z belong to $\{z_1, \ldots, z_n\}$ in Fig. 1. It is also helpful to concatenate backgrounds from the same video clique in the graph to form $B_v^{(i)}$ and all the background cliques to form the $B_v = concat(B_v^{(1)}, ..., B_v^{(N)})$.

The efforts in inferring the latent variables z as well as the parameters ϕ in G-LBM results in maximization of the lower bound $\mathcal{L}$ of the marginal likelihood of the background in video observations [2,20]. Therefore, the total VAE objective for the entire video frames becomes:

$$\log p(v|A_\mathcal{G}, \phi) \geq \tag{3}$$
$$\mathbb{E}_{q_\phi(z|v, A_\mathcal{G})}\big[\log p_\theta(v|z)\big] - KL\big(q_\phi(z|v, A_\mathcal{G})||p(z|A_\mathcal{G})\big) := \mathcal{L}(p(z|A_\mathcal{G}), \phi, \theta).$$

The first term in Eq. (3) can be interpreted as the negative reconstruction error, which encourages the decoder to learn to reconstruct the original input. The second term is the Kullback-Leibler (KL) divergence between prior defined in Eq. (1) and variational posterior of latent process variables defined in Eq. (2), which acts a regularizer to penalize the model complexity. The expectation is taken with respect to the encoder's distribution over the representations given the neighborhood graph adjacency and input video frames. The KL term can be calculated analytically in the case of Gaussian distributions, as indicated by [20].

In proposed G-LBM model, the VAE objective is further constrained to linear dependency of the latent variables corresponding to the input video frames of the same scene (neighbouring data points in graph $\mathcal{G}$). This constraint is imposed by minimizing the rank of the latent manifold coordinates mapped from the video frames in the same clique in the graph $\mathcal{G}$.

$$rank(f_\phi(V^{(i)})) < \delta \quad \forall i \in \{1, ..., N\}, \tag{4}$$

where $f_\phi(V^{(i)})$ is the estimated mean and variance of the latent process z relative to the concatenated input video frames $V^{(i)}$ coming from the same scene (clique in the $\mathcal{G}$). N is the total number of cliques. As schematic illustration presented in Fig. 2 s. For our purpose of background modeling, given the knowledge that moving objects are sparse ouliers to the backgrounds, we extract a motion mask from the video frames. This motion mask is incorporated into the reconstruction loss of the VAE objective in Eq. (3) to provide a motion aware reconstruction loss in G-LBM. Given the motion mask M the VAE objective is updated as follows.

$$\mathcal{L}(p(\mathbf{z}|A_{\mathcal{G}}), \phi, \theta; M) = \tag{5}$$
$$\mathbb{E}_{q_\phi(\mathbf{z}|\mathbf{v}, A_{\mathcal{G}})}\big[\log p_\theta(\mathbf{v}|\mathbf{z}, M)\big] - KL\big(q_\phi(\mathbf{z}|\mathbf{v}, A_{\mathcal{G}})\|p(\mathbf{z}|A_{\mathcal{G}})\big).$$

VAE tries to minimize the reconstruction error between input video frames and estimated backgrounds where there is no foreground object (outlier) given by the motion mask. This minimization is done under an extra constraint to the VAE objective which imposes the sparsity of the outliers/perturbations defined as:

$$\| M^{(i)}(V^{(i)} - B_{\mathbf{v}}^{(i)}) \|_0 < \epsilon \quad \forall i \in \{1, ..., N\}, \tag{6}$$

where $\| \,.\, \|_0$ is the l_0-norm of the difference between concatenated input observations $V^{(i)}$ from each scene and their reconstructed background $B_{\mathbf{v}}^{(i)} = g_\phi(Z^{(i)})$ where motion mask M is present. Putting objective function and constrains together the final optimization problem to train the G-LBM model becomes:

$$\min \ \mathcal{L}(p(\mathbf{z}|A_{\mathcal{G}}), \phi, \theta; M) \tag{7}$$
$$\text{st:} \ \ M^{(i)}(V^{(i)} - B_{\mathbf{v}}^{(i)}) \|_0 < \epsilon \ \ \text{and} \ \ rank(f_\phi(V^{(i)})) < \delta \ \ \forall i \in \{1, ..., N\}.$$

In order to construct the final loss function to be utilized in learning the parameters of the encoder and decoder in G-LBM model, we used Nuclear norm $\| \,.\, \|_\star$ given by the sum of singular values and l_1-norm $\| \,.\, \|_1$ as the tightest convex relaxations of the $rank(.)$ and l_0-norm respectively. Substituting the reconstruction loss and analytical expression of the KL term in Eq. (7) the final loss of G-LBM to be minimized is:

$$\mathcal{L}(\phi, \theta; M, A_{\mathcal{G}}) = \tag{8}$$
$$\sum_{i=1}^{N} BCE(\bar{M}^{(i)}V^{(i)}, B^{(i)}) - \frac{1}{2}\big(tr(\Sigma^{-1}\Pi - I) + \Lambda^T \Sigma^{-1}\Lambda + log\frac{|\Sigma|}{|\Pi|}\big) +$$
$$\beta \sum_{i=1}^{N} \| M^{(i)}(V^{(i)} - B^{(i)}) \|_1 + \alpha \sum_{i=1}^{N} tr(\sqrt{f_\phi(V^{(i)})^T f_\phi(V^{(i)})}).$$

The motion mask is constructed by computing the motion fields using the coarse2fine optical flow [30] between each pair of consecutive frames in the given sequence of frames $V^{(i)}$ from the same scene. Using the motion information we compute a motion mask $M^{(i)}$. Let $\mathbf{v}_i$ and $\mathbf{v}_{i-1}$ be the two consecutive frames in $V^{(i)}$ and $h_{i,k}^x$ and $h_{i,k}^y$ be the horizontal and vertical component of motion vector

$\mathbf{m}_i$ at position k computed between frames $\mathbf{v}_i$ and $\mathbf{v}_{i-1}$ respectively. $\mathbf{m}_i \in \{0,1\}$ is the corresponding vectored motion mask computed as:

$$m_{i,k} = \begin{cases} 1, & \text{if } \sqrt{(h_{i,k}^x)^2 + (h_{i,k}^y)^2} < \tau \\ 0, & \text{otherwise,} \end{cases} \tag{9}$$

where threshold of motion magnitude τ is selected adaptively as a factor of the average of all pixels in motion field such that all pixels in $V^{(i)}$ exhibiting motion larger than τ definitely belong to the foreground not the noise in the background in videos. By concatenating all the motion vectors $\mathbf{m}_i$ computed from input $V^{(i)}$ we construct the $M^{(i)}$. Concatenation of all $M^{(i)}$ is specified as $\mathbf{M}$.

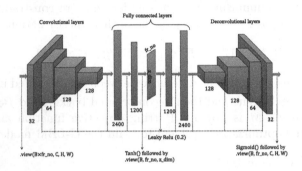

Fig. 3. Network architecture of the G-LBM. Input to the network is a batch of video clips. Each video clip is a sequence of consecutive frames with the same background scene. In order to handle the 4D input videos with 2D convolutions, we squeeze the first two axes of input regarding batch and video frames into one axis. We unsqueeze the first dimension where it is necessary.

2.3 G-LBM Model Architecture and Training Setup

The encoder and decoder parts of the VAE are both implemented using CNN architectures specified in Fig. 3. The encoder takes the video frames as input and outputs the mean and variance of the distribution over underlying low-dimensional latent process. The decoder takes samples drawn from latent distributions as input and outputs the recovered version of the background in original input. We trained G-LBM using the VAE architecture in Fig. 3 by minimizing the loss function defined in Eq. (8). We used Adam optimization to learn the parameters of the encoder and decoder, i.e., θ and ϕ, respectively. We employed learning rate scheduling and gradient clipping in the optimization setup. Training was performed on batches of size 3 video clips with 40 consecutive frames i.e 120 video frames in every input batch for 500 epochs.

3 Experimental Results

In this section, the performance of the proposed G-LBM is evaluated on two publicly available datasets BMC2012 and SBMnet-2016 [19,40]. Both quantitative and qualitative performances compared against state-of-the-art methods are provided in Sect. 3.1 and Sect. 3.2. Results show comparable or better performance against other state-of-the-art methods in background modeling.

3.1 BMC2012 Dataset

We evaluated the performance of our proposed method on the BMC2012 benchmark dataset [40]. We used 9 real-world surveillance videos in this dataset, along with encrypted ground truth (GT) masks of the foreground for evaluations. This dataset focuses on outdoor situations with various weather and illumination conditions making it suitable for performance evaluation of background subtraction (BS) methods in challenging conditions. Since this dataset is designed for the BS task in order to be able to do comparison on this dataset, we further performed BS by utilizing the output of the trained G-LBM model.

To extract the masks of the moving objects in videos, we first trained our model G-LBM using all of the video frames of short videos (with less than 2000 frames) and first 10000 frames of the long videos as explained in Sect. 2.3. After the model was trained, we fed the same frames to the network to estimate the background for each individual frame. Finally, we used the estimated background of each frame to find the mask of the moving objects by thresholding the difference between the original input frame and the estimated background. Table 1 shows the quantitative performance of G-LBM compared to other BS methods including 3TD [27], DP-GMM [12], LSD [23], TVRPCA [7], SRPCA [15], RMAMR [28], LR-FSO [47], GFL [44], MSCL [16]. Fig. 4 shows the estimated backgrounds and extracted masks by G-LBM model on sample video frames in the BMC2012 dataset. Considering that G-LBM is a scene non-specific model of the background and the task of BS is performed by simply thresholding the

Table 1. Comparison of average F_1-score on each video of BMC2012 dataset. Long videos are highlighted in gray.

Video	3TD	DP-GMM	LSD	TVRPCA	SRPCA	RMAMR	LR-FSO	GFL	MSCL	**G-LBM**
001	0.79	0.72	0.79	0.76	0.79	0.78	0.71	0.78	**0.80**	0.73
002	0.76	0.69	0.80	0.67	0.74	0.71	0.66	0.74	0.78	**0.85**
003	0.70	0.75	0.94	0.68	0.83	0.78	0.70	0.61	**0.96**	0.93
004	0.83	0.80	0.88	0.82	0.81	0.79	0.72	0.88	0.86	**0.91**
005	0.79	0.71	0.73	0.77	0.80	0.76	0.66	**0.80**	0.79	0.71
006	0.82	0.68	0.80	0.69	0.69	0.65	0.78	0.74	0.74	**0.85**
007	0.73	0.65	**0.81**	0.71	0.70	0.64	0.54	0.69	0.76	0.70
008	0.81	0.78	0.84	0.79	0.84	0.80	0.80	0.81	**0.89**	0.76
009	0.85	0.79	**0.92**	0.88	0.86	0.82	0.82	0.83	0.86	0.69
Average	0.78	0.73	**0.83**	0.75	0.78	0.74	0.71	0.76	0.82	0.79

difference between the estimated background and the original input frame, it is successful in detecting moving objects and generates acceptable masks of the foreground.

3.2 SBMnet-2016 Dataset

SBMnet dataset [19] provides a diverse set of 79 videos spanning 8 different categories selected to cover a wide range of detection challenges. These categories consist of basic, intermittent motion, clutter, jitter, illumination changes, background motion, very long with more than 3500 frames, and very short with less than 20 frames. The videos are representative of typical indoor and outdoor visual data captured in surveillance, smart environment. Spatial resolutions of the videos vary from 240 × 240 to 800 × 600 and their length varies from 6 to 9370 frames. Following metrics are utilized to measure the performance.

Fig. 4. Visual results of the G-LBM over video sequences of BMC2012. First row is the input video frame, second row is the computed background model by G-LBM, third row is the extracted foreground mask by thresholding the difference between input video frame and the G-LBM background model. Last row is the GT foreground mask.

- Average gray-level error (AGE), average of the gray-level absolute difference between grand truth and the estimated background image.
- Percentage of error pixels (pEPs), percentage of number of pixels in estimated background whose value differs from the corresponding pixel in grand truth by more than a threshold with respect to the total number of pixels.
- Percentage of clustered error pixels (pCEPS), percentage of error pixels whose 4-connected neighbours are also error pixels to the total number of pixels.
- Multi-scale structural similarity index (MSSSIM), estimation of the perceived visual distortion performing at multiple scales.
- Peak signal-to-noise-ratio(PSNR), measuring the image quality defined as $10 \log_{10} \left({(L-1)^2}/{MSE} \right)$ where L is the maximum grey level value 255 and MSE is the mean squared error between GT and estimated background.

- Color image quality measure (CQM), measuring perceptual image quality defined based on the PSNR values in the single YUV bands through $PSNR_Y \times R_w + 0.5C_w(PSNR_U + PSNR_V)$ where PSNR values are in db, R_w and C_w are two coefficients set to 0.9449 and 0.0551 respectively.

The objective of every background model is an accurate estimate of the background which is equivalent to minimizing AGE, pEPs, pCEP with high perceptual quality equivalent to maximizing the PSNR, MSSSIM, and CQM.

Performance Evaluation and Comparison. For comparison, we analyzed the results obtained by the best performing methods reported on the SBMnet-2016 dataset[1] as well as DNN based models compared with our G-LBM model in both quantitative and qualitative manner. Table 2 compares overall performance of our proposed G-LBM model against state-of-the-art background modeling methods including FC-FlowNet [13], BEWiS [8], SC-SOBS-C4 [24], BE-AAPSA [32], MSCL [16], SPMD [46], LaBGen-OF [21], FSBE [9], NExBI [26] with respect to aforementioned metrics. Among deep learning approaches in modeling the background practically, other than G-LBM only FC-FlowNet was fully evaluated on SBMnet-2016 dataset. However, the rank of FC-FlowNet is only 20 (see Table 2) compared with G-LBM and is also outperformed by conventional neural networks approaches like BEWiS, SC-SOBS-C4, and BE-AAPSA. As quantified in Table 3, G-LBM can effectively capture the stochasticity of the background dynamics in its probabilistic model, which outperforms other methods of background modeling in relative challenges of illumination changes, background motion, and jitter. The qualitative comparison of the G-LBM performance with the other methods is shown in Fig. 5 for two samples of jitter

Table 2. Comparison of overall model performance evaluated on SBMnet-2016 dataset in terms of metrics averaged over all categories.

Method	Average ranking	AGE	pEPs	pCEPS	MSSSIM	PSNR	CQM
non-DNN methods							
MSCL	**1.67**	**5.9547**	**0.0524**	**0.0171**	0.9410	**30.8952**	**31.7049**
SPMD	1.83	6.0985	0.0487	0.0154	**0.9412**	29.8439	30.6499
LaBGen-OF	3.00	6.1897	0.0566	0.0232	0.9412	29.8957	30.7006
FSBE	4.33	6.6204	0.0605	0.0217	0.9373	29.3378	30.1777
NExBI	9.33	6.7778	0.0671	0.0227	0.9196	27.9944	28.8810
DNN-based methods							
G-LBM	9.67	6.8779	0.0759	0.0321	0.9181	**28.9336**	**29.7641**
BEWiS	**6.50**	**6.7094**	**0.0592**	0.0266	**0.9282**	28.7728	29.6342
SC-SOBS-C4	12.17	7.5183	0.0711	**0.0242**	0.9160	27.6533	28.5601
FC-FLOWNet	19.83	9.1131	0.1128	0.0599	0.9162	26.9559	27.8767
BE-AAPSA	16.50	7.9086	0.0873	0.0447	0.9127	27.0714	27.9811

---

[1] http://scenebackgroundmodeling.net/.

Table 3. Comparison of G-LBM performance against other approaches evaluated on SBMnet-2016 dataset with respect to averaged AGE over videos in each category.

Method	Basic	Intermittent motion	Clutter	Jitter	Illumination changes	Background motion	Very long	Very short
non-DNN methods								
MSCL	**3.4019**	**3.9743**	5.2695	9.7403	**4.4319**	11.2194	**3.8214**	5.7790
SPMD	3.8141	4.1840	4.5998	9.8095	4.4750	**9.9115**	6.0926	5.9017
LaBGen-OF	3.8421	4.6433	**4.1821**	**9.2410**	8.2200	10.0698	4.2856	**5.0338**
FSBE	3.8960	5.3438	4.7660	10.3878	5.5089	10.5862	6.9832	5.4912
NExBI	4.7466	4.6374	5.3091	11.1301	4.8310	11.5851	6.2698	5.7134
DNN-based methods								
G-LBM	4.5013	7.0859	13.0786	**9.1154**	**3.2735**	9.1644	**2.5819**	6.2223
BEWiS	**4.0673**	**4.7798**	**10.6714**	9.4156	5.9048	9.6776	3.9652	**5.1937**
SC-SOBS-C4	4.3598	6.2583	15.9385	10.0232	10.3591	10.7280	6.0638	5.2953
FC-FLOWNet	5.5856	6.7811	12.5556	10.2805	13.3662	10.0539	7.7727	6.5094
BE-AAPSA	5.6842	6.6997	12.3049	10.1994	7.0447	9.3755	3.8745	8.0857

Input frame   Ground-truth   G-LBM   SC-SOBS-C4   BEWiS   MSCL   FC-FLOWNet

Fig. 5. Visual results of the G-LBM compared against other high performing methods on SBMnet-2016 dataset. First and second rows are sample video frames from illumination changes background and motion categories, respectively.

Jitter/Badminton   BackgroundMotion/advertisementBoard   Jitter/Boulevard   IlluminationChange/CameraParameter   Basic/Blurred   VeryShort/DynamicBackground   Basic/511   VeryShort/CUHK-square   VeryLong/BusStopMorning

Fig. 6. Visual results on categories of SBMnet-2016 that G-LBM successfully models the background with comparable or higher quantitative performance (see Table 3). First row is the input frame, second row is the G-LBM estimated background, and third row is the GT.

Fig. 7. Visualization of G-LBM failure to estimate an accurate model of the background. First row is the input frame, second row is the G-LBM estimated background, and third row is the GT. Since G-LBM is a scene non-specific method it is outperformed by other models that have more specific designs for these challenges.

and background motion categories. However, in clutter and intermittent motion categories that background is heavily filled with clutter or foreground objects are steady for a long time, G-LBM fails in estimating an accurate model compared to other methods that are scene-specific and have special designs for these challenges. Fig. 6 visualizes the qualitative results of G-LBM for different challenges in SBMnet-2016 in which it has comparable or superior performance. Cases that G-LBM fails in providing a robust model for the background are also shown in Fig. 7, which happen mainly in videos with intermittent motion and heavy clutter.

4   Conclusion

Here, we presented our scene non-specific generative low-dimensional background model (G-LBM) using the framework of VAE for modeling the background in videos recorded by stationary cameras. We evaluated the performance of our model in task of background estimation, and showed how well it adapts to the changes of the background on two datasets of BMC2012 and SBMnet-2016. According to the quantitative and qualitative results, G-LBM outperformed other state-of-the-art models specially in categories that stochasticity of the background is the major challenge such as jitter, background motion, and illumination changes.

References

1. Bianco, S., Ciocca, G., Schettini, R.: Combination of video change detection algorithms by genetic programming. IEEE Trans. Evol. Comput. **21**(6), 914–928 (2017)
2. Blei, D.M., Kucukelbir, A., McAuliffe, J.D.: Variational Inference: A Review for Statisticians. ArXiv e-prints, January 2016
3. Bouwmans, T., Garcia-Garcia, B.: Background subtraction in real applications: challenges, current models and future directions. arXiv preprint arXiv:1901.03577 (2019)

4. Bouwmans, T., Javed, S., Sultana, M., Jung, S.K.: Deep neural network concepts for background subtraction: a systematic review and comparative evaluation. Neural Netw. **117**, 8–66 (2019)

5. Bouwmans, T., Zahzah, E.H.: Robust PCA via principal component pursuit: a review for a comparative evaluation in video surveillance. Comput. Vis. Image Underst. **122**, 22–34 (2014)

6. Candès, E.J., Li, X., Ma, Y., Wright, J.: Robust principal component analysis? J. ACM (JACM) **58**(3), 1–37 (2011)

7. Cao, X., Yang, L., Guo, X.: Total variation regularized RPCA for irregularly moving object detection under dynamic background. IEEE Trans. Cybern. **46**(4), 1014–1027 (2015)

8. De Gregorio, M., Giordano, M.: Background estimation by weightless neural networks. Pattern Recogn. Lett. **96**, 55–65 (2017)

9. Djerida, A., Zhao, Z., Zhao, J.: Robust background generation based on an effective frames selection method and an efficient background estimation procedure (FSBE). Sig. Process.: Image Commun. **78**, 21–31 (2019)

10. Dosovitskiy, A., et al.: FlowNet: learning optical flow with convolutional networks. In: Proceedings of the IEEE International Conference on Computer Vision, pp. 2758–2766 (2015)

11. Farnoosh, A., Rezaei, B., Ostadabbas, S.: DeepPBM: deep probabilistic background model estimation from video sequences. arXiv preprint arXiv:1902.00820 (2019)

12. Haines, T.S., Xiang, T.: Background subtraction with Dirichlet process mixture models. IEEE Trans. Pattern Anal. Mach. Intell. **36**(4), 670–683 (2013)

13. Halfaoui, I., Bouzaraa, F., Urfalioglu, O.: CNN-based initial background estimation. In: 2016 23rd International Conference on Pattern Recognition (ICPR), pp. 101–106. IEEE (2016)

14. He, X., Liao, L., Zhang, H., Nie, L., Hu, X., Chua, T.S.: Neural collaborative filtering. In: Proceedings of the 26th International Conference on World Wide Web, pp. 173–182 (2017)

15. Javed, S., Mahmood, A., Bouwmans, T., Jung, S.K.: Spatiotemporal low-rank modeling for complex scene background initialization. IEEE Trans. Circ. Syst. Video Technol. **28**(6), 1315–1329 (2016)

16. Javed, S., Mahmood, A., Bouwmans, T., Jung, S.K.: Background-foreground modeling based on spatiotemporal sparse subspace clustering. IEEE Trans. Image Process. **26**(12), 5840–5854 (2017). https://doi.org/10.1109/TIP.2017.2746268

17. Javed, S., Narayanamurthy, P., Bouwmans, T., Vaswani, N.: Robust PCA and robust subspace tracking: a comparative evaluation. In: 2018 IEEE Statistical Signal Processing Workshop (SSP), pp. 836–840. IEEE (2018)

18. Jiang, S., Lu, X.: WeSamBE: a weight-sample-based method for background subtraction. IEEE Trans. Circ. Syst. Video Technol. **28**(9), 2105–2115 (2017)

19. Jodoin, P.M., Maddalena, L., Petrosino, A., Wang, Y.: Extensive benchmark and survey of modeling methods for scene background initialization. IEEE Trans. Image Process. **26**(11), 5244–5256 (2017)

20. Kingma, D.P., Welling, M.: Auto-Encoding Variational Bayes. ArXiv e-prints, December 2013

21. Laugraud, B., Van Droogenbroeck, M.: Is a memoryless motion detection truly relevant for background generation with LaBGen? In: Blanc-Talon, J., Penne, R., Philips, W., Popescu, D., Scheunders, P. (eds.) ACIVS 2017. Lecture Notes in Computer Science, vol. 10617, pp. 443–454. Springer, Cham (2017). https://doi.org/10.1007/978-3-319-70353-4_38

22. Lim, L.A., Keles, H.Y.: Foreground segmentation using convolutional neural networks for multiscale feature encoding. Pattern Recogn. Lett. **112**, 256–262 (2018)
23. Liu, X., Zhao, G., Yao, J., Qi, C.: Background subtraction based on low-rank and structured sparse decomposition. IEEE Trans. Image Process. **24**(8), 2502–2514 (2015)
24. Maddalena, L., Petrosino, A.: Extracting a background image by a multi-modal scene background model. In: 2016 23rd International Conference on Pattern Recognition (ICPR), pp. 143–148. IEEE (2016)
25. Mondéjar-Guerra, V., Rouco, J., Novo, J., Ortega, M.: An end-to-end deep learning approach for simultaneous background modeling and subtraction. In: British Machine Vision Conference (BMVC), Cardiff (2019)
26. Mseddi, W.S., Jmal, M., Attia, R.: Real-time scene background initialization based on spatio-temporal neighborhood exploration. Multimedia Tools Appl. **78**(6), 7289–7319 (2018). https://doi.org/10.1007/s11042-018-6399-1
27. Oreifej, O., Li, X., Shah, M.: Simultaneous video stabilization and moving object detection in turbulence. IEEE Trans. Pattern Anal. Mach. Intell. **35**(2), 450–462 (2012)
28. Ortego, D., SanMiguel, J.C., Martínez, J.M.: Rejection based multipath reconstruction for background estimation in SBMnet 2016 dataset. In: 2016 23rd International Conference on Pattern Recognition (ICPR), pp. 114–119. IEEE (2016)
29. Papadimitriou, C.H., Raghavan, P., Tamaki, H., Vempala, S.: Latent semantic indexing: a probabilistic analysis. J. Comput. Syst. Sci. **61**(2), 217–235 (2000)
30. Pathak, D., Girshick, R., Dollár, P., Darrell, T., Hariharan, B.: Learning features by watching objects move. In: Proceedings of the IEEE Conference on Computer Vision and Pattern Recognition, pp. 2701–2710 (2017)
31. Qu, Z., Yu, S., Fu, M.: Motion background modeling based on context-encoder. In: 2016 Third International Conference on Artificial Intelligence and Pattern Recognition (AIPR), pp. 1–5. IEEE (2016)
32. Ramirez-Alonso, G., Ramirez-Quintana, J.A., Chacon-Murguia, M.I.: Temporal weighted learning model for background estimation with an automatic re-initialization stage and adaptive parameters update. Pattern Recogn. Lett. **96**, 34–44 (2017)
33. Rezaei, B., Huang, X., Yee, J.R., Ostadabbas, S.: Long-term non-contact tracking of caged rodents. In: 2017 IEEE International Conference on Acoustics, Speech and Signal Processing (ICASSP), pp. 1952–1956. IEEE (2017)
34. Rezaei, B., Ostadabbas, S.: Background subtraction via fast robust matrix completion. In: Proceedings of the IEEE International Conference on Computer Vision Workshops, pp. 1871–1879 (2017)
35. Rezaei, B., Ostadabbas, S.: Moving object detection through robust matrix completion augmented with objectness. IEEE J. Sel. Top. Sig. Process. **12**(6), 1313–1323 (2018)
36. Ronneberger, O., Fischer, P., Brox, T.: U-Net: convolutional networks for biomedical image segmentation. In: Navab, N., Hornegger, J., Wells, W., Frangi, A. (eds.) Medical Image Computing and Computer-Assisted Intervention – CCAI 2015, MICCAI 2015. Lecture Notes in Computer Science, vol. 9351, pp. 234–241. Springer, Cham (2015). https://doi.org/10.1007/978-3-319-24574-4_28
37. St-Charles, P.L., Bilodeau, G.A., Bergevin, R.: A self-adjusting approach to change detection based on background word consensus. In: 2015 IEEE Winter Conference on Applications of Computer Vision, pp. 990–997. IEEE (2015)

38. Sultana, M., Mahmood, A., Javed, S., Jung, S.K.: Unsupervised deep context prediction for background estimation and foreground segmentation. Machine Vis. Appl. **30**(3), 375–395 (2018). https://doi.org/10.1007/s00138-018-0993-0
39. Tao, Y., Palasek, P., Ling, Z., Patras, I.: Background modelling based on generative UNet. In: 2017 14th IEEE International Conference on Advanced Video and Signal Based Surveillance (AVSS), pp. 1–6. IEEE (2017)
40. Vacavant, A., Chateau, T., Wilhelm, A., Lequièvre, L.: A benchmark dataset for outdoor foreground/background extraction. In: Park, J.I., Kim, J. (eds.) ACCV 2012. Lecture Notes in Computer Science, vol. 7728, pp. 291–300. Springer, Heidelberg (2012). https://doi.org/10.1007/978-3-642-37410-4_25
41. Vaswani, N., Bouwmans, T., Javed, S., Narayanamurthy, P.: Robust subspace learning: robust PCA, robust subspace tracking, and robust subspace recovery. IEEE Sig. Process. Mag. **35**(4), 32–55 (2018)
42. Wang, Y., Jodoin, P.M., Porikli, F., Konrad, J., Benezeth, Y., Ishwar, P.: CDnet 2014: an expanded change detection benchmark dataset. In: Proceedings of the IEEE Conference on Computer Vision And Pattern Recognition Workshops, pp. 387–394 (2014)
43. Wang, Y., Luo, Z., Jodoin, P.M.: Interactive deep learning method for segmenting moving objects. Pattern Recogn. Lett. **96**, 66–75 (2017)
44. Xin, B., Tian, Y., Wang, Y., Gao, W.: Background subtraction via generalized fused lasso foreground modeling. In: Proceedings of the IEEE Conference on Computer Vision and Pattern Recognition, pp. 4676–4684 (2015)
45. Xu, P., Ye, M., Li, X., Liu, Q., Yang, Y., Ding, J.: Dynamic background learning through deep auto-encoder networks. In: Proceedings of the 22nd ACM International Conference on Multimedia, pp. 107–116 (2014)
46. Xu, Z., Min, B., Cheung, R.C.: A robust background initialization algorithm with superpixel motion detection. Sig. Process.: Image Commun. **71**, 1–12 (2019)
47. Xue, G., Song, L., Sun, J.: Foreground estimation based on linear regression model with fused sparsity on outliers. IEEE Trans. Circ. Syst. Video Technol. **23**(8), 1346–1357 (2013)
48. Yang, B., Lei, Y., Liu, J., Li, W.: Social collaborative filtering by trust. IEEE Trans. Pattern Anal. Mach. Intell. **39**(8), 1633–1647 (2016)
49. Zheng, W., Wang, K., Wang, F.Y.: A novel background subtraction algorithm based on parallel vision and Bayesian GANs. Neurocomputing **394**, 178–200 (2019)

H3DNet: 3D Object Detection Using Hybrid Geometric Primitives

Zaiwei Zhang$^{(\boxtimes)}$, Bo Sun$^{(\boxtimes)}$, Haitao Yang$^{(\boxtimes)}$, and Qixing Huang$^{(\boxtimes)}$

The University of Texas at Austin, 78710 Austin, TX, USA
zaiweizhang@utexas.edu, bosun0711@gmail.com, yanghtr@outlook.com,
huangqx@cs.utexas.edu

Abstract. We introduce H3DNet, which takes a colorless 3D point cloud as input and outputs a collection of oriented object bounding boxes (or BB) and their semantic labels. The critical idea of H3DNet is to predict a hybrid set of geometric primitives, i.e., BB centers, BB face centers, and BB edge centers. We show how to convert the predicted geometric primitives into object proposals by defining a distance function between an object and the geometric primitives. This distance function enables continuous optimization of object proposals, and its local minimums provide high-fidelity object proposals. H3DNet then utilizes a matching and refinement module to classify object proposals into detected objects and fine-tune the geometric parameters of the detected objects. The hybrid set of geometric primitives not only provides more accurate signals for object detection than using a single type of geometric primitives, but it also provides an overcomplete set of constraints on the resulting 3D layout. Therefore, H3DNet can tolerate outliers in predicted geometric primitives. Our model achieves state-of-the-art 3D detection results on two large datasets with real 3D scans, ScanNet and SUN RGB-D. Our code is open-sourced at here.

Keywords: 3D deep learning · Geometric deep learning · 3D point clouds · 3D bounding boxes · 3D object detection

1 Introduction

Object detection is a fundamental problem in visual recognition. In this work, we aim to detect the 3D layout (i.e., oriented 3D bounding boxes (or BBs) and associated semantic labels) from a colorless 3D point cloud. This problem is fundamentally challenging because of the irregular input and a varying number of objects across different scenes. Choosing suitable intermediate representations

---

B. Sun and H. Yang—Equal Contribution.

Electronic supplementary material The online version of this chapter (https://doi.org/10.1007/978-3-030-58610-2_19) contains supplementary material, which is available to authorized users.

© Springer Nature Switzerland AG 2020
A. Vedaldi et al. (Eds.): ECCV 2020, LNCS 12357, pp. 311–329, 2020.
https://doi.org/10.1007/978-3-030-58610-2_19

Fig. 1. Our approach leverages a hybrid and overcomplete set of geometric primitives to detect and refine 3D object bounding boxes (BBs). Note that red BBs are initial object proposals, green BBs are refined object proposals, and blue surfaces and lines are hybrid geometric primitives. (Color figure online)

to integrate low-level object cues into detected objects is key to the performance of the resulting system. While early works [39,40] classify sliding windows for object detection, recent works [4,17,28–30,33,36,46,52,52,55] have shown the great promise of designing end-to-end neural networks to generate, classify, and refine object proposals.

This paper introduces H3DNet, an end-to-end neural network that utilizes a novel intermediate representation for 3D object detection. Specifically, H3DNet first predicts a hybrid and overcomplete set of geometric primitives (i.e., BB centers, BB face centers, and BB edge centers) and then detects objects to fit these primitives and their associated features. This regression methodology, which is motivated from the recent success of keypoint-based pose regression for 6D object pose estimation [12,20,22,26,27,37], displays two appealing advantages for 3D object detection. First, each type of geometric primitives focuses on different regions of the input point cloud (e.g., points of an entire object for predicting the BB center and points of a planar boundary surface for predicting the corresponding BB face center). Combing diverse primitive types can add the strengths of their generalization behaviors. On new instances, they offer more useful constraints and features than merely using one type of primitives. Second, having an overcomplete set of primitive constraints can tolerate outliers in predicted primitives (e.g., using robust functions) and reduce the influence of individual prediction errors. The design of H3DNet fully practices these two advantages (Fig. 1).

Specifically, H3DNet consists of three modules. The first module computes dense pointwise descriptors and uses them to predict geometric primitives and their latent features. The second module converts these geometric primitives into object proposals. A key innovation of H3DNet is to define a parametric distance function that evaluates the distance between an object BB and the predicted primitives. This distance function can easily incorporate diverse and overcomplete geometric primitives. Its local minimums naturally correspond to object proposals. This method allows us to optimize object BBs continuously and generate high-quality object proposals from imprecise initial proposals.

The last module of H3DNet classifies each object proposal as a detected object or not, and also predicts for each detected object an offset vector of its geometric parameters and a semantic label to fine-tune the detection result. The performance of this module depends on the input. As each object proposal is associated with diverse geometric primitives, H3DNet aggregates latent features associated with these primitives, which may contain complementary semantic and geometric information, as the input to this module. We also introduce a network design that can handle a varying number of geometric primitives.

We have evaluated H3DNet on two popular benchmark datasets ScanNet and SUN RGB-D. On ScanNet, H3DNet achieved 67.2% in mAP (0.25), which corresponded to a 8.5% relative improvement from state-of-the-art methods that merely take the 3D point positions as input. On SUN RGB-D, H3DNet achieved 60.1% in mAP (0.25), which corresponded to a 2.4% relative improvement from the same set of state-of-the-art methods. Moreover, on difficult categories of both datasets (i.e., those with low mAP scores), the performance gains of H3DNet are significant (e.g., from 38.1/47.3/57.1 to 51.9/61.0/75.3/ on window/door/shower-curtain, respectively). We have also performed an ablation study on H3DNet. Experimental results justify the importance of regressing a hybrid and overcomplete set of geometric primitives for generating object proposals and aggregating features associated with matching primitives for classifying and refining detected objects. In summary, the contributions of our work are:

- Formulation of object detection as regressing and aggregating an overcomplete set of geometric primitives
- Predicting multiple types of geometric primitives that are suitable for different object types and scenes
- State-of-the-art results on SUN RGB-D and ScanNet with only point clouds.

2   Related Works

3D Object Detection. From the methodology perspective, there are strong connections between 3D object detection approaches and their 2D counterparts. Most existing works follow the approach of classifying candidate objects that are generated using a sliding window [39,40] or more advanced techniques [28–30,36,46,52,52,55]. Objectness classification involves template-based approaches or deep neural networks. The key differences between 2D approaches and 3D approaches lie in feature representations. For example, [24] leverages a pairwise semantic context potential to guide the proposals' objectness score. [33] uses clouds of oriented gradients (COG) for object detection. [10] utilizes the power of 3D convolution neural networks to identify locations and keypoints of 3D objects. Due to the computational cost in the 3D domain, many methods utilize 2D–3D projection techniques to integrate 2D object detection and 3D data processing. For example, MV3D [4] and VoxelNet [55] represent the 3D input data in a bird's-eye view before proceeding to the rest of the pipeline. Similarly, [14,17,30] first process 2D inputs to identify candidate 3D object proposals.

Point clouds have emerged as a powerful representation for 3D deep learning, particularly for extracting salient geometric features and spatial locations (c.f. [31,32]). Prior usages of point-based neural networks include classification [9,11,16,21,31,32,45,47–49], segmentation [2,7,9,11,16,21,32,41,43,45–49], normal estimation [2], and 3D reconstruction [6,42,50].

There are also growing interests in object detection from point clouds [28, 29,36,46,52,52]. H3DNet is most relevant to [29], which leverages deep neural networks to predict object bounding boxes. The key innovation of H3DNet is that it utilizes an overcomplete set of geometric primitives and a distance function to integrate them for object detection. This strategy can tolerate inaccurate primitive predictions (e.g., due to partial inputs).

Multi-task 3D Understanding. Jointly predicting different types of geometric primitives is related to multi-task learning [3,8,13,19,23,25,28,34,35,53,54,56], where incorporating multiple relevant tasks together boosts the performance of feature learning. In a recent work HybridPose [37], Song et al. show that predicting keypoints, edges between keypoints, and symmetric correspondences jointly lift the prediction accuracies of each type of features. In this paper, we show that predicting BB centers, BB face centers, and BB edge centers together help to improve the generalization behavior of primitive predictions.

Overcomplete Constraints Regression. The main idea of H3DNet is to incorporate an overcomplete set of constraints. This approach achieves considerable performance gains from [29], which uses a single type of geometric primitives. At a conceptual level, similar strategies have been used in tasks of object tracking [44], zero-shot fine-grained classification [1], 6D object pose estimation [37] and relative pose estimation between scans [51], among others. Compared to these works, the novelties of H3DNet lie in designing hybrid constraints that are suitable for object detection, continuous optimization of object proposals, aggregating hybrid features for classifying and fine-tuning object proposals, and end-to-end training of the entire network.

3   Approach

This section describes the technical details of H3DNet. Section 3.1 presents an approach overview. Sect. 3.2 to Sect. 3.5 elaborate on the network design and the training procedure of H3DNet.

3.1   Approach Overview

As illustrated in Fig. 2, the input of H3DNet is a dense set of 3D points (i.e., a point cloud) $S \in \mathbb{R}^{3 \times n}$ ($n = 40000$ in this paper). Such an input typically comes from depth sensors or the result of multi-view stereo matching. The output is given by a collection of (oriented) bounding boxes (or BB) $\mathcal{O}_S \in \overline{\mathcal{O}}$, where $\overline{\mathcal{O}}$ denotes the space of all possible objects. Each object $o \in \overline{\mathcal{O}}$ is given by its class label $l_o \in \mathcal{C}$, where $\mathcal{C}$ is pre-defined, its center $c_o = (c_o^x, c_o^y, c_o^z)^T \in \mathbb{R}^3$ in a world coordinate system, its scales $s_o = (s_o^x, s_o^y, s_o^z)^T \in \mathbb{R}^3$, and its orientation $n_o = (n_o^x, n_o^y)^T \in \mathbb{R}^2$ in the xy-plane of the same world coordinate system (note that the upright direction of an object is always along the z axis).

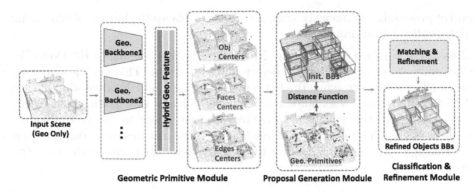

Fig. 2. H3DNet consists of three modules. The first module computes a dense descriptor and predicts three geometric primitives, namely, BB centers, BB face centers, and BB edge centers. The second module converts geometric primitives into object proposals. The third module classifies object proposals and refines the detected objects.

H3DNet consists of three modules, starting from geometric primitive prediction, to proposal generation, to object refinement. The theme is to predict and integrate an overcomplete set of geometric primitives, i.e., BB centers, BB face centers, and BB edge centers. The entire network is trained end-to-end.

Geometric Primitive Module. The first module of H3DNet takes a point cloud S as input and outputs a set of geometric primitives $\mathcal{P}_S$ that predicts locations of BB centers, BB face centers, and BB edge centers of the underlying objects. The network design extends that of [29]. Specifically, it combines a sub-module for extracting dense point-wise descriptors and sub-modules that take point-wise descriptors as input and output offset vectors between input points and the corresponding centers. The resulting primitives are obtained through clustering. In addition to locations, each predicted geometric primitive also possesses a latent feature that is passed through subsequent modules of H3DNet.

In contrast to [29], H3DNet exhibits two advantages. First, since only a subset of predicted geometric primitives is sufficient for object detection, the detected objects are insensitive to erroneous predictions. Second, different types of geometric primitives show complementary strength. For example, BB centers are accurate for complete and solid objects, while BB face centers are suitable for partial objects that possess rich planar structures.

Proposal Generation Module. The second module takes predicted geometric primitives as input and outputs a set of object proposals. A critical innovation of H3DNet is to formulate object proposals as local minimums of a distance function. This methodology is quite flexible in several ways. First, it is easy to incorporate overcomplete geometric primitives, each of which corresponds to an objective term of the distance function. Second, it can handle outlier predictions and mispredictions using robust norms. Finally, it becomes possible to optimize object proposals continuously, and this property relaxes the burden of generating high-quality initial proposals.

Classification and Refinement Module. The last module of H3DNet classifies each object proposal into a detected object or not. This module also computes offset vectors to refine the BB center, BB size, and BB orientation of each detected object, and a semantic label. The key idea of this module is to aggregate features of the geometric primitives that are close to the corresponding primitives of each object proposal. Such aggregated features carry rich semantic information that is unavailable in the feature associated with each geometric primitive.

3.2 Primitive Module

The first module of H3DNet predicts a set of geometric primitives from the input point cloud. Each geometric primitive provides some constraints on the detected objects. In contrast to most prior works that compute a minimum set of primitives, i.e., that is sufficient to determine the object bounding boxes, H3DNet leverages an overcomplete set of geometric primitives, i.e., BB centers, BB face centers, and BB edge centers. In other words, these geometric primitives can provide up-to 19 positional constraints for one BB. As we will see later, they offer great flexibilities in generating, classifying, and refining object proposals.

Similar to [29], the design of this module combines a descriptor sub-module and a prediction sub-module. The descriptor sub-module computes dense pointwise descriptors. Its output is fed into the prediction sub-module, which consists of three prediction branches. Each branch predicts one type of geometric primitives. Below we provide the technical details of the network design.

Descriptor Sub-module. The output of the descriptor sub-module provides semantic information to group points for predicting geometric primitives (e.g., points of the same object for BB centers and points of the same planar boundary faces for BB face centers). Instead of using a single descriptor computation tower [29], H3DNet integrates four separate descriptor computation towers. The resulting descriptors are concatenated together for primitive prediction and subsequent modules of H3DNet. Our experiments indicate that this network design can learn distinctive features for predicting each type of primitives. However, it does not lead to a significant increase in network complexity.

BB Center Prediction. The same as [29], H3DNet leverages a network with three fully connected layers to predict the offset vector between each point and its corresponding object center. The resulting BB centers are obtained through clustering (c.f. [29]). Note that in additional to offset vectors, H3DNet also computes an associated feature descriptor for each BB center. These feature descriptors serve as input feature representations for subsequent modules of H3DNet.

Predictions of BB centers are accurate on complete and rectangular shaped objects. However, there are shifting errors for partial and/or occluded objects, and thin objects, such as pictures or curtains, due to imbalanced weighting for offset prediction. This motivates us to consider centers of BB faces and BB edges.

BB Face Center Prediction. Planar surfaces are ubiquitous in man-made scenes and objects. Similar to BB center, H3DNet uses 3 fully connected layers to perform point-wise predictions. The predicted attributes include a flag that indicates whether a point is close to a BB face or not and if so, an offset vector between that point and its corresponding BB face center. For training, we generate the ground-truth labels by computing the closest BB face for each point. We say a point lies close to a BB face (i.e., a positive instance) if that distance is smaller than 0.2 m. Similar to BB centers, each BB face center prediction also possesses a latent feature descriptor that is fed into the subsequent modules.

Since face center predictions are only affected by points that are close to that face, we found that they are particularly useful for objects with rich planar patches (e.g., refrigerator and shower-curtain) and incomplete objects.

BB Edge Center Prediction. Boundary line features form another type of geometric cues in all 3D scenes and objects. Similar to BB faces, H3DNet employs 3 fully connected layers to predict for each point a flag that indicates whether it is close to a BB edge or not and if so, an offset vector between that point and the corresponding BB edge center. The same as BB face centers, we generate ground-truth labels by computing the closest BB edge for each point. We say a point lies close to a BB edge if the closest distance is smaller than 0.2 m. Again, each BB edge center prediction possesses a latent feature of the same dimension. Compared to BB centers and BB face centers, BB edge centers are useful for objects where point densities are irregular (e.g., with large holes) but BB edges appear to be complete (e.g., window and computer desk).

As analyzed in details in the supplemental material, error distributions of different primitives are largely uncorrelated with each other. Such uncorrelated prediction errors provide a foundation for performance boosting when integrating them together for detecting objects.

3.3   Proposal Module

After predicting geometric primitives, H3DNet proceeds to compute object proposals. Since the predicted geometric primitives are overcomplete, H3DNet converts them into a distance function and generates object proposals as local minimums of this distance function. This approach, which is the crucial contribution of H3DNet, exhibits several appealing properties. First, it automatically incorporates multiple geometric primitives to determine the parameters of each object proposal. Second, the distance function can optimize object proposals continuously. The resulting local minimums are insensitive to initial proposals, allowing us to use simple initial proposal generators. Finally, each local minimum is attached to different types of geometric primitives, which carry potentially complementary semantic information. As discussed in Sect. 3.4, the last module of H3DNet builds upon this property to classify and refine object proposals.

Proposal Distance Function. The proposal distance function $F_S(o)$ measures a cumulative proximity score between the predicted geometric primitives $\mathcal{P}_S$ and the corresponding object primitives of an object proposal o. Recall that $l_o \in \mathcal{C}$,

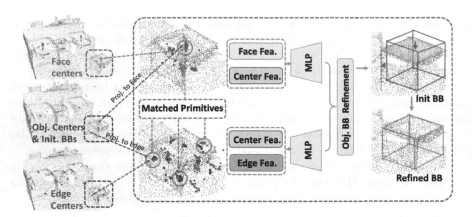

Fig. 3. Illustration of the matching, feature aggregation and refinement process.

$c_o \in \mathbb{R}^3$, $s_o \in \mathbb{R}^3$, and $n_o \in \mathbb{R}^2$ denote the label, center, scales, and orientation of o. With $o = (c_o^T, s_o^T, n_o^T)^T$ we collect all the geometric parameters of o. Note that each object proposal o has 19 object primitives (i.e., one BB center, six BB face centers, and twelve BB edge centers). Let $p_i(o), 1 \leq i \leq 19$ be the location of the i-th primitive of o. Denote $t_i \in \mathcal{T} := \{center, face, edge\}$ as the type of the i-th primitive. Let $\mathcal{P}_{t,S} \subseteq \mathcal{P}_S$ collect all predicted primitives with type $t \in \mathcal{T}$. We define

$$F_S(o) := \sum_{t \in \mathcal{T}} \beta_t \sum_{c \in \mathcal{P}_{t,S}} \min \left(\min_{1 \leq i \leq 19, t_i = t} \|c_i - p_i(o)\|^2 - \delta, 0 \right). \tag{1}$$

In other words, we employ the truncated L2-norm to match predicted primitives and closest object primitives. β_t describes the trade-off parameter for type t. Both β_t and the truncation threshold δ are determined via cross-validation.

Initial Proposals. H3DNet detects object proposals by exploring the local minimums of the distance function from a set of initial proposals. From the perspective of optimization, we obtain the same local minimum from any initial solution that is sufficiently close to that local minimum. This means the initial proposals do not need to be exact. In our experiments, we found that a simple object proposal generation approach is sufficient. Specifically, H3DNet utilizes the method of [29], which initializes an object proposal from each detected BB center.

Proposal Refinement. By minimizing F_S, we refine each initial proposal. Note that different initial proposals may share the same local minimum. The final object proposals only collect distinctive local minimums.

3.4 Classification and Refinement Module

The last module of H3DNet takes the output of the proposal module as input and outputs a collection of detected objects. This module combines a classification sub-module and a refinement sub-module. The classification sub-module determines whether each object proposal is an object or not. The refinement sub-module predicts for each detected object the offsets in BB center, BB size, and BB orientation and a semantic label.

The main idea is to aggregate features associated the primitives (i.e., object centers, edge centers, and face centers) of each object proposal. Such features capture potentially complementary information, yet only at this stage (i.e., after we have detected groups of matching primitives) it becomes possible to fuse them together to determine and fine-tune the detected objects.

As illustrated in Fig. 3, we implement this sub-module by combing four fully connected layers. The input layer concatenates input features of 19 object primitives of an object proposal (i.e., one BB center, six BB face centers, and twelve BB edge centers). Each input feature integrates features associated with primitives that are in the neighborhood of the corresponding object primitive. To address the issue that there is a varying number of neighborhood primitives (e.g., none or multiple), we utilize a variant of the max-pooling layer in Point-Net [31,32] to compute the input feature. Specifically, the input to each max-pooling layer consists of the feature associated with the input object proposal, which addresses the issue of no matching primitives, and 32 feature points that are randomly sampled in the neighborhood of each object primitive. In our implementation, we determine the neighboring primitives via range query, and the radius is 0.05 m.

The output of this module combines the label that indicates objectiveness, offsets in BB center, BB size, and BB orientation, and a semantic label.

3.5   Network Training

Training H3DNet employs a loss function with five objective terms:

$$\min_{\theta_g, \theta_p, \theta_c, \theta_o} \quad \lambda_g l_g(\theta_g) + \lambda_p l_p(\theta_g, \theta_p) + \lambda_f l_f(\theta_g, \theta_p, \theta_o)$$

$$+ \lambda_c l_c(\theta_g, \theta_p, \theta_c) + \lambda_o l_o(\theta_g, \theta_p, \theta_o) \qquad (2)$$

where l_g trains the geometric primitive module θ_g, l_p trains the proposal module θ_p, l_f trains the potential function and refinement sub-network θ_o, l_c trains the classification sub-network θ_c, and l_o trains the refinement sub-network. The trade-off parameters λ_g, λ_p, λ_f, λ_c, and λ_o are determined through 10-fold cross-validation. Intuitively, l_c, l_o and l_f provide end-to-end training of H3DNet, while l_g and l_p offer intermediate supervisions.

Formulation. Formulations of l_g, l_p, l_c, and l_o follow common strategies in the literature. Specifically, both l_g and l_p utilize L2 regression losses and a cross-entropy loss for geometric primitive location and existence flag prediction, and initial proposal generation; l_c applies a cross-entropy loss to train the object classification sub-network; l_o employs L2 regression losses for predicting the shape offset, and a cross-entropy loss for predicting the semantic label. Since these four loss terms are quite standard, we leave the details to the supplemental material.

l_f seeks to match the local minimums of the potential function and the underlying ground-truth objects. Specifically, consider a parametric potential function $f_\Theta(x)$ parameterized by Θ. Consider a local minimum $x_\Theta^\star$ which is a function of Θ. Let x^{gt} be the target location of $x_\Theta^\star$. We define the following alignment potential to pull $x_\Theta^\star$ to close to x^{gt}:

$$l_m(x_\Theta^\star, x^{gt}) := \|x_\Theta^\star - x^{gt}\|^2. \tag{3}$$

The following proposition describes how to compute the derivatives of l_m with respect to Θ. The proof is deferred to the supp. material.

Proposition 1. *The derivatives of l_m with respect to Θ is given by*

$$\frac{\partial l_m}{\partial \Theta} := 2(x_\Theta^\star - x^{gt})^T \cdot \frac{\partial x_\Theta^\star}{\partial \Theta}, \quad \frac{\partial x_\Theta^\star}{\partial \Theta} := -\Big(\frac{\partial^2 f_\Theta(x^\star)}{\partial^2 x}\Big)^{-1} \cdot \frac{\partial^2 f_\Theta(x^\star)}{\partial x \partial \Theta}. \tag{4}$$

We proceed to use l_m to define l_f. For each scene S, we denote the set of ground-truth objects and the set of local minimums of potential function F_S as $\mathcal{O}^{gt}$ and $\mathcal{O}^\star$, respectively. Note that $\mathcal{O}^\star$ depends on the network parameters and hyper-parameters. Let $\mathcal{C}_S \subset \mathcal{O}^{gt} \times \mathcal{O}^\star$ collect the nearest object in $\mathcal{O}^\star$ for each object in $\mathcal{O}^{gt}$. Consider a training set of scenes $\mathcal{S}_{train}$, we define

$$l_f := \sum_{S \in \mathcal{S}_{train}} \sum_{(o^\star, o^{gt}) \in \mathcal{C}_S} l_m(o^\star, o^{gt}). \tag{5}$$

Computing the derivatives of l_f with respect to the network parameters is a straightforward application of Proposition 1.

Training. We train H3DNet end-to-end and from scratch with the Adam optimizer [15]. Please defer to the supplemental material for hyper-parameters used in training, such as learning rate etc.

4   Experimental Results

In this section, we first describe the experiment setup in Sect. 4.2. Then, we compare our method with current state-of-the-art 3D object detection methods quantitatively, and analyze our results in Sect. 4.2, where we show the importance of using geometric primitives and discuss our advantages. Finally, we show ablation results in Sect. 4.3 and qualitative comparison in Figs. 6 and 5. More results and discussions can be found in the supplemental material.

Table 1. 3D object detection results on ScanNet V2 val dataset. We show per-category results of average precision (AP) with 3D IoU threshold 0.25 as proposed by [38], and mean of AP across all semantic classes with 3D IoU threshold 0.25.

	RGB	cab	bed	chair	sofa	tabl	door	wind	bkshf	pic	cntr	desk	curt	fridg	showr	toil	sink	bath	ofurn	mAP
3DSIS-5[10]	✓	19.8	69.7	66.2	71.8	36.1	30.6	10.9	27.3	0.0	10.0	46.9	14.1	53.8	36.0	87.6	43.0	84.3	16.2	40.2
3DSIS[10]	✗	12.8	63.1	66.0	46.3	26.9	8.0	2.8	2.3	0.0	6.9	33.3	2.5	10.4	12.2	74.5	22.9	58.7	7.1	25.4
Votenet[29]	✗	36.3	87.9	88.7	89.6	58.8	47.3	38.1	44.6	7.8	56.1	71.7	47.2	45.4	57.1	94.9	54.7	92.1	37.2	58.7
Ours	✗	**49.4**	**88.6**	**91.8**	**90.2**	**64.9**	**61.0**	**51.9**	**54.9**	**18.6**	**62.0**	**75.9**	**57.3**	**57.2**	**75.3**	**97.9**	**67.4**	**92.5**	**53.6**	**67.2**
w\o refine	✗	37.2	89.3	88.4	88.5	64.4	53.0	44.2	42.2	11.1	51.2	59.8	47.0	54.3	74.3	93.1	57.0	85.6	43.5	60.2

4.1   Experimental Setup

Datasets. We employ two popular datasets ScanNet V2 [5] and SUN RGB-D V1 [38]. ScanNet is a dataset of richly-annotated 3D reconstructions of indoor scenes. It contains 1513 indoor scenes annotated with per-point instance and semantic labels for 40 semantic classes. SUN RGB-D is a single-view RGB-D dataset for 3D scene understanding, which contains 10335 indoor RGB and depth images with per-point semantic labels and object bounding boxes. For both datasets, we use the same training/validation split and BB semantic classes (18 classes for ScanNet and 10 classes for SUN RGB-D) as in VoteNet [29] and sub-sample 40000 points from every scene.

Evaluation Protocol. We use Average Precision (AP) and the mean of AP across all semantic classes (mAP) [38] under different IoU values (the minimum IoU to consider a positive match). Average precision computes the average precision value for recall value over 0 to 1. IoU is given by the ratio of the area of intersection and area of union of the predicted bounding box and ground truth bounding box. Specifically, we use AP/mAP@0.25 and AP/mAP@0.5.

Baseline Methods. We compare H3DNet with STAR approaches: VoteNet [29] is a geometric-only detector that combines deep point set networks and a voting procedure. GSPN [52] uses a generative model for instance segmentation. Both 3D-SIS [10] and DSS [40] extract features from 2D images and 3D shapes to generate object proposals. F-PointNet [30] and 2D-Driven [18] first propose 2D detection regions and project them to 3D frustum for 3D detection. Cloud of gradient(COG) [33] integrates sliding windows with a 3D HoG-like feature.

4.2   Analysis of Results

As shown in Table 2, our approach leads to an average mAP score of 67.2%, with 3D IoU threshold 0.25 (mAP@0.25), on ScanNet V2, which is 8.5% better than the top-performing baseline approach [29]. In addition, our approach is 14.6% better than the baseline approach [29] with 3D IoU threshold 0.5 (mAP@0.5). For SUN RGB-D, our approach gains 2.4% and 6.1% in terms of mAP, with 3D IoU threshold 0.25 and 0.5 respectively. On both datasets, the performance gains of our approach under mAP@0.5 are larger than those under mAP@0.25, meaning our approach offers more accurate predictions than baseline approaches. Such

Table 2. Left: 3D object detection results on ScanNetV2 val set. **Right:** results on SUN RGB-D V1 val set. We show mean of average precision (mAP) across all semantic classes with 3D IoU threshold 0.25 and 0.5.

	Input	mAP@0.25	mAP@0.5		Input	mAP@0.25	mAP@0.5
DSS[40]	Geo + RGB	15.2	6.8	DSS[40]	Geo + RGB	42.1	-
F-PointNet[30]	Geo + RGB	19.8	10.8	COG[33]	Geo + RGB	47.6	-
GSPN[52]	Geo + RGB	30.6	17.7	2D-driven[18]	Geo + RGB	45.1	-
3D-SIS [10]	Geo + 5 views	40.2	22.5	F-PointNet[30]	Geo + RGB	54.0	-
VoteNet [29]	Geo only	58.7	33.5	VoteNet [29]	Geo only	57.7	32.9
Ours	Geo only	**67.2**	**48.1**	Ours	Geo only	**60.1**	**39.0**
w\o refine	Geo only	60.2	37.3	w\o refine	Geo only	58.5	34.2

improvements are attributed to using an overcomplete set of geometric primitives and their associated features for generating and refining object proposals. We can also understand the relative less salient improvements on SUN RGB-D than ScanNet in a similar way, i.e., labels of the former are less accurate than the latter, and the strength of H3DNet is not fully utilized on SUN RGB-D. Except for the classification and refinement module, our approach shares similar computation pipeline and complexity with VoteNet. The computation on multiple descriptor towers and proposal modules can be paralleled, which should not increase computation overhead. In our implementation, our approach requires 0.058 seconds for the last module per scan. Conceptually, our approach requires 50% more time compared to [29] but operates with a higher detection accuracy.

Improvement on Thin Objects. One limitation of the current top-performing baseline [29] is predicting thin objects in 3D scenes, such as doors, windows and pictures. In contrast, with face and edge primitives, H3DNet is able to extract better features for those thin objects. For example, the frames of window or picture provide dense edge feature, and physical texture of curtain or shower-curtain provide dense face/surface feature. As shown in Table 1, H3DNet leads to significant performance gains on thin objects, such as door (13.7%), window (13.8%), picture (10.8%), curtain (10.1%) and shower-curtain (18.2%).

Improvement on Objects with Dense Geometric Primitives. Across the individual object classes in ScanNet in Table 1, other than those thin objects, our approach also leads to significant performance gain on cabinet (13.1%), table (6.1%), bookshelf (10.3%), refrigerator (11.8%), sink (12.7%) and other-furniture (16.4%). One explanation is that the geometric shapes of these object classes

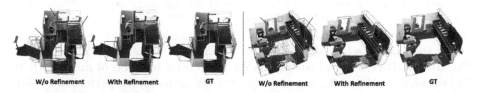

Fig. 4. Effect of geometric primitive matching and refinement.

possess rich planar structures and/or distinct edge structures, which contribute greatly on geometric primitive detection and object refinement.

Effect of Primitive Matching and Refinement. Using a distance function to refine object proposals and aggregating features of matching primitives are crucial for H3DNet. On ScanNet, merely classifying the initial proposals results in a 14.6% drop on mAP 0.5. Figure 4 shows qualitative object detection results, which again justify the importance of optimizing and refining object proposals.

4.3   Ablation Study

Effects of Using Different Geometric Primitives. H3DNet can utilize different groups of geometric primitives for generating, classifying, and refining object proposals. Such choices have profound influences on the detected objects. As illustrated in Fig. 7, when only using BB edge primitives, we can see that objects with prominent edge features, i.e., window, possess accurate predictions. In contrast, objects with dense face/surface features, such as shower curtain, exhibit relative low prediction accuracy. However, these objects can be easily detected by activating BB face primitives. H3DNet, which combines BB centers, BB edge centers, and BB face centers, adds the strength of their generalization behaviors together. The resulting performance gains are salient when compared to using a single set of geometric primitives.

Effects of Proposal Refinement. During object proposal refinement, object center, size, heading angle, semantic and existence are all optimized. As shown

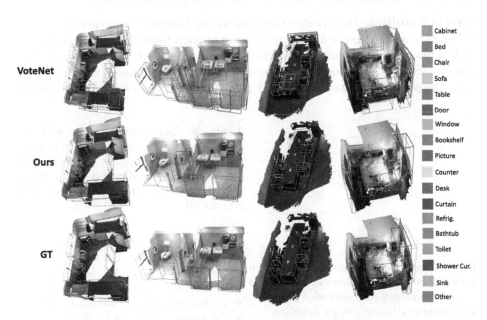

Fig. 5. Qualitative baseline comparisons on ScanNet V2.

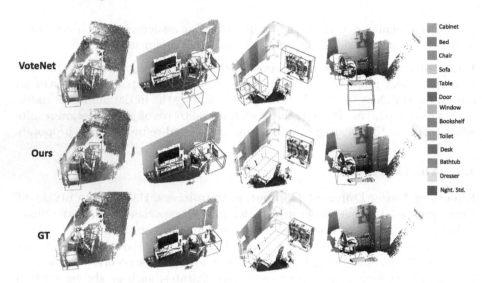

Fig. 6. Qualitative baseline comparisons on SUN RGB-D.

in Table 3, without fine-tuning any of the geometric parameters of the detected objects, the performance drops, which shows the importance of this sub-module.

Effect of Different Truncation Threshold. As shown in Fig. 8, with different truncation values of δ, results with mAP@0.25 and mAP@0.5 remain stable. It shows that our model is robust to different truncation threshold δ.

Effect of Multiple Descriptor Computation Towers. One hyper-parameter of H3DNet is the number of descriptor computation towers. Table 4 shows that adding more descriptor computation towers leads to better results, yet the performance gain of adding more descriptor computation towers quickly drops. Moreover, the performance gain of H3DNet from VoteNet comes from the hybrid set

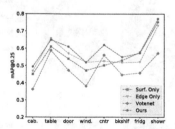

Fig. 7. Quantitative comparisons between VoteNet, our approach, ours with only face primitive and ours with only edge primitive, across sampled categories for ScanNet.

Table 3. Quantitative results without refining predicted center, size, semantic or object existence score for ScanNet, and without refining predicted angle for SUN RGB-D and differences compared with refining all.

	mAP@0.25		mAP@0.5	
w \ o center	66.9	−0.3	46.3	−1.8
w \ o size	65.4	−1.8	44.2	−3.9
w \ o semantic	66.2	−1.0	47.3	−0.8
w \ o existence	65.2	−1.8	45.1	−3.0
w \ o angle	58.6	−1.5	36.6	−2.4

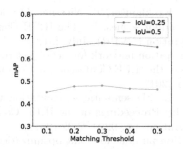

Fig. 8. Quantitative comparisons between different truncation threshold δ for Scan-Net.

Table 4. Quantitative comparisons between different number of descriptor computation towers, among our approach and VoteNet, for ScanNet and SUN RGB-D.

	# of Towers	mAP@0.25	mAP@0.5
Ours	1	64.4	43.4
	2	65.4	46.2
	3	66.0	47.7
	4	67.2	48.3
Vote	4 (Scan)	60.11	37.12
	4 (SUN)	57.5	32.1

of geometric primitives and object proposal matching and refinement. For example, replacing the descriptor computation tower of VoteNet by the four descriptor computation towers of H3DNet only results in modest and no performance gains on ScanNet and SUN RGB-D, respectively (See Table 4).

5   Conclusions and Future Work

In this paper, we have introduced a novel 3D object detection approach that takes a 3D scene as input and outputs a collection of labeled and oriented bounding boxes. The key idea of our approach is to predict a hybrid and overcomplete set of geometric primitives and then fit the detected objects to these primitives and their associated features. Experimental results demonstrate the advantages of this approach on ScanNet and SUN RGB-D. In the future, we would like to apply this approach to other 3D scene understanding tasks such as instance segmentation and CAD model reconstruction. Another future direction is to integrate more geometric primitives, like BB corners, for 3D object detection.

Acknowledgement. We would like to acknowledge the support from NSF DMS-1700234, a Gift from Snap Research, and a hardware donation from NVIDIA.

References

1. Akata, Z., Malinowski, M., Fritz, M., Schiele, B.: Multi-cue zero-shot learning with strong supervision. In: Proceedings of the IEEE Conference on Computer Vision and Pattern Recognition, pp. 59–68 (2016)
2. Atzmon, M., Maron, H., Lipman, Y.: Point convolutional neural networks by extension operators. arXiv preprint arXiv:1803.10091 (2018)
3. Baxter, J.: A Bayesian/information theoretic model of learning to learn via multiple task sampling. Mach. Learn. **28**(1), 7–39 (1997)
4. Chen, X., Ma, H., Wan, J., Li, B., Xia, T.: Multi-view 3D object detection network for autonomous driving. In: Proceedings of the IEEE Conference on Computer Vision and Pattern Recognition, pp. 1907–1915 (2017)

5. Dai, A., Chang, A.X., Savva, M., Halber, M., Funkhouser, T., Niessner, M.: Scan-Net: richly-annotated 3D reconstructions of indoor scenes. In: The IEEE Conference on Computer Vision and Pattern Recognition (CVPR), July 2017
6. Fan, H., Su, H., Guibas, L.J.: A point set generation network for 3D object reconstruction from a single image. In: Proceedings of the IEEE Conference on Computer Vision and Pattern Recognition, pp. 605–613 (2017)
7. Graham, B., Engelcke, M., van der Maaten, L.: 3D semantic segmentation with submanifold sparse convolutional networks. In: Proceedings of the IEEE Conference on Computer Vision and Pattern Recognition, pp. 9224–9232 (2018)
8. Guibas, L.J., Huang, Q., Liang, Z.: A condition number for joint optimization of cycle-consistent networks. In: Advances in Neural Information Processing Systems, pp. 1007–1017 (2019)
9. Hermosilla, P., Ritschel, T., Vázquez, P.P., Vinacua, À., Ropinski, T.: Monte Carlo convolution for learning on non-uniformly sampled point clouds. In: SIGGRAPH Asia 2018 Technical Papers, p. 235. ACM (2018)
10. Hou, J., Dai, A., Niessner, M.: 3D-SIS: 3D semantic instance segmentation of RGB-D scans. In: The IEEE Conference on Computer Vision and Pattern Recognition (CVPR), June 2019
11. Hua, B.S., Tran, M.K., Yeung, S.K.: Pointwise convolutional neural networks. In: Proceedings of the IEEE Conference on Computer Vision and Pattern Recognition, pp. 984–993 (2018)
12. Kendall, A., Cipolla, R.: Geometric loss functions for camera pose regression with deep learning. In: 2017 IEEE Conference on Computer Vision and Pattern Recognition, CVPR 2017, Honolulu, HI, USA, 21–26 July 2017, pp. 6555–6564 (2017). https://doi.org/10.1109/CVPR.2017.694
13. Kendall, A., Gal, Y., Cipolla, R.: Multi-task learning using uncertainty to weigh losses for scene geometry and semantics. In: Proceedings of the IEEE Conference on Computer Vision and Pattern Recognition, pp. 7482–7491 (2018)
14. Kim, B., Xu, S., Savarese, S.: Accurate localization of 3D objects from RGB-D data using segmentation hypotheses. In: 2013 IEEE Conference on Computer Vision and Pattern Recognition, Portland, OR, USA, 23–28 June 2013, pp. 3182–3189 (2013). https://doi.org/10.1109/CVPR.2013.409
15. Kingma, D.P., Ba, J.: Adam: a method for stochastic optimization. In: 3rd International Conference on Learning Representations, Conference Track Proceedings, ICLR 2015, San Diego, CA, USA, 7–9 May 2015 (2015). http://arxiv.org/abs/1412.6980
16. Klokov, R., Lempitsky, V.: Escape from cells: deep Kd-networks for the recognition of 3D point cloud models. In: Proceedings of the IEEE International Conference on Computer Vision, pp. 863–872 (2017)
17. Lahoud, J., Ghanem, B.: 2D-driven 3D object detection in RGB-D images. In: IEEE International Conference on Computer Vision, ICCV 2017, Venice, Italy, 22–29 October 2017, pp. 4632–4640 (2017). https://doi.org/10.1109/ICCV.2017.495
18. Lahoud, J., Ghanem, B.: 2D-driven 3D object detection in RGB-D images. In: The IEEE International Conference on Computer Vision (ICCV), October 2017
19. Lahoud, J., Ghanem, B., Pollefeys, M., Oswald, M.R.: 3D instance segmentation via multi-task metric learning. In: Proceedings of the IEEE International Conference on Computer Vision, pp. 9256–9266 (2019)
20. Lepetit, V., Moreno-Noguer, F., Fua, P.: Epnp: an accurate $O(n)$ solution to the pnp problem. Int. J. Comput. Vis. **81**(2), 155–166 (2009).https://doi.org/10.1007/s11263-008-0152-6

21. Li, Y., Bu, R., Sun, M., Wu, W., Di, X., Chen, B.: PointCNN: convolution on x-transformed points. In: Advances in Neural Information Processing Systems, pp. 820–830 (2018)
22. Li, Y., Wang, G., Ji, X., Xiang, Y., Fox, D.: DeepIM: deep iterative matching for 6D pose estimation. In: Ferrari, V., Hebert, M., Sminchisescu, C., Weiss, Y. (eds.) Computer Vision - ECCV 2018, ECCV 2018. Lecture Notes in Computer Science, vol. 11210, pp. 695–711. Springer, Cham (2018). https://doi.org/10.1007/978-3-030-01231-1_42
23. Liang, M., Yang, B., Chen, Y., Hu, R., Urtasun, R.: Multi-task multi-sensor fusion for 3d object detection. In: Proceedings of the IEEE Conference on Computer Vision and Pattern Recognition, pp. 7345–7353 (2019)
24. Lin, D., Fidler, S., Urtasun, R.: Holistic scene understanding for 3d object detection with RGBD cameras. In: IEEE International Conference on Computer Vision, ICCV 2013, Sydney, Australia, 1–8 December 2013, pp. 1417–1424 (2013). https://doi.org/10.1109/ICCV.2013.179
25. Luvizon, D.C., Picard, D., Tabia, H.: 2D/3D pose estimation and action recognition using multitask deep learning. In: Proceedings of the IEEE Conference on Computer Vision and Pattern Recognition, pp. 5137–5146 (2018)
26. Pavlakos, G., Zhou, X., Chan, A., Derpanis, K.G., Daniilidis, K.: 6-DoF object pose from semantic keypoints. In: 2017 IEEE International Conference on Robotics and Automation, ICRA 2017, Singapore, Singapore, 29 May– 3 June 2017, pp. 2011–2018 (2017). https://doi.org/10.1109/ICRA.2017.7989233
27. Peng, S., Liu, Y., Huang, Q., Zhou, X., Bao, H.: PVNet: pixel-wise voting network for 6DoF pose estimation. In: IEEE Conference on Computer Vision and Pattern Recognition, CVPR 2019, Long Beach, CA, USA, 16–20 June 2019, pp. 4561–4570 (2019). https://doi.org/10.1109/CVPR.2019.00469
28. Pham, Q.H., Nguyen, T., Hua, B.S., Roig, G., Yeung, S.K.: JSIS3D: joint semantic-instance segmentation of 3D point clouds with multi-task pointwise networks and multi-value conditional random fields. In: Proceedings of the IEEE Conference on Computer Vision and Pattern Recognition, pp. 8827–8836 (2019)
29. Qi, C.R., Litany, O., He, K., Guibas, L.J.: Deep hough voting for 3D object detection in point clouds. arXiv preprint arXiv:1904.09664 (2019)
30. Qi, C.R., Liu, W., Wu, C., Su, H., Guibas, L.J.: Frustum PointNets for 3D object detection from rgb-d data. In: Proceedings of the IEEE Conference on Computer Vision and Pattern Recognition, pp. 918–927 (2018)
31. Qi, C.R., Su, H., Mo, K., Guibas, L.J.: PointNet: deep learning on point sets for 3D classification and segmentation. In: Proceedings of the IEEE Conference on Computer Vision and Pattern Recognition, pp. 652–660 (2017)
32. Qi, C.R., Yi, L., Su, H., Guibas, L.J.: PointnNet++: deep hierarchical feature learning on point sets in a metric space. In: Advances in Neural Information Processing Systems, pp. 5099–5108 (2017)
33. Ren, Z., Sudderth, E.B.: Three-dimensional object detection and layout prediction using clouds of oriented gradients. In: The IEEE Conference on Computer Vision and Pattern Recognition (CVPR), June 2016
34. Ruder, S.: An overview of multi-task learning in deep neural networks. CoRR abs/1706.05098 (2017). http://arxiv.org/abs/1706.05098
35. Sener, O., Koltun, V.: Multi-task learning as multi-objective optimization. In: Advances in Neural Information Processing Systems, pp. 527–538 (2018)
36. Shi, S., Wang, X., Li, H.: PointRCNN: 3D object proposal generation and detection from point cloud. In: Proceedings of the IEEE Conference on Computer Vision and Pattern Recognition, pp. 770–779 (2019)

37. Song, C., Song, J., Huang, Q.: HybridPose: 6D object pose estimation under hybrid representations. CoRR abs/2001.01869 (2020). http://arxiv.org/abs/2001.01869
38. Song, S., Lichtenberg, S.P., Xiao, J.: Sun RGB-D: a RGB-D scene understanding benchmark suite. In: Proceedings of the IEEE Conference on Computer Vision and Pattern Recognition, pp. 567–576 (2015)
39. Song, S., Xiao, J.: Sliding shapes for 3D object detection in depth images. In: Fleet, D., Pajdla, T., Schiele, B., Tuytelaars, T. (eds.) ECCV 2014. Lecture Notes in Computer Science, vol. 8694, pp. 634–651. Springer, Cham (2014). https://doi.org/10.1007/978-3-319-10599-4_41
40. Song, S., Xiao, J.: Deep sliding shapes for amodal 3D object detection in RGB-D images. In: The IEEE Conference on Computer Vision and Pattern Recognition (CVPR), June 2016
41. Su, H., et al.: SplatNet: sparse lattice networks for point cloud processing. In: Proceedings of the IEEE Conference on Computer Vision and Pattern Recognition, pp. 2530–2539 (2018)
42. Tatarchenko, M., Dosovitskiy, A., Brox, T.: Octree generating networks: efficient convolutional architectures for high-resolution 3D outputs. In: Proceedings of the IEEE International Conference on Computer Vision, pp. 2088–2096 (2017)
43. Tatarchenko, M., Park, J., Koltun, V., Zhou, Q.Y.: Tangent convolutions for dense prediction in 3D. In: Proceedings of the IEEE Conference on Computer Vision and Pattern Recognition, pp. 3887–3896 (2018)
44. Wang, N., Zhou, W., Tian, Q., Hong, R., Wang, M., Li, H.: Multi-cue correlation filters for robust visual tracking. In: Proceedings of the IEEE Conference on Computer Vision and Pattern Recognition, pp. 4844–4853 (2018)
45. Wang, P.S., Liu, Y., Guo, Y.X., Sun, C.Y., Tong, X.: O-CNN: octree-based convolutional neural networks for 3D shape analysis. ACM Trans. Graph. (TOG) 36(4), 72 (2017)
46. Wang, X., Liu, S., Shen, X., Shen, C., Jia, J.: Associatively segmenting instances and semantics in point clouds. In: Proceedings of the IEEE Conference on Computer Vision and Pattern Recognition, pp. 4096–4105 (2019)
47. Wang, Y., Sun, Y., Liu, Z., Sarma, S.E., Bronstein, M.M., Solomon, J.M.: Dynamic graph CNN for learning on point clouds. arXiv preprint arXiv:1801.07829 (2018)
48. Xie, S., Liu, S., Chen, Z., Tu, Z.: Attentional AhapeContextNet for point cloud recognition. In: Proceedings of the IEEE Conference on Computer Vision and Pattern Recognition, pp. 4606–4615 (2018)
49. Xu, Y., Fan, T., Xu, M., Zeng, L., Qiao, Y.: SpiderCNN: deep learning on point sets with parameterized convolutional filters. In: Ferrari, V., Hebert, M., Sminchisescu, C., Weiss, Y. (eds.) Computer Vision – ECCV 2018. Lecture Notes in Computer Science, vol. 11212, pp. 90–105. Springer, Cham (2018). https://doi.org/10.1007/978-3-030-01237-3_6
50. Yang, Y., Feng, C., Shen, Y., Tian, D.: FoldingNet: point cloud auto-encoder via deep grid deformation. In: Proceedings of the IEEE Conference on Computer Vision and Pattern Recognition, pp. 206–215 (2018)
51. Yang, Z., Yan, S., Huang, Q.: Extreme relative pose network under hybrid representations. CoRR abs/1912.11695 (2019). http://arxiv.org/abs/1912.11695
52. Yi, L., Zhao, W., Wang, H., Sung, M., Guibas, L.J.: GSPN: generative shape proposal network for 3D instance segmentation in point cloud. In: Proceedings of the IEEE Conference on Computer Vision and Pattern Recognition, pp. 3947–3956 (2019)
53. Zhang, Y., Yang, Q.: A survey on multi-task learning. CoRR abs/1707.08114 (2017). http://arxiv.org/abs/1707.08114

54. Zhang, Z., Liang, Z., Wu, L., Zhou, X., Huang, Q.: Path-invariant map networks. In: The IEEE Conference on Computer Vision and Pattern Recognition (CVPR), pp. 11084–11094, June 2019
55. Zhou, Y., Tuzel, O.: VoxelNet: end-to-end learning for point cloud based 3D object detection. In: Proceedings of the IEEE Conference on Computer Vision and Pattern Recognition, pp. 4490–4499 (2018)
56. Zou, Y., Luo, Z., Huang, J.B.: DF-Net: unsupervised joint learning of depth and flow using cross-task consistency. In: Ferrari, V., Hebert, M., Sminchisescu, C., Weiss, Y. (eds.) Computer Vision - ECCV 2018, ECCV 2018. Lecture Notes in Computer Science, vol. 11209, pp. 36–53. Springer, Cham (2018). https://doi.org/10.1007/978-3-030-01228-1_3

Expressive Telepresence via Modular Codec Avatars

Hang Chu[1,2(✉)], Shugao Ma[3], Fernando De la Torre[3], Sanja Fidler[1,2],
and Yaser Sheikh[3]

[1] University of Toronto, Toronto, Canada
chuhang1122@cs.toronto.edu
[2] Vector Institute, Toronto, Canada
[3] Facebook Reality Lab, Pittsburgh, USA

Abstract. VR telepresence consists of interacting with another human in a virtual space represented by an avatar. Today most avatars are cartoon-like, but soon the technology will allow video-realistic ones. This paper aims in this direction, and presents Modular Codec Avatars (MCA), a method to generate hyper-realistic faces driven by the cameras in the VR headset. MCA extends traditional Codec Avatars (CA) by replacing the holistic models with a learned modular representation. It is important to note that traditional person-specific CAs are learned from few training samples, and typically lack robustness as well as limited expressiveness when transferring facial expressions. MCAs solve these issues by learning a modulated adaptive blending of different facial components as well as an exemplar-based latent alignment. We demonstrate that MCA achieves improved expressiveness and robustness w.r.t to CA in a variety of real-world datasets and practical scenarios. Finally, we showcase new applications in VR telepresence enabled by the proposed model.

Keywords: Virtual reality · Telepresence · Codec Avatar

1 Introduction

Telepresence technologies aims to make a person feel as if they were present, to give the appearance of being present, or to have an effect via telerobotics, at a place other than their true location. Telepresence systems can be broadly categorized based on the level of immersiveness. The most basic form of telepresence is video teleconferencing (e.g., Skype, Hangouts, Messenger) that is widely-used, and includes both audio and video transmissions. Recently, a more sophisticated form of telepresence has become available featuring a smart camera that follows a person (e.g., the Portal from Facebook).

Electronic supplementary material The online version of this chapter (https://doi.org/10.1007/978-3-030-58610-2_20) contains supplementary material, which is available to authorized users.

© Springer Nature Switzerland AG 2020
A. Vedaldi et al. (Eds.): ECCV 2020, LNCS 12357, pp. 330–345, 2020.
https://doi.org/10.1007/978-3-030-58610-2_20

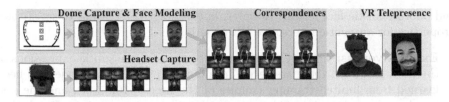

Fig. 1. Train and test pipeline for our VR telepresence system. In the first stage, we capture facial expressions of a user using both a multi-view camera dome and a VR headset (mounted with face-looking cameras). Correspondences between VR headset recording and full face expressions are established using the method described in [1]. Finally, once the person-specific face animation model is learned using these correspondences, a real-time photo-realistic avatar is driven from the VR headset cameras.

This paper addresses a more immersive form of telepresence that utilizes a virtual reality (VR) headset (e.g., Oculus, VIVE headsets). VR telepresence aims to enable a telecommunication system that allows remote social interactions more immersive than any prior media. It is not only a key promise of VR, but also has vast potential socioeconomic impact such as increasing communication efficiency, lowering energy footprint, and such a system could be timely for reducing inter-personal disease transmission [2]. VR telepresence has been an important active area of research in computer vision [1,3–8]. In VR telepresence, the users wear the VR headset, and a 3D face avatar is holographically projected in realtime, as if the user teleports himself/herself into the virtual space. This allows immersive bidirectional face-to-face conversations, facilitating instant interpersonal communication with high fidelity.

Figure 1 illustrates our VR telepresence system that has three main stages: (1) Appearance/shape capture. The person-specific avatar is built by capturing shape and appearance of the person from a multi-camera system (i.e., the dome). The user performs the same set of scripted facial expressions siting in the dome and wearing a headset mounted with face-looking cameras respectively. The 3D faces are reconstructed from the dome-captured multi-view images, and a variational autoencoder (VAE) is trained to model the 3D shape and appearance variability of the person's face. This model is referred to as Codec Avatar (CA) [1,4], since it decodes the 3D face from low-dimensional code. The CA is learned from 10k–14k 3D shapes and texture images. (2) Learning the correspondence between the infra-red (IR) VR cameras and the codec avatar. In the second stage, the CA method establishes the correspondence between the headset cameras and the 3D face model using a image-based synthesis approach [1]. Once a set of IR images (i.e., mouth, eyes) in the VR headset are in correspondence with the CA, we learn a network to map the IR VR headset cameras to the codec avatar codes, that should generalize to unseen situations (e.g. expressions, environments). (3) Realtime inference. Given the input images from the VR headset, and the network learned in step two, we can drive a person-specific and photo-realistic face avatar. However, in this stage the CA has to satisfy two properties for authentic interactive VR experience:

- **Expressiveness:** The VR system needs to transfer the subtle expressions of the user. However, there are several challenges: (1) The CA model has been learned from limited training samples of the user ($\sim$10k–14k), and the system has to precisely interpolate/extrapolate unseen expressions. Recall that is impractical to have a uniform sample of all possible expressions in the training set, because of their long-tail distribution. This requires careful ML algorithms that can learn from few training samples and long-tail distributions. (2) In addition, CA is an holistic model that typically results in rigid facial expression transfers.
- **Robustness:** To enable VR telepresence at scale in realistic scenarios, CAs have to provide robustness across different sources of variability, that includes: (1) iconic changes in the users' appearance (e.g., beard, makeup), (2) variability in the headset camera position (e.g., head strap), (3) different lighting and background from different room environments, (4) hardware variations within manufacturing specification tolerance (e.g., LED intensity, camera placement).

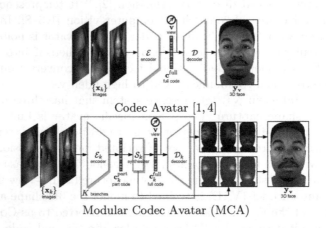

Fig. 2. Model diagrams comparing the previous CA and the proposed MCA. K denote the number of head-mounted cameras. In CA, images of all headset cameras are feed together to the single encoder $\mathcal{E}$ to compute the full face code which is subsequently decoded into 3D face using deocoder $\mathcal{D}$. In MCA, the images of each camera are encoded separately into a modular code $\mathbf{c}_k^{\text{part}}$ with the encoder $\mathcal{E}_k$, which is feed to a synthesizer $\mathcal{S}_k$ to estimate a camera specific full face code $\mathbf{c}_k^{\text{full}}$, and blending weights. Finally all these camera specific full face codes are decoded into 3D faces and blended together to form the final face avatar.

This paper proposes Modular Codec Avatar (MCA) that improves robustness and expressiveness of traditional CA. MCA decomposed the holistic face representation of traditional CA into learned modular local representations. Each local representation corresponds to one headset-mounted camera. MCA learns from data the automatic blending across all the modules. Figure 2 shows a diagram comparing the CA and MCA models. In MCA, a modular encoder first extracts information inside each single headset-mounted camera view. This is followed by a modular synthesizer that estimates a full face expression along with its blending weights from the information extracted within the same modular branch. Finally, multiple estimated 3D faces are aggregated from different modules and blended together to form the final face output.

Our contributions are threefold. First, we present MCA that introduces modularity into CA. Second, we extend MCA to solve the expressivity and robustness issues by learning the blending as well as new constraints in the latent space. Finally, we demonstrate MCA's robustness and expressiveness advantages on a real-world VR dataset.

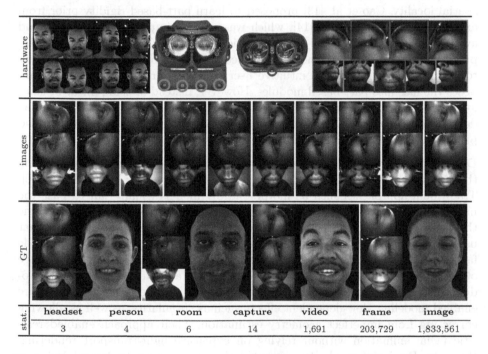

	headset	person	room	capture	video	frame	image
stat.	3	4	6	14	1,691	203,729	1,833,561

Fig. 3. Hardware and dataset examples. First row: capture dome [4], training headset, tracking headset, and sample images [1]. Second row: head-mounted camera examples of left eye, right eye, and mouth from different capture runs. Third row: examples of head-mounted images and ground-truth correspondences. Last row: dataset statistics.

2 Related Work

Morphable 3D Facial Models: Part-based models have been widely used for modeling facial appearance because of their elasticity and the ability to handle occlusion. Cootes et al. [9] introduced Active Appearance Models (AAM) for locating deformable objects such as faces. The facial shape and part-based appearance are iteratively matched to the image, using parameter displacement estimated from residual errors. The 3D morphable face model by Blanz and Vetter [10] decomposes the face into four parts to augment expressiveness, despite the fact that PCA decomposition is still computed holistically on the whole face. Tena et al. [11] proposed region-based face models that use a collection of PCA sub-models with shared boundaries. Their method achieved semantically meaningful expression bases, while generalising better to unseen data compared to the holistic approach. Neumman et al. [12] extended sparse matrix decomposition theory to face mesh sequence processing, and a new way to ensure spatial locality. Cao et al. [13] proposed to learn part-based rigidity prior from existing facial capture data [14], which was used to impose rigid stability and avoid head pose jittery in real-time animation. Recently, Ghafourzadeh et al. [15] presented local PCA-based model combined with anthropometric measurement to ensure expressiveness and intuitive user control. Our approach also decomposes the face into part-based modules. However, instead of using linear or shallow features on the 3D mesh, our modules take place in latent spaces learned by deep neural networks. This enables capturing of complex non-linear effects, and producing facial animation with a new level of realism.

Deep Codec Avatars: Human perception is particularly sensitive to detecting the realism of facial animation, e.g. the well-known Uncanny Valley Effect [16]. Traditional approaches such as morphable 3D models usually fail to pass the uncanny valley. Recent deep codec avatars have brought new hope to overcome this issue. Lombardi et al. [4] introduced Deep Appearance Models (DAM) that use deep Variational Auto-Encoders (VAE) [17] to jointly encode and decode facial geometry and view-dependent appearances into a latent code. The usage of deep VAE enabled capturing complex non-linear animation effects, while producing a smooth and compact latent representation. View-dependent texture enabled modeling view-dependent effects such as specularity, as well as allowing correcting from imperfect geometry estimation. Their approach enabled realistic facial animation without relying on expensive light-transport rendering. Recently, Lombardi et al. [18] extended their prior work to Neural Volumes (NV). They presented a dynamic, volumetric representation learned through encoder-decoder neural networks using a differentiable ray-marching algorithm. Their approach circumvents the difficulties in conventional mesh-based representation and does not require explicit reconstruction or tracking. The work in [4] is an important cornerstone that our work is built upon. Our work differs in that our main focus is producing robust and accurate facial animations in realtime.

VR Telepresence: The previous work that is most closely related to ours is Wei et al. [1]. VR telepresence presents many unique challenges due to its novel hardware setup, e.g. unaccommodating camera views that can only see parts of the face, as well as the image domain gap from head-mounted cameras to realistic avatars. In Thies et al. [6] and Cao et al. [19], a well-posed, unobstructed camera is used to provide image input for realtime reenactment. VR telepresence is different in that clear view of the full face is unavailable because the user has to wear the VR headset. Li et al. [20] presented an augmented VR headset that contains an outreaching arm holding an RGB-D camera. Lombardi et al. [4] used cameras mounted on a commodity VR headset, where telepresence was achieved by image-based rendering to create synthetic head-mounted images, and learning a common representation of real and synthetic images. Wei et al. [1] extended this idea by using CycleGAN [21] to further alleviate the image domain gap. A training headset with 9 cameras was used to establish correspondence, while a standard headset with 3 cameras was used to track the face. Existing work in VR telepresence are based on holistic face modeling. In our work, we revive the classic module-based approach and combine it with codec avatars to achieve expressive and robust VR telepresence.

3 Hardware and Dataset

In this section, we first describe our hardware setup. After that, we provide more details about the dataset that we use to train and evaluate our model.

Gathering high quality facial image data is the foundation of realistic facial animation. For this purpose, we use a large multi-camera dome that contains 40 cameras capturing images at a resolution of 2560×1920 pixels. Cameras lie on the frontal hemisphere of the face at a distance of about 1 m. The cameras are synchronized with a frame rate of 30 fps. 200 LED lights are directed at the face to create uniform illumination. To collect headset images, we used a two-headset design that has a training headset and a tracking headset. The training headset contains 9 cameras, which ensures establishing high-quality expression between head-mounted cameras and the avatar. The tracking headset contains a subset of 3 cameras, it is a consumer-friendly design with minimally intrusive cameras for real-time telepresence. The images are captured by IR cameras with a resolution of 640×480 and frame rate of 30 fps (down-sampled from 90 fps). We refer to [4] and [1] for more details about the camera dome and the VR headsets.

To obtain the high-fidelity facial avatar, we train a person-specific view-dependent VAE using 3D face meshes reconstructed and tracked from dome-captured data similar to [4]. The mesh contains 7306 vertices and a texture map of 1024×1024. We then use the method in [1] to establish correspondences between training headset frames and avatar latent codes. We decode these corresponding latent codes into 3D face meshes. These serve as the ground truth outputs in our dataset. The 9-camera training headset is only used for obtaining accurate ground truth. Both training and evaluation of our model in later sections only uses the subset of 3 cameras on the tracking headset. Note that the method for establishing correspondences provides accurate ground truth output, but it is infeasible for real-time animation due to its expensive computational complexity.

We construct a dataset that covers common variations in practical usage scenarios: varying users, varying headsets and varying environments with different backgrounds and lighting conditions. Figure 3 shows some example data and statistics of the dataset. Specifically, our dataset contains four different users. We used three different headsets, and captured a total of 14 sessions (half an hour for each session) in three different indoor environments: a small room with weak light, an office environment in front of a desk under bright white light, and in front of a large display screen. In the last environment, we capture some sections while playing random high-frequency flashing patterns on the screen, to facilitate the evaluation under extreme lighting condition. Sessions of the same person may be captured on different dates that are months apart, resulting in potential appearance changes in the captured data. For example, for one person, we captured him with heavy beard in some sessions and the other sessions are captured after he shaved.

For each dome capture or headset capture, we recorded a predefined set of 73 unique facial expressions, recitation of 50 phonetically-balanced sentences, two range-of-motion sequences where the user is asked to move jaw or whole face randomly with maximum extend, and 5–10 min conversation. Finally, we split the dataset by assigning one full headset capture session of each person as the testing set. This means the testing set does not have overlap in terms of capture sessions. We use only sequences of sentence reading and conversation for testing as they reflect the usual facial behaviors of a user in social interactions.

4   Method

The goal of our head-only VR telepresence system is to faithfully reconstruct full faces from images captured by the headset-mounted cameras in realtime. In this section, we will first describe the formulation of our model, followed by two important techniques that are important for successfully training the models and lastly implementation details.

4.1   Model Formulation

We denote the images captured by the headset-mounted cameras at each time instance as $\mathbf{X} = \{\mathbf{x}_k \mid k \in \{1, ..., K\}\}$, with K the total number of cameras and $\mathbf{x}_k \in \mathbb{R}^I$ where I is the number of pixels in each image. Note the time subscript t is omitted to avoid notation clutter. We denote $\mathbf{v} \in \mathbb{R}^3$ as the direction from which the avatar is to be viewed in VR. Given $\mathbf{X}$, the telepresence system needs to compute the view-dependent output $\mathbf{y_v}$ which contains the facial geometry $\mathbf{y}^g \in \mathbb{R}^{3G}$ of 3D vertex positions and the facial texture $\mathbf{y}_\mathbf{v}^t \in \mathbb{R}^{3T}$ corresponding to view direction $\mathbf{v}$. G and T are the number of vertices on the facial mesh and number of pixels on the texture map respectively.

The holistic Codec Avatar (CA) [1,4] is formulated as a view-dependent encoder-decoder framework, i.e.

$$\mathbf{y_v} = \mathcal{D}\left(\mathbf{c}, \mathbf{v}\right), \ \mathbf{c} = \mathcal{E}\left(\mathbf{X}\right) \tag{1}$$

where headset cameras' images $\mathbf{X}$ are feed into an image encoder $\mathcal{E}$ to produce the expression code of the whole face, and a decoder $\mathcal{D}$ produces the view-dependent 3D face. $\mathcal{D}$ is trained on dome-captured data to ensure animation quality, and $\mathcal{E}$ is trained using the correspondences between $\mathbf{X}$ and $\mathbf{c}$ that are established using the method in [1]. Because $\mathcal{D}$ is a neural network trained with a limited set of facial expressions, CA has limited expressiveness in out-of-sample expressions due to the long tail distribution nature of human facial expressions. In this work we propose Modular Codec Avatar (MCA), where the 3D face modules are estimated by an image encoder followed by a synthesizer from each headset camera view, and blended together to form the final face. MCA can be formulated as:

$$\mathbf{y_v} = \sum_{k=1}^{K} \mathbf{w}_k \odot \mathcal{D}_k \left(\mathbf{c}_k^{\text{full}}, \mathbf{v} \right) \tag{2}$$

$$\left[\mathbf{c}_k^{\text{full}}, \mathbf{w}_k \right] = \mathcal{S}_k \left(\mathbf{c}_k^{\text{part}} \right), \quad \mathbf{c}_k^{\text{part}} = \mathcal{E}_k \left(\mathbf{x}_k \right) \tag{3}$$

where each camera view $\mathbf{x}_k$ is processed separately. This computation consists of three steps. Firstly, a modular image encoder $\mathcal{E}_k$ estimates the modular expression code $\mathbf{c}_k^{\text{part}}$ which only models the facial part visible in the kth camera view, e.g., left eye. Subsequently, a modular synthesizer $\mathcal{S}_k$ estimates a latent code for the full-face denoted as $\mathbf{c}_k^{\text{full}}$, based only on the information from $\mathbf{c}_k^{\text{part}}$. The synthesizer also estimates blending weights $\mathbf{w}_k$. Lastly, we aggregate the results from all K modules to form the final face: decode the 3D modular faces using $\mathcal{D}_k$, and blend them together using the adaptive weights. $\odot$ represents the element-wise multiplication. In this way, MCA learns part-wise expressions inside each face module, while keeping full flexibility of assembling different modules together.

The objective function for training the MCA model consists of three loss terms for the reconstruction and intermediate latent codes:

$$\mathcal{L}_{\text{MCA}} = \|\mathbf{y_{v_0}} - \hat{\mathbf{y}}_{\mathbf{v_0}}\|_2 + \lambda_1 \sum_{k=1}^{K} \left\| \mathbf{c}_k^{\text{full}} - \hat{\mathbf{c}} \right\|_2 + \lambda_2 \sum_{k=1}^{K} \left\| \mathbf{c}_k^{\text{part}} - \hat{\mathbf{c}}_k^{\text{part}} \right\|_2 \tag{4}$$

where $\hat{\mathbf{y}}_{\mathbf{v_0}}$, $\hat{\mathbf{c}}$, and $\hat{\mathbf{c}}_k^{\text{part}}$ denote different supervision signals. The first term measures the reconstruction error in facial geometry and texture. We set $\hat{\mathbf{y}}_{\mathbf{v_0}} = \mathcal{D}(\hat{\mathbf{c}}, \mathbf{v}_0)$ as the facial geometry and texture decoded from the supervision code $\hat{\mathbf{c}}$ from frontal view direction $\mathbf{v}_0$. Note the supervision $\hat{\mathbf{c}}$ is estimated from the correspondence stage using the method in [1]. We impose equal weights for reconstruction error in geometry and texture. For the texture, we average pool the reconstruction errors of the pixels corresponding to the same closest mesh vertex instead of averaging over all pixels on the texture map. This vertex-based pooling ensures that texture loss has a geometrically uniform impact. The second term encourages each module to produce independent estimation of the correct full-face and the last term directs each encoder to produce correct modular expression code. Section 4.2 describes in detail how we generate the supervision $\hat{\mathbf{c}}_k^{\text{part}}$.

4.2   Exemplar-Based Latent Alignment

The two latent codes $\mathbf{c}_k^{\text{part}}$ and $\mathbf{c}_k^{\text{full}}$ have different purposes. $\mathbf{c}_k^{\text{part}}$ represents information only within its responsible modular region, while $\mathbf{c}_k^{\text{full}}$ further synthesizes a full-face based on the single-module expression information. It is important to have $\mathbf{c}_k^{\text{part}}$, because otherwise the modules will only collectively try to recover the same full-face code through $\mathbf{c}_k^{\text{full}}$, which essentially degrades to the holistic CA. The key to ensure the effectiveness of $\mathbf{c}_k^{\text{part}}$ is through crafting proper supervision signal $\hat{\mathbf{c}}_k^{\text{part}}$. The main challenge is $\hat{\mathbf{c}}_k^{\text{part}}$ resides in an inexplicitly defined latent space. We address this problem with exemplar-based latent alignment.

To obtain $\hat{\mathbf{c}}_k^{\text{part}}$, we first train a separate VAE for each module from dome-captured data. It has a similar architecture as CA, but only uses the region within the module by applying a modular mask on both the VAE input and output. We denote the masked modular VAE decoder as $\mathcal{D}_k^{\text{mask}}$, and the set of codes corresponding to dome-captured modular faces as $\mathbf{C}_k^{\text{mask}}$. The main challenge to obtain $\hat{\mathbf{c}}_k^{\text{part}}$ is the domain gap between dome-captured and headset-captured data. Directly applying the trained modular VAE on masked ground truth often result in spurious code vectors, i.e., the dome code and headset code corresponding to similar expression content do not match. This is caused by lighting and appearance differences between the two drastically different capture setup. Moreover, the domain gap also exist between different headset capture runs. To overcome this mismatch, we use exemplar-based latent alignment that replace the headset-captured codes produced by the decoder by their nearest dome-captured exemplar code. This effectively calibrates $\hat{\mathbf{c}}_k^{\text{part}}$ from different capture runs to a consistent base, i.e.

$$\hat{\mathbf{c}}_k^{\text{part}} = \underset{\mathbf{c} \in \mathbf{C}_k^{\text{mask}}}{\arg\min} \left\| \mathcal{D}_k^{\text{mask}}(\mathbf{c}, \mathbf{v}_0) - \hat{\mathbf{y}}_{\mathbf{v}_0, k}^{\text{mask}} \right\|_2 \tag{5}$$

where $\hat{\mathbf{y}}_{\mathbf{v}_0, k}^{\text{mask}}$ is the modular masked $\hat{\mathbf{y}}_{\mathbf{v}_0}$. This differs from pure image-based synthesis in that result must come from a set of known dome-captured exemplars $\mathbf{C}_k^{\text{mask}}$. The resulting $\hat{\mathbf{c}}_k^{\text{part}}$ is then used in Eq. (4) to train MCA.

4.3   Modulated Adaptive Blending

The blending weights $\mathbf{w}_k$ can be fully learned from data. However, we find automatically learned blending weights lead to animation artifacts. This is because module correlation exists in training data, e.g. left and right eyes often open and close together. Therefore, the dominant module interchanges between left and right eyes across nearby pixels, which results in jigsaw-like artifacts in the eyeball region. To overcome this issue and promote spatial coherence in the modular blending weights, we add a multiplicative and additive modulation signals to the adaptively learned blending weights. This ensures blending weight is constant near the module centroid, and the importance of adaptively learned weights gradually increases, i.e.

$$\mathbf{w}_k = \frac{1}{\mathbf{w}} \odot \left(\mathbf{w}_k^{\mathcal{S}} \odot e^{-\frac{\max\{\|\mathbf{u} - \overline{\mathbf{u}}_k\|^2 - a_k, 0\}}{\sigma^2}} + b_k \mathbb{1}\left\{ \|\mathbf{u} - \overline{\mathbf{u}}_k\|^2 \leq a_k \right\} \right) \tag{6}$$

where $\mathbf{w}_k^{\mathcal{S}}$ denotes the adaptive blending weights produced by the synthesizer $\mathcal{S}_k$, $\mathbf{u}$ denotes the 2D texture map coordinates corresponding to each vertex, $\bar{\mathbf{u}}_k$, a_k, and b_k are constants denoting the module's centroid, area in the texture map, and constant amplitude within its area. $\mathbf{w}$ is computed vertex-wise to normalize the blending weights across all modules.

4.4 Implementation Details

We use $K = 3$ modules following our hardware design, with three head-mounted cameras capturing left eye, right eye, and mouth. Each $\mathbf{x}_k$ is resized to a dimension of 256×256. We use a latent dimension of 256 for both $\hat{\mathbf{c}}_k^{\mathrm{part}}$ and $\hat{\mathbf{c}}$. Similar neural network architectures to [1] are used for our encoder $\mathcal{E}_k$ and decoder $\mathcal{D}_k$. For the decoder, we reduce the feature dimensions to remedy computation cost due to decoding K times. We also set all $\mathcal{D}_k$ to share the same weights to save the capacity requirement for storage and transmission of the model. The synthesizer $\mathcal{S}_k$ consists of three temporal convolution layers [22] (TCN), with connections to a small temporal receptive field of 4 previous frames. The blending weights $\mathbf{w}_k^{\mathcal{S}}$ is predicted through transposed convolution layers with a sigmoid function at the end to fit the texture map resolution. The texture weights were then reordered to produce geometry weights using the correspondence between vertices and texture map coordinates. We also add cross-module skip connections that concatenates the last layer features in $\mathcal{E}_k$ to allow the model to exploit correlations between images from different headset-mounted cameras.

We train the model with the Adam optimizer with both λ_1 and λ_2 set to 1. We augment the headset images by randomly cropping, zooming, and rotating the images. We also add random gaussian noise to the modular code $\hat{\mathbf{c}}_k^{\mathrm{part}}$ with diagonal covariance determined by the distance to the closest neighbour to prevent overfitting. The training is completed using four Tesla V100 GPUs. During test time, the telepresence system using MCA takes in average 21.6 ms to produce one frame in VR, achieving real-time photo-realistic facial animation.

5 Experiments

The first experiment illustrates the advantage of MCA over CA modeling untrained facial expressions. We then provide detailed evaluation results of the full VR telepresence via MCA comparing to CA. Our experiment evaluates the performance from extensive perspectives, including expression accuracy and perceptive quality. We also provide detailed ablation study and discuss failure cases. Finally, we show two extensions of our method which may be useful under certain usage scenarios.

5.1 Facial Expression Modeling: Modular (MCA) vs. Holistic (CA)

In the first experiment, we show the advantage of modeling modular expressions in real world VR telepresence scenarios. Towards this goal, we train a holistic

VAE [1, 4] as well as modular VAEs on dome-captured data. For both approach, we apply agglomerative clustering on the resulting latent codes of the training set data with varying number of clusters to represent the corresponding model's capacity of representing facial expressions. Recall that for headset-captured data, we have their estimated *ground truth* 3D face through the correspondence stage using the method in [1]. We then use these 3D faces to retrieve the closest cluster centers by matching the pixel values within each modular region of the rendered frontal faces, and report the root mean squared error (RMSE) of the pixel values in Fig. 4. For fair comparison, we use $K = 3$ times more clusters for the CA baseline. Figure 4 shows the result.

It can be seen from Fig. 4 that MCA consistently produces lower matching errors than CA throughout all subjects and model capacity levels. Intuitively this implies that MCA generalizes better to untrained expressions. The gap between CA and MCA increases as the number of clusters increases, indicating MCA's advantage increases as the face modeling becomes more fine-grained. It is then clear that by modeling modular expressions in the MCA, we can more truthfully interpolate/extrapolate facial expressions that are not contained in the dome-captured data, achieving better expressiveness in telepresence.

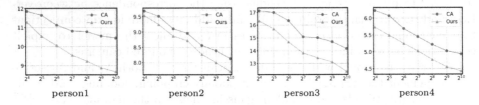

Fig. 4. Comparing holistic versus modular by isolating and evaluating the importance expressiveness. X-axis shows different capacities of the face expressions by the number of clusters. Y-axis shows the RMSE photometric error. Note that CA uses proportionally K times more clusters than each module of MCA for fair comparison.

5.2   Full VR Telepresence

We evaluate the performance of the full VR telepresence system. We feed headset images as input to the model, and evaluate the 3D face output against the ground-truth. Six metrics are used for comprehensive evaluation and the results are reported in Table 1.

Expression Accuracy: The most important metrics are MAE (i.e. Mean Absolute Error) and RMSE on pixels in the rendered frontal view face, as they directly and metrically measure the accuracy of VR telepresence. To further decompose the error into geometry and texture components, we compute RMSE on vertex position (Geo.) and texture map (Tex.) respectively as well. MCA *consistently outperform* CA on all these metrics on almost all identities' test data. These results suggest that MCA can more truthfully recover the 3D face given only the partial views of the face from headset-camera images than CA, leading to more expressive and robust telepresense.

Improvement Consistency: As these metric values are computed from average of all frames, we need to verify that the improvement of MCA is not due to large performance differences on only a small set of frames. To do that, we compute the percentage of frames for which MCA produces more accurate results than CA and vice versa. This is reported as %-better in Table 1. From the results, it is clear that MCA *consistently outperform* CA across frames. For example, for person3, on 99.7% frames MCA produces more accurate results than CA.

Table 1. Quantitative results of our main experiment for evaluating the full VR telepresence system robustness. MCA outperforms CA across various metrics. Please refer to the text for details.

Method	MAE↓		RMSE↓		Geo.↓		Tex.↓		%-better↑		SSIM↑	
	CA	Ours	CA	Ours	CA	Ours	CA	Ours	CA	Ours	CA	Ours
person1	8.82	**8.69**	7.67	**7.47**	1.26	**1.14**	3.40	**3.02**	36.3	**63.7**	0.954	**0.957**
person2	4.44	**4.26**	4.00	**3.84**	1.82	**1.46**	2.05	**2.04**	27.3	**72.7**	0.949	**0.951**
person3	9.09	**6.97**	8.36	**6.66**	1.14	**0.84**	4.58	**3.43**	0.3	**99.7**	0.933	**0.942**
person4	3.33	**3.21**	3.08	**3.04**	**0.54**	0.64	0.86	**0.85**	41.1	**58.9**	0.984	0.984
Overall	6.54	**6.17**	5.81	**5.48**	1.37	**1.17**	2.72	**2.44**	29.3	**70.7**	0.953	**0.956**

Perceptive Quality: We also want to verify that such accuracy improvements align with human perception, i.e. whether a user may actually feel the improvement. Quantitatively, we compute structural similarity index on the grey-level frontal rendering (SSIM) [23] which is a perception-based metric that considers image degradation as perceived change in structural information, while also incorporating important perceptual phenomena [24]. These results are reported in the last column in Table 1: MCA outperforms CA on three persons while is on-par on the last one. This hints that the accuracy improvements of MCA over CA align with perception improvement. Figure 5 shows a qualitative comparison between MCA and CA. It can be seen that MCA is better at handling subtle combined expressions, e.g. mouth open while eyes half-open, showing teeth while eyes shut, and mouth stretched while looking down. Renderings from different viewpoint by varying v are shown in Fig. 6. It can be seen that MCA produces consistent results from different viewing directions.

Ablation Study: Table 2 shows an ablation study of MCA. Ablation factors range from the usage of synthesized blending weights instead of equal weights (blend), training the network end-to-end (end2end), using soft latent part vectors with gaussian noise instead of hard categorical classes (soft-ex.), expanding the dimension of latent codes (dimen.), using skip-module connections so all images are visible to the modular encoder (skip-mod.), and using temporal convolution on the synthesizer (tconv.). It can be seen the former three techniques improve performance significantly, while extra connections such as skip and temporal convolution lead to further improvements. We refer to supplemental material for more details.

Failure Cases: Figure 7 shows example failure cases. Strong background flash is a challenging problem even for MCA, leading to inaccurate output (left of Fig. 7). Although MCA can produce more expressive 3D faces, extreme asymmetric expressions like one pupil in the center while the other roll all the way to the corner as in the right of Fig. 7 still remains challenging to be faithfully reconstructed.

Table 2. An ablation study of our method showing the progression and contribution to the overall performance improvement on the main RMSE metric.

blend	end2end	soft-ex.	dimen.	skip-mod.	tconv.	RMSE	$\Delta_\downarrow$
–	–	–	–	–	–	7.33	–
✓	–	–	–	–	–	6.67	0.66
✓	✓	–	–	–	–	6.10	0.57
✓	✓	✓	–	–	–	5.71	0.39
✓	✓	✓	✓	–	–	5.64	0.07
✓	✓	✓	✓	✓	–	5.52	0.08
✓	✓	✓	✓	✓	✓	5.48	0.04

5.3 Extensive Applications

Flexible Animation: Making funny expressions is part of social interaction. The MCA model can naturally better facilitate this task due to stronger expressiveness. To showcase this, we shuffle the head-mounted image sequences separately for each module, and randomly match them to simulate flexible expressions. It can be seen from Fig. 8 that MCA produces natural flexible expressions, even though such expressions have never been seen holistically in the training set.

Eye Amplification: In practical VR telepresence, we observe users often do not open their eyes to the full natural extend. This maybe due to muscle pressure from the headset wearing, and display light sources near the eyes. We introduce an eye amplification control knob to address this issue. In MCA, this can be simply accomplished by identifying the base $\mathbf{c}_k^{\text{part}}$ that correspond to closed eye, and amplifying the latent space distance by multiplying a user-provided amplification magnitude. Figure 8 shows examples of amplifying by a factor of 2.

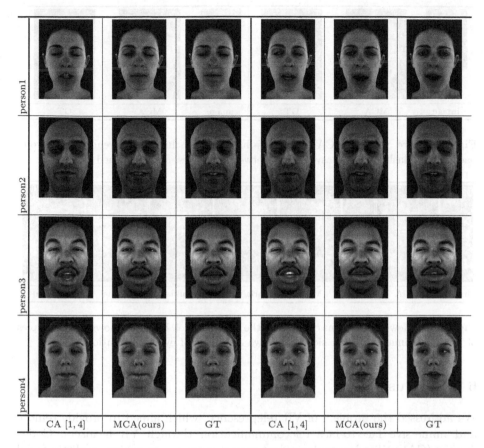

Fig. 5. Qualitative VR telepresence results. Compared to the holistic approach (CA), MCA handles untrained subtle expressions better.

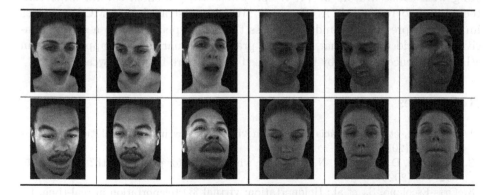

Fig. 6. Qualitative results of MCA from different viewing directions by varying **v**. MCA produces natural expressions that are consistent across viewpoints.

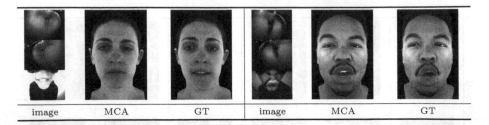

Fig. 7. Failure cases of MCA. Typical failure cases include interference from strong background flash, extreme asymmetry between the eyes, and weakened motion.

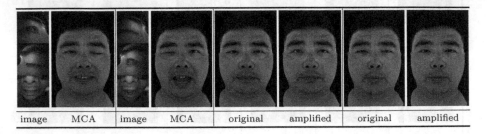

Fig. 8. Two new applications enabled by the MCA model. Left side shows two examples of flexible animation. Right side shows two examples of eye amplification.

6   Conclusion

We addressed the problem of VR telepresence, which aimed to provide remote and immersive face-to-face telecommunication through VR headsets. Codec Avatar (CA) utilized view-dependent neural networks to achieve realistic facial animation. We presented a new formulation of codec avatar named Modular Codec Avatar (MCA). This paper combines classic module-based face modeling with codec avatars in VR telepresence. We presented several important techniques to realize MCA effectively. We demonstrated that MCA achieves improved expressiveness and robustness through experiments on a comprehensive real-world dataset that emulated practical scenarios. New applications in VR telepresence enabled by the proposed model were finally showcased.

References

1. Wei, S.E., et al.: VR facial animation via multiview image translation. In: SIG-GRAPH (2019)
2. Heymann, D.L., Shindo, N.: Covid-19: what is next for public health? Lancet **395**, 542–545 (2020)
3. Orts-Escolano, S., et al.: Holoportation: virtual 3D teleportation in real-time. In: UIST (2016)
4. Lombardi, S., Saragih, J., Simon, T., Sheikh, Y.: Deep appearance models for face rendering. In: SIGGRAPH (2018)

5. Tewari, A., et al.: FML: face model learning from videos. In: CVPR (2019)
6. Thies, J., Zollhofer, M., Stamminger, M., Theobalt, C., Nießner, M.: Face2face: real-time face capture and reenactment of RGB videos. In: CVPR (2016)
7. Elgharib, M., et al.: Egoface: egocentric face performance capture and videorealistic reenactment. arXiv:1905.10822 (2019)
8. Nagano, K., et al.: PaGAN: real-time avatars using dynamic textures. In: SIGGRAPH (2018)
9. Cootes, T.F., Edwards, G.J., Taylor, C.J.: Active appearance models. TPAMI **23**(6), 681–685 (2001)
10. Blanz, V., Vetter, T., et al.: A morphable model for the synthesis of 3D faces. In: SIGGRAPH (1999)
11. Tena, J.R., De la Torre, F., Matthews, I.: Interactive region-based linear 3D face models. In: SIGGRAPH (2011)
12. Neumann, T., Varanasi, K., Wenger, S., Wacker, M., Magnor, M., Theobalt, C.: Sparse localized deformation components. TOG **32**(6), 1–10 (2013)
13. Cao, C., Chai, M., Woodford, O., Luo, L.: Stabilized real-time face tracking via a learned dynamic rigidity prior. TOG **37**(6), 1–11 (2018)
14. Cao, C., Weng, Y., Zhou, S., Tong, Y., Zhou, K.: Facewarehouse: a 3D facial expression database for visual computing. TVCG **20**(3), 413–425 (2013)
15. Ghafourzadeh, D., et al.: Part-based 3D face morphable model with anthropometric local control. In: EuroGraphics (2020)
16. Seyama, J., Nagayama, R.S.: The uncanny valley: effect of realism on the impression of artificial human faces. Presence: Teleoper. Virtual Environ. **16**(4), 337–351 (2007)
17. Kingma, D.P., Welling, M.: Auto-encoding variational bayes. arXiv:1312.6114 (2013)
18. Lombardi, S., Simon, T., Saragih, J., Schwartz, G., Lehrmann, A., Sheikh, Y.: Neural volumes: learning dynamic renderable volumes from images. TOG **38**(4), 65 (2019)
19. Cao, C., Hou, Q., Zhou, K.: Displaced dynamic expression regression for real-time facial tracking and animation. TOG **33**(4), 1–10 (2014)
20. Li, H., et al.: Facial performance sensing head-mounted display. TOG **34**(4), 1–9 (2015)
21. Zhu, J.Y., Park, T., Isola, P., Efros, A.A.: Unpaired image-to-image translation using cycle-consistent adversarial networks. In: ICCV (2017)
22. Bai, S., Kolter, J.Z., Koltun, V.: An empirical evaluation of generic convolutional and recurrent networks for sequence modeling. arXiv:1803.01271 (2018)
23. Wang, Z., Bovik, A., Sheikh, H., Simoncelli, E.: Image quality assessment: from error visibility to structural similarity. TIP **13**(4), 600–612 (2014)
24. Wikipedia: structural similarity. https://en.wikipedia.org/wiki/structural_similarity

Cascade Graph Neural Networks
for RGB-D Salient Object Detection

Ao Luo[1], Xin Li[2], Fan Yang[2], Zhicheng Jiao[3], Hong Cheng[1(✉)], and Siwei Lyu[4]

[1] Center for Robotics, School of Automation Engineering, UESTC, Chengdu, China
aoluo_uestc@hotmail.com, hcheng@uestc.edu.cn
[2] Group 42 (G42), Abu Dhabi, UAE
xinli_uestc@hotmail.com, fanyang_uestc@hotmail.com
[3] University of Pennsylvania, Philadelphia, USA
[4] University at Albany, State University of New York, Albany, USA

Abstract. In this paper, we study the problem of salient object detection (SOD) for RGB-D images using both color and depth information. A major technical challenge in performing salient object detection from RGB-D images is how to fully leverage the two complementary data sources. Current works either simply distill prior knowledge from the corresponding depth map for handling the RGB-image or blindly fuse color and geometric information to generate the coarse depth-aware representations, hindering the performance of RGB-D saliency detectors. In this work, we introduce *Cascade Graph Neural Networks* (CAS-GNN), a unified framework which is capable of comprehensively distilling and reasoning the mutual benefits between these two data sources through a set of cascade graphs, to learn powerful representations for RGB-D salient object detection. CAS-GNN processes the two data sources individually and employs a novel *Cascade Graph Reasoning* (CGR) module to learn powerful dense feature embeddings, from which the saliency map can be easily inferred. Contrast to the previous approaches, the explicitly modeling and reasoning of high-level relations between complementary data sources allows us to better overcome challenges such as occlusions and ambiguities. Extensive experiments demonstrate that CAS-GNN achieves significantly better performance than all existing RGB-D SOD approaches on several widely-used benchmarks. Code is available at https://github.com/LA30/Cas-Gnn.

Keywords: Salient object detection · RGB-D perception · Graph neural networks

1 Introduction

Salient object detection is the crux to dozens of high-level AI tasks such as object detection or classification [52,69,80], weakly-supervised semantic segmentation [30,63], semantic correspondences [77] and others [35,71,72]. An ideal

---

A. Luo and X. Li—Equal contribution

© Springer Nature Switzerland AG 2020
A. Vedaldi et al. (Eds.): ECCV 2020, LNCS 12357, pp. 346–364, 2020.
https://doi.org/10.1007/978-3-030-58610-2_21

solution should identify salient objects of varying shape and appearance, show robustness towards heavy occlusion, various illumination and background. With the development of hardware (sensors and GPU), prediction accuracy of data-driven methods that use deep networks [10,22,37,42,56,67,68,74,79,84,87] have been improved significantly, compared to traditional methods based on hand-crafted features [12,41,81,82]. However, these approaches only take the appearance features from RGB data into consideration, making them unreliable when handling the challenging cases, such as poorly-lighted environments and low-contrast scenes, due to the lack of depth information.

The depth map captured by RGB-D camera preserves important geometry information of the given scene, allowing 2D algorithms to be extend into 3D space. Depth awareness has been proven to be crucial for many applications of scene understanding, e.g., scene parsing [29,61], 6D object pose estimation [27,58] and object detection [24,49], leading to a significant performance enhancement. Recently, there have been a few attempts to take into account the 3D geometric information for salient object detection in the given scene, e.g., by distilling prior knowledge from the depth [51] or incorporating depth information into a SOD framework [21,48,86]. These RGB-D models have achieved better performances than RGB-only models in salient object detection when dealing with challenging cases. However, as we demonstrate empirically, existing RGB-D salient object detection models fall short under heavy occlusions and depth image noise. One primary reason is that these models, which only focus on delivering or gathering information, ignore modeling and reasoning over high-level relations between two data sources. Therefore, it is hard for them to fully exploit the complementary nature of 2D color and 3D depth information for overcoming the ambiguities in complex scenes. These observations inspire us to think about: *How to explicitly reason on high-level relations over 2D appearance (color) and 3D geometry (depth) information for better inferring salient regions?*

Graph neural network (GNN) has been shown to be an optimal way of relation modeling and reasoning [11,54,62,73,88]. Generally, a GNN model propagates messages over a graph, such that the node's representation is not only obtained from its own information but also conditioned on its relations to the neighboring nodes. It has revolutionized deep representation learning and benefitted many computer vision tasks, such as 3D pose estimation [5], action recognition [87], zero-shot learning [70] and language grounding [1], by incorporating graph computation into deep learning frameworks. However, how to design a suitable GNN model for RGB-D based SOD is challenging and, to the best of our knowledge, is still unexplored.

In this paper, we present the first attempt to build a GNN-based model, namely *Cascade Graph Neural Networks* (Cas-Gnn), to explicitly reason about the 2D appearance and 3D geometry information for RGB-D salient object detection. Our proposed deep model including multiple graphs, where each graph is used to handle a specific level of *cross-modality* reasoning. In each graph, two basic types of nodes are contained, i.e., geometry nodes storing depth features and appearance nodes storing RGB-related features, and they are linked to each other by edges. Through message passing, the useful mutual information and high-level relations between two data sources can be gradually distilled for learning the powerful dense feature embeddings, from which the saliency map can be inferred. To further enhance the capability for reasoning over multiple levels of features, we make our CAS-GNN to have these multi-level graphs sequentially chained by coarsening the preceding graph into two domain-specific **guidance nodes** for the following cascade graph. Consequently, each graph in our CAS-GNN (except for the first cascade graph) has three types of nodes in total, and they distill useful information from each other to build powerful feature representations for RGB-D based salient object detection.

Our CAS-GNN is easy to implement and end-to-end learnable. As opposed to prior works which simply fuse features of the two data sources, CAS-GNN is capable of explicitly reasoning about the 2D appearance and 3D geometry information over chained graphs, which is essential to handle heavy occlusions and ambiguities. Extensive experiments show that our CAS-GNN performs remarkably well on 7 widely-used datasets, outperforming state-of-the-art approaches by a large margin. In summary, our major contributions are described below:

1) We are the first to use the graph-based techniques to design network architectures for RGB-D salient object detection. This allows us to fully exploit the mutual benefits between the 2D appearance and 3D geometry information for better inferring salient object(s).
2) We propose a graph-based, end-to-end trainable model, called *Cascade Graph Neural Networks* (CAS-GNN), for RGB-D based SOD, and carefully design *Graph-based Reasoning* (GR) module to distill useful knowledge from different modalities for building powerful feature embeddings.
3) Different from most GNN-based approaches, our CAS-GNN ensembles a set of cascade graphs to reason about relations of the two data sources hierarchically. This cascade reasoning capability ensures the graph-based model to exploit rich, complementary information from multi-level features, which is useful in capturing object details and overcoming ambiguities.
4) We conduct extensive experiments on 7 widely-used datasets and show that our CAS-GNN sets new records, outperforming state-of-the-art approaches.

2   Related Work

This work is related to RGB-D based salient object detection, graph neural network and network cascade. Here, we briefly review these three lines of works.

RGB-D Salient Object Detection. Unlike approaches for RGB-only salient object detection methods [12,17,22,22,23,37,39,41,43,67,75,81,82,84] which only focus on 2D appearance feature learning, RGB-D based SOD approaches [21,48,86] take two different data sources, *i.e.*, 2D appearance (color) and 3D geometry (depth) information, into consideration. Classical approaches extract hand-crafted features from the input RGB-D data and perform cross-modality feature fusion by various strategies, such as random forest regressor [55] and minimum barrier distance [57]. However, with handcrafting of features, classic RGB-D based approaches are limited in the expression ability. Recent works such as CPFP [86] integrates deep feature learning and cross-modality fusion within a unified, end-to-end framework. Piao et al. [48] further enhance the cross-modality feature fusion through a recurrent attention mechanism. Fan et al. [21] introduce a depth-depurator to filter out noises in the depth map for better fusing cross-modality features. These approaches, despite the success, are not able to fully reason the high-order relations of cross-modality data, making them unreliable when handling challenges such as occlusions and ambiguities. In comparison, our CAS-GNN considers a better way to distill the mutual benefit of the two data sources by modeling and reasoning their relations over a set of cascade graphs, and we show that such cross-modality reasoning boosts the performance significantly.

Graph Neural Networks. In recent years, a wide variety of graph neural network (GNN) based models [15,16,33,53] have been proposed for different applications [4,11,44,54,88]. Generally, a GNN can be viewed as a message passing algorithm, where representations for nodes are iteratively computed conditioned on their neighboring nodes through a differentiable aggregation function. Some typical applications in computer vision include semantic segmentation [50], action recognition [65], point cloud classification and segmentation [66], to name a few. In the context of RGB-D based salient object detection – the task that we study in this paper – a key challenge in applying GNNs comes from how the graph model learns high-level relations and low-level details simultaneously. To solve this problem, unlike existing graph models, we ensemble a set of sequentially chained graphs to form a unified, cascade graph reasoning model. Therefore, our CAS-GNN is able to reason about relations across multiple feature levels to capture important hierarchical information for RGB-D based SOD, which is significantly different from all existing GNN based models.

Network Cascade. Network cascade is an effective scheme for a variety of high-level vision applications. Popular examples of cascaded models include DeCaFA for face alignment [14], BDCN for edge detection [26], Bidirectional FCN for object skeleton extraction [76], and Cascade R-CNN for object detection [6], to name a few. The core idea of network cascade is to ensemble a set of models to handle challenging tasks in a *coarse-to-fine* or *easy-to-hard* manner. For salient object detection in RGB-only images, only a few attempts employ the network cascade scheme. Li et al. [36] use a cascade network for gradually integrating saliency prior knowledge from coarse to fine. Wu et al. [67] design a cascaded partial decoder to enhance the learned features for salient object detection. Different from these approaches, our CAS-GNN propagates the knowledge learned from a more global view to assist fine-grained reasoning by chaining multiple graphs, which aids a structured understanding of complex scenes.

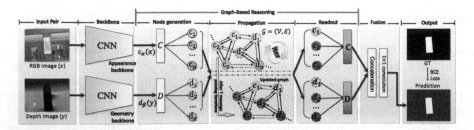

Fig. 1. Overall of our simple cross-modality reasoning model. Our model is built upon two VGG-16 based backbones, and uses a novel graph-based reasoning (GR) module to reason about the high-level relations between the generated 2D appearance and 3D geometry nodes for building more powerful representations. The updated node representations from two modalities are finally fused to infer the salient object regions.

3   Method

The key idea of CAS-GNN is that it enables the fully harvesting of the 2D appearance and 3D geometric information by using a differentiable, cascade module to hierarchically reason about relations between the two data sources. In this section, we elaborate on how to design a graph reasoning module and how to further enhance the capability of graph-based reasoning using the network cascade technique.

3.1   Problem Formulation

The task of RGB-D based salient object detection is to predict a saliency map $z \in \mathcal{Z}$ given an input image $x \in \mathcal{X}$ and its corresponding depth image $y \in \mathcal{Y}$. The input space $\mathcal{X}$ and $\mathcal{Y}$ correspond to the space of images and depths respectively, and the target space $\mathcal{Z}$ consists of only one class. A regression problem is characterized by a continuous target space. In our approach, a graph-based model is defined as a function $f_\Theta : \{\mathcal{X}, \mathcal{Y}\} \mapsto \mathcal{Z}$, parameterized by Θ, which maps an input pair, $i.e.$, $x \in \mathcal{X}$ and $y \in \mathcal{Y}$, to an output $f_\Theta(x, y) \in \mathcal{Z}$. The key challenging is to design a suitable model Θ that can fully exploit useful information from the two data sources (color and depth image) to learn powerful representations so that it can make the mapping more accurately.

3.2   Cross-Modality Reasoning with Graph Neural Networks

We start out with a simple GNN model, which reasons over the cross-modality relations between 2D appearance (color) and 3D geometric (depth) information across multiple scales, for salient object detection, as shown in Fig. 1.

Overview. For RGB-D salient object detection, the key challenge is to fully mine useful information from the two complementary data sources, $i.e.$, the color

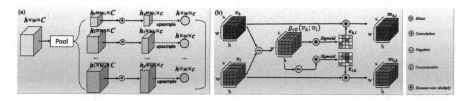

Fig. 2. Detailed illustration of our designs for (a) node embedding and (b) edge embedding. Zoom in for details.

image $x \in \mathcal{X}$ and the depth $y \in \mathcal{Y}$, and learn the mapping function $f_\Theta(x, y)$ which can infer the saliency regions $z \in \mathcal{Z}$. Aiming to achieve this goal, we represent the extracted multi-scale color features $C = \{c_1, \cdots, c_n\}$ and depth features $D = \{d_1, \cdots, d_n\}$ with a directed graph $\mathcal{G} = (\mathcal{V}, \mathcal{E})$, where $\mathcal{V}$ means a finite set of nodes and $\mathcal{E}$ stands for the edges among them. The nodes in the GNN model are naturally grouped into two types: the **geometry nodes** $\mathcal{V}_1 = \{c_1, \cdots, c_n\}$ and the **appearance nodes** $\mathcal{V}_2 = \{d_1, \cdots, d_n\}$, where $\mathcal{V} = \mathcal{V}_1 \cup \mathcal{V}_2$. The edges $\mathcal{E}$ connect i) the nodes from the same modality ($\mathcal{V}_1$ or $\mathcal{V}_2$), and ii) the nodes of the same scale from different modalities, *i.e.*, $c_i \leftrightarrow d_i$ where $i \in \{1, \cdots, n\}$. For each node, c_i or d_i, we learn its updated representation, namely $\mathbf{c}_i^{(t)}$ or $\mathbf{d}_i^{(t)}$, by aggregating the representations of its neighbors. In the end, the updated features are fused to produce the final representations for salient object detection.

Feature Backbones. Before reasoning the cross-modality relations, we first extract the 2D appearance feature C and 3D geometry feature D through the appearance backbone network c_α and geometry backbone network d_β, respectively. Following most of the previous approaches [7,9,25,48,89], we take two VGG-16 networks as the backbones, and use the dilated network technique [78] to ensure that the last two groups of VGG-16 have the same resolution. For the input RGB image x and the corresponding depth image y, we can map them to semantically powerful 2D appearance representations $C = c_\alpha(x) \in \mathbb{R}^{h \times w \times C}$ and 3D geometry representations $D = d_\beta(y) \in \mathbb{R}^{h \times w \times C}$. Rather than directly fusing the extracted features C and D to form the final representations for RGB-D salient object detection, we introduce a *Graph-based Reasoning* (GR) module to reason about the cross-modality, high-order relations between them to build more powerful embeddings, from which the saliency map can be inferred more easily and accurately.

Graph-Based Reasoning Module. The *Graph-based Reasoning* (GR) module g_χ takes the underlying 2D appearance features C and 3D geometry features D as inputs, and outputs powerful embeddings $\mathbf{C}$ and $\mathbf{D}$ after performing cross-modality reasoning: $\{\mathbf{C}, \mathbf{D}\} = g_\chi(C, D)$. We formulate $g_\chi(\cdot, \cdot)$ in a graph-based, end-to-end differentiable way as follows:

1) Graph Construction: Given the 2D appearance features C and 3D geometry features D, we build a graph $\mathcal{G} = (\mathcal{V}, \mathcal{E})$ which has two types of nodes: the **geometry nodes** $\mathcal{V}_1 = \{c_1, \cdots, c_n\}$ and the **appearance nodes** $\mathcal{V}_2 =$

$\{d_1, \cdots, d_n\}$, where $\mathcal{V} = \mathcal{V}_1 \cup \mathcal{V}_2$. Each node c_i or d_i is a feature map for a predefined scale s_i and edges link **i)** the nodes from the same modality but different scales, *i.e.*, $c_i \leftrightarrow c_j$ or $d_i \leftrightarrow d_j$, and **ii)** the nodes of the same scale from different modalities, *i.e.*, $c_i \leftrightarrow d_i$. Next, we show how to parameterize the nodes $\mathcal{V}$, edges $\mathcal{E}$, and message passing functions $\mathcal{M}$ of the graph $\mathcal{G}$ with neural networks.

2) Multi-scale Node Embeddings $\mathcal{V}$: Given the 2D appearance features $\mathcal{C}$ and 3D geometry features $\mathcal{D}$, as shown in Fig. 2(a), we leverage the pyramid pooling module (PPM) [85] followed by a convolution layer and an interpolation layer to extract multi-scale features of the two modalities (n scales) as the initial node representations, resulting in $N = 2 \cdot n$ nodes in total. For the appearance node c_i and geometry node d_i, their initial node representations $\mathbf{c}_i^{(0)} \in \mathbb{R}^{h \times w \times c}$ and $\mathbf{d}_i^{(0)} \in \mathbb{R}^{h \times w \times c}$ can be computed as:

$$\mathbf{c}_i^{(0)} = \mathcal{R}_{h \times w}(Conv(\mathcal{P}(\mathcal{C}; s_i))); \quad \mathbf{d}_i^{(0)} = \mathcal{R}_{h \times w}(Conv(\mathcal{P}(\mathcal{D}; s_i))), \quad (1)$$

where $\mathcal{P}(\cdot\,; s_i)$ means the pyramid pooling operation, which pools the given feature maps to the scale of s_i, and $\mathcal{R}(\cdot)$ is the interpolation operation which ensures multi-scale feature maps to have the same size $h \times w$.

3) Edge Embeddings $\mathcal{E}$: The nodes are linked by edges for information propagation. As mentioned above, in our constructed graph, edges link **i)** the nodes from the same modality but different scales, and **ii)** the nodes of the same scale from different modalities. For simplification, we use v_k and v_l, where $v_k, v_l \in \mathcal{V}$, to represent two nodes linked by the edge[1]. As shown in Fig. 2(b), the edge embedding $\mathbf{e}_{k,l}$ is used to represent the high-level relation on the two sides of the edge from v_k to v_l through a relation function $f_{rel}(\cdot\,;\,\cdot)$:

$$\mathbf{e}_{k,l} = f_{rel}(\mathbf{v}_k; \mathbf{v}_l) = Conv(g_{cb}(\mathbf{v}_k; \mathbf{v}_l)) \in \mathbb{R}^{h \times w \times c}, \quad (2)$$

where $\mathbf{v}_k$ and $\mathbf{v}_l$ are node embeddings for nodes v_k and v_l respectively, $g_{cb}(\cdot\,;\,\cdot)$ is a function that combines the node embeddings $\mathbf{v}_k$ and $\mathbf{v}_l$, and $Conv(\cdot)$ is the convolution operation which learns the relations in an end-to-end manner. For the combination function $g_{cb}(\cdot\,;\,\cdot)$, we follows [66] and model it as: $g_{cb}(\mathbf{v}_k; \mathbf{v}_l) = \mathbf{v}_l - \mathbf{v}_k$. The resulting edge embedding $\mathbf{e}_{k,l}$ for node v_k to v_l is also a c-dimensional feature map with the size of $h \times w$, in which each feature reflects the pixel-wise relationship between linked nodes.

4) Message Passing $\mathcal{M}$: In our GNN model, each node aggregates feature messages from all its neighboring nodes. For the message $\mathbf{m}_{k,l}$ passed from all neighboring nodes v_k to v_l, we define the following message passing function $\mathcal{M}(\cdot\,;\,\cdot)$:

$$\mathbf{m}_{k,l}^{(t)} = \sum_{k \in \mathcal{N}(l)} \mathcal{M}(\mathbf{v}_k^{(t-1)}, \mathbf{e}_{k,l}^{(t-1)}) = \sum_{k \in \mathcal{N}(l)} sigmoid(\mathbf{e}_{k,l}^{(t-1)}) \cdot \mathbf{v}_k^{(t-1)} \in \mathbb{R}^{h \times w \times c}$$

$$(3)$$

---

[1] In our formulation, the edges, message passing function and node-state updating function have no concern with the node types, therefore we simply ignore the node type for more clearly describing the **3) edge embeddings, 4) message passing** and **5) node-state updating**.

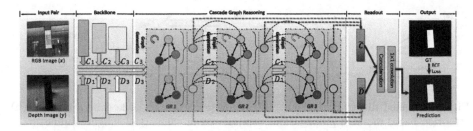

Fig. 3. The overall architecture of our CAS-GNN. Three graph-based reasoning (GR) modules are cascaded in a top-down manner to better distill multi-level information.

where $sigmoid(\cdot)$ is the sigmoid function which maps the edge embedding to link weight. Since our GNN model is designed for a pixel-wise task, the link weight between node is represented by a 2D map.

5) Node-state Updating $\mathcal{F}_{update}$: After the t-th message passing step, each node v_l in our GNN model aggregates information from its neighboring nodes to update its original feature representations. Here, we model the node-state updating process with Gated Recurrent Unit [2],

$$\mathbf{v}_l^{(t)} = \sum_{k \in \mathcal{N}(l)} \mathcal{F}_{update}(\mathbf{v}_l^{(t-1)}, \mathbf{m}_{k,l}^{(t-1)}) = \sum_{k \in \mathcal{N}(l)} \mathcal{U}_{GRU}(\mathbf{v}_l^{(t-1)}, \mathbf{m}_{k,l}^{(t-1)}), \quad (4)$$

where $\mathcal{U}_{GRU}(\cdot \; ; \; \cdot)$ stands for the gated recurrent unit.

6) Saliency Readout $\mathcal{O}$: After T message passing iterations, we upsample all updated node embeddings of each modality to the same size through the interpolation layer $R(\cdot)$, and merge them, *i.e.*, $\mathbf{V}_1 = \{R(\mathbf{c}_i^{(T)})\}_{i=1}^n$ and $\mathbf{V}_2 = \{R(\mathbf{d}_i^{(T)})\}_{i=1}^n$, to form the embeddings:

$$\mathbf{C} = \mathcal{F}_{merge}(\mathbf{V}_1); \quad \mathbf{D} = \mathcal{F}_{merge}(\mathbf{V}_2), \quad (5)$$

where $\mathcal{F}_{merge}(\cdot)$ denotes the merge function which is implemented with a concatenation layer followed by a 3×3 convolution layer. The learned embeddings of each modality can be further fused to form the final representations for RGB-D salient object detection by the following operation:

$$\mathbf{S} = \mathcal{R}_{H \times W}(\mathcal{O}(\mathbf{C}, \mathbf{D})), \quad (6)$$

where $\mathcal{O}(\cdot)$ is the readout function that maps the learned representations to the saliency scores. Here, we implement it with a concatenation layer followed by two 1×1 convolution layers; $\mathcal{R}_{H \times W}(\cdot)$ is used to resize the generated results to the same size of input image $H \times W$ through the interpolation operation.

Overall, all components in our GNN model are formulated in a differentiable manner, and thus can be trained end-to-end. Next, we show how to further enhance the capability of GNN model through network cascade techniques.

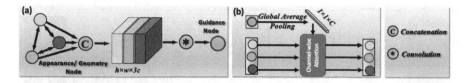

Fig. 4. Detailed illustration of our designs for (a) guidance node generation and (b) attention-based message propagation. Best viewed in color. (Color figure online)

3.3   Cascade Graph Neural Networks

In this part, we further enhance our GNN model for RGB-D salient object detection by using the network cascade technique. As observed by many existing works [20,28,40,59], the deep-layer and shallow-layer features are complementary to each other: the deep layer features encode high-level semantic knowledge while the shallow-layer features capture rich spatial information. Ideally, a powerful deep saliency model should be able to fully explore these multi-level features. Aiming to achieve this, we extend our GNN model to a hierarchical GNN model which is able to perform the reasoning across multiple levels for better inferring the salient object regions.

Hierarchical Reasoning via Multi-level Graphs. A straightforward scheme is to ensemble a set of graphs across multiple levels $\{\mathcal{G}_w\}_{w=1}^W$ to learn the embeddings individually, and then fuse the learned representations to build the final representations. Formally, given the VGG-16 based appearance backbone c_α for RGB image $\mathcal{X}$ and geometry backbone d_β for depth image $\mathcal{Y}$, we follow [28] to map the inputs to W levels of side-output features, $i.e.$, the multi-level appearance features $\tilde{\mathcal{V}}_1 = \{\mathcal{C}_1, \cdots, \mathcal{C}_W\}$ and the multi-level geometry features $\tilde{\mathcal{V}}_2 = \{\mathcal{D}_1, \cdots, \mathcal{D}_W\}$. For the features of each level $w \in [1, W]$, we build a graph $\mathcal{G}_w$ and use our proposed *Graph-based Reasoning* (GR) module $g_\chi(\mathcal{C}_w, \mathcal{D}_w)$ to map them to the corresponding embeddings $\{\mathbf{C}_w, \mathbf{D}_w\}_{i=1}^W$. Then, these multi-level embeddings of each modality, $\tilde{V}_1 = \{\mathbf{C}_1, \cdots, \mathbf{C}_W\}$ and $\tilde{V}_2 = \{\mathbf{D}_1, \cdots, \mathbf{D}_W\}$, can be easily interpolated to have the same resolution through the interpolation layer $R(\cdot)$, $i.e.$, $\tilde{\mathbf{V}}_1 = \{R(\mathbf{C}_1), \cdots, R(\mathbf{C}_W)\}$ and $\tilde{\mathbf{V}}_2 = \{R(\mathbf{D}_1), \cdots, R(\mathbf{D}_W)\}$, and merged by the following function:

$$\tilde{\mathbf{C}} = \mathcal{M}_{cl}(\tilde{\mathbf{V}}_1); \quad \tilde{\mathbf{D}} = \mathcal{M}_{cl}(\tilde{\mathbf{V}}_2) \tag{7}$$

where $\mathcal{M}_{cl}(\cdot)$ is a merge function, which can be either element-wise addition or channel-wise concatenation. Then, the readout function $\mathcal{O}(\tilde{\mathbf{C}}, \tilde{\mathbf{D}})$ can be used to generate the final results.

Generally, this simply hierarchical approach enables the model to perform reasoning across multiple levels. However, as it treats the multi-level reasoning process independently, the mutual benefits are hard to be fully explored.

Cascade Graph Reasoning. To overcome the drawbacks of independent multi-level (graph-based) reasoning, we propose the *Cascade Graph Reasoning* (CGR) module by chaining these graphs $\{\mathcal{G}_w\}_{w=1}^{W}$ for joint reasoning. The resulting model is called *Cascade Graph Neural Networks* (CAS-GNN), as shown in Fig. 3. Specifically, our CAS-GNN includes multi-level graphs $\{\mathcal{G}_w\}_{w=1}^{W}$ which are linked in a top-down manner by coarsening the preceding graph into two domain-specific guidance nodes for the following cascade graph to perform the joint reasoning.

1) Guidance Node: Unlike geometry nodes and appearance nodes, guidance nodes only deliver the guidance information, and will stay fixed during the message passing process. In our formulation, for reasoning the cross-modality relations of the w_th cascade stage, its preceding graph (from the deeper side-output level) is mapped into guidance node embeddings by the following functions:

$$\mathbf{g}_c^w = \mathcal{F}(\mathbf{V}_1^{(w-1)}); \quad \mathbf{g}_d^w = \mathcal{F}(\mathbf{V}_2^{(w-1)}), \tag{8}$$

where $\mathbf{g}_c^w$ and $\mathbf{g}_d^w$ are the guidance node embeddings of cascade stage w, and $\mathcal{F}(\cdot)$ is the graph merging operator, which coarsens the set of learned node embeddings ($\mathbf{V}_1^{(w-1)} = \{\mathbf{c}_i^{(w-1)(T)}\}_{i=1}^{n}$ or $\mathbf{V}_2^{(w-1)} = \{\mathbf{d}_i^{(w-1)(T)}\}_{i=1}^{n}$) of the preceding graph $\mathcal{G}_{(w-1)}$ by firstly concatenating them and then performing the fusion via a 3×3 convolution layer (See Fig. 4(a)).

2) Cascade Message Propagation: Each guidance node, $\mathbf{g}_c^w$ or $\mathbf{g}_d^w$, propagates the guidance information to other nodes of the same domain in the graph $\mathcal{G}_{(w)}$ through the attention mechanism:

$$\check{\mathbf{v}}_c^{w(t)} = \mathbf{v}_c^{w(t)} \odot \mathcal{A}(\mathbf{g}_c^w); \quad \check{\mathbf{v}}_d^{w(t)} = \mathbf{v}_d^{w(t)} \odot \mathcal{A}(\mathbf{g}_d^w) \tag{9}$$

where $\check{\mathbf{v}}_c^{w(t)}$ and $\check{\mathbf{v}}_d^{w(t)}$ denote the updated appearance node embeddings and geometry node embeddings for the cascade stage w after t_th message passing step respectively; $\odot$ means the channel-wise multiplication. $\mathcal{A}(\cdot)$ is the attention function, which can be formulated as:

$$A(\mathbf{g}_c^w) = sigmoid(\mathcal{P}(\mathbf{g}_c^w)); \quad \Lambda(\mathbf{g}_d^w) = sigmoid(\mathcal{P}(\mathbf{g}_d^w)); \tag{10}$$

where $\mathcal{P}(\cdot)$ is the global average pooling operation, and the *sigmoid* is used to map the guidance embeddings of each modality to the channel-wise attention vectors (See Fig. 4(b)). Therefore, the geometry and appearance node embeddings can incorporate important guidance information from previous graph $\mathcal{G}_{(w-1)}$ during performing the joint reasoning over $\mathcal{G}_w$ to create more powerful embeddings.

3) Multi-level Feature Fusion: Through the cascade message propagation, the *Cascade Graph Reasoning* (CGR) learns the embeddings of multi-level features under the guidance information provided by the guidance nodes. Here, we denote these learned multi-level embeddings as $\{\check{\mathbf{C}}_1, \cdots, \check{\mathbf{C}}_W\}$ and $\{\check{\mathbf{D}}_1, \cdots, \check{\mathbf{D}}_W\}$. To fuse them, we rewrite Eq. 7 to create the representations:

$$\check{\mathbf{C}} = \mathcal{M}_{cl}(R(\check{\mathbf{C}}_1), \cdots, R(\check{\mathbf{C}}_W)); \quad \check{\mathbf{D}} = \mathcal{M}_{cl}(R(\check{\mathbf{D}}_1), \cdots, R(\check{\mathbf{D}}_W)); \tag{11}$$

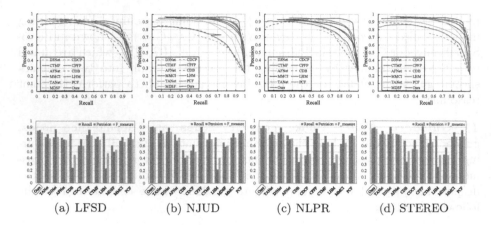

(a) LFSD   (b) NJUD   (c) NLPR   (d) STEREO

Fig. 5. Quantitative comparisons. The PR curves (top) and weighted F-measures (bottom) of the proposed method and state-of-the-art approaches on four datasets.

where $\check{C}$ and $\check{D}$ denote the merged representations for the appearance and geometry domain, respectively. Finally, the saliency readout operation (Eq. 6) is used to produce the final saliency map.

4   Experiments

In this section, we first provide the implementation details of our CAS-GNN. Then, we perform ablation studies to evaluate the effectiveness of each core component of graph-based model. Finally, CAS-GNN is compared with several state-of-the-art RGB-D SOD methods on six widely-used datasets.

Datasets: We conduct our experiments on 7 widely-used datasets: NJUD [31], STEREO [46], NLPR [47], LFSD [34], RGBD135 [13], and SSD [90]. For fair comparison, we follow most SOTAs [7,9,25] to randomly select 1,400 samples from the NJU2K dataset and 650 samples from the NLPR dataset for training, and use all remaining images for evaluation.

Evaluation Metrics: We adopt 5 most-widely used evaluation metrics to comprehensively evaluate the performance of our model, including the mean absolute error (MAE), the precision-recall curve (PR Curve), F-measure (F_β), S-measure (S_α) [18] and E-measure (E_ξ) [19]. Following previous SOTAs [7,9,25], we set β in F_β to 0.3 and α in S_α to 0.5 for fair comparison.

4.1   Implementation Details

Following [7,9,21,25], we utilize two VGG-16 networks as the backbones, where one is used for extracting the 2D appearance (RGB) features and the other for extracting 3D geometric (depth) features. We employ the dilated convolutions

Table 1. Ablation analysis for different graph-related settings.

Methods	Settings		NJUD		RGBD135	
	N	T	F_β	MAE	F_β	MAE
CAS-GNN	2	3	0.887	0.039	0.890	0.033
CAS-GNN	6	3	0.903	0.035	0.906	0.028
CAS-GNN	10	3	0.905	0.035	0.909	0.028
CAS-GNN	6	1	0.881	0.038	0.885	0.031
CAS-GNN	6	3	0.903	0.035	0.906	0.028
CAS-GNN	6	5	0.907	0.034	0.908	0.028

Table 2. Ablation analysis on three widely-used datasets.

Methods	Param.	FLOPs	NJUD [31]		STEREO [46]		RGBD135 [13]	
			F_β	MAE	F_β	MAE	F_β	MAE
Baseline	40.66M	65.64G	0.801	0.073	0.813	0.071	0.759	0.052
Baseline + IL	40.91M	66.21G	0.838	0.065	0.841	0.064	0.788	0.046
Baseline + NL	40.98M	66.86G	0.851	0.059	0.852	0.060	0.807	0.043
Baseline + GR (ours)	41.27M	68.91G	0.874	0.051	0.864	0.048	0.854	0.031
Baseline + CMFS	41.88M	72.63G	0.820	0.068	0.822	0.067	0.780	0.047
Baseline + HR (ours)	42.03M	73.19G	0.886	0.041	0.871	0.045	0.890	0.033
Baseline + CGR (ours)	42.28M	73.62G	0.903	0.035	0.901	0.039	0.906	0.028

to ensure that the last two groups of backbones have the same resolution. In the *Graph-based Reasoning* (GR) module g_χ, three nodes are used in each modality for capturing information of multiple scales, resulting in a graph $\mathcal{G}$ with six nodes in total. $\mathcal{G}$ links all nodes of the same modality. For the nodes of different modalities, the edge only connects those nodes with the same scale. During the construction of the *Cascade Graph Reasoning* (CGR) module, the features from outputs of the second, third and fifth group of each backbone (different resolutions) are used as inputs for performing cascade graph reasoning. Similar to existing approaches [7,9,25], BCE loss is used to train our model.

We implement our CAS-GNN using the Pytorch toolbox. The fully equipped model is trained on a PC with GTX 1080Ti GPU for 40 epochs with the mini-batch size of 8. The input RGB images and depth images are all resized to 256×256. To avoid overfitting, we perform the following data augmentation techniques: random horizontal flip, random rotate and random brightness. We adopt the Adam with a weight decay of 0.0001 to optimize the network parameters. The initial learning rate is set to 0.0001 and the 'poly' policy with the power of 0.9 is used as a mean of adjustment.

4.2 Ablation Analysis

In this section, we perform a series of ablations to evaluate each component in our proposed network.

Conventional Feature Fusion vs. Graph-Based Reasoning. To show the effectiveness of graph-based reasoning, we implement a simple baseline model that directly fuses features from the same multi-modality backbones by first performing the concatenate operation and then learning to fuse the learned features for RGB-D based SOD by two 1×1 convolutions. Clearly, our graph-based reasoning approach (GR module) achieves much more reliable and accurate results.

In addition, we further provide two strong baselines to show the superiority of our proposed graph-based reasoning approach. The first one is designed by using the one-shot induced learner (IL) [3,45] to adapt the learned 3D geometric features to 2D appearance space, making the cross-modality features can be better fused for RGB-D based SOD. The second one uses non-local (NL) module [64] to enable 2D appearance feature map to selectively incorporate useful information from 3D geometric features for building powerful representations. As shown in Table 2, our GR module significantly outperforms these strong baselines. This is

Table 3. Quantitative comparisons with state-of-the-art methods by S-measure (S_α), F-measure (F_β), E-measure (E_ξ) and MAE (M) on 7 widely-used RGB-D datasets.

	Metric	2014-2017					2018-2020									Ours
		LHM [47]	CDB [38]	CDCP [91]	MDSF [55]	CTMF [25]	AFNet [60]	MMCI [9]	PCF [7]	TANet [8]	CPFP [86]	D³Net [21]	DMRA [48]	UCNet [83]	ASIF [32]	
NJUD	$S_\alpha\uparrow$	0.514	0.624	0.669	0.748	0.849	0.772	0.858	0.877	0.878	0.879	0.895	0.886	0.897	0.888	**0.911**
	$F_\beta\uparrow$	0.632	0.648	0.621	0.775	0.845	0.775	0.852	0.872	0.874	0.877	0.889	0.872	0.889	0.900	**0.903**
	$E_\xi\uparrow$	0.724	0.742	0.741	0.838	0.913	0.853	0.915	0.924	0.925	0.926	0.932	0.908	0.903	-	**0.933**
	$M\downarrow$	0.205	0.203	0.180	0.157	0.085	0.100	0.079	0.059	0.060	0.053	0.051	0.051	0.043	0.047	**0.035**
STEREO	$S_\alpha\uparrow$	0.562	0.615	0.713	0.728	0.848	0.825	0.873	0.875	0.871	0.879	0.891	0.886	**0.903**	0.868	0.899
	$F_\beta\uparrow$	0.683	0.717	0.664	0.719	0.831	0.823	0.863	0.860	0.861	0.874	0.881	0.868	0.885	0.893	**0.901**
	$E_\xi\uparrow$	0.771	0.823	0.786	0.809	0.912	0.887	0.927	0.925	0.923	0.925	**0.930**	0.920	0.922	-	**0.930**
	$M\downarrow$	0.172	0.166	0.149	0.176	0.086	0.075	0.068	0.064	0.060	0.051	0.054	0.047	0.040	0.049	**0.039**
RGBD135	$S_\alpha\uparrow$	0.578	0.645	0.709	0.741	0.863	0.770	0.848	0.842	0.858	0.872	0.904	0.901	-	-	**0.905**
	$F_\beta\uparrow$	0.511	0.723	0.631	0.746	0.844	0.728	0.822	0.804	0.827	0.846	0.885	0.857	-	-	**0.906**
	$E_\xi\uparrow$	0.653	0.830	0.811	0.851	0.932	0.881	0.928	0.893	0.910	0.923	0.946	0.945	-	-	**0.947**
	$M\downarrow$	0.114	0.100	0.115	0.122	0.055	0.068	0.065	0.049	0.046	0.038	0.030	0.029	-	-	**0.028**
NLPR	$S_\alpha\uparrow$	0.630	0.629	0.727	0.805	0.860	0.799	0.856	0.874	0.886	0.888	0.906	0.899	0.918	0.884	**0.919**
	$F_\beta\uparrow$	0.622	0.618	0.645	0.793	0.825	0.771	0.815	0.841	0.863	0.867	0.885	0.855	0.890	0.900	**0.904**
	$E_\xi\uparrow$	0.766	0.791	0.820	0.885	0.929	0.879	0.913	0.925	0.941	0.932	0.946	0.942	0.951	-	**0.952**
	$M\downarrow$	0.108	0.114	0.112	0.095	0.056	0.058	0.059	0.044	0.041	0.036	0.034	0.031	**0.025**	0.030	**0.025**
SSD	$S_\alpha\uparrow$	0.566	0.562	0.603	0.673	0.776	0.714	0.813	0.841	0.839	0.807	0.866	0.857	-	-	**0.872**
	$F_\beta\uparrow$	0.568	0.592	0.535	0.703	0.729	0.687	0.781	0.807	0.810	0.766	0.847	0.821	-	-	**0.862**
	$E_\xi\uparrow$	0.717	0.698	0.700	0.779	0.865	0.807	0.882	0.894	0.897	0.852	0.910	0.892	-	-	**0.915**
	$M\downarrow$	0.195	0.196	0.214	0.192	0.099	0.118	0.082	0.062	0.063	0.082	0.058	0.058	-	-	**0.047**
LFSD	$S_\alpha\uparrow$	0.553	0.515	0.712	0.694	0.788	0.738	0.787	0.786	0.801	0.828	0.832	0.847	**0.860**	0.814	0.849
	$F_\beta\uparrow$	0.708	0.677	0.702	0.779	0.787	0.744	0.771	0.775	0.796	0.826	0.819	0.849	0.859	0.858	**0.864**
	$E_\xi\uparrow$	0.763	0.766	0.780	0.819	0.867	0.815	0.839	0.827	0.847	0.863	0.864	**0.899**	0.897	-	0.877
	$M\downarrow$	0.218	0.225	0.172	0.197	0.127	0.133	0.132	0.119	0.111	0.088	0.099	0.075	**0.069**	0.089	0.073
DUT-RGBD	$S_\alpha\uparrow$	0.568	-	0.687	-	0.834	-	0.791	0.801	-	-	-	0.888	-	-	**0.891**
	$F_\beta\uparrow$	0.659	-	0.633	-	0.792	-	0.753	0.760	-	-	-	0.883	-	-	**0.912**
	$E_\xi\uparrow$	0.767	-	0.794	-	0.884	-	0.855	0.858	-	-	-	0.927	-	-	**0.932**
	$M\downarrow$	0.174	-	0.159	-	0.097	-	0.113	0.100	-	-	-	0.048	-	-	**0.042**

because our GR module is capable of explicitly distilling complementary information from 2D appearance (color) and 3D geometry (depth) features while the existing feature fusion approaches fail to reason out high-level relations between them.

The Effectiveness of Cascade Graph Reasoning. A key design of our CAS-GNN is the novel *Cascade Graph Reasoning* module (CGR). To verify the effectiveness of CGR, we use the a common multi-level fusion strategy described in [48] (CMFS) for comparison. As shown in Table 2, our CGR consistently outperforms CMFS across all datasets. Moreover, our CGR is also superior to the hierarchical reasoning (HR) approach without the `guidance nodes` which is described in Sect. 3.3. This indicates that CGR (with the cascade techniques) can better distill and leverage multi-level information than existing strategies.

Node Numbers N. To investigate the impact of node numbers N in the GR module, we report the results of our GR module with different $N = 2 \cdot n$ in Table 1. We observe that when more nodes ($n = 1 \mapsto 3$) in each modality are used, the performance of our model improves accordingly. However, when more nodes are included in each modality ($n = 3 \mapsto 5$), the performance improvements are rather limited. This is caused by the redundant information from generated

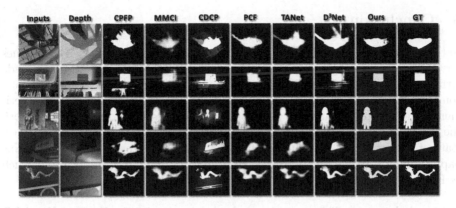

Fig. 6. Qualitative comparisons with state-of-the-art CNNs-based methods.

nodes. Therefore, we believe that setting 3 nodes in each modality ($N = 6$) should be a good balance of the speed and accuracy.

Message Passing Iterations T. We also evaluate the impact of message passing iterations T. As can be seen in Table 1, when more than three message passing iterations are used for graph reasoning, the model can achieve the best performance. Therefore, we set $T = 3$ in our GR module to guarantee a good speed and performance tradeoff.

4.3 Comparison with SOTAs

Quantitative Comparisons. We compare our CAS-GNN with 14 SOTA models on 7 widely-used datasets in Table 3. In general, our CAS-GNN consistently achieves the remarkable performance on all datasets with four evaluation metrics. Clearly, the results demonstrate that explicitly reason and distill mutual beneficial information can help to infer the salient object regions from the clutter images. In addition, we also show the results of widely-used PR curves and weighted F-measure in Fig. 5. As can be seen, our CAS-GNN achieves the best performance on all datasets. All the comparisons with recent SOTAs indicate that mining the high-level relations of multi-modality data sources and perform joint reasoning across multiple feature levels are important, and will largely improve the reliability of deep model for handling cross-modality information.

Qualitative Comparisons. Figure 6 shows some visual samples of results comparing the proposed CAS-GNN with state-of-the-art methods. We observe that our CAS-GNN is good at capturing both of the overall salient object regions and local object/region details. This is because our proposed cascade graph reasoning module is able to take both high-level semantics and low-level local details into consideration to build more powerful embeddings for inferring SOD regions.

5 Conclusion

In this paper, we introduce a novel deep model based on graph-based techniques for RGB-D salient object detection. Besides, we further propose to use cascade structure to enhance our GNN model to make it better take advantages of rich, complementary information from multi-level features. According to our experiments, the proposed CAS-GNN successfully distills useful information from both the 2D (color) appearance and 3D geometry (depth) information, and sets new state-of-the-art records on multiple datasets. We believe the novel designs in this paper is important, and can be used to other cross-modality applications, such as RGB-D based object discover or cross-modality medical image analyse.

Acknowledgement. This research was funded in part by the National Key R&D Progrqam of China (2017YFB1302300) and the NSFC (U1613223).

References

1. Bajaj, M., Wang, L., Sigal, L.: G3raphGround: graph-based language grounding. In: ICCV (2019)
2. Ballas, N., Yao, L., Pal, C., Courville, A.: Delving deeper into convolutional networks for learning video representations (2016)
3. Bertinetto, L., Henriques, J.F., Valmadre, J., Torr, P., Vedaldi, A.: Learning feedforward one-shot learners. In: NIPS (2016)
4. Bi, Y., Chadha, A., Abbas, A., Bourtsoulatze, E., Andreopoulos, Y.: Graph-based object classification for neuromorphic vision sensing. In: ICCV (2019)
5. Cai, Y., et al.: Exploiting spatial-temporal relationships for 3D pose estimation via graph convolutional networks. In: ICCV (2019)
6. Cai, Z., Vasconcelos, N.: Cascade R-CNN: delving into high quality object detection. In: CVPR (2018)
7. Chen, H., Li, Y.: Progressively complementarity-aware fusion network for RGB-D salient object detection. In: CVPR (2018)
8. Chen, H., Li, Y.: Three-stream attention-aware network for RGB-D salient object detection. TIP **28**(6), 2825–2835 (2019)
9. Chen, H., Li, Y., Su, D.: Multi-modal fusion network with multi-scale multi-path and cross-modal interactions for RGB-D salient object detection. Pattern Recogn. **86**, 376–385 (2019)
10. Chen, S., Tan, X., Wang, B., Hu, X.: Reverse attention for salient object detection. In: Ferrari, V., Hebert, M., Sminchisescu, C., Weiss, Y. (eds.) ECCV 2018. LNCS, vol. 11213, pp. 236–252. Springer, Cham (2018). https://doi.org/10.1007/978-3-030-01240-3_15
11. Chen, Y., Rohrbach, M., Yan, Z., Shuicheng, Y., Feng, J., Kalantidis, Y.: Graph-based global reasoning networks. In: CVPR (2019)
12. Cheng, M.M., Mitra, N.J., Huang, X., Torr, P.H., Hu, S.M.: Global contrast based salient region detection. TPAMI **37**(3), 569–582 (2014)
13. Cheng, Y., Fu, H., Wei, X., Xiao, J., Cao, X.: Depth enhanced saliency detection method. In: Proceedings of International Conference on Internet Multimedia Computing and Service (2014)

14. Dapogny, A., Bailly, K., Cord, M.: DeCaFA: deep convolutional cascade for face alignment in the wild. In: ICCV (2019)
15. Defferrard, M., Bresson, X., Vandergheynst, P.: Convolutional neural networks on graphs with fast localized spectral filtering. In: NIPS (2016)
16. Duvenaud, D.K., et al.: Convolutional networks on graphs for learning molecular fingerprints. In: NIPS (2015)
17. Fan, D.-P., Cheng, M.-M., Liu, J.-J., Gao, S.-H., Hou, Q., Borji, A.: Salient objects in clutter: bringing salient object detection to the foreground. In: Ferrari, V., Hebert, M., Sminchisescu, C., Weiss, Y. (eds.) ECCV 2018. LNCS, vol. 11219, pp. 196–212. Springer, Cham (2018). https://doi.org/10.1007/978-3-030-01267-0_12
18. Fan, D.P., Cheng, M.M., Liu, Y., Li, T., Borji, A.: Structure-measure: a new way to evaluate foreground maps. In: CVPR (2017)
19. Fan, D.P., Gong, C., Cao, Y., Ren, B., Cheng, M.M., Borji, A.: Enhanced-alignment measure for binary foreground map evaluation. arXiv preprint arXiv:1805.10421 (2018)
20. Fan, D.P., Ji, G.P., Sun, G., Cheng, M.M., Shen, J., Shao, L.: Camouflaged object detection. In: CVPR (2020)
21. Fan, D.P., Lin, Z., Zhang, Z., Zhu, M., Cheng, M.M.: Rethinking RGB-D salient object detection: models, datasets, and large-scale benchmarks. TNNLS (2020)
22. Fan, D.P., Wang, W., Cheng, M.M., Shen, J.: Shifting more attention to video salient object detection. In: CVPR (2019)
23. Feng, M., Lu, H., Ding, E.: Attentive feedback network for boundary-aware salient object detection. In: CVPR (2019)
24. Gupta, S., Girshick, R., Arbeláez, P., Malik, J.: Learning rich features from RGB-D images for object detection and segmentation. In: Fleet, D., Pajdla, T., Schiele, B., Tuytelaars, T. (eds.) ECCV 2014. LNCS, vol. 8695, pp. 345–360. Springer, Cham (2014). https://doi.org/10.1007/978-3-319-10584-0_23
25. Han, J., Chen, H., Liu, N., Yan, C., Li, X.: CNNs-based RGB-D saliency detection via cross-view transfer and multiview fusion. IEEE Trans. Cybern. 48(11), 3171–3183 (2017)
26. He, J., Zhang, S., Yang, M., Shan, Y., Huang, T.: Bi-directional cascade network for perceptual edge detection. In: CVPR (2019)
27. He, Y., Sun, W., Huang, H., Liu, J., Fan, H., Sun, J.: PVN3D: a deep point-wise 3D keypoints voting network for 6dof pose estimation. arXiv preprint arXiv:1911.04231 (2019)
28. Hou, Q., Cheng, M.M., Hu, X., Borji, A., Tu, Z., Torr, P.H.: Deeply supervised salient object detection with short connections. In: CVPR (2017)
29. Jiao, J., Wei, Y., Jie, Z., Shi, H., Lau, R.W., Huang, T.S.: Geometry-aware distillation for indoor semantic segmentation. In: CVPR (2019)
30. Jin, B., Ortiz Segovia, M.V., Susstrunk, S.: Webly supervised semantic segmentation. In: CVPR (2017)
31. Ju, R., Ge, L., Geng, W., Ren, T., Wu, G.: Depth saliency based on anisotropic center-surround difference. In: ICIP (2014)
32. Li, C., et al.: ASIF-NET: attention steered interweave fusion network for RGB-D salient object detection. TCYB (2020)
33. Li, G., Muller, M., Thabet, A., Ghanem, B.: DeepGCNs: can GCNs go as deep as CNNs? In: ICCV, October 2019
34. Li, N., Ye, J., Ji, Y., Ling, H., Yu, J.: Saliency detection on light field. In: CVPR (2014)
35. Li, X., Chen, L., Chen, J.: A visual saliency-based method for automatic lung regions extraction in chest radiographs. In: ICCWAMTIP (2017)

36. Li, X., Yang, F., Cheng, H., Chen, J., Guo, Y., Chen, L.: Multi-scale cascade network for salient object detection. In: ACM MM (2017)
37. Li, X., Yang, F., Cheng, H., Liu, W., Shen, D.: Contour knowledge transfer for salient object detection. In: Ferrari, V., Hebert, M., Sminchisescu, C., Weiss, Y. (eds.) ECCV 2018. LNCS, vol. 11219, pp. 370–385. Springer, Cham (2018). https://doi.org/10.1007/978-3-030-01267-0_22
38. Liang, F., Duan, L., Ma, W., Qiao, Y., Cai, Z., Qing, L.: Stereoscopic saliency model using contrast and depth-guided-background prior. Neurocomputing **275**, 2227–2238 (2018)
39. Liu, J.J., Hou, Q., Cheng, M.M., Feng, J., Jiang, J.: A simple pooling-based design for real-time salient object detection. In: CVPR (2019)
40. Liu, N., Han, J.: DHSNet: deep hierarchical saliency network for salient object detection. In: CVPR (2016)
41. Liu, T., et al.: Learning to detect a salient object. TPAMI **33**(2), 353–367 (2010)
42. Liu, Y., Zhang, Q., Zhang, D., Han, J.: Employing deep part-object relationships for salient object detection. In: ICCV (2019)
43. Luo, A., Li, X., Yang, F., Jiao, Z., Cheng, H.: Webly-supervised learning for salient object detection. Pattern Recogn. (2020)
44. Luo, A., et al.: Hybrid graph neural networks for crowd counting. In: AAAI (2020)
45. Nie, X., Feng, J., Zuo, Y., Yan, S.: Human pose estimation with parsing induced learner. In: CVPR (2018)
46. Niu, Y., Geng, Y., Li, X., Liu, F.: Leveraging stereopsis for saliency analysis. In: CVPR (2012)
47. Peng, H., Li, B., Xiong, W., Hu, W., Ji, R.: RGBD salient object detection: a benchmark and algorithms. In: Fleet, D., Pajdla, T., Schiele, B., Tuytelaars, T. (eds.) ECCV 2014. LNCS, vol. 8691, pp. 92–109. Springer, Cham (2014). https://doi.org/10.1007/978-3-319-10578-9_7
48. Piao, Y., Ji, W., Li, J., Zhang, M., Lu, H.: Depth-induced multi-scale recurrent attention network for saliency detection. In: The IEEE International Conference on Computer Vision (ICCV), October 2019
49. Qi, C.R., Liu, W., Wu, C., Su, H., Guibas, L.J.: Frustum PointNets for 3D object detection from RGB-D data. In: CVPR (2018)
50. Qi, X., Liao, R., Jia, J., Fidler, S., Urtasun, R.: 3D graph neural networks for RGBD semantic segmentation. In: ICCV (2017)
51. Ren, J., Gong, X., Yu, L., Zhou, W., Ying Yang, M.: Exploiting global priors for RGB-D saliency detection. In: CVPRW (2015)
52. Ren, Z., Gao, S., Chia, L.T., Tsang, I.W.H.: Region-based saliency detection and its application in object recognition. TCSVT **24**(5), 769–779 (2013)
53. Scarselli, F., Gori, M., Tsoi, A.C., Hagenbuchner, M., Monfardini, G.: The graph neural network model. TNN **20**(1), 61–80 (2008)
54. Shen, Y., Li, H., Yi, S., Chen, D., Wang, X.: Person re-identification with deep similarity-guided graph neural network. In: Ferrari, V., Hebert, M., Sminchisescu, C., Weiss, Y. (eds.) ECCV 2018. LNCS, vol. 11219, pp. 508–526. Springer, Cham (2018). https://doi.org/10.1007/978-3-030-01267-0_30
55. Song, H., Liu, Z., Du, H., Sun, G., Le Meur, O., Ren, T.: Depth-aware salient object detection and segmentation via multiscale discriminative saliency fusion and bootstrap learning. TIP **26**(9), 4204–4216 (2017)
56. Su, J., Li, J., Zhang, Y., Xia, C., Tian, Y.: Selectivity or invariance: boundary-aware salient object detection. In: ICCV (2019)
57. Wang, A., Wang, M.: RGB-D salient object detection via minimum barrier distance transform and saliency fusion. SPL **24**(5), 663–667 (2017)

58. Wang, C., et al.: DenseFusion: 6D object pose estimation by iterative dense fusion. In: CVPR (2019)
59. Wang, L., Wang, L., Lu, H., Zhang, P., Ruan, X.: Saliency detection with recurrent fully convolutional networks. In: Leibe, B., Matas, J., Sebe, N., Welling, M. (eds.) ECCV 2016. LNCS, vol. 9908, pp. 825–841. Springer, Cham (2016). https://doi.org/10.1007/978-3-319-46493-0_50
60. Wang, N., Gong, X.: Adaptive fusion for RGB-D salient object detection. IEEE Access **7**, 55277–55284 (2019)
61. Wang, W., Neumann, U.: Depth-aware CNN for RGB-D segmentation. In: Ferrari, V., Hebert, M., Sminchisescu, C., Weiss, Y. (eds.) ECCV 2018. LNCS, vol. 11215, pp. 144–161. Springer, Cham (2018). https://doi.org/10.1007/978-3-030-01252-6_9
62. Wang, W., Lu, X., Shen, J., Crandall, D.J., Shao, L.: Zero-shot video object segmentation via attentive graph neural networks. In: ICCV (2019)
63. Wang, X., You, S., Li, X., Ma, H.: Weakly-supervised semantic segmentation by iteratively mining common object features. In: CVPR (2018)
64. Wang, X., Girshick, R., Gupta, A., He, K.: Non-local neural networks. In: CVPR (2018)
65. Wang, X., Gupta, A.: Videos as space-time region graphs. In: Ferrari, V., Hebert, M., Sminchisescu, C., Weiss, Y. (eds.) ECCV 2018. LNCS, vol. 11209, pp. 413–431. Springer, Cham (2018). https://doi.org/10.1007/978-3-030-01228-1_25
66. Wang, Y., Sun, Y., Liu, Z., Sarma, S.E., Bronstein, M.M., Solomon, J.M.: Dynamic graph CNN for learning on point clouds. TOG **38**(5), 1–12 (2019)
67. Wu, Z., Su, L., Huang, Q.: Cascaded partial decoder for fast and accurate salient object detection. In: CVPR (2019)
68. Wu, Z., Su, L., Huang, Q.: Stacked cross refinement network for edge-aware salient object detection. In: ICCV (2019)
69. Xie, G.S., et al.: Attentive region embedding network for zero-shot learning. In: CVPR (2019)
70. Xie, G.S., et al.: Region graph embedding network for zero-shot learning. In: ECCV (2020)
71. Xie, G.S., et al.: SRSC: selective, robust, and supervised constrained feature representation for image classification. TNNLS (2019)
72. Xu, K., et al.: Show, attend and tell: neural image caption generation with visual attention. In: ICML (2015)
73. Xu, K., Hu, W., Leskovec, J., Jegelka, S.: How powerful are graph neural networks? (2019)
74. Xu, Y., et al.: Structured modeling of joint deep feature and prediction refinement for salient object detection. In: ICCV (2019)
75. Yan, P., et al.: Semi-supervised video salient object detection using pseudo-labels. In: ICCV (2019)
76. Yang, F., Li, X., Cheng, H., Guo, Y., Chen, L., Li, J.: Multi-scale bidirectional FCN for object skeleton extraction. In: AAAI (2018)
77. Yang, F., Li, X., Cheng, H., Li, J., Chen, L.: Object-aware dense semantic correspondence. In: CVPR, July 2017
78. Yu, F., Koltun, V.: Multi-scale context aggregation by dilated convolutions. In: ICLR (2015)
79. Zeng, Y., Zhang, P., Zhang, J., Lin, Z., Lu, H.: Towards high-resolution salient object detection. In: ICCV (2019)
80. Zhang, D., Meng, D., Zhao, L., Han, J.: Bridging saliency detection to weakly supervised object detection based on self-paced curriculum learning. arXiv preprint arXiv:1703.01290 (2017)

81. Zhang, J., Sclaroff, S.: Saliency detection: a Boolean map approach. In: ICCV (2013)
82. Zhang, J., Sclaroff, S., Lin, Z., Shen, X., Price, B., Mech, R.: Minimum barrier salient object detection at 80 FPS. In: ICCV (2015)
83. Zhang, J., et al.: UC-NET: uncertainty inspired RGB-D saliency detection via conditional variational autoencoders. In: CVPR (2020)
84. Zhang, L., Zhang, J., Lin, Z., Lu, H., He, Y.: CapSal: leveraging captioning to boost semantics for salient object detection. In: CVPR (2019)
85. Zhao, H., Shi, J., Qi, X., Wang, X., Jia, J.: Pyramid scene parsing network. In: CVPR (2017)
86. Zhao, J.X., Cao, Y., Fan, D.P., Cheng, M.M., Li, X.Y., Zhang, L.: Contrast prior and fluid pyramid integration for RGBD salient object detection. In: CVPR (2019)
87. Zhao, J.X., Liu, J.J., Fan, D.P., Cao, Y., Yang, J., Cheng, M.M.: EGNet: edge guidance network for salient object detection. In: ICCV (2019)
88. Zhao, L., Peng, X., Tian, Y., Kapadia, M., Metaxas, D.N.: Semantic graph convolutional networks for 3D human pose regression. In: CVPR (2019)
89. Zhu, C., Cai, X., Huang, K., Li, T.H., Li, G.: PDNet: prior-model guided depth-enhanced network for salient object detection. In: ICME (2019)
90. Zhu, C., Li, G.: A three-pathway psychobiological framework of salient object detection using stereoscopic technology. In: CVPRW (2017)
91. Zhu, C., Li, G., Wang, W., Wang, R.: An innovative salient object detection using center-dark channel prior. In: ICCVW, pp. 1509–1515 (2017)

FairALM: Augmented Lagrangian Method for Training Fair Models with Little Regret

Vishnu Suresh Lokhande[1]([✉]) [ID], Aditya Kumar Akash[1] [ID], Sathya N. Ravi[2] [ID], and Vikas Singh[1] [ID]

[1] University of Wisconsin-Madison, Madison, WI, USA
lokhande@cs.wisc.edu, aakash@wisc.edu, vsingh@biostat.wisc.edu
[2] University of Illinois at Chicago, Chicago, IL, USA
sathya@uic.edu

Abstract. Algorithmic decision making based on computer vision and machine learning methods continues to permeate our lives. But issues related to biases of these models and the extent to which they treat certain segments of the population unfairly, have led to legitimate concerns. There is agreement that because of biases in the datasets we present to the models, a fairness-oblivious training will lead to unfair models. An interesting topic is the study of mechanisms via which the *de novo* design or training of the model can be informed by fairness measures. Here, we study strategies to impose fairness concurrently while training the model. While many fairness based approaches in vision rely on training adversarial modules together with the primary classification/regression task, in an effort to remove the influence of the protected attribute or variable, we show how ideas based on well-known optimization concepts can provide a simpler alternative. In our proposal, imposing fairness just requires specifying the protected attribute and utilizing our routine. We provide a detailed technical analysis and present experiments demonstrating that various fairness measures can be reliably imposed on a number of training tasks in vision in a manner that is interpretable.

1 Introduction

Fairness and non-discrimination is a core tenet of modern society. Driven by advances in vision and machine learning systems, algorithmic decision making continues to permeate our lives in important ways. Consequently, ensuring that the decisions taken by an algorithm do not exhibit serious biases is no longer a hypothetical topic, rather a key concern that has started informing legislation [23] (e.g., Algorithmic Accountability act). On one extreme, some types of biases can create inconvenience – a biometric access system could be more error-prone for faces of persons from certain skin tones [9] or a search for homemaker or

---

Electronic supplementary material The online version of this chapter (https://doi.org/10.1007/978-3-030-58610-2_22) contains supplementary material, which is available to authorized users.

A. Vedaldi et al. (Eds.): ECCV 2020, LNCS 12357, pp. 365–381, 2020.
https://doi.org/10.1007/978-3-030-58610-2_22

`programmer` may return gender-stereotyped images [8]. But there are serious ramifications as well – an individual may get pulled aside for an intrusive check while traveling [50] or a model may decide to pass on an individual for a job interview after digesting their social media content [13,25]. Biases in automated systems in estimating recidivism within the criminal judiciary have been reported [38]. There is a growing realization that these problems need to be identified and diagnosed, and then promptly addressed. In the worst case, if no solutions are forthcoming, we must step back and reconsider the trade-off between the benefits versus the harm of deploying such systems, on a case by case basis.

What Leads to Unfair Learning Models? One finds that learning methods in general tend to amplify biases that exist in the training set [46]. While this creates an incentive for the organization training the model to curate datasets that are "balanced" in some sense, from a practical standpoint, it is often difficult to collect data that is balanced along multiple "protected" variables, e.g., gender, race and age. If a protected feature is correlated with the response variable, a learning model can *cheat* and find representations from other features that are collinear or a good surrogate for the protected variable. A thrust in current research is devoted to devising ways to mitigate such shortcuts. If one does not have access to the underlying algorithm, a recent result [24] shows the feasibility of finding thresholds that can impose certain fairness criteria. Such a threshold search can be post-hoc applied to any learned model. But in various cases, because of the characteristics of the dataset, a fairness-oblivious training will lead to biased models. An interesting topic is the study of mechanisms via which the de novo design/training of the model can be informed by fairness measures.

Some General Strategies for Fair Learning. Motivated by the foregoing issues, recent work which may broadly fall under the topic of *algorithmic fairness* has suggested several concepts or measures of fairness that can be incorporated within the learning model. While we will discuss the details shortly, these include demographic parity [40], equal odds and equal opportunities [24], and disparate treatment [42]. In general, existing work can be categorized into a few distinct categories. The *first* category of methods attempts to modify the representations of the data to ensure fairness. While different methods approach this question in different ways, the general workflow involves imposing fairness *before* a subsequent use of standard machine learning methods [10,27]. The *second* group of methods adjusts the decision boundary of an already trained classifier towards making it fair as a *post*-processing step while trying to incur as little deterioration in overall performance as possible [21,22,39]. While this procedure is convenient and fast, it is not always guaranteed to lead to a fair model without sacrificing accuracy. Part of the reason is that the search space for a fair solution in the post-hoc tuning is limited. Of course, we may impose fairness during training directly as adopted in the *third* category of papers such as [4,43], and the approach we take here. Indeed, if we are training the model from scratch and have knowledge of the protected variables, there is little reason not to incorporate this information directly *during* model training. In principle, this strategy provides the maximum control over the model. From the formulation standpoint, it is slightly more involved because it requires satisfying a fairness constraint derived from one or more fairness measure(s) in the literature, while concurrently learning the model parameters. The difficulty varies depending both on the primary task (shallow versus deep model) as well as the specific fairness criteria. For instance, if one were using a deep network for classification, we would need to devise ways to enforce constraints on the *output* of the network, efficiently.

Scope of This Paper and Contributions. Many studies on fairness in learning and vision are somewhat recent and were partly motivated in response to more than a few controversial reports in the news media [17,31]. As a result, the literature on mathematically sound and practically sensible fairness measures that can still be incorporated while training a model is still in a nascent stage. In vision, current approaches have largely relied on training adversarial modules in conjunction with the primary classification or regression task, to remove the influence of the protected attribute. Adversarial training via SGD needs a great deal of care and is not straightforward [36]. In contrast, the **contribution** of our work is to provide a simpler alternative. We show that a number of fairness measures in the literature can be incorporated by viewing them as constraints on the *output* of the learning model. This view allows adapting ideas from constrained optimization, to devise ways in which training can be efficiently performed in a way that at termination, the model parameters correspond to a fair model. For a practitioner, this means that no changes in the architecture or model are needed: imposing fairness only requires specifying the protected attribute, and utilizing our proposed optimization routine.

2   A Primer on Fairness Functions

In this section, we introduce basic notations and briefly review several fairness measures described in the literature.

Basic Notations. We denote classifiers using $h : x \mapsto y$ where x and y are random variables that represent the features and labels respectively. A *protected* attribute is a random variable s on the same probability space as x and y – for example, s may be gender, age, or race. Collectively, a training example would be $z := (x, y, s)$. So, our goal is to learn h (predict y given x) while *imposing fairness-type constraints* over s. We will use $\mathcal{H} = \{h_1, h_2, \ldots, h_N\}$ to denote a set/family of possible classifiers and Δ^N to denote the probability simplex in $\mathbb{R}^N$, i.e., $\Delta := \{q : \sum_{i=1}^N q_i = 1, q_i \geq 0\}$ where q_i is the i-th coordinate of q.

We will assume that the distribution of s has finite support. Unless explicitly specified, we will assume that $y \in \{0, 1\}$. For each $h \in \mathcal{H}$, we will use e_h to denote the misclassification rate of h and $e_{\mathcal{H}} \in \mathbb{R}^N$ to be the vector containing all misclassification rates. We will use superscript to denote conditional expectations. That is, if μ_h corresponds to expectation of some function μ (that depends on $h \in \mathcal{H}$), then the conditional expectation/moment of μ_h with respect to s will be denoted by μ_h^s. With a slight abuse of notation, we will use $\mu_h^{s_0}$ to denote the elementary conditional expectation $\mu_h|(s = s_0)$ whenever it is clear from context. We will use d_h to denote the *difference* between the conditional expectation of the two groups of s, that is, $d_h := \mu_h^{s_0} - \mu_h^{s_1}$. For example, let s be the random variable representing gender, that is, s_0 and s_1 may correspond to male and female. Then, $e_h^{s_i}$ corresponds to the misclassification rate of h on group s_i, and $d_h = e_h^{s_0} - e_h^{s_1}$. Finally, $\mu_h^{s_i,t_j} := \mu_h|(s = s_i, t = t_j)$ denotes the elementary conditional expectation with respect to two random variables s, t.

2.1   Fairness Through the Lens of Confusion Matrix

Recall that a *fairness* constraint corresponds to a performance requirement of a classifier h on subgroups of features x *induced* by a protected attribute s. For instance, say that h predicts the credit-worthiness y of an individual x. Then, we may require that e_h be "approximately" the same across individuals for different races given by s. Does it follow that functions/metrics that are used to evaluate fairness may be written in terms of the error of a classifier e_h *conditioned* on the protected variable s (or in other words e_h^s)? Indeed, it does turn out to be the case. In fact, many widely used functions in practice can be viewed as imposing constraints on the confusion matrix as our intuition suggests. We will now discuss few common fairness metrics to illustrate this idea.

(a) **Demographic Parity (DP)** [40]. A classifier h is said to satisfy Demographic Parity (DP) if $h(x)$ is *independent* of the protected attribute s. Equivalently, h satisfies DP if $d_h = 0$ where we set $\mu_h^{s_i} = e_h^{s_i}$ (using notations introduced above). DP can be seen as equating the total false positives and false negatives between the confusion matrices of the two groups. We denote DDP by the difference of the demographic parity between the two groups.

(b) **Equality of Opportunity (EO)** [24]. A classifier h is said to satisfy EO if $h(x)$ is independent of the protected attribute s for $y \in \{0, 1\}$. Equivalently, h satisfies EO if $d_h^y = 0$ where we set $\mu_h^{s_i} = e_h^{s_i}|(y \in \{0,1\}) =: e_h^{s_i,y_j}$ conditioning on both s and y. Depending on the choice of y in $\mu_h^{s_i}$, we get two different metrics: (i) $y = 0$ corresponds to h with equal *False Positive Rate (FPR)* across s_i [14], whereas (ii) $y = 1$ corresponds to h with equal *False Negative Rate (FNR)* across s_i [14]. Moreover, h satisfies *Equality of Odds* if $d_h^0 + d_h^1 = 0$, i.e., h equalizes both TPR and FPR across s [24]. We denote the difference in EO by DEO.

(c) **Predictive Parity (PP)** [11]. A classifier h satisfies PP if the likelihood of making a misclassification among the positive predictions of the classifier is independent of the protected variable s. Equivalently, h satisfies PP if $d_h^{\hat{y}} = 0$ where we set $\mu_{h_i}^{s_i} = e_h^{s_i}|(\hat{y} = 1)$. It corresponds to matching the False Discovery Rate between the confusion matrices of the two groups.

3   How to Learn Fair Models?

At a high level, the optimization problem that we seek to solve is written as,

$$\min_{h \in \mathcal{H}} \mathbb{E}_{z:(x,y,s) \sim \mathcal{D}} \mathcal{L}(h; (x,y)) \quad \text{subject to} \quad h \in \mathcal{F}_{d_h}, \tag{1}$$

where $\mathcal{L}$ denotes the loss function that measures the accuracy of h in predicting y from x, and $\mathcal{F}_{d_h}$ denotes the set of *fair* classifiers. Our approach to solve (1) *provably efficiently* involves two main steps: (i) first, we reformulate problem (1) to compute a posterior distribution q over $\mathcal{H}$; (ii) second, we incorporate fairness as *soft* constraints on the output of q using the augmented Lagrangian of Problem (1). We assume that we have access to sufficient number of samples to approximate $\mathcal{D}$ and solve the empirical version of Problem (1).

3.1 From Fair Classifiers to Fair Posteriors

The starting point of our development is based on the following simple result that follows directly from the definitions of fairness metrics in Sect. 2:

Observation 1. *Fairness metrics such as DP/EO are linear functions of h, whereas PP takes a linear fractional form due to the conditioning on $\hat{y}$, see [11].*

Observation 1 immediately implies that $\mathcal{F}_{d_h}$ can be represented using linear (fractional) equations in h. To simplify the discussion, we will focus on the case when $\mathcal{F}_{d_h}$ is given by the DP metric. Hence, we can reformulate (1) as,

$$\min_{q \in \Delta} \ \sum_i q_i e_{h_i} \ \text{s.t.} \ q_i(\mu_{h_i}^{s_0} - \mu_{h_i}^{s_1}) = 0 \quad \forall i \in [N], \tag{2}$$

where q represents a distribution over $\mathcal{H}$.

3.2 Imposing Fairness via Soft Constraints

In general, there are two ways of treating the N constraints $q_i d_{h_i} = 0$ in Problem (2) viz., (i) as *hard constraints*; or (ii) as *soft constraints*. Algorithms that can handle explicit constraints efficiently require access to an efficient oracle that can minimize a linear or quadratic function over the feasible set in *each* iteration. Consequently, algorithms that incorporate hard constraints come with high per-iteration computational cost since the number of constraints is (at least) linear in N, and is not applicable in large scale settings. Hence, we propose to use algorithms that incorporate fairness as soft constraints. With these two minor modifications, we will now describe our approach to solve problem (2).

4 Fair Posterior from Proximal Dual

Following the reductions approach in [1], we first write the Lagrangian dual of DP constrained risk minimization problem (2) using dual variables λ as,

$$\max_{\lambda \in \mathbb{R}^N} \min_{q \in \Delta} L(q, \lambda) := \left(\sum_i q_i e_{h_i} \right) + \lambda \left(\sum_i q_i(\mu_{h_i}^{s_0} - \mu_{h_i}^{s_1}) \right) \tag{3}$$

Interpreting the Lagrangian. Problem 3 can be understood as a game between two players: a q-player and a λ-player [16]. We recall an important fact regarding the dual problem (3):

Fact 2. *The objective function of the dual problem* (3) *is* always nonsmooth *with respect to* λ *because of the inner minimization problem in q.*

Technically, there are two main reasons why optimizing nonsmooth functions can be challenging [19]: (i) finding a descent direction in high dimensions N can be difficult; and (ii) subgradient methods can be slow to converge in practice. Due to these difficulties arising from Fact 2, using a first order algorithm such as gradient descent to solve the dual problem in (3) directly can be problematic, and may be suboptimal.

Accelerated Optimization Using Dual Proximal Functions. To overcome the difficulties due to the nonsmoothness of the dual problem, we propose to *augment* the Lagrangian with a proximal term. Specifically, for some λ_T, the augmented Lagrangian function can be written as,

$$L_T(q, \lambda) = \Big(\sum_i q_i e_{h_i}\Big) + \lambda\Big(\sum_i q_i(\mu_{h_i}^{s_0} - \mu_{h_i}^{s_1})\Big) - \frac{1}{2\eta}(\lambda - \lambda_T)^2 \tag{4}$$

Note that, as per our simplified notation, $L_T \equiv L_{\lambda_T}$. The following lemma relates the standard Lagrangian in (3) with its proximal counterpart in (4).

Lemma 1. *At the optimal solution* (q^*, λ^*) *to* L, *we have* $\max_\lambda \min_{q \in \Delta} L = \max_\lambda \min_{q \in \Delta} L_{\lambda^*}$.

This is a standard property of proximal objective functions, where λ^* forms a fixed point of $\min_{q \in \Delta} L_{\lambda^*}(q, \lambda^*)$ (Sect. 2.3 of [32]). Intuitively, Lemma 1 states that L and L_T are not at all different for optimization purposes.

Remark 1. While the augmented Lagrangian L_T still may be nonsmooth, the proximal (quadratic) term can be exploited to design *provably* faster optimization algorithms as we will see shortly.

5 Our Algorithm – FairALM

It is common [1,16,28] to consider the minimax problem in (4) as a zero sum game between the λ-player and the q-player. The Lagrangian(s) L_T (or L) specify the cost which the q-player pays to the λ-player after the latter makes its choice. We update the λ-player by follow-the-leader method [37] which minimizes the cumulative regret. This is distinct from a dual ascent method which relies on a gradient based update scheme. Further, the q-player is updated by following a best response strategy as in [1]. While the q-player's move relies on the availability of an efficient *oracle* to solve the minimization problem, $L_T(q, \lambda)$, being a linear program in q makes it less challenging. We describe our algorithm in Algorithm 1 and call it *FairALM: Linear Classifier*.

5.1 Convergence Analysis

As the game with respect to λ is a maximization problem, we get a reverse regret bound as shown in the following Lemma. Proofs are deferred to the appendix.

Lemma 2. *Let r_t denote the reward at each round of the game. The reward function $f_t(\lambda)$ is defined as $f_t(\lambda) = \lambda r_t - \frac{1}{2\eta}(\lambda - \lambda_t)^2$. We choose λ in round $T + 1$ to maximize the cumulative reward: $\lambda_{T+1} = \mathrm{argmax}_\lambda \sum_{t=1}^T f_t(\lambda)$. Define $L = \max_t |r_t|$. The following bound on the cumulative reward holds, for any λ*

$$\sum_{t=1}^T \left(\lambda r_t - \frac{1}{2\eta}(\lambda - \lambda_t)^2 \right) \leq \sum_{t=1}^T \lambda_t r_t + \eta L^2 \mathcal{O}(\log T) \tag{5}$$

The above lemma indicates that the cumulative reward grows in time as $\mathcal{O}(\log T)$. The proximal term in the augmented Lagrangian gives us a *better* bound than an ℓ_2 or an entropic regularizer (which provides a $\sqrt{T}$ bound [37]).

Next, we evaluate the cost function $L_T(q, \lambda)$ after T rounds of the game. We observe that the average play of both the players converges to a saddle point with respect to $L_T(q, \lambda)$. We formalize this in the following theorem,

Theorem 3. *Recall that d_h represents the difference of conditional means. Assume that $||d_h||_\infty \leq L$ and consider T rounds of the game described above. Let the average plays of the q-player be $\bar{q} = \frac{1}{T}\sum_{t=1}^T q_t$ and the λ-player be $\bar{\lambda} = \frac{1}{T}\sum_{t=1}^T \lambda_t$. Then under the following conditions on q, λ and η, we have $L_T(\bar{q}, \lambda) \leq L_T(q, \bar{\lambda}) + \nu$ and $L_T(\bar{q}, \lambda) \geq L_T(\bar{q}, \lambda) - \nu$*

- *If $\eta = \mathcal{O}(\sqrt{\frac{B^2 T}{L^2(\log T+1)}})$, $\nu = \mathcal{O}(\sqrt{\frac{B^2 L^2(\log T+1)}{T}})$; $\forall |\lambda| \leq B$, $\forall q \in \Delta$*
- *If $\eta = \frac{1}{T}$, $\nu = \mathcal{O}(\frac{L^2(\log T+1)^2}{T})$; $\forall \lambda \in \mathbb{R}$, $\forall q \in \Delta$*

The above theorem indicates that the average play of the q-player and the λ-player reaches a ν-approximate saddle point. Our bounds for $\nu = \frac{1}{T}$ and $\lambda \in \mathbb{R}$ are better than [1].

Algorithm 1 FairALM: Linear Classifier

1: *Notations:* Dual step size η
     $h_t \in \{h_1, h_2, \ldots, h_N\}$.
2: *Input:* Error Vector $e_{\mathcal{H}}$,
     Conditional mean vector $\mu^s_{\mathcal{H}}$
3: *Initializations:* $\lambda_0 = 0$
4: **for** $t = 0, 1, 2, \ldots, T$ **do**
5:     (Primal) $h_t \leftarrow \mathrm{argmin}_i(e_{h_i} + \lambda_t(\mu^{s_0}_{h_i} - \mu^{s_1}_{h_i}))$
6:     (Dual) $\lambda_{t+1} \leftarrow \lambda_t + \eta(\mu^{s_0}_{h_t} - \mu^{s_1}_{h_t})/t$
7: **end for**
8: *Output:* h_T

Algorithm 2 FairALM: DeepNet Classifier

1: *Notations:* Dual step size η, Primal step size τ
2: *Input:* Training Set D
3: *Initializations:* $\lambda_0 = 0$, w_0
4: **for** $t = 0, 1, 2, \ldots, T$ **do**
5:     Sample $z \sim D$
6:     Pick $v_t \in \partial\left(\hat{e}_{h_w}(z) + (\lambda_t + \eta)\hat{\mu}^{s_0}_{h_w}(z) - (\lambda_t - \eta)\hat{\mu}^{s_1}_{h_w}(z)\right)$
7:     (Primal) $w_t \leftarrow w_{t-1} - \tau v_t$
8:     (Dual) $\lambda_{t+1} \leftarrow \lambda_t + \eta(\hat{\mu}^{s_0}_{h_{w_t}}(z) - \hat{\mu}^{s_1}_{h_{w_t}}(z))$
9: **end for**
10: *Output:* w_T

5.2   Can We Train Fair Deep Neural Networks by Adapting Algorithm 1?

The key difficulty from the analysis standpoint we face in extending these results to the deep networks setting is that the number of classifiers $|\mathcal{H}|$ may be exponential in number of nodes/layers. This creates a potential problem in computing Step 5 of Algorithm 1 – if viewed mechanistically, it is not practical since an epsilon net over the family $\mathcal{H}$ (representable by a neural network) is exponential in size. Interestingly, notice that we often use over-parameterized networks for learning. This is a useful fact here because it means that there exists a solution where $\text{argmin}_i(e_{h_i} + \lambda_t d_{h_i})$ is 0. While iterating through all h_is will be intractable, we may still able to obtain a solution via standard stochastic gradient descent (SGD) procedures [45]. The only unresolved question then is if we can do posterior inference and obtain classifiers that are "fair". It turns out that the above procedure provides us an approximation if we leverage two facts: first, SGD can find the minimum of $L(h, \lambda)$ with respect to h and second, recent results show that SGD, in fact, performs variational inference, implying that the optimization can provide an approximate posterior [12]. Having discussed the exponential sized $|\mathcal{H}|$ issue – for which we settle for an approximate posterior – we make three additional adjustments to the algorithm to make it suitable for training deep networks. First, the non-differentiable indicator function $\mathbb{1}[\cdot]$ is replaced with a smooth surrogate function (such as a logistic function). Second, as it is hard to evaluate e_h/μ_h^s due to unavailability of the true data distribution, we instead calculate their empirical estimates $z = (x; y; s)$, and denote it by $\hat{e}_h(z)/\hat{\mu}_h^s(z)$. Third, by exchanging the "max" and "min" in (3), we obtain an objective that *upper-bounds* our current objective in (3). This provides us with a closed-form solution to λ thus reducing the minmax objective to a single simpler minimization problem. We present our *FairALM: DeepNet Classifier* algorithm for deep neural network training in Algorithm 2 (more details are in the supplement).

6   Experiments

A central theme in our experiments is to assess whether our proposed algorithm, FairALM, can indeed obtain meaningful fairness measure scores *without* compromising the test set performance. We evaluate FairALM on a number of problems where the dataset reflects certain inherent societal/stereotypical biases. Our evaluations are also designed with a few additional goals in mind.

Overview. Our **first** experiment on the CelebA dataset seeks to predict the value of a label for a face image while controlling for certain protected attributes (gender, age). We discuss how prediction of some labels is *unfair* in an unconstrained model and contrast with our FairALM. Next, we focus on the label where predictions are the most unfair and present comparisons against methods available in the literature. For our **second** experiment, we use the ImSitu dataset where images correspond to a situation (activities, verb). Expectedly,

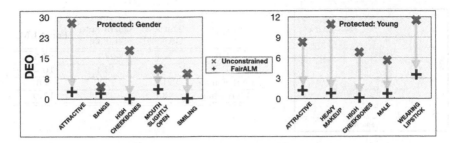

Fig. 1. Identifying unfair labels in CelebA dataset. Using a 3-layers ReLU network, we determine the labels in CelebA dataset that are biased with respect to gender (left) and the attribute young (right). FairALM minimizes the DEO measure, indicated by the green arrow, on these labels while maintaining ±5% precision. (Color figure online)

some activities such as driving or cooking are more strongly associated with a specific gender. We inspect if an unconstrained model is *unfair* when we ask it to learn to predict two gender correlated activities/verbs. Comparisons with baseline methods will help measure FairALM's strengths/weaknesses. We can use heat map visualizations to qualitatively interpret the value of adding fairness constraints. We threshold the heat-maps to get an understanding of a general behavior of the models. Our **third** experiment addresses an important problem in medical/scientific studies. Small sample sizes necessitate pooling data from multiple sites or scanners [49], but introduce a site or scanner specific nuisance variable which must be controlled for – else a deep (also, shallow) model may cheat and use site specific (rather than disease-specific) artifacts in the images for prediction even when the cohorts are age or gender matched [20]. We study one simple setting here: we use FairALM to mitigate site (hospital) specific differences in predicting "tuberculosis" from X-ray images acquired at two hospitals, Shenzhen and Montgomery (and recently made publicly available [26]).

In all the experiments, we impose Difference in Equality of Opportunity (DEO) constraint (defined in Sect. 2.1). We adopt NVP (novel validation procedure) [18] a two-step procedure: first, we search for the hyper-parameters that achieve the best accuracy, and then, we report the minimum fairness measure (DEO) for accuracies within 90% of the highest accuracy.

Remark. Certain attributes such as *attractiveness*, obtained via crowd-sourcing, may have socio-cultural ramifications. Similarly, the gender attribute in the dataset is binary (male versus female) which may be insensitive. We clarify that our goal is to present evidence showing that our algorithm can impose fairness in a sensible way on datasets used in the literature and acknowledge that larger/improved datasets **focused** on societally relevant themes, as they become available, will be much more meaningful.

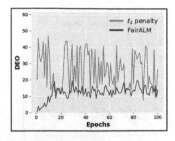

	Fairness GAN[35]	Quadrianto etal[33]	**FairALM**
ERR	26.6	24.1	24.5
DEO	22.5	12.4	**10.4**
FNR Female	21.2	12.8	**6.6**
FNR Male	43.7	25.2	**17.0**

Fig. 2. Quantitative results on CelebA. The target attribute is the label *attractiveness* present in the CelebA dataset and the protected attribute is *gender*. (Left) FairALM has a stable training profile in comparison to naive ℓ_2 penalty. (Right) FairALM attains a lower DEO measure and improves the test set errors (ERR).

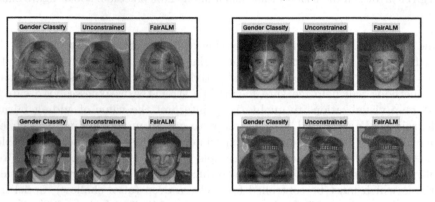

Fig. 3. Interpretable models for CelebA. Unconstrained/FairALM predict the label *attractiveness* present in the CelebA dataset while controlling *gender*. The heatmaps of Unconstrained model overlaps with gender classification task indicating gender leak. FairALM consistently picks non-gender revealing features of the face. Interestingly, these regions are on the left side which appear to agree with psychological studies suggesting that a face's left side is more attractive [6].

6.1   CelebA Dataset

Data and Setup. CelebA [29] consists of 200K celebrity face images from the internet annotated by a group of paid adult participants [7]. There are up to 40 labels available in the dataset, each of which is binary-valued.

Quantitative Results. We begin our analysis by predicting each of the 40 labels with a 3-layer ReLU network. The protected variable, s, we consider are the binary attributes like *Male* and *Young* representing gender and age respectively. We train the SGD algorithm for 5-epochs and select the labels predicted with at least at 70% precision and with a DEO of at least 4% across the protected variables. The biased set of labels thus estimated are shown in Fig. 1. These labels are consistent with other reported results [34]. It is important to bear in mind that the bias in the labels should not be attributed to its relatedness to a specific protected attributed alone. The cause of bias could also be due to the skew in the label distributions. When training a 3-layer ReLU net with FairALM, the precision of the model remained about the same ($\pm 5\%$) while the DEO measure reduced significantly, see Fig. 1. Next, choosing the most unfair label in Fig. 1 (i.e., attractiveness), we train a ResNet18 for a longer duration of about 100 epochs and contrast the performance with a simple ℓ_2-penalty baseline. The training profile is observed to be more stable for FairALM as indicated in Fig. 2. This finding is consistent with the results of [5,30] that discuss the ill-conditioned landscape of non-convex penalties. Comparisons to more recent works such as [33,35] is provided in Fig. 2. Here, we present a new state-of-the-art result for the DEO measure with the label *attractiveness* and protected attribute *gender*.

Qualitatively Assessing Interpretability. While the DEO measure obtained by FairALM is lower, we can ask an interesting question: when we impose the fairness constraint, precisely which aspects of the image are no longer "legal" for the neural network to utilize? This issue can be approached via visualizing activation maps from models such as CAM [48]. As a representative example, our analysis suggests that in general, an unconstrained model uses the entire face image (including the gender-revealing parts). We find some consistency between the activation maps for the label *attractiveness* and activation maps of an unconstrained model trained to predict *gender*! In contrast, when we impose the fairness constraint, the corresponding activation maps turn out to be clustered around specific regions of the face which are *not* gender revealing (Fig. 3). In particular, a surprising finding was that the left regions in the face were far more prominent which turns out to be consistent with studies in psychology [6].

Summary. FairALM minimized the DEO measure without compromising the test error. It has a more stable training profile than an ℓ_2 penalty and is competitive with recent fairness methods in vision. The activation maps in FairALM focus on non-gender revealing features of the face when controlled for gender.

6.2   Imsitu Dataset

Data and Setup. ImSitu [41] is a situation recognition dataset consisting of ~100K color images taken from the web. The annotations for the image is provided as a summary of the activity in the image and includes a verb describing it, the interacting agents and their roles. The protected variable in this experiment is gender. Our objective is to classify a pair of verbs associated with an image. The pair is chosen such that if one of the verbs is biased towards males then the other would be biased towards females. The authors in [47] report the list of labels in the ImSitu dataset that are gender biased: we choose our verb pairs from this list. In particular, we consider the verbs *Cooking vs Driving, Shaving vs Moisturizing, Washing vs Saluting* and *Assembling vs Hanging*. We compare our results against multiple baselines such as **(1)** Unconstrained **(2)** *ℓ_2-penalty*, the penalty applied on the DEO measure **(3)** *Re-weighting*, a weighted loss

376 V. S. Lokhande et al.

Table 1. Quantitative results on ImSitu. Test errors (ERR) and DEO measure are reported in %. The target class that is to be predicted is indicated by a +. FairALM always achieves a zero DEO while remaining competitive in ERR with the best method for a given verb-pair.

	Cooking(+) Driving(-)		Shaving(+) Moisturize(-)		Washing(+) Saluting(-)		Assembling(+) Hanging(-)	
	ERR	DEO	ERR	DEO	ERR	DEO	ERR	DEO
No Constraints	17.9	7.1	23.6	4.2	12.8	25.9	7.5	15.0
ℓ_2 Penalty	14.3	14.0	23.6	1.3	10.9	0.0	5.0	21.6
Reweight	11.9	3.5	19.0	5.3	10.9	0.0	4.9	9.0
Adversarial	4.8	0.0	13.5	11.9	14.6	25.9	6.2	18.3
Lagrangian	2.4	3.5	12.4	12.0	3.7	0.0	5.0	5.8
Proxy-lagrangian	2.4	3.5	12.4	12.0	3.7	0.0	14.9	3.0
FairALM	3.6	0.0	20.0	0.0	7.3	0.0	2.5	0.0

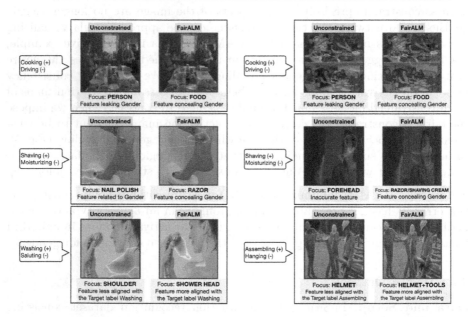

Fig. 4. Interpretability in ImSitu. The activation maps indicate that FairALM conceals gender revealing attributes in an image. Moreover, the attributes are more aligned with label of interest. The target class predicted is indicated by a +. These examples are representative of the general behavior of FairALM on this dataset. More plots in the supplement.

functions where the weights account for the dataset skew (**4**) *Adversarial* [44] (**5**) *Lagrangian* [47] (**6**) *Proxy-Lagrangian* [15]. The supplement includes more details on the baseline methods.

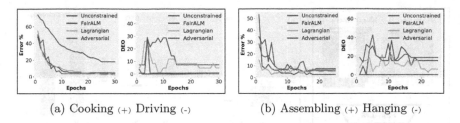

(a) Cooking (+) Driving (-)   (b) Assembling (+) Hanging (-)

Fig. 5. Training profiles. FairALM achieves minimum DEO early in training and remains competitive on testset errors. More plots are available in the supplement.

Quantitative Results. From Fig. 5, it can be seen that FairALM reaches a zero DEO measure very early in training and attains better test errors than an unconstrained model. Within the family of Lagrangian methods such as [15, 47], FairALM performs better on verb pair 'Shaving vs Moisturizing' in both test error and DEO measure as indicated in Table 1. While the results on the other verb pairs are comparable, FairALM was observed to be more stable to different hyper-parameter choices. This finding is in accord with recent studies by [2] who prove that proximal function models are robust to step-size selection. Detailed analysis is provided in the supplement. Turning now to an adversarial method such as [47], results in Table 1 show that the DEO measure is not controlled as competently as FairALM. Moreover, complicated training routines and unreliable convergence [3, 36] makes model-training harder.

Interpretable Models. We again used CAM [48] to inspect the image regions used by the model for target prediction. We observe that the unconstrained model ends up picking features from locations that may not be relevant for the task description but merely co-occur with the verbs in this particular dataset (and are gender-biased). Figure 4 highlights this observation for the selected classification tasks. Overall, we observe that the semantic regions used by the constrained model are more aligned with the action verb present in the image, and this adds to the qualitative advantages of the model trained using FairALM in terms of interpretability.

Limitations. We also note that there are cases where both the unconstrained model and FairALM look at incorrect image regions for prediction, owing to the small dataset sizes. However, the number of such cases are far fewer for FairALM than the unconstrained setup.

Summary. FairALM successfully minimizes the fairness measure while classifying verb/action pairs associated with an image. FairALM uses regions in an image that are more relevant to the target class and less gender revealing.

7 Pooling Multi-site Chest X-Ray Datasets

Data and Setup. The datasets we examine here are publicly available from the U.S. National Library of Medicine [26]. The images come from two sites/sources

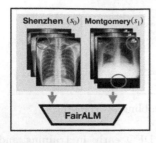 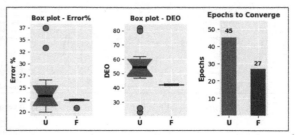

Fig. 6. Datasets pooling with FairALM. (Left:) Data is pooled from two sites/hospitals, Shenzhen s_0 and Montgomery s_1. (Right:) Boxplots indicate a lower variance in testset error and the DEO measure for FairALM. Moreover, FairALM reaches a 20% testset error in fewer epochs.

- first set is collected from patients in Montgomery county, USA and includes 138 X-rays and the second set of 662 images is collected from a hospital in Shenzhen, China. The task is to predict pulmonary tuberculosis (TB) from the X-ray images. Being collected from different X-ray machines with different characteristics, and the images have site-specific markings or artifacts, see Fig. 6. We pool the dataset and set aside 25% of the samples for testing.

Quantitative Results. We treat the site information, Montgomery or Shenzhen, as a nuisance/protected variable and seek to decorrelate it from the TB labels. We train a ResNet18 network and compare an unconstrained model with FairALM model. Our datasets of choice are small in size, and so deep models easily overfit to site-specific biases present in the training data. Our results corroborate this conjecture, the training accuracies reach 100% very early and the test set accuracies for the unconstrained model has a large variance over multiple experimental runs. Conversely, as seen in Fig. 6, a FairALM model not only maintains a lower variance in the test set errors and DEO measure but also attains improved performance on these measures. What stands out in this experiment is that the number of epochs to reach a certain test set error is lower for FairALM indicating that the model generalizes faster compared to an unconstrained model.

Summary. FairALM is effective at learning from datasets from two different sites/sources and minimizes site-specific biases.

8   Conclusion

We introduced FairALM, an augmented Lagrangian framework to impose constraints on fairness measures studied in the literature. On the theoretical side, we provide better bounds: $\mathcal{O}\left(\frac{\log^2 T}{T}\right)$ versus $\mathcal{O}\left(\frac{1}{\sqrt{T}}\right)$, for reaching a saddle point. On the application side, we provide extensive evidence (qualitative and quantitative) on image datasets commonly used in vision to show the benefits of our proposal. Finally, we use FairALM to mitigate site specific differences when performing analysis of pooled medical imaging datasets. In applying deep learning to scientific problems, this is important since sample sizes at individual sites/institutions are often smaller [49]. The overall procedure is simple which we believe will help adoption and follow-up work on this socially relevant topic. The project page is at https://github.com/lokhande-vishnu/FairALM.

Acknowledgments. The authors are grateful to Akshay Mishra for help and suggestions. Research supported by NIH R01 AG062336, NSF CAREER RI#1252725, NSF 1918211, NIH RF1 AG05931201A1, NIH RF1AG05986901, UW CPCP (U54 AI117924) and American Family Insurance. Sathya Ravi was also supported by UIC-ICR start-up funds. Correspondence should be directed to Ravi or Singh.

References

1. Agarwal, A., Beygelzimer, A., Dudík, M., Langford, J., Wallach, H.: A reductions approach to fair classification. arXiv preprint arXiv:1803.02453 (2018)
2. Asi, H., Duchi, J.C.: Stochastic (approximate) proximal point methods: convergence, optimality, and adaptivity. SIAM J. Optim. **29**(3), 2257–2290 (2019)
3. Barnett, S.A.: Convergence problems with generative adversarial networks (GANs). arXiv preprint arXiv:1806.11382 (2018)
4. Bechavod, Y., Ligett, K.: Penalizing unfairness in binary classification. arXiv preprint arXiv:1707.00044 (2017)
5. Bertsekas, D.P.: Constrained Optimization and Lagrange Multiplier Methods. Academic Press, New York (2014)
6. Blackburn, K., Schirillo, J.: Emotive hemispheric differences measured in real-life portraits using pupil diameter and subjective aesthetic preferences. Exp. Brain Res. **219**(4), 447–455 (2012). https://doi.org/10.1007/s00221-012-3091-y
7. Böhlen, M., Chandola, V., Salunkhe, A.: Server, server in the cloud. Who is the fairest in the crowd? arXiv preprint arXiv:1711.08801 (2017)
8. Bolukbasi, T., Chang, K.W., Zou, J.Y., Saligrama, V., Kalai, A.T.: Man is to computer programmer as woman is to homemaker? Debiasing word embeddings. In: Advances in Neural Information Processing Systems, pp. 4349–4357 (2016)
9. Buolamwini, J., Gebru, T.: Gender shades: intersectional accuracy disparities in commercial gender classification. In: Conference on Fairness, Accountability and Transparency, pp. 77–91 (2018)
10. Calmon, F., Wei, D., Vinzamuri, B., Ramamurthy, K.N., Varshney, K.R.: Optimized pre-processing for discrimination prevention. In: Advances in Neural Information Processing Systems, pp. 3992–4001 (2017)
11. Celis, L.E., Huang, L., Keswani, V., Vishnoi, N.K.: Classification with fairness constraints: a meta-algorithm with provable guarantees. In: Proceedings of the Conference on Fairness, Accountability, and Transparency, pp. 319–328. ACM (2019)
12. Chaudhari, P., Soatto, S.: Stochastic gradient descent performs variational inference, converges to limit cycles for deep networks. In: 2018 Information Theory and Applications Workshop (ITA). IEEE (2018)
13. Chin, C.: Assessing employer intent when AI hiring tools are biased, December 2019. https://www.brookings.edu/research/assessing-employer-intent-when-ai-hiring-tools-are-biased/

14. Chouldechova, A.: Fair prediction with disparate impact: a study of bias in recidivism prediction instruments. Big data **5**(2), 153–163 (2017)
15. Cotter, A., Jiang, H., Sridharan, K.: Two-player games for efficient non-convex constrained optimization. arXiv preprint arXiv:1804.06500 (2018)
16. Cotter, A., et al.: Optimization with non-differentiable constraints with applications to fairness, recall, churn, and other goals. arXiv preprint arXiv:1809.04198 (2018)
17. Courtland, R.: Bias detectives: the researchers striving to make algorithms fair. Nature **558**(7710), 357–357 (2018)
18. Donini, M., Oneto, L., Ben-David, S., Shawe-Taylor, J.S., Pontil, M.: Empirical risk minimization under fairness constraints. In: Advances in Neural Information Processing Systems, pp. 2791–2801 (2018)
19. Duchi, J.C., Bartlett, P.L., Wainwright, M.J.: Randomized smoothing for stochastic optimization. SIAM J. Optim. **22**(2), 674–701 (2012)
20. Fawzi, A., Frossard, P.: Measuring the effect of nuisance variables on classifiers, pp. 137.1–137.12, January 2016. https://doi.org/10.5244/C.30.137
21. Fish, B., Kun, J., Lelkes, Á.D.: A confidence-based approach for balancing fairness and accuracy. In: Proceedings of the 2016 SIAM International Conference on Data Mining, pp. 144–152. SIAM (2016)
22. Goh, G., Cotter, A., Gupta, M., Friedlander, M.P.: Satisfying real-world goals with dataset constraints. In: Advances in Neural Information Processing Systems, pp. 2415–2423 (2016)
23. Goodman, B., Flaxman, S.: European union regulations on algorithmic decision-making and a "right to explanation". AI Mag. **38**(3), 50–57 (2017). https://doi.org/10.1609/aimag.v38i3.2741. https://www.aaai.org/ojs/index.php/aimagazine/article/view/2741
24. Hardt, M., Price, E., Srebro, N., et al.: Equality of opportunity in supervised learning. In: Advances in Neural Information Processing Systems, pp. 3315–3323 (2016)
25. Heilweil, R.: Artificial intelligence will help determine if you get your next job, December 2019. https://www.vox.com/recode/2019/12/12/20993665/artificial-intelligence-ai-job-screen
26. Jaeger, S., Candemir, S., Antani, S., Wáng, Y.X.J., Lu, P.X., Thoma, G.: Two public chest x-ray datasets for computer-aided screening of pulmonary diseases. Quant. Imaging Med. Surg. **4**(6), 475 (2014)
27. Kamiran, F., Calders, T.: Classification with no discrimination by preferential sampling. In: Proceedings of the 19th Machine Learning Conference of Belgium and The Netherlands, pp. 1–6. Citeseer (2010)
28. Kearns, M., Neel, S., Roth, A., Wu, Z.S.: Preventing fairness gerrymandering: auditing and learning for subgroup fairness. arXiv preprint arXiv:1711.05144 (2017)
29. Liu, Z., Luo, P., Wang, X., Tang, X.: Large-scale CelebFaces attributes (CelebA) dataset (2018). Accessed 15 Aug 2018
30. Nocedal, J., Wright, S.: Numerical Optimization. Springer, New York (2006). https://doi.org/10.1007/978-0-387-40065-5
31. Obermeyer, Z., Powers, B., Vogeli, C., Mullainathan, S.: Dissecting racial bias in an algorithm used to manage the health of populations. Science **366**(6464), 447–453 (2019)
32. Parikh, N., Boyd, S.: Proximal algorithms. Found. Trends Optim. **1**(3), 127–239 (2014)

33. Quadrianto, N., Sharmanska, V., Thomas, O.: Discovering fair representations in the data domain. In: Proceedings of the IEEE Conference on Computer Vision and Pattern Recognition, pp. 8227–8236 (2019)
34. Ryu, H.J., Adam, H., Mitchell, M.: InclusiveFaceNet: improving face attribute detection with race and gender diversity. arXiv preprint arXiv:1712.00193 (2017)
35. Sattigeri, P., Hoffman, S.C., Chenthamarakshan, V., Varshney, K.R.: Fairness GAN. arXiv preprint arXiv:1805.09910 (2018)
36. Schäfer, F., Anandkumar, A.: Competitive gradient descent. In: Advances in Neural Information Processing Systems, pp. 7625–7635 (2019)
37. Shalev-Shwartz, S., et al.: Online learning and online convex optimization. Found. Trends® Mach. Learn. **4**(2), 107–194 (2012)
38. Ustun, B., Rudin, C.: Learning optimized risk scores from large-scale datasets. Stat **1050**, 1 (2016)
39. Woodworth, B., Gunasekar, S., Ohannessian, M.I., Srebro, N.: Learning non-discriminatory predictors. arXiv preprint arXiv:1702.06081 (2017)
40. Yao, S., Huang, B.: Beyond parity: fairness objectives for collaborative filtering. In: Advances in Neural Information Processing Systems, pp. 2921–2930 (2017)
41. Yatskar, M., Zettlemoyer, L., Farhadi, A.: Situation recognition: visual semantic role labeling for image understanding. In: Conference on Computer Vision and Pattern Recognition (2016)
42. Zafar, M.B., Valera, I., Gomez Rodriguez, M., Gummadi, K.P.: Fairness beyond disparate treatment & disparate impact: learning classification without disparate mistreatment. In: Proceedings of the 26th International Conference on World Wide Web, pp. 1171–1180. International World Wide Web Conferences Steering Committee (2017)
43. Zafar, M.B., Valera, I., Rodriguez, M., Gummadi, K., Weller, A.: From parity to preference-based notions of fairness in classification. In: Advances in Neural Information Processing Systems, pp. 229–239 (2017)
44. Zhang, B.H., Lemoine, B., Mitchell, M.: Mitigating unwanted biases with adversarial learning. In: Proceedings of the 2018 AAAI/ACM Conference on AI, Ethics, and Society, pp. 335–340 (2018)
45. Zhang, C., Bengio, S., Hardt, M., Recht, B., Vinyals, O.: Understanding deep learning requires rethinking generalization. arXiv preprint arXiv:1611.03530 (2016)
46. Zhao, J., Wang, T., Yatskar, M., Cotterell, R., Ordonez, V., Chang, K.W.: Gender bias in contextualized word embeddings. arXiv preprint arXiv:1904.03310 (2019)
47. Zhao, J., Wang, T., Yatskar, M., Ordonez, V., Chang, K.W.: Men also like shopping: reducing gender bias amplification using corpus-level constraints. arXiv preprint arXiv:1707.09457 (2017)
48. Zhou, B., Khosla, A., Lapedriza, À., Oliva, A., Torralba, A.: Learning deep features for discriminative localization. CoRR abs/1512.04150 (2015). http://arxiv.org/abs/1512.04150
49. Zhou, H.H., Singh, V., Johnson, S.C., Wahba, G., Alzheimer's Disease Neuroimaging Initiative, et al.: Statistical tests and identifiability conditions for pooling and analyzing multisite datasets. Proc. Natl. Acad. Sci. **115**(7), 1481–1486 (2018)
50. Zuber-Skerritt, O., Cendon, E.: Critical reflection on professional development in the social sciences: interview results. Int. J. Res. Dev. **5**(1), 16–32 (2014)

Generating Videos of Zero-Shot Compositions of Actions and Objects

Megha Nawhal[1,2]([✉]), Mengyao Zhai[2], Andreas Lehrmann[1], Leonid Sigal[1,3,4,5], and Greg Mori[1,2]

[1] Borealis AI, Vancouver, Canada
[2] Simon Fraser University, Burnaby, Canada
mnawhal@sfu.ca
[3] University of British Columbia, Vancouver, Canada
[4] Vector Institute for AI, Toronto, Canada
[5] CIFAR AI Chair, Toronto, Canada

Abstract. Human activity videos involve rich, varied interactions between people and objects. In this paper we develop methods for generating such videos – making progress toward addressing the important, open problem of video generation in complex scenes. In particular, we introduce the task of generating human-object interaction videos in a zero-shot compositional setting, *i.e.*, generating videos for action-object compositions that are unseen during training, having seen the target action and target object separately. This setting is particularly important for generalization in human activity video generation, obviating the need to observe every possible action-object combination in training and thus avoiding the combinatorial explosion involved in modeling complex scenes. To generate human-object interaction videos, we propose a novel adversarial framework HOI-GAN which includes multiple discriminators focusing on different aspects of a video. To demonstrate the effectiveness of our proposed framework, we perform extensive quantitative and qualitative evaluation on two challenging datasets: EPIC-Kitchens and 20BN-Something-Something v2.

Keywords: Video generation · Compositionality in videos

1 Introduction

Visual imagination and prediction are fundamental components of human intelligence. Arguably, the ability to create realistic renderings from symbolic representations are considered prerequisite for broad visual understanding. Computer vision has seen rapid advances in the field of image generation over the past few years. Existing models are capable of generating impressive results

Electronic supplementary material The online version of this chapter (https://doi.org/10.1007/978-3-030-58610-2_23) contains supplementary material, which is available to authorized users.

© Springer Nature Switzerland AG 2020
A. Vedaldi et al. (Eds.): ECCV 2020, LNCS 12357, pp. 382–401, 2020.
https://doi.org/10.1007/978-3-030-58610-2_23

in this static scenario, ranging from hand-written digits [3,11,19] to realistic scenes [5,29,34,53,78]. Progress on *video generation* [4,25,57,64,66,69,70], on the other hand, has been relatively moderate and remains an open and challenging problem. While most approaches focus on the expressivity and controllability of the underlying generative models, their ability to generalize to unseen scene compositions has not received as much attention. However, such generalizability is an important cornerstone of robust visual imagination as it demonstrates the capacity to reason over elements of a scene.

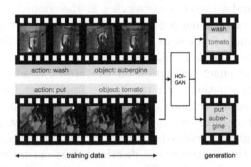

Fig. 1. Generation of zero-shot human-object interactions. Given training examples *"wash aubergine"* and *"put tomato"*, an intelligent agent should be able to imagine action sequences for unseen action-object compositions, *i.e.*, *"wash tomato"* and *"put aubergine"*.

We posit that the domain of human activities constitutes a rich realistic testbed for video generation models. Human activities involve people interacting with objects in complex ways, presenting numerous challenges for generation – the need to (1) render a variety of objects; (2) model the temporal evolution of the effect of actions on objects; (3) understand spatial relations and interactions; and (4) overcome the paucity of data for the complete set of action-object pairings. The last, in particular, is a critical challenge that also serves as an opportunity for designing and evaluating generative models that can generalize to myriad, possibly unseen, action-object compositions. For example, consider Fig. 1. The activity sequences for *"wash aubergine"* (action a_1: "wash"; object o_1: "aubergine") and *"put tomato"*(action a_2: "put"; object o_2: "tomato") are observed in the training data. A robust visual imagination would then allow an agent to imagine videos for *"wash tomato"* (a_1, o_2) and *"put aubergine"* (a_2, o_1).

We propose a novel framework for generating human-object interaction (HOI) videos for unseen action-object compositions. We refer to this task as *zero-shot HOI video generation*. To the best of our knowledge, our work is the first to propose and address this problem. In doing so, we push the envelope on conditional (or controllable) video generation and focus squarely on the model's ability to generalize to unseen action-object compositions. This zero-shot compositional setting verifies that the model is capable of semantic disentanglement of the action and objects in a given context and recreating them separately in other contexts.

The desiderata for performing zero-shot HOI video generation include: (1) mapping the content in the video to the right semantic category, (2) ensuring spatial and temporal consistency across the frames of a video, and (3) producing interactions with the right object in the presence of multiple objects. Based on these observations, we introduce a novel multi-adversarial learning scheme involving multiple discriminators, each focusing on different aspects of an HOI video. Our framework *HOI-GAN* generates a fixed length video clip given an action, an object, and a target scene serving as the context.

Concretely, the conditional inputs to our framework are semantic labels of action and object, and a single start frame with a mask providing the background and location for the object. Then, the model has to create the object, reason over the action, and enact the action on the object (leading to object translation and/or transformation) over the background, thus generating the whole interaction video. During training of the generator, our framework utilizes four discriminators – three pixel-centric discriminators, namely, *frame* discriminator, *gradient* discriminator, *video* discriminator; and one object-centric *relational* discriminator. The three pixel-centric discriminators ensure spatial and temporal consistency across the frames. The novel relational discriminator leverages spatio-temporal scene graphs to reason over the object layouts in videos ensuring the right interactions among objects. Through experiments, we show that our HOI-GAN framework is able to disentangle objects and actions and learns to generate videos with unseen compositions.

In summary, our contributions are as follows:

- We introduce the task of zero-shot HOI video generation. Specifically, given a training set of videos depicting certain action-object compositions, we propose to generate unseen compositions having seen the target action and target object individually, *i.e.*, the target action was paired with a different object and the target object was involved in a different action.
- We propose a novel adversarial learning scheme and introduce our HOI-GAN framework to generate HOI videos in a zero-shot compositional setting.
- We demonstrate the effectiveness of HOI-GAN through empirical evaluation on two challenging HOI video datasets: *20BN-something-something v2* [20] and *EPIC-Kitchens* [9]. We perform both quantitative and qualitative evaluation of the proposed approach and compare with state-of-the-art approaches.

Overall, our work facilitates research in the direction of enhancing generalizability of generative models for complex videos.

2   Related Work

Our paper builds on prior work in: (1) modeling of human-object interactions and (2) GAN-based video generation. In addition, we also discuss literature relevant to HOI video generation in a zero-shot compositional setting.

Modeling Human-Object Interactions. Earlier research attempts to study human-object interactions (HOIs) aimed at studying object affordances [21,38] and semantic-driven understanding of object functionalities [24,62]. Recent work on modeling HOIs in images range from studying semantics and spatial features of interactions between humans and objects [10,18,77] to action information [13,17,76]. Furthermore, there have been attempts to create large scale image and video datasets to study HOI [7,8,20,39]. To model dynamics in HOIs, recent works have proposed methods that jointly model actions and objects in videos [33,35,60]. Inspired by these approaches, we model HOI videos as compositions of actions and objects.

GAN-Based Image & Video Generation. Generative Adversarial Network (GAN) [19] and its variants [3,11,79] have shown tremendous progress in high quality image generation. Built over these techniques, conditional image generation using various forms of inputs to the generator such as textual information [55,75,78], category labels [48,52], and images [29,36,43,80] have been widely studied. This class of GANs allows the generator network to learn a mapping between conditioning variables and the real data distribution, thereby allowing control over the generation process. Extending these efforts to conditional video generation is not straightforward as generating a video involves modeling of both spatial and temporal variations. Vondrick et al. [66] proposed the Video GAN (VGAN) framework to generate videos using a two-stream generator network that decouples foreground and background of a scene. Temporal GAN (TGAN) [57] employs a separate generator for each frame in a video and an additional generator to model temporal variations across these frames. MoCo-GAN [64] disentangles the latent space representations of motion and content in a video to perform controllable video generation using seen compositions of motion and content as conditional inputs. In our paper, we evaluate the extent to which these video generation methods generalize when provided with unseen scene compositions as conditioning variables. Furthermore, promising success has been achieved by recent video-to-video translation methods [4,69,70] wherein video generation is conditioned on a corresponding semantic video. In contrast, our task does not require semantic videos as conditional input.

Video Prediction. Video prediction approaches predict future frames of a video given one or a few observed frames using RNNs [61], variational auto-encoders [67,68], adversarial training [42,46], or auto-regressive methods [32]. While video prediction is typically posed as an image-conditioned (past frame) image generation (future frame) problem, it is substantially different from video generation where the goal is to generate a video clip given a stochastic latent space.

Video Inpainting. Video inpainting/completion refers to the problem of correctly filling up the missing pixels given a video with arbitrary spatio-temporal pixels missing [14,22,50,51,59]. In our setting, however, the model only receives a single static image as input and not a video. Our model is required to go beyond merely filling in pixel values and has to produce an output video with the right visual content depicting the prescribed action upon a synthesized object. In doing so, the background may, and in certain cases should, evolve as well.

Zero-Shot Learning. Zero-shot learning (ZSL) aims to solve the problem of recognizing classes whose instances are not seen during training. In ZSL, external information of a certain form is required to share information between classes to transfer knowledge from seen to unseen classes. A variety of techniques have been used for ZSL ranging from usage of attribute-based information [16,40], word embeddings [71] to WordNet hierarchy [1] and text-based descriptions [15, 23,41,81]. [72] provides a thorough overview of zero-shot learning techniques. Similar to these works, we leverage word embeddings to reason over the unseen compositions of actions and objects in the context of video generation.

Learning Visual Relationships. Visual relationships in the form of scene graphs, *i.e.*, directed graphs representing relationships (edges) between the objects (nodes) have been used for image caption evaluation [2], image retrieval [31] and predicting scene compositions for images [44,49,74]. Furthermore, in a generative setting, [30] aims to synthesize an image from a given scene graph and evaluate the generalizability of an adversarial network to create images with unseen relationships between objects. Similarly, we leverage spatio-temporal scene graphs to learn relevant relations among the objects and focus on the generalizability of video generation models to unseen compositions of actions and objects. However, our task of zero-shot HOI video generation is more difficult as it requires learning to map the inputs to spatio-temporal variations in a video.

Learning Disentangled Representations for Videos. Various methods have been proposed to learn disentangled representations in videos [12,27,64], such as, learning representations by decoupling the content and pose [12], or separating motion from content using image differences [65]. Similarly, our model implicitly learns to disentangle the action and object information of an HOI video.

3   HOI-GAN

Intuitively, for a generated human-object interaction (HOI) video to be realistic, it must: (1) contain the object designated by a semantic label; (2) exhibit the prescribed interaction with that object; (3) be temporally consistent; and (4 – optional) occur in a specified scene. Based on this intuition, we propose an adversarial learning scheme in which we train a generator network $\mathbf{G}$ with a set of 4 discriminators: (1) a frame discriminator $\mathbf{D}_f$, which encourages the generator to learn spatially coherent visual content; (2) a gradient discriminator $\mathbf{D}_g$, which incentivizes $\mathbf{G}$ to produce temporally consistent frames; (3) a video discriminator $\mathbf{D}_v$, which provides the generator with global spatio-temporal context; and (4) a relational discriminator $\mathbf{D}_r$, which assists the generator in producing correct object layouts in a video. We use pretrained word embeddings [54] for semantic representations of actions and objects. All discriminators are conditioned on word embeddings of the action ($\mathbf{s}_a$) and object ($\mathbf{s}_o$) and trained simultaneously in an end-to-end manner. Figure 2 shows an overview of our proposed framework *HOI-GAN*. We now formalize our task and describe each module in detail.

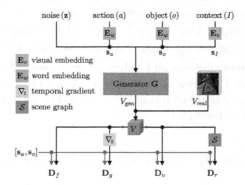

Fig. 2. Architecture overview. The generator network $\mathbf{G}$ is trained using 4 discriminators simultaneously: a frame discriminator $\mathbf{D}_f$, a gradient discriminator $\mathbf{D}_g$, a video discriminator $\mathbf{D}_v$, and a relational discriminator $\mathbf{D}_r$. Given the word embeddings of an action $\mathbf{s}_a$, an object $\mathbf{s}_o$, and a context image $\mathbf{s}_I$, the generator learns to synthesize a video with background I in which the action a is performed on the object o.

3.1 Task Formulation

Let $\mathbf{s}_a$ and $\mathbf{s}_o$ be word embeddings of an action a and an object o, respectively. Furthermore, let I be an image provided as context to the generator. We encode I using an encoder $\mathbf{E}_v$ to obtain a visual embedding $\mathbf{s}_I$, which we refer to as a context vector. Our goal is to generate a video $V = (V^{(i)})_{i=1}^{T}$ of length T depicting the action a performed on the object o with context image I as the background of V. To this end, we learn a function $\mathbf{G} : (\mathbf{z}, \mathbf{s}_a, \mathbf{s}_o, \mathbf{s}_I) \mapsto V$, where $\mathbf{z}$ is a noise vector sampled from a distribution $p_{\mathbf{z}}$, such as a Gaussian distribution.

3.2 Model Description

We describe the elements of our framework below. Overall, the four discriminator networks, *i.e.*, frame discriminator $\mathbf{D}_f$, gradient discriminator $\mathbf{D}_g$, video discriminator $\mathbf{D}_v$, and relational discriminator $\mathbf{D}_r$ are all involved in a zero-sum game with the generator network $\mathbf{G}$. Refer to the supplementary for implementation details.

Frame Discriminator. The frame discriminator network $\mathbf{D}_f$ learns to distinguish between real and generated frames corresponding to the real video V_{real} and generated video $V_{\text{gen}} = \mathbf{G}(\mathbf{z}, \mathbf{s}_a, \mathbf{s}_o, \mathbf{s}_I)$ respectively. Each frame in V_{gen} and V_{real} is processed independently using a network consisting of stacked `conv2d` layers, *i.e.*, 2D convolutional layers followed by spectral normalization [47] and leaky ReLU layers [45] with $a = 0.2$. We obtain a tensor of size $N^{(t)} \times w_0^{(t)} \times h_0^{(t)}$ $(t = 1, 2, \ldots, T)$, where $N^{(t)}$, $w_0^{(t)}$, and $h_0^{(t)}$ are the channel length, width and height of the activation of the last conv2d layer respectively. We concatenate this tensor with spatially replicated copies of $\mathbf{s}_a$ and $\mathbf{s}_o$, which results in a tensor of size $(\dim(\mathbf{s}_a) + \dim(\mathbf{s}_o) + N^{(t)}) \times w_0^{(t)} \times h_0^{(t)}$. We then apply another conv2d layer to obtain a $N \times w_0^{(t)} \times h_0^{(t)}$ tensor. We now perform 1×1 convolutions followed by $w_0^{(t)} \times h_0^{(t)}$ convolutions and a sigmoid to obtain a T-dimensional vector corresponding to the T frames of the video V. The i-th element of the output denotes the probability that the frame $V^{(i)}$ is real. The objective function of the network $\mathbf{D}_f$ is the loss function:

$$L_f = \frac{1}{2T} \sum_{i=1}^{T} [\log(\mathbf{D}_f^{(i)}(V_{\text{real}}; \mathbf{s}_a, \mathbf{s}_o)) + \log(1 - \mathbf{D}_f^{(i)}(V_{\text{gen}}; \mathbf{s}_a, \mathbf{s}_o))], \qquad (1)$$

where $\mathbf{D}_f^{(i)}$ is the i-th element of the output of $\mathbf{D}_f$.

Gradient Discriminator. The gradient discriminator network $\mathbf{D}_g$ enforces temporal smoothness by learning to differentiate between the temporal gradient of a real video V_{real} and a generated video V_{gen}. We define the temporal gradient $\nabla_t V$ of a video V with T frames $V^{(1)}, \ldots, V^{(T)}$ as pixel-wise differences between two consecutive frames of the video. The i-th element of $\nabla_t V$ is defined as:

$$[\nabla_t V]_i = V^{(i+1)} - V^{(i)}, \quad i = 1, 2, \ldots, (T-1). \qquad (2)$$

The architecture of the gradient discriminator $\mathbf{D}_g$ is similar to that of the frame discriminator $\mathbf{D}_f$. The output of $\mathbf{D}_g$ is a $(T-1)$-dimensional vector corresponding to the $(T-1)$ values in gradient $\nabla_t V$. The objective function of $\mathbf{D}_g$ is

$$L_g = \frac{1}{2(T-1)} \sum_{i=1}^{T-1} [\log(\mathbf{D}_g^{(i)}(\nabla_t V_{\text{real}}; \mathbf{s}_a, \mathbf{s}_o)) +$$
$$\log(1 - \mathbf{D}_g^{(i)}(\nabla_t V_{\text{gen}}; \mathbf{s}_a, \mathbf{s}_o))], \qquad (3)$$

where $\mathbf{D}_g^{(i)}$ is the i-th element of the output of $\mathbf{D}_g$.

Video Discriminator. The video discriminator network $\mathbf{D}_v$ learns to distinguish between real videos V_{real} and generated videos V_{gen} by comparing their global spatio-temporal contexts. The architecture consists of stacked conv3d layers, *i.e.*, 3D convolutional layers followed by spectral normalization [47] and leaky ReLU layers [45] with $a = 0.2$. We obtain a $N \times d_0 \times w_0 \times h_0$ tensor, where N, d_0, w_0, and h_0 are the channel length, depth, width, and height of the activation of the last conv3d layer respectively. We concatenate this tensor with spatially replicated copies of $\mathbf{s}_a$ and $\mathbf{s}_o$, which results in a tensor of size $(\dim(\mathbf{s}_a) + \dim(\mathbf{s}_o) + N) \times d_0 \times w_0 \times h_0$, where $\dim(\cdot)$ returns the dimensionality of a vector. We then apply another conv3d layer to obtain a $N \times d_0 \times w_0 \times h_0$

tensor. Finally, we apply a $1 \times 1 \times 1$ convolution followed by a $d_0 \times w_0 \times h_0$ convolution and a sigmoid to obtain the output, which represents the probability that the video V is real. The objective function of the network $\mathbf{D}_v$ is the following loss function:

$$L_v = \frac{1}{2}[\log(\mathbf{D}_v(V_{\text{real}}; \mathbf{s}_a, \mathbf{s}_o)) + \log(1 - \mathbf{D}_v(V_{\text{gen}}; \mathbf{s}_a, \mathbf{s}_o))]. \tag{4}$$

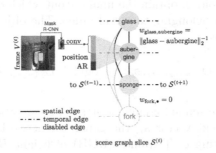

Fig. 3. Relational discriminator. The relational discriminator $\mathbf{D}_r$ leverages a spatiotemporal scene graph to distinguish between object layouts in videos. Each node contains convolutional embedding, position and aspect ratio (AR) of the object crop obtained from MaskRCNN. The nodes are connected in space and time and edges are weighted based on their inverse distance. Edge weights of (dis)appearing objects are 0.

Relational Discriminator. In addition to the three pixel-centric discriminators above, we also propose a novel object-centric discriminator $\mathbf{D}_r$. Driven by a spatio-temporal scene graph, this relational discriminator learns to distinguish between scene layouts of real videos V_{real} and generated videos V_{gen} (Fig. 3).

Specifically, we build a spatio-temporal scene graph $\mathcal{S} = (\mathcal{N}, \mathcal{E})$ from V, where the nodes and edges are represented by $\mathcal{N}$ and $\mathcal{E}$ respectively. We assume one node per object per frame. Each node is connected to all other nodes in the same frame, referred to as spatial edges. In addition, to represent temporal evolution of objects, each node is connected to the corresponding nodes in the adjacent frames that also depict the same object, referred to as temporal edges. To obtain the node representations, we crop the objects in V using Mask-RCNN [26], compute a convolutional embedding for them, and augment the resulting vectors with the aspect ratio (AR) and position of the corresponding bounding boxes. The weights of spatial edges in $\mathcal{E}$ are given by inverse Euclidean distances between the centers of these bounding boxes corresponding to the object appearing in the frame. The weights of the temporal edges is set to 1 by default. When an object is not present in a frame (but appears in the overall video), spatial edges connecting to the object will be absent by design. This is implemented by setting the weights to 0 depicting distance between the objects as ∞. Similarly, if an object does not appear in the adjacent frame, the temporal edge is set to 0.

In case of multiple objects of the same category, the correspondence is established based on the location in the adjacent frames using nearest neighbour data association.

The relational discriminator $\mathbf{D}_r$ operates on this scene graph $\mathcal{S}$ by virtue of a graph convolutional network (GCN) [37] followed by stacking and average-pooling of the resulting node representations along the time axis. We then concatenate this tensor with spatially replicated copies of $\mathbf{s}_a$ and $\mathbf{s}_o$ to result in a tensor of size $(\dim(\mathbf{s}_a) + \dim(\mathbf{s}_o) + N^{(t)}) \times w_0^{(t)} \times h_0^{(t)}$. As before, we then apply convolutions and sigmoid to obtain the final output which denotes the probability of the scene graph belonging to a real video. The objective function of the network $\mathbf{D}_r$ is given by

$$L_r = \frac{1}{2}[\log(\mathbf{D}_r(\mathcal{S}_{\text{real}}; \mathbf{s}_a, \mathbf{s}_o)) + \log(1 - \mathbf{D}_r(\mathcal{S}_{\text{gen}}; \mathbf{s}_a, \mathbf{s}_o))]. \tag{5}$$

Generator. Given the semantic embeddings $\mathbf{s}_a$, $\mathbf{s}_o$ of action and object labels respectively, and context vector $\mathbf{s}_I$, the generator network $\mathbf{G}$ learns to generate video V_{gen} consisting of T frames (RGB) of height H and width W. We concatenate noise $\mathbf{z}$ with the conditions, namely, $\mathbf{s}_a$, $\mathbf{s}_o$, and $\mathbf{s}_I$. We provide this concatenated vector as the input to the network $\mathbf{G}$. The network comprises stacked `deconv3d` layers, *i.e.*, 3D transposed convolution layers each followed by Batch Normalization [28] and leaky ReLU layers [45] with $a = 0.2$ except the last convolutional layer which is followed by a Batch Normalization layer [28] and a `tanh` activation layer. The network is optimized according to the following objective function:

$$L_{gan} = \frac{1}{T} \sum_{i=1}^{T}[\log(1 - \mathbf{D}_f^{(i)}(V_{\text{gen}}; \mathbf{s}_a, \mathbf{s}_o))]+$$
$$\frac{1}{(T-1)} \sum_{i=1}^{T-1}[\log(1 - \mathbf{D}_g^{(i)}(\nabla_t V_{\text{gen}}; \mathbf{s}_a, \mathbf{s}_o))]+ \tag{6}$$
$$\log(1 - \mathbf{D}_v(V_{\text{gen}}; \mathbf{s}_a, \mathbf{s}_o)) + \log(1 - \mathbf{D}_r(\mathcal{S}_{\text{gen}}; \mathbf{s}_a, \mathbf{s}_o)).$$

4   Experiments

We conduct quantitative and qualitative analysis to demonstrate the effectiveness of the proposed framework HOI-GAN for the task of zero-shot generation of human-object interaction (HOI) videos.

4.1   Datasets and Data Splits

We use two datasets for our experiments: EPIC-Kitchens [9] and 20BN-Something-Something V2 [20]. Both of these datasets comprise a diverse set of HOI videos ranging from simple translational motion of objects (*e.g..* push, move) and rotation (*e.g..* open) to transformations in state of objects (*e.g..* cut, fold). Therefore, these datasets, with their wide ranging variety and complexity, provide a challenging setup for evaluating HOI video generation models.

EPIC-Kitchens [9] contains egocentric videos of activities in several kitchens. A video clip V is annotated with action label a and object label o (*e.g.*. open microwave, cut apple, move pan) along with a set of bounding boxes $\mathcal{B}$ (one per frame) for objects that the human interacts with while performing the action. There are around 40k instances in the form of $(V, a, o, \mathcal{B})$ across 352 objects and 125 actions. We refer to this dataset as EPIC hereafter.

20BN-Something-Something V2 [20] contains videos of daily activities performed by humans. A video clip V is annotated with a label l, an action template and object(s) on which the action is applied (*e.g.*. 'hitting ball with racket' has action template 'hitting something with something'). There are 220,847 training instances of the form (V, l) spanning 30,408 objects and 174 action templates. To transform l to action-object label pair (a, o), we use NLTK POS-tagger. We consider the verb tag (after stemming) in l as action label a. We observe that all instances of l begin with the present continuous form of a which is acting upon the subsequent noun. Therefore, we use the noun that appears immediately after the verb as object o. Hereafter, we refer to the transformed dataset in the form of (V, a, o) as SS.

Splitting by Compositions. We believe it is reasonable to only generate combinations that are semantically feasible, and do so by only using action-object pairs seen in the original datasets. We use a subset of action-object pairs as testing pairs – these pairs are not seen during training but are present in the original dataset, hence are semantically feasible. To make the dataset training/testing splits suitable for our zero-shot compositional setting, we first merge the data samples present in the default train and validation sets of the dataset. We then split the combined dataset into training set and test set based on the condition that all the unique object and action labels in appear in the training set, however, any composition of action and object present in the test set is absent in training set and vice versa. We provide further details of the splits for both datasets EPIC and SS in the supplementary.

Generation Scenarios. Recall that the generator network in the HOI-GAN framework (Fig. 2) has 3 conditional inputs, namely, action embedding, object embedding, and context frame I. The context frame serves as the background in the scene. Thus, to provide this context frame during training, we apply a binary mask $M^{(1)}$ corresponding to the first frame $V^{(1)}$ of a real video as $I = (\mathbb{1} - M^{(1)}) \odot V^{(1)}$, where $\mathbb{1}$ represents a matrix of size $M^{(1)}$ containing all ones and $\odot$ denotes elementwise multiplication. This mask $M^{(1)}$ contains ones in regions (either rectangular bounding boxes or segmentation masks) corresponding to the objects (non-*person* classes) detected using MaskRCNN [26] and zeros for other regions. Intuitively, this helps ensure the generator learns to map the action and object embeddings to relevant visual content in the HOI video.

During testing, to evaluate the generator's capability to synthesize the right human-object interactions, we provide a background frame as described above. This background frame can be selected from either the test set or training set,

Table 1. Generation scenarios. Description of the conditional inputs for the two generation scenarios GS1 & GS2 used for evaluation. ✓ denotes 'Yes', ✗ denotes 'No'.

Target conditions	GS1	GS2
Target action a seen during training	✓	✓
Target object o seen during training	✓	✓
Background of target context I seen during training	✗	✓
Object mask in target context I corresponds to target object o	✓	✗
Target action a seen with target context I during training	✗	✓/✗
Target object o seen with target context I during training	✗	✗
Target action-object composition $(a\text{-}o)$ seen during training	✗	✗

and can be suitable or unsuitable for the target action-object composition. To capture these possibilities, we design two different generation scenarios. Specifically, in *Generation Scenario 1 (GS1)*, the input context frame I is the masked first frame of a video from the test set corresponding to the target action-object composition (unseen during training). In *Generation Scenario 2 (GS2)*, I is the masked first frame of a video from the training set which depicts an object other than the target object. The original action in this video could be same or different than the target action. See Table 1 for the contrast between the scenarios.

As such, in GS1, the generator receives a context that it has not seen during training but the context (including object mask) is consistent with the target action-object composition it is being asked to generate. In contrast, in GS2, the generator receives a context frame that it has seen during training but is not consistent with the action-object composition it is being asked to generate. Particularly, the object mask in the context does not correspond to the target object. Although the background is seen, the model has to evolve the background in ways different from training samples to make it suitable for the target composition. Thus, these generation scenarios help illustrate that the generator indeed generalizes over compositions.

4.2   Evaluation Setup

Evaluation of image/video quality is inherently challenging, thus, we use both quantitative and qualitative metrics.

Quantitative Metrics. Inception Score (**I-score**) [58] is a widely used metric for evaluating image generation models. For images x with labels y, I-score is defined as $\exp(\mathbf{KL}(\rho(y|x)||\rho(y)))$ where $\rho(y|x)$ is the conditional label distribution of an ImageNet [56] -pretrained Inception model [63]. We adopted this metric for video quality evaluation. We fine-tune a Kinetics [6]-pretrained video classifier ResNeXt-101 [73] for each of our source datasets and use it for calculating I-score (higher is better). It is based on one of the state-of-the-art video

Scenario	(a, o)	Context	Generated Output				
GS1	take spoon (EPIC)						
GS1	hold cup (SS)						
GS2	move broccoli (EPIC)						
GS2	put apple (SS)						

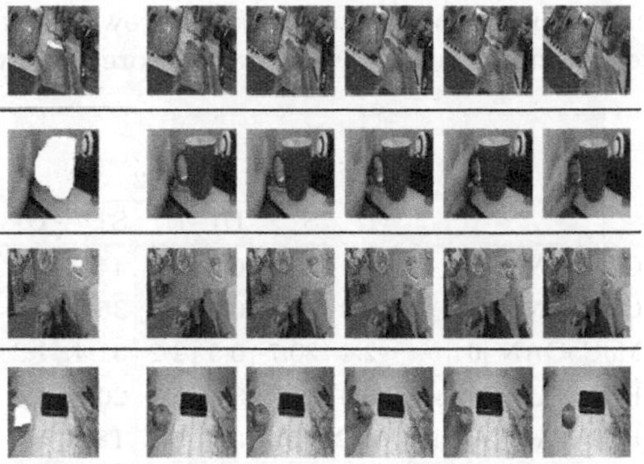

Fig. 4. Qualitative results: Videos generated using our best version of HOI-GAN using embeddings for action (a)-object (o) composition and the context frame. We show 5 frames of the video clip generated for both generation scenarios GS1 and GS2. The context frame in GS1 is obtained from a video in the test set depicting an action-object composition same as the target one. The context frame for GS2 scenarios shown here are from videos depicting *"take carrot"* (for row 3) and *"put bowl"* (for row 4). Refer to supplementary section for additional videos generated using HOI-GAN.

classification architectures. We used the same evaluation setup for the baselines and our model to ensure a fair comparison.

In addition, we believe that measuring realism explicitly is more relevant for our task as the generation process can be conditioned on any context frame arbitrarily to obtain diverse samples. Therefore, in addition to *I-score*, we also analyze the first and second terms of the KL divergence separately. We refer to these terms as: (1) Saliency score or **S-score** (lower is better) to specifically measure realism, and (2) Diversity score or **D-score** (higher is better) to indicate the diversity in generated samples. A smaller value of S-score implies that the generated videos are more realistic as the classifier is very confident in classifying the generated videos. Specifically, the saliency score will have a low value (low is good) only when the classifier is confidently able to classify the generated videos into action-object categories matching the conditional input composition (action-object), thus indicating realistic instances of the required target interaction. In fact, even if a model generates realistic-looking videos but depicts an action-object composition not corresponding to the conditional action-object input, the saliency score will have high values. Finally, a larger value of D-score implies the model generates diverse samples.

Human Preference Score. We conduct a user study for evaluating the quality of generated videos. In each test, we present the participants with two videos generated by two different algorithms and ask which among the two better depicts the given activity, *i.e.*, action-object composition (*e.g.*, lift fork). We evaluate the performance of an algorithm as the overall percentage of tests in which that

Table 2. Quantitative evaluation. Comparison of HOI-GAN with C-VGAN, C-TGAN, and MoCoGAN baselines. We distinguish training of HOI-GAN with bounding boxes (*bboxes*) and segmentation masks (*masks*). Arrows indicate whether lower ($\downarrow$) or higher ($\uparrow$) is better. [I: inception score; S: saliency score; D: diversity score]

Model		EPIC						SS					
		GS1			GS2			GS1			GS2		
		I$\uparrow$	S$\downarrow$	D$\uparrow$	I$\uparrow$	S$\downarrow$	D$\uparrow$	I$\uparrow$	S$\downarrow$	D$\uparrow$	I$\uparrow$	S$\downarrow$	D$\uparrow$
	C-VGAN [66]	1.8	30.9	0.2	1.4	44.9	0.3	2.1	25.4	0.4	1.8	40.5	0.3
	C-TGAN [57]	2.0	30.4	0.6	1.5	35.9	0.4	2.2	28.9	0.6	1.6	39.7	0.5
	MoCoGAN [64]	2.4	30.7	0.5	2.2	31.4	1.2	2.8	17.5	1.0	2.4	33.7	1.4
(Ours)	HOI-GAN (bboxes)	6.0	14.0	3.4	5.7	20.8	4.0	6.6	12.7	3.5	6.0	15.2	2.9
	HOI-GAN (masks)	**6.2**	**13.2**	**3.7**	**5.2**	**18.3**	**3.5**	**8.6**	**11.4**	**4.4**	**7.1**	**14.7**	**4.0**

algorithm's outputs are preferred. This is an aggregate measure over all the test instances across all participants.

Baselines. We compare HOI-GAN with three state-of-the-art video generation approaches: (1) VGAN [66], (2) TGAN, [57] and (3) MoCoGAN [64]. We develop the conditional variants of VGAN and TGAN from the descriptions provided in their papers. We refer to the conditional variants as C-VGAN and C-TGAN respectively. We observed that these two models saturated easily in the initial iterations, thus, we added dropout in the last layer of the discriminator network in both models. MoCoGAN focuses on disentangling motion and content in the latent space and is the closest baseline. We use the code provided by the authors.

4.3 Results

Next, we discuss the results of our qualitative and quantitative evaluation.

Comparison with Baselines. As shown in Table 2, HOI-GAN with different conditional inputs outperforms C-VGAN and C-TGAN by a wide margin in both generation scenarios. In addition, our overall model shows considerable improvement over MoCoGAN, while MoCoGAN has comparable scores to some ablated versions of our models (where gradient discriminator and/or relational discriminator is missing). Furthermore, we varied the richness of the masks in the conditional input context frame ranging from bounding boxes to segmentation masks obtained corresponding to non-*person* classes using MaskRCNN framework [26]. We observe that providing masks during training leads to slight improvements in both scenarios as compared to using bounding boxes (refer to Table 2). We also show the samples generated using the best version of HOI-GAN for the two generation scenarios (Fig. 4). See supplementary for more generated samples and detailed qualitative analysis.

Ablation Study. To illustrate the impact of each discriminator in generating HOI videos, we conduct ablation experiments (refer to Table 3). We observe

Table 3. Ablation study. We evaluate the contributions of our pixel-centric losses (F, G, V) and relational losses (first block vs. second block) by conducting ablation study on HOI-GAN (masks). The last row corresponds to the overall proposed model. [F: frame discriminator $\mathbf{D}_f$; G: gradient discriminator $\mathbf{D}_g$; V: video discriminator $\mathbf{D}_v$; R: relational discriminator $\mathbf{D}_r$]

Model		EPIC						SS					
		GS1			GS2			GS1			GS2		
		I↑	S↓	D↑	I↑	S↓	D↑	I↑	S↓	D↑	I↑	S↓	D↑
−R	HOI-GAN (F)	1.4	44.2	0.2	1.1	47.2	0.3	1.8	34.7	0.4	1.5	39.5	0.3
	HOI-GAN (F+G)	2.3	25.6	0.7	1.9	30.7	0.5	3.0	24.5	0.9	2.7	28.8	0.7
	HOI-GAN (F+G+V)	2.8	21.2	1.3	2.6	29.7	1.7	3.3	18.6	1.2	3.0	20.7	1.0
+R	HOI-GAN (F)	2.4	24.9	0.8	2.2	26.0	0.7	3.1	20.3	1.0	2.9	27.7	0.9
	HOI-GAN (F+G)	5.9	15.4	3.5	4.8	21.3	3.3	7.4	12.1	3.5	5.4	19.2	3.4
	HOI-GAN (F+G+V)	6.2	13.2	3.7	5.2	18.3	3.5	8.6	11.4	4.4	7.1	14.7	4.0

that the addition of temporal information using the gradient discriminator and spatio-temporal information using the video discriminator lead to improvement in generation quality. In particular, the addition of our scene graph based relational discriminator leads to considerable improvement in generation quality resulting in more realistic videos (refer to second block in Table 3). Additional quantitative studies and results are in the supplementary.

Human Evaluation. We recruited 15 sequestered participants for our user study. We randomly chose 50 unique categories and chose generated videos for half of them from generation scenario GS1 and the other half from GS2. For each category, we provided three instances, each containing a pair of videos; one generated using a baseline model and the other using HOI-GAN. For each instance, at least 3 participants (ensuring inter-rater reliability) are asked to choose the video that best depicts the given category. The (aggregate) human preference scores for our model versus the baselines range between 69–84% for both generation scenarios (refer Table 4). These results indicate that HOI-GAN generates more realistic videos than the baselines.

Table 4. Human evaluation. Human preference score (%) for scenarios GS1 and GS2. All the results have p-value less than 0.05 implying statistical significance.

Ours/baseline	GS1	GS2
HOI-GAN/MoCoGAN	**71.7**/28.3	**69.2**/30.8
HOI-GAN/C-TGAN	**75.4**/34.9	**79.3**/30.7
HOI-GAN/C-VGAN	**83.6**/16.4	**80.4**/19.6

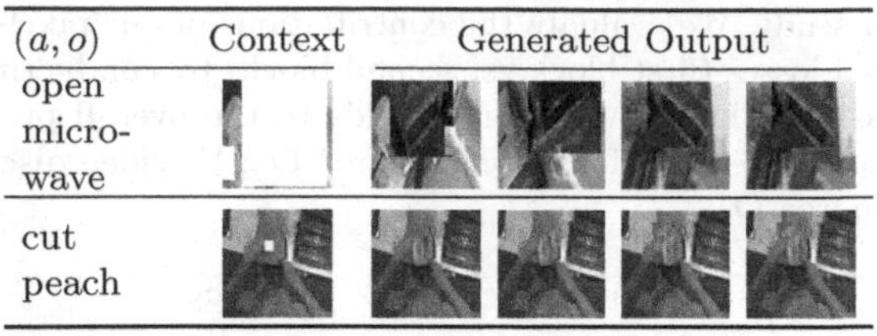

(a, o)	Context	Generated Output			
open micro-wave					
cut peach					

Fig. 5. Failure cases. Videos generated using HOI-GAN corresponding to the given action-object composition (a, o) and the context frame. We show 4 frames of the videos.

Failure Cases. We discuss the limitations of our framework using qualitative examples shown in Fig. 5. For *"open microwave"*, we observe that although HOI-GAN is able to generate conventional colors for a microwave, it shows limited capability to hallucinate such large objects. For *"cut peach"* (Fig. 5), the generated sample shows that our model can learn the increase in count of partial objects corresponding to the action cut and yellow-green color of a peach. However, as the model has not observed the interior of a peach during training (as *cut peach* was not in training set), it is unable to create realistic transformations in the state of *peach* that show the interior clearly. We provide additional discussion on the failure cases in the supplementary.

5 Conclusion

In this paper, we introduced the task of zero-shot HOI video generation, *i.e.*, generating human-object interaction (HOI) videos corresponding to unseen action-object compositions, having seen the target action and target object independently. Towards this goal, we proposed the HOI-GAN framework that uses a novel multi-adversarial learning scheme and demonstrated its effectiveness on challenging HOI datasets. We show that an object-level relational discriminator is an effective means for GAN-based generation of interaction videos. Future work can benefit from our idea of using relational adversaries to synthesize more realistic videos. We believe relational adversaries to be relevant beyond video generation in tasks such as layout-to-image translation.

Acknowledgements. This work was done when Megha Nawhal was an intern at Borealis AI. We would like to thank the Borealis AI team for participating in our user study.

References

1. Akata, Z., Reed, S., Walter, D., Lee, H., Schiele, B.: Evaluation of output embeddings for fine-grained image classification. In: IEEE Conference on Computer Vision and Pattern Recognition (CVPR) (2015)

2. Anderson, P., Fernando, B., Johnson, M., Gould, S.: SPICE: semantic propositional image caption evaluation. In: Leibe, B., Matas, J., Sebe, N., Welling, M. (eds.) ECCV 2016. LNCS, vol. 9909, pp. 382–398. Springer, Cham (2016). https://doi.org/10.1007/978-3-319-46454-1_24

3. Arjovsky, M., Chintala, S., Bottou, L.: Wasserstein generative adversarial networks. In: International Conference on Machine Learning (ICML) (2017)

4. Bansal, A., Ma, S., Ramanan, D., Sheikh, Y.: Recycle-GAN: unsupervised video retargeting. In: Ferrari, V., Hebert, M., Sminchisescu, C., Weiss, Y. (eds.) ECCV 2018. LNCS, vol. 11209, pp. 122–138. Springer, Cham (2018). https://doi.org/10.1007/978-3-030-01228-1_8

5. Brock, A., Donahue, J., Simonyan, K.: Large scale GAN training for high fidelity natural image synthesis. In: International Conference on Learning Representations (ICLR) (2019)

6. Carreira, J., Zisserman, A.: Quo vadis, action recognition? A new model and the kinetics dataset. In: IEEE Conference on Computer Vision and Pattern Recognition (CVPR) (2017)

7. Chao, Y.W., Liu, Y., Liu, X., Zeng, H., Deng, J.: Learning to detect human-object interactions. In: IEEE Winter Conference on Applications of Computer Vision (WACV) (2018)

8. Chao, Y.W., Wang, Z., He, Y., Wang, J., Deng, J.: HICO: a benchmark for recognizing human-object interactions in images. In: IEEE International Conference on Computer Vision (ICCV) (2015)

9. Damen, D., et al.: Scaling egocentric vision: the EPIC-KITCHENS dataset. In: European Conference on Computer Vision (ECCV) (2018)

10. Delaitre, V., Fouhey, D.F., Laptev, I., Sivic, J., Gupta, A., Efros, A.A.: Scene semantics from long-term observation of people. In: Fitzgibbon, A., Lazebnik, S., Perona, P., Sato, Y., Schmid, C. (eds.) ECCV 2012. LNCS, vol. 7577, pp. 284–298. Springer, Heidelberg (2012). https://doi.org/10.1007/978-3-642-33783-3_21

11. Denton, E.L., Chintala, S., Fergus, R., et al.: Deep generative image models using a Laplacian pyramid of adversarial networks. In: Advances in Neural Information Processing Systems (NIPS) (2015)

12. Denton, E.L., et al.: Unsupervised learning of disentangled representations from video. In: Advances in Neural Information Processing Systems (NIPS) (2017)

13. Desai, C., Ramanan, D.: Detecting actions, poses, and objects with relational phraselets. In: Fitzgibbon, A., Lazebnik, S., Perona, P., Sato, Y., Schmid, C. (eds.) ECCV 2012. LNCS, vol. 7575, pp. 158–172. Springer, Heidelberg (2012). https://doi.org/10.1007/978-3-642-33765-9_12

14. Ebdelli, M., Le Meur, O., Guillemot, C.: Video inpainting with short-term windows: application to object removal and error concealment. IEEE Trans. Image Process. 24(10), 3034–3047 (2015)

15. Elhoseiny, M., Saleh, B., Elgammal, A.: Write a classifier: zero-shot learning using purely textual descriptions. In: IEEE International Conference on Computer Vision (ICCV) (2013)

16. Farhadi, A., Endres, I., Hoiem, D., Forsyth, D.: Describing objects by their attributes. In: IEEE Conference on Computer Vision and Pattern Recognition (CVPR) (2009)

17. Fouhey, D.F., Delaitre, V., Gupta, A., Efros, A.A., Laptev, I., Sivic, J.: People watching: human actions as a cue for single view geometry. Int. J. Comput. Vis. (IJCV) 110, 259–274 (2014). https://doi.org/10.1007/s11263-014-0710-z

18. Gkioxari, G., Girshick, R., Dollár, P., He, K.: Detecting and recognizing human-object interactions. In: IEEE Conference on Computer Vision and Pattern Recognition (CVPR) (2018)
19. Goodfellow, I., et al.: Generative adversarial nets. In: Advances in Neural Information Processing Systems (NIPS) (2014)
20. Goyal, R., et al.: The "something something" video database for learning and evaluating visual common sense. In: IEEE International Conference on Computer Vision (ICCV) (2017)
21. Grabner, H., Gall, J., Van Gool, L.: What makes a chair a chair? In: IEEE Conference on Computer Vision and Pattern Recognition (CVPR) (2011)
22. Granados, M., Kim, K.I., Tompkin, J., Kautz, J., Theobalt, C.: Background inpainting for videos with dynamic objects and a free-moving camera. In: Fitzgibbon, A., Lazebnik, S., Perona, P., Sato, Y., Schmid, C. (eds.) ECCV 2012. LNCS, vol. 7572, pp. 682–695. Springer, Heidelberg (2012). https://doi.org/10.1007/978-3-642-33718-5_49
23. Guadarrama, S., et al.: YouTube2Text: recognizing and describing arbitrary activities using semantic hierarchies and zero-shot recognition. In: IEEE International Conference on Computer Vision (ICCV) (2013)
24. Gupta, A., Davis, L.S.: Objects in action: an approach for combining action understanding and object perception. In: IEEE Conference on Computer Vision and Pattern Recognition (CVPR) (2007)
25. He, J., Lehrmann, A., Marino, J., Mori, G., Sigal, L.: Probabilistic video generation using holistic attribute control. In: Ferrari, V., Hebert, M., Sminchisescu, C., Weiss, Y. (eds.) ECCV 2018. LNCS, vol. 11209, pp. 466–483. Springer, Cham (2018). https://doi.org/10.1007/978-3-030-01228-1_28
26. He, K., Gkioxari, G., Dollár, P., Girshick, R.: Mask R-CNN. In: IEEE International Conference on Computer Vision (ICCV) (2017)
27. Hsieh, J.T., Liu, B., Huang, D.A., Fei-Fei, L.F., Niebles, J.C.: Learning to decompose and disentangle representations for video prediction. In: Advances in Neural Information Processing Systems (NIPS) (2018)
28. Ioffe, S., Szegedy, C.: Batch normalization: accelerating deep network training by reducing internal covariate shift. In: International Conference on Machine Learning (ICML) (2015)
29. Isola, P., Zhu, J.Y., Zhou, T., Efros, A.A.: Image-to-image translation with conditional adversarial networks. In: Proceeding of the IEEE Conference on Computer Vision and Pattern Recognition (CVPR) (2017)
30. Johnson, J., Gupta, A., Fei-Fei, L.: Image generation from scene graphs. In: IEEE Conference on Computer Vision and Pattern Recognition (CVPR) (2018)
31. Johnson, J., et al.: Image retrieval using scene graphs. In: IEEE Conference on Computer Vision and Pattern Recognition (CVPR) (2015)
32. Kalchbrenner, N., et al.: Video pixel networks. In: International Conference on Machine Learning (ICML) (2017)
33. Kalogeiton, V., Weinzaepfel, P., Ferrari, V., Schmid, C.: Joint learning of object and action detectors. In: IEEE International Conference on Computer Vision (ICCV). IEEE (2017)
34. Karras, T., Aila, T., Laine, S., Lehtinen, J.: Progressive growing of GANs for improved quality, stability, and variation. In: International Conference on Learning Representations (ICLR) (2018)

35. Kato, K., Li, Y., Gupta, A.: Compositional learning for human object interaction. In: Ferrari, V., Hebert, M., Sminchisescu, C., Weiss, Y. (eds.) Computer Vision – ECCV 2018. LNCS, vol. 11218, pp. 247–264. Springer, Cham (2018). https://doi.org/10.1007/978-3-030-01264-9_15

36. Kim, T., Cha, M., Kim, H., Lee, J.K., Kim, J.: Learning to discover cross-domain relations with generative adversarial networks (2017)

37. Kipf, T.N., Welling, M.: Semi-supervised classification with graph convolutional networks. In: International Conference on Learning Representations (ICLR) (2017)

38. Kjellström, H., Romero, J., Kragić, D.: Visual object-action recognition: Inferring object affordances from human demonstration. Comput. Vis. Image Underst. (CVIU) 115(1), 81–90 (2011)

39. Krishna, R., et al.: Visual genome: connecting language and vision using crowd-sourced dense image annotations. Int. J. Comput. Vis. (IJCV) 123, 32–73 (2017). https://doi.org/10.1007/s11263-016-0981-7

40. Lampert, C.H., Nickisch, H., Harmeling, S.: Learning to detect unseen object classes by between-class attribute transfer. In: IEEE Conference on Computer Vision and Pattern Recognition (CVPR) (2009)

41. Lei Ba, J., Swersky, K., Fidler, S., et al.: Predicting deep zero-shot convolutional neural networks using textual descriptions. In: IEEE International Conference on Computer Vision (2015)

42. Liang, X., Lee, L., Dai, W., Xing, E.P.: Dual motion GAN for future-flow embedded video prediction. In: IEEE International Conference on Computer Vision (ICCV) (2017)

43. Liu, M.Y., Breuel, T., Kautz, J.: Unsupervised image-to-image translation networks. In: Advances in Neural Information Processing Systems (NIPS) (2017)

44. Lu, C., Krishna, R., Bernstein, M., Fei-Fei, L.: Visual relationship detection with language priors. In: Leibe, B., Matas, J., Sebe, N., Welling, M. (eds.) ECCV 2016. LNCS, vol. 9905, pp. 852–869. Springer, Cham (2016). https://doi.org/10.1007/978-3-319-46448-0_51

45. Maas, A.L., Hannun, A.Y., Ng, A.Y.: Rectifier nonlinearities improve neural network acoustic models. In: International Conference on Machine Learning (ICML) (2013)

46. Mathieu, M., Couprie, C., LeCun, Y.: Deep multi-scale video prediction beyond mean square error. In: International Conference on Learning Representations (ICLR) (2016)

47. Miyato, T., Kataoka, T., Koyama, M., Yoshida, Y.: Spectral normalization for generative adversarial networks. In: International Conference on Learning Representations (ICLR) (2018)

48. Miyato, T., Koyama, M.: cGANs with projection discriminator. In: International Conference on Learning Representations (ICLR) (2018)

49. Newell, A., Deng, J.: Pixels to graphs by associative embedding. In: Advances in Neural Information Processing Systems (NeurIPS) (2017)

50. Newson, A., Almansa, A., Fradet, M., Gousseau, Y., Pérez, P.: Video inpainting of complex scenes. SIAM J. Imaging Sci. 7(4), 1993–2019 (2014)

51. Niklaus, S., Mai, L., Liu, F.: Video frame interpolation via adaptive separable convolution. In: IEEE International Conference on Computer Vision (ICCV) (2017)

52. Odena, A., Olah, C., Shlens, J.: Conditional image synthesis with auxiliary classifier GANs. In: International Conference on Machine Learning (ICML) (2017)

53. van den Oord, A., Kalchbrenner, N., Espeholt, L., Vinyals, O., Graves, A., et al.: Conditional image generation with PixelCNN decoders. In: Advances in Neural Information Processing Systems (NIPS) (2016)

54. Pennington, J., Socher, R., Manning, C.: GloVe: global vectors for word representation. In: Conference on Empirical Methods in Natural Language Processing (EMNLP) (2014)
55. Reed, S., Akata, Z., Yan, X., Logeswaran, L., Schiele, B., Lee, H.: Generative adversarial text-to-image synthesis. In: International Conference on Machine Learning (ICML) (2016)
56. Russakovsky, O., et al.: ImageNet large scale visual recognition challenge. Int. J. Comput. Vis. **115**, 211–252 (2015). https://doi.org/10.1007/s11263-015-0816-y
57. Saito, M., Matsumoto, E., Saito, S.: Temporal generative adversarial nets with singular value clipping. In: IEEE International Conference on Computer Vision (ICCV) (2017)
58. Salimans, T., Goodfellow, I., Zaremba, W., Cheung, V., Radford, A., Chen, X.: Improved techniques for training GANs. In: Advances in Neural Information Processing Systems (NIPS) (2016)
59. Shen, Y., Lu, F., Cao, X., Foroosh, H.: Video completion for perspective camera under constrained motion. In: International Conference on Pattern Recognition (ICPR) (2006)
60. Sigurdsson, G.A., Russakovsky, O., Gupta, A.: What actions are needed for understanding human actions in videos? In: IEEE International Conference on Computer Vision (ICCV) (2017)
61. Srivastava, N., Mansimov, E., Salakhudinov, R.: Unsupervised learning of video representations using LSTMs. In: International Conference on Machine Learning (ICML) (2015)
62. Stark, L., Bowyer, K.: Achieving generalized object recognition through reasoning about association of function to structure. IEEE Trans. Pattern Anal. Mach. Intell. (TPAMI) **13**(10), 1097–1104 (1991)
63. Szegedy, C., Vanhoucke, V., Ioffe, S., Shlens, J., Wojna, Z.: Rethinking the inception architecture for computer vision. In: IEEE Conference on Computer Vision and Pattern Recognition (CVPR) (2016)
64. Tulyakov, S., Liu, M.Y., Yang, X., Kautz, J.: MoCoGAN: decomposing motion and content for video generation. In: IEEE Conference on Computer Vision and Pattern Recognition (CVPR) (2018)
65. Villegas, R., Yang, J., Hong, S., Lin, X., Lee, H.: Decomposing motion and content for natural video sequence prediction. In: International Conference on Learning Representations (ICLR) (2017)
66. Vondrick, C., Pirsiavash, H., Torralba, A.: Generating videos with scene dynamics. In: Advances in Neural Information Processing Systems (NIPS) (2016)
67. Walker, J., Doersch, C., Gupta, A., Hebert, M.: An uncertain future: forecasting from static images using variational autoencoders. In: Leibe, B., Matas, J., Sebe, N., Welling, M. (eds.) ECCV 2016. LNCS, vol. 9911, pp. 835–851. Springer, Cham (2016). https://doi.org/10.1007/978-3-319-46478-7_51
68. Walker, J., Marino, K., Gupta, A., Hebert, M.: The pose knows: video forecasting by generating pose futures. In: IEEE International Conference on Computer Vision (ICCV) (2017)
69. Wang, T.C., Liu, M.Y., Tao, A., Liu, G., Kautz, J., Catanzaro, B.: Few-shot video-to-video synthesis. In: Advances in Neural Information Processing Systems (NeurIPS) (2019)
70. Wang, T.C., et al.: Video-to-video synthesis. In: Advances in Neural Information Processing Systems (NeurIPS) (2018)

71. Xian, Y., Lorenz, T., Schiele, B., Akata, Z.: Feature generating networks for zero-shot learning. In: IEEE Conference on Computer Vision and Pattern Recognition (CVPR) (2018)
72. Xian, Y., Schiele, B., Akata, Z.: Zero-shot learning-the good, the bad and the ugly. In: IEEE Conference on Computer Vision and Pattern Recognition (CVPR) (2017)
73. Xie, S., Girshick, R., Dollár, P., Tu, Z., He, K.: Aggregated residual transformations for deep neural networks. In: IEEE Conference on Computer Vision and Pattern Recognition (CVPR) (2017)
74. Xu, D., Zhu, Y., Choy, C.B., Fei-Fei, L.: Scene graph generation by iterative message passing. In: IEEE Conference on Computer Vision and Pattern Recognition (CVPR) (2017)
75. Xu, T., et al.: AttnGAN: fine-grained text to image generation with attentional generative adversarial networks. In: IEEE Conference on Computer Vision and Pattern Recognition (CVPR) (2018)
76. Yao, B., Fei-Fei, L.: Modeling mutual context of object and human pose in human-object interaction activities. In: IEEE Conference on Computer Vision and Pattern Recognition (CVPR) (2010)
77. Zellers, R., Yatskar, M., Thomson, S., Choi, Y.: Neural motifs: scene graph parsing with global context. In: IEEE Conference on Computer Vision and Pattern Recognition (CVPR) (2018)
78. Zhang, H., et al.: StackGAN: text to photo-realistic image synthesis with stacked generative adversarial networks. In: IEEE International Conference on Computer Vision (ICCV) (2017)
79. Zhao, J., Mathieu, M., LeCun, Y.: Energy-based generative adversarial network. In: International Conference on Learning Representations (ICLR) (2017)
80. Zhu, J.Y., Park, T., Isola, P., Efros, A.A.: Unpaired image-to-image translation using cycle-consistent adversarial networks. In: IEEE International Conference on Computer Vision (ICCV) (2017)
81. Zhu, Y., Elhoseiny, M., Liu, B., Peng, X., Elgammal, A.: A generative adversarial approach for zero-shot learning from noisy texts. In: IEEE Conference on Computer Vision and Pattern Recognition (CVPR) (2018)

ViTAA: Visual-Textual Attributes Alignment in Person Search by Natural Language

Zhe Wang[1(✉)], Zhiyuan Fang[2], Jun Wang[1], and Yezhou Yang[2]

[1] Beihang University, Beijing, China
{wangzhewz,wangj203}@buaa.edu.cn
[2] Arizona State University, Tempe, USA
{zfang29,yz.yang}@asu.edu

Abstract. Person search by natural language aims at retrieving a specific person in a large-scale image pool that matches given textual descriptions. While most of the current methods treat the task as a holistic visual and textual feature matching one, we approach it from an attribute-aligning perspective that allows grounding specific attribute phrases to the corresponding visual regions. We achieve success as well as a performance boost by a robust feature learning that the referred identity can be accurately bundled by multiple attribute cues. To be concrete, our Visual-Textual Attribute Alignment model (dubbed as ViTAA) learns to disentangle the feature space of a person into subspaces corresponding to attributes using a light auxiliary attribute segmentation layer. It then aligns these visual features with the textual attributes parsed from the sentences via a novel contrastive learning loss. We validate our ViTAA framework through extensive experiments on tasks of person search by natural language and by attribute-phrase queries, on which our system achieves state-of-the-art performances. Codes and models are available at https://github.com/Jarr0d/ViTAA.

Keywords: Person search by natural language · Person re-identification · Vision and language · Metric learning

1 Introduction

Recently, we have witnessed numerous practical breakthroughs in person modeling related tasks, e.g., pedestrian detection [2,5,47], pedestrian attribute recognition [30,40] and person re-identification [13,26,59]. Person search [15,25] as an

Z. Wang and Z. Fang—Equal contribution. This work was done when Z. Wang was a visiting scholar at Active Perception Group, Arizona State University.

Electronic supplementary material The online version of this chapter (https://doi.org/10.1007/978-3-030-58610-2_24) contains supplementary material, which is available to authorized users.

aggregation of the aforementioned tasks thus gains increasing research attention. Comparing with searching by image queries, person search by natural language [6,24,25,52] makes the retrieving procedure more user-friendly with increased flexibility due to a supporting of open-form natural language queries. Meanwhile, learning robust visual-textual associations becomes increasingly critical, which calls an urgent demand for a representation learning schema that is able to fully exploit both modalities.

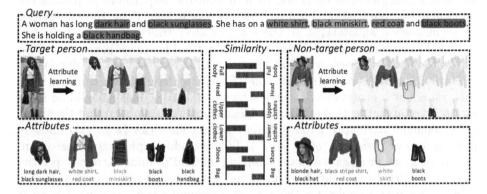

Fig. 1. In a case when two persons exhibit similar appearance attributes, it is hard to discriminate them merely by full-body appearance. Instead of matching the textual descriptions with the images at global-level, we decompose both image and text into attribute components and conduct a fine-grained matching strategy.

Relevant studies in person modeling related research points out the critical role of the discriminative representations, especially of the local fragments in both image and text. For the former, [38,58] learn the pose-related features from the key points map of human, while [20,27] leverage the body-part features by auxiliary segmentation-based supervision. For the latter, [24,25,55] decompose the complex sentences into noun phrases, and [23,52] directly adopt the attribute-specific annotations to learn fine-grained attribute related features. Moving forward, attribute specific features from image and text are even requisite for person search by natural language task, and how to effectively couple them stays an open question. We seek insight from a fatally flawed case that lingers in most of the current visual-language systems in Fig. 1, termed as "malpositioned matching". For example, tasks like textual grounding [33,35], VQA [1], and image-text retrieval [18,36] are measuring the similarities or mutual information across modalities in a holistic fashion by answering: are the feature vectors of image and text match with each other? That way, when users input *"a girl in white shirt and black skirt"* as retrieval query, the model is not able to distinguish the nuances of the two images as shown in Fig. 1, where the false positive one actually shows *"black shirt and white skirt"*. As both the distinct color visual cues (*"white"* and *"black"*) exist in the images, overall matching without the ability of referring them to specific appearance attributes prevents the model from discriminating them as needed. Such cases exist extensively in almost all cross-modal tasks, which pose an indispensable challenge for a system to tackle with the ability of fine-grained interplay between image and text.

Here, we put forward a novel **V**isual-**T**extual **A**ttributes **A**lignment model (ViTAA). For feature extraction, we fully exploit both visual and textual attribute representations. Specifically, we leverage segmentation labels to drive the attribute-aware feature learning from the input image. As shown in Fig. 3, we design multiple local branches, each of which is responsible to predict one particular attribute visual feature. This process is guided by the supervision on segmentation annotations, so the features are intrinsically aligned through the label information. We then use a generic natural language parser to extract attribute-related phrases, which at the same time remove the complex syntax in natural language and redundant non-informative descriptions. Building upon this, we adopt a contrastive learning schema to learn a joint embedding space for both visual and textual attributes. Meanwhile, we also notice that there may exist common attributes across different person identities (*e.g.,* two different persons may wear similar *"black shirt"*). To thoroughly exploits these cases during training, we propagate a novel sampling method to mine surrogate positive examples which largely enriches the sampling space, and also provides us with valid informative samples for the sake of overcoming convergence problem in metric learning.

To this end, we argue and show that the benefits of the attribute alignment to person search model go well beyond the obvious. As the images used for person search tasks often contain a large variance on appearance (*e.g.,* varying poses or viewpoints, with/without occlusion, and with cluttered background), the abstracted attribute-specific features could naturally help to resolve the ambiguity in feature representations. Also, searching by appearance attributes innately brings interpretability for the retrieving task and enables the attribute specific retrieval. It is worth mentioning that, there exist few very recent efforts that attempt to utilize the local fragments in both visual and textual modalities [8,49] and hierarchically align them [3,6]. The pairing schema of visual features and textual phrases in these methods are all based on the same identity, where they neglect the cues that exist across different identities. Comparing with them, ours is a more comprehensive modeling method that fully exploits the identical attributes from different persons thus greatly helps the alignment learning.

To validate these speculations, extensive experiments are conducted to our ViTAA model on the task of 1) person search by natural language and 2) by attribute, showing that our proposed model is capable of linking specific visual cues with specific words/phrases. More concretely, our ViTAA achieves a promising results across all these tasks. Further qualitative analysis verifies that our alignment learning successfully learns the fine-grained level correspondence across the visual and textual attributes. To summarize our contributions:

- We design an attribute-aware representation learning framework to extract and align both visual and textual features for the task of person search by natural language. To the best of our knowledge, this is the first to adopt both semantic segmentation as well as natural language parsing to facilitate a semantically aligned representation learning.
- We design a novel cross-modal alignment learning schema based on contrastive learning which can adaptively highlight the informative samples during the alignment learning. Meanwhile, an unsupervised data sampling method is proposed to facilitate the construction of learning pairs by exploiting more surrogate positive samples across different person identities.
- We evaluate the superiority of ViTAA over other state-of-the-art methods for the person search by natural language task. We also conduct qualitative analysis to demonstrate the interpretability of ViTAA.

2   Related Work

Person Search. Given the form of the querying data, current person search tasks can be categorized into two major thrusts: searching by images (termed as Person Re-Identification), and person search by textual descriptions. Typical person re-identification (re-id) methods [13,26,59] are formulated as retrieving the candidate that shows highest correlation with the query in the image galleries. However, a clear and valid image query is not always available in the real scenario, thus largely impedes the applications of re-id tasks. Recently, researchers alter their attention to re-id by textual descriptions: identifying the target person by using free-form natural languages [3,24,25]. Meanwhile, it also comes with great challenges as requiring the model to deal with the complex syntax from the long and free-form descriptive sentence, and the inconsistent interpretations of low-quality surveillance images. To tackle these, methods like [4,24,25] employ attention mechanism to build relation module between visual and textual representations, while [55,60] propose cross-modal objective functions for joint embedding learning. Dense visual feature is extracted in [32] by cropping the input image for learning a regional-level matching schema. Beyond this, [19] introduces pose estimation information for delicate human body-part parsing.

Attribute Representations. Adopting appropriate feature representations is of crucial importance for learning and retrieving from both image and text. Previous efforts in person search by natural language unanimously use holistic features of the person, which omit the partial visual cues from attributes at the fine-grained level. Multiple re-id systems have focused on the processing of body-part regions for visual feature learning, which can be summarized as: hand-craft horizontal stripes or grid [26,43,46], attention mechanism [37,45], and auxiliary information including keypoints [41,50], human parsing mask [20,27] and dense semantic estimation [56]. Among these methods, the auxiliary information usually provides more accurate partition results on localizing human parts and facilitating body-part attribute representations thanks to the multi-task training or the auxiliary networks. However, only few work [14] pay attention to the accessories (such as the backpack) which could be the potential contextual cues for accurate person retrieval. As the corresponding components to specific visual cues, textual attribute phrases are usually provided as ground-truth labels or can be extracted from sentences through identifying the noun phrases with sentence parsing. Many of them use textual attributes as auxiliary label information to complement the content of image features [23,29,39]. Recently, a few attempts leverage textual attribute as query for person retrieval [6,52]. [52] imposes an attribute-guided attention mechanism to capture the holistic appearance of person. [6] proposes a hierarchical matching model that can jointly learn global category-level and local attribute-level embedding.

Visual-Semantic Embedding. Works in vision and language propagate the notion of visual semantic embedding, with a goal to learn joint feature space for both visual inputs and their correspondent textual annotations [10,53]. Such mechanism plays a core role in a series of cross-modal tasks, *e.g.*, image/video captioning [7,21,51], image retrieval through natural language [8,49,55], and vision question answering [1]. Conventional joint embedding learning framework adopts two-branch architecture [8–10,53,55], where one extracts image features and the other encodes textual descriptions. The extracted cross-modal embedding features are learned through carefully designed objective functions.

3   Our Approach

Our network is composed of an image stream and a language stream (see Fig. 3), with the intention to encode inputs from both modalities for a visual-textual embedding learning. To be specific, given a person image $\mathcal{I}$ and its textual description $\mathcal{T}$, we first use the image stream to extract a global visual representation v_0, and a stack of local visual representations of N_{att} attributes $\{v_1, ..., v_{N_{att}}\}$, $v_i \in \mathbb{R}^d$. Similarly, we follow the language stream to extract overall textual embedding t_0, then decompose the whole sentence using standard natural language parser [22] into a list of the attribute phrases, and encode them as $\{t_1, ..., t_{N_{att}}\}$, $t_i \in \mathbb{R}^d$. Our core contribution is the cross-modal alignment learning that matches each visual component v_a with its corresponding textual phrase t_a, along with the global representation matching $\langle v_0, t_0 \rangle$ for the person search by natural language task.

3.1   The Image Stream

We adopt the sub-network of ResNet-50 (conv1, conv2_x, conv3_x, conv4_x) [17] as the backbone to extract feature maps F from the input image. Then, we introduce a global branch $\mathcal{F}^{glb}$, and multiple local branches $\mathcal{F}^{loc}_a$ to generate global visual features $v_0 = \mathcal{F}^{glb}(F)$, and attribute visual features $\{v_1 ... v_{N_{att}}\}$ respectively, where $v_a = \mathcal{F}^{loc}_a(F)$. The network architectures are shown in Table 1. On the top of all the local branches is an auxiliary segmentation layer supervising each local branch to generate the segmentation map of one specific attribute category (shown in Fig. 3). Intuitively, we argue that the additional auxiliary task acts as a knowledge regulator that diversifies each local branch to present attribute-specific features.

Our segmentation layer utilizes the architecture of a lightweight MaskHead [16] and can be removed during inference phase to reduce the computational cost. The remaining unsolved problem is that parsed annotations are not available in all person search datasets. To address that, we first train a human parsing network with HRNet [42] as an off-the-shelf tool, where the HRNet is jointly trained on multiple human parsing datasets: MHPv2 [57], ATR [28], and VIPeR [44]. We then use the attribute category predictions as our segmentation annotations (illustrated in Fig. 2). With these annotations, local branches receive the supervision needed from the segmentation task to learn attribute-specific features. Essentially, we are distilling the attribute information from a well-trained human parsing networks to the lightweight segmentation layer through joint training[1].

Fig. 2. Attribute annotation generated by human parsing network. Torsos are labeled as background since there is no corresponding textual descriptions.

Discussion. Using attribute feature has the following advantages over the global features. 1) The textual annotations in person search by natural language task describe the person mostly by their dressing/body appearances, where the attribute features perfectly fit the situation. 2) Attribute aligning avoids the "*malpositioned matching*" cases as shown in Fig. 1: using segmentation to regularize feature learning equips the model to be resilient over the diverse human poses or viewpoints, and also robust to the background noises.

3.2 The Language Stream

Given the raw textual description, our language stream first parses and extracts noun phrases *w.r.t.* each attribute through the Stanford POS tagger [31], and then feeds them into a language network to obtain the sentence-level as well as the phrase-level embeddings. We adopt a bi-directional LSTM to generate the global textual embedding t_0 and the local textual embedding. Meanwhile, we adopt a dictionary clustering approach to categorize the novel noun phrases in the sentence to specific attribute phrases as in [8]. Concretely, we manually collect a list of words per attribute category, *e.g.*, "*shirt*", "*jersey*", "*polo*" to

---

[1] More details of our human parsing network and segmentation results can be found in the experimental part and the supplementary materials.

represent the upper-body category, and use the average-pooled word vectors [12] of them as the anchor embedding $\mathbf{d}_a$, and form the dictionary $\mathbf{D} = [\mathbf{d}_1, ..., \mathbf{d}_{N_{att}}]$, where N_{att} is the total number of attributes. Building upon that, we assign the noun phrase to the category that has the highest cosine similarity, and form the local textual embedding $\{t_1 \ldots t_N\}$. Different from previous works like [32,56], we include accessory as one type of attribute as well, which serves as a crucial matching clue in many cases.

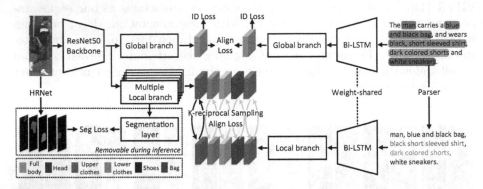

Fig. 3. Illustrative diagram of our ViTAA network, which includes an image stream (left) and a language stream (right). Our image stream first encodes the person image and extract both global and attribute representations. The local branch is additional supervised by an auxiliary segmentation layer where the annotations are acquired by an off-the-shell human parsing network. In the meanwhile, the textual description is parsed and decomposed into attribute atoms, and encoded by a weight-shared Bi-LSTM. We train our ViTAA jointly under global/attribute align loss in an end-to-end manner.

3.3 Visual-Textual Alignment Learning

Once we extract the global and attribute features, the key objective for the next stage is to learn a joint embedding space across the visual and the textual modalities, where the visual cues are tightly matched with the given textual description. Mathematically, we formulate our learning objective as a contrastive learning task that takes input as triplets, *i.e.*, $\langle v^i, t^+, t^- \rangle$ and $\langle t^i, v^+, v^- \rangle$, where i denotes the index of person to identify, and $+/-$ refer to the corresponding feature representations of the person i, and a randomly sampled irrelevant person respectively. We note that features in the triplet can be both at the global-level and the attribute-level. In the following, we discuss the learning schema on $\langle v^i, t^+, t^- \rangle$ which can be extended to $\langle t^i, v^+, v^- \rangle$.

We adopt the cosine similarity as the scoring function between visual and textual features $S = \frac{v^T \cdot t}{\|v\| \cdot \|t\|}$. For a positive pair $\langle v^i, t^+ \rangle$, the cosine similarity S^+ is encouraged to be as large as possible, which we define as *absolute similarity criterion*. While for a negative pair $\langle v^i, t^- \rangle$, enforcing the cosine similarity S^-

to be minimal may yield an arbitrary constraint over the negative samples t^-. Instead, we propose to optimize the deviation between S^- and S^+ to be larger than a preset margin, called *relative similarity criterion*. These criterion can be formulated as:

$$S^+ \to 1 \ and \ S^+ - S^- > m, \tag{1}$$

where m is the least margin that positive and negative similarity should differ and is set to 0.2 in practice.

In contrastive learning, the general form of the basic objective function are either hinge loss $\mathcal{L}(\boldsymbol{x}) = \max\{0, 1 - \boldsymbol{x}\}$ or logistic loss $\mathcal{L}(\boldsymbol{x}) = \log(1 + \exp(-\boldsymbol{x}))$. One crucial drawback of hinge loss is that its derivative *w.r.t.* x is a constant value: $\frac{\partial \mathcal{L}}{\partial x} = -1$. Since the pair-based construction of training data leads to a polynomial growth of training pairs, inevitably we will have a certain part of the randomly sampled negative texts being less informative during training. Treating all the redundant samples equally might raise the risk of a slow convergence and/or even model degeneration for the metric learning tasks. While the derivative of logistic loss *w.r.t.* x is: $\frac{\partial \mathcal{L}}{\partial x} = -\frac{1}{e^x + 1}$, which is related with the input value. Hence, we settle with the logistic loss as our basic objective function.

With the logistic loss, the aforementioned criterion can be further derived and rewritten as:

$$(S^+ - \alpha) > 0, \ -(S^- - \beta) > 0, \tag{2}$$

where $\alpha \to 1$ denotes the lower bound for positive similarity and $\beta = (\alpha - m)$ denotes the upper bound for negative similarity. Together with logistic loss function, our final *Alignment loss* can be unrolled as:

$$\mathcal{L}_{align} = \frac{1}{N} \sum_{i=1}^{N} \left\{ \log\left[1 + e^{-\tau_p(S_i^+ - \alpha)}\right] + \log\left[1 + e^{\tau_n(S_i^- - \beta)}\right] \right\}, \tag{3}$$

where τ_p and τ_n denote the temperature parameters that adjust the slope of gradient. The partial derivatives are calculated as:

$$\frac{\partial \mathcal{L}_{align}}{\partial S_i^+} = \frac{-\tau_p}{1 + e^{\tau_p(S_i^+ - \alpha)}}, \frac{\partial \mathcal{L}_{align}}{\partial S_i^-} = \frac{\tau_n}{1 + e^{\tau_n(\beta - S_i^-)}}. \tag{4}$$

Thus, we show that Eq. 3 outputs continuous gradients and will assign higher weights to more informative samples accordingly.

K-Rreciprocal Sampling. One of the premise of visual-textual alignment is to fully exploit the informative positive and negative samples t^+, t^- to provide valid supervisions. However, most of the current contrastive learning methods [48,54] construct the positive pairs by selecting samples belonging to the same class and simply treat the random samples from other classes as negative. This is viable when using only global information at coarse level during training, but may not be able to handle the case as illustrated in Fig. 1 where a fine-grained level comparison is needed. This practice is largely depending on the average number of samples for each attribute category to provide comprehensive positive samples.

With this insight, we propose to further enlarge the searching space of positive samples from the cross-id incidents.

For instance, as in Fig. 1, though the two ladies are with different identities, they share the extremely alike shoes which can be treated as the positive samples for learning. We term these kinds of samples with identical attributes but belong to different person identities as the "surrogate positive samples". Kindly including the common attribute features of the surrogate positive samples in positive pairs makes much more sense than the reverse. It is worth noting that, this is unique only to our attribute alignment learning phase because attributes can only be compared at the fine-grained level. Now the key question is, how can we dig out the surrogate positive samples since we do not have direct cross-ID attribute annotations? Inspired by the re-ranking techniques in re-id community [11,61], we propose k-reciprocal sampling as an unsupervised method to generate the surrogate labels at the attribute-level. How does the proposed method sample from a batch of visual and textual features? Straightforwardly, for each attribute a, we can extract a batch of visual and textual features from the feature learning network and mine their corresponding surrogate positive samples using our sampling algorithm. Since we are only discussing the input form of $\langle v^i, t^+, t^- \rangle$, our sampling algorithm is actually mining the surrogate positive textual features for each v^i. Note that, if the attribute information in either modality is missing after parsing, we can simply ignore them during sampling.

Algorithm 1: *K-reciprocal* Sampling Algorithm

Input:
    $\mathcal{V}_a = \{v_a^i\}_{i=1}^N$ is a set of visual feature for attribute a
    $\mathcal{T}_a = \{t_a^i\}_{i=1}^N$ is a set of textual feature for attribute a
Output:
    $\mathcal{P}_a$ is a set of surrogate positive sample for attribute a

1 **for** *each* $v_a \in \mathcal{V}_a$ **do**
2   find the top-K nearest neighbours of v_a w.r.t. $\mathcal{T}_a$: $\mathcal{K}_{\mathcal{T}_a}$;
3     $\mathcal{S} \leftarrow \varnothing$;
4   **for** *each* $t_a \in \mathcal{K}_{\mathcal{T}_a}$ **do**
5   find the top-K nearest neighbours of t_a w.r.t. $\mathcal{V}_a$: $\mathcal{K}_{\mathcal{V}_a}$;
6   **if** $v_a \in \mathcal{K}_{\mathcal{V}_a}$ **then**
7           $\mathcal{S} = \mathcal{S} \cup t_a$
8   **end**
9   **end**
10    $\mathcal{P}_a = \mathcal{P}_a \cup \mathcal{S}$
11 **end**

3.4   Joint Training

The entire network is trained in an end-to-end manner. We adopt the widely-used cross-entropy loss (ID Loss) to assist the learning of the discriminative features of each instance, as well as pixel-level cross-entropy loss (Seg Loss) to classify the attribute categories in the auxiliary segmentation task. For the cross-modal alignment learning, we design the Alignment Loss on both the global-level and the attribute-level representations. The overall loss function thus emerges:

$$\mathcal{L} = \mathcal{L}_{id} + \mathcal{L}_{seg} + \mathcal{L}_{align}^{glo} + \mathcal{L}_{align}^{attr}. \qquad (5)$$

Table 1. Detailed architecture of our global and local branches in image stream. *#Branch.* denotes the number of sub-branches.

Layer name	Parameters		Output size	#Branch
$\mathcal{F}^{glb}$	$\begin{bmatrix} 3 \times 3, 2048 \\ 3 \times 3, 2048 \end{bmatrix}$	$\times 2$	24×8	1
	Average pooling		1×1	
$\mathcal{F}^{loc}$	$\begin{bmatrix} 3 \times 3, 256 \\ 3 \times 3, 256 \end{bmatrix}$	$\times 2$	24×8	5
	Max pooling		1×1	

4 Experiment

4.1 Experimental Setting

Datasets. We conduct experiments on the CUHK-PEDES [25] dataset, which is currently the only benchmark for person search by natural language. It contains 40,206 images of 13,003 different persons, where each image comes with two human-annotated sentences. The dataset is split into 11,003 identities with 34,054 images in the training set, 1,000 identities with 3,078 images in validation, and 1,000 identities with 3,074 images in testing set.

Evaluation Protocols. Following the standard evaluation setting, we adopt Recall@K (K = 1, 5, 10) as the retrieval criteria. Specifically, given a text description as query, Recall@K (R@K) reports the percentage of the images where at least one corresponding person is retrieved correctly among the top-K results.

Implementation Details. For the global and local branches in image stream, we use the Basicblock as described in [17], where each branch is randomly initialized (detailed architecture is shown in Table 1). We use horizontally flipping as data augmenting and resize all the images to 384×128. We use the Adam solver as the training optimizer with weight decay set as 4×10^{-5}, and involves 64 image-language pairs per mini-batch. The learning rate is initialized at 2×10^{-4} for the first 40 epochs during training, then decayed by a factor of 0.1 for the remaining 30 epochs. The whole experiment is implemented on a single Tesla V100 GPU machine. The hyperparameters in Eq. 3 are empirically set as: $\alpha = 0.6, \beta = 0.4, \tau_p = 10, \tau_n = 40$.

Pedestrian Attributes Parsing. Based on the analysis of image and natural language annotations in the dataset, we warp both visual and textual attributes into 5 categories: head (including descriptions related to hat, glasses, and face), clothes on the upper body, clothes on the lower body, shoes and bags (including backpack and handbag). We reckon that these attributes are visually distinguishable from both modalities. In Fig. 2, we visualize the segmentation maps generated by our human parsing network, where attribute regions can be properly segmented and associated with correct labels.

4.2   Comparisons with the State-of-The-Arts

Result on CUHK-PEDES Dataset. We summarize the performance of ViTAA and compare it with state-of-the-art methods in Table 2 on the CUHK-PEDES test set. Methods like GNA-RNN [25], CMCE [24], PWM-ATH [4] employ attention mechanism to learn the relation between visual and textual representation, while Dual Path [60], CMPM+CMPC [55] design objective function for better joint embedding learning. These methods only learn and utilize the *"global"* feature representation of both image and text. Moreover, MIA [32] exploits *"region"* information by dividing the input image into several horizontal stripes and extracting noun phrases from the natural language description. Similarly, GALM [19] leverage *"keypoint"* information from human pose estimation as an attention mechanism to assist feature learning and together with a noun phrases extractor implemented on input text. Though the above two utilize the local-level representations, neither of them learns the associations between visual features with textual phrases. From Table 2, we observe that ViTAA shows a consistent lead on all metrics (R@1-10), outperforming the GALM [19] by a margin of 1.85%, 0.39%, 0.55% and claims the new state-of-the-art results. We note that though the performance could be considered as incremental, the shown improvement on the R@1 performance is challenging. It suggests that the alignment learning of ViTAA contributes to the retrieval task directly. We further report the ablation studies on the effect of different components, and exhibit the attribute retrieval results quantitatively and qualitatively.

4.3   Ablation Study

We carry out comprehensive ablations to evaluate the contribution of different components and the training configurations.

Comparisons over Different Component Combinations. To compare the individual contribution of each component, we set the baseline model as the one trained with only ID loss. In Table 3, we report the improvement of the proposed components (segmentation, global-alignment, and attribute-alignment) on the basis of the baseline model. From the table, we have the following observations and analyses: First, using segmentation loss only brings marginal improvement because the visual features are not aligned with their corresponding textual features. Similarly, we observe the same trend when the training is combined

Table 2. Person search results on the CUHK-PEDES test set. Best results are in bold.

Method	Feature	R@1	R@5	R@10
GNA-RNN [25]	global	19.05	–	53.64
CMCE [24]	global	25.94	–	60.48
PWM-ATH [4]	global	27.14	49.45	61.02
Dual Path [60]	global	44.40	66.26	75.07
CMPM+CMPC [55]	global	49.37	–	79.27
MIA [32]	global+region	53.10	75.00	82.90
GALM [19]	global+keypoint	54.12	75.45	82.97
ViTAA	global+attribute	**55.97**	**75.84**	**83.52**

with only attribute-alignment loss where the visual features are not properly segmented, thus can not be associated for retrieval. An incremental gain is obtained by combining these two components. Next, compared with attribute-level, global-level alignment greatly improves the performance under all criteria, which demonstrates the efficiency of the visual-textual alignment schema. The cause of the performance gap is that: the former is learning the attribute similarity across different person identities while the latter is concentrating on the uniqueness of each person. At the end, by combining all the loss terms yields the best performance, validating that our global-alignment and attribute-alignment learning are complimentary with each other.

Table 3. The improvement of components added on baseline model. Glb-Align and Attr-Align represent global-level and attribute-level alignment respectively.

Model component			R@1	R@5	R@10
Segmentation	Attr-Align	Glb-Align			
			29.68	51.84	61.57
✓			30.93	52.71	63.11
	✓		31.40	54.09	63.66
✓	✓		39.26	61.22	68.14
		✓	52.27	73.33	81.61
✓	✓	✓	**55.97**	**75.84**	**83.52**

Visual Attribute Segmentation and Representations. In Fig. 4, we visualize the segmentation maps from the segmentation layer and the feature representations of the local branches. It evidently shows that, even transferred using only a lightweight structure, the auxiliary person segmentation layer produces accurate pixel-wise labels under different human pose. This suggests that person parsing knowledge has been successfully distilled our local branches, which

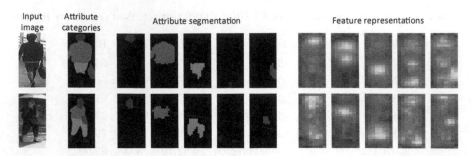

Fig. 4. From left to right, we exhibit the raw input person images, attribute labels generated by the pre-trained HRNet, attribute segmentation result from our segmentation layer, and their corresponded feature maps from the local branches.

is crucial for the precise cross-modal alignment learning. On the right side of Fig. 4, we showcase the feature maps of local branch per attribute.

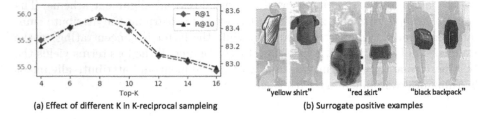

(a) Effect of different K in K-reciprocal sampleing

(b) Surrogate positive examples

Fig. 5. (a) R@1 and R@10 results across different K value in the proposed surrogate positive data sampling method. (b) Some examples of the surrogate positive data with different person identities.

K-Reciprocal Sampling. We investigate how the value of K impacts the pair-based sampling and learning process. We evaluate the R@1 and R@10 performance under different K settings in Fig. 5(a). Ideally, the larger the K is, the more potential surrogate positive samples will be mined, while this also comes with the possibility that more non-relevant examples (false positive examples) might be incorrectly sampled. Result in Fig. 5(a) agrees with our analysis: best R@1 and R@10 is achieved when K is set to 8, and the performances are persistently declining as K goes larger. In Fig. 5(b), we provide visual examinations of the surrogate positive pairs that mined by our sampling method. The visual attributes from different persons serve as valuable positive samples in our alignment learning schema.

Qualitative Analysis. We present the qualitative examples of person retrieval results to provide a more in-depth examination. As shown in Fig. 6, we illustrate the top-10 matching results using the given query. In the successful case (top),

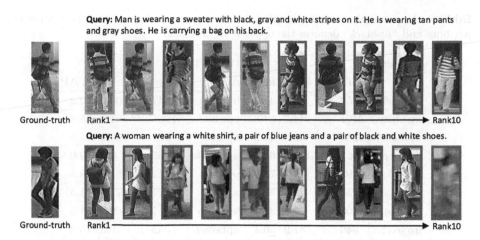

Fig. 6. Examples of person search results on CUHK-PEDES. We indicate the true/false matching results in green/red boxes. (Color figure online)

ViTAA precisely capture all attributes in the target person. It is worth noting that the wrong answers still capture the relevant attributes: "sweater with black, gray and white stripes", "tan pants", and "carrying a bag". For the failure case (bottom), though the retrieved images are incorrect, we observe that all the attributes described in the query are there in almost all retrieved results.

4.4 Extension: Attribute Retrieval

In order to validate the ability of associating the visual attribute with the text phrase, we further conduct attribute retrieval experiment on the datasets of Market-1501 [59] and DukeMTMC [34], where 27 and 23 human related attributes are annotated per image by [29]. In our experiment, we use our pre-trained ViTAA on CUHK-PEDES without any further finetuning, and conduct the retrieval task using the attribute phrase as the query under R@1 and mAP metrics. In our experiment, we simply test on the *upper-body clothing* attribute category, and post the retrieval results in Table 4. We introduce the details of our experiment in the supplementary materials. From Table 4, it clearly shows that ViTAA achieves great performances on almost all sub-attributes. This further strongly supports our argument that ViTAA is able to associate the visual attribute features with textual attribute descriptions successfully.

5 Conclusion

In this work, we present a novel ViTAA model to address the person search by natural language task from the perspective of an attribute-specific alignment learning. In contrast to the existing methods, ViTAA fully exploits the common

Table 4. Upper-body clothing attribute retrieve results. Attr is the short form of attribute and "upblack" denotes the upper-body in black.

Market1501				DukeMTMC			
Attr	#Target	R@1	mAP	Attr	#Target	R@1	m AP
upblack	1759	99.2	44.0	upblack	11047	100.0	82.2
upwhite	3491	44.7	64.8	upwhite	1095	91.3	35.4
upred	1354	92.3	54.8	upred	821	100.0	44.6
uppurple	363	100.0	61.2	uppurple	65	0.0	9.0
upyellow	1195	72.7	75.6	upgray	2012	81.7	29.8
upgray	1755	49.7	55.2	upblue	1577	77.9	31.3
upblue	1100	70.1	32.4	upgreen	417	89.2	24.5
upgreen	949	87.9	50.4	upbrown	345	18.3	15.2

attribute information in both visual and textual modalities across different person identities, and further builds strong association between the visual attribute features and their corresponding textual phrases by using our alignment learning schema. We show that ViTAA achieves state-of-the-art results on the challenging benchmark CUHK-PEDES and demonstrate its promising potential that further advances the person search by natural language domain.

Acknowledgements. Vising scholarship support for Z. Wang from the China Scholarship Council #201806020020 and Amazon AWS Machine Learning Research Award (MLRA) support are greatly appreciated. Any opinions, findings, and conclusion or recommendations expressed in this material are those of the authors and do not necessarily reflect the view of the sponsors.

References

1. Antol, S., et al.: VQA: visual question answering. In: Proceedings of the IEEE International Conference on Computer Vision, pp. 2425–2433 (2015)
2. Benenson, R., Omran, M., Hosang, J., Schiele, B.: Ten years of pedestrian detection, what have we learned? In: Agapito, L., Bronstein, M.M., Rother, C. (eds.) ECCV 2014. LNCS, vol. 8926, pp. 613–627. Springer, Cham (2015). https://doi.org/10.1007/978-3-319-16181-5_47
3. Chen, D., et al.: Improving deep visual representation for person re-identification by global and local image-language association. In: Proceedings of the European Conference on Computer Vision (ECCV), pp. 54–70 (2018)
4. Chen, T., Xu, C., Luo, J.: Improving text-based person search by spatial matching and adaptive threshold. In: 2018 IEEE Winter Conference on Applications of Computer Vision (WACV), pp. 1879–1887, March 2018
5. Dollár, P., Wojek, C., Schiele, B., Perona, P.: Pedestrian detection: a benchmark. In: 2009 IEEE Conference on Computer Vision and Pattern Recognition, CVPR 2009, pp. 304–311. IEEE (2009)

6. Dong, Q., Gong, S., Zhu, X.: Person search by text attribute query as zero-shot learning. In: The IEEE International Conference on Computer Vision (ICCV), October 2019
7. Fang, Z., Gokhale, T., Banerjee, P., Baral, C., Yang, Y.: Video2Commonsense: generating commonsense descriptions to enrich video captioning. arXiv preprint arXiv:2003.05162 (2020)
8. Fang, Z., Kong, S., Fowlkes, C., Yang, Y.: Modularized textual grounding for counterfactual resilience. In: The IEEE Conference on Computer Vision and Pattern Recognition (CVPR), June 2019
9. Fang, Z., Kong, S., Yu, T., Yang, Y.: Weakly supervised attention learning for textual phrases grounding. arXiv preprint arXiv:1805.00545 (2018)
10. Frome, A., et al.: Devise: a deep visual-semantic embedding model. In: Advances in Neural Information Processing Systems, pp. 2121–2129 (2013)
11. Garcia, J., Martinel, N., Micheloni, C., Gardel, A.: Person re-identification ranking optimisation by discriminant context information analysis. In: Proceedings of the IEEE International Conference on Computer Vision, pp. 1305–1313 (2015)
12. Goldberg, Y., Levy, O.: word2vec explained: deriving Mikolov et al.'s negative-sampling word-embedding method. arXiv preprint arXiv:1402.3722 (2014)
13. Gong, S., Cristani, M., Yan, S., Loy, C.C.: Person Re-Identification. Springer, London (2014). https://doi.org/10.1007/978-1-4471-6296-4
14. Guo, J., Yuan, Y., Huang, L., Zhang, C., Yao, J.G., Han, K.: Beyond human parts: dual part-aligned representations for person re-identification. In: The IEEE International Conference on Computer Vision (ICCV), October 2019
15. Han, C., et al.: Re-ID driven localization refinement for person search. In: Proceedings of the IEEE International Conference on Computer Vision, pp. 9814–9823 (2019)
16. He, K., Gkioxari, G., Dollár, P., Girshick, R.: Mask R-CNN. In: Proceedings of the IEEE International Conference on Computer Vision, pp. 2961–2969 (2017)
17. He, K., Zhang, X., Ren, S., Sun, J.: Deep residual learning for image recognition. In: Proceedings of the IEEE Conference on Computer Vision and Pattern Recognition, pp. 770–778 (2016)
18. Jeon, J., Lavrenko, V., Manmatha, R.: Automatic image annotation and retrieval using cross-media relevance models. In: Proceedings of the 26th Annual International ACM SIGIR Conference on Research and Development in Information Retrieval, pp. 119–126 (2003)
19. Jing, Y., Si, C., Wang, J., Wang, W., Wang, L., Tan, T.: Pose-guided joint global and attentive local matching network for text-based person search. arXiv preprint arXiv:1809.08440 (2018)
20. Kalayeh, M.M., Basaran, E., Gökmen, M., Kamasak, M.E., Shah, M.: Human semantic parsing for person re-identification. In: Proceedings of the IEEE Conference on Computer Vision and Pattern Recognition, pp. 1062–1071 (2018)
21. Karpathy, A., Fei-Fei, L.: Deep visual-semantic alignments for generating image descriptions. In: Proceedings of the IEEE Conference on Computer Vision and Pattern Recognition, pp. 3128–3137 (2015)
22. Klein, D., Manning, C.D.: Fast exact inference with a factored model for natural language parsing. In: Advances in Neural Information Processing Systems, pp. 3–10 (2003)
23. Layne, R., Hospedales, T.M., Gong, S.: Attributes-based re-identification. In: Gong, S., Cristani, M., Yan, S., Loy, C.C. (eds.) Person Re-Identification. ACVPR, pp. 93–117. Springer, London (2014). https://doi.org/10.1007/978-1-4471-6296-4_5

24. Li, S., Xiao, T., Li, H., Yang, W., Wang, X.: Identity-aware textual-visual matching with latent co-attention. In: Proceedings of the IEEE International Conference on Computer Vision, pp. 1890–1899 (2017)
25. Li, S., Xiao, T., Li, H., Zhou, B., Yue, D., Wang, X.: Person search with natural language description. In: Proceedings of the IEEE Conference on Computer Vision and Pattern Recognition, pp. 1970–1979 (2017)
26. Li, W., Zhao, R., Xiao, T., Wang, X.: DeepReID: deep filter pairing neural network for person re-identification. In: Proceedings of the IEEE Conference on Computer Vision and Pattern Recognition, pp. 152–159 (2014)
27. Liang, X., Gong, K., Shen, X., Lin, L.: Look into person: joint body parsing & pose estimation network and a new benchmark. IEEE Trans. Pattern Anal. Mach. Intell. **41**(4), 871–885 (2018)
28. Liang, X., et al.: Deep human parsing with active template regression. IEEE Trans. Pattern Anal. Mach. Intell. **12**, 2402–2414 (2015)
29. Lin, Y., et al.: Improving person re-identification by attribute and identity learning. Pattern Recogn. **95**, 151–161 (2019)
30. Liu, X., et al.: HydraPlus-Net: attentive deep features for pedestrian analysis. In: Proceedings of the IEEE International Conference on Computer Vision, pp. 350–359 (2017)
31. Manning, C.D., Surdeanu, M., Bauer, J., Finkel, J., Bethard, S.J., McClosky, D.: The Stanford CoreNLP natural language processing toolkit. In: Association for Computational Linguistics (ACL) System Demonstrations, pp. 55–60 (2014)
32. Niu, K., Huang, Y., Ouyang, W., Wang, L.: Improving description-based person re-identification by multi-granularity image-text alignments. arXiv preprint arXiv:1906.09610 (2019)
33. Plummer, B.A., Wang, L., Cervantes, C.M., Caicedo, J.C., Hockenmaier, J., Lazebnik, S.: Flickr30k entities: collecting region-to-phrase correspondences for richer image-to-sentence models. In: Proceedings of the IEEE International Conference on Computer Vision, pp. 2641–2649 (2015)
34. Ristani, E., Solera, F., Zou, R., Cucchiara, R., Tomasi, C.: Performance measures and a data set for multi-target, multi-camera tracking. In: Hua, G., Jégou, H. (eds.) ECCV 2016. LNCS, vol. 9914, pp. 17–35. Springer, Cham (2016). https://doi.org/10.1007/978-3-319-48881-3_2
35. Rohrbach, A., Rohrbach, M., Hu, R., Darrell, T., Schiele, B.: Grounding of textual phrases in images by reconstruction. In: Leibe, B., Matas, J., Sebe, N., Welling, M. (eds.) ECCV 2016. LNCS, vol. 9905, pp. 817–834. Springer, Cham (2016). https://doi.org/10.1007/978-3-319-46448-0_49
36. Shekhar, R., Jawahar, C.: Word image retrieval using bag of visual words. In: 2012 10th IAPR International Workshop on Document Analysis Systems, pp. 297–301. IEEE (2012)
37. Si, J., et al.: Dual attention matching network for context-aware feature sequence based person re-identification. In: Proceedings of the IEEE Conference on Computer Vision and Pattern Recognition, pp. 5363–5372 (2018)
38. Su, C., Li, J., Zhang, S., Xing, J., Gao, W., Tian, Q.: Pose-driven deep convolutional model for person re-identification. In: Proceedings of the IEEE International Conference on Computer Vision, pp. 3960–3969 (2017)
39. Su, C., Zhang, S., Xing, J., Gao, W., Tian, Q.: Multi-type attributes driven multi-camera person re-identification. Pattern Recog. **75**, 77–89 (2018)
40. Sudowe, P., Spitzer, H., Leibe, B.: Person attribute recognition with a jointly-trained holistic CNN model. In: Proceedings of the IEEE International Conference on Computer Vision Workshops, pp. 87–95 (2015)

41. Suh, Y., Wang, J., Tang, S., Mei, T., Mu Lee, K.: Part-aligned bilinear representations for person re-identification. In: Proceedings of the European Conference on Computer Vision (ECCV), pp. 402–419 (2018)
42. Sun, K., Xiao, B., Liu, D., Wang, J.: Deep high-resolution representation learning for human pose estimation. In: The IEEE Conference on Computer Vision and Pattern Recognition (CVPR), June 2019
43. Sun, Y., Zheng, L., Yang, Y., Tian, Q., Wang, S.: Beyond part models: person retrieval with refined part pooling (and a strong convolutional baseline). In: Proceedings of the European Conference on Computer Vision (ECCV), pp. 480–496 (2018)
44. Tan, Z., Yang, Y., Wan, J., Hang, H., Guo, G., Li, S.Z.: Attention-based pedestrian attribute analysis. IEEE Trans. Image Process. **12**, 6126–6140 (2019)
45. Wang, C., Zhang, Q., Huang, C., Liu, W., Wang, X.: Mancs: a multi-task attentional network with curriculum sampling for person re-identification. In: Proceedings of the European Conference on Computer Vision (ECCV), pp. 365–381 (2018)
46. Wang, G., Yuan, Y., Chen, X., Li, J., Zhou, X.: Learning discriminative features with multiple granularities for person re-identification. In: 2018 ACM Multimedia Conference on Multimedia Conference, pp. 274–282. ACM (2018)
47. Wang, Z., Wang, J., Yang, Y.: Resisting crowd occlusion and hard negatives for pedestrian detection in the wild. arXiv preprint arXiv:2005.07344 (2020)
48. Wen, Y., Zhang, K., Li, Z., Qiao, Y.: A discriminative feature learning approach for deep face recognition. In: Leibe, B., Matas, J., Sebe, N., Welling, M. (eds.) ECCV 2016. LNCS, vol. 9911, pp. 499–515. Springer, Cham (2016). https://doi.org/10.1007/978-3-319-46478-7_31
49. Wu, H., et al.: Unified visual-semantic embeddings: bridging vision and language with structured meaning representations. In: The IEEE Conference on Computer Vision and Pattern Recognition (CVPR), June 2019
50. Xu, J., Zhao, R., Zhu, F., Wang, H., Ouyang, W.: Attention-aware compositional network for person re-identification. In: Proceedings of the IEEE Conference on Computer Vision and Pattern Recognition, pp. 2119–2128 (2018)
51. Xu, K., et al.: Show, attend and tell: neural image caption generation with visual attention. In: International Conference on Machine Learning, pp. 2048–2057 (2015)
52. Yin, Z., et al.: Adversarial attribute-image person re-identification. In: Proceedings of the Twenty-Seventh International Joint Conference on Artificial Intelligence, IJCAI-2018, pp. 1100–1106. International Joint Conferences on Artificial Intelligence Organization, July 2018
53. You, Q., Zhang, Z., Luo, J.: End-to-end convolutional semantic embeddings. In: Proceedings of the IEEE Conference on Computer Vision and Pattern Recognition, pp. 5735–5744 (2018)
54. Zhang, X., Fang, Z., Wen, Y., Li, Z., Qiao, Y.: Range loss for deep face recognition with long-tailed training data. In: Proceedings of the IEEE International Conference on Computer Vision, pp. 5409–5418 (2017)
55. Zhang, Y., Lu, H.: Deep cross-modal projection learning for image-text matching. In: Proceedings of the European Conference on Computer Vision (ECCV), pp. 686–701 (2018)
56. Zhang, Z., Lan, C., Zeng, W., Chen, Z.: Densely semantically aligned person re-identification. In: Proceedings of the IEEE Conference on Computer Vision and Pattern Recognition, pp. 667–676 (2019)

57. Zhao, J., Li, J., Cheng, Y., Sim, T., Yan, S., Feng, J.: Understanding humans in crowded scenes: deep nested adversarial learning and a new benchmark for multi-human parsing. In: 2018 ACM Multimedia Conference on Multimedia Conference, pp. 792–800. ACM (2018)
58. Zheng, L., Huang, Y., Lu, H., Yang, Y.: Pose invariant embedding for deep person re-identification. IEEE Trans. Image Process. **28**(9), 4500–4509 (2019)
59. Zheng, L., Shen, L., Tian, L., Wang, S., Wang, J., Tian, Q.: Scalable person re-identification: a benchmark. In: Proceedings of the IEEE International Conference on Computer Vision, pp. 1116–1124 (2015)
60. Zheng, Z., Zheng, L., Garrett, M., Yang, Y., Shen, Y.D.: Dual-path convolutional image-text embedding with instance loss. arXiv preprint arXiv:1711.05535 (2017)
61. Zhong, Z., Zheng, L., Cao, D., Li, S.: Re-ranking person re-identification with k-reciprocal encoding. In: Proceedings of the IEEE Conference on Computer Vision and Pattern Recognition, pp. 1318–1327 (2017)

Renovating Parsing R-CNN for Accurate Multiple Human Parsing

Lu Yang[1], Qing Song[1($\boxtimes$)], Zhihui Wang[1], Mengjie Hu[1], Chun Liu[1],
Xueshi Xin[1], Wenhe Jia[1], and Songcen Xu[2]

[1] Beijing University of Posts and Telecommunications, Beijing 100876, China
{soeaver,priv,wangzh,mengjie.hu,chun.liu,xinxueshi,
srxhemailbox}@bupt.edu.cn
[2] Noah's Ark Lab, Huawei Technologies, Shenzhen, China
xusongcen@huawei.com

Abstract. Multiple human parsing aims to segment various human parts and associate each part with the corresponding instance simultaneously. This is a very challenging task due to the diverse human appearance, semantic ambiguity of different body parts, and complex background. Through analysis of multiple human parsing task, we observe that human-centric global perception and accurate instance-level parsing scoring are crucial for obtaining high-quality results. But the most state-of-the-art methods have not paid enough attention to these issues. To reverse this phenomenon, we present Renovating Parsing R-CNN (RP R-CNN), which introduces a global semantic enhanced feature pyramid network and a parsing re-scoring network into the existing high-performance pipeline. The proposed RP R-CNN adopts global semantic representation to enhance multi-scale features for generating human parsing maps, and regresses a confidence score to represent its quality. Extensive experiments show that RP R-CNN performs favorably against state-of-the-art methods on CIHP and MHP-v2 datasets. Code and models are available at https://github.com/soeaver/RP-R-CNN.

Keywords: Multiple human parsing · Region-based approach · Global semantic enhanced FPN · Parsing re-scoring network

1 Introduction

Multiple human parsing [8,21,40] is a fundamental task in multimedia and computer vision, which aims to segment various human parts and associate each part with the corresponding instance. It plays a crucial role in applications in human-centric analysis and potential down-stream applications, such as person re-identification [22,27], action recognition [5], human-object interaction [1,7,29], and virtual reality [14].

Due to the successful development of convolutional neural networks [13,33], great progress has been made in multiple human parsing. Current state-of-the-art

© Springer Nature Switzerland AG 2020
A. Vedaldi et al. (Eds.): ECCV 2020, LNCS 12357, pp. 421–437, 2020.
https://doi.org/10.1007/978-3-030-58610-2_25

Fig. 1. Comparison of results between Parsing R-CNN and RP R-CNN on CIHP dataset. The first row is the ground-truth, the second row is the results of Parsing R-CNN, and the third is the predictions of RP R-CNN.

methods can be categorized into bottom-up, one-stage top-down, and two-stage top-down methods. The bottom-up methods [8,9,11] regard multiple human parsing as a fine-grained semantic segmentation task, which predicts the category of each pixel and grouping them into corresponding human instance. This series of methods will have better performance in semantic segmentation metrics, but poor in instance parsing metrics, especially easy to confuse adjacent human instances. Unlike bottom-up methods, the one-stage top-down [30,40] and two-stage top-down methods [16,24,32] locate each instance in the image plane, and then segment each human parts independently. The difference between one-stage and two-stage is whether the detector is trained together with the sub-network used to segment the human part in an end-to-end manner. Compared with the bottom-up, the top-down methods are very flexible, which can easily introduce enhancement modules or train with other human analysis tasks (such as pose estimation [37], dense pose estimation [10,40] or clothing parsing [35]) jointly. So it has become the mainstream research direction of multiple human parsing. But the human parts segmentation of each instance is independent and cannot make full use of context information, so the segmentation of some small scale human parts and human contours still needs to be improved. In addition, it is worth noting that neither bottom-up or top-down methods have a good way to evaluate the quality of predicted instance parsing maps. Resulting in many low-quality results that cannot be filtered.

In this paper, we are devoted to solving the problem of missing global semantic information in top-down methods, and evaluating the quality of predicted instance parsing maps accurately. Therefore, we propose Renovating Parsing R-CNN (RP R-CNN), which introduces a global semantic enhanced feature pyramid network and a parsing re-scoring network to renovate the pipeline of top-down multiple human parsing. The global semantic enhanced feature pyramid network (GSE-FPN) is built on the widely used FPN [23]. We up-sample the multi-scale features generated by FPN to the same scale and fuse them. Using the global human parts segmentation to supervise and generate the global semantic feature, then fusing the global semantic feature with FPN features on the corresponding scales. GSE-FPN encourages the semantic supervision signal to directly propagate to the feature pyramid, so as to strengthen global information of learned multi-scale features. Global semantic enhanced features are passed to the Parsing branch through the RoIAlign [12] operation, ensuring that each independent human instance can still perceive the global semantic information, thereby improving the parsing performance of small targets, human contours, and easily confused categories. On this basis, the parsing re-scoring network (PRSN) is used to sense the quality of instance parsing maps and gives accurate scores. The score of instance parsing map is related to filtering low-quality results and sorting of instances, which is very important in the measurement of method and practical application. However, almost all the top-down methods use the score of detected bounding-box to represents the quality of instance parsing map [32,40]. This will inevitably bring great deviation, because the score of bounding-box can only indicate whether the instance is human or not, while the score of the instance parsing map needs to express the segmentation accuracy of each human part, and there is no direct correlation between them. The proposed PRSN is a very lightweight network, taking the feature map and heat map of each human instance as input, using MSE loss to regress the mean intersection over union (mIoU) between the prediction and ground-truth. During inference, we use the arithmetic square root of predicted mIoU score multiply box classification score as human parsing final score.

Extensive experiments are conducted on two challenging benchmarks, CIHP [8] and MHP-v2 [45], demonstrating that our proposal RP R-CNN significantly outperforms the state-of-the-art for both bottom-up and top-down methods. As shown is Fig. 1, RP R-CNN is more accurate in segmenting small parts and human edges, and the predicted parsing scores can better reflect the quality of instance parsing maps. The main contributions of this work are summarized as follows:

- A novel RP R-CNN is proposed to solve the issue of missing global semantic information and inaccurate scoring of instance parsing maps in top-down multiple human parsing.
- We introduce an effective method to improve the multiple human parsing results by fusing global and instance-level human parts segmentation.
- The proposed RP R-CNN achieves state-of-the-art on two challenging benchmarks. On CIHP val set, RP R-CNN yields 2.0 points mIoU and 7.0 points AP_{50}^p improvements compared with Parsing R-CNN [40]. On MHP-v2 val set, RP R-CNN outperforms Parsing R-CNN by 13.9 points AP_{50}^p and outperforms CE2P [32] by 6.0 points AP_{50}^p, respectively.

Our code and models of RP R-CNN are publicly available.

2   Related Work

Multi-scale Feature Representations. Multi-scale feature is widely used in computer vision tasks [19,23,25,38]. Long *et al.* [25] combine coarse, high layer information with fine, low layer information to generate fine features with high resolution, which greatly promotes the development of semantic segmentation. Lin *et al.* [25] present the feature pyramid network (FPN), and adopt it in object detection, greatly improve the performance of the small object. FPN is a feature pyramid with high-level semantics throughout, through top-down pathway and lateral connections. With the success of FPN, some researches introduce it into other tasks. Panoptic feature pyramid network (PFPN) [19] is proposed by Kirillov *et al.* and applied to panoptic segmentation. PFPN upsamples the feature pyramids and fuse them to the same spatial resolution, then a semantic segmentation branch is attached to generate high-resolution semantic features. However, the computation cost of PFPN is too large, and it only has a single scale semantic feature. Our GSE-FPN solves the above problems well, which adopts a lightweight up-sampling method, and use the global semantic feature to enhance the multi-scale feature.

Instance Scoring. Scoring the predicted instance is a challenging question. The R-CNN series [6,31] of object detection approaches use the object classification score as the confidence of detection results. Recent studies [17,34,46] believe that it cannot accurately reflect the consistency between the predicted bounding-box and the ground-truth. Jiang *et al.* [17] present the IoU-Net, which adopts a IoU-prediction branch to predict the IoU between the predicted bounding box and the corresponding ground truth. Tan *et al.* [34] propose the Learning-to-Rank (LTR) model to produce a ranking score, which is based on IoU to indicate the ranks of candidates during the NMS step. Huang *et al.* [15] consider the difference between classification score and mask quality is greater in instance segmentation. They proposed Mask Scoring R-CNN, which uses a MaskIoU head to predict the quality of mask result. Different from these studies, this work analyzes and solves the inaccurate scoring of instance parsing maps for the first time. The proposed PRSN is concise yet effective, and reducing the gap between score and instance parsing quality.

Multiple Human Parsing. Before the popularity of convolutional neural network, some methods [26,39,41] using hand-crafted visual features and low-level image decompositions have achieved considerable results on single human parsing. However, limited by the representation ability of features, these traditional methods can not be well extended to multiple human parsing. With the successful development of convolutional neural networks [13,20,33] and open source of large-scale multiple human parsing datasets [8,45], some recent researches [8,32,40] have achieved remarkable results in instance-level multiple human parsing. Gong *et al.* [8] present the part grouping network (PGN), which is a typical bottom-up method for instance-level multiple human parsing. PGN reformulates multiple human parsing as semantic part segmentation task and

Table 1. Upper bound analysis of instance-level multiple human parsing via using ground-truth. All models are trained on CIHP `train` set and evaluated on CIHP `val` set. We replace the Bbox branch output with ground-truth box, replace the Parsing branch output with ground-truth segmentation or replace the instance score with ground-truth IoU, respectively. The results suggest that there is still room for improvement in human parsing and scoring.

Backbones	GT-box	GT-parsing	GT-score	mIoU	AP_{50}^P	AP_{vol}^P	PCP_{50}
R50-FPN				56.2	64.6	54.3	60.9
	✓			$58.4_{(+2.2)}$	$58.0_{(-6.6)}$	$51.5_{(-2.8)}$	$62.3_{(+1.4)}$
		✓		$87.8_{(+31.6)}$	$91.4_{(+26.8)}$	$83.6_{(+29.3)}$	$90.6_{(+29.7)}$
			✓	$57.4_{(+1.2)}$	$73.7_{(+9.1)}$	$60.8_{(+6.5)}$	$60.9_{(+0.0)}$

instance-aware edge detection task, the former is used to assign each pixel as human part and the latter is used to group semantic part into different human instances. Ruan et al. [32] rethink and analyze the problems of feature resolution, global context information and edge details in human parsing task, and propose Context Embedding with Edge Perceiving (CE2P) framework for single human parsing. CE2P is a very successful two-stage top-down method, and wins the 1st places on three tracks in the 2018 2nd LIP Challenge. Parsing R-CNN [40] is proposed by Yang et al., which is a one-stage top-down method for multiple human parsing. Based on the in-depth analysis of human appearance characteristics, Parsing R-CNN has made an effective extension on region-based approaches [6,12,23,31] and significantly improved the performance of human parsing. Our work is based on the Parsing R-CNN framework, firstly introducing the global semantic information and reducing the gap between score and instance parsing quality for the top-down methods.

3 Renovating Parsing R-CNN

Our goal is to solve the issue of missing global semantic information and inaccurate scoring of instance parsing map in top-down multiple human parsing pipeline. In this section, we will introduce the motivation, architecture, and components of RP R-CNN in detail.

3.1 Motivation

In the top-down multiple human parsing pipeline, the network outputs three results: bounding-box, instance parsing map and parsing score. The importance of three outputs to the network performance is different. We take Parsing R-CNN [40] as baseline, and make an upper bound analysis of the three outputs. As shown in Table 1, we replace the predicted bounding-box with ground-truth, the multiple human parsing increases by 2.2 points mIoU [25], but AP_{50}^P and AP_{vol}^P [45] decrease. But when we replace the corresponding network output with

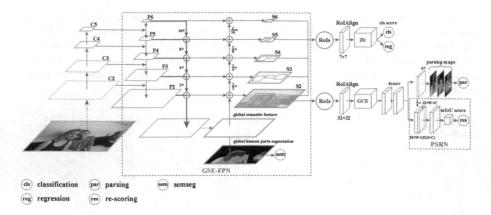

Fig. 2. RP R-CNN architecture. The input image is fed into a backbone with Global Semantic Enhanced FPN [23] to generate RoIs via RPN [31] (not shown in the figure) and RoI features via RoIAlign [12]. The global human parts segmentation is used to supervise and generate the global semantic feature. The BBox branch is standard component of Faster R-CNN which is used to detect human instance. The Parsing branch is mainly composed of GCE module [40] and Parsing Re-Scoring Network for predicting parsing maps and mIoU scores.

the ground-truth of parsing map and mIoU, all the evaluation metrics have significant improvements. In particular, after the adoption of ground-truth parsing map, each evaluation metric has increased by about 30 points. These experimental results show that the accuracy of bounding-box has no significant impact on the multiple human parsing performance.However, the predicted parsing map and score still have a lot of room for improvement and not been paid enough attention by current studies. This is the motivation for our work.

3.2   Architecture

As illustrated in Fig. 2, the proposed RP R-CNN involves four components: Backbone, GSE-FPN, Detector (RPN and BBox branch), and Parsing branch with PRSN. The settings of Backbone and Detector are the same as Parsing R-CNN [40]. The GSE-FPN is attached to the Backbone to generate multi-scale features with global semantic information. The Parsing branch consists of GCE module, parsing map output and Parsing Re-Scoring Network.

3.3   Global Semantic Enhanced Feature Pyramid Network

RoIAlign [12] aims to obtain the features of a specific region on the feature map, so that each instance can be processed separately. However, this makes the instance unable to directly perceive the global (context) information in branch. Global representation is crucial for human parsing, because we not only need to distinguish human body and background, but also give each pixel corresponding

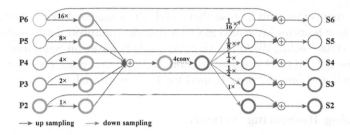

Fig. 3. Global semantic enhanced feature pyramid network (GSE-FPN). Circle is used to represent the feature map, and the circle thickness of circle is used to represent the spatial scale. The semantic segmentation loss is omitted here.

category through understanding the pose and recognizing the clothes the person wears [9]. Therefore, the information about the environment and objects around the human body is helpful for network learning. Some methods [43] perceive more valuable information by changing the area selected by RoIPool/RoIAlign. Different from these, we hope that by explicitly enhancing the global semantic representation of multi-scale features before RoIAlign. As a concrete example, the proposed GSE-FPN are illustrated in Fig. 3, we adopt group normalization [36] and ReLU activation [28] after each convolutional layer.

High-Resolution Feature. For semantic segmentation, high-resolution feature is necessary for generating high-quality results. Dilated convolution is an effective operation, which is adopted by many state-of-the-art semantic segmentation methods [3,42,44]. But dilated convolution substantially increases computation cost, and limits the use of multi-scale features. To keep the efficiency of network, and generate high-resolution features, we extend the multi-scale outputs of FPN [23]. Specifically, we up-sample the FPN generated multi-scale features to the scale of 'P2' level by bilinear interpolation, which is 1/4 resolution of the original image. Each feature map is followed by a 1 × 1 256-d convolutional layer for aligning to the same semantic space, then these feature maps are fused together to generate high-resolution features.

Global Semantic Feature. As shown in Fig. 3, we stack four 3 × 3 256-d convolutional layers after the high-resolution features to generate global semantic feature. Such a design is simple enough, but also can improve the representation ability of the network. In fact, we have tried some popular enhancement modules for semantic segmentation tasks, such as PPM [44] and ASPP [3,4], but experiments show that these modules are not helpful to improve human parsing performance. A 1×1 C-dimension (C is the category number) convolutional layer is attached to the global semantic feature to predict human part segmentation.

Multi-scale Features Fusion. Through the above structure, we can get high-resolution global semantic feature. It is well known that semantic representation can bring performance gains to bounding-box classification and regression [2,12]. Therefore, we down-sample the global semantic feature to the scales of $P3$–$P6$, and use element-wise sum to fuse them with the same scale FPN features. The generated new features are called global semantic enhanced multi-scale features, and denoted as $S2$–$S6$. We follow the Proposals Separation Sampling [40] strategy that the $S2$–$S6$ level features are adopted for extracting region features for BBox branch and only $S2$ level is used for Parsing branch.

3.4   Parsing Re-Scoring Network

Parsing Re-Scoring Network (PRSN) aims to predict accurate mIoU score for each instance parsing map, and can be flexibly integrated into the Parsing branch.

Concise and Lightweight Design. PRSN follows the concise and lightweight design, which will not bring too much computation cost to model training and inference. PRSN receives two inputs, one is the $N \times 512 \times 32 \times 32$ dimension parsing feature map, the other is the $N \times C \times 128 \times 128$ dimension segmentation probability map (N is the number of RoIs, C is the category number). A max pooling layer with stride $= 4$ and kernel $= 4$ is adopted to make the probability map has the same spatial scale with parsing feature map. The down-sampled probability map and parsing feature map are concatenated together, then followed by two 3×3 128-d convolutional layers. A final global average pooling layer, two 256-d fully connected layers, and MSE loss to regress the mIoU between the predicted instance parsing map and ground-truth.

IoU-Aware Ground-Truth. We define the mIoU between the predicted instance parsing map and matched ground-truth as regression target for PRSN. The common Parsing branch can output the segmentation probability map of each human instance, and calculate the loss with the segmentation ground-truth through a cross entropy function. Therefore, the mIoU between them can be calculated directly in the existing framework. It is worth noting that since the instance parsing ground-truth depends on the predicted region of Bbox branch, there is some deviation from the true location of human instance. However, we find that this deviation does not affect the effect of the predicting parsing score, so we do not make corrections to this deviation.

3.5   Training and Inference

As we introduce new supervision into RP R-CNN, there are some changes in the training and inference phases compared with the common methods [30,40].

Training. There are three losses for global human parts segmentation and Parsing branch: $\mathcal{L}_{\text{sem}}$ (segmentation loss), $\mathcal{L}_{\text{par}}$ (parsing loss), $\mathcal{L}_{\text{res}}$ (re-scoring loss). The segmentation loss and parsing loss are computed as a per-pixel cross entropy loss between the predicted segmentation and the ground-truth labels. We use the MSE loss as re-scoring loss. We have observed that the losses from three tasks have different scales and normalization policies. Simply adding them degrades the overall performance. This can be corrected by a simple loss re-weighting

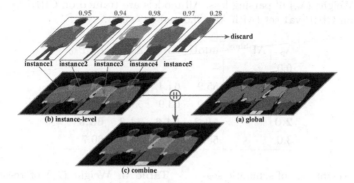

Fig. 4. Combination strategy for generating semantic segmentation results. '$\|$' symbol represents element-wise OR operation.

Table 2. Results of Parsing R-CNN [40] on the CIHP and MHP-v2 datasets. 'our impl.' denotes our implementation of Parsing R-CNN, which uses GN [36] in Parsing branch to stabilize the training.

Dataset	Method	mIoU	AP^p_{50}	AP^p_{vol}	PCP_{50}
CIHP	Parsing R-CNN [40]	56.3	63.7	53.9	60.1
	Parsing R-CNN (our impl.)	56.2	64.6	54.2	60.9
	Δ	-0.1	$+0.9$	$+0.4$	$+0.8$
MHP-v2	Parsing R-CNN [40]	36.2	24.5	39.5	37.2
	Parsing R-CNN (our impl.)	35.5	26.6	40.3	37.9
	Δ	-0.7	$+2.1$	$+0.8$	$+0.7$

strategy. Considering the losses of the detection sub-network, the whole network loss $\mathcal{L}$ can be written as:

$$\mathcal{L} = \mathcal{L}_{rpn} + \mathcal{L}_{bbox} + \lambda_p \mathcal{L}_{par} + \lambda_s \mathcal{L}_{sem} + \lambda_r \mathcal{L}_{res}. \quad (1)$$

The $\mathcal{L}_{rpn}$ and $\mathcal{L}_{bbox}$ are losses of RPN and BBox branch, each of which is composed of classification loss and box regression loss. By tuning λ_p, λ_s and λ_r, it is possible to make the network converge to optimal performance.

Inference. For network inference, we select top 100 candidate bounding-boxes per image from the human detection results. These candidates are fed into Parsing branch to predict instance parsing map and mIoU score. However, the mIoU score is only trained by positive samples, which leads to the lack of the ability to suppress negative samples. So we fuse mIoU score $\mathcal{S}_{iou}$ and classification score $\mathcal{S}_{cls}$ to generate the final parsing score $\mathcal{S}_{parsing} = \sqrt{\mathcal{S}_{cls} * \mathcal{S}_{iou}}$. In addition, we also find that the global human parts segmentation results are complementary to the Parsing branch result, the former has higher recall for each foreground, and the latter has better details. Thus, when generating semantic segmenta-

Table 3. Weight (λ_p) of parsing loss. All models are trained on CIHP `train` set and evaluated on CIHP `val` set (with $\lambda_s = 0.0$ and $\lambda_r = 0.0$).

λ_p	AP^{bbox}	mIoU	AP^p_{50}	AP^p_{vol}	PCP_{50}
0.0	69.1	–	–	–	–
0.5	67.7	55.9	64.4	53.7	59.9
1.0	68.5	55.9	63.9	53.8	60.6
2.0	68.3	**56.2**	**64.6**	**54.3**	**60.9**
3.0	67.8	55.9	**64.6**	54.2	60.7

Table 4. Weight (λ_s) of semantic segmentation loss (with $\lambda_r = 0.0$).

λ_s	AP^{bbox}	mIoU	AP^p_{50}	AP^p_{vol}	PCP_{50}
0.0	69.1	56.2	64.6	54.3	60.9
0.5	67.9	57.0	65.1	54.6	61.1
1.0	67.7	57.8	66.5	55.0	61.7
2.0	67.4	**58.2**	**67.4**	**55.5**	**62.1**
3.0	67.1	58.0	67.4	55.3	61.8
Δ		+2.0	+2.8	+1.2	+1.2

Table 5. Weight (λ_r) of re-scoring loss (with $\lambda_s = 0.0$).

λ_s	AP^{bbox}	mIoU	AP^p_{50}	AP^p_{vol}	PCP_{50}
0.0	69.1	56.2	64.6	54.3	60.9
0.5	68.2	56.3	70.1	57.4	61.1
1.0	68.3	**56.4**	**70.3**	**57.6**	**61.3**
2.0	68.2	56.3	70.2	57.5	**61.3**
3.0	68.1	56.2	**70.3**	57.5	61.1
Δ		+0.2	+5.7	+3.3	+0.4

tion results, we adopt a new combination strategy, as shown in Fig. 4. We filter out the low quality results based on the $S_{parsing}$, and then generate instance-level human parts segmentation (b) *instance-level*. The instance-level (b) was and global human parts segmentation (a) *global* do element-wise OR operation to get the final result (c) *combine*. It is worth noting that if the results of (a) and (b) are different at the same pixel, and both of them are predicted as non-background category, we directly adopt the results of (b). This is because the (b) has a more accurate perception of the human parts inside each instance.

4 Experiments

In this section, we describe experiments on multiple human parsing of RP R-CNN. All experiments are conducted on the CIHP [8] and MHP-v2 [45] datasets. We follow the Parsing R-CNN evaluation protocols. Using mean intersection over union (mIoU) [25] to evaluate the human part segmentation. And using average precision based on part (AP^p) [45] as instance evaluation metric(s).

4.1 Implementation Details

Training Setup. All experiments are based on Pytorch on a server with 8 NVIDIA Titan RTX GPUs. We use 16 batch-size (2 images per GPU) and adopt

Table 6. Semantic segmentation results of different inference methods on CIHP dataset. All models are trained on **train** set and evaluated on **val** set.

Inference methods	mIoU	Pixel acc.	Mean acc.
baseline	56.2	89.3	67.0
(a) semseg	50.2	88.0	61.3
(b) parsing	57.4	89.8	67.1
(c) combine	**58.2**	**90.2**	**69.0**
Δ	*+2.0*	*+0.9*	*+2.0*

Table 7. Ablations of RP R-CNN on CIHP dataset. All models are trained on **train** set and evaluated on **val** set.

Methods	GSE-FPN	PRSN	mIoU	AP_{50}^{P}	AP_{vol}^{P}	PCP_{50}
			56.2	64.6	54.3	60.9
RP R-CNN	✓		$58.2_{(+2.0)}$	$67.4_{(+2.8)}$	$55.5_{(+1.2)}$	$62.1_{(+1.2)}$
		✓	$56.4_{(+0.2)}$	$70.3_{(+5.7)}$	$57.6_{(+3.3)}$	$61.3_{(+0.4)}$
	✓	✓	$58.2_{(+2.0)}$	$71.6_{(+7.0)}$	$58.3_{(+4.0)}$	$62.2_{(+1.3)}$

ResNet50 [13] as backbone. The short side of input image is resized randomly sampled from [512, 864] pixels, and the longer side is limited to 1,400 pixels; inference is on a single scale of 800 pixels. Each image has 512 sampled RoIs for Bbox branch and 16 sampled RoIs for Parsing branch. For CIHP dataset, there are 135,000 iterations (about 75 epochs) of the training process, with a learning rate of 0.02 which is decreased by 10 at the 105,000 and 125,000 iteration. For MHP-v2 dataset, the max iteration is half as long as the CIHP dataset with the learning rate change points scaled proportionally.

Parsing R-CNN Re-implementation. In order to better illustrate the advantages of RP R-CNN, we have re-implemented Parsing R-CNN according to the original paper [40]. We find that the training of Parsing R-CNN is not very stable. We solve this issue by adding group normalization [36] after each convolutional layer of Parsing branch. As shown in Table 2, our re-implemented Parsing R-CNN achieves comparable performance with original version both on CIHP and MHP-v2 datasets. Therefore, this work takes our re-implemented Parsing R-CNN as the baseline.

4.2 Ablation Studies

In this sub-section, we assess the effects of different settings and components on RP R-CNN by details ablation studies.

Table 8. Ablations of RP R-CNN on MHP-v2 dataset. All models are trained on `train` set and evaluated on `val` set.

Methods	GSE-FPN	PRSN	mIoU	AP_{50}^p	AP_{vol}^p	PCP_{50}
RP R-CNN			35.5	26.6	40.3	37.9
	✓		$37.3_{(+1.8)}$	$28.9_{(+2.3)}$	$41.1_{(+0.8)}$	$38.9_{(+1.0)}$
		✓	$35.7_{(+0.2)}$	$39.6_{(+13.0)}$	$44.9_{(+4.6)}$	$38.2_{(+0.3)}$
	✓	✓	$37.3_{(+1.8)}$	$40.5_{(+13.9)}$	$45.2_{(+4.9)}$	$39.2_{(+1.3)}$

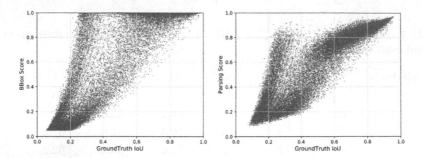

Fig. 5. Comparisons of ground-truth IoU vs. bbox score (**Left**) and ground-truth IoU vs. parsing score (**Right**) on CIHP dataset. All models are trained on `train` set and evaluated on `val` set.

Loss Weights. To combine our GSE-FPN with PRSN in Parsing R-CNN, we need to determine how to train a single, unified network. Previous studies demonstrate that multi-task training is often challenging and can lead to degraded results [18]. We also observe that adding the losses of all tasks directly will not give the best results. But grid searching three hyper-parameters (λ_p, λ_s and λ_r) is very inefficient, so we first determine λ_p, and then determine λ_s and λ_r separately to improve efficiency. As shown in Table 3, we find that the network performance is the best when $\lambda_p = 2.0$. We consider the group normalization layer makes the network convergence more stable, so that a proper large loss weight will bring higher accuracy. Table 4 shows that $\lambda_s = 2.0$ is the proper loss weight for semantic segmentation. Although increasing the weight of global human parts segmentation loss will slightly reduce the accuracy of human detection, the overall performance is improved. Table 5 shows that the re-scoring task is not sensitive to the loss weight, and its loss scale is smaller than other losses, so it has no significant impact on the optimization of other tasks in multi-task training. To sum up, we choose $\lambda_p = 2.0$, $\lambda_s = 2.0$ and $\lambda_r = 1.0$ as the loss weights of Eq. (1).

Inference Methods. The combination inference method proposed in Fig. 4 utilizes the complementarity of global human parts segmentation and Parsing branch results, and we give the detailed results in Table 6. The performance of using (a) *semseg* or (b) *parsing* alone is poor. The combination method (c)

Table 9. Multiple human parsing on the CIHP and MHP-v2 datasets. * denotes longer learning schedule. † denotes using test-time augmentation.

Dataset	Methods	Backbones	Epochs	mIoU	AP^p_{50}	AP^p_{vol}	PCP_{50}
CIHP [8]	Bottom-up						
	PGN† [8]	ResNet101	~80	55.8	34.0	39.0	61.0
	DeepLab v3+ [4]	Xception	100	58.9	–	–	–
	Graphonomy [9]	Xception	100	58.6	–	–	–
	GPM [11]	Xception	100	60.3	–	–	–
	Grapy-ML [11]	Xception	200	60.6	–	–	–
	Two-stage top-down						
	M-CE2P [32]	ResNet101	150	59.5	–	–	–
	BraidNet [24]	ResNet101	150	60.6	–	–	–
	SemaTree [16]	ResNet101	200	60.9	–	–	–
	One-stage top-down						
	Parsing R-CNN [40]	ResNet50	75	56.3	63.7	53.9	60.1
	Parsing R-CNN† [40]	ResNeXt101	75	61.1	71.2	56.5	67.7
	Parsing R-CNN (our impl.)	ResNet50	75	56.2	64.6	54.3	60.9
	Unified [30]	ResNet101	~37	55.2	51.0	48.0	–
	RP R-CNN (ours)	ResNet50	75	58.2	71.6	58.3	62.2
	RP R-CNN (ours)*	ResNet50	150	60.2	74.1	59.5	64.9
	RP R-CNN (ours)*†	**ResNet50**	150	**61.8**	**77.2**	**61.2**	**70.5**
MHP-v2 [45]	Bottom-up						
	MH-Parser [21]	ResNet101	–	–	17.9	36.0	26.9
	NAN [45]	–	~80	–	25.1	41.7	32.2
	Two-stage top-down						
	M-CE2P [32]	ResNet101	150	**41.1**	34.5	42.7	**43.8**
	SemaTree [16]	ResNet101	200	–	34.4	42.5	43.5
	One-stage top-down						
	Mask R-CNN [12]	ResNet50	–	–	14.9	33.8	25.1
	Parsing R-CNN [40]	ResNet50	75	36.2	24.5	39.5	37.2
	Parsing R-CNN (our impl.)	ResNet50	75	35.5	26.6	40.3	37.9
	RP R-CNN (ours)	ResNet50	75	37.3	40.5	45.2	39.2
	RP R-CNN (ours)*	**ResNet50**	150	38.6	**45.3**	**46.8**	**43.8**

combine can significantly improve the metrics of semantic segmentation, and outperforms the baseline by 2 points mIoU.

Ablations on RP R-CNN. In Table 7, we perform the additional ablations of RP R-CNN on CIHP dataset. We observe that GSE-FPN is very helpful to the global human parts segmentation, which yields 2.0 points mIoU improvement. In addition, the global semantic feature also improves the instance metrics, AP^p_{50}, AP^p_{vol} and PCP_{50} increase 2.8, 1.2 and 1.2 points respectively. With PRSN, the improvements of instance metrics are very significant, AP^p_{50} improves 5.7 points, and AP^p_{vol} improves 3.3 points. With GSE-FPN and PRSN, our proposed RP

Fig. 6. Qualitative results of RP R-CNN on the CIHP and MHP-v2 datasets.

R-CNN achieves 58.2 mIoU and 71.6 AP_{50}^p on CIHP. The additional ablations on MHP-v2 dataset is shown in Table 8. Through GSE-FPN and PRSN, the performance of human parsing is also significant improved. Particularly, AP_{50}^p and AP_{vol}^p are raised considerably by PRSN, 13.0 points and 4.6 points, respectively.

4.3 Comparison with State-of-the-Arts

We evaluate RP R-CNN on the CIHP and MHP-v2 datasets and compare the results to state-of-the-art including bottom-up and one-stage/two-stage top-down methods, shown in Table 9. On CIHP dataset, our proposed RP R-CNN achieves 58.2 mIoU and 71.6 AP_{50}^p, which surpasses Parsing R-CNN [40] in all respects. Compared with PGN [8], the performance advantage of RP R-CNN is huge, and AP_{50}^p is even 37.6 points ahead. With longer learning schedule (150 epochs), RP R-CNN achieves compared performance with one-stage top-down methods, e.g. M-CE2P and BraidNet. Even though we have adopted a lighter backbone (ResNet50 vs. ResNet101). Finally, using test-time augmentation, RP R-CNN with RestNet50 achieves state-of-the-art performance on CIHP.

On MHP-v2 dataset, RP R-CNN achieves excellent performance. We can observe that our RP R-CNN outperforms Parsing R-CNN consistently for all the evaluation metrics. And compared with M-CE2P, RP R-CNN yields about 10.8 point AP_{50}^p and 4.1 points AP_{vol}^p improvements. With RestNet50 backbone, it gives new state-of-the-art of 45.3 AP_{50}^p, 46.8 AP_{vol}^p and 43.8 PCP_{50}.

4.4 Analysis and Discussion

Effect of Parsing Re-Scoring Network. Figure 5 shows that the correlation between mIoU of the predicted parsing map with the matched ground-truth and the bbox/parsing score. As shown, the parsing score has better correlation with the ground-truth, especially for high-quality parsing map. However, it is difficult to score low-quality parsing map, which is the source of some false positive detections. Therefore, the evaluation of low-quality prediction is still a problem to be solved.

Qualitative results. We visualize multiple human parsing results of RP R-CNN in Fig. 6. We can observe that RP R-CNN has a good applicability to dense crowds and occlusions. In addition, the parsing score predicted by RP R-CNN reflects the quality of the parsing map.

5 Conclusions

In this paper, we proposed a novel Renovating Parsing R-CNN (RP R-CNN) model for solving the issue of missing global semantic information and inaccurate scoring of parsing result in top-down multiple human parsing. By explicitly introducing global semantic enhanced multi-scale features and learning the mIoU between instance parsing map and matched ground-truth, our RP R-CNN outperforms previous state-of-the-art methods consistently for all the evaluation metrics. In addition, we also adopt a new combination strategy, which improves the results of multiple human parsing by global semantic segmentation and instance-level semantic segmentation. We hope our effective approach will serve as a cornerstone and help the future research in multiple human parsing.

References

1. Chao, Y.W., Wang, Z., He, Y., Wang, J., Deng, J.: HICO: a benchmark for recognizing human-object interactions in images. In: ICCV (2015)
2. Chen, K., et al.: Hybrid task cascade for instance segmentation. In: CVPR (2019)
3. Chen, L., Papandreou, G., Schroff, F., Adam, H.: Rethinking atrous convolution for semantic image segmentation. arXiv:1706.05587 (2017)
4. Chen, L.-C., Zhu, Y., Papandreou, G., Schroff, F., Adam, H.: Encoder-decoder with atrous separable convolution for semantic image segmentation. In: Ferrari, V., Hebert, M., Sminchisescu, C., Weiss, Y. (eds.) ECCV 2018. LNCS, vol. 11211, pp. 833–851. Springer, Cham (2018). https://doi.org/10.1007/978-3-030-01234-2_49
5. Girdhar, R., Ramanan, D.: Attentional pooling for action recognition. In: NIPS (2017)
6. Girshick, R.: Fast R-CNN. In: ICCV (2015)
7. Gkioxari, G., Girshick, R., Dollar, P., He, K.: Detecting and recognizing human-object interactions. In: CVPR (2018)
8. Gong, K., Liang, X., Li, Y., Chen, Y., Yang, M., Lin, L.: Instance-level human parsing via part grouping network. In: Ferrari, V., Hebert, M., Sminchisescu, C., Weiss, Y. (eds.) ECCV 2018. LNCS, vol. 11208, pp. 805–822. Springer, Cham (2018). https://doi.org/10.1007/978-3-030-01225-0_47

9. Gong, K., Gao, Y., Liang, X., Shen, X., Lin, L.: Graphonomy: universal human parsing via graph transfer learning. In: CVPR (2019)
10. Guler, R., Neverova, N., Kokkinos, I.: DensePose: dense human pose estimation in the wild. In: CVPR (2018)
11. He, H., Zhang, J., Zhang, Q., Tao, D.: Grapy-ML: graph pyramid mutual learning for cross-dataset human parsing. In: AAAI (2020)
12. He, K., Gkioxari, G., Dollár, P., Girshick, R.: Mask R-CNN. In: ICCV (2017)
13. He, K., Zhang, X., Ren, S., Sun, J.: Deep residual learning for image recognition. In: CVPR (2016)
14. Hsieh, C.W., Chen, C.Y., Chou, C.L., Shuai, H.H., Liu, J., Cheng, W.H.: FashionOn: semantic-guided image-based virtual try-on with detailed human and clothing information. In: ACM MM (2019)
15. Huang, Z., Huang, L., Gong, Y., Huang, C., Wang, X.: Mask scoring R-CNN. In: CVPR (2019)
16. Ji, R., et al.: Learning semantic neural tree for human parsing. arXiv:1912.09622 (2019)
17. Jiang, B., Luo, R., Mao, J., Xiao, T., Jiang, Y.: Acquisition of localization confidence for accurate object detection. In: Ferrari, V., Hebert, M., Sminchisescu, C., Weiss, Y. (eds.) Computer Vision – ECCV 2018. LNCS, vol. 11218, pp. 816–832. Springer, Cham (2018). https://doi.org/10.1007/978-3-030-01264-9_48
18. Kendall, A., Gal, Y., Cipolla, R.: Multi-task learning using uncertainty to weigh losses for scene geometry and semantics. In: CVPR (2018)
19. Kirillov, A., Girshick, R., He, K., Dollár, P.: Panoptic feature pyramid networks. In: CVPR (2019)
20. Krizhevsky, A., Sutskever, I., Hinton, G.: ImageNet classification with deep convolutional neural networks. In: NIPS (2012)
21. Li, J., et al.: Multi-human parsing in the wild. arXiv:1705.07206 (2017)
22. Li, W., Zhu, X., Gong, S.: Harmonious attention network for person re-identification. In: CVPR (2018)
23. Lin, T., Dollár, P., Girshick, R., He, K., Hariharan, B., Belongie, S.: Feature pyramid networks for object detection. In: CVPR (2017)
24. Liu, X., Zhang, M., Liu, W., Song, J., Mei, T.: BraidNet: braiding semantics and details for accurate human parsing. In: ACM MM (2019)
25. Long, J., Shelhamer, E., Darrell, T.: Fully convolutional networks for semantic segmentation. In: CVPR (2015)
26. Luo, P., Wang, X., Tang, X.: Pedestrian parsing via deep decompositional network. In: ICCV (2013)
27. Miao, J., Wu, Y., Liu, P., Ding, Y., Yang, Y.: Pose-guided feature alignment for occluded person re-identification. In: ICCV (2019)
28. Nair, V., Hinton, G.: Rectified linear units improve restricted Boltzmann machines. In: ICML (2010)
29. Qi, S., Wang, W., Jia, B., Shen, J., Zhu, S.-C.: Learning human-object interactions by graph parsing neural networks. In: Ferrari, V., Hebert, M., Sminchisescu, C., Weiss, Y. (eds.) ECCV 2018. LNCS, vol. 11213, pp. 407–423. Springer, Cham (2018). https://doi.org/10.1007/978-3-030-01240-3_25
30. Qin, H., Hong, W., Hung, W.C., Tsai, Y.H., Yang, M.H.: A top-down unified framework for instance-level human parsing. In: BMVC (2019)
31. Ren, S., He, K., Girshick, R., Sun, J.: Faster R-CNN: towards real-time object detection with region proposal networks. In: NIPS (2015)
32. Ruan, T., et al.: Devil in the details: towards accurate single and multiple human parsing. In: AAAI (2019)

33. Russakovsky, O., et al.: ImageNet large scale visual recognition challenge. IJCV **115**, 211–252 (2015). https://doi.org/10.1007/s11263-015-0816-y
34. Tan, Z., Nie, X., Qian, Q., Li, N., Li, H.: Learning to rank proposals for object detection. In: ICCV (2019)
35. Tangseng, P., Wu, Z., Yamaguchi, K.: Retrieving similar styles to parse clothing. TPAMI (2014)
36. Wu, Y., He, K.: Group normalization. In: Ferrari, V., Hebert, M., Sminchisescu, C., Weiss, Y. (eds.) ECCV 2018. LNCS, vol. 11217, pp. 3–19. Springer, Cham (2018). https://doi.org/10.1007/978-3-030-01261-8_1
37. Xiao, B., Wu, H., Wei, Y.: Simple baselines for human pose estimation and tracking. In: Ferrari, V., Hebert, M., Sminchisescu, C., Weiss, Y. (eds.) ECCV 2018. LNCS, vol. 11210, pp. 472–487. Springer, Cham (2018). https://doi.org/10.1007/978-3-030-01231-1_29
38. Xiao, T., Liu, Y., Zhou, B., Jiang, Y., Sun, J.: Unified perceptual parsing for scene understanding. In: Ferrari, V., Hebert, M., Sminchisescu, C., Weiss, Y. (eds.) ECCV 2018. LNCS, vol. 11209, pp. 432–448. Springer, Cham (2018). https://doi.org/10.1007/978-3-030-01228-1_26
39. Yamaguchi, K., Kiapour, M.H., Ortiz, L.E., Berg, T.L.: Parsing clothing in fashion photographs. In: CVPR (2012)
40. Yang, L., Song, Q., Wang, Z., Jiang, M.: Parsing R-CNN for instance-level human analysis. In: CVPR (2019)
41. Yang, W., Luo, P., Lin, L.: Clothing co-parsing by joint image segmentation and labeling. In: CVPR (2014)
42. Yu, F., Koltun, V.: Multi-scale context aggregation by dilated convolutions. In: ICLR (2016)
43. Zeng, X., et al.: Crafting GBD-net for object detection. TPAMI **40**(9), 2109–2123 (2017)
44. Zhao, H., Shi, J., Qi, X., Wang, X., Jia, J.: Pyramid scene parsing network. In: CVPR (2017)
45. Zhao, J., Li, J., Cheng, Y., Feng, J.: Understanding humans in crowded scenes: deep nested adversarial learning and a new benchmark for multi-human parsing. In: ACM MM (2018)
46. Zhu, B., Song, Q., Yang, L., Wang, Z., Liu, C., Hu, M.: CPM R-CNN: calibrating point-guided misalignment in object detection. arXiv:2003.03570 (2020)

Multi-task Curriculum Framework
for Open-Set Semi-supervised Learning

Qing Yu[1(✉)], Daiki Ikami[1,2], Go Irie[2], and Kiyoharu Aizawa[1]

[1] The University of Tokyo, Bunkyo, Japan
{yu,ikami,aizawa}@hal.t.u-tokyo.ac.jp
[2] NTT Corporation, Chiyoda, Japan
goirie@ieee.org

Abstract. Semi-supervised learning (SSL) has been proposed to leverage unlabeled data for training powerful models when only limited
labeled data is available. While existing SSL methods assume that samples in the labeled and unlabeled data share the classes of their samples,
we address a more complex novel scenario named open-set SSL, where
out-of-distribution (OOD) samples are contained in unlabeled data.
Instead of training an OOD detector and SSL separately, we propose a
multi-task curriculum learning framework. First, to detect the OOD samples in unlabeled data, we estimate the probability of the sample belonging to OOD. We use a joint optimization framework, which updates
the network parameters and the OOD score alternately. Simultaneously,
to achieve high performance on the classification of in-distribution (ID)
data, we select ID samples in unlabeled data having small OOD scores,
and use these data with labeled data for training the deep neural networks to classify ID samples in a semi-supervised manner. We conduct
several experiments, and our method achieves state-of-the-art results by
successfully eliminating the effect of OOD samples.

Keywords: Semi-supervised learning · Out-of-distribution detection ·
Multi-task learning

1 Introduction

After several breakthroughs in deep learning methods, deep neural networks
(DNNs) have achieved impressive results and even outperformed humans on
various machine perception tasks such as image classification [8,26], face recognition [18], and natural language processing [6] with large-scale, annotated training samples. However, creating these large datasets is typically time-consuming
and expensive.

To solve this problem, semi-supervised learning (SSL) is proposed to leverage unlabeled data to improve the performance of a model when only limited
labeled data is available. SSL is able to train large, powerful models when labeling
data is expensive or inconvenient. There is a diverse collection of approaches to

© Springer Nature Switzerland AG 2020
A. Vedaldi et al. (Eds.): ECCV 2020, LNCS 12357, pp. 438–454, 2020.
https://doi.org/10.1007/978-3-030-58610-2_26

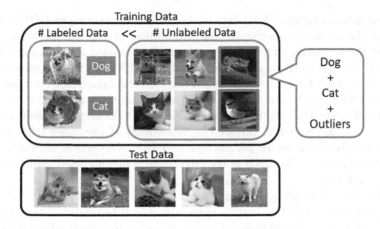

Fig. 1. Problem setting of open-set SSL. Outliers, which do not belong to any class of labeled data, exist in the unlabeled data.

SSL. For example, one approach is consistency regularization [15,24,28], which encourages a model to produce the same prediction when the input is perturbed. Another approach, entropy minimization [7], encourages the model to produce high-confidence predictions. The recent state-of-the-art method, MixMatch [2], combines the aforementioned techniques in a unified loss function and achieves strong performance on a variety of image classification benchmarks.

These existing SSL methods assume that the labeled and unlabeled data have the same distribution, meaning that they share the classes of their samples, and there is no outlier sample in unlabeled data. However, in the real world, it is hard to ensure that the unlabeled data does not contain any out-of-distribution (OOD) sample that is drawn from different distributions. Oliver et al. [20] have shown that adding unlabeled data from a mismatched set of classes can actually damage the performance of SSL.

Hence, we consider a new, realistic setting called "Open-Set Semi-supervised Learning", as shown in Fig. 1. Outliers, which do not belong to the classes of labeled data, exist in the unlabeled data, and the model should be trained on labeled and unlabeled data by eliminating the effect of these outliers. To the best of our knowledge, our study is the first to tackle the problem of open-set SSL.

Although there are many algorithms for detecting OOD samples [9,16,17,29], these methods are trained on a large number of labeled in-distribution (ID) samples with class labels. In the setting of SSL, the number of labeled data is very limited. Hence, the previous methods cannot achieve high performance of detection and are not suitable for open-set SSL. Therefore, we propose a method that uses multi-task curriculum learning, which is a multi-task framework aiming to solve OOD detection and SSL simultaneously.

First, we detect OOD samples in the unlabeled data. We propose a new OOD detection method by a joint optimization framework, which can utilize the unlabeled data containing OOD data in the process of training an OOD detector. We train the network to estimate the probability of the sample belonging to OOD. At the beginning of training, we treat all unlabeled samples as OOD and all labeled samples as ID by assigning an initial OOD score to each sample (0 for labeled data and 1 for unlabeled data). Next, we train the model to classify the sample as OOD or ID. Since unlabeled data also contains a reasonable amount of ID samples, treating all unlabeled samples as OOD samples would result in incorrect label assignments. Inspired by a solution of the noisy label problem [27], we update the network parameters and the OOD scores alternately as a joint optimization to clean the noisy OOD scores of unlabeled samples, which ranges from 0 to 1.

At the same time, while training the network for OOD detection, we also train the network to classify ID samples correctly, which forms multi-task learning. Since ID samples in the unlabeled data are expected to have smaller OOD scores than the real OOD samples, we use curriculum learning that excludes the samples with higher OOD scores in unlabeled data. Then we combine remaining ID unlabeled samples with labeled data for training the CNN to classify ID samples correctly by any SSL method, where MixMatch [2] is used in this paper.

We evaluate our method on a diverse set of open-set SSL settings. In many settings, our method outperforms existing methods by a large margin. We summarize the contributions of this paper as the following:

- We propose a novel experimental setting and training methodology for open-set SSL.
- We propose a multi-task curriculum learning framework that detects OOD samples by alternate optimization and classifies ID samples by applying SSL according to the results of OOD detection.
- We evaluate our method across several open-set SSL tasks and outperforms state-of-the-art by a considerable margin. Our approach successfully eliminates the effect of OOD samples in the unlabeled data.

2   Related Work

At present, there are several different methods of SSL and OOD detection. We will briefly explain some important studies in this section.

2.1   Semi-supervised Learning

Although there are many studies on SSL techniques, such as transductive models [10,11], graph-based methods [33] and generative modeling [13,23,25], we focus mainly on the recent state-of-the-art methods, based on consistency regularization [15,24,28].

In general supervised learning, data augmentation is a common regularization technique. In image classification, it is common to add some noise to an input image to change the pixel values of an image but keep its label [4], which means data augmentation is able to artificially increase the size of a training set by generating new modified data.

Consistency regularization is a method that applies data augmentation to SSL. It imposes a constraint in the form of regularization so that the classification result of each unlabeled sample does not change before and after augmentation.

In the simplest method, Laine and Aila [15] proposed π-model, which applies two different stochastic augmentations to an unlabeled data to generate two inputs and minimize the distance of the two network outputs of these two inputs.

Mean Teacher [28] used an exponential moving average of network parameter values to generate a more stable output on one of the two inputs, instead of generating two outputs for the two inputs by the same network, to improve the effectiveness of their method.

The state-of-the-art method for SSL is MixMatch [2], which works by guessing low-entropy labels for data-augmented unlabeled examples and mixing labeled and unlabeled data using MixUp [32]. MixMatch [2] then uses π-model to train a model using the mixed labeled and unlabeled data. We refer the reader to their paper [2] for further details.

However, these methods assume that the labeled and unlabeled data share the classes of their samples. When some OOD samples are contained in the unlabeled data, Oliver et al. [20] showed that the existing methods achieved bad performance, which is even lower than the performance of supervised learning trained only by limited labeled data in some cases. The motivation of this research is to solve this problem.

2.2 Out-of-Distribution Detection

There are also some methods for OOD detection. In the simplest method, Hendrycks & Gimpel [9] used the predicted softmax class probability to detect OOD samples. They observed that the prediction probability of incorrect and OOD samples tends to be lower than that of the correct samples. However, they also found that a pre-trained neural network can still classify some OOD samples overconfidently, which limits its performance.

To improve the effectiveness of Hendrycks & Gimpel's method [9], Liang et al. [17] applied temperature scaling and input preprocessing to detect OOD samples, called Out-of-DIstribution detector for Neural networks (ODIN). They found that the difference between the largest logit (the outputs of which are not normalized by softmax) and the remaining logits is larger for ID samples than for OOD samples if the logits are scaled by a large constant (temperature scaling). They showed that the separation of the softmax scores between the ID and OOD samples could be increased by temperature scaling. They also found that the addition of small perturbations to the input (through the loss gradient) increases the maximum predicted softmax score. As a result, the ID samples show a greater increase in score than the OOD samples. Using these techniques, their method outperformed the baseline method [9].

In another method using the predicted probability of the network, Bendale & Boult [1] calculated the score for an unknown class by taking the weighted average of all other classes obtained from a Weibull distribution, named openMax layer.

However, all the methods described earlier need a large number of labeled ID samples to achieve stable results and they are unable to utilize any unlabeled data. In open-set SSL, the number of labeled ID samples is small but we have access to a huge amount of unlabeled data containing some OOD samples. Our method aims at training a model to not only detect OOD samples with limited labeled and plenty of unlabeled data, but also achieve high recognition performance on the classification of ID samples.

3 Method

3.1 Problem Statement

We assume that an ID image-label pair, $\{x_l, y_l\}$, drawn from a set of labeled ID images $\{X_l, Y_l\}$, as well as an unlabeled image, x_{ul}, drawn from a set of unlabeled images X_{ul}, is accessible. The labeled ID sample $\{x_l, y_l\}$ can be classified into one of K classes denoted by $\{c_1, \ldots, c_K\}$, meaning that $y_l \in \{c_1, \ldots, c_K\}$. Besides ID samples, outliers (OOD samples) also exist in the unlabeled data, which signifies that the true class of some unlabeled data x_{ul} is not in $\{c_1, \ldots, c_K\}$.

The goal of our method is to train a model that can correctly classify ID samples into $\{c_1, \ldots, c_K\}$ on a combination of labeled ID samples and unlabeled samples under semi-supervised setting. Our technique achieves this by distinguishing whether the image x_{ul} is from in-distribution to eliminate the negative effect of OOD samples during the training.

3.2 Overall Concept

The most challenging part of this task is the detection of OOD samples when only limited labeled ID samples are available. As mentioned in Sect. 2, traditional OOD detection methods [9,17] assumed that there is a large number of labeled ID samples for training the recognition model and these methods did not utilize unlabeled data in training. Thus, these methods cannot achieve high performance OOD detection in SSL.

We propose a multi-task curriculum learning framework for open-set SSL, which aims to solve OOD detection and SSL simultaneously as a multi-task framework.

Since the number of labeled ID samples is limited in SSL, we use a joint optimization framework inspired by [27], which updates DNN parameters and estimates the probability of the sample belonging to OOD alternately. First, we assign an initial pseudo label representing the probability of the sample belonging to OOD, named OOD score, to all the data. For labeled samples, since they are ID, we initialize the OOD scores as 0 and for unlabeled samples, we initialize the OOD scores as 1. Since some amount of ID samples are present in unlabeled

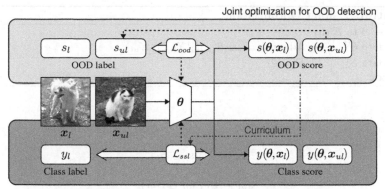

Fig. 2. An overview of our framework. Noisy OOD scores of unlabeled data are reassigned to the outputs of OOD scores by the DNN. The network parameters and OOD scores are alternately updated for each epoch. The unlabeled samples used to calculate semi-supervised loss are selected by their OOD scores.

data, we can consider the binary classification of OOD as a noisy label problem. [27] showed that a DNN trained on noisy labeled datasets does not memorize noisy labels under a high learning rate. Thus, the noisy label of a sample can be corrected by reassigning the probability output of the DNN to the sample as a new label. We utilize this property to increase the number of ID samples during the training by cleaning the OOD scores of unlabeled data. To achieve this, the network parameters of the DNN and the OOD scores of unlabeled samples are alternately updated for each epoch and the OOD scores of unlabeled samples are reassigned to the estimation of OOD scores by DNN.

At the same time, to achieve high performance on the classification of ID samples, we select ID samples in unlabeled data having low OOD scores and combine them with labeled ID samples for training the DNN to classify ID samples correctly in a semi-supervised manner. Although our method can be applied to any SSL method, we choose the state-of-the-art SSL method, MixMatch [2], in this paper.

Instead of training OOD detection and ID classification separately, we further propose a multi-task training framework combining all the steps to formulate an end-to-end trainable network that can detect OOD samples and classify ID samples simultaneously. The overview of our framework is shown in Fig. 2.

3.3   Training Procedure

Noisy Label Optimization for OOD Detection. First, we assign an initial OOD score s to all the data, which makes an ID image-label pair, $\{\boldsymbol{x}_l, y_l\} \in \{X_l, Y_l\}$ become $\{\boldsymbol{x}_l, y_l, s_l\} \in \{X_l, Y_l, S_l\}$, and an unlabeled image, $\boldsymbol{x}_{ul} \in X_{ul}$ become $\{\boldsymbol{x}_{ul}, s_{ul}\} \in \{X_{ul}, S_{ul}\}$. We initialize the OOD scores S_l as 0 for the labeled (ID) samples, whereas the unlabeled samples are assigned the OOD

Algorithm 1. Joint Optimization

---

    for $t \leftarrow 0$ to E **do**

        update $\boldsymbol{\theta}^{(t+1)}$ by Adam on $\mathcal{L}_{ood}(\boldsymbol{\theta}^t, S_{ul}^t | X_l, S_l, X_{ul})$

        update $S_{ul}^{(t+1)}$ by Eq. (6)

    end for

---

score S_{ul} of 1. This is denoted by:

$$s_l = 0 \ (\forall s_l \in S_l), \tag{1}$$

$$s_{ul} = 1 \ (\forall s_{ul} \in S_{ul}). \tag{2}$$

As a binary classification of OOD, parameters of the network $\boldsymbol{\theta}$ can be optimized as follows:

$$\min_{\boldsymbol{\theta}} \mathcal{L}_{ood}(\boldsymbol{\theta} | X_l, S_l, X_{ul}, S_{ul}), \tag{3}$$

$$\mathcal{L}_{ood} = -\frac{1}{|X_l|} \sum_{i=1}^{|X_l|} (s_{l_i} \log s(\boldsymbol{\theta}, \boldsymbol{x}_{l_i}) + (1 - s_{l_i}) \log(1 - s(\boldsymbol{\theta}, \boldsymbol{x}_{l_i})))$$

$$-\frac{1}{|X_{ul}|} \sum_{i=1}^{|X_{ul}|} (s_{ul_i} \log s(\boldsymbol{\theta}, \boldsymbol{x}_{ul_i}) + (1 - s_{ul_i}) \log(1 - s(\boldsymbol{\theta}, \boldsymbol{x}_{ul_i}))), \tag{4}$$

where $\mathcal{L}_{ood}$ denotes the cross entropy loss under the supervision of OOD score and $s(\boldsymbol{\theta}, \boldsymbol{x}_{l_i})$ (or $s(\boldsymbol{\theta}, \boldsymbol{x}_{ul_i})$) denotes the predicted OOD score of the image $\boldsymbol{x}_{l_i}$ (or $\boldsymbol{x}_{ul_i}$) with the network parameters being $\boldsymbol{\theta}$.

However, although the OOD scores S_{ul} are initialized to 1 for all unlabeled samples, the existence of some ID samples in the unlabeled data leads to the classification of OOD as a noisy label problem. Tanaka et al. [27] showed that a network trained with a high learning rate is less likely to overfit to noisy labels, which means the loss Eq. (4) is high for noisy labels and low for clean labels. So we obtain clean OOD scores by updating the OOD scores in the direction to decrease Eq. (4) Hence, we formulate the problem as the joint optimization of the network parameters and OOD scores as follows:

$$\min_{\boldsymbol{\theta}, S_{ul}} \mathcal{L}_{ood}(\boldsymbol{\theta}, S_{ul} | X_l, S_l, X_{ul}), \tag{5}$$

Alternately updating the network parameters $\boldsymbol{\theta}$ and OOD scores of unlabeled data S_{ul} is achieved via joint optimization [27] by repeating the following two steps:

Updating $\boldsymbol{\theta}$ with fixed S_{ul}: Since all terms in the loss function Eq. (4) are differentiable with respect to $\boldsymbol{\theta}$, we update $\boldsymbol{\theta}$ by the Adam optimizer [12] on Eq. (4).

Updating S_{ul} with fixed $\boldsymbol{\theta}$: Considering the update of S_{ul}, we need to minimize $\mathcal{L}_{ood}$ with fixed $\boldsymbol{\theta}$ to correct OOD scores. $\mathcal{L}_{ood}$ can be minimized when the

---

Algorithm 2. Multi-task curriculum learning

---

for $t \leftarrow 0$ to E **do**

    Select ID samples in unlabeled data X_{ul}^{id} by Eq. (9)

    update $\theta^{(t+1)}$ by Adam on $\mathcal{L}_{ssl}(\theta^t|X_l, Y_l, X_{ul}^{id}) + \mathcal{L}_{ood}(\theta^t, S_{ul}^t|X_l, S_l, X_{ul})$

    update $S_{ul}^{(t+1)}$ by Eq. (6)

end for

---

predicted OOD scores of the network equals S_{ul}. As a result, S_{ul} is updated as follows:

$$s_{ul_i} \leftarrow s(\theta, \boldsymbol{x}_{ul_i}). \tag{6}$$

The whole algorithm of joint optimization is shown in Algorithm 1.

Multi-task Curriculum Learning for Open-Set SSL. As general SSL, the optimization problem of the network parameters θ is formulated as follows:

$$\min_{\theta} \mathcal{L}_{ssl}(\theta|X_l, Y_l, X_{ul}), \tag{7}$$

where $\mathcal{L}_{ssl}$ denotes a loss function such as the sum of cross-entropy loss on labeled data and L2 loss on unlabeled data in [15].

During the training of the network for OOD detection, we also train the network to classify ID data by SSL. As a result, our method is a multi-task learning problem formulated as follows:

$$\min_{\theta, S_{ul}} \mathcal{L}_{ssl}(\theta|X_l, Y_l, X_{ul}) + \mathcal{L}_{ood}(\theta, S_{ul}|X_l, S_l, X_{ul}). \tag{8}$$

The inclusion of the training for ID data classification is helpful to OOD detection because the network can learn more discriminative features. However, while optimizing θ on the semi-supervised part $\mathcal{L}_{ssl}(\theta^t|X_l, Y_l, X_{ul})$ in Eq. (8), the existence of OOD samples in X_{ul} is detrimental to the training for the classification of ID samples. This problem is solved by using curriculum learning that picks up ID samples X_{ul}^{id} from unlabeled data X_{ul} according to the OOD scores of unlabeled data S_{ul}.

Although we can simply sample top $\eta\%$ samples from X_{ul} in ascending order of OOD scores S_{ul}, we implement Otsu thresholding [21] to decide the threshold th_{otsu} automatically, which reduces one hyper-parameter. Then, the selected unlabeled samples are denoted as follows:

$$X_{ul}^{id} = \{\boldsymbol{x}_{ul_i}|s(\theta, \boldsymbol{x}_{ul_i}) < th_{otsu}, 1 \leq i \leq N\}, \tag{9}$$

which converts the total loss function to:

$$\min_{\theta, S_{ul}} \mathcal{L}_{ssl}(\theta|X_l, Y_l, X_{ul}^{id}) + \mathcal{L}_{ood}(\theta, S_{ul}|X_l, S_l, X_{ul}). \tag{10}$$

The entire algorithm of our method is as shown in Algorithm 2. It is to be noted that the semi-supervised loss $\mathcal{L}_{ssl}(\theta^t|X_l, Y_l, X'_{ul})$ is a general loss of SSL, which implies that our method is applicable to any SSL method. In this paper, we use MixMatch [2] as SSL.

4   Experiments

In this section, we discuss our experimental settings and results. We demonstrate the effectiveness of our method on a diverse set of in- and out-of-distribution dataset pairs for open-set SSL. We found that our method outperformed the current state-of-the-art methods by a considerable margin. We used PyTorch 1.1.0 [22] to run all the experiments.

4.1   Neural Network Architecture

Following [2,20], we implemented our network based on Wide ResNet (WRN) [31]. We first trained the model only on the OOD classification loss for 100 epochs and the update of the OOD scores was set to start at the 10th epoch, to achieve stable performance. The model was then trained on the total loss function in Algorithm 2 for 1,024 epochs and 1,024 iterations of each epoch, which is the same as [2]. We used the Adam [12] optimizer and the learning rate was set as 0.002. 64 samples each from the labeled and unlabeled data are sampled for a batch. We report the average test accuracy of the last 10 checkpoints.

4.2   In-Distribution Datasets

CIFAR-10 [14] and SVHN [19] (each containing 10 classes) were used as in-distribution datasets. A total of 5,000 samples were split from the original training data as validation data, and all original test samples were used for testing. We further split the remaining training samples (45,000 for CIFAR-10 and 68,257 for SVHN) into labeled and unlabeled data. Following [2,20], we used {250, 1000, 4000} samples as labeled data and remaining samples as unlabeled data.

4.3   Out-of-Distribution Datasets

As the OOD data are mixed in the unlabeled data, we added 10,000 samples from the following four datasets for each setting:

1. **TinyImageNet (TIN).** The Tiny ImageNet dataset [5] contains 10,000 test images from 200 different classes, which are drawn from the original 1,000 classes of ImageNet [5]. All samples are downsampled by resizing the original image to a size of 32×32.
2. **LSUN.** The Large-scale Scene Understanding dataset (LSUN) consists of 10,000 test images from 10 different scene categories [30]. Similar to TinyImageNet, all samples are downsampled by resizing the original image to a size of 32×32.
3. **Gaussian.** The synthetic Gaussian noise dataset contains 10,000 random 2D Gaussian noise images, where each RGB value of every pixel is sampled from an independent and identically distributed Gaussian distribution with mean 0.5 and unit variance. We further clip each pixel value into the range $[0, 1]$.

4. **Uniform.** The synthetic uniform noise dataset contains 10,000 images where each RGB value of every pixel is independently and identically sampled from a uniform distribution on $[0, 1]$.

Examples of these datasets are shown in Fig. 3. The experimental setting is summarized in Table 1.

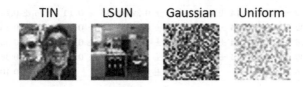

Fig. 3. Examples of Out-of-Distribution Datasets. The samples from TinyImageNet and LSUN are resized to 32×32.

Table 1. The number and type of labeled and unlabeled samples in the experimental setting.

#Labeled	#Unlabeled	#Unlabeled outlier	#Valid	#Test
CIFAR-10				
250	44,750	10,000	5,000	10,000
1,000	44,000			
4,000	41,000			
SVHN				
250	68,007	10,000	5,000	26,032
1,000	67,257			
4,000	64,257			

4.4 Results

The results for the CIFAR-10 dataset are summarized in the upper part of Table 2, which shows the comparison of our method and the baseline. We used the original MixMatch [2] without any OOD detection as the baseline method. Table 2 clearly shows that our approach significantly outperforms the baseline by eliminating the effect of OOD samples in unlabeled data. Compared to TIN and LSUN, which are natural images, synthetic datasets (Gaussian and Uniform) are more harmful to the performance of SSL. Our technique has successfully enabled SSL methods to achieve stable performance on these outliers by detecting them.

In Fig. 4, we show test accuracy vs. the number of epochs. We observe that our method continuously improves the performance of the model during the

Table 2. Accuracy (%) for CIFAR-10/SVHN and OOD dataset pairs. We report the averages and the standard deviations of the scores obtained from three trials. Bold values represent the highest accuracy in each setting. *Clean* shows the upper limit of the model when the unlabeled data contains no OOD data.

OOD dataset	250 labeled		1000 labeled		4000 labeled	
	Baseline	Ours	Baseline	Ours	Baseline	Ours
CIFAR-10						
TIN	82.42 ± 0.70	**86.44 ± 0.64**	88.03 ± 0.22	**89.85 ± 0.11**	91.25 ± 0.13	**93.03 ± 0.05**
LSUN	76.32 ± 4.19	**86.65 ± 0.41**	87.03 ± 0.41	**90.19 ± 0.47**	91.18 ± 0.33	**92.91 ± 0.03**
Gaussian	75.76 ± 3.49	**87.34 ± 0.13**	85.71 ± 1.14	**89.80 ± 0.26**	91.51 ± 0.35	**92.53 ± 0.08**
Uniform	72.90 ± 0.96	**85.54 ± 0.11**	84.49 ± 1.06	**89.87 ± 0.08**	90.47 ± 0.38	**92.83 ± 0.04**
Clean	87.65 ± 0.29		90.67 ± 0.29		93.30 ± 0.10	
SVHN						
TIN	94.66 ± 0.14	**95.21 ± 0.27**	95.58 ± 0.38	**96.65 ± 0.14**	96.73 ± 0.05	**97.01 ± 0.03**
LSUN	94.98 ± 0.23	**95.40 ± 0.17**	95.46 ± 0.05	**96.51 ± 0.16**	96.75 ± 0.01	**97.15 ± 0.02**
Gaussian	93.42 ± 1.09	**95.23 ± 0.04**	95.85 ± 0.33	**96.50 ± 0.11**	96.97 ± 0.02	**97.07 ± 0.07**
Uniform	94.78 ± 0.25	**95.07 ± 0.12**	95.62 ± 0.50	**96.47 ± 0.24**	96.86 ± 0.12	**97.04 ± 0.02**
Clean	96.04 ± 0.39		96.84 ± 0.06		97.23 ± 0.05	

latter half of the training process and its performance is more stable compared to the baseline method during the training. At the beginning of the training process, our method is observed to converge slower than the baseline method. We consider this to be due to the multi-task learning, where our method also learns to detect OOD samples.

The lower part of Table 2 shows the comparison of our method and the baseline for the SVHN dataset. Compared to the case where CIFAR-10 is used as the ID dataset, the effect of outliers on the model is much smaller in this case, possibly because the classification of SVHN is a comparatively easier task. In this situation, our method continues to exhibit a higher and more stable performance than the baseline.

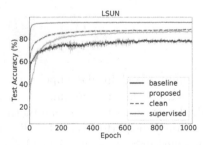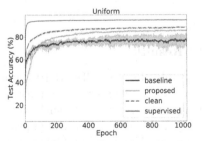

Fig. 4. Test accuracy vs. the number of epochs using CIFAR-10 as ID and other datasets as OOD when 250 labeled samples are used. *Clean* shows the upper limit of the model when the unlabeled data contains no OOD data and *supervised* shows the performance when all the samples are labeled and no OOD data is contained.

Table 3. Performance (%) of OOD detection by our proposed method. We report the averages of the scores obtained from three trials.

OOD dataset	250 labeled		1000 labeled		4000 labeled	
	Recall	Precision	Recall	Precision	Recall	Precision
CIFAR-10						
TIN	99.22	98.48	99.48	97.11	100.00	99.30
LSUN	99.48	99.38	99.95	98.95	100.00	99.64
Gaussian	100.00	100.00	100.00	100.00	100.00	100.00
Uniform	100.00	100.00	100.00	100.00	100.00	100.00
SVHN						
TIN	84.28	99.93	98.4	99.83	99.59	99.87
LSUN	88.55	99.98	98.28	99.97	99.70	99.98
Gaussian	87.52	100.00	99.28	100.00	99.76	100.00
Uniform	82.67	100.00	99.21	100.00	99.78	100.00

We also studied the performance of OOD detection by the proposed method. Since we use the threshold calculated by Otsu thresholding to select ID samples in unlabeled data, we evaluate the performance of OOD detection by precision and recall. Precision is calculated by the percentage of ID samples in the selected samples by curriculum learning. Recall is calculated by the percentage of selected ID samples among all ID samples in unlabeled data. Higher precision and recall indicate better OOD detection – the precision and recall of a perfect detector are both 1. Table 3 shows the results. We find that our method achieves high precision and recall in all the cases, indicating that our method successfully in selecting ID samples from unlabeled data for semi-supervised training.

4.5   Ablation Studies

We further analyzed the effects of the following factors:

The Number of OOD Samples in the Unlabeled Data. We used LSUN as OOD and we changed the number of OOD samples in X_{ul}. The result is summarized in Table 4, which shows that our proposed method works under different OOD conditions except for a case with few outliers in unlabeled data. When there are more OOD samples in the unlabeled data, the performance of the baseline model is lower while our method can achieve stable performance.

The Performance of OOD Detection Compared to Existing OOD Detection Methods. As mentioned in Sect. 2, the existing OOD detection methods cannot achieve high performance when the labeled data is limited and the unlabeled data cannot be utilized in training. We show the results of applying the existing OOD detection method in the setting of open-set semi-supervised learning in Table 5. We choose the most challenging cases when only 250 labeled

Table 4. Accuracy (%) for CIFAR-10 as ID and LSUN as OOD on different numbers of OOD samples when 250 labeled samples are used.

#OOD samples	2000		5000		10000		20000	
	Baseline	Ours	Baseline	Ours	Baseline	Ours	Baseline	Ours
Accuracy (%)	**88.16**	84.82	82.98	**86.20**	76.32	**86.65**	70.3	**85.83**

samples are available. We report the AUROC (the area under the false positive rate against the true positive rate curve) of the OOD detection score of each method. The AUROC of a perfect detector is 1. Table 5 shows that our approach significantly outperforms other OOD detection methods. It is also interesting that the AUROC of previous methods is less than 50% in some cases, which means the model shows more confidence on the predictions of OOD samples than those of ID samples (and this observation conflicts with the idea of [9,17]).

Table 5. The comparison of AUROC (%) in the task of OOD Detection.

ID dataset	OOD dataset	Hendrycks & Gimpel [9]	ODIN [17]	Ours
CIFAR-10	TIN	50.92	54.54	**98.86**
	LSUN	54.34	58.02	**99.82**
	Gaussian	32.41	37.49	**100.00**
	Uniform	45.43	51.05	**100.00**
SVHN	TIN	50.48	57.09	**99.57**
	LSUN	51.44	53.68	**99.84**
	Gaussian	21.20	1.87	**99.98**
	Uniform	2.79	8.31	**99.97**

The Performance of Otsu Thresholding. We use Otsu thresholding to calculate the threshold for splitting ID and OOD samples for the curriculum learning of semi-supervised learning. Figure 5 shows the histogram of OOD scores and the threshold calculated using Otsu thresholding. We find that both ID and OOD samples can be successfully separated by the threshold.

4.6   Discussion

Limitations. As shown in Table 4, our method fails to improve the baseline if there are few outliers in the unlabeled data. This failure mainly comes from the wrong threshold calculated by Otsu thresholding, since the number of ID samples and OOD samples is extremely imbalanced. This problem can be solved

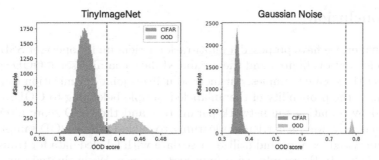

Fig. 5. The histogram of OOD scores and the threshold (the green line) calculated by Otsu thresholding when 250 labeled samples are used. (Color figure online)

Table 6. Performance (%) of UASD [3] and our proposed method.

OOD dataset	Test accuracy			Detection recall		Detection precision	
	Baseline	UASD [3]	Ours	UASD [3]	Ours	UASD [3]	Ours
TIN	82.42	83.53	**86.44**	66.47	**99.22**	96.84	**98.48**
LSUN	76.32	80.87	**86.65**	63.88	**99.48**	96.50	**99.38**

by changing the OOD threshold, which means we can introduce a new parameter to control the number of unlabeled samples selected as ID data.

"Similar" Outliers. The outlier datasets used in Sect. 4 are still quite different from the original training datasets CIFAR-10 and SVHN. Including similar outliers in the unlabeled data is a more complicated scenario. We tried using the animal classes in CIFAR-10 as ID and other classes as OOD and found that these similar outliers are not as harmful as dissimilar outliers, which leads to a 3% decrease in test accuracy. Our method can still reach 2% higher than the test accuracy of the baseline with 94% precision and 98% recall of OOD detection.

Additional Comparisons. Chen et al. [3] works on a close setting to our paper at around the same time as our paper, but there are two significant differences between [3] and our work. First, the experimental setting is different. [3] defines the mismatch of class distribution as some known classes are not contained in the unlabeled dataset and some unknown classes are contained in the unlabeled dataset. Second, the method of utilizing OOD samples in unlabeled data is different. In [3], the OOD samples are simply ignored by filtering the confidence score, which means they mainly train an SSL model and just use the output of the model directly.

For the comparison, we implemented [3] for filtering OOD in MixMatch and show the comparison in the setting of CIFAR-10 with 250 labeled samples. The results are summarized in Table 6 and it shows our method has better performance not only in SSL but also in OOD detection, because we explicitly train an OOD detector by unlabeled OOD samples together with the training of SSL model, which enables our method to achieve higher OOD detection performance.

5 Conclusion

In this paper, we have proposed a multi-task curriculum for open-set SSL, where the labeled data is limited and the unlabeled data contains some OOD samples. To detect these OOD samples, our method utilizes joint optimization framework to estimate the probability of the unlabeled sample belonging to OOD, which is achieved by updating the network parameters and the OOD score alternately. Simultaneously, we use curriculum learning to exclude these OOD samples from semi-supervised learning and utilize these data with labeled data for training the model to classify ID samples with high performance. We evaluated our method on several open-set semi-supervised benchmarks and proved that our method achieves state-of-the-art performance by detecting the OOD samples with high accuracy.

Acknowledgements. This work was supported by JSPS KAKENHI Grant Number 18H03254 and JST CREST Grant Number JPMJCR1686, Japan.

References

1. Bendale, A., Boult, T.E.: Towards open set deep networks. In: Proceedings of the IEEE Conference on Computer Vision and Pattern Recognition (2016)
2. Berthelot, D., Carlini, N., Goodfellow, I., Papernot, N., Oliver, A., Raffel, C.: Mixmatch: a holistic approach to semi-supervised learning. In: Advances in Neural Information Processing Systems (2019)
3. Chen, Y., Zhu, X., Li, W., Gong, S.: Semi-supervised learning under class distribution mismatch. In: AAAI Conference on Artificial Intelligence (2020)
4. Cubuk, E.D., Zoph, B., Mane, D., Vasudevan, V., Le, Q.V.: Autoaugment: learning augmentation policies from data. In: Proceedings of the IEEE Conference on Computer Vision and Pattern Recognition (2019)
5. Deng, J., Dong, W., Socher, R., Li, L.J., Li, K., Fei-Fei, L.: Imagenet: a large-scale hierarchical image database. In: Proceedings of the IEEE Conference on Computer Vision and Pattern Recognition (2009)
6. Devlin, J., Chang, M.W., Lee, K., Toutanova, K.: Bert: pre-training of deep bidirectional transformers for language understanding. In: Proceedings of the Conference of the North American Chapter of the Association for Computational Linguistics (2019)
7. Grandvalet, Y., Bengio, Y.: Semi-supervised learning by entropy minimization. In: Advances in Neural Information Processing Systems (2005)
8. He, K., Zhang, X., Ren, S., Sun, J.: Deep residual learning for image recognition. In: Proceedings of the IEEE Conference on Computer Vision and Pattern Recognition (2016)
9. Hendrycks, D., Gimpel, K.: A baseline for detecting misclassified and out-of-distribution examples in neural networks. In: International Conference on Learning Representations (2017)
10. Joachims, T.: Transductive inference for text classification using support vector machines. In: International Conference on Machine Learning (1999)
11. Joachims, T.: Transductive learning via spectral graph partitioning. In: International Conference on Machine Learning (2003)

12. Kingma, D.P., Ba, J.: Adam: a method for stochastic optimization. In: International Conference on Learning Representations (2014)

13. Kingma, D.P., Mohamed, S., Rezende, D.J., Welling, M.: Semi-supervised learning with deep generative models. In: Advances in Neural Information Processing Systems (2014)

14. Krizhevsky, A.: Learning multiple layers of features from tiny images. Technical report (2009)

15. Laine, S., Aila, T.: Temporal ensembling for semi-supervised learning. In: International Conference on Learning Representations (2017)

16. Lee, K., Lee, H., Lee, K., Shin, J.: Training confidence-calibrated classifiers for detecting out-of-distribution samples. In: International Conference on Learning Representations (2018)

17. Liang, S., Li, Y., Srikant, R.: Enhancing the reliability of out-of-distribution image detection in neural networks. In: International Conference on Learning Representations (2018)

18. Liu, W., Wen, Y., Yu, Z., Li, M., Raj, B., Song, L.: Sphereface: deep hypersphere embedding for face recognition. In: Proceedings of the IEEE Conference on Computer Vision and Pattern Recognition (2017)

19. Netzer, Y., Wang, T., Coates, A., Bissacco, A., Wu, B., Ng, A.Y.: Reading digits in natural images with unsupervised feature learning. In: NeurIPS Workshop on Deep Learning and Unsupervised Feature Learning (2011)

20. Oliver, A., Odena, A., Raffel, C.A., Cubuk, E.D., Goodfellow, I.: Realistic evaluation of deep semi-supervised learning algorithms. In: Advances in Neural Information Processing Systems (2018)

21. Otsu, N.: A threshold selection method from gray-level histograms. IEEE Trans. Syst. Man Cybern. **9**, 62–66 (1979)

22. Paszke, A., et al.: Pytorch: an imperative style, high-performance deep learning library. In: Advances in Neural Information Processing Systems (2019)

23. Pu, Y., et al.: Variational autoencoder for deep learning of images, labels and captions. In: Advances in Neural Information Processing Systems (2016)

24. Sajjadi, M., Javanmardi, M., Tasdizen, T.: Regularization with stochastic transformations and perturbations for deep semi-supervised learning. In: Advances in Neural Information Processing Systems (2016)

25. Salimans, T., Goodfellow, I., Zaremba, W., Cheung, V., Radford, A., Chen, X.: Improved techniques for training GANs. In: Advances in Neural Information Processing Systems (2016)

26. Tan, M., Le, Q.: Efficientnet: rethinking model scaling for convolutional neural networks. In: International Conference on Machine Learning (2019)

27. Tanaka, D., Ikami, D., Yamasaki, T., Aizawa, K.: Joint optimization framework for learning with noisy labels. In: Proceedings of the IEEE Conference on Computer Vision and Pattern Recognition (2018)

28. Tarvainen, A., Valpola, H.: Mean teachers are better role models: weight-averaged consistency targets improve semi-supervised deep learning results. In: Advances in Neural Information Processing Systems (2017)

29. Vyas, A., Jammalamadaka, N., Zhu, X., Das, D., Kaul, B., Willke, T.L.: Out-of-distribution detection using an ensemble of self supervised leave-out classifiers. In: Ferrari, V., Hebert, M., Sminchisescu, C., Weiss, Y. (eds.) ECCV 2018. LNCS, vol. 11212, pp. 560–574. Springer, Cham (2018). https://doi.org/10.1007/978-3-030-01237-3_34

30. Yu, F., Seff, A., Zhang, Y., Song, S., Funkhouser, T., Xiao, J.: Construction of a large-scale image dataset using deep learning with humans in the loop. arXiv preprint arXiv: 1506.03365 (2015)
31. Zagoruyko, S., Komodakis, N.: Wide residual networks. In: Proceedings of the British Machine Vision Conference (2016)
32. Zhang, H., Cisse, M., Dauphin, Y.N., Lopez-Paz, D.: mixup: beyond empirical risk minimization. In: International Conference on Learning Representations (2018)
33. Zhu, X., Ghahramani, Z., Lafferty, J.D.: Semi-supervised learning using Gaussian fields and harmonic functions. In: International Conference on Machine Learning (2003)

Gradient-Induced Co-Saliency Detection

Zhao Zhang[1] (iD), Wenda Jin[2] (iD), Jun Xu[1] (iD), and Ming-Ming Cheng[1 (✉)] (iD)

[1] TKLNDST, CS, Nankai University, Tianjin, China
zzhang@mail.nankai.edu.cn, cmm@nankai.edu.cn
[2] College of Intelligence and Computing, Tianjin University, Tianjin, China

Abstract. Co-saliency detection (Co-SOD) aims to segment the common salient foreground in a group of relevant images. In this paper, inspired by human behavior, we propose a gradient-induced co-saliency detection (GICD) method. We first abstract a consensus representation for a group of images in the embedding space; then, by comparing the single image with consensus representation, we utilize the feedback gradient information to induce more attention to the discriminative co-salient features. In addition, due to the lack of Co-SOD training data, we design a jigsaw training strategy, with which Co-SOD networks can be trained on general saliency datasets without extra pixel-level annotations. To evaluate the performance of Co-SOD methods on discovering the co-salient object among multiple foregrounds, we construct a challenging *CoCA* dataset, where each image contains at least one extraneous foreground along with the co-salient object. Experiments demonstrate that our GICD achieves state-of-the-art performance. Our codes and dataset are available at https://mmcheng.net/gicd/.

Keywords: Co-saliency detection · New dataset · Gradient inducing · Jigsaw training

1 Introduction

Co-Saliency Detection (Co-SOD) aims to discover the common and salient objects by exploring the inherent connection of multiple relevant images. It is a challenging computer vision task due to complex variations on the co-salient objects and backgrounds. As a useful task for understanding correlations in multiple images, Co-SOD is widely employed as a pre-processing step for many vision tasks, such as weakly-supervised semantic segmentation [44,47], image surveillance [15,30], and video analysis [18,19], *etc.*

Previous researches study the Co-SOD problem from different aspects [6, 20,23]. At the early stage, researchers explored the consistency among a group of relevant images using handcrafted features, *e.g.*, SIFT [5,18], color and texture [13,24], or multiple cues fusion [4], *etc.* These shallow features are not

Electronic supplementary material The online version of this chapter (https://doi.org/10.1007/978-3-030-58610-2_27) contains supplementary material, which is available to authorized users.

© Springer Nature Switzerland AG 2020
A. Vedaldi et al. (Eds.): ECCV 2020, LNCS 12357, pp. 455–472, 2020.
https://doi.org/10.1007/978-3-030-58610-2_27

discriminative enough to separate co-salient objects in real-world scenarios. Recently, learning-based methods achieve encouraging Co-SOD performance by exploring the semantic connection within a group of images, via deep learning [23,43], self-paced learning [17,52], metric learning [16], or graph learning [20,55], *etc.* However, these methods suffer from the inherent discrepancy in features, due to varying viewpoints, appearance, and positions of the common objects. How to better utilize the connections of relevant images is worth deeper investigation.

Fig. 1. Human behavior inspired GICD. GIM is the gradient inducing module.

How do humans segment co-salient objects from a group of images? Generally, humans first browse the group of images, summarize the shared attributes of the co-salient objects with "general knowledge" [33], and then segment the common objects in each image with these attributes. This process is shown in Fig. 1. Inspired by human behavior, we design an end-to-end network with corresponding two stages. To obtain the shared attributes of the common objects as humans do, we calculate the consensus representation of multiple relevant images in a high-dimensional space with a learned embedding network. Once the consensus representation is obtained, for each image, we propose a Gradient Inducing Module (GIM) to imitate the human behavior of comparing a specific scene with the consensus description to feedback matching information.

In GIM, the similarity between the single and consensus representations can be measured first. As high-level convolutional kernels with different semantic awareness [36,56], we can find out the kernels that are more related to the consensus representation and enhance them to detect co-salient objects. To this end, by partially back-propagating, we calculate the gradients of the similarity with respect to the top convolution layer as the feedback information. High gradient values mean corresponding kernels have a positive impact on the similarity results; thus, by assigning more weight to these kernels, the model will be induced to focus on the co-salient related features. Moreover, to better discriminate the co-salient object in each level of the top-down decoder, we propose an Attention Retaining Module (ARM) to connect the corresponding encoder-decoder pairs of our model. We call this two-stage framework with GIM and ARM as Gradient-Induced Co-saliency Detection (GICD) network. Experiments on benchmark datasets demonstrate the advantages of our GICD over previous Co-SOD methods.

Without sufficient labels, existing Co-SOD networks [23,43,48] are trained with semantic segmentation datasets, *e.g.*, Microsoft COCO [27]. However, the annotated objects in segmentation datasets are not necessarily salient. In this paper, we introduce a novel jigsaw strategy to extend existing salient object detection (SOD) datasets, without extra pixel-level annotating, for training Co-SOD networks.

In addition, to better evaluate the Co-SOD methods' ability of discovering co-salient object(s) among multiple foregrounds, most images in an evaluation dataset should contain at least one unrelated salient foreground except for the co-salient object(s). As can be seen in Fig. 3, this is ignored by the current Co-SOD datasets [2,11,45,51]. To alleviate the problem, we meticulously construct a more challenging dataset, named *C*ommon *C*ategory *A*ggregation (*CoCA*).

In summary, our major contributions are as follows:

- **We propose a gradient-induced co-saliency detection (GICD) network for Co-SOD**. Specifically, we propose a gradient inducing module (GIM) to pay more attention to the discriminative co-salient features, and an attention retaining module (ARM) to keep the attention during the top-down decoding.
- **We present a jigsaw strategy to train Co-SOD models on general SOD datasets without extra pixel-level annotations**, to alleviate the problem of lacking Co-SOD training data.
- **We construct a challenging *CoCA* dataset with meticulous annotations**, providing practical scenarios to better evaluate Co-SOD methods.
- Experiments on the *CoSal2015* [51] and our *CoCA* datasets demonstrate that our GICD outperforms previous Co-SOD methods. Extensive ablation studies validate the effectiveness of our contributions.

2   Related Works

2.1   Co-Saliency Object Detection (Co-SOD)

Different from traditional salient object detection (SOD) task [8,12,14], Co-SOD aims to automatically segment the common salient objects in a group of relevant images. Early Co-SOD methods assume that the co-salient objects in multiple images share low-level consistency [52]. For instance, Li *et al.* [24] introduced a co-multi-layer graph by exploring color and texture properties. Fu *et al.* [13] explored the contrast, spatial, and corresponding cues to enforce global association constraint by clustering. Cao *et al.* [4] integrated multiple saliency cues by a self-adaptive weighting manner. Tsai *et al.* [39] extracted co-salient objects by solving an energy minimization problem over a graph.

Recently, many deep learning-based methods have been proposed to explore high-level features for the Co-SOD task [17,51,53]. These methods can be divided into two categories. One is a natural deep extension from traditional low-level consistency. It explores the high-level similarity to enhance the similar candidate

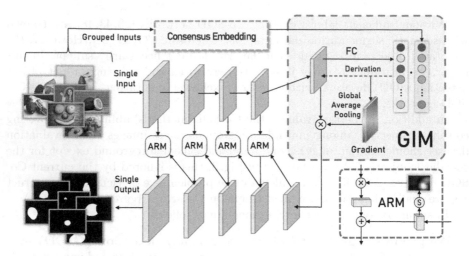

Fig. 2. Pipeline of our Gradient-Induced Co-saliency Detection (GICD) method. GIM denotes the Gradient Inducing Module, while ARM means the Attention Retaining Module. "•", "⊗", "⊕", and "Ⓢ" represent the inner product, element-wise production, element-wise addition, and the sigmoid function, respectively.

regions among multiple images. For example, Zhang *et al.* [51] jointly investigated inter-group separability and intra-group consistency depending on high-level CNN features. Hsu *et al.* [17] proposed an unsupervised method by maximizing the similarity among multiple foregrounds and minimizing the similarity between foregrounds and backgrounds with graphical optimization. Jiang *et al.* [20] explored the superpixel-level similarity by intra- and inter-graph learning using the graph convolution network. Zhang *et al.* [53] proposed a mask-guided network to obtain coarse Co-SOD results and then refined the results by multi-label smoothing. The second category of deep methods is based on joint feature extracting. They often extract the common feature for a group of images, and then fuse it with each single image feature. For instance, Wei *et al.* [43] learn a shared feature for every five images with a group learning branch, then concatenate the shared feature with every single feature to get the final prediction. Li *et al.* [23] extend this idea with a sequence model to process variable length input. Wang *et al.* [40] and Zha *et al.* [48] learn a category vector for an image group to concatenate with each spatial position of a single image feature on multiple levels.

2.2   Co-SOD Datasets

Current Co-SOD datasets include mainly *MSRC* [45], *iCoseg* [2], *CoSal2015* [51], and *CoSOD3k* [11], *etc.* In Fig. 3, we show some examples of these datasets and our *CoCA* dataset. *MSRC* [45] is mainly for recognizing objects from images. In [13,51], they select 233 images of seven groups from *MSRC-v1* for evaluating detection accuracy. *iCoseg* [2] contains 643 images of 38 groups in invariant

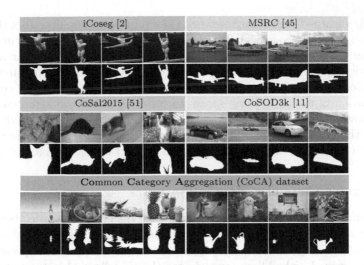

Fig. 3. Examples of current popular datasets and our proposed *CoCA* dataset. In *CoCA*, except for the co-salient object(s), each image contains at least one extraneous salient object, which enables the dataset to better evaluate the models' ability of discovering co-salient object(s) among multiple foregrounds.

scenes. In the above datasets, the co-salient objects are mostly the same in similar scenes, and consistent in appearance. *CoSal2015* [51] and *CoSOD3k* [11] are two large-scale datasets containing 2015 and 3316 images, respectively. In the two datasets, some target objects belong to the same category differ greatly in appearance, which makes them more challenging datasets. However, these above datasets are not well-designed for evaluating the Co-SOD algorithms because they only have a single salient object in most images. Taking the athlete of the *iCoseg* in Fig. 3 as an example, although the athlete is co-salient in different images, these data can be easily processed by a SOD method because there is no other extraneous salient foreground interference. Although this awkwardness has been avoided in some groups in *CoSal2015* and *CoSOD3k*, it is not guaranteed in most cases. As discovering the co-salient object(s) among multiple foregrounds is the primary pursuit of a Co-SOD method in real-world applications [50], to evaluate this ability better, we construct a challenging *CoCA* dataset, where each image contains at least one extraneous salient object.

3   Proposed Method

Figure 2 shows the flowchart of our gradient-induced co-saliency detection (GICD) network. Our backbone is the widely used Feature Pyramid Network (FPN) [26]. For the Co-SOD task, we incorporate it with two proposed modules: the gradient inducing module (GIM), and the attention retaining module (ARM). GICD detects co-salient objects in two stages. It first receives a group of images as input for exploring a consensus representation in a high-dimensional space with a learned embedding network. The representation describes the common patterns of the co-salient objects within the group. Then, it turns back to segment the co-salient object(s) for each sample. In this stage, for inducing the attention of the model on co-salient regions, we utilize GIM to enhance the features closely related to co-salient object by comparing single and consensus representation in the embedding space. In order to retain the attention during the top-down decoding, we use ARM to connect each encoder-decoder pairs. We train the GICD network with jigsaw training strategy, where the Co-SOD models can be trained on SOD dataset without extra pixel-level annotations.

3.1   Learning Consensus Representation

Given a group of images $\mathcal{I} = \{I_n\}_{n=1}^N$, to locate the co-salient object(s) in each image, we should first know what patterns the co-salient objects have based on prior knowledge. To this end, we propose to learn a consensus representation with a pre-trained embedding network, for the co-salient objects of the image group $\mathcal{I}$. Deep classifiers can be naturally utilized for representation learning [34], where the prior knowledge of semantic attribute can be transformed from the parameters pre-trained on ImageNet [7]. In this case, we employ a pre-trained classified network $\mathcal{F}(\cdot)$, such as VGG-16, as our embedding network by removing the softmax layer. It first extracts the representation $e_n = \mathcal{F}(I_n) \in \mathbb{R}^d$ of each image I_n, where d is the dimension of the last full connection layer. The consensus representation $e^\dagger$ can be calculated by $e^\dagger = \mathtt{Softmax}\left(\sum_{n=1}^N e_n\right)$, to describe the common attributes of this image group.

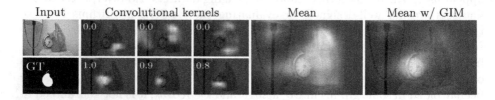

Fig. 4. Visualization of high-level features induced by GIM. In the six small images on the left, the kernels above are not sensitive to the target object, while the kernels below are related to the target object. Their corresponding important values based on gradients are marked with **green** numbers on the top-left corners. The mean values of F_n^5 and induced $\tilde{F}_n^5$ are shown in the form of orange heat maps. (Color figure online)

3.2   Gradient Inducing Module

After obtaining the consensus representation $e^\dagger$ of the group $\mathcal{I}$, for each image, we focus on how to find the discriminative features that match the consensus description. As demonstrated in [36,56], high-level convolutional layers naturally possess semantic-specific spatial information. We denote the five convolutional feature blocks of $\mathcal{F}(\cdot)$ as $\{F^1, F^2, \ldots, F^5\}$. In Fig. 4, we show the feature maps of the last convolutional layer F^5. The input image (1st column) contains a pocket watch and blue gloves, and the convolutional kernels focus on different regions (2nd to 4th columns). If assigning more importance to these kernels which closely concern about the co-salient objects, the model will tend to segment the co-salient objects (pocket watch) by decoding the induced features. As indicated in [36], the discriminability of features in neural networks can be measured by the gradient obtained by optimizing objectives. Therefore, we propose a gradient inducing module (GIM) for enhancing the discriminative feature by exploring the feedback gradient information. As the encoder of our FPN backbone shares the fixed parameters with the consensus embedding network, it can also embed each image into the same space as consensus representation $e^\dagger$. For the extracted representation e_n of the n-th image, the similarity c_n between e_n and its consensus representation $e^\dagger$ can be defined by inner product, i.e., $c_n = e_n^\top e^\dagger$. Then we compute the positive gradient G_n flowing back into the last convolutional layer $F^5 \in \mathbb{R}^{w \times h \times c}$ to select discriminative features in F_n^5, specifically,

$$G_n = \mathrm{ReLU}\left(\frac{\partial c_n}{\partial F_n^5}\right) \in \mathbb{R}^{w \times h \times c}. \tag{1}$$

In this partial backpropagation, the positive gradient G_n reflects the sensitivity of the corresponding position to the final similarity score; that is, increasing activation value with a larger gradient will make the specific representation e_n more consistent with the consensus one $e^\dagger$. Therefore, the importance of a convolution kernel for a particular object can be measured by the mean of its feature gradients. Specifically, the channel-wise importance values can be calculated by global average pooling (GAP), namely $w_n = \mathtt{GAP}(G_n) = \frac{1}{wh}\sum_i \sum_j G_n$, where $i = 1, ..., w$ and $j = 1, ..., h$. Once obtaining the weight, we can induce the high-level feature F_n^5 by assigning the importance value to each kernel $\tilde{F}_n^5 = F_n^5 \otimes w_n$, where $\otimes$ denotes the element-wise production. As shown in Fig. 4, we visualize the mean heat-maps of F_n^5 and $\tilde{F}_n^5$. without our GIM module, the kernels will averagely focus on both objects. One can see that the kernels more relevant to the co-salient category have higher gradient weights (marked with green numbers), and the attention of the network has shifted to the co-salient object(s) after gradient inducing.

3.3   Attention Retaining Module

In GIM, the high-level features have been induced by the gradient. However, top-down decoder is built upon the bottom-up backbone, and the induced high-level features will be gradually diluted when transmitted to lower layers. To this end, we propose an attention retaining module (ARM) to connect the corresponding encoder-decoder pairs of our GICD network. As shown in Fig. 2, for each ARM, the feature of encoder used for skip-connection is guided by the higher-level prediction. Through top-down iterative reminding, the network will focus the detail recovery of the co-salient regions without being interfered by other irrelevant objects. We take the channel-wise mean of $\tilde{F}_n^5$ as the first low-resolution guide map S_n^5, and reduce $\tilde{F}_n^5$ to feature P_n^5 containing 64 channels. The decoding process with ARM is as follows:

$$\begin{cases} \tilde{F}_n^i = \left(S_n^{i+1}\right) \uparrow \odot F_n^i \\ P_n^i = \mathcal{E}^i \left(\left(P_n^{i+1}\right) \uparrow + \mathcal{R}^i \left(\tilde{F}_n^i\right)\right), \ i \in \{4,3,2,1\}, \\ S_n^i = \mathcal{D}^i \left(P_n^i\right), \end{cases} \quad (2)$$

where $(\cdot) \uparrow$ is the up-sampling operation. $\mathcal{R}^i(\cdot)$ consists of two convolutional layers, and reduces the enhanced features $\tilde{F}_n^i$ to 64 channels. $\mathcal{E}^i(\cdot)$ is the corresponding two convolutional layers, with 64 kernels, in decoder. $\mathcal{D}^i(\cdot)$ is applied for deep supervision, and outputs a prediction by two convolutional layers followed by a sigmoid layer. The last S_n^1 is the final output.

To validate the effectiveness of our ARM, in Fig. 5, we show the intermediate features in different levels of the decoder with ARM (1st row) and without ARM (2nd row). We observe that, through GIM, both locate the co-salient object (*i.e.*, Teddy bear) successfully, while our GICD w/o ARM is gradually interfered by other salient objects during upsampling and produces inaccurate detection results. These results show that our ARM can effectively hold the attention on the co-salient objects in relevant images.

Image   w/ ARM |<- - - - - - - - - Top-Down Features w/ ARM - - - - - - - -|

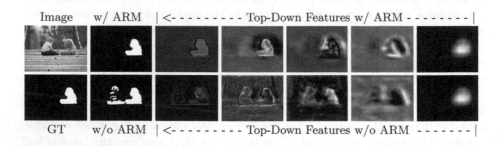

GT   w/o ARM |<- - - - - - - - - Top-Down Features w/o ARM - - - - - - -|

Fig. 5. Visualization of attention retaining with ARM. The first row shows the multiple level intermediate features with (w/) our ARM, and the second row shows salient maps without (w/o) ARM. The prediction w/ ARM (second column, up) is more accurate than that w/o ARM (second column, down), since our ARM pays stronger attention to the co-salient regions.

3.4   Jigsaw Training Strategy

Strategy. One important problem in Co-SOD task is that current SOD datasets, *e.g.*, DUTS [41] and MSRA-B [29], are not suitable for the training of Co-SOD networks. The reasons are two-fold: 1) they do not have class information, so it is impossible to train models in groups; 2) most samples in them only contain one salient foreground object. It is difficult to enable the network to distinguish the co-salient object(s) among multiple foreground objects. Recent Co-SOD methods [23,40,43] are trained on semantic segmentation datasets [27]. This suffers from two problems: 1) the label of such a semantic segmentation dataset is relatively rough, so the ability of recovery details of the trained network is not ideal, which cannot meet the accuracy requirements of downstream tasks; 2) the objects in such datasets are not necessarily salient. To alleviate these problems, we design a jigsaw strategy to transform SOD datasets into suitable training data for Co-SOD models: Step 1: we employ a classifier [31] to classify every SOD dataset into multiple categories since it has no category information. Step 2: we splice the samples of one category with the samples of other ones to form a new jigsaw, as shown in Fig. 6. This step is to ensure that an input image contains not only a co-salient foreground but also extraneous foreground objects. Through the above steps, existing SOD datasets can be seamlessly utilized to train Co-SOD networks without additional pixel-level annotations.

Loss Function. Considering the most important goal for co-saliency detection is to find the position of the common foreground objects correctly, we employ the soft intersection over union (IoU) loss [25,35] for GICD, specifically,

$$\mathcal{L}(S,G) = 1 - \frac{\sum_c S(c) G(c)}{\sum_c [S(c) + G(c) - S(c) G(c)]}, \tag{3}$$

where S is the prediction, and G denotes ground-truth. c represents each pixel position in the image. The loss function of our model can be expressed as

$$L_{total} = \sum_{n=1}^{N} \sum_{i=1}^{4} \mathcal{L}\left(S_n^i, G_n\right). \tag{4}$$

Fig. 6. Demo of jigsaw training. A sample (cat), together with the samples from other categories, constitute new jigsaws for training.

4   Proposed *CoCA* Dataset

Construction Guidelines. We construct our *CoCA* dataset under three guidelines. **G1**: each image should contain at least one extraneous foreground, excluding the co-salient object(s). **G2**: in each image group, the co-salient objects are better to be different. **G3**: the dataset needs to be misaligned with the categories of the common training set, to explore the ability of the model on handling unseen categories. The guideline **G1** reflects whether or not the model can detect the co-salient objects, rather than only segmenting the foreground and background. **G2** can evaluate whether the model is robust to the intra-group differences. **G3** ensures that the model can be evaluated for its ability to detect co-salient objects from unknown categories robustly.

Construction Procedures. With the above guidelines, we collect images from pixabay[1]. We divide them into 80 categories, covering everyday indoor and outdoor scenes. It is worth noting that these categories are outright staggered with Microsoft COCO [27], which is often used for the training of co-saliency models [23, 40, 43]. Most importantly, with manual screening, the images in our dataset include at least one extraneous salient object, excluding the co-salient object(s). We provide four levels of annotations: class level, bounding box level, object level, and instance level. The high-quality object-level annotations are applicable to the co-saliency detection task in this paper. Different levels of annotations of our dataset corresponds to different tasks, such as co-localization [21, 38], few-shot object segmentation [49, 54], and instance co-segmentation [37].

Dataset Statistics. Our *CoCA* dataset consists of 80 categories with 1295 images. As shown in Fig. 3, these images are challenging in occlusion, clutter background, extraneous object interference, *etc.* The number of images in each category is different, varying from 8 to 40. This diversity is helpful in evaluating the ability of the model for different image set sizes. The number of co-salient instances in an image is also diverse. 336 images have more than two co-salient instances. The diversity of the number of instances can help to evaluate the robustness of the model to multi-object scenarios.

5   Experiments

5.1   Implementation Details

We train our GICD network on the training set of *DUTS* [41] with our jigsaw training strategy. These samples are classified into 291 groups, which contains 8250 images with removing the noisy samples. Each sample will be combined with others to form three jigsaws as supplementary samples, as shown in Fig. 6; thus,

---

[1] https://pixabay.com.

the candidate training data is quadrupled. In each training epoch, we randomly select at most 20 samples from each group. The Adam optimizer [22] is used with an initial learning rate of 0.0001, $\beta_1 = 0.9$, and $\beta_2 = 0.99$. The learning rate is divided by 10 at the 50-th epoch. We train our GICD for 100 epochs in total. To accommodate the input images with our FPN backbone (VGG network), we resize them to 224×224 during the training and test stage, and the output saliency maps are resized back to the original size for evaluating. Our GICD is implemented in PyTorch [32], and runs at ~ 55 FPS on an NVIDIA GeForce RTX 2080Ti.

5.2 Evaluation Datasets and Metrics

Datasets. We employ two challenging datasets to evaluate the performance of various methods. The first dataset is *CoSal2015* [51]. In some image groups, *e.g.* baseball, it is challenging in the interference of extraneous salient objects. The other is our *CoCA*, where most images possess more than one irrelevant salient objects besides the co-salient target.

Metrics. We employ five widely used metrics as suggested by [17,50,53]: mean F-measure (F_{avg}) [1], maximum F-measure (F_{max}) [3], Precision-Recall (PR) curve, S-measure (S_α) [9], and mean E-measure (E_ξ) [10].

Table 1. Quantitative comparisons of mean F-measure [1] (F_{avg}), maximum F-measure [3] (F_{max}), S-measure [9] (S_α), and mean E-measure [10] (E_ξ) by our GICD and other methods on the *CoSal2015* [51] and *CoCA* datasets. "↑" means that the higher the numerical value, the better the model performance.

	Metric	CBCD [13]	GW [42]	CSMG [53]	RCAN [23]	BASNet [35]	PoolNet [28]	SCRN [46]	GICD Ours
CoSal2015	F_{avg} ↑	0.378	0.639	0.721	0.670	**0.778**	0.768	0.755	0.835
	F_{max} ↑	0.547	0.706	0.787	0.764	**0.791**	0.785	0.783	0.844
	S_α ↑	0.550	0.744	0.776	0.779	0.822	**0.823**	0.817	0.844
	E_ξ ↑	0.516	0.727	0.763	0.742	**0.841**	0.836	0.822	0.883
CoCA	F_{avg} ↑	0.230	0.358	0.390	0.360	**0.398**	0.394	0.394	0.504
	F_{max} ↑	0.313	0.408	**0.499**	0.422	0.408	0.404	0.413	0.513
	S_α ↑	0.523	0.602	**0.627**	0.616	0.592	0.602	0.612	0.658
	E_ξ ↑	0.535	0.615	0.606	0.614	0.600	0.616	**0.625**	0.701

5.3 Comparison with State-of-the-Arts

Comparison Methods. We compare our GICD with seven state-of-the-art methods, including four Co-SOD method: RCAN [23], CSMG [53], GW [42], and CBCD [13], as well as three SOD methods: BASNet (ResNet-34) [35], PoolNet (ResNet-50) [28], and SCRN (ResNet-50) [46].

Quantitative Evaluation. In the Table 1, we illustrate the quantitative results of our GICD and other state-of-the-art methods on the *CoSal2015* and our *CoCA* datasets. As can be seen, our GICD achieves better performance. The results show some interesting phenomena. On the *CoSal2015*, the SOD methods outperform most Co-SOD methods except GICD. The reason is that a large part of the images in *CoSal2015* have only one salient object, which can be solved by SOD algorithms. The advantages of Co-SOD algorithms cannot be fully reflected on this data, and these detail-oriented SOD methods easily surpass their performance. However, in our newly proposed *CoCA* dataset, this phenomenon is no longer obvious, because the salient objects in an image contain many objects that are not co-salient. This is why our *CoCA* dataset is more suitable for evaluating Co-SOD algorithms. Nevertheless, our GICD still surpasses the SOD methods on *CoSal2015*. It brings 11.4% improvement in terms of mean F-measure compared with the best Co-SOD method, 5.7% improvement compared with the SOD method. In our *CoCA* dataset, GICD brings 3.1% improvement in terms of S-measure compared with the best Co-SOD method, 4.6% improvement compared with the SOD method. Seen from Fig. 7, our method also outperforms other methods on the PR curve and F-measure curve. The trend of the curves demonstrates our method is less affected by the threshold, because it predicts the result with high confidence. This can avoid the problem of how to select the appropriate threshold in the subsequent practical applications.

Qualitative Results. In Fig. 8, we show some saliency maps produced by GICD and other compared methods for intuitive comparison. The samples we illustrate are challenging because the salient objects in each input include not only co-salient object(s) but also interference from other extraneous foreground(s). This is also reflected in the prediction results of SOD algorithms, which over segmented many unrelated regions. From the overall results, our GICD has high confidence in the prediction maps, even at the edge, while most of the other methods suffer from uncertain regions. Back to specific examples, baseball is the most challenging subset of the *CoSal2015* [51], because it varies greatly in size across images and is interfered by other salient objects. Results show that our method successfully handles the tiny size and the occlusions. In *CoCA*, boots class faces the interference of background color, and strawberry class has multiple segmentation targets. Nevertheless, GICD locates the target object accurately.

5.4   Ablation Study

In order to explore the contribution and mechanism of gradient inducing module (GIM), attention retaining module (ARM), and jigsaw training (JT) to our GICD network, we evaluated all possible combinations of the three candidates. Note that, these three candidates are interdependent and are not recommended to be used alone. As shown in Table 2, "A" is the baseline without JT, GIM, and ARM. It is actually a SOD model because it does not take into account any relationship between images.

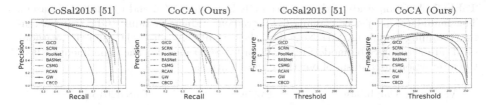

Fig. 7. Precision-Recall (PR) and F-measure curves of our GICD and seven state-of-the-art methods on the *CoSal2015* and *CoCA* datasets. The node on each PR curve denotes the precision and recall value used for calculating maximum F-measure.

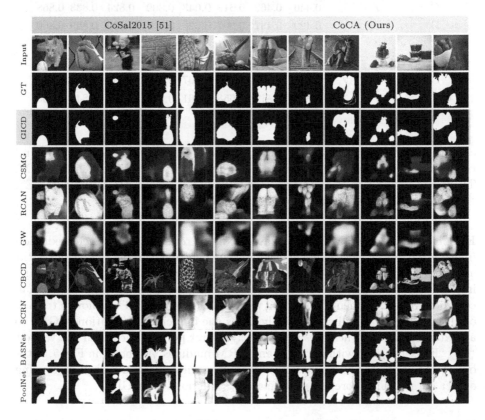

Fig. 8. Visual comparison of our GICD with 7 state-of-the-arts (4 Co-SOD methods and 3 SOD methods) on the *CoSal2015* [51] and our *CoCA* datasets.

Effectiveness of GIM. GIM is the core module of our GICD. With GIM, the variants of "C", "E", "G", and our GICD can be regarded as a Co-SOD network; while, without GIM, the variants of "A", "B", "D", and "F" have limited ability on discovering co-salient objects. By directly applying GIM, the variant "C" shifts the attention to co-salient objects in the high-level features, and achieves a certain performance improvement compared to the baseline. However, in this

Table 2. Ablation study of the proposed GICD on the *CoCA* and *CoSal2015* datasets. The candidates are jigsaw training (JT), gradient inducing module (GIM), and attention retaining module (ARM). Note that, the variants "A", "B", "D", and "F", without GIM, are actually SOD models rather than Co-SOD models. The experiments reflect the interaction mechanism of our three contributions.

Variant	Candidate			CoCA				CoSal2015 [51]			
	JT	GIM	ARM	$F_{avg}\uparrow$	$F_{max}\uparrow$	$S_\alpha\uparrow$	$E_\xi\uparrow$	$F_{avg}\uparrow$	$F_{max}\uparrow$	$S_\alpha\uparrow$	$E_\xi\uparrow$
A				0.420	0.430	0.601	0.627	0.788	0.800	0.818	0.852
B	✓			0.424	0.430	0.602	0.655	0.750	0.759	0.782	0.821
C		✓		0.446	0.462	0.618	0.643	0.809	0.824	0.833	0.868
D			✓	0.429	0.437	0.607	0.628	0.800	0.809	0.829	0.860
E	✓	✓		0.470	0.478	0.631	0.689	0.795	0.803	0.808	0.850
F	✓		✓	0.436	0.442	0.612	0.654	0.762	0.770	0.795	0.832
G		✓	✓	0.471	0.480	0.636	0.667	0.826	0.835	0.845	0.879
GICD	✓	✓	✓	0.504	0.513	0.658	0.701	0.835	0.844	0.844	0.883

case, the training set is ill-posed for co-saliency detection task without JT, and the attention will be disturbed during the decoding without the help of ARM. These factors limit its performance. By introducing JT or ARM (variants "E", "G" and our GICD), the effect of GIM is further enhanced.

Effectiveness of ARM. ARM plays a role in retaining high-level prediction information during the top-down decoding. As shown in variant "D", using ARM alone does not improve Co-SOD performance. The reason is that, without inducing by GIM, the prediction in high-level is actually the salient objects rather than co-salient objects. When cooperating with GIM in variant "G", although trained on ill-posed data, it still compulsorily keeps the inducing information of GIM to an extend; thus, "G" achieves significantly better performance than the variant "C". "E" is a variant with our GIM and ARM modules. As shown in Fig. 5, without ARM, it is easy to be interfered by irrelevant foreground when recovering object details. Therefore, its performance is inferior to our GICD.

Effectiveness of JT. The jigsaw training (JT) helps turn SOD datasets into Co-SOD ones, and serves as a useful strategy for training Co-SOD networks. In Table 2, without GIM, the variant model "B" and "F" are SOD models, not Co-SOD ones. Since no interactive cue between images is considered, a SOD model trained on Co-SOD dataset is unable to discover group-wise co-salient connections, and the generated JT labels will bring meaningless predictions in this ill-posed scene; therefore, the JT does not work in these cases. When working with GIM in variant "E", JT improves the effect in the challenging *CoCA* dataset. Similarly, this improvement can also be seen through the comparison between our GICD and variant "G".

In summary, our three contributions of the GIM, ARM, and JT candidates are mutually reinforced for better co-saliency detection performance, as validated through comprehensive experiments.

6   Conclusions

In this paper, inspired by the mechanism of how human behaves on the Co-SOD task, we proposed an end-to-end Gradient-Induced Co-saliency Detection (GICD) method. In GICD, the gradient information, which highlights the discrimination of features, is generated from the comparison between single and consensus representations. Induced by the gradient, GICD pays more attention to discriminative convolutional kernels, enabling our model to locate the co-salient regions. Due to the lack of Co-SOD training data, we designed a novel jigsaw training strategy, with which we trained Co-SOD models on a general SOD dataset without extra pixel-level annotations. In addition, we constructed a challenging *CoCA* dataset for Co-SOD evaluation, to prosper the subsequent research on exploring real-world Co-SOD scenarios.

Acknowledgements. Zhao Zhang and Wenda Jin are the joint first authors. This research was supported by Major Project for New Generation of AI under Grant No. 2018AAA0100400, NSFC (61922046), Tianjin Natural Science Foundation (18ZXZNGX00110), and the Fundamental Research Funds for the Central Universities, Nankai University (63201169).

References

1. Achanta, R., Hemami, S., Estrada, F., Susstrunk, S.: Frequency-tuned salient region detection. In: CVPR, pp. 1597–1604 (2009)
2. Batra, D., Kowdle, A., Parikh, D., Luo, J., Chen, T.: iCoseg: interactive co-segmentation with intelligent scribble guidance. In: CVPR, pp. 3169–3176. IEEE (2010)
3. Borji, A., Cheng, M.M., Jiang, H., Li, J.: Salient object detection: a benchmark. IEEE TIP **24**(12), 5706–5722 (2015)
4. Cao, X., Tao, Z., Zhang, B., Fu, H., Feng, W.: Self-adaptively weighted co-saliency detection via rank constraint. IEEE TIP **23**(9), 4175–4186 (2014)
5. Chang, K.Y., Liu, T.L., Lai, S.H.: From co-saliency to co-segmentation: an efficient and fully unsupervised energy minimization model. In: CVPR 2011, pp. 2129–2136. IEEE (2011)
6. Chen, H.T.: Preattentive co-saliency detection. In: ICIP, pp. 1117–1120. IEEE (2010)
7. Deng, J., Dong, W., Socher, R., Li, L.J., Li, K., Fei-Fei, L.: Imagenet: a large-scale hierarchical image database. In: CVPR, pp. 248–255. IEEE (2009)
8. Fan, D.-P., Cheng, M.-M., Liu, J.-J., Gao, S.-H., Hou, Q., Borji, A.: Salient objects in clutter: bringing salient object detection to the foreground. In: Ferrari, V., Hebert, M., Sminchisescu, C., Weiss, Y. (eds.) ECCV 2018. LNCS, vol. 11219, pp. 196–212. Springer, Cham (2018). https://doi.org/10.1007/978-3-030-01267-0_12

9. Fan, D.P., Cheng, M.M., Liu, Y., Li, T., Borji, A.: Structure-measure: a new way to evaluate foreground maps. In: ICCV, pp. 4548–4557 (2017)
10. Fan, D.P., Gong, C., Cao, Y., Ren, B., Cheng, M.M., Borji, A.: Enhanced-alignment measure for binary foreground map evaluation. In: IJCAI, pp. 698–704 (2018)
11. Fan, D.P., Lin, Z., Ji, G.P., Zhang, D., Fu, H., Cheng, M.M.: Taking a deeper look at the co-salient object detection. In: CVPR (2020)
12. Fan, D.P., Zhai, Y., Borji, A., Yang, J., Shao, L.: BBS-Net: RGB-D salient object detection with a bifurcated backbone strategy network. In: Vedaldi, A., et al. (eds.) ECCV 2020. LNCS, vol. 12357, pp. 275–292. Springer, Heidelberg (2020)
13. Fu, H., Cao, X., Tu, Z.: Cluster-based co-saliency detection. IEEE TIP **22**(10), 3766–3778 (2013)
14. Gao, S.H., Tan, Y.Q., Cheng, M.M., Lu, C., Chen, Y., Yan, S.: Highly efficient salient object detection with 100k parameters. In: ECCV (2020)
15. Gao, Z., Xu, C., Zhang, H., Li, S., de Albuquerque, V.H.C.: Trustful internet of surveillance things based on deeply-represented visual co-saliency detection. IEEE Internet Things J. **7**, 4092–4100 (2020)
16. Han, J., Cheng, G., Li, Z., Zhang, D.: A unified metric learning-based framework for co-saliency detection. IEEE TCSVT **28**(10), 2473–2483 (2017)
17. Hsu, K.-J., Tsai, C.-C., Lin, Y.-Y., Qian, X., Chuang, Y.-Y.: Unsupervised CNN-based co-saliency detection with graphical optimization. In: Ferrari, V., Hebert, M., Sminchisescu, C., Weiss, Y. (eds.) ECCV 2018. LNCS, vol. 11209, pp. 502–518. Springer, Cham (2018). https://doi.org/10.1007/978-3-030-01228-1_30
18. Jerripothula, K.R., Cai, J., Yuan, J.: CATS: co-saliency activated tracklet selection for video co-localization. In: Leibe, B., Matas, J., Sebe, N., Welling, M. (eds.) ECCV 2016. LNCS, vol. 9911, pp. 187–202. Springer, Cham (2016). https://doi.org/10.1007/978-3-319-46478-7_12
19. Jerripothula, K.R., Cai, J., Yuan, J.: Efficient video object co-localization with co-saliency activated tracklets. IEEE TCSVT **29**(3), 744–755 (2018)
20. Jiang, B., Jiang, X., Zhou, A., Tang, J., Luo, B.: A unified multiple graph learning and convolutional network model for co-saliency estimation. In: ACM Multimedia, pp. 1375–1382 (2019)
21. Joulin, A., Tang, K., Fei-Fei, L.: Efficient image and video co-localization with Frank-Wolfe algorithm. In: Fleet, D., Pajdla, T., Schiele, B., Tuytelaars, T. (eds.) ECCV 2014. LNCS, vol. 8694, pp. 253–268. Springer, Cham (2014). https://doi.org/10.1007/978-3-319-10599-4_17
22. Kingma, D.P., Ba, J.L.: Adam: a method for stochastic optimization. In: ICLR (2015)
23. Li, B., Sun, Z., Tang, L., Sun, Y., Shi, J.: Detecting robust co-saliency with recurrent co-attention neural network. In: IJCAI, pp. 818–825 (2019)
24. Li, H., Ngan, K.N.: A co-saliency model of image pairs. IEEE TIP **20**(12), 3365–3375 (2011)
25. Li, Z., Chen, Q., Koltun, V.: Interactive image segmentation with latent diversity. In: CVPR, pp. 577–585 (2018)
26. Lin, T.Y., Dollár, P., Girshick, R., He, K., Hariharan, B., Belongie, S.: Feature pyramid networks for object detection. In: CVPR, pp. 2117–2125 (2017)
27. Lin, T.-Y., et al.: Microsoft COCO: common objects in context. In: Fleet, D., Pajdla, T., Schiele, B., Tuytelaars, T. (eds.) ECCV 2014. LNCS, vol. 8693, pp. 740–755. Springer, Cham (2014). https://doi.org/10.1007/978-3-319-10602-1_48
28. Liu, J.J., Hou, Q., Cheng, M.M., Feng, J., Jiang, J.: A simple pooling-based design for real-time salient object detection. In: CVPR, pp. 3917–3926 (2019)

29. Liu, T., et al.: Learning to detect a salient object. IEEE TPAMI **33**(2), 353–367 (2010)
30. Luo, Y., Jiang, M., Wong, Y., Zhao, Q.: Multi-camera saliency. IEEE TPAMI **37**(10), 2057–2070 (2015)
31. Mahajan, D., et al.: Exploring the limits of weakly supervised pretraining. In: Ferrari, V., Hebert, M., Sminchisescu, C., Weiss, Y. (eds.) ECCV 2018. LNCS, vol. 11206, pp. 185–201. Springer, Cham (2018). https://doi.org/10.1007/978-3-030-01216-8_12
32. Paszke, A., et al.: PyTorch: an imperative style, high-performance deep learning library. In: NeurIPS, pp. 8024–8035 (2019)
33. Plaut, D.C.: Graded modality-specific specialisation in semantics: a computational account of optic aphasia. Cognit. Neuropsychol. **19**(7), 603–639 (2002)
34. Qi, H., Brown, M., Lowe, D.G.: Low-shot learning with imprinted weights. In: CVPR, pp. 5822–5830 (2018)
35. Qin, X., Zhang, Z., Huang, C., Gao, C., Dehghan, M., Jagersand, M.: Basnet: boundary-aware salient object detection. In: CVPR, pp. 7479–7489 (2019)
36. Selvaraju, R.R., Cogswell, M., Das, A., Vedantam, R., Parikh, D., Batra, D.: Grad-CAM: visual explanations from deep networks via gradient-based localization. In: ICCV, pp. 618–626 (2017)
37. Sun, H., Zhen, X., Zheng, Y., Yang, G., Yin, Y., Li, S.: Learning deep match kernels for image-set classification. In: CVPR, pp. 6240–6249 (2017)
38. Tang, K., Joulin, A., Li, L.J., Fei-Fei, L.: Co-localization in real-world images. In: CVPR, pp. 1464–1471 (2014)
39. Tsai, C.C., Li, W., Hsu, K.J., Qian, X., Lin, Y.Y.: Image co-saliency detection and co-segmentation via progressive joint optimization. IEEE TIP **28**(1), 56–71 (2018)
40. Wang, C., Zha, Z.J., Liu, D., Xie, H.: Robust deep co-saliency detection with group semantic. In: AAAI, vol. 33, pp. 8917–8924 (2019)
41. Wang, L., Lu, H., Wang, Y., Feng, M., Wang, D., Yin, B., Ruan, X.: Learning to detect salient objects with image-level supervision. In: CVPR (2017)
42. Wei, L., Zhao, S., Bourahla, O.E.F., Li, X., Wu, F.: Group-wise deep co-saliency detection. In: IJCAI, pp. 3041–3047 (2017)
43. Wei, L., Zhao, S., Bourahla, O.E.F., Li, X., Wu, F., Zhuang, Y.: Deep group-wise fully convolutional network for co-saliency detection with graph propagation. IEEE TIP **28**, 5052–5063 (2019)
44. Wei, Y., et al.: Stc: a simple to complex framework for weakly-supervised semantic segmentation. IEEE TPAMI **39**(11), 2314–2320 (2016)
45. Winn, J., Criminisi, A., Minka, T.: Object categorization by learned universal visual dictionary. In: ICCV, pp. 1800–1807 (2005)
46. Wu, Z., Su, L., Huang, Q.: Stacked cross refinement network for edge-aware salient object detection. In: ICCV, pp. 7264–7273 (2019)
47. Zeng, Y., Zhuge, Y., Lu, H., Zhang, L.: Joint learning of saliency detection and weakly supervised semantic segmentation. In: ICCV, pp. 7223–7233 (2019)
48. Zha, Z., Wang, C., Liu, D., Xie, H., Zhang, Y.: Robust deep co-saliency detection with group semantic and pyramid attention. IEEE TNNLS **31**(7), 2398–2408 (2020)
49. Zhang, C., Lin, G., Liu, F., Yao, R., Shen, C.: CANet: class-agnostic segmentation networks with iterative refinement and attentive few-shot learning. In: CVPR, pp. 5217–5226 (2019)
50. Zhang, D., Fu, H., Han, J., Borji, A., Li, X.: A review of co-saliency detection algorithms: fundamentals, applications, and challenges. ACM TIST **9**(4), 1–31 (2018)

51. Zhang, D., Han, J., Li, C., Wang, J., Li, X.: Detection of co-salient objects by looking deep and wide. IJCV **120**(2), 215–232 (2016). https://doi.org/10.1007/s11263-016-0907-4

52. Zhang, D., Meng, D., Han, J.: Co-saliency detection via a self-paced multiple-instance learning framework. IEEE TPAMI **39**(5), 865–878 (2016)

53. Zhang, K., Li, T., Liu, B., Liu, Q.: Co-saliency detection via mask-guided fully convolutional networks with multi-scale label smoothing. In: CVPR, pp. 3095–3104 (2019)

54. Zhang, X., Wei, Y., Yang, Y., Huang, T.: SG-One: similarity guidance network for one-shot semantic segmentation (2018)

55. Zheng, X., Zha, Z.J., Zhuang, L.: A feature-adaptive semi-supervised framework for co-saliency detection. In: ACM Multimedia, pp. 959–966 (2018)

56. Zhou, B., Khosla, A., Lapedriza, A., Oliva, A., Torralba, A.: Learning deep features for discriminative localization. In: CVPR, pp. 2921–2929 (2016)

Nighttime Defogging Using High-Low Frequency Decomposition and Grayscale-Color Networks

Wending Yan[1]([✉])[iD], Robby T. Tan[1,2][iD], and Dengxin Dai[3][iD]

[1] National University of Singapore, Singapore, Singapore
{eleyanw,robby.tan}@nus.edu.sg
[2] Yale-NUS College, Singapore, Singapore
[3] ETH Zurich, Zürich, Switzerland
dai@vision.ee.ethz.ch

Abstract. We address the problem of nighttime defogging from a single image by introducing a framework consisting of two modules: grayscale and color modules. Given an RGB foggy nighttime image, our grayscale module takes the grayscale version of the image as input, and decomposes it into high and low frequency layers. The high frequency layers contain the scene texture information, which is less affected by fog. While the low frequency layers contain the scene layout/structure information including fog and glow. Our grayscale module then enhances the visibility of the textures in the high frequency layers, and removes the presence of glow and fog in the low frequency layers. Having processed the high/low frequency information, it fuses the two layers to obtain a grayscale defogged image. Our second module, the color module, takes the original RGB image, and process it similarly to what the grayscale module does. However, to obtain fog-free high and low frequency information, the module is guided by the grayscale module. The reason of doing this is because grayscale images are less affected by multiple colors of atmospheric light, which are commonly present in nighttime scenes. Moreover, having the grayscale module allows us to have consistency losses between the outputs of the two modules, which is critical to our framework, since we do not have paired ground-truths for our real data. Our extensive experiments on real foggy nighttime images show the effectiveness of our method.

1 Introduction

Fog in nighttime can significantly degrades visibility. Few methods have been proposed to address this degradation problem, e.g.: [17,26,27]. Unfortunately, due to the complexity of illumination colors, the presence of multiple light sources and the strong glow, these methods still perform suboptimally as shown in Fig. 1. To our knowledge, there is no deep-learning based nighttime defogging method

© Springer Nature Switzerland AG 2020
A. Vedaldi et al. (Eds.): ECCV 2020, LNCS 12357, pp. 473–488, 2020.
https://doi.org/10.1007/978-3-030-58610-2_28

Fig. 1. Top left: Input image. Top right: Our result. Bottom left: Li et al.'s result [17]. Bottom right: Zhang et al.'s result [26]. These methods are the state of the art in nighttime defogging. Zoom-in for better visualization.

that has been proposed so far. The major reason is because collecting pairs of foggy nighttime images and their corresponding clear nighttime images, which are necessary for training a network, is significantly intractable. Rendering synthetic foggy nighttime images is a possible solution, however to render them physically realistic for many different nighttime scenes and atmospheric conditions is challenging, since various information is required (such as depths, light source locations, particle distributions, how lights with different colors and intensities interact, etc).

Unlike foggy daytime images, foggy nighttime images suffer from low light, strong noise, multiple light sources, non-uniform distribution of atmospheric light intensity, multiple colors of lights, strong glow, etc. Because of these problems, directly applying existing daytime defogging methods to foggy nighttime images will not be effective. One of the reasons is that most of the methods (e.g. [3,7, 12,19,22,24]) assume a single atmospheric light (thus, a uniformly ambient light color), which is inapplicable to most of foggy nighttime images, since there are usually many man-made light sources with different colors in a single image.

In this paper, we introduce a framework that consists of grayscale and color modules. Given an RGB input image, the grayscale module takes the grayscale version of the input RGB image as input. The grayscale module decomposes the image into high and low frequency layers. These layers are then processed by two separate networks: a network for the high frequency layers that enhances the textures, and a network for the low frequency layers that removes glow and fog. Subsequently, fusing the processed high and low frequency information will produce a defogged grayscale image. Our second module, the color module, takes the original RGB image as input. The operations in the color module are similar to those in the grayscale module. Unlike the grayscale module, however, we train the module based on the outputs of the grayscale module. The reason of doing this is because we found defogging a grayscale image is more effective than defogging the RGB version of the image alone. This is mainly because, the grayscale image is less complicated by multiple colors of atmospheric light (see the detailed discussion in Sect. 4). Moreover, having the grayscale module, we can have consistency losses between the outputs of the two modules, which is critical to our framework, since we do not have paired ground-truths for our real data. To train our networks, we employ a few foggy-nighttime synthetic images, which come with ground-truths, and real foggy-nighttime images that have no paired ground-truths.

As a summary, our contributions are as follows:

- We introduce a nighttime defogging method based on the grayscale and color modules. Our method works for both grayscale and RGB input image.
- We propose the use of the low and high frequency layer decomposition. The high frequency layer contains the scene texture information, which is less affected by fog and ambient light colors; while, the low frequency layer contains the scene structure information affected by fog and glow.
- We introduce new consistency losses between the outputs of the grayscale and color modules, which are useful to strengthen the training process, particularly when we do not have paired ground-truths.

2 Related Work

Many methods have been proposed to deal with fog or haze in daytime (e.g., [3,6,7,12,14,15,19,22–25]). Tan [24] proposes a method by maximizing contrast. Fattal [7] estimates the transmission map by assuming that it is independent on the surface shading. He et al. [12] propose with the dark channel prior. Meng et al. [19] predict the transmission map with minimum boundaries and contextual regularization. Berman et al. [3] present a haze-line prior. In the era of deep learning, Ren et al. [22] design a multi-scale network for estimating the transmission map. Li et al. [14] propose all-in-one network, which predicted defogged images without estimating the transmission map and atmospheric light. Li et al. [15] apply a conditional GAN and Engin et al. [6] applied a cycle GAN to their method. More detailed discussion can be found in [18]. However, all these existing methods do not have good performance in nighttime fog removal. They have problem in estimating the atmospheric light. Even CNN-based methods have not been shown to be able to handle nighttime complex ambient lights, due to the lack of paired ground-truths.

There are few methods that specifically address nighttime fog removal. Pei and Lee [20] transfer the colors of a foggy nighttime image into the foggy daytime style, and apply the dark channel prior. Zhang et al. [27] propose a method using varying illumination compensation and post-processing. This technique brings more realistic results, but also color artifacts. Li et al. [17] apply glow

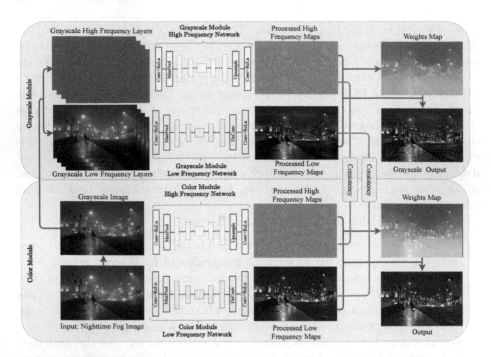

Fig. 2. The pipeline of our framework that consists of two main modules: grayscale module (top) and the color module (bottom). For the images, zoom-in for better visualization. (Color figure online)

removal before fog removal. The glow is decomposed from input image by a layer separation method. Then, the deglowed image is further defogged using the dark channel prior. Ancuti et al. [1] provide a multi-scale fusion approach to enhance the nighttime fog visibility. They compute the atmospheric light component on an image patch level and derive three components, which will be weighed and fused to obtain the final output. Zhang et al. [26] propose a prior named maximum reflectance prior. They claim that the ambient illumination can be estimated by this prior. Like most of the nighttime defogging methods, the method employs the dark prior channel.

3   Proposed Method

Figure 2 shows the pipeline of our method, which consists of two modules: grayscale and color modules. Given an RGB input image, the grayscale module takes the grayscale version of the input rgb image as input. It then decomposes the input into high and low frequency layers. A network dedicated to process the high frequency layers boosts the texture information of the layers. Another network dedicated to process the low frequency layers removes glow and fog.

From the low and high frequency layers, a weight map is computed to suppress the noise of the processed high frequency information. Finally, the grayscale defogged output is obtained by fusing the low/high frequency information. The color module does similar processes, and thus we can create losses that can keep the outputs of the grayscale and color modules to be consistent. Note that, for real training data, we do not have paired ground-truths. The details of the discussion on the two modules are as follows.

3.1   Grayscale Module

There are two main reasons of having the grayscale module in our solution. First, in nighttime, ambient lights with multiple colors are common. This is due to the fact that there are many man-made light sources with various colors. Unlike nighttime colored images, nighttime grayscale images are less complicated by the colors of multiple light (more discussion in Sect. 4). Second, for real foggy nighttime images, we do not have their clean fog-free ground-truths. Hence, by having the grayscale module, we can have additional constraints by enforcing the consistency between the outputs of the grayscale and color modules.

High/Low Frequency Layers and Networks. The grayscale module decomposes the input image into a few high and low frequency layers. Any state of the art decomposition techniques can be used (e.g. [11]). In our implementation, we produces 8 high frequency layers, and their 8 corresponding low frequency layers using different decomposition parameters. The reason of doing this is to reduce missing information possibly caused by inappropriate choice of parameters with respect to the input image. The high frequency layers contain the scene textures, and the low frequency layers contain the scene layout/structures. Due to their smoothness, fog and glow will go to the low frequency layers, and consequently the high frequency layers is not much affected by fog and glow (see Fig. 2). We separate the high and low frequency is because they represent different aspects of a foggy scene, and hence they should be processed differently. Otherwise, when removing fog, we may also remove genuine textures of the scene.

After decomposition, we process the high frequency layers to boost the texture information. The main reason of why we want to boost the textures is because in foggy nighttime images, the contrast is low, and thus the textures can be weak. We design an autoencoder network to handle this task, and we call it high-frequency network. The input of this network is the concatenation of the high frequency layers. Like the high frequency part, for the low frequency layers, we create a low-frequency network to process them. Our intention of creating the network is to suppress the presence of fog and glow. We call the outputs of the high/low frequency networks high/low frequency maps, respectively.

Fusing High/Low Frequency Maps. Having obtained the high/low frequency maps, we can directly fuse them. However, since boosting the textures in the high frequency layers also means boosting the noise, prior fusing the two maps, we compute a weight map. The weights in this map represents the noise levels of the high frequency map. Thus, multiplying the weight map with the high frequency map will reduce noise. Note that, foggy nighttime images suffer significantly from noise; this is because many parts of foggy nighttime images are regions with low light. We adopt an existing technique [16] to compute the weight map, where we first generate a coarse map from the discrete cosine transform (DCT) coefficients of the high frequency map. Larger DCT coefficients represent more texture details. Thus, we set the sum of coefficients as the coarse weight map. However, since the DCT is computed patch by patch, the weight map may looks coarse. As suggested in [16], we utilize the low frequency map to smoothen the weights. This refined weight map has high values on texture regions and low values on uniform regions like sky regions. Figure 2, the top right image, shows a sample of our weight map.

In the grayscale module, we have a few losses: smoothness, pair and discriminative. The smoothness loss is applied only to the low frequency network; while the grayscale pair and discriminative losses are applied to the final defogged image. Hence, the backpropagated errors from the grayscale pair and discriminative losses will go to both the high and low frequency networks.

– **Grayscale Smoothness Loss.** The loss is defined as:

$$\mathcal{L}_{smooth}^{gray} = \mathbb{E}\big[(J_{gray}^{f} - LF_{gray}[J_{gray}^{f}]) * f_2\big], \tag{1}$$

where $LF[.]$ is the low frequency network, and J_{gray}^{f} is the grayscale foggy nighttime input (the superscript f stands for fog). f_2 is the second order Laplacian filter and $*$ is the convolution operator. The loss enforces that the removed fog and glow by the network must be smooth spatially. The subtraction is done pixel by pixel. The operator $\mathbb{E}[S]$, where S is any matrix, implies adding every element of the matrix, and dividing the total sum with the size of the matrix.

– **Grayscale Pair Loss.** Our synthetic nightime foggy data has paired clean ground-truths. Thus, we can employ the pair loss:

$$\mathcal{L}_{pair}^{gray} = \mathbb{E}\big[\|M_{gray}[J_{gray}^{f}] - J_{gray}^{gt}\|\big], \tag{2}$$

where J_{gray}^{f} and J_{gray}^{gt} are paired synthetic grayscale fog image and clean ground-truth image, respectively. $M_{gray}[.]$ is our grayscale module.

– **Grayscale Discriminative Loss.** We apply a discriminative loss only to real images that do not have paired ground-truths. We define our discriminative loss as:

$$\mathcal{L}_{GAN}^{gray} = -\log(\sigma[D_{gray}[M_{gray}[J_{gray}^{f}]]]), \tag{3}$$

where $\sigma[.]$ is the sigmoid function, and D_{gray} is the discriminator [9].

Overall, the total loss functions of the grayscale module is:

$$\mathcal{L}_{total}^{gray} = \mathcal{L}_{pair}^{gray} + \lambda_1^{gray}\mathcal{L}_{GAN}^{gray} + \lambda_2^{gray}\mathcal{L}_{smooth}^{gray}, \tag{4}$$

Grayscale Input   Grayscale Defogged   Grayscale Input   Grayscale Defogged
Output   Output

Fig. 3. The inputs and outputs of our grayscale module. Two left images show a pair of our input and output. Two right images show the same. Zoom-in for better visualization.

where λ_1^{gray} and λ_2^{gray} are the weighting factors. Empirically in our experiments, we set $\lambda_1^{gray} = 0.1$ and $\lambda_2^{gray} = 1$.

Figure 3 shows some defogging results using the grayscale module. The glow and fog in input images are significantly removed, while the textures of the background are still retained. Note that, fog-free nighttime images without glow and sufficient lighting usually suffer from the low light problem. This is because glow is in fact light scattered by particles in the atmosphere. Hence, after we successfully remove fog and thus glow, it is natural that the image looks dimmer than the input image. We will discuss about this further in the later part of the following section.

3.2 Color Module

In the color module, we take the original RGB input image as input. The operations in this module are similar to those of the grayscale module. However, in this module, we do not apply the high and low frequency decomposition technique like in the grayscale module. Instead, we input the RGB foggy nighttime image to the low frequency network directly, and the grayscale version of the input image to the high frequency network. The high and low frequency networks in this module are then guided by the counterparts in the grayscale module.

High/Low Frequency Networks. Since the high frequency map is independent from colors, the input of the high frequency network is the grayscale version of the input image. Like its counterpart in the grayscale module, the task of the high frequency network is to boost textures. The high frequency network learns to boost textures largely from the consistency loss that measures the consistency between its output and the outputs of the high frequency network in the grayscale module. Note that, unlike the high frequency network in the grayscale module that takes high frequency layers as input, in this module, the high frequency module takes the grayscale input image. Hence, we cannot simply use the one in the grayscale module.

Similarly, the low frequency network is also trained using the consistency with respect to the output of the low frequency counterpart in the grayscale module. However, different from the low frequency network in the grayscale module, the low frequency network in the color module produces a colored image. Hence, we convert the colored low frequency map to grayscale before applying the consistency loss. Doing this process, however, allows the color in the low frequency map to be shifted, since for real data, we do not have their paired ground-truths. To tackle this problem, we need another loss that ensures the correctness of the colors of the low frequency map. Here are all the losses for training the high and low frequency networks in the color module. Note that, after all the networks in the grayscale module have been trained, we freeze them.

- **High Frequency Consistency Loss.** We define it as:

$$\mathcal{L}_{HF}^{rgb} = \mathbb{E}\big[\|[HF_{gray}[J_{gray}^f]] - HF_{rgb}[J_{gray}^f]\|\big], \tag{5}$$

where $HF_{gray}[.]$ and $HF_{rgb}[.]$ are the low frequency networks from the grayscale and color modules, respectively.
- **Low Frequency Consistency Loss.** We define it as:

$$\mathcal{L}_{LF}^{rgb} = \mathbb{E}\big[\|LF_{gray}[J_{gray}^f] - G\left(LF_{rgb}[J_{rgb}^f]\right)\|\big], \tag{6}$$

where $LF_{gray}[.]$ and $LF_{rgb}[.]$ are the low frequency networks from the grayscale and color modules, respectively. Function $G(.)$ is to convert an rgb image to its grayscale version.
- **Hue Loss.** This loss is to ensure there is no color shift in the output of the low frequency network. The basic idea is that if we can estimate the color (or chromaticity) of the atmospheric light of the input RGB image, then we can normalize the image such that the atmospheric light colors will be canceled. Once the input image is normalized, it is known that the hue values of the input fog image will be the same as those of the normalized clean fog-free image (e.g. [13]). Hence, we define the loss as:

$$\mathcal{L}_{hue}^{rgb} = \mathbb{E}\big[\|Hue[CN[LF_{rgb}[J_{rgb}^f], A[x_{rgb}^f]]]$$
$$-Hue[CN[J_{rgb}^f, A[J_{rgb}^f]]]\|\big], \tag{7}$$

where $Hue[.]$ are a function to compute the hue values of an image. $CN[.]$ is the function that normalizes an image based on the atmospheric light chromaticity. $A[.]$ is the function to estimate the atmospheric light chromaticity. We can use any existing technique to estimate the atmospheric light chromaticity for every pixel such as that described in [17,26].

Fusing High and Colored Low Frequency Maps. Having obtained the grayscale high and colored low frequency maps, we multiple the grayscale high frequency map with the weight map we obtained in the grayscale module to suppress noise. Subsequently, we fuse the weighted high frequency map and the colored low frequency map. To ensure good quality of outputs, we also impose a few loss functions to train the networks in this color module:

- **Color Pair Loss.** This loss is only for our paired synthetic data. The loss is defined as:

$$\mathcal{L}_{pair}^{rgb} = \mathbb{E}\big[\|M_{rgb}[J_{rgb}^{f}] - J_{rgb}^{gt}\|\big], \tag{8}$$

where x_{rgb}^{f} and x_{rgb}^{gt} are paired synthetic fog image and clean ground-truth image, respectively.
- **Color Discriminative Loss.** We define the loss as:

$$\mathcal{L}_{GAN}^{rgb} = -\log(\sigma[D_{rgb}[M_{rgb}[J_{rgb}^{f}]]]), \tag{9}$$

where x_{rgb}^{f} is a real colored foggy nighttime image.
- **Image Metric Loss.** We include an unsupervised loss based on contrast and acutance [2], which are defined as:

$$\mathcal{L}_{IQ}^{rgb} = \mathbb{E}\big[(1 - C[M_{rgb}[J_{rgb}^{f}]] + C[J_{rgb}^{f}]) \\ + (1 - Ac[M_{rgb}[J_{rgb}^{f}]] + Ac[J_{rgb}^{f}])\big], \tag{10}$$

where the $C[.]$ and $Ac[.]$ are the functions to obtain the contrast and acutance of an image.

Overall, the total loss functions for color module is:

$$\mathcal{L}_{total}^{rgb} = \mathcal{L}_{pair}^{rgb} + \lambda_1^{rgb}\mathcal{L}_{GAN}^{rgb} + \lambda_2^{rgb}\mathcal{L}_{LF}^{rgb} \\ + \lambda_3^{rgb}\mathcal{L}_{HF}^{rgb} + \lambda_4^{rgb}\mathcal{L}_{hue}^{rgb} + \lambda_5^{rgb}\mathcal{L}_{IQ}^{rgb}, \tag{11}$$

where $\lambda_i^{rgb}, i = \{1...5\}$ are the weighting factors. Empirically in our experiments, we set $\lambda_1^{rgb} = 0.1$, $\lambda_2^{rgb} = 1$, $\lambda_3^{rgb} = 1$, $\lambda_4^{rgb} = 0.5$ and $\lambda_5^{rgb} = 1$.

Intensity Boosting. Degradation in foggy nighttime images is caused not only by fog but also by other nighttime conditions such as low light and noise. Hence, successfully removing fog does not mean that we can have an output that is bright and noise free, since low light and noise are inherent even in clear nighttime images. In fact, foggy nighttime images look relatively brighter than clear nighttime images mainly due to the presence of glow (= the light scattered into the atmosphere). Hence, when we successfully remove fog and thus glow, implying removing the scattered light, it is natural that the images look darker.

Nevertheless, we can boost the intensity of our output to improve the visibility quality using our current framework. Although we have to note that, the

Without Boosting   With Boosting   Without Boosting   With Boosting

Fig. 4. After modified the low frequency consistency loss, the image brightness is boosted.

issue of nighttime visibility enhancement is still an open problem (e.g. [4,10]), and thus effective nighttime/low-light visibility enhancement is actually beyond the scope of our defogging paper. In our intensity boosting, we simply modify the low frequency consistency loss in Eq. (6), namely by brightening up the reference grayscale low-frequency map; since, the low frequency map in the color module influences the tone and brightness of the final output. Our modified loss is expressed as:

- **Modified Low Frequency Consistency Loss:**

$$\mathcal{L}_{LF}^{boost} = \mathbb{E}\big[\|B\big([LF_{gray}[J_{gray}^{f}]]]\big) - G\big(LF_{rgb}[J_{rgb}^{f}]\big)\|\big], \tag{12}$$

where $B(.)$ is the intensity boosting function, where we use a gamma function. $G(S)$, where S is any matrix, means for each element in S we power it to γ. In other words: S_{ij}^{γ}, where γ in our experiment is 0.5. As a result of this, In Eq. (11), the loss $\mathcal{L}_{LF}^{rgb}$ is replaced by this loss $\mathcal{L}_{LF}^{boost}$.

Figure 4 shows the results of our intensity boosting.

4   Discussion: Multiple Light Colors

In this section, we discuss why using grayscale images (and thus our grayscale module) can be more effective in dealing with multiple-light colors, which are commonly present in the nighttime images.

Most daytime defogging methods (e.g., [8,12,24]) assume the atmospheric light is uniform across the input image. Which is a reasonable assumption, since in foggy daytime, the dominant light source is only the skylight. This assumption can thus simplify the problem of defogging, since we do not need to estimate the atmospheric light **A** for every pixel locally, reducing the complexity of the already ill-posed problem. Unfortunately, this assumption cannot be applied to most foggy nighttime images. Since in foggy nighttime, the skylight is usually much dimmer than the man-made light sources; and, there are multiple intensities/colors of these light sources, and their potential combination. This implies, that we need to estimate the atmospheric light locally for every pixel in the input image, which render a more complication of the problem. Existing nighttime defogging methods (e.g. [1,17,26]) propose a few techniques to estimate the atmospheric light locally, based on the patches of the input image. The basic assumption is that in a small patch, there is

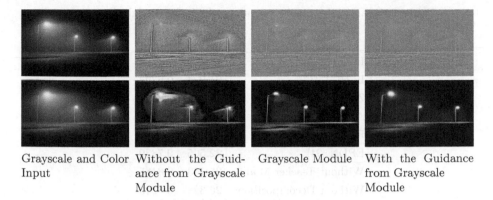

| Grayscale and Color Input | Without the Guidance from Grayscale Module | Grayscale Module | With the Guidance from Grayscale Module |

Fig. 5. Comparison of the high and low frequency maps: The color module without guidance, the grayscale module and the color module with guidance from the grayscale module. (Color figure online)

a uniform atmospheric light. While to some extent the proposed techniques work, their accuracy relies on the decided patch size. If the size is too large, the accuracy will drop, since the patch can actually contains multiple light colors. And, if the size is too small, the accuracy will also drop, since the background intensity/color can be uniform and mistakenly estimated as the light intensity/color.

In our proposed method, to deal with the problem, we leverage the grayscale version of the RGB input image. Our underlying idea is that grayscale images do not have color information, and thus the problem of multiple colors of the atmospheric light is not present. This reduces the complexity of the problem, and our graysacle module can focus on extracting the background properties, i.e., high and low frequencies, without worrying the correctness of their colors. Figure 5 show the results of the high and low frequency from an RGB image (processing them directly without the guidance from its grayscale images), from a grayscale version of the image, and from our color module guided by our grayscale module. As can be observed, the grayscale high/low frequency can reveal more about the background properties. Moreover, the one without the guidance shows more artifacts, fogginess and color shifts in some regions.

5   Experimental Results

In our experiments, we compare our method with the following baselines: Li et al. [17], Ancuti et al. [1], and Zhang et al. [26], which are the three state-of-art nighttime defogging methods. Besides, we also compare our results with the state-of-art daytime defogging methods Berman et al. [3] and EPDN [21]. In our training process, we combine synthetic data with ground-truths and real foggy nighttime data without ground-truths. The synthetic data is generated by a video game engine, GTA5 [5], and our real foggy nighttime data are collected from the Internet.

Table 1. Quantitative results on our synthetic foggy nighttime data.

	PSNR	SSIM
Input Image	18.987	0.6764
Li et al. [17]	21.024	0.6394
Zhang et al. [26]	20.921	0.6461
Ancuti et al. [1]	20.585	0.6233
Berman et al. [3]	19.085	0.5373
EPDN [21]	22.565	0.7330
Without Teacher Module	26.163	0.8185
Without Decomposition	26.328	0.8219
Without Weights Map	26.892	0.8327
Our Result	**26.997**	**0.8499**

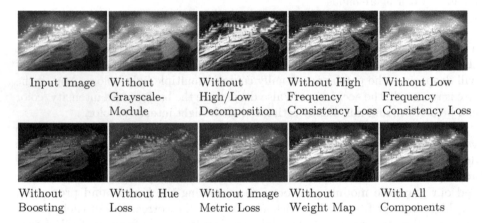

Input Image | Without Grayscale-Module | Without High/Low Decomposition | Without High Frequency Consistency Loss | Without Low Frequency Consistency Loss

Without Boosting | Without Hue Loss | Without Image Metric Loss | Without Weight Map | With All Components

Fig. 6. Ablation studies on the high/low frequency decomposition, grayscale/color modules, loss functions and weight map. Zoom-in for better visualization.

For the quantitative evaluation, we use 200 pairs of synthetic data, which are generated from GTA5. The quantitative evaluation results are shown in Table 1, where our method shows better performance on both PSNR and SSIM by significant margins compared to all the baseline methods. Figure 7 shows the qualitative evaluation results on real foggy nighttime images. As one can notice that our method qualitatively provides better defogging results compared with the results of the other baseline methods.

5.1   Ablation Studies

To show the effectiveness of our grayscale module in helping the color module, we train our color module without the losses that involve the grayscale module. In other words, we cut off the color module from the grayscale module. The network

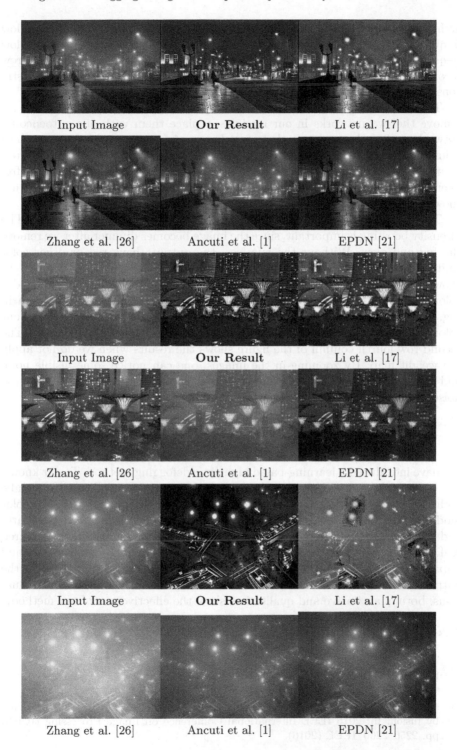

Fig. 7. Qualitative comparisons with the state of the art methods on real images. Zoom-in for better visualization. (Color figure online)

architecture in this case is similar to the grayscale module, but the input of the low frequency network is the RGB low frequency layers (instead of the grayscale low frequency layers), and the same decomposition technique is applied. The first row second column of Fig. 6 shows the defogging results. As can be seen, the fog is still considerably noticeable.

To show the effectiveness of the high/low frequency decomposition, we remove the two networks in our modules, replace them with one autoencoder network, and apply our losses that are relevant. As shown in the first row third column of Fig. 6, there are some regions still affected by fog, as the network is mistaken it with genuine textures. The ablation studies on the high/low frequency grayscale guidance (i.e. the high/low frequency consistency losses) are shown in the last two columns and first row of Fig. 6.

The second row of Fig. 6 shows that the effectiveness of other losses. The intensity boosting is important, since the image becomes dimmer after we remove the fog and glow. Without the hue loss, the outputs have red or other color shifting. Our hue loss can constrain the color shifting of the RGB outputs. To show the effectiveness of our image metric loss, we remove this loss from our color module. The results are shown in the second row and third column in Fig. 6. After adding this unsupervised image metric loss, the outputs of our color module are constrained to have higher contrast and higher fidelity. The second row fourth column of the figure shows the results when we do not apply the weight map. By zooming-in the images, one can notice some visible noise and artifacts particularly in the dark regions. By applying the weight map, the noise/artifact will be suppressed, as shown in the last column of the figure.

6   Conclusion

We have introduced a learning-based nighttime defogging method. To our knowledge, this is the first time, a deep learning-based method is dedicated to handle nighttime defogging problem. To achieve our goal, we design grayscale and color modules, which rely mainly on the high/low frequency layers to enhance textures and at the same time suppress glow, fog and noise. Due to the lack of paired real ground-truths, our training process employs both paired synthetic data and unpaired real data. For this, we introduce new consistency losses between the outputs of the grayscale and color modules. Experimental results and evaluations, both quantitative and qualitative, show the effectiveness of our method.

Acknowledgment. This work is supported by MOE2019-T2-1-130.

References

1. Ancuti, C., Ancuti, C.O., De Vleeschouwer, C., Bovik, A.C.: Night-time dehazing by fusion. In: 2016 IEEE International Conference on Image Processing (ICIP), pp. 2256–2260. IEEE (2016)

2. Barbosa, W.V., Amaral, H.G., Rocha, T.L., Nascimento, E.R.: Visual-quality-driven learning for underwater vision enhancement. In: 2018 25th IEEE International Conference on Image Processing (ICIP), pp. 3933–3937. IEEE (2018)

3. Berman, D., Avidan, S., et al.: Non-local image dehazing. In: Proceedings of the IEEE Conference on Computer Vision and Pattern Recognition, pp. 1674–1682 (2016)

4. Chen, C., Chen, Q., Xu, J., Koltun, V.: Learning to see in the dark. In: Proceedings of the IEEE Conference on Computer Vision and Pattern Recognition, pp. 3291–3300 (2018)

5. Doan, A.D., Jawaid, A.M., Do, T.T., Chin, T.J.: G2D: from GTA to Data. arXiv preprint arXiv:1806.07381, pp. 1–9 (2018)

6. Engin, D., Genc, A., Kemal Ekenel, H.: Cycle-dehaze: enhanced cyclegan for single image dehazing. In: Proceedings of the IEEE Conference on Computer Vision and Pattern Recognition Workshops, pp. 825–833 (2018)

7. Fattal, R.: Single image dehazing. ACM Trans. Graph. (TOG) **27**(3), 72 (2008)

8. Fattal, R.: Dehazing using color-lines, **34**(1), 13 (2014)

9. Goodfellow, I., et al.: Generative adversarial nets. In: Advances in Neural Information Processing Systems, pp. 2672–2680 (2014)

10. Guo, X., Li, Y., Ling, H.: Lime: low-light image enhancement via illumination map estimation. IEEE Trans. Image Process **26**(2), 982–993 (2016)

11. He, K., Sun, J., Tang, X.: Guided image filtering. In: Daniilidis, K., Maragos, P., Paragios, N. (eds.) ECCV 2010. LNCS, vol. 6311, pp. 1–14. Springer, Heidelberg (2010). https://doi.org/10.1007/978-3-642-15549-9_1

12. He, K., Sun, J., Tang, X.: Single image haze removal using dark channel prior. IEEE Trans. Pattern Anal. Mach. Intell. **33**(12), 2341–2353 (2011)

13. Jiang, Y., Sun, C., Zhao, Y., Yang, L.: Image dehazing using adaptive bi-channel priors on superpixels. Comput. Vis. Image Underst. **165**, 17–32 (2017)

14. Li, B., Peng, X., Wang, Z., Xu, J., Feng, D.: Aod-net: All-in-one dehazing network. In: Proceedings of the IEEE International Conference on Computer Vision, pp. 4770–4778 (2017)

15. Li, R., Pan, J., Li, Z., Tang, J.: Single image dehazing via conditional generative adversarial network. In: Proceedings of the IEEE Conference on Computer Vision and Pattern Recognition, pp. 8202–8211 (2018)

16. Li, Yu., Guo, F., Tan, R.T., Brown, M.S.: A contrast enhancement framework with JPEG artifacts suppression. In: Fleet, D., Pajdla, T., Schiele, B., Tuytelaars, T. (eds.) ECCV 2014. LNCS, vol. 8690, pp. 174–188. Springer, Cham (2014). https://doi.org/10.1007/978-3-319-10605-2_12

17. Li, Y., Tan, R.T., Brown, M.S.: Nighttime haze removal with glow and multiple light colors. In: Proceedings of the IEEE International Conference on Computer Vision, pp. 226–234 (2015)

18. Li, Y., You, S., Brown, M.S., Tan, R.T.: Haze visibility enhancement: a survey and quantitative benchmarking. Comput. Vis. Image Underst. **165**, 1–16 (2017)

19. Meng, G., Wang, Y., Duan, J., Xiang, S., Pan, C.: Efficient image dehazing with boundary constraint and contextual regularization. In: Proceedings of the IEEE International Conference on Computer Vision, pp. 617–624 (2013)

20. Pei, S.C., Lee, T.Y.: Nighttime haze removal using color transfer pre-processing and dark channel prior. In: 2012 19th IEEE International Conference on Image Processing, pp. 957–960. IEEE (2012)

21. Qu, Y., Chen, Y., Huang, J., Xie, Y.: Enhanced pix2pix dehazing network. In: Proceedings of the IEEE Conference on Computer Vision and Pattern Recognition, pp. 8160–8168 (2019)

22. Ren, W., Liu, S., Zhang, H., Pan, J., Cao, X., Yang, M.-H.: Single image dehazing via multi-scale convolutional neural networks. In: Leibe, B., Matas, J., Sebe, N., Welling, M. (eds.) ECCV 2016. LNCS, vol. 9906, pp. 154–169. Springer, Cham (2016). https://doi.org/10.1007/978-3-319-46475-6_10
23. Ren, W., et al.: Gated fusion network for single image dehazing. In: Proceedings of the IEEE Conference on Computer Vision and Pattern Recognition, pp. 3253–3261 (2018)
24. Tan, R.T.: Visibility in bad weather from a single image. In: 2008 IEEE Conference on Computer Vision and Pattern Recognition, pp. 1–8. IEEE (2008)
25. Zhang, H., Patel, V.M.: Densely connected pyramid dehazing network. In: Proceedings of the IEEE Conference on Computer Vision and Pattern Recognition, pp. 3194–3203 (2018)
26. Zhang, J., Cao, Y., Fang, S., Kang, Y., Wen Chen, C.: Fast haze removal for nighttime image using maximum reflectance prior. In: Proceedings of the IEEE Conference on Computer Vision and Pattern Recognition, pp. 7418–7426 (2017)
27. Zhang, J., Cao, Y., Wang, Z.: Nighttime haze removal based on a new imaging model, pp. 4557–4561. IEEE (2014)

SegFix: Model-Agnostic Boundary Refinement for Segmentation

Yuhui Yuan[1,2,4], Jingyi Xie[3], Xilin Chen[1,2], and Jingdong Wang[4(✉)]

[1] Key Lab of Intelligent Information Processing of Chinese Academy of Sciences
(CAS), Institute of Computing Technology, CAS, Beijing, China
[2] University of Chinese Academy of Sciences, Beijing, China
xlchen@ict.ac.cn
[3] University of Science and Technology of China, Hefei, China
hsfzxjy@mail.ustc.edu.cn
[4] Microsoft Research Asia, Beijing, China
{yuhui.yuan,jingdw}@microsoft.com

Abstract. We present a model-agnostic post-processing scheme to improve the boundary quality for the segmentation result that is generated by any existing segmentation model. Motivated by the empirical observation that the label predictions of interior pixels are more reliable, we propose to replace the originally unreliable predictions of boundary pixels by the predictions of interior pixels. Our approach processes only the input image through two steps: (i) localize the boundary pixels and (ii) identify the corresponding interior pixel for each boundary pixel. We build the correspondence by learning a direction away from the boundary pixel to an interior pixel. Our method requires no prior information of the segmentation models and achieves nearly real-time speed. We empirically verify that our SegFix consistently reduces the boundary errors for segmentation results generated from various state-of-the-art models on Cityscapes, ADE20K and GTA5. Code is available at: https://github.com/openseg-group/openseg.pytorch.

Keywords: Semantic segmentation · Instance segmentation · Boundary refinement · Model agnostic

1 Introduction

The task of semantic segmentation is formatted as predicting the semantic category for each pixel in an image. Based on the pioneering fully convolutional network [46], previous studies have achieved great success as reflected by increasing the performance on various challenging semantic segmentation benchmarks [7,16,68].

Y. Yuan and J. Xie—Equal contribution.

Electronic supplementary material The online version of this chapter (https://doi.org/10.1007/978-3-030-58610-2_29) contains supplementary material, which is available to authorized users.

© Springer Nature Switzerland AG 2020
A. Vedaldi et al. (Eds.): ECCV 2020, LNCS 12357, pp. 489–506, 2020.
https://doi.org/10.1007/978-3-030-58610-2_29

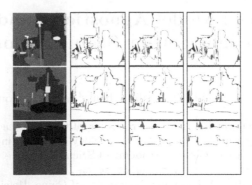

Fig. 1. Qualitative analysis of the segmentation error maps. The 1st column presents the ground-truth segmentation maps, and the 2nd/3rd/4th column presents the error maps of DeepLabv3/HRNet/Gated-SCNN separately. These examples are cropped from Cityscapes `val` set. We can see that there exist many errors along the thin boundary for all three methods.

Most of the existing works mainly addressed semantic segmentation through (i) increasing the resolution of feature maps [12,13,54], (ii) constructing more reliable context information [23,27,28,39,62–65] and (iii) exploiting boundary information [5,9,44,55]. In this work, we follow the 3rd line of work and focus on improving segmentation result on the pixels located within the thinning boundary[1] via an effective model-agnostic boundary refinement mechanism.

Our work is mainly motivated by the observation that *most of the existing state-of-the-art segmentation models fail to deal well with the error predictions along the boundary.* We illustrate some examples of the segmentation error maps with DeepLabv3 [12], Gated-SCNN [55] and HRNet [54] in Fig. 1. More specifically, we illustrate the statistics on the numbers of the error pixels *vs.* the distances to the object boundaries in Fig. 2. We can observe that, for all three methods, the number of error pixels significantly decrease with larger distances to the boundary. In other words, predictions of the interior pixels are more reliable.

We propose a novel model-agnostic post-processing mechanism to reduce boundary errors by replacing labels of boundary pixels with the labels of corresponding interior pixels for a segmentation result. We estimate the pixel correspondences by processing the input image (without exploring the segmentation result) with two steps. The first step aims to localize the pixels along the object boundaries. We follow the contour detection methods [2,4,21] and simply use a convolutional network to predict a binary mask indicating the boundary pixels. In the second step, we learn a direction away from the boundary pixel to an interior pixel and identify the corresponding interior pixel by moving from the

---

[1] In this paper, we treat the pixels with neighboring pixels belonging to different categories as the boundary pixels. We use the distance transform to generate the ground-truth boundary map with any given width in our implementation.

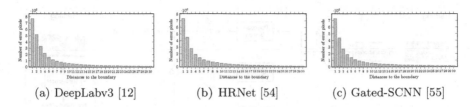

(a) DeepLabv3 [12]   (b) HRNet [54]   (c) Gated-SCNN [55]

Fig. 2. Histogram statistics of errors: the number of error pixels *vs.* their (Euclidean) distances to the boundaries on Cityscapes `val` based on DeepLabv3/HRNet/Gated-SCNN. We can see that pixels with larger distance tend to be well-classified with higher probability and there exist many errors distributing within ~ 5 pixels width along the boundary.

boundary pixel along the direction by a certain distance. Especially, our SegFix can reach nearly real-time speed with high resolution inputs.

Our SegFix is a general scheme that consistently improves the performance of various segmentation models across multiple benchmarks without any prior information. We evaluate the effectiveness of SegFix on multiple semantic segmentation benchmarks including Cityscapes, ADE20K and GTA5. We also extend SegFix to instance segmentation task on Cityscapes. According to the Cityscapes leaderboard, "HRNet + OCR + SegFix" and "PolyTransform + Seg-Fix" achieve 84.5% and 41.2%, which rank the 1st and 2nd place on the semantic and instance segmentation track separately by the ECCV 2020 submission deadline.

2   Related Work

Distance/Direction Map for Segmentation: Some recent work [3,25,56] performed distance transform to compute distance maps for instance segmentation task. For example, [3,25] proposed to train the model to predict the truncated distance maps within each cropped instance. The other work [6,10,17,51] proposed to regularize the semantic or instance segmentation predictions with distance map or direction map in a multi-task mechanism. Compared with the above work, the key difference is that our approach does not perform any segmentation predictions and instead predicts the direction map from only the image, and then we refine the segmentation results of the existing approaches.

Level Set for Segmentation: Many previous efforts [8,31,50] have used the level set approach to address the semantic segmentation problem before the era of deep learning. The most popular formulation of level set is the signed distance function, with all the zero values corresponding to predicted boundary positions. Recent work [1,14,33,56] extended the conventional level-set scheme to deep network for regularizing the boundaries of predicted segmentation map. Instead of representing the boundary with a level set function directly, we implicitly

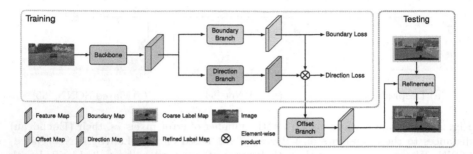

Fig. 3. Illustrating the SegFix framework: In the training stage, we first send the input image into a backbone to predict a feature map. Then we apply a boundary branch to predict a binary boundary map and a direction branch to predict a direction map and mask it with the binary boundary map. We apply boundary loss and direction loss on the predicted boundary map and direction map separately. In the testing stage, we first convert the direction map to offset map and then refine the segmentation results of any existing methods according to the offset map.

encode the relative distance information of the boundary pixels with a boundary map and a direction map.

DenseCRF for Segmentation: Previous work [11,41,58,67] improved their segmentation results with the DenseCRF [36]. Our approach is also a kind of general post processing scheme while being simpler and more efficient for usage. We empirically show that our approach not only outperforms but also is complementary with the DenseCRF.

Refinement for Segmentation: Extensive studies [22,24,29,37,38] have proposed various mechanisms to refine the segmentation maps from coarse to fine. Different from most of the existing refinement approaches that depend on the segmentation models, to the best of our knowledge, our approach is the first model-agnostic segmentation refinement mechanism that can be applied to refine the segmentation results of any approach without any prior information.

Boundary for Segmentation: Some previous efforts [1,45,59,60] focused on localizing semantic boundaries. Other studies [5,18–20,32,43,44,55] also exploited the boundary information to improve the segmentation. For example, BNF [5] introduced a global energy model to consider the pairwise pixel affinities based on the boundary predictions. Gated-SCNN [55] exploited the duality between the segmentation predictions and the boundary predictions with a two-branch mechanism and a regularizer.

These methods [5,18,32,55] are highly dependent on the segmentation models and require careful re-training or fine-tuning. Different from them, SegFix does not perform either segmentation prediction or feature propagation and we instead refine the segmentation maps with an offset map directly. In other words, we only need to train a single unified SegFix model once w/o any further fine-tuning the different segmentation models (across multiple different datasets).

We also empirically verify that our approach is complementary with the above methods, e.g., Gated-SCNN [55] and Boundary-Aware Feature Propagation [18].

Guided Up-Sampling Network: The recent work [47, 48] performed a segmentation guided offset scheme to address boundary errors caused by the bilinear up-sampling. The main difference is that they do not apply any explicit supervision on their offset maps and require re-training for different models, while we apply explicit semantic-aware supervision on the offset maps and our offset maps can be applied to various approaches directly without any re-training. We also empirically verify the advantages of our approach.

3   Approach

3.1   Framework

The overall pipeline of SegFix is illustrated in Fig. 3. We first train a model to pick out boundary pixels (with the boundary maps) and estimate their corresponding interior pixels (with offsets derived from the direction maps) from only the image. We do not perform segmentation directly during training. We apply this model to generate offset maps from the images and use the offsets to get the corresponding pixels which should mostly be the more confident interior pixels, and thereby refine segmentation results from any segmentation model. We mainly describe SegFix scheme for semantic segmentation and we illustrate the details for instance segmentation in the Appendix E.

Training Stage. Given an input image $\mathbf{I}$ of shape $H \times W \times 3$, we first use a backbone network to extract a feature map $\mathbf{X}$, and then send $\mathbf{X}$ in parallel to (1) the *boundary branch* to predict a binary map $\mathbf{B}$, with 1 for the boundary pixels and 0 for the interior pixels, and (2) the *direction branch* to predict a direction map $\mathbf{D}$ with each element storing the direction pointing from the boundary pixel to the interior pixel. The direction map $\mathbf{D}$ is then masked by the binary map $\mathbf{B}$ to yield the input for the offset branch.

For model training, we use a binary cross-entropy loss as the boundary loss on $\mathbf{B}$ and a categorical cross-entropy loss as the direction loss on $\mathbf{D}$ separately.

Testing Stage. Based on the predicted boundary map $\mathbf{B}$ and direction map $\mathbf{D}$, we apply the *offset branch* to generate a offset map $\Delta\mathbf{Q}$. A coarse label map $\mathbf{L}$ output by any semantic segmentation model will be refined as:

$$\widetilde{\mathbf{L}}_{\mathbf{p}_i} = \mathbf{L}_{\mathbf{p}_i + \Delta\mathbf{q}_i}, \tag{1}$$

where $\widetilde{\mathbf{L}}$ is refined label map, $\mathbf{p}_i$ represents the coordinate of the boundary pixel i, $\Delta\mathbf{q}_i$ is the generated offset vector pointing to an interior pixel, which is indeed an element of $\Delta\mathbf{Q}$. $\mathbf{p}_i + \Delta\mathbf{q}_i$ is the position of the identified interior pixel.

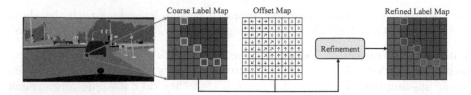

Fig. 4. Illustrating the refinement mechanism of our approach: we refine the coarse label map based on the offset map by replacing the labels of boundary pixels with the labels of (more) reliable interior pixels. We represent different offset vectors with different arrows. We mark the error positions in the coarse label map with □ and the corresponding corrected positions in the refined label map with ▫. For example, the top-most error pixel (class road) in the coarse label map is associated with a direction →. We use the label (class car) of the updated position based on offset $(1, 0)$ as the refined label. Only one-step shifting based on the offset map already refines several boundary errors.

Considering that there might be some "fake" interior pixels [2] when the boundary is thick, we propose two different schemes as following: (i) re-scaling all the offsets by a factor, e.g., 2. (ii) iteratively applying the offsets (of the "fake" interior pixels) until finding an interior pixel. We choose (i) by default for simplicity as their performance is close.

During testing stage, we only need to generate the offset maps on test set *for once*, and could apply the same offset maps to refine the segmentation results from any existing segmentation model without requiring any prior information. In general, our approach is agnostic to any existing segmentation models.

3.2   Network Architecture

Backbone. We adopt the recently proposed high resolution network (HRNet) [54] as backbone, due to its strengths at maintaining high resolution feature maps and our need to apply full-resolution boundary maps and direction maps to refine full-resolution coarse label maps. Besides, we also modify HRNet through applying a 4×4 deconvolution with stride 2 on the final output feature map of HRNet to increase the resolution by $2\times$, which is similar to [15], called Higher-HRNet. We directly perform the boundary branch and the direction branch on the output feature map with the highest resolution. The resolution is $\frac{H}{s} \times \frac{W}{s} \times D$, where $s = 4$ for HRNet and $s = 2$ for Higher-HRNet. We empirically verify that our approach consistently improves the coarse segmentation results for all variations of our backbone choices in Sect. 4.2, e.g., HRNet-W18 and HRNet-W32.

---

[2] We use "fake" interior pixels to represent pixels (after offsets) that still lie on the boundary when the boundary is thick. Notably, we identify an pixel as interior pixel/boundary pixel if its value in the predicted boundary map **B** is 0/1.

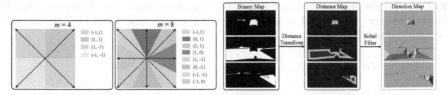

(a) Illustrating the directions → offsets. (b) Ground-truth generation procedure.

Fig. 5. (a) We divide the entire direction value range $[0°, 360°)$ to m partitions or categories (marked with different colors), For example, when $m = 4$, we have $[0°, 90°)$, $[90°, 180°)$, $[180°, 270°)$ and $[270°, 360°)$ correspond to 4 different categories separately. The above 4 direction categories correspond to offsets $(1, 1)$, $(-1, 1)$, $(-1, -1)$ and $(1, -1)$ respectively. The situation for $m = 8$ is similar. (b) Binary maps → Distance maps → Direction maps. The ground-truth binary maps are of category car, road and side-walk. We first apply distance transform on each binary map to compute the ground-truth distance maps. Then we use Sobel filter on the distance maps to compute the ground-truth direction maps. We choose different colors to represent different distance values or the direction values. (Color figure online)

Boundary Branch/Loss. We implement the boundary branch as $1×1$ Conv → BN → ReLU with 256 output channels. We then apply a linear classifier ($1 × 1$ Conv) and up-sample the prediction to generate the final boundary map **B** of size $H × W × 1$. Each element of **B** records the probability of the pixel belonging to the boundary. We use binary cross-entropy loss as the boundary loss.

Direction Branch/Loss. Different from the previous approaches [1,3] that perform regression on continuous directions in $[0°, 360°)$ as the ground-truth, our approach directly predicts discrete directions by evenly dividing the entire direction range to m partitions (or categories) as our ground-truth ($m = 8$ by default). In fact, we empirically find that our discrete categorization scheme outperforms the regression scheme, e.g., mean squared loss in the angular domain [3], measured by the final segmentation performance improvements. We illustrate more details for the discrete direction map in Sect. 3.3.

We implement the direction branch as $1 × 1$ Conv → BN → ReLU with 256 output channels. We further apply a linear classifier ($1 × 1$ Conv) and up-sample the classifier prediction to generate the final direction map **D** of size $H × W × m$. We mask the direction map **D** by multiplying by the (binarized) boundary map **B** to ensure that we only apply direction loss on the pixels identified as boundary by the boundary branch. We use the standard category-wise cross-entropy loss to supervise the discrete directions in this branch.

Offset Branch. The offset branch is used to convert the predicted direction map **D** to the offset map $\Delta\mathbf{Q}$ of size $H × W × 2$. We illustrate the mapping mechanism in Fig. 5 (a). For example, the "upright" direction category (corresponds to the value within range $[0°, 90°)$) will be mapped to offset $(1, 1)$ when $m = 4$.

Table 1. Improvements with ground-truth boundary offset on Cityscapes **val**. We report both the segmentation performance mIoU and the boundary performance F-score (1px width).

Metric	Method	w/o SegFix	w/ SegFix	
			$m = 4$	$m = 8$
mIoU	DeepLabv3 (Our impl.)	79.5	82.6 (+3.1)	82.4 (+2.9)
	HRNet-W48 (Our impl.)	81.1	84.1 (+3.0)	84.1 (+3.0)
	Gated-SCNN (Our impl.)	81.0	84.2 (+3.2)	84.1 (+3.1)
F-score	DeepLabv3 (Our impl.)	56.6	68.6 (+12.0)	68.4 (+11.8)
	HRNet-W48 (Our impl.)	62.4	73.8 (+11.4)	73.8 (+11.4)
	Gated-SCNN (Our impl.)	61.4	72.3 (+10.9)	72.3 (+10.9)

Last, we generate the refined label map through shifting the coarse label map with the grid-sample scheme [30]. The process is shown in Fig. 4.

3.3 Ground-Truth Generation and Analysis

There may exist many different mechanisms to generate ground-truth for the boundary maps and the direction maps. In this work, we mainly exploit the conventional distance transform [34] to generate ground-truth for both semantic segmentation task and the instance segmentation task.

We start from the ground-truth segmentation label to generate the ground-truth distance map, followed by boundary map and direction map. Figure 5 (b) illustrates the overall procedure.

Distance Map. For each pixel, our distance map records its minimum (Euclidean) distance to the pixels belonging to other object category. We illustrate how to compute the distance map as below.

First, we decompose the ground-truth label into K binary maps associated with different semantic categories, e.g., car, road, sidewalk. The kth binary map records the pixels belonging to the kth semantic category as 1 and 0 otherwise. Second, we perform distance transform [34][3] on each binary map independently to compute the distance map. The element of kth distance map encodes the distance from a pixel belonging to k^{th} category to the nearest pixel belonging to other categories. Such distance can be treated as the distance to the object boundary. We compute a fused distance map through aggregating all the K distance maps.

Note that the values in our distance map are (always positive) different from the conventional signed distances that represent the interior/exterior pixels with positive/negative distances separately.

---

[3] We use `scipy.ndimage.morphology.distance_transform_textttedt` in implementation.

Boundary Map. As the fused distance map represents the distances to the object boundary, we can construct the ground-truth boundary map through setting all the pixels with distance value smaller than a threshold γ as boundary[4]. We empirically choose small γ value, e.g., $\gamma = 5$, as we are mainly focused on the thin boundary refinement.

Direction Map. We perform the Sobel filter (with kernel size 9×9) on the K distance maps independently to compute the corresponding K direction maps respectively. The Sobel filter based direction is in the range $[0°, 360°)$, and each direction points to the interior pixel (within the neighborhood) that is furthest away from the object boundary. We divide the entire direction range to m categories (or partitions) and then assign the direction of each pixel to the corresponding category. We illustrate two kinds of partitions in Fig. 5 (a) and we choose the $m = 8$ partition by default. We apply the evenly divided direction map as our ground-truth for training. Besides, we also visualize some examples of direction map in Fig. 5 (b).

Empirical Analysis. We apply the generated ground-truth on the segmentation results of three state-of-the-art methods including DeepLabv3 [12], HRNet [54] and Gated-SCNN [55] to investigate the potential of our approach. Specifically, we first project the ground-truth direction map to offset map and then refine the segmentation results on Cityscapes val based on our generated ground-truth offset map. Table 1 summarizes the related results. We can see that our approach significantly improves both the overall mIoU and the boundary F-score. For example, our approach ($m = 8$) improves the mIoU of Gated-SCNN by 3.1%. We may achieve higher performance through re-scaling the offsets for different pixels adaptively, which is not the focus of this work.

Discussion. The key condition for ensuring the effectiveness of our approach is that *segmentation predictions of the interior pixels are more reliable empirically*. Given accurate boundary maps and direction maps, we could always improve the segmentation performance in expectation. In other words, the segmentation performance ceiling of our approach is also determined by the interior pixels' prediction accuracy.

4   Experiments: Semantic Segmentation

4.1   Datasets and Implementation Details

Cityscapes. [16] is a real-world dataset that consists of $2,975/500/1,525$ images with resolution 2048×1024 for training/validation/testing respectively, which contains 19/8 semantic categories for semantic/instance segmentation task.

ADE20K. [68] is a very challenging benchmark consisting of around $20,000/2,000$ images for training/validation respectively. The dataset contains 150 fine-grained semantic categories.

---

[4] We define the boundary pixels and interior pixels based on their distance values.

Table 2. Influence of backbones. The runtime is tested with an input image of resolution 2048 × 1024 on a single V100 GPU (PyTorch1.4 + TensorRT). SegFix reaches real-time speed with light-weight backbone, e.g., HRNet-W18 or HRNet-W32.

Backbone	#param (M)	Runtime (ms)	Mask F-score	Direction accuracy	mIoU△	F-score△
HRNet-W18	9.6	16	71.44	64.44	+0.8	+3.7
HRNet-W32	29.4	20	72.24	65.10	+0.9	+3.9
Higher-HRNet	47.3	69	73.67	66.87	+1.0	+4.4

GTA5. [52] is a synthetic dataset that consists of 12, 402/6, 347/6, 155 images with resolution 1914 × 1052 for training/validation/testing respectively. The dataset contains 19 semantic categories which are compatible with Cityscapes.

Implementation Details. We perform the same training and testing settings on Cityscapes and GTA5 benchmarks as follow. We set the initial learning rate as 0.04, weight decay as 0.0005, crop size as 512 × 512 and batch size as 16, and train for 80K iterations. For the ADE20K benchmark, we set the initial learning as 0.02 and all the other settings are kept the same as on Cityscapes. We use "poly" learning rate policy with power = 0.9. For data augmentation, we all apply random flipping horizontally, random cropping and random brightness jittering within the range of [−10, 10]. Besides, we all apply syncBN [53] across multiple GPUs to stabilize the training. We simply set the loss weight as 1.0 for both the boundary loss and direction loss without tuning.

Notably, our approach does not require extra training or fine-tuning any semantic segmentation models. We only need to predict the boundary mask and the direction map for all the test images *in advance* and refine the segmentation results of any existing approaches accordingly.

Evaluation Metrics. We use two different metrics including: *mask F-score* and top-1 *direction accuracy* to evaluate the performance of our approach during the training stage. Mask F-score is performed on the predicted binary boundary map and direction accuracy is performed on the predicted direction map. Especially, we only measure the direction accuracy within the regions identified as boundary by the boundary branch.

To verify the effectiveness of our approach for semantic segmentation, we follow the recent Gated-SCNN [55] and perform two quantitative measures including: *class-wise mIoU* to measure the overall segmentation performance on regions; *boundary F-score* to measure the boundary quality of predicted mask with a small slack in distance. In our experiments, we measure the boundary F-score using thresholds 0.0003, 0.0006 and 0.0009 corresponding to 1, 2 and 3 pixels respectively. We mainly report the performance with threshold as 0.0003 for most of our ablation experiments.

Table 3. Influence of the boundary width and direction number. SegFix is robust to boundary width and direction number. We choose $\gamma = 5$ and $m = 8$ according to their F-scores.

	Boundary width				# directions		
	$\gamma = 3$	$\gamma = 5$	$\gamma = 10$	$\gamma = \infty$	$m = 4$	$m = 8$	$m = 16$
mIoU△	+0.94	+0.96	+0.95	+0.84	+0.97	+0.96	+0.96
F-score△	+4.1	+4.2	+4.1	+3.6	+4.1	+4.2	+4.2

Fig. 6. Qualitative results of our direction branch predictions. The 1st and 3rd columns represent the ground-truth segmentation map. The 2nd and 4th columns illustrate the predicted directions with the segmentation map of HRNet as the background. We mark the directions that fix errors with blue arrow and directions that lead to extra errors with red arrow. Our predicted directions addresses boundary errors for various object categories such as bicycle, traffic light and traffic sign. (Better viewed zoom in) (Color figure online)

4.2 Ablation Experiments

We conduct a group of ablations to analyze the influence of various factors within SegFix. We report the improvements over the segmentation baseline DeepLabv3 (mIoU/F-score is 79.5%/56.6%) if not specified.

Backbone. We study the performance of our SegFix based on three different backbones with increasing complexities, i.e., HRNet-W18, HRNet-W32 and Higher-HRNet. We apply the same training/testing settings for all three backbones. According to the comparisons in Table 2, our SegFix consistently improves both the segmentation performance and the boundary quality with different backbone choices. We choose Higher-HRNet in the following experiments if not specified as it performs best. Besides, we also report their running time in Table 2.

Boundary Branch. We verify that SegFix is robust to the choice of hyper-parameter γ within the boundary branch and illustrate some qualitative results.

☐ **Boundary Width:** Table 3 shows the performance improvements based on boundary with different widths. We choose different γ values to control the boundary width, where smaller γ leads to thinner boundaries. We also report the performance with $\gamma = \infty$, which means all pixels is identified as boundary. We find their improvements are close and we choose $\gamma = 5$ by default.

Table 4. Comparison with GUM [47]. SegFix not only outperforms GUM but also is complementary with GUM.

	Baseline	GUM (Our impl.)	SegFix	GUM+SegFix
mIoU	79.5	79.8 (+0.3)	80.5 (+1.0)	80.6 (+1.1)
F-score	56.6	57.7 (+1.1)	60.9 (+4.3)	61.6 (+5.0)

Table 5. Comparison with DenseCRF [36]. SegFix achieves comparable F-score improvements and much larger mIoU gains.

	Baseline	DenseCRF	SegFix	DenseCRF+SegFix
mIoU	79.5	79.7 (+0.2)	80.5 (+1.0)	80.5 (+1.0)
F-score	56.6	60.9 (+4.3)	61.0 (+4.4)	64.1 (+7.5)

Table 6. Comparison with Gated-SCNN [55]. The result of Gated-SCNN is based on multi-scale testing.

	Gated-SCNN	Gated-SCNN+SegFix
mIoU	81.0	81.5 (+0.5)
F-score	61.4	63.1 (+1.7)

Table 7. DeepLabv3 with SegFix on ADE20K and GTA5. We all choose DeepLabv3 as the baseline.

	ADE20K		GTA5	
	Baseline	+SegFix	Naseline	+SegFix
mIoU	44.8	45.4(+0.6)	77.8	80.6(+2.8)
F-score	16.4	19.3(+2.9)	50.2	61.7(+11.5)

☐ **Qualitative Results:** We show the qualitative results with our boundary branch in the Appendix G. We find that the predicted boundaries are of high quality. Besides, we also compute the F-scores between the boundary computed from the segmentation map of the existing approaches, e.g., Gated-SCNN and HRNet, and the predicted boundary from our boundary branch. The F-scores are around 70%, which (in some degree) means that their boundary maps are well aligned and ensures that more accurate direction predictions bring larger performance gains.

Direction Branch. We analyze the influence of the direction number m and then present some qualitative results of our predicted directions.

☐ **Direction Number:** We choose different direction numbers to perform different direction partitions and control the generated offset maps that are used to refine the coarse label map. We conduct the experiments with $m = 4$, $m = 8$ and $m = 16$. According to the reported results on the right 3 columns in Table 3, we find different direction numbers all lead to significant improvements and we choose $m = 8$ if not specified as our SegFix is less sensitive to the choice of m.

☐ **Qualitative Results:** In Fig. 6, we show some examples to illustrate that our predicted boundary directions improve the errors. Overall, the improved pixels (marked with blue arrow) are mainly distributed along the very thin boundary.

Comparison with GUM. We compare SegFix with the previous model-dependent guided up-sampling mechanism [47,48] based on DeepLabv3 as the baseline. We report the related results in Table 4. It can be seen that our approach significantly outperforms GUM measured by both mIoU and F-score. We achieve higher performance through combining GUM with our approach, which achieves 5.0% improvements on F-score compared to the baseline.

Comparison with DenseCRF. We compare our approach with the conventional well-verified DenseCRF [36] based on the DeepLabv3 as our baseline. We fine-tune the hyper-parameters of DenseCRF and set them empirically following [11]. According to Table 5, our approach not only outperforms DenseCRF but also is complementary with DenseCRF. The possible reasons for the limited mIoU improvements of DenseCRF might be that it brings more extra errors on the interior pixels.

Application to Gated-SCNN. Considering that Gated-SCNN [55] introduced multiple components to improve the performance, it is hard to compare our approach with Gated-SCNN fairly to a large extent. To verify the effectiveness of our approach to some extent, we first take the open-sourced Gated-SCNN (multi-scale testing) segmentation results on Cityscapes validation set as the coarse segmentation maps, then we apply the SegFix offset maps to refine the results. We report the results in Table 6 and SegFix improves the boundary F-score by 1.7%, suggesting that SegFix is complementary with the strong baseline that also focuses on improving the segmentation boundary quality. Besides, we also report the detailed category-wise improvements measured by both mIoU and boundary F-score in the Table 2 and Table 3 of Appendix.

4.3   Application to State-of-the-art

We generate the boundary maps and the direction maps in advance and apply them to the segmentation results of various state-of-the-art approaches without extra training or fine-tuning.

Cityscapes val: We first apply our approach on various state-of-the-art approaches (on Cityscapes val) including DeepLabv3, Gated-SCNN and HRNet. We report the category-wise mIoU improvements in Table 2 of Appendix. It can be seen that our approach significantly improves the segmentation quality along the boundaries of all the evaluated approaches. We provides some qualitative examples of the improvements with our approach along the thin boundaries based on both DeepLabv3 and HRNet in the Fig. 1 of Appendix.

Cityscapes Test: We further apply our approach on several recent state-of-the-art methods on Cityscapes test including PSANet [66], DANet [23], BFP [18], HRNet [54], Gated-SCNN [55], VPLR [69] and HRNet + OCR [61]. We directly apply the same model that are trained with only the 2,975 training images without any other tricks, e.g., training with validation set or Mapillary Vistas [49], online hard example mining.

Notably, the state-of-the-art methods have applied various advanced techniques, e.g., multi-scale testing, multi-grid, performing boundary supervision or utilizing extra training data such as Mapillary Vistas or Cityscapes video, to improve their results. In Table 8, our model-agnostic boundary refinement

Table 8. Results on Cityscapes Semantic Segmentation task. Category-wise improvements of SegFix based on various state-of-the-art methods on Cityscapes `test`. Notably, "HRNet + OCR + SegFix" ranks the first place on the Cityscapes semantic segmentation leaderboard by the ECCV 2020 submission deadline.

method	road	sidewalk	building	wall	fence	pole	traffic light	traffic sign	vegetation	terrain	sky	person	rider	car	truck	bus	train	motorcycle	bicycle	mean
PSANet	98.7	87.0	93.5	58.9	62.5	67.8	76.0	80.0	93.7	72.6	95.4	86.9	73.0	96.2	79.3	91.2	84.9	71.1	77.9	81.4
+ SegFix	98.7	87.4	93.7	59.3	62.8	69.5	77.6	81.4	93.9	73.0	95.6	88.0	73.9	96.5	79.6	91.5	85.1	71.8	78.6	82.0
DANet	98.6	86.1	93.5	56.1	63.3	69.7	77.3	81.3	93.9	72.9	95.7	87.3	72.9	96.2	76.8	89.4	86.5	72.2	78.2	81.5
+ SegFix	98.7	86.6	93.7	56.5	63.5	71.4	78.7	82.4	94.1	73.2	95.9	88.2	73.7	96.5	77.0	89.7	86.8	72.8	78.8	82.0
BFP	98.7	87.0	93.5	59.8	63.4	68.9	76.8	80.9	93.7	72.8	95.5	87.0	72.1	96.0	77.6	89.0	86.9	69.2	77.6	81.4
+ SegFix	98.7	87.5	93.7	60.2	63.7	71.1	78.4	82.4	94.0	73.2	95.7	88.1	72.9	96.3	77.8	89.3	87.2	69.9	78.4	82.0
HRNet	98.8	87.5	93.7	55.6	62.3	71.8	79.3	81.8	94.0	73.1	95.8	88.5	76.1	96.5	72.2	86.5	84.7	73.8	79.4	81.8
+ SegFix	98.8	87.9	93.9	56.0	62.5	73.6	80.7	83.2	94.1	73.4	95.9	89.3	76.7	96.6	72.4	86.7	85.0	74.3	80.2	82.2
VPLR	98.8	87.8	94.2	64.1	65.0	72.4	79.0	82.8	94.2	74.0	96.1	88.2	75.4	96.5	78.8	94.0	91.6	73.8	79.0	83.5
+ SegFix	98.8	88.0	94.3	64.4	65.3	73.3	80.0	83.5	94.3	74.3	96.2	89.0	76.2	96.7	79.0	94.2	92.0	74.4	79.7	83.9
HRNet + OCR	98.9	88.3	94.3	66.8	66.6	73.6	80.3	83.7	94.3	74.4	96.0	88.7	75.4	96.6	82.5	94.0	90.8	73.8	79.7	84.2
+ SegFix	98.9	88.3	94.4	68.0	67.8	73.6	80.6	83.9	94.4	74.5	96.1	89.2	75.9	96.8	83.6	94.2	91.3	74.0	80.1	84.5

Table 9. Results on Cityscapes Instance Segmentation task. Our SegFix significantly improves the mask AP of Mask-RCNN [26], PointRend [35], PANet [42] and Poly-Transform [40] on Cityscapes `test` (w/ COCO pre-training). Notably, "PolyTransform + SegFix" ranks the second place on the Cityscapes instance segmentation leaderboard by the ECCV 2020 submission deadline.

method	person	rider	car	truck	bus	train	motorcycle	bicycle	mean (%)
Mask-RCNN	36.0	28.8	51.6	30.0	38.7	27.3	23.9	19.4	32.0
+ SegFix	37.9	30.3	54.1	31.0	40.0	27.9	25.1	20.5	33.3 (+1.3)
PointRend	36.6	29.7	53.7	29.9	40.4	33.3	23.6	19.6	33.3
+ SegFix	38.7	31.1	56.2	31.1	41.6	34.1	24.6	20.7	34.8 (+1.5)
PANet	41.5	33.6	58.2	31.8	45.3	28.7	28.2	24.1	36.4
+ SegFix	43.3	34.9	60.4	32.9	47.0	30.1	29.1	24.7	37.8 (+1.4)
PolyTransform	42.4	34.8	58.5	39.8	50.0	41.3	30.9	23.4	40.1
+ SegFix	44.3	35.9	60.5	40.5	51.2	41.6	31.7	24.1	41.2 (+1.1)

scheme consistently improves all the evaluated approaches. For example, with our SegFix, "HRNet + OCR" achieves 84.5% on Cityscapes `test`. The improvements of our SegFix is in fact already significant considering the baseline is already very strong and the performance gap between top ranking methods is just around 0.1%–0.3%. We believe that lots of other advanced approaches might also benefit from our approach.

We also apply the SegFix scheme on two other challenging semantic segmentation benchmarks including ADE20K and GTA5. Table 7 reports the results and SegFix achieves significant performance improvements along the boundary on both benchmarks. e.g., the boundary F-score of DeepLabv3 gains 2.9%/11.5% on ADE20K `val`/GTA5 `test` separately. Besides, we propose a unified SegFix model and compare our SegFix to model ensemble in Appendix C and D.

5   Experiments: Instance Segmentation

In Table 9, we illustrate the results of SegFix on Cityscapes instance segmentation task. We can find that the SegFix consistently improves the mean AP scores over Mask-RCNN [26], PANet [42], PointRend [35] and PolyTransform [40]. For example, with SegFix scheme, PANet gains 1.4% points on the Cityscapes `test` set. We also apply our SegFix on the very recent PointRend and PolyTransform. Our SegFix consistently improves the performance of PointRend and PolyTransform by 1.5% and 1.1% separately, which further verifies the effectiveness of our method.

We use the public available checkpoints from Dectectron2[5] and PANet[6] to generate the predictions of Mask-RCNN, PointRend and PANet. Besides, we use the segmentation results of PolyTransform directly. More training/testing details of SegFix on Cityscapes instance segmentation task are illustrated in the Appendix E. We believe that SegFix can be used to improve various other state-of-the-art instance segmentation methods directly w/o any prior requirements.

Notably, the improvements on the instance segmentation tasks (+1.1%–1.5%) are more significant than the improvements on semantic segmentation task (+0.3%–0.5%). We guess the main reason is that the instance segmentation evaluation (on Cityscapes) only considers 8 object categories without including the stuff categories. The performance of stuff categories is less sensitive to the boundary errors due to that their area is (typically) larger than the area of object categories. In summary, our SegFix achieves larger improvements on object categories than stuff categories.

6   Conclusion

In this paper, we have proposed a novel model-agnostic approach to refine the segmentation maps predicted by an unknown segmentation model. The insight is that the predictions of the interior pixels are more reliable. We propose to replace the predictions of the boundary pixels using the predictions of the corresponding interior pixels. The correspondence is learnt only from the input image. The main advantage of our method is that SegFix generalizes well on various strong segmentation models. Empirical results show that the effectiveness of our approach for both semantic segmentation and instance segmentation tasks. We hope our SegFix scheme can become a strong baseline for more accurate segmentation results along the boundary.

Acknowledgement. This work is partially supported by Natural Science Foundation of China under contract No. 61390511, and Frontier Science Key Research Project CAS No. QYZDJ-SSW-JSC009.

References

1. Acuna, D., Kar, A., Fidler, S.: Devil is in the edges: learning semantic boundaries from noisy annotations. In: CVPR (2019)

---

[5] Detectron2: https://github.com/facebookresearch/detectron2.
[6] PANet: https://github.com/ShuLiu1993/PANet.

2. Arbelaez, P., Maire, M., Fowlkes, C., Malik, J.: Contour detection and hierarchical image segmentation. PAMI **33**, 898–916 (2010)
3. Bai, M., Urtasun, R.: Deep watershed transform for instance segmentation. In: CVPR (2017)
4. Bertasius, G., Shi, J., Torresani, L.: High-for-low and low-for-high: efficient boundary detection from deep object features and its applications to high-level vision. In: ICCV (2015)
5. Bertasius, G., Shi, J., Torresani, L.: Semantic segmentation with boundary neural fields. In: CVPR (2016)
6. Bischke, B., Helber, P., Folz, J., Borth, D., Dengel, A.: Multi-task learning for segmentation of building footprints with deep neural networks. In: ICIP (2019)
7. Caesar, H., Uijlings, J., Ferrari, V.: Coco-stuff: thing and stuff classes in context. In: CVPR (2018)
8. Caselles, V., Kimmel, R., Sapiro, G.: Geodesic active contours. IJCV **22**, 61–79 (1997). https://doi.org/10.1023/A:1007979827043
9. Chen, L.C., Barron, J.T., Papandreou, G., Murphy, K., Yuille, A.L.: Semantic image segmentation with task-specific edge detection using CNNS and a discriminatively trained domain transform. In: CVPR (2016)
10. Chen, L.C., Hermans, A., Papandreou, G., Schroff, F., Wang, P., Adam, H.: Masklab: instance segmentation by refining object detection with semantic and direction features. In: CVPR (2018)
11. Chen, L.C., Papandreou, G., Kokkinos, I., Murphy, K., Yuille, A.L.: Deeplab: semantic image segmentation with deep convolutional Nets, Atrous convolution, and fully connected CRFs. PAMI **40**, 834–848 (2017)
12. Chen, L.C., Papandreou, G., Schroff, F., Adam, H.: Rethinking Atrous convolution for semantic image segmentation. arXiv:1706.05587 (2017)
13. Chen, L.-C., Zhu, Y., Papandreou, G., Schroff, F., Adam, H.: Encoder-decoder with Atrous separable convolution for semantic image segmentation. In: Ferrari, V., Hebert, M., Sminchisescu, C., Weiss, Y. (eds.) ECCV 2018. LNCS, vol. 11211, pp. 833–851. Springer, Cham (2018). https://doi.org/10.1007/978-3-030-01234-2_49
14. Chen, X., Williams, B.M., Vallabhaneni, S.R., Czanner, G., Williams, R., Zheng, Y.: Learning active contour models for medical image segmentation. In: CVPR (2019)
15. Cheng, B., Xiao, B., Wang, J., Shi, H., Huang, T.S., Zhang, L.: Bottom-up higher-resolution networks for multi-person pose estimation. arXiv preprint arXiv:1908.10357 (2019)
16. Cordts, M., et al.: The cityscapes dataset for semantic urban scene understanding. In: CVPR (2016)
17. Dangi, S., Yaniv, Z., Linte, C.: A distance map regularized CNN for cardiac cine MR image segmentation. arXiv:1901.01238 (2019)
18. Ding, H., Jiang, X., Liu, A.Q., Thalmann, N.M., Wang, G.: Boundary-aware feature propagation for scene segmentation. In: ICCV (2019)
19. Ding, H., Jiang, X., Shuai, B., Liu, A.Q., Wang, G.: Semantic correlation promoted shape-variant context for segmentation. In: CVPR (2019)
20. Ding, H., Jiang, X., Shuai, B., Qun Liu, A., Wang, G.: Context contrasted feature and gated multi-scale aggregation for scene segmentation. In: CVPR (2018)
21. Dollár, P., Zitnick, C.L.: Fast edge detection using structured forests. ArXiv (2014)
22. Fieraru, M., Khoreva, A., Pishchulin, L., Schiele, B.: Learning to refine human pose estimation. In: CVPRW (2018)
23. Fu, J., et al.: Dual attention network for scene segmentation. In: CVPR (2019)

24. Gidaris, S., Komodakis, N.: Detect, replace, refine: deep structured prediction for pixel wise labeling. In: CVPR (2017)
25. Hayder, Z., He, X., Salzmann, M.: Boundary-aware instance segmentation. In: CVPR (2017)
26. He, K., Gkioxari, G., Dollár, P., Girshick, R.: Mask R-CNN. In: ICCV (2017)
27. Huang, L., Yuan, Y., Guo, J., Zhang, C., Chen, X., Wang, J.: Interlaced sparse self-attention for semantic segmentation. arXiv preprint arXiv:1907.12273 (2019)
28. Huang, Z., Wang, X., Huang, L., Huang, C., Wei, Y., Liu, W.: CCNet: criss-cross attention for semantic segmentation. In: ICCV (2019)
29. Islam, M.A., Naha, S., Rochan, M., Bruce, N., Wang, Y.: Label refinement network for coarse-to-fine semantic segmentation. arXiv:1703.00551 (2017)
30. Jaderberg, M., Simonyan, K., Zisserman, A., et al.: Spatial transformer networks. In: NIPS (2015)
31. Kass, M., Witkin, A., Terzopoulos, D.: Snakes: active contour models. IJCV 1, 321–331 (1988). https://doi.org/10.1007/BF00133570
32. Ke, T.-W., Hwang, J.-J., Liu, Z., Yu, S.X.: Adaptive affinity fields for semantic segmentation. In: Ferrari, V., Hebert, M., Sminchisescu, C., Weiss, Y. (eds.) ECCV 2018. LNCS, vol. 11205, pp. 605–621. Springer, Cham (2018). https://doi.org/10.1007/978-3-030-01246-5_36
33. Kim, Y., Kim, S., Kim, T., Kim, C.: CNN-based semantic segmentation using level set loss. In: WACV (2019)
34. Kimmel, R., Kiryati, N., Bruckstein, A.M.: Sub-pixel distance maps and weighted distance transforms. JMIV 6, 223–233 (1996)
35. Kirillov, A., Wu, Y., He, K., Girshick, R.: Pointrend: image segmentation as rendering. arXiv:1912.08193 (2019)
36. Krähenbühl, P., Koltun, V.: Efficient inference in fully connected CRFs with gaussian edge potentials. In: NIPS (2011)
37. Kuo, W., Angelova, A., Malik, J., Lin, T.Y.: Shapemask: learning to segment novel objects by refining shape priors. In: ICCV (2019)
38. Li, K., Hariharan, B., Malik, J.: Iterative instance segmentation. In: CVPR (2016)
39. Li, X., Zhong, Z., Wu, J., Yang, Y., Lin, Z., Liu, H.: Expectation-maximization attention networks for semantic segmentation. In: ICCV (2019)
40. Liang, J., Homayounfar, N., Ma, W.C., Xiong, Y., Hu, R., Urtasun, R.: Polytransform: Deep polygon transformer for instance segmentation. arXiv:1912.02801 (2019)
41. Lin, G., Milan, A., Shen, C., Reid, I.: Refinenet: multi-path refinement networks for high-resolution semantic segmentation. In: CVPR (2017)
42. Liu, S., Qi, L., Qin, H., Shi, J., Jia, J.: Path aggregation network for instance segmentation. In: CVPR (2018)
43. Liu, S., De Mello, S., Gu, J., Zhong, G., Yang, M.H., Kautz, J.: Learning affinity via spatial propagation networks. In: NIPS (2017)
44. Liu, T., et al.: Devil in the details: towards accurate single and multiple human parsing. arXiv:1809.05996 (2018)
45. Liu, Y., Cheng, M.M., Hu, X., Wang, K., Bai, X.: Richer convolutional features for edge detection. In: CVPR (2017)
46. Long, J., Shelhamer, E., Darrell, T.: Fully convolutional networks for semantic segmentation. In: CVPR (2015)
47. Mazzini, D.: Guided upsampling network for real-time semantic segmentation. arXiv preprint arXiv:1807.07466 (2018)
48. Mazzini, D., Schettini, R.: Spatial sampling network for fast scene understanding. In: CVPRW (2019)

49. Neuhold, G., Ollmann, T., Rota Bulo, S., Kontschieder, P.: The mapillary vistas dataset for semantic understanding of street scenes. In: ICCV (2017)
50. Osher, S., Sethian, J.A.: Fronts propagating with curvature-dependent speed: algorithms based on Hamilton-Jacobi formulations. J. Comput. Phys. **79**, 12–49 (1988)
51. Papandreou, G., Zhu, T., Chen, L.-C., Gidaris, S., Tompson, J., Murphy, K.: PersonLab: person pose estimation and instance segmentation with a bottom-up, part-based, geometric embedding model. In: Ferrari, V., Hebert, M., Sminchisescu, C., Weiss, Y. (eds.) Computer Vision – ECCV 2018. LNCS, vol. 11218, pp. 282–299. Springer, Cham (2018). https://doi.org/10.1007/978-3-030-01264-9_17
52. Richter, S.R., Vineet, V., Roth, S., Koltun, V.: Playing for data: ground truth from computer games. In: Leibe, B., Matas, J., Sebe, N., Welling, M. (eds.) ECCV 2016. LNCS, vol. 9906, pp. 102–118. Springer, Cham (2016). https://doi.org/10.1007/978-3-319-46475-6_7
53. Rota Bulò, S., Porzi, L., Kontschieder, P.: In-place activated batchnorm for memory-optimized training of DNNs. In: CVPR (2018)
54. Sun, K., et al.: High-resolution representations for labeling pixels and regions. arXiv:1904.04514 (2019)
55. Takikawa, T., Acuna, D., Jampani, V., Fidler, S.: Gated-SCNN: gated shape CNNs for semantic segmentation. In: ICCV (2019)
56. Wang, Z., Acuna, D., Ling, H., Kar, A., Fidler, S.: Object instance annotation with deep extreme level set evolution. In: CVPR (2019)
57. Wu, Y., Kirillov, A., Massa, F., Lo, W.Y., Girshick, R.: Detectron2. https://github.com/facebookresearch/detectron2 (2019)
58. Yu, F., Koltun, V.: Multi-scale context aggregation by dilated convolutions. arXiv preprint arXiv:1511.07122 (2015)
59. Yu, Z., Feng, C., Liu, M.Y., Ramalingam, S.: CASENet: deep category-aware semantic edge detection. In: CVPR (2017)
60. Yu, Z., et al.: Simultaneous edge alignment and learning. In: Ferrari, V., Hebert, M., Sminchisescu, C., Weiss, Y. (eds.) ECCV 2018. LNCS, vol. 11207, pp. 400–417. Springer, Cham (2018). https://doi.org/10.1007/978-3-030-01219-9_24
61. Yuan, Y., Chen, X., Wang, J.: Object-contextual representations for semantic segmentation. arXiv preprint arXiv:1909.11065 (2019)
62. Yuan, Y., Wang, J.: OCNet: object context network for scene parsing. arXiv:1809.00916 (2018)
63. Zhang, H., et al.: Context encoding for semantic segmentation. In: CVPR (2018)
64. Zhang, H., Zhang, H., Wang, C., Xie, J.: Co-occurrent features in semantic segmentation. In: CVPR (2019)
65. Zhao, H., Shi, J., Qi, X., Wang, X., Jia, J.: Pyramid scene parsing network (2017)
66. Zhao, H., et al.: PSANet: point-wise spatial attention network for scene parsing. In: Ferrari, V., Hebert, M., Sminchisescu, C., Weiss, Y. (eds.) ECCV 2018. LNCS, vol. 11213, pp. 270–286. Springer, Cham (2018). https://doi.org/10.1007/978-3-030-01240-3_17
67. Zheng, S., et al.: Conditional random fields as recurrent neural networks. In: ICCV (2015)
68. Zhou, B., Zhao, H., Puig, X., Fidler, S., Barriuso, A., Torralba, A.: Scene parsing through ade20k dataset. In: CVPR (2017)
69. Zhu, Y., et al.: Improving semantic segmentation via video propagation and label relaxation. In: CVPR (2019)

Spatio-Temporal Graph Transformer Networks for Pedestrian Trajectory Prediction

Cunjun Yu[1], Xiao Ma[1,2(✉)], Jiawei Ren[1], Haiyu Zhao[1], and Shuai Yi[1]

[1] SenseTime Research, Beijing, China
yucunjun@sensetime.com
[2] National University of Singapore, Singapore, Singapore
xiao-ma@comp.nus.edu.sg

Abstract. Understanding crowd motion dynamics is critical to real-world applications, e.g., surveillance systems and autonomous driving. This is challenging because it requires effectively modeling the socially aware crowd spatial interaction and complex temporal dependencies. We believe attention is the most important factor for trajectory prediction. In this paper, we present *STAR*, a Spatio-Temporal grAph tRansformer framework, which tackles trajectory prediction by only attention mechanisms. STAR models intra-graph crowd interaction by *TGConv*, a novel Transformer-based graph convolution mechanism. The inter-graph temporal dependencies are modeled by separate temporal Transformers. STAR captures complex spatio-temporal interactions by interleaving between spatial and temporal Transformers. To calibrate the temporal prediction for the long-lasting effect of disappeared pedestrians, we introduce a read-writable external memory module, consistently being updated by the temporal Transformer. We show that with only attention mechanism, STAR achieves the state-of-the-art performance on 5 commonly used real-world pedestrian prediction datasets (code available at https://github.com/Majiker/STAR).

Keywords: Trajectory prediction · Transformer · Graph neural networks

1 Introduction

Crowd trajectory prediction is of fundamental importance to both the computer vision [1,16,21,22,52] and robotics [33,34] community. This task is challenging because 1) human-human interactions are multi-modal and extremely hard to

---

C. Yu and X. Ma—Equal contribution, listed in alphabetical order.

Electronic supplementary material The online version of this chapter (https://doi.org/10.1007/978-3-030-58610-2_30) contains supplementary material, which is available to authorized users.

A. Vedaldi et al. (Eds.): ECCV 2020, LNCS 12357, pp. 507–523, 2020.
https://doi.org/10.1007/978-3-030-58610-2_30

Fig. 1. STAR successfully models spatio-temporal crowd dynamics with only a strong Transformer-based attention mechanism. STAR produces more accurate prediction trajectories compared to the state-of-the-art model, SR-LSTM.

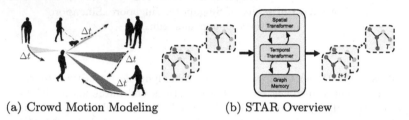

(a) Crowd Motion Modeling   (b) STAR Overview

Fig. 2. (a) People decide their future motions by paying different attentions (light yellow for less attention and dark red for more attention) to the potential future motions of their neighbors up to a certain time interval (Δt). (b) STAR models the crowd as a graph and learns spatio-temporal interaction of the crowd motion by interleaving between a graph-based spatial Transformer and a temporal Transformer. An external read-writable graph memory module is applied to improve the smoothness of the temporal predictions.

capture, e.g., strangers would avoid intimate contact with others, while fellows tend to walk in group [52]; 2) the complex temporal prediction is coupled with the spatial human-human interaction, e.g., humans condition their motions on the history and future motion of their neighbors [21].

Classic models capture human-human interaction by handcrafted energy-functions [18,19,34], which require significant feature engineering effort and normally fail to build crowd interactions in crowded spaces [21]. With the recent advances in deep neural networks, Recurrent Neural Networks (RNNs) have been extensively applied to trajectory prediction and demonstrated promising performance [1,16,21,22,52]. RNN-based methods capture pedestrian motion by their latent state and model the human-human interaction by merging latent states of spatially proximal pedestrians. Social-pooling [1,16] treat pedestrians in a neighborhood area equally and merge their latent state by a pooling mechanism. Attention mechanisms [21,22,52] relax this assumption and weigh pedestrians according to a learned function, which encodes unequal importance of neighboring pedestrians for trajectory prediction. However, existing predictors have two shared limitations: 1) the attention mechanisms used are still simple, which fails to fully model the human-human interaction, 2) RNNs normally have difficulty modeling complex temporal dependencies [43] (Fig. 1).

Recently, Transformer networks have made ground-breaking progress in Natural Language Processing domains (NLP) [10,26,43,49,51]. Transformers discard the sequential nature of language sequences and model temporal dependencies with only the powerful self-attention mechanism. The major benefit of Transformer architecture is that self-attention significantly improves temporal modeling, especially for horizon sequences, compared to RNNs [43]. Nevertheless, Transformer-based models are restricted to normal data sequences and it is hard to generalize them to more structured data, e.g., graph sequences.

In this paper, we introduce the Spatio-Temporal grAph tRansformer (STAR) framework, a novel framework for spatio-temporal trajectory prediction based purely on self-attention mechanism. We believe that learning the temporal, spatial and temporal-spatial attentions is the key to accurate crowd trajectory prediction, and Transformers provide a neat and efficient solution to this task. STAR captures the human-human interaction with a novel spatial graph Transformer. In particular, we introduce *TGConv*, a Transformer-based graph convolution mechanism. TGConv improves the attention-based graph convolution [44] by self-attention mechanism with Transformers and can capture more complex social interactions. Specifically, TGConv tends to improve more on datasets with higher pedestrian densities (ZARA1, ZARA2, UNIV). We model pedestrian motions with separate temporal Transformers, which better captures temporal dependencies compared to RNNs. STAR extracts spatio-temporal interaction among pedestrians by interleaving between spatial Transformer and temporal Transformer, a simple yet effective strategy. Besides, as Transformers treat a sequence as a bag of words, they normally have problem modeling time series data where strong temporal consistency is enforced [29]. We introduce an additional read-writable graph memory module that continuously performs smoothing over the embeddings during prediction. An overview of STAR is given by Fig. 2(b)

We experimented on 5 commonly used real-world pedestrian trajectory prediction datasets. With only attention mechanism, STAR achieves the state-of-the-art on all 5 datasets. We conduct extensive ablation studies to better understand each proposed component.

2 Background

2.1 Self-Attention and Transformer Networks

Transformer networks have achieved great success in the NLP domain, such as machine translation, sentiment analysis, and text generation [10]. Transformer networks follow the famous encoder-decoder structure widely used in the RNN seq2seq models [3,6].

The core idea of Transformer is to replace the recurrence completely by multi-head self-attention mechanism. For embeddings $\{h_t\}_{t=1}^T$, the self-attention of Transformers first learns the query matrix $Q = f_Q(\{h_t\}_{t=1}^T)$, key matrix $K = f_K(\{h_t\}_{t=1}^T)$ and a corresponding value matrix $V = f_V(\{h_t\}_{t=1}^T)$ of all embeddings from $t = 1$ to T. It computes the attention by

$$Att(Q, K, V) = \frac{\text{Softmax}(QK^{\text{T}})}{\sqrt{d_k}} V \tag{1}$$

where d_k is the dimension of each query. The $1/\sqrt{d_k}$ implements the scaled-dot product term for numerical stability for attentions. By computing the self-attention between embeddings across different time steps, the self-attention mechanism is able to learn temporal dependencies over long time horizon, in contrast to RNNs that remember the history with a single vector with limited memory. Besides, decoupling attention into the query, key and value tuples allows the self-attention mechanism to capture more complex temporal dependencies.

Multi-head attention mechanism learns to combine multiple hypotheses when computing attentions. It allows the model to jointly attend to information from different representations at different positions. With k heads, we have

$$\text{MultiHead}(Q, K, V) = f_O([\text{head}_i]_{i=1}^k)$$
$$\text{where head}_i = Att_i(Q, K, V) \tag{2}$$

where f_O is a fully connected layer merging the output from k heads and $Att_i(Q, K, V)$ denote the self-attention of the i-th head. Additional positional encoding is used to add positional information to the Transformer embeddings. Finally, Transformer outputs the updated embeddings by a fully connected layer with two skip connections.

However, one major limitation of current Transformer-based models is they only apply to non-structured data sequences, e.g., word sequences. STAR extends Transformers to more structured data sequences, as a first step, graph sequences, and apply it to trajectory prediction.

2.2 Related Works

Graph Neural Networks. Graph Neural Networks (GNNs) are powerful deep learning architectures for graph-structured data. Graph convolutions [9,15,24, 27,47] have demonstrated significant improvement on graph machine learning tasks, e.g., modeling physical systems [4,28], drug prediction [31] and social recommendation systems [11]. In particular, Graph Attention Networks (GAT) [44] implement efficient weighted message passing between nodes and achieved state-of-the-art results across multiple domains. From the sequence prediction perspective, temporal graph RNNs allow learning spatio-temporal relationship in graph sequences [8,17]. Our STAR improves GAT with TGConv, a transformer boosted attention mechanism and tackles the graph spatio-temporal modeling with transformer architecture.

Sequence Prediction. RNNs and its variants, e.g., LSTM [20] and GRU [7], have achieved great success in sequence prediction tasks, e.g., speech recognition [39,46], robot localization [14,36], robot decision making [23,37], and etc.

RNNs have been also successfully applied to model the temporal motion pattern of pedestrians [1,16,21,22,52]. RNNs-based predictors make predictions with a Seq2Seq structure [41]. Additional structure, e.g., social pooling [1,16], attention mechanism [22,45,48] and graph neural networks [21,52], are used to improve the trajectory prediction with social interaction modeling.

Transformer networks have dominated Natural Language Processing domains in recent years [10,26,43,49,51]. Transformer models completely discard the recurrence and focus on the attention across time steps. This architecture allows long-term dependency modeling and large-batch parallel training. Transformer architecture has also been applied to other domains with success, e.g., stock prediction [30], robot decision making [12] etc. STAR applies the idea of Transformer to the graph sequences. We demonstrate it on a challenging crowd trajectory prediction task, where we consider crowd interaction as a graph. STAR is a general framework and could be applied to other graph sequence prediction tasks, e.g., event prediction in social networks [35] and physical system modeling [28]. We leave this for future study.

Crowd Interaction Modeling. As the pioneering work, Social Force models [19,32], has been proven effective in various applications, e.g., crowd analysis [18] and robotics [13]. They assume the pedestrians are driven by virtual forces for goal navigation and collision avoidance. Social Force models work well on interaction modeling while performing poorly on trajectory prediction [25]. Geometry based methods, e.g., ORCA [42] and PORCA [34], consider the geometry of the agent and convert the interaction modeling into an optimization problem. One major limitation of classic approaches is that they rely on hand-crafted features, which is non-trivial to tune and hard to generalize.

Deep learning based models achieve automatic feature engineering by directly learning the model from data. Behavior CNNs [50] capture crowd interaction by CNNs. Social-Pooling [1,16] further encodes the proximal pedestrian states by a pooling mechanism that approximates the crowd interaction. Recent works consider crowd as a graph and merge information of the spatially proximal pedestrians with attention mechanisms [22,45,48]. Attention mechanism models pedestrians with importance compared to the pooling methods. Graph neural networks are also applied to address crowd modeling [21,52]. Explicit message passing allows the network to model more complex social behaviors.

3   Method

3.1   Overview

In this section, we introduce the proposed spatio-temporal graph Transformer based trajectory prediction framework, STAR. We believe attention is the most important factor for effective and efficient trajectory prediction.

STAR decomposes the spatio-temporal attention modeling into temporal modeling and spatial modeling. For temporal modeling, STAR considers each

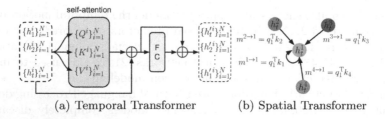

(a) Temporal Transformer   (b) Spatial Transformer

Fig. 3. STAR has two main components, Temporal Transformer and Spatial Transformer. (a) Temporal Transformer treats each pedestrians independently and extracts the temporal dependencies by Transformer model (h is the embedding of pedestrian positions, Q, K and V are the query, key, value matrix in Transformers). (b) Spatial Transformer models the crowd as a graph, and applies TGConv, a Transformer-based message passing graph convolution, to model the social interactions ($m^{i \to j}$ is the message from node i to j represented by Transformer attention)

pedestrian independently and applies a standard temporal Transformer network to extract the temporal dependencies. The temporal Transformer provides a better temporal dependency modeling protocol compared to RNNs, which we validate in our ablation studies. For spatial modeling, we introduce *TGConv*, a Transformer-based message passing graph convolution mechanism. TGConv improves the state-of-the-art graph convolution methods with a better attention mechanism and gives a better model for complex spatial interactions. In particular, TGConv tends to improve more on datasets with higher pedestrian densities (ZARA1, ZARA2, UNIV) and complex interactions. We construct two encoder modules, each including a pair of spatial and temporal Transformers, and stack them to extract spatio-temporal interactions.

3.2   Problem Setup

We are interested in the problem of predicting future trajectories starting at time step $T_{obs} + 1$ to T of total N pedestrians involved in a scene, given the observed history during time steps 1 to T_{obs}. At each time step t, we have a set of N pedestrians $\{p_t^i\}_{i=1}^N$, where $p_t^i = (x_t^i, y_t^i)$ denotes the position of the pedestrian in a top-down view map. We assume the pedestrian pairs (p_t^i, p_t^j) with distance less than d would have an undirected edge (i, j). This leads to an *interaction graph* at each time step t: $G_t = (V_t, E_t)$, where $V_t = \{p_t^i\}_{i=1}^N$ and $E_t = \{(i, j) \mid i, j \text{ is connected at time } t\}$. For each node i at time t, we define its neighbor set as $Nb(i, t)$, where for each node $j \in Nb(i, t)$, $e_t(i, j) \in E_t$.

3.3   Temporal Transformer

The temporal Transformer block in STAR uses a set of pedestrian trajectory embeddings $\{h_1^i\}_{i=1}^N, \{h_2^i\}_{i=1}^N, \ldots, \{h_t^i\}_{i=1}^N$ as input, and output a set of updated embeddings $\{h'_1^i\}_{i=1}^N, \{h'_2^i\}_{i=1}^N, \ldots, \{h'_t^i\}_{i=1}^N$ with temporal dependencies as output, considering each pedestrian independently.

The structure of a temporal Transformer block is given by Fig. 3(a). The self-attention block first learns the query matrices $\{Q^i\}_{i=1}^N$, key matrix $\{K^i\}_{i=1}^N$ and the value matrix $\{V^i\}_{i=1}^N$ given the inputs. For i-th pedestrian, we have

$$Q^i = f_Q(\{h_j^i\}_{j=1}^t), \quad K^i = f_K(\{h_j^i\}_{j=1}^t), \quad V^i = f_V(\{h_j^i\}_{j=1}^t) \tag{3}$$

where f_Q, f_K and f_V are the corresponding query, key and value functions shared by pedestrians $i = 1, \ldots, N$. We could parallel the computation for all pedestrians, benefiting from the GPU acceleration.

We compute the attention for each single pedestrian separately, following Eq. 1. Similarly, we have the multi-head attention (k heads) for pedestrian i represented as

$$Att(Q^i, K^i, V^i) = \frac{\mathrm{Softmax}(Q^i K^{iT})}{\sqrt{d_k}} V^i \tag{4}$$

$$\mathrm{MultiHead}(Q^i, K^i, V^i) = f_O([head_j]_{j=1}^k) \tag{5}$$

$$\text{where } head_j = Att_j(Q^i, K^i, V^i) \tag{6}$$

where f_O is a fully connected layer that merges the k heads and Att_j indexes the j-th head. The final embedding is generated by two skip connections and a final fully connected layers, as shown in Fig. 3(a).

The temporal Transformer is a simple generalization of Transformer networks to a data sequence set. We demonstrate in our experiment that Transformer based architecture provides better temporal modeling.

3.4 Spatial Transformer

The spatial Transformer block extracts the spatial interaction among pedestrians. We propose a novel Transformer based graph convolution, TGConv, for message passing on a graph.

Our key observation is that the self-attention mechanism can be regarded as message passing on an undirected fully connected graph. For a feature vector h_i of feature set $\{h_i\}_{i=1}^n$, we can represent its corresponding query vector as $q_i = f_Q(h_i)$, key vector as $k_i = f_K(h_i)$ and value vector as $v_i = f_V(h_i)$. We define the message from node j to i in the fully connected graph as

$$m^{j \to i} = q_i^T k_j \tag{7}$$

and the attention function (Eq. 1) can be rewritten as

$$Att(Q, K, V) = \frac{\mathrm{Softmax}\left([m^{j \to i}]_{i,j=1:n}\right)}{\sqrt{d_k}} [v_i]_{i=1}^n \tag{8}$$

Built upon the above insight, we introduce *Transformer-based Graph Convolution (TGConv)*. TGConv is essentially an attention-based graph convolution mechanism, similar to GATConv [44], but with a better attention mechanism powered by Transformers. For an arbitrary graph $G = (V, E)$ where

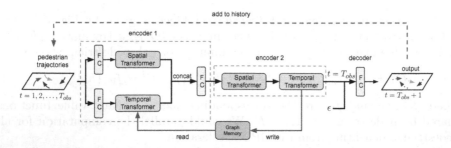

Fig. 4. Network structure of STAR with application to trajectory prediction. In STAR, trajectory prediction is achieved completely by attention mechanisms. STAR interleaves spatial Transformer and temporal Transformer in two encoder blocks to extract spatio-temporal pedestrian dependencies. An external read-writable graph memory module helps to smooth the graph embeddings and improve the consistency of temporal predictions. The prediction at $T_{obs} + 1$ is added back to history to predict the pedestrian poses at $T_{obs} + 2$.

$V = \{1, 2, \ldots, n\}$ is the node set and $E = \{(i, j) \mid i, j \text{ is connected}\}$. Assume each node i is associated with an embedding h_i and a neighbor set $Nb(i)$. The graph convolution operation for node i is written as

$$Att(i) = \frac{\text{Softmax}\left(\left[m^{j \to i}\right]_{j \in Nb(i) \bigcup \{i\}}\right)}{\sqrt{d_k}} \left[v_j\right]^{\text{T}}_{j \in Nb(i) \bigcup \{i\}} + h_i \tag{9}$$

$$h'_i = f_{out}(Att(i)) + Att(i) \tag{10}$$

where f_{out} is the output function, in our case, a fully connected layer, and h'_i is the updated embedding of node i by TGConv. We summarize the TGConv function for node i by $TGConv(h_i)$. In a Transformer structure, we would normally apply layer normalization [2] after each skip connection in the above equations. We ignored them in the equations for a clean notation.

The spatial Transformer, as shown in Fig. 3(b), can be easily implemented by the TGConv. A TGConv with shared weights is applied to each graph G_t separately. We believe TGConv is general and can be applied to other tasks and we leave it for future study.

3.5   Spatio-Temporal Graph Transformer

In this section, we introduce the Spatio-Temporal grAph tRansformer (STAR) framework for pedestrian trajectory prediction.

Temporal transformer can model the motion dynamics of each pedestrian separately, but fails to incorporate spatial interactions; spatial Transformer tackles crowd interaction with TGConv but can be hard to generalize to temporal sequences. One major challenge of pedestrian prediction is modeling coupled spatio-temporal interaction. The spatial and temporal dynamics of a pedestrian is tightly dependent on each other. For example, when one decides her next action, one would first predict the future motions of her neighbors, and choose an action that avoids collision with others in a time interval Δt.

STAR addresses the coupled spatio-temporal modeling by interleaving the spatial and temporal Transformers in a single framework. Figure 4 shows the network structure of STAR. STAR has two encoder modules and a simple decoder module. The input to the network is the pedestrian position sequences from $t = 1$ to $t = T_{obs}$, where the pedestrian positions at time step t is denoted by $\{p_t^i\}_{i=1}^N$ with $p_t^i = (x_t^i, y_t^i)$. In the first encoder, we embed the positions by two separate fully connected layers and pass the embeddings to spatial Transformer and Temporal Transformer, to extract independent spatial and temporal information from the pedestrian history. The spatial and temporal features are then merged by a fully connected layer, which gives a set of new features with spatio-temporal encodings. To further model spatio-temporal interaction in the feature space, we perform post-processing of the features with the second encoder module. In encoder 2, spatial Transformer models spatial interaction with temporal information; the temporal Transformer enhances the output spatial embeddings with temporal attentions. STAR predicts the pedestrians positions at $t = T_{obs}+1$ using a simple fully connected layer with the $t = T_{obs}$ embeddings from the second temporal Transformer as input, concatenated with a random Gaussian noise to generate various future predictions [21]. We construct $G_{T_{obs}+1}$ by connecting the nodes with distance smaller than d according to the predicted positions. The prediction is added to the history for the next step prediction.

The STAR architecture significantly improves the spatio-temporal modeling ability compared to naively combining spatial and temporal Transformers.

3.6   External Graph Memory

Although Transformer networks improve long-horizon sequence modeling by self-attention mechanism, it would potentially have difficulties handling continuous time-series data which requires a strong temporal consistency [29]. Temporal consistency, however, is a strict requirement for trajectory prediction, because pedestrian positions normally would not change sharply during a short period.

We introduce a simple external *graph memory* to tackle this dilemma. A graph memory $M_{1:T}$ is read-writable and learnable, where $M_t(i)$ has the same size with h_t^i and memorizes the embeddings of pedestrian i. At time step t, in encoder 1, the temporal Transformer first reads from memory M the past graph embeddings with function $\{\tilde{h}_1^i, \tilde{h}_2^i, \ldots, \tilde{h}_{t-1}^i\}_{i=1}^N = f_{read}(M)$ and concatenate it with the current graph embedding $\{h_t^i\}_{i=1}^N$. This allows the Temporal Transformers to condition current embeddings on the previous embedding for a consistent prediction. In encoder 2, we write the output $\{h'^i_1, h'^i_2, \ldots, h'^i_t\}_{i=1}^N$ of Temporal Transformer to the graph memory by function $M' = f_{write}(\{h'^i_1, h'^i_2, \ldots, h'^i_t\}_{i=1}^N, M)$, which performs a smoothing over the time series data. For any $t' < t$, the embeddings will be updated by the information from $t'' > t$, which gives temporally smoother embeddings for a more consistent trajectory.

For implementing f_{read} and f_{write}, many potential function forms could be adopted. In this paper, we only consider a very simple strategy

$$\{\tilde{h}_1^i, \tilde{h}_2^i, \ldots, \tilde{h}_{t-1}^i\}_{i=1}^N = f_{read}(M) = \{M_1(i), M_2(i), \ldots, M_{t-1}(i)\}_{i=1}^N \tag{11}$$

$$M' = f_{write}(\{h'^i_1, h'^i_2, \ldots, h'^i_t\}_{i=1}^N, M) = \{h'^i_1, h'^i_2, \ldots, h'^i_t\}_{i=1}^N \tag{12}$$

that is, we directly replace the memory with the embeddings and copy the memory to generate the output. This simple strategy works well in practice. More complicated functional form of f_{read} and f_{write} could be considered, e.g., fully connected layers or RNNs. We leave this for future study.

4 Experiments

In this section, we first report our results on five pedestrian trajectory datasets which serve as the major benchmark for the task of trajectory prediction: ETH (ETH and HOTEL) and UCY (ZARA1, ZARA2, and UNIV) datasets. We compare STAR to 9 trajectory predictors, including the SOTA model, SR-LSTM [52]. We follow the leave-one-out cross-validation evaluation strategy which is commonly adopted by previous works. We also perform extensive ablation studies to understand the effect of each proposed component and try to provide deeper insights for model design in the trajectory prediction task.

As a brief conclusion, we show that: 1) STAR outperforms the SOTA model on 4 out of 5 datasets and have a comparable performance to the SOTA model on the other dataset; 2) the spatial Transformer improves crowd interaction modeling compared to existing graph convolution methods; 3) the temporal Transformer generally improves the LSTM; 4) the graph memory gives a smoother temporal prediction and a better performance.

4.1 Experiment Setup

We follow the same data prepossessing strategy as SR-LSTM [52] for our method. The origin of all the input is shifted to the last observation frame. Random rotation is adopted for data augmentation.

- Average Displacement Error (ADE): the mean square error (MSE) overall estimated positions in the predicted trajectory and ground-truth trajectory.
- Final Displacement Error (FDE): the distance between the predicted final destination and the ground-truth final destination.

We take 8 frames (3.2s) as an sequence and 12 frames(4.8s) as the target sequence for prediction to have a fair comparison with all the existing works.

4.2   Implementation Details

Coordinates as input would be first encoded into a vector in size of 32 by a fully connected layer followed with ReLU activation. The dropout ratio at 0.1 is applied when processing the input data. All the transformer layers accept input with feature size at 32. Both spatial transformer and temporal transformer consists of encoding layers with 8 heads. We performed a hyper-parameter search over the learning rate, from 0.0001 to 0.004 with interval 0.0001 on a smaller network and choose the best-performed learning rate (0.0015) to train all the other models. As a result, we train the network using Adam optimizer with a learning rate of 0.0015 and batch size 16 for 300 epochs. Each batch contains around 256 pedestrians in different time windows indicated by an attention mask to accelerate the training and inference process.

4.3   Baselines

We compare STAR with a wide range of baselines, including: 1) LR: A simple temporal linear regressor; 2) LSTM: a vanilla temporal LSTM; 3) S-LSTM [1]: each pedestrian is modeled with an LSTM, and the hidden state is pooled with neighbors at each time-step; 4) Social Attention [45]: it models the crowd as a spatio-temporal graph and uses two LSTMs to capture spatial and temporal dynamics; 5) CIDNN [48]: a modularized approach for spatio-temporal crowd trajectory prediction with LSTMs; 6) SGAN [16]: a stochastic trajectory predictor with GANs; 7) SoPhie [40]: one of the SOTA stochastic trajectory predictors with LSTMs. 8) TrafficPredict [38]: LSTM-based motion predictor for heterogeneous traffic agents. Note that TrafficPredict in [38] reports isometrically normalized results. We scale them back for a consistent comparison; 9) SR-LSTM: the SOTA trajectory predictor with motion gate and pair-wise attention to refine the hidden state encoded by LSTM to obtain social interactions.

4.4   Quantitative Results and Analyses

We compare STAR with state-of-the-art approaches as mentioned in Sect. 4.3. All the stochastic method samples 20 times and reports the best-performed sample.

The main results are presented in Table 1. We observe that STAR-D outperforms SOTA deterministic models on the overall performance, and the stochastic STAR significantly outperforms all SOTA models by a large margin.

One interesting finding is that the simple model LR significantly outperforms many deep learning approaches including the SOTA model, SR-LSTM, in the HOTEL scene, which mostly contains straight-line trajectories and is relatively less crowded. This indicates that these complex models might overfit to those complex scenes like UNIV. Another example is that STAR significantly outperforms SR-LSTM on ETH and HOTEL, but is only comparable to SR-LSTM on UNIV, where the crowd density is high. This can potentially be explained by that SR-LSTM has a well-designed gated-structure for message passing on the graph, but has a relatively weak temporal model, a single LSTM. The design of

SR-LSTM potentially improves spatial modeling but might also lead to overfitting. In contrast, our approach performs well in both simple and complex scenes. We then will further demonstrate this in Sect. 4.5 with visualized results.

Table 1. Comparison with baselines models. STAR-D denotes the deterministic version of STAR. †: The results marked with † are calculated on 20 samples since they are stochastic models. $*$: SoPhie takes extra image input.

Deterministic	Performance (ADE/FDE)					
	ETH	HOTEL	ZARA1	ZARA2	UNIV	AVERAGE
LR	1.33/2.94	0.39/0.72	0.62/1.21	0.77/1.48	0.82/1.59	0.79/1.59
LSTM	1.13/2.39	0.69/1.47	0.64/1.43	0.54/1.21	0.73/1.60	0.75/1.62
S-LSTM [1]	0.77/1.60	0.38/0.80	0.51/1.19	0.39/0.89	0.58/1.28	0.53/1.15
CIDNN [48]	1.25/2.32	1.31/1.86	0.90/1.28	0.50/1.04	**0.51/1.07**	0.89/1.73
SocialAttention [45]	1.39/2.39	2.51/2.91	1.25/2.54	1.01/2.17	0.88/1.75	1.41/2.35
TrafficPredict [38]	5.46/9.73	2.55/3.57	4.32/8.00	3.76/7.20	3.31/6.37	3.88/6.97
SR-LSTM [52]	0.63/1.25	0.37/0.74	**0.41/0.90**	0.32/**0.70**	**0.51**/1.10	0.45/0.94
STAR-D	**0.56/1.11**	**0.26/0.50**	**0.41/0.90**	**0.31**/0.71	0.52/1.15	**0.41/0.87**
Stochastic	ETH	HOTEL	ZARA1	ZARA2	UNIV	AVERAGE
SGAN† [16]	0.81/1.52	0.72/1.61	0.34/0.69	0.42/0.84	0.60/1.26	0.58/1.18
SoPhie$*^\dagger$ [40]	0.70/1.43	0.76/1.67	0.30/0.63	0.38/0.78	0.54/1.24	0.54/1.15
STGAT† [21]	0.65/1.12	0.35/0.66	0.34/0.69	0.29/0.60	0.52/1.10	0.43/0.83
STAR†	**0.36/0.65**	**0.17/0.36**	**0.26/0.55**	**0.22/0.46**	**0.31/0.62**	**0.26/0.53**

4.5 Qualitative Results and Analyses

We present our qualitative results in Fig. 5 and Fig. 6.

- *STAR is able to predict temporally consistent trajectories.* In Fig. 5(a), STAR successfully captures the intention and velocity of the single pedestrian, where no social interaction exists.
- *STAR successfully extracts the social interaction of the crowd.* We visualize the attention values of the second spatial Transformer in Fig. 6. We notice that pedestrians are paying high attention to themselves and the neighbors who might potentially collide with them, e.g., Fig. 6(c) and (d); less attention is paid to spatially far away pedestrians and pedestrians without conflict of intentions, e.g., Fig. 6(a) and (b).
- *STAR is able to capture spatio-temporal interaction of the crowd.* In Fig. 5(b), we can see that the prediction of pedestrian considers the future motions of their neighbors. In addition, STAR better balances the spatial modeling and temporal modeling, compared to SR-LSTM. SR-LSTM potentially overfits on the spatial modeling and often tends to predict curves even when pedestrians are walking straight. This also corresponds to our findings in the quantitative

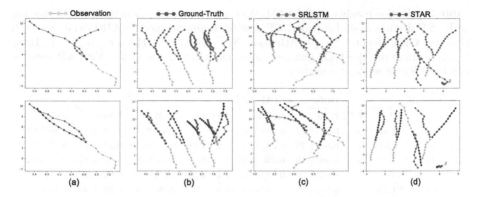

Fig. 5. Trajectory visualization. STAR successfully models the spatio-temporal interaction of the crowd and makes better predictions than the SOTA model, SR-LSTM. (a) STAR accurately extracts the temporal dynamics of the agent; (b, c, d) STAR is able to model crowd interaction and spatio-temporal interactions.

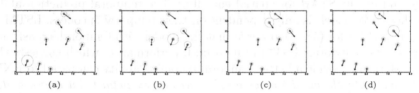

Fig. 6. Attention visualization of the spatial Transformer in encoder 2. We visualize the attention of all pedestrians with respect to the red dotted pedestrian. The size of circles represents the attention value and bigger circles indicate higher attention. STAR learns reasonable spatial attention, the pedestrians have higher attentions over themselves and their neighbors.

analyses section, that deep predictors overfits onto complex datasets. STAR better alleviates this issue with the spatial-temporal Transformer structure.

– *Auxiliary information is required for more accurate trajectory prediction.* Although STAR achieves the SOTA results, prediction can be still inaccurate occasionally, e.g., Fig. 5(d). The pedestrian takes a sharp turn, which makes it impossible to predict future trajectory purely based on the history of locations. For future work, additional information, e.g., environment setup or map, should be used to provide extra information for prediction.

4.6   Ablation Studies

We conduct extensive ablation studies on all 5 datasets to understand the influence of each STAR component. Specifically, we choose deterministic STAR to remove the influence of random sample and focus on the effect of the proposed components. The results are presented in Table 2.

– *The temporal Transformer improves the temporal modeling of pedestrian dynamics compared to RNNs.* In (4) and (5), we remove the graph mem-

520   C. Yu et al.

Table 2. Ablation Study on SR-LSTM. We replace components in STAR with existing works. **SP** denotes spaital encoder. **TP** denotes temporal encoder. **GM** denotes Graph Memory. **GAT** denotes Graph Attention Network [44], **MHA** denotes Multi-Head Additive attention [5].**STAR** denotes components in original STAR. **VSTAR** denotes simplified STAR without encoder2.

Components			Performance (ADE/FDE)					
SP	TP	GM	ETH	HOTEL	ZARA1	ZARA2	UNIV	AVG
(1) GCN	STAR	✓	3.06/5.57	0.99/1.80	2.49/4.58	1.37/2.52	1.38/2.47	1.86 /3.34
(2) GAT	STAR	✓	0.64/1.25	0.34/0.72	0.47/1.09	0.37/0.86	0.55/1.19	0.48/1.02
(3) MHA	STAR	✓	0.58/1.15	**0.25/0.48**	0.50/0.98	0.35/0.76	0.60/1.24	0.56/0.92
(4) STAR	LSTM	-	0.66/1.29	0.34/0.68	0.45/0.96	0.34/0.74	0.60/1.29	0.48/0.99
(5) STAR	STAR	×	0.60/1.18	0.28/0.60	0.53/1.13	0.36/0.76	0.57/1.20	0.47/0.97
(6) VSTAR	VSTAR	✓	0.61/1.18	0.29/0.56	0.48/1.00	0.36/0.76	0.58/1.24	0.46/0.95
(7) STAR	STAR	✓	**0.56/1.11**	0.26/0.50	**0.41/0.90**	**0.31/0.71**	**0.52/1.15**	**0.41/0.87**

ory and fix the STAR for spatial encoding. The temporal prediction ability of these two models is only dependent on their temporal encoders, LSTM for (4) and STAR for (5). We observe that the model with temporal Transformer encoding outperforms LSTM in its overall performance, which suggests that Transformers provide a better temporal modeling ability compared to RNNs.

- *TGConv outperforms the other graph convolution methods on crowd motion modeling.* In (1), (2), (3) and (7), we change the spatial encoders and compare the spatial Transformer by TGConv (7) with the GCN [24], GATConv [44] and the multi-head additive graph convolution [5]. We observe that TGConv, under the scenario of crowd modeling, achieves higher performance gain compared to the other two alternative attention-based graph convolutions.
- *Interleaving spatial and temporal Transformer is able to better extract spatio-temporal correlations.* In (6) and (7), we observe that the two encoder structures proposed in the STAR framework (7), generally outperforms the single encoder structure (6). This empirical performance gain potentially suggests that interleaving the spatial and temporal Transformers is able to extract more complex spatio-temporal interactions of pedestrians.
- *Graph memory gives a smoother temporal embedding and improves performance.* In (5) and (7), we verify the embedding smoothing ability of the graph memory module, where (5) is the STAR variant without GM. We first noticed that graph memory improves the performance of STAR on all datasets. In addition, we noticed that on ZARA1, where the spatial interaction is simple and temporal consistency prediction is more important, graph memory improves (6) to (7) by the largest margin. According to the empirical evidence, we can conclude that the embedding smoothing of graph memory is able to improve the overall temporal modeling for STAR.

5   Conclusion

We have introduced STAR, a framework for spatio-temporal crowd trajectory prediction with only attention mechanisms. STAR consists of two encoder modules, composed of spatial Transformers and temporal Transformers. We also have introduced TGConv, a novel powerful Transformer based graph convolution mechanism. STAR, using only attention mechanisms, achieves SOTA performance on 5 commonly used datasets.

STAR makes prediction only with the past trajectories, which might fail to detect the unpredictable sharp turns. Additional information, e.g., environment configuration, could be incorporated into the framework to solve this issue.

STAR framework and TGConv are not limited to trajectory prediction. They can be applied to any graph learning task. We leave it for future study.

References

1. Alahi, A., Goel, K., Ramanathan, V., Robicquet, A., Fei-Fei, L., Savarese, S.: Social LSTM: Human trajectory prediction in crowded spaces. In: CVPR (2016)
2. Ba, J.L., Kiros, J.R., Hinton, G.E.: Layer normalization. arXiv preprint arXiv:1607.06450 (2016)
3. Bahdanau, D., Cho, K., Bengio, Y.: Neural machine translation by jointly learning to align and translate. arXiv preprint arXiv:1409.0473 (2014)
4. Battaglia, P., Pascanu, R., Lai, M., Rezende, D.J., et al.: Interaction networks for learning about objects, relations and physics. In: Advances in Neural Information Processing Systems (2016)
5. Chen, B., Barzilay, R., Jaakkola, T.: Path-augmented graph transformer network (2019). https://doi.org/10.26434/chemrxiv.8214422
6. Cho, K., et al.: Learning phrase representations using RNN encoder-decoder for statistical machine translation. In: Proceedings of the 2014 Conference on Empirical Methods in Natural Language Processing (2014)
7. Chung, J., Gulcehre, C., Cho, K., Bengio, Y.: Empirical evaluation of gated recurrent neural networks on sequence modeling. arXiv preprint arXiv:1412.3555 (2014)
8. Cui, Z., Henrickson, K., Ke, R., Wang, Y.: Traffic graph convolutional recurrent neural network: A deep learning framework for network-scale traffic learning and forecasting. IEEE Trans. Intell. Transp. Syst. (2019)
9. Defferrard, M., Bresson, X., Vandergheynst, P.: Convolutional neural networks on graphs with fast localized spectral filtering. In: Advances in Neural Information Processing Systems (2016)
10. Devlin, J., Chang, M.W., Lee, K., Toutanova, K.: Bert: pre-training of deep bidirectional transformers for language understanding. arXiv preprint arXiv:1810.04805 (2018)
11. Fan, W., et al.: Graph neural networks for social recommendation. In: WWW (2019)
12. Fang, K., Toshev, A., Fei-Fei, L., Savarese, S.: Scene memory transformer for embodied agents in long-horizon tasks. In: CVPR (2019)
13. Ferrer, G., Garrell, A., Sanfeliu, A.: Robot companion: a social-force based approach with human awareness-navigation in crowded environments. In: IROS (2013)

14. Förster, A., Graves, A., Schmidhuber, J.: RNN-based learning of compact maps for efficient robot localization. In: ESANN (2007)
15. Gilmer, J., Schoenholz, S.S., Riley, P.F., Vinyals, O., Dahl, G.E.: Neural message passing for quantum chemistry. In: ICML (2017)
16. Gupta, A., Johnson, J., Fei-Fei, L., Savarese, S., Alahi, A.: Social Gan: socially acceptable trajectories with generative adversarial networks. In: CVPR (2018)
17. Hajiramezanali, E., Hasanzadeh, A., Narayanan, K., Duffield, N., Zhou, M., Qian, X.: Variational graph recurrent neural networks. In: Advances in Neural Information Processing Systems (2019)
18. Helbing, D., Buzna, L., Johansson, A., Werner, T.: Self-organized pedestrian crowd dynamics: experiments, simulations, and design solutions. Transp. Sci. **39**, 1–24 (2005)
19. Helbing, D., Molnar, P.: Social force model for pedestrian dynamics. Phys. Rev. E **51**, 4282 (1995)
20. Hochreiter, S., Schmidhuber, J.: Long short-term memory. Neural Comput. (1997)
21. Huang, Y., Bi, H., Li, Z., Mao, T., Wang, Z.: Stgat: modeling spatial-temporal interactions for human trajectory prediction. In: ICCV (2019)
22. Ivanovic, B., Pavone, M.: The trajectron: probabilistic multi-agent trajectory modeling with dynamic spatiotemporal graphs. In: ICCV (2019)
23. Karkus, P., Ma, X., Hsu, D., Kaelbling, L.P., Lee, W.S., Lozano-Pérez, T.: Differentiable algorithm networks for composable robot learning. arXiv preprint arXiv:1905.11602 (2019)
24. Kipf, T.N., Welling, M.: Semi-supervised classification with graph convolutional networks. arXiv preprint arXiv:1609.02907 (2016)
25. Kuderer, M., Kretzschmar, H., Sprunk, C., Burgard, W.: Feature-based prediction of trajectories for socially compliant navigation. In: RSS (2012)
26. Lan, Z., Chen, M., Goodman, S., Gimpel, K., Sharma, P., Soricut, R.: Albert: a lite bert for self-supervised learning of language representations. arXiv preprint arXiv:1909.11942 (2019)
27. Li, Y., Tarlow, D., Brockschmidt, M., Zemel, R.: Gated graph sequence neural networks. arXiv preprint arXiv:1511.05493 (2015)
28. Li, Y., Wu, J., Tedrake, R., Tenenbaum, J.B., Torralba, A.: Learning particle dynamics for manipulating rigid bodies, deformable objects, and fluids. arXiv preprint arXiv:1810.01566 (2018)
29. Lim, B., Arik, S.O., Loeff, N., Pfister, T.: Temporal fusion transformers for interpretable multi-horizon time series forecasting. arXiv preprint arXiv:1912.09363 (2019)
30. Liu, J., et al.: Transformer-based capsule network for stock movement prediction. In: Proceedings of the First Workshop on Financial Technology and Natural Language Processing (2019)
31. Liu, K., et al.: Chemi-Net: a molecular graph convolutional network for accurate drug property prediction. Int. J. Mol. Sci. **20**, 3389 (2019)
32. Löhner, R.: On the modeling of pedestrian motion. Appl. Math. Model. **34**, 366–382 (2010)
33. Luo, Y., Cai, P.: Gamma: A general agent motion prediction model for autonomous driving. arXiv preprint arXiv:1906.01566 (2019)
34. Luo, Y., Cai, P., Bera, A., Hsu, D., Lee, W.S., Manocha, D.: Porca: modeling and planning for autonomous driving among many pedestrians. IEEE Robot. Autom. Lett. **3**, 3418–3425 (2018)
35. Ma, X., Gao, X., Chen, G.: Beep: a Bayesian perspective early stage event prediction model for online social networks. In: ICDM (2017)

36. Ma, X., Karkus, P., Hsu, D., Lee, W.S.: Particle filter recurrent neural networks. arXiv preprint arXiv:1905.12885 (2019)
37. Ma, X., Karkus, P., Hsu, D., Lee, W.S., Ye, N.: Discriminative particle filter reinforcement learning for complex partial observations. arXiv preprint arXiv:2002.09884 (2020)
38. Ma, Y., Zhu, X., Zhang, S., Yang, R., Wang, W., Manocha, D.: Trafficpredict: trajectory prediction for heterogeneous traffic-agents. In: AAAI (2019)
39. Miao, Y., Gowayyed, M., Metze, F.: EESEN: End-to-end speech recognition using deep RNN models and WFST-based decoding. In: ASRU (2015)
40. Sadeghian, A., Kosaraju, V., Sadeghian, A., Hirose, N., Rezatofighi, H., Savarese, S.: Sophie: an attentive Gan for predicting paths compliant to social and physical constraints. In: CVPR (2019)
41. Sutskever, I., Vinyals, O., Le, Q.V.: Sequence to sequence learning with neural networks. In: Advances in Neural Information Processing Systems (2014)
42. Van Den Berg, J., Guy, S.J., Lin, M., Manocha, D.: Reciprocal n-body collision avoidance. In: Pradalier, C., Siegwart, R., Hirzinger, G. (eds.) Robotics Research. Springer Tracts in Advanced Robotics, vol. 70, pp. 3–19. Springer, Heidelberg (2011). https://doi.org/10.1007/978-3-642-19457-3_1
43. Vaswani, A., et al.: Attention is all you need. In: Advances in Neural Information Processing Systems (2017)
44. Veličković, P., Cucurull, G., Casanova, A., Romero, A., Lio, P., Bengio, Y.: Graph attention networks. arXiv preprint arXiv:1710.10903 (2017)
45. Vemula, A., Muelling, K., Oh, J.: Social attention: modeling attention in human crowds. In: ICRA (2018)
46. Xiong, W., Wu, L., Alleva, F., Droppo, J., Huang, X., Stolcke, A.: The Microsoft 2017 conversational speech recognition system. In: Proceedings of the IEEE International Conference on Acoustics, Speech and Signal Processing (2018)
47. Xu, K., Hu, W., Leskovec, J., Jegelka, S.: How powerful are graph neural networks? arXiv preprint arXiv:1810.00826 (2018)
48. Xu, Y., Piao, Z., Gao, S.: Encoding crowd interaction with deep neural network for pedestrian trajectory prediction. In: CVPR (2018)
49. Yang, Z., Dai, Z., Yang, Y., Carbonell, J., Salakhutdinov, R.R., Le, Q.V.: Xlnet: generalized autoregressive pretraining for language understanding. In: Advances in Neural Information Processing Systems (2019)
50. Yi, S., Li, H., Wang, X.: Pedestrian behavior understanding and prediction with deep neural networks. In: Leibe, B., Matas, J., Sebe, N., Welling, M. (eds.) ECCV 2016. LNCS, vol. 9905, pp. 263–279. Springer, Cham (2016). https://doi.org/10.1007/978-3-319-46448-0_16
51. Young, T., Hazarika, D., Poria, S., Cambria, E.: Recent trends in deep learning based natural language processing. IEEE Comput. Intel. Mag. **13**, 55–75 (2018)
52. Zhang, P., Ouyang, W., Zhang, P., Xue, J., Zheng, N.: SR-LSTM: state refinement for LSTM towards pedestrian trajectory prediction. In: CVPR (2019)

Fast Bi-Layer Neural Synthesis of One-Shot Realistic Head Avatars

Egor Zakharov[1,2], Aleksei Ivakhnenko[1], Aliaksandra Shysheya[1,3],
and Victor Lempitsky[1,2(✉)]

[1] Samsung AI Center – Moscow, Moscow, Russia
v.lempitsky@samsung.com
[2] Skolkovo Institute of Science and Technology, Moscow, Russia
[3] University of Cambridge, Cambridge, UK

Abstract. We propose a neural rendering-based system that creates head avatars from a single photograph. Our approach models a person's appearance by decomposing it into two layers. The first layer is a pose-dependent coarse image that is synthesized by a small neural network. The second layer is defined by a pose-independent texture image that contains high-frequency details. The texture image is generated offline, warped and added to the coarse image to ensure a high effective resolution of synthesized head views. We compare our system to analogous state-of-the-art systems in terms of visual quality and speed. The experiments show significant inference speedup over previous neural head avatar models for a given visual quality. We also report on a real-time smartphone-based implementation of our system.

Keywords: Neural avatars · Talking heads · Neural rendering · Head synthesis · Head animation

1 Introduction

Personalized head avatars driven by keypoints or other mimics/pose representation is a technology with manifold applications in telepresence, gaming, AR/VR applications, and special effects industry. Modeling human head appearance is a daunting task, due to complex geometric and photometric properties of human heads including hair, mouth cavity and surrounding clothing. For at least two decades, creating head avatars (talking head models) was done with computer graphics tools using mesh-based surface models and texture maps. The resulting systems fall into two groups. Some [4] are able to model specific people with very high realism after significant acquisition and design efforts are spent on those particular people. Others [18] are able to create talking head models from as little as a single photograph, but do not aim to achieve photorealism.

Electronic supplementary material The online version of this chapter (https://doi.org/10.1007/978-3-030-58610-2_31) contains supplementary material, which is available to authorized users.

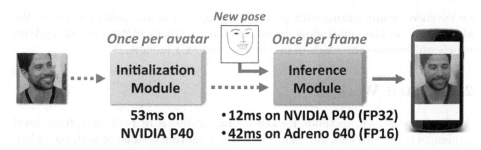

Fig. 1. Our new architecture creates photorealistic neural avatars in one-shot mode and achieves considerable speed-up over previous approaches. Rendering takes just 42 milliseconds on Adreno 640 (Snapdragon 855) GPU, FP16 mode.

In recent years, *neural talking heads* have emerged as an alternative to classic computer graphics pipeline striving to achieve both high realism and ease of acquisition. The first works required a video [25,38] or even multiple videos [27,34] to create a neural network that can synthesize talking head view of a person. Most recently, several works [12,16,32,32,35,36,40] presented systems that create neural head avatars from a handful of photographs (*few-shot* setting) or a single photograph (*one-shot* setting), causing both excitement and concerns about potential misuse of such technology.

Existing few-shot neural head avatar systems achieve remarkable results. Yet, unlike some of the graphics-based avatars, the neural systems are too slow to be deployed on mobile devices and require a high-end desktop GPU to run in real-time. We note that most application scenarios of neural avatars, especially those related to telepresence, would benefit highly from the capability to run in real-time on a mobile device. While in theory neural architectures within state-of-the-art approaches can be scaled down in order to run faster, we show that such scaling down results in a very unfavourable speed-realism tradeoff.

In this work, we address the speed limitations of one-shot neural head avatar systems, and develop an approach that can run much faster than previous models. To achieve this, we adopt a *bi-layer representation*, where the image of an avatar in a new pose is generated by summing two components: a coarse image directly predicted by a rendering network, and a warped texture image. While the warping itself is also predicted by the rendering network, the texture is estimated at the time of avatar creation and is static at runtime. To enable the few-shot capability, we use the meta-learning stage on a dataset of videos, where we (meta)-train the inference (rendering) network, the embedding network, as well as the texture generation network.

The separation of the target frames into two layers allows us both to improve the effective resolution and the speed of neural rendering. This is because we can use off-line avatar generation stage to synthesize high-resolution texture, while at test time both the first component (coarse image) and the warping of the texture need not contain high frequency details and can therefore be predicted by a relatively small rendering network. These advantages of our system are validated by extensive comparisons with previously proposed neural avatar systems. We also report on the smartphone-based real-time implementation of our system, which was beyond the reach of previously proposed models.

2 Related Work

As discussed above, methods for the neural synthesis of realistic talking head sequences can be divided into many-shot (i.e. requiring a video or multiple videos of the target person for learning the model) [20,25,27,38] and a more recent group of few-shot/singe-shot methods capable of acquiring the model of a person from a single or a handful photographs [16,32,35,36,39,40]. Our method falls into the latter category as we focus on the one-shot scenario (modeling from a single photograph).

Along another dimension, these methods can be divided according to the architecture of the generator network. Thus, several methods [25,35,38,40] use generators based on *direct synthesis*, where the image is generated using a sequence of convolutional operators, interleaved with elementwise non-linearities, and normalizations. Person identity information may be injected into such architecture, either with a lengthy learning process (in the many-shot scenario) [25,38] or by using adaptive normalizations conditioned on person embeddings [12,35,40]. The method [40] effectively combines both approaches by injecting identity through adaptive normalizations, and then fine-tuning the resulting generator on the few-shot learning set. The direct synthesis approach for human heads can be traced back to [34] that generated lips of a famous person in the talking head sequence, and further towards first works on conditional convolutional neural synthesis of generic objects such as [10].

The alternative to the direct image synthesis is to use differentiable warping [21] inside the architecture. The X2Face approach [39] applies warping twice, first from the source image to a standardized image (texture), and then to the target image. The Codec Avatar system [27] synthesizes a pose-dependent texture for a simplified mesh geometry. The MarioNETte system [16] applies warping to the intermediate feature representations. The Few-shot Vid-to-Vid system [36] combines direct synthesis with the warping of the previous frame in order to obtain temporal continuity. The First Order Motion Model [32] learns to warp the intermediate feature representation of the generator based on keypoints that are learned from data. Beyond heads, differentiable warping/texturing have recently been used for full body re-rendering [29,31]. Earlier, DeepWarp system [13] used neural warping to alter the appearance of eyes for the purpose of gaze redirection, and [42] also used neural warping for the resynthesis of generic scenes. Our method combines direct image synthesis with warping in a new way, as we obtain the fine layer by warping an RGB *pose-independent* texture, while the coarse-grained *pose-dependent* RGB component is synthesized by a neural network directly.

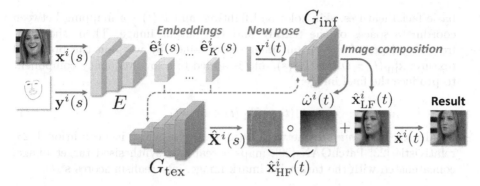

Fig. 2. During training, we first encode a source frame into the embeddings, then we initialize adaptive parameters of both inference and texture generators, and predict a high-frequency texture. These operations are only done once per avatar. Target keypoints are then used to predict a low-frequency component of the output image and a warping field, which, applied to the texture, provides the high-frequency component. Two components are then added together to produce an output.

3 Methods

We use video sequences annotated with keypoints and, optionally, segmentation masks, for training. We denote t-th frame of the i-th video sequence as $\mathbf{x}^i(t)$, corresponding keypoints as $\mathbf{y}^i(t)$, and segmentation masks as $\mathbf{m}^i(t)$ We will use an index t to denote a target frame, and s – a source frame. Also, we mark all tensors, related to generated images, with a hat symbol, ex. $\hat{\mathbf{x}}^i(t)$. We assume the spatial size of all frames to be constant and denote it as $H \times W$. In some modules, input keypoints are encoded as an RGB image, which is a standard approach in a large body of previous works [16,36,40]. In this work, we will call it a *landmark* image. But, contrary to these approaches, at test-time we input the keypoints into the inference generator directly as a vector. This allows us to significantly reduce the inference time of the method.

3.1 Architecture

In our approach, the following networks are trained in an end-to-end fashion:

- The *embedder* network $E\big(\mathbf{x}^i(s), \mathbf{y}^i(s)\big)$ encodes a concatenation of a source image and a landmark image into a stack of embeddings $\{\hat{\mathbf{e}}_k^i(s)\}$, which are used for initialization of the adaptive parameters inside the generators.
- The *texture generator* network $G_{\text{tex}}\big(\{\hat{\mathbf{e}}_k^i(s)\}\big)$ initializes its adaptive parameters from the embeddings and decodes an inpainted high-frequency component of the source image, which we call a texture $\hat{\mathbf{X}}^i(s)$.
- The *inference generator* network $G\big(\mathbf{y}^i(t), \{\hat{\mathbf{e}}_k^i(s)\}\big)$ maps target poses into a predicted image $\hat{\mathbf{x}}^i(t)$. The network accepts vector keypoints as an input and outputs a low-frequency layer of the output image $\hat{\mathbf{x}}_{\text{LF}}^i(t)$, which encodes

basic facial features, skin color and lighting, and $\hat{\omega}^i(t)$ – a mapping between coordinate spaces of the texture and the output image. Then, the high-frequency layer of the output image is obtained by warping the predicted texture: $\hat{\mathbf{x}}^i_{\mathrm{HF}}(t) = \hat{\omega}^i(t) \circ \hat{\mathbf{X}}^i(s)$, and is added to a low-frequency component to produce the final image:

$$\hat{\mathbf{x}}^i(t) = \hat{\mathbf{x}}^i_{\mathrm{LF}}(t) + \hat{\mathbf{x}}^i_{\mathrm{HF}}(t). \tag{1}$$

– Finally, the *discriminator* network $D(\mathbf{x}^i(t), \mathbf{y}^i(t))$, which is a conditional [28] relativistic [23] PatchGAN [20], maps a real or a synthesised target image, concatenated with the target landmark image, into realism scores $\mathbf{s}^i(t)$.

During training, we first input a source image $\mathbf{x}^i(s)$ and a source pose $\mathbf{y}^i(s)$, encoded as a landmark image, into the embedder. The outputs of the embedder are K tensors $\hat{\mathbf{e}}^i_k(s)$, which are used to predict the adaptive parameters of the texture generator and the inference generator. A high-frequency texture $\hat{\mathbf{X}}^i(s)$ of the source image is then synthesized by the texture generator. Next, we input corresponding target keypoints $\mathbf{y}^i(t)$ into the inference generator, which predicts a low-frequency component of the output image $\hat{\mathbf{x}}^i_{\mathrm{LF}}(t)$ directly and a high-frequency component $\hat{\mathbf{x}}^i_{\mathrm{HF}}(t)$ by warping the texture with a predicted field $\hat{\omega}^i(t)$. Finally, the output image $\hat{\mathbf{x}}^i(t)$ is obtained as a sum of these two components.

It is important to note that while the texture generator is manually forced to generate only a high-frequency component of the image via the design of the loss functions, which is described in the next section, we do not specifically constrain it to perform texture inpainting for occluded head parts. This behavior is emergent from the fact that we use two different images with different poses for initialization and loss calculation.

3.2   Training Process

We use multiple loss functions for training. The main loss function responsible for the realism of the outputs is trained in an adversarial way [15]. We also use pixelwise loss to preserve source lightning conditions and perceptual [22] loss to match the source identity in the outputs. Finally, a regularization of the texture mapping adds robustness to the random initialization of the model.

Pixelwise and Perceptual Losses ensure that the predicted images match the ground truth, and are respectively applied to low- and high-frequency components of the output images. Since usage of pixelwise losses assumes independence of all pixels in the image, the optimization process leads to blurry images [20], which is suitable for the low-frequency component of the output. Thus the pixelwise loss is calculated by simply measuring mean L_1 distance between the target image and the low-frequency component:

$$\mathcal{L}^G_{\mathrm{pix}} = \frac{1}{HW} ||\hat{\mathbf{x}}^i_{\mathrm{LF}}(t) - \mathbf{x}^i(t)||_1. \tag{2}$$

On the contrary, the optimization of the perceptual loss leads to crisper and more realistic images [22], which we utilize to train the high-frequency component. To calculate the perceptual loss, we use the stop-gradient operator SG, which allows us to prevent the gradient flow into a low-frequency component. The input generated image is, therefore, calculated as following:

$$\tilde{\mathbf{x}}^i(t) = \text{SG}\big(\hat{\mathbf{x}}^i_{\text{LF}}(t)\big) + \hat{\mathbf{x}}^i_{\text{HF}}(t). \tag{3}$$

Following [16] and [40], our variant of the perceptual loss consists of two components: features evaluated using an ILSVRC (ImageNet) pre-trained VGG19 network [33], and the VGGFace network [30], trained for face recognition. If we denote the intermediate features of these networks as $\mathbf{f}^i_{k,\text{IN}}(t)$ and $\mathbf{f}^i_{k,\text{face}}(t)$, and their spatial size as $H_k \times W_k$, the objectives can be written as follows:

$$\mathcal{L}^G_{\text{IN}} = \frac{1}{K} \sum_k \frac{1}{H_k W_k} ||\tilde{\mathbf{f}}^i_{k,\text{IN}}(t) - \mathbf{f}^i_{k,\text{IN}}(t)||_1, \tag{4}$$

$$\mathcal{L}^G_{\text{face}} = \frac{1}{K} \sum_k \frac{1}{H_k W_k} ||\tilde{\mathbf{f}}^i_{k,\text{face}}(t) - \mathbf{f}^i_{k,\text{face}}(t)||_1. \tag{5}$$

Texture Mapping Regularization is proposed to improve the stability of the training. In our model, the coordinate space of the texture is learned implicitly, and there are two degrees of freedom that can mutually compensate each other: the position of the face in the texture, and the predicted warping. If, after initial iterations, the major part of the texture is left unused by the model, it can easily compensate that with a more distorted warping field. This artifact of an initialization is not fixed during training, and clearly is not the behavior we need, since we want all the texture to be used to achieve the maximum effective resolution in the outputs. We address the problem by regularizing the warping in the first iterations to be close to an identity mapping:

$$\mathcal{L}^G_{\text{reg}} = \frac{1}{HW} ||\omega^i(t) - \mathcal{I}||_1. \tag{6}$$

Adversarial Loss is optimized by both generators, the embedder and the discriminator networks. Usually, it resembles a binary classification loss function between real and fake images, which discriminator is optimized to minimize, and generators – maximize [15]. We follow a large body of previous works [6, 16, 36, 40] and use a hinge loss as a substitute for the original binary cross entropy loss. We also perform relativistic realism score calculation [23], following its recent success in tasks such as super-resolution [38] and denoising [24]. Additionally, we use PatchGAN [20] formulation of the adversarial learning. The discriminator is trained only with respect to its adversarial loss $\mathcal{L}^D_{\text{adv}}$, while the generators and the embedder are trained via the adversarial loss $\mathcal{L}^G_{\text{adv}}$, and also a feature matching loss $\mathcal{L}_{\text{FM}}$ [37]. The latter is introduced for better stability of the training.

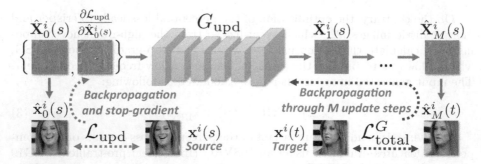

Fig. 3. Texture enhancement network (updater) accepts the current state of the texture and the guiding gradients to produce the next state. The guiding gradients are obtained by reconstructing the source image from the current state of the texture and matching it to the ground-truth via a lightweight updater loss. These gradients are only used as inputs and are detached from the computational graph. This process is repeated M times. The final state of the texture is then used to obtain a target image, which is matched to the ground-truth via the same loss as the one used during training of the main model. The gradients from this loss are then backpropagated through all M copies of the updater network.

3.3   Texture Enhancement

To minimize the identity gap, [40] suggested to fine-tune the generator weights to the few-shot training set. Training on a person-specific source data leads to significant improvement in realism and identity preservation of the synthesized images [40], but is computationally expensive. Moreover, when the source data is scarce, like in one-shot scenario, fine-tuning may lead to over-fitting and performance degradation, which is observed in [40]. We address both of these problems by using a learned gradient descend (LGD) method [5] to optimize only the synthesized texture $\hat{\mathbf{X}}^i(s)$. Optimizing with respect to the texture tensor prevents the model from overfitting, while the LGD allows us to perform optimization with respect to computationally expensive objectives by doing forward passes through a pre-trained network.

Specifically, we introduce a lightweight loss function $\mathcal{L}_{\text{upd}}$ (we use a sum of squared errors), that measures the distance between a generated image and a ground-truth in the pixel space, and a *texture updating* network G_{upd}, that uses the current state of the texture and the gradient of $\mathcal{L}_{\text{upd}}$ with respect to the texture to produce an update $\Delta\hat{\mathbf{X}}^i(s)$. During fine-tuning we perform M update steps, each time measuring the gradients of $\mathcal{L}_{\text{upd}}$ with respect to an updated texture. The visualization of the process can be seen in Fig. 3. More formally, each update is computed as:

$$\hat{\mathbf{X}}_{m+1}^i(s) = \hat{\mathbf{X}}_m^i(s) + G_{\text{upd}}\left(\hat{\mathbf{X}}_m^i(s), \frac{\partial\mathcal{L}_{\text{upd}}}{\partial\hat{\mathbf{X}}_m^i(s)}\right), \qquad (7)$$

where $m \in \{0, \ldots, M-1\}$ denotes an iteration number, with $\hat{\mathbf{X}}_0^i(s) \equiv \hat{\mathbf{X}}^i(s)$.

The network G_{upd} is trained by back-propagation through all M steps. For training, we use the same objective $\mathcal{L}_{\mathrm{total}}^G$ that was used during the training of the base model. We evaluate it using a target frame $\mathbf{x}^i(t)$ and a generated frame

$$\hat{\mathbf{x}}_M^i(t) = \hat{\mathbf{x}}_{\mathrm{LF}}^i(t) + \hat{\omega}^i(t) \circ \hat{\mathbf{X}}_M^i(s). \tag{8}$$

It is important to highlight that $\mathcal{L}_{\mathrm{upd}}$ is not used for training of G_{upd}, but simply guides the updates to the texture. Also, the gradients with respect to this loss are evaluated using the source image, while the objective in Eq. 8 is calculated using the target image, which implies that the network has to produce updates for the whole texture, not just a region "visible" on the source image. Lastly, while we do not propagate any gradients into the generator part of the base model, we keep training the discriminator using the same objective $\mathcal{L}_{\mathrm{adv}}^D$. Even though training the updater network jointly with the base generator is possible, and can lead to better quality (following the success of model agnostic meta-learning [11] method), we resort to two-stage training due to memory constraints.

3.4 Segmentation

The presence of static background leads to a certain degradation of our model for two reasons. Firstly, part of the capacity of the texture and the inference generators has to be spent on modeling high variety of background patterns. Secondly, and more importantly, the static nature of backgrounds in most training videos biases the warping towards identity mapping. We therefore, have found it advantageous to include background segmentation into our model.

We use a state-of-the-art face and body segmentation model [14] to obtain the ground truth masks. Then, we add the mask prediction output $\hat{\mathbf{m}}^i(t)$ to our inference generator alongside with its other outputs, and train it via a binary cross-entropy loss $\mathcal{L}_{\mathrm{seg}}$ to match the ground truth mask $\mathbf{m}^i(t)$. To filter out the training signal, related to the background, we have explored multiple options. Simple masking of the gradients that are fed into the generator leads to severe overfitting of the discriminator. We also could not simply apply the ground truth masks to all the images in the dataset, since the model [14] works so well that it produces a sharp border between the foreground and the background, leading to border artifacts that emerge after adversarial training.

Instead, we have found out that masking the ground truth images that are fed to the discriminator with the predicted masks $\hat{\mathbf{m}}^i(t)$ works well. Indeed, these masks are smooth and prevent the discriminator from overfitting to the lack of background, or sharpness of the border. We do not backpropagate the signal from the discriminator and from perceptual losses to the generator via the mask pathway (i.e. we use stop gradient/detach operator $\mathrm{SG}\big(\hat{\mathbf{m}}^i(t)\big)$ before applying the mask). The stop-gradient operator also ensures that the training does not converge to a degenerate state (empty foreground).

3.5   Implementation Details

All our networks consist of pre-activation residual blocks [17] with LeakyReLU activations. We set a minimum number of features in these blocks to 64, and a maximum to 512. By default, we use half the number of features in the inference generator, but we also evaluate our model with full- and quater-capacity inference part, with the results provided in the experiments section.

We use batch normalization [19] in all the networks except for the embedder and the texture updater. Inside the texture generator, we pair batch normalization with adaptive SPADE layers [36]. We modify these layers to predict pixelwise scale and bias coefficients using feature maps, which are treated as model parameters, instead of being input from a different network. This allows us to save memory by removing additional networks and intermediate feature maps from the optimization process, and increase the batch size. Also, following [36], we predict weights for all 1×1 convolutions in the network from the embeddings $\{\hat{\mathbf{e}}_k^i(s)\}$, which includes the scale and bias mappings in AdaSPADE layers, and skip connections in the residual upsampling blocks. In the inference generator, we use standard adaptive batch normalization layers [6], but also predict weights for the skip connections from the embeddings.

We do simultaneous gradient descend on parameters of the generator networks and the discriminator using Adam [26] with a learning rate of $2 \cdot 10^{-4}$. We use 0.5 weight for adversarial losses, and 10 for all other losses, except for the VGGFace perceptual loss (Eq. 5), which is set to 0.01. The weight of the regularizer (Eq. 6) is then multiplicatively reduced by 0.9 every 50 iterations. We train our models on 8 NVIDIA P40 GPUs with the batch size of 48 for the base model, and a batch size of 32 for the updater model. We set unrolling depth M of the updater to 4 and use a sum of squared errors as the lightweight objective. Batch normalization statistics are synchronized across all GPUs during training. During inference they are replaced with "standing" statistics, similar to [6], which significantly improves the quality of the outputs, compared to the usage of running statistics. Spectral normalization is also applied in all linear and convolutional layers of all networks.

Please refer to the supplementary material for a detailed description of our model's architecture, as well as the discussion of training and architectural features that we have adopted.

4   Experiments

We perform evaluation in multiple scenarios. First, we use the original Vox-Celeb2 [8] dataset to compare with state-of-the-art systems. To do that, we annotated this dataset using an off-the-shelf facial landmarks detector [7]. Overall, the dataset contains 140697 videos of 5994 different people. We also use a high-quality version of the same dataset, additionally annotated with the segmentation masks (which were obtained using a model [14]), to measure how the performance of our model scales with a dataset of a significantly higher quality.

We obtained this version by downloading the original videos via the links provided in the VoxCeleb2 dataset, and filtering out the ones with low resolution. This dataset is, therefore, significantly smaller and contains only 14859 videos of 4242 people, with each video having at most 250 frames (first 10 s). Lastly, we do ablation studies on both VoxCeleb2 and VoxCeleb2-HQ, and report on a smartphone-based implementation of the method. For comparisons and ablation studies we show the results qualitatively and also evaluate the following metrics:

- Learned perceptual image patch similarity [41] (LPIPS), which measures overall predicted image similarity to ground truth.
- Cosine similarity between the embedding vectors of a state-of-the-art face recognition network [9] (CSIM), calculated using the synthesized and the target images. This metric evaluates the identity mismatch.
- Normalized mean error of the head pose in the synthesized image (NME). We use the same network [7], which was used for the annotation of the dataset, to evaluate the pose of the synthesized image. We normalize the error, which is a mean euclidean distance between the predicted and the target points, by the distance between the eyes in the target pose, multiplied by 10.
- Multiply-accumulate operations (MACs), which measure the complexity of each method. We exclude from the evaluation initialization steps, which are calculated only once per avatar.

The test set in both datasets does not intersect with the train set in terms of videos or identities. For evaluation, we use a subset of 50 test videos with different identities (for VoxCeleb2, it is the same as in [40]). The first frame in each sequence is used as a source. Target frames are taken sequentially at 1 FPS.

We only discuss most important results in the main paper. For additional qualitative results and comparisons please refer to the supplementary materials.

4.1   Comparison with the State-of-the-art Methods

We compare against three state-of-the-art systems: Few-shot Talking Heads [40], Few-shot Vid-to-Vid [36] and First Order Motion Model [32]. The first system is a problem-specific model designed for avatar creation. Few-shot Vid-to-Vid is a state-of-the-art video-to-video translation system, which has also been successfully applied to this problem. First Order Motion Model (FOMM) is a general motion transfer system that does not use precomputed keypoints, but can also be used as an avatar. We believe that these models are representative of the most recent and successful approaches to one-shot avatar generation. We also acknowledge the work of [16], but do not compare to them extensively due to unavailability of the source code, pretrained models or pre-calculated results. A small-scale qualitative comparison is provided in the supplementary materials. Additionally, their method is limited to the usage of 3D keypoints, while our method does not have such restriction. Lastly, since Few-shot Vid-to-Vid is an autoregressive model, we use a full test video sequence for evaluation (25 FPS) and save the predicted frames at 1 FPS.

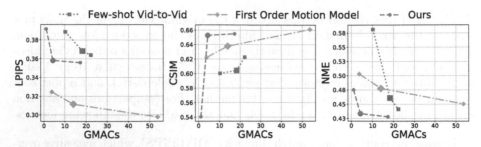

Fig. 4. In order to evaluate a quality against performance trade off, we train a family of models with varying complexity for each of the compared methods. For quality metrics, we have compared synthesized images to their targets using a perceptual image similarity (LPIPS ↓), identity preservation metric (CSIM ↑), and a normalized pose error (NME ↓). We highlight a model which was used for the comparison in Fig. 5 with a bold marker. We observe that our model outperforms the competitors in terms of identity preservation (CSIM) and pose matching (NME) in the settings, when models' complexities are comparable. In order to better compare with FOMM, we did a user study, where users have preferred the image generated by our model to FOMM 59.6% of the time.

Importantly, the base models in these approaches have a lot of computational complexity, so for each method we evaluate a family of models by varying the number of parameters. The performance comparison for each family is reported in Fig. 4 (with Few-shot Talking Heads being excluded from this evaluation, since their performance is much worse than the compared methods). Overall, we can see that our model's family outperforms competing methods in terms of pose error and identity preservation, while being, on average, up to an order of magnitude faster. To better compare with FOMM in terms of image similarity, we have performed a user study, where we asked crowd-sourced users which generated image better matches the ground truth. In total, 361 users evaluated 1600 test pairs of images, with each one seeing on average 21 pairs. In 59.6% of comparisons, the result of our medium model was preferred to a medium sized model of FOMM.

Another important note is on how the complexity was evaluated. In Few-shot Vid-to-Vid we have additionally excluded from the evaluation parts that are responsible for the temporal consistency, since other compared methods are evaluated frame-by-frame and do not have such overhead. Also, in FOMM we have excluded the keypoints extractor network, because this overhead is shared implicitly by all the methods via usage of the precomputed keypoints.

We visualize the results for medium-sized models of each of the compared methods in Fig. 5. Since all methods perform similarly in case when source and target images have marginal differences, we have shown the results where a source and a target have different head poses. In this extrapolation setting, our method has a clear advantage, while other methods either introduce more artifacts or more blurriness.

| | | Few-shot | Few-shot | | |
| Source | Target | T. Heads | Vid-to-Vid | FOMM | Ours |

Fig. 5. Comparison on a VoxCeleb2 dataset. The task is to reenact a **target** image, given a **source** image and target keypoints. The compared methods are Few-shot Talking Heads [40], Few-shot Vid-to-Vid [36], First Order Motion Model (FOMM) [32] and our proposed Bi-layer Model. For each method, we used the models with a similar number of parameters, and picked source and target images to have diverse poses and expressions, in order to highlight the differences between the compared methods.

Evaluation on High-Quality Images. Next, we evaluate our method on the high-quality dataset and present the results in Fig. 6. Overall, in this case, our method is able to achieve a smaller identity gap, compared to the dataset with the background. We also show the decomposition between the texture and a low frequency component in Fig. 7. Lastly, in Fig. 8, we show that our texture enhancement pipeline allows us to render small person-specific features like wrinkles and moles on out-of-domain examples. For more qualitative examples, as well as reenactment examples with a driver of a different person, please refer to the supplementary materials.

Smartphone-Based Implementation. We train our model using PyTorch [1] and then port it to smartphones with Qualcomm Snapdragon chips. There are several frameworks which provide APIs for mobile inference on such devices. From our experiments, we measured the Snapdragon Neural Processing Engine (SNPE) [2] to be about 1.5 times faster than PyTorch Mobile [1] and up to two times faster than TensorFlow Lite [3]. The medium-sized model ported to the

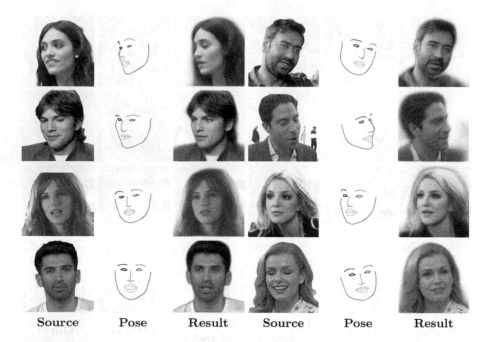

Source   Pose   Result   Source   Pose   Result

Fig. 6. High quality synthesis results. We can see that our model is both capable of viewpoint extrapolation and low identity gap synthesis. The architecture in this experiment has the same number of parameters as the medium architecture in the previous comparison.

Snapdragon 855 (Adreno 640 GPU, FP16 mode) takes 42 ms per frame, which is sufficient for real-time performance, given that the keypoint tracking is being run in parallel, e.g. on a mobile CPU.

Ablation Study. Finally, we evaluate the contribution of individual components. First, we evaluate the contribution of adaptive SPADE layers in the texture generator (by replacing them with adaptive batch normalization and per-pixel biases) and adaptive skip-connections in both generators. A model with these features removed makes up our baseline. Lastly, we evaluate the contribution of the updater network. The results can be seen in Table 1 and Fig. 9. We evaluate the baseline approach only on a VoxCeleb2 dataset, while the full models with and without the updater network are evaluated on both low- and high-quality datasets. Overall, we see a significant contribution of each component with respect to all metrics, which is particularly noticeable in the high-quality scenario. In all ablation comparisons, medium-sized models were used.

| Source | Texture | LF | Result | | Source | Target | Pose | Ours |

Fig. 7. Detailed results on the generation process of the output image. **LF** denotes a low-frequency component.

Fig. 8. Our method can preserve a lot of details in the facial features, like the famous Marylin's mole.

Table 1. Ablation studies of our approach. We first evaluate the baseline method without AdaSPADE or adaptive skip connections. Then we add these layers, following [36], and observe significant quality improvement. Finally, our updater network provides even more improvement across all metrics, especially noticeable in the high-quality scenario.

Method	LPIPS ↓	CSIM ↑	NME ↓
VoxCeleb2			
Baseline	0.377	0.547	0.447
Ours	0.370	0.595	0.441
+Updater	**0.358**	**0.653**	**0.433**
VoxCeleb2-HQ			
Ours	0.313	0.432	0.476
+Updater	**0.298**	**0.649**	**0.456**

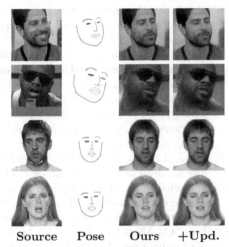

| Source | Pose | Ours | +Upd. |

Fig. 9. Examples from the ablation study on VoxCeleb2 (first two rows) and VoxCeleb2-HQ (last two rows).

5   Conclusion

We have proposed a new neural rendering-based system that creates head avatars from a single photograph. Our approach models person appearance by decomposing it into two layers. The first layer is a pose-dependent coarse image that is synthesized by a small neural network. The second layer is defined by a pose-independent texture image that contains high-frequency details and is generated offline. During test-time it is warped and added to the coarse image to ensure high effective resolution of synthesized head views. We compare our system to analogous state-of-the-art systems in terms of visual quality and speed. The experiments show up to an order of magnitude inference speedup over previous neural head avatar models, while achieving state-of-the-art quality. We also report on a real-time smartphone-based implementation of our system.

References

1. PyTorch homepage. https://pytorch.org
2. SNPE homepage. https://developer.qualcomm.com/sites/default/files/docs/snpe
3. TensorFlow Lite homepage. https://www.tensorflow.org/lite
4. Alexander, O., et al.: The Digital Emily project: achieving a photorealistic digital actor. IEEE Comput. Graph. Appl. **30**(4), 20–31 (2010)
5. Andrychowicz, M., et al.: Learning to learn by gradient descent by gradient descent. In: Advances in Neural Information Processing Systems 29: Annual Conference on Neural Information Processing Systems (2016)
6. Brock, A., Donahue, J., Simonyan, K.: Large scale GAN training for high fidelity natural image synthesis. In: 7th International Conference on Learning Representations, ICLR 2019 (2019)
7. Bulat, A., Tzimiropoulos, G.: How far are we from solving the 2D & 3D face alignment problem? (and a dataset of 230, 000 3D facial landmarks). In: IEEE International Conference on Computer Vision, ICCV 2017, Venice, Italy, 22–29 October 2017, pp. 1021–1030 (2017)
8. Chung, J.S., Nagrani, A., Zisserman, A.: Voxceleb2: deep speaker recognition. In: Interspeech 2018, 19th Annual Conference of the International Speech Communication Association (2018)
9. Deng, J., Guo, J., Niannan, X., Zafeiriou, S.: Arcface: Additive angular margin loss for deep face recognition. In: CVPR (2019)
10. Dosovitskiy, A., Tobias Springenberg, J., Brox, T.: Learning to generate chairs with convolutional neural networks. In: Proceedings of the IEEE Conference on Computer Vision and Pattern Recognition, pp. 1538–1546 (2015)
11. Finn, C., Abbeel, P., Levine, S.: Model-agnostic meta-learning for fast adaptation of deep networks. In: Proceedings of the 34th International Conference on Machine Learning, ICML 2017 (2017)
12. Fu, C., Hu, Y., Wu, X., Wang, G., Zhang, Q., He, R.: High fidelity face manipulation with extreme pose and expression. arXiv preprint arXiv:1903.12003 (2019)
13. Ganin, Y., Kononenko, D., Sungatullina, D., Lempitsky, V.: DeepWarp: photorealistic image resynthesis for gaze manipulation. In: Leibe, B., Matas, J., Sebe, N., Welling, M. (eds.) ECCV 2016. LNCS, vol. 9906, pp. 311–326. Springer, Cham (2016). https://doi.org/10.1007/978-3-319-46475-6_20
14. Gong, K., Gao, Y., Liang, X., Shen, X., Wang, M., Lin, L.: Graphonomy: universal human parsing via graph transfer learning. In: IEEE Conference on Computer Vision and Pattern Recognition, CVPR (2019)
15. Goodfellow, I.J., et al.: Generative adversarial nets. In: Advances in Neural Information Processing Systems 27: Annual Conference on Neural Information Processing Systems 2014 (2014)
16. Ha, S., Kersner, M., Kim, B., Seo, S., Kim, D.: Marionette: Few-shot face reenactment preserving identity of unseen targets. CoRR abs/1911.08139 (2019)
17. He, K., Zhang, X., Ren, S., Sun, J.: Identity mappings in deep residual networks. In: Leibe, B., Matas, J., Sebe, N., Welling, M. (eds.) ECCV 2016. LNCS, vol. 9908, pp. 630–645. Springer, Cham (2016). https://doi.org/10.1007/978-3-319-46493-0_38

18. Hu, L., et al.: Avatar digitization from a single image for real-time rendering. ACM Trans. Graph. **36**(6), 195:1–195:14 (2017)
19. Ioffe, S., Szegedy, C.: Batch normalization: accelerating deep network training by reducing internal covariate shift. In: Proceedings of the 32nd International Conference on Machine Learning, ICML 2015 (2015)
20. Isola, P., Zhu, J., Zhou, T., Efros, A.A.: Image-to-image translation with conditional adversarial networks. In: 2017 IEEE Conference on Computer Vision and Pattern Recognition, CVPR 2017 (2017)
21. Jaderberg, M., Simonyan, K., Zisserman, A., Kavukcuoglu, K.: Spatial transformer networks. In: Advances in Neural Information Processing Systems 28: Annual Conference on Neural Information Processing Systems (2015)
22. Johnson, J., Alahi, A., Fei-Fei, L.: Perceptual losses for real-time style transfer and super-resolution. In: Leibe, B., Matas, J., Sebe, N., Welling, M. (eds.) ECCV 2016. LNCS, vol. 9906, pp. 694–711. Springer, Cham (2016). https://doi.org/10.1007/978-3-319-46475-6_43
23. Jolicoeur-Martineau, A.: The relativistic discriminator: a key element missing from standard GAN. In: 7th International Conference on Learning Representations, ICLR (2019)
24. Kim, D., Chung, J.R., Jung, S.: GRDN: grouped residual dense network for real image denoising and GAN-based real-world noise modeling. In: IEEE Conference on Computer Vision and Pattern Recognition Workshops, CVPR Workshops 2019 (2019)
25. Kim, H., et al.: Deep video portraits. arXiv preprint arXiv:1805.11714 (2018)
26. Kingma, D.P., Ba, J.: Adam: a method for stochastic optimization. CoRR abs/1412.6980 (2014)
27. Lombardi, S., Saragih, J., Simon, T., Sheikh, Y.: Deep appearance models for face rendering. ACM Trans. Graph. (TOG) **37**(4), 68 (2018)
28. Mirza, M., Osindero, S.: Conditional generative adversarial nets. CoRR abs/1411.1784 (2014)
29. Neverova, N., Alp Güler, R., Kokkinos, I.: Dense pose transfer. In: Ferrari, V., Hebert, M., Sminchisescu, C., Weiss, Y. (eds.) ECCV 2018. LNCS, vol. 11207, pp. 128–143. Springer, Cham (2018). https://doi.org/10.1007/978-3-030-01219-9_8
30. Parkhi, O.M., Vedaldi, A., Zisserman, A.: Deep face recognition. In: Proceedings of the British Machine Vision Conference 2015, BMVC 2015 (2015)
31. Shysheya, A., et al.: Textured neural avatars. In: IEEE Conference on Computer Vision and Pattern Recognition, CVPR (2019)
32. Siarohin, A., Lathuilière, S., Tulyakov, S., Ricci, E., Sebe, N.: First order motion model for image animation. In: Advances in Neural Information Processing Systems 32: Annual Conference on Neural Information Processing Systems 2019, NeurIPS 2019 (2019)
33. Simonyan, K., Zisserman, A.: Very deep convolutional networks for large-scale image recognition. CoRR abs/1409.1556 (2014). http://arxiv.org/abs/1409.1556
34. Suwajanakorn, S., Seitz, S.M., Kemelmacher-Shlizerman, I.: Synthesizing Obama: learning lip sync from audio. ACM Trans. Graph. (TOG) **36**(4), 95 (2017)
35. Tripathy, S., Kannala, J., Rahtu, E.: ICface: interpretable and controllable face reenactment using GANs. CoRR abs/1904.01909 (2019). http://arxiv.org/abs/1904.01909
36. Wang, T., Liu, M., Tao, A., Liu, G., Catanzaro, B., Kautz, J.: Few-shot video-to-video synthesis. In: Advances in Neural Information Processing Systems 32: Annual Conference on Neural Information Processing Systems 2019, NeurIPS 2019 (2019)

37. Wang, T., Liu, M., Zhu, J., Tao, A., Kautz, J., Catanzaro, B.: High-resolution image synthesis and semantic manipulation with conditional GANs. In: 2018 IEEE Conference on Computer Vision and Pattern Recognition, CVPR 2018 (2018)
38. Wang, T., et al.: Video-to-video synthesis. In: Advances in Neural Information Processing Systems 31: Annual Conference on Neural Information Processing Systems 2018, NeurIPS 2018, 3–8 December 2018, Montréal, Canada (2018)
39. Wiles, O., Koepke, A.S., Zisserman, A.: X2Face: a network for controlling face generation using images, audio, and pose codes. In: Ferrari, V., Hebert, M., Sminchisescu, C., Weiss, Y. (eds.) ECCV 2018. LNCS, vol. 11217, pp. 690–706. Springer, Cham (2018). https://doi.org/10.1007/978-3-030-01261-8_41
40. Zakharov, E., Shysheya, A., Burkov, E., Lempitsky, V.S.: Few-shot adversarial learning of realistic neural talking head models. In: IEEE International Conference on Computer Vision, ICCV 2019 (2019)
41. Zhang, R., Isola, P., Efros, A.A., Shechtman, E., Wang, O.: The unreasonable effectiveness of deep features as a perceptual metric. In: 2018 IEEE Conference on Computer Vision and Pattern Recognition, CVPR (2018)
42. Zhou, T., Tulsiani, S., Sun, W., Malik, J., Efros, A.A.: View synthesis by appearance flow. In: Leibe, B., Matas, J., Sebe, N., Welling, M. (eds.) ECCV 2016. LNCS, vol. 9908, pp. 286–301. Springer, Cham (2016). https://doi.org/10.1007/978-3-319-46493-0_18

Neural Geometric Parser for Single Image Camera Calibration

Jinwoo Lee[1], Minhyuk Sung[2], Hyunjoon Lee[3], and Junho Kim[1(✉)]

[1] Kookmin University, Seoul, South Korea
junho@kookmin.ac.kr
[2] Adobe Research, San Jose, USA
[3] Intel Korea, Seoul, South Korea

Abstract. We propose a neural geometric parser learning single image camera calibration for man-made scenes. Unlike previous neural approaches that rely only on semantic cues obtained from neural networks, our approach considers both semantic and geometric cues, resulting in significant accuracy improvement. The proposed framework consists of two networks. Using line segments of an image as geometric cues, the first network estimates the zenith vanishing point and generates several candidates consisting of the camera rotation and focal length. The second network evaluates each candidate based on the given image and the geometric cues, where prior knowledge of man-made scenes is used for the evaluation. With the supervision of datasets consisting of the horizontal line and focal length of the images, our networks can be trained to estimate the same camera parameters. Based on the Manhattan world assumption, we can further estimate the camera rotation and focal length in a weakly supervised manner. The experimental results reveal that the performance of our neural approach is significantly higher than that of existing state-of-the-art camera calibration techniques for single images of indoor and outdoor scenes.

Keywords: Single image camera calibration · Neural geometric parser · Horizon line · Focal length · Vanishing Points · Man-made scenes

1 Introduction

This paper deals with the problem of inferring camera calibration parameters from a single image. It is used in various applications of computer vision and graphics, including image rotation correction [11], perspective control [19], camera rotation estimation [35], metrology [8], and 3D vision [14, 21] (Fig. 1). Due to its importance, single image camera calibration has been revisited in various ways.

Electronic supplementary material The online version of this chapter (https://doi.org/10.1007/978-3-030-58610-2_32) contains supplementary material, which is available to authorized users.

© Springer Nature Switzerland AG 2020
A. Vedaldi et al. (Eds.): ECCV 2020, LNCS 12357, pp. 541–557, 2020.
https://doi.org/10.1007/978-3-030-58610-2_32

Fig. 1. Applications of the proposed framework: (a) image rotation correction, (b) perspective control, and virtual object insertions with respect to the (c) horizon and (d) VPs; before (top) and after (bottom).

Conventional approaches focus on reasoning vanishing points (VPs) in images by assembling geometric cues in the images. Most methods find straight line segments in the images using classic image processing techniques [2,13] and then estimate the VPs by carefully selecting parallel or orthogonal segments in the 3D scene as geometric cues [19]. In practice, however, line segments found in images contain a large amount of noisy data, and it is therefore important to carefully select an inlier set of line segments for the robust detection of VPs [12,26]. Because the accuracy of the inlier set is an important performance indicator, the elapsed time may exponentially increase if stricter criteria are applied to draw the inlier set.

Recently, several studies have proposed estimating camera intrinsic parameters using semantic cues obtained from deep neural networks. It has been investigated [16,29,30] that well-known backbone networks, such as ResNet [15] and U-Net [23], can be used to estimate the focal length or horizon line of an image without significant modifications of the networks. In these approaches, however, it is difficult to explain which geometric interpretation inside of the networks infers certain camera parameters. In several studies [31,35], neural networks were designed to infer geometric structures; however, they required a new convolution operator [35] or 3D supervision datasets [31].

In this paper, we propose a novel framework for single image camera calibration that combines the advantages of both conventional and neural approaches. The basic idea is for our network to leverage line segments to reason camera parameters. We specifically focus on calibrating camera parameters from a single image of a man-made scene. By training with image datasets annotated with horizon lines and focal lengths, our network infers pitch, roll, and focal lengths (3DoF) and can further estimate camera rotations and focal lengths through three VPs (4DoF).

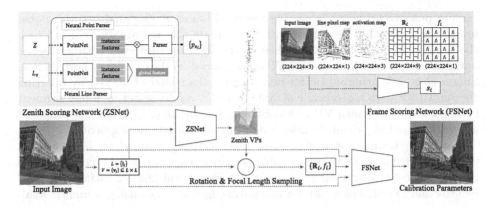

Fig. 2. Overview of the proposed neural geometric parser.

The proposed framework consists of two networks. The first network, the Zenith Scoring Network (ZSNet), takes line segments detected from the input image and deduces reliable candidates of parallel world lines along the zenith VP. Then, from the lines directed at the zenith VP, we generate candidate pairs consisting of a camera rotation and a focal length as inputs of the following step. The second network, the Frame Scoring Network (FSNet), evaluates the score of its input in conjunction with the given image and line segment information. Here, geometric cues from the line segments are used as prior knowledge about the man-made scenes in our network training. This allows us to obtain significant improvement over previous neural methods that only use semantic cues [16, 30]. Furthermore, it is possible to estimate camera rotation and focal length in a *weakly supervised* manner based on the Manhattan world assumption, as we reason camera parameters with pairs consisting of a camera rotation and a focal length. It should be noted that the ground truth for our supervisions is readily available with Google Street View [1] or with consumer-level devices possessing a camera and an inertial measurement unit (IMU) sensor, in contrast to the method in [31], which requires 3D supervision datasets.

2   Related Work

Projective geometry [14, 21] has historically stemmed from the study on imaging of perspective distortions occurring in the human eyes when one observes man-made architectural scenes [3]. In this regard, conventional methods of single image camera calibration [7, 10, 18, 19, 24, 27, 28, 32] involve extracting line segments from an image, inferring the combinations of world parallel or orthogonal lines, identifying two or more VPs, and finally estimating the rotation and focal length of the camera. LSD [13] or EDLine [2] were commonly used as effective line segment detectors, and RANSAC [12] or J-Linkage [26] were adopted to identify VPs describing as many of the extracted line segments as possible. Lee *et al.* [19] proposed robust estimation of camera parameters and automatic adjustment of camera poses to achieve perspective control. Zhai et al. [34] analyzed the global image context with a neural network to estimate the probability field in which the horizon line was formed. In their work, VPs were inferred with geometric optimization, in which horizontal VPs were placed on the estimated horizon line. Simon et al. [25] achieved better performance than Zhai et al. [34] by inferring the zenith VP with a geometric algorithm and carefully selecting a line segment orthogonal to the zenith VP to identify the horizon line. Li et al. [20] proposed a quasi-optimal algorithm to infer VPs from annotated line segments.

Recently, neural approaches have been actively studied to infer camera parameters from a single image using semantic cues learned by convolutional neural networks. Workman et al. proposed DeepFocal [29] for estimating focal lengths and DeepHorizon [30] for estimating horizon lines using semantic analyses of images with neural networks. Hold-Geoffroy et al. [16] trained a neural classifier that jointly estimates focal lengths and horizons. They demonstrated that their joint estimation leads to more accurate results than those produced by independent estimations [29,30]. Although they visualized how the convolution filters react near the edges (through the method proposed by Zeiler and Fergus [33]), it is difficult to intuitively understand how the horizon line is geometrically determined through the network. Therefore, [16,29,30] have a common limitation that it is non-trivial to estimate VPs from the network inference results. Zhou et al. [35] proposed NeurVPS that infers VPs with conic convolutions for a given image. However, NeurVPS [35] assumes normalized focal lengths and does not estimate focal lengths.

Inspired by UprightNet [31], which takes geometric cues into account, we propose a neural network that learns camera parameters by leveraging line segments. Our method can be compared to Lee et al. [19] and Zhai et al. [34], where line segments are used to infer the camera rotation and focal length. However, in our proposed method, the entire process is designed with neural networks. Similar to Workman et al. [30] and Hold-Geoffroy et al. [16], we utilize semantic cues from neural networks but our network training differs in that line segments are utilized as prior knowledge about man-made scenes. Unlike UprightNet [16], which requires the supervisions of depth and normal maps for learning roll/pitch (2DoF), the proposed method learns the horizon and focal length (3DoF) with supervised learning and the camera rotation and focal length (4DoF) with weakly supervised learning.

The relationship between our proposed method and the latest neural RANSACs [5,6,17] is described below. Our ZSNet is related to neural-guided RANSAC [6] in that it updates the line features with backpropagation when learning to sample zenith VP candidates. In addition, our FSNet is related to DSAC [5] in that it evaluates each input pair consisting of a camera rotation and focal length based on the hypothesis on man-made scenes. Our work differs from CONSAC [17], which requires the supervision of all VPs, as we focus on learning single image camera calibrations from the supervision of horizons and focal lengths.

3 Neural Geometric Parser for Camera Calibration

From a given input image, our network estimates up to four camera intrinsic and extrinsic parameters; the focal length f and three camera rotation angles ψ, θ, ϕ. Then, a 3D point $(P_x, P_y, P_z)^T$ in the world coordinate is projected onto the image plane as follows:

$$\begin{bmatrix} p_x \\ p_y \\ p_w \end{bmatrix} = (\mathbf{KR}) \begin{bmatrix} P_x \\ P_y \\ P_z \end{bmatrix}, \text{ where } \mathbf{K} = \begin{bmatrix} f & 0 & c_u \\ 0 & f & c_v \\ 0 & 0 & 1 \end{bmatrix} \text{ and } \mathbf{R} = \mathbf{R}_\psi \mathbf{R}_\theta \mathbf{R}_\phi, \quad (1)$$

where $(p_x, p_y, p_w)^T$ represents the mapped point in the image space, and $\mathbf{R}_\psi$, $\mathbf{R}_\theta$, $\mathbf{R}_\phi$ represent the rotation matrices along x-, y-, and z-axes, with rotation angles ψ, θ, ϕ, respectively. The principal point is assumed to be on the image center such that $c_u = W/2$ and $c_v = H/2$, where W and H represent the width and height of the image, respectively.

Under the Manhattan world assumption, calibration can be done once we obtain the *Manhattan directions*, which are three VPs corresponding to x-, y-, and z-directions in 3D [7]. In Sect. 3.1, we describe how to extract a set of candidate VPs along the zenith direction. Then, in Sect. 3.2, we present our weakly supervised method for estimating all three directions and calibrating the camera parameters.

We use LSD [13] as a line segment detector in our framework. A line segment is represented by a pair of points in the image space. Before estimating the focal length in Sect. 3.2, we assume that each image is transformed into a pseudo camera space as $\mathbf{p} = \mathbf{K}_p^{-1} (p_x, p_y, p_w)^T$, where $\mathbf{K}_p$ represents a pseudo camera intrinsic matrix of $\mathbf{K}$, built by assuming f as $\min(W, H)/2$.

3.1 Zenith Scoring Network (ZSNet)

We first explain our ZSNet, which is used to estimate the zenith VP (see Fig. 2 top-left). Instead of searching for a single zenith VP, we extract a set of candidates that are sufficiently close to the ground truth.

Similar to PointNet [22], ZSNet takes sets of unordered vectors in 2D homogeneous coordinates - line equations and VPs - as inputs. Given a line segment, a line equation $\mathbf{l}$ can be computed as a cross product of its two endpoints:

$$\mathbf{l} = [\mathbf{p}_0]_\times \mathbf{p}_1, \quad (2)$$

where $[\cdot]_\times$ represents a skew-symmetric matrix of a vector. A candidate VP $\mathbf{v}$ can then be computed as an intersection point of the two lines:

$$\mathbf{v} = [\mathbf{l}_0]_\times \mathbf{l}_1. \quad (3)$$

Motivated by [25], we sample a set of line equations roughly directed to the zenith $L_z = \{\mathbf{l}_0, \ldots, \mathbf{l}_{|L_z|}\}$ from the line segments, using the following equation:

$$\left| \tan^{-1} \left(-\frac{a}{b} \right) \right| > \tan^{-1} (\delta_z), \quad (4)$$

where $\mathbf{l} = (a, b, c)^T$ represents a line equation as in Eq. (2) and the angle threshold δ_z is set to $67.5°$ as recommended in [25]. Then, we randomly select pairs of line segments from L_z and compute their intersection points as in Eq. (3) to extract a set of zenith VP candidates $Z = \{\mathbf{z}_0, \dots, \mathbf{z}_{|Z|}\}$. Finally, we feed L_z and Z to ZSNet. We set both the number of samples, $|L_z|$ and $|Z|$, 256 in the experiments.

The goal of our ZSNet is to score each zenith candidate in Z; 1 if a candidate is sufficient close to the ground truth zenith, and 0 otherwise. Fig. 2 top-left shows the architecture of our ZSNet.

In the original PointNet [22], each point is processed independently, except for transformer blocks, to generate point-wise features. A global max pooling layer is then applied to aggregate all the features and generate a global feature. The global feature is concatenated to each point-wise feature, followed by several neural network blocks to classify/score each point.

In our network, we also feed the set of zenith candidates Z to the network, except that we do not compute the global feature from Z. Instead, we use another network, feeding the set of line equations L_z, to extract the global feature of L_z that is then concatenated with each point-wise feature of Z (Fig. 2, top-left).

Let $h_z(\mathbf{z}_i)$ be a point-wise feature of the point $\mathbf{z}_i$ in Z, where $h_z(\cdot)$ represents a PointNet feature extractor. Similarly, let $h_l(L_z) = \{h_l(\mathbf{l}_0), \dots, h_l(\mathbf{l}_{|L_z|})\}$ be a set of features of L_z. A global feature $\mathbf{g}_l$ of $h_l(L_z)$ is computed via a global max-pooling operation (gpool), and is concatenated to $h_z(\mathbf{z}_i)$ as follows:

$$\mathbf{g}_l = \text{gpool}\,(h_l(L_z)) \tag{5}$$

$$h'_z(\mathbf{z}_i) = \mathbf{g}_l \otimes h_z(\mathbf{z}_i), \tag{6}$$

where $\otimes$ represents the concatenation operation.

Finally, the concatenated features are fed into a scoring network computing $[0, 1]$ scores such that:

$$p_{z_i} = \text{sigmoid}\,(s_z(h'_z(\mathbf{z}_i))), \tag{7}$$

where p_{z_i} represents the computed scores of each zenith candidate. The network $s_z(\cdot)$ in Eq. (7) consists of multiple MLP layers, similar to the latter part of the PointNet [22] segmentation architecture.

To train the network, we assign a ground truth label y_i to each zenith candidate $\mathbf{z}_i$ using the following equation:

$$y_i = \begin{cases} 1 & \text{if } \text{cossim}(\mathbf{z}_i, \mathbf{z}_{gt}) > \cos(\delta_p) \\ 0 & \text{if } \text{cossim}(\mathbf{z}_i, \mathbf{z}_{gt}) < \cos(\delta_n) \end{cases}, \tag{8}$$

where $\text{cossim}(x, y) = \frac{|x \cdot y|}{\|x\|\|y\|}$ and $\mathbf{z}_{gt}$ represents the ground truth zenith. The two angle thresholds δ_p and δ_n are empirically selected as $2°$ and $5°$, respectively, from our experiments. The zenith candidates each of which y_i is undefined are not used in the training. The cross entropy loss is used to train the network as follows:

$$\mathcal{L}_{cls} = \frac{1}{N} \sum_i^N -y_i \log(p_{z_i}). \tag{9}$$

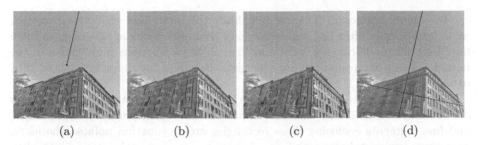

    (a)   (b)   (c)   (d)

Fig. 3. Sampling horizontal line segments into two groups: (a) Based on a zenith VP representative (green), we want to classify the line segments (blue) of a given image. (b) Line segments are classified as follows: vanishing lines of the zenith VP (green) and the remaining lines (red). (c) Junction points (cyan) are computed as the intersections of spatially adjacent line segments that are classified differently; line segments whose endpoints are close to junction points are selected. (d) Using a pseudo-horizon (dotted line), we divide horizontal line segments into two groups (magenta and cyan). (Color figure online)

To better train our ZSNet we use another loss in addition to the cross entropy. Specifically, we constrain the weighted average of zenith candidates close to the ground truth, where estimated scores p_{z_i} are used as weights. To average the zenith candidates, which represent vertical directions of the scene, we use structure tensors of l_2 normalized 2D homogeneous points. Given a 2D homogeneous point $\mathbf{v} = (v_x, v_y, v_w)^T$, a structure tensor of the normalized point is computed as follows:

$$\mathrm{ST}(\mathbf{v}) = \frac{1}{(v_x^2 + v_y^2 + v_w^2)} \begin{bmatrix} v_x^2 & v_x v_y & v_x v_w \\ v_x v_y & v_y^2 & v_y v_w \\ v_x v_w & v_y v_w & v_w^2 \end{bmatrix}, \tag{10}$$

The following loss is used in our network:

$$\mathcal{L}_{loc} = \left\| \mathrm{ST}(\mathbf{z}_{gt}) - \overline{\mathrm{ST}}(\mathbf{z}) \right\|_F, \quad \overline{\mathrm{ST}}(\mathbf{z}) = \frac{\sum_i p_{z_i} \mathrm{ST}(\mathbf{z}_i)}{\sum_i p_{z_i}} \tag{11}$$

where $\| \cdot \|_F$ represents the Frobenius norm. Finally, we select zenith candidates whose scores p_{z_i} are larger than δ_c to the set of zenith candidates, as:

$$Z_c = \{ \mathbf{z}_i \mid p_{z_i} > \delta_c \}, \tag{12}$$

where $\delta_c = 0.5$ in our experiments. The set Z_c is then used in our FSNet.

3.2 Frame Scoring Network (FSNet)

After we extract a set of zenith candidates, we estimate the remaining two horizontal VPs taking into account the given set of zenith VP candidates. We first generate a set of hypotheses on all three VPs. Each hypothesis is then scored by our FSNet.

To sample horizontal VPs, we first filter the input line segments. However, we cannot simply filter line segments using their directions in this case, as there may be multiple horizontal VPs, and lines in any directions may vanish in the horizon. As a workaround, we use a heuristic based on the characteristics of most urban scenes.

Many man-made structures contain a large number of rectangles (e.g., facades or windows of a building) that are useful for calibration parameter estimation, and line segments enclosing these rectangles create junction points. Therefore, we sample horizontal direction line segments by only using their endpoints when they are close to the endpoints of the estimated vertical vanishing lines.

Figure 3 illustrates the process of sampling horizontal line segments into two groups. Let $\mathbf{z}_{est} = (z_x, z_y, z_w)$ be a representative of the estimated zenith VPs, which is computed as the eigenvector with the largest eigenvalue of $\overline{\mathrm{ST}}(\mathbf{z})$ in Eq. (11). We first draw a pseudo-horizon by using $\mathbf{z}_{est}$ and then compute the intersection points between each sampled line segment and the pseudo-horizon. Finally, using a line connecting $\mathbf{z}_{est}$ and the image center as a pivot, we divide horizontal line segments into two groups; one that intersects the pseudo-horizon on the left side of the pivot and the other that intersects the pseudo-horizon on the right side of the pivot. The set of horizontal VP candidates is composed of intersection points by randomly sampling pairs of horizontal direction line segments in each group. We sample an equal number of candidates for both groups.

Once the set of horizontal VP candidates is sampled, we sample candidates of Manhattan directions. To sample each candidate, we draw two VPs; one from zenith candidates and the other from either set of horizontal VP candidates. The calibration parameters for the candidate can then be estimated by solving Eq. (1) with the two VPs, assuming that the principal point is on the image center [18].

We design our FSNet for inferring camera calibration parameters to utilize all the available data, including VPs, lines, and the original raw image (Fig. 2, top-right). ResNet [15] is adapted to our FSNet to handle raw images, appending all the other data as additional color channels. To append the information of the detected line segments, we rasterize line segments as a binary line segment map whose width and height are the same as those of the input image, as follows:

$$\mathbf{L}(u, v) = \begin{cases} 1 & \text{if a line } \mathbf{l} \text{ passes through } (u, v) \\ 0 & \text{otherwise} \end{cases}, \qquad (13)$$

where (u, v) represents a pixel location of the line segment map. We also append the information of vanishing line segments (i.e., expanding lines are close to a VP) as a weighted line segment map for all three VPs of a candidate, where weights are computed using the closeness between the line segments and VPs. For a given VP $\mathbf{v}$ and the line equation $\mathbf{l}$ of a line segment, we compute the closeness between $\mathbf{v}$ and $\mathbf{l}$ using the conventional line-point distance as follows:

$$\text{closeness}(\mathbf{l}, \mathbf{v}) = 1 - \frac{|\mathbf{l} \cdot \mathbf{v}|}{\|\mathbf{l}\| \|\mathbf{v}\|}. \qquad (14)$$

Three activation maps are drawn for each candidate (x-, y- and z-directions), as:

$$\mathbf{A}_{\{x|y|z\}}(u, v) = \begin{cases} \text{closeness}(\mathbf{l}, \mathbf{v}_{\{x|y|z\}}) & \text{if a line } \mathbf{l} \text{ passes through } (u, v) \\ 0 & \text{otherwise} \end{cases}. \quad (15)$$

All the maps are appended to the original image as illustrated in Fig. 2. Finally, we append the Manhattan directions and the estimated focal length to each pixel of the concatenated map so that the input to the scoring network have size of $(height, width, channel) = (224, 224, 17)$ (Fig. 2, top-right).

To train FSNet, we assign GT score to each candidate by measuring similarities between horizon and zenith of each candidate and those of the GT. For the zenith, we measure the cosine similarities of GT zenith and that of candidate as follows:

$$s_{z_i} = \text{cossim}(\mathbf{z}_{gt}, \mathbf{z}_i), \quad (16)$$

where $\mathbf{z}_{gt}$ and $\mathbf{z}_i$ represent the GT and candidate zenith. For the horizon, we adapt the distance metric proposed in [4]. For this, we compute the intersection points between the GT and candidate horizons and left/right image boundaries. Let $\mathbf{h}_l$ and $\mathbf{h}_r$ be intersection points of the predicted horizon and left/right border of the image. Similarly, we compute $\mathbf{g}_l$ and $\mathbf{g}_r$ using the ground truth horizon. Inspired by [4], the similarity between the GT and a candidate is computed as:

$$s_{h_i} = \exp\left(-\max\left(\|\mathbf{h}_{l_i} - \mathbf{g}_l\|_1, \|\mathbf{h}_{r_i} - \mathbf{g}_r\|_1\right)^2\right). \quad (17)$$

Our scoring network h_{score} is then trained with the cross entropy loss, defined as:

$$\mathcal{L}_{score} = \sum_i -h_{score}(\mathbf{R}_i) \log(c_i) \quad (18)$$

$$c_i = \begin{cases} 0 & \text{if } s_{vh_i} < \delta_s \\ 1 & \text{otherwise} \end{cases} \quad (19)$$

$$s_{vh_i} = \exp\left(-\frac{\left(\frac{s_{h_i}+s_{v_i}}{2} - 1.0\right)^2}{2\sigma^2}\right), \quad (20)$$

where $\sigma = 0.1$ and $\delta_s = 0.5$ in our experiments.

Robust Score Estimation Using the Manhattan World Assumption. Although our FSNet is able to accurately estimate camera calibration parameters in general, it can sometimes be noisy and unstable. In our experiments, we found that incorporating with the Manhattan world assumption increased the robustness of our network. Given a line segment map and three closeness maps (Eqs. (14) and (15)), we compute the extent to which a candidate follows the Manhattan world assumption using Eq. (21):

$$m_i = \frac{\sum_u \sum_v \max\left(\mathbf{A}_x(u, v), \mathbf{A}_y(u, v), \mathbf{A}_z(u, v)\right)}{\sum_u \sum_v \mathbf{L}(u, v)}, \quad (21)$$

and the final score of a candidate is computed as:

$$s_i = s_{vh_i} \cdot m_i. \tag{22}$$

Once all the candidates are scored, we estimate the final focal length and zenith by averaging those of top-k high score candidates such that:

$$f_{est} = \frac{\sum_i s_i f_i}{\sum_i s_i} \quad \text{and} \quad \overline{ST}_{est} = \frac{\sum_i s_i ST(z_i)}{\sum_i s_i}, \tag{23}$$

where $ST(\mathbf{z}_i)$ represents the structure tensor of a zenith candidate (Eq. (10)). We set $k = 8$ in the experiments. $\mathbf{z}_{est}$ can be estimated from $\overline{ST}_{est}$, and the two camera rotation angles ψ and ϕ in Eq. (1) can be computed from f_{est} and $\mathbf{z}_{est}$. For the rotation angle θ, we simply take the value from the highest score candidate, as there may be multiple pairs of horizontal VPs that are not close to each other, particularly when the scene does not follow the Manhattan world assumption but the Atlanta world assumption. Note that they still share similar zeniths and focal lengths [4,19].

3.3   Training Details and Runtime

In training, we train ZSNet first and then train FSNet with the outputs of ZSNet. Both networks are trained with Adam optimizer with initial learning rates of 0.001 and 0.0004 for ZSNet and FSNet, respectively. Both learning rates are decreased by half for every 5 epochs. The mini-batch sizes of ZSNet and FSNet are set to 16 and 2, respectively. The input images are always downsampled to 224×224, and the LSD [13] is computed on these low-res images.

At test time, it takes ~0.08 s for data preparation with Intel Core i7-7700K CPU and another ~0.08 s for ZSNet/FSNet per image with Nvidia GeForce GTX 1080 Ti GPU.

4   Experiments

We provide the experimental results with Google Street View [1] and HLW [30] datasets. Refer to the supplementary material for more experimental results with the other datasets and more qualitative results.

Google Street View [1] **Dataset.** It provides panoramic images of outdoor city scenes for which the Manhanttan assumption is satisfied. For generating training and test data, we first divide the scenes for each set and rectify and crop randomly selected panoramic images by sampling FoV, pitch, and roll in the ranges of $40 \sim 80°$, $-30 \sim 40°$, and $-20 \sim 20°$, respectively. 13,214 and 1,333 images are generated for training and test sets, respectively.

HLW [30] **Dataset.** It includes images only with the information of horizon line but no other camera intrinsic parameters. Hence, we use this dataset only for verifying the generalization capability of methods at test time.

Evaluation Metrics. We measure the accuracy of the output camera up vector, focal length, and horizon line with several evaluation metrics. For camera up

Table 1. Supervision and output characteristics of baseline methods and ours. The first two are unsupervised methods not leveraging neural networks, and the others are deep learning methods. Ours is the only network-based method predicting all four outputs.

Method	Supervision				Output			
	Horizon Line	Focal Length	Camera Rotation	Per-Pixel Normal	Horizon Line	Focal Length	Camera Rotation	Up Vector
Upright [19]	N/A				✓	✓	✓	✓
A-Contrario [25]					✓	✓	✓	✓
DeepHorizon [30]	✓				✓			
Perceptual [16]	✓	✓			✓	✓		✓
UprightNet [31]			✓	✓				✓
Ours	✓	✓			✓	✓	✓	✓

Table 2. Quantitative evaluations with Google Street View dataset [1]. See **Evaluation Metrics** in Sect. 4 for details. Bold is the best result, and underscore is the second-best result. Note that, for DeepHorizon [30]*, we use GT FoV to calculate the camera up vector (angle, pitch, and roll errors) from the predicted horizon line. Also, for UprightNet [31]**, we use a pretrained model on ScanNet [9] due to the lack of required supervision in the Google Street View dataset.

Method	Angle (°) ↓		Pitch (°) ↓		Roll (°) ↓		FoV (°) ↓		AUC
	Mean	Med.	Mean	Med.	Mean	Med.	Mean	Med.	(%) ↑
Upright [19]	3.05	1.92	2.90	1.80	6.19	**0.43**	9.47	4.42	77.43
A-Contrario [25]	3.93	<u>1.85</u>	3.51	<u>1.64</u>	13.98	0.52	-	-	74.25
DeepHorizon [30]*	3.58	3.01	2.76	2.12	1.78	1.67	-	-	80.29
Perceptual [16]	2.73	2.13	2.39	1.78	0.96	0.66	**4.61**	<u>3.89</u>	80.40
Perceptual [16] +L	<u>2.66</u>	2.10	<u>2.31</u>	1.80	<u>0.92</u>	0.93	<u>5.27</u>	3.99	80.40
UprightNet [31]**	28.20	26.10	26.56	24.56	6.22	4.33	-	-	-
Ours	**2.12**	**1.61**	**1.92**	**1.38**	**0.75**	<u>0.47</u>	6.01	**3.72**	**83.12**

vector, we measure the difference of angle, pitch, and roll with the GT. For the focal length, we first convert the output focal length to FoV and measure the angle difference with the GT. Lastly, for the horizon line, analogous to our similarity definition in Eq. (17), we measure the distances between the predicted and GT lines at the left/right boundary of the input image (normalized by the image height) and take the maximum of the two distances. We also report the area under curve (AUC) of the cumulative distribution with the x-axis of the distance and the y-axis of the percentage, as introduced in [4]. The range of x-axis is $[0, 0.25]$.

4.1 Comparisons

We compare our method with six baseline methods, where Table 1 presents the required supervision characteristics and outputs. Upright [19] and A-Contrario

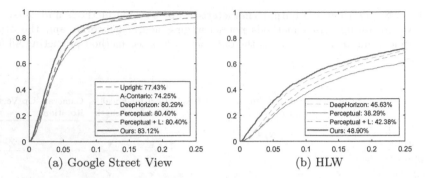

(a) Google Street View   (b) HLW

Fig. 4. Comparison of the cumulative distributions of the horizon line error and their AUCs tested on (a) Google Street View and (b) HLW. Note that, in (b), neural approaches are trained with the Google Street View training dataset and to demonstrate the generalization capability. The AUCs in (a) are also reported in Table 2.

Detection [25] are non-neural-net methods based on line detection and RANSAC. We use the authors' implementations in our experiments. For Upright [19], the evaluation metrics are applied after optimizing the Manhattan direction per image, assuming the principal point as the image center. A-Contrario Detection [25] often fails to estimate focal length when horizon VP candidates are insufficient in the input. Thus, we exclude [25] in the evaluation of FoV. The AUC of [25] is measured, regardless of the failures in focal length estimations, with the horizon lines estimated by the method proposed in [25]. The other metrics of [25] are evaluated only for the cases that focal lengths are successfully estimated. DeepHorizon [30] and Perceptual Measure [16] are neural-network-based methods directly taking the global feature of an image and performing classifications in discretized camera parameter spaces. For fair comparisons, we use ResNet [15] as a backbone architecture in the implementations of these methods. Note that DeepHorizon [30] does not predict focal length, and thus we use ground truth focal length in the estimation of the camera up vector. Additionally, we train Perceptual Measure [16] by feeding it both the input image and the line map used in our FSNet (Sect. 3.2), and we assess whether the extra input improves the performance. UprightNet [31] is another deep learning method that requires additional supervision in training, such as camera extrinsic parameters and per-pixel normals in the 3D space. Due to the lack of such supervision in our datasets, in our experiments, we use the author's pretrained model on ScanNet [9].

The quantitative results with Google Street View dataset [1] are presented in Table 2. The results demonstrate that our method outperforms all the baseline methods in most evaluation metric. Upright [19] provides a slightly lower median roll than ours, although its mean roll is much greater than the median, meaning that it completely fails in some test cases. In addition, Perceptual Measure [16] gives a slightly smaller mean FoV error; however, the median FoV error is higher than ours. When Perceptual Measure [16] is trained with the additional line map

■ ■ ■ ■ Ground Truth ■ ■ ■ ■ Upright [19]   ■ ■ ■ ■ A-Contario [25]   ■ ■ ■ ■ DeepHorizon [30]
■ ■ ■ ■ Perceptual [16] ■ ■ ■ ■ Perceptual [16] +**L** ■ ■ ■ ■ Ours

Fig. 5. Examples of horizon line prediction on the Google Street View test set (top two rows) and on the HLW test set (bottom two rows). Each example also shows the Manhattan direction of the highest score candidate.

input, the result indicates that it does not lead to a meaning difference in performance. As mentioned earlier, a pretrained model is used for UprightNet [31] (trained on ScanNet [9]) due to the lack of required supervision in Google Street View; thus the results are much poorer than others.

Figure 5 visualizes several examples of horizon line predictions as well as our weakly-supervised Manhattan directions. Recall that we do not use *full* supervision of the Manhattan directions in training; we only use the supervision of horizon lines and focal lengths. In each example in Fig. 5, we illustrate the Manhattan direction of the highest score candidate.

To evaluate the generalization capability of neural-network-based methods, we also take the network models trained on Google Street View training dataset and test them on the HLW dataset [30]. Because the HLW dataset only has the GT horizon line, Fig. 4(b) only reports the cumulative distributions of the horizon prediction errors. As shown in the figure, our method provides the largest AUC with a significant margin compared with the other baselines. Interestingly, Perceptual Measure [16] shows improvement when trained with the additional line map, meaning that the geometric interpretation helps more when parsing *unseen* images in network training.

Table 3. Evaluation of ZSNet.

	Angle (°) ↓	
	Mean	Med.
ZSNet (Ours)	**2.53**	**1.52**
Upright [19]	3.15	2.11
A-Contrario [25]	3.06	2.38

Fig. 6. Visualizations of FSNet focus: (left) input; (right) feature highlight. (Color figure online)

We conduct an experiment comparing the outputs of ZSNet with the outputs of Upright [19] and A-Contrario [25]. Using the weighted average of the zenith candidates (in Eq. (11)), we measured the angle to the GT zenith $\text{cossim}(z_i, z_{gt})$, as provided in Eq. (8). Table 3 show that our ZSNet computes the zenith VP more accurately than the other non-neural-net methods.

Figure 6 visualizes the weights of the second last convolution layer in FSNet (the layer in the ResNet backbone); red means high, and blue means low. It can be seen that our FSNet focused on the areas with many line segments, such as buildings, window frames, and pillars. The supplementary material contains more examples.

4.2 Ablation Study

We conduct an ablation study using Google Street View dataset to demonstrate the effect of each component in our framework. All results are reported in Table 4, where the last row shows the result of our final version framework.

We first evaluate the effect of the entire ZSNet by ablating it in the training. When sampling zenith candidates in FSNet (Sect. 3.2), the score p_{z_i} in Eq. (7) is not predicted but set uniformly. The first row of Table 4 indicates that the performance significantly decreases in all evaluation metrics; e.g., the AUC decreased by 9%. This indicates that ZSNet plays an important role in finding inlier Zenith VPs that satisfy the Manhattan/Atlanta assumption. In ZSNet, we also evaluate the effect of global line feature $\mathbf{g}_l$ in Eq. (5) by not concatenating it with the point feature $h_z(\mathbf{z}_i)$ in Eq. (6). Without the line feature, ZSNet is still able to prune the outlier zeniths in some extent, as indicated in the second row, but the performance is far inferior to that of our final framework (the last row). This result indicates that the equation of a horizon line is much more informative than the noisy coordinates of the zenith VP.

In FSNet, we first ablate some parts of the input fed to the network per frame. When we do not provide the given image but the rest of the input (the third row of Table 4), the performance decreases somewhat; however, the change is less significant than when omitting the line map $\mathbf{L}$ (Eq. (13)) and the activation map $\mathbf{A}$ (Eq. (15)) in the input (the fourth row). This demonstrates that FSNet learns more information from the line map and activation map, which contain explicit geometric interpretations of the input image. The combination of the

Table 4. Ablation study results. Bold is the best result. See Sect. 4.2 for details.

	Angle (°) ↓		Pitch (°) ↓		Roll (°) ↓		FoV (°) ↓		AUC
	Mean	Med.	Mean	Med.	Mean	Med.	Mean	Med.	(%) ↑
w/o ZSNet	3.00	2.04	2.81	1.98	1.62	0.95	8.42	4.47	74.01
$h'_z(\mathbf{z}_i) = h_z(\mathbf{z}_i)$ (Eq. (6))	4.34	1.96	3.91	1.76	1.64	0.59	7.88	4.16	77.65
FSNet−Image	2.45	1.78	2.19	1.52	**0.68**	**0.47**	6.71	4.35	80.20
FSNet−L − A	3.74	2.22	3.09	1.91	1.68	0.66	8.26	5.40	74.31
$s_i = s_{vh_i}$ (Eq. (22))	2.32	1.80	2.09	1.57	0.72	0.54	6.06	4.12	80.85
Ours	**2.12**	**1.61**	**1.92**	**1.38**	0.75	**0.47**	**6.01**	**3.72**	**83.12**

two maps (our final version) produces the best performance. In addition, the results get worse when the activation map score m_i is not used in the final score of candidates—i.e., $s_i = s_{vh_i}$ in Eq. (22) (the fifth row).

5   Conclusion

In this paper, we introduced a neural method that predicts camera calibration parameters from a single image of a man-made scene. Our method fully exploits line segments as prior knowledge of man-made scenes, and in our experiments, it exhibited better performance than that of previous approaches. Furthermore, compared to previous neural approaches, our method demonstrated a higher generalization capability to unseen data. In future work, we plan to investigate neural camera calibration that considers a powerful but small number of geometric cues through analyzing image context, as humans do.

Acknowledgements. This research was supported by the National Research Foundation of Korea (NRF) funded by the Ministry of Education (2017R1D1A1B03034907).

References

1. Google Street View Images API. https://developers.google.com/maps/
2. Akinlar, C., Topal, C.: EDLines: a real-time line segment detector with a false detection control. Pattern Recogn. Lett. **32**(13), 1633–1642 (2011)
3. Alberti, L.B.: Della Pittura (1435)
4. Barinova, O., Lempitsky, V., Tretiak, E., Kohli, P.: Geometric image parsing in man-made environments. In: Daniilidis, K., Maragos, P., Paragios, N. (eds.) ECCV 2010. LNCS, vol. 6312, pp. 57–70. Springer, Heidelberg (2010). https://doi.org/10.1007/978-3-642-15552-9_5
5. Brachmann, E., et al.: DSAC – differentiable RANSAC for camera localization. In: Proceedings of CVPR, pp. 6684–6692 (2017)
6. Brachmann, E., Rother, C.: Neural-guided RANSAC: learning where to sample model hypotheses. In: Proceedings of ICCV, pp. 4322–4331 (2019)
7. Coughlan, J.M., Yuille, A.L.: Manhattan world: compass direction from a single image by Bayesian inference. In: Proceedings of ICCV, pp. 941–947 (1999)

8. Criminisi, A., Reid, I., Zisserman, A.: Single view metrology. Int. J. Comput. Vis. **40**(2), 123–148 (2000)
9. Dai, A., Chang, A.X., Savva, M., Halber, M., Funkhouser, T., Nießner, M.: Scan-Net: richly-annotated 3D reconstructions of indoor scenes. In: Proceedings of CVPR, pp. 5828–5839 (2017)
10. Denis, P., Elder, J.H., Estrada, F.J.: Efficient edge-based methods for estimating manhattan frames in urban imagery. In: Forsyth, D., Torr, P., Zisserman, A. (eds.) ECCV 2008. LNCS, vol. 5303, pp. 197–210. Springer, Heidelberg (2008). https://doi.org/10.1007/978-3-540-88688-4_15
11. Fischer, P., Dosovitskiy, A., Brox, T.: Image orientation estimation with convolutional networks. In: Gall, J., Gehler, P., Leibe, B. (eds.) GCPR 2015. LNCS, vol. 9358, pp. 368–378. Springer, Cham (2015). https://doi.org/10.1007/978-3-319-24947-6_30
12. Fischler, M.A., Bolles, R.C.: Random sample consensus: a paradigm for model fitting with applications to image analysis and automated cartography. Commun. ACM **24**(6), 381–395 (1981)
13. von Gioi, R.G., Jakubowicz, J., Morel, J.M., Randall, G.: LSD: a fast line segment detector with a false detection control. IEEE Trans. Pattern Anal. Mach. Intell. **32**(4), 722–732 (2010)
14. Hartley, R., Zisserman, A.: Multiple View Geometry in Computer Vision, 2 edn. Cambridge University Press, Cambridge (2003)
15. He, K., Zhang, X., Ren, S., Sun, J.: Deep residual learning for image recognition. In: Proceedings of CVPR, pp. 770–778 (2016)
16. Hold-Geoffroy, Y., et al.: A perceptual measure for deep single image camera calibration. In: Proceedings of CVPR, pp. 2354–2363 (2018)
17. Kluger, F., Brachmann, E., Ackermann, H., Rother, C., Yang, M.Y., Rosenhahn, B.: CONSAC: robust multi-model fitting by conditional sample consensus. In: Proceedings of CVPR, pp. 4633–4642 (2020)
18. Košecká, J., Zhang, W.: Video compass. In: Heyden, A., Sparr, G., Nielsen, M., Johansen, P. (eds.) ECCV 2002. LNCS, vol. 2353, pp. 476–490. Springer, Heidelberg (2002). https://doi.org/10.1007/3-540-47979-1_32
19. Lee, H., Shechtman, E., Wang, J., Lee, S.: Automatic upright adjustment of photographs with robust camera calibration. IEEE Trans. Pattern Anal. Mach. Intell. **36**(5), 833–844 (2014)
20. Li, H., Zhao, J., Bazin, J.C., Chen, W., Liu, Z., Liu, Y.H.: Quasi-globally optimal and efficient vanishing point estimation in manhattan world. In: Proceedings of ICCV, pp. 1646–1654 (2019)
21. Ma, Y., Soatto, S., Kosecka, J., Sastry, S.S.: An Invitation to 3-D Vision: From Images to Geometric Models. Springer, Heidelberg (2004). https://doi.org/10.1007/978-0-387-21779-6
22. Qi, C.R., Su, H., Mo, K., Guibas, L.J.: PointNet: deep learning on point sets for 3D classification and segmentation. In: Proceedings of CVPR, pp. 652–660 (2017)
23. Ronneberger, O., Fischer, P., Brox, T.: U-Net: convolutional networks for biomedical image segmentation. In: Navab, N., Hornegger, J., Wells, W.M., Frangi, A.F. (eds.) MICCAI 2015. LNCS, vol. 9351, pp. 234–241. Springer, Cham (2015). https://doi.org/10.1007/978-3-319-24574-4_28
24. Schindler, G., Dellaert, F.: Atlanta world: an expectation maximization framework for simultaneous low-level edge grouping and camera calibration in complex man-made environments. In: Proceedings of CVPR (2004)

25. Simon, G., Fond, A., Berger, M.-O.: *A-Contrario* horizon-first vanishing point detection using second-order grouping laws. In: Ferrari, V., Hebert, M., Sminchisescu, C., Weiss, Y. (eds.) ECCV 2018. LNCS, vol. 11214, pp. 323–338. Springer, Cham (2018). https://doi.org/10.1007/978-3-030-01249-6_20

26. Tardif, J.P.: Non-iterative approach for fast and accurate vanishing point detection. In: Proceedings of ICCV, pp. 1250–1257 (2009)

27. Tretyak, E., Barinova, O., Kohli, P., Lempitsky, V.: Geometric image parsing in man-made environments. Int. J. Comput. Vis. **97**(3), 305–321 (2012)

28. Wildenauer, H., Hanbury, A.: Robust camera self-calibration from monocular images of manhattan worlds. In: Proceedings of CVPR, pp. 2831–2838 (2012)

29. Workman, S., Greenwell, C., Zhai, M., Baltenberger, R., Jacobs, N.: DEEPFOCAL: a method for direct focal length estimation. In: Proceedings of ICIP, pp. 1369–1373 (2015)

30. Workman, S., Zhai, M., Jacobs, N.: Horizon lines in the wild. In: Proceedings of BMVC, pp. 20.1–20.12 (2016)

31. Xian, W., Li, Z., Fisher, M., Eisenmann, J., Shechtman, E., Snavely, N.: UprightNet: geometry-aware camera orientation estimation from single images. In: Proceedings of ICCV, pp. 9974–9983 (2019)

32. Xu, Y., Oh, S., Hoogs, A.: A minimum error vanishing point detection approach for uncalibrated monocular images of man-made environments. In: Proceedings of CVPR, pp. 1376–1383 (2013)

33. Zeiler, M.D., Fergus, R.: Visualizing and understanding convolutional networks. In: Fleet, D., Pajdla, T., Schiele, B., Tuytelaars, T. (eds.) ECCV 2014. LNCS, vol. 8689, pp. 818–833. Springer, Cham (2014). https://doi.org/10.1007/978-3-319-10590-1_53

34. Zhai, M., Workman, S., Jacobs, N.: Detecting vanishing points using global image context in a non-manhattan world. In: Proceedings of CVPR, pp. 5657–5665 (2016)

35. Zhou, Y., Qi, H., Huang, J., Ma, Y.: NeurVPS: neural vanishing point scanning via conic convolution. In: Proceedings of NeurIPS (2019)

Learning Flow-Based Feature Warping for Face Frontalization with Illumination Inconsistent Supervision

Yuxiang Wei[1], Ming Liu[1], Haolin Wang[1], Ruifeng Zhu[2,4], Guosheng Hu[3],
and Wangmeng Zuo[1,5(✉)]

[1] School of Computer Science and Technology, Harbin Institute of Technology,
Harbin, China
yuxiang.wei.cs@gmail.com, csmliu@outlook.com, Why_cs@outlook.com,
wmzuo@hit.edu.cn
[2] University of Burgundy Franche-Comté, Besançon, France
reefing.z@gmail.com
[3] Anyvision, Belfast, UK
huguosheng100@gmail.com
[4] University of the Basque Country, Eibar, Spain
[5] Peng Cheng Lab, Shenzhen, China

Abstract. Despite recent advances in deep learning-based face frontalization methods, photo-realistic and illumination preserving frontal face synthesis is still challenging due to large pose and illumination discrepancy during training. We propose a novel Flow-based Feature Warping Model (FFWM) which can learn to synthesize photo-realistic and illumination preserving frontal images with illumination inconsistent supervision. Specifically, an Illumination Preserving Module (IPM) is proposed to learn illumination preserving image synthesis from illumination inconsistent image pairs. IPM includes two pathways which collaborate to ensure the synthesized frontal images are illumination preserving and with fine details. Moreover, a Warp Attention Module (WAM) is introduced to reduce the pose discrepancy in the feature level, and hence to synthesize frontal images more effectively and preserve more details of profile images. The attention mechanism in WAM helps reduce the artifacts caused by the displacements between the profile and the frontal images. Quantitative and qualitative experimental results show that our FFWM can synthesize photo-realistic and illumination preserving frontal images and performs favorably against the state-of-the-art results. Our code is available at https://github.com/csyxwei/FFWM.

Keywords: Face frontalization · Illumination preserving · Optical flow · Guided filter · Attention mechanism

Electronic supplementary material The online version of this chapter (https://doi.org/10.1007/978-3-030-58610-2_33) contains supplementary material, which is available to authorized users.

A. Vedaldi et al. (Eds.): ECCV 2020, LNCS 12357, pp. 558–574, 2020.
https://doi.org/10.1007/978-3-030-58610-2_33

1   Introduction

Face frontalization aims to synthesize the frontal view face from a given profile. Frontalized faces can be directly used for general face recognition methods without elaborating additional complex modules. Apart from face recognition, generating photo-realistic frontal face is beneficial for a series of face-related tasks, including face reconstruction, face attribute analysis, facial animation, *etc.*

Traditional methods address this problem through 2D/3D local texture warping [6,36] or statistical modeling [22]. Recently, GAN-based methods have been proposed to recover a frontal face in a data-driven manner [1,8,10,28,31,32, 34,35]. For instance, Yin *et al.* [32] propose DA-GAN to capture the long-displacement contextual information from illumination discrepancy images under large poses. However, it recovers inconsistent illumination on the synthesized image. Flow-based method [33] predicts a dense pixel correspondence between the profile and frontal image and uses it to deform the profile face to the frontal view. However, deforming the profile face in the image space directly leads to obvious artifacts and missing pixels should be addressed under large poses.

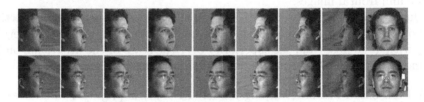

Fig. 1. $\pm 45°$, $\pm 60°$, $\pm 75°$ and $\pm 90°$ images of the two persons in the Multi-PIE. Each row images have the same flash in the recording environment.

The existing methods do not consider the illumination inconsistency between the profile and ground-truth frontal image. Taking the widely used benchmark Multi-PIE [4] as an example, the visual illumination conditions on several poses are significantly different from the ground-truth frontal images as shown in Fig. 1. Except $\pm 90°$, the other face images are produced by the same camera type. The variation in camera types causes obvious illumination inconsistency between the $\pm 90°$ images and the ground-truth frontal image. Although efforts have been made to manually color-balance those same type cameras, the illumination of resulting images within $\pm 75°$ (except $0°$) still look visually distinguishable with the ground-truth frontal image. Since the existing methods minimize pixel-wise loss between the synthesized image and the illumination inconsistent ground-truth, they tend to change both the pose and the illumination of the profile face image, while the latter actually is not acceptable in face editing and synthesis.

To address the above issue, this paper proposes a novel Flow-based Feature Warping Model (FFWM) which can synthesize photo-realistic and illumination preserving frontal image from illumination inconsistent image pairs. In particular, FFWM incorporates the flow estimation with two modules: Illumination Preserving Module (IPM) and Warp Attention Module (WAM). Specifically, we estimate the optical flow fields from the given profile: the reverse and forward flow fields are predicted to warp the front face to the profile view and vice versa, respectively. The estimated flow fields are fed to IPM and WAM to conduct face frontalization.

The IPM is proposed to synthesize illumination preserving images with fine facial details from illumination inconsistent image pairs. Specifically, IPM contains two pathways: (1) Illumination Preserving Pathway and (2) Illumination Adaption Pathway. For (1), an illumination preserving loss equipped with the reverse flow field is introduced to constrain the illumination consistency between synthesized images and the profiles. For (2), guided filter [7] is introduced to further eliminate the illumination discrepancy and learns frontal view facial details from illumination inconsistent ground-truth image. The WAM is introduced to reduce the pose discrepancy in the feature level. It uses the forward flow field to align the profile features to the frontal view. This flow provides an explicit and accurate supervision signal to guide the frontalization. The attention mechanism in WAM helps to reduce the artifacts caused by the displacements between the profile and frontal images.

Quantitative and qualitative experimental results demonstrate the effectiveness of our FFWM on synthesizing photo-realistic and illumination preserving faces with large poses and the superiority over the state-of-the-art results on the testing benchmarks. Our contributions can be summarized as:

- A Flow-based Feature Warping Model (FFWM) is proposed to address the challenging problem in face frontalization, i.e. photo-realistic and illumination preserving image synthesis.
- Illumination Preserving Module (IPM) equipped with guided filter and flow field is proposed to achieve illumination preserving image synthesis. Warp Attention Module (WAM) uses the attention mechanism to effectively reduce the pose discrepancy in the feature level under the explicit and effective guidance from flow estimation.
- Quantitative and qualitative results demonstrate that the proposed FFWM outperforms the state-of-the-art methods.

2   Related Work

2.1   Face Frontalization

Face frontalization aims to synthesize the frontal face from a given profile. Traditional methods address this problem through 2D/3D local texture warping [6,36] or statistical modeling [22]. Hassner et al. [6] employ a mean 3D model for face normalization. A statistical model [22] is used for frontalization and landmark detection by solving a constrained low-rank minimization problem.

Benefiting from deep learning, many GAN-based methods [8,10,27,28,32,33] are proposed for face frontalization. Huang *et al.* [10] use a two-pathway GAN architecture for perceiving global structures and local details simultaneously. Domain knowledge such as symmetry and identity information of face is used to make the synthesized faces photo-realistic. Zhao *et al.* [34] propose PIM with introducing a domain adaptation strategy for pose invariant face recognition. 3D-based methods [1,2,31,35] attempt to combine prior knowledge of 3D face with face frontalization. Yin *et al.* [31] incorporate 3D face model into GAN to solve the problem of large pose face frontalization in the wild. HF-PIM [1] combines the advantages of 3D and GAN-based methods and frontalizes profile images via a novel texture warping procedure. In addition to supervised learning, Qian *et al.* [17] propose a novel Face Normalization Model (FNM) for unsupervised face generation with unpaired face images in the wild. Note that FNM focuses on face normalization, without considering preserving illumination.

Instead of learning function to represent the frontalization procedure, our method gets frontal warped feature by flow field and reconstructs illumination preserving and identity preserving frontal view face.

2.2   Optical Flow

Optical flow estimation has many applications, *e.g.*, action recognition, autonomous driving and video editing. With the progress in deep learning, FlowNet [3], FlowNet2 [13] and others achieve good results by end-to-end supervised learning. While SpyNet [18], PWC-Net [26] and LiteFlowNet [11] also use coarse-to-fine strategery to refine the initial flow. It is worth mentioning that PWC-Net and LiteFlowNet have smaller size and are easier to train. Based on weight sharing and residual subnetworks, Hur and Roth [12] learn bi-directional optical flow and occlusion estimation jointly. Bilateral refinement of flow and occlusion address blurry estimation, particularly near motion boundaries. By the global and local correlation layers, GLU-Net [29] can resolve the challenges of large displacements, pixel-accuracy, and appearance changes.

In this work, we estimate bi-directional flow fields to represent dense pixel correspondence between the profile and frontal faces, which are then exploited to obtain frontal view features and preserve illumination condition, respectively.

3   Proposed Method

Let $\{I, I^{gt}\}$ be a pair of profile and frontal face image of the same person. Given a profile image I, our goal is to train a model $\mathcal{R}$ to synthesize the corresponding frontal face image $\hat{I} = \mathcal{R}(I)$, which is expected to be photo-realistic and illumination preserving. To achieve this, we propose the Flow-based Feature Warping Model (FFWM). As shown in Fig. 2, FFWM takes U-net [20] as the backbone and incorporates with the Illumination Preserving Module (IPM) and the Warp Attention Module (WAM) to synthesize $\hat{I}$. In addition, FFWM uses optical flow fields which are fed to IPM and WAM to conduct frontalization. Specifically, we

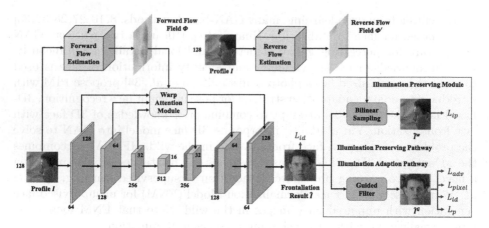

Fig. 2. The architecture of our FFWM. Illumination Preserve Module is incorporated to facilitate synthesized frontal image $\hat{I}$ to be illumination preserving and facial details preserving in two independent pathways. Based on the skip connection, the Warp Attention Module helps synthesize frontal image effectively. Losses are shown in red color, which $\hat{I}^w$ is the synthesized image $\hat{I}$ warped by Φ' and the $\hat{I}^G$ is the guided filter output. (Color figure online)

compute the forward and reverse flow fields to warp the profile to the frontal view and vice versa, respectively.

In this section, we first introduce the bi-directional flow fields estimation in Sect. 3.1. IPM and WAM are introduced in Sect. 3.2 and Sect. 3.3. Finally, the loss functions are detailed in Sect. 3.4.

3.1 Bi-directional Flow Fields Estimation

Face frontalization can be viewed as the face rotation transformation, and the flow field can model this rotation by establishing the pixel-level correspondence between the profile and frontal faces. Traditional optical flow methods [3,13] take two frames as the input. However, we only use one profile image as the input. In this work, we adopt the FlowNetSD in FlowNet2 [13] as our flow estimation network, and change the input channel from 6 (two frames) to 3 (one image). For preserving illumination and frontalization, we estimate the reverse flow field Φ' and the forward flow field Φ from the profile image, respectively.

Reverse Flow Field. Given the profile image I, reverse flow estimation network $\mathcal{F}'$ predicts the reverse flow field Φ' which can warp the ground-truth frontal image I^{gt} to the profile view as I.

$$\Phi' = \mathcal{F}'(I; \Theta_{\mathcal{F}'}), \tag{1}$$

$$I^{w'} = \mathcal{W}(I^{gt}, \Phi'), \tag{2}$$

Where $\Theta_{\mathcal{F}'}$ denotes the parameters of $\mathcal{F}'$, and $\mathcal{W}(\cdot)$ [14] is the bilinear sampling operation. To learn an accurate reverse flow field, $\mathcal{F}'$ is pretrained with the landmark loss [15], sampling correctness loss [19] and the regularization term [19].

Forward Flow Field. Given the profile image I, forward flow estimation network $\mathcal{F}$ predicts the forward flow field Φ which can warp I to the frontal view.

$$\Phi = \mathcal{F}(I; \Theta_{\mathcal{F}}), \tag{3}$$

$$I^w = \mathcal{W}(I, \Phi), \tag{4}$$

Where $\Theta_{\mathcal{F}}$ denotes the parameters of $\mathcal{F}$. To learn an accurate forward flow field, $\mathcal{F}$ is pretrained with the same losses as $\mathcal{F}'$.

Then two flow fields Φ' and Φ are used for the IPM and WAM to generate illumination preserving and photo-realistic frontal images.

3.2 Illumination Preserving Module

Without considering inconsistent illumination in the face datasets, the existing frontalization methods potentially overfit to the wrong illumination. To effectively decouple the illumination and the facial details, hence to synthesize illumination preserving faces with fine details, we propose the Illumination Preserving Module (IPM). As shown in Fig. 2, IPM consists of two pathways. Illumination preserving pathway ensures that the illumination condition of the synthesized image $\hat{I}$ is consistent with the profile I. Illumination adaption pathway ensures that the facial details of the synthesized image $\hat{I}$ are consistent with the ground-truth I^{gt}.

Illumination Preserving Pathway. Because the illumination condition is diverse and cannot be quantified as a label, it is hard to learn reliable and independent illumination representation from face images. Instead of constraining the illumination consistency between the profile and the synthesized image in the feature space, we directly constrain it in the image space. As shown in Fig. 2, in the illumination preserving pathway, we firstly use the reverse flow field Φ' to warp the synthesized image $\hat{I}$ to the profile view,

$$\hat{I}^w = \mathcal{W}(\hat{I}, \Phi'). \tag{5}$$

Then an illumination preserving loss is defined on the warped synthesized image $\hat{I}^w$ to constrain the illumination consistency between the synthesized image $\hat{I}$ and the profile I. By minimizing it, FFWM can synthesize illumination preserving frontal images.

564   Y. Wei et al.

Illumination Adaption Pathway. Illumination preserving pathway cannot ensure the consistency of facial details between the synthesized image $\hat{I}$ and the ground-truth I^{gt}, so we constrain it in the illumination adaption pathway. Since the illumination of profile I is inconsistent with the ground-truth I^{gt} under large poses, adding constraints directly between $\hat{I}$ and I^{gt} eliminates the illumination consistency between $\hat{I}$ and I. So a guided filter layer [7] is firstly used to transfer the illumination of images. Specifically, the guided filter takes I^{gt} as the input image and $\hat{I}$ as the guidance image,

$$\hat{I}^G = \mathcal{G}(\hat{I}, I^{gt}), \tag{6}$$

where $\mathcal{G}(\cdot)$ denotes the guided filter, and we set the radius of filter as the quarter of the image resolution. After filtering, the guided filter result $\hat{I}^G$ has the same illumination with I^{gt} while keeping the same facial details with $\hat{I}$. Then the illumination-related losses (*e.g.*, pixel-wise loss, perceptual loss) are defined on $\hat{I}^G$ to facilitate our model synthesize $\hat{I}$ with much finer details. By this means, $\hat{I}$ can become much more similar to I^{gt} in facial details without changing the illumination consistency between $\hat{I}$ and I.

Note that the guided filter has no trainable parameters and potentially cause our model trap into local minima during training. So we apply the guided filter after several iterations, providing stable and robust initialization to our model.

3.3   Warp Attention Module

The large pose discrepancy makes it difficult to synthesize correct facial details in the synthesized images. To reduce the pose discrepancy between the profile and frontal face, Warp Attention Module (WAM) is proposed to align the profile face to the frontal one in the feature space. We achieve this by warping the profile features guided by the forward flow field Φ. The architecture of our WAM is illustrated in Fig. 3. It contains two steps: flow-guided feature warping and feature attention.

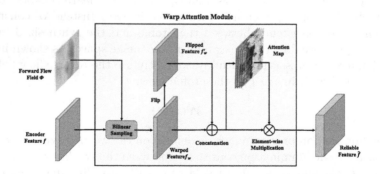

Fig. 3. The architecture of Warp Attention Module. Considering the symmetry prior of human face, WAM also contains flipped warped feature.

Flow-Guided Feature Warping. Because the profile and frontal face have different visible areas, the forward flow field Φ cannot establish a complete pixel-level correspondence between them. Hence, warping profile face directly leads to artifacts. Here we incorporate Φ with bilinear sampling operation $\mathcal{W}(\cdot)$ to warp the profile face to the frontal one in the feature space. Additionally, we use the symmetry prior of human face, and take both warped features and its horizontal flip to guide the frontal image synthesis.

$$f_w = \mathcal{W}(f, \Phi), \tag{7}$$

Where f denotes the encoder feature of the profile. Let $f_w{}'$ denotes the horizontal flip of f_w, and $(f_w \oplus f_w{}')$ denotes the concatenation of f_w and $f_w{}'$.

Feature Attention. After warping, the warped feature encodes the backgrounds and self-occlusion artifacts, which leads to degraded frontalization performance. To eliminate above issue and extract reliable frontal feature, an attention mechanism is then used to adaptively focus on the critical parts of $(f_w \oplus f_w{}')$. The warped feature $(f_w \oplus f_w{}')$ is firstly fed into a Conv-BN-ReLU-ResidualBlock Layer to generate an attention map A, which has the same height, width and channel size with $(f_w \oplus f_w{}')$. Then the reliable frontalized feature $\hat{f}$ is obtained by,

$$\hat{f} = A \otimes (f_w \oplus f_w{}'), \tag{8}$$

where $\otimes$ denotes element-wise multiplication. $\hat{f}$ is then skip connected to the decoder to help generate photo-realistic frontal face image $\hat{I}$.

3.4   Loss Functions

In this section, we formulate the loss functions used in our work. The background of images is masked to make the loss functions focus on the facial area.

Pixel-Wise Loss. Following [8,10], we employ a multi-scale pixel-wise loss on the guided filter result $\hat{I}^G$ to constrain the content consistency,

$$\mathcal{L}_{pixel} = \sum_{s=1}^{S} \left\| \hat{I}_s^G - I_s^{gt} \right\|_1, \tag{9}$$

Where S denotes the number of scales. In our experiments, we set $S = 3$, and the scales are 32×32, 64×64 and 128×128.

Perceptual Loss. Pixel-wise loss tends to generate over-smoothing results. To alleviate this, we introduce the perceptual loss defined on the VGG-19 network [25] pre-trained on ImageNet [21],

$$\mathcal{L}_p = \sum_i w_i \left\| \phi_i(\hat{I}^G) - \phi_i(I^{gt}) \right\|_1, \tag{10}$$

where $\phi_i(\cdot)$ denotes the output of the i-th VGG-19 layer. In our implementation, we use Conv1-1, Conv2-1, Conv3-1, Conv4-1 and Conv5-1 layer, and set $w = \{1, 1/2, 1/4, 1/4, 1/8\}$. To improve synthesized imagery in the particular facial regions, we also use the perceptual loss on the facial regions like eyes, nose and mouth.

Adversarial Loss. Following [24], we adpot a multi-scale discriminator and adversarial learning to help synthesize photo-realistic images.

$$\mathcal{L}_{adv} = \min_R \max_D \mathbb{E}_{I^{gt}}[\log D(I^{gt})] - \mathbb{E}_{\hat{I}^G}[\log(1 - D(\hat{I}^G))]. \tag{11}$$

Illumination Preserving Loss. To preserve the illumination of profile I on synthesized image $\hat{I}$, we define the illumination preserving loss on the warped synthesized image $\hat{I}^w$ at different scales,

$$\mathcal{L}_{ip} = \sum_{s=1}^{S} \left\| \hat{I}_s^w - I_s \right\|_1, \tag{12}$$

Where S denotes the number of scales, and the scale setting is same as Eq. (9).

Identity Preserving Loss. Following [8,10], we present an identity preserving loss to preserve the identity information of the synthesized image $\hat{I}$,

$$\mathcal{L}_{id} = \left\| \psi_{fc2}(\hat{I}) - \psi_{fc2}(I^{gt}) \right\|_1 + \left\| \psi_{pool}(\hat{I}) - \psi_{pool}(I^{gt}) \right\|_1, \tag{13}$$

Where $\psi(\cdot)$ denotes the pretrained LightCNN-29 [30]. $\psi_{fc2}(\cdot)$ and $\psi_{pool}(\cdot)$ denote the outputs of the last pooling layer and the fully connected layer respectively. To preserve the identity information, we add the identity loss on both $\hat{I}$ and $\hat{I}^G$.

Overall Losses. Finally, we combine all the above losses to give the overall model objective,

$$\mathcal{L} = \lambda_0 \mathcal{L}_{pixel} + \lambda_1 \mathcal{L}_p + \lambda_2 \mathcal{L}_{adv} + \lambda_3 \mathcal{L}_{ip} + \lambda_4 \mathcal{L}_{id}, \tag{14}$$

Where λ_* denotes the different losses tradeoff parameters.

4   Experiments

To illustrate our model can synthesize photo-realistic and illumination preserving images while preserving identity, we evaluate our model qualitatively and quantitatively under both controlled and in the wild settings. In the following subsections, we begin with an introduction of datasets and implementation details. Then we demonstrate the merits of our model on qualitative synthesis results and quantitative recognition results over the state-of-the-art methods. Lastly, we conduct an ablation study to demonstrate the benefits from each part of our model.

4.1   Experimental Settings

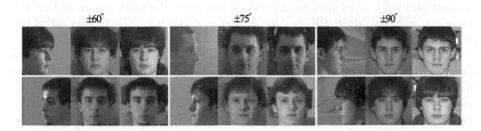

Fig. 4. Synthesis results of the Multi-PIE dataset by our model under large poses and illumination inconsistent conditions. Each pair presents profile (left), synthesized frontal face (middle) and ground-truth frontal face (right).

Datasets. We adopt the Multi-PIE dataset [4] as our training and testing set. Multi-PIE is widely used for evaluating face synthesis and recognition in the controlled setting. It contains 754,204 images of 337 identities from 15 poses and 20 illumination conditions. In this paper, the face images with neutral expression under 20 illuminations and 13 poses within $\pm 90°$ are used. For a fair comparison, we follow the test protocols in [8] and utilize two settings to evaluate our model. The first setting (Setting 1) only contains images from Session 1. The training set is composed of all the first 150 identities images. For testing, one gallery image with frontal view and normal illumination is used for the remaining 99 identities. For the second setting (Setting 2), we use neutral expression images from all four sessions. The first 200 identities and the remaining 137 identities are used for training and testing, respectively. Each testing identity has one gallery image with frontal view and normal illumination from the first appearance.

LFW [9] contains 13,233 face images collected in unconstrained environment. It will be used to evaluate the frontalization performance in uncontrolled settings.

Implementation Details. All images in our experiments are cropped and resized to 128×128 according to facial landmarks, and image intensities are linearly scaled to the range of $[0, 1]$. The LightCNN-29 [30] is pretrained on MS-Celeb-1M [5] and fine-tuned on the training set of Multi-PIE.

In all our experiments, we empirically set $\lambda_0 = 5, \lambda_1 = 1, \lambda_2 = 0.1, \lambda_3 = 15, \lambda_4 = 1$. The learning rate is initialized by 0.0004 and the batch size is 8. The flow estimation networks $\mathcal{F}$ and $\mathcal{F}'$ are pre-trained and then all networks are end-to-end trained by minimizing the objective $\mathcal{L}$ with setting lr = 0.00005 for $\mathcal{F}$ and $\mathcal{F}'$.

4.2   Qualitative Evaluation

In this subsection, we qualitatively compare the synthesized results of our model against state-of-the-art face frontalization methods. We train our model on the training set of the Multi-PIE Setting 2, and evaluate it on the testing set of the Multi-PIE Setting 2 and the LFW [9].

Figure 4 shows the face synthesized results under large poses, and it is obvious that our model can synthesize photo-realistic images. To demonstrate the illumination preserving strength of our model, we choose the profiles with obvious inconsistent illumination. As shown in Fig. 4, the illumination of profile faces can be well preserved in the synthesized images. More synthesized results are provided in the supplementary material.

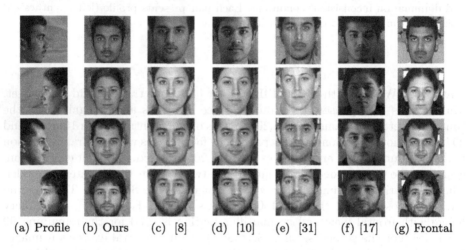

(a) Profile   (b) Ours   (c) [8]   (d) [10]   (e) [31]   (f) [17]   (g) Frontal

Fig. 5. Face frontalization comparison on the Multi-PIE dataset under the pose of 90° (first two rows) and 75° (last two rows).

Figure 5 illustrates the comparison with the state-of-the-art face frontalization methods [8,10,17,31] on the Multi-PIE dataset. In the large pose cases, existing methods are disable to preserve the illumination of profiles on the synthesized results. Face shape and other face components (e.g., eyebrows, mustache and nose) also occur deformation. The reason is those methods are less able to preserve reliable details from the profiles. Compared with the existing methods, our method produces more identity preserving results while keeping the facial details of the profiles as much as possible. In particular, under large poses, our model can recover photo-realistic illumination conditions of the profiles, which is important when frontalized images are used for some other face-related tasks, such as face editing, face pose transfer and face-to-face synthesis.

We further qualitatively compare face frontalization results of our model on the LFW dataset with [6,10,17,28,34]. As shown in Fig. 6, the existing methods fail to recover clear global structures and fine facial details. Also they cannot

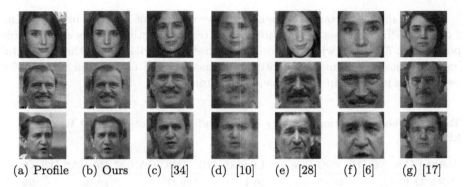

(a) Profile (b) Ours (c) [34] (d) [10] (e) [28] (f) [6] (g) [17]

Fig. 6. Face frontalization comparison on the LFW dataset. Our method is trained on Mulit-PIE and tested on LFW.

Table 1. Rank-1 recognition rates (%) across poses under Setting 2 of the Multi-PIE. The best two results are highlighted by **bold** and <u>underline</u> respectively.

Method	±15°	±30°	±45°	±60°	±75°	±90°	Avg
Light CNN [30]	98.59	97.38	92.13	62.09	24.18	5.51	63.31
DR-GAN [28]	94.90	91.10	87.20	84.60	–	–	89.45
FF-GAN [31]	94.60	92.50	89.70	85.20	77.20	61.20	83.40
TP-GAN [10]	98.68	98.06	95.38	87.72	77.43	64.64	86.99
CAPG-GAN [8]	99.82	99.56	97.33	90.63	83.05	66.05	89.41
PIM [34]	99.30	99.00	98.50	98.10	95.00	86.50	96.07
3D-PIM [35]	99.64	99.48	98.81	98.37	95.21	86.73	96.37
DA-GAN [32]	<u>99.98</u>	<u>99.88</u>	99.15	97.27	93.24	81.56	95.18
HF-PIM [1]	**99.99**	**99.98**	**99.98**	**99.14**	<u>96.40</u>	<u>92.32</u>	<u>97.97</u>
Ours	99.86	99.80	<u>99.37</u>	<u>98.85</u>	**97.20**	**93.17**	**98.04**

preserve the illumination of the profiles. Though FNM [17] generates high qualitative images, it is still disable to preserve identity. It is worth noting that our method produces more photo-realistic faces with identity and illumination well-preserved, which also demonstrates the generalizability of our model in the uncontrolled environment. More results under large poses are provided in the supplementary material.

4.3 Quantitative Evaluation

In this subsection, we quantitatively compare the proposed method with other methods in terms of recognition accuracy on Multi-PIE and LFW. The recognition accuracy is calculated by firstly extracting deep features with LightCNN-29 [30] and then measuring similarity between features with a cosine-distance metric.

Table 1 shows the Rank-1 recognition rates of different methods under Setting 2 of Multi-PIE. Our method has advantages over competitors, especially at large poses (*e.g.*, 75°, 90°), which demonstrates that our model can synthesize frontal images while preserving the identity information. The recognition rates under Setting 1 is provided in the supplementary material.

Table 2. Face verification accuracy (ACC) and area-under-curve (AUC) results on LFW.

Method	FaceNet [23]	VGG Face [16]	FF-GAN [31]	CAPG-GAN [8]	DA-GAN [32]	**Ours**
ACC(%)	99.63	98.95	96.42	99.37	99.56	**99.65**
AUC(%)	–	–	99.45	99.90	99.91	**99.92**

Table 2 compares the face verification performance (ACC and AUC) of our method with other state-of-the-arts [8,16,23,31,32] on the LFW. Our method achieves 99.65 on accuracy and 99.92 on AUC, which is also comparable with other state-of-the-art methods. The above quantitative results prove that our method is able to preserve the identity information effectively.

4.4   Ablation Study

In this subsection, we analyze the respective roles of the different modules and loss functions in frontal view synthesis. Both qualitative perceptual performance (Fig. 7) and face recognition rates (Table 3) are reported for comprehensive comparison under the Multi-PIE Setting 2. We can see that our FFWM exceeds all its variants in both quantitative and qualitative evaluations.

Effects of the Illumination Preserving Module (IPM). Although without IPM the recognition rates drop slightly (as shown in Table 3), the synthesized results cannot preserve illumination and are approximate to the inconsistent ground-truth illumination (as shown in Fig. 7). We also explore the contributions

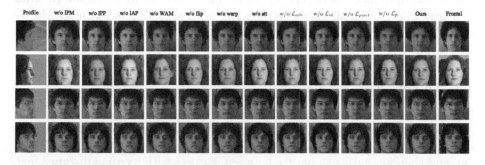

Fig. 7. Model comparsion: synthesis results of our model and its variants on Multi-PIE

Table 3. Incomplete variants analysis: Rank-1 recognition rates (%) across poses under Setting 2 of the Multi-PIE dataset. IAP and IPP denote the illumination adaption pathway and illumination preserving pathway in the Illumination Preserving Module (IPM). Warp, flip and att denote the three variants in Warp Attention Module (WAM).

Method	±15°	±30°	±45°	±60°	±75°	±90°	Avg
w/o IPM	99.83	99.77	99.35	98.74	97.18	93.15	98.00
IPM w/o IPP	99.84	99.74	99.36	98.47	96.73	91.56	97.62
IPM w/o IAP	99.83	99.76	99.30	98.70	97.11	92.83	97.92
w/o WAM	99.84	99.46	98.91	97.27	93.18	86.23	95.81
WAM w/o flip	99.84	99.69	99.27	98.10	96.57	92.65	97.69
WAM w/o warp	99.83	99.64	99.16	97.83	94.60	88.16	96.54
WAM w/o att	99.85	99.79	99.36	98.71	96.81	93.09	97.94
w/o $\mathcal{L}_{adv}$	99.83	99.72	99.28	98.57	97.09	93.11	97.93
w/o $\mathcal{L}_{id}$	99.85	99.62	99.12	97.42	93.93	86.05	96.00
w/o $\mathcal{L}_{pixel}$	99.83	99.77	99.35	98.79	97.05	92.85	97.94
w/o $\mathcal{L}_p$	99.81	99.75	99.33	98.62	97.13	93.10	97.96
Ours	**99.86**	**99.80**	**99.37**	**98.85**	**97.20**	**93.17**	**98.04**

of illumination adaption pathway (IAP) and illumination preserving pathway (IPP) in the IPM. As shown in Fig. 7, without IPP, the illumination of synthesized images tend to be inconsistent with the profiles and ground-truth images. And without IAP, the illumination of synthesized images tends to be a tradeoff between the profiles and the illumination inconsistent ground-truth images. Only integrating IPP and IAP together, our model can achieve illumination preserving image synthesis. Furthermore, our model archives a lower recognition rate when removing the IPP, which demonstrates that the IPP prompts the synthesized results to keep reliable information of the profiles.

Effects of the Warp Attention Module (WAM). We can see that without WAM, the synthesized results tend to be smooth and distorted in the self-occlusion parts (as shown in Fig. 7). As shown in Table 3, without WAM, the recognition rates drop significantly, which proves that WAM dominates in preserving identity information. Moreover, we explore the contributions of three components in the WAM, including taking flipped warped feature as additional input (w/o flip), feature warping (w/o warp) and feature attention (w/o att). As shown in Fig. 7, taking flip feature as additional input has benefits on recovering the self-occlusion parts on the synthesized images. Without the feature attention mechanism, there are artifacts on the synthesized images. Without feature warping, the synthesized results get worse visual performance. These results above suggest that each component in WAM is essential for synthesizing identity preserving and photo-realistic frontal images.

Effects of the Losses. As shown in Table 3, the recognition rates decrease if one loss function is removed. Particularly, the rates drop significantly for all poses if the $\mathcal{L}_{id}$ loss is not adapted. We also report the qualitative visualization results in Fig. 7. Without $\mathcal{L}_{adv}$ loss, the synthesized images tend to be blurry, suggesting the usage of adversarial learning. Without $\mathcal{L}_{id}$ and $\mathcal{L}_{pixel}$, our model cannot promise the visual performance on the local textures (*e.g.*, eyes). Without $\mathcal{L}_p$, the synthesized faces present artifacts at the edges (*e.g.*, face and hair).

5   Conclusion

In this paper, we propose a novel Flow-based Feature Warping Model (FFWM) to effectively address the challenging problem in face frontalization, photo-realistic and illumination preserving image synthesis with illumination inconsistent supervision. Specifically, an Illumination Preserve Module is proposed to address the illumination inconsistent issue. It helps FFWM to synthesize photo-realistic frontal images while preserving the illumination of profile images. Furthermore, the proposed Warp Attention Module reduces the pose discrepancy in the feature space and helps to synthesize frontal images effectively. Experimental results demonstrate that our method not only synthesizes photo-realistic and illumination preserving results but also outperforms state-of-the-art methods on face recognition across large poses.

Acknowledgement. This work is partially supported by the National Natural Science Foundation of China (NSFC) under Grant Nos. 61671182 and U19A2073.

References

1. Cao, J., Hu, Y., Zhang, H., He, R., Sun, Z.: Learning a high fidelity pose invariant model for high-resolution face frontalization. In: Advances in Neural Information Processing Systems, pp. 2867–2877 (2018)
2. Deng, J., Cheng, S., Xue, N., Zhou, Y., Zafeiriou, S.: UV-GAN: adversarial facial UV map completion for pose-invariant face recognition. In: Proceedings of the IEEE Conference on Computer Vision and Pattern Recognition, pp. 7093–7102 (2018)
3. Dosovitskiy, A., et al.: FlowNet: learning optical flow with convolutional networks. In: Proceedings of the IEEE International Conference on Computer Vision, pp. 2758–2766 (2015)
4. Gross, R., Matthews, I., Cohn, J., Kanade, T., Baker, S.: Multi-pie. Image Vis. Comput. **28**(5), 807–813 (2010)
5. Guo, Y., Zhang, L., Hu, Y., He, X., Gao, J.: MS-CELEB-1M: a dataset and benchmark for large-scale face recognition. In: Leibe, B., Matas, J., Sebe, N., Welling, M. (eds.) ECCV 2016. LNCS, vol. 9907, pp. 87–102. Springer, Cham (2016). https://doi.org/10.1007/978-3-319-46487-9_6
6. Hassner, T., Harel, S., Paz, E., Enbar, R.: Effective face frontalization in unconstrained images. In: Proceedings of the IEEE Conference on Computer Vision and Pattern Recognition, pp. 4295–4304 (2015)

7. He, K., Sun, J., Tang, X.: Guided image filtering. In: Daniilidis, K., Maragos, P., Paragios, N. (eds.) ECCV 2010. LNCS, vol. 6311, pp. 1–14. Springer, Heidelberg (2010). https://doi.org/10.1007/978-3-642-15549-9_1

8. Hu, Y., Wu, X., Yu, B., He, R., Sun, Z.: Pose-guided photorealistic face rotation. In: Proceedings of the IEEE Conference on Computer Vision and Pattern Recognition, pp. 8398–8406 (2018)

9. Huang, G.B., Ramesh, M., Berg, T., Learned-Miller, E.: Labeled faces in the wild: a database for studying face recognition in unconstrained environments. Technical report 07-49, University of Massachusetts, Amherst, October 2007

10. Huang, R., Zhang, S., Li, T., He, R.: Beyond face rotation: global and local perception GAN for photorealistic and identity preserving frontal view synthesis. In: Proceedings of the IEEE International Conference on Computer Vision, pp. 2439–2448 (2017)

11. Hui, T.W., Tang, X., Change Loy, C.: LiteFlowNet: a lightweight convolutional neural network for optical flow estimation. In: Proceedings of the IEEE Conference on Computer Vision and Pattern Recognition, pp. 8981–8989 (2018)

12. Hur, J., Roth, S.: Iterative residual refinement for joint optical flow and occlusion estimation. In: Proceedings of the IEEE Conference on Computer Vision and Pattern Recognition, pp. 5754–5763 (2019)

13. Ilg, E., Mayer, N., Saikia, T., Keuper, M., Dosovitskiy, A., Brox, T.: FlowNet 2.0: evolution of optical flow estimation with deep networks. In: Proceedings of the IEEE Conference on Computer Vision and Pattern Recognition, pp. 2462–2470 (2017)

14. Jaderberg, M., Simonyan, K., Zisserman, A., et al.: Spatial transformer networks. In: Advances in Neural Information Processing Systems, pp. 2017–2025 (2015)

15. Li, X., Liu, M., Ye, Y., Zuo, W., Lin, L., Yang, R.: Learning warped guidance for blind face restoration. In: Proceedings of the European Conference on Computer Vision, pp. 272–289 (2018)

16. Parkhi, O.M., Vedaldi, A., Zisserman, A.: Deep face recognition (2015)

17. Qian, Y., Deng, W., Hu, J.: Unsupervised face normalization with extreme pose and expression in the wild. In: Proceedings of the IEEE Conference on Computer Vision and Pattern Recognition, pp. 9851–9858 (2019)

18. Ranjan, A., Black, M.J.: Optical flow estimation using a spatial pyramid network. In: Proceedings of the IEEE Conference on Computer Vision and Pattern Recognition, pp. 4161–4170 (2017)

19. Ren, Y., Yu, X., Chen, J., Li, T.H., Li, G.: Deep image spatial transformation for person image generation. In: Proceedings of the IEEE/CVF Conference on Computer Vision and Pattern Recognition, pp. 7690–7699 (2020)

20. Ronneberger, O., Fischer, P., Brox, T.: U-Net: convolutional networks for biomedical image segmentation. In: Navab, N., Hornegger, J., Wells, W.M., Frangi, A.F. (eds.) MICCAI 2015. LNCS, vol. 9351, pp. 234–241. Springer, Cham (2015). https://doi.org/10.1007/978-3-319-24574-4_28

21. Russakovsky, O., et al.: ImageNet large scale visual recognition challenge. Int. J. Comput. Vis. 115(3), 211–252 (2015)

22. Sagonas, C., Panagakis, Y., Zafeiriou, S., Pantic, M.: Robust statistical face frontalization. In: Proceedings of the IEEE International Conference on Computer Vision, pp. 3871–3879 (2015)

23. Schroff, F., Kalenichenko, D., Philbin, J.: FaceNet: a unified embedding for face recognition and clustering. In: Proceedings of the IEEE Conference on Computer Vision and Pattern Recognition, pp. 815–823 (2015)

24. Shocher, A., Bagon, S., Isola, P., Irani, M.: InGAN: capturing and remapping the "DNA" of a natural image. arXiv preprint arXiv:1812.00231 (2018)
25. Simonyan, K., Zisserman, A.: Very deep convolutional networks for large-scale image recognition. arXiv preprint arXiv:1409.1556 (2014)
26. Sun, D., Yang, X., Liu, M.Y., Kautz, J.: PWC-NET: CNNs for optical flow using pyramid, warping, and cost volume. In: Proceedings of the IEEE Conference on Computer Vision and Pattern Recognition, pp. 8934–8943 (2018)
27. Tian, Y., Peng, X., Zhao, L., Zhang, S., Metaxas, D.N.: CR-GAN: learning complete representations for multi-view generation. arXiv preprint arXiv:1806.11191 (2018)
28. Tran, L., Yin, X., Liu, X.: Disentangled representation learning GAN for pose-invariant face recognition. In: Proceedings of the IEEE Conference on Computer Vision and Pattern Recognition, pp. 1415–1424 (2017)
29. Truong, P., Danelljan, M., Timofte, R.: GLU-NeT: global-local universal network for dense flow and correspondences. arXiv preprint arXiv:1912.05524 (2019)
30. Wu, X., He, R., Sun, Z., Tan, T.: A light CNN for deep face representation with noisy labels. IEEE Trans. Inf. Forensics Secur. **13**(11), 2884–2896 (2018)
31. Yin, X., Yu, X., Sohn, K., Liu, X., Chandraker, M.: Towards large-pose face frontalization in the wild. In: Proceedings of the IEEE International Conference on Computer Vision, pp. 3990–3999 (2017)
32. Yin, Y., Jiang, S., Robinson, J.P., Fu, Y.: Dual-attention GAN for large-pose face frontalization. arXiv preprint arXiv:2002.07227 (2020)
33. Zhang, Z., Chen, X., Wang, B., Hu, G., Zuo, W., Hancock, E.R.: Face frontalization using an appearance-flow-based convolutional neural network. IEEE Trans. Image Process. **28**(5), 2187–2199 (2018)
34. Zhao, J., et al.: Towards pose invariant face recognition in the wild. In: Proceedings of the IEEE Conference on Computer Vision and Pattern Recognition, pp. 2207–2216 (2018)
35. Zhao, J., et al.: 3D-aided deep pose-invariant face recognition. In: IJCAI, vol. 2, p. 11 (2018)
36. Zhu, X., Lei, Z., Yan, J., Yi, D., Li, S.Z.: High-fidelity pose and expression normalization for face recognition in the wild. In: Proceedings of the IEEE Conference on Computer Vision and Pattern Recognition, pp. 787–796 (2015)

Learning Architectures for Binary Networks

Dahyun Kim[1], Kunal Pratap Singh[2], and Jonghyun Choi[1(✉)]

[1] GIST (Gwangju Institute of Science and Technology), Gwangju, South Korea
killawhale@gm.gist.ac.kr, jhc@gist.ac.kr
[2] Indian Institute of Technology (IIT) Roorkee, Roorkee, India
ksingh@ee.iitr.ac.in
https://github.com/gistvision/bnas

Abstract. Backbone architectures of most binary networks are well-known floating point (FP) architectures such as the ResNet family. Questioning that the architectures designed for FP networks might not be the best for binary networks, we propose to search architectures for binary networks (BNAS) by defining a new search space for binary architectures and a novel search objective. Specifically, based on the cell based search method, we define the new search space of binary layer types, design a new cell template, and rediscover the utility of and propose to use the *Zeroise* layer instead of using it as a placeholder. The novel search objective *diversifies early search* to learn better performing binary architectures. We show that our method searches architectures with stable training curves despite the quantization error inherent in binary networks. Quantitative analyses demonstrate that our searched architectures outperform the architectures used in state-of-the-art binary networks and outperform or perform *on par* with state-of-the-art binary networks that employ various techniques other than architectural changes.

Keywords: Binary networks · Backbone architecture · Architecture search

1 Introduction

Increasing demand for deploying high performance visual recognition systems encourages research on efficient neural networks. Approaches include pruning [12], efficient architecture design [14,15,42], low-rank decomposition [16], network quantization [6,20,32] and knowledge distillation [13,37]. Particularly,

---

D. Kim and K. P. Singh—Indicates equal contribution. This work is done while KPS is at GIST for internship.

Electronic supplementary material The online version of this chapter (https://doi.org/10.1007/978-3-030-58610-2_34) contains supplementary material, which is available to authorized users.

© Springer Nature Switzerland AG 2020
A. Vedaldi et al. (Eds.): ECCV 2020, LNCS 12357, pp. 575–591, 2020.
https://doi.org/10.1007/978-3-030-58610-2_34

network quantization, especially binary or 1-bit CNNs, are known to provide extreme computational and memory savings. The computationally expensive floating point convolutions are replaced with computationally efficient XNOR and bit-count operations, which significantly speeds up inference [32]. Hence, binary networks are incomparable with efficient floating point networks due to the extreme computational and memory savings.

Current binary networks, however, use architectures designed for floating point weights and activations [25,28,32,34]. We hypothesize that the backbone architectures used in current binary networks may not be optimal for binary parameters as they were designed for floating point ones. Instead, we may learn better binary network architectures by exploring the space of binary networks.

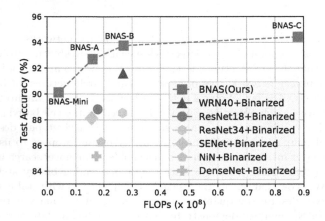

Fig. 1. Test accuracy (%) *vs* FLOPs on CIFAR10 using the XNOR-Net binarization scheme [32]. Our searched architectures outperform the binarized floating point architectures. Note that our BNAS-Mini, which has much less FLOPs, outperforms all other binary networks except the one based on WideResNet40 (WRN40)

To discover better performing binary networks, we first apply one of the widely used binarization schemes [32] to the searched architectures from floating point NAS which use cell based search and gradient based search algorithms [9,27,40]. We then train the resulting binary networks on CIFAR10. Disappointingly, the binarized searched architectures do not perform well (Sect. 3). We hypothesize two reasons for the failure of binarized searched floating point architectures. First, the search space used in the floating point NAS is not necessarily the best one for binary networks. For example, separable convolutions will have large quantization error when binarized, since nested convolutions increase quantization error (Sect. 4.1). Additionally, we discover that the *Zeroise* layer, which was only used as a placeholder in floating point NAS, improves the accuracy of binary networks when kept in the final architecture (Sect. 4.1). Second, the cell template used for floating point cell based NAS methods is not well suited for the binary domain because of unstable gradients due to quantization error (Sect. 4.2).

Based on the above hypotheses and empirical observations, we formulate a cell based search space explicitly defined for binary networks and further propose a novel search objective with the diversity regularizer. The proposed regularizer encourages exploration of diverse layer types in the early stages of search, which is particularly useful for discovering better binary architectures. We call this method as Binary Network Architecture Search or *BNAS*. We show that the new search space and the diversity regularizer in BNAS helps in searching better performing binary architectures (Sect. 5).

Given the same binarization scheme, we compare our searched architectures to several handcrafted architectures including the ones shown in the Fig. 1. Our searched architectures clearly outperforms the architectures used in the state-of-the-art binary networks, indicating the prowess of our search method in discovering better architectures for binary networks.

We summarize our contributions as follows:

- We propose the first architecture search method for binary networks. The searched architectures are adjustable to various computational budgets (in FLOPs) and outperform backbone architectures used in state-of-the-art binary networks on both CIFAR10 and ImageNet dataset.
- We define a new search space for binary networks that is more robust to quantization error; a new cell template and a new set of layers.
- We propose a new search objective aimed to diversify early stages of search and demonstrate its contribution in discovering better performing binary networks.

2   Related Work

2.1   Binary Neural Networks

There have been numerous proposals to improve the accuracy of binary (1-bit) precision CNNs whose weights and activations are all binary valued. We categorize them into binarization schemes, architectural modifications and training methods.

Binarization Schemes. As a pioneering work, [6] proposed to use the sign function to binarize the weights and achieved compelling accuracy on CIFAR10. [7] binarized the weights and the activations by the sign function and use the straight through estimator (STE) to estimate the gradient. [32] proposed XNOR-Net which uses the sign function with a scaling factor to binarize the weights and the activations. They showed impressive performance on a large scale dataset (ImageNet ILSVRC 2012) and that the computationally expensive floating point convolution operations can be replaced by highly efficient XNOR and bit counting operations. Many following works including recent ones [25,28] use the binarization scheme of XNOR-Net as do we. [22] approximated both weights and activations as a weighted sum of multiple binary filters to improve performance.

Very recently, new binarization schemes have been proposed [3,10]. [10] uses projection convolutional layers while [3] improves upon the analytically calculated scaling factor in XNOR-Net.

These different binarization schemes do not modify the backbone architecture while we focus on finding better backbone architectures given a binarization scheme. A newer binarization scheme can be incorporated into our search framework but that was not the focus of this work.

Architectural Advances. It has been shown that appropriate modifications to the backbone architecture can result in great improvements in accuracy [25,28,32]. [32] proposed XNOR-Net which shows that changing the order of batch normalization (BN) and the sign function is crucial for the performance of binary networks. [28] connected the input floating point activations of consecutive blocks through identity connections before the sign function. They aimed to improve the representational capacity for binary networks by adding the floating point activation of the current block to the consequent block. They also introduced a better approximation of the gradient of the sign function for back-propagation. [25] used circulant binary convolutions to enhance the representational capabilities of binary networks. [31] proposed a modified version of separable convolutions to binarize the MobileNetV1 architecture. However, we observe that the modified separable convolution modules do not generalize to architectures other than MobileNet. These methods do not alter the connectivity or the topology of the network while we search for entirely new network architectures.

Training Methods. There have been a number of methods proposed for training binary networks. [44] showed that quantized networks, when trained progressively from higher to lower bit-width, do not get trapped in a local minimum. [11] proposed a training method for binary networks using two new losses; Bayesian kernel loss and Bayesian feature loss. Recently, [18] proposed to pretrain the network with ternary activation which are later decoupled to binary activations for fine-tuning. The training methods can be used in our searched networks as well, but we focus on the architectural advances.

Recent Works on Binary Architecture Search. Recently, [34] performed a hyper-parameter search (*e.g.*, number of channels) using an evolutionary algorithm to efficiently increase the FLOPs of a binarized ResNet backbone. However, they trade more computation cost for better performance, reducing their inference speed up ($\sim 2.74\times$) to be far smaller than other binary networks ($\sim 10\times$). Concurrently in this conference, [2] propose to search for binary networks but using a different search space and search strategy than ours. However, the reported accuracy was difficult to reproduce with the given configuration details. We expect further progress in this field with reproducible public codes.

2.2 Efficient Neural Architecture Search

We search architectures for binary networks by adopting ideas from neural architecture search (NAS) methods for floating point networks [27,30,40,45,46]. To reduce the severe computation cost of NAS methods, there are numerous proposals focused on accelerating the NAS algorithms [1,4,5,8,9,21,23,24,26,27,29, 30,38–41,43]. We categorize these attempts into cell based search and gradient based search algorithms.

Cell Based Search. Pioneered by [46], many NAS methods [1,5,8,9,21,23, 24,26,27,29,38–41,43] have used the cell based search, where the objective of the NAS algorithm is to search for a cell, which will then be stacked to form the final network. The cell based search reduces the search space drastically from the entire network to a cell, significantly reducing the computational cost. Additionally, the searched cell can be stacked any number of times given the computational budget. Although the scalability of the searched cells to higher computational cost is a non-trivial problem [36], it is not crucial to our work because binary networks focus more on smaller computational budgets.

Gradient Based Search Algorithms. In order to accelerate the search, methods including [5,9,27,38,40] relax the discrete sampling of child architectures to be differentiable so that the gradient descent algorithm can be used. The relaxation involves taking a weighted sum of several layer types during the search to approximate a single layer type in the final architecture. [27] uses softmax of learnable parameters as the weights, while other methods [9,38,40] use the Gumbel-softmax [17] instead, both of which allow seamless back-propagation by gradient descent. Coupled with the use of the cell based search, certain work has been able to drastically reduce the search complexity [9].

We make use of both the cell based search and gradient based search algorithms but propose a novel search space along with a modified cell template and a new regularized search objective to search binary networks.

3   Binarizing Searched Architectures by NAS

It is well known that architecture search results in better performing architecture than the hand-crafted ones. To obtain better binary networks, we first binarize the searched architectures by cell based gradient search methods. Specifically, we apply the binarization scheme of XNOR-Net along with their architectural modifications [32] to architectures searched by DARTS [27], SNAS [40] and GDAS [9]. We show the learning curves of the binarized searched floating point architectures on CIFAR10 dataset in Fig. 2.

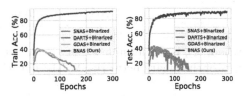

Fig. 2. Train (left) and test (right) accuracy of binarized searched architectures on CIFAR10. The XNOR-Net's binarization scheme and architectural modifications are applied in all cases. Contrasting to our BNAS, the binarized searched architectures fail to train well

Disappointingly, GDAS and SNAS reach around 40% test accuracy and quickly plummet while DARTS did not train at all. This implies that floating point NAS methods are not trivially extended to search binary networks. We investigate the failure modes in training and find two issues; 1) the search space is not well suited for binary networks, *e.g.*, using separable convolutions accumulates the quantization error repetitively and 2) the cell template does not propagate the gradients properly, due to quantization error. To search binary networks, the search space and the cell template should be redesigned to be robust to quantization error.

4   Approach

To search binary networks, we first write the problem of cell-based architecture search in general as:

$$\alpha^* = \underset{\alpha \in A(\mathcal{S},T)}{\mathrm{argmin}} \; \mathcal{L}_S(D; \theta_\alpha), \tag{1}$$

where A is the feasible set of final architectures, $\mathcal{S}$ is the search space (a set of layer types to be searched), T is the cell template which is used to create valid networks from the chosen layer types, L_S is the search objective, D is the dataset, θ_α is the parameters of the searched architecture α which contain both architecture parameters (used in the continuous relaxation [27], Eq. 6) and the network weights (the learnable parameters of the layer types, Eq. 6), and α^* is the searched final architecture. Following [27], we solve the minimization problem using SGD.

Based on the observation in Sect. 3, we propose a new search space ($\mathcal{S}_B$), cell template (T_B) and a new search objective $\widetilde{\mathcal{L}}_S$ for binary networks which have binary weights and activations. The new search space and the cell template are more robust to quantization error and the new search objective $\widetilde{\mathcal{L}}_S$ promotes diverse search which is important when searching binary networks (Sect. 4.3). The problem of architecture search for binary network α_B^* can be rewritten as:

$$\alpha_B^* = \underset{\alpha_B \in A_B(\mathcal{S}_B,T_B)}{\mathrm{argmin}} \; \widetilde{\mathcal{L}}_S(D; \theta_{\alpha_B}), \tag{2}$$

where A_B is the feasible set of binary network architectures and θ_{α_B} is parameters of the binary networks. We detail each proposal in the following subsections.

4.1   Search Space for Binary Networks ($\mathcal{S}_B$)

Unlike the search space used in floating point NAS, the search space used for binary networks should be robust to quantization error. Starting from the search space popularly used in floating point NAS [9,27,40,46], we investigate the robustness of various convolutional layers to quantization error and selectively define the space for the binary networks. Note that the quantization error depends on the binarization scheme and we use the scheme proposed in [32].

Convolutions and Dilated Convolutions. To investigate the convolutional layers' resilience to quantization error, we review the binarization scheme we use [32]. Let $\mathbf{W}$ be the weights of a floating point convolution layer with dimension $c \cdot w \cdot h$ (number of channels, width and height of an input) and $\mathbf{A}$ be an input activation. The floating point convolution can be approximated by binary parameters, $\mathbf{B}$, and the binary input activation, $\mathbf{I}$ as:

$$\mathbf{W} * \mathbf{A} \approx \beta \mathbf{K} \odot (\mathbf{B} * \mathbf{I}), \tag{3}$$

where $*$ denotes the convolution operation, $\odot$ is the Hadamard product (element wise multiplication), $\mathbf{B} = sign(\mathbf{W})$, $\mathbf{I} = sign(\mathbf{A})$, $\beta = \frac{1}{n}\|\mathbf{W}\|_1$ with $n = c \cdot w \cdot h$, $\mathbf{K} = \mathbf{D} * \mathbf{k}$, $\mathbf{D} = \frac{\sum |A_{i,:,:}|}{c}$ and $\mathbf{k}_{ij} = \frac{1}{w \cdot h}$ $\forall ij$. Dilated convolutions are identical to convolutions in terms of quantization error.

Since both convolutions and dilated convolutions show tolerable quantization error in binary networks [32] (and our empirical study in Table 1), we include the standard convolutions and dilated convolutions in our search space.

Separable Convolutions. Separable convolutions [35] have been widely used to construct efficient network architectures for floating point networks [14] in both hand-crafted and NAS methods. Unlike floating point networks, we argue that the separable convolution is not suitable for binary networks due to large quantization error. It uses nested convolutions to approximate a single convolution for computational efficiency. The nested convolution are approximated to binary convolutions as:

$$Sep(\mathbf{W} * \mathbf{A}) \approx \beta_2(\mathbf{B}_2 * \mathbf{A}_2) \approx \beta_1\beta_2(\mathbf{B}_2 * (\mathbf{K}_1 \odot (\mathbf{B}_1 * \mathbf{I}_1))), \tag{4}$$

where $Sep(\mathbf{W} * \mathbf{A})$ denotes the separable convolution, $\mathbf{B}_1$ and $\mathbf{B}_2$ are the binary weights for the first and second convolution operation in the separable convolution layer, $\mathbf{I}_1 = sign(\mathbf{A})$, $\mathbf{A}_2 = \beta_1\mathbf{K}_1 \odot (\mathbf{B}_1 * \mathbf{I}_1)$ and $\beta_1, \beta_2, \mathbf{K}_1$ are the scaling factors for their respective binary weights and activations. Since every scaling factor induces quantization error, the nested convolutions in separable convolutions will result in more quantization error.

To empirically investigate how the quantization error affects training for different convolutional layer types, we construct small networks formed by repeating each kind of convolutional layers three times, followed by three fully connected layers. We train these networks on CIFAR10 in floating point and binary domain and summarize the results in Table 1.

When binarized, both convolution and dilated convolution layers show only a reasonable drop in accuracy, while the separable convolution layers show performance equivalent to random guessing (10% for CIFAR10). The observations in Table 1 imply that the accumulated quantization error by the nested convolutions fails binary networks in training. This also partly explains why the binarized architecture searched by DARTS in Fig. 2 does not train as it selects a large number of separable convolutions.

Table 1. Test Accuracy (%) of a small CNN composed of each layer type only, in floating point (FP Acc.) and in binary domain (Bin. Acc) on CIFAR10. *Conv, Dil. Conv* and *Sep. Conv* refer to the convolutions, dilated convolutions and separable convolutions, respectively. Separable convolutions show a drastically low performance on the binary domain

Layer Type	Conv		Dil. Conv		Sep. Conv	
Kernel Size	3×3	5×5	3×3	5×5	3×3	5×5
FP Acc. (%)	61.78	60.14	56.97	55.17	56.38	57.00
Bin. Acc. (%)	46.15	42.53	41.02	37.68	10.00	10.00

Zeroise. The *Zeroise* layer outputs all zeros irrespective of the input [27]. It was originally proposed to model the lack of connections. Further, in the authors' implementation of [27][1], the final architecture excludes the *Zeroise* layers and replaces it with the second best layer type, even if the search picks the *Zeroise* layers. Thus, the *Zeroise* layers are not being used as they were originally proposed but simply used as a placeholder for a different and sub-optimal layer type. Such replacement of layer types effectively removes all architectures that have *Zeroise* layers from the feasible set of final architectures.

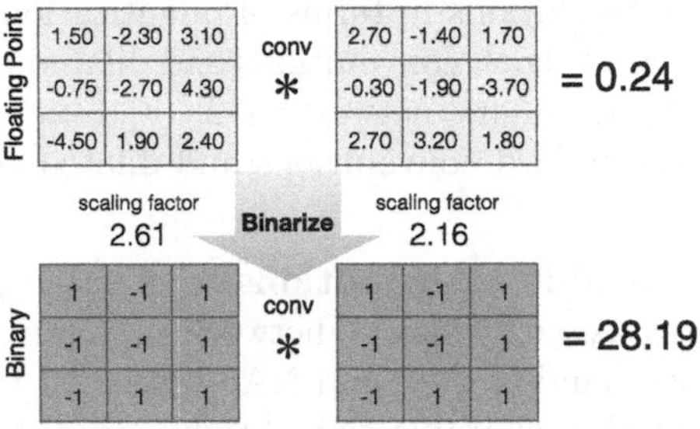

Fig. 3. An example when the *Zeroise* layer is beneficial for binary networks. Since the floating point convolution is close to zero but the binarized convolution is far greater than 0, if the search selects the *Zeroise* layer instead of the convolution layer, the quantization error reduces significantly

In contrast, we use the *Zeroise* layer for *reducing the quantization error* and are the first to keep it in the *final* architectures instead of using it as a placeholder for other layer types. As a result, our feasible set is different from that of [27] not only in terms of precision (binary), but also in terms of the network topology it contains.

As the exclusion of the *Zeroise* layers is not discussed in [27], we compare the accuracy with and without the *Zeroise* layer for DARTS in the DARTS column of Table 2 and empirically verify that the *Zeroise* layer is not particularly useful for floating point networks. However, we observe that the *Zeroise* layer improve the accuracy by a meaningful margin in binary networks as shown in the table. We argue that the *Zeroise* layer can reduce quantization error in binary networks as an example in Fig. 3. Including the *Zeroise* layer in the final architecture is particularly beneficial when the situation similar to Fig. 3 happens frequently as the quantization error reduction is significant. But the degree of benefit may

---

[1] https://github.com/quark0/darts.

Table 2. DARTS and BNAS w/ and w/o the *Zeroise* layers in the final architecture on CIFAR10. *Zeroise Layer* indicates whether the *Zeroise* layers were kept (✓) or not (✗). The test accuracy of DARTS drops by 3.02% when you include the *Zeroise* layers and the train accuracy drops by 63.54% and the training stagnates. In constrast, the *Zeroise* layers improves BNAS in both train and test accuracy

Precision Zeroise Layer	Floating Point (DARTS)			Binary (BNAS)		
	✗	✓	Gain	✗	✓	Gain
Train Acc. (%)	99.18	35.64	-63.54%	93.41	97.46	+4.05%
Test Acc. (%)	97.45	94.43	-3.02%	89.47	92.70	+3.23%

differ from dataset to dataset. As the dataset used for search may differ from the dataset used to train and evaluate the searched architecture, we propose to tune the probability of including the *Zeroise* layer. Specifically, we propose a generalized layer selection criterion to adjust the probability of including the *Zeroise* layer by a transferability hyper-parameter γ as:

$$p^* = \max\left[\frac{p_z}{\gamma}, p_{op_1}, ..., p_{op_n}\right], \tag{5}$$

where p_z is the architecture parameter corresponding to the *Zeroise* layer and p_{op_i} are the architecture parameters corresponding to the ith layer other than *Zeroise*. Larger γ encourages to pick the *Zeroise* layer only if it is substantially better than the other layers.

With the separable convolutions and the *Zeroise* layer type considered, we summarize the defined search space for BNAS ($\mathcal{S}_B$) in Table 3.

4.2 Cell Template for Binary Networks (T_B)

With the defined search space, we now learn a network architecture with the convolutional cell template proposed in [46]. However, the learned architecture still suffers from unstable gradients in the binary domain as shown in Fig. 4-(a) and (b). Investigating the reasons for the unstable gradients, we observe that the skip-connections in the cell template proposed in [46] are confined to be inside a single convolutional cell, *i.e.*, intra-cell skip-connections. The intra-cell skip-connections do not propagate the gradients outside the cell, forcing the cell to aggregate outputs that always have quantization error created inside the cell. To help convey information without the cumulative quantization error through multiple cells, we propose to add skip-connections between multiple cells as illustrated in Fig. 5.

The proposed cell template with inter-cell skip-connections help propagate gradients with less quantization error throughout the network, stabilizing the training curve. We empirically validate the usefulness of the inter-cell skip connections in Sect. 5.4.

Table 3. Proposed search space for BNAS. *Bin Conv, Bin Dil. Conv, MaxPool and AvgPool* refer to the binary convolution, binary dilated convolution, max pooling and average pooling layers, respectively

Layer Type	Bin Conv.		Bin Dil. Conv.		MaxPool	AvgPool	Zeroise
Kernel Size	3×3	5×5	3×3	5×5	3×3	3×3	N/A

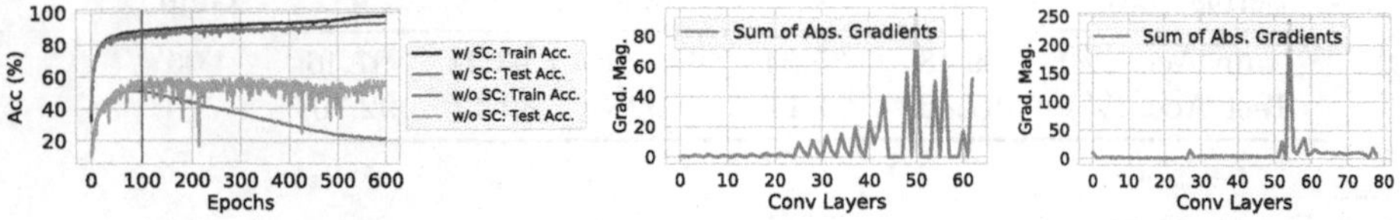

(a) Learning curve   (b) Gradients w/o SC   (c) Gradients w/ SC

Fig. 4. Unstable gradients in the binary domain. 'w/o SC' indicates the cell template of [46] (Fig. 5-(a)). 'w/ SC' indicates the proposed cell template (Fig. 5-(b)). The gradient magnitudes are taken at epoch 100. With the proposed cell template ('w/ SC'), the searched network trains well (a). The proposed cell template shows far less spiky gradients along with generally larger gradient magnitudes ((b) vs. (c)), indicating that our template helps to propagate gradients more effectively in the binary domain

4.3   Search Objective with Diversity Regularizer ($\widetilde{\mathcal{L}}_S$)

With the feasible set of binary architectures (A_B) defined by S_B and T_B, we solve the optimization problem similar to [27]. However, the layers with learnable parameters (*e.g.*, convolutional layers) are not selected as often early on as the layers requiring no learning, because the parameter-free layers are more favorable than the under-trained layers. The problem is more prominent in the binary domain because binary layers train slower than the floating point counterparts [7]. To alleviate this, we propose to use an exponentially annealed entropy based regularizer in the search objective to promote

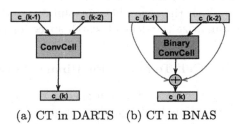

(a) CT in DARTS   (b) CT in BNAS

Fig. 5. Cell templates (CT) of (a) DARTS and (b) BNAS. Red lines in BNAS indicate inter-cell skip connections. ConvCell indicates the convolutional cell. $c\_(k)$ indicates the output of the kth cell

selecting diverse layers and call it the *diversity regularizer*. Specifically, we subtract the entropy of the architecture parameter distribution from the search objective as:

$$\widetilde{\mathcal{L}}_S(D; \theta_{\alpha_B}) = \mathcal{L}_S(D; \theta, p) - \lambda H(p)e^{(-t/\tau)}, \tag{6}$$

where $\mathcal{L}_S(\cdot)$ is the search objective of [27], which is a cross-entropy, θ_{α_B} is the parameters of the sampled binary architecture, which is split into the architecture parameters p and the network weights θ, $H(\cdot)$ is the entropy, λ is a balancing hyper-parameter, t is the epoch, and τ is an annealing hyper-parameter. This will encourage the architecture parameter distribution to be closer to uniform in the early stages, allowing the search to explore diverse layer types.

Using the proposed diversity regularizer, we observed a 16% relative increase in the average number of learnable layer types selected in the first 20 epochs of the search. More importantly, we empirically validate the benefit of the diversity regularizer with the test accuracy on the CIFAR10 dataset in Table 4 and in Sect. 5.4. While the accuracy improvement from the diversity regularizer in the floating point NAS methods such as DARTS [27] is marginal (+0.2%), the improvement in our binary network is more meaningful (+1.75%).

Table 4. Effect of searching diversity on CIFAR10. *Diversity* refers to whether diversity regularization was applied (✓) or not (✗) during the search. DARTS only gains 0.20% test accuracy while BNAS gains 1.75% test accuracy

Precision	Floating Point (DARTS)			Binary (BNAS)		
Diversity	✗	✓	Gain	✗	✓	Gain
Test Acc. (%)	96.53	96.73	+0.20	90.95	92.70	+1.75

5   Experiments

5.1   Experimental Setup

Datasets. We use CIFAR10 [19] and ImageNet (ILSVRC 2012) [33] datasets to evaluate the image classification accuracy. We follow [27] in splitting the datasets for search and training. Please refer to the supplementary material for additional details regarding the search and final evaluation settings.

Details on Comparison with Other Binary Networks. For XNOR-Net with different backbone architectures, we use the floating point architectures from torchvision or a public source[2] and apply the binarization scheme of XNOR-Net. Following previous work [10,28] on comparing ABC-Net with a single base [22], we compare PCNN with a single projection kernel for both CIFAR10 and ImageNet. Please refer to the supplementary material for a qualitative comparison between our searched cells and other non-searched networks. The code and learned models will be available in our repository.

5.2   Comparisons on Backbone Architectures for Binary Networks

We quantitatively compare our searched architectures to various backbone architectures that have been used in the state-of-the-art binary networks with the

---

[2] https://github.com/kuangliu/pytorch-cifar.

Table 5. Comparison of different backbone architectures for binary networks with XNOR-Net binarization scheme [32] in various FLOPs budgets. Bi-Real* indicates Bi-Real Net's method with only the architectural modifications. We refer to [18] for the FLOPs of CBCN. CBCN* indicates the highest accuracy for CBCN with the ResNet18 backbone as [25] report multiple different accuracy for the same network configuration. Additionally, [25] does not report the exact FLOPs of their model, hence we categorized them conservatively into the '~ 0.27' bracket

Dataset	FLOPs ($\times 10^8$)	Model (Backbone Arch.)	Top-1 Acc. (%)	Top-5 Acc. (%)
CIFAR10	~ 0.16	XNOR-Net (ResNet18)	88.82	-
		XNOR-Net (DenseNet)	85.16	-
		XNOR-Net (NiN)	86.28	-
		XNOR-Net (SENet)	88.12	-
		BNAS-A	**92.70**	-
	~ 0.27	XNOR-Net (ResNet34)	88.54	-
		XNOR-Net (WRN40)	91.58	-
		CBCN* (ResNet18) [25]	91.91	-
		BNAS-B	**93.76**	-
	~ 0.90	XNOR-Net (ResNext29-64)	84.27	-
		BNAS-C	**94.43**	-
ImageNet	~ 1.48	XNOR-Net (ResNet18)	51.20	73.20
		BNAS-D	**57.69**	**79.89**
	~ 1.63	Bi-Real* (Bi-Real Net18) [28]	32.90	56.70
		BNAS-E	**58.76**	**80.61**
	~ 1.78	XNOR-Net (ResNet34)	56.49	79.13
		BNAS-F	**58.99**	**80.85**
	~ 1.93	Bi-Real* (Bi-Real Net34) [28]	53.10	76.90
		BNAS-G	**59.81**	**81.61**
	~ 6.56	CBCN (Bi-Real Net18) [25]	61.40	82.80
		BNAS-H	**63.51**	**83.91**

binarization scheme of XNOR-Net [32] in Table 5. The comparisons differ only in the backbone architecture, allowing us to isolate the effect of our searched architectures on the final accuracy, *i.e.*, the comparison with XNOR-Net with different backbone architectures for various FLOPs and newer binary networks with the architectural contributions only. To single out the architectural contributions of Bi-Real Net, we used Table 1 in [28] to excerpt the ImageNet classification accuracy with using only the Bi-Real Net architecture. Note that CBCN is based on the Bi-Real Net architecture with the convolutions being changed to circulant convolutions[3]. Additionally, as mentioned in Sect. 2, we do not compare with [34] as the inference speed-up is significantly worse than other binary networks (~ 2.7× compared to ~ 10×), which makes the comparison less meaningful.

---

[3] They mention that center loss and gaussian gradient update is also used but they are not elaborated and not the main focus of CBCN's method.

As shown in Table 5, our searched architectures outperform other architectures used in binary networks in all FLOPs brackets and on both CIFAR10 and ImageNet. Notably, comparing XNOR-Net with the ResNet18 and ResNet34 backbone to BNAS-D and BNAS-F, we gain +6.49% or +2.50% top-1 accuracy and +6.69% or +1.72% top-5 accuracy on ImageNet.

Furthermore, BNAS retains the accuracy much better at lower FLOPs, showing that our searched architectures are better suited for efficient binary networks. Additionally, comparing CBCN to BNAS-H, we gain +2.11% top-1 accuracy and +1.11% top-5 accuracy, showing that our architecture can scale to higher FLOPs budgets better than CBCN. In sum, replacing the architectures used in current binary networks to our searched architectures can greatly improve the performance of binary networks.

Table 6. Comparison of other binary networks in various FLOPs budgets. The binarization schemes are: *'Sign + Scale'*: using fixed scaling factor and the sign function [32], *'Sign'*: using the sign function [7], *'Clip + Scale'*: using clip function with shift parameter [22], *'Sign + Scale*'*: using learned scaling factor and the sign function [3], *'Projection'*: using projection convolutions[10], *'Bayesian'*: using a learned scaling factor from the Bayesian losses [11] and the sign function, and *'Decoupled'*: decoupling ternary activations to binary activations [18]

Dataset	FLOPs (×10⁸)	Method (Backbone Arch.)	Binarization Scheme	Pretraining	Top-1 Acc. (%)	Top-5 Acc. (%)
CIFAR10	~0.04	PCNN($i=16$) (ResNet18) [10]	Projection	✗	89.16	-
		BNAS-Mini	Sign + Scale	✗	**90.12**	-
	~0.16	BinaryNet (ResNet18) [7]	Sign	✗	89.95	-
		BNAS-A	Sign + Scale	✗	**92.70**	-
	~0.27	PCNN($i=64$) (ResNet18) [10]	Projection	✓	**94.31**	-
		BNAS-B	Sign + Scale	✗	93.76	-
ImageNet	~1.48	BinaryNet (ResNet18) [7]	Sign	✗	42.20	67.10
		ABC-Net (ResNet18) [22]	Clip + Sign	✗	42.70	67.60
		BNAS-D	Sign + Scale	✗	**57.69**	**79.89**
	~1.63	Bi-Real (Bi-Real Net18) [28]	Sign + Scale	✓	56.40	79.50
		XNOR-Net++ (ResNet18) [3]	Sign + Scale*	✗	57.10	79.90
		PCNN (ResNet18) [10]	Projection	✓	57.30	80.00
		BONN (Bi-Real Net18) [11]	Bayesian	✗	59.30	81.60
		BinaryDuo (ResNet18) [18]	Decoupled	✓	**60.40**	**82.30**
		BNAS-E	Sign + Scale	✗	58.76	80.61
	~1.78	ABC-Net (ResNet34) [22]	Clip + Scale	✗	52.40	76.50
		BNAS-F	Sign+Scale	✗	**58.99**	**80.85**
	~1.93	Bi-Real (Bi-Real Net34) [28]	Sign + Scale	✓	**62.20**	**83.90**
		BNAS-G	Sign + Scale	✗	59.81	81.61

5.3 Comparison with Other Binary Networks

As we focus on improving binary networks by architectural benefits only, comparison to other binary network methods is not of our interest. However, it is still intriguing to compare gains from a pure architectural upgrade to gains from new binarization schemes or new training methods. As shown in Table 6, our searched architectures outperform other methods in more than half the FLOPs brackets spread across CIFAR10 and ImageNet. Moreover, the state-of-the-art methods that focus on discovering better training schemes are complementary to our searched architectures, as these training methods were not designed exclusively for a fixed network topology.

Note that, with the same backbone of ResNet18 or ResNet34, Bi-Real, PCNN, XNOR-Net++ and BONN have higher FLOPs than ABC-Net, XNOR-Net and BinaryNet. The higher FLOPs are from unbinarizing the downsampling convolutions in the ResNet architecture.[4]

5.4   Ablation Studies

We perform ablation studies on the proposed components of our method. We use the CIFAR10 dataset for the experiments with various FLOPs budgets and summarize the results in Table 7.

Table 7. Classification acc. (%) of ablated models on CIFAR10. *Full* refers to the proposed method with all components. *No Skip*, *No Zeroise*, and *No Div* refers to our method without the inter-cell skip connections, with explicitly discarding the *Zeroise* layers, or without the diversity regularizer respectively.

Model	Full	No Skip	No Zeroise	No Div
BNAS-A	**92.70**	61.23	89.47	90.95
BNAS-B	**93.76**	67.15	91.69	91.55
BNAS-C	**94.43**	70.58	88.74	92.66

All components have decent contributions to the accuracy, with the inter-cell skip connection in the new cell template contributing the most; without it, the models eventually collapsed to very low training and test accuracy and exhibited unstable gradient issues as discussed in Sect. 4.2. Comparing *No Div* with *Full*, the searched cell with the diversity regularizer has a clear gain over the searched cell without it in all the model variants. Interestingly, the largest model (BNAS-C) without *Zeroise* layers performs worse than BNAS-A and BNAS-B, due to excess complexity. Please refer to the supplementary material for more discussion.

6   Conclusion

To design better performing binary network architectures, we propose a method to search the space of binary networks, called BNAS. BNAS searches for a cell that can be stacked to generate networks for various computational budgets. To configure the feasible set of binary architectures, we define a new search

---

[4] We have confirmed with the authors of [32] that their results were reported without unbinarizing the downsampling convolutions.

space of binary layer types and a new cell template. Specifically, we propose to exclude separable convolution layer and include *Zeroise* layer type in the search space for less quantization error. Further, we propose a new search objective with the diversity regularizer and show that it helps in obtaining better binary architectures. The learned architectures outperform the architectures used in the state-of-the-art binary networks on both CIFAR-10 and ImageNet.

Acknowledgement. This work was partly supported by the National Research Foundation of Korea (NRF) grant funded by the Korea government (MSIT) (No.2019R1 C1C1009283), Institute of Information & communications Technology Planning & Evaluation (IITP) grant funded by the Korea government (MSIT) (No.2019-0-01842, Artificial Intelligence Graduate School Program (GIST) and No.2019-0-01351, Development of Ultra Low-Power Mobile Deep Learning Semiconductor With Compression/Decompression of Activation/Kernel Data), "GIST Research Institute(GRI) GIST-CNUH research Collaboration" grant funded by the GIST in 2020, and a study on the "HPC Support" Project, supported by the 'Ministry of Science and ICT' and NIPA.

The authors would like to thank Dr. Mohammad Rastegari for valuable comments and training details of XNOR-Net and Dr. Chunlei Liu and other authors of [25] for sharing their code.

References

1. Bender, G., Kindermans, P.J., Zoph, B., Vasudevan, V., Le, Q.: Understanding and simplifying one-shot architecture search. In: ICML (2018)
2. Bulat, A., Martínez, B., Tzimiropoulos, G.: Bats: binary architecture search. ArXiv preprint arXiv:2003.01711 abs/2003.01711 (2020)
3. Bulat, A., Tzimiropoulos, G.: XNOR-Net++: improved binary neural networks. In: BMVC (2019)
4. Cai, H., Zhu, L., Han, S.: ProxylessNAS: direct neural architecture search on target task and hardware. In: ICLR (2019). https://openreview.net/forum?id=HylVB3AqYm
5. Chen, X., Xie, L., Wu, J., Tian, Q.: Progressive differentiable architecture search: bridging the depth gap between search and evaluation. In: ICCV, pp. 1294–1303 (2019)
6. Courbariaux, M., Bengio, Y., David, J.P.: Binaryconnect: training deep neural networks with binary weights during propagations. In: NIPS (2015)
7. Courbariaux, M., Hubara, I., Soudry, D., El-Yaniv, R., Bengio, Y.: Binarized neural networks: training deep neural networks with weights and activations constrained to +1 or −1. arXiv preprint arXiv:1602.02830 (2016)
8. Dong, J.-D., Cheng, A.-C., Juan, D.-C., Wei, W., Sun, M.: DPP-Net: device-aware progressive search for pareto-optimal neural architectures. In: Ferrari, V., Hebert, M., Sminchisescu, C., Weiss, Y. (eds.) ECCV 2018. LNCS, vol. 11215, pp. 540–555. Springer, Cham (2018). https://doi.org/10.1007/978-3-030-01252-6_32
9. Dong, X., Yang, Y.: Searching for a robust neural architecture in four GPU hours. In: CVPR (2019)
10. Gu, J., et al.: Projection convolutional neural networks for 1-bit CNNs via discrete back propagation. In: AAAI (2019)

11. Gu, J., et al.: Bayesian optimized 1-bit CNNs. In: CVPR (2019)
12. Han, S., Pool, J., Tran, J., Dally, W.: Learning both weights and connections for efficient neural network. In: NIPS (2015)
13. Hinton, G., Vinyals, O., Dean, J.: Distilling the knowledge in a neural network. arXiv preprint arXiv:1503.02531 (2015)
14. Howard, A.G., et al.: Mobilenets: efficient convolutional neural networks for mobile vision applications. arXiv preprint arXiv:1704.04861 (2017)
15. Iandola, F.N., Han, S., Moskewicz, M.W., Ashraf, K., Dally, W.J., Keutzer, K.: SqueezeNet: alexnet-level accuracy with 50x fewer parameters and <0.5 mb model size. arXiv:1602.07360 (2016)
16. Jaderberg, M., Vedaldi, A., Zisserman, A.: Speeding up convolutional neural networks with low rank expansions. arXiv preprint arXiv:1405.3866 (2014)
17. Jang, E., Gu, S., Poole, B.: Categorical reparameterization with gumbel-softmax. In: ICLR (2017). https://arxiv.org/abs/1611.01144
18. Kim, H., Kim, K., Kim, J., Kim, J.J.: Binaryduo: reducing gradient mismatch in binary activation network by coupling binary activations. In: ICLR (2020). https://openreview.net/forum?id=r1x0lxrFPS
19. Krizhevsky, A.: Learning multiple layers of features from tiny images. Technical report (2009)
20. Li, F., Zhang, B., Liu, B.: Ternary weight networks. arXiv preprint arXiv:1605.04711 (2016)
21. Li, L., Talwalkar, A.: Random search and reproducibility for neural architecture search. arXiv preprint arXiv:1902.07638 (2019)
22. Lin, X., Zhao, C., Pan, W.: Towards accurate binary convolutional neural network. In: NIPS (2017)
23. Liu, C., et al.: Auto-deeplab: hierarchical neural architecture search for semantic image segmentation. In: CVPR (2019)
24. Liu, C., et al.: Progressive neural architecture search. In: Ferrari, V., Hebert, M., Sminchisescu, C., Weiss, Y. (eds.) ECCV 2018. LNCS, vol. 11205, pp. 19–35. Springer, Cham (2018). https://doi.org/10.1007/978-3-030-01246-5_2
25. Liu, C., et al.: Circulant binary convolutional networks: enhancing the performance of 1-bit DCNNs with circulant back propagation. In: CVPR (2019)
26. Liu, H., Simonyan, K., Vinyals, O., Fernando, C., Kavukcuoglu, K.: Hierarchical representations for efficient architecture search. In: ICLR (2018). https://openreview.net/forum?id=BJQRKzbA-
27. Liu, H., Simonyan, K., Yang, Y.: DARTS: differentiable architecture search. In: ICLR (2019). https://openreview.net/forum?id=S1eYHoC5FX
28. Liu, Z., Wu, B., Luo, W., Yang, X., Liu, W., Cheng, K.-T.: Bi-Real Net: enhancing the performance of 1-Bit CNNs with improved representational capability and advanced training algorithm. In: Ferrari, V., Hebert, M., Sminchisescu, C., Weiss, Y. (eds.) ECCV 2018. LNCS, vol. 11219, pp. 747–763. Springer, Cham (2018). https://doi.org/10.1007/978-3-030-01267-0_44
29. Luo, R., Tian, F., Qin, T., Chen, E., Liu, T.Y.: Neural architecture optimization. In: NIPS (2018)
30. Pham, H., Guan, M., Zoph, B., Le, Q., Dean, J.: Efficient neural architecture search via parameters sharing. In: ICML (2018)
31. Phan, H., Huynh, D., He, Y., Savvides, M., Shen, Z.: Mobinet: a mobile binary network for image classification. arXiv preprint arXiv:1907.12629 (2019)

32. Rastegari, M., Ordonez, V., Redmon, J., Farhadi, A.: XNOR-Net: ImageNet classification using binary convolutional neural networks. In: Leibe, B., Matas, J., Sebe, N., Welling, M. (eds.) ECCV 2016. LNCS, vol. 9908, pp. 525–542. Springer, Cham (2016). https://doi.org/10.1007/978-3-319-46493-0_32

33. Russakovsky, O., et al.: ImageNet large scale visual recognition challenge. IJCV **115**(3), 211–252 (2015)

34. Shen, M., Han, K., Xu, C., Wang, Y.: Searching for accurate binary neural architectures. In: ICCV Workshop (2019)

35. Sifre, L., Mallat, S.: Rigid-motion scattering for image classification (2014)

36. Tan, M., Le, Q.V.: Efficientnet: rethinking model scaling for convolutional neural networks. In: ICML (2019)

37. Tan, S., Caruana, R., Hooker, G., Koch, P., Gordo, A.: Learning global additive explanations for neural nets using model distillation. arXiv preprint arXiv:1801.08640 (2018)

38. Wu, B., et al.: FBNet: hardware-aware efficient convnet design via differentiable neural architecture search. In: CVPR (2019)

39. Xie, S., Kirillov, A., Girshick, R., He, K.: Exploring randomly wired neural networks for image recognition. arXiv preprint arXiv:1904.01569 (2019)

40. Xie, S., Zheng, H., Liu, C., Lin, L.: SNAS: stochastic neural architecture search. In: ICLR (2019). https://openreview.net/forum?id=rylqooRqK7

41. Zhang, C., Ren, M., Urtasun, R.: Graph hypernetworks for neural architecture search. In: ICLR (2019). https://openreview.net/forum?id=rkgW0oA9FX

42. Zhang, X., Zhou, X., Lin, M., Sun, J.: Shufflenet: an extremely efficient convolutional neural network for mobile devices. In: CVPR (2018)

43. Zhou, Y., Ebrahimi, S., Arık, S.Ö., Yu, H., Liu, H., Diamos, G.: Resource-efficient neural architect. arXiv preprint arXiv:1806.07912 (2018)

44. Zhuang, B., Shen, C., Tan, M., Liu, L., Reid, I.: Towards effective low-bitwidth convolutional neural networks. In: CVPR (2018)

45. Zoph, B., Le, Q.V.: Neural architecture search with reinforcement learning. In: ICLR (2017). https://openreview.net/forum?id=r1Ue8Hcxg

46. Zoph, B., Vasudevan, V., Shlens, J., Le, Q.V.: Learning transferable architectures for scalable image recognition. In: CVPR (2018)

Semantic View Synthesis

Hsin-Ping Huang[1]([✉]), Hung-Yu Tseng[2], Hsin-Ying Lee[2], and Jia-Bin Huang[3]

[1] UT Austin, Austin, USA
hsinping@cs.utexas.edu
[2] University of California, Merced, USA
[3] Virginia Tech, Blacksburg, USA

Abstract. We tackle a new problem of semantic view synthesis—generating free-viewpoint rendering of a synthesized scene using a semantic label map as input. We build upon recent advances in semantic image synthesis and view synthesis for handling photographic image content generation and view extrapolation. Direct application of existing image/view synthesis methods, however, results in severe ghosting/blurry artifacts. To address the drawbacks, we propose a two-step approach. First, we focus on synthesizing the color and depth of the visible surface of the 3D scene. We then use the synthesized color and depth to impose explicit constraints on the multiple-plane image (MPI) representation prediction process. Our method produces sharp contents at the original view and geometrically consistent renderings across novel viewpoints. The experiments on numerous indoor and outdoor images show favorable results against several strong baselines and validate the effectiveness of our approach.

1 Introduction

Visual content creation using generative models has been gaining increasing attention. Driving by the advances in generative models, recent work has demonstrated impressive performance on a wide range of tasks, including image generation from various contexts (e.g., noises [12,24], images [1,20,22,26,56], text [43, 51], and audio [28]), view interpolation and extrapolation [8,15,41,44,55], and image editing [2,5,42]. These algorithms greatly help unleash human imagination and support creative processes. In this paper, we introduce a new form of visual content creation task by integrating (1) semantic image synthesis and (2) novel view synthesis.

Semantic image synthesis [3,35,37,46] is a specific form of image-to-image translation task that aims to generate photorealistic images from semantic label maps. Such an application is intuitive as users can easily draw and refine the

Electronic supplementary material The online version of this chapter (https://doi.org/10.1007/978-3-030-58610-2_35) contains supplementary material, which is available to authorized users.

A. Vedaldi et al. (Eds.): ECCV 2020, LNCS 12357, pp. 592–608, 2020.
https://doi.org/10.1007/978-3-030-58610-2_35

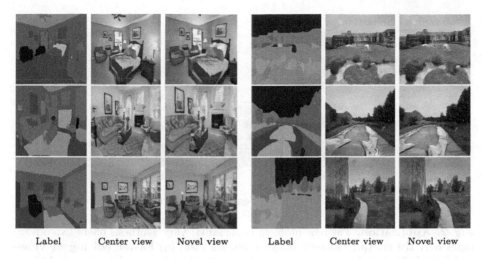

| Label | Center view | Novel view | Label | Center view | Novel view |

Fig. 1. Semantic view synthesis. We introduce a new visual synthesis problem, semantic view synthesis—synthesizing a *photorealistic* image that supports *free-viewpoint rendering* given a single semantic label map. To achieve such visual effects, we build a two-step inference pipeline upon recent advances in semantic view synthesis and novel view synthesis. We show that our model learns to generate scene representations for rendering geometrically consistent and semantically meaningful novel views. We demonstrate the efficacy of our method using a wide variety of indoor (*left*) and outdoor (*right*) scenes.

semantic map on a digital canvas and then use the algorithm to synthesize *2D images* with plausible appearances. As these algorithms produce only 2D outputs, it is challenging for users to manipulate the viewpoints of the synthesized image in a geometrically consistent manner.

View synthesis, on the other hand, takes a sparse set of real images (captured at different viewpoints) as inputs and synthesizes novel views of the same scene [7,15,41,44,55]. This is achieved by explicitly or implicitly modeling the *3D structure* of the scene. However, these methods are applicable only to real images.

In this paper, we propose to tackle a new problem: *semantic view synthesis*—generating free-viewpoint rendering of a synthesized scene using a semantic label map as input (Fig. 1). Compared to the existing semantic image synthesis task, the semantic view synthesis problem offers two unique advantages (Fig. 2). First, it allows the users to easily manipulate the viewpoints of the synthesized image with minimal effort. Second, it supports temporally and geometrically consistent rendering of 3D fly-through effects.

To enable this new application, we develop a two-step method, drawing inspirations from the recent advances in semantic image synthesis and view synthesis algorithms. First, given the input semantic label map, we leverage a state-of-the-art image synthesis model, SPADE [35], to generate a photorealistic color image and the corresponding disparity map. The synthesized color/disparity images

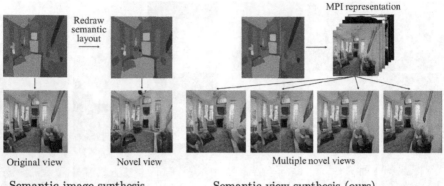

Fig. 2. Application. The new problem of semantic view synthesis offers two advantages over the existing semantic image synthesis task. (a) **Faster editing of viewpoints.** (*Left*) To refine the viewpoint of a synthesized image, the users would have to *redraw* the semantic layout of the scene and apply the image synthesis algorithm on the new semantic layout again to produce the desired view. (*Right*) Taking a single semantic layout as input, our method produces an MPI representation that naturally supports fast, free-form novel view rendering. (b) **Consistent rendering over viewpoints.** (*Left*) As novel view images are *independently* generated, the synthesized contents may not be consistent. (*Right*) Our semantic view synthesis, in contrast, enables 3D fly-through effects with plausible motion parallax.

capture the appearance and structure of the *visible surface* of the scene. Second, to handle the dis-occluded contents (which become visible at novel views), we infer a multiplane images (MPI) representation [55] using the synthesized color/disparity as constraints. The resulting output of our method is an MPI representation that naturally supports view synthesis at any viewpoints. We conduct extensive quantitative and visual comparisons on three datasets (ADE20K [53], ADE20k-outdoor [37], and NYUv2 [33]) covering various indoor and outdoor scenes. Our results demonstrate clear improvement over several strong baseline methods and alternative designs.

In summary, we make the following contributions:

- We introduce a new *semantic view synthesis* task that aims to synthesize images of free-viewpoint from semantic masks.
- We propose a novel two-step training and inference pipeline: (1) color and disparity image synthesis for the visible surface and (2) MPI prediction with explicit constraints from the first step (Sect. 3).
- We build several baseline approaches for this new problem and validate the efficacy of our proposed framework on a wide variety of indoor and outdoor scenes (Sect. 4).

2   Related Work

Monocular depth prediction aims to estimate the depth of a scene from a *single-view* RGB image. It is a challenging problem due to the difficulty of obtaining explicit 3D cue from the single-view RGB image without additional information (e.g., stereo pair). To conquer the problem, several supervised learning schemes [9,19,25,48] utilize the ground-truth depth notation in the RGB-D dataset and train fully-convolutional networks (e.g., [31]) to capture the image prior. However, these approaches require large and diverse annotated data for the training. Numerous self-supervised approaches [10,11,49,54,59] have been proposed to avoid the labor-intensive annotating process. For instances, training with stereo videos [10], monocular videos [11], incorporating the information of camera poses or optical flow [49,54,58,59]. Nevertheless, these supervised and unsupervised methods often train their models using data from specific domains (e.g., driving scenes from the KITTI dataset) and therefore have difficulty in generalizing to diverse scenes in the wild. On the other hand, a line of approaches uses multi-view internet photos [30], MannequinChallenge [29] or 3D movies [38,45] as the source of data. In particular, training with mixed datasets from different sources achieves strong generality on unseen scenes. Our work leverage the pre-trained single-view depth estimation model from MiDaS [38] to obtain (pseudo) ground truth of depth/disparity maps for images in our training dataset.

Novel view synthesis aims to generate novel views based on single or multiple images. Earlier learning-based approaches [8,23] take multiple posed images as input and produce the target views by blending the warped input images. Such approaches, however, only *interpolate* among the given viewpoints and do not handle dis-occlusions. Recent advances explore generating novel view through a 3D scene representations, such as multi-plane images [7,32,41,44,55], layered depth images [6], mesh representations [14,40], and point clouds [47]. The multi-plane image representation [7,32,41,44,55] is a set of RGBA layers at discrete disparity levels. The novel views are rendered by homographic projection and alpha blending of the MPI layers. The layered depth image approach [6] represents 3D images as a foreground RGBD image and a background RGBD image. To generate the novel views, the RGB image is warped by the depth image, then composite by a predicted visibility mask. This approach requires supervision of the background image and only works for synthetic scenes. 3D photography [14,40] focuses on generating 3D effects for real-world photos; they represent 3D images as a multi-layer 3D mesh. These methods generate scene representation at the reference (original) viewpoint. The novel view images can be rendered by projecting the scene representation to the desired viewpoint.

Our work also produces an MPI representation as our output for supporting novel view synthesis. Our problem setting, however, differs significantly from prior MPI-based methods. Prior methods often require (at least) two images as inputs, which consist of the appearance of visible surfaces, cues of scene depth, and some content of the occluded background. In contrast, the input to our method is one semantic label map. Our experimental results show that direct application of prior MPI-based methods leads to severe blurry ghosting artifacts when rendered at novel views. Our two-step approach substantially reduces these artifacts via imposing explicit constraints on the MPI representation during training and testing time.

Image-to-image translation Aims to learn the mapping between two image domains [1,20,22,27,56,57]. These techniques demonstrate a wide range of applications such as image inpainting, image super-resolution, domain adaptation [4,18], and semantic image synthesis [35,46]. In particular, semantic image synthesis learns to generate photo-realistic images conditioned on semantic label maps. Pix2pix [22] adopts a U-Net architecture to synthesize low-resolution images from a semantic map. To operate in high-resolution settings, Pix2pixHD [46] introduces the multi-scale generator and discriminator network structure to enhance the quality of the generated images. SPADE [35] further improves Pix2pixHD with the spatially-adaptive normalization layers. Different from the semantic image synthesis frameworks, we aim to synthesize 3D representation of a scene from a *single-view* semantic segmentation layout.

Cross-modal distillation Transfers the knowledge between different modalities. Existing works [13,17] use learned representation from a large labeled dataset of the source modality as a supervised signal to train tasks of target modality with limited data. For example, the method in [13] utilize ImageNet-pretrained model to train new representations for optical flow and depth images. To address the problem of collecting a large indoor/outdoor dataset of semantic map to depth image pairs, our work also incorporates the idea of cross-modal distillation. Specifically, We transfer the knowledge of monocular depth prediction model (predicting depth maps from images) and semantic segmentation (predicting semantic layouts from images) to our *semantic depth synthesis* (predicting depth from semantic layouts). To this end, we present a two-branch version of a SPADE network [35] to predict both color and depth from a single semantic map.

3   Method

3.1   Overview

Our goal is to learn to synthesize novel-view color images from a given a semantic label map. As shown in Fig. 3, our scene representation generation process consists of (1) image and disparity generation module and (2) MPI prediction module. With the generated MPI, we can project and blend the MPI to produce the desired target views. In this section, we first describe the data preparation in Sect. 3.2. We then detail the training procedure of scene representation generation including image and disparity generation and the MPI prediction in Sect. 3.3. Finally, we introduce the novel view synthesis procedure at test time in Sect. 3.4.

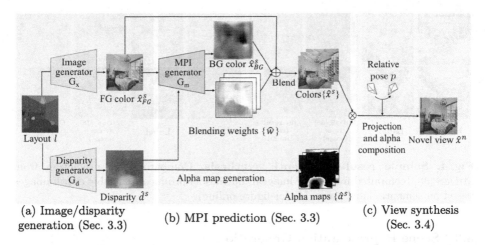

(a) Image/disparity generation (Sec. 3.3)

(b) MPI prediction (Sec. 3.3)

(c) View synthesis (Sec. 3.4)

Fig. 3. Method overview. Our method first produces an MPI-based scene representation via a two-step approach (a) (b). (a) Our first step focuses on synthesizing the color and disparity image from the given semantic label map as the *visible surface*. Here, we present a Y-shaped network with partially shared color/depth decoder architecture to ensure consistency between the synthesized color and depth maps. (b) We then infer the MPI representation that captures the color and structure of both the visible surface and the dis-occluded surfaces. With only one single RGB image as input, it is challenging to learn MPI with high-quality view renderings. This is because the network needs to predict both the appearances at multiple depth levels as well as the alpha (transparency) maps. To address this issue, we directly generate the alpha maps using the synthesized depth map from step (a), and we use the synthesized depth map for modulating the activations in normalization layers [35] in our MPI generator. Such an approach imposes effective constraints and results in improved MPI prediction. (c) Given target camera poses, we can then project and blend the generated MPI representation for rendering images at novel views. (Color figure online)

3.2   Data Preparation

We build a dataset from the RealEstate10K dataset [55], which consists of 80,000 indoor/outdoor YouTube video clips with camera poses for each frame. To extract training pairs of the semantic layout and the corresponding disparity map, we adopt the idea of cross-modal distillation (Fig. 5a). Specifically, we apply PSPNet [52] (pretrained on the ADE20K [53]) to obtain segmentation map annotation. Similar, we apply the pre-trained MiDaS [38] monocular depth estimation network to estimate the corresponding disparity map. Since MiDaS predicts the *relative* disparity with unknown scale/shift, we use the absolute depth prediction from DPSNet [21] to estimate the scale and shift for each training image. The relative disparity images are then transformed into absolute disparity images that serve as the (pseudo) ground-truth images for training. We collect training pairs from each frame in the RealEstate10K dataset. While existing Habitat [39] framework also provides semantic layouts, disparity maps and multi-view images with camera poses, we did not use it as the dataset contains indoor scenes only.

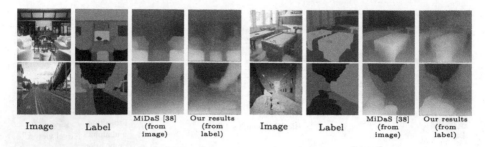

| Image | Label | MiDaS [38] (from image) | Our results (from label) | Image | Label | MiDaS [38] (from image) | Our results (from label) |

Fig. 4. Sample results of depth synthesis. Comparing the prediction from MiDaS [38] (computed from color images), our model produces plausible depth images based on semantic label maps. (Color figure online)

3.3 Scene Representation Generation

We adopt a two-step prediction strategy due to the difficulty of predicting MPI representation in one step. First, our image and disparity generator takes the semantic layout l as input and learns to synthesize the corresponding color image $\hat{x}_{FG}^s$ and disparity image $\hat{d}^s$ of the visible surface. Second, the MPI generator uses the synthesized color image and the disparity as input and predicts an MPI representation $\hat{m}^s$ of the scene.

Image and Disparity Generation. Image and disparity generator aims to synthesize the color $\hat{x}_{FG}^s$ and disparity image $\hat{d}^s$ of *visible surface* of the scene (Fig. 5b). To this end, we modify the SPADE [35] model into two-stream generators (with the color generator G_x and the disparity generator G_d). The two-stream generators G_x and G_d share the first three SPADE-style ResNet blocks. Using the training pairs of semantic layout l and disparity image d, we use the losses in SPADE [35] for training the color stream and an $\ell 1$ reconstruction loss for training the disparity stream. Figure 4 shows sample results of disparity prediction from a semantic label map.

MPI Prediction. For simple scenes (e.g., there is no apparent occluded region in the input image), using a single image with the associated disparity map will suffice for modeling the 3D scene. However, synthesizing novel-view images with only color and disparity map inevitably induce visible artifacts, particularly in the dis-occluded regions, thereby failing to render general scenes where multiple depth layers exist. We therefore use an MPI representation [55] for handling the depth-complex scenarios. An MPI [55] $m = \{(x_k, \alpha_k)\}_{k=1}^K$ is a collection of RGBA images, where K is the number of depth planes. Each layer k is an image plane placed at a fixed depth with respect to a virtual reference camera. The color images x_k at each depth plane indicate the visible view, while the alpha image α_k represents the visibility, which has a range between 0 and 1.

However, we find that predicting the MPI using only a single color image results in poor visual quality. The primary reason is that without depth cues (e.g., stereo pair in [55]), it is challenging to predict accurate alpha (transparency) maps for compositing multi-plane images. To tackle this issue, we

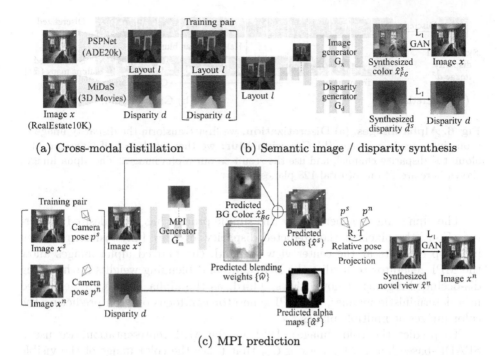

(a) Cross-modal distillation   (b) Semantic image / disparity synthesis

(c) MPI prediction

Fig. 5. Training pipeline. Our model training process consists of the following steps. (a) **Cross-modal distillation**: We generate *pseudo* training pairs for training the semantic image/depth generation by applying the pre-trained depth estimation model [38] and semantic segmentation PSPNet [52] on training images. (b) **Semantic depth and image synthesis**: Using the generated training pairs from the cross-modal distillation step, we use a two-stream (color and disparity) SPADE network to generate the visible surface. We train the color stream using the losses in [35] and the disparity stream with an $\ell 1$ loss (based on the normalized disparity values). (c) **MPI prediction**: We use training pairs of source/target images with relative pose annotations (provided by [55]). We train the MPI generator to produce colors at multiple depth levels and use $\ell 1$ and GAN loss to enforce the consistency between the projected image and the target image. Note that the MPI generator does *not* need to predict the alpha (transparency) maps. (Color figure online)

directly compute and constrain the alpha images from the synthesized disparity map $\hat{d}^s$. Since the synthesized disparity map $\hat{d}^s$ provides a strong prior for the scene visibility at different depth layers, we transform it into the alpha images $\{\hat{\alpha}_k^s\}$ in our MPI representation (Fig. 6). Specifically, we first transform the disparity image into a *one-hot representation* with K disparity channels, according to the inverse depth. Then, we apply a half Gaussian blur along the disparity channel, which produces blurring effect only *behind* the predicted disparity and has a peak value at the predicted disparity. The blurred one-hot disparity images are then used as the alpha images in our MPI representation.

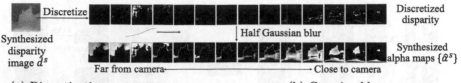

Fig. 6. Alpha images. (a) **Discretization:** we first transform the disparity image to a one-hot encoding image. (b) **Gaussian blur:** we then apply a half Gaussian blur along the disparity channel, and use the result as our alpha images. The alpha images shown here are 14 out of total 128 planes.

The alpha images generated by this simple process has three desired properties. First, the pixels at the predicted disparity level are fully visible, resulting in sharp contents at the center view. Second, the blurred alpha images allow the MPI generator to predict the BG colors and blending weights for handling dis-occluded regions at novel views. Third, as the alpha images are generated in a deterministic manner, the MPI generator can focus only on predicting the color images at multiple planes.

To predict the color images, $\{\hat{x}_k^s\}$ in the MPI representation, we use a SPADE-based [35] MPI generator G_m that takes the color image of the visible surface $\hat{x}_{FG}^s$ as main input, and uses the disparity image $\hat{d}^s$ for modulating the activations in normalization layers. The MPI generator synthesizes a background color image $\hat{x}_{BG}^s$ and a set of blending weights $\{\hat{w}_k\}$. The color images $\{\hat{x}_k^s\}$ are calculated as the weighted sum of the foreground $\hat{x}_{FG}^s$ and the background $\hat{x}_{BG}^s$:

$$\hat{x}_k^s = \hat{w}_k \odot \hat{x}_{FG}^s + (1 - \hat{w}_k) \odot \hat{x}_{BG}^s \qquad (1)$$

We refer the reader to Zhou et al. [55] for more details on synthesizing novel view images using an MPI representation.

Training MPI Generator. Figure 5c illustrates the training process of MPI prediction. We use the data sampling strategy in [55] to sample the training image pair $(x^s, x^n) = (x_{FG}^s, x^n)$ (note that x^s is equivalent to x_{FG}^s) with corresponding camera poses (p^s, p^n), as well as the disparity image d^s, where the notation s and n indicate the *source* and *novel* view, respectively. Our MPI generator predicts the color images $\{\hat{x}_k^s\}$ from the source color image x_{FG}^s. We transform the disparity image d^s into alpha images $\{\alpha_k^s\}$.

With the predicted MPI representation $\hat{m}^s = (\{\hat{x}_k^s\}, \{\alpha_k^s\})$, we can use the warped multi-plane images according to the relative pose p^{n-s} between the source pose p^s and novel pose p^n. Given the warped MPIs, we then use the over-composited approach [36] to composite the novel view $\hat{x}^n$. We train the MPI generator using an $\ell 1$ loss and a GAN loss of weight 0.01 between the generated and the ground-truth color image at the novel view x^n.

3.4 Novel View Synthesis

Similar to the training process, at test time, we follow the two-step approach for generating an MPI. First, we generate color $\hat{x}_{FG}^s$ and disparity image $\hat{d}^s$ from input semantic layout l. We then use both color $\hat{x}_{FG}^s$ and disparity image $\hat{d}^s$ to predict the MPI representation $\hat{m}^s = (\{\hat{x}_k^s\}, \{\hat{\alpha}_k^s\})$. Given a relative camera pose, we can warp and over-composite the predicted MPI and obtain the novel view image $\hat{x}^n$.

4   Experimental Results

4.1   Experimental Setup

Datasets. We validate our method on three datasets.

- **ADE20K** [53] is a dataset of diverse indoor and outdoor scenes. It consists of 2,000 testing images with 150 semantic classes.
- **ADE20K-outdoor** [37] is a subset of outdoor scenes in ADE20K dataset. It consists of 1,035 testing images with 150 semantic classes.
- **NYU** [33] is an indoor dataset. It consists of 249 testing images with 13 semantic classes.

Implementation Details. We implement our system in PyTorch and use the Adam optimizer with $\beta_1 = 0$, $\beta_2 = 0.9$ for all network training. All the experiments are conducted on an NVIDIA GTX 1080. The color module, the disparity module and the MPI module are trained for 600k/300k/300k iterations respectively. We use a batch size of one with a learning rate of 0.0002. We use $K = 128$ image planes for our MPI representations. We set the disparity of each alpha map equally distributed from 0.01 m to 1 m, according to the inverse depth. The Gaussian blur we use for the alpha images has a peak 1, window 31, and the σ value of 10. We set the size of the target synthesized images as 384×384 for all the models. Our source code and the pre-trained models are available on the project website.

Baselines. We compare our methods with four baseline methods.

- (a) **Direct (U-Net)** synthesizes the multi-plane images directly from the semantic layout using a fully-convolutional encoder-decoder architecture [55].
- (b) **Direct (SPADE)** also synthesizes the multi-plane images directly from the semantic layout, but uses a generator with spatially-adaptive normalization [35].
- (c) **Cascade (MPI)** first synthesizes a color image from the semantic layout using SPADE [35], then apply an MPI predictor using the synthesized image as input. Here, we modify the original MPI generation model in [55] so that it takes a single image as input.
- (d) **Cascade (KB)** first synthesizes a color image from the semantic layout using SPADE [35], then apply a recent single-image view synthesis method (3D Ken Burns [34]).

Training and testing details of the baseline models can be found in the supplementary material.

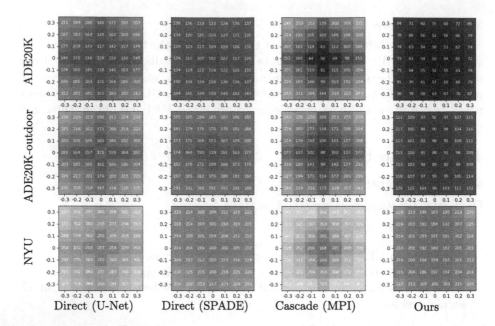

Fig. 7. Quantitative evaluation. We compare the results of three alternative approaches for semantic view synthesis and our model on the ADE20K, ADE20K-outdoor, and NYU datasets. Each table shows the FID score of generated novel view images at 7×7 grids of target viewpoints. A lower FID score is better. Using Cascade (MPI) and Direct (U-Net) MPI architectures is unable to produce sharp, photorealistic contents (therefore high FID scores). The Direct (SPADE) method can synthesize detailed contents at the center view due to the use of SPADE [35]. However, its performance degrades rapidly when the camera viewpoints move away from the center view. Our two-step generation preserves the detailed content in the front layer while maintaining photorealism under novel views. We were not able to include Cascade (KB) due to different camera movements.

4.2   Quantitative Evaluation

We use the Fréchet Inception Distance (FID) [16] to measure the distance between the distribution of generated images and real images. We use ADE20K images as real images. For measuring the realism of novel view synthesis, we evaluate the FID scores of generating novel views at 7×7-grid viewpoints on x-y planes with camera movement from $-0.3\,\text{m}$ to $0.3\,\text{m}$ across both axes. The center view with camera movement $(0,0)$ shows the performance of semantic image synthesis. As shown in Fig. 7, all the baselines, and our model produce the lowest FID score at the center view, and the FID score gradually increases when the camera movement becomes larger. The trend is similar across different datasets. We discuss the results based on the ADE20K dataset below.

Results at the Center View. Comparing methods directly synthesizing MPIs from layouts, Direct (SPADE) performs better than Direct (U-Net) (102 vs. 128) due to the use of the SPADE architecture. Comparing methods that both employ

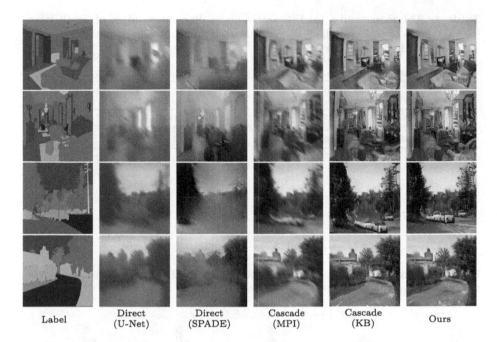

| Label | Direct (U-Net) | Direct (SPADE) | Cascade (MPI) | Cascade (KB) | Ours |

Fig. 8. Visual comparisons. We compare the generated novel view images of four other baselines and our model among ADE20K and ADE20K-outdoor datasets. The left column shows the input label at the center viewpoint.

the SPADE generator, Cascade (MPI) performs better than Direct (SPADE) (50 vs. 102), suggesting the difficulty of directly predicting MPI from semantic layout. Our method achieves the same FID score 50 when compared with Cascade (MPI) at the center view as the input (synthesized color image) is the same.

Results at the Novel Views. When evaluating the results at a novel view (e.g., $(0.3, 0.3)$ m away from the center), we observe that while the Cascade (MPI) method performs well at the center view, it produces significantly inferior to the methods that directly predict MPI. In contrast, our method produces lowest FID scores among the competing baselines.

4.3   Visual Comparisons

Figure 8 compares the generated novel view images of four baselines and our model. Two-step methods, Cascade (MPI), Cascade (3D Ken Burns) and Ours, produce images with sharper contents. Direct (U-Net) and Direct (SPADE) tend to produce blurry and less plausible contents. In particular, the results of Cascade (MPI) suffer from blurry due to the difficulty of generating alpha images when no depth cues (e.g., multiple images, plane sweep volume) are available. The Cascade (KB) inpaints the dis-occluded region at only *one* novel viewpoint. Such a method supports 3D Ken Burns effect with a simple camera trajectory such as zooming in, but not free-viewpoint rendering.

Table 1. Ablation study. (a) FID scores under different numbers of depth layers. (b) FID scores of replacing the MPI prediction with per-frame background inpainting. We use NYU dataset for this experiment.

(a) Number of depth layers.

	Camera movement		
	(0, 0)	(0.1, 0.1)	(0.2, 0.2)
128	**188.83**	**191.04**	207.70
64	190.70	193.71	210.06
32	190.60	194.73	**205.59**

(b) Handling dis-occlusion.

	Camera movement		
	(0, 0)	(0.1, 0.1)	(0.2, 0.2)
Ours	**188.83**	**191.04**	**207.70**
Diffusion	190.63	192.55	210.16
GatedConv	190.67	192.83	210.00

K=32   K=64   K=128   K=32   K=64   K=128

Fig. 9. Number of depth layers. Increasing the number of depth levels improves the rendered quality.

Foreground only   Diffusion   GatedConv [50]   KB [34]   Ours

Fig. 10. Disocclusion handling. The purple regions (left) are the dis-occluded region. *Diffusion* and *GatedConv* produce artifacts. The 3D Ken Burns method [34] generates blurry and unnatural dis-occluded contents. Our model hallucinates visually appealing results. (Color figure online)

4.4   Ablation Study

Number of Depth Layers. Table 1a shows the results of having a different number of depth layers in our MPI. At $(0.2, 0.2)$, the model with $K = 32$ achieves better FID. At $(0, 0)$ and $(0.1, 0.1)$, the model with $K = 128$ achieves better FID. We conclude that more MPI planes lead to slightly blurrier results for large camera movement. Figure 9 illustrates that the novel view synthesized with 32 depth layers show more artifacts than 64 or 128 depth layers.

Background Inpainting. We explore alternative methods for handling the dis-occluded regions when rendering at novel views. We use the standard backward warping to project the synthesized color image using disparity image to render the novel views. We then inpaint the missing pixels using either simple diffusion (implemented in OpenCV) or a learning-based image inpainting model (GatedConv [50]).

Table 1b shows that our method achieves lower FID scores at three viewpoints. Note that as all the novel view images are processed independently, *Diffusion* and *GatedConv* approaches do not retain the consistency across different viewpoints. We refer the readers to the supplementary materials for video results. Figure 10 shows that while our method produces slightly blurry foreground (due to the over-composition of multi-plane images), our MPI representation hallucinates plausible dis-occluded regions.

4.5 User Study

We conducted a perceptual user study to quantify the user preference over the proposed method and the six baseline approaches. For each test during the study, we present two novel view videos of the same scene generated by two different methods with circular camera motion (in randomized order). We then ask the participant to select his/her preferred result.

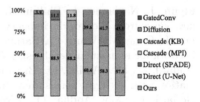

Fig. 11. User study. We show the user preference between the proposed method and baselines.

There are 120 videos (60 pairwise comparisons) generated from the layouts in ADE20K, ADE20K-outdoor, and NYU datasets used. We conduct the study with 47 participants (2820 binary votes). The results shown in Fig. 11 validate that the proposed method synthesizes more realistic novel view videos compared to the baseline approaches.

5   Conclusions

We have introduced a new problem called semantic view synthesis. The problem aims to generate a photorealistic image from a given semantic label map that supports novel view rendering. The new form of visual content creation offers significantly more immersive experience than the conventional 2D image synthesis task. This is technically achieved by carefully integrating techniques from semantic image synthesis and view synthesis. Our core idea is to model the 3D scene by first modeling the visible surface then further inferring the full 3D scene representation. We conduct an extensive experimental evaluation to validate our model design and show favorable results over several baseline methods.

References

1. AlBahar, B., Huang, J.B.: Guided image-to-image translation with bi-directional feature transformation. In: ICCV (2019)
2. Bau, D., et al.: Semantic photo manipulation with a generative image prior. ACM Trans. Graph. (TOG) **38**(4), 1–11 (2019)
3. Chen, Q., Koltun, V.: Photographic image synthesis with cascaded refinement networks. In: ICCV (2017)
4. Chen, Y.C., Lin, Y.Y., Yang, M.H., Huang, J.B.: Crdoco: pixel-level domain transfer with cross-domain consistency. In: CVPR (2019)
5. Cheng, Y.C., Lee, H.Y., Sun, M., Yang, M.H.: Controllable image synthesis via SegVAE. In: Vedaldi, A., et al. (eds.) ECCV 2020. LNCS, vol. 12352. Springer, Heidelberg (2020)
6. Dhamo, H., Tateno, K., Laina, I., Navab, N., Tombari, F.: Peeking behind objects: layered depth prediction from a single image. In: Pattern Recognition Letters (2018)
7. Flynn, J., et al.: DeepView: view synthesis with learned gradient descent. In: CVPR (2015)
8. Flynn, J., Neulander, I., Philbin, J., Snavely, N.: DeepStereo: learning to predict new views from the world's imagery. In: CVPR (2016)
9. Fu, H., Gong, M., Wang, C., Batmanghelich, K., Tao, D.: Deep ordinal regression network for monocular depth estimation. In: CVPR (2018)
10. Godard, C., Mac Aodha, O., Brostow, G.J.: Unsupervised monocular depth estimation with left-right consistency. In: CVPR (2017)
11. Godard, C., Mac Aodha, O., Firman, M., Brostow, G.J.: Digging into self-supervised monocular depth prediction. In: ICCV (2019)
12. Goodfellow, I., et al.: Generative adversarial nets. In: NIPS (2014)
13. Gupta, S., Hoffman, J., Malik, J.: Cross modal distillation for supervision transfer. In: CVPR (2016)
14. Hedman, P., Kopf, J.: Instant 3D photography. In: SIGGRAPH (2018)
15. Hedman, P., Philip, J., Price, T., Frahm, J.M., Drettakis, G., Brostow, G.: Deep blending for free-viewpoint image-based rendering. ACM Trans. Graph. (TOG) **37**(6), 1–15 (2018)
16. Heusel, M., Ramsauer, H., Unterthiner, T., Nessler, B., Hochreiter, S.: GANs trained by a two time-scale update rule converge to a local Nash equilibrium. In: NIPS (2017)
17. Hoffman, J., Gupta, S., Darrell, T.: Learning with side information through modality hallucination. In: CVPR (2016)
18. Hoffman, J., et al.: Cycada: cycle-consistent adversarial domain adaptation. In: ICML (2018)
19. Hu, J., Ozay, M., Zhang, Y., Okatani, T.: Revisiting single image depth estimation: toward higher resolution maps with accurate object boundaries. In: WACV (2019)
20. Huang, X., Liu, M.-Y., Belongie, S., Kautz, J.: Multimodal unsupervised image-to-image translation. In: Ferrari, V., Hebert, M., Sminchisescu, C., Weiss, Y. (eds.) ECCV 2018. LNCS, vol. 11207, pp. 179–196. Springer, Cham (2018). https://doi.org/10.1007/978-3-030-01219-9_11
21. Im, S., Jeon, H.G., Lin, S., Kweon, I.S.: DPSNet: end-to-end deep plane sweep stereo. In: ICLR (2019)
22. Isola, P., Zhu, J.Y., Zhou, T., Efros, A.A.: Image-to-image translation with conditional adversarial networks. In: CVPR (2017)

23. Kalantari, N.K., Wang, T.C., Ramamoorthi, R.: Learning-based view synthesis for light field cameras. In: SIGGRAPH Asia (2016)

24. Karras, T., Laine, S., Aittala, M., Hellsten, J., Lehtinen, J., Aila, T.: Analyzing and improving the image quality of StyleGAN. In: CoRR. vol. abs/1912.04958 (2019)

25. Laina, I., Rupprecht, C., Belagiannis, V., Tombari, F., Navab, N.: Deeper depth prediction with fully convolutional residual networks. In: 3DV (2016)

26. Lee, H.-Y., Tseng, H.-Y., Huang, J.-B., Singh, M., Yang, M.-H.: Diverse image-to-image translation via disentangled representations. In: Ferrari, V., Hebert, M., Sminchisescu, C., Weiss, Y. (eds.) ECCV 2018. LNCS, vol. 11205, pp. 36–52. Springer, Cham (2018). https://doi.org/10.1007/978-3-030-01246-5_3

27. Lee, H.Y., et al.: Drit++: diverse image-to-image translation via disentangled representations. IJCV, 1–16 (2020)

28. Lee, H.Y., et al.: Dancing to music. In: NeurIPS (2019)

29. Li, Z., et al.: Learning the depths of moving people by watching frozen people. In: CVPR (2019)

30. Li, Z., Snavely, N.: MegaDepth: learning single-view depth prediction from internet photos. In: CVPR (2018)

31. Long, J., Shelhamer, E., Darrell, T.: Fully convolutional networks for semantic segmentation. In: CVPR (2015)

32. Mildenhall, B., et al.: Local light field fusion: practical view synthesis with prescriptive sampling guidelines. In: SIGGRAPH (2019)

33. Silberman, N., Hoiem, D., Kohli, P., Fergus, R.: Indoor segmentation and support inference from RGBD images. In: Fitzgibbon, A., Lazebnik, S., Perona, P., Sato, Y., Schmid, C. (eds.) ECCV 2012. LNCS, vol. 7576, pp. 746–760. Springer, Heidelberg (2012). https://doi.org/10.1007/978-3-642-33715-4_54

34. Niklaus, S., Mai, L., Yang, J., Liu, F.: 3D ken burns effect from a single image. ACM Trans. Graph. **38**, 1–15 (2019)

35. Park, T., Liu, M.Y., Wang, T.C., Zhu, J.Y.: Semantic image synthesis with spatially-adaptive normalization. In: CVPR (2019)

36. Porter, T., Duff, T.: Compositing digital images. In: SIGGRAPH (1984)

37. Qi, X., Chen, Q., Jia, J., Koltun, V.: Semi-parametric image synthesis. In: CVPR (2018)

38. Ranftl, R., Lasinger, K., Hafner, D., Schindler, K., Koltun, V.: Towards robust monocular depth estimation: Mixing datasets for zero-shot cross-dataset transfer. arXiv:1907.01341 (2019)

39. Savva, M., et al.: Habitat: a platform for embodied AI research. In: ICCV (2019)

40. Shih, M.L., Su, S.Y., Kopf, J., Huang, J.B.: 3D photography using context-aware layered depth inpainting. In: CVPR (2020)

41. Srinivasan, P.P., Tucker, R., Barron, J.T., Ramamoorthi, R., Ng, R., Snavely, N.: Pushing the boundaries of view extrapolation with multiplane images. In: CVPR (2019)

42. Tseng, H.Y., Fisher, M., Lu, J., Li, Y., Kim, V., Yang, M.H.: Modeling artistic workflows for image generation and editing. In: Vedaldi, A., et al. (eds.) ECCV 2020. LNCS. Springer, Heidelberg (2020)

43. Tseng, H.Y., Lee, H.Y., Jiang, L., Yang, W., Yang, M.H.: RetrieveGAN: image synthesis via differentiable patch retrieval. In: Vedaldi, A., et al. (eds.) ECCV 2020. LNCS, vol. 12353. Springer, Heidelberg (2020)

44. Tucker, R., Snavely, N.: Single-view view synthesis with multiplane images. In: CVPR (2020)

45. Wang, C., Lucey, S., Perazzi, F., Wang, O.: Web stereo video supervision for depth prediction from dynamic scenes. In: 3DV (2019)

46. Wang, T.C., Liu, M.Y., Zhu, J.Y., Tao, A., Kautz, J., Catanzaro, B.: High-resolution image synthesis and semantic manipulation with conditional GANs. In: CVPR (2018)
47. Wiles, O., Gkioxari, G., Szeliski, R., Johnson, J.: SynSin: end-to-end view synthesis from a single image. In: CVPR (2020)
48. Xu, D., Ricci, E., Ouyang, W., Wang, X., Sebe, N.: Multi-scale continuous CRFs as sequential deep networks for monocular depth estimation. In: CVPR (2017)
49. Yin, Z., Shi, J.: GeoNet: unsupervised learning of dense depth, optical flow and camera pose. In: CVPR (2018)
50. Yu, J., Lin, Z., Yang, J., Shen, X., Lu, X., Huang, T.S.: Free-form image inpainting with gated convolution. In: ICCV (2019)
51. Zhang, H., et al.: StackGAN: text to photo-realistic image synthesis with stacked generative adversarial networks. In: ICCV (2017)
52. Zhao, H., Shi, J., Qi, X., Wang, X., Jia, J.: Pyramid scene parsing network. In: CVPR (2017)
53. Zhou, B., Zhao, H., Puig, X., Fidler, S., Barriuso, A., Torralba, A.: Scene parsing through ADE20K dataset. In: CVPR (2017)
54. Zhou, T., Brown, M., Snavely, N., Lowe, D.G.: Unsupervised learning of depth and ego-motion from video. In: CVPR (2017)
55. Zhou, T., Tucker, R., Flynn, J., Fyffe, G., Snavely, N.: Stereo magnification: learning view synthesis using multiplane images. In: SIGGRAPH (2018)
56. Zhu, J.Y., Park, T., Isola, P., Efros, A.A.: Unpaired image-to-image translation using cycle-consistent adversarial networks. In: ICCV (2017)
57. Zhu, J.Y., et al.: Toward multimodal image-to-image translation. In: NIPS (2017)
58. Zou, Y., Ji, P., Tran, Q.H., Huang, J.B., Chandraker, M.: Learning monocular visual odometry via self-supervised long-term modeling. In: Vedaldi, A., et al. (eds.) ECCV 2020. LNCS, vol. 12359. Springer, Heidelberg (2020)
59. Zou, Y., Luo, Z., Huang, J.-B.: DF-Net: unsupervised joint learning of depth and flow using cross-task consistency. In: Ferrari, V., Hebert, M., Sminchisescu, C., Weiss, Y. (eds.) ECCV 2018. LNCS, vol. 11209, pp. 38–55. Springer, Cham (2018). https://doi.org/10.1007/978-3-030-01228-1_3

An Analysis of Sketched IRLS
for Accelerated Sparse Residual
Regression

Daichi Iwata$^{(\boxtimes)}$, Michael Waechter, Wen-Yan Lin, and Yasuyuki Matsushita

Graduate School of Information Science and Technology,
Osaka University, Osaka, Japan
{iwata.daichi,waechter.michael,lin.daniel,yasumat}@ist.osaka-u.ac.jp

Abstract. This paper studies the problem of sparse residual regression, *i.e.*, learning a linear model using a norm that favors solutions in which the residuals are sparsely distributed. This is a common problem in a wide range of computer vision applications where a linear system has a lot more equations than unknowns and we wish to find the maximum feasible set of equations by discarding unreliable ones. We show that one of the most popular solution methods, iteratively reweighted least squares (IRLS), can be significantly accelerated by the use of matrix sketching. We analyze the convergence behavior of the proposed method and show its efficiency on a range of computer vision applications. The source code for this project can be found at https://github.com/Diwata0909/Sketched_IRLS.

Keywords: Sparse residual regression · ℓ_1 minimization · Randomized algorithm · Matrix sketching

1 Introduction

We consider the problem of residual minimization, where we wish to learn a linear model that minimizes the residuals that deviate from the model in some distance metric. For a linear model we have a matrix $\mathbf{A} \in \mathbb{R}^{n \times d}$ (we consider tall matrices, *i.e.*, the strongly over-determined case with $n \gg d$), a vector $\mathbf{b} \in \mathbb{R}^{n \times 1}$, $\mathbf{b} \notin \mathcal{R}(\mathbf{A})$ (the range of $\mathbf{A}$), and we seek to find

$$\mathbf{x}^* = \underset{\mathbf{x}}{\mathrm{argmin}} \left\| \mathbf{A}\mathbf{x} - \mathbf{b} \right\|_p^p \tag{1}$$

for some p-norm. In this paper, we consider this linear model and call it an ℓ_p (residual) minimization problem. For the more general case including non-linear residual minimization we refer the reader to, *e.g.*, Aftab and Hartley [1] or Kiani and Drummond [36].

One of the most popular methods for ℓ_p residual minimization with $1 \le p < 2$ is iteratively reweighted least squares (IRLS), in which weighted ℓ_2 minimization is repeatedly computed, eventually converging to the ℓ_p solution. It is generally understood that $p = 2$ is optimal for Gaussian distributed errors whereas

© Springer Nature Switzerland AG 2020
A. Vedaldi et al. (Eds.): ECCV 2020, LNCS 12357, pp. 609–626, 2020.
https://doi.org/10.1007/978-3-030-58610-2_36

the 1-norm leads to solutions with sparser residuals (or Laplacian distributed errors). For *sparse* residual regression, the ℓ_0 pseudo-norm is appropriate to consider; however, due to its computational complexity, a convex ℓ_1 relaxation is often employed. Because of its capability of ignoring large outliers and reaching approximate solutions that yield sparse residuals, ℓ_1 residual minimization has become an important tool for various computer vision tasks. It can be speculated that the core reason of why ℓ_2 is still widely preferred over ℓ_1 despite ℓ_1's better suitability for many applications, is simply that ℓ_2 residual minimization can be solved efficiently in closed form while ℓ_1 residual minimization cannot. Therefore, acceleration of ℓ_1 residual minimization is wanted, especially when computing budget is limited, *e.g.*, in end-user applications on mobile devices.

In recent years, randomized algorithms are gaining attention because they offer immense speed up potential with small failure probability in various applications. In linear algebra, matrix operations can be made efficient with randomized *matrix sketching* [31, 39,58]. It is based on the idea of approximating an input matrix by multiplying a randomly generated *sketching matrix* to obtain a much smaller matrix that still preserves important properties of the input matrix. In particular, certain sketching matrices **S** fulfill the *subspace embedding property* [11, Sec. 1.2]

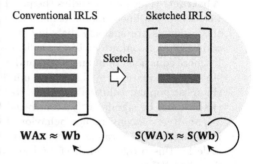

Fig. 1. Illustration of sketched IRLS: Unlike canonical IRLS, we suggest performing an approximate computation in sketched lower dimensions, yielding significant speed-up while retaining accuracy.

$$\frac{1}{\gamma}\|\mathbf{Ax}\|_2^2 \leq \|\mathbf{SAx}\|_2^2 \leq \gamma\|\mathbf{Ax}\|_2^2 \qquad (2)$$

for some $\gamma > 1$ with high probability, meaning that the ℓ_2 subspace embedding **S** preserves the lengths of all vectors **Ax** within the bounds specified by γ, indicating that we can preserve **A**'s range via sketching even though the sketched matrix **SA** lives in lower dimensions. This allows us to accelerate ℓ_2 regression.

In this paper, we show that ℓ_1 residual regression with IRLS can be significantly accelerated by matrix sketching, making it much more useful in practical applications. We call the new method *sketched IRLS*. The key idea is to speed up the internal computation block by projecting the linear system to lower dimensions using matrix sketching as illustrated in Fig. 1. Through comparisons with other robust techniques such as RANSAC, we also show that our method is versatile. This paper's contributions are summarized as follows:

– We propose accelerating IRLS for ℓ_1 minimization with matrix sketching,
– we analyze the error bound of sketched IRLS compared to canonical IRLS,

 – and we provide an analysis of our proposed method's effectiveness on common computer vision problems using both synthetic and real-world data.

The proposed method yields the important benefit that the fundamental computer vision task of regression can be executed in an outlier-robust manner with the flexible tool of IRLS without excessive computational burden, allowing it to be used in diverse applications with large datasets.

2 Related Works

The problem we study in this paper is related to sparse regression and matrix sketching. Here we briefly review these two subject areas.

Sparse Regression. The field of sparse regression has recently developed rapidly, and its usefulness due to its outlier and noise robustness has been recognized in many tasks such as face recognition [38,59], image denoising [23], general signal processing [40], and background estimation [20]. Elad [22] and Zhang [61] give technical overviews on sparse regression. The goal of sparse regression is to obtain solutions where many of the unknowns become zero. A typical setting aims at minimizing the ℓ_0 norm of the solution with an under-determined constraint as

$$\min_{\mathbf{x}} \|\mathbf{x}\|_0 \quad \text{s.t.} \quad \mathbf{A}\mathbf{x} = \mathbf{b}. \tag{3}$$

Similar to this problem is the problem of minimizing the ℓ_0 norm of residuals (called robust sensing in [35]) with over-determined $\mathbf{A}\mathbf{x} \simeq \mathbf{b}$, expressed as

$$\min_{\mathbf{r}} \|\mathbf{r}\|_0 \quad \text{s.t.} \quad \mathbf{r} = \mathbf{b} - \mathbf{A}\mathbf{x}. \tag{4}$$

Both of these look for *sparse* unknowns via minimization of the ℓ_0-norm of a vector, which is NP-hard. To address this issue, various optimization strategies have been developed. One of them is based on greedy pursuit and tries to find an approximate solution for ℓ_0 minimization directly in an iterative manner [15, 41,43,47,54]. Greedy methods are simple and easy to implement, but they are only guaranteed to work under very restricted conditions [52].

On the other hand, relaxation methods cast the NP-hard ℓ_0 minimization into, for example, ℓ_1 minimization, which is a convex surrogate for ℓ_0 minimization. Gorodnitsky and Rao proposed the FOCUSS algorithm [28], in which they approximate ℓ_1 minimization with the 2-norm using IRLS. It is also understood that ℓ_1 minimization can be solved by linear programming (LP) [26]. Several methods have been proposed to solve the ℓ_1 minimization problem, for example, one based on the interior point method [33], Least Angle Regression (LARS) [21] which is a homotopy algorithm, and the Iterative Shrinkage Thresholding Algorithm (ISTA) [13], to name a few. An excellent summary of recent ℓ_1 minimization methods is given by Yang *et al.* [60].

Among the ℓ_1 minimization methods, IRLS is simple and easily implemented as it only iterates weighted ℓ_2 minimization, and it is known for its fast convergence, *i.e.*, only requiring a small number of iterations to achieve an accurate solution [9,14,46]. Despite the simplicity of the algorithm, ℓ_1 minimization with IRLS shows good performance in recent applications [20,38,42]. Also, IRLS can be naturally extended to more general ℓ_p minimization and minimization with other robust functions such as Huber and the Pseudo-Huber functions [1]. To handle nonlinear problems, generalized IRLS has also been proposed and analyzed theoretically [44,49]. Due to this simplicity and versatility, IRLS has been one of the most popular methods for ℓ_1 minimization and related problems.

Matrix Sketching. Randomized algorithms are attracting attention recently as methods to speed up fundamental computation. In linear algebra, matrix sketching is a randomized algorithm for matrix operations. Woodruff [58], Mahoney [39], Halko *et al.* [31], and Drineas *et al.* [16] give technical overviews on matrix sketching and randomized algorithms. In recent years, randomized algorithms have been widely applied to some specialized linear and algebraic problems, *e.g.*, *consistent* linear systems [29], *symmetric diagonally dominant* linear systems [12], and matrix inversion [30]. Matrix sketching is also used for accelerating several types of matrix decomposition, *e.g.*, singular value decomposition [31], QR decomposition [19], and dynamic mode decomposition [24].

As mentioned, matrix sketching is based on the idea that the range of an input matrix can be well approximated with high probability by random projection. Such a sketched matrix can then for example be used in regression problems as they fulfill the subspace embedding property of Eq. (2). One of the earliest sketching matrices (projectors) were random Gaussian matrices [27]: Let $\mathbf{S} \in \mathbb{R}^{s \times n}$ be a matrix randomly sampled from a Gaussian distribution. Such an $\mathbf{S}$ fulfills the subspace embedding property with $\gamma = 1 + \varepsilon$ and a small $\varepsilon \geq 0$. This is a consequence of the Johnson-Lindenstrauss lemma [32]. However, with a dense $\mathbf{S}$ sketching takes $\mathcal{O}(nds)$ time which is very costly. To overcome this, various methods with other sketching matrices have been proposed. They can roughly be divided into sampling-based and projection-based methods.

Sampling-based methods extract rows of the matrix. The simple intuition is that, in strongly overdetermined systems, the row vectors are mostly linearly dependent and a subset of them is sufficient to maintain their span. Selecting rows of the matrix is equivalent to subsampling the data points. Row sampling could be achieved by premultiplying the input matrix with a binary matrix that picks the desired rows, but for efficiency this is in practice implemented with simply selecting those rows in a streaming fashion. The simplest sampling method is *uniform sampling* that selects rows with uniform probability. Drineas *et al.* [17] devised leverage score sampling, which samples based on precomputed leverage scores that unfortunately require computing the singular value decomposition (SVD), making leverage score sampling somewhat impractical.

Projection-based methods premultiply the input matrix with more general matrices so that the codomain bases are linear combinations of multiple of $\mathbf{A}$'s domain bases and not simply selections of $\mathbf{A}$'s domain bases. Ailon and Chazelle [3] proposed the Fast Johnson-Lindenstrauss Transform (FJLT), and Tropp [53] and Drineas et $al.$ [18] proposed the Subsampled Randomized Hadamard Transform (SRHT) as an extension, which take $\mathcal{O}(nd\log s)$ time. Clarkson and Woodruff further proposed the much faster $CountSketch$ method [11], which was originally used in data streaming [8]. CountSketch takes only $\mathcal{O}(\mathrm{nnz}(\mathbf{A}))$ time, where $\mathrm{nnz}(\mathbf{A})$ is the number of non-zero entries of $\mathbf{A}$.

3   Proposed Method: Sketched IRLS

For ℓ_p minimization $(1 \leq p < 2)$ with linear models, IRLS converges to Eq. (1)'s solution by iteratively minimizing

$$\mathbf{x}^{(t+1)} = \underset{\mathbf{y}}{\mathrm{argmin}} \left\| \mathbf{W}^{(t)}\mathbf{A}\mathbf{y} - \mathbf{W}^{(t)}\mathbf{b} \right\|_2^2 \qquad (5)$$

with a diagonal weight matrix $\mathbf{W}^{(t)}$ at the $(t)^{\mathrm{th}}$ iteration that is initialized with $\mathbf{W}^{(0)} = \mathbf{I}_n$. This is called the ℓ_p Weiszfeld algorithm [2,56,57]. For ℓ_1 minimization, where $p = 1$, the weights are updated as

$$\mathbf{W}_{i,i}^{(t)} = \left| \mathbf{A}_{i,*}\mathbf{x}^{(t)} - b_i \right|^{-\frac{1}{2}},$$

where $\mathbf{A}_{i,*}$ is matrix $\mathbf{A}$'s i^{th} row, and $\mathbf{W}_{i,i}$ is weight matrix $\mathbf{W}$'s i^{th} diagonal element. Solving Eq. (5) requires $\mathcal{O}(nd^2 + d^3)$ arithmetic operations and is thus expensive for large matrices. Further, since it is repeatedly solved with updated $\mathbf{W}$ in IRLS, it is the computational bottleneck for IRLS-based ℓ_p minimization.

The key idea of the proposed method is to accelerate this computation block, the weighted ℓ_2 minimization of Eq. (5), with matrix sketching. Via matrix sketching, we reduce n-dimensional to much smaller s-dimensional vectors so that the computational complexity is significantly reduced. Specifically, Eq. (5)'s weighted ℓ_2 minimization is modified as

$$\min_{\mathbf{x}} \left\| \mathbf{W}\mathbf{A}\mathbf{x} - \mathbf{W}\mathbf{b} \right\|_2^2 \xrightarrow{\text{sketch}} \min_{\mathbf{x}} \left\| \widetilde{\mathbf{W}\mathbf{A}}\mathbf{x} - \widetilde{\mathbf{W}\mathbf{b}} \right\|_2^2,$$

where $\mathbf{W}\mathbf{A} \in \mathbb{R}^{n \times d}$ and $\widetilde{\mathbf{W}\mathbf{A}} \in \mathbb{R}^{s \times d}$, $s \ll n$, with retaining the solution's accuracy. For adopting matrix sketching in IRLS, there are two aspects to consider; (1) when to sketch in the algorithm, and (2) the choice of the sketching method.

3.1   When to Sketch?

There are two possible points in time for applying matrix sketching in IRLS; (1) sketch only once before all IRLS iterations (named $sketch$-$once$), and (2) sketch in every iteration of the weighted ℓ_2 minimization (named $sketch$-$iteratively$). We analyze the behavior of these two strategies in this paper.

Sketch-Once. Not only the ℓ_2 minimization but also the sketching contributes to the algorithm's overall runtime. One way to reduce the sketching costs is to

---

Algorithm 1. IRLS with sketch-once
***

Input: $\mathbf{A} \in \mathbb{R}^{n \times d}$, $\mathbf{b} \in \mathbb{R}^n$, sketching size s, #iterations T, threshold δ for termination
Output: Approximate solution $\mathbf{x}$
  $\widetilde{\mathbf{W}}^{(0)} \leftarrow$ Identity matrix $\mathbf{I}_s$
  $\mathbf{x}^{(0)} \leftarrow \mathbf{0}$
  $\widetilde{\mathbf{A}}, \widetilde{\mathbf{b}} \leftarrow \text{sketch}(\mathbf{A}, \mathbf{b}, s)$
  for $t = 0 : T$ **do**
    $\mathbf{x}^{(t+1)} \leftarrow \text{solve\_linear\_least\_squares}(\widetilde{\mathbf{W}}^{(t)}\widetilde{\mathbf{A}}, \widetilde{\mathbf{W}}^{(t)}\widetilde{\mathbf{b}})$
    if $\left\| \mathbf{x}^{(t+1)} - \mathbf{x}^{(t)} \right\|_2 < \delta$ **then break end if**
    $\widetilde{\mathbf{W}}_{i,i}^{(t+1)} \leftarrow \left(\max\{\epsilon, |\widetilde{\mathbf{A}}_{i,*}\mathbf{x}^{(t+1)} - \widetilde{b}_i|\} \right)^{-\frac{1}{2}}$
  end for

Algorithm 2. IRLS with sketch-iteratively
***

Input: $\mathbf{A} \in \mathbb{R}^{n \times d}$, $\mathbf{b} \in \mathbb{R}^n$, sketching size s, #iterations T, threshold δ for termination
Output: Approximate solution $\mathbf{x}$
  $\mathbf{W}^{(0)} \leftarrow$ Identity matrix $\mathbf{I}_n$
  $\mathbf{x}^{(0)} \leftarrow \mathbf{0}$
  for $t = 0 : T$ **do**
    $\widetilde{\mathbf{WA}}^{(t)}, \widetilde{\mathbf{Wb}}^{(t)} \leftarrow \text{sketch}(\mathbf{W}^{(t)}\mathbf{A}, \mathbf{W}^{(t)}\mathbf{b}, s)$
    $\mathbf{x}^{(t+1)} \leftarrow \text{solve\_linear\_least\_squares}(\widetilde{\mathbf{WA}}^{(t)}, \widetilde{\mathbf{Wb}}^{(t)})$
    if $\left\| \mathbf{x}^{(t+1)} - \mathbf{x}^{(t)} \right\|_2 < \delta$ **then break end if**
    $\mathbf{W}_{i,i}^{(t+1)} \leftarrow \left(\max\{\epsilon, |\mathbf{A}_{i,*}\mathbf{x}^{(t+1)} - b_i|\} \right)^{-\frac{1}{2}}$
  end for

---

only perform it once at the beginning of IRLS, outside of IRLS's iteration loop. Algorithm 1 shows pseudo code for IRLS with sketch-once.

Sketch-Iteratively. Unlike sketch-once, sketch-iteratively performs sketching within the IRLS iteration loop. In each IRLS iteration, we generate an $\mathbf{S} \in \mathbb{R}^{s \times n}$ with $s \ll n$ and instead of Eq. (5) we solve

$$\mathbf{x}^{(t+1)} = \underset{\mathbf{x}}{\text{argmin}} \left\| \underbrace{\mathbf{SW}^{(t)}\mathbf{A}}_{=\widetilde{\mathbf{WA}}^{(t)}} \mathbf{x} - \underbrace{\mathbf{SW}^{(t)}\mathbf{b}}_{=\widetilde{\mathbf{Wb}}^{(t)}} \right\|_2^2.$$

While sketch-iteratively requires more computation time than sketch-once because of the sketching operation in the loop, it is expected to be more stable because the linear system is sketched differently in each iteration. Pseudo code for IRLS with sketch-iteratively is shown in Algorithm 2.

3.2 Sketching Method Choice

In Sect. 2, we discussed various matrix sketching methods. In this study, we mainly focus on *uniform sampling* and *CountSketch*, because of their computational efficiency. Both make the sketching matrix sparse, resulting in sketching computation costs of only $\mathcal{O}(\text{nnz}(\mathbf{A}))$ time. In practice, they can be implemented in a streaming fashion without explicitly forming the sketching matrices $\mathbf{S}$.

Uniform Sampling. samples rows of a tall matrix $[\mathbf{A}|\mathbf{b}]$ or $\mathbf{W}[\mathbf{A}|\mathbf{b}]$ with uniform probability so that the row-dimension can be reduced. Uniform sampling of a massively over-determined system has been employed in the past in practical applications, in a similar manner to the sketch-once approach described in the previous section. In an explicit matrix form, the sketching matrix $\mathbf{S}$ is a sparse matrix in which each row has only one element that is 1 while the rest are zeros.

CountSketch. [11] is similar to uniform sampling, but instead of subsampling rows of matrix $\mathbf{A}$, it uses a sketching matrix $\mathbf{S}$ in which each column has a single randomly chosen non-zero entry (typically, 1 or -1).

4   Theoretical Analysis

We now analyze the errors in IRLS solutions obtained with and without sketching. To this end, we will show that the proposed sketched IRLS can reliably derive a solution close to the non-sketched case. As mentioned, sketching can approximate a matrix's range with a certain accuracy. Since "sketch-iteratively" uses sketching repeatedly, we consider the errors introduced in each iteration. The goal is to reveal the relationship between $\left\| \mathbf{A}\widetilde{\mathbf{x}} - \mathbf{b} \right\|_1$ and $\left\| \mathbf{A}\mathbf{x}^* - \mathbf{b} \right\|_1$ for IRLS, where $\mathbf{x}^*$ is the optimal solution, and $\widetilde{\mathbf{x}}$ is the approximate solution obtained with sketched IRLS. We derive it by combining the error bounds of sketching and IRLS's convergence rate. Let $\mathbf{x}^{*(t)}$ and $\widetilde{\mathbf{x}}^{(t)}$ be the solutions of canonical and sketched IRLS, resp., after t iterations, and $\widetilde{\mathbf{x}}^{*(t+1)}$ be the solution obtained by solving Eq. (5) without sketching and using $\mathbf{W}$ based on $\widetilde{\mathbf{x}}^{(t)}$.

4.1   Condition for the Residual to Decrease

Before considering error bounds, we first show what condition must be fulfilled so that the residual in sketched IRLS decreases monotonically, *i.e.*, $\left\| \mathbf{A}\widetilde{\mathbf{x}}^{(t+1)} - \mathbf{b} \right\|_1 \leq \left\| \mathbf{A}\widetilde{\mathbf{x}}^{(t)} - \mathbf{b} \right\|_1$. For matrix sketching with ℓ_2 regression problems, several error bounds are known [17,39,48]. In a general form, we can write them as

$$\left\| \mathbf{A}\widetilde{\mathbf{x}} - \mathbf{b} \right\|_2 \leq (1 + \varepsilon) \left\| \mathbf{A}\mathbf{x}^* - \mathbf{b} \right\|_2, \tag{6}$$

$$\left\| \widetilde{\mathbf{x}} - \mathbf{x}^* \right\|_2 = \left\| \Delta\mathbf{x} \right\|_2 \leq \varepsilon_x, \tag{7}$$

using $\widetilde{\mathbf{x}} - \mathbf{x}^* := \Delta\mathbf{x}$ and small errors ε and ε_x, which depend on the sketching method and sketching size. The error decay rate is known for canonical IRLS, *e.g.*, linear and super-linear convergence rates [14,49]. According to Daubechies [14, Sec. 6.1], for sparse variable $\mathbf{x}$ problems (Eq. (3)) we have

$$\left\| \widetilde{\mathbf{x}}^{*(t+1)} - \mathbf{x}^* \right\|_1 \leq \mu \left\| \widetilde{\mathbf{x}}^{(t)} - \mathbf{x}^* \right\|_1 \tag{8}$$

with a constant $\mu \leq 1$. It is known that sparse variable problems (Eq. (3)) and sparse residual problems (Eq. (4)) can be treated as the same problem under $\mathbf{x} = \mathbf{r}$ [7], and for sparse residual problems we have

$$\left\| (\mathbf{A}\widetilde{\mathbf{x}}^{*(t+1)} - \mathbf{b}) - (\mathbf{A}\mathbf{x}^* - \mathbf{b}) \right\|_1 \leq \mu \left\| (\mathbf{A}\widetilde{\mathbf{x}}^{(t)} - \mathbf{b}) - (\mathbf{A}\mathbf{x}^* - \mathbf{b}) \right\|_1. \tag{9}$$

616   D. Iwata et al.

From these, the residuals $\left\|A\widetilde{x}^{(t+1)} - b\right\|_1$ and $\left\|A\widetilde{x}^{(t)} - b\right\|_1$ satisfy Eq. 9

$$\left\|(A\widetilde{x}^{(t+1)} - b)\right\|_1 - \left\|(Ax^* - b)\right\|_1 \leq \left\|(A\widetilde{x}^{(t+1)} - b) - (Ax^* - b)\right\|_1$$
$$= \left\|(A\widetilde{x}^{*(t+1)} - b) + A\Delta x - (Ax^* - b)\right\|_1$$
$$\leq \left\|(A\widetilde{x}^{*(t+1)} - b) - (Ax^* - b)\right\|_1 + \left\|A\Delta x\right\|_1$$
$$\leq \mu\left\|(A\widetilde{x}^{(t)} - b) - (Ax^* - b)\right\|_1 + \left\|A\Delta x\right\|_1$$
$$\leq \mu\left\|(A\widetilde{x}^{(t)} - b)\right\|_1 + \mu\left\|(Ax^* - b)\right\|_1 + \left\|A\Delta x\right\|_1. \quad (10)$$

From Eq. (10), we finally have

$$\left\|(A\widetilde{x}^{(t+1)} - b)\right\|_1 \leq \mu\left\|(A\widetilde{x}^{(t)} - b)\right\|_1 + (1+\mu)\left\|(Ax^* - b)\right\|_1 + \left\|A\Delta x\right\|_1, \quad (11)$$

and know that when $\left\|A\Delta x\right\|_1 \leq (1-\mu)\left\|(A\widetilde{x}^{(t)} - b)\right\|_1 - (1+\mu)\left\|(Ax^* - b)\right\|_1$ holds, then $\left\|A\widetilde{x}^{(t+1)} - b\right\|_1 \leq \left\|A\widetilde{x}^{(t)} - b\right\|_1$. We remark that Δx satisfies Eq. (7), and when the error of sketching Δx is sufficiently close to 0, then sketched IRLS lets the objective decrease monotonically.

4.2 Worst-Case Residual Bound

In this section, we show the relationship between the residuals of canonical IRLS $\left\|Ax^{*(t)} - b\right\|_1$ and sketched IRLS $\left\|A\widetilde{x}^{(t)} - b\right\|_1$ after t IRLS iterations using norm relations. For canonical IRLS, after t iterations we have

$$\eta\left\|Ax^{*(0)} - b\right\|_1 = \left\|Ax^{*(t)} - b\right\|_1, \quad (12)$$

where, assuming that t is big enough for $x^{*(t)}$ to converge to x^*, $\eta \in [0,1]$ is the ratio between the ℓ_1 residuals of the solutions for Eq. (1) with $p=1$ and $p=2$. For sketched IRLS, under Sect. 4.1's condition, after t iterations we have

$$\left\|A\widetilde{x}^{(t)} - b\right\|_1 \leq \left\|A\widetilde{x}^{(0)} - b\right\|_1. \quad (13)$$

Further, we have the norm relations

$$\left\|x\right\|_2 \leq \left\|x\right\|_1 \leq \sqrt{n}\left\|x\right\|_2, \quad (14)$$

where n is the number of dimensions of x. We can now derive a bound on the solution error due to the sketching after t IRLS iterations (Eqs. 6, 12, 13 and 14):

$$\left\|A\widetilde{x}^{(t)} - b\right\|_1 \leq \left\|A\widetilde{x}^{(0)} - b\right\|_1 \leq \sqrt{n}\left\|A\widetilde{x}^{(0)} - b\right\|_2 \leq \sqrt{n}(1+\varepsilon)\left\|Ax^{*(0)} - b\right\|_2$$
$$\leq \sqrt{n}(1+\varepsilon)\left\|Ax^{*(0)} - b\right\|_1 \leq \sqrt{n}(1+\varepsilon)\eta^{-1}\left\|Ax^{*(t)} - b\right\|_1. \quad (15)$$

This bound may not be very tight if n is large; however, we next show that a tighter bound can be expected in practice.

4.3 Expected Residual Bound

The reason why the bound shown in Eq. (15) looks somewhat loose is mainly due to the right side expression of Eq. (14). The bound expresses the worst case and if that case rarely occurs, the bound does not have strong implications. Therefore, it is reasonable to consider the expected value for the ℓ_1-norm. Here, let $\mathbf{x} \in \mathbb{R}^n$ be an arbitrary vector with $\|\mathbf{x}\|_2 = r$ and the expected value for the ℓ_1-norm be $\mathbb{E}[\|\mathbf{x}\|_1]$. The expectation of the ℓ_1-norm of a vector $\mathbf{x}$ becomes $\mathbb{E}[\|\mathbf{x}\|_1] = \mathbb{E}[|x_1|] + \ldots + \mathbb{E}[|x_n|]$. In polar coordinates, $x_1, \ldots, x_n$ are

$$
\begin{cases}
x_1 &= r\cos\theta_1, \\
x_2 &= r\sin\theta_1\cos\theta_2, \\
& \quad \ldots \\
x_{n-1} &= r\sin\theta_1\sin\theta_2\ldots\sin\theta_{n-2}\cos\theta_{n-1}, \\
x_n &= r\sin\theta_1\sin\theta_2\ldots\sin\theta_{n-2}\sin\theta_{n-1}.
\end{cases}
$$

Here, we assume that arbitrary random vectors are generated by uniformly distributed θ, which means that a probability density function $p(\theta)$ is a constant. In this case, the probability density function for x_1 depends on the arcsine distribution. Namely, for $-r \le x_1 \le r$, the probability density function is $p(x_1) = \frac{1}{\pi\sqrt{r^2 - x_1^2}}$. Therefore, the expectation $\mathbb{E}[|x_1|]$ becomes

$$
\begin{aligned}
\mathbb{E}[|x_1|] &= \int_{-r}^{r} |x_1|\, p(x_1)\, \mathrm{d}x_1 = \int_{-r}^{r} |x_1| \frac{1}{\pi\sqrt{r^2 - x_1^2}}\, \mathrm{d}x_1 = \frac{2}{\pi}\int_0^r \frac{x_1}{\sqrt{r^2 - x_1^2}}\, \mathrm{d}x_1 \\
&= \frac{2}{\pi}\left[-\sqrt{r^2 - x_1^2} \right]_0^r = \frac{2}{\pi}r.
\end{aligned} \tag{16}
$$

We can write x_2 as $x_2 = r_1\cos\theta_2$ with $r_1 = r\sin\theta_1$, and we can obtain $\mathbb{E}[|x_2|]$ by using Eq. (16) recursively, $i.e.$, $\mathbb{E}[|x_2|] = \int_{-\mathbb{E}[r_1]}^{\mathbb{E}[r_1]} |x_2| p(x_2)\, \mathrm{d}x_2 = \frac{2}{\pi}\mathbb{E}[r_1] = \left(\frac{2}{\pi}\right)^2 r$. Finally, the expected value $\mathbb{E}[\|\mathbf{x}\|_1]$ becomes

$$
\begin{aligned}
\mathbb{E}[\|\mathbf{x}\|_1] &= \mathbb{E}[|x_1|] + \mathbb{E}[|x_2|] + \ldots + \mathbb{E}[|x_{n-1}|] + \mathbb{E}[|x_n|] \\
&= \frac{2}{\pi}r + \left(\frac{2}{\pi}\right)^2 r + \ldots + \left(\frac{2}{\pi}\right)^{n-1} r + \left(\frac{2}{\pi}\right)^{n-1} r \\
&= \sum_{i=1}^{n-1}\left(\frac{2}{\pi}\right)^i r + \left(\frac{2}{\pi}\right)^{n-1} r = \frac{\frac{2}{\pi} + \left(\frac{2}{\pi}\right)^{n-1} - 2\left(\frac{2}{\pi}\right)^n}{1 - \frac{2}{\pi}} r.
\end{aligned}
$$

We thus have an expected bound of $\left\|\mathbf{A}\widetilde{\mathbf{x}}^{(t)} - \mathbf{b}\right\|_1 \le \frac{\frac{2}{\pi} + \left(\frac{2}{\pi}\right)^{n-1} - 2\left(\frac{2}{\pi}\right)^n}{1 - \frac{2}{\pi}}\, (1 + \varepsilon)\, \eta^{-1}\left\|\mathbf{A}\mathbf{x}^{*(t)} - \mathbf{b}\right\|_1$. With $n \to \infty$, the fraction expression converges to $\frac{2}{\pi - 2} \simeq 1.75$. The expected value is considerably smaller than the worst case value $\sqrt{n}$, indicating that when n is large, Sect. 4.2's worst case bound is hardly relevant. Although it is still far from a strict bound due to the dependency on η, as we will see in the following sections, sketched IRLS yields highly accurate approximations in practice in various settings.

5   Performance Evaluation

We first evaluate the proposed method's performance in comparison to canonical IRLS using synthetic data in common problems, namely, in residual minimization and low-rank approximation.

5.1   Residual Minimization

In ℓ_1 residual minimization, given $\mathbf{A} \in \mathbb{R}^{n \times d}$, $n > d$ and $\mathbf{b} \in \mathbb{R}^n$, we wish to solve

$$\min_{\mathbf{x}} \|\mathbf{A}\mathbf{x} - \mathbf{b}\|_1$$

for $\mathbf{x} \in \mathbb{R}^d$. To form a highly over-determined linear system, we set the size of matrix $\mathbf{A}$ to $(n, d) = (10^6, 40)$. For assessing the performance variations with respect to the outlier distribution, we formed matrix $\mathbf{A}$ in three distinct ways: (a) $\mathbf{A}$ drawn from a uniform distribution in $[0, 10]$ and flipped signs of 20 % elements to create outliers (called *uniform data 20%*, hereafter), (b) $\mathbf{A}$ created like (a) but with 60% outliers (called *uniform data 60%*, hereafter), and (c) $\mathbf{A}$ created like (a) but further corrupted by adding a large value ($=10^3$) to randomly selected 0.1% of rows (called *biased data*, hereafter). The ground truth $\mathbf{x}^*$ is created, and based on $\mathbf{x}^*$, $\mathbf{b}$ is pre-computed before adding outliers. We evaluate the *error* defined as $\frac{1}{d}\|\mathbf{x}^* - \mathbf{x}^{(t)}\|_2$, where $\mathbf{x}^{(t)}$ is the solution after the tth iteration. We also compare the accuracy with RANSAC [25] to assess the robustness of ℓ_1 minimization against outliers. Unlike ℓ_1 minimization, RANSAC requires adjusting a few parameters. We changed the number of samplings, RANSAC's most important parameter, and chose the best result from $\{d+1, d \times 10, d \times 10^2\}$.

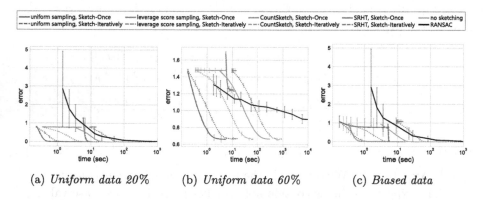

(a) *Uniform data 20%*   (b) *Uniform data 60%*   (c) *Biased data*

Fig. 2. Averages of the error $\frac{1}{d}\|\mathbf{x}^* - \mathbf{x}^{(t)}\|_2$ and the standard deviations over time in residual minimization with various matrix sketching methods and RANSAC on uniform synthetic data with a lower outlier rate (*left*), with a higher outlier rate (*center*), and with biased synthetic data (*right*).

Figures 2a, 2b and 2c show the results of *uniform data 20%*, *uniform data 60%*, and *biased data*, resp. All results are averages of 10 trials with different random seeds. The plots show averages of the *error*, error bars indicate standard deviations. From Figs. 2a and 2b we can observe that ℓ_1 minimization works well for problems with relatively few outliers, and RANSAC shows slow convergences. Both methods need to solve least squares minimization many times, IRLS can be expected to improve the solution for each loop, but RANSAC does not necessarily do so. RANSAC is further known to fail at high-dimensional problems. Regarding sketching methods, while sampling-based methods with sketch-once work well for *uniform data* as shown in Fig. 2a, their accuracies become lower and variances become larger on the biased data as shown in Fig. 2c. Projection-based methods work well for either data with low variances, especially CountSketch with sketch-iteratively. For leverage score sampling, the comparison was done using the known leverage scores while it is usually unknown a priori. From here on out, we will focus only on uniform sampling and CountSketch as the chosen sketching methods because of their efficiency as discussed in Sect. 3.2.

5.2 Low-Rank Approximation

Next, we evaluate the proposed method on low-rank approximation with ℓ_1 singular value decomposition. Given a matrix $\mathbf{M} \in \mathbb{R}^{m \times n}$, its rank-$r$ approximation ($r < \min(m, n)$) with ℓ_1-norm can be written as

$$\min_{\mathbf{U},\mathbf{V}} \|\mathbf{M} - \mathbf{U}\mathbf{V}^\top\|_1,$$

where $\mathbf{U}$ and $\mathbf{V}^\top$ are $m \times r$ and $r \times n$ matrices, respectively. The solution method involves ℓ_1 minimization subproblems that are computed iteratively [34] as

$$\mathbf{U} \leftarrow \operatorname*{argmin}_{\mathbf{U}} \|\mathbf{M} - \mathbf{U}\mathbf{V}^\top\|_1, \quad \mathbf{V} \leftarrow \operatorname*{argmin}_{\mathbf{V}} \|\mathbf{M} - \mathbf{U}\mathbf{V}^\top\|_1,$$

starting from a random initialization of $\mathbf{V}$.

We generated an $\mathbf{M} \in \mathbb{R}^{10^4 \times 10^4}$ with rank($\mathbf{M}$) = 40, flipped signs of 10% and 40% of the elements. We set the sketching size s as $1,600$ and $3,200$. We also compare the accuracy with robust principal component analysis (R-PCA) [6]. R-PCA decomposes a corrupted matrix $\mathbf{M}$ into a low-rank matrix $\mathbf{A}$ and a sparse corruption matrix $\mathbf{E}$, *i.e.*, $\mathbf{M} = \mathbf{A} + \mathbf{E}$. To solve R-PCA, the Augmented Lagrange Multiplier method (ALM) [37] is a known effective approach. Also a previous study accelerated ALM by fast randomized singular value thresholding (FRSVT) [45]. We used native ALM and ALM with FRSVT as comparison. These methods also require tuning hyper-parameter λ, and we chose the best result from $\lambda \in \{10^{-1}, 10^{-2}, 10^{-3}\}$.

To assess the accuracy, we define the *error* as $\frac{1}{mn}\|\mathbf{M}^* - \mathbf{M}^{(t)}\|_F$, where $\mathbf{M}^*$ is the uncorrupted original matrix and $\mathbf{M}^{(t)}$ is the estimated low-rank matrix after the tth iteration. Figures 3a and 3b show the results. In this setting, ℓ_1 minimization achieves high accuracy in both datasets. R-PCA converges quickly

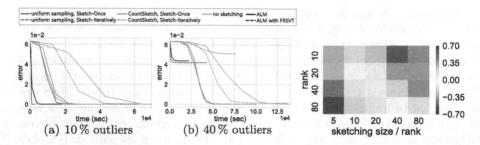

Fig. 3. Error over time in ℓ_1 singular value decomposition on synthetic data with lower outlier rate, (*left*), with higher outlier rate (*right*)

Fig. 4. Comparison between sketch-iteratively and sketch-once. Colors indicate faster convergence for **sketch-iteratively** or *sketch-once*.

but does not work well for the dataset with many outliers. The sketch-once strategy shows about 3 times faster convergence compared to canonical IRLS in Fig. 3a, and also uniform sampling strategy shows fast convergence in Fig. 3b.

5.3   When to Sketch?

In the following, we show an experiment that serves to give a first intuition for when a user should pick the sketch-once or the sketch-iteratively sketching regime. We generated $\mathbb{R}^{10^4 \times 10^4}$ matrices with 10% noise, we picked the ranks from $\{10, 20, 40, 80\}$ and sketching sizes from $\{\text{rank} \times 5, \times 10, \times 20, \times 40, \times 80\}$, and conducted a low-rank approximation experiment. We checked the times t_{SO} and t_{SI} until sketch-once and sketch-iteratively achieved an accuracy of 10^{-5}. Figure 4 shows the values of $\ln(t_{SO}/t_{SI})$. For each rank, the larger the sketching size, the faster sketch-once converges. As the matrix rank increases, sketch-iteratively shows faster convergence at larger sketching sizes. The sketching size for ideal approximations increases faster than $O(r)$. When the sketching size is not big enough, the risk of not making ideal approximations becomes higher. Especially, at small sketching sizes, sketch-once is susceptible to making bad approximations, whereas sketch-iteratively alleviates this effect by repeated sketching and shows faster convergence for smaller sketching sizes.

6   Applications

Countless computer vision tasks are built upon robust regression. In practical scenarios, the input is typically quite large and noisy, either due to a noisy capturing process or a noisy precomputation step, *e.g.*, incorrect feature matches. In this section, we demonstrate the proposed sketched IRLS's effectiveness on such real-world applications. We adapted our method to problems with over-determined system: homography estimation and point cloud alignment.

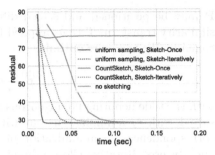

(a) *Left:* Input images. *Right:* ℓ_1 and ℓ_2 stitching results.

(b) ℓ_1 residual variation over time

Fig. 5. Performance of image stitching with ℓ_1 and ℓ_2 homography estimation

6.1   Homography Estimation

Homography estimation is an important building block for image stitching and plane-to-plane alignment. Given correspondences in two images, the image transformation can be written by a 3×3 homography $\mathbf{H}$ with 8 DoF as

$$\lambda[x', y', 1]^\top = \mathbf{H}[x, y, 1]^\top,$$

where λ is a scaling factor, and (x, y) and (x', y') are corresponding points in the left and right images, respectively [50]. Given a set of correspondences, one can estimate $\mathbf{H}$ by solving a homogeneous system [51, Chap. 6.1]. The set of correspondences may contain many wrong correspondences from erroneous feature matching. Therefore, conventional methods use robust estimation techniques, such as RANSAC or ℓ_1 regression.

Result. We obtained point correspondences by matching AKAZE [4] features in the two images shown in the left column of Fig. 5a. From $17{,}076$ obtained point correspondences we estimated the homography. The second column of Fig. 5a shows the stitching results of ℓ_1 and ℓ_2 minimization, respectively. ℓ_1 minimization successfully joined the two images whereas ℓ_2 minimization produced a strongly erroneous result due to feature mismatches. In this experiment, we sketched the matrix to $17{,}076 \times 2 \times 1\% \simeq 341$ equations (note that each point pair gives two equations), and confirmed that the solution did not differ from the direct ℓ_1 minimization without sketching. In Fig. 5b, we can see that the proposed method (uniform sampling + sketch-once) converges about $5\times$ faster than canonical IRLS while maintaining the same accuracy.

6.2   Point Cloud Alignment

Consider the two point sets $\mathbf{P} = [\mathbf{p}_1, \ldots, \mathbf{p}_l]$ and $\mathbf{Q} = [\mathbf{q}_1, \ldots, \mathbf{q}_l]$ captured from different viewpoints around an object. For each i, the points $\mathbf{p}_i \in \mathbb{R}^3$ and $\mathbf{q}_i \in \mathbb{R}^3$ approximately correspond to the same surface point on the 3D object. Since

$\mathbf{P}$ may be positioned and scaled differently in space than $\mathbf{Q}$, we search for a similarity transform $\mathbf{T}$ that, applied to all points in $\mathbf{P}$, makes $\mathbf{P}$ and $\mathbf{Q}$ roughly coincide. We therefore optimize

$$\min_{\mathbf{T}} \sum_i \left\| \mathbf{T}\tilde{\mathbf{p}}_i - \tilde{\mathbf{q}}_i \right\|, \tag{17}$$

with the tilde denoting homogeneous representations. This problem is commonly solved with ℓ_2 minimization but with large outliers, $e.g.$, due to wrong point correspondences, ℓ_1 minimization may be superior. In the veteran but still frequently used Iterative Closest Point (ICP) algorithm [5,10], the point sets are first manually pre-aligned, then it searches for the nearest neighbor in $\mathbf{Q}$ for each point in $\mathbf{P}$, optimize Eq. (17), and iterate.

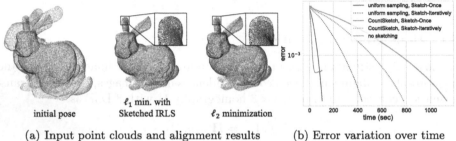

(a) Input point clouds and alignment results   (b) Error variation over time

initial pose ℓ_1 min. with Sketched IRLS ℓ_2 minimization

Fig. 6. Performance of point cloud alignment via ICP with ℓ_1 and ℓ_2 similarity transform estimation

Result. In this task, we used the Stanford Bunny [55]. The two input point clouds are shown in Fig. 6a (left). Each set consists of 10^5 points, viewed from different viewpoints in different scalings. We applied the proposed sketched IRLS to perform ICP with ℓ_1 residual minimization. The second and third figures of Fig. 6a show the results of ℓ_1 minimization with sketched IRLS and conventional ℓ_2 minimization. It is observed that ℓ_1 minimization results in an accurate alignment of the two point clouds, unaffected by inaccurate correspondences.

For this experiment we set the sketching size s as 10% of the original problem, and gradually transformed the point set $\mathbf{Q}$ with fixed $\mathbf{P}$. Since we have the ground truth here, we evaluate the error by $\frac{1}{l}\|\mathbf{Q}^* - \mathbf{Q}^{(t)}\|_F$, where $\mathbf{Q}^{(t)}$ is the result after the tth iteration. The evaluation of the error variation $w.r.t.$ time only for computing the transformation (excluding matchings) is summarized in Fig. 6b. While CountSketch + sketch-once did not show good convergence, the other sketching methods find a good alignment with significant speed up compared to the conventional method.

7   Discussion

Our experiments showed that sketching is effective in reducing IRLS's runtime and ℓ_1 minimization with IRLS works well on a wide range of computer vision problems. Other robust methods such as RANSAC and RPCA are certainly good at specific problems, but IRLS ℓ_1 minimization is versatile without requiring parameter tuning and its convergence is demonstrably superior in some tasks.

Regarding when to sketch, sketch-once is superior if the rank r of the design matrix $\mathbf{A}$ is very small and the sketching is not aggressive, $i.e.$, $s \gg r$. However, if the rank is high or we sketch aggressively, $e.g.$, $s < 5r$, then it is likely that $\mathbf{A}$'s range will not be preserved, and we need to perform sketch-iteratively to be able to recover from badly chosen samples.

If one naïvely performs subsampling on the input data, this would be equal to sketch-once sketching, basically treating IRLS as a black box that can never be touched again after the data has initially been subsampled. The experiments showed that in applications where sketch-iteratively performs better, we want to open that black box and perform subsampling in every iteration.

Acknowledgments. This work is supported by JSPS CREST Grant Number JPMJCR1764, Japan. Michael Waechter was supported through a postdoctoral fellowship by the Japan Society for the Promotion of Science (JP17F17350).

References

1. Aftab, K., Hartley, R.: Convergence of iteratively re-weighted least squares to robust M-estimators. In: Winter Conference on Applications of Computer Vision (WACV) (2015)
2. Aftab, K., Hartley, R., Trumpf, J.: Generalized Weiszfeld algorithms for Lq optimization. Trans. Pattern Anal. Mach. Intell. (PAMI) **37**(4), 728–745 (2015)
3. Ailon, N., Chazelle, B.: Approximate nearest neighbors and the fast Johnson-Lindenstrauss transform. In: Symposium on Theory of Computing (2006)
4. Alcantarilla, P.F., Nuevo, J., Bartoli, A.: Fast explicit diffusion for accelerated features in nonlinear scale spaces. In: British Machine Vision Conference (BMVC) (2013)
5. Besl, P.J., McKay, N.D.: A method for registration of 3-D shapes. Trans. Pattern Anal. Mach. Intell. (PAMI) **14**(2), 239–256 (1992)
6. Candès, E.J., Li, X., Ma, Y., Wright, J.: Robust principal component analysis? J. ACM **58**(3), 11:1–11:37 (2011)
7. Candes, E.J., Tao, T.: Decoding by linear programming. IEEE Trans. Inf. Theory **51**(12), 4203–4215 (2005)
8. Charikar, M., Chen, K., Farach-Colton, M.: Finding frequent items in data streams. In: Widmayer, P., Eidenbenz, S., Triguero, F., Morales, R., Conejo, R., Hennessy, M. (eds.) ICALP 2002. LNCS, vol. 2380, pp. 693–703. Springer, Heidelberg (2002). https://doi.org/10.1007/3-540-45465-9_59
9. Chen, C., He, L., Li, H., Huang, J.: Fast iteratively reweighted least squares algorithms for analysis-based sparse reconstruction. Med. Image Anal. **49**, 141–152 (2018)
10. Chen, Y., Medioni, G.: Object modeling by registration of multiple range images. In: International Conference on Robotics and Automation (ICRA) (1991)
11. Clarkson, K.L., Woodruff, D.P.: Low rank approximation and regression in input sparsity time. In: Symposium on Theory of Computing (2013)

12. Cohen, M.B., et al.: Solving SDD linear systems in nearly $m \log^{1/2} n$ time. In: Symposium on Theory of Computing (2014)
13. Daubechies, I., Defrise, M., De Mol, C.: An iterative thresholding algorithm for linear inverse problems with a sparsity constraint. Commun. Pure Appl. Math. **57**(11), 1413–1457 (2004)
14. Daubechies, I., DeVore, R., Fornasier, M., Güntürk, C.S.: Iteratively reweighted least squares minimization for sparse recovery. Commun. Pure Appl. Math. **63**(1), 1–38 (2010)
15. Donoho, D.L., Tsaig, Y., Drori, I., Starck, J.L.: Sparse solution of underdetermined systems of linear equations by stagewise orthogonal matching pursuit. Trans. Inf. Theory **58**(2), 1094–1121 (2012)
16. Drineas, P., Mahoney, M.W.: RandNLA: randomized numerical linear algebra. Commun. ACM **59**(6), 80–90 (2016)
17. Drineas, P., Mahoney, M.W., Muthukrishnan, S.: Sampling algorithms for ℓ_2 regression and applications. In: ACM-SIAM Symposium on Discrete Algorithm (2006)
18. Drineas, P., Mahoney, M.W., Muthukrishnan, S., Sarlós, T.: Faster least squares approximation. Numer. Math. **117**(2), 219–249 (2011)
19. Duersch, J., Gu, M.: Randomized QR with column pivoting. SIAM J. Sci. Comput. **39**(4), 263–291 (2017)
20. Dutta, A., Richtárik, P.: Online and batch supervised background estimation via L1 regression. In: Winter Conference on Applications of Computer Vision (WACV) (2019)
21. Efron, B., Hastie, T., Johnstone, I., Tibshirani, R.: Least angle regression. Ann. Stat. **32**(2), 407–499 (2004)
22. Elad, M.: Sparse and Redundant Representations: From Theory to Applications in Signal and Image Processing, 1st edn. Springer, Heidelberg (2010). https://doi.org/10.1007/978-1-4419-7011-4
23. Elad, M., Aharon, M.: Image denoising via sparse and redundant representations over learned dictionaries. Trans. Image Process. **15**(12), 3736–3745 (2006)
24. Erichson, N., Mathelin, L., Brunton, S., Kutz, J.: Randomized dynamic mode decomposition. SIAM J. Appl. Dyn. Syst. **18**(4), 1867–1891 (2017)
25. Fischler, M.A., Bolles, R.C.: Random sample consensus: a paradigm for model fitting with applications to image analysis and automated cartography. Commun. ACM **24**(6), 381–395 (1981)
26. Gentle, J.E.: Matrix Algebra: Theory, Computations, and Applications in Statistics. Springer, Heidelberg (2007). https://doi.org/10.1007/978-3-319-64867-5
27. Goldstine, H.H., von Neumann, J.: Numerical inverting of matrices of high order. ii. Am. Math. Soc. **2**, 188–202 (1951)
28. Gorodnitsky, I.F., Rao, B.D.: Sparse signal reconstruction from limited data using FOCUSS: a re-weighted minimum norm algorithm. Trans. Signal Process. **45**(3), 600–616 (1997)
29. Gower, R., Richtárik, P.: Randomized iterative methods for linear systems. SIAM J. Matrix Anal. Appl. **36**(4), 1660–1690 (2015)
30. Gower, R., Richtárik, P.: Randomized quasi-Newton updates are linearly convergent matrix inversion algorithms. SIAM J. Matrix Anal. Appl. **38**(4), 1380–1409 (2016)
31. Halko, N., Martinsson, P.G., Tropp, J.A.: Finding structure with randomness: probabilistic algorithms for constructing approximate matrix decompositions. SIAM Rev. **53**(2), 217–288 (2011)
32. Johnson, W.B., Lindenstrauss, J.: Extensions of Lipschitz mappings into a Hilbert space. Contemp. Math. **26**, 189–206 (1984)

33. Karmarkar, N.: A new polynomial-time algorithm for linear programming. Combinatorica **4**(4), 373–395 (1984)
34. Ke, Q., Kanade, T.: Robust L_1 norm factorization in the presence of outliers and missing data by alternative convex programming. In: Conference on Computer Vision and Pattern Recognition (CVPR) (2005)
35. Kekatos, V., Giannakis, G.B.: From sparse signals to sparse residuals for robust sensing. IEEE Trans. Signal Process. **59**(7), 3355–3368 (2011)
36. Kiani, K.A., Drummond, T.: Solving robust regularization problems using iteratively re-weighted least squares. In: Winter Conference on Applications of Computer Vision (WACV) (2017)
37. Lin, Z., Chen, M., Wu, L., Ma, Y.: The augmented Lagrange multiplier method for exact recovery of corrupted low-rank matrices. Technical report UILU-ENG-09-2215, Coordinated Science Laboratory, University of Illinois at Urbana-Champaign (2010)
38. Lu, C., Lin, Z., Yan, S.: Smoothed low rank and sparse matrix recovery by iteratively reweighted least squares minimization. Trans. Image Process. **24**(2), 646–654 (2015)
39. Mahoney, M.W.: Randomized algorithms for matrices and data. Found. Trends® Mach. Learn. **3**(2), 123–224 (2011)
40. Mallat, S.G.: A Wavelet Tour of Signal Processing: The Sparse Way, 3rd edn. Academic Press Inc., Cambridge (2008)
41. Mallat, S.G., Zhang, Z.: Matching pursuits with time-frequency dictionaries. Trans. Signal Process. **41**(12), 3397–3415 (1993)
42. Millikan, B., Dutta, A., Rahnavard, N., Sun, Q., Foroosh, H.: Initialized iterative reweighted least squares for automatic target recognition. In: Military Communications Conference (2015)
43. Needell, D., Tropp, J.A.: CoSaMP: iterative signal recovery from in complete and inaccurate samples. Appl. Comput. Harmonic Anal. **26**(3), 301–321 (2009)
44. Ochs, P., Dosovitskiy, A., Brox, T., Pock, T.: On iteratively reweighted algorithms for non-smooth non-convex optimization in computer vision. SIAM J. Imaging Sci. **8**(1), 331–372 (2015)
45. Oh, T.H., Matsushita, Y., Tai, Y.W., Kweon, I.S.: Fast randomized singular value thresholding for low-rank optimization. Trans. Pattern Anal. Mach. Intell. (PAMI) **40**(2), 376–391 (2018)
46. Osborne, M.R.: Finite Algorithms in Optimization and Data Analysis. Wiley, Hoboken (1985)
47. Pati, Y.C., Rezaiifar, R., Krishnaprasad, P.S.: Orthogonal matching pursuit: Recursive function approximation with applications to wavelet decomposition. In: Asilomar Conference on Signals, Systems, and Computers (1993)
48. Sarlós, T.: Improved approximation algorithms for large matrices via random projections. In: Symposium on Foundations of Computer Science, pp. 143–152 (2006)
49. Sigl, J.: Nonlinear residual minimization by iteratively reweighted least squares. Comput. Optim. Appl. **64**(3), 755–792 (2016)
50. Szeliski, R.: Video mosaics for virtual environments. Comput. Graph. Appl. **16**(2), 22–30 (1996)
51. Szeliski, R.: Computer Vision - Algorithms and Applications. Texts in Computer Science. Springer, New York (2011). https://doi.org/10.1007/978-1-84882-935-0
52. Tropp, J.A.: Greed is good: algorithmic results for sparse approximation. Trans. Inf. Theory **50**(10), 2231–2242 (2004)
53. Tropp, J.A.: Improved analysis of the subsampled randomized Hadamard transform. Adv. Adapt. Data Anal. **3**, 115–126 (2011)

54. Tropp, J.A., Gilbert, A.C., Strauss, M.J.: Algorithms for simultaneous sparse approximation. Part I. Signal Process. **86**(3), 572–588 (2006)
55. Turk, G., Levoy, M.: Zippered polygon meshes from range images. In: SIGGRAPH (1994)
56. Weiszfeld, E.: Sur le point pour lequel la somme des distances de n points donnés est minimum. Tohoku Math. J. **43**, 355–386 (1937)
57. Weiszfeld, E., Plastria, F.: On the point for which the sum of the distances ton given points is minimum. Ann. Oper. Res. **167**, 7–41 (2009)
58. Woodruff, D.P.: Sketching as a tool for numerical linear algebra. Found. Trends® Theor. Comput. Sci. **10**(1–2), 1–157 (2014)
59. Wright, J., Yang, A.Y., Ganesh, A., Sastry, S.S., Ma, Y.: Robust face recognition via sparse representation. Trans. Pattern Anal. Mach. Intell. (PAMI) **31**(2), 210–227 (2009)
60. Yang, A., Zhou, Z., Ganesh Balasubramanian, A., Sastry, S., Ma, Y.: FastL1-minimization algorithms for robust face recognition. Trans. Image Process. **22**(8), 3234–3246 (2013)
61. Zhang, Z., Xu, Y., Yang, J., Li, X., Zhang, D.: A survey of sparse representation: algorithms and applications. IEEE Access **3**, 490–530 (2015)

Relative Pose from Deep Learned Depth and a Single Affine Correspondence

Ivan Eichhardt[1,3]([✉]) [iD] and Daniel Barath[1,2] [iD]

[1] Machine Perception Research Laboratory, SZTAKI, Budapest, Hungary
{eichhardt.ivan,barath.daniel}@sztaki.mta.hu
[2] VRG, Faculty of Electrical Engineering,
Czech Technical University in Prague, Prague, Czechia
[3] Faculty of Informatics, University of Debrecen, Debrecen, Hungary

Abstract. We propose a new approach for combining deep-learned non-metric monocular depth with affine correspondences (ACs) to estimate the relative pose of two calibrated cameras from a single correspondence. Considering the depth information and affine features, two new constraints on the camera pose are derived. The proposed solver is usable within 1-point RANSAC approaches. Thus, the processing time of the robust estimation is linear in the number of correspondences and, therefore, orders of magnitude faster than by using traditional approaches. The proposed 1AC+D (Source code: ⓞ https://github.com/eivan/one-ac-pose) solver is tested both on synthetic data and on 110395 publicly available real image pairs where we used an off-the-shelf monocular depth network to provide up-to-scale depth per pixel. The proposed 1AC+D leads to similar accuracy as traditional approaches while being significantly faster. When solving large-scale problems, *e.g.* pose-graph initialization for Structure-from-Motion (SfM) pipelines, the overhead of obtaining ACs and monocular depth is negligible compared to the speed-up gained in the pairwise geometric verification, *i.e.*, relative pose estimation. This is demonstrated on scenes from the 1DSfM dataset using a state-of-the-art global SfM algorithm.

Keywords: Pose estimation · Minimal solver · Depth prediction · Affine correspondences · Global structure from motion · Pose graph initialization

1 Introduction

This paper investigates the challenges and viability of combining deep-learned monocular depth with affine correspondences (ACs) to create minimal pose solvers. Estimating pose is a fundamental problem in computer vision [1, 8, 17,

Electronic supplementary material The online version of this chapter (https://doi.org/10.1007/978-3-030-58610-2_37) contains supplementary material, which is available to authorized users.

A. Vedaldi et al. (Eds.): ECCV 2020, LNCS 12357, pp. 627–644, 2020.
https://doi.org/10.1007/978-3-030-58610-2_37

(a) Vienna Cathedral

(b) Madrid Metropolis

Fig. 1. Scenes from the 1DSfM dataset [56]. From left to right: color image from the dataset, predicted monocular depth [30], and 3D reconstruction using global SfM [50]. Pose graphs were initialized by the proposed solver, 1AC+D, as written in Sect. 4.3. (Color figure online)

21, 22, 25–29, 36, 37, 42, 43, 48, 49, 55], enabling reconstruction algorithms such as simultaneous localization and mapping [34,35] (SLAM) as well as structure-from-motion [44,45,51,57] (SfM). Also, there are several other applications, *e.g.*, in 6D object pose estimation [23,31] or in robust model fitting [2,7,12,15,40,52], where the accuracy and runtime highly depends on the minimal solver applied. The proposed approach uses monocular depth predictions together with ACs to estimate the relative camera pose from a single correspondence. Exploiting ACs for geometric model estimation is a nowadays popular approach with a number of solvers proposed in the recent years. For instance, for estimating the epipolar geometry of two views, Bentolila and Francos [10] proposed a method using three correspondences interpreting the problem via conic constraints. Perdoch *et al.* [38] proposed two techniques for approximating the pose using two and three correspondences by converting each affine feature to three points and applying the standard point-based estimation techniques. Raposo *et al.* [41] proposed a solution for essential matrix estimation from two correspondences. Baráth *et al.* [3,6] showed that the relationship of ACs and epipolar geometry is linear and geometrically interpretable. Eichhardt *et al.* [14] proposed a method that uses two ACs for relative pose estimation based on general central-projective views. Recently, Hajder *et al.* [19] and Guan *et al.* [18] proposed minimal solutions for relative pose from a single AC when the camera is mounted on a moving vehicle. Homographies can also be estimated from two ACs as it was first shown by Köser [24]. In the case of known epipolar geometry, a single correspondence [5] is enough to recover the parameters of the homography, or the underlying surface normal [4,24]. Pritts *et al.* [39] used affine features for the simultaneous estimation of affine image rectification and lens distortion.

In this paper, we propose to use affine correspondences together with deep-learned priors derived directly into minimal pose solvers – exploring new ways of utilizing these priors for pose estimation. We use an off-the-shelf deep monocular depth estimator [30] to provide a 'relative' (non-metric) depth for each pixel (examples are in Fig. 1) and use the depth together with affine correspondences for estimating the relative pose from a single correspondence. Aside from the fact that predicted depth values are far from being perfect, they provide a sufficiently strong prior about the underlying scene geometry. This helps in reducing the degrees-of-freedom of the relative pose estimation problem. Consequently, fewer correspondences are needed to determine the pose which is highly favorable in robust estimation, *i.e.*, robustly determining the camera motion when the data is contaminated by outliers and noise.

We propose a new minimal solver which estimates the general relative camera pose, *i.e.*, 3D rotation and translation, and the scale of the depth map from a single AC. We show that it is possible to use the predicted imperfect depth signal to robustly estimate the pose parameters. The depth and the AC together constrain the camera geometry so that the relative motion and scale can be determined as the closed-form solution of the implied least-squares optimization problem. The proposed new constraints are derived from general central projection and, therefore, they are valid for arbitrary pairs of central-projective views. It is shown that the proposed solver has significantly lower computational cost, compared to state-of-the-art pose solvers. The imperfections of the depth signal and the AC are alleviated through using the solver with a modern robust estimator [2], providing state-of-the-art accuracy at exceptionally low run-time.

The reduced number of data points required for the model estimation leads to linear time complexity when combined with robust estimators, *e.g.* RANSAC [15]. The resulting 1-point RANSAC has to check only n model hypotheses (where n is the point number) instead of, *e.g.*, $\binom{n}{5}$ which the five-point solver implies. This improvement in efficiency has a significant positive impact on solving large-scale problems, *e.g.*, on view-graph construction (*i.e.*, a number of 2-view matching and geometric verification) [45,46,51] which operates over thousands of image pairs. We provide an evaluation of the proposed solver both on synthetic and real-world data. It is demonstrated, on 110395 image pairs from the 1DSFM dataset [56], that the proposed methods are similarly robust to image noise while being up to 2 orders of magnitude faster, when applied within Graph-Cut RANSAC [2], than the traditional five-point algorithm. Also, it is demonstrated that when using the resulting pose-graph in the global SfM pipeline [11,56] as implemented in the Theia library [50], the accuracy of the obtained reconstruction is similar to when the five-point algorithm is used.

2   Constraints from Relative Depths and Affine Frames

Let us denote a local affine frame (LAF) as $(\mathbf{x}, \mathbf{M})$, where $\mathbf{x} \in \mathbb{R}^2$ is an image point and $\mathbf{M} \in \mathbb{R}^{2 \times 2}$ is a linear transformation describing the local coordinate system of the associated image region. The following expression maps points $\mathbf{y}$ from a local coordinate system onto the image plane, in the vicinity of $\mathbf{x}$.

$$\mathbf{x}(\mathbf{y}) \approx \mathbf{x} + \mathbf{M}\mathbf{y}. \tag{1}$$

LAFs typically are acquired from images by affine-covariant feature extractors [33]. In practice, they can easily be obtained by, $e.g.$, the VLFeat library [54].

Two LAFs $(\mathbf{x}_1, \mathbf{M}_1)$ and $(\mathbf{x}_2, \mathbf{M}_2)$ form an affine correspondence, namely, $\mathbf{x}_1$ and $\mathbf{x}_2$ are corresponding points and $\mathbf{M}_2\mathbf{M}_1^{-1}$ is the linear transformation mapping points between the infinitesimal vicinities of $\mathbf{x}_1$ and $\mathbf{x}_2$ [13].

2.1   Local Affine Frames and Depth Through Central Projection

Let $q : \mathbb{R}^2 \to \mathbb{R}^3$ be a function that maps image coordinates to bearing vectors. $\mathbf{R} \in \mathrm{SO}(3)$ and $\mathbf{t} \in \mathbb{R}^3$ are the relative rotation and translation of the camera coordinate system, such that $\mathbf{x} \in \mathbb{R}^2$ is the projection of $\mathbf{X} \in \mathbb{R}^3$, as follows:

$$\exists! \lambda : \mathbf{X} = \lambda \mathbf{R}q(\mathbf{x}) + \mathbf{t}. \tag{2}$$

That is, a unique $depth$ λ corresponds to $\mathbf{X}$, along the bearing vector $q(\mathbf{x})$.

When image coordinates $\mathbf{x}$ are locally perturbed, the corresponding $\mathbf{X}$ and λ are also expected to change as per Eq. (2). The first order approximation of the projection expresses the connection between local changes of $\mathbf{x}$ and $\mathbf{X}$, $i.e.$, it is an approximate linear description how local perturbations would relate the two. Differentiating Eq. (2) along image coordinates $k = u, v$ results in an expression for the first order approximation of projection, as follows:

$$\exists! \lambda, \partial_k \lambda : \partial_k \mathbf{X} = \mathbf{R}\left(\partial_k \lambda q(\mathbf{x}) + \lambda \partial_k q(\mathbf{x}) \partial_k \mathbf{x}\right). \tag{3}$$

Observe that $\mathbf{t}$ is eliminated from the expression above. Note that, considering the k-th image coordinate, $\partial_k \mathbf{X}$ is the Jacobian of the 3D point $\mathbf{X}$, and $\partial_k \lambda$ is the Jacobian of the depth λ. Using the differential operator $\nabla = [\partial_u, \partial_v]$ it is more convenient to use a single compact expression, that includes the local affine frame $\mathbf{M}$, as $\nabla \mathbf{x} = [\partial_u \mathbf{x}, \partial_v \mathbf{x}] = \mathbf{M}$.

$$\exists! \lambda, \partial_u \lambda, \partial_v \lambda : \nabla \mathbf{X} = \mathbf{R}\left(q(\mathbf{x}) \nabla \lambda + \lambda \nabla q(\mathbf{x}) \mathbf{M}\right). \tag{4}$$

All in all, Eqs. (2) and (4) together describe all constraints on the 3D point $\mathbf{X}$ and on its vicinity $\nabla \mathbf{X}$, imposed by a LAF $(\mathbf{x}, \mathbf{M})$.

2.2   Affine Correspondence and Depth Constraining Camera Pose

Assume two views observing $\mathbf{X}$ under corresponding LAFs $(\mathbf{x}_1, \mathbf{M}_1)$ and $(\mathbf{x}_2, \mathbf{M}_2)$. Without the loss of generality, we define the coordinate system of the first view as the identity – thus putting it in the origin – while that of the second one is described by the relative rotation and translation, $\mathbf{R}$ and $\mathbf{t}$, respectively. In this two-view system, Eqs. (2) and (4) lead to two new constraints on the camera pose, depth and local affine frames, as follows:

$$\underbrace{\lambda_1 q_1\left(\mathbf{x}_1\right)}_{\mathbf{a}} = \mathbf{R}\underbrace{\lambda_2 q_2\left(\mathbf{x}_2\right)}_{\mathbf{b}} + \mathbf{t},$$

$$\underbrace{q_1\left(\mathbf{x}_1\right)\nabla\lambda_1 + \lambda_1\nabla q_1\left(\mathbf{x}_1\right)\mathbf{M}_1}_{\mathbf{A}} = \mathbf{R}\underbrace{\left(q_2\left(\mathbf{x}_2\right)\nabla\lambda_2 + \lambda_2\nabla q_2\left(\mathbf{x}_2\right)\mathbf{M}_2\right)}_{\mathbf{B}},$$

(5)

where the projection functions $q_1\left(\mathbf{x}_1\right)$ and $q_2\left(\mathbf{x}_2\right)$ assign bearing vectors to image points $\mathbf{x}_1$ and $\mathbf{x}_2$ of the first and second view, respectively. The first equation comes from Eq. (2) through the point correspondence, and the second one from Eq. (4) through the AC, by equating $\mathbf{X}$ and $\nabla\mathbf{X}$, respectively. Note that depths λ_1, λ_2 and their Jacobians $\nabla\lambda_1, \nabla\lambda_2$ are intrinsic to each view. Also observe that 3D point $\mathbf{X}$ and its Jacobian $\nabla\mathbf{X}$ are now eliminated from these constraints.

In the rest of the paper, we are resorting to the more compact notations using $\mathbf{a}, \mathbf{b}, \mathbf{A}$ and $\mathbf{B}$, as highlighted above, i.e., the two lines of Eq. (5) take the simplified forms of $\mathbf{a} = \mathbf{Rb} + \mathbf{t}$ and $\mathbf{A} = \mathbf{RB}$, respectively.

Relative Depth. In this paper, as monocular views are used to provide relative depth predictions, λ_1 and λ_2 are only known up to a common scale Λ, so that Eq. (5), with the simplified notation, is modified as follows:

$$\mathbf{a} = \Lambda\mathbf{Rb} + \mathbf{t}, \quad \mathbf{A} = \Lambda\mathbf{RB}. \tag{6}$$

These constraints describe the relationship of relative camera pose and ACs in the case of known relative depth.

3   Relative Pose and Scale from a Single Correspondence

Given a single affine correspondence, the optimal estimate for the relative pose and scale is given as the solution of the following optimization problem.

$$\min_{\mathbf{R},\mathbf{t},\Lambda} \underbrace{\frac{1}{2}\|\mathbf{a} - (\Lambda\mathbf{Rb} + \mathbf{t})\|_2^2 + \frac{1}{2}\|\mathbf{A} - \Lambda\mathbf{RB}\|_F^2}_{f(\Lambda,\mathbf{R},\mathbf{t})} \tag{7}$$

To solve the problem, the first order optimality conditions have to be investigated. At optimality, differentiating $f\left(\Lambda, \mathbf{R}, \mathbf{t}\right)$ by $\mathbf{t}$ gives:

$$\nabla_\mathbf{t} f\left(\Lambda, \mathbf{R}, \mathbf{t}\right) = \mathbf{a} - \Lambda\mathbf{Rb} - \mathbf{t} \overset{!}{=} 0, \tag{8}$$

that is, $\mathbf{t}$ can only be determined once the rest of the unknowns, $\mathbf{R}$ and Λ, are determined. This also means that above optimization problem can be set free of the translation, by performing the following substitution $\mathbf{t} \leftarrow \Lambda\mathbf{Rb} - \mathbf{a}$.

The optimization problem, where the translation is replaced as previously described, only contains the rotation and scale as unknowns, as follows:

$$\min_{\mathbf{R}, \Lambda} \underbrace{\frac{1}{2} \|\mathbf{A} - \Lambda \mathbf{R} \mathbf{B}\|_F^2}_{g(\mathbf{R}, \Lambda)}. \tag{9}$$

Below, different approaches for determining the unknown scale and rotation are described, solving Eq. (9) exactly or, in some manner, approximately.

1AC+D (Umeyama): An SVD Solution. In the least-squares sense, the optimal rotation and scale satisfying Eq. (9) can be acquired by the singular value decomposition of the covariance matrix of $\mathbf{A}$ and $\mathbf{B}$, as follows:

$$\mathbf{U}\mathbf{S}\mathbf{V}^\mathsf{T} = \mathbf{A}\mathbf{B}^\mathsf{T}. \tag{10}$$

Using these matrices, the optimal rotation and scale are expressed as

$$\mathbf{R} = \mathbf{U}\mathbf{D}\mathbf{V}^\mathsf{T}, \quad \Lambda = \frac{\mathrm{tr}\,(\mathbf{S}\mathbf{D})}{\mathrm{tr}\,(\mathbf{B}^\mathsf{T}\mathbf{B})}, \tag{11}$$

where $\mathbf{D}$ is a diagonal matrix with its diagonal being $[1, 1, \det(\mathbf{U}\mathbf{V}^\mathsf{T})]$, to constrain $\det \mathbf{R} = 1$. Given $\mathbf{R}$ and Λ, the translation is expressed as $\mathbf{t} = \mathbf{a} - \Lambda \mathbf{R} \mathbf{b}$.

This approach was greatly motivated by Umeyama's method [53], where a similar principle is used for solving point cloud alignment.

1AC+D (Proposed): Approximate Relative Orientation Solution. Matrices $\mathbf{A}$ and $\mathbf{B}$, together with their 1D left-nullspaces, define two non-orthogonal coordinate systems. By orthogonalizing the respective basis vectors, one can construct matrices $\mathbf{R_A}$ and $\mathbf{R_B}$, corresponding to $\mathbf{A}$ and $\mathbf{B}$. The relative rotation between these two coordinate systems can be written as follows:

$$\mathbf{R} = \mathbf{R_A}^\mathsf{T} \mathbf{R_B}. \tag{12}$$

$\mathbf{R_A}$ and $\mathbf{R_B}$ can be determined, *e.g.*, using Gram-Schmidt orthonormalization [47]. For our special case, a faster approach is shown in Algorithm 1.

Using this approach, the solution is biased towards the first columns of $\mathbf{A}$ and $\mathbf{B}$, providing perfect alignment of those two axes. In our experiments, this was rather a favorable property than an issue. The first axis of a LAF represents the orientation of the feature while the second one specifies the affine shape, as long as the magnitude of $\nabla \lambda_i$ is negligible compared to λ_i, which is usually the case. The orientation of a LAF is usually more reliable than its shape.

As $\mathbf{R}$ is now known and $\mathbf{t}$ has been ruled out, Λ is to be determined. Differentiating $g(\mathbf{R}, \Lambda)$ by Λ gives:

$$\nabla_\Lambda g(\mathbf{R}, \Lambda) = \mathrm{tr}\,(\mathbf{A}^\mathsf{T}\mathbf{R}\mathbf{B}) + \Lambda \mathrm{tr}\,(\mathbf{B}^\mathsf{T}\mathbf{B}), \tag{13}$$

Table 1. The theoretical computational complexity of the solvers used in RANSAC. The reported properties are: the number of operations of each solver (steps; 1st row); the computational complexity of one estimation (1 iter; 2nd); the number of correspondences required for the estimation (m; 3rd); possible outlier ratios ($1 - \mu$; 4th); the number of iterations needed for RANSAC with the required confidence set to 0.99 (# iters; 5th); and computational complexity of the full procedure (# comps; 6th). The major operations are: singular value decomposition (SVD), eigenvalue decomposition (EIG), LU factorization (LU), and QR decomposition (QR).

	Point-based	AC-based	AC+depth	
	5PT [48]	2AC [6]	1AC+D (Umeyama)	1AC+D (Proposed)
Steps	59 SVD + 1020 LU + 1010 EIG	69 SVD + 109 QR	33 SVD + cov. + trans.	4 × (cross + norm) + trans. + etc.
1 iter	$5^2 * 9 + 10^2 * 20 + 10^2 = 2325$	$6^2 * 9 + 10^3 + 10^2 = 1424$	$11 * 3^3 + 27 + 21 = 345$	$72 + 21 + 40 = 133$
m	5	2	1	1
1-μ	0.50 0.75 0.90	0.50 0.75 0.90	0.50 0.75 0.90	0.50 0.75 0.90
# iters	145 4713 $\sim 10^6$	16 71 458	7 16 44	7 16 44
# comps	$\sim 10^5$ $\sim 10^7$ $\sim 10^9$	$\sim 10^4$ $\sim 10^5$ $\sim 10^5$	$\sim 10^3$ $\sim 10^3$ $\sim 10^4$	$\sim 10^2$ $\sim 10^3$ $\sim 10^3$

that is, once $\mathbf{R}$ is known, Λ can be determined as follows:

$$\Lambda = -\frac{\mathrm{tr}\left(\mathbf{A}^{\mathsf{T}}\mathbf{R}\mathbf{B}\right)}{\mathrm{tr}\left(\mathbf{B}^{\mathsf{T}}\mathbf{B}\right)}. \tag{14}$$

Finally, the translation parameters are expressed as $\mathbf{t} = \mathbf{a} - \Lambda\mathbf{R}\mathbf{b}$.

The complete algorithm is shown in Algorithm 1. Note that, although, we have tested various approaches to computing the relative rotation between the frames $\mathbf{A}$ and $\mathbf{B}$, such as the above described SVD approach or the Gram-Schmidt process [47], the one introduced in Algorithm 1 proved to be the fastest with no noticeable deterioration in accuracy.

Algorithm 1. The 1AC+D (Proposed) algorithm for relative pose computation.

1: **procedure** 1AC+D (PROPOSED)$(\mathbf{a}, \mathbf{b}, \mathbf{A}, \mathbf{B})$ ▷ Computes the relative camera pose.
2:     $\mathbf{R_A} \leftarrow$ ORTHONORM$(\mathbf{A})$
3:     $\mathbf{R_B} \leftarrow$ ORTHONORM$(\mathbf{B})$
4:     $\mathbf{R} \leftarrow \mathbf{R_A^{\mathsf{T}}} \mathbf{R_B}$ ▷ Applying Eq. (12).
5:     $\Lambda \leftarrow -\mathrm{tr}\left(\mathbf{A}^{\mathsf{T}}\mathbf{R}\mathbf{B}\right)/\mathrm{tr}\left(\mathbf{B}^{\mathsf{T}}\mathbf{B}\right)$ ▷ Applying Eq. (14).
6:     $\mathbf{t} \leftarrow \mathbf{a} - \Lambda\mathbf{R}\mathbf{b}$
7:     **return** $\mathbf{R}, \mathbf{t}, \Lambda$
8: **function** ORTHONORM$(\mathbf{Y})$ ▷ Quick orthonormalization of $\mathbf{Y}$.
9:     $\mathbf{r}_x \leftarrow$ normalize $\left(\mathbf{Y}_{(:,1)} \times \mathbf{Y}_{(:,2)}\right)$ ▷ $\mathbf{r}_x$ is a normal to the underlying plane.
10:     $\mathbf{r}_z \leftarrow$ normalize $\left(\mathbf{Y}_{(:,2)}\right)$
11:     **return** $\left[\mathbf{r}_x, \mathbf{r}_z \times \mathbf{r}_x, \mathbf{r}_z\right]$ ▷ Return a 3 × 3 rotation matrix.

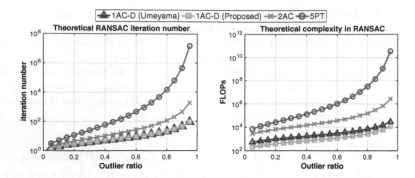

Fig. 2. *Left:* the theoretical number of RANSAC iterations – calculated as in [20]. *Right:* the number of floating point operations (horizontal axes) plotted as the function of the outlier ratio (vertical), displayed on logarithmic scales.

4   Experimental Results

In this section, experiments and complexity analyses compare the performance of the two versions of the proposed 1AC+D solver, *i.e.*, 1AC+D (Umeyama) and 1AC+D (Proposed), 2AC [6] and 5PT [48] methods for relative pose estimation.

4.1   Processing Time

In Table 1, we compare the computational complexity of state-of-the-art pose solvers used in our evaluation. The first row consists of the major steps of each solver. For instance, 5×9 SVD $+ 10 \times 20$ EIG means that the major steps are: the SVD decomposition of a 5×9 matrix and eigendecomposition of a 10×20 matrix. In the second row, the implied computational complexities are summed. In the third one, the number of correspondences required for the solvers are written. The fourth row lists example outlier ratios. In the fifth one, the theoretical numbers of iterations of RANSAC [20] are written with confidence set to 0.99. The last row reports the total complexity, *i.e.*, the complexity of one iteration multiplied by the number of RANSAC iterations.

The proposed methods have significantly smaller computational requirements than the five-point solver, 5PT, or the affine correspondence-based 2AC method [6] when included in RANSAC. Note that while the reported values for the 1AC+D methods are the actual FLOPs, for 5PT and 2AC, it is in practice about an order of magnitude higher due to the iterative manner of various linear algebra operations, *e.g.*, SVD.   In Fig. 2, the theoretical number of RANSAC iterations – calculated as in [20] – and the number of floating point operations are plotted as the function of the outlier ratio (horizontal axes), displayed on logarithmic scales (vertical axes). The proposed solvers lead to fewer iterations compared to the solvers solely based on point or affine correspondences.

In summary of the above analysis of the computational complexity, the proposed approach speeds up the robust estimation procedure in three ways. First, it requires at least an order of magnitude fewer operations to complete a single iteration of RANSAC than by using the five-point algorithm. Second, it leads to significantly fewer iterations due to the fact that 1AC+D takes only a single correspondence to propose a solution to pose estimation. Third, 1AC+D returns a single solution from a minimal sample, in contrast to the five-point algorithm. Therefore, each RANSAC iteration requires the validation of a single model.

4.2 Synthetic Evaluation

The synthetic evaluation was carried out in a setup consisting of two cameras with randomized poses represented by rotation $\mathbf{R}_i \in SO(3)$ and translation $\mathbf{t}_i \in \mathbb{R}^3$ ($i = 1, 2$). The cameras were placed around the origin at a distance sampled from $[1.0, 2.0]$, oriented towards a random point sampled from $[-0.5, 0.5]^3$. Both cameras had a common intrinsic camera matrix $\mathbf{K}$ with focal length $f = 600$ and principal point $[300, 300]$, representing pinhole projection $\pi : \mathbb{R}^3 \to \mathbb{R}^2$. For generating LAFs $(\mathbf{x}_i, \mathbf{M}_i)$, depths λ_i and their derivatives $\nabla\lambda_i$, randomized 3D points $\mathbf{X} \sim \mathcal{N}(\mathbf{0}, \mathbf{I}_{3\times3})$ with corresponding normals $\mathbf{n} \in \mathbb{R}^3$, $\|\mathbf{n}\|_2 = 1$ were projected into the two image planes using Eqs. (15) and (16), for $i = 1, 2$.

$$\mathbf{x}_i = \pi(\mathbf{R}_i\mathbf{X} + \mathbf{t}_i), \qquad\qquad \mathbf{M}_i = \nabla\pi(\mathbf{R}_i\mathbf{X} + \mathbf{t}_i)\mathbf{R}_i\nabla\mathbf{X}, \qquad (15)$$

$$\lambda_i = \mathbf{R}_i|_{(3,:)}\mathbf{X} + \mathbf{t}_i|_{(3)}, \qquad\qquad \nabla\lambda_i = \mathbf{R}_i|_{(3,:)}\nabla\mathbf{X}, \qquad (16)$$

where $\mathbf{R}_i|_{(3,:)}$ is the 3rd row of $\mathbf{R}_i$, $\nabla\mathbf{X} = \text{nulls}(\mathbf{n})$ simulates the local frame of the surface as the nullspace of the surface normal $\mathbf{n}$. Note that [14] proposed the approach for computing local frames $\mathbf{M}_i$, i.e., Eq. (15). Their synthetic experiment are also similar in concept, but contrast to them, we also had to account for the depth in Eq. (16), and we used the nullspace of $\mathbf{n}$ to express the Jacobian $\nabla\mathbf{X}$. The 2nd part of Eq. (16), the expression for the Jacobian of the depth, $\nabla\lambda_i$, was derived from the 1st part by differentiation. In this setup λ_i is the projective depth, a key element of perspective projection. Note that the depth and its derivatives are different for various other camera models. Finally, zero-mean Gaussian-noise was added to the coordinates/components of $\mathbf{x}_i$, $\mathbf{M}_i$, λ_i and $\nabla\lambda_i$, with σ, $\sigma_\mathbf{M}$, σ_λ and $\sigma_{\nabla\lambda}$ standard deviation (STD), respectively.

The reported errors for rotation and translation both were measured in degrees (°) in this evaluation. The rotation error was calculated as follows: $\epsilon_\mathbf{R} = \cos^{-1}\left(\frac{1}{2}\left(\text{tr}(\hat{\mathbf{R}}\mathbf{R}^\mathsf{T}) - 1\right)\right)$, where $\hat{\mathbf{R}}$ is the measured and $\mathbf{R}$ is the ground truth rotation matrix. Translation error $\epsilon_\mathbf{t}$ is the angular difference between the estimated and ground truth translations. For some of the tests, we also show the avg. Sampson error [20] of the implied essential matrix.

Solver Stability. We randomly generated 30000 instances of the above described synthetic setup, except for adding any noise to the LAFs and depths, in order to evaluate the stability of the various solver in a noise-free environment. Figures 3(a), 3(b) and 3(c) show the distribution of the Sampson, rotation and translation errors on logarithmic scales. The figures show that the proposed method is one of the best performers given that its distribution of errors is

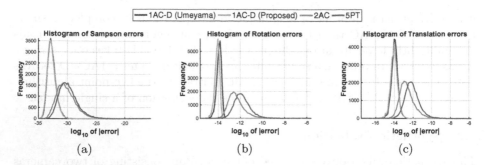

Fig. 3. Stability study of four different relative pose solvers: two versions of the proposed 1AC+D method, 1AC+D (Umeyama) and 1AC+D (Proposed), 2AC [6] solver using two affine correspondences and the five point solver 5PT [48]. The three plots show the distribution of various errors on logarithmic scales as histograms, namely (a) Sampson error, (b) rotation and (c) translation errors in degrees.

shifted lower compared to the other methods. However, all methods are quite stable having no peaks on the right side of the figures.

Noise Study. We added noise to the 30 000 instances of the synthetic setup. The estimated poses were evaluated to determine how sensitive the solvers are to the noise in the data, *i.e.*, image coordinates, depth, and LAFs. As expected, the noise in the depth or parameters of LAFs has no effect on the 5PT [48] solver. This can be seen in Fig. 5(b), where 5PT [48] is only affected by image noise. Increasing noise on either axes has a negative effect on the output of the proposed method as seen on Fig. 5(a). The effect of noise on rotation and translation estimated by 1AC+D and 5PT [48] are visualized on diagrams Fig. 4(a)-(f). The curves show that the effect of image noise – on useful scales, such as 2.5 px – in itself has a less significant effect on the degradation of rotation and translation estimates of the proposed method, compared to 5PT [48]. However, adding realistic scales of depth or affine noise, *e.g.* Fig. 4(a), has an observable negative effect on the accuracy of the proposed ones. Although 2AC [6] is unaffected by the noise on depth, it is moderately affected by noise on the affinities of LAFs. Its rotation estimates are worse than that of any other methods in the comparison. The provided translation vectors are of acceptable quality.

Summarizing the synthetic evaluation, we can state that, on realistic scales of noise, the accuracy of the rotation and translation estimates of 5PT [48] and both versions of 1AC+D are comparable. 2AC is the worst performer among all.

4.3 Real-World Evaluation

We tested the proposed solver on the 1DSfM dataset [56][1]. It consists of 13 scenes of landmarks with photos of varying sizes collected from the internet.

---

[1] http://www.cs.cornell.edu/projects/1dsfm/.

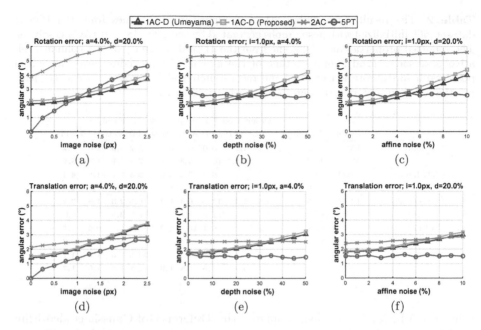

Fig. 4. Synthetic evaluation – the effect of different levels of noise on various relative pose solvers. Plots (a–f) compare the two proposed 1AC+D solvers, 2AC [6] and 5PT [48]. In the 1st row (a–c), the rotation error is shown. In the 2nd one (d–f) translation errors are plotted. All errors are angular errors in degrees. In each column, different setups are shown, where we fixed two of the three sources of noise – image, affine or depth noise – to analyse the negative effect of the third as its level increases.

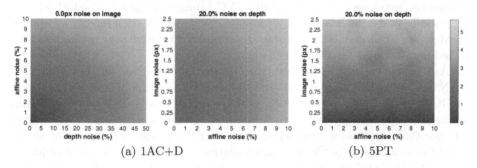

Fig. 5. Synthetic evaluation – the effect of image, affine and depth noise on the proposed 1AC+D and the point-based 5PT [48] – rotation errors, in degrees, displayed as heat-maps. As expected, 1AC+D is affected by all three noise types, as seen on (a). Being a point-based relative pose solver, 5PT [48] is affected only by the image noise (b).

1DSfM provides 2-view matches with epipolar geometries and a reference reconstruction from incremental SfM (computed with Bundler [45,46]) for measuring error. We iterated through the provided 2-view matches, detected ACs [32]

Table 2. The results of a global SfM [50] algorithm, on scenes from the 1DSfM dataset [56], initialized with pose-graphs generated by the 5PT [48] and 1AC+D solvers. The reported properties are: the scene from the 1DSfM dataset [56] (1st column), relative pose solver (2nd), total runtime of the global SfM procedure given an initial pose-graph (3rd), rotation error of the reconstructed global poses in degrees (4th), position error in meters (5th) and focal length errors (6th).

Scene	Solver	Runtime (s)	Orientation (°)			Position (m)			Focal len. (×e − 2)		
			AVG	MED	STD	AVG	MED	STD	AVG	MED	STD
Piccadilly	1AC+D	51,2	**6,1**	4,3	**8,0**	6,5	**3,4**	**7,7**	**2,3**	1.6	**3,2**
	5PT	**48,4**	6,8	**2,5**	10,1	**4,8**	3,5	7,8	2,4	1,7	3,3
NYC Library	1AC+D	10,4	6,0	2,0	**6,3**	**3,4**	3,6	**3,3**	**2,7**	1,5	**4,1**
	5PT	**5,9**	**5,9**	1,9	6,7	4,2	3,7	5,5	2,9	1,5	4,6
Vienna Cathedral	1AC+D	**56,5**	**4,5**	**1,7**	**11,6**	**6,1**	1,4	**10,3**	**2,3**	1,3	**3,7**
	5PT	81,4	4,6	2,4	12,1	8,4	1,6	12,5	**2,3**	1,3	4,0
Madrid Metropolis	1AC-D	**14,6**	**5,2**	**2,4**	**6,1**	13,6	**1,3**	**15,5**	**1,2**	**0,5**	2,4
	5PT	20,5	6,9	4,7	7,2	18,3	3,6	21,1	**1,2**	**0,5**	**2,3**
Ellis Island	1AC-D	9,0	4,4	**5,6**	7,2	11,2	2,6	11,2	**1,6**	1,1	**1,7**
	5PT	**5,7**	**3,1**	6,0	**3,9**	11,8	**1,9**	12,0	**1,6**	1,0	**1,7**

using the VLFeat library [54], applying the Difference-of-Gaussians algorithm combined with the affine shape adaptation procedure as proposed in [9]. In our experiments, affine shape adaptation only had a small ∼10% extra time demand, *i.e.*, (0.31 ± 0.25) s per view, over regular feature extraction. This extra overhead is negligible, compared to the more time-consuming feature matching. Matches were filtered by the standard ratio test [32]. We did not consider image pairs in the evaluation with fewer than 20 corresponding features between them. For the evaluation, we chose scenes Piccadilly, NYC Library, Vienna Cathedral, Madrid Metropolis, and Ellis Island. To get monocular depth for each image, we applied the detector of Li *et al.* [30]. In total, the compared relative pose estimators were tested on a total of 110395 image pairs.

As a robust estimator, we chose the Graph-Cut RANSAC [2] algorithm (GC-RANSAC) since it is state-of-the-art and has publicly available implementation[2]. In GC-RANSAC, and other locally optimized RANSACs, two different solvers are applied: (i) one for fitting to a minimal sample and (ii) one for fitting to a larger-than-minimal sample when improving the model parameters by fitting to all found inliers or in the local optimization step. For (i), the main objective is to solve the problem using as few points as possible since the overall wall-clock time is a function of the point number required for the estimation. For (ii), the goal is to estimate an accurate model from the provided set of points. In practice, step (ii) is applied rarely and, therefore, its processing time is not so crucial for achieving efficient robust estimation. We used the normalized eight-point algorithm followed by a rank-2 projection to estimate the essential matrix from a larger-than-minimal sample. We applied GC-RANSAC with a confidence set to 0.99 and inlier-outlier threshold set to be the 0.05% of the image diagonal size. For the other parameters, the default values were used.

---

[2] https://github.com/danini/graph-cut-ransac.

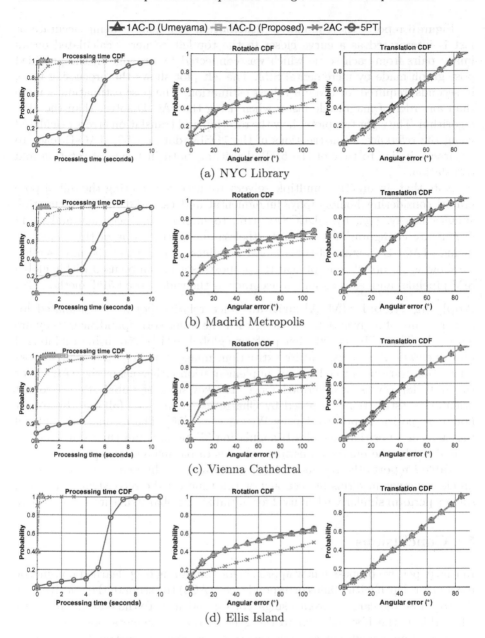

Fig. 6. Relative pose estimation on a total of 110395 image pairs from the 1DSfM dataset. The cumulative distribution functions are shown for the processing time (in seconds), angular error of the estimated rotations and translations (in degrees). Being accurate or fast is interpreted as a curve close to the top-left corner.

Figure 6 reports the cumulative distribution functions – being accurate or fast is interpreted as a curve closer to the top-left corner – calculated on all image pairs from each scene which was connected by an edge in the provided pose graph made by incremental SfM. The left plot shows the processing time, in seconds, required for the full robust estimation. The pose estimation is more than an order of magnitude faster when using the 1AC-D solver than the 5PT algorithm. The right two plots show the rotation and translation errors calculated using the reference reconstructions of the 1DSfM dataset [56]. 1AC-D leads to accuracy similar to that of the 5PT algorithm, both in terms of rotation and translation.

Note that, in practice, multiple solvers are used for creating the initial pose graph, considering homography, fundamental and essential matrix estimation simultaneously, *e.g.*, by QDEGSAC [16], to improve the accuracy and avoid degenerate configurations. It is nevertheless out of the scope of this paper, to include 1AC+D in state-of-the-art SfM pipelines. However, in the next section we show as a proof-of-concept that the proposed solver can be used within global SfM pipelines and leads to similar accuracy as the widely-used 5PT method [48].

Applying Global SfM Algorithm. Once relative poses are estimated for camera pairs of a given dataset, along with the inlier correspondences, they are fed to the Theia library [50] that performs global SfM [11,56] using its internal implementation. That is, feature extraction, image matching and relative pose estimation were performed by our implementation either using the 1AC+D or the 5PT [48] solvers, as described above. The key steps of global SfM are robust orientation estimation, proposed by Chatterjee *et al.* [11], followed by robust nonlinear position optimization by Wilson *et al.* [56]. The estimation of global rotations and positions enables triangulating 3D points, and the reconstruction is finalized by the bundle adjustment of camera parameters and points.

Table 2 reports the results of Theia initialized by different solvers. There is no clear winner in terms of accuracy or run-time (of the reconstruction). Both solvers perform similarly when used for initializing global structure-from-motion.

5   Conclusions

In this paper, we propose a new approach for combining deep-learned non-metric monocular depth with affine correspondences (ACs) to estimate the relative pose of two calibrated cameras from a single correspondence. To the best of our knowledge, this is the first solution to the general relative camera pose estimation problem, from a single correspondence. Two new general constraints are derived interpreting the relationship of camera pose, affine correspondences and relative depth. Since the proposed solver requires a single correspondence, robust estimation becomes significantly faster, compared to traditional techniques, with speed depending linearly on the number of correspondences.

The proposed 1AC+D solver is tested both on synthetic data and on 110395 publicly available real image pairs from the 1DSfM dataset. It leads to an accuracy similar to traditional approaches while being significantly faster. When solving large-scale problems, *e.g.*, pose-graph initialization for Structure-from-Motion (SfM) pipelines, the overhead of obtaining affine correspondences and monocular depth is negligible compared to the speed-up gain in the pairwise geometric verification. As a proof-of-concept, it is demonstrated on scenes from the 1DSfM dataset, via using a state-of-the-art global SfM algorithm, that acquiring the initial pose-graph by the proposed method leads to reconstruction of similar accuracy to the commonly used five-point solver.

Acknowledgement. Supported by Exploring the Mathematical Foundations of Artificial Intelligence (2018-1.2.1-NKP-00008), -Intensification of the activities of HU-MATHS-IN–Hungarian Service Network of Mathematics for Industry and Innovation' under grant number EFOP-3.6.2-16-2017-00015, the Ministry of Education OP VVV project CZ.02.1.01/0.0/0.0/16 019/0000765 Research Center for Informatics, and the Czech Science Foundation grant GA18-05360S.

References

1. Albl, C., Kukelova, Z., Fitzgibbon, A., Heller, J., Smid, M., Pajdla, T.: On the two-view geometry of unsynchronized cameras. In: Computer Vision and Pattern Recognition, July 2017
2. Barath, D., Matas, J.: Graph-cut RANSAC. In: Computer Vision and Pattern Recognition, pp. 6733–6741 (2018)
3. Baráth, D., Tóth, T., Hajder, L.: A minimal solution for two-view focal-length estimation using two affine correspondences. In: Computer Vision and Pattern Recognition (2017)
4. Barath, D., Eichhardt, I., Hajder, L.: Optimal multi-view surface normal estimation using affine correspondences. IEEE Trans. Image Process. **28**(7), 3301–3311 (2019)
5. Barath, D., Hajder, L.: A theory of point-wise homography estimation. Pattern Recogn. Lett. **94**, 7–14 (2017)
6. Barath, D., Hajder, L.: Efficient recovery of essential matrix from two affine correspondences. IEEE Trans. Image Process. **27**(11), 5328–5337 (2018)
7. Barath, D., Matas, J., Noskova, J.: MAGSAC: marginalizing sample consensus. In: Computer Vision and Pattern Recognition, pp. 10197–10205 (2019)
8. Batra, D., Nabbe, B., Hebert, M.: An alternative formulation for five point relative pose problem. In: Workshop on Motion and Video Computing, p. 21. IEEE (2007)
9. Baumberg, A.: Reliable feature matching across widely separated views. In: Computer Vision and Pattern Recognition, vol. 1, pp. 774–781. IEEE (2000)
10. Bentolila, J., Francos, J.M.: Conic epipolar constraints from affine correspondences. Comput. Vis. Image Underst. **122**, 105–114 (2014)
11. Chatterjee, A., Madhav Govindu, V.: Efficient and robust large-scale rotation averaging. In: Proceedings of International Conference on Computer Vision, pp. 521–528 (2013)
12. Chum, O., Matas, J.: Matching with PROSAC-progressive sample consensus. In: Computer Vision and Pattern Recognition. IEEE (2005)
13. Eichhardt, I., Barath, D.: Optimal multi-view correction of local affine frames. In: British Machine Vision Conf. (September 2019)

14. Eichhardt, I., Chetverikov, D.: Affine correspondences between central cameras for rapid relative pose estimation. In: Ferrari, V., Hebert, M., Sminchisescu, C., Weiss, Y. (eds.) ECCV 2018. LNCS, vol. 11210, pp. 488–503. Springer, Cham (2018). https://doi.org/10.1007/978-3-030-01231-1_30
15. Fischler, M.A., Bolles, R.C.: Random sample consensus: a paradigm for model fitting with applications to image analysis and automated cartography. Commun. ACM **24**, 381–395 (1981)
16. Frahm, J.M., Pollefeys, M.: RANSAC for (quasi-) degenerate data (QDEGSAC). In: Computer Vision and Pattern Recognition, pp. 453–460. IEEE (2006)
17. Fraundorfer, F., Tanskanen, P., Pollefeys, M.: A minimal case solution to the calibrated relative pose problem for the case of two known orientation angles. In: Daniilidis, K., Maragos, P., Paragios, N. (eds.) ECCV 2010. LNCS, vol. 6314, pp. 269–282. Springer, Heidelberg (2010). https://doi.org/10.1007/978-3-642-15561-1_20
18. Guan, B., Zhao, J., Li, Z., Sun, F., Fraundorfer, F.: Minimal solutions for relative pose with a single affine correspondence. In: Computer Vision and Pattern Recognition (2020)
19. Hajder, L., Baráth, D.: Relative planar motion for vehicle-mounted cameras from a single affine correspondence. In: Proceedings of International Conference of Robotics and Automation (2020)
20. Hartley, R., Zisserman, A.: Multiple View Geometry in Computer Vision. Cambridge University Press, Cambridge (2003)
21. Hartley, R., Li, H.: An efficient hidden variable approach to minimal-case camera motion estimation. IEEE Trans. Pattern Anal. Mach. Intell. **34**(12), 2303–2314 (2012)
22. Hesch, J.A., Roumeliotis, S.I.: A direct least-squares (DLS) method for PnP. In: Proceedings of International Conference on Computer Vision, pp. 383–390. IEEE (2011)
23. Hodaň, T., et al.: BOP: benchmark for 6D object pose estimation. In: Ferrari, V., Hebert, M., Sminchisescu, C., Weiss, Y. (eds.) ECCV 2018. LNCS, vol. 11214, pp. 19–35. Springer, Cham (2018). https://doi.org/10.1007/978-3-030-01249-6_2
24. Köser, K.: Geometric estimation with local affine frames and free-form surfaces. Shaker (2009)
25. Kukelova, Z., Bujnak, M., Pajdla, T.: Polynomial eigenvalue solutions to minimal problems in computer vision. IEEE Trans. Pattern Anal. Mach. Intell. **34**(7), 1381–1393 (2011)
26. Larsson, V., Kukelova, Z., Zheng, Y.: Camera pose estimation with unknown principal point. In: Computer Vision and Pattern Recognition, June 2018
27. Larsson, V., Sattler, T., Kukelova, Z., Pollefeys, M.: Revisiting radial distortion absolute pose. In: Proceedings of International Conference on Computer Vision, October 2019
28. Li, B., Heng, L., Lee, G.H., Pollefeys, M.: A 4-point algorithm for relative pose estimation of a calibrated camera with a known relative rotation angle. In: International Conference on Intelligent Robots and Systems, pp. 1595–1601. IEEE (2013)
29. Li, H., Hartley, R.: Five-point motion estimation made easy. In: Proceedings of International Conference on Pattern Recognition, vol. 1, pp. 630–633. IEEE (2006)
30. Li, Z., Snavely, N.: MegaDepth: learning single-view depth prediction from internet photos. In: Computer Vision and Pattern Recognition, June 2018
31. Li, Z., Wang, G., Ji, X.: CDPN: coordinates-based disentangled pose network for real-time RGB-based 6-DoF object pose estimation. In: Proceedings of International Conference on Computer Vision, pp. 7678–7687 (2019)

32. Lowe, D.G.: Object recognition from local scale-invariant features. In: Proceedings of International Conference on Computer Vision. IEEE (1999)
33. Mikolajczyk, K., Schmid, C.: Comparison of affine-invariant local detectors and descriptors. In: European Signal Processing Conference, pp. 1729–1732. IEEE (2004)
34. Mur-Artal, R., Montiel, J.M.M., Tardos, J.D.: ORB-SLAM: a versatile and accurate monocular SLAM system. IEEE Trans. Robot. **31**(5), 1147–1163 (2015)
35. Mur-Artal, R., Tardós, J.D.: ORB-SLAM2: an open-source SLAM system for monocular, stereo, and RGB-D cameras. IEEE Trans. Robot. **33**(5), 1255–1262 (2017)
36. Nakano, G.: A versatile approach for solving PnP, PnPf, and PnPfr problems. In: Leibe, B., Matas, J., Sebe, N., Welling, M. (eds.) ECCV 2016. LNCS, vol. 9907, pp. 338–352. Springer, Cham (2016). https://doi.org/10.1007/978-3-319-46487-9_21
37. Nistér, D.: An efficient solution to the five-point relative pose problem. IEEE Trans. Pattern Anal. Mach. Intell. **26**, 756–770 (2004)
38. Perdoch, M., Matas, J., Chum, O.: Epipolar geometry from two correspondences. In: Proceedings of International Conference on Computer Vision (2006)
39. Pritts, J., Kukelova, Z., Larsson, V., Lochman, Y., Chum, O.: Minimal solvers for rectifying from radially-distorted conjugate translations. arXiv preprint arXiv:1911.01507 (2019)
40. Raguram, R., Chum, O., Pollefeys, M., Matas, J., Frahm, J.M.: USAC: a universal framework for random sample consensus. IEEE Trans. Pattern Anal. Mach. Intell. **35**, 2022–2038 (2013)
41. Raposo, C., Barreto, J.P.: Theory and practice of structure-from-motion using affine correspondences. In: Computer Vision and Pattern Recognition, pp. 5470–5478 (2016)
42. Saurer, O., Pollefeys, M., Lee, G.H.: A minimal solution to the rolling shutter pose estimation problem. In: International Conference on Intelligent Robots and Systems, pp. 1328–1334. IEEE (2015)
43. Scaramuzza, D.: 1-point-RANSAC structure from motion for vehicle-mounted cameras by exploiting non-holonomic constraints. Int. J. Comput. Vis. **95**(1), 74–85 (2011)
44. Schonberger, J.L., Frahm, J.M.: Structure-from-motion revisited. In: Computer Vision and Pattern Recognition, pp. 4104–4113 (2016)
45. Snavely, N., Seitz, S.M., Szeliski, R.: Photo tourism: exploring photo collections in 3D. In: ACM Transmission Graphics, vol. 25, pp. 835–846. ACM (2006)
46. Snavely, S., Seitz, S.M., Szeliski, R.: Modeling the world from internet photo collections. Int. J. Comput. Vis. **80**(2), 189–210 (2008)
47. Solomon, J.: Numerical Algorithms: Methods for Computer Vision, Machine Learning, and Graphics. AK Peters/CRC Press, London (2015)
48. Stewenius, H., Engels, C., Nistér, D.: Recent developments on direct relative orientation. J. Photogrammetry Remote Sens. **60**(4), 284–294 (2006)
49. Stewénius, H., Nistér, D., Kahl, F., Schaffalitzky, F.: A minimal solution for relative pose with unknown focal length. In: Computer Vision and Pattern Recognition, vol. 2, pp. 789–794. IEEE (2005)
50. Sweeney, C.: Theia multiview geometry library. http://theia-sfm.org
51. Sweeney, C., Sattler, T., Hollerer, T., Turk, M., Pollefeys, M.: Optimizing the viewing graph for structure-from-motion. In: Proceedings of International Conference on Computer Vision, pp. 801–809 (2015)
52. Torr, P.H.S., Zisserman, A.: MLESAC: a new robust estimator with application to estimating image geometry. Comput. Vis. Image Underst. **78**, 138–156 (2000)

53. Umeyama, S.: Least-squares estimation of transformation parameters between two point patterns. IEEE Trans. Pattern Anal. Mach. Intell. **13**(4), 376–380 (1991)
54. Vedaldi, A., Fulkerson, B.: VLFeat: an open and portable library of computer vision algorithms (2008). http://www.vlfeat.org/
55. Ventura, J., Arth, C., Reitmayr, G., Schmalstieg, D.: A minimal solution to the generalized pose-and-scale problem. In: Computer Vision and Pattern Recognition, pp. 422–429 (2014)
56. Wilson, K., Snavely, N.: Robust global translations with 1DSfM. In: Fleet, D., Pajdla, T., Schiele, B., Tuytelaars, T. (eds.) ECCV 2014. LNCS, vol. 8691, pp. 61–75. Springer, Cham (2014). https://doi.org/10.1007/978-3-319-10578-9_5
57. Wu, C.: Towards linear-time incremental structure from motion. In: International Conference on 3D Vision, pp. 127–134. IEEE (2013)

Video Super-Resolution with Recurrent Structure-Detail Network

Takashi Isobe[1,2], Xu Jia[2(✉)], Shuhang Gu[3], Songjiang Li[2], Shengjin Wang[1(✉)], and Qi Tian[2]

[1] Department of Electronic Engineering, Tsinghua University, Beijing, China
jbj18@mails.tsinghua.edu.cn,wgsgj@tsinghua.edu.cn
[2] Noah's Ark Lab, Huawei Technologies, Shenzhen, China
{x.jia,songjiang.li,tian.qi1}@huawei.com
[3] School of Eie, The University of Sydney, Sydney, Australia
shuhanggu@gmail.com

Abstract. Most video super-resolution methods super-resolve a single reference frame with the help of neighboring frames in a temporal sliding window. They are less efficient compared to the recurrent-based methods. In this work, we propose a novel recurrent video super-resolution method which is both effective and efficient in exploiting previous frames to super-resolve the current frame. It divides the input into structure and detail components which are fed to a recurrent unit composed of several proposed two-stream structure-detail blocks. In addition, a hidden state adaptation module that allows the current frame to selectively use information from hidden state is introduced to enhance its robustness to appearance change and error accumulation. Extensive ablation study validate the effectiveness of the proposed modules. Experiments on several benchmark datasets demonstrate superior performance of the proposed method compared to state-of-the-art methods on video super-resolution. Code is available at https://github.com/junpan19/RSDN.

Keywords: Video super-resolution · Recurrent neural network · Two-stream block

1 Introduction

Super-resolution is one of the fundamental problem in image processing, which aims at reconstructing a high resolution (HR) image from a single low-resolution (LR) image or a sequence of LR images. According to the number of input frames, the field of SR can be divided into two categories, *i.e.*, single image super-resolution (SISR) and multi-frame super-resolution (MFSR). For SISR, the key issue is to exploit natural image prior for compensating missing details; while for MFSR, how to take full advantage from additional temporal information is of pivotal importance. In this work, we focus on the video super-resolution (VSR)

T. Isobe—The work was done in Noah's Ark Lab, Huawei Technologies.

© Springer Nature Switzerland AG 2020
A. Vedaldi et al. (Eds.): ECCV 2020, LNCS 12357, pp. 645–660, 2020.
https://doi.org/10.1007/978-3-030-58610-2_38

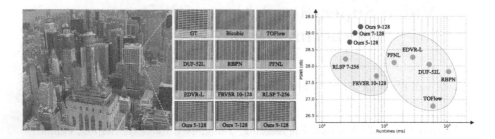

Fig. 1. VSR results on the **City** sequence in Vid4. Our method produces finer details and stronger edges with better balance between speed and performance than both temporal sliding window based [7,12,26,27,29] and recurrent based methods [4,23]. Blue box represents recurrent-based and green box represents sliding window based methods. Runtimes (ms) are calculated on an HR image of size 704 × 576. (Color figure online)

task which belongs to MFSR. It draws much attention in both research and industrial communities because of its great value on computational photography and surveillance (Fig. 1).

In the last several years, great attempts have been made to exploit multi-frame information for VSR. One category of approaches utilize multi-frame information by conducting explicit motion compensation. These approaches [1,13,21,25,27] firstly compute optical flow between a reference frame and neighboring frames and then employ the aligned observations to reconstruct the high-resolution reference frame. However, estimating dense optical flow itself is a challenging and time-consuming task. Inaccurate flow estimation often leads to unsatisfactory artifacts in the SR results of these flow-based VSR approaches. In addition, the heavy computational burden also impedes the application of these applications in resource-constrained devices and time-sensitive scenarios. In order to avoid explicit motion compensation, another category of methods propose to exploit the motion information in an implicit manner. The dynamic upsampling filters [12] and the progressive fusion residual blocks [29] are designed to explore flow-free motion compensation. With these flexible compensation strategies, [12,29] not only avoid heavy motion estimation step but also achieve highly competitive VSR performance. However, they still suffer from the redundant computation for several neighboring frames within a temporal window and need to cache several frames in advance to conduct VSR. Recently, for the pursuit of efficiency, there is an emerging trend of applying recurrent connection to address the VSR task.

These approaches [4,23] make use of recurrent connection to conduct video super-resolution in a streaming way, that is, output or hidden state of previous time steps is used to help super-resolve future frames. In addition, they are able to exploit temporal information from many frames in the past. By simply propagating output and hidden state of previous steps with a recurrent unit, they achieve promising VSR performance with considerably less processing time.

In this paper, we propose a novel recurrent network for efficient and effective video super-resolution. Instead of simply concatenating consecutive three frames with previous hidden state as in [4], we propose to decompose each frame of a sequence into components of structure and detail and aggregate both current and previous structure and detail information to super-resolve each frame. Such a strategy not only allows our method to address different difficulties in the structure and detail components, but also able to impose flexible supervision to recover high-frequency details and strengthen edges in the reconstruction.

In addition, we observe that hidden state in a recurrent network captures different typical appearances of a scene over time. To make full use of temporal information in hidden state, we treat the hidden state as a historical dictionary and compute correlation between the reference frame and each channel in hidden state. This allows the current frame to highlight the potentially helpful information and suppress outdated information such that information fusion would be more robust to appearance change and accumulated errors. Extensive ablation study demonstrates the effectiveness of the proposed method. It performs very favorably against state-of-the-art methods on several benchmark datasets, in both super-resolution performance and speed.

2 Related Work

2.1 Single Image Super-Resolution

Traditional SISR methods include interpolation-based methods and dictionary learning-based methods. However, since the rise of deep learning, most traditional methods are outperformed by deep learning based methods. A simple three-layer CNN is proposed by Dong [2], showing great potential of deep learning in super-resolution for the first time. Since then, plenty of new network architectures [6,14,18,19,30,31] have been designed to explore power of deep learning to further improve performance of SISR. In addition, researchers also investigate the role of losses for better perceptual quality. More discussions can be found in a recent survey [28]. A very relevant work is the DualCNN method proposed by Pan *et al.* [22], where authors proposed a network with two parallel branches to reconstruct structure and detail components of an image, respectively. However, different from that work, our method aims at addressing the video super-resolution task. It decomposes the input frames into structure and detail components and propagates them with a recurrent unit that is composed of two interleaved branches to reconstruct the high-resolution targets. It is motivated by the assumption that structure and detail components not only suffer from different difficulties in high-resolution reconstruction but also take benefit from other frames in different ways.

2.2 Video Super-Resolution

Although SISR methods can also be used to address the video super-resolution task, they are not very effective because they only learn to explore natural prior and self-similarity within an image and ignore rich temporal information in a sequence. The key to video super-resolution is to make full use of complementary information across frames. Most video super-resolution methods can be roughly divided into two categories according to whether they conduct motion compensation in an explicit way or not.

Explicit Motion Compensation. Most methods with explicit motion compensation follow a pipeline of motion estimation, motion compensation, information fusion and upsampling. VESPCN [1] presents a joint motion compensation and video super-resolution with a coarse-to-fine spatial transformer module. Tao *et al.* [25] proposed an SPMC module for sub-pixel motion compensation and used a ConvLSTM to fuse information across aligned frames. Xue *et al.* [27] proposed a task-oriented flow module that is trained together with a video processing network for video denoising, deblock or super-resolution. In [23], Sajjadi *et al.* proposed to super-resolve a sequence of frames in a recurrent manner, where the result of previous frame is warped to the current frame and two frames are concatenated for video super-resolution. Haris *et al.* [7] proposed to use a recurrent encoder-decoder module to exploit explicitly estimated inter-frame motion. Wang *et al.* [26] proposed to align multiple frames to a reference frame in feature space with a deformable convolution based module and fuse aligned frames with a temporal and spatial attention module. However, the major drawback of such methods is the heavy computational load introduced by motion estimation and motion compensation.

Implicit Motion Compensation. As for methods with implicit motion compensation [4,7,12,29], they do not estimate motion between frames and align them to a reference frame, but focus on designing an advanced fusion module such that it can make full use of complementary information across frames. Jo *et al.* [12] proposed to use a 3D CNN to exploit spatial-temporal information and predict a dynamic upsampling filter to reconstruct HR images. In [29], Yi *et al.* proposed to fuse spatial-temporal information across frames in a progressive way and use a non-local module to avoid explicit motion compensation. Video super-resolution with implicit motion can also be done with recurrent connection. Huang *et al.* [9] proposed a bidirectional recurrent convolutional network to model temporal information across multiple frames for efficient video super-resolution. In [4], Fuoli *et al.* proposed to conduct temporal information propagation with a recurrent architecture in feature space. Our method also adopts the recurrent way to conduct video super-resolution without explicit motion compensation. However, different from the above methods, we proposed to decompose a frame into two components of structure and detail and propagate them separately. In addition, we also compute correlation between the current frame and the hidden state to adaptively use the history information in the hidden state for better performance and less risk of error accumulation.

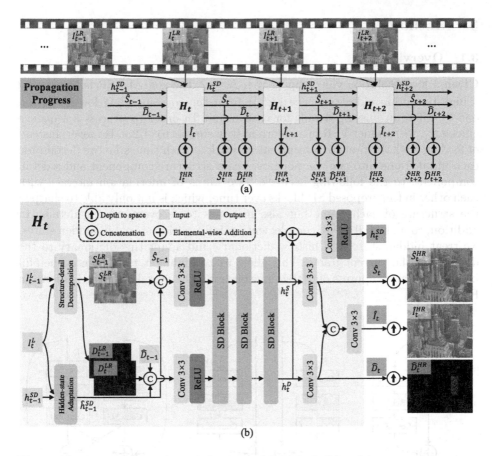

Fig. 2. (a) The overall pipeline of the proposed method; (b) architecture of the recurrent structure-detail unit.

2.3 Recurrent Networks for Video-Based Tasks

Recurrent networks have been widely used in different video recognition tasks. Donahue *et al.* [15] proposed a class of recurrent convolutional architectures which combine convolutional layers and long-range temporal information for action recognition and image captioning. In [24], a bi-directional LSTM is applied after a multi-stream CNN to fully explore temporal information in a sequence for fine-grained action detection. Du *et al.* [3] proposed a recurrent network with a pose attention mechanism which exploits spatial-temporal evolution of human pose to assist action recognition. Recurrent networks are capable of processing sequential information by integrating information from each frame in their hidden states. They can not only be used for high-level video recognition tasks but are also suitable for low-level video processing tasks.

3 Method

3.1 Overview

Given a low-resolution video clip $\{I_{1:N}^{LR}\}$, $N \geq 2$, the goal of VSR is to produce a high-resolution video sequence $\{\hat{I}_{1:N}^{HR}\}$ from the corresponding low-resolution one by filling in missing details for each frame. In order to process a sequence efficiently, we conduct VSR in a recurrent way similar to [4,23]. However, instead of feeding a whole frame to a recurrent network at each time step, we decompose each input frame into two components, i.e., a structure component and a detail component, to the following network. Two kinds of information interact with each other in the proposed SD blocks over time, which is not only able to sharpen the structure of each frame but also manages to recovers missing details. In addition, to make full use of complementary information stored in hidden states, we treat hidden state as a history dictionary and adapt this dictionary to the demand of the current frame. This allow us to highlight the potential helpful information and suppress outdated information. The overall pipeline is shown in Fig. 2(a).

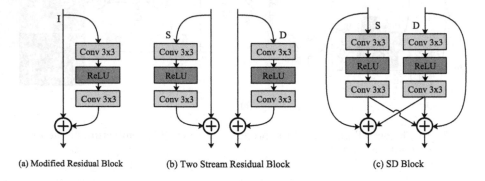

(a) Modified Residual Block   (b) Two Stream Residual Block   (c) SD Block

Fig. 3. Variant design for structure-detail block.

3.2 Recurrent Structure-Detail Network

Recurrent Unit. Each frame can be decomposed into a structure component and a detail component. The structure component models low-frequency information in an image and motion between frames. While the detail component captures fine high-frequency information and slight change in appearance. These two components suffer from different difficulty in high-resolution reconstruction and take different benefit from other frames, hence should be processed separately.

In this work, we simply apply a pair of bicubic downsampling and upsampling operations to extract structural information from a frame I_t^{LR}, which is denoted as S_t^{LR}. The detail component D_t^{LR} can be then computed as the difference between the input frame I_t^{LR} and the structure component S_t^{LR}. In fact, we can also use other ways such as low-pass filtering and high-pass filtering to get these two components. For simplicity, we adopt a symmetric architecture for the two components in the recurrent unit, as shown in Fig. 2 (b). Here we only take D-branch at time step t as an example to explain its architecture design. Detail components of the previous and current frames $\{D_{t-1}^{LR}, D_t^{LR}\}$ are concatenated with the previously estimated detail map $\hat{D}_{t-1}$ and hidden state $\hat{h}_{t-1}^{SD}$ along the channel axis. Such information is further fused by one 3×3 convolutional layer and several structure-detail (SD) blocks. In this way, this recurrent unit manages to integrate together information from two consecutive input frames, output of the previous time step and historical information stored in the hidden state. h_t^D denotes the feature computed after several SD blocks. It goes through another 3×3 convolutional layer and an upsampling layer to produce the high resolution detail component $\hat{D}_t^{HR}$. The S-branch is designed in a similar way. h_t^S and h_t^D are combined to produce the final high resolution image $\hat{I}_t^{HR}$ and new hidden state h_t^{SD}. The D-branch focuses on extracting complementary details from past frames for the current frame while the S-branch focuses on enhancing existed edges and textures in the current frame.

Structure-Detail Block. Residual block [18] and dense block [8] are widely used in both high-level and low-level computer vision tasks because of their effectiveness in mining and propagating information. In this section, we compare several variants of blocks in propagating information in a recurrent structure-detail unit. For comparison, we also include a modified residual block as shown in Fig. 3(a), which only has one branch and takes the whole frames as input. To adapt it to address two branches, the easiest way is to have two modified residual blocks that process two branches separately, as shown in Fig. 3(b). However, in this way each branch only sees the component-specific information and can not makes full use of all information in the input frames. Therefore, we propose a new module called structure-detail (SD) block, as shown in Fig. 3(c). The two components are first fed to two individual branches and then combined with an addition operation. In this way, it not only specializes on each component but also promotes information exchange between structure and detail components. Its advantage over the other two variants is validated in the experiment section (Fig. 4).

3.3 Hidden State Adaptation

In a recurrent neural network, hidden state at time step t would summarize past information in the previous frames. When applying a recurrent neural network to the video super-resolution task, hidden state is expected to model how a scene's appearance evolves over time, including both structure and detail. The previous recurrent-based VSR method [4] directly concatenates previous hidden state and two input frames and feeds it to several convolutional layers. However, for each LR frame to be super-resolved, it has distinct appearance and is expected

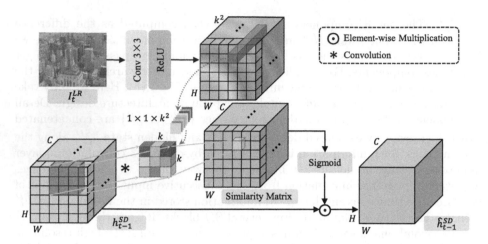

Fig. 4. Design of hidden state adaptation module.

to borrow complementary information from previous frames in different ways. Applying the same integration manner to all frames is not optimal and could hurt the final performance. As shown in Fig. 5, different channels in hidden state describe different scene appeared in the past. They should make different contribution to different positions of different frames, especially when there are occlusion and large deformation with some channels of the hidden state.

In this work, we proposed the Hidden State Adaptation (HSA) module to adapt a hidden state to the appearance of the current frame. As for each unit in hidden state, it should be highlighted if it has similar appearance as the current frame; otherwise, it should be suppressed if it looks very different. With this module the proposed method carefully chooses only useful information in previous frames, hence alleviate the influence of drastic appearance change and error accumulation. Since response of a filter models correlation between the filter and a neighborhood on an image, here we take similar way to compute correlation between an input frame and hidden state. Inspired by [11], we generate spatially variant and sample specific filters for each position in the current frame and use those filters to compute their correlation with the corresponding positions in each channel of hidden state. Specifically, these spatially variant filters $F_t^\theta \in R^{H \times W \times (k \times k)}$ are obtained by feeding the current frame $I_t^{LR} \in R^{H \times W \times 3}$ into a convolutional layer with ReLU activation function [5], where H and W are respectively height and width of the current frame, and k denotes the size of filters. Then, each filter $F_t^\theta(i,j)$ are applied to a $k \times k$ window of h_{t-1}^{SD} centered at position (i,j) to conduct spatially variant filtering. This process can be formulated as:

$$M_t(i,j,c) = \sum_{u=-\lceil k/2 \rceil}^{\lfloor k/2 \rfloor} \sum_{v=-\lfloor k/2 \rfloor}^{\lfloor k/2 \rfloor} F_t^\theta(i,j,u,v) \times h_{t-1}^{SD}(i+u,j+v,c), \qquad (1)$$

where $M_t(i, j, c)$ represents correlation between the current frame and the c-th channel of hidden state at position (i, j). It is further fed to a sigmoid activation function $\sigma(\cdot)$ that transforms it into a similarity value in range $[0, 1]$. Finally, the adapted hidden state $\hat{h}_{t-1}^{SD}$ is computed by:

$$\hat{h}_{t-1}^{SD} = M_t \odot h_{t-1}^{SD}, \tag{2}$$

where '$\odot$' denotes element-wise multiplication.

3.4 Loss Functions

Since the proposed recurrent network has two streams, the trade-off between supervision on structure and detail during training is very important. Imbalanced supervision on structure and detail might produce either sharpened frames but with less details or frames with many weak edges and details. Therefore, we propose to train the proposed network with three loss terms as shown in Eq. 3, one for structure component, one for detail component, and one for the whole frame. α, β and γ

Fig. 5. Four channels in hidden state at a certain time step are selected for visualization. Yellow arrow denotes the difference in appearance among these four channels. Zoom in for better visualization. (Color figure online)

are hyper-parameters to balance the trade-off of these three terms. The loss to train an N-frame sequence is formulated as:

$$\mathcal{L} = \frac{1}{N} \sum_{t=1}^{N} (\alpha \mathcal{L}_t^{\mathcal{S}} + \beta \mathcal{L}_t^{\mathcal{D}} + \gamma \mathcal{L}_t^{\mathcal{I}}). \tag{3}$$

Similar to [17], we use Charbonnier loss function to compute the difference between reconstruction and high-resolution targets. Hence, we have $\mathcal{L}_t^{\mathcal{S}} = \sqrt{\|S_t^{HR} - \hat{S}_t^{HR}\|^2 + \varepsilon^2}$ for structure component, $\mathcal{L}_t^{\mathcal{D}} = \sqrt{\|D_t^{HR} - \hat{D}_t^{HR}\|^2 + \varepsilon^2}$ for detail component, and $\mathcal{L}_t^{\mathcal{I}} = \sqrt{\|I_t^{HR} - \hat{I}_t^{HR}\|^2 + \varepsilon^2}$ for the whole frame. The effectiveness of these loss functions is validated in the experiment section.

4 Experiments

In this section, we first explain the experiment datasets and implementation details of the proposed method. Then extensive ablation study is conducted to analyze the effectiveness of the proposed SD block and hidden state adaptation module. Furthermore, the proposed method is compared with state-of-the-art video super-resolution methods in terms of both effectiveness and efficiency.

Table 1. Ablation study on different network architecture.

Method	One stream 7-256		Two stream 7-128			SD block 7-128		
Model	Model 1	Model 2	Model 3	Model 4	Model 5	Model 6	Model 7	Model 8
HSA?	w/o	w/	w/	w/o	w/	w/	w/o	w/
Input	Image	Image	Image	S & D	S & D	Image	S & D	S & D
PSNR/SSIM	27.58/0.8410	27.65/0.8444	27.70/0.8452	27.64/0.8404	27.68/0.8429	27.73/0.8460	27.76/0.8463	27.79/0.8474

4.1   Implementation Details

Datasets. Some works [23,29] collect private training data from youtube on their own, which is not suitable for fair comparison with other methods. In this work, we adopt a widely used video processing dataset Vimeo-90K to train video super-resolution models. Vimeo-90K is a recent proposed large dataset for video processing tasks, which contains about $90K$ 7-frame video clips with various motions and diverse scenes. About $7K$ video clips select out of $90K$ as the test set, termed as Vimeo-90K-T. To train our model, we crop patches of size 256×256 from HR video sequences as the target. Similar to [4,12,23,29], the corresponding low-resolution patches are obtained by applying Gaussian blur with $\sigma = 1.6$ to the target patches followed by $\times 4$ times downsampling.

To validate the effectiveness of the proposed method, we evaluate our models on several popular benchmark datasets, including Vimeo-90K-T [27], Vid4 [20] and UDM10 [29]. As mentioned above, Vimeo-90K-T contains a lot of video clips, but each clip has only 7 frames. Vid4 and UDM10 are long sequences with diverse scenes, which is suitable to evaluate the effectiveness of recurrent-based method in information accumulation [4,23].

Training Details. The base model of our method consists of 5 SD blocks where each convolutional layer has 128 channels, *i.e.*, RSDN 5-128. By adding more SD blocks, we can obtain RSDN 7-128 and RSDN 9-128. The performance can be further boosted with only small increase on computational cost and runtime. We adopt $K = 3$ for HSA module for efficiency. To fully utilize all given frames, we pad each sequence by reflecting the second frame at the beginning of the sequence. When dealing with the first frame of a sequence, the previous estimated detail $\hat{D}_{t-1}$, structure $\hat{S}_{t-1}$ and hidden state feature h_{t-1}^{SD} are all initialized with zeros. The model training is supervised with Charbonnier penalty loss function and is optimized with Adam optimizer [16] with $\beta_1 = 0.9$ and $\beta_2 = 0.999$. Each mini-batch consists of 16 samples. The learning rate is initially set to 1×10^{-4} and is later down-scaled by a factor of 0.1 every 60 epoch till 70 epochs. The training data is augmented by standard flipping and rotating. All experiments are conducted on a server with Python 3.6.4, PyTorch 1.1 and Nvidia Tesla V100 GPU.

Recurrent Unit. We compare three kinds of blocks for information flow in the recurrent unit, *i.e.*, the three blocks shown in Fig. 3. For fair comparison among these blocks, we keep these three networks with almost the same parameters by setting the channel of convolutional layers in model 1 to 256, and setting the one in model 4 and 7 to 128.

4.2   Ablation Study

In this section, we conduct several ablation studies to analyze the effectiveness of the proposed SD block and the hidden state adaptation module. In addition, we also investigate the influence of different supervision on structure and detail components on the reconstruction performance. As shown in Table 1, model 1 and model 4 achieves similar performance, with model 1 a little higher SSIM and model 4 a little higher PSNR. This implies that simply dividing the input into structure and detail components and processing each one individually does not work well. Although it seems that having two branches to process each component divides a difficult task into two easier ones, it makes each one blind to the other and can not make full use of the information in the input to reconstruct either component.

By introducing informa-
tion exchange between structure and detail components,
model 7 obtains better performance than model 1 and 4 in
both PSNR and SSIM. Similar
result can also found in comparison among model 2, 5 and
8. In addition, we experiment
with taking the whole frames

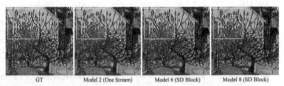

Fig. 6. Qualitative comparison between different network structures. Zoom in to see better visualization.

as input of both branches, that is, model 3 and model 6. By comparing model 3 and model 5 (and also model 6 and model 8), we show that the improvement comes not only from architecture of the two-stream block itself but also indeed from the decomposition into structure and detail components. The network with the proposed SD block allows each branch to explicitly focus on reconstructing a single component, which is easier than reconstructing a mixture of multiple components. Each branch makes use of the other one such that it can obtain enough information to reconstruct the high-resolution version for that component. The advantage of the proposed SD blocks can also be observed in the qualitative comparison as shown in Fig. 6.

In addition, we show in Table 1 that each model can gain further boost in performance with the proposed HSA module, about 0.04 dB in PSNR and 0.002 in SSIM on average. This module does not only work for the proposed network with SD blocks but also helps improve the performance for the ones with one-stream and two-stream residual blocks. The hidden state adaptation module allows the model to selectively use the history information stored in hidden state, which makes it robust to appearance change and error accumulation to some extent.

Influence of Different Components. The above experiment shows that decomposing the input into two components and processing them with the proposed SD blocks brings much improvement. We also investigate the relative importance of these two components by imposing different levels of supervision on the reconstruction of two components. It implies that the relative

Table 2. Ablation study on influence of different loss items.

(α, β, γ)	$(1, 0.5, 1)$	$(0.5, 1, 1)$	$(1, 1, 0)$	$(1, 1, 1)$
PSNR/SSIM	27.56/0.8440	27.77/0.8459	27.73/0.8453	27.79/0.8474

Table 3. Quantitative comparison (PSNR (dB) and SSIM) on **Vid4** for 4× video super-resolution. Red text indicates the best and blue text indicates the second best performance. Y and RGB indicate the luminance and RGB channels, respectively. FLOPs (MAC) are calculated on an HR image of size 720 × 480. '†' means the values are either taken from paper or calculated using provided models.

Vid4	#Frame	FLOPs	#Param.	Calendar (Y)	City (Y)	Foliage (Y)	Walk (Y)	Average (Y)	Average (RGB)
Bicubic	1	N/A	N/A	18.83/0.4936	23.84/0.5234	21.52/0.4438	23.01/0.7096	21.80/0.5426	20.37/0.5106
SPMC † [25]	3	-	-	-/-	-/-	-/-	-/-	25.52/0.76	-/-
Liu† [21]	5	-	-	21.61/-	26.29/-	24.99/-	28.06/-	25.23/-	-/-
TOFlow [27]	7	0.81T	1.41M	22.29/0.7273	26.79/0.7446	25.31/0.7118	29.02/0.8799	25.85/0.7659	24.39/0.7438
DUF-52L [12]	7	0.62T	5.82M	24.17/0.8161	28.05/0.8235	26.42/0.7758	30.91/0.9165	27.38/0.8329	25.91/0.8166
RBPN [7]	7	9.30T	12.2M	24.02/0.8088	27.83/0.8045	26.21/0.7579	30.62/0.9111	27.17/0.8205	25.65/0.7997
EDVR-L† [26]	7	0.93T	20.6M	24.05/0.8147	28.00/0.8122	26.34/0.7635	31.02/0.9152	27.35/0.8264	25.83/0.8077
PFNL† [29]	7	0.70T	3.00M	23.56/0.8232	28.11/0.8366	26.42/0.7761	30.55/0.9103	27.16/0.8365	25.67/0.8189
TGA [10]	7	0.23T	5.87M	24.50/0.8285	28.50/0.8442	26.59/0.7795	30.96/0.9171	27.63/0.8423	26.14/0.8258
FRVSR 10-128 [23]	recurrent (2)	0.14T	5.05M	22.67/0.7844	27.70/0.8063	25.83/0.7541	29.72/0.8971	26.48/0.8104	25.01/0.7917
RLSP 7-256 [4]	recurrent (3)	0.09T	4.21M	24.36/0.8235	28.22/0.8362	26.66/0.7821	30.71/0.9134	27.48/0.8388	25.69/0.8153
RSDN 5-128	recurrent (2)	0.08T	3.83M	24.34/0.8242	28.73/0.8374	26.66/0.7842	30.73/0.9149	27.61/0.8402	26.13/0.8238
RSDN 7-128	recurrent (2)	0.10T	5.01M	24.46/0.8305	29.01/0.8480	26.78/0.7921	30.92/0.9189	27.79/0.8474	26.30/0.8314
RSDN 9-128	recurrent (2)	0.13T	6.19M	24.60/0.8355	29.20/0.8527	26.84/0.7931	31.04/0.9210	27.92/0.8505	26.43/0.8349

Table 4. Quantitative comparison (PSNR(dB) and SSIM) on **UDM10** and **Vimeo-90K-T** for 4× video super-resolution, respectively. Flops and runtimes are calculated on an HR image size of 1280 × 720 and 448 × 256 for UDM10 and Vimeo-90K-T, respectively. Red text indicates the best and blue text indicates the second best performance. Y and RGB indicate the luminance and RGB channels, respectively. '†' means the values are either taken from paper or calculated using provided models.

UDM10	Bicubic	TOFlow [27]	DUF-52L [12]	RBPN [7]	PFNL† [29]	FRVSR 10-128 [23]	RLSP 7-256 [4]	RSDN 7-128	RSDN 9-128
FLOPs [TMAC]	N/A	2.17	1.65	24.81	1.88	0.36	0.24	0.28	0.35
Runtime [ms]	N/A	1693	1413	3567	295	137	49	79	94
Average (Y)	28.47/0.8523	36.26/0.9438	38.48/0.9605	38.66/0.9596	38.74/0.9627	37.09/0.9522	38.48/0.9606	39.13/0.9645	39.35/0.9653
Average (RGB)	27.05/0.8267	34.46/0.9298	36.78/0.9514	36.53/0.9462	36.78/0.9514	35.39/0.9403	36.39/0.9465	37.26/0.9548	37.46/0.9557
Vimeo-90K-T	Bicubic	TOFlow [27]	DUF-52L [12]	RBPN [7]	EDVR-L† [26]	FRVSR 10-128 [23]	RLSP 7-256 [4]	RSDN 7-128	RSDN 9-128
FLOPs [TMAC]	N/A	0.27	0.20	3.08	0.30	0.04	0.03	0.03	0.04
Runtime [ms]	N/A	215	167	470	99	28	11	13	15
Average (Y)	31.30/0.8687	34.62/0.9212	36.87/0.9447	37.20/0.9458	37.61/0.9489	35.64/0.9319	36.49/0.9403	37.05/0.9454	37.23/0.9471
Average (RGB)	29.77/0.8490	32.78/0.9040	34.96/0.9313	35.39/0.9340	35.79/0.9374	33.96/0.9192	34.56/0.9274	35.14/0.9325	35.32/0.9344

supervision strength applied to different components also plays an important role in the super-resolution performance. As shown in Table 2, when the weights for structure component, detail component and the whole frame are set to $(\alpha, \beta, \gamma) = (1, 1, 1)$, it achieves a good performance of 27.79/0.8474 in PSNR/SSIM. The performance degrades when the weigh for structure component more than the weight for detail component (*i.e.* $(\alpha, \beta, \gamma) = (1, 0.5, 1)$), and verse vise (*i.e.* $(\alpha, \beta, \gamma) = (0.5, 1, 1)$). The result of $(1, 1, 0)$ is 0.06 dB lower than that of $(1, 1, 1)$, which means applying additional supervision on the combined image helps the training of the model.

4.3 Comparison with State-of-the-Arts

In this section, we compare our methods with several state-of-the-art VSR approaches, including SPMC [25], TOFlow [27], Liu [21], DUF [7], EDVR [26], PFNL [29], TGA [10], FRVSR [23] and RLSP [4]. The first seven methods super-resolve a single reference within a temporal sliding window. Among these methods, SPMC, TOFlow, Liu, RBPN and EDVR need to explicitly estimate the motion between the reference frame and other frames within the window, which requires redundant computation for several frames. DUF, PFNL and TGA skip the motion estimation process and partially ameliorate this issue. The last two methods FRVSR and RLSP super-resolve each frame in a recurrent way and are more efficient. We carefully implement most of these methods either on our own or by running the publicly available code, and manage to reproduce the results in their paper. The quantitative result of state-of-the-art methods on Vid4 is shown in Table 3, where the number is either reported in the original papers or computed with our implementation. In addition, we also include the number of parameters and FLOPs for most methods when super-resolution is conducted on an LR image of size 112×64 in Table 3.

On Vid4, our model with only 5 SD block achieves 27.61dB PSNR in Y channel and 26.13dB PSNR in RGB channels, which already outperforms most of the previous methods by a large margin. By increasing the number of SD block to 7 and 9, our methods respectively gain another 0.18dB and 0.31dB PSNR in Y channel while with only a little increase in FLOPs. We also evaluate our method on other three popular test sets. The quantitative results on UDM10 [29] and Vimeo-90K-T [27] two datasets are reported in Table 4. Our method achieves a very good balance between reconstruction performance and speed on these datasets. On UDM10 test set, RSDN 9-128 achieves new state-of-the-art, and is about 15 and 37 times faster than DUF and RBPN, respectively. RSDN 9-128 outperforms the recent proposed PFNL, where this dataset is proposed by 0.61 dB in PSNR in Y channel while being 3 times faster. The proposed method is also evaluated on Vimeo-90K-T, which only contains 7-frame in each sequence. In this case, although our method can not take full of its advantage because of the short length of the sequence, it only lags behind the large model EDVR-L but is 6 times faster.

We also show the qualitative comparison with other state-of-the-art methods. As shown in Fig. 7, our method produces higher quality high-resolution images on all three datasets, including finer details and sharper edges. Other methods are either prone to generate some artifacts (e.g., wrong stripes in an image) or can not recover missing details (e.g., small windows of the building). We also examine temporal consistency of the video super-resolution results in Fig. 8, which is produced by extracting a horizontal row of pixels at the same position from consecutive frames and stacking them vertically. The temporal profile produced by our method is not only temporally smoother but also much sharper, satisfying both requirements of the video super-resolution task.

658   T. Isobe et al.

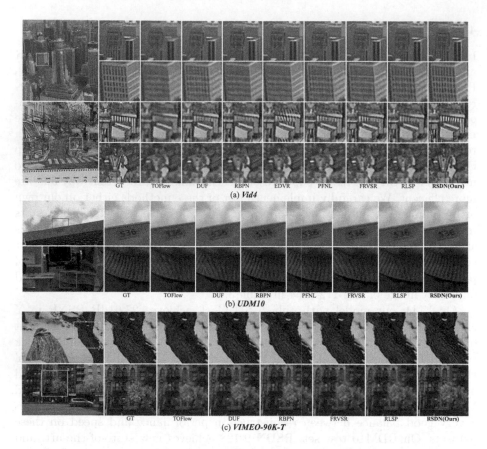

Fig. 7. Qualitative comparison on **Vid4**, **UDM10** and **Vimeo-90K-T** test set for 4×
SR. Zoom in for better visualization.

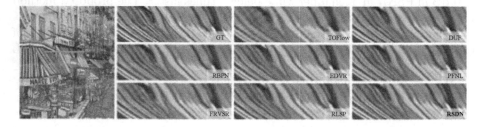

Fig. 8. Visualization of temporal profile for the green line on the calendar sequence.

5   Conclusion

In this work we have presented an effective and efficient recurrent network to
super-resolve a video in a streaming manner. The input is decomposed into
structure and detail components and fed to two interleaved branches to respec-

tively reconstruct the corresponding components of high-resolution frames. Such a strategy allows our method to address different difficulties in the structure and detail components and to enjoy flexible supervision applied to each components for good performance. In addition we find that hidden state in a recurrent network captures different typical appearance of a scene over time and selectively using information from hidden state can enhance its robustness to appearance change and error accumulation. Extensive experiments on several benchmark datasets demonstrate its superiority in terms of both effectiveness and efficiency.

References

1. Caballero, J., et al.: Real-time video super-resolution with spatio-temporal networks and motion compensation. In: CVPR (2017)
2. Dong, C., Loy, C.C., He, K., Tang, X.: Learning a deep convolutional network for image super-resolution. In: Fleet, D., Pajdla, T., Schiele, B., Tuytelaars, T. (eds.) ECCV 2014. LNCS, vol. 8692, pp. 184–199. Springer, Cham (2014). https://doi.org/10.1007/978-3-319-10593-2_13
3. Du, W., Wang, Y., Qiao, Y.: RPAN: an end-to-end recurrent pose-attention network for action recognition in videos. In: CVPR (2017)
4. Fuoli, D., Gu, S., Timofte, R.: Efficient video super-resolution through recurrent latent space propagation. CoRR abs/1909.08080 (2019)
5. Glorot, X., Bordes, A., Bengio, Y.: Deep sparse rectifier neural networks. In: AISTATS (2011)
6. Haris, M., Shakhnarovich, G., Ukita, N.: Deep back-projection networks for super-resolution. In: CVPR (2018)
7. Haris, M., Shakhnarovich, G., Ukita, N.: Recurrent back-projection network for video super-resolution. In: CVPR (2019)
8. Huang, G., Liu, Z., Van Der Maaten, L., Weinberger, K.Q.: Densely connected convolutional networks. In: CVPR (2017)
9. Huang, Y., Wang, W., Wang, L.: Bidirectional recurrent convolutional networks for multi-frame super-resolution. In: NeurIPS (2015)
10. Isobe, T., et al.: Video super-resolution with temporal group attention. In: CVPR (2020)
11. Jia, X., De Brabandere, B., Tuytelaars, T., Gool, L.V.: Dynamic filter networks. In: NeurIPS (2016)
12. Jo, Y., Wug Oh, S., Kang, J., Joo Kim, S.: Deep video super-resolution network using dynamic upsampling filters without explicit motion compensation. In: CVPR (2018)
13. Kappeler, A., Yoo, S., Dai, Q., Katsaggelos, A.K.: Video super-resolution with convolutional neural networks. IEEE Trans. Comput. Imaging 2(2), 109–122 (2016)
14. Kim, J., Lee, J.K., Lee, K.M.: Accurate image super-resolution using very deep convolutional networks. In: CVPR (2016)
15. Kim, J., Lee, J.K., Lee, K.M.: Deeply-recursive convolutional network for image super-resolution. In: CVPR (2016)
16. Kingma, D.P., Ba, J.: Adam: a method for stochastic optimization. In: ICLR (2015)
17. Lai, W.S., Huang, J.B., Ahuja, N., Yang, M.H.: Fast and accurate image super-resolution with deep laplacian pyramid networks. IEEE Trans. Pattern Anal. Mach. Intell. 41(11), 2599–2613 (2018)

18. Ledig, C., et al.: Photo-realistic single image super-resolution using a generative adversarial network. In: CVPR (2017)
19. Lim, B., Son, S., Kim, H., Nah, S., Mu Lee, K.: Enhanced deep residual networks for single image super-resolution. In: CVPR Workshops (2017)
20. Liu, C., Sun, D.: On bayesian adaptive video super resolution. IEEE Trans. Pattern Anal. Mach. Intell. **36**(2), 346–360 (2013)
21. Liu, D., et al.: Robust video super-resolution with learned temporal dynamics. In: ICCV (2017)
22. Pan, J., et al.: Learning dual convolutional neural networks for low-level vision. In: CVPR (2018)
23. Sajjadi, M.S., Vemulapalli, R., Brown, M.: Frame-recurrent video super-resolution. In: CVPR (2018)
24. Singh, B., Marks, T.K., Jones, M., Tuzel, O., Shao, M.: A multi-stream bi-directional recurrent neural network for fine-grained action detection. In: CVPR (2016)
25. Tao, X., Gao, H., Liao, R., Wang, J., Jia, J.: Detail-revealing deep video super-resolution. In: ICCV (2017)
26. Wang, X., Chan, K.C., Yu, K., Dong, C., Change Loy, C.: EDVR: Video restoration with enhanced deformable convolutional networks. In: CVPR Workshops (2019)
27. Xue, T., Chen, B., Wu, J., Wei, D., Freeman, W.T.: Video enhancement with task-oriented flow. Int. J. Comput. Vis. **127**(8), 1106–1125 (2019)
28. Yang, W., Zhang, X., Tian, Y., Wang, W., Xue, J.H., Liao, Q.: Deep learning for single image super-resolution: a brief review. IEEE Trans. Multimed. **21**(12), 3106–3121 (2019)
29. Yi, P., Wang, Z., Jiang, K., Jiang, J., Ma, J.: Progressive fusion video super-resolution network via exploiting non-local spatio-temporal correlations. In: ICCV (2019)
30. Zhang, Y., Li, K., Li, K., Wang, L., Zhong, B., Fu, Y.: Image super-resolution using very deep residual channel attention networks. In: Ferrari, V., Hebert, M., Sminchisescu, C., Weiss, Y. (eds.) ECCV 2018. LNCS, vol. 11211, pp. 294–310. Springer, Cham (2018). https://doi.org/10.1007/978-3-030-01234-2_18
31. Zhang, Y., Tian, Y., Kong, Y., Zhong, B., Fu, Y.: Residual dense network for image super-resolution. In: CVPR (2018)

Shape Adaptor: A Learnable Resizing Module

Shikun Liu[1]([✉]), Zhe Lin[2], Yilin Wang[2], Jianming Zhang[2], Federico Perazzi[2], and Edward Johns[1]

[1] Department of Computing, Imperial College London, London, UK
{shikun.liu17,e.johns}@imperial.ac.uk
[2] Adobe Research, San Jose, USA
{zlin,yilwang,jianmzha,perazzi}@adobe.com

Abstract. We present a novel resizing module for neural networks: *shape adaptor*, a drop-in enhancement built on top of traditional resizing layers, such as pooling, bilinear sampling, and strided convolution. Whilst traditional resizing layers have fixed and deterministic reshaping factors, our module allows for a learnable reshaping factor. Our implementation enables shape adaptors to be trained end-to-end without any additional supervision, through which network architectures can be optimised for each individual task, in a fully automated way. We performed experiments across seven image classification datasets, and results show that by simply using a set of our shape adaptors instead of the original resizing layers, performance increases consistently over human-designed networks, across all datasets. Additionally, we show the effectiveness of shape adaptors on two other applications: network compression and transfer learning.

Keywords: Automated Machine Learning · Resizing layer · Neural architecture search

1 Introduction

Deep neural networks have become popular for many machine learning applications, since they provide simple strategies for end-to-end learning of complex representations. However, success can be highly sensitive to network architectures, which places a great demand on manual engineering of architectures and hyper-parameter tuning.

A typical human-designed convolutional neural architecture is composed of two types of computational modules: i) a *normal layer*, such as a stride-1 convolution or an identity mapping, which maintains the spatial dimension of incoming

---

Electronic supplementary material The online version of this chapter (https://doi.org/10.1007/978-3-030-58610-2_39) contains supplementary material, which is available to authorized users.

A. Vedaldi et al. (Eds.): ECCV 2020, LNCS 12357, pp. 661–677, 2020.
https://doi.org/10.1007/978-3-030-58610-2_39

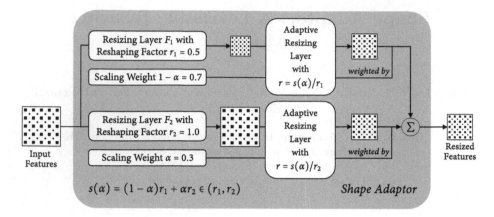

Fig. 1. Left: visualisation of a shape adaptor module build on top of two resizing layers. Right: Different network shapes in the same network architecture ResNet-50 can result a significantly different performance.

feature maps; ii) a *resizing layer*, such as max/average pooling, bilinear sampling, or stride-2 convolution, which reshapes the incoming feature map into a different spatial dimension. We hereby define the *shape* of a neural network as the composition of the feature dimensions in all network layers, and the *architecture* as the overall structure formed by stacking multiple normal and resizing layers.

To move beyond the limitations of human-designed network architectures, there has been a growing interest in developing Automated Machine Learning (AutoML) algorithms [9] for automatic architecture design, known as Neural Architecture Search (NAS) [2,16,17,26]. However, whilst this has shown promising results in discovering powerful network architectures, these methods still rely heavily on human-designed network shapes, and focus primarily on learning connectivities between layers. Typically, reshaping factors of 0.5 (down-sampling) and 2 (up-sampling) are chosen, and the total number of reshaping layers is defined manually, but we argue that network shape is an important inductive bias which should be directly optimised.

For example, Fig. 1 Right shows three networks with the exact same design of network structure, but different shapes. For the two human-designed networks [8], we see that a ResNet-50 model designed specifically for CIFAR-100 dataset (Human Designed B) leads to a 15% performance increase over a ResNet-50 model designed for ImageNet dataset (Human Designed A). The performance can be further improved with the network shape designed by the shape adaptors we will later introduce. Therefore, by learning network shapes rather than manually designing them, a more optimal network architecture can be found.

To this end, we propose *Shape Adaptor*, a novel resizing module which can be dropped into any standard neural network to learn task-specific network shape. A shape adaptor module (see Fig. 1 Left) takes in an input feature, and reshapes it into two intermediate features. Each reshaping operation is done using a standard resizing layer $F_i(x, r_i), i = 1, 2$, where each resizing layer has a different, pre-defined reshaping factor r_i to reshape feature map x. Finally, the two intermediate features are softly combined with a scalar weighting $1 - \alpha$ and α respectively (for $\alpha \in (0, 1)$), after reshaping them into the same spatial dimension via a learned reshaping factor in the search space $s(\alpha) \in (r_1, r_2)$, assuming $r_1 < r_2$. The module's output represents a mixed combination over these two intermediate features, and the scalar α can be learned solely based on the task-specific training loss with stochastic gradient descent, without any additional supervision. Thus, by simply optimising these scaling weights for every shape adaptor, the entire neural architecture is differential and we are able to learn network shape in an automated, end-to-end manner.

2 Related Work and Background

Neural Architecture Search. Neural architecture search (NAS) presents an interesting research direction in AutoML, in automatically discovering an optimal neural structure for a particular dataset, alleviating the hand design of neural architectures which traditionally involves tedious trial-and-error. NAS approaches can be highly computationally demanding, requiring hundreds of thousands of GPU days of search time, due to intensive techniques such as reinforcement learning [37] and evolutionary search [27]. Several approaches have been proposed to speed up the search, based on parameter sharing [26], hypernetworks [1], and gradient-based optimisation [17]. But despite their promising performance, these approaches come with controversial debate questioning the lack of reproducibility, and sensitivity to initialisations [15,33].

Architecture Pruning and Compression. Network pruning is another direction towards achieving optimal network architectures. But instead of searching from scratch as in NAS, network pruning is applied to existing human-designed networks and removes redundant neurons and connectivities. Such methods can be based on $\mathcal{L}_0$ regularisation [19], batch-norm scaling parameters [18], and weight quantization [7]. As with our shape adaptors, network pruning does not require the extensive search cost of NAS, and can be performed alongside regular training. Our shape adaptors can also be formulated as a pruning algorithm, by optimising the network shape within a bounded search space. We provide a detailed explanation of this in Sect. 5.1.

Design of Resizing Modules. A resizing module is one of the essential components in deep convolutional network design, and has seen continual modifications to improve performance and efficiency. The most widely used resizing modules are max pooling, average pooling, bilinear sampling, and strided convolutions, which are deterministic, efficient, and simple. But despite their benefits in increasing computational efficiency and providing regularisation, there are two issues with current designs: i) *lack of spatial invariance*, and ii) *fixed scale*. Prior works focus on improving spatial robustness with a learnable combination between max and average pooling [14,32], and with anti-aliased low-pass filters [35]. Other works impose regularisation and adjustable inference by stochastically inserting pooling layers [13,34], and sampling different network shapes [36].

In contrast, shape adaptors solve both problems simultaneously, with a learnable mixture of features in different scales, and with which reshaping factors can be optimised automatically based on the training objective.

3   Shape Adaptors

In this section, we introduce the details of the proposed Shape Adaptor module. We discuss the definition of these modules, and the optimisation strategy used to train them.

3.1   Formulation of Shape Adaptors

A visual illustration of a shape adaptor module is presented in Fig. 1 Left. It is a two-branch architecture composed of two different resizing layers $F_i(x, r_i)_{i=1,2}$, assuming $r_1 < r_2$, taking the same feature map x as the input. A resizing layer F_i can be any classical sampling layer, such as max pooling, average pooling, bilinear sampling, or strided convolution, with a fixed reshaping factor r_i. Each resizing layer reshapes the input feature map by this factor, which represents the ratio of spatial dimension between the output and input feature maps, and outputs an *intermediate feature*. An adaptive resizing layer G with a learnable reshaping factor is then used to reshape these intermediate features into the same spatial dimension, and combine them with a weighted average to compute the module's output.

Each module has a learnable parameter $\alpha \in (0, 1)$, parameterised by a sigmoid function, which is the only extra learnable parameter introduced by shape adaptors. The role of α is to optimally combine two intermediate features after reshaping them by an adaptive resizing layer G. To enable a non-differential reshaping factor in G to be learned, we use a monotone function s, which monotonically maps from α into the search space $s(\alpha) \in \mathcal{R} = (r_1, r_2)$, representing the scaling ratio of the module's reshaping operation. With this formulation, a learnable reshaping factor $s(\alpha)$ allows a shape adaptor to reshape at any scale between r_1 and r_2, rather than being restricted to a discrete set of scales as with typical manually-designed network architectures.

Using this formulation, a shape adaptor module can be expressed as a function:

$$\texttt{ShapeAdaptor}(x, \alpha, r_{1,2}) = (1-\alpha) \cdot G\left(F_1(x, r_1), \frac{s(\alpha)}{r_1}\right) + \alpha \cdot G\left(F_2(x, r_2), \frac{s(\alpha)}{r_2}\right),$$
(1)

with reshaping factor $s(\alpha)$, a monotonic mapping which satisfies:

$$\lim_{\alpha \to 0} s(\alpha) = r_1, \quad \text{and} \quad \lim_{\alpha \to 1} s(\alpha) = r_2.$$
(2)

We choose our adaptive resizing layer G to be a bilinear interpolation function, which allows feature maps to be resized into any shape. We design module's learnable reshaping factor $s(\alpha) = (r_2 - r_1)\alpha + r_1$, a convex combination over these pre-defined reshaping factors, assuming having no prior knowledge on the network shape.

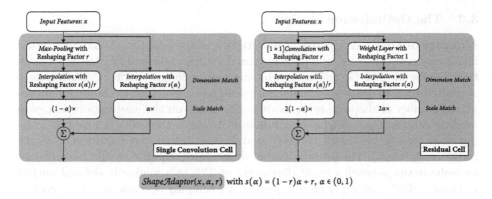

Fig. 2. Visualisation of a down-sampling shape adaptor built on a single convolutional cell and a residual cell with a reshaping factor in the range $\mathcal{R} = (r, 1)$.

Each shape adaptor is arranged as a soft and learnable operator to search the optimal reshaping factor $s(\alpha^*) = r^* \in \mathcal{R}$ over a combination of intermediate reshaped features $F_i(x, r_i)$. Thus, it can also be easily coupled with a continuous approximate of categorical distribution, such as Gumbel SoftMax [10,20], to control the softness. This technique is commonly used in gradient-based NAS methods [17], where a categorical distribution is learned over different operations.

The overall shape adaptor module ensures that its reshaping factor $s(\alpha)$ can be updated through the updated scaling weights. Thus, we enable differentiability of $s(\alpha)$ in a shape adaptor module as an approximation from the mapping of the derivative of its learnable scaling weight: $\nabla s(\alpha) \approx s(\nabla \alpha)$. This formulation enables shape adaptors to be easily trained with standard back-propagation and end-to-end optimisation.

In our implementation, we use one resizing layer to maintain the incoming feature dimension (an identity layer), and the other resizing layer to change the dimension. If F_2 is the layer which maintains the dimension with $r_2 = 1$, then a shape adaptor module acts as a learnable down-sampling layer when $0 < r_1 < 1$, and a learnable up-sampling layer when $r_1 > 1$.

In Fig. 2, we illustrate our learnable down-sampling shape adaptor in two commonly used computational modules: a single convolutional cell in VGG-like neural networks [30], and a residual cell in ResNet-like neural networks [8]. To seamlessly insert shape adaptors into human-designed networks, we build shape adaptors on top of the same sampling functions used in the original network design. For example, in a single convolutional cell, we apply max pooling as the down-sampling layer, and the identity layer is simply an identity mapping. And in a residual cell, we use the 'shortcut' $[1 \times 1]$ convolutional layer as the down-sampling layer, and the weight layer stacked with multiple convolutional layers as the identity layer. In the ResNet design, we double the scaling weights in the residual cell, in order to match the same feature scale as in the original design.

3.2   The Optimisation Recipe

Number of Shape Adaptors. Theoretically, shape adaptors should be inserted into every network layer, to enable maximal search space and flexibility. In practice, we found that beyond a certain number of shape adaptors, performance actually began to degrade. We therefore designed a heuristic to choose an appropriate number of shape adaptor modules N, based on the assumption that each module contributes a roughly equal amount towards the network's overall resizing effect. Let us consider that each module resizes its input feature map in the range (r_{min}, r_{max}). The overall number of modules required should be sufficient to reshape the network's input dimension of $\mathcal{D}^{in}$ to a manually defined output dimension $\mathcal{D}^{last}$, by applying a sequence of reshaping operations, where each is $\sim r_{min}$. As such, the optimal number of modules can be expressed as a logarithmic function of the overall ratio between the network's input and output, based on the scale of the reshaping operation in each module:

$$N = \left\lfloor \log_{1/r_{min}}(\mathcal{D}^{in}/\mathcal{D}^{last}) \right\rfloor \tag{3}$$

Initialisations in Shape Adaptors. As with network weights, a good initialisation for shape adaptors, i.e. the initial values for α, is important. Again, assuming we have every shape adaptor designed in the same search space $\mathcal{R} = (r_{min}, r_{max})$ with the reshaping factor $s(\alpha) = (r_{max} - r_{min})\alpha + r_{min}$, we propose a formula to automatically compute the initialisations such that the output feature dimension of the initialised shape would map to the user-defined dimension $\mathcal{D}^{out}$. Assuming we want to initialise the raw scaling parameters $\bar{\alpha}$ before sigmoid function $\alpha = \sigma(\bar{\alpha})$, we need to solve the following equation:

$$\mathcal{D}^{in} \cdot s(\sigma(\bar{\alpha}))^N = \mathcal{D}^{out}. \tag{4}$$

Suppose we use N as defined in Eq. 3, then Eq. 4 is only solvable when $D^{last} \leq D^{out}$. Otherwise, we then initialise the smallest possible shape when encountering the case for $D^{last} > D^{out}$. This eventually derives the following:

$$\bar{\alpha} = \begin{cases} \ln\left(-\dfrac{\sqrt[N]{\mathcal{D}^{out}/\mathcal{D}^{in}}-r_{min}}{\sqrt[N]{\mathcal{D}^{out}/\mathcal{D}^{in}}-r_{max}} + \epsilon\right) & \text{if } \mathcal{D}^{last} \leq \mathcal{D}^{out} \\ \ln(\epsilon) & \text{otherwise} \end{cases}, \quad \epsilon = 10^{-4}. \tag{5}$$

where ϵ is a small value to avoid encountering $\pm\infty$ values.

Shape Adaptors with Memory Constraint. During experiments, we observed that shape adaptors tend to converge to a larger shape than the human designed network, which may then require very large memory. For practical applications, it is desirable to have a constrained search space for learning the optimal network shape given a user-defined memory limit. For any layer designed with down-sampling shape adaptors, the spatial dimension of which is guaranteed to be smaller than the one from the previous layers. We thus again use the final feature dimension to approximate the memory usage for the network shape.

Suppose we wish to constrain the network shape with the final feature dimension to be no greater than D^{limit}. We then limit the scaling factors in shape adaptors by use of a penalty value ρ, which is applied whenever the network's final feature dimension after the current update D^{cout} is greater than the defined limit, i.e. when $D^{cout} > D^{limit}$. When this occurs, the penalty term ρ is applied on every shape adaptor module, and we compute ρ dynamically for every iteration so that we make sure $D^{cout} \leq D^{limit}$ in the entire training stage. The penalised scaling parameter α_ρ is then defined as follows,

$$\alpha_\rho = \alpha \cdot \rho + \frac{r_{min}}{r_{max} - r_{min}}(\rho - 1). \tag{6}$$

Then the penalised module's reshaping factor $s(\alpha_\rho)$ becomes,

$$s(\alpha_\rho) = (r_{max} - r_{min})\alpha_\rho + r_{min} = s(\alpha)\rho. \tag{7}$$

Using Eq. 4, we can compute ρ as,

$$\rho = \sqrt[N]{\frac{D^{limit}}{D^{cout}}}. \tag{8}$$

Iterative Optimisation Strategy. To optimise a neural network equipped with shape adaptor modules, there are two sets of parameters to learn: the weight parameters $w = \{w_i\}$, and the shape parameters $\alpha = \{\alpha_i\}$. Unlike NAS algorithms which require optimisation of network weights and structure parameters on separate datasets, shape adaptors are optimised on the same dataset and require no re-training.

Since the parameter space for the network shape is significantly smaller than the network weight, we update the shape parameters less frequently than the weight parameters, at a rate of once every α_s steps. The entire optimisation for a network equipped with shape adaptors is illustrated in Algorithm 1.

4   Experiments

In this section, we present experimental results to evaluate shape adaptors on image classification tasks.

4.1   Experimental Setup

Datasets. We evaluated on seven different image classification datasets, with varying sizes and complexities, to fully assess the robustness and generalisation of shape adaptors. These seven datasets are divided into three categories: i) small (resolution) datasets: CIFAR-10/100 [12], SVHN [5]; ii) fine-grained classification datasets: FGVC-Aircraft (Aircraft) [21], CUBS-200-2011 (Birds) [31], Stanford Cars (Cars) [11]; and iii) ImageNet [3]. Small datasets are in resolution [32 × 32], and fine-grained classification and ImageNet datasets are in resolution [224×224].

Algorithm 1: Optimisation for Shape Adaptor Networks

---

1 **Define:** shape adaptors: $\alpha_s, r_{min}, r_{max}, D^{last}, D^{out}, D^{limit}$
2 **Define:** network architecture $f_{\alpha,w}$ defined with shape and network parameters
3 **Initialise:** shape parameters: $\alpha = \{\alpha_i\}$ with Eq. 3, and Eq. 5
4 **Initialise:** weight parameters: $w = \{w_i\}$
5 **Initialise:** learning rate: λ_1, λ_2
6 **while** *not converged* **do**
7 | **for** *each training iteration i* **do**
8 | | $(x_{(i)}, y_{(i)}) \in (x, y)$ ▷ *fetch one batch of training data*
9 | | **if** *requires memory constraint* **then**
10 | | | **Compute:** ρ using Eq. 8
11 | | **else**
12 | | | **Define:** $\rho = 1$
13 | | **end**
14 | | **if** *in α_s step* **then**
15 | | | **Update:** $\alpha \leftarrow \lambda_1 \nabla_\alpha \mathcal{L}(f_{\alpha_\rho, w}(x_{(i)}), y_{(i)})$ ▷ *update shape parameters*
16 | | **end**
17 | | **Update:** $w \leftarrow \lambda_2 \nabla_w \mathcal{L}(f_{\alpha_\rho, w}(x_{(i)}), y_{(i)})$ ▷ *update weight parameters*
18 | **end**
19 **end**

Baselines. We ran experiments with three widely-used networks: VGG-16 [30], ResNet-50 [8], and MobileNetv2 [29]. The baseline *Human* represents the original human-designed networks, which require manually adjusting the number of resizing layers according to the resolution of each dataset. For smaller $[32 \times 32]$ datasets, human-designed VGG-16, ResNet-50 and MobileNetv2 networks were equipped with 4, 3, 3 resizing layers respectively, and for $[224 \times 224]$ datasets, all human designed networks have 5 resizing layers.

Implementation of Shape Adaptors. For all experiments in this section, since we assume no prior knowledge of the optimal network architecture, we inserted shape adaptors uniformly into the network layers (except for the last layer). We initialised shape adaptors with $D^{last} = 2, D^{out} = 8$, which we found to work well across all datasets and network choices. All shape adaptors use the search space $\mathcal{R} = (0.5, 1)$ with the design in Fig. 2. (Other choices of the search space are discussed in the supplementary material.) We applied the memory constraint on shape adaptors so that the network shape can grow no larger than the running GPU memory. We optimised shape adaptors every $\alpha_s = 20$ steps for non-ImageNet datasets, and every $\alpha_s = 1500$ steps for ImageNet. The full hyper-parameter choices are provided in the supplementary material.

4.2 Results on Image Classification Datasets

First, we compared networks built with shape adaptors to the original human-designed networks, to test whether shape adaptors can improve performances solely by finding a better network shape, without using any additional parameter space. To ensure fairness, all network weights in the human-designed and shape adaptor networks were optimised using the same hyper-parameters, optimiser, and scheduler. Table 1 shows the test accuracies of shape adaptor and human-designed networks, with each accuracy averaged over three individual runs. We see that in nearly all cases, shape adaptor designed networks outperformed human-designed networks by a significant margin, despite both methods using exactly the same parameter space. We also see that performance of shape adaptor designed networks are stable, with a relatively low variance across different runs. This is similar to the human-designed networks, showing stability and robustness of our method without needing the domain knowledge that is required for human-designed networks.

Table 1. Top-1 test accuracies on different datasets for networks equipped with human-designed resizing layers and with shape adaptors. We present the results with the range of three independent runs. Best results are in bold.

Dataset	VGG-16		ResNet-50		MobileNetv2	
	Human	Shape adaptor	Human	Shape adaptor	Human	Shape adaptor
CIFAR-10	$94.11_{\pm0.17}$	$\mathbf{95.35_{\pm0.06}}$	$\mathbf{95.50_{\pm0.09}}$	$95.48_{\pm0.17}$	$93.71_{\pm0.25}$	$\mathbf{93.86_{\pm0.23}}$
CIFAR-100	$75.39_{\pm0.11}$	$\mathbf{79.16_{\pm0.23}}$	$78.53_{\pm0.11}$	$\mathbf{80.29_{\pm0.10}}$	$73.80_{\pm0.17}$	$\mathbf{75.74_{\pm0.31}}$
SVHN	$96.26_{\pm0.03}$	$\mathbf{96.89_{\pm0.07}}$	$96.74_{\pm0.20}$	$\mathbf{96.84_{\pm0.13}}$	$96.50_{\pm0.08}$	$\mathbf{96.86_{\pm0.14}}$
Aircraft	$85.28_{\pm0.09}$	$\mathbf{86.95_{\pm0.29}}$	$81.57_{\pm0.51}$	$\mathbf{85.60_{\pm0.32}}$	$77.64_{\pm0.23}$	$\mathbf{83.00_{\pm0.30}}$
Birds	$73.37_{\pm0.35}$	$\mathbf{74.86_{\pm0.50}}$	$68.62_{\pm0.10}$	$\mathbf{71.02_{\pm0.48}}$	$60.37_{\pm1.12}$	$\mathbf{68.53_{\pm0.21}}$
Cars	$89.30_{\pm0.21}$	$\mathbf{90.13_{\pm0.11}}$	$87.23_{\pm0.48}$	$\mathbf{89.67_{\pm0.20}}$	$80.86_{\pm0.13}$	$\mathbf{84.62_{\pm0.38}}$
ImageNet	$\mathbf{73.92_{\pm0.12}}$	$73.53_{\pm0.09}$	$77.18_{\pm0.04}$	$\mathbf{78.74_{\pm0.12}}$	$71.72_{\pm0.02}$	$\mathbf{73.32_{\pm0.07}}$

4.3 Ablative Analysis and Visualisations

In this section, we perform an ablative analysis on CIFAR-100 and Aircraft to understand the behaviour of shape adaptors with respect to the number and initialisation of shape adaptors. We observed that conclusions are consistent across different networks, thus we performed experiments in two networks only: VGG-16 and MobileNetv2. All results are averaged over two independent runs.

Number of Shape Adaptors. We first evaluate the performance by varying different number of shape adaptors used in the network, whilst fixing all other hyper-parameters used in Sect. 4.2. In Table 2, we show that the performance of shape adaptor networks is consistent across the number of shape adaptors used. Notably, performance is always better than networks with human-designed resizing layers, regardless of the number of shape adaptors used. This again shows the ability of shape adaptors to automatically learn optimal shapes without requiring domain knowledge. The optimal number of shape adaptor modules given by our heuristic in Eq. 3 is highlighted in teal, and we can therefore see that this is a good approximation to the optimal number of modules.

Table 2. Test accuracies of VGG-16 on CIFAR-100 and MobileNetv2 on Aircraft, when different numbers of shape adaptors are used. Best results are in bold. The number produced in Eq. 3 is highlighted in teal.

CIFAR-100	Human	SA (with number of)					Aircraft	Human	SA (with number of)				
		3	4	5	6	8			5	6	7	8	10
VGG-16	75.39	79.03	**79.16**	78.56	78.43	78.16	VGG-16	85.28	84.80	**86.95**	86.44	86.72	85.76
MobileNetv2	73.80	75.39	**75.74**	75.22	74.92	74.86	MobileNetv2	77.64	81.12	83.00	**83.02**	82.43	80.36

In Fig. 3, we present visualisations of network shapes in human-designed and shape adaptor designed networks. We can see that the network shapes designed by our shape adaptors are visually similar when different numbers of shape adaptor modules are used. In Aircraft dataset, we see a narrower shape with MobileNetv2 due to inserting an excessive number of 10 shape adaptors, which eventually converged to a local minima and lead to a degraded performance.

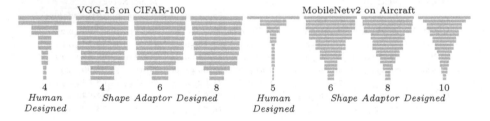

Fig. 3. Visualisation of human-designed and shape adaptor designed network shapes. The number on the second row represents the number of resizing layers (or shape adaptors) applied in the network.

Initialisations in Shape Adaptors. Here, we evaluate the robustness of shape adaptors by varying initialisation of α. Initialisation with a "wide" shape (large α) causes high memory consumption and a longer training time, whereas initialisation with a "narrow" shape (small α) results in weaker gradient signals and a more likely convergence to a non-optimal local minima. In Table 3, we can see that the performance is again consistently better than the human-designed architecture, across all tested initialisations.

In Fig. 4, we present the learning dynamics for each shape adaptor module across the entire training stage. We can observe that shape adaptors are learning in an almost identical pattern across different initialisations in the CIFAR-100

Table 3. Test accuracies of VGG-16 on CIFAR-100 and MobileNetv2 on Aircraft datasets in shape adaptors with different initialisations. Best results are in bold. The initialisation produced in Eq. 5 is highlighted in teal.

CIFAR-100	Human	SA (with $s(\alpha)$ initialised)				Aircraft	Human	SA (with $s(\alpha)$ initialised)			
		0.60	0.70	0.80	0.90			0.52	0.58	0.62	0.68
VGG-16	75.39	**79.21**	79.16	78.79	78.53	VGG-16	85.28	83.90	**86.95**	86.36	86.54
MobileNetv2	73.80	75.16	**75.74**	74.89	74.74	MobileNetv2	77.64	79.64	**83.00**	82.51	81.56

dataset, with nearly no variance. For the larger resolution Aircraft dataset, different initialised shape adaptors converged to a different local minimum. They still follow a general trend, for which the reshaping factor of a shape adaptor inserted in the deeper layers would converge into a smaller scale.

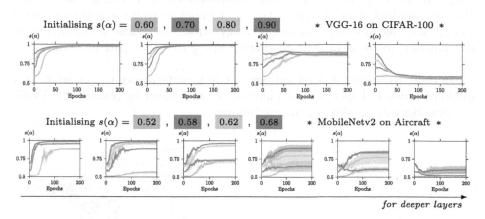

Fig. 4. Visualisation of learning dynamics for every shape adaptor module across the entire training stage.

4.4 A Detailed Study on Neural Shape Learning

In this section, we propose a study to analyse different shape learning strategies, and the transferability of learned shapes. Likewise, all results are averaged over two independent runs.

We evaluate different neural shape learning strategies by running shape adaptors in three different versions. *Standard*: the standard implementation from previous sections; *Fix (Final)*: a network retrained with a fixed optimal shape obtained from shape adaptors; and *Fix (Large)*: a network retrained with a fixed largest possible shape in the current running GPU memory. The Fix (Final) baseline is designed to align with the training strategy from NAS algorithms [17,26,37]. The Fix (Large) baseline is to test whether naively increasing network computational cost can give improved performance.

Table 4. Test accuracies and MACs (the number of multiply-adds) on CIFAR-100 and Aircraft datasets trained with different shape learning strategies.

CIFAR-100	Human	Shape adaptors			Aircraft	Human	Shape Adaptors		
		Standard	Fix (Final)	Fix (Large)			Standard	Fix (Final)	Fix (Large)
VGG-16	75.39	79.16	78.62	78.51	VGG-16	85.28	86.95	86.27	84.49
	314M	5.21G	5.21G	9.46G		15.4G	50.9G	50.9G	97.2G
MobileNetv2	73.80	75.74	75.54	75.46	MobileNetv2	77.64	83.00	82.26	81.18
	94.7M	923M	923M	1.35G		326M	9.01G	9.01G	12.0G

In Table 4, we can observe that our standard version achieves the best performance among all shape learning strategies. In addition, we found that just having a large network would not guarantee an improved performance (VGG-16 on Aircraft). This validates that shape adaptors are truly learning the optimal shape, rather than naively increasing computational cost. Finally, we can see that our original shape learning strategy without re-training performs better than a NAS-like two-stage training strategy, which we assume is mainly due to dynamically updating of network shape helping to learn spatial-invariant features.

In order to further understand how network performance is correlated with different network shapes, we ran a large-scale experiment by training 200 VGG-16 networks with randomly generated shapes.

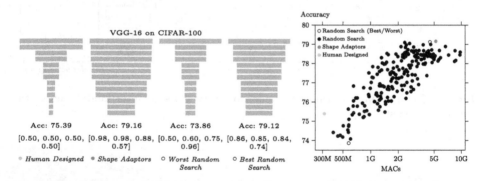

Fig. 5. Visualisation and test accuracies of VGG-16 on CIFAR-100 in 200 randomly generated shapes. The second row represents the precise reshaping factor in each resizing layer.

In Fig. 5, we visualise the randomly generated network shapes with the best and the worst performance, and compare them to the network shapes in human designed and shape adaptor networks.

First, we can see that the best randomly searched shape obtains a very similar performance as well as a similar structure of shape compared to the ones learned from shape adaptors. Second, the reshaping factors in the worst searched shape are arranged from small to large, which is the direct opposite trend to the reshaping factors automatically learned by our shape adaptors. Third, human-designed networks are typically under-sized, and just by increasing network memory cost is not able to guarantee an improved performance. Finally, we can see a clear correlation between memory cost and performance, where a higher memory cost typically increases performance. However, this correlation ceases after 5G of memory consumption, after which point we see no improved performance. Interestingly, the memory cost of shape adaptors lies just on the edge of this point, which again shows the shape adaptor's ability to learn optimal design.

5 Other Applications

5.1 Automated Shape Compression

In previous sections, we have shown that shape adaptors are able to improve performance by finding the optimal network shapes, but with a cost of a huge memory requirement of the learned network. In AutoSC, we show that shape adaptors can also achieve strong results, when automatically finding optimal memory-bounded network shapes based on an initial human design. Instead of the original implementation of shape adaptors where these are assumed to be the only resizing layers in the network, with AutoSC we attach down-sampling shape adaptors only on top of the non-resized layers of the human-designed architecture, whilst keeping the original human-designed resizing layers unchanged. We initialise shape adaptors so that the network shape is identical to the human-designed architecture, and thus the down-sampling shape adaptors can only learn to compress the network shape.

In Table 5, we present AutoSC built on MobileNetv2, an efficient network design for mobile applications. We evaluate AutoSC on three datasets: CIFAR-100, Aircraft and ImageNet. During training of MobileNetv2, we initialised a small width multiplier on the network's channel dimension to slightly increase the parameter space (if applicable). By doing this, we ensure that this "wider" network after compression would have a similar memory consumption as the human-designed MobileNetv2, for a fair comparison. In all three datasets, we can observe that shape adaptors are able to improve performance, despite having similar memory consumption compared to human-designed networks.

Table 5. Test accuracies for AutoSC and human-designed MobileNetv2 on CIFAR-100, Aircraft, and ImageNet. × represents the applied width multiplier.

200/300M MobileNetv2	Params	MACs	Acc.
Human 0.75×	2.6M	233M	69.8
AutoSC 0.85×	2.9M	262M	70.7
Human 1.0×	3.5M	330M	71.8
AutoSC 1.1×	4.0M	324M	72.3

Plain MobileNetv2	Params	MACs	Acc.
Human 1.0×	2.3M	94.7M	73.80
AutoSC 1.0×	2.3M	91.5M	74.81
Human 1.0×	2.3M	330M	77.64
AutoSC 1.0×	2.3M	326M	78.95

(a) Results on ImageNet

(b) Results on CIFAR-100 (up) and Aircraft (down)

5.2   Automated Transfer Learning

In this section, we present how shape adaptors can be used to perform transfer learning in an architectural level. In AutoTL, we directly replace the original human-designed resizing layers with shape adaptors, and initialise them with the reshaping factors designed in the original human-defined architecture, to match the spatial dimension of each pre-trained network layer. During fine-tuning, the network is then fine-tuning with network weights along with network shapes, thus improving upon the standard fine-tuning in a more flexible manner.

The results for Auto TL and other state-of-the-art transfer learning methods are listed in Table 6, for which we outperform 4 out of 5 datasets. The most related methods to our approach are standard fine-tuning and Spot-Tune [6], which opti-

Table 6. Test accuracies of transfer Learning methods built on ResNet-50 on fine-grained datasets. Best results are in bold.

	Birds [31]	Cars [11]	Flowers [25]	WikiArt [28]	Sketch [4]
PackNet [23]	80.41	86.11	93.04	69.40	76.17
PiggyBack [22]	81.59	89.62	94.77	71.33	79.91
NetTailor [24]	82.52	90.56	95.79	72.98	80.48
Fine-tune [6]	81.86	89.74	93.67	75.60	79.58
SpotTune [6]	84.03	92.40	**96.34**	75.77	80.20
AutoTL	**84.29**	**93.66**	96.22	**77.47**	**80.74**

mise the entire network parameters for each dataset. Other approaches like Pack-Net [23], Piggyback [22], and NetTailor [24] focus on efficient transfer learning by updating few task-specific weights. We design AutoTL with standard fine-tuning, as the simplest setting to show the effectiveness of shape adaptors. In practise, AutoTL can be further improved, and integrated into other efficient transfer learning techniques.

6   Conclusions and Future Directions

In this paper, we presented *shape adaptor*, a learnable resizing module to enhance existing neural networks with task-specific network shapes. With shape adaptors, the learned network shapes can further improve performances compared to human-designed architectures, without requiring an increase in parameter space. In future work, we will investigate improving shape adaptors in a multi-branch design, where the formulation provided in this paper can be extended to integrating more than two resizing layers in each shape adaptor module. In addition, we will also study use of shape adaptors for more applications, such as neural architecture search, and multi-task learning.

References

1. Brock, A., Lim, T., Ritchie, J., Weston, N.: SMASH: One-shot model architecture search through hypernetworks. In: International Conference on Learning Representations (2018). https://openreview.net/forum?id=rydeCEhs-
2. Cai, H., Zhu, L., Han, S.: ProxylessNAS: direct neural architecture search on target task and hardware. In: International Conference on Learning Representations (2019). https://openreview.net/forum?id=HylVB3AqYm
3. Deng, J., Dong, W., Socher, R., Li, L.J., Li, K., Fei-Fei, L.: ImageNet: a large-scale hierarchical image database. In: 2009 IEEE Conference on Computer Vision and Pattern Recognition, pp. 248–255. IEEE (2009)
4. Eitz, M., Hays, J., Alexa, M.: How do humans sketch objects? ACM Trans. Graph. (TOG) **31**(4), 1–10 (2012)
5. Goodfellow, I.J., Bulatov, Y., Ibarz, J., Arnoud, S., Shet, V.: Multi-digit number recognition from street view imagery using deep convolutional neural networks. arXiv preprint arXiv:1312.6082 (2013)
6. Guo, Y., Shi, H., Kumar, A., Grauman, K., Rosing, T., Feris, R.: Spottune: transfer learning through adaptive fine-tuning. In: Proceedings of the IEEE Conference on Computer Vision and Pattern Recognition, pp. 4805–4814 (2019)
7. Han, S., Mao, H., Dally, W.J.: Deep compression: compressing deep neural networks with pruning, trained quantization and huffman coding. In: International Conference on Learning Representations (2016)
8. He, K., Zhang, X., Ren, S., Sun, J.: Deep residual learning for image recognition. In: Proceedings of the IEEE Conference on Computer Vision and Pattern Recognition, pp. 770–778 (2016)
9. Hutter, F., Kotthoff, L., Vanschoren, J.: Automated Machine Learning-Methods, Systems, Challenges. Springer, Heidelberg (2019)
10. Jang, E., Gu, S., Poole, B.: Categorical reparameterization with gumbel-softmax. In: International Conference on Learning Representations (2017)
11. Krause, J., Stark, M., Deng, J., Fei-Fei, L.: 3D object representations for fine-grained categorization. In: 4th International IEEE Workshop on 3D Representation and Recognition (3dRR-13), Sydney, Australia (2013)
12. Krizhevsky, A., Hinton, G.: Learning multiple layers of features from tiny images. Technical report (2009)
13. Kuen, J., et al.: Stochastic downsampling for cost-adjustable inference and improved regularization in convolutional networks. In: Proceedings of the IEEE Conference on Computer Vision and Pattern Recognition, pp. 7929–7938 (2018)
14. Lee, C.Y., Gallagher, P.W., Tu, Z.: Generalizing pooling functions in convolutional neural networks: mixed, gated, and tree. In: Artificial Intelligence and Statistics, pp. 464–472 (2016)
15. Li, L., Talwalkar, A.: Random search and reproducibility for neural architecture search. arXiv preprint arXiv:1902.07638 (2019)
16. Liu, C., et al.: Progressive neural architecture search. In: Ferrari, V., Hebert, M., Sminchisescu, C., Weiss, Y. (eds.) ECCV 2018. LNCS, vol. 11205, pp. 19–35. Springer, Cham (2018). https://doi.org/10.1007/978-3-030-01246-5_2
17. Liu, H., Simonyan, K., Yang, Y.: Darts: differentiable architecture search. In: International Conference on Learning Representations (2019). https://openreview.net/forum?id=S1eYHoC5FX
18. Liu, Z., Li, J., Shen, Z., Huang, G., Yan, S., Zhang, C.: Learning efficient convolutional networks through network slimming. In: Proceedings of the IEEE International Conference on Computer Vision, pp. 2736–2744 (2017)

19. Louizos, C., Welling, M., Kingma, D.P.: Learning sparse neural networks through l_0 regularization. In: International Conference on Learning Representations (2018). https://openreview.net/forum?id=H1Y8hhg0b
20. Maddison, C.J., Mnih, A., Teh, Y.W.: The concrete distribution: a continuous relaxation of discrete random variables. In: International Conference on Learning Representations (2017)
21. Maji, S., Kannala, J., Rahtu, E., Blaschko, M., Vedaldi, A.: Fine-grained visual classification of aircraft. Technical report (2013)
22. Mallya, A., Davis, D., Lazebnik, S.: Piggyback: adapting a single network to multiple tasks by learning to mask weights. In: Ferrari, V., Hebert, M., Sminchisescu, C., Weiss, Y. (eds.) ECCV 2018. LNCS, vol. 11208, pp. 72–88. Springer, Cham (2018). https://doi.org/10.1007/978-3-030-01225-0_5
23. Mallya, A., Lazebnik, S.: Packnet: adding multiple tasks to a single network by iterative pruning. In: Proceedings of the IEEE Conference on Computer Vision and Pattern Recognition, pp. 7765–7773 (2018)
24. Morgado, P., Vasconcelos, N.: Nettailor: tuning the architecture, not just the weights. In: Proceedings of the IEEE Conference on Computer Vision and Pattern Recognition, pp. 3044–3054 (2019)
25. Nilsback, M.E., Zisserman, A.: Automated flower classification over a large number of classes. In: 2008 Sixth Indian Conference on Computer Vision, Graphics & Image Processing, pp. 722–729. IEEE (2008)
26. Pham, H., Guan, M., Zoph, B., Le, Q., Dean, J.: Efficient neural architecture search via parameters sharing. In: Dy, J., Krause, A. (eds.) Proceedings of the 35th International Conference on Machine Learning. Proceedings of Machine Learning Research, vol. 80, pp. 4095–4104. PMLR, Stockholmsmässan, Stockholm Sweden, 10–15 July 2018. http://proceedings.mlr.press/v80/pham18a.html
27. Real, E., Aggarwal, A., Huang, Y., Le, Q.V.: Regularized evolution for image classifier architecture search. In: Proceedings of the AAAI Conference on Artificial Intelligence, vol. 33, pp. 4780–4789 (2019)
28. Saleh, B., Elgammal, A.: Large-scale classification of fine-art paintings: learning the right metric on the right feature. Int. J. Digit. Art Hist. (2016)
29. Sandler, M., Howard, A., Zhu, M., Zhmoginov, A., Chen, L.C.: Mobilenetv 2: inverted residuals and linear bottlenecks. In: Proceedings of the IEEE Conference on Computer Vision and Pattern Recognition, pp. 4510–4520 (2018)
30. Simonyan, K., Zisserman, A.: Very deep convolutional networks for large-scale image recognition. In: International Conference on Learning Representations (2015)
31. Wah, C., Branson, S., Welinder, P., Perona, P., Belongie, S.: The Caltech-UCSD Birds-200-2011 Dataset. Technical report CNS-TR-2011-001, California Institute of Technology (2011)
32. Yu, D., Wang, H., Chen, P., Wei, Z.: Mixed pooling for convolutional neural networks. In: Miao, D., Pedrycz, W., Ślęzak, D., Peters, G., Hu, Q., Wang, R. (eds.) RSKT 2014. LNCS (LNAI), vol. 8818, pp. 364–375. Springer, Cham (2014). https://doi.org/10.1007/978-3-319-11740-9_34
33. Yu, K., Sciuto, C., Jaggi, M., Musat, C., Salzmann, M.: Evaluating the search phase of neural architecture search. In: International Conference on Learning Representations (2020). https://openreview.net/forum?id=H1loF2NFwr
34. Zeiler, M., Fergus, R.: Stochastic pooling for regularization of deep convolutional neural networks. In: Proceedings of the International Conference on Learning Representation (2013)

35. Zhang, R.: Making convolutional networks shift-invariant again. In: International Conference on Machine Learning, pp. 7324–7334 (2019)

36. Zhu, Y., Zhang, X., Yang, T., Sun, J.: Resizable neural networks (2020). https://openreview.net/forum?id=BJe_z1HFPr

37. Zoph, B., Le, Q.V.: Neural architecture search with reinforcement learning. In: International Conference on Learning Representations (2017)

Shuffle and Attend: Video Domain Adaptation

Jinwoo Choi[1]([✉]), Gaurav Sharma[2], Samuel Schulter[2], and Jia-Bin Huang[1]

[1] Virginia Tech, Blacksburg, VA 24060, USA
{jinchoi,jbhuang}@vt.edu
[2] NEC Labs America, San Jose, CA 95110, USA

Abstract. We address the problem of domain adaptation in videos for the task of human action recognition. Inspired by image-based domain adaptation, we can perform video adaptation by aligning the features of frames or clips of source and target videos. However, equally aligning all clips is sub-optimal as not all clips are informative for the task. As the first novelty, we propose an attention mechanism which focuses on more discriminative clips and directly optimizes for video-level (cf. clip-level) alignment. As the backgrounds are often very different between source and target, the source background-corrupted model adapts poorly to target domain videos. To alleviate this, as a second novelty, we propose to use the clip order prediction as an auxiliary task. The clip order prediction loss, when combined with domain adversarial loss, encourages learning of representations which focus on the humans and objects involved in the actions, rather than the uninformative and widely differing (between source and target) backgrounds. We empirically show that both components contribute positively towards adaptation performance. We report state-of-the-art performances on two out of three challenging public benchmarks, two based on the UCF and HMDB datasets, and one on Kinetics to NEC-Drone datasets. We also support the intuitions and the results with qualitative results.

1 Introduction

Recent computer vision-based methods have reached very high performances in supervised tasks [2,17,18,22,51] and many real-world applications have been made possible such as image search, face recognition, automatic video tagging etc. The two main ingredients for success are (i) high capacity network design with an associated practical learning method, and (ii) large amounts of *annotated* data. While the first aspect is scalable, in terms of deployment to multiple novel

J. Choi—Part of this work was done when Jinwoo Choi was an intern at NEC Labs America.

Electronic supplementary material The online version of this chapter (https://doi.org/10.1007/978-3-030-58610-2_40) contains supplementary material, which is available to authorized users.

A. Vedaldi et al. (Eds.): ECCV 2020, LNCS 12357, pp. 678–695, 2020.
https://doi.org/10.1007/978-3-030-58610-2_40

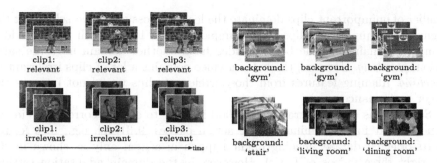

Fig. 1. Motivation. We do video domain adaptation and introduce the following two key components: (*Left*): Clip attention. The top video and the lower video have the same action punching. However, the lower video has only one relevant punching clip, while the top video has three relevant punching clips. Our proposed attention suppresses features from irrelevant clips, improving the feature alignment across domains. (*Right*): Clip order prediction. The top and bottom videos are from different domains, but all capture the action fencing. However, the backgrounds are different: the top domain has a gym as a background, and the lower domain has a dining room or a living room or a stair as a background. Predicting the order of clip encourages the model to focus more on the humans, not the background, as the background is uninformative for predicting temporal order.

scenarios, the second aspect becomes the limiting factor. The annotation issue is even more complicated in video-related tasks, as we need temporal annotation, i.e., we need to specify the start and end of actions in long videos. Domain adaptation has emerged as an important and popular problem in the community to address this issue. The applications of domain adaptation have ranged from simple classification [13,33,40,47,48,56] to more complex tasks like semantic segmentation [5,7,20,46,49,57] and object detection [1,6,19,21,26,60]. However, the application on video tasks e.g., action recognition is still limited [3,10,23].

We address this less studied but challenging and practically important task of video domain adaptation for human action recognition. We work in an unsupervised domain adaptation setting. That is, we have annotated data for the source domain and only *unannotated* data for the target domain. Examples domains that we use in experiments include (human) actions from movies, unconstrained actions from sports videos, YouTube videos, and even videos taken from drones.

We exploit two insights related to the problem and propose two novel adaptation components inspired by them. First, we note that the existing domain adaptation methods, when applied directly to the video adaptation task, sample frames or clips [3,23], depending on whether the video encoding is based on a 2D network, e.g., temporal relation network [58] or a 3D network, e.g., C3D [44]. We sample clips (or frames) and then average the final outputs from multiple clips at test time, following the video classification networks they are built upon. Performing domain adaptation by aligning features for all sampled clips is suboptimal, as a lot of network capacity is wasted on aligning clips that are not crucial for the task. In the worst case, it can even be detrimental if a large number of unimportant clips dominate the learning loss and adversely affect the alignment of important clips. For example, in Fig. 1 left, both the top video from one domain and the bottom video from another domain have the same action, *punching*. However, the bottom video contains a lot of clips irrelevant to *punching*. Aligning features from those irrelevant clips would not improve the target performance much.

Second, this clip-wise training method is likely to exploit correlations in the scene context for discriminating the action classes [9,31,32], e.g., in a formal sports-oriented dataset fencing might happen in a gym only as shown in the top right three videos of Fig. 1. However, in the domain adaptation setting, the target domain might have vastly different scene contexts, e.g., the same fencing might happen in a living room or dining room, as shown in the bottom right three videos of Fig. 1. When the source model uses the correlated gym information to predict a fencing action, it may perform poorly on the same class in the target domain, which does not have a gym scene. Similar scene context corruption issues have been identified for transfer learning, and few works have addressed the problem of *debiasing* the representations explicitly [9,52].

Based on the above insights, we propose **Shuffle and Attend: Video domain Adaptation** (SAVA) with two novel components. First, we propose to identify and align *important* (which we define as *discriminative*) clips in source and target videos via an attention mechanism. The attention mechanism leads to the suppression of temporal background clips, which helps us focus on aligning only the important clips. Such attention is learned jointly for video-level adaptation and classification. We estimate the clip's importance by employing an auxiliary network and derive the video feature as the weighted combination of the identified important clip features.

Second, we propose to learn *spatial-background invariant human action representations* by employing a self-supervised clip order prediction task. While there could be some correlation between the scene context/background and the action class, e.g., soccer field for 'kicking the ball' action, the scene context is not sufficient for predicting the temporal clip order. In contrast, the actual human actions are indicative of the temporal order, e.g., for 'kicking the ball' action the clip order follows roughly the semantics of 'approaching the ball', 'swinging the leg' and 'kicking'; if we shuffle the clips, the actual human action representation would be able to recover the correct order, but the scene context based representation would be likely to fail.

Thus using the clip order prediction based loss helps us counter the scene context corruption in the action representations and improves adaptation performance. We employ the self-supervised clip order prediction task for both source and target. As this auxiliary task is self-supervised, it does not require any annotation (which we do not have for target videos).

We provide extensive empirical evaluations to demonstrate the benefits of the proposed method on three challenging video domain adaptation benchmark settings. We also give qualitative results to highlight the benefits of our system.

In summary, our contributions are as follows.

- We propose to learn to align important (discriminative) clips to achieve improved representation for the target domain.
- We propose to employ a self-supervised task which encourages a model to focus more on actual action and suppresses the scene context information, to learn representations more robust to domain shifts. The self-supervised task does not require extra annotations.
- We obtain state-of-the-art results on the HMDB to UCF adaptation benchmark, and Kinetics to NEC-Drone benchmarks.

2   Related Work

Action Recognition. Action recognition using deep neural networks has shown quick progress recently, starting from two-stream networks [42] to 3D [2,44,51] or 2D and 1D separable CNNs [45,53] have performed very well on the task. More recent advances in action recognition model long-term temporal contexts [11,51]. However, most models still rely on target supervised data when finetuning on target datasets. In contrast, we are interested in unsupervised domain adaptation, where we do not have access to target labels during training.

Unsupervised Domain Adaptation for Images. Based on adversarial learning, domain adaptation methods have been proposed for image classification [13,33,40,47,48,56], object detection [1,6,19,21,26,60], semantic segmentation [5,7,20,46,49,57], and low-level vision tasks [39]. We also build upon adversarial learning. However, we work with videos and not still images.

Unsupervised Domain Adaptation for Videos. Unlike image-related tasks, there are only a few works on video domain adaptation [3,10,23]. We also use the basic adversarial learning framework but improve upon it by adding auxiliary tasks that depend on the temporal order in videos, (i) to encourage suppression of spatial-background, and (ii) to focus on important clips in the videos to align.

Self-supervision. Image based self-supervised methods work with spatial context, e.g., by solving jigsaw puzzle [36], image inpainting [38], image colorization [29], and image rotation [15] to learn more generalizable image representation. In contrast, video based self-supervised methods exploit temporal context, e.g., by order verification [34], frame sorting [30], and clip sorting [54]. Recent video domain adaptation methods employ self-supervised domain sequence prediction [4], or self-supervised RGB/flow modality correspondence prediction [35].

We make a connection between the self-supervised task of clip order prediction [54] and learning a robust spatial-background decoupled representation for action recognition. We hypothesize (see Sect. 1) that, in combination with adversarial domain adaptation loss, this leads to suppression of domain correlated background, and simultaneous enhancement of the task correlated human part in the final representation leading to better domain adaptation performance.

Attention. There are numerous methods employing attention model for image [14,24] and video tasks [12,16,25,27,37,41,50,55]. The most closely related

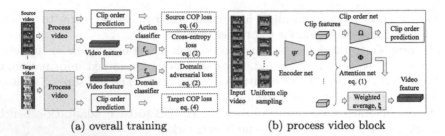

(a) overall training   (b) process video block

Fig. 2. Overview of SAVA. We employ standard domain adversarial loss along with two novel components. The first component is the self-supervised clip order prediction loss. The second is a clip attention based feature alignment mechanism. We predict attention weights for the uniformly sampled clips from the videos and construct the video feature as a weighted average of the clip features. Then we align the source and target video features. Best viewed with zoom and color. (Color figure oline)

work is the that by Chen et al. [3]. While both the proposed method and Chen et al. [3] are based on attention, the main difference is in *what* they attend to. Chen et al. [3] attends to temporal relation features (proposed by Zhou et al. [58]) with *larger domain gaps*. In contrast, our proposed method attends to *discriminative* clip features. The clips in the same video may have different discriminative content, e.g., leg swinging (more discriminative) vs. background clips (less so) in a video of 'kicking a ball' class. The proposed method attends to more discriminative clips and focuses on aligning them. Chen et al. [3] samples 2−5 frames relation features and attends to the ones with a larger domain gap measured by the entropy of the domain classifiers. However, the relation feature with a larger domain gap might come from frames irrelevant to the action, aligning them would be suboptimal. The proposed method addresses this problem. In another closely related work Pan et al. [37] temporally align the source and target features using temporal co-attention and match their distributions. In contrast, the proposed method argues that human-focused representation is more robust to domain shifts, and captures it via self-supervised clip order prediction.

3   Method

We work in an unsupervised domain adaptation setting. We have (i) *annotated source data* $(\mathbf{x}_s, \mathbf{y}_s) \in \mathbf{X}^s \times \mathbf{Y}^s$, where $\mathbf{X}^s$ is the set of videos containing human-centered videos and $\mathbf{Y}^s$ is the actions label set, and (ii) *unannotated target data* $\mathbf{x}_t \in \mathbf{X}^t$. The task is to train a model, which performs well on the target data. Since the source data distribution, e.g., actions in movies, is expected to be very different from the target data distribution, e.g., actions in sports videos, the model trained on the source data only does not work well on target videos. The challenge is to design methods that can adapt a model to work on the target data, using both annotated source data and unannotated target data. The proposed method has three main components for adaptation: domain adversarial loss, clip order prediction losses, and an attention module for generating video features.

Figure 2 gives an overview of the proposed method, which we call Shuffle and Attend Video domain Adaptation (SAVA). We start with uniformly sampling N clips, with L frames, from an arbitrary length input video, as shown in Figure 2 (b). We encode source and target clips into clip features by an encoder network $\Psi(\cdot)$; which can be either the same for both or different. Here we assume it is the same for the brevity of notation. Then we use the clip features for (i) the clip order prediction network $\Omega(\cdot)$, and (ii) constructing the video-level features using the attention network $\Phi(\cdot)$. The video-level features obtained after the attention network, are then used with (i) a linear classifier, for source videos only, and (ii) a domain classifier, for both source and target videos, as shown in Fig. 2 (a).

In total, there are three types of losses that we optimize, (i) domain adversarial loss, (ii) clip order prediction loss for both source and target, and (iii) classification loss for source only. The clip order prediction loss works with clip level features, while the other two work on video-level features. As discussed in Sect. 1, the clip order prediction loss helps a model to learn a representation that is less reliant on correlated source data background. The attention network gives us the final video feature by focusing on important clips. The domain adversarial loss helps a model to align video-level features between source and target videos. All these are jointly learned and hence lead to a trained system that gives aligned representations and achieves higher action classification performance than the baselines. We now describe each of our proposed components individually in detail in the following subsections.

3.1 Clip Order Prediction

As shown on Fig. 1 (right), the source videos of the same class may have correlations with similar background context [32], and the target videos of the same class might have a background which is vastly different from the source background. While the source model might benefit from learning representation, which is partially dependent on the correlated background, this would lead to poor target classification. To address this problem, we propose to employ clip order prediction (COP) to enable better generalization of the representation. COP would not be very accurate if a model focuses on the background as the background might not change significantly over time. However, the temporal evolution of the clip depends more on the humans performing actions, and possibly the objects. Thus, if we employ the COP, the representation would focus more on the relevant humans and objects, while relying less on the background.

We build our COP module
upon the work by Xu et al.
[54]. We show the illustration
of the COP network Ω in
Fig. 3. We incorporate an auxiliary network, taking clip features as input, to predict the
correct order of shuffled clips
of an input video. We sample
M clips, with L frames each,
from an input video and shuffle them. The task of the module is to predict the order of
the shuffled clips. We formulate
the COP task as a classification

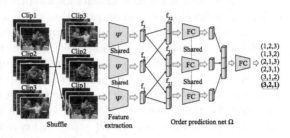

Fig. 3. Clip order prediction network Ω (the layers after Ψ).

task with $M!$ classes, corresponding to all permutation tuples of the clips, and
consider the correct order tuple as the ground truth class. We concatenate clip
features pairwise and pass them to a fully connected layer with ReLU activation
followed by a dropout layer. Then we concatenate all of the output features and
use a final linear classifier to predict the order of the input clips. Since this is
a self-supervised task and requires no extra annotation, we can use the task for
the videos from source, target, or both; we evaluate this empirically in Sect. 4.3.

3.2 Clip-Attention Based Video-Level Features

As shown in the left side of Fig. 1, all clips are not equally *important* (*discriminative* or *relevant*) for predicting the action. Aligning the irrelevant clip features
is suboptimal, and it might even degrade performance if they dominate the loss
cf. the important clips. Focusing on and aligning the important clips would lead
to better adaptation and classification performance. To achieve such focus on
important clips, we propose a clip attention module. The attention module takes
N number of clip features as inputs, and outputs N softmax scores indicating
the importance of each of them. The final video-level feature is obtained by the
weighted average of the clip features. Formally, given $\mathbf{x}_1,\ldots,\mathbf{x}_N$ as the N clips
from an input video $\mathbf{x}$, we obtain the video-level feature $\mathbf{x}_v$ as

$$\mathbf{w} = \Phi(\Psi(\mathbf{x}_1),\ldots,\Psi(\mathbf{x}_N)), \quad \mathbf{x}^v = \xi(\mathbf{w},\Psi(\mathbf{x}_1),\ldots,\Psi(\mathbf{x}_N)) = \sum_{i=1}^{N} w_i\Psi(\mathbf{x}_i), \quad (1)$$

where, $\xi(\cdot)$ is the weighted average function.

The attention module $\Phi(\cdot)$ is a network that takes N clip features with D
dimension as an input. It outputs the importance vector $\mathbf{w} \in R^N$, which is used
for weighted averaging to obtain the video-level feature. Thus, we can train the
model end-to-end with the full domain adaptation system.

There can be multiple valid choices for the architecture of the attention
module, e.g., a standard feed-forward network which takes concatenation of the

clip features as input, or a recurrent network that consumes the clip features one by one. We explore two specific choices in an ablation experiment in Sect. 4.3, (i) Multi Layer Perceptron (MLP) similar to Kar et al. [25], and (ii) Gated Recurrent Units (GRU).

3.3 Training

We pre-train the attention module with standard binary cross-entropy loss, where we get the ground truth attention vector as follows. The ground truth label is 1 if the clip is correctly classified by the baseline clip-based classification network, and has confidence higher than a threshold c_{th}, and 0 otherwise. The pre-training makes the attention module to start from good local optima, mimicking the baseline classifier. Once pre-trained, the attention module can then either be fixed or can be trained end-to-end with the rest of the network. Please note that we train the attention module only on the source dataset as the training requires the ground truth action labels.

For the feature distribution alignment, we follow the well-known adversarial domain adaptation framework of ADDA [48]. We define our losses as,

$$
L_{\mathrm{CE}} = -\mathbb{E}_{(\mathbf{x}_s, \mathbf{y}_s) \sim (\mathbf{X}^s, \mathbf{Y}^s)} \sum_{k=1}^{K} [y_{s,k} \log f_C(\mathbf{x}_s^v)],
$$

$$
L_{\mathrm{ADV}_{f_D}} = -\mathbb{E}_{\mathbf{x}_s \sim \mathbf{X}^s} [\log f_D(\mathbf{x}_s^v)] - \mathbb{E}_{\mathbf{x}_t \sim \mathbf{X}^t} [\log(1 - f_D(\mathbf{x}_t^v)],
$$

$$
L_{\mathrm{ADV}_{\psi_t}} = -\mathbb{E}_{\mathbf{x}_t \sim \mathbf{X}^t} [\log f_D(\mathbf{x}_t^v)], \tag{2}
$$

where f_C is the linear source classifier and f_D is the domain classifier. The video feature $\mathbf{x}^v = \xi(\mathbf{w}, \Psi(\mathbf{x}_1) \ldots, \Psi(\mathbf{x}_N))$ is the weighted average of clip level features, with weights $\mathbf{w} = \Phi(\Psi(\mathbf{x}_1), \ldots, \Psi(\mathbf{x}_N))$ obtained from the attention module. Then our optimization objective is as follows,

$$
\theta_s^*, \theta_{f_C}^*, \theta_\Phi^* = \operatorname*{argmin}_{\theta_s, \theta_{f_C}} L_{\mathrm{CE}, \theta_\Phi}, \theta_{f_D}^* = \operatorname*{argmin}_{\theta_{f_D}} L_{\mathrm{ADV}_{f_D}}, \theta_t^* = \operatorname*{argmin}_{\theta_t} L_{\mathrm{ADV}_{\psi_t}}, \tag{3}
$$

where θ_s is the parameter of the source encoder $\Psi_s(\cdot)$, θ_{f_C} is the parameter of the source classifier $f_C(\cdot)$, θ_t is the parameter of the target encoder $\Psi_t(\cdot)$, and θ_{f_D} is the parameter of the domain classifier $f_D(\cdot)$.

We optimize this objective function in a stage-wise fashion [48]. We first optimize the source cross-entropy loss L_{CE} over the source parameters θ_s and θ_{f_C} with the annotated source data. Then we freeze source model parameters θ_s and θ_{f_C}, and optimize the domain classification loss $L_{\mathrm{ADV}_{f_D}}$ over the domain classifier parameter θ_{f_D}, and the inverted GAN loss $L_{\mathrm{ADV}_{\psi_t}}$ over the target encoder parameter θ_t with both the labeled source and the unlabeled target data.

Clip Order Prediction. We define the COP loss as follows.

$$
L_{\mathrm{COP}} = -\mathbb{E}_{(\mathbf{x}, \mathbf{y}) \sim (\mathbf{X}, \mathbf{Y})} \sum_{k=1}^{M!} [y_k \log f_O(\phi)]. \tag{4}
$$

Here, f_O is the linear classification function for the COP, $\phi = \Omega(\Phi(\mathbf{x}_1),$ $..., \Phi(\mathbf{x}_M))$ is the ReLU activation of the MLP which takes M clip features as input. We can employ the L_{COP} for both source and target. We optimize the loss L_{COP} over the source encoder parameter θ_s, target encoder parameter θ_t, COP MLP parameter θ_Ω, and clip order cliassifier parameter θ_{f_O}.

3.4   Inference

At inference time, we remove the domain discriminator and clip order prediction network. We divide the input video into N clips and extract clip features. These features are then weight averaged with weights obtained using the attention network. The action classifier predicts the action using the video-level feature.

4   Experimental Results

4.1   Datasets

We show results on the publicly available benchmark based on the UCF [43] and HMDB [28] datasets. We further show the result in a more challenging setting where the source dataset is part of the Kinetics dataset [2], and the target dataset is drone-captured action dataset [10]. In the following, the direction of the arrow indicates the source (arrow start) to target (arrowhead).

UCF↔HMDB. Chen et al. [3] released the UCF-HMDB dataset for studying video domain adaptation. This dataset has $3,209$ videos with 12 action classes. All the videos come from the original UCF [43] and HMDB [28] datasets. They subsampled overlapping 12 classes out of $101/51$ classes from the UCF/HMDB, respectively. There are two settings of interest, UCF $\rightarrow$ HMDB, and the other is HMDB $\rightarrow$ UCF. We show the performance of our method in both of the two settings. We use the official split provided by the authors [3].

Kinetics→NEC-Drone. We also test our method on a more challenging target dataset captured by drones [10]. The dataset contains $5K$ videos with 16 classes in total, while the domain adaptation subset used contains 994 videos from 7 classes, which overlap with Kinetics dataset. We use the official train/-val/test split provided by Choi et al. [10]. We conduct domain adaptation experiments with Kinetics→NEC-Drone setting, which is more challenging than UCF↔HMDB as there is a more significant domain gap between source and target domains.

In all three settings, we report top-1 accuracy on the target dataset and compare it to other methods.

4.2   Implementation Details

We implement our method with the PyTorch library. We use the I3D [2] network as our clip feature encoder architecture for both source and target. The source

Table 1. Ablation experiments on the COP loss, on Kinetics→NEC-Drone.

Method	COP on		Top-1 acc (%)	Δ
	Source	Target		
Clip DA + COP	✓	✓	28.5	+11.3
Clip DA + COP	✓	×	25.9	+8.7
Clip DA + COP	×	✓	22.4	+5.2
Clip DA only	×	×	23.7	+6.5
Supervised source only	×	×	17.2	Reference

Table 2. Ablation experiments on the clip attention on Kinetics→NEC-Drone.

Method	Align	Clip attention	Top-1 acc (%)
SAVA (ours)	Video-level	✓	31.6
SAVA (ours) w/o. clip attention	Video-level	×	30.3
Clip-level align	Clip-level	×	28.5

and target encoders are different from each other and do not share parameters. Both are initialized with the Kinetics pre-trained model weights and then trained further as appropriate. Such pre-training on a large dataset is common in domain adaptation, e.g. for images (ImageNet) [6,13,46,48] and videos (Sports-1M [23], Kinetics [4,35]). The input to the clip feature encoder is a 3 channels × 16 frames × 224 × 224 pixels clip. We set the number of clips per video to $N = 4$ via validation. During testing, we sample the same $N = 4$ number of clips. COP module is a 2-layer MLP with 512 hidden units. We sample $M = 3$ clips per video for the COP task by following Xu [54].

By using attention, we compute the weighted average of the clip-level softmax score as our final video-level softmax score. We evaluate two types of networks for the attention module. One is 4-layer MLP with 1024 hidden units in each layer, and the other is a GRU [8] with 1024 hidden units. We found GRU to be better in two out of the three cases (Sect. 4.3), so we report all results with GRU. We set the attention module's confidence threshold c_{th} as 0.96 for the UCF and HMDB and 0.8 for Kinetics by validation on the source dataset. We use 4-layer MLP with 4096 hidden units in each layer as our domain classifier.

We set the batch size to 72. The learning rate starts from 0.01, and we divide the learning rate by 10 after two epochs and ten epochs. We train models for 40 epochs. We set the weight decay to 10^{-7}. We use stochastic gradient descent with momentum 0.9 as our optimizer.

We follow the 'pre-train then adapt' training procedure similar to previous work [48]. (i) We train the feature extractor $\Psi(\cdot)$ with the COP loss (4). We train our feature extractor $\Psi(\cdot)$ on both source and target datasets as we do not require any labels. (ii) Given the trained feature extractor $\Psi(\cdot)$, we further train it on the labeled source and unlabeled target datasets with a domain classifier

$f_D(\cdot)$ attached. We also train the attention module $\Phi(\cdot)$ on the labeled source dataset, given the trained feature extractor $\Psi(\cdot)$ from step 1. (iii) Given the feature extractor $\Psi(\cdot)$ and the attention module $\Phi(\cdot)$, we train our full model with the labeled source dataset and unlabeled target dataset.

4.3 Ablation Study

We perform several ablation experiments to analyze the proposed domain adaptation method. We conduct the experiments on more challenging Kinetics→NEC-Drones setting except the attention module design choice experiment in Table 3, which we performed on the UCF, HMDB, and Kinetics datasets.

Effect of Clip Order Prediction. Table 1 gives the results showing the effect of COP. Here, the source only is the I3D network trained on the source dataset, which we directly test on the target dataset without any adaptation. Clip-level domain adaptation (Clip DA) is the baseline where we randomly sample clips and align features of the clips without any attention. On top of the clip DA, we can optionally use the COP losses for either source or target or both.

Table 3. Effect of using different attention implementation. We show the attention module accuracy (%) on the Kinetics, UCF, and HMDB datasets.

Method	No. params	Kinetics	UCF	HMDB
MLP	$6.3M$	72.2	86.1	75.4
GRU	$6.3M$	78.0	78.9	76.6

The clip DA only (without COP) improves performance over the supervised source only baseline by 6.5%p. More interestingly, the results show that using both source and target COP improves performance significantly compared to the clip DA only baseline by 4.8%p. We also observe that the source COP is more crucial compared to the target COP. This is because the target NEC-Drone dataset (i) contains similar background appearance across all videos (a high school gym), (ii) has a limited number of training videos ($\sim$1 K), and (iii) has the main activities occurring with small spatial footprint (as the actors are small given the videos were captured by drones). Thus, applying COP on the NEC-Drone dataset does not lead to improved results. However, applying COP on the source or both source/target produces large improvements over the baseline.

Clip Attention Performance. We evaluate the two different design choices, MLP and GRU, for our attention module in this experiment. We show the clip attention accuracy on the three source datasets in Table 3. We get the attention accuracy by comparing the ground truth importance label (see Sect. 3.3 for the details) and the predicted importance. We compute the clip attention performance on the three source datasets Kinetics, UCF, and HMDB. Using such curated ground truth ensures that the attention module starts from good local minima, which is in tune with the base I3D encoder network.

The GRU shows a higher attention performance in two out of the three cases, while it has a similar number of parameters to the MLP-based attention. Thus, we employ the GRU-based attention module on all experiments in this paper.

Effect of Attention Module. We show the effect of the attention module in the overall method in Table 2. Here, all the methods are pre-trained using source and target COP losses turned on. We train our domain adaptation network with three settings, (i) video-level alignment with clip attention (our full model), (ii) video-level alignment without clip attention (using temporal average pooling instead), and finally (iii) clip-level alignment.

The results show that video-level alignment gives an improvement over random clip sampling alignment, 30.3% vs. 28.5%. Our full model with clip attention alignment further improves the performance to 31.6%, over video-level alignment without attention. The video-level alignment without attention treats every clip equally. Hence, if there are some non informative clips, e.g., temporal background, equally aligning those clips is a waste of the network capacity. Our *discriminative* clip attention alignment is more effective in determining more discriminative clips and doing alignment based on those.

Table 4. Results on UCF↔HMDB.

Method	Encoder	UCF→HMDB	HMDB→UCF
Supervised source only [3]	ResNet-101-based TRN	73.1 (71.7)	73.9 (73.9)
TA^{3}N [3]	ResNet-101-based TRN	75.3 (78.3)	79.3 (81.8)
Supervised target only [3]	ResNet-101-based TRN	90.8 (82.8)	95.6 (94.9)
Supervised source only [3]	I3D-based TRN	80.6	88.8
TA^{3}N [3]	I3D-based TRN	81.4	90.5
Supervised target only [3]	I3D-based TRN	93.1	97.0
TCoN [37]	ResNet-101-based TRN	87.2	89.1
Supervised source only	I3D	80.3	88.8
SAVA (ours)	I3D	82.2	91.2
Supervised target only	I3D	95.0	96.8

4.4   Comparison with Other Methods

Methods Compared. The methods reported are (i) 'supervised source only': the network trained with supervised source data (a lower bound for adaptation methods), (ii) 'supervised target only': the network trained with supervised target data (an upper bound for the adaptation methods), and (iii) different unsupervised domain adaptation methods. For the TA^{3}N, we compare with the latest results obtained by running the public code[1] provided by the authors [3] and not the results in the paper (given in brackets for reference, in Table 4). While the original TA^{3}N [3] works with 2D features based temporal relation network (TRN) [58], we go beyond and integrate the TA^{3}N with stronger I3D [2] based TRN features. This allows a fair comparison with our method when all other

---

[1] https://github.com/cmhungsteve/TA3N.

factors (backbone, computational complexity, etc) are similar. For TCoN, we report the numbers from the paper [37] as code is not publicly available.

For the Kinetics → NEC-Drone setting, we implement video versions of the DANN [13] and ADDA [48], which align the clip-level I3D [2] features and show the results. We also compare with both unsupervised and semi-supervised methods of Choi et al. [10].

UCF → HMDB. We compare our method with existing methods in Table 4. The first three blocks contain the results of the method with TRN-based encoder [58]. The fourth block shows the results of our SAVA with domain adaptation as well as the source only I3D [2] baseline. We also show the result of fully supervised finetuning of the I3D network on the target dataset as an upper bound.

SAVA with the I3D-based encoder shows 82.2% top-1 accuracy on the HMDB dataset, in this setting. SAVA improves the performance of the strong I3D encoder, 80.3%, which in itself obtains better results than the TRN-based adaptation results, 75.3% with TA^{3}N. Our SAVA is closer to the upper bound (82.2% vs. 95.0%), than the gap between TA^{3}N and its upper bound (75.3% vs. 90.8%). Furthermore, SAVA outperforms TA^{3}N with I3D-based TRN features, 81.4%.

HMDB → UCF. Table 4 gives the comparison of our method with existing methods in this setting. We achieve state-of-the-art results in this setting while the other trend is similar to the UCF→HMDB setting. SAVA achieves 91.2% accuracy on the target dataset with domain adaptation and without using any target labels. The baseline source only accuracy of the I3D network is already quite strong cf. the existing best adaptation method, i.e., 88.8% vs. 90.5% for TA^{3}N with I3D-based TRN features. We improve this to 91.2%. SAVA is quite close to the upper bound of 96.8%, which strongly supports the proposed method. In contrast, TA^{3}N is still far behind its upper bound (79.3% vs. 95.6%).

Table 5. Results on the Kinetics→NEC-Drone.

Method	Encoder	Target labels used (%)	Top-1 acc (%)
Supervised source only [3]	ResNet-101-based TRN	None	15.8
TA^{3}N [3]	ResNet-101-based TRN	None	25.0
Supervised source only [3]	I3D-based TRN	None	15.8
TA^{3}N [3]	I3D-based TRN	None	28.1
Supervised source only	I3D	None	17.2
DANN [13]	I3D	None	22.3
ADDA [48]	I3D	None	23.7
Choi et al. [10] (on val set)	I3D	None	15.1
SAVA (ours)	I3D	None	31.6
Choi et al. [10]	I3D	6	32.0
Supervised target only	I3D	100	81.7

Kinetics → NEC-Drone. This setting is more challenging, as the domain gap is larger, i.e., the gap between the source only and target finetuned classifiers is 64.5% cf. 14.7% for UCF→HMDB.

Table 5 gives the results. The first block uses the TRN with ResNet-101 features, and the second block uses the TRN with I3D features while the others use I3D features. We observe that similar to previous cases, SAVA outperforms all methods, e.g., DANN (42% relative), ADDA (33% relative), TA³N (26.4% relative), TA³N with I3D features (12.4% relative), and Choi et al. (the unsupervised domain adaptation case). It is very close to the semi-supervised result of Choi et al. (31.6 vs. 32.0), where they use 5 target labeled examples per class.

While the improvements achieved by SAVA are encouraging in this challenging setting, the gap is still significant, 31.6% with adaptation vs. 81.7% with the model finetuned with the target labels. The gap highlights the challenging nature of the dataset, and the large margin for improvement in the future, for video-based domain adaptation methods (Fig. 5).

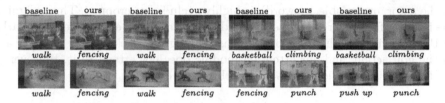

Fig. 4. Class activation maps (CAM) on the UCF (first row) and HMDB (second row) datasets. The actions green are correct predictions, and those in red are incorrect predictions. Here the baseline is ADDA without COP, and ours is ADDA with COP. Note how the COP encourages the model to focus more on human action instead of scene context. (Color figure online)

Fig. 5. Attention visualization on center frames of 4 clips from 4 videos. The frames with green borders are given more importance by our attention module cf. those with red borders. Note that our attention module can attend to relevant clips where the action is clearly visible, while the baseline without attention would align all clips equally, even those where the actor is missing or highly occluded. (Color figure online)

4.5 Qualitative Evaluation

Clip Order Prediction. To better understand the effect of the proposed COP module, we show class activation maps (CAM) [59] of target videos in Fig. 4. We compute the CAM of the center (8th) frame of a 16 frames long clip. We show CAMs from models with and without COP (baseline/ours). The baseline without COP tends to focus more on the scene context. However, the proposed model with COP focuses more on the actual human action (typically around the actors). As the model with COP focuses more on the actual action, it generalizes better to a new domain with a completely different scene cf. the model without COP, which is heavily biased by the scene context.

Clip Attention. We show the center frames of 4 clips per video with the clip attention module based selection. The videos demonstrate how the proposed clip attention module focuses more on the action class relevant clips and less on the irrelevant clips with either highly occluded actors or mainly background. E.g., in the fencing video in the second row, first and the fourth clips are not informative as the actor, or the object (sword), is highly occluded or cropped. Thus, aligning the features from the relevant second and the third clips is encouraged. Similarly, in the golf video of the first row, the last clip (green background) is irrelevant to the golf action, and our attention module does not attend to it. However, a model without attention treats all the clips equally.

5　Conclusion

We proposed **Shuffle and Attend: Video domain Adaptation** (SAVA), a novel video domain adaptation method with self-supervised clip order prediction and clip attention based feature alignment. We showed that both of the two components contribute to the performance. We achieved state-of-the-art performance on the publicly available HMDB→UCF and Kinetics→Drone datasets. We showed extensive ablation studies to show the impact of different aspects of the method. We also validated the intuitions for designing the method with qualitative results for both the contributions.

Acknowledgment. This work was supported in part by NSF under Grant No. 1755785 and a Google Faculty Research Award. We thank NVIDIA Corporation for the GPU donation.

References

1. Cai, Q., Pan, Y., Ngo, C.W., Tian, X., Duan, L., Yao, T.: Exploring object relation in mean teacher for cross-domain detection. In: CVPR (2019)
2. Carreira, J., Zisserman, A.: Quo vadis, action recognition? A new model and the kinetics dataset. In: CVPR (2017)
3. Chen, M.H., Kira, Z., AlRegib, G., Woo, J., Chen, R., Zheng, J.: Temporal attentive alignment for large-scale video domain adaptation. In: ICCV (2019)

4. Chen, M.H., Li, B., Bao, Y., AlRegib, G., Kira, Z.: Action segmentation with joint self-supervised temporal domain adaptation. In: CVPR (2020)
5. Chen, M., Xue, H., Cai, D.: Domain adaptation for semantic segmentation with maximum squares loss. In: ICCV (2019)
6. Chen, Y., Li, W., Sakaridis, C., Dai, D., Van Gool, L.: Domain adaptive faster r-cnn for object detection in the wild. In: CVPR (2018)
7. Chen, Y.C., Lin, Y.Y., Yang, M.H., Huang, J.B.: Crdoco: pixel-level domain transfer with cross-domain consistency. In: CVPR (2019)
8. Cho, K., et al.: Learning phrase representations using rnn encoder-decoder for statistical machine translation. In: EMNLP (2014)
9. Choi, J., Gao, C., Messou, J.C., Huang, J.B.: Why can't i dance in the mall? learning to mitigate scene bias in action recognition. In: NeurIPS (2019)
10. Choi, J., Sharma, G., Chandraker, M., Huang, J.B.: Unsupervised and semi-supervised domain adaptation for action recognition from drones. In: WACV (2020)
11. Feichtenhofer, C., Fan, H., Malik, J., He, K.: Slowfast networks for video recognition. In: ICCV (2019)
12. Gaidon, A., Harchaoui, Z., Schmid, C.: Temporal localization of actions with actoms. TPAMI **35**(11), 2782–2795 (2013)
13. Ganin, Y., Lempitsky, V.: Unsupervised domain adaptation by backpropagation. In: ICML (2015)
14. Gao, C., Zou, Y., Huang, J.B.: ican: instance-centric attention network for human-object interaction detection. In: BMVC (2018)
15. Gidaris, S., Singh, P., Komodakis, N.: Unsupervised representation learning by predicting image rotations. In: ICLR (2018)
16. Girdhar, R., Ramanan, D.: Attentional pooling for action recognition. In: NeurIPS (2017)
17. He, K., Gkioxari, G., Dollár, P., Girshick, R.: Mask R-CNN. In: ICCV (2017)
18. He, K., Zhang, X., Ren, S., Sun, J.: Deep residual learning for image recognition. In: CVPR (2016)
19. He, Z., Zhang, L.: Multi-adversarial faster-rcnn for unrestricted object detection. In: ICCV (2019)
20. Hoffman, J., Wang, D., Yu, F., Darrell, T.: Fcns in the wild: Pixel-level adversarial and constraint-based adaptation. arXiv preprint arXiv:1612.02649 (2016)
21. Hsu, H.K., et al.: Progressive domain adaptation for object detection. In: WACV (2020)
22. Huang, G., Liu, Z., Van Der Maaten, L., Weinberger, K.Q.: Densely connected convolutional networks. In: CVPR (2017)
23. Jamal, A., Namboodiri, V.P., Deodhare, D., Venkatesh, K.: Deep domain adaptation in action space. In: BMVC (2018)
24. Jetley, S., Lord, N.A., Lee, N., Torr, P.H.: Learn to pay attention. In: ICLR (2018)
25. Kar, A., Rai, N., Sikka, K., Sharma, G.: Adascan: adaptive scan pooling in deep convolutional neural networks for human action recognition in videos. In: CVPR (2017)
26. Khodabandeh, M., Vahdat, A., Ranjbar, M., Macready, W.G.: A robust learning approach to domain adaptive object detection. In: ICCV (2019)
27. Korbar, B., Tran, D., Torresani, L.: Scsampler: sampling salient clips from video for efficient action recognition. In: ICCV (2019)
28. Kuehne, H., Jhuang, H., Garrote, E., Poggio, T., Serre, T.: HMDB: a large video database for human motion recognition. In: ICCV (2011)
29. Larsson, G., Maire, M., Shakhnarovich, G.: Colorization as a proxy task for visual understanding. In: CVPR (2017)

30. Lee, H.Y., Huang, J.B., Singh, M., Yang, M.H.: Unsupervised representation learning by sorting sequences. In: ICCV (2017)
31. Li, Y., Vasconcelos, N.: Repair: removing representation bias by dataset resampling. In: CVPR (2019)
32. Li, Y., Li, Y., Vasconcelos, N.: Resound: towards action recognition without representation bias. In: ECCV (2018)
33. Luo, Z., Zou, Y., Hoffman, J., Fei-Fei, L.F.: Label efficient learning of transferable representations across domains and tasks. In: NeurIPS (2017)
34. Misra, I., Zitnick, C.L., Hebert, M.: Shuffle and learn: unsupervised learning using temporal order verification. In: Leibe, B., Matas, J., Sebe, N., Welling, M. (eds.) ECCV 2016. LNCS, vol. 9905, pp. 527–544. Springer, Cham (2016). https://doi.org/10.1007/978-3-319-46448-0_32
35. Munro, J., Damen, D.: Multi-modal domain adaptation for fine-grained action recognition. In: CVPR (2020)
36. Noroozi, M., Favaro, P.: Unsupervised learning of visual representations by solving jigsaw puzzles. In: Leibe, B., Matas, J., Sebe, N., Welling, M. (eds.) ECCV 2016. LNCS, vol. 9910, pp. 69–84. Springer, Cham (2016). https://doi.org/10.1007/978-3-319-46466-4_5
37. Pan, B., Cao, Z., Adeli, E., Niebles, J.C.: Adversarial cross-domain action recognition with co-attention. In: AAAI (2020)
38. Pathak, D., Krahenbuhl, P., Donahue, J., Darrell, T., Efros, A.A.: Context encoders: Feature learning by inpainting. In: CVPR (2016)
39. Ren, Z., Jae Lee, Y.: Cross-domain self-supervised multi-task feature learning using synthetic imagery. In: CVPR (2018)
40. Saito, K., Watanabe, K., Ushiku, Y., Harada, T.: Maximum classifier discrepancy for unsupervised domain adaptation. In: CVPR (2018)
41. Sikka, K., Sharma, G.: Discriminatively trained latent ordinal model for video classification. TPAMI **40**(8), 1829–1844 (2017)
42. Simonyan, K., Zisserman, A.: Two-stream convolutional networks for action recognition in videos. In: NeurIPS (2014)
43. Soomro, K., Zamir, A.R., Shah, M.: UCF101: A dataset of 101 human actions classes from videos in the wild. arXiv preprint arXiv:1212.0402 (2012)
44. Tran, D., Bourdev, L., Fergus, R., Torresani, L., Paluri, M.: Learning spatiotemporal features with 3d convolutional networks. In: ICCV (2015)
45. Tran, D., Wang, H., Torresani, L., Ray, J., LeCun, Y., Paluri, M.: A closer look at spatiotemporal convolutions for action recognition. In: CVPR (2017)
46. Tsai, Y.H., Sohn, K., Schulter, S., Chandraker, M.: Domain adaptation for structured output via discriminative representations. In: ICCV (2019)
47. Tzeng, E., Hoffman, J., Darrell, T., Saenko, K.: Simultaneous deep transfer across domains and tasks. In: ICCV (2015)
48. Tzeng, E., Hoffman, J., Saenko, K., Darrell, T.: Adversarial discriminative domain adaptation. In: CVPR (2017)
49. Vu, T.H., Jain, H., Bucher, M., Cord, M., Pérez, P.: Dada: Depth-aware domain adaptation in semantic segmentation. In: ICCV (2019)
50. Wang, J., Wang, W., Huang, Y., Wang, L., Tan, T.: M3: multimodal memory modelling for video captioning. In: CVPR (2018)
51. Wang, X., Girshick, R., Gupta, A., He, K.: Non-local neural networks. In: CVPR (2018)
52. Wang, Y., Hoai, M.: Pulling actions out of context: explicit separation for effective combination. In: CVPR (2018)

53. Xie, S., Sun, C., Huang, J., Tu, Z., Murphy, K.: Rethinking spatiotemporal feature learning for video understanding. In: ECCV (2018)
54. Xu, D., Xiao, J., Zhao, Z., Shao, J., Xie, D., Zhuang, Y.: Self-supervised spatiotemporal learning via video clip order prediction. In: CVPR (2019)
55. Xu, D., et al.: Video question answering via gradually refined attention over appearance and motion. In: ACM MM (2017)
56. Zhang, J., Li, W., Ogunbona, P.: Joint geometrical and statistical alignment for visual domain adaptation. In: CVPR (2017)
57. Zhang, Q., Zhang, J., Liu, W., Tao, D.: Category anchor-guided unsupervised domain adaptation for semantic segmentation. In: NeurIPS (2019)
58. Zhou, B., Andonian, A., Oliva, A., Torralba, A.: Temporal relational reasoning in videos. In: ECCV (2018)
59. Zhou, B., Khosla, A., Lapedriza, A., Oliva, A., Torralba, A.: Learning deep features for discriminative localization. In: CVPR (2016)
60. Zhu, X., Pang, J., Yang, C., Shi, J., Lin, D.: Adapting object detectors via selective cross-domain alignment. In: CVPR (2019)

DRG: Dual Relation Graph
for Human-Object Interaction Detection

Chen Gao$^{(\boxtimes)}$, Jiarui Xu, Yuliang Zou, and Jia-Bin Huang

Virginia Tech, Blacksburg, USA
{chengao,jiaruixu,ylzou,jbhuang}@vt.edu

Abstract. We tackle the challenging problem of human-object interaction (HOI) detection. Existing methods either recognize the interaction of each human-object pair in isolation or perform joint inference
based on complex appearance-based features. In this paper, we leverage an abstract spatial-semantic representation to describe each human-
object pair and aggregate the contextual information of the scene via a
dual relation graph (one *human-centric* and one *object-centric*). Our proposed dual relation graph effectively captures discriminative cues from
the scene to resolve ambiguity from local predictions. Our model is conceptually simple and leads to favorable results compared to the state-of-
the-art HOI detection algorithms on two large-scale benchmark datasets.

1 Introduction

Detecting individual persons and objects in isolation often does not provide sufficient information for understanding complex human activities. Moving beyond
detecting/recognizing individual objects, we aim to detect persons, objects, and
recognize their interaction relationships (if any) in the scene. This task, known
as human-object interaction (HOI) detection, can produce rich semantic information with visual grounding.

State-of-the-art HOI detection methods often use appearance features from
the detected human/object instances as well as their relative spatial layout
for predicting the interaction relationships [1,3,9,12–14,21,24,25,32,40,41,51].
These methods, however, often predict the interaction relationship between each
human-object pair *in isolation*, thereby ignoring the contextual information
in the scene. In light of this, several methods have been proposed to capture
the contextual cues through iterative message passing [34,43] or attentional
graph convolutional networks [44]. However, existing approaches rely on complex
appearance-based features to encode the human-object relation (e.g., deep features extracted from a union of two boxes) and do not exploit the informative
spatial cues. In addition, the contexts are aggregated via a *densely connected*
graph (where the nodes represent all the detected objects).

Electronic supplementary material The online version of this chapter (https://
doi.org/10.1007/978-3-030-58610-2_41) contains supplementary material, which is
available to authorized users.

© Springer Nature Switzerland AG 2020
A. Vedaldi et al. (Eds.): ECCV 2020, LNCS 12357, pp. 696–712, 2020.
https://doi.org/10.1007/978-3-030-58610-2_41

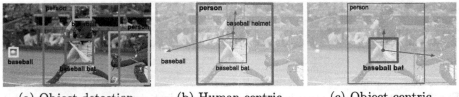

(a) Object detection   (b) Human-centric   (c) Object-centric

Fig. 1. Human-object interaction (HOI) detection using dual relation graph. Predicting each HOI in isolation is ambiguous due to the lack of context. In this work, we propose to leverage a dual relation graph. For each human node h, we obtain a *human-centric* subgraph where all object nodes are connected to h. Similarly, we can obtain an *object-centric* subgraph for each object node o. The human subgraph helps to adjust single HOI's prediction based on the same person's other HOIs. For example, knowing a person is wearing a baseball helmet and hitting a baseball suggests that the person may be holding a **baseball bat**. Similarly, the object subgraph helps to refine the HOI's prediction based on other HOIs associated with the same object. For example, knowing a **baseball bat** is held by a person lowers the chance that it is held by another person. Our method exploits such cues for improving HOI detection.

In this paper, we first propose to use spatial-semantic representation to describe each human-object pair. Specifically, our spatial-semantic representation encodes (1) the relative spatial layout between a person and an object and (2) the semantic word embedding of the object category. Using spatial-semantic representation for HOI prediction has two main advantages: First, it is invariant to complex appearance variations. Second, it enables knowledge transfer among object classes and helps with rare interaction during training and inference.

While such representations are informative, predicting HOI in isolation fails to leverage the contextual cues. In the example of Fig. 1, a model might struggle to recognize that the person (in the red box) is hitting the baseball, by using only the spatial-semantic features from this particular human-object pair. Such ambiguity, however, may be alleviated if given the relation among different HOIs *from the same person*, e.g., this person is wearing a baseball helmet and holding a baseball bat. Similarly, we can exploit the relations among different HOIs *from the same object*. For example, a model may recognize both persons are holding the same baseball bat when making prediction independently. Knowing that the person (red box) is more likely to hold the baseball bat reduces the probability of another person (blue box) holding the same baseball bat. Inspired by these observations, we construct a *human-centric* and an *object-centric* HOI subgraph and apply attentional graph convolution to encode and aggregate the contextual information. We refer to our method as Dual Relation Graph (DRG).

Our Contributions

- We propose Dual Relation Graph, an effective method to capture and aggregate contextual cues for improving HOI predictions.

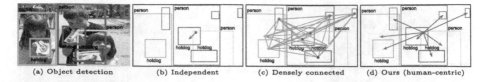

(a) Object detection   (b) Independent   (c) Densely connected   (d) Ours (human-centric)

Fig. 2. Leveraging contextual information. Given object detections in the scene (a), existing HOI detection algorithms only perform *independent* prediction for each Human-object pair (b), ignoring the rich contextual cues. Recent methods in visual relationship detection (or scene graph generation) perform *joint inference* on a densely connected graph (c). While being general, the large number of relations among the dense connections makes the learning and inference on such a graph challenging. In contrast, our work leverages the human/object-centric graph to focus only on relevant contexts for improved HOI detection (d).

- We demonstrate that using the proposed spatial-semantic representation alone (without using appearance features) can achieve competitive performance compared to the state-of-the-art.
- We conduct extensive ablation study of our model, identifying contributions from individual components and exploring different model design choices.
- We achieve competitive results compared with the state-of-the-art on the VCOCO and HICO-DET datasets.

2   Related Work

Human-Object Interaction Detection. The task of human-object interaction detection aims to localize persons, object instances, as well as recognize the interactions (if any) between each pair of a person and an object. State-of-the-art HOI detection algorithms generally rely on two types of visual cues: (1) appearance features of the detected persons and objects (e.g., using the ROI pooling features extracted from a ConvNet) and (2) the spatial relationship between each human-object pair (e.g., using the bounding box transformation between the agent and the object [12–14], a two-channel interaction pattern [3,9], or modeling the mutual contexts of human pose and object [14,24,46]). Recent advances focus on incorporating *contexts* to resolve potential ambiguity in interaction prediction based on independent human-object pairs, including pairwise body-parts [7,40] or object-parts [51], instance-centric attention [9,41], or message passing on a graph [34]. Our work shares similar spirits with these recent efforts as we also aim to capture contextual cues. The key difference lies in that the above approaches learn to aggregate contextual information from the other objects, body parts, or the scene background, while our method exploits *relations among different HOIs* to refine the predictions.

Inspired by the design of two-stage object detectors [35], recent works also show that filtering out candidate pairs with no relations using a relation proposal network [44] or an interactiveness network [24] improves the performance. Our method does not train an additional network for pruning unlikely relations. We believe that incorporating such a strategy may lead to further improvement.

Recent advances in HOI detection focus on tackling the long-tailed distributions of HOI classes. Examples include transferring knowledge from seen categories to unseen ones by an analogy transformation [31], performing data augmentation of semantically similar objects [1], or leveraging external knowledge graph [20]. While we do not explicitly address rare HOI classes, our method shows a small performance gap between rare and non-rare classes. We attribute this to the use of our abstract spatial-semantic representation.

Visual Relationship Detection. Many recent efforts have been devoted to detecting visual relationships [2,5,17,22,31,50,52]. Unlike object detection, the number of relationship classes can be prohibitively large due to the compositionality of object pairs, predicates, and limited data samples. To overcome this issue, some forms of language prior have been applied [27,33]. Our focus in this work is on one particular class of relationship: human-centric interactions. Compared with other object classes, the possible interactions (the predicate) between a person and objects are significantly more fine-grained.

Scene Graph. A scene graph is a graphical structure representation of an image where objects are represented as nodes, and the relationships between objects are represented as edges [30,43–45,49]. As the scene graph captures richer information beyond categorizing scene types or localizing object instances, it has been successfully applied to image retrieval [19], captioning [23], and generation [18]. Recent advances in scene graph generation leverage the idea of iterative message passing to capture contextual cues and produce a holistic interpretation of the scene [34,43,44,49]. Our work also exploits contextual information but has the following distinctions: (1) Unlike existing methods that apply message passing to update *appearance features* (e.g., the appearance feature extracted from the union of human-object pair) at each step, we use an abstract spatial-semantic representation with an *explicit* encoding of relative spatial layout. (2) In contrast to prior works that use a single densely connected graph structure where edges connecting all possible object pairs, we operate on human-centric and object-centric subgraphs to focus on relevant contextual information specifically for HOI. Figure 2 highlights the differences between methods that capture contextual cues.

The mechanisms for dynamically capturing contextual cues for resolving ambiguity in local predictions have also been successfully applied to sequence prediction [38], object detection [16], action recognition [10,37,42], and HOI detection [9]. Our dual relation graph shares a similar high-level idea with these approaches but with a focus on exploiting the contexts of spatial-semantic representations.

Visual Abstraction. The use of visual abstraction helps direct the focus to study the semantics of an image [53]. Visual abstraction has also been applied to learn common sense [39], forecasting object dynamics [8], and visual question

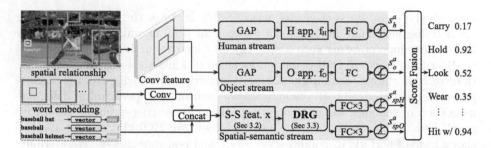

Fig. 3. Overview of the proposed model. Our network consists of three streams (human, object, and spatial-semantic). The human and object stream leverage the appearance feature $\mathbf{f}_h$ and $\mathbf{f}_o$. The spatial-semantic stream makes a prediction from the abstract spatial-semantic feature $\mathbf{x}$. We apply our proposed dual relation graph (DRG) to this stream. The three streams predict the scores s_h^a, s_o^a, s_{spH}^a and s_{spO}^a, which are fused to form final prediction.

answering [47]. Our work leverages the contexts of an abstract representation between human-object pairs for detecting HOIs.

Spatial-Semantic Representation. Spatial-semantic representation has also been applied in other problem domains such as image search [28], multi-class object detection [6], and image captioning [48].

3   Method

In this section, we present our network for HOI detection (Fig. 3). We start with an overview of our network (Sect. 3.1). We then introduce the spatial-semantic representation (Sect. 3.2) and describe how we can leverage the proposed Dual Relation Graph (DRG) to propagate contextual information (Sect. 3.3). Finally, we outline our inference (Sect. 3.4) and the training procedure (Sect. 3.5).

3.1   Algorithm Overview

Figure 3 provides a high-level overview of our HOI detection network. We decompose the HOI detection problem into two steps: (1) object detection and (2) HOI prediction. Following Gao et al. [9], we first apply an off-the-shelf object detector Faster R-CNN [35] to detect all the human/object instances in an image. We denote $\mathbb{H}$ as the set of human detections, and $\mathbb{O}$ as the set of object detections. Note that "person" is also an object category. We denote b_h as the detected bounding box for a person and b_o for an object instance. We use s_h and s_o to denote the confidence scores produced by the object detector for a detected person b_h and an object b_o, respectively. Given the detected bounding boxes b_h and b_o, we first extract the ROI pooled features and pass them into the human and object stream. We then pass the detected bounding boxes as well as the object category information to the spatial-semantic stream. We apply the proposed

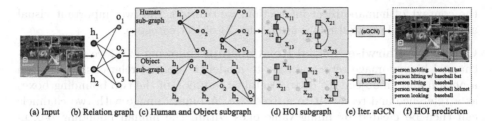

(a) Input   (b) Relation graph (c) Human and Object subgraph   (d) HOI subgraph   (e) Iter. aGCN   (f) HOI prediction

Fig. 4. HOI detection using Dual Relation Graph. (a) The input to our model are the detected objects in the given image. We denote $\mathbb{H}$ as the set of human detections, and $\mathbb{O}$ as the set of object detections. (b) We construct a *relation graph* from the detections where the two sets are $\mathbb{H}$ and $\mathbb{O}$. (c) For each human node h in $\mathbb{H}$, we obtain a human-centric sub-graph where all nodes in $\mathbb{O}$ are connected to h. Similarly, we can obtain an object-centric sub-graph for each object node o in $\mathbb{O}$. Note that "person" is also an object category. For simplicity, we do not show it in the figure. (d) In order to predict HOIs, we need to construct the HOI graph explicitly. Taking human sub-graph for example, we insert an HOI node x between human node h and object node o. We then connect all the HOI nodes and obtain the *human-centric* HOI sub-graph and the *object-centric* HOI sub-graph. (e) We iteratively update the HOI node feature via a trainable attentional graph convolutional network. This helps to aggregate the contextual information. (f) We fuse the scores from both sub-graphs and make the final HOI prediction.

Dual Relation Graph (DRG) in the spatial-semantic stream. Lastly, we fuse the action scores from the three streams (human, object, and spatial-semantic) to produce our final predictions.

Human and Object Stream. Our human/object streams follow the standard object detection pipeline for feature extraction and classification. For each ROI pooled human/object feature, we pass it into a one-layer MLP followed by global average pooling and obtain the human appearance feature $\mathbf{f}_h$ and the object appearance feature $\mathbf{f}_o$. We then apply a standard classification layer to obtain the A-dim action scores s_h^a (from human stream) and s_o^a (from object stream).

Spatial-Semantic Stream. Our inputs to this stream are the spatial-semantic features (described in Sect. 3.2). In an image, we pair all the detected persons in $\mathbb{H}$ with all the objects in $\mathbb{O}$, and extract spatial-semantic features for each human-object pair. We then pass all the features into our proposed Dual Relation Graph (Sect. 3.3) to aggregate the contextual information and produce updated features for each human-object pair. Our dual relation graph consists of a human-centric subgraph and an object-centric subgraph. These two subgraphs produce the action scores, s_{spH}^a and s_{spO}^a.

3.2 Spatial-Semantic Representation

We leverage the *abstract visual representation* of human-object pair for HOI prediction. The visual abstraction of a human-object pair allows us to construct representations that are invariant to intra-class appearance variations. In the context of human-object interaction, we consider the two important visual abstractions: (1) spatial relationship and (2) object category.[1]

Capturing Pairwise Spatial Relationship. Following [3,9], we use the two-channel binary image representation to model the spatial relationship between a person and an object. To do so, we take the union of the two bounding boxes as a reference and rescale it to a fixed size. A binary image with two channels can then be created by filling value ones within the human bounding box in the first channel and filling value ones within the object bounding box in the second channel. The remaining locations are then filled with value 0. We then feed these two-channel binary images into a simple two-layer ConvNet to extract the spatial relation feature.

Capturing Object Semantics. We find that using spatial features by itself leads to poor results in predicting the interaction. To address this issue, we augment the spatial feature with the word embedding of each object's category, $vector(o)$, using fastText [29]. Let $\mathbf{x}_{ij}$ denote the *spatial-semantic feature* between the i-th person and the j-th object. We construct $\mathbf{x}_{ij} \in \mathbb{R}^{5708}$ by concatenating (1) the spatial feature and (2) the 300-dimensional word embedding vector.

3.3   Dual Relation Graph

Here, we introduce the Dual Relation Graph for aggregating the spatial-semantic features. Figure 4 illustrates the overall process.

Relation Graph. Given the object detection results, i.e., instances in $\mathbb{H}$ and $\mathbb{O}$, we construct a relation graph (Fig. 4b). There are two types of nodes in this graph, $\mathbb{H}$ (human) and $\mathbb{O}$ (object). For each node h in $\mathbb{H}$, we connect it to all the other nodes in $\mathbb{O}$. Similarly, for each node o in $\mathbb{O}$, we connect it to all nodes in $\mathbb{H}$.

Human-Centric Subgraph and Object-Centric Subgraph. Unlike previous methods [34,44], we do not use the densely connected graphs. To exploit the relation among different HOIs performed by *the same person*, we construct a *human-centric* subgraph. Similarly, we construct an *object-centric* subgraph for the HOIs performed on *the same object* (Fig. 4c). So far, each node stands for an object instance detection. To explicitly represent the HOI, we insert an HOI node x_{ij} between each paired human node h_i and object node o_j. We then connect all the HOI nodes and obtain *human-centric* HOI subgraph and *object-centric* HOI subgraph (Fig. 4d). We use the before mentioned *spatial-semantic feature* $\mathbf{x}_{ij}$ to encode each HOI node between the i-th person and the j-th object.

Contextual Feature Aggregation. With these two HOI subgraphs, we follow a similar procedure for propagating and aggregating features as in relation

---

[1] Other types of abstracted representation such as the pose of the person, the attribute of the person/object can also be incorporated into our formulation. We leave this to future work.

network [16], non-local neural network [42], and attentional graph convolutional network [44].

Human-Centric HOI Subgraph. To update node x_{ij}, we aggregate all the spatial-semantic feature of the nodes involving the same i-th person $\{x_{ij'}|j' \in \mathcal{N}(j)\}$. The feature aggregation can be written as:

$$\mathbf{x}_{ij}^{(l+1)} = \sigma \left(\mathbf{x}_{ij}^{(l)} + \sum_{j' \in \mathcal{N}(j)} \alpha_{jj'} W \mathbf{x}_{ij'}^{(l)} \right), \tag{1}$$

where $W \in \mathbb{R}^{5708 \times 5708}$ is a learned linear transformation that projects features into the embedding space, $\alpha_{ij'}$ is a learned attention weight.

We can rewrite this equation compactly with matrix operation:

$$\mathbf{x}_{ij}^{(l+1)} = \sigma \left(W X^{(l)} \boldsymbol{\alpha}_j \right) \tag{2}$$

$$u_{jj'} = \left(W_q \mathbf{x}_{ij'}^{(l)} \right)^{\top} \left(W_k \mathbf{x}_{ij}^{(l)} \right) / \sqrt{d_k} \tag{3}$$

$$\boldsymbol{\alpha}_j = \text{softmax}(\boldsymbol{u}_j), \tag{4}$$

where $W_q, W_k \in \mathbb{R}^{1024 \times 5708}$ are linear projections that project feature into a query and a key embedding space. Following [38], we calculate the attention weights using scaled dot-product, normalized by $\sqrt{d_k}$ where $d_k = 1024$ is the dimension of the key embedding space. We do not directly use the aggregated feature $\sigma \left(W X^{(l)} \boldsymbol{\alpha}_j \right)$ as our output updated feature. Instead, we add it back to the original spatial-semantic feature $\mathbf{x}_{ij}^{(l)}$. We then pass the addition through a LayerNorm to get the final aggregated feature on the human-centric subgraph.

$$\mathbf{x}_{ij}^{(l+1)} = \text{LayerNorm} \left(\mathbf{x}_{ij}^{(l)} + \sigma \left(W X^{(l)} \boldsymbol{\alpha}_j \right) \right). \tag{5}$$

The linear transformation W does not change the dimension of the input feature, thus the output $\mathbf{x}_{ij}^{(l+1)}$ has the same size as input $\mathbf{x}_{ij}^{(l)}$. As a result, we can perform several iterations of feature aggregation (Fig. 4e). We explore the effectiveness of more iteration in Table 3(a).

Object-Centric HOI Subgraph. Similarly, to update node x_{ij}, we aggregate all the spatial-semantic feature of the nodes which involved the same object $\{x_{i'j}|i' \in \mathcal{N}(i)\}$. The two subgraphs have independent weights and aggregate contextual information independently.

3.4   Inference

For each human-object bounding box pair (b_h, b_o) in image I, we predict the score $S_{h,o}^a$ for each action $a \in \{1, \cdots, A\}$, where A denotes the total number of possible actions. The final score $S_{h,o}^a$ depends on (1) the confidence for the individual object detections (s_h and s_o), (2) the prediction score from the appearance of the

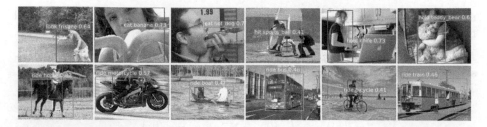

Fig. 5. Sample HOI detections on V-COCO (first row) and HICO-DET (second row) *test* set.

person s_h^a and the object s_o^a, and (3) the prediction score based on the aggregated spatial-semantic feature, using human-centric and object-centric subgraph, s_{spH}^a and s_{spO}^a. We compute the HOI score $S_{h,o}^a$ for the human-object pair (b_h, b_o) as

$$S_{h,o}^a = s_h \cdot s_o \cdot s_h^a \cdot s_o^a \cdot s_{spH}^a \cdot s_{spO}^a \tag{6}$$

Note that we are not able to obtain the action scores using object s_o^a or the spatial-semantic stream for some classes of actions as they do not involve any objects (e.g., walk, smile). For those cases, we use only the score s_h^a from the human stream. For those actions, our final scores are $s_h \cdot s_h^a$.

3.5 Training

HOI detection is a multi-label classification problem because a person can simultaneously perform different actions on different objects, e.g., sitting on a chair and reading a book. Thus, we minimize the cross-entropy loss *for each individual action class* between the ground-truth action label and the score produced from each stream. The total loss is the summation of the loss at each stream.

4 Experimental Results

In this section, we first outline our experimental setup, including datasets, metrics, and implementation details. We then report the quantitative results on two large-scale HOI benchmark datasets and compare the performance with the state-of-the-art HOI detection algorithms. Next, we show sample visual results on HOI detection. We conduct a detailed ablation study to quantify the contributions from individual components and validate our design choices. More results can be found in the supplementary material. We will make the source code and pre-trained models publicly available to foster future research.

4.1 Experimental Setup

Datasets. V-COCO dataset [13] is constructed by augmenting the COCO dataset [26] with additional human-object interaction annotations. Each person

Table 1. Comparison with the state-of-the-art on the V-COCO *test* set. The best performance is in **bold** and the second best is underscored. Character * indicates that the method uses both VCOCO and HICO-DET training data. "S-S only" shows the performance of our spatial-semantic stream.

Method	Use human pose	Feature backbone	AP_{role}
VSRL [13]	–	ResNet-50-FPN	31.8
InteractNet [12]	–	ResNet-50-FPN	40.0
BAR-CNN [21]	–	Inception-ResNet	41.1
GPNN [34]	–	ResNet-101	44.0
iCAN [9]	–	ResNet-50	45.3
Wang et al. [41]	–	ResNet-50	47.3
RPNN [51]	✓	ResNet-50	47.5
$RP_{T2}C_D$* [24]	✓	ResNet-50	48.7
PMFNet [40]	–	ResNet-50-FPN	48.6
PMFNet [40]	✓	ResNet-50-FPN	**52.0**
Ours (S-S only)	–	–	47.1
Ours	–	ResNet-50-FPN	51.0

is annotated with a binary label vector for 29 different action categories (five of them do not involve associated objects). **HICO-DET** [4] is a larger dataset containing 600 HOI categories over 80 object categories (same as [26]) with more 150 K annotated instances of human-object pairs. For applying our method on the HICO-DET dataset, we disentangle the 600 HOI categories into 117 object-agnostic action categories and train our network over these 117 action categories. At test time, we then combine the predicted action and the detected object and convert them back to the original 600 HOI classes. Note that the evaluation for the HICO-DET dataset remains the same.

Evaluation Metrics. To evaluate the performance of our model, we adopt the commonly used role mean average precision (role mAP) [13] for both V-COCO and HICO datasets. The goal is to detect and correctly predict the ⟨ human, verb, object ⟩ triplet. We consider a triplet as true positive if and only if it localizes the human and object accurately (i.e., with IoUs ≥ 0.5 w.r.t the ground truth annotations) and predicts the action correctly.

Implementation Details. We build our network with the publicly available PyTorch framework. Following Gao et al. [9], we use the Detectron [11] with a feature backbone of ResNet-50 to generate human and object bounding boxes. For VCOCO, we conduct an ablation study on the validation split to determine the best threshold. We keep the detected human boxes with scores s_h higher than 0.8 and object boxes with scores s_o higher than 0.1. For HICO-DET, since there is no validation split available, we follow the setting in [32]. We use the

Table 2. Comparison with the state-of-the-art on HICO-DET *test* set. The best performance is in bold and the second best is underscored. Character * indicates that the method uses both VCOCO and HICO-DET training data. For the object detector, "COCO" means that the detector is trained on COCO, while "HICO-DET" means that the detector is first pre-trained on COCO and then further fine-tuned on HICO-DET.

Method	Detector	Use human pose	Feature backbone	Default			Known object		
				Full	Rare	Non Rare	Full	Rare	Non Rare
Shen et al. [36]	COCO	–	VGG-19	6.46	4.24	7.12	–	–	–
HO-RCNN [3]	COCO	–	CaffeNet	7.81	5.37	8.54	10.41	8.94	10.85
InteractNet [12]	COCO	–	ResNet-50-FPN	9.94	7.16	10.77	–	–	–
GPNN [34]	COCO	–	ResNet-101	13.11	9.34	14.23	–	–	–
iCAN [9]	COCO	–	ResNet-50	14.84	10.45	16.15	16.26	11.33	17.73
Wang et al. [41]	COCO	–	ResNet-50	16.24	11.16	17.75	17.73	12.78	19.21
Bansal et al. [1]	COCO	–	ResNet-101	16.96	11.73	18.52	–	–	–
$RP_D C_D$ [24]	COCO	✓	ResNet-50	17.03	13.42	18.11	19.17	15.51	20.26
$RP_{T2} C_D$* [24]	COCO	✓	ResNet-50	17.22	13.51	18.32	19.38	15.38	20.57
no-frills [14]	COCO	✓	ResNet-152	17.18	12.17	18.68	–	–	–
RPNN [51]	COCO	✓	ResNet-50	17.35	12.78	18.71	–	–	–
PMFNet [40]	COCO	–	ResNet-50-FPN	14.92	11.42	15.96	18.83	15.30	19.89
PMFNet [40]	COCO	✓	ResNet-50-FPN	17.46	15.65	18.00	20.34	17.47	21.20
Peyre et al. [32]	COCO	–	ResNet-50-FPN	19.40	14.63	20.87	–	–	–
Ours (S-S only)	COCO	–	–	12.45	9.84	13.23	15.77	12.76	16.66
Ours	COCO	–	ResNet-50-FPN	19.26	17.74	19.71	23.40	21.75	23.89
Bansal et al. [1]	HICO-DET	–	ResNet-101	21.96	16.43	23.62	–	–	–
Ours	HICO-DET	–	ResNet-50-FPN	24.53	19.47	26.04	27.98	23.11	29.43

score threshold 0.6 to filter out unreliable human boxes and threshold 0.4 to filter out unconfident object boxes. To augment the training data, we apply random spatial jitterring to the human and object bounding boxes and ensure that the IOU with the ground truth bounding box is greater than 0.7. We pair all the detected human and objects, and regard those who are not ground truth as negative training examples. We keep the negative to positive ratio to three.

We initialize our appearance feature backbone with the COCO pre-trained weight from Mask R-CNN [15]. We perform two iterations of feature aggregation on both *human-centric* and *object-centric* subgraphs. We train the three streams (human appearance, object appearance, and spatial-semantic) using the V-COCO *train* set. We use early stopping criteria by monitoring the validation loss. We train our network with a learning rate of 0.0025, a weight decay of 0.0001, and a momentum of 0.9 on both the V-COCO *train* set and HICO-DET *train* set. Training our network takes 14 h on a single NVIDIA P100 GPU on V-COCO and 24 h on HICO-DET. At test time, our model runs at 3.3 fps for VCOCO and 5 fps for HICO-DET.

4.2 Quantitative Evaluation

We report the main quantitative results in terms of AP_{role} on V-COCO in Table 1 and HICO-DET in Table 2. For the V-COCO dataset, our method com-

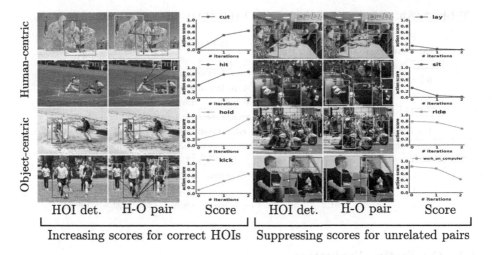

Fig. 6. More iteration of feature aggregation leads to a more accurate prediction. The human-centric and object-centric subgraph in the spatial-semantic stream propagates contextual information to produce increasingly accurate HOI predictions.

pares favorably against state-of-the-art algorithms [24,41,51] except PMFNet [40], which uses human pose as an additional feature. Since pose estimation required additional training data (with pose annotations), we expect to see performance gain using human pose. PMFNet [40] also reports the AP_{role} *without* human pose, which is 2.4 mAP lower to our method. We also note that the spatial-semantic stream alone *without* using any visual features achieves a competitive performance (47.1 mAP) when compared with the state-of-the-art. This highlights the effectiveness of the abstract spatial-semantic representation and contextual information. Compared with methods that perform joint inference on densely connected graph [34], our approach produces significant performance gains.

For the HICO-DET dataset, our method also achieves competitive performance with state-of-the-art methods [14,24,31,40]. Our method achieves the best performance for the *rare categories*, showing that our method handles the long-tailed distributions of HOI classes well.

We note that the current best performing model [1] uses an object detector which is fine-tuned on HICO-DET *train* set using the annotated object bounding boxes. For a fair comparison, we also fine-tune our object detector on HICO-DET and report our result. Note that we do *not* re-train our model, but only replace the object detector at the test time.

Here, the large performance gain from fine-tuning the object detector may not reflect the progress on the HOI detection task. This is because the objects (and the associated HOIs) in the HICO-DET dataset are *not* fully annotated. Using a fine-tuned object detector can thus improve the HOI detection performance by exploiting such annotation biases.

Table 3. Ablation study on the V-COCO *val* set. We show the role mAP AP_{role}.

(a) **More message passing iters.**

iter.	H graph	O graph	H + O
0-iter.	48.78	47.47	50.14
1-iter.	48.83	47.35	50.74
2-iter.	50.20	47.87	51.37

(b) **Feature used in DRG**

	mAP
App. feature (entire image)	35.69
App. feature (H-O union box)	46.93
Word2vec embedding	37.36
Spatial-semantic feature (ours)	51.37

(c) **Different subgraph**

H graph	O graph	mAP
-	-	50.14
✓	-	51.10
-	✓	50.78
✓	✓	51.37

(d) **Effectiveness of O subgraph**

	1-3	4-6	7+	all
H graph	57.89	52.77	50.96	51.10
H graph + O graph	58.28	53.75	51.06	51.37
Margin	+0.39	+0.98	+0.10	+0.27
% of testing images	68%	12%	20%	100%

4.3 Qualitative Evaluation

HOI Detection Results. Here we show sample results on the V-COCO dataset and the HICO-DET dataset in Fig. 5. We highlight the detected person and the associated object with red and green bounding boxes, respectively.

Visualizing the Effect of the Dual Relation Graph. In Fig. 6, we show the effectiveness of the proposed DRG. In the first two rows, we show that by aggregating contextual information, using the *human-centric* subgraph produces more accurate HOI predictions. Another example in the top right image indicates that the *human-centric* subgraph can also suppress the scores for unrelated human-object pairs. In the bottom two rows, we show four examples of how the *object-centric* subgraph propagates contextual information in each step to produce increasingly more accurate HOI predictions. For instance, the bottom right images show that for a person and an object without interaction, our model learns to suppress the predicted score by learning from the relationship of other HOI pairs associated with this particular object (laptop). In this example, the model starts with predicting a high score for the woman working on a computer. By learning from the relationship between the man and the computer, our model suppresses the score in each iteration.

4.4 Ablation Study

We examine several design choices using the V-COCO *val* set.

More Iteration of Feature Aggregation. Table 3(a) shows the performance using different iterations of feature aggregation. For either *human-centric* or *object-centric* subgraph, using more iterations of feature aggregation improves the overall performance. This highlights the advantages of exploiting contextual information among different HOIs. Performing feature aggregation on *both* subgraphs further improves the final performance.

Incorrect object   Incorrect action

Fig. 7. Failure cases of our method.

Effectiveness of Each Subgraph. To validate the effectiveness of the proposed subgraph, we show different variants of our model in Table 3(c). Adding only *human-centric* subgraph improves upon the baseline model (without using any subgraph) by 0.96 absolute mAP, while adding only *object-centric* subgraph gives a 0.64 absolute mAP. More importantly, our results show that the performance gain of each subgraph is complementary to each other. To further validate the effectiveness of the *object-centric* subgraph, we show in Table 3(d) the breakdown of Table 3(c) in terms of the number of persons in the scene. The *object-centric* subgraph is less effective for cases with few people. For example, if there is only one person, the *object-centric* subgraph has no effect. For images with a moderate amount of persons (4–6), however, our *object-centric* subgraph shows a clear 0.98 mAP gain. As the number of persons getting larger (7+), the *object-centric* subgraph shows a relatively smaller improvement due to clutter. Among the 2,867 testing images, 68% of them have only 1–3 persons. As a result, we do not see significant overall improvement.

Spatial-Semantic Representation. To demonstrate the advantage and effectiveness of the use of the abstract spatial-semantic representation, we show in Table 3(b) the comparison with alternative features, e.g., word2vec (as used in [27]) or appearance-based features. By using our spatial-semantic representation in the dual relation graph, we achieve 51.37 mAP. This shows a clear margin over the other alternative options, highlighting the contribution of spatial-semantic representation.

4.5   Limitations

While we demonstrated improved performance, our model is far from perfect. Below, we discuss two main limitations of our approach, with examples in Fig. 7.

First, we leverage the off-the-shelf object detector to detect object instances in an image. The object detection does *not* benefit from the rich contextual cues captured by our method. We believe that a joint end-to-end training approach may help reduce this type of errors.

Second, our model may be confused by plausible spatial configuration and predicts incorrect action. In the third image, our model predicts that the person is sitting on a bench even though our model confidently predicts this person is standing and catching a Frisbee. Capturing the statistics of co-occurring actions may resolve such mistakes.

5   Conclusions

In this paper, we present a Dual Relation Graph network for HOI detection. Our core idea is to exploit the *global object layout* as contextual cues and use a *human-centric* as well as an *object-centric* subgraph to propagate and integrate rich relations among individual HOIs. We validate the efficacy of our approach on two large-scale HOI benchmark datasets and show our model achieves a sizable performance boost over the state-of-the-art algorithms. We also find that using the abstract spatial-semantic representation alone (i.e., without the appearance features extracted from a deep CNN) yields competitive accuracy, demonstrating a promising path of activity understanding through visual abstraction.

Acknowledgements. We thank the support from Google Faculty Award.

References

1. Bansal, A., Rambhatla, S.S., Shrivastava, A., Chellappa, R.: Detecting human-object interactions via functional generalization. In: AAAI (2020)
2. Bilen, H., Vedaldi, A.: Weakly supervised deep detection networks. In: CVPR (2016)
3. Chao, Y.W., Liu, Y., Liu, X., Zeng, H., Deng, J.: Learning to detect human-object interactions. In: WACV (2017)
4. Chao, Y.W., Wang, Z., He, Y., Wang, J., Deng, J.: HICO: A benchmark for recognizing human-object interactions in images. In: CVPR (2015)
5. Dai, B., Zhang, Y., Lin, D.: Detecting visual relationships with deep relational networks. In: CVPR (2017)
6. Desai, C., Ramanan, D., Fowlkes, C.C.: Discriminative models for multi-class object layout. IJCV **95**(1), 1–12 (2011)
7. Fang, H.-S., Cao, J., Tai, Y.-W., Lu, C.: Pairwise body-part attention for recognizing human-object interactions. In: Ferrari, V., Hebert, M., Sminchisescu, C., Weiss, Y. (eds.) ECCV 2018. LNCS, vol. 11214, pp. 52–68. Springer, Cham (2018). https://doi.org/10.1007/978-3-030-01249-6_4
8. Fouhey, D.F., Zitnick, C.L.: Predicting object dynamics in scenes. In: CVPR (2014)
9. Gao, C., Zou, Y., Huang, J.B.: iCAN: instance-centric attention network for human-object interaction detection. In: BMVC (2018)
10. Girdhar, R., Carreira, J., Doersch, C., Zisserman, A.: Video action transformer network. In: CVPR (2019)
11. Girshick, R., Radosavovic, I., Gkioxari, G., Dollár, P., He, K.: Detectron (2018). https://github.com/facebookresearch/detectron
12. Gkioxari, G., Girshick, R., Dollár, P., He, K.: Detecting and recognizing human-object interactions. In: CVPR (2018)
13. Gupta, S., Malik, J.: Visual semantic role labeling. arXiv preprint arXiv:1505.04474 (2015)
14. Gupta, T., Schwing, A., Hoiem, D.: No-frills human-object interaction detection: factorization, appearance and layout encodings, and training techniques. In: ICCV (2019)
15. He, K., Gkioxari, G., Dollár, P., Girshick, R.: Mask R-CNN. In: ICCV (2017)

16. Hu, H., Gu, J., Zhang, Z., Dai, J., Wei, Y.: Relation networks for object detection. In: CVPR (2018)
17. Hu, R., Rohrbach, M., Andreas, J., Darrell, T., Saenko, K.: Modeling relationships in referential expressions with compositional modular networks. In: CVPR (2017)
18. Johnson, J., Gupta, A., Fei-Fei, L.: Image generation from scene graphs. In: CVPR (2018)
19. Johnson, J., et al.: Image retrieval using scene graphs. In: CVPR (2015)
20. Kato, K., Li, Y., Gupta, A.: Compositional learning for human object interaction. In: Ferrari, V., Hebert, M., Sminchisescu, C., Weiss, Y. (eds.) Computer Vision – ECCV 2018. LNCS, vol. 11218, pp. 247–264. Springer, Cham (2018). https://doi.org/10.1007/978-3-030-01264-9_15
21. Kolesnikov, A., Lampert, C.H., Ferrari, V.: Detecting visual relationships using box attention. In: ICCV (2019)
22. Li, Y., Ouyang, W., Wang, X., Tang, X.: VIP-CNN: visual phrase guided convolutional neural network. In: CVPR (2017)
23. Li, Y., Ouyang, W., Zhou, B., Wang, K., Wang, X.: Scene graph generation from objects, phrases and region captions. In: ICCV (2017)
24. Li, Y.L., et al.: Transferable interactiveness prior for human-object interaction detection. In: CVPR (2019)
25. Liao, Y., Liu, S., Wang, F., Chen, Y., Qian, C., Feng, J.: PPDM: parallel point detection and matching for real-time human-object interaction detection. In: CVPR (2020)
26. Lin, T.Y., et al.: Microsoft COCO: common objects in context. In: Fleet, D., Pajdla, T., Schiele, B., Tuytelaars, T. (eds.) ECCV 2014. LNCS, vol. 8693, pp. 740–755. Springer, Cham (2014). https://doi.org/10.1007/978-3-319-10602-1_48
27. Lu, C., Krishna, R., Bernstein, M., Fei-Fei, L.: Visual relationship detection with language priors. In: Leibe, B., Matas, J., Sebe, N., Welling, M. (eds.) ECCV 2016. LNCS, vol. 9905, pp. 852–869. Springer, Cham (2016). https://doi.org/10.1007/978-3-319-46448-0_51
28. Mai, L., Jin, H., Lin, Z., Fang, C., Brandt, J., Liu, F.: Spatial-semantic image search by visual feature synthesis. In: CVPR (2017)
29. Mikolov, T., Grave, E., Bojanowski, P., Puhrsch, C., Joulin, A.: Advances in pretraining distributed word representations. In: LREC (2018)
30. Newell, A., Deng, J.: Pixels to graphs by associative embedding. In: NeurIPS (2017)
31. Peyre, J., Laptev, I., Schmid, C., Sivic, J.: Weakly-supervised learning of visual relations. In: ICCV (2017)
32. Peyre, J., Laptev, I., Schmid, C., Sivic, J.: Detecting rare visual relations using analogies. In: ICCV (2019)
33. Plummer, B.A., Mallya, A., Cervantes, C.M., Hockenmaier, J., Lazebnik, S.: Phrase localization and visual relationship detection with comprehensive linguistic cues. In: ICCV (2017)
34. Qi, S., Wang, W., Jia, B., Shen, J., Zhu, S.-C.: Learning human-object interactions by graph parsing neural networks. In: Ferrari, V., Hebert, M., Sminchisescu, C., Weiss, Y. (eds.) ECCV 2018. LNCS, vol. 11213, pp. 407–423. Springer, Cham (2018). https://doi.org/10.1007/978-3-030-01240-3_25
35. Ren, S., He, K., Girshick, R., Sun, J.: Faster R-CNN: towards real-time object detection with region proposal networks. In: NeurIPS (2015)
36. Shen, L., Yeung, S., Hoffman, J., Mori, G., Fei-Fei, L.: Scaling human-object interaction recognition through zero-shot learning. In: WACV (2018)

37. Sun, C., Shrivastava, A., Vondrick, C., Murphy, K., Sukthankar, R., Schmid, C.: Actor-centric relation network. In: Ferrari, V., Hebert, M., Sminchisescu, C., Weiss, Y. (eds.) ECCV 2018. LNCS, vol. 11215, pp. 335–351. Springer, Cham (2018). https://doi.org/10.1007/978-3-030-01252-6_20
38. Vaswani, A., et al.: Attention is all you need. In: NeurIPS (2017)
39. Vedantam, R., Lin, X., Batra, T., Lawrence Zitnick, C., Parikh, D.: Learning common sense through visual abstraction. In: ICCV (2015)
40. Wan, B., Zhou, D., Zhou, Y., Li, R., He, X.: Pose-aware multi-level feature network for human object interaction detection. In: ICCV (2019)
41. Wang, T., et al.: Deep contextual attention for human-object interaction detection. In: ICCV (2019)
42. Wang, X., Girshick, R., Gupta, A., He, K.: Non-local neural networks. In: CVPR (2018)
43. Xu, D., Zhu, Y., Choy, C.B., Fei-Fei, L.: Scene graph generation by iterative message passing. In: CVPR (2017)
44. Yang, J., Lu, J., Lee, S., Batra, D., Parikh, D.: Graph R-CNN for scene graph generation. In: Ferrari, V., Hebert, M., Sminchisescu, C., Weiss, Y. (eds.) ECCV 2018. LNCS, vol. 11205, pp. 690–706. Springer, Cham (2018). https://doi.org/10.1007/978-3-030-01246-5_41
45. Yang, X., Zhang, H., Cai, J.: Shuffle-then-assemble: learning object-agnostic visual relationship features. In: Ferrari, V., Hebert, M., Sminchisescu, C., Weiss, Y. (eds.) ECCV 2018. LNCS, vol. 11216, pp. 38–54. Springer, Cham (2018). https://doi.org/10.1007/978-3-030-01258-8_3
46. Yao, B., Fei-Fei, L.: Modeling mutual context of object and human pose in human-object interaction activities. In: CVPR (2010)
47. Yi, K., Wu, J., Gan, C., Torralba, A., Kohli, P., Tenenbaum, J.B.: Neural-symbolic VQA: disentangling reasoning from vision and language understanding. In: NeurIPS (2018)
48. Yin, X., Ordonez, V.: Obj2text: generating visually descriptive language from object layouts. In: EMNLP (2017)
49. Zellers, R., Yatskar, M., Thomson, S., Choi, Y.: Neural motifs: scene graph parsing with global context. In: CVPR (2018)
50. Zhang, H., Kyaw, Z., Yu, J., Chang, S.F.: PPR-FCN: weakly supervised visual relation detection via parallel pairwise R-FCN. In: ICCV (2017)
51. Zhou, P., Chi, M.: Relation parsing neural network for human-object interaction detection. In: ICCV (2019)
52. Zhuang, B., Liu, L., Shen, C., Reid, I.: Towards context-aware interaction recognition for visual relationship detection. In: ICCV (2017)
53. Zitnick, C.L., Parikh, D.: Bringing semantics into focus using visual abstraction. In: CVPR (2013)

Flow-edge Guided Video Completion

Chen Gao[1($\boxtimes$)], Ayush Saraf[2], Jia-Bin Huang[1], and Johannes Kopf[2]

[1] Virginia Tech, Blacksburg, USA
chengao@vt.edu
[2] Facebook, Seattle, USA

Abstract. We present a new flow-based video completion algorithm. Previous flow completion methods are often unable to retain the sharpness of motion boundaries. Our method first extracts and completes motion edges, and then uses them to guide piecewise-smooth flow completion with sharp edges. Existing methods propagate colors among *local* flow connections between adjacent frames. However, not all missing regions in a video can be reached in this way because the motion boundaries form impenetrable barriers. Our method alleviates this problem by introducing *non-local* flow connections to temporally distant frames, enabling propagating video content over motion boundaries. We validate our approach on the DAVIS dataset. Both visual and quantitative results show that our method compares favorably against the state-of-the-art algorithms.

1 Introduction

Video completion is the task of filling a given space-time region with newly synthesized content. It has many applications, including restoration (removing scratches), video editing and special effects workflows (removing unwanted objects), watermark and logo removal, and video stabilization (filling the exterior after shake removal instead of cropping). The newly generated content should embed seamlessly in the video, and the alteration should be as imperceptible as possible. This is challenging because we need to ensure that the result is temporally coherent (does not flicker) and respects dynamic camera motion as well as complex object motion in the video.

Up until a few years ago, most methods used patch-based synthesis techniques [14,26,39]. These methods are often slow and have limited ability to synthesize new content because they can only remix existing patches in the video. Recent learning-based techniques achieve more plausible synthesis [5,38], but due to the high memory requirements of video, methods employing 3D spatial-temporal kernels suffer from resolution issues. The most successful methods to date [14,42] are flow-based. They synthesize color and flow jointly and propagate color along

---

Electronic supplementary material The online version of this chapter (https://doi.org/10.1007/978-3-030-58610-2_42) contains supplementary material, which is available to authorized users.

A. Vedaldi et al. (Eds.): ECCV 2020, LNCS 12357, pp. 713–729, 2020.
https://doi.org/10.1007/978-3-030-58610-2_42

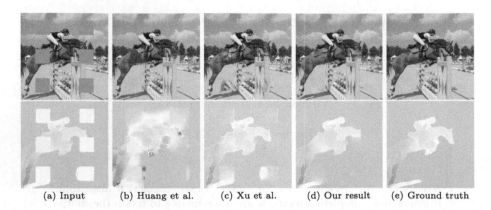

(a) Input   (b) Huang et al.   (c) Xu et al.   (d) Our result   (e) Ground truth

Fig. 1. Flow-edge guided video completion. Our new flow-based video completion method synthesizes sharper motion boundaries than previous methods and can propagate content across motion boundaries using non-local flow connections.

flow trajectories to improve temporal coherence, which alleviates memory problems and enables high-resolution output. Our method also follows this general approach.

The key to achieving good results with the flow-based approach is accurate flow completion, in particular, synthesizing sharp *flow edges* along the object boundaries. However, the aforementioned methods are not able to synthesize sharp flow edges and often produce over-smoothed results. While this still works when removing *entire* objects in front of *flat* backgrounds, it breaks down in more complex situations. For example, existing methods have difficulty in completing *partially* seen *dynamic* objects well (Fig. 1b–c). Notably, this situation is ubiquitous when completing *static screen-space masks*, such as logos or watermarks. In this work, we improve the flow completion by explicitly completing flow edges. We then use the completed flow edges to guide the flow completion, resulting in *piecewise-smooth* flow with sharp edges (Fig. 1d).

Another limitation of previous flow-based methods is that chained flow vectors between adjacent frames can only form *continuous* temporal constraints. This prevents constraining and propagating to many parts of a video. For example, considering the situation of the periodic leg motion of a walking person: here, the background is repeatedly visible between the legs, but the sweeping motion prevents forming continuous flow trajectories to reach (and fill) these areas. We alleviate this problem by introducing additional flow constraints to a set of *non-local* (i.e., temporally distant) frames. This creates short-cuts across flow barriers and propagates color to more parts of the video.

Finally, previous flow-based methods propagate color values directly. However, the color often subtly changes over time in a video due to effects such as lighting changes, shadows, lens vignetting, auto exposure, and white balancing, which can lead to visible color seams when combining colors propagated from

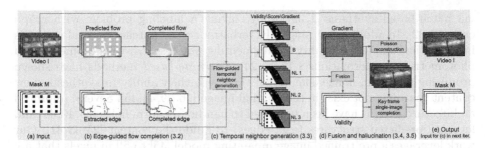

Fig. 2. Algorithm overview. (a) The input to our video completion method is a color video and a binary mask video that indicates which parts need to be synthesized. (b) We compute forward and backward flow between adjacent frames as well as a set of non-adjacent frames, extract and complete flow edges, and then use the completed edges to guide a piecewise-smooth flow completion (Sect. 3.2). (c) We follow the flow trajectories to compute a set of candidate pixels for each missing pixel. For each candidate, we estimate a confidence score as well as a binary validity indicator (Sect. 3.3). (d) We fuse the candidates in the gradient domain for each missing pixel using a confidence-weighted average. We pick a frame with most missing pixels and fill it with image inpainting (Sect. 3.4). (e) The result will be passed into the next iteration until there is no missing pixel (Sect. 3.5).

different frames. Our method reduces this problem by operating in the gradient domain.

In summary, our method alleviates the limitations of existing flow-based video completion algorithms through the following key contributions:

1. **Flow edges:** By explicitly completing flow edges, we obtain piecewise-smooth flow completion.
2. **Non-local flow:** We handle regions that cannot be reached through transitive flow (e.g., periodic motion, such as walking) by leveraging non-local flow.
3. **Seamless blending:** We avoid visible seams in our results through operating in the gradient domain.
4. **Memory efficiency:** Our method handles videos with up 4K resolution, while other methods fail due to excessive GPU memory requirements.

We validate the contribution of individual components to our results and show clear improvement over the prior methods in both quantitative evaluation and the quality of visual results.

2   Related Work

Image completion aims at filling missing regions in images with plausibly synthesized content. *Example-based* methods exploit the redundancy in natural images and transfer patches or segments from known regions to unknown (missing) regions [7,9]. These methods find correspondences for content transfer either via patch-based synthesis [1,39] or by solving a labeling problem with graph cuts [12,32]. In addition to using only verbatim copied patches, several methods improve the completion quality by augmenting patch search with geometric and photometric transformations[8,13,15,24]. *Learning-based* methods have shown promising results in image completion thanks to their ability to synthesize new content that may not exist in the original image [16,29,43,45]. Several improved architecture designs have been proposed to handle free-form holes [23,40,44] and leverage predicted structures (e.g., edges) to guide the content [25,33,41]. Our work leverages a pre-trained image inpainting model [45] to fill in pixels that are not filled through temporal propagation.

Video completion inherits the challenges from the image completion problems and introduces new ones due to the additional time dimension. Below, we only discuss the video completion methods that are most relevant to our work. We refer the readers to a survey [17] for a complete map of the field.

Patch-based synthesis techniques have been applied to video completion by using 3D (spatio-temporal) patches as the synthesis unit [26,39]. It is, however, challenging to handle dynamic videos (e.g., captured with a hand-held camera) with 3D patches, because they cannot adapt to deformations induced by camera motion. For this reason, several methods choose to fill the hole using 2D spatial patches and enforce temporal coherence with homography-based registration [10, 11] or explicit flow constraints [14,34,36]. In particular, Huang et al. [14] propose an optimization formulation that alternates between optical flow estimation and flow-guided patch-based synthesis. While the impressive results have been shown, the method is computationally expensive. Recent work [3,28] shows that the speed can be substantially improved by (1) decoupling the flow completion step from the color synthesis step and (2) removing patch-based synthesis (i.e., relying solely on flow-based color propagation). These flow-based methods, however, are unable to infer sharp flow edges in the missing regions and thus have difficulties synthesizing dynamic object boundaries. Our work focuses on overcoming the limitations of these flow-based methods.

Driven by the success of learning-based methods for visual synthesis, recent efforts have focused on developing CNN-based approaches for video completion. Several methods adopt 3D CNN architectures for extracting features and learning to reconstruct the missing content [5,38]. However, the use of 3D CNNs substantially limits the spatial (and temporal) resolution of the videos one can process due to the memory constraint. To alleviate this issue, the methods in [20,22,27] sample a small number of nearby frames as references. These methods, however, are unable to transfer temporally distant content due to the fixed temporal windows used by the method. Inspired by flow-based methods [3,14,28], Xu et al. [42] explicitly predict and complete dense flow field to facilitate propagating content from potentially distant frames to fill the missing regions. Our method builds upon the flow-based video completion formulation and makes several technical contributions to substantially improve the visual quality of completion, including completing edge-preserving flow fields, leveraging non-local flow, and gradient-domain processing for seamless results.

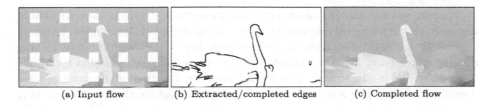

(a) Input flow (b) Extracted/completed edges (c) Completed flow

Fig. 3. Flow completion. (a) Optical flow estimated on the input video. Missing regions tend to have zero value (white). (b) Extracted and completed flow edges. (c) Piecewise-smooth completed flow, using the edges as guidance.

Gradient-domain processing techniques are indispensable tools for a wide variety of applications, including image editing [2,31], image-based rendering [21], blending stitched panorama [37], and seamlessly inserting moving objects in a video [6]. In the context of video completion, Poisson blending could be applied as a post-processing step to blend the synthesized content with the original video and hide the seams along the hole boundary. However, such an approach would not be sufficient because the propagated content from multiple frames may introduce visible seams *within* the hole that cannot be removed via Poisson blending. Our method alleviates this issue by propagating gradients (instead of colors) in our flow-based propagation process.

3 Method

3.1 Overview

The input to our video completion method is a color video and a binary mask video indicating which parts need to be synthesized (Fig. 2a). We refer to the masked pixels as the *missing* region and the others as the *known* region. Our method consists of the following three main steps. **(1) Flow completion:** We first compute forward and backward flow between adjacent frames as well as a set of non-adjacent ("non-local") frames, and complete the missing region in these flow fields (Sect. 3.2). Since edges are typically the most salient features in flow maps, we extract and complete them first. We then use the completed edges to produce piecewise-smooth flow completion (Fig. 2b). **(2) Temporal propagation:** Next, we follow the flow trajectories to propagate a set of candidate pixels for each missing pixel (Sect. 3.3). We obtain two candidates from chaining forward and backward flow vectors until a known pixel is reached. We obtain three additional candidates by checking three temporally distant frames with the help of non-local flow vectors. For each candidate, we estimate a confidence score as well as a binary validity indicator (Fig. 2c). **(3) Fusion:** We fuse the candidates for each missing pixel with at least one valid candidate using a confidence-weighted average (Sect. 3.4). We perform the fusion in the gradient domain to avoid visible color seams (Fig. 2d).

If there are still missing pixels after this procedure, it means that they could not be filled via temporal propagation (e.g., being occluded throughout the entire video). To handle these pixels, we pick a single key frame (with most remaining missing pixels) and fill it completely using a single-image completion technique (Sect. 3.5). We use this result as input for another iteration of the same process described above. The spatial completion step guarantees that we are making progress in each iteration, and its result will be propagated to the remainder of the video for enforcing temporal consistency in the next iteration. In the following sections, we provide more details about each of these steps.

3.2   Edge-Guided Flow Completion

The first step in our algorithm is to compute optical flow between adjacent frames as well as between several non-local frames (we explain how we choose the set of non-local connections in Sect. 3.3) and to complete the missing regions in the flow fields in an edge-guided manner.

Flow Computation. Let I_i and M_i be the color and mask of the i-th frame, respectively (we drop the subscript i if it is clear from the context), with $M(p) = 1$ if pixel p is missing, and 0 otherwise.

We compute the flow between adjacent frames i and j using the pretrained FlowNet2 [18] network $\mathcal{F}$:

$$F_{i \to j} = \mathcal{F}(I_i, I_j), \quad |i - j| = 1. \tag{1}$$

Note that we set the missing pixels in the color video to black, but we do not treat them in any special way except during flow computation. In these missing regions, the flow is typically estimated to be zero (white in visualizations, e.g., in Fig. 3a).

We notice that the flow estimation is *substantially degraded* or even *fails* in the presence of large motion, which frequently occurs in non-local frames. To alleviate this problem, we use a homography warp $\mathcal{H}_{j \to i}$ to compensate for the large motion between frame i and frame j (e.g., from camera rotation) before estimating the flow:

$$F_{i \to j} = \mathcal{F}(I_i, \mathcal{H}_{j \to i}(I_j)) + H_{i \to j}, \quad |i - j| > 1. \tag{2}$$

Since we are not interested in the flow between the homography-aligned frames but between the original frames, we add back the flow field $H_{i \to j}$ of the *inverse* homography transformation, i.e., mapping each flow vector back to the original pixel location in the unaligned frame j. We estimate the aligning homography using RANSAC on ORB feature matches [35]. This operation takes about 3% of the total computational time.

Flow Edge Completion. After estimating the flow fields, our next goal is to replace missing regions with plausible completions. We notice that the influence of missing regions extends slightly outside the masks (see the bulges in the white regions in Fig. 3a). Therefore, we dilate the masks by 15 pixels for flow completion. As can be seen in numerous examples throughout this paper, flow fields are generally piecewise-smooth, i.e., their gradients are small except along distinct motion boundaries, which are the most salient features in these maps. However, we observed that many prior flow-based video completion methods are unable to preserve sharp boundaries. To improve this, we first extract and complete the flow edges, and then use them as guidance for a piecewise-smooth completion of the flow values.

We use the Canny edge detector [4] to extract a flow edge map $E_{i \to j}$ (Fig. 3b, black lines). Note that we remove the edges of missing regions using the masks. We follow *EdgeConnect* [25] and train a flow edge completion network (See Sect. 4.1 for details). At inference time, the network predicts a completed edge map $\tilde{E}_{i \to j}$ (Fig. 3b, red lines).

Flow Completion. Now that we have hallucinated flow *edges* in the missing region, we are ready to complete the actual flow *values*. Since we are interested in a smooth completion except at the edges, we solve for a solution that minimizes the gradients everywhere (except at the edges). We obtain the completed flow $\tilde{F}$ by solving the following problem:

$$\operatorname*{argmin}_{\tilde{F}} \sum_{p \mid \tilde{E}(p)=1} \left\| \Delta_x \tilde{F}(p) \right\|_2^2 + \left\| \Delta_y \tilde{F}(p) \right\|_2^2,$$

$$\text{subject to} \ \ \tilde{F}(p) = F(p) \mid M(p) = 0, \ (3)$$

where Δ_x and Δ_y respectively denote the horizontal and vertical finite forward difference operator. The summation is over all non-edge pixels, and the boundary condition ensures a smooth continuation of the flow outside the mask. The solution to Eq. 3 is a set of sparse linear equations, which we solve using a standard linear least-squares solver. Figure 3c shows an example of flow completion.

3.3   Local and Non-local Temporal Neighbors

Now we can use the *completed* flow fields to guide the completion of the color video. This proceeds in two steps: for each missing pixel, we (1) find a set of known *temporal neighbor* pixels (this section), and (2) resolve a color by *fusing* the candidates using weighted averaging (Sect. 3.4).

The flow fields establish a connection between related pixels across frames, which are leveraged to guide the completion by propagating colors from known pixels through the missing regions along flow trajectories. Instead of *push*-propagating colors to the missing region (and suffering from repeated resampling), it is more desirable to transitively follow the forward and backward flow links for a given missing pixel, until known pixels are reached, and *pull* their colors.

We check the validity of the flow by measuring the forward-backward cycle consistency error,

$$\tilde{D}_{i \to j}(p) = \left\| F_{i \to j}(p) + F_{j \to i}\big(p + F_{i \to j}(p)\big) \right\|_2^2, \tag{4}$$

Non-local frame 1   Current frame   Non-local frame 3   space-time

Fig. 4. Non-local completion candidates. The right figure shows a *space-time* visualization for the highlighted scanlines in the left images. Green regions are missing. The yellow, orange and brown line in the right subfigure represents the scanline at the first non-local frame, the current frame and the third non-local frame, respectively. The figure illustrates the completion candidates for the red and **blue** pixels (large discs on the orange line). By following the flow trajectories (dashed black lines) until the edge of the missing region, we obtain *local* candidates for the **blue** pixel (small discs), but not for the red pixel, because the sweeping legs of the person form impassable flow barriers. With the help of the non-local flow that connects to the temporally distant frames, we obtain extra *non-local* neighbors for the red pixel (red discs on the yellow and brown line). As a result, we can reveal the true background that is covered by the sweeping legs. (Color figure online)

and stop the tracing if we encounter an error of more than $\tau = 5$ pixels. We call the known pixels that can be reached in this manner *local* temporal neighbors because they are computed by chaining flow vector between adjacent frames.

Sometimes, we might not be able to reach a local known pixel, either because the missing region extends to the end of the video, because of invalid flow, or because we encounter a *flow barrier*. Flow barriers occur at every major motion boundary because the occlusion/dis-occlusion breaks the forward/backward cycle consistency there. A typical example is shown in Fig. 4. Barriers can lead to large regions of *isolated* pixels without local temporal neighbors. Previous methods relied on hallucinations to generate content in these regions. However, hallucinations are more artifact-prone than propagation.

In particular, even if the synthesized content is plausible, it will most likely be different from the actual content visible *across* the barrier, which would lead to temporarily inconsistent results.

We alleviate this problem by introducing *non-local* temporal neighbors, i.e., computing flow to a set of temporally distant frames that short-cut across flow barriers, which dramatically reduces the number of isolated pixels and the need for hallucination. For every frame, we compute non-local flow to three additional frames using the homography-aligned method (Eq. 2). For simplicity, we always select the first, middle, and last frames of the video as non-local neighbors. Figure 5 shows an example.

Forward neighbor   Backward neighbor   Non-local neighbor 1   Non-local neighbor 2   Non-local neighbor 3

Fig. 5. Temporal neighbors. Non-local temporal neighbors (the second non-local neighbor in this case) are useful when correct known pixels cannot be reached with local flow chains due to flow barriers (red: invalid neighbors.) (Color figure online)

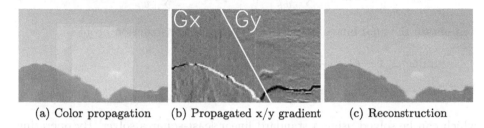

(a) Color propagation   (b) Propagated x/y gradient   (c) Reconstruction

Fig. 6. Gradient domain reconstruction. Previous methods operate directly in the color domain, which results in seams (a). We propagate in the gradient domain (b), and reconstruct the results by Poisson reconstruction (c).

Discussion: We experimented with adaptive schemes for non-local neighbor selection, but found that the added complexity was hardly justified for the relatively short video sequences we worked with in this paper. When working with longer videos, it might be necessary to resort to more sophisticated schemes, such as constant frame offsets, and possibly adding additional non-local frames.

3.4   Fusing Temporal Neighbors

Now that we have computed temporal neighbors for the missing pixels, we are ready to fuse them to synthesize the completed color values. For a given missing pixel p, let $k \in N(p)$ be the set of valid local and non-local temporal neighbors (we reject neighbors with flow error exceeding τ, and will explain how to deal with pixels that have no neighbors in Sect. 3.5). We compute the completed color as a weighted average of the candidate colors c_k,

$$\tilde{I}(p) = \frac{\sum_k w_k c_k}{\sum_k w_k}. \tag{5}$$

The weights, w_k are computed from the flow cycle consistency error:

$$w_k = \exp(-d_k/T), \tag{6}$$

where d_k is the consistency error $\tilde{D}_{i \to j}(p)$ for *non-local* neighbors, and the maximum of these errors along the chain of flow vectors for *local* neighbors. We set $T = 0.1$ to strongly down-weigh neighbors with large flow error.

Gradient-Domain Processing. We observed that directly propagating color values often yields visible seams, even with the correct flow. This is because of subtle color shifts in the input video (Fig. 6a). These frequently occur due to effects such as lighting changes, shadows, lens vignetting, auto exposure, and white balancing, etc. We address this issue by changing Eq. 5 to compute a weighted average of color *gradients*, rather than color values,

$$\tilde{G}_x(p) = \frac{\sum_k w_k \, \Delta_x c_k}{\sum_k w_k}, \quad \tilde{G}_y(p) = \frac{\sum_k w_k \, \Delta_y c_k}{\sum_k w_k}, \tag{7}$$

and obtain the final image by solving a Poisson reconstruction problem,

$$\underset{\tilde{I}}{\operatorname{argmin}} \left\| \Delta_x \tilde{I} - \tilde{G}_x \right\|_2^2 + \left\| \Delta_y \tilde{I} - \tilde{G}_y \right\|_2^2,$$

$$\text{subject to } \tilde{I}(p) = I(p) \mid M(p) = 0, \tag{8}$$

which can be solved using a standard linear least-squares solver. By operating in the gradient domain (Fig. 6b), the color seams are suppressed (Fig. 6c).

3.5   Iterative Completion

In each iteration, we propagate color gradients and obtain (up to) five candidate gradients. Then we fuse all candidate gradients and obtain missing pixel color values by solving a Poisson reconstruction problem (Eq. 8). This will fill all the missing pixels that have valid temporal neighbors. Some missing pixels might not have any valid temporal neighbors, even with the non-local flow, which, for example, happens when the pixel is occluded in all non-local frames, or when the flow is incorrectly estimated. Similar to past work [14], we formulate this problem as a single-image completion task, and solve it with Deepfill [45]. However, if we would complete the remaining missing regions in *all* frames with this single-image method, the result would not be temporally coherent. Instead, we select only *one* frame with the most remaining missing pixels and complete it with the single-image method. Then, we feed the inpainting result as input to another iteration of our whole pipeline (with the notable exception of flow computation, which does not need to be recomputed). In this subsequent iteration, the single-image completed frame is treated as a known region, and its color gradients are coherently propagated to the surrounding frames.

The iterative completion process ends when there is no missing pixel. In practice, our algorithm needs around 5 iterations to fill all missing pixels in the video sequences we have tried. We have included the pseudo-code in the supplementary material, which summarizes the entire pipeline.

Table 1. Video completion results with two types of synthetic masks. We report the average PSNR, SSIM and LPIPS results with comparisons to existing methods on DAVIS dataset. The best performance is in **bold** and the second best is <u>underscored</u>. Missing entries indicate the method fails at the respective resolution.

| | 720 × 384 resolution | | | | | | 960 × 512 resolution | | | | | |
| | Stationary masks | | | Object masks | | | Stationary masks | | | Object masks | | |
	PSNR ↑	SSIM ↑	LPIPS ↓	PSNR ↑	SSIM ↑	LPIPS ↓	PSNR ↑	SSIM ↑	LPIPS ↓	PSNR ↑	SSIM ↑	LPIPS ↓
Kim et al. [20]	25.19	0.8229	0.301	28.07	0.8673	0.283	-	-	-	-	-	-
Newson et al. [26]	27.50	0.9070	<u>0.067</u>	32.65	0.9648	0.023	-	-	-	-	-	-
Xu et al. [42]	27.69	0.9264	0.077	<u>39.67</u>	<u>0.9894</u>	<u>0.008</u>	27.17	<u>0.9216</u>	<u>0.085</u>	<u>38.88</u>	<u>0.9882</u>	<u>0.009</u>
Lee et al. [22]	28.47	0.9170	0.111	35.76	0.9819	0.021	<u>28.08</u>	0.9141	0.117	35.34	0.9814	0.022
Huang et al. [14]	28.72	0.9256	0.070	34.64	0.9725	0.018	-	-	-	-	-	-
Oh et al. [27]	<u>30.28</u>	<u>0.9279</u>	0.082	33.78	0.9630	0.058	-	-	-	-	-	-
Ours	**31.38**	**0.9592**	**0.042**	**42.72**	**0.9917**	**0.007**	**30.91**	**0.9564**	**0.048**	**41.89**	**0.9910**	**0.007**

Fig. 7. Qualitative results. We show the results of stationary screen-space inpainting task (first three columns) and object removal task (last three columns).

4 Experimental Results

4.1 Experimental Setup

Scenarios. We consider two application scenarios for video completion: (1) screen-space mask inpainting and (2) object removal. For the inpainting setting, we generate a stationary mask with a uniform grid of 5 × 4 square blocks (see an example in Fig. 7). This setting simulates the tasks of watermark or subtitle removal. Recovering content from such holes is particularly challenging because it often requires synthesizing *partially* visible dynamic objects over their background. For the object removal setting, we aim at recovering the missing content from a dynamically moving mask that covers the *entire* foreground object. This task is relatively easier because, typically, the dominant dynamic object is removed entirely. Results in the object removal setting, however, are difficult to compare and evaluate due to the lack of ground truth content behind the masked object. For this reason, we introduce a further *synthetic* object mask inpainting task. Specifically, we take a collection of free-form object masks and randomly pair them with other videos, pretending there is an object occluding the scene.

Evaluation Metrics. For tasks where the ground truth is available (stationary mask inpainting and object mask inpainting), we quantify the quality of the completed video using PSNR, SSIM, and LPIPS [46]. For LPIPS, we follow the

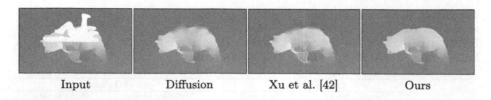

Input   Diffusion   Xu et al. [42]   Ours

Fig. 8. Flow completion. Comparing different methods for flow completion. Our method has better ability to retain the piecewise-smooth nature of flow fields (sharp motion boundaries, smooth everywhere else) than the other two methods.

default setting; we use Alexnet as the backbone, and we add a linear calibration on top of intermediate features.

Dataset. We evaluate our method on the DAVIS dataset [30], which contains a total of 150 video sequences. Following the evaluation protocol in [42], we use the 60 sequences in `2017-test-dev` and `2017-test-challenge` for training our flow edge completion network. We use the 90 sequences in `2017-train` and `2017-val` for testing the stationary mask inpainting task. For the object removal task, we test on the 29 out of the 90 sequences for which refined masks provided by Huang et al.[14] are available (these masks include shadows cast by the foreground object). For the object mask inpainting task, we randomly pair these 29 video sequences with mask sequences from the same set that have the same or longer duration. We resize the object masks by a uniform random factor in [0.8, 1], and trim them to match the number of frames. We resize all sequences to 960×512.

Implementation Details. We build our flow edge completion network upon the publicly available official implementation of EdgeConnect [25][1]. We use the following parameters for the Canny edge detector [4]: Gaussian $\sigma = 1$, low threshold 0.1, high threshold 0.2. We run the Canny edge detector on the flow magnitude image. In addition to the mask and edge images, EdgeConnect takes a "grayscale" image as additional input; we substitute the flow magnitude image for it. We load weights pretrained on the Places2 dataset [47], and then fine-tune on 60 sequences in DAVIS `2017-test-dev` and `2017-test-challenge` for 3 epochs. We adopt masks from NVIDIA Irregular Mask Dataset testing split[2]. During training, we first crop the edge images and corresponding flow magnitude images to 256×256 patches. Then we corrupt them with a randomly chosen mask, which is resized to 256×256. We use the ADAM optimizer with a learning rate of 0.001. Training our network takes 12 h on a single NVIDIA P100 GPU.

4.2 Quantitative Evaluation

We report quantitative results under the stationary mask inpainting and object mask inpainting setting in Table 1. Because not all methods were able to

---

[1] https://github.com/knazeri/edge-connect.
[2] https://www.dropbox.com/s/01dfayns9s0kevy/test_mask.zip?dl=0.

| Without non-local neighbors | With non-local neighbors |

Fig. 9. Non-local temporal neighbor ablation. Video completion results *with* and *without* non-local temporal neighbors. The result without non-local neighbors (left) does not recover well from the lack of well-propagated content.

Table 2. Ablation study. We report the average scores on DAVIS.

(a) **Domain and non-local**

Gradient	Non-local	Stationary masks			Object masks		
		PSNR ↑	SSIM ↑	LPIPS ↓	PSNR ↑	SSIM ↑	LPIPS ↓
-	-	28.28	0.9451	0.067	39.29	0.9893	0.009
-	✓	28.47	0.9469	0.069	39.67	0.9897	0.009
✓	-	30.78	0.9552	0.049	41.55	0.9907	0.007
✓	✓	30.91	0.9564	0.048	41.89	0.9910	0.007

(b) **Flow completion methods**

	Stationary masks				Object masks			
	Flow EPE ↓	PSNR ↑	SSIM ↑	LPIPS ↓	Flow EPE ↓	PSNR ↑	SSIM ↑	LPIPS ↓
Diffusion	1.79	30.18	0.9526	0.049	0.04	41.12	0.9902	0.008
Xu et al. [42]	2.01	27.17	0.9216	0.085	0.26	38.88	0.9882	0.009
Ours	1.63	30.91	0.9564	0.048	0.03	41.89	0.9910	0.007

handle the full 960×512 resolution due to memory constraint, we downscaled all scenes to 720×384 and reported numbers for both resolutions. Our method substantially improves the performance over state-of-the-art algorithms [14,20,22,26,27,42] on the three metrics. Following [14], we also show the detailed running time analysis of our method in the supplementary material. We report the time for each component of our method on the "CAMEL" video sequence under the object removal setting. Our method runs at 7.2 frames per minute.

4.3 Qualitative Evaluation

Figure 7 shows sample completion results for a diverse set of sequences. In all these cases, our method produces temporally coherent and visually plausible content. Please refer to the supplementary video results for extensive qualitative comparison to the methods listed in Table 1.

4.4 Ablation Study

In this section, we validate the effectiveness of our design choices.

Gradient Domain Processing. We compare the proposed gradient propagation process with color propagation (used in [14,42]). Figure 6 shows a visual comparison. When filling the missing region with directly propagated colors, the result contains visible seams due to color differences in different source frames (Fig. 6a), which are removed when operating in the gradient domain (Fig. 6c). Table 2(a) analyzes the contribution of the gradient propagation quantitatively.

Non-local Temporal Neighbors. We study the effectiveness of the non-local temporal neighbors. Table 2(a) shows the quantitative comparisons. The overall

Large stationary hole   Semantic structure   Fast motion

Fig. 10. Failure cases. Left, middle: hallucinated content in large missing regions (i.e., not filled by propagation) is sometimes not plausible. Right: fast motion might lead to poorly estimated flow, which results in a poor color completion.

quantitative improvement is somewhat subtle because, in many simple scenarios, the forward/backward flow neighbors are sufficient for propagating the correct content. In challenging cases, the use of non-local neighbors helps substantially reduce artifacts when both forward and backward (transitively connected) flow neighbors are incorrect due to occlusion or not available. Figure 9 shows such an example. Using non-local neighbors enables us to transfers the correct contents from temporally distant frames.

Edge-Guided Flow Completion. We evaluate the performance of completing the flow field with different methods. In Fig. 8, we show two examples of flow completion results using diffusion (essentially Eq. 3 without edge guidance), a trained flow completion network [42], and our proposed edge-guided flow completion. The diffusion-based method maximizes smoothness in the flow field everywhere and thus cannot create motion boundaries. The learning-based flow completion network [42] fails to predict a smooth flow field and sharp flow edges. In contrast, the proposed edge-guided flow completion fills the missing region with a piecewise-smooth flow and no visible seams along the hole boundary. Table 2(b) reports the endpoint error (EPE) between the pseudo ground truth flow (i.e., flow computed from the original, uncorrupted videos using FlowNet2) and the completed flow. The results show that the proposed flow completion achieves significantly lower EPE errors than diffusion and the trained flow completion network [42]. As a result, our proposed flow completion method helps improve the quantitative results.

4.5   Limitations

Failure Results. Video completion remains a challenging problem. We show and explain several failure cases in Fig. 10.

Processing Speed. Our method runs at 0.12 fps, which is comparable to other flow-based methods. End-to-end models are relatively faster, e.g., Lee et al.[22]

runs at 0.405 fps, but with worse performance. We acknowledge our slightly slower running time to be a weakness.

4.6   Negative Results

We explored several alternatives to our design choices to improve the quality of our video completion results. Unfortunately, these changes either ended up degrading performance or not producing clear improvement.

Flow Completion Network. As many CNN-based methods have shown impressive results on the task of image completion, using a CNN for flow completion seems a natural approach. We modified and experimented with several inpainting architectures, including partial conv [23] and EdgeConnect [25] for learning to complete the missing flow (by training on flow fields extracted from a large video dataset [19]). However, we found that in both cases, the network fails to generalize to unseen video sequences and produce visible seams along the hole boundaries.

Learning-Based Fusion. We explored using a U-Net based model for learning the weights for fusing the candidate (Sect. 3.4). Our model takes a forward-backward consistency error maps and the validity mask as inputs and predict the fusion weights so that the fused gradients are as similar to the ground truth gradients as possible. However, we did not observe a clear improvement from this learning-based method over the hand-crafted weights.

References

1. Barnes, C., Shechtman, E., Finkelstein, A., Goldman, D.B.: PatchMatch: a randomized correspondence algorithm for structural image editing. In: ACM TOG (Proceedings of the SIGGRAPH), vol. 28, p. 24 (2009)
2. Bhat, P., Zitnick, C.L., Cohen, M.F., Curless, B.: GradientShop: a gradient-domain optimization framework for image and video filtering. ACM TOG (Proc. SIGGRAPH) **29**(2), 10-1 (2010)
3. Bokov, A., Vatolin, D.: 100+ times faster video completion by optical-flow-guided variational refinement. In: ICIP (2018)
4. Canny, J.: A computational approach to edge detection. IEEE Trans. Pattern Anal. Mach. Intell. 679–698 (1986)
5. Chang, Y.L., Liu, Z.Y., Hsu, W.: Free-form video inpainting with 3D gated convolution and temporal PatchGAN. In: ICCV (2019)
6. Chen, T., Zhu, J.Y., Shamir, A., Hu, S.M.: Motion-aware gradient domain video composition. TIP **22**(7), 2532–2544 (2013)
7. Criminisi, A., Perez, P., Toyama, K.: Object removal by exemplar-based inpainting. In: CVPR (2003)
8. Darabi, S., Shechtman, E., Barnes, C., Goldman, D.B., Sen, P.: Image melding: combining inconsistent images using patch-based synthesis. ACM TOG (Proc. SIGGRAPH) **31**(4), 82-1 (2012)
9. Drori, I., Cohen-Or, D., Yeshurun, H.: Fragment-based image completion. In: ACM TOG (Proceedings of the SIGGRAPH), vol. 22, pp. 303–312 (2003)

10. Gao, C., Moore, B.E., Nadakuditi, R.R.: Augmented robust PCA for foreground-background separation on noisy, moving camera video. In: 2017 IEEE Global Conference on Signal and Information Processing (GlobalSIP) (2017)
11. Granados, M., Kim, K.I., Tompkin, J., Kautz, J., Theobalt, C.: Background inpainting for videos with dynamic objects and a free-moving camera. In: Fitzgibbon, A., Lazebnik, S., Perona, P., Sato, Y., Schmid, C. (eds.) ECCV 2012. LNCS, vol. 7572, pp. 682–695. Springer, Heidelberg (2012). https://doi.org/10.1007/978-3-642-33718-5_49
12. He, K., Sun, J.: Image completion approaches using the statistics of similar patches. TPAMI 36(12), 2423–2435 (2014)
13. Huang, J.B., Kang, S.B., Ahuja, N., Kopf, J.: Image completion using planar structure guidance. ACM TOG (Proc. SIGGRAPH) 33(4), 129 (2014)
14. Huang, J.B., Kang, S.B., Ahuja, N., Kopf, J.: Temporally coherent completion of dynamic video. ACM Trans. Graph. (TOG) (2016)
15. Huang, J.B., Kopf, J., Ahuja, N., Kang, S.B.: Transformation guided image completion. In: ICCP (2013)
16. Iizuka, S., Simo-Serra, E., Ishikawa, H.: Globally and locally consistent image completion. ACM TOG (Proc. SIGGRAPH) 36(4), 107 (2017)
17. Ilan, S., Shamir, A.: A survey on data-driven video completion. Comput. Graph. Forum 34, 60–85 (2015)
18. Ilg, E., Mayer, N., Saikia, T., Keuper, M., Dosovitskiy, A., Brox, T.: FlowNet 2.0: evolution of optical flow estimation with deep networks. In: CVPR (2017)
19. Kay, W., et al.: The kinetics human action video dataset. arXiv preprint arXiv:1705.06950 (2017)
20. Kim, D., Woo, S., Lee, J.Y., Kweon, I.S.: Deep video inpainting. In: CVPR (2019)
21. Kopf, J., Langguth, F., Scharstein, D., Szeliski, R., Goesele, M.: Image-based rendering in the gradient domain. ACM TOG (Proc. SIGGRAPH) 32(6), 199 (2013)
22. Lee, S., Oh, S.W., Won, D., Kim, S.J.: Copy-and-paste networks for deep video inpainting. In: ICCV (2019)
23. Liu, G., Reda, F.A., Shih, K.J., Wang, T.-C., Tao, A., Catanzaro, B.: Image inpainting for irregular holes using partial convolutions. In: Ferrari, V., Hebert, M., Sminchisescu, C., Weiss, Y. (eds.) ECCV 2018. LNCS, vol. 11215, pp. 89–105. Springer, Cham (2018). https://doi.org/10.1007/978-3-030-01252-6_6
24. Mansfield, A., Prasad, M., Rother, C., Sharp, T., Kohli, P., Van Gool, L.J.: Transforming image completion. In: BMVC (2011)
25. Nazeri, K., Ng, E., Joseph, T., Qureshi, F., Ebrahimi, M.: EdgeConnect: generative image inpainting with adversarial edge learning. In: ICCVW (2019)
26. Newson, A., Almansa, A., Fradet, M., Gousseau, Y., Pérez, P.: Video inpainting of complex scenes. SIAM J. Imaging Sci. (2014)
27. Oh, S.W., Lee, S., Lee, J.Y., Kim, S.J.: Onion-peel networks for deep video completion. In: ICCV (2019)
28. Okabe, M., Noda, K., Dobashi, Y., Anjyo, K.: Interactive video completion. IEEE Comput. Graph. Appl. (2019)
29. Pathak, D., Krahenbuhl, P., Donahue, J., Darrell, T., Efros, A.A.: Context encoders: feature learning by inpainting. In: CVPR (2016)
30. Perazzi, F., Pont-Tuset, J., McWilliams, B., Van Gool, L., Gross, M., Sorkine-Hornung, A.: A benchmark dataset and evaluation methodology for video object segmentation. In: CVPR (2016)
31. Pérez, P., Gangnet, M., Blake, A.: Poisson image editing. ACM TOG (Proc. SIGGRAPH) 22(3), 313–318 (2003)

32. Pritch, Y., Kav-Venaki, E., Peleg, S.: Shift-map image editing. In: ICCV (2009)
33. Ren, Y., Yu, X., Zhang, R., Li, T.H., Liu, S., Li, G.: StructureFlow: image inpainting via structure-aware appearance flow. In: CVPR, pp. 181–190 (2019)
34. Roxas, M., Shiratori, T., Ikeuchi, K.: Video completion via spatio-temporally consistent motion inpainting. IPSJ Trans. Comput. Vis. Appl. (2014)
35. Rublee, E., Rabaud, V., Konolige, K., Bradski, G.: Orb: an efficient alternative to sift or surf. In: ICCV (2011)
36. Strobel, M., Diebold, J., Cremers, D.: Flow and color inpainting for video completion. In: Jiang, X., Hornegger, J., Koch, R. (eds.) GCPR 2014. LNCS, vol. 8753, pp. 293–304. Springer, Cham (2014). https://doi.org/10.1007/978-3-319-11752-2_23
37. Szeliski, R., Uyttendaele, M., Steedly, D.: Fast poisson blending using multi-splines. In: ICCP, pp. 1–8 (2011)
38. Wang, C., Huang, H., Han, X., Wang, J.: Video inpainting by jointly learning temporal structure and spatial details. In: AAAI (2019)
39. Wexler, Y., Shechtman, E., Irani, M.: Space-time completion of video. TPAMI **3**, 463–476 (2007)
40. Xie, C., et al.: Image inpainting with learnable bidirectional attention maps. In: ICCV (2019)
41. Xiong, W., et al.: Foreground-aware image inpainting. In: CVPR (2019)
42. Xu, R., Li, X., Zhou, B., Loy, C.C.: Deep flow-guided video inpainting. In: CVPR (2019)
43. Yan, Z., Li, X., Li, M., Zuo, W., Shan, S.: Shift-Net: image inpainting via deep feature rearrangement. In: Ferrari, V., Hebert, M., Sminchisescu, C., Weiss, Y. (eds.) Computer Vision – ECCV 2018. LNCS, vol. 11218, pp. 3–19. Springer, Cham (2018). https://doi.org/10.1007/978-3-030-01264-9_1
44. Yu, J., Lin, Z., Yang, J., Shen, X., Lu, X., Huang, T.S.: Free-form image inpainting with gated convolution. arXiv preprint arXiv:1806.03589 (2018)
45. Yu, J., Lin, Z., Yang, J., Shen, X., Lu, X., Huang, T.S.: Generative image inpainting with contextual attention. In: CVPR (2018)
46. Zhang, R., Isola, P., Efros, A.A., Shechtman, E., Wang, O.: The unreasonable effectiveness of deep features as a perceptual metric. In: CVPR (2018)
47. Zhou, B., Lapedriza, A., Khosla, A., Oliva, A., Torralba, A.: Places: a 10 million image database for scene recognition. IEEE Trans. Pattern Anal. Mach. Intell. **40**(6), 1452–1464 (2017)

End-to-End Trainable Deep Active Contour Models for Automated Image Segmentation: Delineating Buildings in Aerial Imagery

Ali Hatamizadeh[✉], Debleena Sengupta, and Demetri Terzopoulos

Computer Science Department, University of California,
Los Angeles, CA 90095, USA
{ahatamiz,debleenas,dt}@cs.ucla.edu

Abstract. The automated segmentation of buildings in remote sensing imagery is a challenging task that requires the accurate delineation of multiple building instances over typically large image areas. Manual methods are often laborious and current deep-learning-based approaches fail to delineate all building instances and do so with adequate accuracy. As a solution, we present Trainable Deep Active Contours (TDACs), an automatic image segmentation framework that intimately unites Convolutional Neural Networks (CNNs) and Active Contour Models (ACMs). The Eulerian energy functional of the ACM component includes per-pixel parameter maps that are predicted by the backbone CNN, which also initializes the ACM. Importantly, both the ACM and CNN components are fully implemented in TensorFlow and the entire TDAC architecture is end-to-end automatically differentiable and backpropagation trainable without user intervention. TDAC yields fast, accurate, and fully automatic simultaneous delineation of arbitrarily many buildings in the image. We validate the model on two publicly available aerial image datasets for building segmentation, and our results demonstrate that TDAC establishes a new state-of-the-art performance.

Keywords: Computer vision · Image segmentation · Active contour models · Convolutional neural networks · Building delineation

1 Introduction

The delineation of buildings in remote sensing imagery [24] is a crucial step in applications such as urban planning [29], land cover analysis [35], and disaster relief response [28], among others. Manual or semi-automated approaches can be very slow, laborious, and sometimes imprecise, which can be detrimental to

Electronic supplementary material The online version of this chapter (https://doi.org/10.1007/978-3-030-58610-2_43) contains supplementary material, which is available to authorized users.

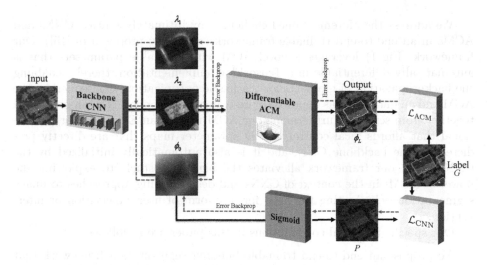

Fig. 1. TDAC is a fully-automated, end-to-end automatically differentiable and back-propagation trainable ACM and backbone CNN framework.

the prompt, accurate extraction of situational information from high-resolution aerial and satellite images.

Convolutional Neural Networks (CNNs) and deep learning have been broadly applied to various computer vision tasks, including semantic and instance segmentation of natural images in general [6,7] and particularly to the segmentation of buildings in remote sensing imagery [3,28]. However, building segmentation challenges CNNs. First, since CNN architectures often include millions of trainable parameters, successful training relies on large, accurately-annotated datasets; but creating such datasets from high-resolution imagery with possibly many building instances is very tedious. Second, CNNs rely on a filter learning approach in which edge and texture features are learned together, which adversely impacts the ability to properly delineate buildings and capture the details of their boundaries [11,16].

One of the most influential computer vision techniques, the Active Contour Model (ACM) [19], has been successfully employed in various image analysis tasks, including segmentation. In most ACM variants, the deformable curves of interest dynamically evolve according to an iterative procedure that minimizes a corresponding energy functional. Since the ACM is a model-based formulation founded on geometric and physical principles, the segmentation process relies mainly on the content of the image itself, not on learning from large annotated image datasets with hours or days of training and extensive computational resources. However, the classic ACM relies to some degree on user input to specify the initial contour and tune the parameters of the energy functional, which undermines its usefulness in tasks requiring the automatic segmentation of numerous images.

We address the aforementioned challenges by intimately uniting CNNs and ACMs in an end-to-end trainable framework (originally proposed in [15]). Our framework (Fig. 1) leverages a novel ACM with trainable parameters that is automatically differentiable in a TensorFlow implementation, thereby enabling the backpropagation of gradients for stochastic optimization. Consequently, the ACM and an untrained, as opposed to pre-trained, backbone CNN can be trained together from scratch. Furthermore, our ACM utilizes an Eulerian energy functional that affords local control via 2D parameter maps that are directly predicted by the backbone CNN, and it is also automatically initialized by the CNN. Thus, our framework alleviates the biggest obstacle to exploiting the power of ACMs in the context of CNNs and deep learning approaches to image segmentation—eliminating the need for any form of user supervision or intervention.

Our specific technical contributions in this paper are as follows:

- We propose an end-to-end trainable building segmentation framework that establishes a tight merger between the ACM and any backbone CNN in order to delineate buildings and accurately capture the fine-grained details of their boundaries.
- To this end, we devise an implicit ACM formulation with pixel-wise parameter maps and differentiable contour propagation steps for each term of the associated energy functional, thus making it amenable to TensorFlow implementation.
- We present new state-of-the-art benchmarks on two popular publicly available building segmentation datasets, *Vaihingen* and *Bing Huts*, with performance surpassing the best among competing methods [9,25].

2   Related Work

2.1   CNN-Based Building Segmentation Models

Audebert *et al.* [1] leveraged CNN-based models for building segmentation by applying SegNet [2] with multi-kernel convolutional layers at three different resolutions. Subsequently, Wang *et al.* [31] applied ResNet [17], first to identify the instances, followed by an MRF to refine the predicted masks. Some methods combine CNN-based models with classical optimization methods. Costa *et al.* [10] proposed a two-stage model in which they detect roads and intersections with a Dual-Hop Generative Adversarial Network (DH-GAN) at the pixel level and then apply a smoothing-based graph optimization to the pixel-wise segmentation to determine a best-covering road graph. Wu *et al.* [33] employed a U-Net encoder-decoder architecture with loss layers at different scales to progressively refine the segmentation masks. Xu *et al.* [34] proposed a cascaded approach in which pre-processed hand-crafted features are fed into a Residual U-Net to extract building locations and a guided filter refines the results.

In an effort to address the problem of poor boundary predictions by CNN models, Bischke *et al.* [3] proposed a cascaded multi-task loss function to simultaneously predict the semantic masks and distance classes. Recently, Rudner *et al.* [28] proposed a method to segment flooded buildings using multiple streams of encoder-decoder architectures that extract spatiotemporal information from medium-resolution images and spatial information from high-resolution images along with a context aggregation module to effectively combine the learned feature map.

2.2   CNN/ACM Hybrid Models

Hu *et al.* [18] proposed a model in which the network learns a level-set function for salient objects; however, the authors predefined a fixed scalar weighting parameter λ, which will not be optimal for all cases in the analyzed set of images. Hatamizadeh *et al.* [14] connected the output of a CNN to an implicit ACM through spatially-varying functions for the λ parameters. Le *et al.* [22] proposed a framework for the task of semantic segmentation of natural images in which level-set ACMs are implemented as RNNs. There are three key differences between that effort and our TDAC: (1) TDAC does not reformulate ACMs as RNNs, which makes it more computationally efficient. (2) TDAC benefits from a novel, locally-parameterized energy functional, as opposed to constant weighted parameters (3) TDAC has an entirely different pipeline—we employ a single CNN that is trained from scratch along with the ACM, as opposed to requiring two pre-trained CNN backbones. The dependence of [22] on pre-trained CNNs limits its applicability.

Marcos *et al.* [25] proposed Deep Structured Active Contours (DSAC), an integration of ACMs with CNNs in a structured prediction framework for building instance segmentation in aerial images. There are three key differences between that work and our TDAC: (1) TDAC is fully automated and runs without any external supervision, as opposed to depending heavily on the manual initialization of contours. (2) TDAC leverages the Eulerian ACM, which naturally segments multiple building instances simultaneously, as opposed to a Lagrangian formulation that can handle only a single building at a time. (3) Our approach fully automates the direct back-propagation of gradients through the entire TDAC framework due to its automatically differentiable ACM implementation.

Cheng *et al.* [9] proposed the Deep Active Ray Network (DarNet), which uses a polar coordinate ACM formulation to prevent the problem of self-intersection and employs a computationally expensive multiple initialization scheme to improve the performance of the proposed model. Like DSAC, DarNet can handle only single instances of buildings due to its explicit ACM formulation. Our approach is fundamentally different from DarNet, as (1) it uses an implicit ACM formulation that handles multiple building instances and (2) leverages a CNN to automatically and precisely initialize the implicit ACM.

Wang *et al.* [32] proposed an interactive object annotation framework for instance segmentation in which a backbone CNN and user input guide the evolution of an implicit ACM. Recently, Gur *et al.* [12] introduced Active Contours via Differentiable Rendering Network (ACDRNet) in which an explicit ACM

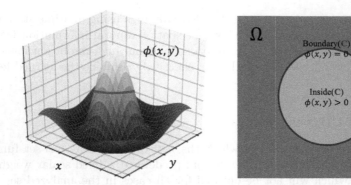

Fig. 2. Boundary C represented as the zero level-set of implicit function $\phi(x,y)$.

is represented by a "neural renderer" and a backbone encoder-decoder U-Net predicts a shift map to evolve the contour via edge displacement.

Some efforts have also focused on deriving new loss functions that are inspired by ACM principles. Inspired by the global energy formulation of [5], Chen *et al.* [8] proposed a supervised loss layer that incorporated area and size information of the predicted masks during training of a CNN and tackled a medical image segmentation task. Similarly, Gur *et al.* [13] presented an unsupervised loss function based on morphological active contours without edges [26].

3   The TDAC Model

3.1   Localized Level-Set ACM with Trainable Parameters

Our ACM formulation allows us to create a differentiable and trainable active contour model. Instead of working with a parametric contour that encloses the desired area to be segmented [19], we represent the contour(s) as the zero level-set of an implicit function. Such so-called "level-set active contours" evolve the segmentation boundary by evolving the implicit function so as to minimize an associated Eulerian energy functional.

The most well-known approaches that utilize this implicit formulation are geodesic active contours [4] and active contours without edges [5]. The latter, also known as the Chan-Vese model, relies on image intensity differences between the interior and exterior regions of the level set. Lankton and Tannenbaum [21] proposed a reformulation in which the energy functional incorporates image properties in the local region near the level set, which more accurately segments objects with heterogeneous features.[1]

---

[1] These approaches numerically solve the PDE that governs the evolution of the implicit function. Interestingly, Márquez-Neila *et al.* [26] proposed a morphological approach that approximates the numerical solution of the PDE by successive application of morphological operators defined on the equivalent binary level set.

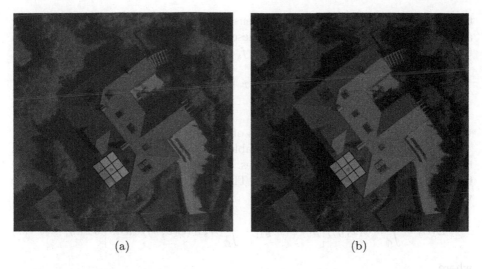

(a)   (b)

Fig. 3. The filter is divided by the contour into interior and exterior regions. The point x is represented by the red dot and the interior (a) and exterior (b) regions are shaded yellow. (Color figure online)

Let I represent an input image and $C = \{(x,y)|\phi(x,y) = 0\}$ be a closed contour in $\Omega \in R^2$ represented by the zero level set of the signed distance map $\phi(x,y)$ (Fig. 2). The interior and exterior of C are represented by $\phi(x,y) > 0$ and $\phi(x,y) < 0$, respectively. Following [5], we use a smoothed Heaviside function

$$H(\phi(x,y)) = \frac{1}{2} + \frac{1}{\pi} \arctan\left(\frac{\phi(x,y)}{\epsilon}\right) \tag{1}$$

to represent the interior as $H(\phi)$ and exterior as $(1 - H(\phi))$. The derivative of $H(\phi(x,y))$ is

$$\frac{\partial H(\phi(x,y))}{\partial \phi(x,y)} = \frac{1}{\pi}\frac{\epsilon}{\epsilon^2 + \phi(x,y)^2} = \delta(\phi(x,y)). \tag{2}$$

In TDAC, we evolve C to minimize an energy function according to

$$E(\phi) = E_{\text{length}}(\phi) + E_{\text{image}}(\phi), \tag{3}$$

where

$$E_{\text{length}}(\phi) = \int_{\Omega} \mu\delta(\phi(x,y))|\nabla\phi(x,y)|\,dx\,dy \tag{4}$$

penalizes the length of C whereas

$$E_{\text{image}}(\phi) = \int_{\Omega} \delta(\phi(x,y))\Big[H(\phi(x,y))(I(x,y) - m_1)^2 + (1 - H(\phi(x,y)))(I(x,y) - m_2)^2\Big]\,dx\,dy \tag{5}$$

takes into account the mean image intensities m_1 and m_2 of the regions interior and exterior to C [5]. We compute these local statistics using a characteristic function W_s with local window of size f_s (Fig. 3), as follows:

$$W_s = \begin{cases} 1 & \text{if } x - f_s \leq u \leq x + f_s, \quad y - f_s \leq v \leq y + f_s; \\ 0 & \text{otherwise,} \end{cases} \quad (6)$$

where x, y and u, v are the coordinates of two independent points.

To make our level-set ACM trainable, we associate parameter maps with the foreground and background energies. These maps, $\lambda_1(x, y)$ and $\lambda_2(x, y)$, are functions over the image domain Ω. Therefore, our energy function may be written as

$$E(\phi) = \int_\Omega \delta(\phi(x, y)) \left[\mu |\nabla\phi(x, y)| + \int_\Omega W_s F(\phi(u, v)) \, du \, dv \right] dx \, dy, \quad (7)$$

where

$$F(\phi) = \lambda_1(x, y)(I(u, v) - m_1(x, y))^2 (H(\phi(x, y)) \\ + \lambda_2(x, y)(I(u, v) - m_2(x, y))^2 (1 - H(\phi(x, y)). \quad (8)$$

The variational derivative of E with respect to ϕ yields the Euler-Lagrange PDE[2]

$$\frac{\partial\phi}{\partial t} = \delta(\phi) \left[\mu \text{div} \left(\frac{\nabla\phi}{|\nabla\phi|} \right) + \int_\Omega W_s \nabla_\phi F(\phi) \, dx \, dy \right] \quad (9)$$

with

$$\nabla_\phi F = \delta(\phi) \big(\lambda_1(x, y)(I(u, v) - m_1(x, y))^2 - \lambda_2(x, y)(I(u, v) - m_2(x, y))^2 \big). \quad (10)$$

To avoid numerical instabilities during the evolution and maintain a well-behaved $\phi(x, y)$, a distance regularization term [23] can be added to (9).

It is important to note that our formulation enables us to capture the fine-grained details of boundaries, and our use of *pixel-wise* parameter maps $\lambda_1(x, y)$ and $\lambda_2(x, y)$ allows them to be directly predicted by the backbone CNN along with an initialization map $\phi_0(x, y)$. Thus, not only does the implicit ACM propagation now become fully automated, but it can also be directly controlled by a CNN through these learnable parameter maps.

3.2   CNN Backbone

For the backbone CNN, we use a standard encoder-decoder with convolutional layers, residual blocks, and skip connections between the encoder and decoder. Each 3×3 convolutional layer is followed by ReLU activation and batch normalization. Each residual block consists of two 3×3 convolutional layers and an additive identity skip connection. As illustrated in Fig. 4, the first stage of the encoder comprises two 3×3 convolutional layers and a max pooling operation.

---

[2] The derivation is found in the supplemental document.

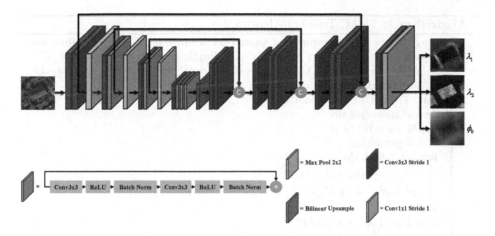

Fig. 4. TDAC's CNN backbone employs a standard encoder-decoder architecture.

Its second and third stages are comprised of a residual block followed by a max pooling operation. Each stage of the decoder performs a bilinear upsampling followed by two convolutional layers. The encoder is connected to the decoder via three residual blocks as well as skip connections at every stage. The output of the decoder is connected to a 1×1 convolution with three output channels for predicting the $\lambda_1(x, y)$ and $\lambda_2(x, y)$ parameter maps as well as the initialization map $\phi_0(x, y)$.[3]

3.3 Differentiable ACM

The ACM is evolved according to (9) in a differentiable manner in TensorFlow. The first term is computed according to the surface curvature expression:

$$\text{div}\left(\frac{\nabla\phi}{|\nabla\phi|}\right) = \frac{\phi_{xx}\phi_y^2 - 2\phi_{xy}\phi_x\phi_y + \phi_{yy}\phi_x^2}{(\phi_x^2 + \phi_y^2)^{3/2}}, \tag{11}$$

where the subscripts denote the spatial partial derivatives of ϕ, which are approximated using central finite differences. For the second term, convolutional operations are leveraged to efficiently compute $m_1(x, y)$ and $m_2(x, y)$ in (8) within image regions interior and exterior to C. Finally, $\partial\phi/\partial t$ in (9) is evaluated and $\phi(x, y)$ updated according to

$$\phi^t = \phi^{t-1} + \Delta t \frac{\partial\phi^{t-1}}{\partial t}, \tag{12}$$

where Δt is the size of the time step.

---

[3] Additional details about the CNN backbone are found in the supplemental document.

Algorithm 1: TDAC Training Algorithm

Data: I: Image; G: Corresponding ground truth label; g: ACM energy function
with parameter maps λ_1, λ_2; ϕ: ACM implicit function; L: Number of
ACM iterations; W: CNN with weights w; P: CNN prediction; $\mathcal{L}$: Total
loss function; η: Learning rate

Result: Trained TDAC model

while *not converged* **do**

    $\lambda_1, \lambda_2, \phi_0 = W(I)$

    $P = \text{Sigmoid}(\phi_0)$

    for $t = 1$ *to* L **do**

        $\frac{\partial \phi_{t-1}}{\partial t} = g(\phi_{t-1}; \lambda_1, \lambda_2, I)$

        $\phi^t = \phi^{t-1} + \Delta t \frac{\partial \phi^{t-1}}{\partial t}$

    end

    $\mathcal{L} = \mathcal{L}_{\text{ACM}}(\phi_L) + \mathcal{L}_{\text{CNN}}(P)$

    Compute $\frac{\partial \mathcal{L}}{\partial w}$ and backpropagate the error

    Update the weights of W: $w \leftarrow w - \eta \frac{\partial \mathcal{L}}{\partial w}$

end

3.4 TDAC Training

Referring to Fig. 1, we simultaneously train the CNN and level-set components
of TDAC in an end-to-end manner with no human intervention. The CNN guides
the ACM by predicting the $\lambda_1(x, y)$ and $\lambda_2(x, y)$ parameter maps, as well as an
initialization map $\phi_0(x, y)$ from which $\phi(x, y)$ evolves through the L layers of the
ACM in a differentiable manner, thus enabling training error backpropagation.
The $\phi_0(x, y)$ output of the CNN is also passed into a Sigmoid activation function
to produce the prediction P. Training optimizes a loss function that combines
binary cross entropy and Dice losses:

$$\hat{\mathcal{L}}(X) = -\frac{1}{N} \sum_{j=1}^{N} [X_j \log G_j + (1 - X_j) \log(1 - G_j)] + 1 - \frac{\sum_{j=1}^{N} 2X_j G_j}{\sum_{j=1}^{N} X_j + \sum_{j=1}^{N} G_j},$$

$$(13)$$

where X_j denotes the output prediction and G_j the corresponding ground truth
at pixel j, and N is the total number of pixels in the image. The total loss of
the TDAC model is

$$\mathcal{L} = \mathcal{L}_{\text{ACM}} + \mathcal{L}_{\text{CNN}},$$

$$(14)$$

where $\mathcal{L}_{\text{ACM}} = \hat{\mathcal{L}}(\phi_L)$ is the loss computed for the output ϕ_L from the final
ACM layer and $\mathcal{L}_{\text{CNN}} = \hat{\mathcal{L}}(P)$ is the loss computed over the prediction P of the
backbone CNN. Algorithm 1 presents the details of the TDAC training proce-
dure.

3.5 Implementation Details

We have implemented the TDAC architecture and training algorithm entirely
in TensorFlow. Our ACM implementation benefits from the automatic

differentiation utility of TensorFlow and has been designed to enable the back-propagation of the error gradient through the L layers of the ACM. We set $L = 60$ iterations in the ACM component of TDAC since, as will be discussed in Sect. 4.3, the performance does not seem to improve significantly with additional iterations. We set a filter size of $f = 5$, as discussed in Sect. 4.3. The training was performed on an Nvidia Titan RTX GPU, and an Intel® Core™ i7-7700K CPU @ 4.20 GHz. The size of the training minibatches for both datasets is 2. All the training sessions employ the Adam optimization algorithm [20] with an initial learning rate of $\alpha_0 = 0.001$ decreasing according to [27]

$$\alpha = \alpha_0 \left(1 - e/N_e\right)^{0.9} \tag{15}$$

with epoch counter e and total number of epochs N_e.

4   Empirical Study

4.1   Datasets

Vaihingen: The Vaihingen buildings dataset consists of 168 aerial images of size 512×512 pixels. Labels for each image were generated by using a semi-automated approach. We used 100 images for training and 68 for testing, following the same data partition as in [25]. In this dataset, almost all the images include multiple instances of buildings, some of which are located at image borders.

Bing Huts: The Bing Huts dataset consists of 605 aerial images of size 64×64 pixels. We followed the same data partition used in [25], employing 335 images for training and 270 images for testing. This dataset is especially challenging due the low spatial resolution and contrast of the images.

4.2   Evaluation Metrics

To evaluate TDAC's performance, we utilized four different metrics—Dice, mean Intersection over Union (mIoU), Boundary F (BoundF) [9], and Weighted Coverage (WCov) [30].

Given the prediction X and ground truth mask G, the Dice (F1) score is

$$\text{Dice}(X, G) = \frac{2 \sum_{i=1}^{N} X_i G_i}{\sum_{i=1}^{N} X_i + \sum_{i=1}^{N} G_i}, \tag{16}$$

where N is the number of image pixels and G_i and X_i denote pixels in G and X.

Similarly, the IoU score measures the overlap of two objects by calculating the ratio of intersection over union, according to

$$\text{IoU}(X, G) = \frac{|X \cap G|}{|X \cup G|}. \tag{17}$$

Table 1. Model Evaluations: Single-Instance Segmentation.

Model		Vaihingen				Bing Huts			
Method	Backbone	Dice	mIoU	WCov	BoundF	Dice	mIoU	WCov	BoundF
FCN	ResNet	84.20	75.60	77.50	38.30	79.90	68.40	76.14	39.19
FCN	Mask R-CNN	86.00	76.36	81.55	36.80	77.51	65.03	76.02	65.25
FCN	UNet	87.40	78.60	81.80	40.20	77.20	64.90	75.70	41.27
FCN	Ours	90.02	81.10	82.01	44.53	82.24	74.09	73.67	42.04
FCN	DSAC	–	81.00	81.40	64.60	–	69.80	73.60	30.30
FCN	DarNet	–	87.20	86.80	76.80	–	74.50	77.50	37.70
DSAC	DSAC	–	71.10	70.70	36.40	–	38.70	44.60	37.10
DSAC	DarNet	–	60.30	61.10	24.30	–	57.20	63.00	15.90
DarNet	DarNet	93.66	88.20	88.10	75.90	85.21	75.20	77.00	38.00
TDAC-const λs	Ours	91.18	83.79	82.70	73.21	84.53	73.02	74.21	48.25
TDAC	Ours	**94.26**	**89.16**	**90.54**	**78.12**	**89.12**	**80.39**	**81.05**	**53.50**

Table 2. Model Evaluations: Multiple-Instance Segmentation.

Model		Vaihingen				Bing Huts			
Method	Backbone	Dice	mIoU	WCov	BoundF	Dice	mIoU	WCov	BoundF
FCN	UNet	81.00	69.10	72.40	34.20	71.58	58.70	65.70	40.60
FCN	ResNet	80.10	67.80	70.50	32.50	74.20	61.80	66.59	39.48
FCN	Mask R-CNN	88.35	79.42	80.26	41.92	76.12	63.40	70.51	41.97
FCN	Ours	89.30	81.00	82.70	49.80	75.23	60.31	72.41	41.12
TDAC-const λs	Ours	90.80	83.30	83.90	47.20	81.19	68.34	75.29	44.61
TDAC	Ours	**95.20**	**91.10**	**91.71**	**69.02**	**83.24**	**71.30**	**78.45**	**48.49**

BoundF computes the average of Dice scores over 1 to 5 pixels around the boundaries of the ground truth segmentation.

In WConv, the maximum overlap output is selected and the IoU between the ground truth segmentation and best output is calculated. IoUs for all instances are summed up and weighted by the area of the ground truth instance. Assuming that $S_G = \{r_1^{S_G}, \ldots, r_{|S_G|}^{S_G}\}$ is a set of ground truth regions and $S_X = \{r_1^{S_X}, \ldots, r_{|S_X|}^{S_X}\}$ is a set of prediction regions for single image, and $|r_j^{S_G}|$ is the number of pixels in $r_j^{S_G}$, the weighted coverage can be expressed as

$$\text{WCov}(S_X, S_G) = \frac{1}{N} \sum_{j=1}^{|S_G|} |r_j^{S_G}| \max_{k=1\ldots|S_X|} \text{IoU}(r_k^{S_X}, r_j^{S_G}). \tag{18}$$

4.3 Experiments and Ablation Studies

Single-Instance Segmentation: Although most of the images in the Vaihingen dataset depict multiple instances of buildings, the DarNet and DSAC models can deal only with a single building instance at a time. For a fair comparison against

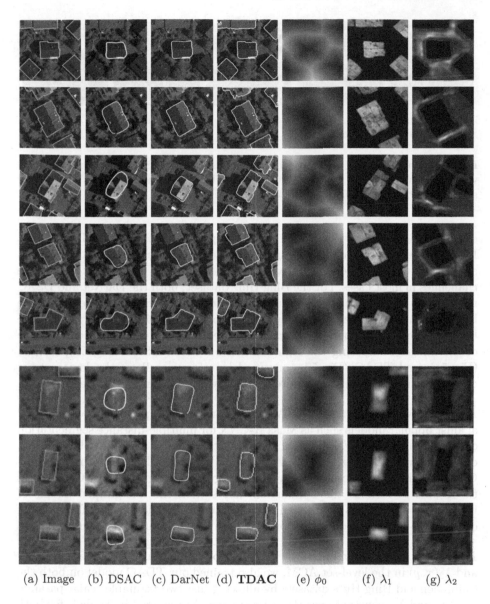

(a) Image (b) DSAC (c) DarNet (d) **TDAC**   (e) ϕ_0 (f) λ_1 (g) λ_2

Fig. 5. Comparative visualization of the labeled image and the outputs of DSAC, Dar-Net, and our TDAC for the Vaihingen (top) and Bing Huts (bottom) datasets. (a) Image labeled with (green) ground truth segmentation. (b) DSAC output. (c) Dar-Net output. (d) TDAC output. (e) TDAC's learned initialization map $\phi_0(x, y)$ and parameter maps (f) $\lambda_1(x, y)$ and (g) $\lambda_2(x, y)$. (Color figure online)

these models, we report single-instance segmentation results in the exact same manner as [25] and [9]. As reported in Table 1, our TDAC model outperforms both DarNet and DSAC in all metrics on both the Vaihingen and Bing Huts

742   A. Hatamizadeh et al.

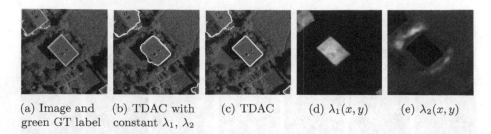

(a) Image and   (b) TDAC with   (c) TDAC   (d) $\lambda_1(x,y)$ (e) $\lambda_2(x,y)$
green GT label   constant λ_1, λ_2

Fig. 6. (a) Image labeled with (green) ground truth segmentation. (b) Output of TDAC with constant λ_1 and λ_2. (c) TDAC output and learned parameter maps (d) $\lambda_1(x,y)$ and (e) $\lambda_2(x,y)$. (Color figure online)

datasets. Figure 5 shows that with the Vaihingen dataset, both the DarNet and DSAC models have difficulty coping with the topological changes of the buildings and fail to appropriately capture sharp edges, while TDAC overcomes these challenges in most cases. For the Bing Huts dataset, both the DarNet and DSAC models are able to localize the buildings, but they inaccurately delineate the buildings in many cases. This may be due to their inability to distinguish the building from the surrounding terrain because of the low contrast and small size of the image. Comparing the segmentation output of DSAC (Fig. 5b), DarNet (Fig. 5c), and TDAC (Fig. 5d), our model performs well on the low contrast dataset, delineating buildings more accurately than the earlier models.

Multiple-Instance Segmentation: We next compare the performance of TDAC against popular models such as Mask R-CNN for multiple-instance segmentation of all buildings in the Vaihingen and Bing Huts datasets. As reported in Table 2, our extensive benchmarks confirm that the TDAC model outperforms Mask R-CNN and the other methods by a wide margin. Although Mask R-CNN seems to be able to localize the building instances well, the fine-grained details of boundaries are lost, as is attested by the BoundF metric. The performance of other CNN-based approaches follow the same trend in our benchmarks.

Parameter Maps: To validate the contribution of the parameter maps $\lambda_1(x,y)$ and $\lambda_2(x,y)$ in the level-set ACM, we also trained our TDAC model on both the Vaihingen and Bing Huts datasets by allowing just two trainable scalar parameters, λ_1 and λ_2, constant over the entire image. As reported in Table 1, for both the Vaihingen and Bing Huts datasets, this "constant-λ" formulation (i.e., the Chan-Vese model [5,21]) still outperforms the baseline CNN in most evaluation metrics for both single-instance and multiple-instance buildings, thus establishing the effectiveness of the end-to-end training of the TDAC. Nevertheless, our TDAC with its full $\lambda_1(x,y)$ and $\lambda_2(x,y)$ maps outperforms this constant-λ version by a wide margin in all experiments and metrics. A key metric of interest in this comparison is the BoundF score, which elucidates that our formulation captures the details of the boundaries more effectively by locally adjusting the inward and outward forces on the contour. Figure 6 shows that our TDAC has

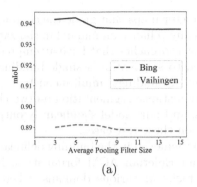 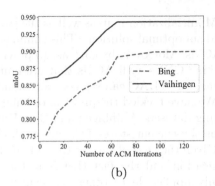

(a)   (b)

Fig. 7. The effects on mIoU of (a) varying the convolutional filter size and (b) varying the number L of ACM iterations.

well delineated the boundaries of the building instances, compared to the TDAC hobbled by the constant-λ formulation.

Convolutional Filter Size: The filter size of the convolutional operation is an important hyper-parameter for the accurate extraction of localized image statistics. As illustrated in Fig. 7a, we have investigated the effect of the convolutional filter size on the overall mIoU for both the Vaihingen and Bing datasets. Our experiments indicate that filter sizes that are too small are sub-optimal while excessively large sizes defeat the benefits of the localized formulation. Hence, we recommend a filter size of $f = 5$ for the TDAC.

Number of Iterations: The direct learning of an initialization map $\phi_0(x, y)$ as well as its efficient TensorFlow implementation have enabled the TDAC to require substantially fewer iterations to converge with a better chance of avoiding undesirable local minima. As shown in Fig. 7b, we have investigated the effect of the number of iterations on the overall mIoU for both Vaihingen and Bing datasets and our results reveal that TDAC exhibits a robust performance after a certain threshold. Therefore, we have chosen a fixed number of iterations (i.e., ACM layers) for optimal performance, $L = 60$, yielding a runtime of less than 1 sec in TensorFlow.

5   Conclusions and Future Work

We have introduced a novel image segmentation framework, called Trainable Deep Active Contour Models (TDACs), which is a full, end-to-end merger of ACMs and CNNs. To this end, we proposed a new, locally-parameterized, Eulerian ACM energy model that includes pixel-wise learnable parameter maps that can adjust the contour to precisely delineate the boundaries of objects of interest in the image. Our model is fully automatic, as its backbone CNN learns the

ACM initialization map as well as the parameter maps that guide the contour to avoid suboptimal solutions. This eliminates any reliance on manual initialization of ACMs. Moreover, by contrast to previous approaches that have attempted to combine CNNs with ACMs that are limited to segmenting a single building at a time, our TDAC can segment any number of buildings simultaneously.

We have tackled the problem of building instance segmentation on two challenging datasets, Vaihingen and Bing Huts, and our model significantly outperforms the current state-of-the-art methods on these test cases.

Given the level of success that TDAC has achieved in the building delineation application and the fact that it features an Eulerian ACM formulation, it is readily applicable to other segmentation tasks in various domains, wherever purely CNN filter-based approaches can benefit from the versatility and precision of ACMs to accurately delineate object boundaries in images.

References

1. Audebert, N., Le Saux, B., Lefèvre, S.: Semantic segmentation of earth observation data using multimodal and multi-scale deep networks. In: Lai, S.-H., Lepetit, V., Nishino, K., Sato, Y. (eds.) ACCV 2016. LNCS, vol. 10111, pp. 180–196. Springer, Cham (2017). https://doi.org/10.1007/978-3-319-54181-5_12

2. Badrinarayanan, V., Kendall, A., Cipolla, R.: SegNet: a deep convolutional encoder-decoder architecture for image segmentation. IEEE Trans. Pattern Anal. Mach. Intell. **39**(12), 2481–2495 (2017)

3. Bischke, B., Helber, P., Folz, J., Borth, D., Dengel, A.: Multi-task learning for segmentation of building footprints with deep neural networks. In: 2019 IEEE International Conference on Image Processing (ICIP), pp. 1480–1484. IEEE (2019)

4. Caselles, V., Kimmel, R., Sapiro, G.: Geodesic active contours. Int. J. Comput. Vis. **22**(1), 61–79 (1997)

5. Chan, T.F., Vese, L.A.: Active contours without edges. IEEE Trans. Image Process. **10**(2), 266–277 (2001)

6. Chen, K., et al.: Hybrid task cascade for instance segmentation. In: Proceedings of the IEEE Conference on Computer Vision and Pattern Recognition, pp. 4974–4983 (2019)

7. Chen, L.C., Papandreou, G., Kokkinos, I., Murphy, K., Yuille, A.L.: DeepLab: semantic image segmentation with deep convolutional nets, atrous convolution, and fully connected CRFs. IEEE Trans. Pattern Anal. Mach. Intell. **40**(4), 834–848 (2018)

8. Chen, X., Williams, B.M., Vallabhaneni, S.R., Czanner, G., Williams, R., Zheng, Y.: Learning active contour models for medical image segmentation. In: Proceedings of the IEEE Conference on Computer Vision and Pattern Recognition, pp. 11632–11640 (2019)

9. Cheng, D., Liao, R., Fidler, S., Urtasun, R.: DARNet: deep active ray network for building segmentation. In: Proceedings of the IEEE Conference on Computer Vision and Pattern Recognition, pp. 7431–7439 (2019)

10. Costea, D., Marcu, A., Slusanschi, E., Leordeanu, M.: Creating roadmaps in aerial images with generative adversarial networks and smoothing-based optimization. In: Proceedings of the IEEE International Conference on Computer Vision (ICCV) Workshops, October 2017

11. Geirhos, R., Rubisch, P., Michaelis, C., Bethge, M., Wichmann, F.A., Brendel, W.: ImageNet-trained CNNs are biased towards texture; increasing shape bias improves accuracy and robustness. In: International Conference on Learning Representations (ICLR) (2019)
12. Gur, S., Shaharabany, T., Wolf, L.: End to end trainable active contours via differentiable rendering. In: Proceedings of the International Conference on Learning Representations (ICLR) (2019)
13. Gur, S., Wolf, L., Golgher, L., Blinder, P.: Unsupervised microvascular image segmentation using an active contours mimicking neural network. In: Proceedings of the IEEE International Conference on Computer Vision, pp. 10722–10731 (2019)
14. Hatamizadeh, A., et al.: Deep active lesion segmentation. In: Suk, H.-I., Liu, M., Yan, P., Lian, C. (eds.) MLMI 2019. LNCS, vol. 11861, pp. 98–105. Springer, Cham (2019). https://doi.org/10.1007/978-3-030-32692-0_12
15. Hatamizadeh, A., Sengupta, D., Terzopoulos, D.: End-to-end deep convolutional active contours for image segmentation. arXiv preprint arXiv:1909.13359 (2019)
16. Hatamizadeh, A., Terzopoulos, D., Myronenko, A.: End-to-end boundary aware networks for medical image segmentation. In: Suk, H.-I., Liu, M., Yan, P., Lian, C. (eds.) MLMI 2019. LNCS, vol. 11861, pp. 187–194. Springer, Cham (2019). https://doi.org/10.1007/978-3-030-32692-0_22
17. He, K., Zhang, X., Ren, S., Sun, J.: Deep residual learning for image recognition. In: Proceedings of the IEEE Conference on Computer Vision and Pattern Recognition, pp. 770–778 (2016)
18. Hu, P., Shuai, B., Liu, J., Wang, G.: Deep level sets for salient object detection. In: Proceedings of the IEEE Conference on Computer Vision and Pattern Recognition (CVPR) (2017)
19. Kass, M., Witkin, A., Terzopoulos, D.: Snakes: active contour models. Int. J. Comput. Vis. 1(4), 321–331 (1988)
20. Kingma, D.P., Ba, J.: Adam: a method for stochastic optimization. arXiv preprint arXiv:1412.6980 (2014)
21. Lankton, S., Tannenbaum, A.: Localizing region-based active contours. IEEE Trans. Image Process. 17(11), 2029–2039 (2008)
22. Le, T.H.N., Quach, K.G., Luu, K., Duong, C.N., Savvides, M.: Reformulating level sets as deep recurrent neural network approach to semantic segmentation. IEEE Trans. Image Process. 27(5), 2393–2407 (2018)
23. Li, C., Xu, C., Gui, C., Fox, M.D.: Distance regularized level set evolution and its application to image segmentation. IEEE Trans. Image Process. 19(12), 3243 (2010)
24. Lillesand, T., Kiefer, R.W., Chipman, J.: Remote Sensing and Image Interpretation. Wiley, Hoboken (2015)
25. Marcos, D., et al.: Learning deep structured active contours end-to-end. In: Proceedings of the IEEE Conference on Computer Vision and Pattern Recognition (CVPR), pp. 8877–8885 (2018)
26. Marquez-Neila, P., Baumela, L., Alvarez, L.: A morphological approach to curvature-based evolution of curves and surfaces. IEEE Trans. Pattern Anal. Mach. Intell. 36(1), 2–17 (2013)
27. Myronenko, A., Hatamizadeh, A.: Robust semantic segmentation of brain tumor regions from 3D MRIs. In: Crimi, A., Bakas, S. (eds.) BrainLes 2019. LNCS, vol. 11993, pp. 82–89. Springer, Cham (2020). https://doi.org/10.1007/978-3-030-46643-5_8

28. Rudner, T.G., et al.: Multi3Net: segmenting flooded buildings via fusion of multiresolution, multisensor, and multitemporal satellite imagery. In: Proceedings of the AAAI Conference on Artificial Intelligence, vol. 33, pp. 702–709 (2019)
29. Shrivastava, N., Rai, P.K.: Remote-sensing the urban area: automatic building extraction based on multiresolution segmentation and classification. Geogr. Malays. J. Soc. Space **11**(2) (2017)
30. Silberman, N., Sontag, D., Fergus, R.: Instance segmentation of indoor scenes using a coverage loss. In: Fleet, D., Pajdla, T., Schiele, B., Tuytelaars, T. (eds.) ECCV 2014. LNCS, vol. 8689, pp. 616–631. Springer, Cham (2014). https://doi.org/10.1007/978-3-319-10590-1_40
31. Wang, S., et al.: TorontoCity: seeing the world with a million eyes. arXiv preprint arXiv:1612.00423 (2016)
32. Wang, Z., Acuna, D., Ling, H., Kar, A., Fidler, S.: Object instance annotation with deep extreme level set evolution. In: Proceedings of the IEEE Conference on Computer Vision and Pattern Recognition, pp. 7500–7508 (2019)
33. Wu, G., et al.: Automatic building segmentation of aerial imagery using multiconstraint fully convolutional networks. Remote Sens. **10**(3), 407 (2018)
34. Xu, Y., Wu, L., Xie, Z., Chen, Z.: Building extraction in very high resolution remote sensing imagery using deep learning and guided filters. Remote Sens. **10**(1), 144 (2018)
35. Zhang, P., Ke, Y., Zhang, Z., Wang, M., Li, P., Zhang, S.: Urban land use and land cover classification using novel deep learning models based on high spatial resolution satellite imagery. Sensors **18**(11), 3717 (2018)

Towards End-to-End Video-Based Eye-Tracking

Seonwook Park[(✉)], Emre Aksan, Xucong Zhang, and Otmar Hilliges

Department of Computer Science, ETH Zurich, Zürich, Switzerland
{seonwook.park,emre.aksan,xucong.zhang,otmar.hilliges}@inf.ethz.ch

Abstract. Estimating eye-gaze from images alone is a challenging task, in large parts due to un-observable person-specific factors. Achieving high accuracy typically requires labeled data from test users which may not be attainable in real applications. We observe that there exists a strong relationship between what users are looking at and the appearance of the user's eyes. In response to this understanding, we propose a novel dataset and accompanying method which aims to explicitly learn these semantic and temporal relationships. Our video dataset consists of time-synchronized screen recordings, user-facing camera views, and eye gaze data, which allows for new benchmarks in temporal gaze tracking as well as label-free refinement of gaze. Importantly, we demonstrate that the fusion of information from visual stimuli as well as eye images can lead towards achieving performance similar to literature-reported figures acquired through supervised personalization. Our final method yields significant performance improvements on our proposed EVE dataset, with up to 28% improvement in Point-of-Gaze estimates (resulting in 2.49° in angular error), paving the path towards high-accuracy screen-based eye tracking purely from webcam sensors. The dataset and reference source code are available at https://ait.ethz.ch/projects/2020/EVE.

Keywords: Eye tracking · Gaze estimation · Computer vision dataset

1 Introduction

The task of gaze estimation from a single low-cost RGB sensor is an important topic in Computer Vision and Machine Learning. It is an essential component in intelligent user interfaces [4,13], user state awareness [15,20], and serves as input modality to Computer Vision problems such as zero-shot learning [23], object referral [2], and human attention estimation [9]. Un-observable person-specific differences inherent in the problem are challenging to tackle and as such high accuracy general purpose gaze estimators are hard to attain. In response, person-specific adaptation techniques [29,30,37] have seen much attention, albeit at the

Electronic supplementary material The online version of this chapter (https://doi.org/10.1007/978-3-030-58610-2_44) contains supplementary material, which is available to authorized users.

© Springer Nature Switzerland AG 2020
A. Vedaldi et al. (Eds.): ECCV 2020, LNCS 12357, pp. 747–763, 2020.
https://doi.org/10.1007/978-3-030-58610-2_44

cost of requiring test-user-specific labels. We propose a dataset and accompanying method which holistically combine multiple sources of information explicitly. This novel approach yields large performance improvements without needing ground-truth labels from the final target user. Our large-scale dataset (EVE) and network architecture (GazeRefineNet) effectively showcase the newly proposed task and demonstrate up to 28% in performance improvements.

The human gaze can be seen as a closed-loop feedback system, whereby the appearance of target objects or regions (or visual stimuli) incur particular movements in the eyes. Many works consider this interplay in related but largely separate strands of research, for instance in estimating gaze from images of the user (bottom-up, e.g. [50]) or post-hoc comparison of the eye movements with the visual distribution of the presented stimuli (top-down, e.g. [42]). Furthermore, gaze estimation is often posed as a frame-by-frame estimation problem despite its rich temporal dynamics. In this paper, we suggest that by taking advantage of the interaction between user's eye movements and what they are looking at, significant improvements in gaze estimation accuracy can be attained even in the *absence of labeled samples* from the final target. This can be done without explicit gaze estimator personalization. We are not aware of existing datasets that would allow for the study of these semantic relations and temporal dynamics. Therefore, we introduce a novel dataset designed to facilitate research on the joint contributions of dynamic eye gaze and visual stimuli. We dub this dataset the EVE dataset (**E**nd-to-end **V**ideo-based **E**ye-tracking). EVE is collected from 54 participants and consists of 4 camera views, over 12 million frames and 1327 unique visual stimuli (images, video, text), adding up to approximately 105 h of video data in total.

Accompanying the proposed EVE dataset, we introduce a novel bottom-up-and-top-down approach to estimating the user's point of gaze. The Point-of-Gaze (PoG) refers to the actual target of a person's gaze as measured on the screen plane in metric units or pixels. In our method, we exploit the fact that more visually salient regions on a screen often coincide with the gaze. Unlike previous methods which adopt and thus depend on pre-trained models of visual saliency [6,42,43], we define our task as that of online and conditional PoG refinement. In this setting a model takes raw screen content and an initial gaze estimate as explicit conditions, to predict the final and refined PoG. Our final architecture yields significant improvements in predicted PoG accuracy on the proposed dataset. We achieve a mean test error of 2.49 degrees in gaze direction or 2.75cm (95.59 pixels) in screen-space Euclidean distance. This is an improvement of up to 28% compared to estimates of gaze from an architecture that does not consider screen content. We thus demonstrate a meaningful step towards the proliferation of screen-based eye tracking technology.

In summary, we propose the following contributions:

- A new task of online point-of-gaze (PoG) refinement, which combines bottom-up (eye appearance) and top-down (screen content) information to allow for a truly end-to-end learning of human gaze,

Table 1. Comparison of EVE with existing screen-based datasets. EVE is the first to provide natural eye movements (free-viewing, without specific instructions) synchronized with full-frame user-facing video and screen content

Name	Region	# Subjects	# Samples	Temporal data	Natural eye movements	Screen content video	Publicly available
Columbia Gaze [41]	Frame	56	5,800	–	N	N	Y
EYEDIAP [16]	Frame	16	62,500	30 Hz	N*	N	Y
UT Multiview [44]	Eyes	50	64,000	–	N	N	Y
MPIIGaze [53]	Eyes	15	213,659	–	N	N	Y
TabletGaze [21]	Frame	51	1,785	–	N	N	Y
GazeCapture [25]	Frame	1,474	2,129,980	–	N	N	Y
Deng and Zhu [11]	Eyes	200	240,000	–	N	N	N
MPIIFaceGaze [54]	Face	15	37,639	–	N	N	Y
DynamicGaze [47]	Eyes	20	645,000	∼30 Hz	Y	N	N
EVE (Ours)	Frame	54	12,308,334	30 Hz, 60 Hz	Y	30 Hz	Y

*Only smooth pursuits eye movements are available.

- EVE, a large-scale video dataset of over 12 million frames from 54 participants consisting of 4 camera views, natural eye movements (as opposed to following specific intructions or smoothly moving targets), pupil size annotations, and screen content video to enable the new task (Table 1),
- a novel method for eye gaze refinement which exploits the complementary sources of information jointly for improved PoG estimates, in the absence of ground-truth annotations from the user.

In combination these contributions allow us to demonstrate a gaze estimator performance of 2.49° in angular error, comparing favorably with supervised person-specific model adaptation methods [7,29,37].

2 Related Work

In our work we consider the task of remote gaze estimation from RGB, where a monocular camera is located away from and facing a user. We outline here recent approaches, proposed datasets, and relevant methods for refining gaze estimates.

2.1 Remote Gaze Estimation

Remote gaze estimation from unmodified monocular sensors is challenging due to the lack of reference features such as reflections from near infra-red light sources. Recent methods have increasingly used machine learning methods to tackle this problem [3,32,36] with extensions to allow for greater variations in head pose [11,31,40]. The task of cross-person gaze estimation is defined as one where a model is evaluated on a previously unseen set of participants. Several extensions have been proposed for this challenging task in terms of self-assessed uncertainty [8], novel representations [38,39,52], and Bayesian learning [49,50].

Novel datasets have contributed to the progress of gaze estimation methods and the reporting of their performance, notably in challenging illumination settings [25,53,54], or at greater distances from the user [14,16,24] where image details are lost. Screen-based gaze estimation datasets have had a particular focus [11,16,21,25,33,53,54] due to the practical implications in modern times, with digital devices being used more frequently. Very few existing datasets include videos, and even then often consist of participants gazing at points [21] or following smoothly moving targets only (via smooth pursuits) [16]. While the RT-GENE dataset includes natural eye movement patterns such as fixations and saccades, it is not designed for the task of screen-based gaze estimation [14]. The recently proposed DynamicGaze dataset [47] includes natural eye movements from 20 participants gazing upon video stimuli. However, it is yet to be publicly released and it is unclear if it will contain screen-content synchronization. We are the first to provide a video dataset with full camera frames and associated eye gaze and pupil size data, in conjunction with screen content. Furthermore, EVE includes a large number of participants (=54) and frames (12.3M) over a large set of visual stimuli (1004 images, 161 videos, and 162 wikipedia pages).

2.2   Temporal Models for Gaze Estimation

Temporal modelling of eye gaze is an emerging research topic. An initial work demonstrates the use of a recurrent neural network (RNN) in conjunction with a convolutional neural network (CNN) for feature extraction [35]. While no improvements are shown for gaze estimates in the screen-space, results on smooth pursuits sequence of the EYEDIAP dataset [16] are encouraging. In [47], a top-down approach for gaze signal filtering is presented, where a probabilistic estimate of state (fixation, saccade, or smooth pursuits) is initially made, and consequently a state-specific linear dynamical system is applied to refine the initially predicted gaze. Improvements in gaze estimation performance are demonstrated on a custom dataset. As one of our evaluations, we re-confirm previous findings that a temporal gaze estimation network can improve on a static gaze estimation network. We demonstrate this on our novel video dataset, which due to its diversity of visual stimuli and large number of participants should allow for future works to benchmark their improvements well.

2.3   Refining Gaze Estimates

While eye gaze direction (and subsequent Point-of-Gaze) can be predicted just from images of the eyes or face of a given user, an initial estimate can be improved with additional data. Accordingly, various methods have been proposed to this end. A primary example is that of using few samples of labeled data - often dubbed "person-specific gaze estimation" - where a pre-trained neural network is fine-tuned or otherwise adapted on very few samples of a target test person's data, to yield performance improvements on the final test data from the same person. Building on top of initial works [25,39], more recent works have

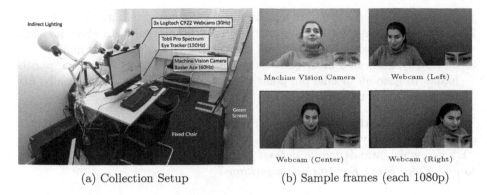

(a) Collection Setup   (b) Sample frames (each 1080p)

Fig. 1. EVE data collection setup and example of (undistorted) frames collected from the 4 camera views with example eye patches shown as insets.

demonstrated significant performance improvements with as few as 9 calibration samples or less [7,29,30,37,51]. Although the performance improvements are impressive, all such methods still require labeled samples from the final user.

Alternative approaches to refining gaze estimates in the screen-based setting, consider the predicted visual saliency of the screen content. Given a sufficient time horizon, it is possible to align estimates for PoG so-far, with an estimate for visual saliency [1,6,43,45,48]. However, visual saliency estimation methods can over-fit to presented training data. Hence, methods have been suggested to merge estimates of multiple saliency models [42] or use face positions as likely gaze targets [43]. We propose an alternate and direct approach, which formulates the problem of gaze refinement as one that is conditioned explicitly on screen content and an initial gaze estimate.

3   The EVE Dataset

To study the semantic relations and temporal dynamics between eye gaze and visual content, we identify a need for a new gaze dataset that:

1. allows for the training and evaluation of temporal models on natural eye movements (including fixations, saccades, and smooth pursuits),
2. enables the training of models that can process full camera frame inputs to yield screen-space Point-of-Gaze (PoG) estimates,
3. and provide a community-standard benchmark for a good understanding of the generalization capabilities of upcoming methods.

Furthermore, we consider the fact that the distribution of visual saliency on a computer screen at a given time is indicative of likely gaze positions. In line with this observation, prior work reports difficulty in generalization when considering saliency estimation and gaze estimation as separate components [42,43]. Thus, we define following further requirements for our new dataset:

1. a video of the screen content synchronized with eye gaze data,
2. a sufficiently large set of visual stimuli must be presented to allow for algorithms to generalize better without over-fitting to a few select stimuli,
3. and lastly, gaze data must be collected over time without instructing participants to gaze at specific pin-point targets such that they act naturally, like behaviours in a real-world setting.

We present in this section the methodologies we adopt to construct such a dataset, and briefly describe its characteristics. We call our proposed dataset "EVE", which stands for *"a dataset for enabling progress towards truly End-to-end Video-based Eye-tracking algorithms"*.

3.1 Captured Data

The minimum requirements for constructing our proposed dataset is the captured video from a webcam, gaze ground truth data from a commercial eye tracker, and screen frames from a given display. Furthermore, we:

- use the Tobii Pro Spectrum eye tracker, which reports high accuracy and precision in predicted gaze[1] even in the presence of natural head movements,
- add a high performance Basler Ace acA1920-150uc machine vision camera with global shutter, running at 60 Hz,
- install three Logitech C922 webcams (30 Hz) for a wider eventual coverage of head orientations, assuming that the final user will not only be facing the screen in a fully-frontal manner (see Fig. 1b),
- and apply MidOpt BP550 band-pass filters to all webcams and machine vision camera to remove reflections and glints on eyeglass and cornea surfaces due to the powerful near-infra-red LEDs used by the Tobii eye tracker.

All video camera frames are captured at 1920×1080 pixels resolution, but the superior timestamp-reliability and image quality of the Basler camera is expected to yield better estimates of gaze compared to the webcams.

The data captured by the Tobii Pro Spectrum eye tracker can be of very high quality which is subject to participant and environment effects. Hence to ensure data quality and reliability, an experiment coordinator is present during every data collection session to qualitatively assess eye tracking data via a live-stream of camera frames and eye movements. Additional details on our hardware setup and steps we take to ensure the best possible eye tracking calibration and subsequent data quality are described in the supplementary materials.

3.2 Presented Visual Stimuli

A large variety of visual stimuli are presented to our participants. Specifically, we present image, video, and wikipedia page stimuli (shown later in Fig. 4).

---

[1] See https://www.tobiipro.com/pop-ups/accuracy-and-precision-test-report-spectrum/?v=1.1.

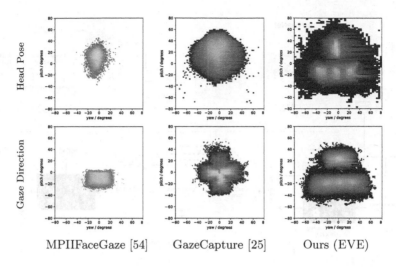

Fig. 2. Head orientation and gaze direction distributions are compared with existing screen-based gaze datasets [25,54]. We capture a larger range of parameter space due to a multi-view camera setup and 25-inch display. 2D histogram plot values are normalized and colored with log-scaling.

For static image stimuli, we select the widely used MIT1003 dataset [22] originally created for the task of image saliency estimation. Most images in the dataset span 1024 pixels in either horizontal or vertical dimensions. We randomly scale the image between 1320 and 1920 pixels in width or 480 to 1080 pixels in height, to be displayed on our 25-inch screen (with a resolution of 1080p).

All video stimuli are displayed in 1080p resolution (to span the full display), and taken from the DIEM [34], VAGBA [27], and Kurzhals et al. [26] datasets. These datasets consist of 720p, 1080p, and 1080p videos respectively, and thus are of high-resolution compared to other video-based saliency datasets. DIEM consists of various videos sampled from public repositories such as trailers and documentaries. VAGBA includes human movement or interactions in everyday scenes, and Kurzhals et al. contain purposefully designed video sequences with intentionally-salient regions. To increase the variety of the final set of video stimuli further, we select 23 videos from Wikimedia (at 1080p resolution).

Wikipedia pages are randomly selected on-the-fly by opening the following link in a web browser: https://en.m.wikipedia.org/wiki/Special:Random#/random and participants are then asked to freely view and navigate the page, as well as to click on links. Links leading to pages outside of Wikipedia are automatically removed using the GreaseMonkey web browser extension.

In our data collection study, we randomly sample the image and video stimuli from the mentioned datasets. We ensure that each participant observes 60 image stimuli (for three seconds each), at least 12 min of video stimuli, and six minutes of wikipedia stimulus (three 2-min sessions). At the conclusion of data collection,

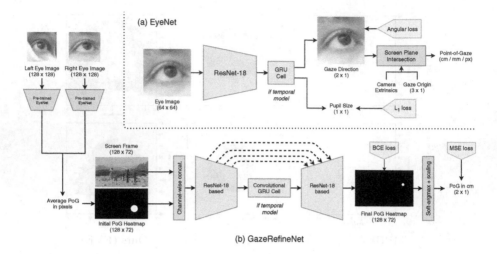

Fig. 3. We adopt (a) a simple EyeNet architecture for gaze direction and pupil size estimation with an optional recurrent component, and propose (b) a novel GazeRefineNet architecture for label-free PoG refinement using screen content.

we found that each image stimulus has been observed 3.35 times ($SD = 0.73$), and each video stimulus has been observed 9.36 times ($SD = 1.28$).

3.3 Dataset Characteristics

The final dataset is collected from 54 participants (30 male, 23 female, 1 unspecified). The details of responses to our demographics questionnaire can be found in our supplementary materials along with how we pre-process the dataset. We ensure that the subjects in both training and test sets exhibit diverse gender, age, and ethnicity, some with and some without glasses.

In terms of gaze direction and head orientation distributions, EVE compares favorably to popular screen-based datasets such as MPIIFaceGaze [54] and Gaze-Capture [25]. Figure 2 shows that we cover a larger set of gaze directions and head poses. This is likely due to the 4 camera views that we adopt, together with a large screen size of 25 in. (compared to the other datasets).

4 Method

We now discuss a novel architecture designed to exploit the various sources of information in datasets and to serve as baseline for follow-up work. We first introduce a simple eye gaze estimation network (EyeNet$_{\text{static}}$) and its recurrent counterparts (EyeNet$_{\text{RNN}}$, EyeNet$_{\text{LSTM}}$, EyeNet$_{\text{GRU}}$) for the task of per-frame or temporal gaze and pupil size estimation (see Fig. 3a). As the EVE dataset contains synchronized visual stimuli, we propose a novel technique to process these initial eye-gaze predictions further by taking the raw screen content directly into consideration. To this end, we propose the GazeRefineNet architecture (Fig. 3b), and describe its details in the second part of this section.

4.1 EyeNet Architecture

Learning-based eye gaze estimation models typically output their predictions as a unit direction vector or in Euler angles in the spherical coordinate system. The common metric to evaluate the goodness of predicted gaze directions is via an angular distance error metric in degrees. Assuming that the predicted gaze direction is represented by a 3-dimensional unit vector $\hat{\mathbf{g}}$, the calculation of the angular error loss when given ground-truth $\mathbf{g}$ is then:

$$\mathcal{L}_{\text{gaze}}(\mathbf{g}, \hat{\mathbf{g}}) = \frac{1}{NT} \sum_{}^{N} \sum_{}^{T} \frac{180}{\pi} \arccos\left(\frac{\mathbf{g} \cdot \hat{\mathbf{g}}}{\|\mathbf{g}\|\|\hat{\mathbf{g}}\|}\right) \tag{1}$$

where a mini-batch consists of N sequences each of length T.

To calculate PoG, the predicted gaze direction must first be combined with the 3D gaze origin position $\mathbf{o}$ (determined during data pre-processing), yielding a gaze ray with 6 degrees of freedom. We can then intersect this ray with the screen plane to calculate the PoG by using the camera transformation with respect to the screen plane. Pixel dimensions (our 1920×1080 screen 553 mm wide 311 mm tall) can be used to convert the PoG to pixel units for an alternative interpretation. We denote the predicted PoG in centimeters as $\hat{\mathbf{s}}$.

Assuming that the pupil size can be estimated, we denote it as $\hat{\mathbf{p}}$ and define an ℓ_1 loss given ground-truth $\mathbf{p}$ as:

$$\mathcal{L}_{\text{pupil}}(\mathbf{p}, \hat{\mathbf{p}}) = \frac{1}{NT} \sum_{}^{N} \sum_{}^{T} \|\mathbf{p} - \hat{\mathbf{p}}\|_1 \tag{2}$$

The two values of gaze direction and pupil size are predicted by a ResNet-18 architecture [17]. To make the network recurrent, we optionally incorporate a RNN [46], LSTM [18], or GRU [10] cell.

4.2 GazeRefineNet Architecture

Given the left and right eye images $\mathbf{x}_l$ and $\mathbf{x}_r$ of a person, we hypothesize that incorporating the corresponding screen content can improve the initial PoG estimate. Provided that an initial estimate of PoG $\tilde{\mathbf{s}} = f(\mathbf{x})$ can be made for the left and right eyes $\tilde{\mathbf{s}}_l$ and $\tilde{\mathbf{s}}_r$ respectively, we first take the average of the predicted PoG values with $\tilde{\mathbf{s}} = \frac{1}{2}(\tilde{\mathbf{s}}_l + \tilde{\mathbf{s}}_r)$ to yield a single estimate of gaze. Here f denotes the previously described EyeNet. We define and learn a new function, $\mathbf{s} = g(\mathbf{x}_S, \tilde{\mathbf{s}})$, to refine the EyeNet predictions by incorporating the screen content and temporal information. The function g is parameterized by a fully convolutional neural network (FCN) to best preserve spatial information. Following the same line of reasoning, we represent our initial PoG estimate $\tilde{\mathbf{s}}$ as a confidence map. More specifically, we use an isotropic 2D Gaussian function centered at the estimated gaze position on the screen. The inputs to the FCN are concatenated channel-wise.

To allow the model to better exploit the temporal information, we use an RNN cell in the bottleneck. Inspired by prior work in video-based saliency estimation, we adopt a convolutional recurrent cell [28] and evaluate RNN [46], LSTM [18], and GRU [10] variants.

The network optionally incorporate concatenative skip connections between the encoder and decoder layers, as this is shown to be helpful in FCNs. We train the GazeRefineNet by using pixel-wise binary cross-entropy loss on the output heatmap and MSE loss on the final numerical estimate of the PoG. It is calculated in a differentiable manner via a soft-argmax layer [5, 19]. The PoG is converted to centimeters to keep the loss term from exploding (due to its magnitude). Please refer to Fig. 3b for the full architecture diagram, and our supplementary materials for implementation details.

Offset Augmentation. In the task of cross-person gaze estimation, it is common to observe high discrepancies between the training and validation objectives. This is not necessarily due to overfitting or non-ideal hyperparameter selections but rather due to the inherent nature of the problem. Specifically, every human has a person-specific offset between their optical and visual axes in each eye, often denoted by a so-called Kappa parameter. While the optical axis can be observed by the appearance of the iris, the visual axis cannot be observed at all as it is defined by the position of the fovea at the back of the eyeball.

During training, this offset is absorbed into the neural network's parameters, limiting generalization to unseen people. Hence, prior work typically incur a large error increase in cross-person evaluations ($\sim 5°$) in comparison to person-specific evaluations ($\sim 3°$). Our insight is that we are now posing a gaze refinement problem, where an initially incorrect assessment of offset could actually be corrected by additional signals such as that of screen content. This is in contrast with the conventional setting, where no such corrective signal is made available. Therefore, the network should be able to learn to overcome this offset when provided with randomly sampled offsets to a given person's gaze.

This randomization approach can intuitively be understood as learning to undo all possible inter-personal differences rather than learning the corrective parameters for a specific user, as would be the case in traditional supervised personalization (e.g., [37]). We dub our training data augmentation approach as an *"offset augmentation"*, and provide further details of its implementation in our supplementary materials.

5   Results

In this section, we evaluate the variants of EyeNet and find that temporal modelling can aid in gaze estimation. Based on a pre-trained EyeNet$_{GRU}$, we then evaluate the effects of our contributions in refining an initial estimate of PoG using variants of GazeRefineNet. We demonstrate large and consistent performance improvements even across camera views and visual stimulus types.

Table 2. Cross-person gaze estimation and pupil size errors of EyeNet variants, evaluated on the test set of EVE. The GRU variant performs best in terms of both gaze and pupil size estimates

Model	Left Eye				Right Eye			
	Gaze Dir. (°)	PoG (cm)	PoG (px)	Pupil size (mm)	Gaze Dir. (°)	PoG (cm)	PoG (px)	Pupil size (mm)
EyeNet$_{static}$	4.54	5.10	172.7	**0.29**	4.75	5.29	181.0	**0.29**
EyeNet$_{RNN}$	4.33	4.86	166.7	**0.29**	4.91	5.48	186.5	**0.28**
EyeNet$_{LSTM}$	4.17	4.66	161.0	0.32	**4.71**	**5.25**	**180.5**	0.33
EyeNet$_{GRU}$	**4.11**	**4.60**	**158.5**	**0.28**	4.80	5.33	183.9	**0.29**

Table 3. An ablation study of our contributions in GazeRefineNet, where a frozen and pre-trained EyeNet$_{GRU}$ is used for initial gaze predictions. Temporal modelling and our novel offset augmentation both yield large gains in performance.

Model	Screen Content	Offset Augmen.	Skip Conn.	Gaze Dir. (°)	PoG (cm)	PoG (px)
Baseline (EyeNet$_{GRU}$)				3.48	3.85	132.56
GazeRefineNet$_{static}$	o			3.33	3.67	127.59
	o	o		2.80	3.09	107.42
	o	o	o	2.87	3.16	109.85
GazeRefineNet$_{RNN}$	o	o		2.67	2.95	102.36
	o	o	o	2.57	2.83	98.38
GazeRefineNet$_{LSTM}$	o	o		**2.49**	**2.75**	**95.43**
	o	o	o	2.53	2.79	96.97
GazeRefineNet$_{GRU}$	o	o		2.51	2.77	96.24
	o	o	o	**2.49**	**2.75**	**95.59**

5.1 Eye Gaze Estimation

We first consider the task of eye gaze estimation purely from a single eye image patch. Table 2 shows the performance of the static EyeNet$_{static}$ and its temporal variants (EyeNet$_{RNN}$, EyeNet$_{LSTM}$, EyeNet$_{GRU}$) on predicting gaze direction, PoG, and pupil size. The networks are trained on the training split of EVE. Generally, we find our gaze direction error values to be in line with prior works in estimating gaze from single eye images [53], and see that the addition of recurrent cells improve gaze estimation performance modestly. This makes a case for training gaze estimators on temporal data, using temporally-aware models, and corroborates observations from a prior learning-based gaze estimation approach on natural eye movements [47].

Pupil size errors are presented in terms of mean absolute error. Considering that the size of pupils in our dataset vary 2 mm to 4 mm, the presented errors of 0.3 mm should allow for meaningful insights to be made in fields such as the

Table 4. Improvement in PoG prediction (in px) of our method in comparison with two saliency-based alignment methods, as evaluated on the EVE dataset.

Method \ Stimulus Type	Image	Video	Wikipedia
Saliency-based (scale + bias)	$78.4^{\downarrow 36.3\%}$	$116.7^{\downarrow 12.0\%}$	$198.3^{\uparrow 43.6\%}$
Saliency-based (kappa)	$75.0^{\downarrow 39.2\%}$	$110.9^{\downarrow 17.0\%}$	$258.0^{\uparrow 84.4\%}$
GazeRefineNet$_{\mathrm{GRU}}$ (Ours)	$\mathbf{48.7}^{\downarrow 60.4\%}$	$\mathbf{96.7}^{\downarrow 27.1\%}$	$\mathbf{116.3}^{\downarrow 15.8\%}$

Table 5. Final gaze direction errors (in degrees, lower is better) from the output of GazeRefineNet$_{\mathrm{GRU}}$, evaluated on the EVE test set in cross-stimuli settings. Indicated improvements are with respect to initial PoG predictions (mean of left+right) from EyeNet$_{\mathrm{GRU}}$ trained on specified source stimuli types.

Source \ Target	Images	Videos	Wikipedia
Images	$\mathbf{1.30}^{\downarrow 60.55\%}$	$3.60^{\uparrow 4.10\%}$	$4.74^{\uparrow 30.13\%}$
Videos	$1.97^{\downarrow 40.09\%}$	$\mathbf{2.60}^{\downarrow 24.88\%}$	$3.71^{\uparrow 1.94\%}$
Wikipedia	$2.12^{\downarrow 35.75\%}$	$3.32^{\downarrow 3.84\%}$	$\mathbf{3.04}^{\downarrow 16.62\%}$

cognitive sciences. We select the GRU variant (EyeNet$_{\mathrm{GRU}}$) for the next steps as it shows consistently good performance for both eyes.

5.2 Screen Content Based Refinement of PoG

GazeRefineNet consists of a fully-convolutional architecture which takes as input a screen content frame, and an offset augmentation procedure at training time. Our baseline performance for this experiment is different to Table 2 as gaze errors are improved when averaging the PoG from the left and right eyes, with according adjustments to the label (averaged in screen space). Even with the new competitive baseline from PoG averaging, we find in Table 3 that each of our additional contributions yield large performance improvements, amounting to a 28% improvement in gaze direction error, reducing it to 2.49°. While not directly comparable due to differences in setting, this value is lower even than recently reported performances of supervised few-shot adaptation approaches on in-the-wild datasets [29,37]. Specifically, we find that the offset augmentation procedure yields the greatest performance improvements, with temporal modeling further improving performance. Skip connections between the encoder and decoder do not necessarily help (except in the case of GazeRefineNet$_{\mathrm{RNN}}$), presumably because the output relies mostly on information processed at the bottleneck. We present additional experiments of GazeRefineNet in the following paragraphs, and describe their setup details in our supplementary materials.

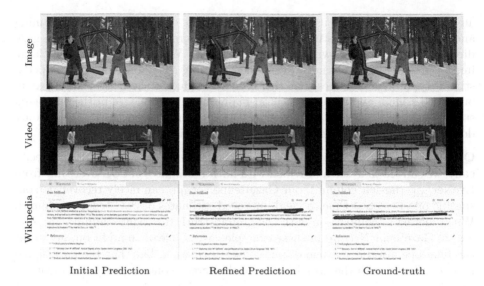

Fig. 4. Qualitative results of our gaze refinement method on our test set, where PoG over time are colored from blue-to-red (old-to-new). It can be seen that GazeRefineNet corrects offsets between the initial prediction and ground-truth.

Comparison to Saliency-Based Methods. In order to assess how our GazeRefineNet approach compares with existing saliency-based methods, we implement two up-to-date methods loosely based on [1] and [48]. First, we use the state-of-the-art UNISAL approach [12] to attain high quality visual saliency predictions. We accumulate these predictions over time for the full exposure duration of each visual stimulus in EVE (up to 2 min), which should provide the best context for alignment (as opposed to our online approach, which is limited to 3 seconds of history). Standard back propagation is then used to optimize for either scale and bias in screen-space (similar to [1]) or the visual-optical axis offset, kappa (similar to [48]) using a KL-divergence objective between accumulated visual saliency predictions and accumulated heatmaps of refined gaze estimates in the screen space. Table 4 shows that while both saliency-based baselines perform respectably on the well-studied image and video stimuli, they fail completely on wikipedia stimuli despite the fact that the saliency estimation model was provided with full 1080p frames (as opposed to the 128×72 input used by GazeRefineNet$_{\text{GRU}}$). Furthermore, our direct approach takes raw screen pixels and gaze estimations up to the current time-step as explicit conditions and thus is a simpler yet explicit solution for live gaze refinement that can be learned end-to-end. Both the training of our approach and its large-scale evaluation is made possible by the EVE, which should allow for insightful comparisons in the future.

Cross-Stimuli Evaluation. We study if our method generalizes to novel stimuli types, as this has previously been raised as in issue for saliency-based gaze alignment methods (such as in [42]). In Table 5, we confirm that indeed training and testing on the same stimulus type yields the greatest improvements in gaze direction estimation (shown in diagonal of table). We find in general that large improvements can be observed even when training solely on video or wikipedia stimuli types. One assumes that this is the case due to the existence of text in our video stimuli and the existence of small amounts of images in the wikipedia stimulus. In contrast, we can see that training a model on static images only does not lead to good generalization on the stimuli types.

Qualitative Results. We visualize our results qualitatively in Fig. 4. Specifically, we can see that when provided with initial estimates of PoG over time from EyeNet$_{GRU}$ (far-left column), our GazeRefineNet$_{GRU}$ can nicely recover person-specific offsets at test time to yield improved estimates of PoG (center column). When viewed in comparison with the ground-truth (far-right column), the success of GazeRefineNet$_{GRU}$ in these example cases is clear. In addition, note that the final operation is not one of pure offset-correction, but that the gaze signal is more aligned with the visual layout of the screen content post-refinement.

6   Conclusion

In this paper, we introduced several effective steps towards increasing screen-based eye-tracking performance even in the absence of labeled samples or eye-tracker calibration from the final target user. Specifically, we identified that eye movements and the change in visual stimulus have a complex interplay which previous literature have considered in a disconnected manner. Subsequently, we proposed a novel dataset (EVE) for evaluating temporal gaze estimation models and for enabling a novel online PoG-refinement task based on raw screen content. Our GazeRefineNet architecture performs this task effectively, and demonstrates large performance improvements of up to 28%. The final reported angular gaze error of 2.49° is achieved without labeled samples from the test set.

The EVE dataset is made publicly available[2], with a public web server implemented for consistent test metric calculations. We provide the dataset and accompanying training and evaluation code in hopes of further progress in the field of remote webcam-based gaze estimation. Comprehensive additional information regarding the capture, pre-processing, and characteristics of the dataset is made available in our supplementary materials.

Acknowledgements. We thank the participants of our dataset for their contributions, our reviewers for helping us improve the paper, and Jan Wezel for helping with the hardware setup. This project has received funding from the European Research Council (ERC) under the European Union's Horizon 2020 research and innovation programme grant agreement No. StG-2016-717054.

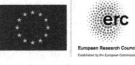

---

[2] https://ait.ethz.ch/projects/2020/EVE.

References

1. Alnajar, F., Gevers, T., Valenti, R., Ghebreab, S.: Calibration-free gaze estimation using human gaze patterns. In: ICCV, December 2013
2. Balajee Vasudevan, A., Dai, D., Van Gool, L.: Object referring in videos with language and human gaze. In: CVPR, pp. 4129–4138 (2018)
3. Baluja, S., Pomerleau, D.: Non-intrusive gaze tracking using artificial neural networks. In: NeurIPS, pp. 753–760 (1993)
4. Biedert, R., Buscher, G., Schwarz, S., Hees, J., Dengel, A.: Text 2.0. In: ACM CHI EA (2010)
5. Chapelle, O., Wu, M.: Gradient descent optimization of smoothed information retrieval metrics. Inf. Retrieval **13**(3), 216–235 (2010)
6. Chen, J., Ji, Q.: Probabilistic gaze estimation without active personal calibration. In: CVPR, pp. 609–616 (2011)
7. Chen, Z., Shi, B.: Offset calibration for appearance-based gaze estimation via gaze decomposition. In: WACV, March 2020
8. Cheng, Y., Lu, F., Zhang, X.: Appearance-based gaze estimation via evaluation-guided asymmetric regression. In: Ferrari, V., Hebert, M., Sminchisescu, C., Weiss, Y. (eds.) ECCV 2018. LNCS, vol. 11218, pp. 105–121. Springer, Cham (2018). https://doi.org/10.1007/978-3-030-01264-9_7
9. Chong, E., Ruiz, N., Wang, Y., Zhang, Y., Rozga, A., Rehg, J.M.: Connecting gaze, scene, and attention: generalized attention estimation via joint modeling of gaze and scene saliency. In: Ferrari, V., Hebert, M., Sminchisescu, C., Weiss, Y. (eds.) ECCV 2018. LNCS, vol. 11209, pp. 397–412. Springer, Cham (2018). https://doi.org/10.1007/978-3-030-01228-1_24
10. Chung, J., Gulcehre, C., Cho, K., Bengio, Y.: Empirical evaluation of gated recurrent neural networks on sequence modeling. In: NeurIPS Workshop on Deep Learning (2014)
11. Deng, H., Zhu, W.: Monocular free-head 3D gaze tracking with deep learning and geometry constraints. In: ICCV, pp. 3143–3152 (2017)
12. Droste, R., Jiao, J., Noble, J.A.: Unified image and video saliency modeling. In: ECCV (2020)
13. Feit, A.M., et al.: Toward everyday gaze input: accuracy and precision of eye tracking and implications for design. In: ACM CHI, pp. 1118–1130 (2017)
14. Fischer, T., Chang, H.J., Demiris, Y.: RT-GENE: real-time eye gaze estimation in natural environments. In: Ferrari, V., Hebert, M., Sminchisescu, C., Weiss, Y. (eds.) ECCV 2018. LNCS, vol. 11214, pp. 339–357. Springer, Cham (2018). https://doi.org/10.1007/978-3-030-01249-6_21
15. Fridman, L., Reimer, B., Mehler, B., Freeman, W.T.: Cognitive load estimation in the wild. In: ACM CHI (2018)
16. Funes Mora, K.A., Monay, F., Odobez, J.M.: EYEDIAP: a database for the development and evaluation of gaze estimation algorithms from RGB and RGB-D cameras. In: ACM ETRA. ACM, March 2014
17. He, K., Zhang, X., Ren, S., Sun, J.: Delving deep into rectifiers: surpassing human-level performance on imagenet classification. In: ICCV (2015)
18. Hochreiter, S., Schmidhuber, J.: Long short-term memory. Neural Comput. **9**(8), 1735–1780 (1997)
19. Honari, S., Molchanov, P., Tyree, S., Vincent, P., Pal, C., Kautz, J.: Improving landmark localization with semi-supervised learning. In: CVPR (2018)

20. Huang, M.X., Kwok, T.C., Ngai, G., Chan, S.C., Leong, H.V.: Building a personalized, auto-calibrating eye tracker from user interactions. In: ACM CHI, pp. 5169–5179. ACM, New York (2016)
21. Huang, Q., Veeraraghavan, A., Sabharwal, A.: TabletGaze: dataset and analysis for unconstrained appearance-based gaze estimation in mobile tablets. Mach. Vis. Appl. **28**(5–6), 445–461 (2017)
22. Judd, T., Ehinger, K., Durand, F., Torralba, A.: Learning to predict where humans look. In: ICCV, pp. 2106–2113. IEEE (2009)
23. Karessli, N., Akata, Z., Schiele, B., Bulling, A.: Gaze embeddings for zero-shot image classification. In: CVPR (2017)
24. Kellnhofer, P., Recasens, A., Stent, S., Matusik, W., Torralba, A.: Gaze360: physically unconstrained gaze estimation in the wild. In: ICCV, October 2019
25. Krafka, K., et al.: Eye tracking for everyone. In: CVPR (2016)
26. Kurzhals, K., Bopp, C.F., Bässler, J., Ebinger, F., Weiskopf, D.: Benchmark data for evaluating visualization and analysis techniques for eye tracking for video stimuli. In: Proceedings of the Fifth Workshop on Beyond Time and Errors: Novel Evaluation Methods for Visualization, pp. 54–60 (2014)
27. Li, Z., Qin, S., Itti, L.: Visual attention guided bit allocation in video compression. Image Vis. Comput. **29**(1), 1–14 (2011)
28. Linardos, P., Mohedano, E., Nieto, J.J., O'Connor, N.E., Giro-i Nieto, X., McGuinness, K.: Simple vs complex temporal recurrences for video saliency prediction. In: BMVC (2019)
29. Lindén, E., Sjostrand, J., Proutiere, A.: Learning to personalize in appearance-based gaze tracking. In: ICCVW (2019)
30. Liu, G., Yu, Y., Mora, K.A.F., Odobez, J.: A differential approach for gaze estimation with calibration. In: BMVC (2018)
31. Lu, F., Okabe, T., Sugano, Y., Sato, Y.: A head pose-free approach for appearance-based gaze estimation. In: BMVC (2011)
32. Lu, F., Sugano, Y., Okabe, T., Sato, Y.: Inferring human gaze from appearance via adaptive linear regression. In: ICCV (2011)
33. Martinikorena, I., Cabeza, R., Villanueva, A., Porta, S.: Introducing I2Head database. In: PETMEI, pp. 1–7 (2018)
34. Mital, P.K., Smith, T.J., Hill, R.L., Henderson, J.M.: Clustering of gaze during dynamic scene viewing is predicted by motion. Cogn. Comput. **3**(1), 5–24 (2011)
35. Palmero, C., Selva, J., Bagheri, M.A., Escalera, S.: Recurrent CNN for 3D gaze estimation using appearance and shape cues. In: BMVC (2018)
36. Papoutsaki, A., Sangkloy, P., Laskey, J., Daskalova, N., Huang, J., Hays, J.: WebGazer: scalable webcam eye tracking using user interactions. In: IJCAI, pp. 3839–3845 (2016)
37. Park, S., Mello, S.D., Molchanov, P., Iqbal, U., Hilliges, O., Kautz, J.: Few-shot adaptive gaze estimation. In: ICCV (2019)
38. Park, S., Spurr, A., Hilliges, O.: Deep pictorial gaze estimation. In: Ferrari, V., Hebert, M., Sminchisescu, C., Weiss, Y. (eds.) ECCV 2018. LNCS, vol. 11217, pp. 741–757. Springer, Cham (2018). https://doi.org/10.1007/978-3-030-01261-8_44
39. Park, S., Zhang, X., Bulling, A., Hilliges, O.: Learning to find eye region landmarks for remote gaze estimation in unconstrained settings. In: ACM ETRA (2018)
40. Ranjan, R., Mello, S.D., Kautz, J.: Light-weight head pose invariant gaze tracking. In: CVPRW (2018)
41. Smith, B., Yin, Q., Feiner, S., Nayar, S.: Gaze locking: passive eye contact detection for human-object interaction. In: ACM UIST, pp. 271–280, October 2013

42. Sugano, Y., Bulling, A.: Self-calibrating head-mounted eye trackers using egocentric visual saliency. In: ACM UIST, pp. 363–372. ACM, New York (2015)
43. Sugano, Y., Matsushita, Y., Sato, Y.: Calibration-free gaze sensing using saliency maps. In: CVPR, pp. 2667–2674 (2010)
44. Sugano, Y., Matsushita, Y., Sato, Y.: Learning-by-synthesis for appearance-based 3D gaze estimation. In: CVPR (2014)
45. Sugano, Y., Matsushita, Y., Sato, Y., Koike, H.: An incremental learning method for unconstrained gaze estimation. In: Forsyth, D., Torr, P., Zisserman, A. (eds.) ECCV 2008. LNCS, vol. 5304, pp. 656–667. Springer, Heidelberg (2008). https://doi.org/10.1007/978-3-540-88690-7_49
46. Sutskever, I., Vinyals, O., Le, Q.V.: Sequence to sequence learning with neural networks. In: NeurIPS, pp. 3104–3112 (2014)
47. Wang, K., Su, H., Ji, Q.: Neuro-inspired eye tracking with eye movement dynamics. In: CVPR, pp. 9831–9840 (2019)
48. Wang, K., Wang, S., Ji, Q.: Deep eye fixation map learning for calibration-free eye gaze tracking. In: ACM ETRA, pp. 47–55. ACM, New York (2016)
49. Wang, K., Zhao, R., Ji, Q.: A hierarchical generative model for eye image synthesis and eye gaze estimation. In: CVPR (2018)
50. Wang, K., Zhao, R., Su, H., Ji, Q.: Generalizing eye tracking with Bayesian adversarial learning. In: CVPR, pp. 11907–11916 (2019)
51. Yu, Y., Liu, G., Odobez, J.M.: Improving few-shot user-specific gaze adaptation via gaze redirection synthesis. In: CVPR, pp. 11937–11946 (2019)
52. Yu, Y., Odobez, J.M.: Unsupervised representation learning for gaze estimation. In: CVPR, June 2020
53. Zhang, X., Sugano, Y., Fritz, M., Bulling, A.: Appearance-based gaze estimation in the wild. In: CVPR (2015)
54. Zhang, X., Sugano, Y., Fritz, M., Bulling, A.: It's written all over your face: full-face appearance-based gaze estimation. In: CVPRW (2017)

Generating Handwriting via Decoupled Style Descriptors

Atsunobu Kotani$^{(\boxtimes)}$ ⓘ, Stefanie Tellex ⓘ, and James Tompkin ⓘ

Brown University, Providence, USA
Atsunobu_Kotani@brown.edu

Abstract. Representing a space of handwriting stroke styles includes the challenge of representing both the style of each character and the overall style of the human writer. Existing VRNN approaches to representing handwriting often do not distinguish between these different style components, which can reduce model capability. Instead, we introduce the Decoupled Style Descriptor (DSD) model for handwriting, which factors both character- and writer-level styles and allows our model to represent an overall greater space of styles. This approach also increases flexibility: given a few examples, we can generate handwriting in new writer styles, and also now generate handwriting of new characters across writer styles. In experiments, our generated results were preferred over a state of the art baseline method 88% of the time, and in a writer identification task on 20 held-out writers, our DSDs achieved 89.38% accuracy from a single sample word. Overall, DSDs allows us to improve both the quality and flexibility over existing handwriting stroke generation approaches.

1 Introduction

Producing computational models of handwriting is a deeply *human* and *personal* topic—most people can write, and each writer has a unique style to their script. Capturing these styles flexibly and accurately is important as it determines the space of descriptive expression of the model; in turn, these models define the usefulness of our recognition and generation applications. For deep-learning-based models, our concern is how to architecture the neural network such that we can represent the underlying stroke characteristics of the styles of writing.

Challenges in handwriting representation include reproducing fine detail, generating unseen characters, enabling style interpolation and transfer, and using human-labeled training data efficiently. Across these, one foundational problem is how to succinctly represent both the style variation of each character and the overall style of the human writer—to capture both the variation within an 'h' letterform and the overall consistency with other letterform for each writer.

Electronic supplementary material The online version of this chapter (https://doi.org/10.1007/978-3-030-58610-2_45) contains supplementary material, which is available to authorized users.

© Springer Nature Switzerland AG 2020
A. Vedaldi et al. (Eds.): ECCV 2020, LNCS 12357, pp. 764–780, 2020.
https://doi.org/10.1007/978-3-030-58610-2_45

As handwriting strokes can be modeled as a sequence of points over time, supervised deep learning methods to handwriting representation can use recurrent neural networks (RNNs) [2,17]. This allows consistent capture of style features that are distant in time and, with the use of variational RNNs (VRNNs), allows the diverse generation of handwriting by drawing from modeled distributions. However, the approach of treating handwriting style as a 'unified' property of a sequence can limit the representation of both character- and writer-level features. This includes specific character details being averaged out to maintain overall writer style, and an reduced representation space of writing styles.

Instead, we explicitly represent 1) writer-, 2) character- and 3) writer-character-level style variations within an RNN model. We introduce a method of Decoupled Style Descriptors (DSD) that models style variations such that character style can still depend on writer style. Given a database of handwriting strokes as timestamped sequences of points with character string labels [32], we learn a representation that encodes three key factors: writer-independent character representations ($\mathbf{C}_h$ for character h, $\mathbf{C}_{his}$ for the word his), writer-dependent character-string style descriptors ($\mathbf{w}_h$ for character h, $\mathbf{w}_{his}$ for the word his), and writer-dependent global style descriptors ($\mathbf{w}$ per writer). This allows new sequence generation for existing writers (via new $\mathbf{w}_{she}$), new writer generation via style transfer and interpolation (via new $\mathbf{w}$), and new character generation in the style of existing writers (via new $\mathbf{C}_2$, from only a few samples of character 2 from any writer). Further, our method helps to improve generation quality as more samples are provided for projection, rather than tending towards average letterforms in existing VRNN models.

In a qualitative user study, our model's generations were preferred 88% of the time over an existing baseline [2]. For writer classification tasks on a held-out 20-way test, our model achieves accuracy of 89.38% from a single word sample, and 99.70% from 50 word-level samples. In summary, we contribute:

- Decoupled Style Descriptors as a way to represent latent style information;
- An architecture with DSDs to model handwriting, with demonstration applications in generation, recognition, and new character adaptation; and
- A new database—BRUSH (BRown University Stylus Handwriting)—of handwritten digital strokes in the Latin alphabet, which includes 170 writers, 86 characters, 488 common words written by all writers, and 3668 rarer words written across writers.

Our dataset, code, and model will be open source at http://dsd.cs.brown.edu.

2   Related Work

Handwriting modeling methods either handle images, which capture writing appearance, or handle the underlying strokes collected via digital pens. Each may be online, where observation happens along with writing, or offline. Offline methods support historical document analysis, but cannot capture the motion of writing. We consider an online stroke-based approach, which avoids the stroke

Fig. 1. Illustrating synthesis approaches. Given test sample *his* for reference, we wish to generate *she* in the same style. Left: Pixels of *h* and *s* are copied from input with a slight modification [19]; however, this fails to synthesize *e* as it is missing in the reference. Middle: A global latent writer style is inferred from *his* and used as the initial state for LSTM generation [2]. Right: Our approach infers both character and writer style vectors to improve quality and flexibility.

extraction problem and allows us to focus on modelling style variation. Work also exists in the separate problem of typeface generation [5,12,24,31,41].

General Style Transfer Methods. Current state-of-the-art style transfer works use a part of the encoded reference sample as a style component, e.g., the output of a CNN encoder for 2D images [26,29], or the last output of an LSTM for speech audio [34]. These can be mixed to allocate parts of a conditioning style vector to disentangled variation [25]. Common style representations often cannot capture small details, with neural networks implicitly filtering out this content information, because the representations fail to structurally decouple style from content in the style reference source. Other approaches [27,38] tackle this problem by making neural networks predict parameters for a transformation model (an idea that originates from neuroevolution [18,36]); our **C** prediction is related.

Recent Image-Based Offline Methods. Haines et al. produced a system to synthesize new sentences in a specific style inferred from source images [19]. Some human intervention is needed during character segmentation, and the model can only recreate characters that were in the source images. Alonso et al. addressed the labeling issue with a GAN-based approach [3,16]; however, their model presents an image quality trade-off and struggles to generate new characters. There are also studies on typeface generation from few reference data [4,37]: Baluja generates typefaces for Latin alphabets [6], and Lian et al. for Chinese [30]. Our method does not capture writing implement appearance, but does

Table 1. Property comparison of state-of-the-art handwriting generation models.

Method	Style transfer?	No human segmentation?	Infinite variations?	Synthesize mis-sing samples?	Benefit from more samples?	Smooth interpolation?	Learn new characters?
Graves (2013)	No	Yes	Yes	No	No	No	No
Berio et al. (2017)	Yes	Yes	Yes	No	No	Sort of	No
Haines et al. (2017)	Yes	No	Sort of	No	Yes	No	No
Aksan et al. (2018)	Yes	Sort of	Yes	Yes	No	Sort of	No
Ours	Yes	Yes	Yes	Yes	Yes	Yes	Yes

provides underlying stroke generation and synthesizes new characters from few examples.

Stroke-Based Online Methods. Deep learning methods, such as Graves' work, train RNN models conditioned on target characters [13,17,40]. The intra-variance of a writer's style was achieved with Mixture Density Networks (MDN) as the final synthesis layer [10]. Berio et al. use recurrent-MDN for graffiti style transfer [9]. However, these methods cannot learn to represent handwriting styles per writer, and so cannot perform writer style transfer.

State-of-the-art models can generate characters in an inferred style [2]. Aksan et al.'s DeepWriting model uses Variational Recurrent Neural Networks (VRNN) [15] and assumes a latent vector z that controls writer handwriting style. Across writers, this method tends to average out specific styles and so reduces detail. Further, while sample efficient, VRNN models have trouble exploiting an abundance of inference samples because the style representation is only the last hidden state of an LSTM. We avoid this limitation by extracting character-dependent style vectors from samples and querying them as needed in generation.

Sequence Methods Beyond Handwriting. Learning-based architectures for sequences were popularized in machine translation [14], where the core idea is to encode sequential inputs into a fixed-length latent representation. Likewise, text-to-speech processing has been improved by sequence models [33,35], with extensions to style representation for speech-related tasks like speaker verification and voice conversion. Again, one approach is to use the (converted) last output of an LSTM network as a style representation [21]. Other approaches [22,23] models multiple stylistic latent variables in a hierarchical manner and introduces an approach to transfer styles within a standard VAE setting [28].

Broadly, variational RNN approaches [2,15,22] have the drawback that they are incapability of improving generation performance with more inference samples. While VRNNs are sample efficient when only a few samples are available for style inference, a system should also generate better results as more inference samples are provided (as in [19]). Our method attempts to be scalable and sample efficient through learning decoupled underlying generation factors.

We compare properties of four state of the art handwriting synthesis models (Table 1), and illustrate two of their different approaches (Fig. 1).

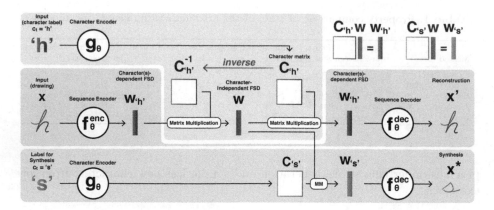

Fig. 2. High-level architecture. Circles are parametrized function approximators, and rectangles/squares are variables. *Blue region*: Encoder-decoder architecture. *Orange region*: Character-conditioned layers. *Green region*: Synthesis procedure. (Color figure online)

3 Method

Input, Preprocess, and Output. A stroke sequence $\mathbf{x} = (p_1, \ldots, p_N)$ has each p_t store the change in x- and y-axis from the previous timestep ($\Delta x_t = x_t - x_{t-1}$, $\Delta y_t = y_t - y_{t-1}$), and a binary termination flag for the 'end of stroke' ($eos = \{0, 1\}$). This creates an $(N, 3)$ matrix. A character sequence $\mathbf{s} = (\mathbf{c}_1, \ldots, \mathbf{c}_M)$ contains character vectors $\mathbf{c}_t$ where each is a one-hot vector of length equal to the total number of characters considered. This similarly is an (M, Q) matrix.

The IAM dataset [32] and our stroke dataset were collected by asking participants to naturally write character sequences or words, which often produces cursive writing. As such, we must solve a segmentation problem to attribute stroke points to specific characters in $\mathbf{s}$. This is complex; we defer explanation to our supplemental. For now, it is sufficient to say that we use unsupervised learning to train a segmentation network $k_\theta(\mathbf{x}, \mathbf{s})$ to map regions in $\mathbf{x}$ to characters, and to demark 'end of character' labels ($eoc = \{0, 1\}$) for each point.

As output, we wish to predict $\mathbf{x}'$ comprised of $\mathbf{p}'_t$ with: 1) coefficients for Mixture Density Networks [10] ($\pi_t, \mu_x, \mu_y, \sigma_x, \sigma_y, \rho$), which provide variation in output by sampling Δx_t and Δy_t from these distributions at runtime; 2) 'end of stroke' eos probability; and 3) 'end of character' eoc probability. This lets us generate cursive writing when eos probability is low and eoc probability is high.

Decoupled Style Descriptors (DSD). We begin with the encoder-decoder architecture proposed by Cho et al. [14] (Fig. 2, blue region). Given a supervised database $\mathbf{x}, \mathbf{s}$ and a target string c_t, to represent handwriting style we train a parameterized encoder function f_θ^{enc} to learn writer-dependent character-dependent latent vectors $\mathbf{w}_{c_t}$. Then, given $\mathbf{w}_{c_t}$, we simultaneously train a parameterized decoder function f_θ^{dec} to predict the next point p'_t given all past points

$p'_{1:t-1}$. Both encoder and decoder f_θ are RNNs such as LSTM models:

$$p'_t = f_\theta^{\text{dec}}(p'_{1:t-1}|\mathbf{w}_{c_t}). \tag{1}$$

This method does not factor character-independent writer style; yet, we have no way of explicitly describing this property via supervision and so we must devise a construction to learn it implicitly. Thus, we add a layer of abstraction (Fig. 2, orange region) with three assumptions:

1. If two stroke sequences $\mathbf{x}_1$ and $\mathbf{x}_2$ are written by the same writer, then consistency in their writing style is manifested by a character-independent writer-dependent latent vector $\mathbf{w}$.
2. If two character sequences $\mathbf{s}_1$ and $\mathbf{s}_2$ are written by different writers, then consistency in their stroke sequences is manifested by a character-dependent writer-independent latent matrix $\mathbf{C}$. $\mathbf{C}$ can be estimated via a parameterized encoder function g_θ, which is also an RNN such as an LSTM:

$$\mathbf{C}_{c_t} = g_\theta(\mathbf{s}, c_t). \tag{2}$$

3. $\mathbf{C}_{c_t}$ instantiates a writer's style $\mathbf{w}$ to draw a character via $\mathbf{w}_{c_t}$, such that $\mathbf{C}_{c_t}$ and $\mathbf{w}$ are latent factors:

$$\mathbf{w}_{c_t} = \mathbf{C}_{c_t}\mathbf{w}, \tag{3}$$

$$\mathbf{w} = \mathbf{C}_{c_t}^{-1}\mathbf{w}_{c_t}. \tag{4}$$

This method assumes that $\mathbf{C}_{c_t}$ is invertible, which we will demonstrate in Sect. 4. Intuitively, the multiplication of writer-dependent character vectors $\mathbf{w}_{c_t}$ with the inverse of character-DSD $\mathbf{C}_{c_t}^{-1}$ (Eq. 4) factors out character-dependent information from writer-dependent information in $\mathbf{w}_{c_t}$ to extract a writer style representation $\mathbf{w}$. Likewise, Eq. 3 restores writer-dependent character $\mathbf{w}_{c_t}$ by multiplying the writer-specific style $\mathbf{w}$ with a relevant character-DSD $\mathbf{C}_{c_t}$.

We use this property in synthesis (Fig. 2, green region). Given a target character c_t, we use encoder g_θ to generate a $\mathbf{C}$ matrix. Then, we multiply $\mathbf{C}_{c_t}$ by a desired writer style $\mathbf{w}$ to generate $\mathbf{w}_{c_t}$. Finally, we use trained decoder f_θ^{dec} to create a new point p'_t given previous points $p'_{1:t-1}$:

$$p'_t = f_\theta^{\text{dec}}(p'_{1:t-1}|\mathbf{w}_{c_t}), \text{ where } \mathbf{w}_{c_t} = \mathbf{C}_{c_t}\mathbf{w}. \tag{5}$$

Interpreting the Linear Factors. Equation 3 states a linear relationship between $\mathbf{C}_{c_t}$ and $\mathbf{w}$. This exists at the latent representation level: $\mathbf{w}_{c_t}$ and $\mathbf{C}_{c_t}$ are separately approximated by independent neural networks f_θ^{enc} and g_θ, which themselves are nonlinear function approximators [27,38]. As $\mathbf{C}_{c_t}$ maps a vector $\mathbf{w}$ to another vector $\mathbf{w}_{c_t}$, we can consider $\mathbf{C}_{c_t}$ to be a fully-connected neural network layer (without bias). However, unlike standard layers, $\mathbf{C}_{c_t}$'s weights are not implicitly learned through backpropagation but are predicted by a neural network g_θ in Eq. 2. A further interpretation of $\mathbf{C}_{c_t}$ and $\mathbf{C}_{c_t}^{-1}$ as two layers of a network is that they respectively share a set of weights and their inverse.

Explicitly forming $\mathbf{C}_{c_t}$ in this linear way makes it simple to estimate $\mathbf{C}_{c_t}$ for *new* characters that are not in the training dataset, given few sample pairs of $\mathbf{w}_{c_t}$ and $\mathbf{w}$, using standard linear least squares methods (Sect. 4).

Mapping Character and Stroke Sequences with f_θ and g_θ. Next, we turn our attention to how we map sequences of characters and strokes within our function approximators. Consider the LSTM f_θ^{enc}: Given a character sequence $\mathbf{s}$ as size of (M, Q) where M is the number of characters, and a stroke sequence $\mathbf{x}$ of size $(N, 3)$ where N is the number of points, our goal is to obtain a style vector for each character $\mathbf{w}_{c_t}$ in that sequence. The output of our segmentation network k_θ preprocess defines 'end of character' bits, and so we know at which point in $\mathbf{x}$ that a character switch occurs, e.g., from h to e in *hello*.

First, we encode $\mathbf{x}$ using f_θ^{enc} to obtain a $\mathbf{x}^*$ of size (N, L), where L is the latent feature dimension size (we use 256). Then, from $\mathbf{x}^*$, we extract M vectors at these switch indices—these are our writer-dependent character-dependent DSDs $\mathbf{w}_{c_t}$. As f_θ^{enc} is an LSTM, the historical sequence data up to that index is encoded within the vector at that index (Fig. 3, top). For instance, for *his*, $\mathbf{x}^*$ at switch index 2 represents how the writer writes the first two characters *hi*, i.e., $\mathbf{w}_{hi}$. We refer to these $\mathbf{w}_{c_t}$ as 'writer-character-DSDs'.

Likewise, LSTM g_θ takes a character sequence $\mathbf{s}$ of size (M, Q) and outputs an array of $\mathbf{C}$ matrices that forms a tensor of size (M, L, L) and preserves sequential dependencies between characters: The i-th element of the tensor $\mathbf{C}_{c_i}$ is a matrix of size (L, L)—that is, it includes information about previous characters up to and including the i-th character. Similar to $\mathbf{x}^*$, for *his*, the second character matrix $\mathbf{C}_{c_2}$ contains information about the first two characters *hi*—$\mathbf{C}$ is really a character sequence matrix. Multiplying character information $\mathbf{C}_{c_t}$ with writer style vector $\mathbf{w}$ creates a writer-character-DSD $\mathbf{w}_{c_t}$.

Estimating $\mathbf{w}$. When we encode a stroke sequence $\mathbf{x}$ that draws $\mathbf{s}$ characters via f_θ^{enc}, we extract M character(s)-dependent DSDs $\mathbf{w}_{c_t}$ (e.g., $\mathbf{w}_h$, $\mathbf{w}_{hi}$ and $\mathbf{w}_{his}$, *right*). Via Eq. 4, we obtain M distinct candidates for writer-DSDs $\mathbf{w}$. To overcome this, for each sample, we simply take the mean to form $\mathbf{w}$:

$$\overline{\mathbf{w}} = \frac{1}{M} \sum_{t=1}^{M} \mathbf{C}_{c_t}^{-1} \mathbf{w}_{c_t}. \tag{6}$$

Generation Approaches via $\mathbf{w}_{c_t}$. Consider the synthesis task in Fig. 1: given our trained model, generate how a new writer would write *she* given a reference sample of them writing *his*. From the *his* sample, we can extract 1) segment-level writer-character-DSDs ($\mathbf{w}_h$, $\mathbf{w}_i$, $\mathbf{w}_s$), and 2) the global $\overline{\mathbf{w}}$. To synthesize *she*, our model must predict three writer-character-DSDs ($\mathbf{w}_s$, $\mathbf{w}_{sh}$, $\mathbf{w}_{she}$) as input to the decoder f_θ^{dec}. We introduce two methods to estimate $\mathbf{w}_{c_t}$:

$$\text{Method } \alpha : \mathbf{w}_{c_t}^\alpha = \mathbf{C}_{c_t} \overline{\mathbf{w}} \tag{7a}$$

$$\text{Method } \beta : \mathbf{w}_{c_t}^\beta = h_\theta([\mathbf{w}_{c_1}, \dots, \mathbf{w}_{c_t}]) \tag{7b}$$

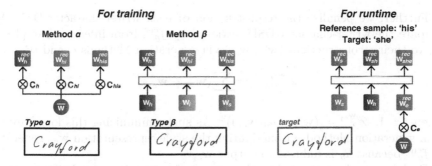

Fig. 3. Reconstruction methods to produce writer-character-DSD $\mathbf{w}_{c_t}$, with training sample $\mathbf{s}, \mathbf{x}$ of *his* and test sample $\mathbf{s}$ of *she*. Green rectangle is h_θ as defined in Eq. 7b. *Training*: Method α multiplies writer style $\overline{\mathbf{w}}$ with each character string matrix $\mathbf{C}_{c_t}$. Method β restore temporal dependencies of segment-level writer-character-DSDs ($\mathbf{w}_h$, $\mathbf{w}_i$, $\mathbf{w}_s$) via an LSTM, which produces higher-quality results that are preferred by users (Sect. 4). Target test image is in red. *Runtime*: Both prediction model Method α and β are combined to synthesize a new sample given contents within the reference sample. (Color figure online)

where h_θ is an LSTM that restore dependencies between temporally-separated writer-character-DSDs as illustrated in Fig. 3, green rectangle. We train our model to reconstruct $\mathbf{w}_{c_t}$ both ways. This allows us to use method α when test reference samples do not include target characters, e.g., *his* is missing an *e* for *she*, and so we can reconstruct $\mathbf{w}_e$ via $\mathbf{w}$ and $\mathbf{C}_e$ (Fig. 3, right). It also allows us to use Method β when test reference samples include relevant characters that, via f_θ^{enc}, provide writer-character-DSDs, e.g., *his* contains s and h in *she* and so we can estimate $\mathbf{w}_s$ and $\mathbf{w}_h$. As these characters could come from any place in the reference samples, h_θ restores the missing sequence dependencies.

3.1   Training Losses

We defer full architecture details for our supplemental material, and here explain our losses. We begin with a point location loss $\mathscr{L}^{loc}$ on predicted shifts in x, y coordinates, $(\Delta x, \Delta y)$. As we employ mixture density networks as a final prediction layer in f_θ^{dec}, we try to maximize the probability for the target shifts $(\Delta x^*, \Delta y^*)$ as explained by Graves et al. [17]:

$$\mathscr{L}^{loc} = -\sum_t \log \Big(\sum_j \pi_t^j \mathcal{N}(\Delta x_t^*, \Delta y_t^* | \mu_{xt}^j, \mu_{yt}^j, \sigma_{xt}^j, \sigma_{yt}^j, \rho_t^j) \Big).$$

Further, we consider 'end of sequence' flags *eos* and 'end of character' flags *eoc* by computing binary cross-entropy losses $\mathscr{L}^{eos}, \mathscr{L}^{eoc}$ for each.

Next, we consider consistency in predicting writer-DSD $\mathbf{w}$ from different writer-character-DSDs $\mathbf{w}_{c_t}$. We penalize a loss $\mathscr{L}^{\mathbf{w}}$ that minimizes the variance in $\mathbf{w}_t$ in Eq. 6:

$$\mathscr{L}^{\mathbf{w}} = \sum_t (\overline{\mathbf{w}} - \mathbf{w}_t)^2 \tag{8}$$

Further, we penalize the reconstruction of each writer-character-DSD. We compare the writer-character-DSD retrieved by f_θ^{enc} from inference samples as $\mathbf{w}_{c_t}$ to their reconstructions $(\mathbf{w}_{c_t}^\alpha, \mathbf{w}_{c_t}^\beta)$ via generation Methods α and β:

$$\mathscr{L}_{A \in (\alpha, \beta)}^{\mathbf{w}_{c_t}} = \sum_t (\mathbf{w}_{c_t} - \mathbf{w}_{c_t}^A)^2 \tag{9}$$

When $t = 1$, $\mathscr{L}_\beta^{\mathbf{w}_{c_1}} = (\mathbf{w}_{c_1} - h_\theta(\mathbf{w}_{c_1}))^2$. As such, minimizing this loss prevents h_θ in generation Method β from diluting the style representation $\mathbf{w}_{c_1}$ generated by f_θ^{enc} because h_θ is induced to output $\mathbf{w}_{c_1}$.

Each loss can be computed for three types of writer-character-DSD $\mathbf{w}_{c_t}$: those predicted by f_θ^{enc}, Method α, and Method β. These losses can also be computed at character, word, and sentence levels, e.g., for words:

$$\mathscr{L}_{word} = \sum_{A \in (f_\theta^{\text{enc}}, \alpha, \beta)} \left(\mathscr{L}_A^{loc} + \mathscr{L}_A^{eos} + \mathscr{L}_A^{eoc} + \mathscr{L}_A^{\mathbf{w}} + \mathscr{L}_A^{\mathbf{w}_{c_t}} \right). \tag{10}$$

Thus, the total loss is: $\mathscr{L}_{total} = \mathscr{L}_{char} + \mathscr{L}_{word} + \mathscr{L}_{sentence}$. $\mathscr{L}_{f_\theta^{\text{enc}}}^{\mathbf{w}_{c_t}} = 0$ by construction from Eq. 9; we include it here for completeness.

Sentence-level losses help to make the model predict spacing between words. While our model could train just with character- and word-level losses, this would cause a problem if we ask the model to generate a sentence from a reference sample of a single word. Training with $\mathscr{L}_{sentence}$ lets our model predict how a writer would space words based on their writer-DSD $\mathbf{w}$.

Implicit **C** *Inverse Constraint.* Finally, we discuss how $\mathscr{L}^{\mathbf{w}_{c_t}}$ at the character level implicitly constrains character-DSD **C** to be invertible. If we consider a single character sample, then mean $\overline{\mathbf{w}}$ in Eq. 6 is equal to $\mathbf{C}_{c_1}^{-1}\mathbf{w}_{c_1}$. In this case, as $\mathbf{w}_{c_t}^\alpha = \mathbf{C}_{c_t}\overline{\mathbf{w}}$ (Eq. 7a), $\mathscr{L}_\alpha^{\mathbf{w}_{c_t}}$ becomes:

$$\mathscr{L}_\alpha^{\mathbf{w}_{c_t}} = (\mathbf{w}_{c_1} - \mathbf{C}_{c_1}\mathbf{C}_{c_1}^{-1}\mathbf{w}_{c_1})^2 \tag{11}$$

This value becomes nonzero when **C** is singular ($\mathbf{CC}^{-1} \neq \mathbf{I}$), and so our model avoids non-invertible **C**s.

Training Through Inverses. As we train our network end-to-end, our model must backpropagate through $\mathbf{C}_{c_t}$ and $\mathbf{C}_{c_t}^{-1}$. As derivative of matrix inverses can be obtained with $\frac{d\mathbf{C}^{-1}}{dx} = -\mathbf{C}^{-1}\frac{d\mathbf{C}}{dx}\mathbf{C}^{-1}$, our model can train.

4   Experiments

Dataset. Our new dataset—BRUSH—provides characteristics that other online English handwriting datasets do not, including the typical online English handwriting dataset IAM [32]. First, we explicitly display a baseline in every drawing box during data collection. This enables us to create handwriting samples whose initial action is the x, y shift from the baseline to the starting point. This additional information might also help improve performance in recognition tasks.

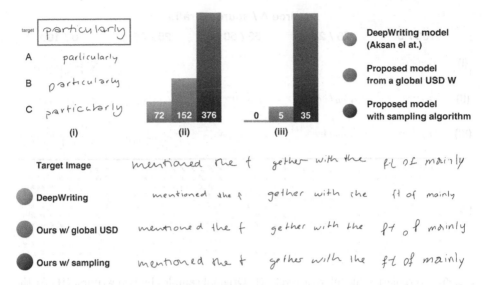

Fig. 4. Comparison of our proposed model vs. the state-of-the-art model [2]. *Top:* (i) Example writing similarity ordering task assigned to MTurk workers. (ii) Counts of most similar results with the target image. (iii) Sample-level vote. *Bottom:* Three examples of task orderings; see supplemental for all 40. The model of Aksan et al. [2] typically over-smooths the style and loses key details.

Second, our 170 individuals wrote 488 words *in common* across 192 sentences. This helps to evaluate handwriting models and observe whether **w** and **C** are decoupled: given a sample that failed to generate, we can compare the generated results of the same word across writers. If writer A failed but B succeeded, then it is likely that the problem is not with **C** representations but with either **w** or $\mathbf{w}_{c_t}$. If both A and B failed to draw the word but succeeding in generating other words, it is likely that **C** or $\mathbf{w}_{c_t}$ representations are to blame. We provide further details about our dataset and collection process in our supplemental material.

Third, for DeepWriting [2] comparisons, we use their training and test splits on IAM that mix writer identities—i.e., in training, we see some data from every writer. For all other experiments, we use our dataset, where we split between writers—our 20 test writers have never been seen writing *anything* in training.

Invertibility of **C**. To compute **w** in Eq. 4, we must invert the character-DSD **C**. Our network is designed to make **C** invertible as training proceeds by penalizing a reconstruction loss for $\mathbf{w}_{c_t}$ and $\mathbf{C}_{c_t}\mathbf{C}_{c_t}^{-1}\mathbf{w}_{c_t}$ (Sect. 3.1). To test its success, we compute **C**s from our model for all single characters (86 characters) and character pairs ($86^2 = 7,396$ cases), and found **C** to have full rank in each case. Next, we test all possible 3-character-strings ($86^3 = 636,056$ cases). Here, there were 37 rare cases with non-invertible **C**s, such as *1Zb* and *6ak*. In these cases, we can still extract two candidate **w** from the first two characters (e.g., *1* and *1Z* in the *1Zb* sample) to complete generation tasks.

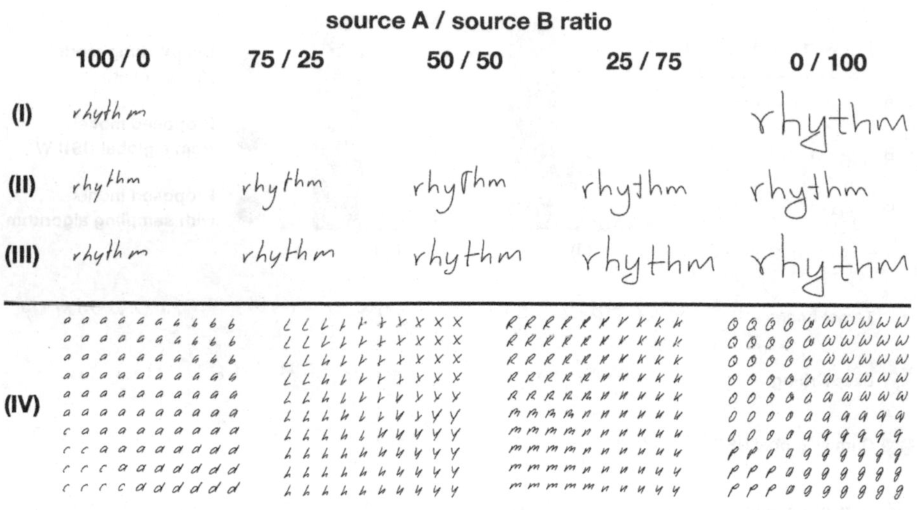

774   A. Kotani et al.

source A / source B ratio

	100 / 0	75 / 25	50 / 50	25 / 75	0 / 100
(I)	rhythm				rhythm
(II)	rhythm	rhythm	rhythm	rhythm	rhythm
(III)	rhythm	rhythm	rhythm	rhythm	rhythm

(IV) *(handwriting character interpolation grid)*

Fig. 5. Interpolation at different levels. (I) Original samples by two writers. (II) At the writer-DSD **w** level. (III) At the writer-character-DSD $\mathbf{w}_{c_t}$ level. (IV) At **C** level. *Left to right:* Characters used are *abcd, Lxhy, Rkmy, QWPg.*

Qualitative Evaluation with Users. We use Amazon Mechanical Turk to asked 25 participants to rank generated handwriting similarity to a target handwriting (Fig. 4 (i)). We randomly selected 40 sentence-level target handwriting samples from the validation set of IAM dataset [32]. Each participant saw randomly-shuffled samples; in total, 600 assessments were made. We compared the abilities of three models to generate the same handwriting style without seeing the actual target sample. We compare to the state-of-the-art DeepWriting model [2], which uses a sample from the same writer (but of a different character sequence) for style inference. We test both Methods α and β from our model. Method α uses the same sample to predict **w** and to generated a new sample. Method β randomly samples 10 sentence-level drawing by the target writer and creates a sample with the algorithm discussed in Sect. 3. DeepWriting cannot take advantage of any additional character samples at inference time because it estimates only a single character-independent style vector.

Figure 4 (ii) displays how often each model was chosen as the most similar to the target handwriting; our model with sampling algorithm was selected 5.22 × as often as Aksan et al.'s model. Figure 4 (iii) displays which model was preferred across the 40 cases: of the 15 assessments per case, we count the number of times each model was the most popular. We show all cases in supplemental material.

Interpolation of **w**, $\mathbf{w}_{c_t}$, *and* **C***.* Figure 5 demonstrates that our method can interpolate (II) at the writer-DSD **w** level, (III) at the writer-character-DSD $\mathbf{w}_{c_t}$ level, and (IV) at the character-DSD **C** level. Given two samples of the same word by two writers $\mathbf{x}^A$ and $\mathbf{x}^B$, we first extract writer-character-DSDs from each sample (e.g., $\mathbf{w}^A_{rhy}$, $\mathbf{w}^B_{rhythm}$), then we derive writer-DSDs $\overline{\mathbf{w}^A}$ and $\overline{\mathbf{w}^B}$

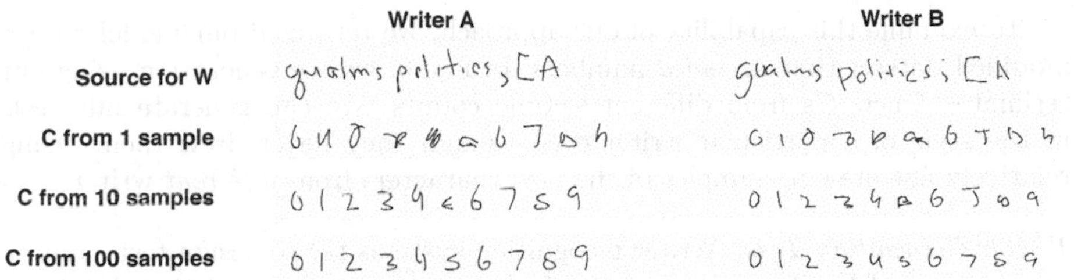

	Writer A	Writer B
Source for W	*qualms politics, LA*	*qualms politics; LA*
C from 1 sample	*6 4 0 x 4 a b 7 o h*	*6 1 0 0 p a 6 J b h*
C from 10 samples	*0 1 2 3 4 6 6 7 s 9*	*0 1 2 3 4 a 6 J o 9*
C from 100 samples	*0 1 2 3 4 s 6 7 8 9*	*0 1 2 3 4 s 6 7 8 9*

Fig. 6. Predicting $\mathbf{C}$ from new character samples, given a version of our model that is not trained on numbers. As we increase the number of samples used to estimate $\mathbf{C}$, the better the stylistic differences are preserved when multiplying with ws from different writers A and B. *Note:* neither writers provided numeral samples; by our construction, samples can come from any writer.

as in Sect. 3. To interpolate by γ between two writers, we compute the weighted average $\overline{\mathbf{w}^C} = \gamma \overline{\mathbf{w}^A} + (1 - \gamma) \overline{\mathbf{w}^B}$. Finally, we reconstruct writer-character-DSDs from $\overline{\mathbf{w}^C}$ (e.g., $\mathbf{w}_{rhy}^C = \mathbf{C}_{rhy} \overline{\mathbf{w}^C}$) and feed this into f_θ^{dec} to generate a new sample. For (III), we simply interpolate at the sampled character-level (e.g., $\mathbf{w}_{rhy}^A$ and $\mathbf{w}_{rhy}^B$). For (IV), we bilinearly interpolate four character-DSDs $\mathbf{C}_{c_t}$ placed at the corners of each image: $\overline{\mathbf{C}} = (r_A \times \mathbf{C}_A + r_B \times \mathbf{C}_B + r_C \times \mathbf{C}_C + r_D \times \mathbf{C}_D)$, where all r sum to 1. From $\overline{\mathbf{C}}$, we compute a writer-character-DSD as $\mathbf{w}_{\overline{c}} = \overline{\mathbf{C}} W$ and synthesize a new sample. In each case (II-IV), our representations are smooth.

Synthesis of New Characters. Our approach allows us to generate handwriting for new characters from a few samples from any writer. Let us assume that writer A produces a new character sample *3* that is not in our dataset. To make *3* available for generation in other writer styles, we need to recover the character-DSD $\mathbf{C}_3$ that represents the shape of the character *3*. Given $\mathbf{x}$ for newly drawn character *3*, encoder f_θ^{enc} first extracts the writer-character-DSD $\mathbf{w}_3$. Assuming that writer A provided other non-*3* samples in our dataset, we can compute multiple writer-DSD $\mathbf{w}$ for A. This lets us solve for $\mathbf{C}_3$ using least squares methods. We form matrices $\mathbf{Q}, \mathbf{P}_3$ where each column of $\mathbf{Q}$ is one specific instance of $\mathbf{w}$, and where each column of $\mathbf{P}_3$ is one specific instance of $\mathbf{w}_3$. Then, we minimize the sum of the squared error, which is the Frobenius norm $\|\mathbf{C}_3 \mathbf{Q} - \mathbf{P}_3\|_F^2$, e.g., via $\mathbf{C}_3 = \mathbf{P}_3 \mathbf{Q}^+$.

As architectured (and detailed in supplemental), g_θ actually has two parts: an LSTM encoder g_θ^{LSTM} that generates a 256×1 character representation vector $\mathbf{c}_{c_t}^{raw}$ for a substring c_t, and a fully-connected layer g_θ^{FC2} that expands $\mathbf{c}_{c_t}^{raw}$ and reshapes it into a 256×256 matrix $\mathbf{C}_{c_t} = g_\theta^{FC2}(\mathbf{c}_{c_t}^{raw})$. Further, as the output of an LSTM, we know that $\mathbf{c}_{c_t}^{raw}$ should be constrained to values $[-1, +1]$. Thus, for this architecture, we directly optimize the (smaller set of) parameters of the latent vector $\mathbf{c}_{c_t}^{raw}$ to create $\mathbf{C}_{c_t}$ given the pre-trained fully-connected layer weights, using a constrained non-linear optimization algorithm (L-BFGS-B) via the objective $f(\mathbf{c}_{c_t}^{raw}) = \|\mathbf{P}_3 - g_\theta^{FC2}(\mathbf{c}_{c_t}^{raw})\mathbf{Q}\|_F^2$.

To examine this capability of our approach, we retrained our model with a modified dataset that *excluded* numbers. In Fig. 6, we see generation using our estimate of new Cs from different sample counts. We can generate numerals in the style of a particular writer even though they never drew them, using relatively few drawing samples of the new characters from *different* writers.

Writer Recognition Task. Writer recognition systems try to assign test samples (e.g., a page of handwriting) to a particular writer given an existing database. Many methods use codebook approaches [7,8,11,20] to catalogue characteristic patterns such as graphemes, stroke junctions, and key-points from offline handwriting images and compare them to test samples. Zhang et al. [39] extend this idea to online handwriting, and Adak et al. study idiosyncratic character style per person and extract characteristic patches to identify the writer [1].

To examine how well our model might represent the latent distribution of handwriting styles, we perform a writer recognition task on our trained model on the randomly-selected 20-writer hold out set from our dataset. First, we compute 20 writer DSDs $\overline{\mathbf{w}}_i^{writer}$ from 10 sentence-level samples—this is our offline 'codebook' representing the style of each writer. Then, for testing, we sample from 1–50 new word-level stroke sequences per writer (using words with at least 5 characters), and calculate the corresponding writer DSDs ($N = 1,000$ in total). With the vector L of true writer labels, we compute prediction accuracy:

$$A = \frac{1}{N} \sum_{i=1}^{N} I(L_i, \arg\min_j (\overline{\mathbf{w}}_i^{word} - \overline{\mathbf{w}}_j^{writer})^2), \quad I(x,y) = \begin{cases} 1, & \text{if } x = y \\ 0, & \text{otherwise} \end{cases} \quad (12)$$

We repeated the random sampling of 1–50 words over 100 trials and compute mean accuracy and standard error. When multiple test samples are provided, we predict writer identity for each word and average their predictions. Random accuracy performance is 5%. Our test prediction accuracy rises from $89.20\% \pm 6.23$ for one word sample, to $97.85\% \pm 2.57$ for ten word samples, to $99.70\% \pm 1.18$ for 50 word samples. Increasing the number of test samples per writer increases accuracy because some words may not be as characteristic as others (e.g., 'tick' vs. 'anon'). Overall, while our model was not trained to perform this classification task, we can still achieved promising accuracy results from few samples—this is an indication that our latent space is usefully descriptive.

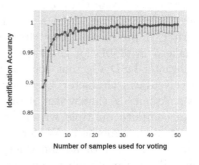

Additional Experiments. In our supplemental material, along with more architecture, model training procedure, and sampling algorithm details, we also: 1) compare to two style extraction pipelines, a stacked FC+ReLU layers and AdaIN, and find our approach more capable; 2) demonstrate the importance of learning style and content of character-DSD $\mathbf{C}$ by comparing with a randomly-initialized version; 3) ablate parts of our loss function, and illustrate key components; 4) experimentally show that our model is more efficient than DeepWriting by comparing generation given the same number of model parameters.

5 Discussion

While users preferred our model in our study, it still sometimes fails to generate readable letters or join cursive letters. One issue here is the underlying inconsistency in human writers, which we only partially capture in our data and represent in our model (e.g., cursive inconsistency). Another issue is collecting high-quality data with digital pens in a crowdsourced setting, which can still be a challenge and requires careful cleaning (see supplemental for more details).

Decoupling Additional Styles. Our model could potentially scale to more styles. For instance, we might create an age matrix $\mathbf{A}$ from a numerical age value a as $\mathbf{C}$ is constructed from c_t, and extract character-independent age-independent style descriptor as $\mathbf{w}^* = \mathbf{A}^{-1}\mathbf{C}_{c_t}^{-1}\mathbf{w}_{c_t}$. Introducing a new age operator $\mathbf{A}$ invites our model to find latent-style similarities across different age categories (e.g., between a child and a mature writer). Changing the age value and thus $\mathbf{A}$ may predict how a child's handwriting changes as s/he becomes older. However, training multiple additional factors in this way is likely to be challenging.

Alternatives to Linear $\mathbf{C}$ Multiplication Operator. Our model can generate new characters by approximating a new $\mathbf{C}$ matrix from few pairs of $\mathbf{w}$ and $\mathbf{w}_{c_t}$ thanks to their linear relationship. However, one might consider replacing our matrix multiplication 'operator' on $\mathbf{C}$ with parametrized nonlinear function approximators, such as autoencoders. Multiplication by $\mathbf{C}^{-1}$ would become an encoder, with multiplication by $\mathbf{C}$ being a decoder; in this way, g_θ would be tasked with predicting encoder weights given some predefined architecture. Here, consistency with $\mathbf{w}$ must still be retained. We leave this for future work.

6 Conclusion

We introduce an approach to online handwriting stroke representation via the Decoupled Style Descriptor (DSD) model. DSD succeeds in generating drawing samples which are preferred more often in a user study than the state-of-the-art model. Further, we demonstrate the capabilities of our model in interpolating samples at different representation levels, recovering representations for new characters, and achieving a high writer-identification accuracy, despite not being trained explicitly to perform these tasks. Online handwriting synthesis is still challenging, particularly when we infer a stylistic representation from few numbers of samples and try to generate new samples. However, we show that decoupling style factors has potential, and believe it could also apply to style-related tasks like transfer and interpolation in other sequential data domains, such as in speech synthesis, dance movement prediction, and musical understanding.

Acknowledgements. This work was supported by the Sloan Foundation and the National Science Foundation under award number IIS-1652561. We thank Kwang In Kim for fruitful discussions and for being our matrix authority. We thank Naveen Srinivasan and Purvi Goel for the ECCV deadline snack delivery service. Finally, we thank all anonymous writers who contributed to our dataset.

References

1. Adak, C., Chaudhuri, B.B., Lin, C.T., Blumenstein, M.: Intra-variable handwriting inspection reinforced with idiosyncrasy analysis (2019)
2. Aksan, E., Pece, F., Hilliges, O.: DeepWriting: making digital ink editable via deep generative modeling. In: SIGCHI Conference on Human Factors in Computing Systems, CHI 2018. ACM, New York (2018)
3. Alonso, E., Moysset, B., Messina, R.O.: Adversarial generation of handwritten text images conditioned on sequences. arXiv abs/1903.00277 (2019)
4. Azadi, S., Fisher, M., Kim, V.G., Wang, Z., Shechtman, E., Darrell, T.: Multi-content GAN for few-shot font style transfer. In: Proceedings of the IEEE Conference on Computer Vision and Pattern Recognition, pp. 7564–7573 (2018)
5. Balashova, E., Bermano, A.H., Kim, V.G., DiVerdi, S., Hertzmann, A., Funkhouser, T.: Learning a stroke-based representation for fonts. Comput. Graph. Forum **38**(1), 429–442 (2019). https://doi.org/10.1111/cgf.13540. https://onlinelibrary.wiley.com/doi/abs/10.1111/cgf.13540
6. Baluja, S.: Learning typographic style: from discrimination to synthesis. Mach. Vis. Appl. **28**(5–6), 551–568 (2017)
7. Bennour, A., Djeddi, C., Gattal, A., Siddiqi, I., Mekhaznia, T.: Handwriting based writer recognition using implicit shape codebook. Forensic Sci. Int. **301**, 91–100 (2019)
8. Bensefia, A., Nosary, A., Paquet, T., Heutte, L.: Writer identification by writer's invariants. In: Proceedings Eighth International Workshop on Frontiers in Handwriting Recognition, pp. 274–279. IEEE (2002)
9. Berio, D., Akten, M., Leymarie, F.F., Grierson, M., Plamondon, R.: Calligraphic stylisation learning with a physiologically plausible model of movement and recurrent neural networks. In: Proceedings of the 4th International Conference on Movement Computing, pp. 1–8 (2017)
10. Bishop, C.M.: Mixture density networks (1994)
11. Bulacu, M., Schomaker, L.: Text-independent writer identification and verification using textural and allographic features. IEEE Trans. Pattern Anal. Mach. Intell. **29**(4), 701–717 (2007)
12. Campbell, N.D., Kautz, J.: Learning a manifold of fonts. ACM Trans. Graph. (TOG) **33**(4), 1–11 (2014)
13. Carter, S., Ha, D., Johnson, I., Olah, C.: Experiments in handwriting with a neural network. Distill (2016). https://doi.org/10.23915/distill.00004. http://distill.pub/2016/handwriting
14. Cho, K., et al.: Learning phrase representations using RNN encoder-decoder for statistical machine translation. In: Proceedings of the 2014 Conference on Empirical Methods in Natural Language Processing (EMNLP), pp. 1724–1734. Association for Computational Linguistics, Doha, Qatar, October 2014. https://doi.org/10.3115/v1/D14-1179. https://www.aclweb.org/anthology/D14-1179

15. Chung, J., Kastner, K., Dinh, L., Goel, K., Courville, A.C., Bengio, Y.: A recurrent latent variable model for sequential data. In: Cortes, C., Lawrence, N.D., Lee, D.D., Sugiyama, M., Garnett, R. (eds.) Advances in Neural Information Processing Systems 28, pp. 2980–2988. Curran Associates, Inc. (2015). http://papers.nips.cc/paper/5653-a-recurrent-latent-variable-model-for-sequential-data.pdf
16. Goodfellow, I., et al.: Generative adversarial nets. In: Advances in Neural Information Processing Systems, pp. 2672–2680 (2014)
17. Graves, A.: Generating sequences with recurrent neural networks. arXiv preprint arXiv:1308.0850 (2013)
18. Ha, D., Dai, A., Le, Q.V.: Hypernetworks. arXiv preprint arXiv:1609.09106 (2016)
19. Haines, T.S.F., Mac Aodha, O., Brostow, G.J.: My text in your handwriting. ACM Trans. Graph. **35**(3) (2016). https://doi.org/10.1145/2886099. https://doi.org/10.1145/2886099
20. He, S., Wiering, M., Schomaker, L.: Junction detection in handwritten documents and its application to writer identification. Pattern Recogn. **48**(12), 4036–4048 (2015)
21. Heigold, G., Moreno, I., Bengio, S., Shazeer, N.: End-to-end text-dependent speaker verification. In: 2016 IEEE International Conference on Acoustics, Speech and Signal Processing (ICASSP), pp. 5115–5119. IEEE (2016)
22. Hsu, W.N., Glass, J.: Scalable factorized hierarchical variational autoencoder training. arXiv preprint arXiv:1804.03201 (2018)
23. Hsu, W.N., Zhang, Y., Glass, J.: Unsupervised learning of disentangled and interpretable representations from sequential data. In: Advances in Neural Information Processing Systems (2017)
24. Hu, C., Hersch, R.D.: Parameterizable fonts based on shape components. IEEE Comput. Graphics Appl. **21**(3), 70–85 (2001)
25. Hu, Q., Szabó, A., Portenier, T., Favaro, P., Zwicker, M.: Disentangling factors of variation by mixing them. In: The IEEE Conference on Computer Vision and Pattern Recognition (CVPR), June 2018
26. Huang, X., Belongie, S.: Arbitrary style transfer in real-time with adaptive instance normalization. In: Proceedings of the IEEE International Conference on Computer Vision (ICCV), October 2017
27. Jia, X., De Brabandere, B., Tuytelaars, T., Gool, L.V.: Dynamic filter networks. In: Advances in Neural Information Processing Systems, pp. 667–675 (2016)
28. Kingma, D.P., Welling, M.: Auto-encoding variational bayes. arXiv preprint arXiv:1312.6114 (2013)
29. Kotovenko, D., Sanakoyeu, A., Lang, S., Ommer, B.: Content and style disentanglement for artistic style transfer. In: Proceedings of the IEEE International Conference on Computer Vision, pp. 4422–4431 (2019)
30. Lian, Z., Zhao, B., Chen, X., Xiao, J.: EasyFont: a style learning-based system to easily build your large-scale handwriting fonts. ACM Trans. Graph. (TOG) **38**(1), 1–18 (2018)
31. Lopes, R.G., Ha, D., Eck, D., Shlens, J.: A learned representation for scalable vector graphics. In: The IEEE International Conference on Computer Vision (ICCV), October 2019
32. Marti, U.V., Bunke, H.: The IAM-database: an English sentence database for offline handwriting recognition. Int. J. Doc. Anal. Recogn. **5**(1), 39–46 (2002)
33. van den Oord, A., et al.: WaveNet: a generative model for raw audio. arXiv preprint arXiv:1609.03499 (2016)

34. Qian, K., Zhang, Y., Chang, S., Yang, X., Hasegawa-Johnson, M.: AUTOVC: zero-shot voice style transfer with only autoencoder loss. In: International Conference on Machine Learning, pp. 5210–5219 (2019)
35. Shen, J., et al.: Natural TTS synthesis by conditioning WaveNet on MEL spectrogram predictions. In: 2018 IEEE International Conference on Acoustics, Speech and Signal Processing (ICASSP), pp. 4779–4783. IEEE (2018)
36. Stanley, K.O., D'Ambrosio, D.B., Gauci, J.: A hypercube-based encoding for evolving large-scale neural networks. Artif. Life **15**(2), 185–212 (2009)
37. Suveeranont, R., Igarashi, T.: Example-based automatic font generation. In: Taylor, R., Boulanger, P., Krüger, A., Olivier, P. (eds.) SG 2010. LNCS, vol. 6133, pp. 127–138. Springer, Heidelberg (2010). https://doi.org/10.1007/978-3-642-13544-6_12
38. Wang, H., Liang, X., Zhang, H., Yeung, D.Y., Xing, E.P.: ZM-NET: real-time zero-shot image manipulation network. arXiv preprint arXiv:1703.07255 (2017)
39. Zhang, X.Y., Xie, G.S., Liu, C.L., Bengio, Y.: End-to-end online writer identification with recurrent neural network. IEEE Trans. Hum.-Mach. Syst. **47**(2), 285–292 (2016)
40. Zhang, X.Y., Yin, F., Zhang, Y.M., Liu, C.L., Bengio, Y.: Drawing and recognizing Chinese characters with recurrent neural network. IEEE Trans. Pattern Anal. Mach. Intell. **40**(4), 849–862 (2017)
41. Zongker, D.E., Wade, G., Salesin, D.H.: Example-based hinting of true type fonts. In: Proceedings of the 27th Annual Conference on Computer Graphics and Interactive Techniques, pp. 411–416 (2000)

LEED: Label-Free Expression Editing via Disentanglement

Rongliang Wu📷 and Shijian Lu(✉)📷

Nanyang Technological University, Singapore, Singapore
ronglian001@e.ntu.edu.sg, shijian.lu@ntu.edu.sg

Abstract. Recent studies on facial expression editing have obtained very promising progress. On the other hand, existing methods face the constraint of requiring a large amount of expression labels which are often expensive and time-consuming to collect. This paper presents an innovative label-free expression editing via disentanglement (LEED) framework that is capable of editing the expression of both frontal and profile facial images without requiring any expression label. The idea is to disentangle the identity and expression of a facial image in the expression manifold, where the neutral face captures the identity attribute and the displacement between the neutral image and the expressive image captures the expression attribute. Two novel losses are designed for optimal expression disentanglement and consistent synthesis, including a mutual expression information loss that aims to extract pure expression-related features and a siamese loss that aims to enhance the expression similarity between the synthesized image and the reference image. Extensive experiments over two public facial expression datasets show that LEED achieves superior facial expression editing qualitatively and quantitatively.

Keywords: Facial expression editing · Image synthesis · Disentangled representation learning

1 Introduction

Facial expression editing (FEE) allows users to edit the expression of a face image to a desired one. Compared with facial attribute editing which only considers appearance modification of specific facial regions [31,42,58], FEE is much more challenging as it often involves large geometrical changes and requires to modify multiple facial components simultaneously. FEE has attracted increasing interest due to the recent popularity of digital and social media and a wide spectrum of applications in face animations, human-computer interactions, etc.

Until very recently, this problem was mainly addressed from a graphical perspective in which a 3D Morphable Model (3DMM) was first fitted to the image

Electronic supplementary material The online version of this chapter (https://doi.org/10.1007/978-3-030-58610-2_46) contains supplementary material, which is available to authorized users.

© Springer Nature Switzerland AG 2020
A. Vedaldi et al. (Eds.): ECCV 2020, LNCS 12357, pp. 781–798, 2020.
https://doi.org/10.1007/978-3-030-58610-2_46

and then re-rendered with a different expression [17]. Such methods typically involve tracking and optimization to fit a source video into a restrictive set of facial poses and expression parametric space [54]. A desired facial expression can be generated by combining the graphical primitives [30]. Unfortunately, 3DMMs can hardly capture all subtle movements of face with the pre-defined parametric model and often produce blurry outputs due to the Gaussian assumption [17].

Inspired by the recent success of Generative Adversarial Nets (GANs) [19], a number of networks [10,12,16,17,39,41,45,53] have been developed and achieved very impressive FEE effects. Most of these networks require a large amount of training images with different types of expression labels/annotations, e.g. discrete labels [10,12], action units intensity [39,50,53] and facial landmarks [16,41,45], whereas labelling a large amount of facial expression images is often expensive and time-consuming which has impeded the advance of relevant research on this task. At the other end, the ongoing research [17,39,41,45,53] is largely constrained on the expression editing of frontal faces due to the constraint of existing annotations, which limits the applicability of FEE in many tasks.

This paper presents a novel label-free expression editing via disentanglement (LEED) framework that can edit both frontal and profile expressions without requiring any expression label or annotation by humans. Inspired by the manifold analysis of facial expressions [7,14,24] that different persons have analogous expression manifolds, we design an innovative disentanglement network that is capable of separating the identity and expression of facial images of different poses. The label-free expression editing is thus accomplished by fusing the identity of an input image with an arbitrary expression and the expression of a reference image. Two novel losses are designed for optimal identity-expression disentanglement and identity-preserving expression editing in training the proposed method. The first loss is a mutual expression information loss that guides the network to extract pure expression-related features from the reference image. The second loss is a siamese loss that enhances the expression similarity between the synthesized image and the reference image. Extensive experiments show that our proposed LEED even outperforms supervised expression editing networks qualitatively and quantitatively.

The contributions of this work are threefold. First, we propose a novel label-free expression editing via disentanglement (LEED) framework that is capable of editing expressions of frontal and profile facial images without requiring any expression label and annotation by humans. Second, we design a mutual expression information loss and a siamese loss that help extract pure expression-related features and enhance the expression similarity between the edited and reference facial images effectively. Third, extensive experiments show that the proposed LEED is capable of generating high-fidelity facial expression images and even outperforms many supervised networks.

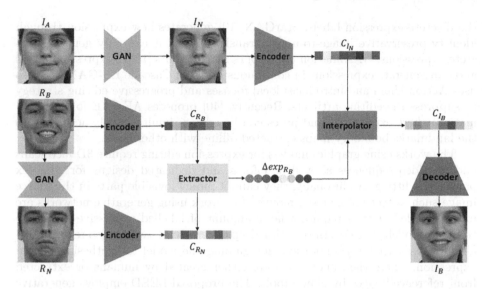

Fig. 1. The framework of LEED: given an input image I_A and a reference image R_B, the corresponding neutral faces I_N and R_N are first derived by a pre-trained GAN. The *Encoder* maps I_N, R_N, and R_B to latent codes C_{I_N}, C_{R_N}, and C_{R_B} in an embedded space which capture the identity attribute of I_A, the identity attribute of R_B, and the identity and expression attributes of R_B, respectively. The *Extractor* then extracts the expression attribute of R_B (Δexp_{R_B}) from C_{R_B} and C_{R_N}, and the *Interpolator* further generates the target latent code C_{I_B} from Δexp_{R_B} and C_{I_N}. Finally, the *Decoder* projects C_{I_B} to the image space to generate the edited image I_B.

2   Related Work

Facial Expression Editing: FEE is a challenging task and existing works can be broadly grouped into two categories. The first category is more conventional which exploits graphic models for expression editing. A typical approach is to first fit a 3D Morphable Model to a face image and then re-render it with a different expression. A pioneering work of Blanz and Vetter [5] presents the first public 3D Morphable Model. Vlasic et al. [49] proposes a video based multilinear model to edit the facial expressions. Cao et al. [6] introduces a video-to-image facial retargeting application that requires user interaction for accurate editing. Thies et al. [47] presents Face2Face for video-to-video facial expression retargeting which assumes the target video contains sufficient visible expression variation.

The second category exploits deep generative networks [19,28]. For example, warp-guided GAN [16] and paGAN [36] are presented to edit the expression of frontal face images with neutral expression. G2-GAN [45] and GCGAN [41] adopt facial landmarks as geometrical priors to control the generated expressions, where ground-truth images are essential for extracting the geometrical information. [17] proposes a model that combines 3DMM and GAN to synthesize expressions on RGB-D images. ExprGAN [12] introduces an expression controller to control the intensity of the generated expressions conditioned on the discrete expression labels. StarGAN [10] generates new expression through identity-preservative image-to-image translation and it can only generate discrete expressions. GANimation [39] adopts Action Units [15] as expression labels and can generate expressions in continuous domain. Cascade EF-GAN [53] also uses Action Units and introduces local focuses and progressive editing strategy to suppress the editing artifacts. Recently, [40] proposes AF-VAE for face animation where expressions and poses can be edited simultaneously according to the landmarks boundary maps extracted offline with other tools.

The works using graphics models for expression editing require 3D face scans and/or video sequences as well as efforts and dedicated designs for complex parametric fitting. Additionally, they cannot model invisible parts in the source image such as teeth of a closed mouth. The work using generative networks are more flexible but they require a large amount of labelled expressive images to train the models. Besides, the existing deep generative networks also require suitable expression labels/annotations for guiding the model to synthesize desired expression, where the annotations are either created by humans or extracted from reference images by offline tools. The proposed LEED employs generative networks which can hallucinate missing parts of input face images and it just requires a single photo for the input image and reference image which makes it much simpler to implement. At the same time, it can edit the expression of both frontal facial images and profile images without requiring any expression label/annotation by either humans or other tools.

Disentangled Representations: The key of learning disentangled representation is to model the distinct, informative factors of variations in the data [4]. Such representations have been applied successfully to image editing [24,37,43], image-to-image translation [57] and recognition tasks [38,51,55]. However, previous works achieve disentangled learning by training a multi-task learning model [24,51,57], where labels for each disentangled factors are essential. Recently, the unsupervised setting has been explored [8,23]. InfoGAN [8] achieves disentanglement by maximizing the mutual information between latent variables and data variation and β-VAE [23] learns the independent data generative factors by introducing an adjustable hyper-parameter β to the original VAE objective function. But these methods suffer from the lack of interpretability, and the meaning of each learned factor is uncontrollable. Based on the expression manifold analysis [7], our proposed method seeks another way to disentangle the identity and expression attributes from the facial images.

3 Proposed Method

3.1 Overview

Our idea of label-free expression editing via disentanglement (LEED) is inspired by the manifold analysis of facial expressions [7,14,24] that the expression manifold of different individuals is analogous. On the expression manifold, similar expressions are points in the local neighborhood with a 'neutral' face as the central reference point. Each individual has its neutral expression that corresponds to the original point in its own expression manifold and represents the identity attribute. The displacement of an expressive face and its neutral face gives the expression attribute.

Our proposed method achieves label-free expression editing by learning to disentangle the identity and expression attributes and fusing the identity of the input image and the expression of the reference image for synthesizing the desired expression images. As illustrated in Fig. 1, our network has five major components: an extractor for extracting expression attribute; an interpolator for fusing the extracted expression attribute and the identity attribute of the input image; an encoder for mapping the facial images into a compact expression and identity embedded space; a decoder for projecting the interpolated code to image space and a pre-trained GAN for synthesizing the neutral faces.

3.2   Extractor and Interpolator

Learning Expression Attribute Extractor: Given an input image with arbitrary expression A (denoted as I_A) and a reference image with desired expression B (denoted as R_B), our goal is to synthesize a new image I_B that combines the identity attribute of I_A and expression attribute of R_B. Without the expression labels, our proposed method needs to address two key challenges: 1) how to extract identity attribute from the input image and expression attribute from the reference image, and 2) how to combine the extracted identity and expression attributes properly to synthesize the desired expression images. We address the two challenges by learning an expression attribute extractor $\mathcal{X}$ and an interpolator $\mathcal{I}$, more details to be shared in the following texts.

The label-free expression editing is achieved by disentangling the identity and expression attributes. Given I_A and R_B, LEED first employs a pre-trained GAN to generate their corresponding neutral faces I_N and R_N. An encoder E is then employed to map all the images to a latent space, producing C_{I_A}, C_{I_N}, C_{R_B} and C_{R_N}, where C_{I_A}/C_{R_B} and C_{I_N}/C_{R_N} are the latent codes of the input/reference image and its neutral face, respectively. More details of the pre-trained GAN and E are to be discussed in Sect. 3.4.

According to [7], the latent code of the neutral face (C_{I_N}) represents the identity attribute of the input image, and the displacement between C_{R_B} and C_{R_N} represents the expression attribute of the reference image:

$$\Delta exp^*_{R_B} = C_{R_B} - C_{R_N}. \tag{1}$$

On the other hand, $\Delta exp^*_{R_B}$ depends on the embedded space, and the residual between C_{R_B} and C_{R_N} may contain expression-unrelated information such as head-poses variations that could lead to undesired changes in the synthesized images. We therefore propose to learn the expression attribute with an extractor $\mathcal{X}$ rather than directly using $\Delta exp^*_{R_B}$.

Formally, we train an expression extractor $\mathcal{X}$ to extract the expression Δexp_{R_B} from C_{R_B} and C_{R_N} with $\Delta exp^*_{R_B}$ as the pseudo label:

$$\min_{\mathcal{X}} \mathcal{L}_{exp} = \|\Delta exp_{R_B} - \Delta exp^*_{R_B}\|^2, \tag{2}$$

where $\Delta exp_{R_B} = \mathcal{X}(C_{R_B}, C_{R_N})$.

In addition, we design a mutual expression information loss to encourage the extractor to extract pure expression-related information. Specifically, we first use a pre-trained facial expression classification model Ψ, i.e. the ResNet [21] pre-trained on Real-world Affective Faces Database [32], to extract the features from R_B. As such a model is trained for classification task, the features of the last layers contain rich expression-related information [48]. We take the features from penultimate layer as the representation of the expression attribute of R_B and denote it as F_{R_B}, where $F_{R_B} = \Psi(R_B)$. As Ψ is used for extracting the features, we do not update its parameters in the training process.

In information theory, the mutual information between A and B measures the reduction of uncertainty in A when B is observed. If A and B are related by a deterministic, invertible function, the maximal mutual information is attained [8]. By maximizing the mutual information between Δexp_{R_B} and F_{R_B}, the extractor will be encouraged to extract pure expression-related features and ignore expression-unrelated information. However, directly maximizing the mutual information is hard as it requires access to the posterior distribution. We follow [8,12] to impose a regularizer Q on top of the extractor to approximate it by maximizing its derived lower bound [3]:

$$\min_{Q,\mathcal{X}} \mathcal{L}_Q = -\mathbb{E}[\log(Q(\Delta exp_{R_B}|F_{R_B})], \tag{3}$$

By combining Eqs. (2) and (3), the overall objective function of $\mathcal{X}$ is

$$\mathcal{L}_{\mathcal{X}} = \mathcal{L}_{exp} + \lambda_Q \mathcal{L}_Q, \tag{4}$$

where λ_Q is the hyper-parameter to balance the terms.

Learning Interpolator: With the identity attribute C_{I_N} of the input image and the expression attribute Δexp_{R_B} of the reference image, we can easily obtain the latent code C_{I_B} for the target image through linear interpolation

$$C^*_{I_B} = C_{I_N} + \Delta exp_{R_B}. \tag{5}$$

On the other hand, the linearly interpolated latent code may not reside on the manifold of real facial images and lead to weird editing (e.g. ghost faces) while projected back to the image space. Hence, we train an interpolator $\mathcal{I}$ to generate interpolated codes and impose an adversarial regularization term on it (details of the regularization term to be discussed in Sect. 3.4) as follows:

$$\min_{\mathcal{I}} \mathcal{L}_{interp} = \mathcal{L}_{adv_E,\mathcal{I}} + \|\mathcal{I}(C_{I_N}, \alpha\Delta exp_{R_B}) - (C_{I_N} + \alpha\Delta exp_{R_B})\|^2, \tag{6}$$

where $\alpha \in [0, 1]$ is the interpolated factor that controls the expression intensity of the synthesized image. We can obtain a smooth transition sequences of different expressions by simply changing the value of α once the model is trained.

In addition, the interpolator should be able to recover the original latent code of the input image given his/her identity attribute and the corresponding expression attribute. The loss term can be formulated as follows:

$$\min_{\mathcal{I}} \mathcal{L}_{idt} = \|\mathcal{I}(C_{I_N}, \Delta exp_{I_A}) - C_{I_A}\|^2, \tag{7}$$

where $\Delta exp_{I_A} = \mathcal{X}(C_{I_A}, C_{I_N})$.

The final objective function for the interpolator $\mathcal{I}$ is

$$\mathcal{L}_{\mathcal{I}} = \mathcal{L}_{interp} + \mathcal{L}_{idt}. \tag{8}$$

3.3 Expression Similarity Enhancement

To further enhance the expression similarity between the synthesized image I_B and the reference image R_B, we introduce a siamese network to encourage the synthesized images to share similar semantics with the reference image. The idea of siamese network is first introduced in natural language processing applications [18] that learns a space where the vector that transforms the word *man* to the word *woman* is similar to the vector that transforms *hero* to *heroine* [1]. In our problem, we define the difference between an expression face and its corresponding neutral face as the expression transform vector. And we minimize the difference between the expression transform vector of R_B and R_N and that of I_B and I_N. The intuition is that the transformation that turns a similar expressive face into neutral face should be analogous for different identities, which is aligned with the analysis of expression manifold [7].

Specifically, given reference image with expression B (R_B), its corresponding neutral face (R_N), synthesized image with expression B (I_B) and its corresponding neutral face (I_N), we first map them into a latent space by the siamese network S and obtain the transform vectors:

$$v_R = S(R_B) - S(R_N), \tag{9}$$

$$v_I = S(I_B) - S(I_N), \tag{10}$$

then we minimize the difference between v_R and v_I:

$$\min_{S} \mathcal{L}_S = Dist(v_R, v_I), \tag{11}$$

where $Dist$ is a distance metric. We adopt cosine similarity as the distance measurement and incorporate the siamese loss in learning the encoder.

3.4   Encoder, Decoder and GAN

Learning Encoder and Decoder: Given a collection of facial images I, we train an encoder to map them to a compact expression and identity embedded space to facilitate the disentanglement. We aim to obtain a flattened latent space so as to generate smooth transition sequences of different expressions by changing the interpolated factor (Sect. 3.2). This is achieved by minimizing the Wasserstein distance between the latent codes of real samples and the interpolated ones.

Specifically, a discriminator $\mathcal{D}$ is learned to distinguish the real samples and the interpolated ones and the encoder E and interpolator $\mathcal{I}$ are trained to fool the discriminator. We adopt the WGAN-GP [20] to learn the parameters. The adversarial loss functions are formulated as

$$\min_{\mathcal{D}} \mathcal{L}_{adv_{\mathcal{D}}} = \mathbb{E}_{\hat{C} \sim P_{\hat{I}}}[\log \mathcal{D}(\hat{C})] - \mathbb{E}_{C \sim P_{data}}[\log \mathcal{D}(C)]$$
$$+ \lambda_{gp} \mathbb{E}_{\tilde{C} \sim P_{\tilde{C}}}[(\|\nabla_{\tilde{C}} \mathcal{D}(\tilde{C})\|_2 - 1)^2], \tag{12}$$

$$\min_{E, \mathcal{I}} \mathcal{L}_{adv_{E, \mathcal{I}}} = -\mathbb{E}_{\hat{C} \sim P_{\hat{I}}}[\log \mathcal{D}(\hat{C})], \tag{13}$$

where $C = E(I)$ stands for the code generated by the encoder, $\hat{C}$ the interpolated code generated by the interpolator $\mathcal{I}$, P_{data} the data distribution of the codes of real images, $P_{\hat{I}}$ the distribution of the interpolated ones and $P_{\tilde{C}}$ the random interpolation distribution introduced in [20].

The model may suffer from 'mode collapse' problem if we simply optimize the parameters with Eqs. (12) and (13). The encoder learns to map all images to a small latent space where the real and interpolated codes are closed that yields a small Wasserstein distance. To an extreme, the Wasserstein distance could be 0 if the encoder maps all images to a single point [9]. To avoid this trivial solution, we train a decoder D to project the latent codes back to the image space. We follow [9,33] to train the decoder with perceptual loss [25] as Eq. (14), and impose an reconstruction constraint on the encoder as Eq. (15).

$$\min_{D} \mathcal{L}_{D} = \mathbb{E}(\|\Phi(D(C)) - \Phi(I)\|^2), \tag{14}$$

$$\min_{E} \mathcal{L}_{recon} = \mathbb{E}(\|\Phi(D(E(I))) - \Phi(I)\|^2), \tag{15}$$

where Φ is the VGG network [44] pre-trained on ImageNet [11].

The final objective function of the encoder can thus be derived as follows:

$$\mathcal{L}_{E} = \mathcal{L}_{GAN_{E, \mathcal{I}}} + \lambda_{recon} \mathcal{L}_{recon} + \lambda_{S} \mathcal{L}_{S}, \tag{16}$$

where λ_{recon} and λ_{S} are the hyper-parameters. E and S are updated in an alternative manner.

Pre-training GAN: We generate the neutral face of the input and reference images by using a pre-trained GAN which can be adapted from many existing image-to-image translation models [10,26,56,59]. In our experiment, we adopt the StarGAN [10] and follow the training strategy in [10] to train the model. The parameters are fixed once the GAN is trained.

Table 1. Quantitative comparison with state-of-the-art methods on datasets RaFD and CFEED by using FID (lower is better) and SSIM (higher is better).

	RaFD		CFEED	
	FID ↓	SSIM ↑	FID ↓	SSIM ↑
StarGAN [10]	62.51	0.8563	42.39	0.8011
GANimation [39]	45.55	0.8686	29.07	0.8088
Ours	**38.20**	**0.8833**	**23.60**	**0.8194**

Table 2. Quantitative comparison with state-of-the-art methods on datasets RaFD and CFEED by using facial expression classification accuracy (higher is better).

Dataset	Method	R	G	R + G
RaFD	StarGAN [10]	92.21	82.37	88.48
	GANimation [39]		84.36	92.31
	Ours		**88.67**	**93.25**
CFEED	StarGAN [10]	88.23	77.80	81.87
	GANimation [39]		79.46	84.42
	Ours		**84.35**	**90.06**

4 Experiments

4.1 Dataset and Evaluation Metrics

Our experiments are conducted on two public datasets including Radboud Faces Database (RaFD) [29] and Compound Facial Expressions of Emotions Database (CFEED) [13]. RaFD consists of 8,040 facial expression images collected from 67 participants. CFEED [13] contains 5,060 compound expression images collected from 230 participants. We randomly sample 90% images for training and the rest for testing. All the images are center cropped and resized to 128 × 128.

We evaluate and compare the quality of the synthesized facial expression images with different metrics, namely, Fréchet Inception Distance (FID) [22], structural similarity (SSIM) index [52], expression classification accuracy and the Amazon Mechanical Turk (AMT) user study results. The FID scores are calculated between the final average pooling features of a pre-trained inception model [46] of the real faces and the synthesized faces, and the SSIM is computed over synthesized expressions and corresponding expressions of the same identity.

4.2 Implementation Details

Our model is trained using Adam optimizer [27] with $\beta_1 = 0.5$, $\beta_2 = 0.999$. The detailed network architecture is provided in the supplementary materials. For a fair comparison, we train StarGAN [10] and GANimation [39] using the implementations provided by the authors. In all the experiments, StarGAN [10]

Table 3. Quantitative comparison with state-of-the-art methods on RaFD and CFEED by Amazon-Mechanical-Turk based user studies (higher is better for both metrics).

	RaFD		CFEED	
	Real or Fake	Which's More Real	Real or Fake	Which's More Real
Real	78.82	–	72.50	–
StarGAN [10]	31.76	7.06	14.37	8.75
GANimation [39]	47.06	15.29	31.87	9.38
Ours	**74.12**	**77.65**	**70.63**	**81.87**

is trained with discrete expression labels provided in the two public datasets, while GANimation [39] is trained with AU intensities extracted by OpenFace toolkit [2]. Our network does not use any expression annotation in training.

4.3 Quantitative Evaluation

We evaluate and compare our expression editing technique with state-of-the-art StarGAN [10] and GANimation [39] quantitatively by using FID, SSIM, expression classification accuracy and user study evaluation on RaFD and CFEED.

FID and SSIM: Table 1 shows the evaluation results of all compared methods on the datasets RaFD and CFEED by using the FID and SSIM. As Table 1 shows, our method outperforms the state-of-the-art methods by a large margin in FID, with a 7.35 improvement on RaFD and a 5.47 improvement on CFEED. The achieved SSIMs are also higher than the state-of-the-art by 1.5% and 1.1% for the two datasets. All these results demonstrate the superior performance of our proposed LEED in synthesizing high fidelity expression images.

Expression Classification: We perform quantitative evaluations with expression classification as in StarGAN [10] and ExprGAN [12]. Specifically, we first train expression editing models on the training set and perform expression editing on the corresponding testing set. The edited images are then evaluated by expression classification – a higher classification accuracy means more realistic expression editing. Two classification tasks are designed: 1) train expression classifiers by using the training set images (real) and evaluate them over the edited images; 2) train classifiers by combining the real and edited images and evaluate them over the test set images. The first task evaluates whether the edited images lie in the manifold of natural expressions, and the second evaluates whether the edited images help train better classifiers.

Table 2 shows the classification accuracy results (only seven primary expressions evaluated for CFEED). Specifically, **R** trains classifier with the original training set images and evaluates on the corresponding testing set images. **G** applies the same classifier (in **R**) to the edited images. **R + G** trains classifiers by combining the original training images and the edited ones, and evaluates on the same images in **R**. As Table 2 shows, LEED outperforms the state-of-the-art by 4.31% on RaFD and 4.89% on CFEED, respectively. Additionally, the LEED

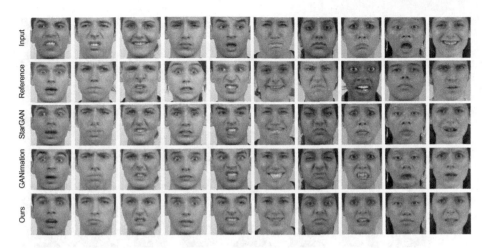

Fig. 2. Expression editing by LEED and state-of-the-art methods: Columns 1–5 show the editing of RaFD images, and columns 6–10 show the editing of CFEED images. Our method produces more realistic editing with better details and less artifacts.

edited images help to train more accurate classifiers while incorporated in training, where the accuracy is improved by 1.04% on RaFD and 1.83% on CFEED, respectively. They also outperform StarGAN and GANimation edited images, the latter even degrade the classification probably due to the artifacts within the edited images as illustrated in Fig. 2. The two experiments demonstrate the superiority of LEED in generating more realistic expression images.

User Studies: We also evaluate and benchmark the LEED edited images by conducting two Amazon-Mechanical-Turk (AMT) user studies under two evaluation metrics: 1) Real or Fake: subjects are presented with a set of expression images including real ones and edited ones by LEED, GANimation, and Star-GAN, and tasked to identify whether the images are real or fake; 2) Which's More Real: subjects are presented by three randomly-ordered expression images edited by the three methods, and are tasked to identify the most real one. Table 3 shows experimental results, where LEED outperforms StarGAN and GANimation significantly under both evaluation metrics. The two user studies further demonstrate the superior perceptual fidelity of the LEED edited images.

4.4 Qualitative Evaluation

Figure 2 shows qualitative experimental results with images from RaFD (cols 1–5) and CFEED (cols 6–10). Each column shows an independent expression editing, including an input image and a reference image as well as editing by StarGAN [10], GANimation [39] and our proposed LEED.

As Fig. 2 shows, StarGAN [10] and GANimation [39] tend to generate blurs and artifacts and even corrupted facial regions (especially around eyes and mouths). LEED can instead generate more realistic facial expressions with much

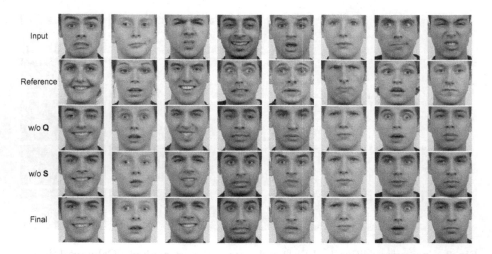

Fig. 3. Ablation study of LEED over RaFD: From top to bottom: input image, reference image, editing without mutual expression information loss, editing without siamese loss, final result. The graphs show the effectiveness of our designed losses.

less blurs and artifacts, and the generated images are also clearer and sharper. In addition, LEED preserves the identity information well though it does not adopt any identity preservation loss, largely due to the identity disentanglement which encodes the identity information implicitly.

4.5 Ablation Study

We study the two designed losses by training three editing networks on RaFD: 1) a network without the mutual expression information loss (regularizer Q) as labelled by 'w/o Q'; 2) a network without the siamese loss as labelled by 'w/o S'; and 3) a network with both losses as labelled by 'Final' in Fig. 3. As Fig. 3 shows, the mutual expression information loss guides the extractor to extract pure expression-relevant features. When it is absent, the extracted expression is degraded by expression-irrelevant information which leads to undesired editing such as eye gazing direction changes (column 1), head pose changes (columns 2, 3, 5 and 8), and identity attribute changes (missing mustache in column 4). The siamese loss enhances the expression similarity of the edited and reference images without which the expression intensity of the edited images becomes lower than that of the reference image (columns 2, 3, 6 and 7) as illustrated in Fig. 3.

4.6 Discussion

Feature Visualization: We use t-SNE [34] to show that LEED learns the right expression features via disentanglement. Besides the *Extractor features*, we also show the *Encoder features* (i.e. the dimension reduced representation of original

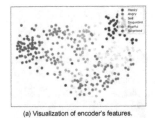 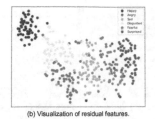 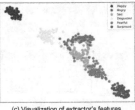

(a) Visualization of encoder's features.　　(b) Visualization of residual features.　　(c) Visualization of extractor's features.

Fig. 4. Expression feature Visualization with t-SNE: The *Extractor* learns much more compact clusters for expression features of different classes. Best view in colors.

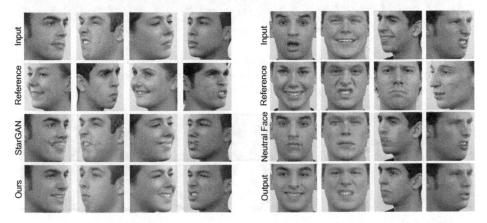

Fig. 5. LEED can 'transfer' expressions across profile images of different poses whereas state-of-the-art Star-GAN tends to produce clear artifacts.

Fig. 6. Though the GAN-generated neutral faces may not be perfect, LEED still generate sharp and clear expression thanks to our adopted adversarial loss.

image) and the *Residual features* (i.e. the difference between the Encoder features of expressive and neutral faces) as illustrated in Fig. 4 (learnt from the RaFD images). As Fig. 4 shows, the *Encoder features* and *Residual features* cannot form compact expression clusters as the former learns entangled features and the latter contains expression-irrelevant features such as head-poses variations. As a comparison, the *Extractor features* cluster each expression class compactly thanks to the mutual expression information loss.

Expression Editing on Profile Images: LEED is capable of 'transferring' expression across profile images of different poses. As illustrated in Fig. 5, LEED produces realistic expression editing with good detail preservation whereas Star-GAN introduces lots of artifacts (GANimation does not work as OpenFace cannot extract AUs accurately from profile faces). The capability of handling profile images is largely attributed to the mutual expression information loss that helps extract expression related features in the reference image. Note AF-VAE [40]

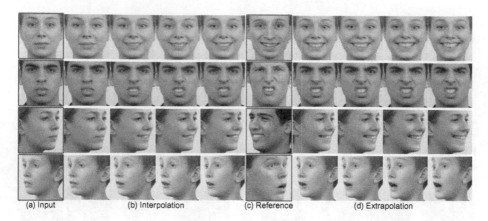

(a) Input   (b) Interpolation   (c) Reference   (d) Extrapolation

Fig. 7. Expression editing via interpolation/extrapolation: Given input images in (a) and reference images in (c), LEED can edit expressions by either interpolation ($\alpha <$ 1) or extrapolation ($\alpha > 1$) as shown in (b) and (d).

Fig. 8. Facial expression editing by LEED on wild images: In each triplet, the first column is input facial image, the second column is the image with desired expression and the last column is the synthesized result.

can also work with non-frontal profile images but it can only transfer expressions across facial images of the same pose.

Robustness to Imperfect Neutral Expression Images: LEED uses a pretrain GAN to generate neutral expression images for the disentanglement but the generated neural face may not be perfect as illustrated in Fig. 6 (row 3). LEED is tolerant to such imperfection as shown in Fig. 6 (row 4), largely because of the adversarial loss that is included into the interpolated latent code. However, the imperfect neutral face of reference image may contain residual of the original expression and lead to lower expression intensity in the output. This issue could be mitigated by adopting a stronger expression normalization model.

Continuous Editing: Our method can generate continuous expression sequences by changing the interpolated factor α (Sect. 3.2) as shown in Fig. 7. Besides interpolation, we show that the extrapolation can generate extreme expressions. This shows our method could uncover the structure of natural expression manifolds.

Facial Expression Editing on Wild Images: FEE for wild images is much more challenging as the images have more variations in complex background, uneven lighting, etc. LEED can adapt to handle wild images well as illustrated in Fig. 8, where the model is trained on expressive images sampled from Affect-Net [35]. As Fig. 8 shows, LEED can transform the expressions successfully while maintaining the expression-unrelated information unchanged.

5 Conclusion

We propose a novel label-free expression editing via disentanglement (LEED) framework for realistic expression editing of both frontal and profile facial images without any expression annotation. Our method disentangles the identity and expression of facial images and edits expressions by fusing the identity of the input image and the expression of the reference image. Extensive experiments over two public datasets show that LEED achieves superior expression editing as compared with the state-of-the-art techniques. We expect that LEED will inspire new insights and attract more interests for better FEE in the near future.

Acknowledgement. This work is supported by Data Science & Artificial Intelligence Research Centre, NTU Singapore.

References

1. Amodio, M., Krishnaswamy, S.: TraVeLGAN: image-to-image translation by transformation vector learning. In: Proceedings of the IEEE Conference on Computer Vision and Pattern Recognition, pp. 8983–8992 (2019)
2. Baltrusaitis, T., Zadeh, A., Lim, Y.C., Morency, L.P.: OpenFace 2.0: facial behavior analysis toolkit. In: 2018 13th IEEE International Conference on Automatic Face & Gesture Recognition (FG 2018), pp. 59–66. IEEE (2018)
3. Barber, D., Agakov, F.V.: The IM algorithm: a variational approach to information maximization. In: Advances in Neural Information Processing Systems (2003)
4. Bengio, Y., Courville, A., Vincent, P.: Representation learning: a review and new perspectives. IEEE Trans. Pattern Anal. Mach. Intell. **35**(8), 1798–1828 (2013)
5. Blanz, V., Vetter, T., et al.: A morphable model for the synthesis of 3D faces. In: SIGGRAPH, vol. 99, pp. 187–194 (1999)
6. Cao, C., Weng, Y., Zhou, S., Tong, Y., Zhou, K.: FaceWarehouse: a 3D facial expression database for visual computing. IEEE Trans. Vis. Comput. Graph. **20**(3), 413–425 (2013)
7. Chang, Y., Hu, C., Feris, R., Turk, M.: Manifold based analysis of facial expression. Image Vis. Comput. **24**(6), 605–614 (2006)
8. Chen, X., Duan, Y., Houthooft, R., Schulman, J., Sutskever, I., Abbeel, P.: InfoGAN: interpretable representation learning by information maximizing generative adversarial nets. In: Advances in Neural Information Processing Systems, pp. 2172–2180 (2016)
9. Chen, Y.C., Xu, X., Tian, Z., Jia, J.: Homomorphic latent space interpolation for unpaired image-to-image translation. In: Proceedings of the IEEE Conference on Computer Vision and Pattern Recognition, pp. 2408–2416 (2019)

10. Choi, Y., Choi, M., Kim, M., Ha, J.W., Kim, S., Choo, J.: StarGAN: unified generative adversarial networks for multi-domain image-to-image translation. In: Proceedings of the IEEE Conference on Computer Vision and Pattern Recognition, pp. 8789–8797 (2018)
11. Deng, J., Dong, W., Socher, R., Li, L.J., Li, K., Fei-Fei, L.: ImageNet: a large-scale hierarchical image database. In: 2009 IEEE Conference on Computer Vision and Pattern Recognition, pp. 248–255. IEEE (2009)
12. Ding, H., Sricharan, K., Chellappa, R.: ExprGAN: facial expression editing with controllable expression intensity. In: Thirty-Second AAAI Conference on Artificial Intelligence (2018)
13. Du, S., Tao, Y., Martinez, A.M.: Compound facial expressions of emotion. Proc. Nat. Acad. Sci. **111**(15), E1454–E1462 (2014)
14. Ekman, P., Friesen, W., Hager, J.: Facial action coding system (FACS) a human face. Salt Lake City (2002)
15. Friesen, E., Ekman, P.: Facial action coding system: a technique for the measurement of facial movement. Palo Alto **3** (1978)
16. Geng, J., Shao, T., Zheng, Y., Weng, Y., Zhou, K.: Warp-guided GANs for single-photo facial animation. ACM Trans. Graph. (TOG) **37**(6), 1–12 (2018)
17. Geng, Z., Cao, C., Tulyakov, S.: 3D guided fine-grained face manipulation. In: Proceedings of the IEEE Conference on Computer Vision and Pattern Recognition, pp. 9821–9830 (2019)
18. Goldberg, Y., Levy, O.: word2vec explained: deriving Mikolov et al'.s negative-sampling word-embedding method. arXiv preprint arXiv:1402.3722 (2014)
19. Goodfellow, I., et al.: Generative adversarial nets. In: Advances in Neural Information Processing Systems, pp. 2672–2680 (2014)
20. Gulrajani, I., Ahmed, F., Arjovsky, M., Dumoulin, V., Courville, A.C.: Improved training of Wasserstein GANs. In: Advances in Neural Information Processing Systems, pp. 5767–5777 (2017)
21. He, K., Zhang, X., Ren, S., Sun, J.: Deep residual learning for image recognition. In: Proceedings of the IEEE Conference on Computer Vision and Pattern Recognition, pp. 770–778 (2016)
22. Heusel, M., Ramsauer, H., Unterthiner, T., Nessler, B., Hochreiter, S.: GANs trained by a two time-scale update rule converge to a local Nash equilibrium. In: Advances in Neural Information Processing Systems, pp. 6626–6637 (2017)
23. Higgins, I., Matthey, L., Pal, A., Burgess, C., Glorot, X., Botvinick, M., Mohamed, S., Lerchner, A.: β-VAE: learning basic visual concepts with a constrained variational framework. ICLR **2**(5), 6 (2017)
24. Jiang, Z.H., Wu, Q., Chen, K., Zhang, J.: Disentangled representation learning for 3d face shape. arXiv preprint arXiv:1902.09887 (2019)
25. Johnson, J., Alahi, A., Fei-Fei, L.: Perceptual losses for real-time style transfer and super-resolution. In: Leibe, B., Matas, J., Sebe, N., Welling, M. (eds.) ECCV 2016. LNCS, vol. 9906, pp. 694–711. Springer, Cham (2016). https://doi.org/10.1007/978-3-319-46475-6_43
26. Kim, T., Cha, M., Kim, H., Lee, J.K., Kim, J.: Learning to discover cross-domain relations with generative adversarial networks. In: Proceedings of the 34th International Conference on Machine Learning, vol. 70, pp. 1857–1865. JMLR.org (2017)
27. Kingma, D.P., Ba, J.: Adam: a method for stochastic optimization. arXiv preprint arXiv:1412.6980 (2014)
28. Kingma, D.P., Welling, M.: Auto-encoding variational bayes. arXiv preprint arXiv:1312.6114 (2013)

29. Langner, O., Dotsch, R., Bijlstra, G., Wigboldus, D.H., Hawk, S.T., Van Knippenberg, A.: Presentation and validation of the radboud faces database. Cogn. Emot. **24**(8), 1377–1388 (2010)
30. Li, H., Weise, T., Pauly, M.: Example-based facial rigging. ACM Trans. Graph. (TOG) **29**(4), 1–6 (2010)
31. Li, M., Zuo, W., Zhang, D.: Deep identity-aware transfer of facial attributes. arXiv preprint arXiv:1610.05586 (2016)
32. Li, S., Deng, W., Du, J.: Reliable crowdsourcing and deep locality-preserving learning for expression recognition in the wild. In: 2017 IEEE Conference on Computer Vision and Pattern Recognition (CVPR), pp. 2584–2593. IEEE (2017)
33. Li, Y., Fang, C., Yang, J., Wang, Z., Lu, X., Yang, M.H.: Universal style transfer via feature transforms. In: Advances in Neural Information Processing Systems, pp. 386–396 (2017)
34. van der Maaten, L., Hinton, G.: Visualizing data using t-SNE. J. Mach. Learn. Res. **9**(Nov), 2579–2605 (2008)
35. Mollahosseini, A., Hasani, B., Mahoor, M.H.: AffectNet: a database for facial expression, valence, and arousal computing in the wild. IEEE Trans. Affect. Comput. **10**(1), 18–31 (2017)
36. Nagano, K., et al.: paGAN: real-time avatars using dynamic textures. ACM Trans. Graph. (TOG) **37**(6), 1–12 (2018)
37. Narayanaswamy, S., et al.: Learning disentangled representations with semi-supervised deep generative models. In: Advances in Neural Information Processing Systems, pp. 5925–5935 (2017)
38. Peng, X., Yu, X., Sohn, K., Metaxas, D.N., Chandraker, M.: Reconstruction-based disentanglement for pose-invariant face recognition. In: Proceedings of the IEEE International Conference on Computer Vision, pp. 1623–1632 (2017)
39. Pumarola, A., Agudo, A., Martinez, A.M., Sanfeliu, A., Moreno-Noguer, F.: GANimation: anatomically-aware facial animation from a single image. In: Proceedings of the European Conference on Computer Vision (ECCV), pp. 818–833 (2018)
40. Qian, S., et al.: Make a face: towards arbitrary high fidelity face manipulation. In: Proceedings of the IEEE International Conference on Computer Vision, pp. 10033–10042 (2019)
41. Qiao, F., Yao, N., Jiao, Z., Li, Z., Chen, H., Wang, H.: Geometry-contrastive GAN for facial expression transfer. arXiv preprint arXiv:1802.01822 (2018)
42. Shen, W., Liu, R.: Learning residual images for face attribute manipulation. In: Proceedings of the IEEE Conference on Computer Vision and Pattern Recognition, pp. 4030–4038 (2017)
43. Shu, Z., Yumer, E., Hadap, S., Sunkavalli, K., Shechtman, E., Samaras, D.: Neural face editing with intrinsic image disentangling. In: Proceedings of the IEEE Conference on Computer Vision and Pattern Recognition, pp. 5541–5550 (2017)
44. Simonyan, K., Zisserman, A.: Very deep convolutional networks for large-scale image recognition. arXiv preprint arXiv:1409.1556 (2014)
45. Song, L., Lu, Z., He, R., Sun, Z., Tan, T.: Geometry guided adversarial facial expression synthesis. In: 2018 ACM Multimedia Conference on Multimedia Conference, pp. 627–635. ACM (2018)
46. Szegedy, C., Ioffe, S., Vanhoucke, V., Alemi, A.A.: Inception-v4, inception-resnet and the impact of residual connections on learning. In: Thirty-First AAAI Conference on Artificial Intelligence (2017)
47. Thies, J., Zollhofer, M., Stamminger, M., Theobalt, C., Nießner, M.: Face2Face: real-time face capture and reenactment of RGB videos. In: Proceedings of the IEEE Conference on Computer Vision and Pattern Recognition, pp. 2387–2395 (2016)

48. Upchurch, P., et al.: Deep feature interpolation for image content changes. In: Proceedings of the IEEE Conference on Computer Vision and Pattern Recognition, pp. 7064–7073 (2017)
49. Vlasic, D., Brand, M., Pfister, H., Popovic, J.: Face transfer with multilinear models. In: ACM SIGGRAPH 2006 Courses, p. 24-es (2006)
50. Wang, J., Zhang, J., Lu, Z., Shan, S.: DFT-NET: disentanglement of face deformation and texture synthesis for expression editing. In: 2019 IEEE International Conference on Image Processing (ICIP), pp. 3881–3885. IEEE (2019)
51. Wang, Y., et al.: Orthogonal deep features decomposition for age-invariant face recognition. In: Proceedings of the European Conference on Computer Vision (ECCV), pp. 738–753 (2018)
52. Wang, Z., Bovik, A.C., Sheikh, H.R., Simoncelli, E.P., et al.: Image quality assessment: from error visibility to structural similarity. IEEE Trans. Image Process. **13**(4), 600–612 (2004)
53. Wu, R., Zhang, G., Lu, S., Chen, T.: Cascade EF-GAN: progressive facial expression editing with local focuses. In: Proceedings of the IEEE/CVF Conference on Computer Vision and Pattern Recognition, pp. 5021–5030 (2020)
54. Wu, W., Zhang, Y., Li, C., Qian, C., Change Loy, C.: ReenactGAN: learning to reenact faces via boundary transfer. In: Proceedings of the European Conference on Computer Vision (ECCV), pp. 603–619 (2018)
55. Wu, X., Huang, H., Patel, V.M., He, R., Sun, Z.: Disentangled variational representation for heterogeneous face recognition. In: Proceedings of the AAAI Conference on Artificial Intelligence, vol. 33, pp. 9005–9012 (2019)
56. Xiao, T., Hong, J., Ma, J.: ELEGANT: exchanging latent encodings with GAN for transferring multiple face attributes. In: Proceedings of the European Conference on Computer Vision (ECCV), pp. 168–184 (2018)
57. Yang, L., Yao, A.: Disentangling latent hands for image synthesis and pose estimation. In: Proceedings of the IEEE Conference on Computer Vision and Pattern Recognition, pp. 9877–9886 (2019)
58. Zhang, G., Kan, M., Shan, S., Chen, X.: Generative adversarial network with spatial attention for face attribute editing. In: Proceedings of the European Conference on Computer Vision (ECCV), pp. 417–432 (2018)
59. Zhu, J.Y., Park, T., Isola, P., Efros, A.A.: Unpaired image-to-image translation using cycle-consistent adversarial networks. In: Proceedings of the IEEE International Conference on Computer Vision, pp. 2223–2232 (2017)

Correction to: Single Image Super-Resolution via a Holistic Attention Network

Ben Niu, Weilei Wen, Wenqi Ren, Xiangde Zhang, Lianping Yang,
Shuzhen Wang, Kaihao Zhang, Xiaochun Cao, and Haifeng Shen

Correction to:
Chapter "Single Image Super-Resolution via a Holistic Attention Network" in: A. Vedaldi et al. (Eds.): *Computer Vision – ECCV 2020*, LNCS 12357, https://doi.org/10.1007/978-3-030-58610-2_12

In the originally published version of chapter 12, the first affiliation stated a wrong city and country. This has been corrected.

The updated version of this chapter can be found at
https://doi.org/10.1007/978-3-030-58610-2_12

Author Index

Printed in the United States
By Bookmasters